Introduction to Multiple Antenna Communications and Reconfigurable Surfaces

Emil Björnson and
Özlem Tuğfe Demir

now
the essence of knowledge

Published, sold and distributed by:
now Publishers Inc.
PO Box 1024
Hanover, MA 02339
United States
Tel. +1-781-985-4510
www.nowpublishers.com
sales@nowpublishers.com

Outside North America:
now Publishers Inc.
PO Box 179
2600 AD Delft
The Netherlands
Tel. +31-6-51115274

ISBN: 978-1-63828-314-0
E-ISBN: 978-1-63828-315-7
DOI: 10.1561/9781638283157

Suggested citation: Emil Björnson and Özlem Tuğfe Demir. (2024). *Introduction to Multiple Antenna Communications and Reconfigurable Surfaces.* Boston–Delft: Now Publishers

Supplementary material, such as simulation code that reproduces all numerical results in this work, is available at https://github.com/emilbjornson/mimobook and is delivered under a different license

Contents

Preface

The writing of this book started in 2018 as a small compendium written for the course "Multiple Antenna Communications" at Linköping University. The initial goal was to cover a few crucial aspects not included in the course book *Fundamentals of Massive MIMO*. The principle in the writing was to explain the fundamentals of the topic with as simple mathematics as possible while including all the practical insights we gathered as researchers in the field. For each year that passed, the compendium became 50 pages longer. We added a recap of the theoretical foundations that the topic builds on, practical aspects often overlooked by academia (e.g., polarization), and additional concepts needed in a prolonged version of the course given to doctoral students. During the COVID-19 pandemic, lecture recordings from the course were uploaded to YouTube, receiving thousands of views and many positive reviews. Hence, when we both moved to the KTH Royal Institute of Technology in 2021-2022 and stopped teaching the original course, we did not want to bury the compendium in a digital folder. Instead, we decided to turn it into a complete textbook that can be shared with an international audience.

As the original course's syllabus no longer limited us, we could focus on writing the definitive introductory book on multiple-input multiple-output (MIMO) communications. A key motivation for us is that with the advent of fifth-generation (5G) mobile networks, MIMO technology is everywhere: each base station and mobile phone is equipped with antenna arrays capable of transmitting/receiving signals with controllable directivity. This feature leads to stronger signals, robustness against channel fading, and spatial multiplexing that can drastically raise data rates. This is only the beginning of the MIMO saga because larger antenna arrays and higher frequency bands that can accommodate more antennas in the same enclosure are envisioned for future network generations. The MIMO technology affects the physical-layer transmissions and changes how resource allocation and network optimization are done. The same methodology also underpins emerging technologies such as reconfigurable intelligent surfaces (RIS) and integrated sensing and communication (ISAC). Hence, we believe that anyone who will research or develop future wireless communication systems must understand the fundamentals of multiple antenna communications. The first textbooks on the topic were written 25 years ago, and the basic theory remains valid; yet many recent insights and methodologies are not covered in classic textbooks, new terminologies and hardware architectures have arisen, and some old concepts are outdated.

This incentivized us to spend two years finalizing this textbook, including adding new chapters and numerous examples, exercises, and simulations that can be reproduced using MATLAB code available on the book's website.

How to Use This Book

This book is primarily written as the course material for a first-year graduate-level course and builds on undergraduate courses on signals and systems, linear algebra, probability theory, and digital communications. We believe the book should also appeal to wireless engineers and researchers who want to broaden their knowledge base and learn specific methods and algorithms.

Chapter 1 provides a high-level introduction and motivation to multiple antenna communications. To ensure that the reader remembers the essential results from the mentioned undergraduate courses, Chapter 2 summarizes the theoretical foundations used in later chapters. The basics of point-to-point MIMO communications between two transceivers equipped with multiple antennas are provided in Chapter 3. The theory is then expanded for static line-of-sight (LOS) channels in Chapter 4 and random non-LOS channels in Chapter 5. Next, we consider multi-user MIMO channels in Chapter 6, where a base station with multiple antennas serves multiple user devices. These chapters constitute the core of the book and should be included when it is used for teaching a course. If these chapters are too extensive, one can omit Section 4.5 on planar antenna arrays, Section 4.6 on polarization, Section 5.5 on block-fading channels, and Section 5.6 on sparse multipath propagation.

The last three chapters are mostly independent and cover three different topics. Chapter 7 extends the theory to wideband MIMO channels with orthogonal frequency-division multiplexing (OFDM). The chapter also describes hybrid analog-digital implementation architectures and MIMO terminology that one might encounter elsewhere. Chapter 8 covers the basics of direction-of-arrival estimation, localization, and radar sensing using antenna arrays. We explain how these array signal processing topics connect to the MIMO communication theory from previous chapters. The book ends with Chapter 9, which covers reconfigurable surfaces consisting of multiple antenna-like elements that can reflect signals in desirable ways to enhance communication channels. The basic theory borrows much from that described in previous chapters but comes with its characteristics and constraints.

We recommend solving exercises while reading the book. The answers are available online, and a solution manual is provided to instructors who use the book in their teaching—contact us to retrieve it.

This is an introductory book, so there are more advanced methodologies and applications to learn. If you want to dig deeper into the topic, we recommend the textbooks *Massive MIMO Networks* [1], *Foundations of User-Centric Cell-Free Massive MIMO* [2], and *Fundamentals of Massive MIMO* [3].

Acknowledgments

First and foremost, we would like to thank our families for supporting us throughout the seemingly neverending journey of writing this book.

The know-how that we share has been developed through scientific conversations and collaborations with Erik G. Larsson, Thomas Marzetta, Luca Sanguinetti, Jakob Hoydis, Björn Ottersten, Mats Bengtsson, Petar Popovski, Merouane Debbah, and many other excellent researchers. A special thanks go to Daniel Verenzuela, Marcus Karlsson, Giovanni Interdonato, Özgecan Özdogan, and Nikolaos Kolomvakis, who have influenced the material by being great teaching assistants in the course "Multiple Antenna Communications".

Many colleagues and students have read drafts of the book and provided detailed feedback, which enables us to polish off the rough edges and improve the technical rigor. Some of these are Alva Kosasih, Amna Irshad, Eren Berk Kama, Mert Alıcıoğlu, Morteza Tavana, Parisa Ramezani, Salih Gümüşbuğa, Sarvendranath Rimalapudi, Unnikrishnan Kunnath Ganesan, Yasaman Khorsandmanesh, and Zakir Hussain Shaik. We apologize to everyone we have forgotten since we did not make notes of contributors until recently. We would also like to thank Daniel Aronsson for helping us resolve MATLAB issues and everyone who has asked questions related to our videos and blog posts, which guided us in what concepts to explain in the book.

Finally, we would like to thank KTH Royal Institute of Technology, Linköping University, and TOBB University of Economics and Technology for giving us the time to write this book, as well as the Swedish Research Council, Swedish Foundation for Strategic Research, Knut and Alice Wallenberg Foundation, VINNOVA, and the Scientific and Technological Research Council of Türkiye that have funded our research efforts in these areas in recent years.

The authors, November 2023

Chapter 1

Introduction and Motivation

The basic scenario in wireless communications is that of a transmit antenna that radiates an electromagnetic waveform that spreads out and eventually is measured by a receive antenna located at another geographical location. The transmitted waveform is designed to carry information that can be extracted by the receiver from its measured received signal. A combination of digital modulation and channel coding is used to generate the waveform and encode information into it, which is done in such a way that the receiver can extract it even if the signal is attenuated and distorted.

There are many wireless technologies currently in use, such as the IEEE 802.11 technology family for WiFi, the IEEE 802.15.1 family for Bluetooth, the 3GPP family with GSM/UMTS/LTE/NR for cellular (mobile) communications [4], [5], and the competing but somewhat outdated 3GPP2 family with IS-95/CDMA2000/EV-DO. These technologies are based on open standards, created in collaboration between companies that jointly decide on the basic features but compete in building and selling commercial implementations. Some standards are designed to replace previous standards, targeting the same use cases. Other standards are optimized for different use cases—for example, long-range versus short-range communications, high data rate versus low power, or operation in licensed versus unlicensed frequency bands.

This chapter first introduces the fundamental concepts of signal power, channel gain, and antenna directivity. Then the use of multiple antennas will be motivated by outlining three main benefits this technology can provide.

1.1 Transmitted and Received Signal Power

In the technologies mentioned above, the transmit power P varies substantially with the type of device, signal bandwidth, technology, and use case. The cellular base stations deployed on rooftops and towers might transmit tens of watts; for example, 40 W per 10 MHz of bandwidth is typical in 4G LTE systems [6]. Base stations deployed closer to the potential users might only transmit a few hundred milliwatts; for example, 0.1 W is typical for WiFi access points, and 0.4 W is a limit for local-area cellular base stations in 5G NR [7]. A cell (mobile) phone typically radiates up to 0.1 W, and a short-range Bluetooth transmitter might operate at only 1 mW = 0.001 W. The power of a transmitter connected to an electrical grid is often limited by national regulations, selected to enable coexistence between different wireless systems and limit human exposure to strong electromagnetic fields. There are also regulations on battery-powered user devices; however, the devices are also subject to more practical limitations, such as keeping the power down to alleviate the need for active cooling and make the battery last longer. While the numbers mentioned above are the maximum power, battery-powered devices can purposely reduce their power during transmission and turn the transceivers on/off with time to save energy, especially when the data rate the system supports is higher than the device requires for the moment.

Due to the large transmit power variations, a decibel scale is often used to report the power numbers conveniently. In particular, the unit dBm is used to report the ratio between the signal power and 1 mW in decibels (dB):

$$10\log_{10}\left(\frac{\text{Signal power}}{1\,\text{mW}}\right)\ \text{dBm}, \tag{1.1}$$

where $\log_{10}(\cdot)$ is the base-10 logarithm. This means that 1 mW is equal to 0 dBm, 0.1 W is 20 dBm, and 40 W is 46 dBm. We note that $10\log_{10}(2) \approx 3$, $10\log_{10}(4) \approx 6$, and $10\log_{10}(8) \approx 9$. These approximations are often treated as being exact in the communication literature. Hence, doubling the signal power equals a 3 dB increase.

Example 1.1. The decibel scale is generally used to measure the relative size of two power values. Compare $P_1 = 8$ W and $P_2 = 1$ W using the dBm unit.

A direct computation based on (1.1) yields $P_1 \approx 39$ dBm and $P_2 = 30$ dBm because $10\log_{10}(8/10^{-3}) \approx 39$ and $10\log_{10}(1/10^{-3}) = 30$. The ratio P_1/P_2 is equal to 8, which can be expressed in decibels as

$$10\log_{10}\left(\frac{P_1}{P_2}\right) = 10\log_{10}\left(\frac{8}{1}\right) \approx 9\,\text{dB}. \tag{1.2}$$

This ratio can also be computed as P_1 [dBm] $-$ P_2 [dBm] $\approx 39 - 30 = 9$ dB, by first converting both numbers to dBm and then computing their difference. Note that the difference between 39 dBm and 30 dBm is expressed in dB, although their individual units are dBm. While dBm measures an absolute power value compared to 1 mW, dB is used to measure the relative ratio between two specific power values. In this example, we can say that P_1 is 9 dB larger than P_2, or that P_1 is 8 times larger than P_2.

A transmit antenna radiates an electromagnetic signal waveform that travels in all directions at the speed of light. The signal power is quickly

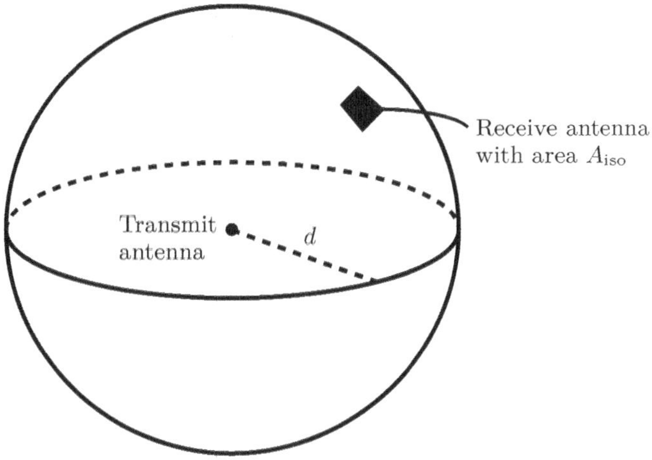

Figure 1.1: An isotropic transmit antenna radiates a signal that spreads like an inflatable sphere. At a propagation distance d in free space, the surface area of a sphere with radius d is $4\pi d^2$. This area is typically huge compared to the area A_{iso} of an isotropic receive antenna; thus, the receiver only captures a tiny fraction of the signal.

dispersed over the surrounding environment; thus, the power measured by a receiving device is incredibly much smaller than the transmit power. One can picture this as if the signal power exists on the surface of a balloon. As we blow up the balloon, the radius of the balloon grows, and the surface area becomes larger and larger, but the surface material also becomes thinner and thinner. When the signal waveform has traveled a distance d in free space, the signal power exists on a sphere with radius d, as illustrated in Figure 1.1. The surface area is $4\pi d^2$. If the power is equally distributed over the sphere's surface, the transmit antenna is said to be *isotropic*. This is also called a *point source*. Isotropic antennas are impossible to build[1] but are used for theoretical analysis and as a benchmark for other antennas by measuring how close to isotropic a practical antenna is radiating its signals.

An elementary kind of signal waveform is the sinusoid illustrated in Figure 1.2. This is an oscillating periodic function of time with a *frequency* denoted by f in this figure. The frequency represents the number of repeated periods per second observed at a specific location and is measured in Hertz (Hz). The *period* can be measured between two adjacent peaks observed in time and is $1/f$ seconds. When a sinusoidal electromagnetic wave propagates at the speed of light $c\,\text{m/s}$, at any given time instance, each period will cover a spatial interval of length c/f meters. This quantity is very important when

[1]The radiated field from an antenna must satisfy the Helmholtz wave equation, which originates from Maxwell's equations. One can prove that an isotropic field does not do that. Even if one could build an isotropic antenna, it is not practically useful since it must be connected to transceiver hardware that generates wireless signals. This connection would block the wave propagation in some directions.

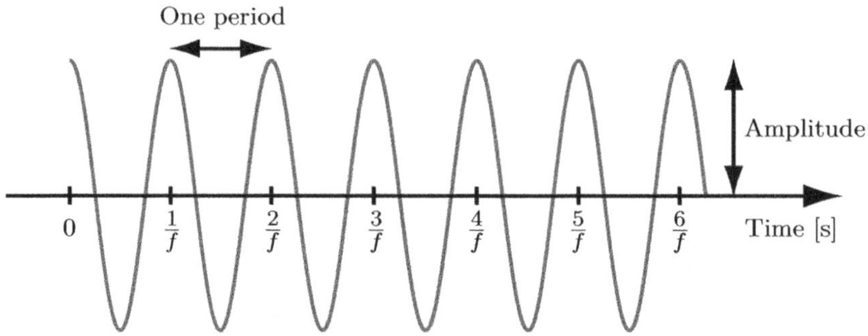

Figure 1.2: A sinusoid is a signal waveform characterized by its amplitude and frequency f [Hz]. The time period between two peaks is $1/f$ seconds.

analyzing how the wave interacts with objects in the surroundings, including antennas. It is called the signal's *wavelength* and will be denoted as $\lambda = c/f$. The speed of light is 299 792 458 m/s in free space (vacuum), but we will use the close approximation $c = 3\cdot 10^8$ m/s throughout this book to enable a simple conversion between frequencies and wavelengths; for example, $f = 3\,\text{GHz}$ gives $\lambda = 0.1\,\text{m}$.

The receive antenna converts the impinging electromagnetic waves into an electric current and can thereby be used to collect signal power. The power-capturing ability of an antenna is quantified by its *effective area*. It is defined as the ratio of the power that the antenna can collect (in W) to the *power flux density* of the incident wave (in W/m^2) [8]. It can be proved that a hypothetical lossless isotropic antenna must have the effective area

$$A_{\text{iso}} = \frac{\lambda^2}{4\pi}, \tag{1.3}$$

where λ is the wavelength of the type of waveform the antenna was built for. Since $\lambda = c/f$, the effective area in (1.3) can be equivalently expressed as

$$A_{\text{iso}} = \frac{c^2}{4\pi f^2}. \tag{1.4}$$

This means that the higher the signal's frequency, the smaller the area of the matching isotropic antenna. The word "effective" in the term "effective area" refers to the following: Suppose a planar waveform travels in a given direction, and you place a surface perpendicular to that direction to block a part of the signal. The antenna captures power proportional to what would pass through the surface if it has the specified effective area. This does not mean a practical antenna must have that specific area, but it depends on the hardware implementation and deployment.[2] For example, if the antenna is not

[2]The effective area of an aperture-type antenna is always less than or equal to its physical area. The *aperture efficiency*, which is the ratio of the maximum effective area (over all directions) to the physical area of an antenna, is an essential metric in antenna design [8].

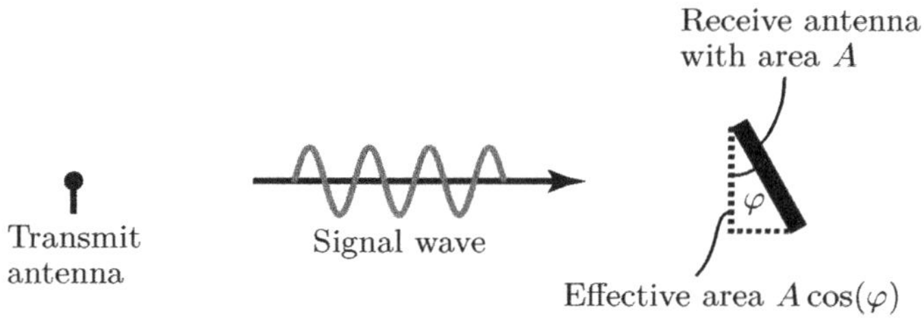

Figure 1.3: The effective area of a receive antenna is generally smaller than the antenna's physical area. The physical area is A in this figure. However, the effective area $A\cos(\varphi)$ perpendicular to the direction that the signal propagates determines the received signal power. Any non-isotropic antenna has a varying effective area for different angular directions φ. The maximum effective area among all rotations is used as the reference value when comparing practical antennas of different kinds.

perpendicular to the direction in which the wave travels, the effective area is smaller than the physical area of the antenna. This is illustrated in Figure 1.3, where the receive antenna has the physical area A. Since the antenna is not deployed perpendicularly to the direction that the signal is traveling, the effective area is the projection of the physical antenna area in that direction. In the figure, the antenna is rotated by an angle $\varphi \in [-\pi/2, \pi/2]$; thus, the effective area is $A\cos(\varphi)$, which is smaller or equal to the physical area.

Example 1.2. Consider a lossless isotropic antenna designed for the wavelength $\lambda = 0.1\,\text{m}$ ($f = 3\,\text{GHz}$). What is the power captured by this antenna if the power flux density of the incident electromagnetic wave is $50\,\mu\text{W/m}^2$?

The answer is the product of the effective area and the power flux density:

$$A_{\text{iso}} \cdot 50 \cdot 10^{-6} = \frac{\lambda^2}{4\pi} \cdot 50 \cdot 10^{-6} \approx 3.98 \cdot 10^{-8}\,\text{W}. \tag{1.5}$$

Suppose a so-called *short dipole* replaces the isotropic antenna. This non-isotropic antenna captures different amounts of power depending on its rotation with respect to the incident wave. The maximum effective area among all rotations is used as the reference value when analyzing such an antenna. If we measure the received power over different rotations and notice that $5.96 \cdot 10^{-8}\,\text{W}$ is the maximum value, what is the maximum effective area?

The effective area A_{eff} is the ratio of the captured power to the power flux density. In this case, it becomes

$$A_{\text{eff}} = \frac{5.96 \cdot 10^{-8}}{50 \cdot 10^{-6}} \approx 0.00119\,\text{m}^2, \tag{1.6}$$

which is approximately 1.5 times larger than A_{iso}.

The black area in Figure 1.1 represents an isotropic receive antenna placed on the surface area of the sphere; that is, perpendicular to the direction that the transmitted waveform is traveling outwards from the origin. If the receive antenna is located at the distance d from the transmitter, its area A_{iso} in (1.3) should be compared with the total surface area $A_{\text{sphere}}(d) = 4\pi d^2$ of a sphere with radius d. If $A_{\text{sphere}}(d) \geq A_{\text{iso}}$, the fraction of the transmit power that reaches the receive antenna is

$$\frac{A_{\text{iso}}}{A_{\text{sphere}}(d)} = \frac{\frac{\lambda^2}{4\pi}}{4\pi d^2} = \frac{\lambda^2}{(4\pi)^2}\frac{1}{d^2}. \tag{1.7}$$

The factor $\lambda^2/(4\pi)^2$ is determined only by the wavelength, while the second factor is inversely proportional to the square of the propagation distance. This means that the signal power captured by the receive antenna decays rapidly with the distance d. Note that this example assumes so-called free-space propagation, which means there are no objects inside (or outside) the sphere in Figure 1.1 that interact with the radiated waveform to increase or decrease the received power. We will use this as the basic scenario in this book but also cover some other scenarios. The expression in (1.7) is a special case of the classical Friis' transmission formula for free-space propagation [9], which also applies to other types of antennas than isotropic.

The ratio in (1.7) is called the *channel gain*, while its inverse is called the *pathloss*.[3] In this book, we often let the parameter β denote the channel gain. This is a dimensionless parameter computed as the ratio between two areas. To get a sense of the typical size of the channel gain, Figure 1.4 shows its value as a function of the distance d for three different frequencies that are relevant for wireless communications:

- $f = 1\,\text{GHz}$ with wavelength $\lambda = 0.3\,\text{m}$;
- $f = 3\,\text{GHz}$ with wavelength $\lambda = 0.1\,\text{m}$;
- $f = 30\,\text{GHz}$ with wavelength $\lambda = 0.01\,\text{m}$.

Since the channel gains are generally tiny, they are presented in the decibel scale in Figure 1.4; that is, the vertical axis presents

$$10\log_{10}\left(\frac{\lambda^2}{(4\pi)^2}\frac{1}{d^2}\right) = 10\log_{10}\left(\frac{\lambda^2}{(4\pi)^2}\right) - 20\log_{10}(d)\,\text{dB}. \tag{1.8}$$

The curves start at a $1\,\text{m}$ distance, where the channel gain is $-42\,\text{dB}$ at the $3\,\text{GHz}$ frequency. When increasing the distance by a factor of 10, from $1\,\text{m}$ to $10\,\text{m}$, the channel gain reduces by $20\,\text{dB}$ to $-62\,\text{dB}$. Hence, if we divide the transmit power into (roughly) one million parts, only one reaches the receive antenna. As seen from the last term in (1.8), the channel gain reduces by $20\,\text{dB}$ every time the distance increases by 10 times. Hence, another $20\,\text{dB}$ is lost when the distance increases from $10\,\text{m}$ to $100\,\text{m}$.

[3] It also happens that (1.7) is called the pathloss in the communication literature, so it is vital to know the dimensionality of this type of term to understand which definition is used in a particular text. Importantly, a wireless channel can only attenuate signals, so the channel gain must be smaller than or equal to 1, while its inverse must be greater than or equal to 1.

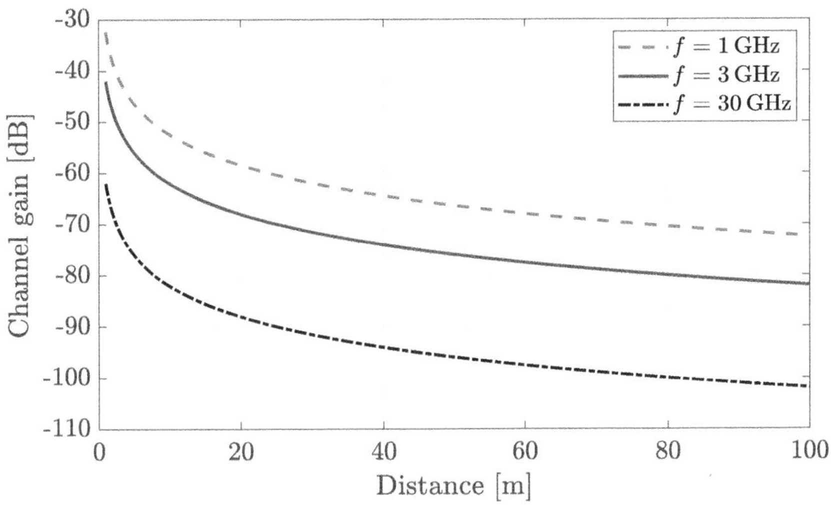

Figure 1.4: The channel gain in (1.7) depends on the propagation distance d and the frequency f of the waveform, assuming that different matching isotropic antennas are used when communicating at each of the considered frequencies. The channel gain is reported using the decibel scale since the variations are huge.

Compared to communications at the 3 GHz frequency, the channel gain in Figure 1.4 is larger when using the lower frequency 1 GHz and smaller when using the higher frequency 30 GHz. This is purely due to the differences in the effective area in (1.3) for the corresponding isotropic receive antennas, which is proportional to λ^2. The waveforms are attenuated identically when propagating in free space irrespective of the frequency; that is, the power flux density is constant at the receiver location but is multiplied by different effective areas depending on the frequency band. In particular, it is only the first term in (1.8) that depends on the wavelength, while the distance-dependent second term is the same for any wavelength.

Example 1.3. The channel gain with an isotropic receive antenna at $f = 3$ GHz and the distance $d = 10$ m is -62 dB, as shown in Figure 1.4. What is the corresponding channel gain if we replace the isotropic receive antenna with another antenna whose effective area is twice as large? What is the channel gain with this new antenna at a 100 m distance?

The channel gain is proportional to the effective area, as can be seen from (1.7) where the effective area of an isotropic antenna is divided by the area of a sphere. If we double the effective area, the channel gain is doubled, and in the decibel scale, it becomes $-62 + 3 = -59$ dB at the 10 m distance.

For the considered channel gain model in (1.8), there is a 20 dB gain reduction each time the distance increases by 10 times. Hence, the channel gain with the new antenna at a 100 m distance is $-59 - 20 = -79$ dB.

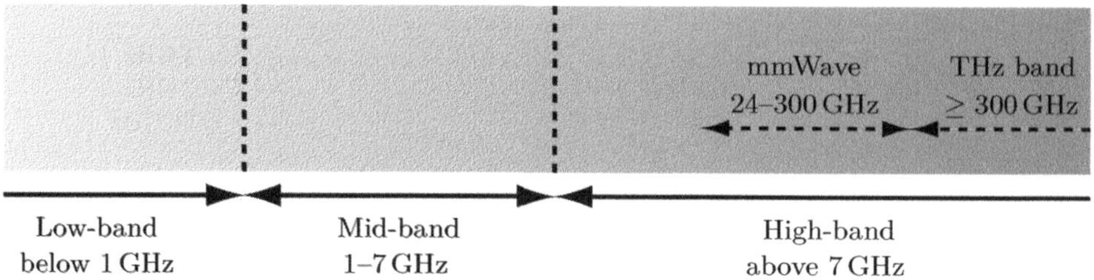

Figure 1.5: The radio frequency spectrum ranges from 3 kHz to 3000 GHz (i.e., 3 THz) and is used for many different services. The spectrum used for wireless communications is commonly divided into the low-band, mid-band, and high-band, as indicated in this figure. The high-band range 24–300 GHz is referred to as the mmWave band since the wavelength ranges from 12 to 1 mm. The range 300–3000 GHz is called the THz band.

The three exemplified frequencies were selected to represent the three specific bands considered in 5G NR [10]. Most wireless communication systems operate in the part of the electromagnetic frequency spectrum called the *radio spectrum*, even if there are exceptions.[4] The radio spectrum ranges have changed with time as the applications and hardware have evolved. According to the 2020 regulations from the International Telecommunication Union (ITU) [11], the radio spectrum consists of all frequencies from 3 kHz to 3000 GHz. In the context of 5G NR, the spectrum is further divided into the *low-band* containing carrier frequencies up to 1 GHz, the *mid-band* in the range 1–7 GHz, and the *high-band* with frequencies above 7 GHz, as illustrated in Figure 1.5.[5] The millimeter-wave (mmWave) band is a particularly prominent part of the high-band spectrum and, strictly speaking, covers 30–300 GHz, where the wavelength is between 10 and 1 mm. For practical reasons, the mmWave band is typically said to start at 24 GHz since spectrum is available from that frequency in some countries. Moreover, only mmWave bands below 100 GHz are considered in 5G NR; thus, the range 100–300 GHz is often called the sub-THz band by researchers who want to differentiate future technologies from existing 5G solutions [13]. Finally, the range 300–3000 GHz is called the THz band since this range can also be expressed as 0.3–3 THz.

It is commonly stated that the maximum coverage range of a wireless communication system is longer in the low-band than in the high-band. This statement is often correct, but it is not caused by the phenomenon illustrated in Figure 1.4. Recall that we considered a free-space propagation model without objects between the transmitter and receiver, where the power flux density is independent of frequency. The differences in the free-space channel gains in Figure 1.4 can be fully compensated for by increasing the effective area of the

[4]Two notable exceptions are free-space optical communication that uses visible or near-visible light and sonic communication that uses audio waves.

[5]The convention of whether a frequency band is considered low or high shifts with time and application; in particular, the low-band for cellular communications is known as the ultra-high frequency band for radar, and some other radio applications [12].

receive antenna; thus, it is the same irrespective of the signal's frequency. In particular, the channel gain definition in (1.7) becomes frequency-independent if the area in the numerator is constant instead of proportional to λ^2. Since the effective area of a single receive antenna reduces with increasing frequency, a fair comparison between two frequency bands requires antenna configurations with the same effective area in both bands. One way to achieve this in practice is by using multiple receive antennas in the higher band so that their collective effective area sums up to the same value as in the lower band. We will consider this in detail later in this book.

The main reason low-band frequencies generally have a longer coverage range is the signal behaviors in scenarios other than free-space propagation. In terrestrial communications, there are many objects in the environment around and between the transmitter and receiver. Signals with a lower frequency range propagate better through and around such objects and are reflected off walls more favorably. The signal absorption by atmospheric gases in the air also increases with the frequency. For these reasons, base stations for wide-area coverage typically use the low-band, while medium-range and local-area networks use the mid-band. Short-range networks might use the mmWave spectrum (or even the THz spectrum) in the high-band. Nevertheless, satellites commonly use the high-band spectrum to communicate with the ground over incredibly long distances. This works well if no blocking objects exist and the antennas have large effective areas.

Despite the reduced range, there are two good reasons why new wireless communication systems are gradually supporting higher frequency bands. Firstly, large parts of the low-band and mid-band are already occupied by existing wireless services, making it hard to launch new services there. Secondly, there is generally more bandwidth available at higher carrier frequencies, and we will see later that the data rates increase with the bandwidth. To give some indicative numbers, a network operator might have licenses for 20 MHz in the low-band, 100 MHz in the mid-band, and 1 GHz in the high-band.

The channel gain depends on the propagation distance d in typical terrestrial communication scenarios, where the transmitting base station might be deployed on a rooftop and the receiving user device is located in an urban city. In that case, there is no unequivocal channel gain model because the wave propagation depends on the exact geographical locations of buildings and other large-scale objects. However, we can describe the average propagation conditions by fitting a parametric channel gain model of the kind

$$\beta = \Upsilon \left(\frac{1\,\mathrm{m}}{d} \right)^{\alpha} \tag{1.9}$$

to real-world channel measurements. The parameter α is called the *pathloss exponent* while Υ is the channel gain at a 1 m reference distance. This parametric model is inspired by the free-space model in (1.7), which is obtained by $\alpha = 2$ and $\Upsilon = \left(\frac{\lambda/(1\,\mathrm{m})}{4\pi}\right)^2 = \left(\frac{0.3\,\mathrm{GHz}}{4\pi f}\right)^2$ because $c/(1\,\mathrm{m}) = 0.3\,\mathrm{GHz}$.

Example 1.4. The 3GPP technical report [6] presents channel gain models for several propagation scenarios typical in cellular communications. For example, in the non-line-of-sight urban microcell (UMi) scenario [6, Table B.1.2.1-1], the channel gain is modeled (in decibel) as

$$\beta_{\text{UMi}} = -36.7 \log_{10}\left(\frac{d}{1\,\text{m}}\right) - 22.7 - 26 \log_{10}\left(\frac{f}{1\,\text{GHz}}\right) \text{ dB}. \tag{1.10}$$

This model can be used for distances d in the range 10–2000 m and frequencies f in the range 2–6 GHz. What are the values of α and Υ in this case, and how does it differ from the free-space propagation case?

The distance-dependent term in (1.10) is $-36.7\log_{10}(d) = -10\log_{10}(d^{3.67})$; thus, the pathloss exponent for this UMi channel is $\alpha = 3.67$. Since the exponent is larger than in free-space propagation ($\alpha = 2$), the channel gain decays more rapidly with the distance. This represents the fact that the wireless signals must interact with objects in the environment to reach the receiver.

The channel gain Υ at the reference distance of 1 m is given by the last two terms in (1.10) and becomes $\Upsilon = 10^{-2.27}(\frac{1\,\text{GHz}}{f})^{2.6}$. This parameter is valid for specifying the pathloss model even if the UMi model should only be used for $d \geq 10$ m. We notice that Υ decays with the frequency as $f^{-2.6}$. This is faster than the f^{-2} behavior in free-space propagation, which is caused by the isotropic receive antenna assumption. The extra decay describes how the wireless signals interact less favorably with objects as the frequency increases.

Apart from the scaling behaviors, we can compare the channel gains obtained at the minimum values $d = 10$ m and $f = 2$ GHz. The channel gain is -67.2 dB with the UMi model and -58.5 dB in free-space propagation; thus, the UMi model consistently gives lower gains at all distances and frequencies.

1.1.1 Signal-to-Noise Ratio

Although the channel gains are typically tiny in wireless communications, many existing systems operate efficiently. This is possible because what matters is not the absolute amount of signal power received but its relative size compared to the noise power in the receiver hardware (and the interference power received from other concurrent transmissions).

We let σ^2 denote the noise power. It is computed as the product $\sigma^2 = N_0 B$ of the *noise power spectral density* N_0 W/Hz and the *signal bandwidth* B Hz. The intuition behind this model is that the thermal noise in the receiver is a white random process with the constant power spectral density N_0 over all frequencies, but the receiver hardware filters out the noise that lies outside the signal band, thereby making the total noise power equal to N_0 times the signal bandwidth B. We will return to these modeling assumptions in

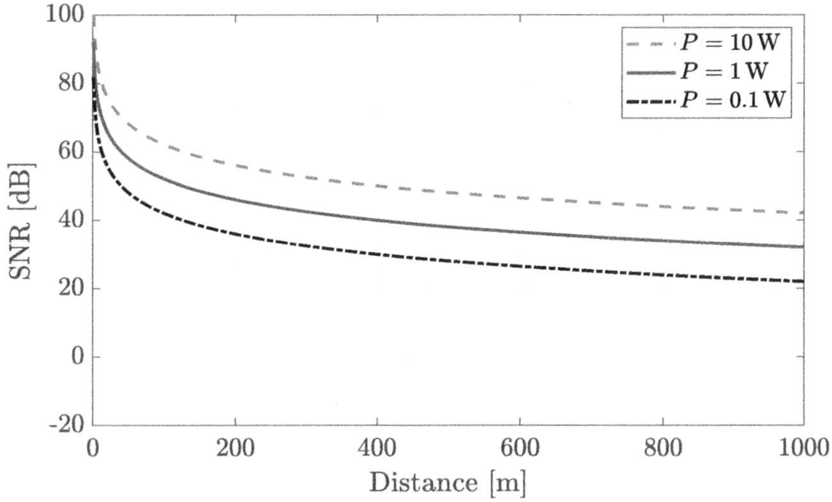

(a) The SNR when using the free-space channel gain model in (1.7).

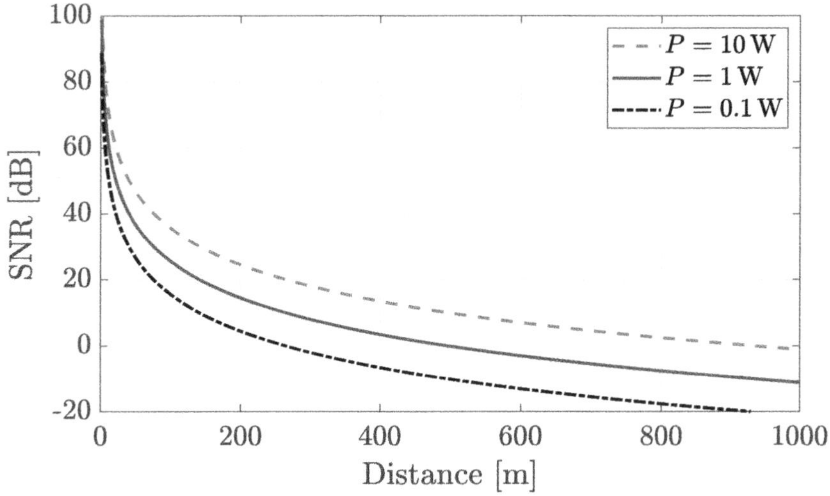

(b) The SNR when using the UMi channel gain model in (1.10).

Figure 1.6: The SNR in (1.13) as a function of the propagation distance d for two different channel gain models: free-space propagation and the non-line-of-sight UMi model. The setup is defined by $f = 3\,\text{GHz}$, $B = 10\,\text{MHz}$, and either $P = 10\,\text{W}$, $P = 1\,\text{W}$, or $P = 0.1\,\text{W}$.

Section 2.3.2. The noise power spectral density depends on the temperature, but the variations are small in most use cases. Therefore it is common to take the number at room temperature (i.e., 20°C) and treat it as a constant:[6]

$$N_0 = 10^{-20.4} \text{ W/Hz}. \tag{1.11}$$

[6]The actual noise power spectral density in wireless receivers is normally larger than the number in (1.11) since the receiver hardware is amplifying the thermal noise. For example, the practical noise power might be 4-8 dB higher than the theoretical lower limit in (1.11).

When reporting noise powers in the decibel scale, using dBm, the formula is

$$\sigma^2 = 10\log_{10}\left(\frac{N_0 B}{1\,\text{mW}}\right) = -174 + 10\log_{10}(B)\ \text{dBm}. \tag{1.12}$$

The *signal-to-noise ratio (SNR)* is defined as

$$\text{SNR} = \frac{P\beta}{\sigma^2} = \frac{P\beta}{N_0 B}, \tag{1.13}$$

where we recall that P is the transmit power, β is the channel gain, and $\sigma^2 = N_0 B$ is the noise power. The SNR is a dimensionless variable since it is computed as the ratio of two powers. To get a sense of what the practical range of SNR values is, Figure 1.6 shows the SNR in (1.13) in the decibel scale as a function of the propagation distance. We consider a bandwidth of $B = 10\,\text{MHz}$ around the frequency $f = 3\,\text{GHz}$ and use either the free-space channel gain model in (1.7) or the UMi channel gain model in (1.10). The SNR can be many tens of decibels for very short distances (e.g., inside a room). For practical distances in outdoor scenarios, we can expect an SNR below 40 dB, particularly when using the non-line-of-sight UMi model, where the channel gain decays more rapidly with the distance. If we reduce the transmit power, the SNR curve is shifted downwards accordingly. Many other phenomena affect the SNR, but as a rule-of-thumb, the SNR in a wireless communication system is between $-10\,\text{dB}$ and $+40\,\text{dB}$.

Example 1.5. Consider a communication setup where the SNR is 30 dB at a 400 m distance from the transmitter when using the free-space channel gain in (1.7) with $f = 3\,\text{GHz}$ (i.e., $\lambda = 0.1\,\text{m}$). What will be the new SNR at that distance if we switch to using the UMi channel gain model in (1.10)?

Due to the linear relation between SNR and the channel gain in (1.13), the SNR in the modified UMi setup is

$$\text{SNR}_{\text{UMi}} = \frac{P\beta_{\text{UMi}}}{N_0 B} = \frac{P\beta}{N_0 B}\frac{\beta_{\text{UMi}}}{\beta} = 30 + 10\log_{10}\left(\frac{\beta_{\text{UMi}}}{\beta}\right)\ \text{dB}, \tag{1.14}$$

where β is the free-space channel gain from (1.7) and β_{UMi} was defined in (1.10). By inserting numbers into this expression, we obtain

$$\begin{aligned}\text{SNR}_{\text{UMi}} &= 30 - 36.7\log_{10}(400) - 22.7 - 26\log_{10}(3) - 20\log_{10}\left(\frac{0.1}{4\pi\cdot 400}\right)\\ &\approx -6.58\ \text{dB}.\end{aligned} \tag{1.15}$$

This new SNR is 36.58 dB smaller (i.e., 4550 times smaller), which shows that the SNR can vary greatly with the propagation conditions. Such large variations can hardly be compensated for by increasing the transmit power.

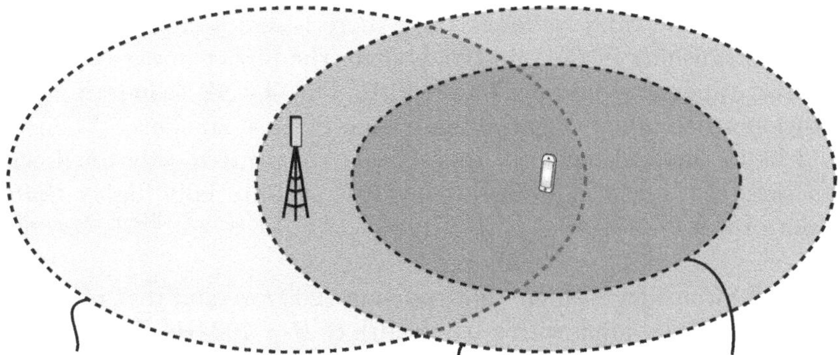

Figure 1.7: Since the base station (to the left) and the phone (to the right) have different transmit powers, the areas where the SNR is above the minimum threshold that enables successful communications will be different in the downlink and uplink. One way to deal with this problem is to reduce the bandwidth in the uplink so that the SNR becomes the same as in the downlink.

As mentioned earlier in this chapter, the transmit power can vary significantly between different devices, including those communicating with each other using the same communication standard. In a cellular network, the base station might transmit with 40 W, while the cell phone uses 0.1 W. This is a difference of $40/0.1 = 400 \approx 26$ dB, which implies that the SNR is 26 dB better when transmitting in the *downlink* (from the base station to the phone) than when transmitting in the *uplink* (from the phone to the base station) over the same frequency band. It is necessary to communicate in both directions to keep a cellular network operational, which makes the uplink transmission the weakest link. A practical solution to this problem is to utilize only a fraction of the bandwidth when the user transmits, which increases the SNR since the noise power reduces. In other words, we put all the signal power into a narrower range of frequencies. This principle is illustrated in Figure 1.7 by showing the geographical area where a receiver would get an SNR above a certain threshold required for successful communication (e.g., -10 dB). The yellow area for the downlink transmission with $B = 10$ MHz contains the phone; thus, the downlink transmission will be successful. However, the red area for the uplink transmission is substantially smaller and does not contain the base station. The yellow and red areas use the same bandwidth of 10 MHz in the uplink and downlink. However, if the phone only uses 10 MHz/400 = 25 kHz of bandwidth, the blue uplink area is obtained, and it is as large as the yellow downlink area. In practice, the bandwidth that is used by the phone can be varied dynamically depending on how far from the base station the user is.

Another solution is to use different frequency bands in the uplink and downlink. Suppose the base station and phone can use both the low-band and the mid-band. It is then possible to let the phone transmit its signals in the low-band where the range is longer and there is less bandwidth, while the base station transmits in the mid-band where the higher power compensates for a shorter range and broader bandwidth. The 5G NR standard supports this solution to enhance the coverage range of base stations. When wider bandwidths in the mid-band (or high-band) are utilized only for downlink transmission, the downlink data rates will be substantially higher than the uplink data rates.

Example 1.6. Consider a phone that transmits 200 mW and that is connected to a communication system with a bandwidth of $B = 20\,\text{MHz}$. When using the entire bandwidth, the uplink SNR is $-30\,\text{dB}$. Suppose the uplink SNR must be at least $-10\,\text{dB}$ for the system to be operational. How much bandwidth can the phone use?

The phone must reduce the uplink bandwidth so that the SNR increases by $-10-(-30) = 20\,\text{dB}$, which is 100 times more. Hence, at most, it can use an uplink bandwidth of $20\,\text{MHz}/100 = 200\,\text{kHz}$.

1.1.2 Fraunhofer Distance

The analysis has thus far been based on isotropic antennas, which is a hypothetical concept, as noted earlier. This book is not focused on antenna design or detailed modeling of individual antennas but on the phenomena, benefits, and challenges that occur when having multiple antennas. However, we will briefly describe a few fundamental antenna properties essential to understanding the connection between fixed directive antennas and the adaptive directivity obtained using multiple antennas.

When we derived the channel gain equation for free-space propagation, we used Figure 1.1, where the receive antenna is located on the surface of a sphere because the transmitted signal spreads out as a sphere with an increasing radius. This implies that the receive antenna must be curved to fit on the surface area; otherwise, the transmitted signal will reach different parts of the antenna at different times. Practical antennas are generally flat, creating a mismatch that we will now analyze in detail. Figure 1.8 shows a flat receive antenna perpendicular to the direction of the propagating wave. When the spherical wavefront of the transmitted signal reaches the center of the receive antenna, it has not yet reached its edges. As a result, the impinging electric field will vary in phase and amplitude over the antenna surface. This has consequences for the intercepted signal power, which can typically be computed by integrating the power flux density of the impinging electric field over the receive antenna's surface. The maximum power is intercepted when the impinging electric field is constant over the antenna, which happens in the ideal case when the wavefront is planar and impinges perpendicularly.

When the propagation distance is sufficiently large compared to the antenna

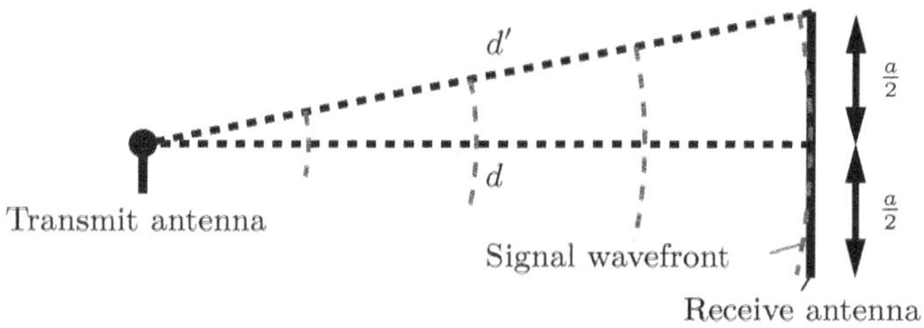

Figure 1.8: When a spherical wavefront approaches a flat receive antenna, there will be a delay between when the wave reaches the antenna's center and edge. This delay (or difference in propagation distance) turns into a phase-shift. The phase-shift is small or large depending on the relation between the distance d, the width a of the receive antenna, and the wavelength λ.

size, the spherical wavefront can be locally approximated as planar when considering the power the antenna intercepts. If the distance is d from the transmitter to the antenna's center and the antenna's width is a, then we can compute the distance d' from the transmitter to the antenna's edges using the Pythagorean theorem as

$$d' = \sqrt{d^2 + \left(\frac{a}{2}\right)^2} = d\sqrt{1 + \left(\frac{a}{2d}\right)^2}. \tag{1.16}$$

When a sinusoidal signal with the wavelength λ needs to travel an extra distance $d' - d$ to reach the edge, then there will be a phase difference of[7]

$$\frac{2\pi}{\lambda}(d' - d) = \frac{2\pi}{\lambda}\left(d\sqrt{1 + \left(\frac{a}{2d}\right)^2} - d\right) \approx \frac{2\pi}{\lambda}\left(d + \frac{a^2}{8d} - d\right) = \frac{\pi a^2}{4\lambda d} \text{ [rad]} \tag{1.17}$$

between the signal received at the edge and the center. The simplified expression in (1.17) is obtained by using the Taylor approximation $\sqrt{1+x^2} \approx 1 + \frac{x^2}{2}$ which is tight (the error is less than 0.05%) for $0 \leq x \leq 0.25$. Since $x = a/(2d)$ in this case, $x \leq 0.25$ implies we need to consider distances $d \geq 2a$. The phase difference in (1.17) will never be zero, but it will be close to zero when the propagation distance d is much larger than the width a of the antenna. It is common to assume (somewhat arbitrarily) that the phase variations over the antenna can be neglected if the maximum difference in (1.17) is no larger than $\pi/8$ radians (22.5 degrees) [14]. By following this convention, we get the relation

$$\frac{\pi}{8} \geq \frac{\pi a^2}{4\lambda d} \quad \Rightarrow \quad d \geq \frac{2a^2}{\lambda}. \tag{1.18}$$

The impinging wavefront also varies in amplitude between the center and the edge since the received signal amplitude is inversely proportional to the

[7]Suppose the signal $\sin(2\pi f t)$ is transmitted in Figure 1.8. The signal reaching the center of the antenna is $\sin(2\pi f(t - d/c))$, while the signal reaching the edge of the antenna is $\sin(2\pi f(t - d'/c))$. The phase difference between these signals is $2\pi f(d' - d)/c = 2\pi(d' - d)/\lambda$.

distance. The relative difference is d'/d and this ratio is between 0.97 and 1 for distances $d \geq 2a$, because $d = 2a$ gives

$$\frac{d}{d'} = \frac{d}{\sqrt{d^2 + \left(\frac{a}{2}\right)^2}} \geq \frac{2a}{\sqrt{4a^2 + \left(\frac{a}{2}\right)^2}} = \sqrt{\frac{16}{17}} \approx 0.97. \tag{1.19}$$

Hence, if the distance between the transmitter and receiver is simultaneously greater than $2a^2/\lambda$ and $2a$, we can neglect the spherical shape of the waveform (when considering both the phase and amplitude) and compute channel gains in the way previously described. In other words, we can treat the impinging wave as a *plane wave* traveling in one angular direction and only depends on time and the location along that direction; at any time instance, the wave is constant within any given plane perpendicular to the direction of travel.[8] The impinging wave is only approximately plane at the local level, observable at the receiver, but remains spherical at the global level. This is similar to how Earth appears flat to an observer on the ground, although it is curved.

The minimum distance in (1.18) is called the *Fraunhofer distance* and is named after Joseph von Fraunhofer, who studied many electromagnetic phenomena. It is occasionally also called the *Rayleigh distance*. The region that lies beyond the Fraunhofer distance is known as the *far-field* of the antenna. The Fraunhofer distance was derived based on two approximations but is known to be a good rule-of-thumb. When the propagation distance d is either smaller than $2a^2/\lambda$ or $2a$, we are in the *near-field* of the antenna. The near-field can be divided into two parts. The *radiative* near-field is an intermediate region where the propagation distance to the receiver is too short to neglect the phase and/or amplitude variations over the receive antenna but large enough to avoid direct hardware interaction between the transmitter and receiver. The *reactive* near-field is closest to the transmitter and includes additional electromagnetic effects such as evanescent waves and magnetic induction. These are examples of electric and magnetic field components that can only be observed near the transmitter, typically up to a maximum distance of $\lambda/(2\pi)$. Specific standards exist for near-field communication (NFC) that are commonly used by smartphones and cards to enable short-range payments and identification. This book, which focuses on radiated electromagnetic waves, will not cover these technologies.

To shed light on how far away the far-field is, suppose the receive antenna in Figure 1.8 has the length $a = \lambda$ for which $2a^2/\lambda$ and $2a$ are both equal to 2λ. Hence, if the receive antenna is at least 1 m from the isotropic transmit antenna, we are guaranteed to be in the far-field for any frequency band of interest in wireless communications (because the wavelength is typically shorter than 0.5 m). This condition is almost always satisfied.

[8]An ideal plane wave fills the infinitely large three-dimensional world (i.e., $\mathbb{R}^3$) and, thus, cannot exist in practice. However, the impinging wave observed over an antenna of finite width a will be perceived as being a finite-sized portion of a plane wave when $d \geq 2a^2/\lambda$.

The Fraunhofer distance in (1.18) is truly wavelength-dependent, in contrast to the free-space channel gain in (1.7) whose wavelength-dependence was caused by the assumption of having an isotropic receive antenna. The distances d and d' in Figure 1.8 are computed based on geometrical arguments that do not involve the wavelength λ. However, when the wave travels the extra distance $d' - d$ to reach the edge, the wavelength determines how large the resulting phase-shift is. For a fixed-sized antenna, the Fraunhofer distance in (1.18) is inversely proportional to λ, making it larger in the high-band than in the low-band. However, suppose the antenna size is proportional to the wavelength. In that case, we get the opposite behavior, as shown by the fact that $a = \lambda$ gives the Fraunhofer distance 2λ proportional to λ.

Example 1.7. What is the Fraunhofer distance when considering a rectangular receive antenna with width a and height b?

Suppose d is the distance from the transmitter to the center of the antenna. Following the same steps as before, we can compute the distance d' to the antenna's corners as

$$d' = \sqrt{d^2 + \left(\frac{a}{2}\right)^2 + \left(\frac{b}{2}\right)^2} = d\sqrt{1 + \left(\frac{D}{2d}\right)^2} \tag{1.20}$$

where we have defined $D = \sqrt{a^2 + b^2}$ as the length of the diagonal of the rectangular antenna. The difference $d' - d$ leads to the phase difference

$$\frac{2\pi}{\lambda}(d' - d) = \frac{2\pi}{\lambda}\left(d\sqrt{1 + \left(\frac{D}{2d}\right)^2} - d\right) \approx \frac{2\pi}{\lambda}\left(d + \frac{D^2}{8d} - d\right) = \frac{\pi D^2}{4\lambda d} \text{ [rad]} \tag{1.21}$$

between the signals captured at the center and the corners, using the same Taylor approximation as in (1.17). We recall that the Fraunhofer distance is obtained when the phase difference is $\pi/8$. Solving $\frac{\pi D^2}{4\lambda d} = \frac{\pi}{8}$ for d yields $2D^2/\lambda$. The only difference from (1.18) is that D has replaced a. Generally speaking, for any antenna shape, the Fraunhofer distance is $2D^2/\lambda$ by letting D be the largest distance between any two points on the antenna.

1.1.3 Antenna Directivity Gains

We will now move beyond isotropic antennas and provide the basic characterization of antenna directivity. Practical transmit antennas radiate a larger fraction of their power in some angular directions than others. The transmitted signal will still propagate as a sphere with an expanding radius, as illustrated in Figure 1.1, but the signal power is unequally distributed over the surface area. We need a spherical coordinate system to specify the power distribution over the sphere. There are different ways to define spherical coordinates. We use the definition in Figure 1.9 where a point at a distance d from the origin is characterized by the azimuth angle $\varphi \in [-\pi, \pi)$ in the xy-plane and the elevation angle $\theta \in [-\pi/2, \pi/2]$. Any point in the three-dimensional world can be uniquely described using either conventional Cartesian coordinates (x, y, z) or the spherical coordinates (d, φ, θ). The one-to-one mapping between these coordinate systems can be defined as

$$\begin{bmatrix} x \\ y \\ z \end{bmatrix} = d \begin{bmatrix} \cos(\varphi)\cos(\theta) \\ \sin(\varphi)\cos(\theta) \\ \sin(\theta) \end{bmatrix}. \tag{1.22}$$

This relation makes it easy to compute the Cartesian coordinates (x, y, z) of a point that is specified in spherical coordinates. The opposite transformation involves inverse trigonometric functions, and we must be careful when computing the azimuth angle so it is not shifted incorrectly by $\pm\pi$.

Example 1.8. How can the point with the Cartesian coordinates $(x, y, z) = (3, 4, 5)$ be expressed using the spherical coordinates (d, φ, θ)?

Using the relations in (1.22), we first obtain that

$$x^2 + y^2 + z^2 = d^2\Big(\underbrace{\cos^2(\varphi)\cos^2(\theta) + \sin^2(\varphi)\cos^2(\theta)}_{\cos^2(\theta)} + \sin^2(\theta)\Big) = d^2. \tag{1.23}$$

Hence, we have $d = \sqrt{x^2 + y^2 + z^2} = \sqrt{3^2 + 4^2 + 5^2} = 5\sqrt{2}$. By using (1.22), we can further notice that

$$\frac{y}{x} = \frac{d\sin(\varphi)\cos(\theta)}{d\cos(\varphi)\cos(\theta)} = \tan(\varphi). \tag{1.24}$$

We know that $\varphi \in [-\pi/2, \pi/2]$ since x is positive; thus, we obtain $\varphi = \arctan(4/3)$ radians when solving for φ. Lastly, we note that

$$\frac{z}{\sqrt{x^2 + y^2}} = \frac{d\sin(\theta)}{d\sqrt{\cos^2(\varphi)\cos^2(\theta) + \sin^2(\varphi)\cos^2(\theta)}} = \tan(\theta), \tag{1.25}$$

where we have utilized that $\cos(\theta) \geq 0$ for $\theta \in [-\pi/2, \pi/2]$. By solving for θ, we obtain $\theta = \arctan(5/5) = \arctan(1) = \pi/4$ radians. In summary, the spherical coordinates of the given point are $(d, \varphi, \theta) = (5\sqrt{2}, \arctan(4/3), \pi/4)$.

When transmitting with power P, the signal intensity at the point (d, φ, θ) is determined by the general power flux density function $U(P, d, \varphi, \theta)$ measured in $\mathrm{W/m^2}$. We will only consider the far-field (i.e., d larger than the Fraunhofer distance) because then the angular distribution over the sphere is approximately constant when we change the radius. This is not the case in the near-field for various electromagnetic reasons. In the far-field, we can

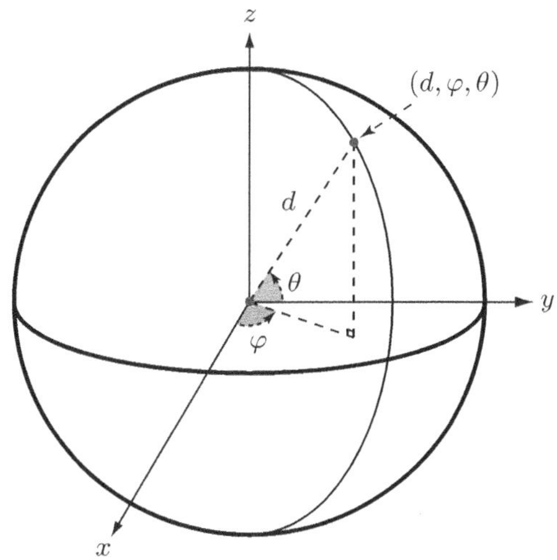

Figure 1.9: The directivity gain of an antenna is described using spherical coordinates. A location on the surface area is determined by the distance d, the azimuth angle $\varphi \in [-\pi, \pi)$, and the elevation angle $\theta \in [-\pi/2, \pi/2]$.

decompose the power flux density function as

$$U(P, d, \varphi, \theta) = \underbrace{\frac{P}{4\pi d^2}}_{\text{Average power density}} \cdot \underbrace{G(\varphi, \theta)}_{\text{Antenna gain}}, \tag{1.26}$$

where the first term is the average power flux density at the given distance d in $\mathrm{W/m^2}$ (i.e., the transmit power divided by the surface area) and $G(\varphi, \theta)$ is the *antenna gain* function. The antenna gain function describes how the radiated power is distributed over azimuth angles $\varphi \in [-\pi, \pi)$ and elevation angles $\theta \in [-\pi/2, \pi/2]$. A lossless isotropic antenna is represented by $G(\varphi, \theta) = 1$ for all angles, often reported using the decibel scale as 0 dBi, where dBi stands for decibels-isotropic (i.e., the gain relative to an isotropic antenna).

Any practical antenna has a varying antenna gain function larger than 0 dBi for some angles and smaller for others. However, the average antenna gain is identical to an isotropic antenna. This implies that all antenna gain functions for lossless antennas must satisfy the condition[9]

$$\frac{1}{4\pi} \int_{-\pi}^{\pi} \int_{-\pi/2}^{\pi/2} G(\varphi, \theta) \cos(\theta) \partial\theta \partial\varphi = 1, \tag{1.27}$$

where 4π is the surface area of the unit sphere and $\cos(\theta)\partial\theta\partial\varphi$ is the area of a surface element in the direction (φ, θ) that appears when integrating over a sphere using spherical coordinates. The cosine-term represents the fact that there is less area near the north/south poles than along the equator.

[9]Power losses appear in practical antennas, in which case the left-hand side of (1.27) becomes smaller than one.

Example 1.9. To examine how the formula (1.27) is derived, we consider an isotropic lossless antenna in the origin that transmits with power P. What is the total power reaching the surface of a sphere with radius d?

The power flux density function is $U(P, d, \varphi, \theta) = \frac{P}{4\pi d^2}$ with an isotropic lossless transmit antenna. By integrating over the surface area of a sphere with radius d, we obtain the total power as

$$P_{\text{tot}}(d) = \iiint_{\sqrt{x^2+y^2+z^2}=d} \frac{P}{4\pi(x^2+y^2+z^2)} \partial x \partial y \partial z. \tag{1.28}$$

It is convenient first to transform the Cartesian coordinates into spherical coordinates to evaluate the integral. The integral in (1.28) then becomes

$$P_{\text{tot}}(d) = \frac{P}{4\pi} \int_{-\pi}^{\pi} \int_{-\pi/2}^{\pi/2} \frac{|J(d, \varphi, \theta)|}{d^2} \partial\theta \partial\varphi, \tag{1.29}$$

where there is no integral with respect to the distance since all points on the sphere have the same distance d. The Jacobian determinant $J(d, \varphi, \theta)$ appears due to the change of variables and is computed based on (1.22) as

$$\begin{aligned}
J(d, \varphi, \theta) &= \det\left(\begin{bmatrix} \frac{\partial d\cos(\varphi)\cos(\theta)}{\partial d} & \frac{\partial d\cos(\varphi)\cos(\theta)}{\partial \varphi} & \frac{\partial d\cos(\varphi)\cos(\theta)}{\partial \theta} \\ \frac{\partial d\sin(\varphi)\cos(\theta)}{\partial d} & \frac{\partial d\sin(\varphi)\cos(\theta)}{\partial \varphi} & \frac{\partial d\sin(\varphi)\cos(\theta)}{\partial \theta} \\ \frac{\partial d\sin(\theta)}{\partial d} & \frac{\partial d\sin(\theta)}{\partial \varphi} & \frac{\partial d\sin(\theta)}{\partial \theta} \end{bmatrix}\right) \\
&= \det\left(\begin{bmatrix} \cos(\varphi)\cos(\theta) & -d\sin(\varphi)\cos(\theta) & -d\cos(\varphi)\sin(\theta) \\ \sin(\varphi)\cos(\theta) & d\cos(\varphi)\cos(\theta) & -d\sin(\varphi)\sin(\theta) \\ \sin(\theta) & 0 & d\cos(\theta) \end{bmatrix}\right) \\
&= d^2\Big(\underbrace{\cos^2(\varphi)\cos^3(\theta) + \sin^2(\varphi)\cos^3(\theta)}_{\cos^3(\theta)} \\
&\quad + \underbrace{\sin^2(\varphi)\cos(\theta)\sin^2(\theta) + \cos^2(\varphi)\cos(\theta)\sin^2(\theta)}_{\cos(\theta)\sin^2(\theta)}\Big) \\
&= d^2\cos(\theta).
\end{aligned} \tag{1.30}$$

After inserting $J(d, \varphi, \theta)$ into the integral in (1.29), we obtain

$$P_{\text{tot}}(d) = \frac{P}{4\pi} \int_{-\pi}^{\pi} \int_{-\pi/2}^{\pi/2} \cos(\theta) \partial\theta \partial\varphi = P. \tag{1.31}$$

This is equivalent to (1.27) for $G(\varphi, \theta) = 1$, which is the gain of a lossless isotropic antenna. If we consider an arbitrary lossless antenna, its gain function $G(\varphi, \theta)$ also appears inside the integral, and we thereby obtain the general condition in (1.27) for preserving the total transmit power.

The antenna gain function $G(\varphi, \theta)$ provides a complete description of the angular variations in antenna gain. However, if the antenna is rotated perfectly towards the receiver, it is sufficient to know the maximum gain

$$G_{\text{max}} = \max_{\varphi,\theta} G(\varphi, \theta). \tag{1.32}$$

This value is typically used when categorizing and comparing practical antennas. It is particularly common to represent the maximum gain in decibel scale as

$$10 \log_{10} (G_{\text{max}}) = \max_{\varphi,\theta} 10 \log_{10} (G(\varphi, \theta)) \quad [\text{dBi}]. \tag{1.33}$$

A simple example of a non-isotropic antenna gain function is

$$G(\varphi, \theta) = \begin{cases} 4\cos(\varphi)\cos(\theta), & \text{if } \varphi \in [-\pi/2, \pi/2], \theta \in [-\pi/2, \pi/2], \\ 0, & \text{elsewhere.} \end{cases} \tag{1.34}$$

This antenna concentrates the radiated power in the direction $\varphi = \theta = 0$ where the maximum antenna gain is $G_{\text{max}} = 4$, which is usually reported as $10 \log_{10}(4) \approx 6\,\text{dBi}$. When varying the azimuth angle, the gain reduces as $\cos(\varphi)$ and reaches zero at $\varphi = \pm\pi/2$. The gain value is zero for $\varphi \in [-\pi, -\pi/2]$ and $\varphi \in [\pi/2, \pi]$, which effectively means that the antenna only radiates into one half-space. The gain variations are similar in the elevation domain. In practice, this behavior can be achieved by a microstrip patch antenna, consisting of a metal patch printed on a substrate that acts as a reflecting ground plane. The maximum gain is then obtained perpendicularly to the patch while there is (ideally) no signal radiated at the backside. Patch antennas are extensively used in both mobile phones and base stations, thanks to their compact size and weight. Exact antenna gain models can be found in textbooks on antenna theory [8, Ch. 14], but (1.34) serves as a basic abstraction that we call the *cosine antenna*.

Example 1.10. Verify that the cosine antenna satisfies the lossless antenna condition in (1.27).

Direct computation based on the antenna gain expression in (1.34) yields

$$\begin{aligned} \frac{1}{4\pi} \int_{-\pi}^{\pi} \int_{-\pi/2}^{\pi/2} G(\varphi, \theta) \cos(\theta) \partial\theta \partial\varphi &= \frac{1}{4\pi} \int_{-\pi/2}^{\pi/2} \int_{-\pi/2}^{\pi/2} 4\cos(\varphi)\cos^2(\theta) \partial\theta \partial\varphi \\ = \frac{1}{\pi} \underbrace{\int_{-\pi/2}^{\pi/2} \cos(\varphi) \partial\varphi}_{=2} \underbrace{\int_{-\pi/2}^{\pi/2} \cos^2(\theta) \partial\theta}_{=\pi/2} &= 1. \end{aligned} \tag{1.35}$$

The antenna gain function of the cosine antenna is illustrated in Figure 1.10, where its values are plotted over the surface of a unit sphere. The pattern illustrates how the radiated power is distributed over different angular

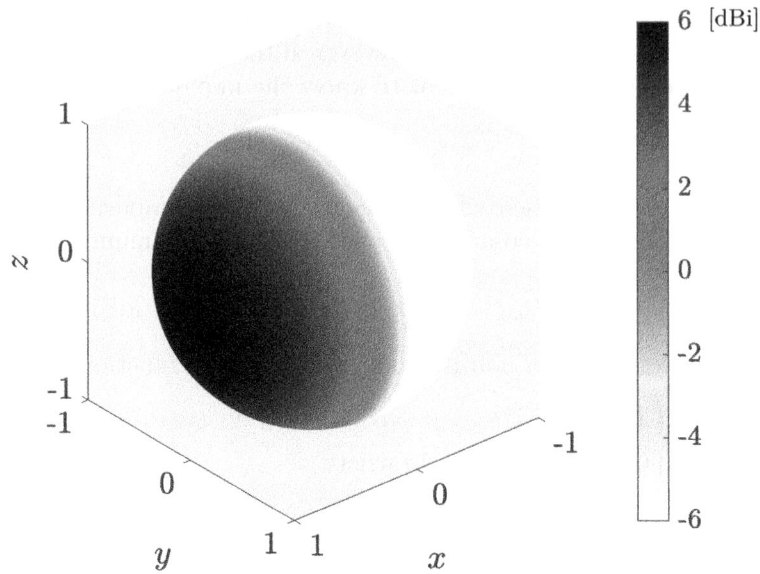

Figure 1.10: The antenna gain of the cosine antenna in (1.34) is plotted over the unit sphere. The pattern shows how the radiated power is distributed unequally over the angular directions, with the maximum appearing at $(x, y, z) = (1, 0, 0)$. The color shows the antenna gain in the decibel scale compared to an isotropic antenna.

directions. The power is concentrated over half of the sphere and maximized at its center. The maximum value is 6 dBi, while the average value is 0 dBi, as is the case for all lossless antennas.

Figure 1.11 compares the antenna gain functions of a cosine antenna and an isotropic antenna for $\theta = 0$ and different values of the azimuth angle φ. The isotropic antenna has a constant gain value of 0 dBi, while the gain of the cosine antenna ranges from 6 dBi to zero ($-\infty$ dBi). The total transmit power is the same for both types of antennas, but the cosine antenna concentrates the radiated power in specific directions. This means that a receiver located in that direction will receive a stronger signal than when using an isotropic antenna. Receivers in other directions will receive less power, and those at the backside of the antenna receive nothing. Hence, depending on the receiver's location, the antenna gain variations can be either a benefit or a drawback. Receivers located in directions where the curved solid curve in Figure 1.11 is above the dashed line will experience signal amplification compared to an isotropic transmit antenna.

Antennas are reciprocal by nature, which means that the same antenna gain is achieved when transmitting to a receiver in the direction (φ, θ) and when receiving a signal from that direction. Recall that the antenna gain describes how much stronger/weaker the signal power is compared to the reference case with an isotropic antenna. We stated in (1.3) that the effective

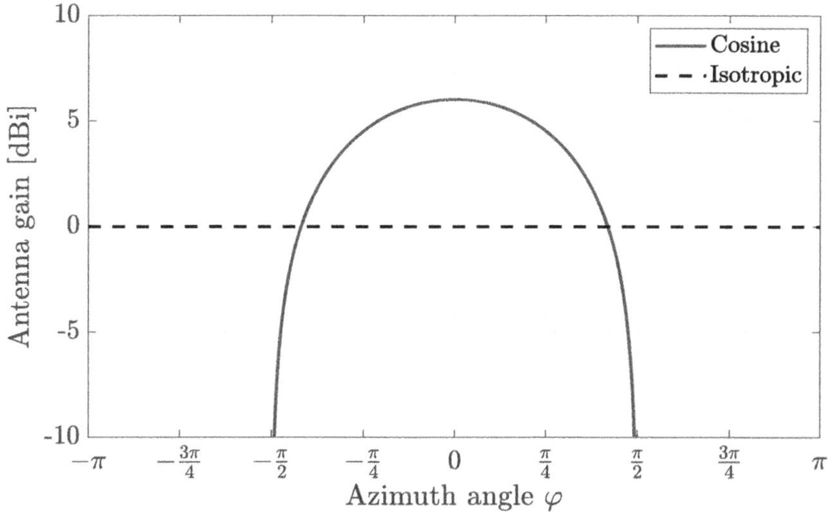

Figure 1.11: The antenna gains observed in different azimuth angles $\varphi \in [-\pi, \pi)$ for the elevation angle $\theta = 0$, when using the cosine antenna from (1.34) or an isotropic antenna.

area of a receiving isotropic antenna is $\lambda^2/(4\pi)$. Hence, if the antenna gain function is $G(\varphi, \theta)$ for another type of antenna, its effective area will be

$$A(\varphi, \theta) = \frac{\lambda^2}{4\pi} G(\varphi, \theta) \tag{1.36}$$

when receiving a signal from direction (φ, θ).

To emphasize the relation between the antenna gain and effective area, we return to Figure 1.3, which considered a receive antenna with the physical area A that receives a signal from the azimuth angle φ. We previously concluded that its effective area is $A\cos(\varphi)$ for $\varphi \in [-\pi/2, \pi/2]$, but we implicitly assumed the elevation angle was zero. When considering both angles, the effective area becomes $A(\varphi, \theta) = A\cos(\varphi)\cos(\theta)$ by the same arguments. If we further assume (for the sake of argument) that the physical antenna area is $A = 4A_{\text{iso}} = \frac{\lambda^2}{\pi}$, then the relation in (1.36) between the effective area and antenna gain becomes

$$\frac{\lambda^2}{\pi}\cos(\varphi)\cos(\theta) = \frac{\lambda^2}{4\pi} G(\varphi, \theta) \quad \Rightarrow \quad G(\varphi, \theta) = 4\cos(\varphi)\cos(\theta) \tag{1.37}$$

for $\varphi \in [-\pi/2, \pi/2]$ and $\theta \in [-\pi/2, \pi/2]$. This result coincides with the cosine antenna in (1.34). Hence, we have found a way to tie the concepts together: A patch antenna with a physical area that is 4 times larger than A_{iso} has a 4 times higher maximum gain. The gain function varies according to a cosine pattern since the patch looks smaller from non-perpendicular viewing angles.

Example 1.11. There are many other cosine-type radiation patterns in the field of antenna design than the one defined in (1.34). As an example, consider the gain function

$$G(\varphi,\theta) = \begin{cases} c\cos(3\varphi)\cos(\theta), & \text{if } \varphi \in [-\pi/6, \pi/6], \theta \in [-\pi/2, \pi/2], \\ 0, & \text{elsewhere.} \end{cases} \tag{1.38}$$

If this antenna is known to be lossless, what should be the value of the scalar $c > 0$? What is the maximum antenna gain?

The left-hand side of the lossless antenna condition in (1.27) becomes

$$\frac{1}{4\pi}\int_{-\pi}^{\pi}\int_{-\pi/2}^{\pi/2} G(\varphi,\theta)\cos(\theta)\partial\theta\partial\varphi = \frac{c}{4\pi}\int_{-\pi/6}^{\pi/6}\int_{-\pi/2}^{\pi/2}\cos(3\varphi)\cos^2(\theta)\partial\theta\partial\varphi$$
$$= \frac{c}{4\pi}\underbrace{\int_{-\pi/6}^{\pi/6}\cos(3\varphi)\partial\varphi}_{=2/3}\underbrace{\int_{-\pi/2}^{\pi/2}\cos^2(\theta)\partial\theta}_{=\pi/2} = \frac{c}{12}. \tag{1.39}$$

We notice that this value only becomes 1 if $c = 12$. The maximum antenna gain is achieved in the direction $\varphi = \theta = 0$ and is $G_{\text{max}} = c = 12$.

1.1.4 Revisiting the Signal-to-Noise Ratio

We will now revisit the SNR calculation and consider arbitrary antenna gains. The SNR was defined in (1.13) as $\text{SNR} = \frac{P\beta}{N_0 B}$ and depends on the channel gain β. The channel gain in free-space propagation with isotropic transmit and receive antennas was computed in (1.7) as $\frac{\lambda^2}{(4\pi d)^2}$, where d is the distance. We can generalize this expression for arbitrary antennas as [9]

$$\beta = \frac{\lambda^2}{(4\pi d)^2}G_{\text{t}}(\varphi_{\text{t}},\theta_{\text{t}})G_{\text{r}}(\varphi_{\text{r}},\theta_{\text{r}}), \tag{1.40}$$

where $G_{\text{t}}(\varphi,\theta)$ is the antenna gain function of the transmitter and $G_{\text{r}}(\varphi,\theta)$ is the antenna gain function of the receiver. These functions are defined for an arbitrary azimuth angle φ and elevation angle θ, but the functions are evaluated in (1.40) for the angles $(\varphi_{\text{t}},\theta_{\text{t}})$ at the transmitter that lead to the receiver and the angles $(\varphi_{\text{r}},\theta_{\text{r}})$ at the receiver that lead to the transmitter. Figure 1.12 illustrates this setup and, particularly, makes the point that the transmitter and receiver measure the angles based on their local coordinate systems. The antenna gain functions can then have their peak values at $\varphi = \theta = 0$, irrespective of how the transmitter and receiver are rotated with respect to each other.

It might seem strange to call (1.40) the *channel gain* when it also contains the antenna gains at the transmitter and receiver. However, this is unavoidable

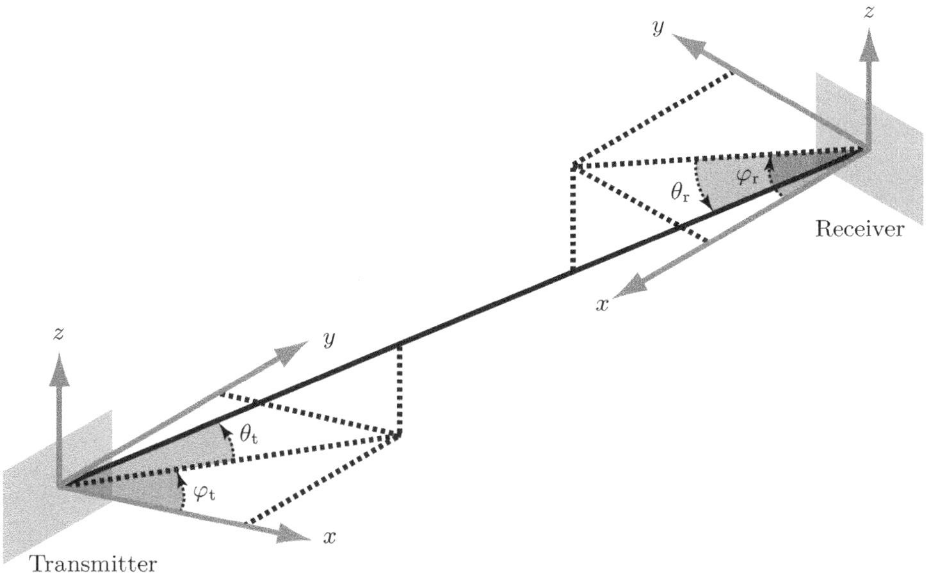

Figure 1.12: The transmitter sees the receiver in the angular direction $(\varphi_{\mathrm{t}}, \theta_{\mathrm{t}})$, measured using the transmitter's local coordinate system. The receiver sees the transmitter in the angular direction $(\varphi_{\mathrm{r}}, \theta_{\mathrm{r}})$, measured using the receiver's local coordinate system. These angles can be used when evaluating the antenna gains in (1.40).

since the effective area of the receiver always determines the fraction of the transmit power that is received, even when the transmitter is isotropic. One must always make assumptions regarding the antenna gains to compute a channel gain. Hence, the channel starts at the input to the transmit antenna and ends at the output from the receive antenna.

The channel gain in (1.40) is an increasing function of the antenna gains $G_{\mathrm{t}}(\varphi_{\mathrm{t}}, \theta_{\mathrm{t}})$ and $G_{\mathrm{r}}(\varphi_{\mathrm{r}}, \theta_{\mathrm{r}})$, which gives the impression that it is preferable to have strongly directive antennas in wireless communications. This is a valid conclusion for fixed wireless links where the person that deploys the transmitter and receiver can rotate the antennas so that the maximum gains are achieved precisely at the angles $(\varphi_{\mathrm{t}}, \theta_{\mathrm{t}})$ and $(\varphi_{\mathrm{r}}, \theta_{\mathrm{r}})$. This is the case for links between a geostationary satellite and receivers on the ground (e.g., using parabolic dish antennas to receive television broadcasts) or for fixed wireless broadband links where the customer has a fixed receive antenna at the outside of its house pointing towards the nearest base station.

The situation is more complicated in mobile communications, as illustrated in Figure 1.13, where a rooftop-mounted base station serves Receiver 1 and Receiver 2. The receivers are mobile phones, and it is not reasonable to require the users to hold their phones in precisely the right directions all the time. Hence, nearly isotropic antennas are utilized in mobile devices to ensure that almost the same SNR is achieved irrespective of how the device is rotated. The transmitter in Figure 1.13 emits a signal with an antenna gain function that

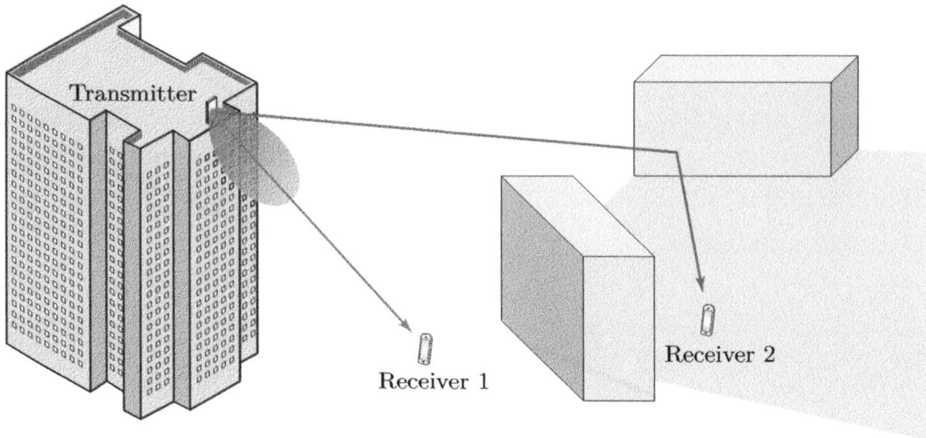

Figure 1.13: An example of a mobile communication scenario where the transmitting base station has a directive antenna. The maximum antenna gain is achieved in the direction leading to Receiver 1. The path leading to Receiver 2 experiences a weak antenna gain.

is illustrated to resemble that of a cosine antenna. Receiver 1 happens to be located in the direction with the maximum antenna gain. In contrast, Receiver 2 is located behind a building and can only be reached if the wireless signals are reflected off another building, as indicated in the figure. This receiver will experience a low antenna gain since the transmitter's gain function is low in the angular direction leading to the receiver. This example pinpoints the practical tradeoff between having a large maximum antenna gain and having a wide coverage area (wide enough to cover all prospective users) when selecting the antenna to be used at a base station.

Ideally, we would like to rotate the antenna gain function depending on the receiver's location, so we can always provide the maximum antenna gain. This could be achieved by mechanically rotating the base station antenna, but it is quite impractical since receivers can move rapidly. The preferred practical solution is to use multiple antennas to rotate the directivity of transmitted signals using the theory developed in later chapters of this book.

The free-space channel gain in (1.40) can also be expressed in terms of the effective areas $A_{\mathrm{t}}(\varphi_{\mathrm{t}}, \theta_{\mathrm{t}})$ and $A_{\mathrm{r}}(\varphi_{\mathrm{r}}, \theta_{\mathrm{r}})$. By using the relation stated in (1.36), an equivalent version of (1.40) is

$$\beta = \frac{A_{\mathrm{t}}(\varphi_{\mathrm{t}}, \theta_{\mathrm{t}}) A_{\mathrm{r}}(\varphi_{\mathrm{r}}, \theta_{\mathrm{r}})}{(d\lambda)^2}. \tag{1.41}$$

The impact of the antenna design and wavelength on the free-space channel gain can be understood by inspecting (1.40) and (1.41). If the antenna gains in (1.40) are constant as we reduce the wavelength λ (i.e., increase the carrier frequency), then the channel gain β will reduce proportionally to λ^2. This reduces the SNR because the effective receive antenna area is reduced, so the receiver captures less power. This is manifested by the relationship between area and gain in (1.36). On the other hand, if the effective areas in (1.41) are constant as we reduce the wavelength, then the channel gain β will instead increase proportionally to λ^{-2} when λ is reduced. This results in an SNR improvement because the antenna gains are increased; that is, the antennas become more directive. This is beneficial if the transmit and receive antennas are aligned to deliver the maximum antenna gains to the communication system. In other words, the high-band can provide better channel conditions in free-space propagation than the low-band, if we compare two systems with equal-sized antennas that are perfectly aligned. This is one of the features that fixed wireless links rely on (e.g., communication with geostationary satellites). Using the high-band spectrum for mobile communications, where the physical directions of the devices' antennas change over time, requires that the directivity can change accordingly to keep them directed toward the base station. We will explore how this is achieved using multiple antennas.

Example 1.12. How does the SNR in free-space propagation depend on the wavelength λ if the base station has a fixed wavelength-independent effective antenna area $A_{\mathrm{t}}(\varphi_{\mathrm{t}}, \theta_{\mathrm{t}})$ while the user device has an isotropic antenna?

The effective area of the isotropic receive antenna is $A_{\mathrm{r}}(\varphi_{\mathrm{r}}, \theta_{\mathrm{r}}) = \frac{\lambda^2}{4\pi}$. We can compute the SNR using (1.41) as

$$\mathrm{SNR} = \frac{P\beta}{N_0 B} = \frac{P}{N_0 B} \frac{A_{\mathrm{t}}(\varphi_{\mathrm{t}}, \theta_{\mathrm{t}}) A_{\mathrm{r}}(\varphi_{\mathrm{r}}, \theta_{\mathrm{r}})}{(d\lambda)^2} = \frac{P}{N_0 B} \frac{A_{\mathrm{t}}(\varphi_{\mathrm{t}}, \theta_{\mathrm{t}})}{4\pi d^2}. \tag{1.42}$$

This expression is independent of λ since the two wavelength-dependent effects are canceling out. The area of the receiver is proportional to λ^2, while the gain of the transmit antenna is obtained from (1.36) as $G_{\mathrm{t}}(\varphi, \theta) = \frac{4\pi}{\lambda^2} A_{\mathrm{t}}(\varphi, \theta)$, which is inversely proportional to λ^2 when the area is fixed. Hence, if the wavelength shrinks, the receiver becomes physically smaller but captures the same signal power since the transmit antenna becomes more directive. The same principle applies when the device transmits, but then the radiated signal is isotropic and induces a frequency-independent power flux density on the fixed-area receive antenna.

A general parametric channel gain model was defined in (1.9), as a function of the pathloss exponent α and the channel gain Υ at a 1 m reference distance. The parameter values are normally stated for isotropic antennas but can be used along with other antennas by multiplying with the antenna gains:

$$\beta = \Upsilon \left(\frac{1\,\mathrm{m}}{d} \right)^{\alpha} G_{\mathrm{t}}(\varphi_{\mathrm{t}}, \theta_{\mathrm{t}}) G_{\mathrm{r}}(\varphi_{\mathrm{r}}, \theta_{\mathrm{r}}). \tag{1.43}$$

1.2 Three Main Benefits of Having Multiple Antennas

This book will cover how using multiple antennas can improve the operation of wireless communication systems. We have already provided some hints of what the benefits could be in the context of mobile communications, where the location and rotation of the transmitter/receiver change with time. In this section, we will describe the three main categories of benefits that multiple antenna communication systems have over conventional systems with a single antenna at the transmitter and receiver. These benefits have been given several different names over the years. In this book, we call them:

1. Beamforming gain;

2. Spatial multiplexing;

3. Spatial diversity.

These benefits will be introduced below, including a short historical expose, and then covered in further detail in later chapters.

1.2.1 Beamforming Gain

The wireless telegraph was invented in the 1890s as the first system for wireless communications. The technology used *Morse code* to transfer words encoded as a sequence of "dots" and "dashes", represented by transmitting sinusoidal signal pulses of two different durations. The wireless telegraph played an essential role during the First World War since it allowed for direct communication between continents [15]. The distance from North America to Europe is more than 5000 km; thus, if the channel gain is computed as in (1.7), it would be much smaller than the values shown in Figure 1.4. To reach over the oceans, the radio stations had to broadcast their signals with very high transmit power (tens of kilowatts). Therefore, researchers started to look for ways to achieve directive transmission and reception to reduce the transmit power or to reach even further distances with the same power. This was where multiple antenna communications appeared as a solution (in addition to using directive antennas). Guglielmo Marconi made the first transatlantic transmission in 1901 using two tall antenna poles in the United Kingdom [16]. Karl Ferdinand Braun did an experiment using three antennas in 1905, which he described publicly when he and Marconi shared the Nobel Prize in Physics in 1909 [17]. Ernst F. W. Alexanderson filed a patent application in 1917 describing the first practical implementation of radio communications [18]. The patent did not use the term *beamforming* but outlined all the same benefits as will be described in this section. The implementation was analog then, while current systems are digitally controlled. Some early field trials for mobile communications in the 1990s are described in [19], [20].

To exemplify the basic phenomenon that was discovered and utilized in the early 1900s, we consider the transmission of a time-limited sinusoidal pulse

$$p(t) = \begin{cases} \sqrt{2}\sin(2\pi f t), & \text{if } t \in [0, T], \\ 0, & \text{otherwise}, \end{cases} \tag{1.44}$$

where f is the frequency and the time duration is $T = l/f$, for some integer $l > 0$. This means the pulse consists of l full periods of the sine wave. The power of this pulse is

$$\begin{aligned} \frac{1}{T}\int_0^T p^2(t)\partial t &= \frac{2}{T}\int_0^T \sin^2(2\pi f t)\partial t \\ &= \frac{2}{T}\left(\int_0^T \frac{1}{2}\partial t - \int_0^T \frac{\cos(4\pi f t)}{2}\partial t\right) = 1, \end{aligned} \tag{1.45}$$

where we utilize the trigonometric identity $\sin^2(x) = (1 - \cos(2x))/2$ and notice that the last integral is zero since we integrate over $2l$ periods.

The Morse code is transmitted using *on-off keying*, which means we switch between transmitting the sinusoidal pulse $Ap(t)$ with an amplitude $A > 0$ and being silent. If we transmit the pulse with amplitude A, then the transmitted signal power is computed as

$$\frac{1}{T}\int_0^T (Ap(t))^2\,\partial t = A^2\frac{1}{T}\int_0^T p^2(t)\partial t = A^2. \tag{1.46}$$

We notice that the signal power is proportional to the square of the pulse's amplitude. The received signal at some destination will be $\sqrt{\beta}Ap(t)$, where the channel gain β represents the signal propagation loss and can, for example, be computed as described in (1.7) for free-space propagation with isotropic antennas or in (1.43) for arbitrary antennas and propagation modeling. In any case, the received signal power is βA^2, which is also proportional to A^2.

Suppose the received signal is too weak for the receiver to decode the Morse code accurately. If we want to increase the received signal power by 100 times (i.e., 20 dB), we can increase the signal amplitude by a factor of 10, from A to $10A$. The transmitted signal power will then instead be

$$\frac{1}{T}\int_0^T (10Ap(t))^2\,\partial t = (10A)^2\frac{1}{T}\int_0^T p^2(t)\partial t = 100A^2. \tag{1.47}$$

This means we need to spend 100 times more transmit power to receive the signal $\sqrt{\beta}10Ap(t)$ that contains 100 times more power.

An alternative solution is to generate the original signal $Ap(t)$ at 10 different transmit antennas. Each signal has a power of A^2 so this approach requires a total transmit power of

$$10\frac{1}{T}\int_0^T (Ap(t))^2\,\partial t = 10A^2. \tag{1.48}$$

We can then radiate these signals simultaneously from the multiple antennas and let them add/superimpose constructively over the air. In this way, the received signal will also be $\sqrt{\beta}10Ap(t)$, but we only need to spend 10 times more power, instead of 100 times more as in the single-antenna case. In other words, if the destination requires a specific received signal power level to decode the information successfully, we can satisfy that requirement using only 1/10 of the power when using 10 transmit antennas instead of one.

In general, if we compare single-antenna transmission with transmission from M antennas, we can reduce the total transmit power by a factor of $1/M$ while keeping the received signal power constant. How is this possible? It might seem that additional signal power is "magically" created when the M transmitted signals are combined in the air. The simple yet physically accurate explanation is that the transmission becomes spatially directed toward the receiver. In other words, when observed at a distant receiver, the combination of M transmitted signals looks like the signal emitted from a single "virtual" antenna with high directivity; that is, a virtual antenna having an M times higher antenna gain than the individual physical antennas had.

Figure 1.14 shows an array with $M = 4$ isotropic antennas deployed on a line. The adjacent antennas are separated by half-a-wavelength: $\lambda/2 = c/(2f)$. If all the antennas transmit the signal $Ap(t)$ simultaneously, then each of the emitted signal components will radiate as in the single-antenna case described earlier. A superposition of the M signal components can be observed at every point in space. The components have, generally, traveled different distances to reach the considered point and, thus, are time-delayed differently.

Let us consider points many wavelengths away from the array (i.e., in its far-field) so that the propagation distance is much larger than the distance between the individual antennas. For any such point on the horizontal axis in Figure 1.14, the distances to each antenna will be roughly the same. This can be understood by considering the triangle in Figure 1.15, which has corners at two different antennas and the considered receiver location along the horizontal axis. Hence, the M signal components will be approximately time-synchronized, and the received signal becomes $M\sqrt{\beta}Ap(t)$. This is the *constructive interference* behavior that we are looking for. However, for any point on the vertical axis in Figure 1.14, the distances to the antennas differ by integer multiples of $\lambda/2$. This distance difference remains even if the considered point of the receiver is far away. The corresponding time delay difference between two adjacent antennas is an integer multiple of $\tau = \frac{\lambda}{2c} = \frac{1}{2f}$, which corresponds to a half period of the sine wave:

$$\sin(2\pi f(t - \tau)) = \sin(2\pi ft - \pi) = -\sin(2\pi ft). \tag{1.49}$$

Hence, the signals emitted from two adjacent antennas cancel out along the vertical direction, called *destructive interference*. The horizontal and vertical axes represent the extreme cases, while partially constructive or destructive interference can be observed elsewhere, as indicated in Figure 1.14.

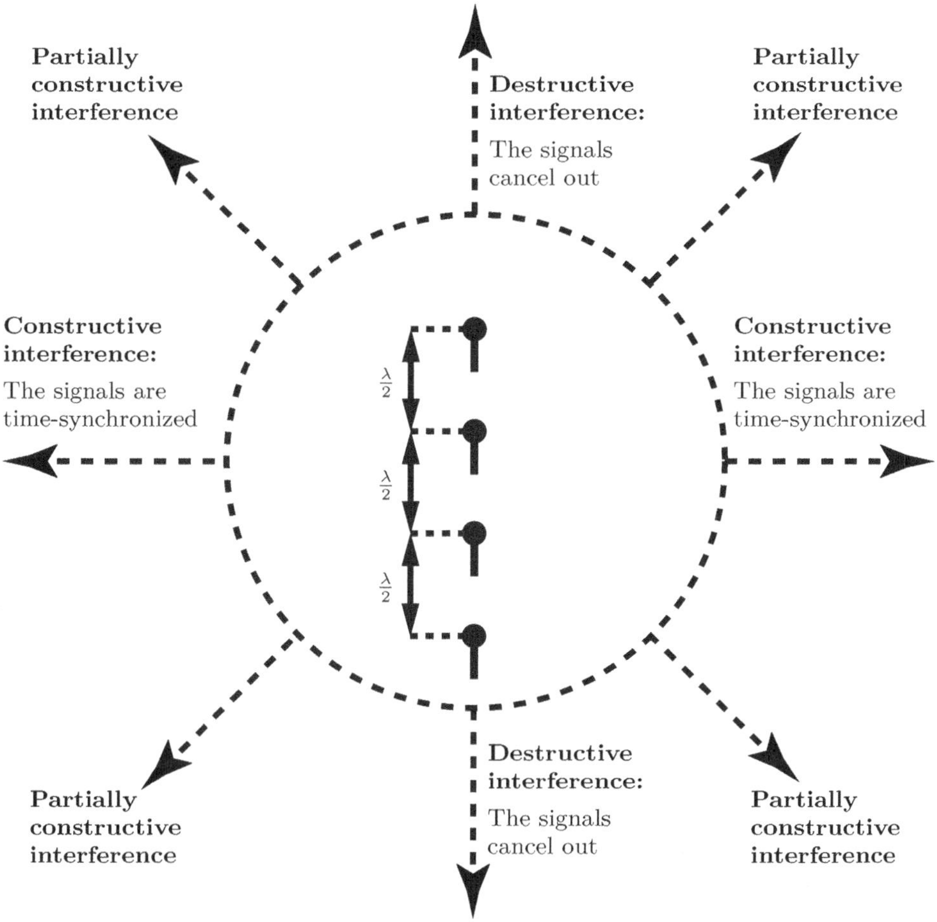

Figure 1.14: When transmitting the same signal from all the antennas in a one-dimensional array, the signal components will propagate time-synchronously in the direction perpendicular to the array, leading to constructive interference in the horizontal direction in this figure. On the other hand, the signals will propagate non-synchronously in other directions leading to partially constructive or fully destructive interference.

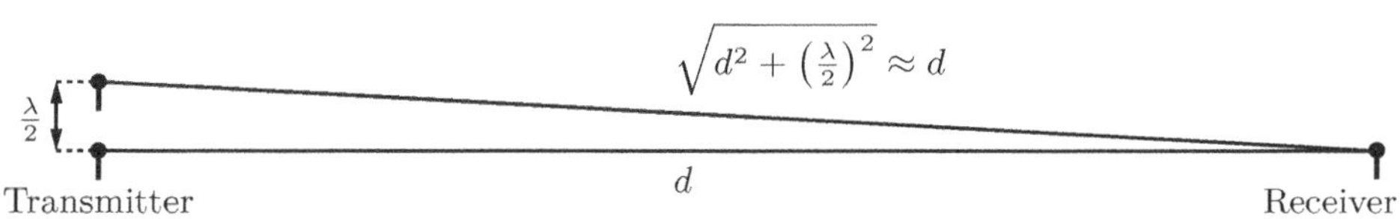

Figure 1.15: If the distance $\lambda/2$ between the two transmit antennas is much smaller than the propagation distances to the receive antenna, then we have approximately the same distance d from both transmit antennas. If the antennas transmit the same signal, constructive interference will occur at the receiver.

Example 1.13. An array with $M = 2$ isotropic antennas is located at the Cartesian coordinates $(0, +\lambda/4, 0)$ and $(0, -\lambda/4, 0)$, where λ is the wavelength. The sinusoidal pulse $p(t)$ in (1.44) is transmitted from both antennas with the amplitude $A/\sqrt{2}$, so the total transmit power is A^2. What is the received power at a point with spherical coordinates $(d, \varphi, 0)$, assuming that $d \gg \lambda$ and the channel gain β is the same from both transmit antennas to the receiver?

We let d_1 and d_2 denote the distances to the receiver from the antennas at $(0, +\lambda/4, 0)$ and $(0, -\lambda/4, 0)$, respectively. The received signal becomes

$$\sqrt{\beta} A \sin\left(\frac{2\pi c}{\lambda}\left(t - \frac{d_1}{c}\right)\right) + \sqrt{\beta} A \sin\left(\frac{2\pi c}{\lambda}\left(t - \frac{d_2}{c}\right)\right). \tag{1.50}$$

By using the trigonometric identity $\sin(a) + \sin(b) = 2\sin\left(\frac{a+b}{2}\right)\cos\left(\frac{a-b}{2}\right)$, (1.50) can be expressed as

$$2\sqrt{\beta} A \sin\left(\frac{2\pi c}{\lambda} t - \frac{\pi}{\lambda}(d_1 + d_2)\right) \cos\left(\frac{\pi}{\lambda}(d_2 - d_1)\right). \tag{1.51}$$

To determine its power, we need d_1 and d_2. The Cartesian coordinates of the receiver is $(d\cos(\varphi), d\sin(\varphi), 0)$. Since $d \gg \lambda$, d_1 can be approximated as

$$\begin{aligned} d_1 &= \sqrt{(d\cos(\varphi) - 0)^2 + \left(d\sin(\varphi) - \frac{\lambda}{4}\right)^2} = \sqrt{d^2 - \frac{d\lambda\sin(\varphi)}{2} + \frac{\lambda^2}{4^2}} \\ &= d\sqrt{1 - \frac{\lambda\sin(\varphi)}{2d} + \frac{\lambda^2}{16d^2}} \approx d - \frac{\lambda\sin(\varphi)}{4} + \frac{\lambda^2}{32d} \approx d - \frac{\lambda\sin(\varphi)}{4} \end{aligned} \tag{1.52}$$

by using that $\sqrt{1+x} \approx 1 + \frac{x}{2}$ for $0 \leq x \ll 1$. Similarly, d_2 can be approximated as $d_2 \approx d + \frac{\lambda\sin(\varphi)}{4}$. We can now approximate (1.51) as

$$\underbrace{\sqrt{2\beta} A \sin\left(\frac{2\pi c}{\lambda}\left(t - \frac{d}{c}\right)\right)}_{\substack{\text{Received signal at distance } d \\ \text{with a single antenna}}} \underbrace{\sqrt{2}\cos\left(\frac{\pi}{2}\sin(\varphi)\right)}_{\text{Angle-dependent multiplicative factor}}, \tag{1.53}$$

which is the product of the signal received with a single transmit antenna and an angle-dependent factor that describes the constructive/destructive interference. By integrating the square of (1.53) over one signal period and utilizing that $\int_0^1 2\sin^2(2\pi t)\partial t = 1$, we obtain the received signal power

$$P(\varphi) = 2\cos^2\left(\frac{\pi}{2}\sin(\varphi)\right)\beta A^2. \tag{1.54}$$

The largest power $2\beta A^2$ is achieved if $\varphi = 0$ or $\varphi = \pi$, as in Figure 1.14, which is twice the received power compared to a single antenna using the same total power. Destructive interference occurs when $\varphi = \pm\pi/2$ since $\cos(\pm\pi/2) = 0$, while half of the maximum power is received when $\varphi = \pm\pi/6$.

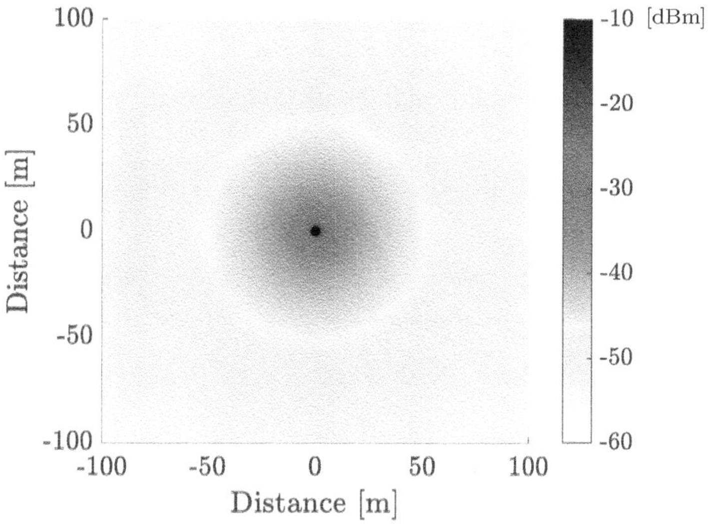

Figure 1.16: The received signal power in different directions and distances when transmitting 1 W from an isotropic antenna. The color shows the received signal power in dBm when using (1.7) to compute the channel gain in free-space propagation with $f = 3$ GHz as the frequency.

We will now illustrate the constructive, partially constructive, and destructive interference behavior when using multiple antennas and compare it to the single-antenna transmission case. Figure 1.16 shows how the signal power from a single isotropic transmit antenna spreads out over an area of 200×200 m. The transmitter is located in the origin and transmits a signal with 1 W of power. The color shows the received signal power in dBm, and we use the free-space model in (1.7) to compute the channel gain β at different distances. We notice that the signal power spreads out identically in all directions and decays with distance. If we rotate the figure around the origin, the pattern remains the same, as expected when using an isotropic transmit antenna.

In contrast, a transmitter with an array of $M = 10$ isotropic antennas is considered in Figure 1.17. The antennas are deployed along the vertical axis with $\lambda/2$ antenna spacing, as illustrated in Figure 1.14, and are centered around the origin. Exactly the same signal is simultaneously transmitted from all the antennas. Figure 1.17(a) considers the case when the total transmit power is 0.1 W (i.e., scaled down as $1/M$), which leads to 0.01 W per antenna. Figure 1.17(b) considers the case when the total transmit power is 1 W (i.e., the same as in the single-antenna case); thus, the power per antenna is 0.1 W. Although each antenna radiates its signal isotropically, the figure shows that the combined effect is a directive signal in the two horizontal directions. Hence, we create constructive and destructive interference patterns aligned with the previous discussion related to Figure 1.14.

The constructive interference pattern in Figure 1.17 takes the shape of a *beam* (also known as a *lobe*), and the antenna array is therefore said to

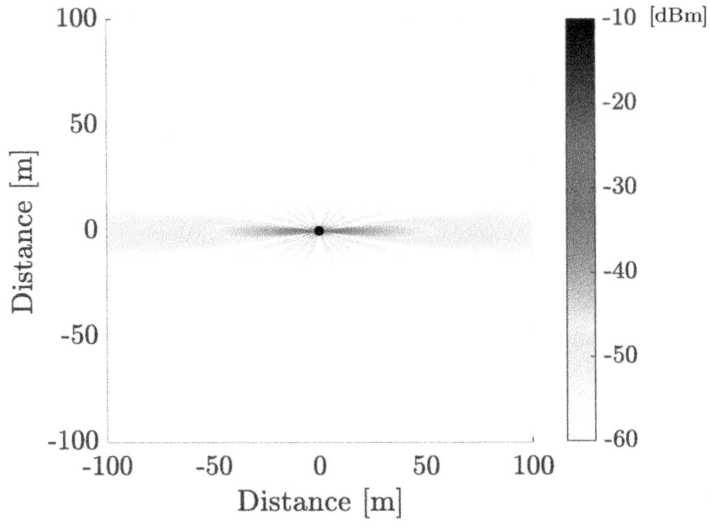

(a) $M = 10$ transmit antennas with a total transmit power of $0.1\,\text{W}$ ($0.01\,\text{W}$ per antenna).

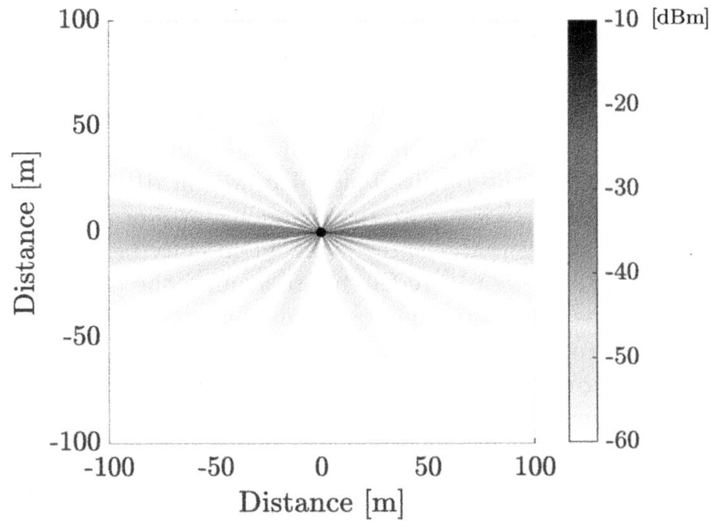

(b) $M = 10$ transmit antennas with a total transmit power of $1\,\text{W}$ ($0.1\,\text{W}$ per antenna).

Figure 1.17: The received signal power in different directions and at different distances, when transmitting the *same signal* from $M = 10$ isotropic antennas located in the origin. The color shows the received signal power in dBm when using (1.7) to compute the channel gain in free-space propagation with $f = 3\,\text{GHz}$ as the signal frequency.

perform *beamforming*. There is a strong main beam along the horizontal axis, but also several *side-beams* pointing in other directions, usually referred to as *side-lobes*. By comparing Figure 1.16 and Figure 1.17(a), we can notice that a receiver located in the direction of the beam (i.e., along the horizontal axis) will receive the same power in both cases. However, the transmit power has been reduced with a factor $1/M$ in Figure 1.17(a) so we can deliver the same wireless communication service but save power. This is called an M-times *beamforming gain* or *array gain*. Receivers in other directions will receive less power when using multiple antennas because there is no magical appearance of signal power but only a power redistribution from some angular directions to other directions. In particular, no signal power is observed along the vertical axis. Hence, beamforming can be both a blessing and a curse—it is helpful if the main beam points in the direction preferred by the receiver and can be detrimental otherwise. This issue resembles that of using directive antennas (described earlier), but there is a crucial difference: an individual antenna has a fixed antenna gain function, while the direction of the beam from an antenna array can be controlled when using beamforming. The ability to change the direction is often seen as an inherent part of the beamforming concept but it has also been called *adaptive beamforming*. Various methods to point beams toward the desired receivers are developed later in this book.

In Figure 1.17(b), the total transmit power is the same as in the single-antenna case. The received signal power for a user located along the horizontal axis is then M times stronger than in the single-antenna case. Hence, the beamforming gain provides a stronger received signal for users that the beam is pointed toward. There are many directions outside the main beam where less power is received than in the single-antenna case.

The fact that beamforming distributes the transmit power unequally between different angular directions is illustrated in Figure 1.18, where a sphere is centered around the array. The color illustrates the received power level at different points on the sphere relative to the maximum value. The x-axis corresponds to the horizontal axis in the previous figures, the y-axis corresponds to the vertical axis, while the z-axis was not visible before. As M increases, the black stripe where the received signal power is high will contain a larger and larger fraction of the transmit power but also become narrower. If one would integrate over the sphere to sum up all the power, it would always be equal to the total transmit power irrespective of the value of M.

In summary, the beamforming gain can be utilized to achieve an M times higher SNR than in the single-antenna case using the same transmit power, or it can be used to achieve the same SNR using M times less transmit power. Although the example above considers transmission from M antennas to a single-antenna receiver, the same gains can be achieved when transmitting with one antenna to M receive antennas. We will study this in detail later in this book. The beamforming distributes the transmit power unequally over

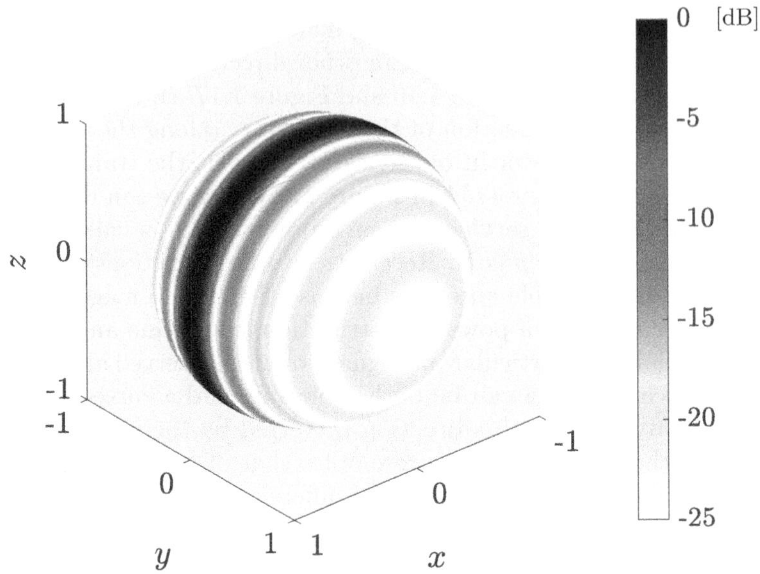

Figure 1.18: The normalized received power on different parts of a sphere centered around an array with $M = 10$ isotropic antennas, in the same setup as in Figure 1.17. The color shows the normalized received power in dB-scale where the maximum value is 0 dB. All distances are normalized.

different angular directions, similar to a single directive antenna (compare Figure 1.10 and Figure 1.18) but with the vital difference that the directivity of an antenna array can be changed, as described next.

1.2.2 Spatial Multiplexing

Many wireless systems have more than one user and must multiplex their communication services on the shared wireless channel. Traditionally, the users are multiplexed by assigning non-overlapping time-frequency resources; for example, different time intervals and/or frequency bands. The reason for this system design is to avoid interference. If two signals are radiated with equal power from an isotropic antenna at the same time and frequency, each signal will propagate isotropically as illustrated in Figure 1.16. At every point in space, a superposition of the two signals will be observed where the signals remain equally strong. Each receiver is only interested in one of the two signals. When measuring the corresponding communication quality, the ratio between the desired signal's power and the summation of the interfering signal's power plus the noise power is a common performance metric. This is known as the *signal-to-interference-plus-noise ratio (SINR)* and is a generalization of the SNR metric to situations with interference:

$$\mathrm{SINR} = \frac{\text{Received signal power}}{\text{Received interference power} + \text{Noise power}}. \tag{1.55}$$

The SINR is always smaller than or equal to the SNR because we obtain the SNR by removing the interference from the denominator in (1.55). The interference is problematic when the SINR is much smaller than the SNR and might severely limit communication performance. For example, when the signal and interference powers in (1.55) are equally large, the SINR cannot surpass 1. In contrast, the SNR values exemplified in Figure 1.6 can be many orders-of-magnitude larger than one (e.g., 30 dB is 1000). This issue cannot be addressed using a directive transmit antenna since both signals will be radiated with the same directivity. The following example proves this mathematically.

Example 1.14. Consider an isotropic antenna that transmits to two receivers with the same channel gain $\beta \in (0, 1]$. It assigns power $P_1 \geq 0$ to receiver 1 and power $P_2 \geq 0$ to receiver 2. Suppose an SINR of 1 (i.e., 0 dB) is needed for reliable communication. Is it possible to select the powers P_1 and P_2 so that the transmitter can communicate to both receivers reliably?

If we let $\sigma^2 > 0$ denote the noise power, then we can use (1.55) to obtain the SINR achieved by the first receiver:

$$\text{SINR}_1 = \frac{P_1 \beta}{P_2 \beta + \sigma^2} = \frac{P_1}{P_2 + \frac{\sigma^2}{\beta}}. \tag{1.56}$$

Similarly, the SINR achieved by the second receiver is

$$\text{SINR}_2 = \frac{P_2 \beta}{P_1 \beta + \sigma^2} = \frac{P_2}{P_1 + \frac{\sigma^2}{\beta}}. \tag{1.57}$$

For jointly reliable communication to the two receivers, both SINR_1 and SINR_2 must be greater than or equal to 1, which is equivalent to the conditions

$$P_1 \geq P_2 + \frac{\sigma^2}{\beta}, \tag{1.58}$$

$$P_2 \geq P_1 + \frac{\sigma^2}{\beta}. \tag{1.59}$$

Since both inequalities require one power to be strictly larger than the other one, they cannot be satisfied simultaneously. This happens even if the channel gain is large, so σ^2/β is small. Only in the hypothetical noise-free case of $\sigma^2/\beta = 0$ can reliable communication be guaranteed for both receivers. Even in that case, the common SINR cannot surpass 1. This is why single-antenna systems avoid interference by letting the users take turns communicating.

Using multiple antennas fundamentally changes the situation since each radiated signal can have a unique spatial directivity. Recall that Figure 1.17 illustrated a situation where a signal is focused along the horizontal axis, so the signal vanishes entirely along the vertical axis. Hence, a device that is located in that direction will not observe any interference at all. With this phenomenon in mind, the concept of *spatial multiplexing*, also known as *space-division multiple access (SDMA)*, was conceived in the late 1980s and early 1990s [21]–[24]. The key idea was to equip the base stations in cellular networks with multiple antennas and exploit beamforming to suppress interference between the users, thereby enabling efficient communications where multiple users are using the same time and frequency resource. The SDMA concept had been considered for satellite systems decades earlier [25].

Suppose $p(t)$ is the signal transmitted to the receiver. When all the transmit antennas emit this signal simultaneously, a particular pattern of constructive and destructive interference is created, as exemplified in Figure 1.14. Other patterns can be generated by emitting different signals from the antennas; in particular, we can transmit a time-shifted copy of $p(t)$, where we adapt the time-shift to obtain constructive interference in any direction or at any point of choice. The methodology of adaptive beamforming is to:

1. Measure the propagation time delays $\tau_1, \ldots, \tau_M$ from each of the M transmit antennas to the intended receiver.

2. Compensate for the time delays by transmitting the signal $p(t)$ earlier from the more distant antennas in the array: $x_m(t) = p(t + \tau_m)$ is the signal transmitted from the mth antenna.

3. All the signal components arrive at exactly the same time at the intended receiver since the received signal is an attenuated version of

$$\begin{aligned} x_1(t-\tau_1) + \ldots + x_M(t-\tau_M) &= p(t+\tau_1-\tau_1) + \ldots + p(t+\tau_M-\tau_M) \\ &= Mp(t). \end{aligned} \tag{1.60}$$

Suppose we want to direct the signal towards a user located on the vertical axis in Figure 1.14 instead of the horizontal axis. Since the antennas are separated by a distance $\lambda/2$, the geometry implies that each transmitted signal becomes time-shifted by half a period compared to the signal from the adjacent antenna. Hence, if we emit a signal already shifted by half a period, the two effects cancel out at every point on the vertical axis. The result is shown in Figure 1.19, where the main beams point along the vertical axis, while the signal components cancel out along the horizontal axis. Apart from the angular rotation of the beamforming, the general behavior is the same as before: the beamforming gain from the M antennas can be either utilized to achieve the same SNR as in the single-antenna case using M times less total transmit power (as in Figure 1.19(a)) or achieve M times higher SNR (as in Figure 1.19(b)) using the same total power.

The beamforming gain is once again achieved by redistributing the transmit power between different angular directions. Figure 1.20 illustrates the received power level at different points on a sphere centered around the array. The

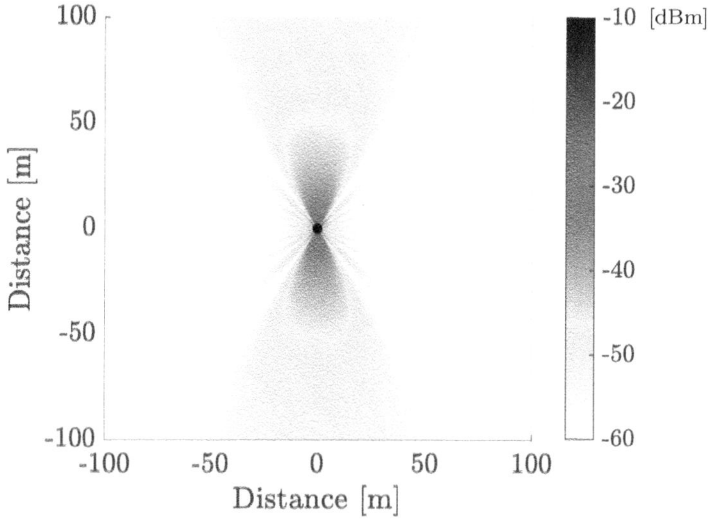

(a) $M = 10$ transmit antennas with a total transmit power of 0.1 W (0.01 W per antenna).

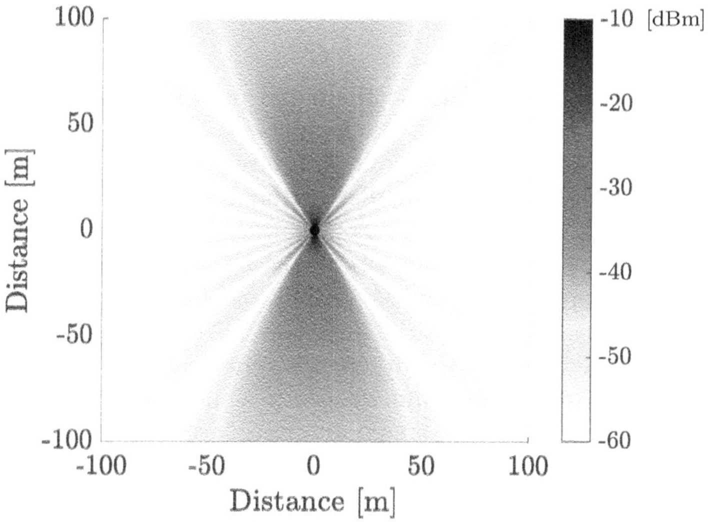

(b) $M = 10$ transmit antennas with a total transmit power of 1 W (0.1 W per antenna).

Figure 1.19: The received signal power in different directions and at different distances, when transmitting *time-shifted signals* from $M = 10$ isotropic antennas located in the origin. The time shifts are selected to achieve constructive interference along the vertical axis. The color shows the received signal power in dBm when using (1.7) to compute the channel gain in free-space propagation, and $f = 3$ GHz is the signal frequency.

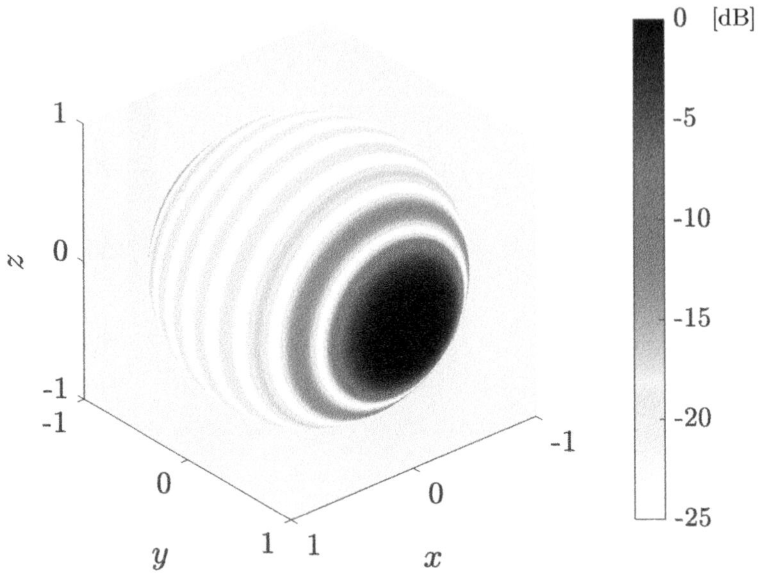

Figure 1.20: The normalized received power on different parts of a sphere centered around an array with $M = 10$ isotropic antennas, in the same setup as in Figure 1.19. The color shows the normalized received power in dB-scale where the maximum value is 0 dB.

power is focused in one direction (the same pattern appears at the back of the sphere that is not visible). As M increases, the black dot where the received signal power is high will contain a larger and larger fraction of the transmit power but also become smaller. Although the pattern on the sphere differs from Figure 1.18, we can always obtain the original transmit power by integrating over the sphere, to sum up all the radiated power.

How is this example related to spatial multiplexing? Suppose two users are located in sufficiently different spatial directions. There will be low interference if each user is located outside the other user's main beam. Hence, these users can be served at the same time and frequency while achieving a decent SINR (much higher than that in the single-antenna case). Ideally, the data rate becomes proportional to the number of users. If K users are served by spatial multiplexing, then K times more data can be transmitted compared to the single-user case if the beamforming deals with the interference. A basic setup of spatial multiplexing is illustrated in Figure 1.21.

The term "interference" has two different meanings in this context. The physical phenomenon of constructive/destructive interference determines how the signal copies emitted from multiple antennas superimpose over the air to form a directive beam. Moreover, when a signal reaches an unintended receiver, it is called interference for different reasons, and the interfering signal's power is included in the denominator of the SINR. In the remainder of this book, we will only use the term in the latter sense.

Example 1.15. Two isotropic transmit antennas are deployed with a $\lambda/2$ separation and transmit to two single-antenna receivers, located as in Figure 1.21. The one-bit data intended for receiver k is represented by $s_k \in \{-1, 1\}$, for $k = 1, 2$. It is multiplied by the sinusoidal pulse in (1.44) before transmission. Suppose both receivers need an SINR of 1 (i.e., 0 dB) to reliably decode their data and that $\sigma^2/\beta = 10^{-1}$ W. Is it possible to select the transmit powers P_1 and P_2 to enable reliable communication to both receivers simultaneously?

Since the distance to receiver 1 is identical for the two transmit antennas, we can focus a beam towards this receiver by transmitting $\sqrt{P_1/2}s_1p(t)$ from both antennas, where the power P_1 is divided equally. To focus a beam on receiver 2, the two antennas can transmit $\sqrt{P_2/2}s_2p(t)$ and $\sqrt{P_2/2}s_2p\left(t + \frac{\lambda}{2c}\right)$, where the delay is selected to compensate for the propagation delay difference of $\frac{\lambda}{2c}$. The received signal at receiver 1 then becomes

$$
\begin{aligned}
y_1(t) &= \underbrace{2\sqrt{\frac{P_1}{2}}\beta s_1 p(t-\tau_1)}_{\text{Desired signal}} + \underbrace{\sqrt{\frac{P_2}{2}}\beta s_2\left(p(t-\tau_1) + p\left(t + \frac{\lambda}{2c} - \tau_1\right)\right)}_{\text{Interference from the second signal}} + \underbrace{n_1(t)}_{\text{Noise}} \\
&= \sqrt{2P_1\beta}s_1 p(t-\tau_1) + n_1(t),
\end{aligned}
\tag{1.61}
$$

where τ_1 is propagation delay and $n_1(t)$ is the noise. The second equality follows from that $p\left(t + \frac{\lambda}{2c} - \tau_1\right) = -p(t - \tau_1)$, as stated in (1.49).

The received signal at receiver 2 becomes

$$
\begin{aligned}
y_2(t) =& \sqrt{\frac{P_2}{2}}\beta s_2 \Bigg(\underbrace{p(t-\tau_{2,1}) + p\left(t + \frac{\lambda}{2c} - \tau_{2,2}\right)}_{=2p(t-\tau_{2,1})} \Bigg) \\
&+ \sqrt{\frac{P_1}{2}}\beta s_1 \Bigg(\underbrace{p(t-\tau_{2,1}) + p(t-\tau_{2,2})}_{=0} \Bigg) + \underbrace{n_2(t)}_{\text{Noise}},
\end{aligned}
\tag{1.62}
$$

where $n_2(t)$ is the noise while $\tau_{2,1}$ and $\tau_{2,2}$ are the propagation delays from the first and second transmit antenna, respectively. The interference vanishes since $\tau_{2,2} = \tau_{2,1} + \frac{\lambda}{2c}$ and $p(t - \tau_{2,2}) = p\left(t - \frac{\lambda}{2c} - \tau_{2,1}\right) = -p(t - \tau_{2,1})$.

Since there is no interference and $s_1^2 = s_2^2 = 1$, the SINRs at receiver 1 and receiver 2 respectively become

$$
\text{SINR}_1 = \frac{2P_1\beta}{\sigma^2} = 20P_1, \quad \text{SINR}_2 = \frac{2P_2\beta}{\sigma^2} = 20P_2. \tag{1.63}
$$

For jointly reliable communication to the two receivers, we need $\text{SINR}_1 \geq 1$ and $\text{SINR}_2 \geq 1$, which is equivalent to $20P_1 \geq 1$ and $20P_2 \geq 1$. We notice that P_1 and P_2 can be selected independently and that both conditions are satisfied if the powers are greater than or equal to 50 mW.

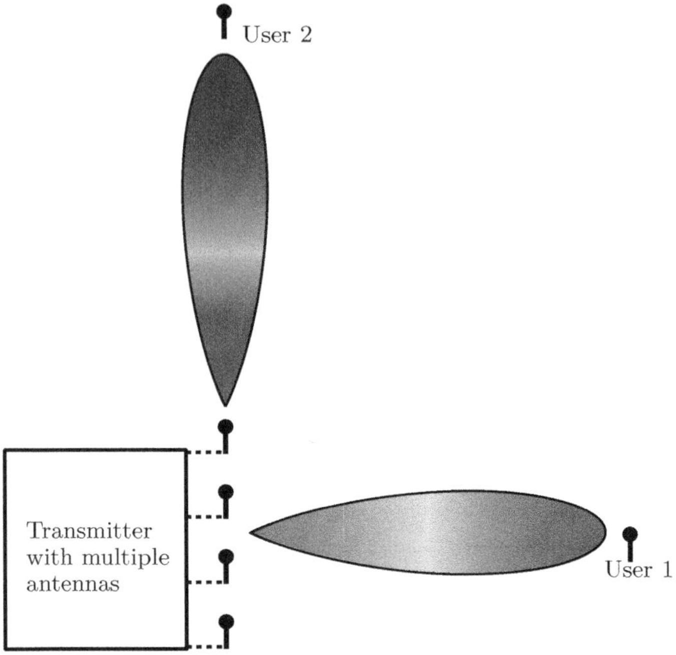

Figure 1.21: Schematic illustration of spatial multiplexing where two users are served at the same time and frequency, but their signals are transmitted using different beamforming. The true beam patterns are those shown in Figure 1.17 and Figure 1.19.

The previous example substantiated the claim that we can communicate simultaneously with two receivers thanks to the use of multiple antennas. This was impossible in the single-antenna case analyzed in Example 1.14. In the considered geometrical setup, the beamforming towards the receivers simultaneously maximizes their SINRs and SNRs since the receivers in Figure 1.21 are located in ideal perpendicular directions. When considering other receiver locations, the beamforming that maximizes the SINR must balance achieving a high SNR and avoiding interference. This corresponds to not directing each main beam exactly onto its intended receiver but fine-tuning the beamforming to balance between high signal power and low interference. These factors are analyzed in detail later in this book.

Adaptive beamforming from an antenna array is a much more flexible solution than using a single directive antenna. When serving a single user, adaptive beamforming can steer the emitted signal precisely toward the receiver, wherever it is. This is achieved electrically by time-delaying the signals emitted by the individual antennas. The same effect could be achieved by mechanically rotating a directive antenna. However, this is only an alternative in free-space propagation where there is only a single path and not in the more complicated non-line-of-sight propagation environments that often occur in practice. Moreover, an antenna array can simultaneously spatially multiplex several users with different beamforming, while a directive antenna can only transmit with one directivity at a time. The spatial multiplexing feature was used in a few commercial networks in Southeast Asia in the early 2000s [26, Example 10.1]. It is nowadays a widely supported feature in WiFi 5 (802.11ac) [27] and 5G NR [5]. It will likely be a core feature also in future systems.

Spatial multiplexing was conceived when cellular networks were used for voice communications. Hence, the network performance was characterized by the number of user connections that could be multiplexed and how good the network coverage was; the latter is the fraction of all spatial locations for which the SNR is above the threshold required for the voice quality to be acceptable. Both criteria could be improved by beamforming and spatial multiplexing of users. When wireless technology began to transmit data packets primarily, the *data rate* (bit/s) achieved by each user also became an important performance metric. The spatial multiplexing concept was then extended to setups where a single user device has multiple antennas [28]–[31], in which case one can assign multiple beams to the same device, and send several parallel layers of data to increase the data rate. The current wireless standards support a combination of these single-user and multi-user features [5], [27]: spatial multiplexing of many user devices and a few layers per device.

1.2.3 Spatial Diversity

In addition to increasing the SNR, using multiple antennas can improve the reliability of a wireless communication system. Thus far, we have mainly considered the free-space propagation scenario in Figure 1.1, in which there are no reflections or scattering: the only signal component that reaches the receiver is the one that has traveled along the *direct path* between the transmitter and receiver. This can be a good channel model for wireless communications in outer space but not for terrestrial systems where many reflecting/scattering objects might exist. This leads to so-called *multipath propagation*.

To exemplify the basic impact of multipath propagation, suppose the transmitter emits a pure sinusoidal signal $x(t) = \sin(2\pi f t)$, where f is the frequency. We consider the setup in Figure 1.22, where the signal reaches the receive antenna via two paths: the direct path has a distance d_1, and the reflected path has a distance d_2. Since electromagnetic waves travel at the speed of light c, the two distances correspond to the propagation time delays

$$\tau_i = \frac{d_i}{c} = \frac{d_i}{\lambda f} \quad \text{for } i = 1, 2, \tag{1.64}$$

where $\lambda = c/f$ is the wavelength. For the sake of argument, we disregard that the two paths will have (slightly) different channel gains and omit the channel gain parameters in this example to simplify the notation. Disregarding the

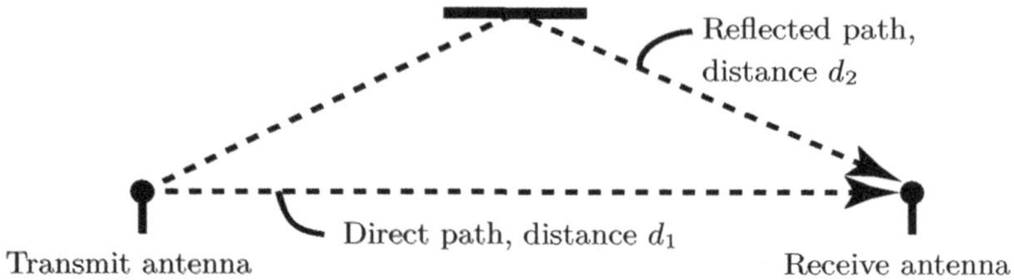

Figure 1.22: A basic multipath channel with a direct and reflected path.

additive noise, the received signal $y(t)$ can be expressed as

$$y(t) = x(t - \tau_1) + x(t - \tau_2) = \sin(2\pi f(t - \tau_1)) + \sin(2\pi f(t - \tau_2)), \quad (1.65)$$

where each term is called a *multipath component*. Depending on the relationship between the time delays τ_1, τ_2, the two multipath components in (1.65) can either reinforce or cancel each other. Using a trigonometric identity[10] we can rewrite (1.65) as

$$y(t) = \underbrace{2\cos(\pi f(\tau_1 - \tau_2))}_{\text{Amplitude}} \underbrace{\sin\left(2\pi f\left(t - \frac{\tau_1 + \tau_2}{2}\right)\right)}_{\text{Delayed version of the signal}}, \quad (1.66)$$

where $2\cos(\pi f(\tau_1 - \tau_2))$ is the constant amplitude of the received signal and $\sin(2\pi f(t - \frac{\tau_1+\tau_2}{2})) = x(t - \frac{\tau_1+\tau_2}{2})$ is a version of the transmitted signal with the average time delay. The constant amplitude can be rewritten as

$$2\cos(\pi f(\tau_1 - \tau_2)) = 2\cos\left(\pi f\left(\frac{d_1}{\lambda f} - \frac{d_2}{\lambda f}\right)\right) = 2\cos\left(\pi \frac{d_1 - d_2}{\lambda}\right) \quad (1.67)$$

by utilizing (1.64). We notice that this amplitude has a sign and can take any value between -2 and $+2$ depending on the argument of the cosine function. By comparing this amplitude with the unit amplitude achieved with only the direct path, we notice that multipath propagation can be either a blessing or a curse. In particular, if $(d_1 - d_2)/\lambda$ is an integer, then (1.67) becomes ± 2, and we benefit from having two paths by getting twice the amplitude. This happens because the signals received over the two paths have identical phases. On the other hand, (1.67) is zero when d_1 and d_2 differ by $\lambda/2$ ($\pm$ any integer number of wavelengths), because then the signals received over the two paths have opposite phases and their multipath components cancel out. When this happens (exactly or approximately), the channel is said to be in a *deep fade*.

This phenomenon is illustrated in Figure 1.23, where the signless amplitude $2\left|\cos\left(\pi \frac{d_1 - d_2}{\lambda}\right)\right|$ is shown. The key message is that a small change in the distance difference $d_1 - d_2$ can make the amplitude of the received signal either

[10]We use the fact that $\sin(\phi) + \sin(\psi) = 2\cos\left(\frac{\phi - \psi}{2}\right)\sin\left(\frac{\phi + \psi}{2}\right)$.

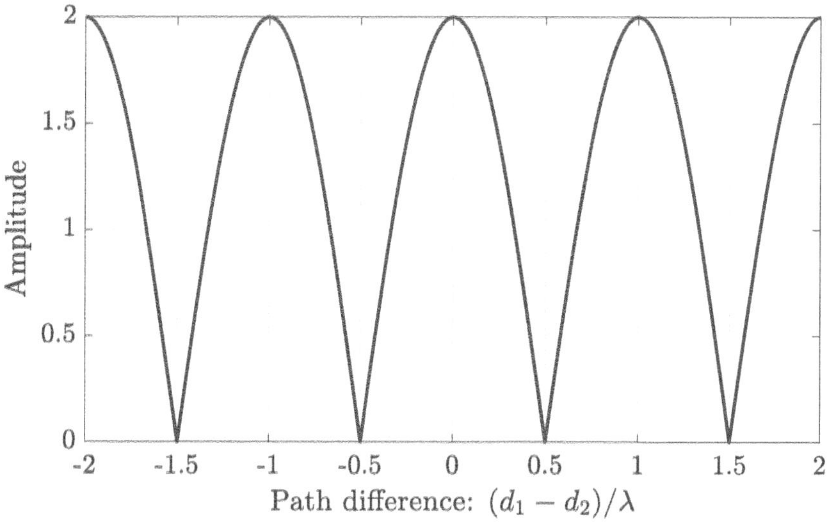

Figure 1.23: In the multipath example in Figure 1.22, the amplitude $2\left|\cos\left(\pi\frac{d_1-d_2}{\lambda}\right)\right|$ of the received signal varies rapidly as the difference in path distances changes.

grow or fade away. The interpretation of "small" is that the change happens when the transmitter and/or receiver move a fraction of the wavelength; for example, $\lambda/4$ is the change in the path difference $d_1 - d_2$ that is needed to move from the peak amplitude 2 to $\sqrt{2}$ in Figure 1.23, which corresponds to losing half the signal power (i.e., the channel gain reduces from $2^2 = 4$ to $\sqrt{2}^2 = 2$). That distance is 2.5 cm if $f = 3$ GHz and 2.5 mm if $f = 30$ GHz. These rapid channel changes are called *multipath fading* or *small-scale fading*.

The described two-path scenario resembles the behavior that appeared when transmitting from two different antennas in free-space propagation: the emitted signals are received along two paths with different time delays. The core difference is that with multiple transmit antennas, we can compensate for the time delays at the transmitter side (this is what we call adaptive beamforming). This is impossible in single-antenna multipath propagation since the two signal copies originate from the same transmit antenna.

Since a slight movement of the transmitting and receiving devices can lead to huge SNR fluctuations, multipath fading is a problematic phenomenon that makes wireless communications fundamentally unreliable. When transmitting a data packet, we must select a particular digital modulation and channel coding scheme in advance. Based on this selection, the receiver needs a particular SNR level during the transmission to decode the packet correctly. This level cannot always be fulfilled when the SNR fluctuates after the modulation/coding has been selected. When a packet cannot be decoded due to the channel being in a deep fade, we say an *outage* has occurred. Interestingly, multiple receive antennas can be used to protect communication against outages. Pioneering research on this topic was performed already in the 1930s by [32], [33], and the mathematical analysis presented in this book dates back to the 1950s [34]. Multiple transmit antennas can also protect against fading, but this requires more complex techniques, first developed in the 1990s [35]–[37]. In this section, we will introduce the basic concepts and then return to the topic in Chapter 5.

Suppose a random variable models the channel between the transmit and receive antennas: the communication works flawlessly with probability $1-p$, while it breaks down entirely with probability p. Hence, an outage occurs with probability p. This means that whenever you want to transmit a data packet, the *outage probability* is p.

If we instead make use of two receive antennas and the channel to each one of them is described by an independent random variable with the same distribution as above, three random events can occur:

1. Both antennas have good channels, happening with probability $(1-p)^2$;
2. One antenna has a good channel, and the other antenna experiences an outage, which happens with probability $(1-p)p + p(1-p) = 2(1-p)p$;
3. Both antennas experience outages, which happens with probability p^2.

It is only in the third case that the receiver cannot decode the data packet. Hence, the outage probability is p^2 in this two-antenna setup.

By following the same logic, if we have M receive antennas and each one experiences an outage with probability p, then the probability that all the antennas are simultaneously experiencing outages is p^M (assuming that the outage events occur independently for every antenna). This means that the reliability of the communication system rapidly improves as we add more receive antennas, known as *spatial diversity*. The name suggests that, at every time instance, we utilize the spatial domain to combat fading; for example, by only using those antennas that are located at spatial locations that currently experience good channel conditions. The argument above applies when using any antenna type. Directive antennas can be used to improve the SNR, but they cannot be used to obtain spatial diversity; multiple antennas are needed for that.[11] However, the antenna array can be actively designed to extract as much diversity as possible in a given propagation environment. This can be achieved by deploying the antennas far apart, rotating their antenna gains differently, and making them sensitive to waves with different polarization; the overarching goal with this is to ensure that the antennas experience outage events nearly independently so that the maximum diversity can be achieved. The term *antenna diversity* is sometimes used to describe how spatial diversity and antenna design are utilized jointly to achieve reliable communications.

[11] Directive antennas might reduce the impact of multipath propagation, compared to isotropic antennas since some multipath components can "disappear" because there are low antenna gains in their directions. However, active exploitation of spatial diversity requires multiple antennas.

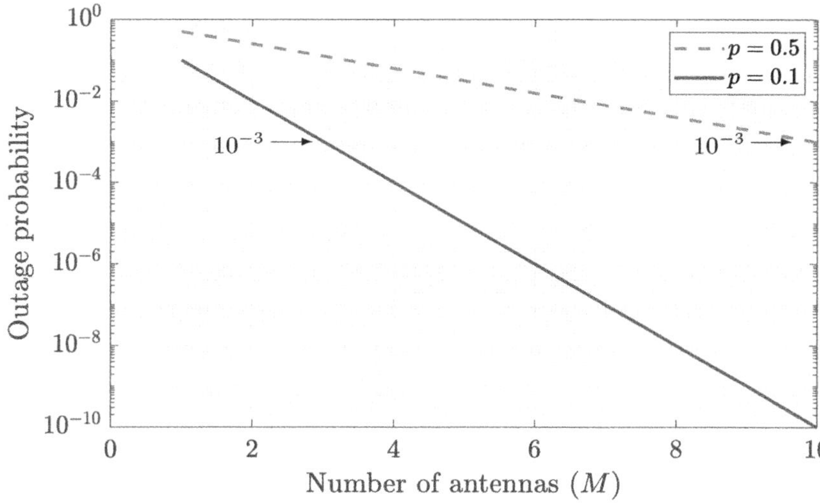

Figure 1.24: The outage probability p^M as a function of the number of antennas for different values of p, which is the probability that an arbitrary antenna observes an outage.

The benefit of spatial diversity is illustrated in Figure 1.24 for $p = 0.5$ and $p = 0.1$. The vertical axis shows the outage probability using a logarithmic scale, while the horizontal axis shows the number of antennas on a linear scale. The figure demonstrates that the outage probability reduces rapidly when the number of antennas increases. The slope of the curve becomes steeper when p is smaller since the outage probability is $10\log_{10}(p^M) = 10M\log_{10}(p)$ dB. Hence, a single-antenna system that has a noticeable outage probability can be greatly improved by adding additional antennas. We can never achieve a zero-valued outage probability when $p > 0$, but suppose $10^{-3} = 0.001$ is acceptable in a practical system. The figure shows that it can be achieved using 3 antennas if $p = 0.1$ and 10 antennas if $p = 0.5$. The latter represents a very unreliable channel, but it can be turned into a very reliable communication system by using the spatial diversity provided by having many antennas.

Spatial diversity can also be utilized in the opposite scenario where the transmitter has multiple antennas while the receiver has a single antenna. We then must be mindful of both outage events for the channels between each transmit antenna and the receiver and the risk that the signals emitted from the antennas cancel over the air. A simple way to alleviate the latter issue is to transmit from the antennas at different times or frequencies and then let the receiver jointly process the received signals to retrieve the information without any outage. This is inefficient since the same signal must be repeated several times before the next signal can be transmitted. There are more efficient solutions called *space-time codes* where multiple signals are repeated at the same time in an intricate way that enables the receiver to exploit diversity. We will describe these methods in Section 5.3.

Example 1.16. There might be a correlation between the channel conditions experienced at the different antennas of an array. In this example, we consider a single-antenna transmitter and a receiver with an array of M antennas. We let B_m denote an outage event at receive antenna m, and it occurs with the *marginal* probability $\Pr\{B_m\} = p$, for $m = 1, \dots, M$, as previously in this section. The outage probability of the channel is the *joint* probability that all antennas are experiencing an outage simultaneously: $\Pr\{B_1, \dots, B_M\}$. It equals the product p^M of the marginal probabilities when the outage events are independent between the antennas but not if the events are correlated.

We assume that if one antenna experiences an outage, the conditional probability for the other antennas changes to $\varrho \in [0, 1]$. Hence, $\Pr\{B_1\} = p$ but $\Pr\{B_2 | B_1\} = \varrho$, which can be larger or smaller than p depending on the value of ϱ. The typical situation in practice is that $\varrho \geq p$ so that an outage at antenna 1 increases the probability of an outage at the other antennas. What is the outage probability $\Pr\{B_1, \dots, B_M\}$ of this channel?

Based on the assumed correlation model, an outage event at antenna m, given the information that the antennas $1, \dots, m-1$ experience outage, is

$$\Pr\{B_m | B_1, \dots, B_{m-1}\} = \varrho. \tag{1.68}$$

We can then use the chain rule[a] for random events to compute

$$\begin{aligned} \Pr\{B_1, \dots, B_M\} &= \Pr\{B_1\}\Pr\{B_2, \dots, B_M | B_1\} \\ &= \Pr\{B_1\}\Pr\{B_2 | B_1\}\Pr\{B_3, \dots, B_M | B_1, B_2\} \\ &= \dots = \Pr\{B_1\} \prod_{m=2}^{M} \Pr\{B_m | B_1, \dots, B_{m-1}\} = p\varrho^{M-1}. \end{aligned} \tag{1.69}$$

If $\varrho = 1$, so that an outage event at one antenna guarantees outages on all other antennas, we get $\Pr\{B_1, \dots, B_M\} = p$. There is no spatial diversity benefit from having multiple antennas in this extreme case, but having the extra antennas does not hurt. However, whenever $\varrho < 1$, the outage probability will decay as ϱ^{M-1} when increasing the number of antennas. On a decibel scale, the outage probability behaves as $10\log_{10}(p\varrho^{M-1}) = 10M\log_{10}(\varrho) + 10\log_{10}(p/\varrho)$, similar to the case with independent outage events. If we would add a new curve to Figure 1.24 that represents this new scenario with correlated outages, it will decay similarly to the existing curves, but the slope depends on the correlation ϱ rather than the marginal probability p. The key conclusion is that the spatial diversity brought by having multiple antennas helps lower the outage probability compared to the single-antenna case, even if the outage events are correlated between the antennas.

[a]If A and B are two random events, then the chain rule says that $\Pr\{A, B\} = \Pr\{A\}\Pr\{B|A\}$. The rule can be expanded by including more than two events and can then be applied repeatedly, as done in (1.69).

1.3 Exercises

Exercise 1.1. Since the power levels in wireless communications can be extremely different, it is convenient to use decibel scales.

(a) What is 1 mW expressed in dBm?

(b) What is 30 dBm expressed in Watt?

(c) Suppose we transmit a signal with power $P_{\text{tx}} = 20\,\text{dBm}$ and that 90 dB is lost on the way to the receiver. What is the received signal power P_{rx}? Express the answer in both dBm and mW.

(d) Suppose the noise power is $N_0 B = -100\,\text{dBm}$. What is the SNR $P_{\text{rx}}/(N_0 B)$ at the receiver? What is the unit of the SNR?

Exercise 1.2. The SNR determines how much data can be transmitted per modulation symbol in a wireless communication system. The system is not operational if the SNR is below a specific value, in which case we are out-of-coverage. In this exercise, we consider a system that is operational when the SNR is equal to or larger than $-10\,\text{dB}$.

(a) A single-antenna base station communicates with a single-antenna user device. The base station transmits with 10 W and the device with 0.1 W. The channel gain is $-110\,\text{dB}$, the bandwidth is 10 MHz, and the noise power spectral density is $10^{-17}\,\text{W/Hz}$. Compute the SNRs achieved in the uplink and downlink.

(b) The computation in (a) reveals that the uplink SNR is below $-10\,\text{dB}$. Hence, the system is not operational, even if the downlink SNR is above $-10\,\text{dB}$. This is a common issue that can be resolved using multiple antennas at the base station. How many antennas are needed in this case, if the SNR is proportional to the number of antennas?

(c) Can we instead change how much bandwidth that is used? If yes, explain how and what the consequences will be. If no, explain why.

Exercise 1.3. The parametric channel gain model in (1.9) is entirely determined by the channel gain values at two different distances. Suppose the channel gain is $-100\,\text{dB}$ at $d = 100\,\text{m}$ and $-135\,\text{dB}$ at $d = 1000\,\text{m}$.

(a) What are the values of the pathloss exponent α and the constant Υ? Assume that the measurements were made using isotropic antennas.

(b) Suppose the measurements were made using short dipoles with antenna gains of 1.5 at the transmitter and the receiver. What are the values of the pathloss exponent α and the constant Υ?

Exercise 1.4. Consider a (hypothetical) antenna with the gain function

$$G(\varphi, \theta) = \begin{cases} c\cos(4\varphi + \pi)\cos(\theta), & \text{if } \varphi \in [-3\pi/8, -\pi/8], \theta \in [-\pi/2, \pi/2], \\ c\cos(3\varphi - \pi)\cos(\theta), & \text{if } \varphi \in [\pi/6, \pi/2], \theta \in [-\pi/2, \pi/2], \\ 0, & \text{elsewhere}, \end{cases} \tag{1.70}$$

where $c > 0$ is a constant.

(a) If the antenna is lossless, what should be the value of c?

(b) What is the maximum effective area of this antenna, and for which angles (φ, θ) is it achieved?

Exercise 1.5. A single-antenna transmitter located at the point with Cartesian coordinates $(300, 400, 0)$ m communicates in free space with a single-antenna receiver located in the origin. The transmit power is 20 dBm, the carrier frequency is $f = 3$ GHz, and the bandwidth is $B = 10$ MHz. Due to noise amplification in the receiver hardware, the noise power spectral density is $N_0 F$, where N_0 is given in (1.11) and $F = 4$ dB is called the noise figure.

(a) What is the SNR if isotropic antennas are used?

(b) What is the SNR if antennas with the cosine gain function in (1.34) are used? The transmit antenna achieves its maximum gain in the azimuth plane in the negative y-axis direction, while the receive antenna achieves its maximum gain in the positive y-axis direction. This setup is shown to the left in the figure below.

(c) If the transmit antenna in (b) is rotated clockwise by $\pi/2$ radians in the azimuth plane, what is the SNR? This setup is shown to the right in the figure below. Note that the antenna gain pattern in (1.34) should be rotated accordingly.

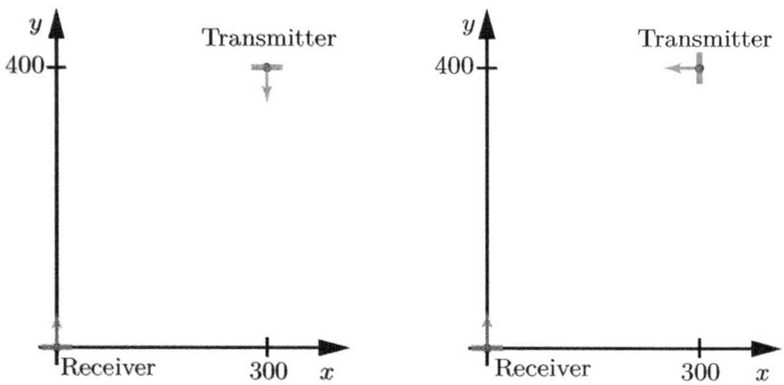

Exercise 1.6. Consider an isotropic transmit antenna and a flat receive antenna having the width a. For simplicity, the antennas are located in the same two-dimensional plane, and the geometry is similar to Figure 1.8 but rotated. The transmitter is located at the origin $(0, 0)$. The receive antenna covers the line segment from $(\sqrt{3}d/2, d/2 - a/2)$ to $(\sqrt{3}d/2, d/2 + a/2)$, where d is the propagation distance to the center $(\sqrt{3}d/2, d/2)$ of the receive antenna.

(a) Suppose $d \gg a$ and the transmitted signal has wavelength λ. What are the approximate phase differences between the signal received at the center and the signals received at the two corners?

(b) Suppose $d = \frac{2a^2}{\lambda}$, which is the Fraunhofer distance defined in (1.18). What is the maximum phase difference between two points on the receive antenna? Is this value in line with the definition of the Fraunhofer distance?

Exercise 1.7. Consider a transmitter with an array of $M = 3$ isotropic antennas located at the Cartesian coordinates $(\lambda/4, 0, -\lambda/2)$, $(0, -\lambda/3, 0)$, and $(0, 0, \lambda/2)$, where λ is the wavelength. The transmitted signal from the mth antenna is $x_m(t) = Ap(t + \tau_m)$ where $p(t)$ is the sine pulse in (1.44). We want to maximize the received signal power at the spherical coordinate (d, φ, θ), where $d \gg \lambda$. What values of the delays τ_1, τ_2, and τ_3 can be selected? Is the solution unique?

Exercise 1.8. A transmitter equipped with $M = 2$ antennas communicates with a single-antenna receiver. The propagation time delays from the first and second antennas are denoted by τ_1 and τ_2, respectively. The transmitter compensates for these delays by transmitting the signal $x_m(t) = A_m p(t + \tau_m)$ from the mth antenna, for $m = 1, 2$, where $A_m \geq 0$ is the amplitude and $p(t)$ is the sine pulse in (1.44). The total transmit power is $A_1^2 + A_2^2$.

(a) Suppose the channel gains to the receiver are the same for both antennas and denoted by β. If the total transmit power must be equal to P, what values of A_1 and A_2 maximize the received signal power?

(b) Suppose the channel gains from the first and second antenna are denoted by β_1 and β_2, respectively. If the total transmit power must be equal to P, what values of A_1 and A_2 should be selected to maximize the received signal power?

(c) If $\beta_1 > \beta_2$, which antenna will transmit with the highest power according to the answer in (b)?

Exercise 1.9. Consider a transmitter with two isotropic antennas that emit the same signal $Ap(t)$, where A is the amplitude and $p(t)$ is the sine pulse in (1.44). The antennas are located at the Cartesian coordinates $(0, y_0, 0)$ and $(0, -y_0, 0)$, for some value of $y_0 \geq 0$. We are interested in receiver locations with spherical coordinates $(d, \varphi, 0)$ that are at a large but fixed distance $d \gg y_0$ from the transmitter but have varying azimuth angle φ. The channel gain β is the same from both antennas to any of these points.

(a) What is the minimum value of y_0 for which destructive interference occurs at $(d, \varphi, 0)$ for at least one $\varphi \in [-\pi, \pi)$?

(b) For what range of y_0 values will constructive interference occur at $(d, \varphi, 0)$ for six different values of $\varphi \in [-\pi, \pi)$?

Exercise 1.10. Consider a transmitter array with two isotropic antennas having the Cartesian coordinates $(0, \lambda/4, 0)$ and $(0, -\lambda/4, 0)$, respectively. These antennas jointly transmit signals to two receivers located at the spherical coordinates $(d, 0, 0)$ and $(d, \pi/3, 0)$, respectively. The distance d is large, so the channel gain β is the same between any transmit antenna and receive antenna. The time-limited pulse in (1.44) is used to carry the two symbols $s_1, s_2 \in \{-1, 1\}$ intended for the two receivers. To beamform towards the first receiver, both transmit antennas send $\sqrt{P_1} s_1 p(t)$ using some power P_1. Moreover, to beamform towards the second receiver, the two antennas transmit $\sqrt{P_2} s_2 p(t)$ and $\sqrt{P_2} s_2 p(t + \frac{\lambda \sin(\pi/3)}{2c})$, respectively, using some power P_2. Suppose that $\sigma^2/\beta = 10^{-1}$ W. Is it possible to select the powers P_1 and P_2 so that both receivers achieve an SINR of 10 dB? If yes, give an example of how it can be done.

Exercise 1.11. Communication systems that operate in the mmWave and sub-THz bands are sensitive to signal blockage by the human body, which might lower the received power by more than 20 dB. To circumvent this issue, a handheld device can be built with antennas at different sides (e.g., at the top and on the right side) to make it unlikely that they are all blocked simultaneously by the user.

(a) Consider a device with two antennas. The outage probability of one antenna is p. However, if one antenna is in outage, the outage probability of the other antenna reduces to $\varrho < p$ thanks to the antenna placement. What is the probability that both antennas are in outage simultaneously?

(b) How much larger is the outage probability with independent outage events compared to the probability in (a)?

Chapter 2

Theoretical Foundations

This book is dedicated to analyzing multiple antenna communication systems, and we will rely on methods from linear algebra, probability theory, signal processing, and information theory. This chapter will describe the key results from these fields that we will utilize in later chapters, using the notation and terminology used in the remainder of this book. The reader is expected to be familiar with these general topics since the chapter mainly summarizes essential results, and we refer to other textbooks for an in-depth introduction. The focus is on complex numbers and how they enter into the aforementioned theory when developing concise models of communication systems.

2.1 Complex Numbers and Algebra

Complex numbers naturally appear when analyzing communication systems, for example, since the frequency representation of signals and systems is generally complex. The fundamental component of complex numbers is the imaginary unit, which we denote $\mathrm{j} = \sqrt{-1}$. Note that the letter "j" is used in electrical engineering instead of the letter "i" commonly used in the mathematical literature to not confuse it with the letter used for electrical currents.

We let $\mathbb{C}$ denote the set of all complex numbers. Any complex number $c \in \mathbb{C}$ can be decomposed as

$$c = a + \mathrm{j}b \tag{2.1}$$

for some real numbers $a, b \in \mathbb{R}$. In this case, a is the real part of c, while b is the imaginary part of c. We will let $\Re(\cdot)$ be the function that outputs the real part of its input, while $\Im(\cdot)$ is the function that outputs the imaginary part. Hence, if $c = a + \mathrm{j}b$, it follows that $\Re(c) = a$ and $\Im(c) = b$.

The representation in (2.1) is called the *Cartesian form.* Instead of decomposing a complex number c in its real and imaginary part, we can use the *polar form* to decompose it using the magnitude and argument. More precisely,

$$c = |c|e^{\mathrm{j}\arg(c)}, \tag{2.2}$$

where $|c| = \sqrt{a^2+b^2} \geq 0$ is the magnitude (also known as absolute value), describing the length of the vector $[a,b]^{\mathrm{T}}$, and the argument $\arg(c) \in [-\pi, \pi)$ is the angle of that vector in $\mathbb{R}^2$. The polar form contains *Euler's number* $e \approx 2.71828$ and makes use of the *complex exponential* function

$$e^{\mathrm{j}x} = \cos(x) + \mathrm{j}\sin(x), \tag{2.3}$$

where the real and imaginary parts contain the cosine and sine of the argument $x \in \mathbb{R}$, respectively. This relation is known as *Euler's formula.*

The different ways to represent a complex number are illustrated in Figure 2.1. From the definition of the sine and cosine functions, we can also establish the relation

$$c = \underbrace{|c|\cos\left(\arg(c)\right)}_{=a} + \mathrm{j}\underbrace{|c|\sin\left(\arg(c)\right)}_{=b} \tag{2.4}$$

between the Cartesian and polar forms. Hence, when considering signals, $|c|$ can represent the magnitude/amplitude while $\arg(c)$ can represent the phase.

The *complex conjugate* is a vital operation when considering complex numbers. The complex conjugate of $c = a + \mathrm{j}b$ is denoted as c^* and computed by switching the sign of the imaginary part: $c^* = a - \mathrm{j}b$. This is equivalent to switching the sign of the argument: $c^* = |c|e^{-\mathrm{j}\arg(c)}$. Note that

$$cc^* = (a+\mathrm{j}b)(a-\mathrm{j}b) = a^2 + \mathrm{j}ab - \mathrm{j}ab - \mathrm{j}^2b^2 = a^2 + b^2 = |c|^2. \tag{2.5}$$

Hence, we can compute the squared magnitude of a complex number by multiplying it with its complex conjugate. We can also extract the real and imaginary parts by adding c and c^* with different scaling factors:

$$\frac{1}{2}(c + c^*) = \frac{1}{2}(a + \mathrm{j}b + a - \mathrm{j}b) = a = \Re(c), \tag{2.6}$$

$$\frac{1}{\mathrm{j}2}(c - c^*) = \frac{1}{\mathrm{j}2}(a + \mathrm{j}b - a + \mathrm{j}b) = b = \Im(c). \tag{2.7}$$

The complex exponential function is the essential building block to create sinusoids oscillating at a specified frequency f_{c}. If x is replaced by $2\pi f_{\mathrm{c}}t$ in (2.3), we obtain the complex exponential $e^{\mathrm{j}2\pi f_{\mathrm{c}}t} = \cos(2\pi f_{\mathrm{c}}t) + \mathrm{j}\sin(2\pi f_{\mathrm{c}}t)$, where t represents time. By following the procedure in (2.6) and (2.7), we can extract the real and imaginary parts as

$$\cos(2\pi f_{\mathrm{c}}t) = \Re\left(e^{\mathrm{j}2\pi f_{\mathrm{c}}t}\right) = \frac{1}{2}e^{\mathrm{j}2\pi f_{\mathrm{c}}t} + \frac{1}{2}e^{-\mathrm{j}2\pi f_{\mathrm{c}}t}, \tag{2.8}$$

$$\sin(2\pi f_{\mathrm{c}}t) = \Im\left(e^{\mathrm{j}2\pi f_{\mathrm{c}}t}\right) = \frac{1}{\mathrm{j}2}e^{\mathrm{j}2\pi f_{\mathrm{c}}t} - \frac{1}{\mathrm{j}2}e^{-\mathrm{j}2\pi f_{\mathrm{c}}t}. \tag{2.9}$$

The unique aspect of the complex exponential is that it only contains the frequency f_{c}, and no other frequencies. Since the cosine and sine functions

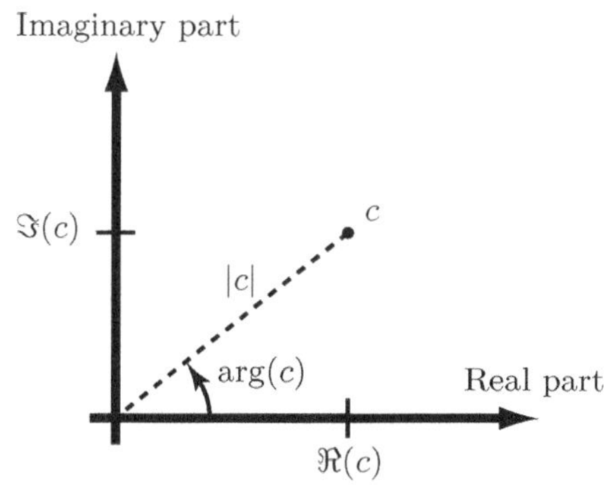

Figure 2.1: The complex number c can be equivalently represented by two real-valued numbers. The Cartesian form is $c = \Re(c) + \mathrm{j}\Im(c)$, with the real part $\Re(c)$ and imaginary part $\Im(c)$. The polar form is $c = |c|e^{\mathrm{j}\arg(c)}$, where $|c|$ is the magnitude and $\arg(c)$ is the argument.

are created as linear combinations of both $e^{\mathrm{j}2\pi f_c t}$ and $e^{-\mathrm{j}2\pi f_c t}$, these signals are said to contain both the positive frequency f_c and the negative frequency $-f_c$. Any real-valued signal contains a range of positive frequencies and the corresponding negative ones. We will continue to study the frequency representation of signals in Sections 2.3 and 2.8.

Example 2.1. Let $c = a + \mathrm{j}b$ be an arbitrary complex number. Show that the sinusoid $a\cos(t) + b\sin(t)$ with the time variable t can be written as a single cosine function, using the polar form $c = |c|e^{\mathrm{j}\arg(c)}$.

The sinusoid can be rewritten as

$$\begin{aligned} a\cos(t) + b\sin(t) &= \Re\left((\cos(t) + \mathrm{j}\sin(t))\,(a - \mathrm{j}b)\right) \\ &= \Re\left(e^{\mathrm{j}t}c^*\right) = \Re\left(e^{\mathrm{j}t}|c|e^{-\mathrm{j}\arg(c)}\right) = |c|\Re\left(e^{\mathrm{j}(t-\arg(c))}\right) \\ &= |c|\cos\left(t - \arg(c)\right). \end{aligned} \tag{2.10}$$

This shows how the amplitude and phase of a sinusoid can be represented by a complex number, which is a primary reason for using them in communications.

2.1.1 Vector Analysis

Vectors and matrices are commonly used when describing systems with multiple antennas, where each entry is related to one of the antennas. The entries will be complex in most of the chapters of this book. Thus, we will briefly review the foundational linear algebra results in the complex domain.

An M-dimensional vector containing the complex entries $x_1, \ldots, x_M \in \mathbb{C}$ can be expressed as

$$\mathbf{x} = \begin{bmatrix} x_1 \\ \vdots \\ x_M \end{bmatrix}. \tag{2.11}$$

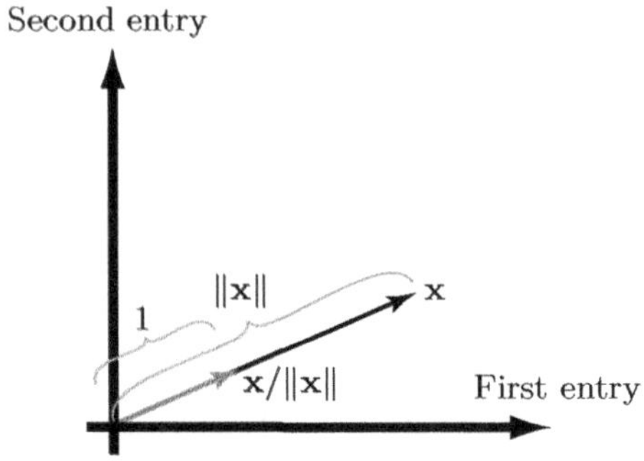

Figure 2.2: The complex vector $\mathbf{x}$ has a length determined by the norm $\|\mathbf{x}\|$. The unit-length vector $\mathbf{x}/\|\mathbf{x}\|$ points in the same direction as $\mathbf{x}$.

We denote vectors using lower-case bold-faced letters, such as $\mathbf{x}$. The entries are expressed using the same letter and a subscript indicating the entry number, such as x_m for the mth entry of $\mathbf{x}$. Since the vector belongs to the M-dimensional complex vector space $\mathbb{C}^M$, we can write that $\mathbf{x} \in \mathbb{C}^M$.

A vector $\mathbf{x}$ has a norm that describes the distance between the origin and the point $\mathbf{x}$ in the vector space. Since it describes the length, it can be viewed as the generalization of the magnitude to vectors. The *Euclidean norm* is denoted by $\|\mathbf{x}\|$ and is computed as

$$\|\mathbf{x}\| = \sqrt{|x_1|^2 + \ldots + |x_M|^2} = \sqrt{\sum_{m=1}^{M} |x_m|^2}. \tag{2.12}$$

By using the norm, we can decompose the vector as

$$\mathbf{x} = \underbrace{\|\mathbf{x}\|}_{\text{Length}} \cdot \underbrace{\frac{\mathbf{x}}{\|\mathbf{x}\|}}_{\text{Direction}}, \tag{2.13}$$

where the second term is the length-one vector pointing in the same direction as $\mathbf{x}$. Figure 2.2 illustrates how an arbitrary vector $\mathbf{x}$ is described by its length/norm $\|\mathbf{x}\|$ and the direction $\mathbf{x}/\|\mathbf{x}\|$. There will be occasions in this book where we want to select two vectors that point in the same direction but have different norms, in which case we can utilize this decomposition.

All the vectors in this book are column matrices, meaning they have one column and multiple rows. When dealing with matrices, one can switch the meaning of rows and columns using the operation called *transpose*. The transpose of an arbitrary vector $\mathbf{x}$ is denoted as $\mathbf{x}^{\mathrm{T}}$. For example, the transpose of (2.11) is

$$\mathbf{x}^{\mathrm{T}} = \begin{bmatrix} x_1 & \ldots & x_M \end{bmatrix}, \tag{2.14}$$

which is a row matrix containing the same entries.

When dealing with complex vectors, there is another type of transpose that also includes the complex conjugate operation:

$$\mathbf{x}^{\mathrm{H}} = \begin{bmatrix} x_1^* & \dots & x_M^* \end{bmatrix}. \tag{2.15}$$

This will be called the *conjugate transpose* in this book but is also known as the *Hermitian transpose*, which explains the letter ^{H}. A third operation that we will use is the complex conjugation of a vector (or matrix), which is defined as taking the complex conjugate of the individual entries:

$$\mathbf{x}^* = \begin{bmatrix} x_1^* \\ \vdots \\ x_M^* \end{bmatrix}. \tag{2.16}$$

The conjugate transpose is simply a combination of the conventional transpose and the conjugation, $\mathbf{x}^{\mathrm{H}} = (\mathbf{x}^{\mathrm{T}})^*$, but it is so commonly occurring in complex vector analysis that it deserves its own notation.

The inner product (or dot product) between two M-dimensional complex vectors $\mathbf{x}$ and $\mathbf{y} = [y_1, \dots, y_M]^{\mathrm{T}}$ is defined using the conjugate transpose as

$$\mathbf{x}^{\mathrm{H}}\mathbf{y} = \sum_{m=1}^{M} x_m^* y_m. \tag{2.17}$$

The magnitude $|\mathbf{x}^{\mathrm{H}}\mathbf{y}|$ of the inner product becomes larger the more similar the directions of the two vectors are and smaller when the directions are very different. This statement can be quantified by the *Cauchy-Schwarz inequality*, which states that

$$|\mathbf{x}^{\mathrm{H}}\mathbf{y}| \leq \|\mathbf{x}\|\|\mathbf{y}\| \tag{2.18}$$

with equality if and only if $\mathbf{x}$ and $\mathbf{y}$ are parallel (i.e., $\mathbf{x} = c\mathbf{y}$ for some non-zero $c \in \mathbb{C}$). The upper bound is the product of the lengths of the two vectors. Figure 2.3 illustrates how the inner product varies depending on the directions of the vectors, with the parallel vectors $\mathbf{x}, \mathbf{y}_1$ achieving the upper bound in the Cauchy-Schwarz inequality and orthogonal vectors $\mathbf{x}, \mathbf{y}_3$ having an inner product equal to zero. The latter vectors span a two-dimensional plane in the M-dimensional vector space and are separated by 90° in that plane.

Example 2.2. Suppose we are given a vector $\mathbf{x} \in \mathbb{C}^M$ and can select the vector $\mathbf{y} \in \mathbb{C}^M$ freely. Which selections will maximize or minimize $\frac{|\mathbf{x}^{\mathrm{H}}\mathbf{y}|}{\|\mathbf{y}\|}$?

The minimum is 0 and achieved for any vector $\mathbf{y}$ that is orthogonal to $\mathbf{x}$. The Cauchy-Schwarz inequality implies that the maximum is obtained for $\mathbf{y} = c\mathbf{x}$ for any non-zero $c \in \mathbb{C}$.

When one of the vectors has a unit length, the inner product can also be interpreted as an orthogonal projection onto that vector. Suppose $\mathbf{x}$ has unit

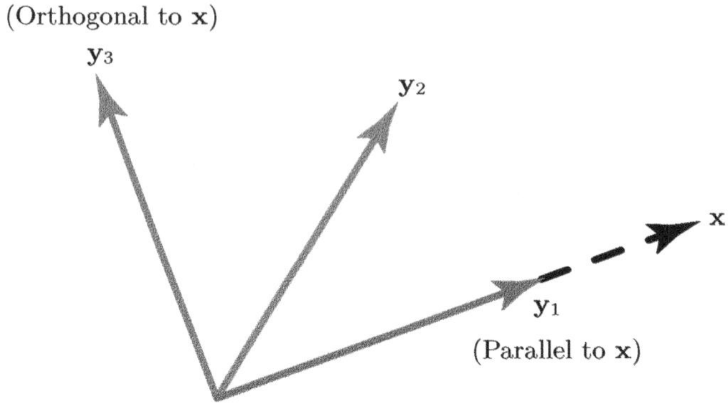

Figure 2.3: The magnitude of the inner product $|\mathbf{x}^{\mathrm{H}}\mathbf{y}_i|$ between two vectors depends on how similar their directions are. Parallel vectors give the largest value and achieve the upper bound in the Cauchy-Schwarz inequality in (2.18): $|\mathbf{x}^{\mathrm{H}}\mathbf{y}_1| = \|\mathbf{x}\|\|\mathbf{y}_1\|$. Orthogonal vectors give $\mathbf{x}^{\mathrm{H}}\mathbf{y}_3 = 0$, while other vectors give a number in between zero and the upper bound.

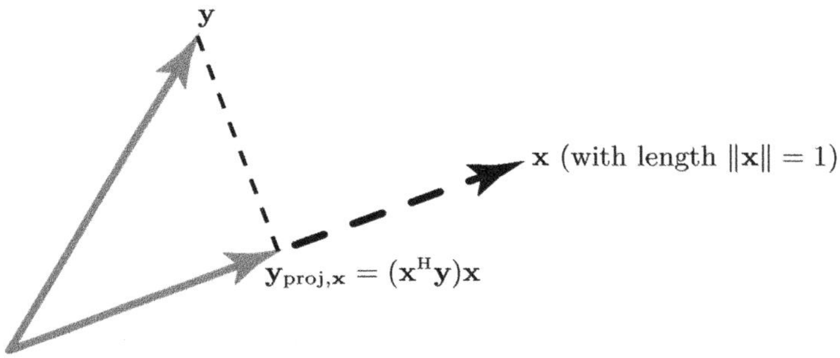

Figure 2.4: If $\mathbf{x}$ is a unit-length vector, the inner product $\mathbf{x}^{\mathrm{H}}\mathbf{y}$ is tightly connected to the orthogonal projection of $\mathbf{y}$ onto $\mathbf{x}$. The orthogonal projection is $\mathbf{y}_{\mathrm{proj},\mathbf{x}} = (\mathbf{x}^{\mathrm{H}}\mathbf{y})\mathbf{x}$ and has the length $|\mathbf{x}^{\mathrm{H}}\mathbf{y}|$.

length (i.e., $\|\mathbf{x}\| = 1$) and let $\mathbf{y}$ be any other vector of the same dimension. The magnitude $|\mathbf{x}^{\mathrm{H}}\mathbf{y}|$ of their inner product is also the length of the vector

$$\mathbf{y}_{\mathrm{proj},\mathbf{x}} = (\mathbf{x}^{\mathrm{H}}\mathbf{y})\mathbf{x}, \tag{2.19}$$

which is the orthogonal projection of $\mathbf{y}$ onto the direction pointed out by $\mathbf{x}$. This projection is illustrated in Figure 2.4. From this example, we can notice that only the part of $\mathbf{y}$ that is parallel to $\mathbf{x}$ will affect the inner product; thus, there are many different vectors $\mathbf{y}$ that have the same inner product with $\mathbf{x}$. It can also be proved that $\mathbf{y}_{\mathrm{proj},\mathbf{x}}$ is orthogonal to $\mathbf{y} - \mathbf{y}_{\mathrm{proj},\mathbf{x}}$.

A special case where the upper bound is achieved is when the inner product is computed between $\mathbf{x}$ and itself:

$$\mathbf{x}^{\mathrm{H}}\mathbf{x} = \sum_{m=1}^{M} x_m^* x_m = \sum_{m=1}^{M} |x_m|^2 = \|\mathbf{x}\|^2, \tag{2.20}$$

where the last equality follows from (2.12). Hence, the squared norm of a vector $\mathbf{x}$ can be computed using the inner product. This is a generalization of (2.5), where we computed the squared magnitude of a complex number by multiplying it with its complex conjugate.

By utilizing (2.20), the squared norm of the summation of two arbitrary vectors $\mathbf{x}$ and $\mathbf{y}$ (of the same dimension) can be expanded as

$$\begin{aligned}\|\mathbf{x}+\mathbf{y}\|^2 &= \mathbf{x}^{\mathrm{H}}\mathbf{x}+\mathbf{y}^{\mathrm{H}}\mathbf{y}+\mathbf{x}^{\mathrm{H}}\mathbf{y}+\mathbf{y}^{\mathrm{H}}\mathbf{x} \\ &= \|\mathbf{x}\|^2+\|\mathbf{y}\|^2+2\Re(\mathbf{x}^{\mathrm{H}}\mathbf{y}) \end{aligned} \tag{2.21}$$

by utilizing the fact that $\mathbf{x}^{\mathrm{H}}\mathbf{y}$ and $\mathbf{y}^{\mathrm{H}}\mathbf{x}$ have the same real part but imaginary parts with opposite signs.

Example 2.3. Consider a set of K unit-length vectors $\mathbf{x}_1,\ldots,\mathbf{x}_K \in \mathbb{C}^M$ that are mutually orthogonal, where $K \leq M$. Compute the squared norm of the vector $\mathbf{y}=\sum_{k=1}^{K} c_k\mathbf{x}_k$, where $c_1,\ldots,c_K \in \mathbb{C}$ are scalar coefficients.

From the provided information, we have $\|\mathbf{x}_k\|=1$, for $k=1,\ldots,K$, and $\mathbf{x}_k^{\mathrm{H}}\mathbf{x}_m=0$, for $k\neq m$. We use these properties to expand the squared norm as

$$\begin{aligned}\|\mathbf{y}\|^2 = \mathbf{y}^{\mathrm{H}}\mathbf{y} = \sum_{k=1}^{K} c_k^*\mathbf{x}_k^{\mathrm{H}} \sum_{m=1}^{K} c_m\mathbf{x}_m &= \sum_{k=1}^{K} |c_k|^2 \underbrace{\mathbf{x}_k^{\mathrm{H}}\mathbf{x}_k}_{=\|\mathbf{x}_k\|^2=1} + \sum_{k=1}^{K}\sum_{\substack{m=1\\ m\neq k}}^{K} c_k^* c_m \underbrace{\mathbf{x}_k^{\mathrm{H}}\mathbf{x}_m}_{=0} \\ &= \sum_{k=1}^{K} |c_k|^2. \end{aligned} \tag{2.22}$$

We notice that $\|\mathbf{y}\|^2$ is the summation of the squared coefficients, which determine the length of $\mathbf{y}$ in each of the K orthogonal directions $\mathbf{x}_1,\ldots,\mathbf{x}_K$.

The summation of vectors, multiplied by scalar coefficients, is known as a *linear combination.* If $\mathbf{x}_1,\ldots,\mathbf{x}_K \in \mathbb{C}^M$ are K vectors and $c_1,\ldots,c_K \in \mathbb{C}$ are K scalar coefficients, then the linear combination of the vectors using those coefficients is

$$c_1\mathbf{x}_1 + c_2\mathbf{x}_2 + \ldots + c_K\mathbf{x}_K = \sum_{k=1}^{K} c_k\mathbf{x}_k. \tag{2.23}$$

This concept is helpful in making geometrical comparisons of vectors in high-dimensional situations where we cannot draw them on paper.

Definition 2.1. The vectors $\mathbf{x}_1,\mathbf{x}_2,\ldots,\mathbf{x}_K$ are said to be *linearly independent* if the system of equations

$$c_1\mathbf{x}_1 + c_2\mathbf{x}_2 + \ldots + c_K\mathbf{x}_K = \mathbf{0} \tag{2.24}$$

only has the solution $c_1=\ldots=c_K=0$. If additional non-zero solutions exist, the vectors are said to be *linearly dependent*.

Any two vectors are linearly independent except if they are entirely parallel; thus, linear independence is a broader condition than orthogonality. For example, we can pick any two of $\mathbf{y}_1$, $\mathbf{y}_2$, and $\mathbf{y}_3$ in Figure 2.3 and get a set of linearly independent vectors. However, the set of all three vectors in the figure is linearly dependent because $\mathbf{y}_2$ points partially in the direction of $\mathbf{y}_1$ and partially in the direction of $\mathbf{y}_3$. This is a typical situation when considering two-dimensional vectors, as in the figure: if we pick more than two vectors, they must always be linearly dependent because they share the same two dimensions. More generally, any set of more than M vectors that are M-dimensional must be linearly dependent, but we can find a set with exactly M linearly independent vectors. Moreover, any set of pairwise orthogonal vectors can be shown to be linearly independent.

Example 2.4. Consider the vector $\mathbf{y} = \sum_{k=1}^{M} c_k \mathbf{x}_k$, constructed using the mutually orthogonal unit-length vectors $\mathbf{x}_1, \ldots, \mathbf{x}_M \in \mathbb{C}^M$ and scalar coefficients $c_1, \ldots, c_M \in \mathbb{C}$. Let $\mathbf{y}_{\text{proj},\mathbf{x}_m}$ denote the orthogonal projection of $\mathbf{y}$ onto $\mathbf{x}_m$, which is the mth of the unit-length vectors. . What are the squared norms of $\mathbf{y}_{\text{proj},\mathbf{x}_m}$ and the residual vector $\mathbf{y} - \mathbf{y}_{\text{proj},\mathbf{x}_m}$?

The vector $\mathbf{y}_{\text{proj},\mathbf{x}_m}$ is computed similarly to (2.19) as

$$\mathbf{y}_{\text{proj},\mathbf{x}_m} = (\mathbf{x}_m^{\text{H}} \mathbf{y})\, \mathbf{x}_m = \left(\sum_{k=1}^{M} c_k \underbrace{\mathbf{x}_m^{\text{H}} \mathbf{x}_k}_{=\begin{cases} 1, & m = k \\ 0, & m \neq k \end{cases}} \right) \mathbf{x}_m = c_m \mathbf{x}_m. \tag{2.25}$$

Hence, we obtain $\|\mathbf{y}_{\text{proj},\mathbf{x}_m}\|^2 = |c_m|^2$, which is the squared coefficient associated with $\mathbf{x}_m$. The squared norm of the residual $\mathbf{y} - \mathbf{y}_{\text{proj},\mathbf{x}_m}$ becomes

$$\left\|\mathbf{y} - \mathbf{y}_{\text{proj},\mathbf{x}_m}\right\|^2 = \left\| \sum_{k=1}^{M} c_k \mathbf{x}_k - c_m \mathbf{x}_m \right\|^2 = \left\| \sum_{\substack{k=1 \\ k\neq m}}^{M} c_k \mathbf{x}_k \right\|^2 = \sum_{\substack{k=1 \\ k\neq m}}^{M} |c_k|^2, \tag{2.26}$$

which is the sum of all the other squared coefficients.

We notice that $\|\mathbf{y}_{\text{proj},\mathbf{x}_m}\|^2 + \left\|\mathbf{y} - \mathbf{y}_{\text{proj},\mathbf{x}_m}\right\|^2 = \|\mathbf{y}\|^2$, which is a consequence of the fact that $\mathbf{y}_{\text{proj},\mathbf{x}_m}$ is orthogonal to $\mathbf{y} - \mathbf{y}_{\text{proj},\mathbf{x}_m}$.

An *orthonormal basis* in $\mathbb{C}^M$ is a set of M vectors $\mathbf{b}_1, \ldots, \mathbf{b}_M$ that satisfies the following two conditions:

1. The vectors are mutually orthogonal, so that their inner products are $\mathbf{b}_i^{\text{H}} \mathbf{b}_j = 0$ for any choice of $i, j \in \{1, \ldots, M\}$ such that $i \neq j$;

2. The vectors have length one so that their norm is $\|\mathbf{b}_i\| = 1$ for all $i \in \{1, \ldots, M\}$.

There are many examples of orthonormal bases. One way of constructing it is to let $\mathbf{b}_i$ be 1 in entry i and zeros elsewhere. For $M = 4$, this results in

$$\mathbf{b}_1 = \begin{bmatrix} 1 \\ 0 \\ 0 \\ 0 \end{bmatrix}, \quad \mathbf{b}_2 = \begin{bmatrix} 0 \\ 1 \\ 0 \\ 0 \end{bmatrix}, \quad \mathbf{b}_3 = \begin{bmatrix} 0 \\ 0 \\ 1 \\ 0 \end{bmatrix}, \quad \mathbf{b}_4 = \begin{bmatrix} 0 \\ 0 \\ 0 \\ 1 \end{bmatrix}. \tag{2.27}$$

A common reason for defining an orthonormal basis is that any other vector $\mathbf{x} \in \mathbb{C}^M$ can be written as a linear combination of the M basis vectors: $\mathbf{x} = \sum_{i=1}^{M} c_i \mathbf{b}_i$ for some coefficients $c_1, \ldots, c_M$. This follows from the fact that any set of $M + 1$ vectors is linearly dependent in $\mathbb{C}^M$.

2.1.2 Matrix Analysis

A vector is a special case of a matrix. An $M \times K$ matrix has M rows and K columns, and contains MK entries. Let $h_{m,k} \in \mathbb{C}$ denote the entry at the mth row in the kth column. The full matrix can then be expressed as

$$\mathbf{H} = \begin{bmatrix} h_{1,1} & \ldots & h_{1,K} \\ \vdots & \ddots & \vdots \\ h_{M,1} & \ldots & h_{M,K} \end{bmatrix}. \tag{2.28}$$

We denote matrices using upper-case bold-faced letters, such as $\mathbf{H}$. The space of all complex matrices of size $M \times K$ is denoted as $\mathbb{C}^{M \times K}$; thus, we can write that $\mathbf{H} \in \mathbb{C}^{M \times K}$. The transpose and conjugate transpose are computed as

$$\mathbf{H}^{\mathrm{T}} = \begin{bmatrix} h_{1,1} & \ldots & h_{M,1} \\ \vdots & \ddots & \vdots \\ h_{1,K} & \ldots & h_{M,K} \end{bmatrix}, \quad \mathbf{H}^{\mathrm{H}} = \begin{bmatrix} h_{1,1}^* & \ldots & h_{M,1}^* \\ \vdots & \ddots & \vdots \\ h_{1,K}^* & \ldots & h_{M,K}^* \end{bmatrix}, \tag{2.29}$$

respectively. Note that $\mathbf{H}^{\mathrm{T}}$ is obtained by flipping the matrix over its diagonal, while $\mathbf{H}^{\mathrm{H}}$ is obtained by both flipping the matrix and replacing each entry by its complex conjugate. Both operations change the dimensions of the matrix: $\mathbf{H}^{\mathrm{T}}$ and $\mathbf{H}^{\mathrm{H}}$ belong to the space $\mathbb{C}^{K \times M}$ with all complex $K \times M$ matrices. Only in the square matrix case of $M = K$ is the dimensionality unchanged.

The columns of a matrix are important when analyzing its properties. Let $\mathbf{h}_1, \ldots, \mathbf{h}_K$ denote the K columns of an $M \times K$ matrix $\mathbf{H}$. We notice that each column is an M-dimensional vector. The *matrix-vector product* between the matrix $\mathbf{H}$ and a K-dimensional vector $\mathbf{c} = [c_1, \ldots, c_K]^{\mathrm{T}}$ is denoted as $\mathbf{Hc}$ and is an M-dimensional vector computed as

$$\mathbf{Hc} = \begin{bmatrix} h_{1,1}c_1 + \ldots + h_{1,K}c_K \\ \vdots \\ h_{M,1}c_1 + \ldots + h_{M,K}c_K \end{bmatrix} = c_1\mathbf{h}_1 + \ldots + c_K\mathbf{h}_K. \tag{2.30}$$

This is the linear combination of the column vectors of $\mathbf{H}$ using the corresponding entries of $\mathbf{c}$ as coefficients. Hence, the directions of the columns will determine which directions the vector $\mathbf{Hc}$ can have. In particular, we can never get a vector that is orthogonal to all the columns of $\mathbf{H}$.

A square matrix where all the off-diagonal entries are zero is called a *diagonal matrix*. If the diagonal of an $M \times M$ diagonal matrix $\mathbf{D}$ contains the entries $d_1, \ldots, d_M$, then the matrix is

$$\mathbf{D} = \begin{bmatrix} d_1 & 0 & \ldots & 0 \\ 0 & d_2 & \ddots & \vdots \\ \vdots & \vdots & \ddots & 0 \\ 0 & \ldots & 0 & d_M \end{bmatrix} \tag{2.31}$$

and will be written in short form as $\mathbf{D} = \operatorname{diag}(d_1, \ldots, d_M)$.

A diagonal matrix with only ones on the diagonal is known as an *identity matrix*. We will denote the $M \times M$ identity matrix as $\mathbf{I}_M$. The columns of an identity matrix are an orthonormal basis in $\mathbb{C}^M$, as exemplified in (2.27).

Non-diagonal square matrices can be transformed into diagonal matrices by a process known as diagonalization. We will summarize this process because it reveals several key properties of matrices, starting with the eigenvalues.

Definition 2.2. Consider an $M \times M$ matrix $\mathbf{A}$ and a non-zero vector $\mathbf{u} \in \mathbb{C}^M$. If

$$\mathbf{Au} = \lambda \mathbf{u} \tag{2.32}$$

for some scalar $\lambda \in \mathbb{C}$, then $\mathbf{u}$ is an *eigenvector* of $\mathbf{A}$ with λ being the corresponding *eigenvalue*.

The output of the matrix-vector product $\mathbf{Au}$ is generally a rotated and stretched version of $\mathbf{u}$. The unique property of an eigenvector $\mathbf{u}$ is that it is only stretched by the scalar factor λ (the eigenvalue). Two different matrices generally have different eigenvectors and eigenvalues.

Each M-dimensional matrix has M eigenvalues, which can be denoted as $\lambda_1, \ldots, \lambda_M$. There are two matrix operations that directly expose the eigenvalues. The first operation is the *trace* $\operatorname{tr}(\mathbf{A})$ that is defined as the sum of the diagonal entries of $\mathbf{A}$, but also has the property

$$\operatorname{tr}(\mathbf{A}) = \sum_{m=1}^{M} \lambda_m. \tag{2.33}$$

The second operation is the *determinant* $\det(\mathbf{A})$, which has a complicated definition and can be computed in multiple ways but satisfies the property

$$\det(\mathbf{A}) = \lambda_1 \cdot \lambda_2 \cdot \ldots \cdot \lambda_M. \tag{2.34}$$

Hence, the trace and determinant are the sum and product of the eigenvalues, respectively. The determinant is zero whenever one of the eigenvalues is zero.

The eigenvalue definition $\mathbf{Au} = \lambda \mathbf{u}$ in (2.32) is equivalent to $(\mathbf{A} - \lambda \mathbf{I}_M)\mathbf{u} = \mathbf{0}$, which means that $\mathbf{A} - \lambda \mathbf{I}_M$ must have a zero-valued eigenvalue. Hence, $\det(\mathbf{A} - \lambda \mathbf{I}_M) = 0$ and we can use this equation to identify the eigenvalue λ.

More generally, the *characteristic polynomial* of a matrix $\mathbf{A}$ is expressed as

$$\det(\mathbf{A} - \lambda \mathbf{I}_M) = (\lambda_1 - \lambda)(\lambda_2 - \lambda) \cdots (\lambda_M - \lambda), \tag{2.35}$$

where λ is the variable and the determinant plays an essential role. All the M eigenvalues are roots of the characteristic polynomial and vice versa. The same eigenvalue can appear multiple times in the characteristic polynomial.

Example 2.5. What are the eigenvalues of the 2×2 matrix

$$\mathbf{A} = \begin{bmatrix} 4 & -2 \\ 5 & -3 \end{bmatrix}? \tag{2.36}$$

The characteristic polynomial of this matrix is

$$\begin{aligned} \det(\mathbf{A} - \lambda \mathbf{I}_2) &= \det\left(\begin{bmatrix} 4-\lambda & -2 \\ 5 & -3-\lambda \end{bmatrix}\right) = (4-\lambda)(-3-\lambda) - 5(-2) \\ &= \lambda^2 - \lambda - 2 = (\lambda+1)(\lambda-2), \end{aligned} \tag{2.37}$$

where we utilized the property that the determinant of a 2×2 is the product of the diagonal entries minus the product of the off-diagonal entries. The roots to the characteristic polynomial are $\lambda_1 = -1$ and $\lambda_2 = 2$, which are also the eigenvalues of $\mathbf{A}$.

The *rank* of an $M \times K$ matrix equals the maximum number of linearly independent columns the matrix has. The rank is also equal to the maximum number of linearly independent rows. The rank can take any value between 0 and $\min(M, K)$; that is, the minimum of M and K. In the case of an $M \times M$ square matrix, the rank is greater than or equal to the number of non-zero eigenvalues. In fact, the rank is usually equal to the number of non-zero eigenvalues for the square matrices appearing in communications, but one can create counterexamples where this is not the case. Later in this section, we will provide additional conditions that guarantee equivalence.

Recall from (2.30) that the matrix-vector product $\mathbf{Hc}$ is computed as a linear combination of the columns of $\mathbf{H}$ with coefficients from $\mathbf{c}$. Suppose we want to create a set $\mathbf{Hc}_1, \mathbf{Hc}_2, \ldots$ of linearly independent vectors (or even mutually orthogonal vectors) by multiplying $\mathbf{H}$ by different vectors $\mathbf{c}_1, \mathbf{c}_2, \ldots$ The rank of $\mathbf{H}$ limits how many such vectors we can create. The rank property will be utilized in later chapters to quantify how many parallel data streams we can transmit over a communication channel, where the matrix dimensions represent antennas and/or frequency bands.

Example 2.6. Let $\mathbf{c}_1, \ldots, \mathbf{c}_K \in \mathbb{C}^K$ be K linearly independent vectors. For an arbitrary matrix $\mathbf{H} \in \mathbb{C}^{K \times K}$ that has rank r satisfying $r < K$, show that $\mathbf{H}\mathbf{c}_1, \ldots, \mathbf{H}\mathbf{c}_K$ cannot be linearly independent.

Since the rank of $\mathbf{H}$ is $r < K$ and the number of non-zero eigenvalues is smaller than or equal to the rank, $\mathbf{H}$ must have at least $K - r$ zero-valued eigenvalues. Consequently, there must exist an eigenvector $\mathbf{x} \neq \mathbf{0}$ satisfying $\mathbf{H}\mathbf{x} = \mathbf{0}$. Since $\mathbf{c}_1, \ldots, \mathbf{c}_K$ are linearly independent, any such non-zero $\mathbf{x} \in \mathbb{C}^K$ can be expressed as $\sum_{k=1}^{K} \alpha_k \mathbf{c}_k$ for some selection of the coefficients, with not all α_k being zero. Inserting $\mathbf{x} = \sum_{k=1}^{K} \alpha_k \mathbf{c}_k$ into $\mathbf{H}\mathbf{x} = \mathbf{0}$, we obtain

$$\mathbf{H}\left(\sum_{k=1}^{K} \alpha_k \mathbf{c}_k\right) = \sum_{k=1}^{K} \alpha_k \mathbf{H}\mathbf{c}_k = \mathbf{0}. \tag{2.38}$$

According to Definition 2.1, $\mathbf{H}\mathbf{c}_1, \ldots, \mathbf{H}\mathbf{c}_K$ are linearly independent if and only if the above linear system of equations (with respect to $\alpha_1, \ldots, \alpha_K \in \mathbb{C}$) only has the solution $\alpha_1 = \ldots = \alpha_K = 0$. However, for a non-zero eigenvector $\mathbf{x}$, we should have at least one non-zero α_k, which implies $\mathbf{H}\mathbf{c}_1, \ldots, \mathbf{H}\mathbf{c}_K$ cannot be linearly independent if the rank of $\mathbf{H}$ is strictly less than K.

Square matrices can be factorized and diagonalized using the eigenvalues and eigenvectors. For brevity, we will only present this *eigendecomposition* in the special case of symmetric matrices, which are defined as follows.

Definition 2.3. A matrix $\mathbf{A}$ is *Hermitian* if $\mathbf{A} = \mathbf{A}^{\mathrm{H}}$.

Only square matrices can be Hermitian, and the condition $\mathbf{A} = \mathbf{A}^{\mathrm{H}}$ implies a specific symmetry: the entries at the opposite sides of the diagonals have the same real part, while the imaginary parts have the same magnitude but opposite signs. The symmetry implies that any eigenvalue of $\mathbf{A}$ must satisfy $\lambda = \lambda^*$, which only holds if the imaginary part is zero. Hence, all the eigenvalues of Hermitian matrices must be real-valued. One common type of matrix that satisfies the Hermitian property is covariance matrices, which will be described later in this chapter. Before considering the eigendecomposition of Hermitian matrices, we will define one more type of matrix.

Definition 2.4. A matrix $\mathbf{U} \in \mathbb{C}^{M \times M}$ is *unitary* if $\mathbf{U}^{\mathrm{H}}\mathbf{U} = \mathbf{I}_M$ and $\mathbf{U}\mathbf{U}^{\mathrm{H}} = \mathbf{I}_M$. The former implies that the columns of $\mathbf{U}$ are mutually orthogonal, while the latter implies that the rows are mutually orthogonal.

A unitary matrix's column vectors are an orthonormal basis in $\mathbb{C}^M$. We notice that the conjugate transpose $\mathbf{U}^{\mathrm{H}}$ of a unitary matrix $\mathbf{U}$ acts as a *matrix inverse* because their multiplication results in an identity matrix. This is the matrix extension of how $1/u$ is the inverse of the scalar u because their multiplication is 1. If the eigenvectors of a Hermitian matrix are placed as the columns of a matrix, it will be a unitary matrix.

Lemma 2.1. Any Hermitian $M \times M$ matrix $\mathbf{A}$ can be factorized as

$$\mathbf{A} = \mathbf{U}\mathbf{D}\mathbf{U}^{\mathrm{H}}, \tag{2.39}$$

where $\mathbf{U}$ is a unitary $M \times M$ matrix containing the unit-length eigenvectors as columns and $\mathbf{D} = \mathrm{diag}(\lambda_1, \ldots, \lambda_M)$ is a diagonal matrix containing the corresponding real-valued eigenvalues.

The factorization in (2.39) is known as the eigendecomposition. For a Hermitian matrix, the rank is exactly equal to the number of non-zero eigenvalues. If we let $\mathbf{u}_1, \ldots, \mathbf{u}_M$ denote the columns of $\mathbf{U}$ (i.e., the eigenvectors), then we can also express (2.39) as

$$\mathbf{A} = \sum_{m=1}^{M} \lambda_m \mathbf{u}_m \mathbf{u}_m^{\mathrm{H}}. \tag{2.40}$$

Hence, the matrix is the summation of the eigenvalues multiplied by the respective eigenvectors. This property can be utilized to diagonalize the matrix. More precisely, we can rearrange (2.39) as

$$\mathbf{U}^{\mathrm{H}}\mathbf{A}\mathbf{U} = \mathbf{D} \tag{2.41}$$

by utilizing the properties of unitary matrices. This shows how the Hermitian matrix $\mathbf{A}$ can be transformed into the diagonal matrix $\mathbf{D}$ with eigenvalues by multiplying with the matrix $\mathbf{U}$ containing the eigenvectors.

Non-Hermitian square matrices can also be diagonalized, but the notation is more complicated, and one can find special cases where it is not possible. Since we will not utilize those results, we will not cover them here.

If all the eigenvalues of a Hermitian matrix $\mathbf{A}$ are non-zero, then the matrix is invertible. This implies that there exists a matrix denoted as $\mathbf{A}^{-1}$ with the property that $\mathbf{A}\mathbf{A}^{-1} = \mathbf{A}^{-1}\mathbf{A} = \mathbf{I}_M$. By utilizing the eigendecomposition in (2.39), we can notice that the inverse can be computed as

$$\mathbf{A}^{-1} = \mathbf{U}\mathbf{D}^{-1}\mathbf{U}^{\mathrm{H}}, \tag{2.42}$$

where $\mathbf{D}^{-1} = \mathrm{diag}(\lambda_1^{-1}, \ldots, \lambda_M^{-1})$. Hence, the inverse matrix has the same eigenvectors but reciprocal eigenvalues.

If all the eigenvalues of a Hermitian matrix $\mathbf{A}$ are non-negative, then the matrix is said to be *positive semi-definite*. The reason is that $\mathbf{x}^{\mathrm{H}}\mathbf{A}\mathbf{x} \geq 0$ for all vectors $\mathbf{x}$ of matching dimension, because (2.40) implies that $\mathbf{x}^{\mathrm{H}}\mathbf{A}\mathbf{x} = \sum_{m=1}^{M} \lambda_m |\mathbf{u}_m^{\mathrm{H}}\mathbf{x}|^2$ which only has non-negative terms. For such matrices, we can define the square root of the matrix as follows.

Lemma 2.2. Any Hermitian $M \times M$ matrix $\mathbf{A}$ that is also positive semi-definite has a *square root* defined as

$$\mathbf{A}^{1/2} = \mathbf{U}\mathbf{D}^{1/2}\mathbf{U}^{\mathrm{H}}, \tag{2.43}$$

using the notation from Lemma 2.1 with $\mathbf{D}^{1/2} = \mathrm{diag}(\sqrt{\lambda_1}, \ldots, \sqrt{\lambda_M})$. The square root $\mathbf{A}^{1/2}$ is also Hermitian and satisfies the property $\mathbf{A}^{1/2}\mathbf{A}^{1/2} = \mathbf{A}$.

If all the eigenvalues of the Hermitian matrix $\mathbf{A}$ are strictly positive, then the matrix is said to be *positive definite*. In this case, both the matrix and its square root are invertible. The inverse square root is denoted as

$$\mathbf{A}^{-1/2} = \mathbf{U}\mathbf{D}^{-1/2}\mathbf{U}^{\mathrm{H}}, \tag{2.44}$$

where $\mathbf{D}^{-1/2} = \mathrm{diag}(1/\sqrt{\lambda_1}, \ldots, 1/\sqrt{\lambda_M})$.

Example 2.7. Consider a Hermitian matrix $\mathbf{A} \in \mathbb{C}^{M \times M}$ with the eigendecomposition

$$\mathbf{A} = \mathbf{U}\mathbf{D}\mathbf{U}^{\mathrm{H}}. \tag{2.45}$$

What is the eigendecomposition of $\mathbf{B} = \mathbf{A} + \epsilon\mathbf{I}_M$ if $\epsilon > 0$?

Since $\mathbf{U}$ is a unitary matrix (i.e., $\mathbf{U}\mathbf{U}^{\mathrm{H}} = \mathbf{I}_M$), we can express $\mathbf{B}$ as

$$\mathbf{B} = \mathbf{A} + \epsilon\mathbf{I}_M = \mathbf{U}\mathbf{D}\mathbf{U}^{\mathrm{H}} + \epsilon\mathbf{U}\mathbf{U}^{\mathrm{H}} = \mathbf{U}\left(\mathbf{D} + \epsilon\mathbf{I}_M\right)\mathbf{U}^{\mathrm{H}}, \tag{2.46}$$

which has the correct structure to be its eigendecomposition. Hence, adding a scaled identity matrix to $\mathbf{A}$ does not change the eigenvectors, but all the eigenvalues are increased by the scaling factor ϵ.

The following *matrix inversion lemma* can be helpful when analyzing expressions containing invertible matrices.

Lemma 2.3. Consider the matrices $\mathbf{A} \in \mathbb{C}^{M \times M}$, $\mathbf{B} \in \mathbb{C}^{M \times N}$, $\mathbf{C} \in \mathbb{C}^{N \times N}$, and $\mathbf{D} \in \mathbb{C}^{N \times M}$. The following identity holds if all the involved inverses exist:

$$(\mathbf{A} + \mathbf{B}\mathbf{C}\mathbf{D})^{-1} = \mathbf{A}^{-1} - \mathbf{A}^{-1}\mathbf{B}(\mathbf{D}\mathbf{A}^{-1}\mathbf{B} + \mathbf{C}^{-1})^{-1}\mathbf{D}\mathbf{A}^{-1}. \tag{2.47}$$

A special case of this lemma, known as the *rank-one update formula*, is obtained when $\mathbf{A}$ is an invertible Hermitian matrix, $\mathbf{C} = 1$, $\mathbf{B} = \mathbf{x} \in \mathbb{C}^M$ is a vector, and $\mathbf{D} = \mathbf{x}^{\mathrm{H}}$:

$$(\mathbf{A} + \mathbf{x}\mathbf{x}^{\mathrm{H}})^{-1} = \mathbf{A}^{-1} - \frac{1}{1 + \mathbf{x}^{\mathrm{H}}\mathbf{A}^{-1}\mathbf{x}}\mathbf{A}^{-1}\mathbf{x}\mathbf{x}^{\mathrm{H}}\mathbf{A}^{-1}. \tag{2.48}$$

If we multiply the expression in (2.48) by $\mathbf{x}$ from the right, we obtain

$$(\mathbf{A} + \mathbf{x}\mathbf{x}^{\mathrm{H}})^{-1}\mathbf{x} = \frac{1}{1 + \mathbf{x}^{\mathrm{H}}\mathbf{A}^{-1}\mathbf{x}}\mathbf{A}^{-1}\mathbf{x}, \tag{2.49}$$

which shows that the vectors $\mathbf{A}^{-1}\mathbf{x}$ and $(\mathbf{A}+\mathbf{x}\mathbf{x}^{\mathrm{H}})^{-1}\mathbf{x}$ are equal except for the scaling factor $1+\mathbf{x}^{\mathrm{H}}\mathbf{A}^{-1}\mathbf{x}$. This property will be utilized in this book when analyzing different signal processing methods.

Consider two matrices $\mathbf{A} \in \mathbb{C}^{M\times K}$ and $\mathbf{B} \in \mathbb{C}^{K\times M}$ having opposite dimensions, which means that it is feasible to compute both the matrix products $\mathbf{AB}$ and $\mathbf{BA}$. A matrix identity similar to the matrix inversion lemma is

$$\begin{aligned}(\mathbf{AB}+\mathbf{I}_M)^{-1}\mathbf{A} &= (\mathbf{AB}+\mathbf{I}_M)^{-1}\mathbf{A}(\mathbf{BA}+\mathbf{I}_K)(\mathbf{BA}+\mathbf{I}_K)^{-1}\\ &= (\mathbf{AB}+\mathbf{I}_M)^{-1}(\mathbf{AB}+\mathbf{I}_M)\mathbf{A}(\mathbf{BA}+\mathbf{I}_K)^{-1}\\ &= \mathbf{A}(\mathbf{BA}+\mathbf{I}_K)^{-1},\end{aligned} \tag{2.50}$$

where the matrix $\mathbf{A}$ is moved from one side of the inverse to the other side. The content of the inverse is also changing and, interestingly, the identity matrix changes dimension. There is a deeper matrix algebraic property enabling this result. The eigenvalues of $\mathbf{AB}$ and $\mathbf{BA}$ are always the same, except that the bigger of these matrices has $|M-K|$ extra eigenvalues that are equal to zero. This can be proved as follows. We let $\mathbf{u}$ denote an arbitrary eigenvector of $\mathbf{AB}$ associated with the eigenvalue λ, so that $\mathbf{ABu}=\lambda\mathbf{u}$. It then follows that $\mathbf{Bu}$ is an eigenvector of $\mathbf{BA}$ with the same eigenvalue λ because

$$\lambda\mathbf{Bu} = \mathbf{B}(\lambda\mathbf{u}) = \mathbf{B}(\mathbf{ABu}) = \mathbf{BA}(\mathbf{Bu}). \tag{2.51}$$

One can further prove that the eigenvalue multiplicity is the same. A consequence is that we can switch the matrix order in the trace function as

$$\mathrm{tr}(\mathbf{AB}) = \mathrm{tr}(\mathbf{BA}) \tag{2.52}$$

because the sum of the eigenvalues is the same for $\mathbf{AB}$ and $\mathbf{BA}$. Another consequence is *Sylvester's determinant theorem*

$$\det(\mathbf{AB}+\mathbf{I}_M) = \det(\mathbf{BA}+\mathbf{I}_K), \tag{2.53}$$

which holds because the identity matrix adds one to all the eigenvalues, and the determinant then multiplies them together. The matrix identities in (2.52) and (2.53) will be used repeatedly in this book.

Consider the two vectors $\mathbf{x} = [x_1, x_2, \ldots, x_M]^{\mathrm{T}}$ and $\mathbf{y} = [y_1, y_2, \ldots, y_K]^{\mathrm{T}}$, which might have different dimensions. The *Kronecker product* between these vectors is defined as

$$\mathbf{x}\otimes\mathbf{y} = \begin{bmatrix} x_1\mathbf{y}\\ x_2\mathbf{y}\\ \vdots\\ x_M\mathbf{y}\end{bmatrix}, \tag{2.54}$$

which is an MK-dimensional vector. The first K entries contain x_1 multiplied by each of the entries of $\mathbf{y}$, the next K entries contain x_2 multiplied by each of the entries of $\mathbf{y}$, etc. The Kronecker product is closely related to the outer product $\mathbf{y}\mathbf{x}^{\mathrm{T}}$ between the same vectors. One obtains the Kronecker product by stacking the columns of $\mathbf{y}\mathbf{x}^{\mathrm{T}}$ into a single vector.

2.2 Probability Theory

This book will use random variables to describe signals, noise, and communication channels. Any continuous random variable x is entirely determined by its probability density function (PDF), which we denote by $f_{\mathsf{x}}(x)$. This function determines how the probability mass is distributed over all possible realizations. The realizations of the random variable take values in some *sample set* Ω, which is typically the real space $\mathbb{R}$ or the complex space $\mathbb{C}$. The probability of obtaining a realization in a subset $\mathcal{A} \subset \Omega$ of the sample set is the integral of the PDF over that subset:

$$\Pr\{x \in \mathcal{A}\} = \int_{\mathcal{A}} f_{\mathsf{x}}(x) \partial x. \tag{2.55}$$

When considering complex random variables, the integral in (2.55) should be interpreted as a double-integral over the real and imaginary parts. The PDF $f_{\mathsf{x}}(x)$ is non-negative for all $x \in \Omega$ and the total probability is one: $\int_{\Omega} f_{\mathsf{x}}(x) \partial x = 1$. Hence, the probability $\Pr\{x \in \mathcal{A}\}$ is between zero and one.

Based on the PDF, we can compute the (arithmetic) *mean*

$$\mathbb{E}\{x\} = \int_{\Omega} x f_{\mathsf{x}}(x) \partial x, \tag{2.56}$$

which is also known as the expected value, first moment, and average. The variability is often measured by computing the squared deviation $|x - \mathbb{E}\{x\}|^2$ from the mean and taking its mean. It is denoted $\mathbb{V}\mathrm{ar}\{x\}$ and computed as

$$\mathbb{V}\mathrm{ar}\{x\} = \mathbb{E}\{|x - \mathbb{E}\{x\}|^2\} = \mathbb{E}\{|x|^2\} - |\mathbb{E}\{x\}|^2. \tag{2.57}$$

This is known as the *variance* or second moment, and it measures how large variations from the mean we can expect to observe when generating many realizations. It is essential to use magnitudes in (2.57) when the random variable takes complex values. If the random variable has zero mean, then (2.57) shows that the variance coincides with the quadratic mean computed as

$$\mathbb{E}\{|x|^2\} = \int_{\Omega} |x|^2 f_{\mathsf{x}}(x) \partial x. \tag{2.58}$$

It is common in the probability theory literature to use a different notation for the random variable and its realizations; for example, x for the variable and x as the realization. In this book, we have instead chosen to use the same notation but write out what is considered in each context.

Definition 2.5. The random variables x and y are *statistically independent* if their *joint PDF* $f_{\mathsf{x},\mathsf{y}}(x, y)$ can be factorized as

$$f_{\mathsf{x},\mathsf{y}}(x, y) = f_{\mathsf{x}}(x) f_{\mathsf{y}}(y), \tag{2.59}$$

where $f_{\mathsf{x}}(x)$ and $f_{\mathsf{y}}(y)$ are their individual PDFs, called *marginal PDFs*.

This independence concept is entirely different from the linear independence of vectors in Definition 2.1. Statistical independence of random variables implies that the realization of x will not affect the realization of y whatsoever, which happens in practice when the variables are associated with different sources of randomness. For example, in communications, the variable representing random data from the transmitter is typically independent of the variable representing random thermal noise in the receiver hardware.

We will now consider L independent realizations of the same random variable, which can be thought of as having L independent and identically distributed random variables (i.e., with the same marginal PDF), and generate one realization from each of them. Suppose we compute the arithmetic average of these realizations. In that case, we will obtain a value close to the mean in (2.56), at least under the technical condition that the variance is finite. This result can be formalized mathematically as the following *law of large numbers*.

Lemma 2.4. Let $x_1, \ldots, x_L$ be a sequence of L independent and identically distributed random variables with mean $\mathbb{E}\{x_i\} = \mu$ and finite variance σ^2 for $i = 1, \ldots, L$. The arithmetic sample average $\frac{1}{L}\sum_{i=1}^{L} x_i$ satisfies

$$\lim_{L \to \infty} \frac{1}{L} \sum_{i=1}^{L} x_i = \mu. \tag{2.60}$$

We will utilize this lemma when studying the impact of random variables on communication performance and also as a way to approximate an unknown mean value using multiple realizations from the random variable.

The variance measures the average squared deviation, which has a different unit than the original variable (i.e., it is squared). The square root $\sqrt{\mathbb{V}\text{ar}\{x\}}$ of the variance can be utilized to understand better how large deviations from the mean are likely to occur. This measure is called the *standard deviation*, and whenever the variance is finite, most random realizations will occur within a few standard deviations from the mean. The exact characteristics depend on the distribution of the random variable, but the following worst-case result known as *Chebyshev's inequality* can be established.

Lemma 2.5. Consider a random variable x with mean $\mathbb{E}\{x\} = \mu$ and finite standard deviation $\sigma = \sqrt{\mathbb{V}\text{ar}\{x\}}$. For any constant $k > 0$, it holds that

$$\Pr\{|x - \mu| \geq k\sigma\} \leq \frac{1}{k^2}. \tag{2.61}$$

Suppose we insert $k = 2$ or $k = 3$ into Lemma 2.5. In that case, the inequality says that the probability of obtaining realizations that are more than two or three standard deviations from the mean is smaller than 0.25 and

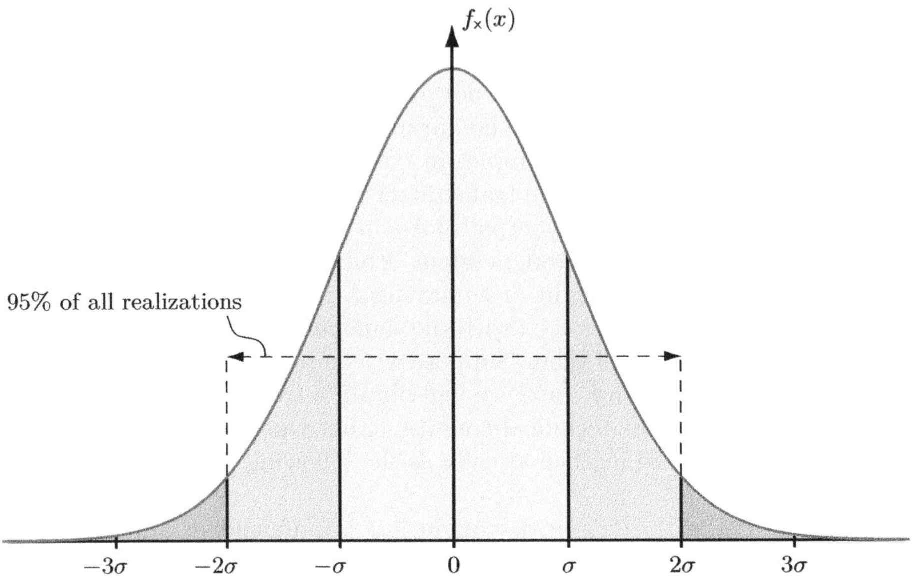

Figure 2.5: The PDF of the zero-mean Gaussian distribution $x \sim \mathcal{N}(0, \sigma^2)$ with the standard deviation indicated. If another mean value is considered, the PDF is shifted to be centered around it. 95% of all realizations occur between -2σ and 2σ.

0.11, respectively:

$$\Pr\{|x - \mu| \geq k\sigma\} \leq \begin{cases} 0.25 & \text{if } k = 2, \\ 0.11 & \text{if } k = 3. \end{cases} \tag{2.62}$$

Since Chebyshev's inequality provides an upper bound on the probability of obtaining realizations further than k standard deviations from the mean, most random distributions have a much smaller probability than that. In other words, Chebyshev's inequality characterizes the worst-case situation of having a distribution with a high probability of realizations far from the mean.

2.2.1 Gaussian Distribution

A common example is a *Gaussian random variable*, which is denoted as $x \sim \mathcal{N}(\mu, \sigma^2)$ and has the PDF

$$f_x(x) = \frac{1}{\sqrt{2\pi\sigma^2}} e^{-\frac{(x-\mu)^2}{2\sigma^2}}. \tag{2.63}$$

This distribution has the mean $\mathbb{E}\{x\} = \mu$, variance $\mathbb{V}\text{ar}\{x\} = \mathbb{E}\{(x-\mu)^2\} = \sigma^2$, and standard deviation $\sqrt{\mathbb{V}\text{ar}\{x\}} = \sigma$. The PDF is illustrated in Figure 2.5 and is symmetric around the mean value. When the mean is zero, and the variance is one, we have a *standard Gaussian distribution*.

The Gaussian distribution is also known as the *normal distribution* since it has become the norm to utilize it as an approximation of other random distributions. A contributing factor is the following classical result, called the *central limit theorem*.

Lemma 2.6. Let $x_1, \ldots, x_L$ be a sequence of L real-valued independent and identically distributed random variables with zero mean and finite variance σ^2. As $L \to \infty$, the distribution of

$$\frac{1}{\sqrt{L\sigma^2}} \sum_{i=1}^{L} x_i \tag{2.64}$$

converges to a standard Gaussian distribution $\mathcal{N}(0, 1)$.

The interpretation of this theorem is that the summation of a set of independent and identically distributed random variables tends to be approximately Gaussian distributed, with the approximation error being smaller the more variables are considered. This property is often used in communications to motivate that the noise in the receiver hardware is Gaussian distributed (because the random motion of many electrons creates it) and that wireless channels behave as Gaussian distributed when they contain many propagation paths, which will be considered later in this book.

The scaling factor $1/\sqrt{L\sigma^2}$ in (2.64) was selected so that the variance of the quantity becomes one, instead of going to zero or infinity when adding L terms and letting $L \to \infty$. However, any scaling factor can be utilized along with Lemma 2.6 if the central limit theorem is merely used to motivate that the summation of a finite number of independent random variables is approximately Gaussian distributed. For example, the law of large numbers in Lemma 2.4 considered the sample average and when combined with Lemma 2.6, we obtain

$$\frac{1}{L} \sum_{i=1}^{L} x_i \mathrel{\dot\sim} \mathcal{N}\left(\mu, \frac{\sigma^2}{L}\right), \tag{2.65}$$

where the notation $\dot\sim$ means *approximately distributed as.* The variance in (2.65) goes to zero as $L \to \infty$, which implies that the sample average converges to the mean μ, as previously stated in the law of large numbers. The added benefit of (2.65) is that it also suggests that the variance goes to zero as $1/L$ and that the deviation from the mean is approximately Gaussian distributed.

The Gaussian distribution has unbounded support (i.e., we can get arbitrarily large positive or negative realizations), but the probability mass is concentrated around the mean. In fact, it is much more concentrated than the worst-case situation determined by Chebyshev's inequality. The probabilities of obtaining realizations that are beyond one, two, or three standard deviations away from the mean value are

$$\Pr\{|x-\mu| \geq k\sigma\} = 1 - \int_{-k\sigma}^{k\sigma} \frac{1}{\sqrt{2\pi\sigma^2}} e^{-\frac{x^2}{2\sigma^2}} \partial x \approx \begin{cases} 0.32 & \text{if } k=1, \\ 0.05 & \text{if } k=2, \\ 0.003 & \text{if } k=3. \end{cases} \tag{2.66}$$

Hence, only 5% of all realizations are beyond two standard deviations from the mean, while (2.62) states that it can be the case for up to 25% of all realizations when considering an arbitrary random distribution. Figure 2.5 illustrates that 95% of all realizations appear from -2σ to 2σ.

2.2.2 Complex Gaussian Distribution

We will now consider the complex generalization of the Gaussian distribution. Suppose $a, b \sim \mathcal{N}(0, \sigma^2/2)$ are two independent Gaussian variables, each having zero mean and variance $\sigma^2/2$. The complex variable $x = a + jb$ will then have a complex Gaussian distribution. We denote it as $x \sim \mathcal{N}_{\mathbb{C}}(0, \sigma^2)$ and the PDF is

$$f_{\mathsf{x}}(x) = \frac{1}{\pi\sigma^2} e^{-\frac{|x|^2}{\sigma^2}}. \tag{2.67}$$

This distribution has the mean $\mathbb{E}\{x\} = 0$ and variance

$$\mathbb{V}\text{ar}\{x\} = \mathbb{E}\{|x|^2\} = \mathbb{E}\{a^2 + b^2\} = \sigma^2, \tag{2.68}$$

where the real and imaginary parts each contribute with $\sigma^2/2$. The PDF in (2.67) is illustrated in Figure 2.6 and has the classical shape of a Gaussian distribution but in two dimensions. There are other types of complex Gaussian distributions than the one described above. To be precise, we have defined what is known as the *circularly symmetric complex Gaussian distribution*. The circular symmetry refers to the fact that if $x \sim \mathcal{N}_{\mathbb{C}}(0, \sigma^2)$, then $xe^{\mathrm{j}\psi}$ has the same distribution for any value of $\psi \in \mathbb{R}$. In other words, the distribution does not change when applying a phase-shift. This property can be proved by noticing that $f_{\mathsf{x}}(x) = f_{\mathsf{x}}(xe^{\mathrm{j}\psi})$ for the PDF in (2.67). The circular symmetry implies that we can rotate the PDF in the complex plane without changing its shape, as seen from Figure 2.6. Looking at the mean value, the circular symmetry implies $\mathbb{E}\{x\} = \mathbb{E}\{xe^{\mathrm{j}\psi}\} = e^{\mathrm{j}\psi}\mathbb{E}\{x\}$, which only holds for all $\psi \in \mathbb{R}$ if $\mathbb{E}\{x\} = 0$. Hence, all circularly symmetric distributions have zero means. The circular symmetry follows from the assumptions of having independent and identically distributed real and imaginary parts. One can define other complex Gaussian distributions that do not satisfy these conditions, but these are not considered in this book. We will refer to the circularly symmetric complex Gaussian distribution as the *complex Gaussian* distribution for brevity.

Multiplying a complex Gaussian distribution with a constant $c \in \mathbb{C}$ will change the variance but not the shape of the distribution. Suppose $x \sim \mathcal{N}_{\mathbb{C}}(0, \sigma^2)$ and recall that the variance can be computed as $\mathbb{E}\{|x|^2\} = \sigma^2$ since

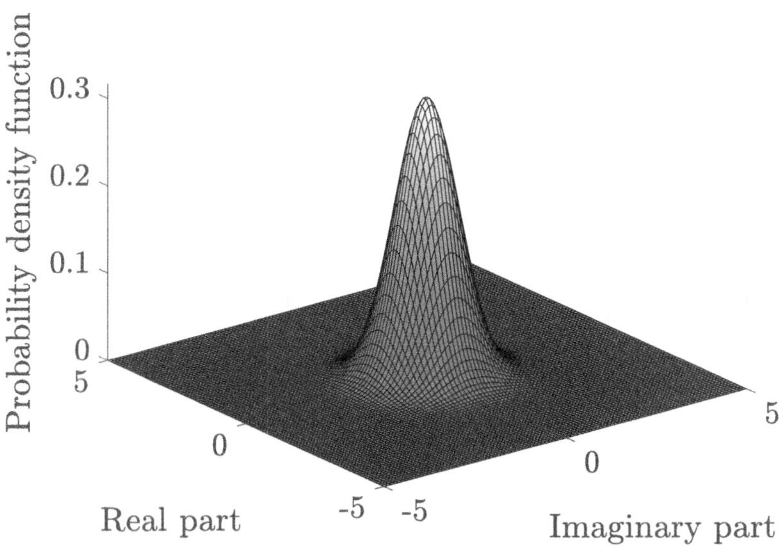

Figure 2.6: The PDF of the circularly symmetric complex Gaussian distribution $x \sim \mathcal{N}_{\mathbb{C}}(0, 1)$. The real and imaginary parts are statistically independent and jointly Gaussian distributed with identical variance.

the variable has zero mean. The random variable cx will, therefore, have the variance $\mathbb{E}\{|cx|^2\} = |c|^2\mathbb{E}\{|x|^2\} = |c|^2\sigma^2$. This implies that $cx \sim \mathcal{N}_{\mathbb{C}}(0, |c|^2\sigma^2)$.

2.2.3 Covariance and Conditional Distribution

Multiple random variables can affect a communication system, some independent (see Definition 2.5) and others statistically dependent. Consider the two independent random variables $v \sim \mathcal{N}_{\mathbb{C}}(0, \sigma_v^2)$ and $w \sim \mathcal{N}_{\mathbb{C}}(0, \sigma_w^2)$. The summation of these variables is also complex Gaussian distributed and has a variance that is the summation of the individual variances:

$$z = v + w \sim \mathcal{N}_{\mathbb{C}}(0, \sigma_v^2 + \sigma_w^2). \tag{2.69}$$

Although v and w are independent variables, z is clearly dependent on both.

The variance concept can be extended to measure the *covariance* between two random variables. For two arbitrary random variables z and v, the covariance is defined as

$$\mathbb{E}\{(z - \mathbb{E}\{z\})(v - \mathbb{E}\{v\})^*\} = \mathbb{E}\{zv^*\} - \mathbb{E}\{z\}\mathbb{E}\{v^*\}, \tag{2.70}$$

where the complex conjugate is important when the variables are complex. The variables are said to be *uncorrelated* if the covariance is zero, while a non-zero covariance measures how strongly the random realization of one variable affects the realization of the other variable. Independent random variables are always uncorrelated, but the converse might not hold: uncorrelated variables can still influence each others' realizations but in more subtle ways.

The covariance in (2.70) can be both positive and negative, and takes values between $-\sqrt{\mathbb{V}\text{ar}\{z\}\mathbb{V}\text{ar}\{v\}}$ and $\sqrt{\mathbb{V}\text{ar}\{z\}\mathbb{V}\text{ar}\{v\}}$. The bounds are achieved when the two variables are equal except for a negative/positive scaling factor.

Example 2.8. What is the covariance between z and v, defined in (2.69)?

Direct computation based on the covariance definition in (2.70) yields

$$\mathbb{E}\{(z-\mathbb{E}\{z\})(v-\mathbb{E}\{v\})^*\} = \mathbb{E}\{zv^*\} = \mathbb{E}\{vv^*\} + \mathbb{E}\{wv^*\} = \sigma_v^2, \tag{2.71}$$

where the last equality follows from the fact that $\mathbb{E}\{wv^*\} = \mathbb{E}\{w\}\mathbb{E}\{v^*\} = 0$ since w and v are independent. The non-zero covariance demonstrates that z and v are dependent random variables and implies that their realizations are statistically connected, which is logical since $z = v + w$.

Suppose we can observe z but want to know the value of v. We are then interested in the conditional PDF $f_{\mathsf{v}|\mathsf{z}}(v|z)$ of v given the realization of z. If we know the opposite conditional PDF $f_{\mathsf{z}|\mathsf{v}}(z|v)$, we can compute $f_{\mathsf{v}|\mathsf{z}}(v|z)$ using *Bayes' theorem*:

$$f_{\mathsf{v}|\mathsf{z}}(v|z) = \frac{f_{\mathsf{z}|\mathsf{v}}(z|v) f_{\mathsf{v}}(v)}{f_{\mathsf{z}}(z)}. \tag{2.72}$$

This rule says that $f_{\mathsf{v}|\mathsf{z}}(v|z)$ and $f_{\mathsf{z}|\mathsf{v}}(z|v)$ are equal up to the scaling factor $f_{\mathsf{v}}(v)/f_{\mathsf{z}}(z)$. We can compute this factor using the marginal PDFs of z and v.

Example 2.9. Determine the conditional PDFs $f_{\mathsf{z}|\mathsf{v}}(z|v)$ and $f_{\mathsf{v}|\mathsf{z}}(v|z)$ that relate the random variables v and z that were defined in (2.69).

If we know v, then $z - v = w \sim \mathcal{N}_{\mathbb{C}}(0, \sigma_w^2)$. This implies that

$$f_{\mathsf{z}|\mathsf{v}}(z|v) = \frac{1}{\pi\sigma_w^2} e^{-\frac{|z-v|^2}{\sigma_w^2}}. \tag{2.73}$$

We can now compute $f_{\mathsf{v}|\mathsf{z}}(v|z)$ using Bayes' theorem in (2.72):

$$\begin{aligned} f_{\mathsf{v}|\mathsf{z}}(v|z) &= \frac{\frac{1}{\pi\sigma_w^2} e^{-\frac{|z-v|^2}{\sigma_w^2}} \frac{1}{\pi\sigma_v^2} e^{-\frac{|v|^2}{\sigma_v^2}}}{\frac{1}{\pi(\sigma_v^2+\sigma_w^2)} e^{-\frac{|z|^2}{\sigma_v^2+\sigma_w^2}}} = \frac{\sigma_v^2+\sigma_w^2}{\pi\sigma_v^2\sigma_w^2} e^{-\frac{|z-v|^2}{\sigma_w^2} - \frac{|v|^2}{\sigma_v^2} + \frac{|z|^2}{\sigma_v^2+\sigma_w^2}} \\ &= \frac{\sigma_v^2+\sigma_w^2}{\pi\sigma_v^2\sigma_w^2} e^{-\frac{\sigma_v^2+\sigma_w^2}{\sigma_v^2\sigma_w^2}\left|v - \frac{\sigma_v^2}{\sigma_v^2+\sigma_w^2} z\right|^2}. \end{aligned} \tag{2.74}$$

This conditional PDF resembles that of the complex Gaussian distribution. In particular, $v - \frac{\sigma_v^2}{\sigma_v^2+\sigma_w^2} z \sim \mathcal{N}_{\mathbb{C}}\left(0, \frac{\sigma_v^2\sigma_w^2}{\sigma_v^2+\sigma_w^2}\right)$ when z is known.

2.2.4 Multivariate Complex Gaussian Distribution

A random vector can be created by taking a collection of M random scalar variables $x_1, \ldots, x_M$ and collecting them in a vector

$$\mathbf{x} = \begin{bmatrix} x_1 \\ \vdots \\ x_M \end{bmatrix}. \tag{2.75}$$

This is also known as a multivariate random variable, and the mean is denoted as $\mathbb{E}\{\mathbf{x}\}$. The variance of the individual entries and the covariance between any pair of entries is captured by the *covariance matrix* $\mathrm{Cov}\{\mathbf{x}\}$ defined as

$$\mathrm{Cov}\{\mathbf{x}\} = \mathbb{E}\{(\mathbf{x} - \mathbb{E}\{\mathbf{x}\})(\mathbf{x} - \mathbb{E}\{\mathbf{x}\})^{\mathrm{H}}\}. \tag{2.76}$$

If we take the conjugate (Hermitian) transpose of this expression, we will get the same expression, which shows that all covariance matrices are Hermitian matrices (see Definition 2.3). The covariance matrix is also positive semi-definite because, for any deterministic $\mathbf{y} \in \mathbb{C}^M$, it holds that

$$\mathbf{y}^{\mathrm{H}}\mathrm{Cov}\{\mathbf{x}\}\mathbf{y} = \mathbb{E}\{\mathbf{y}^{\mathrm{H}}(\mathbf{x} - \mathbb{E}\{\mathbf{x}\})(\mathbf{x} - \mathbb{E}\{\mathbf{x}\})^{\mathrm{H}}\mathbf{y}\} = \mathbb{E}\left\{|\mathbf{y}^{\mathrm{H}}(\mathbf{x} - \mathbb{E}\{\mathbf{x}\})|^2\right\} \geq 0. \tag{2.77}$$

The *correlation matrix* is similarly defined as $\mathbb{E}\{\mathbf{x}\mathbf{x}^{\mathrm{H}}\}$ without subtracting the mean. This implies that a deterministic vector $\mathbf{x}$ has a zero-valued covariance matrix but $\mathbf{x}\mathbf{x}^{\mathrm{H}}$ as its correlation matrix. Hence, the covariance matrix is a better measure of the amount of randomness in the considered vector.

Suppose the M variables are independent and identically distributed complex Gaussian variables with variance σ^2; that is, $x_m \sim \mathcal{N}_{\mathbb{C}}(0, \sigma^2)$ for $m = 1, \ldots, M$. The mean value is $\mathbb{E}\{\mathbf{x}\} = \mathbf{0}$ since each of the individual variables has a zero mean. Moreover, the covariance matrix is

$$\mathrm{Cov}\{\mathbf{x}\} = \mathbb{E}\{(\mathbf{x} - \mathbb{E}\{\mathbf{x}\})(\mathbf{x} - \mathbb{E}\{\mathbf{x}\})^{\mathrm{H}}\} = \mathbb{E}\{\mathbf{x}\mathbf{x}^{\mathrm{H}}\} = \sigma^2\mathbf{I}_M, \tag{2.78}$$

where the diagonal entries are the variances of the individual entries and the zero-valued off-diagonal entries represent that the independent variables have zero covariance. This multivariate complex Gaussian distribution with independent entries is denoted as

$$\mathbf{x} \sim \mathcal{N}_{\mathbb{C}}(\mathbf{0}, \sigma^2\mathbf{I}_M). \tag{2.79}$$

This distribution is often utilized to model receiver noise in communication systems. It is then referred to as *white* Gaussian noise, where the color (or lack thereof) indicates the independence of the entries. Following Definition 2.5, the PDF of $\mathbf{x}$ is the product of M marginal PDFs of the kind in (2.67):

$$f_{\mathbf{x}}(\mathbf{x}) = \prod_{m=1}^{M} \frac{1}{\pi\sigma^2} e^{-\frac{|x_m|^2}{\sigma^2}} = \frac{1}{(\pi\sigma^2)^M} e^{-\frac{\|\mathbf{x}\|^2}{\sigma^2}}. \tag{2.80}$$

In this book, we will also consider a complex Gaussian random vector with correlated entries, in which case the covariance matrix is not an identity matrix. We can create such a matrix by starting from a K-length unit-variance complex Gaussian random vector with independent entries $\tilde{\mathbf{x}} \sim \mathcal{N}_{\mathbb{C}}(\mathbf{0}, \mathbf{I}_K)$ and an $M \times K$ deterministic matrix $\mathbf{A}$ with $M \leq K$. We can then create an M-length complex Gaussian random vector $\mathbf{x}$ by computing the product

$$\mathbf{x} = \mathbf{A}\tilde{\mathbf{x}}, \tag{2.81}$$

irrespective of whether M and K are the same or different. This new random vector has zero mean since

$$\mathbb{E}\{\mathbf{x}\} = \mathbf{A} \underbrace{\mathbb{E}\{\tilde{\mathbf{x}}\}}_{=\mathbf{0}} = \mathbf{0}. \tag{2.82}$$

The covariance matrix can be computed as

$$\begin{aligned} \mathrm{Cov}\{\mathbf{x}\} &= \mathbb{E}\{(\mathbf{x} - \mathbb{E}\{\mathbf{x}\})(\mathbf{x} - \mathbb{E}\{\mathbf{x}\})^{\mathrm{H}}\} = \mathbb{E}\{\mathbf{x}\mathbf{x}^{\mathrm{H}}\} \\ &= \mathbf{A} \underbrace{\mathbb{E}\{\tilde{\mathbf{x}}\tilde{\mathbf{x}}^{\mathrm{H}}\}}_{=\mathbf{I}_K} \mathbf{A}^{\mathrm{H}} = \mathbf{A}\mathbf{A}^{\mathrm{H}}. \end{aligned} \tag{2.83}$$

Hence, the random vector created by the product in (2.81) is distributed as $\mathbf{x} \sim \mathcal{N}_{\mathbb{C}}(\mathbf{0}, \mathbf{A}\mathbf{A}^{\mathrm{H}})$. This example shows how we can create a correlated complex Gaussian vector $\mathbf{x}$ from a complex Gaussian vector $\tilde{\mathbf{x}}$ with independent entries by multiplying with a matrix, which will happen later in this book.

Example 2.10. Show that if $\tilde{\mathbf{x}} \sim \mathcal{N}_{\mathbb{C}}(\mathbf{0}, \mathbf{I}_K)$, then $\mathbf{x} = \mathbf{U}\tilde{\mathbf{x}}$ has the same distribution if $\mathbf{U} \in \mathbb{C}^{K \times K}$ is a unitary matrix.

The vector $\mathbf{x}$ is created as in (2.81) with $\mathbf{A} = \mathbf{U}$. The corresponding covariance matrix is computed in (2.83) and becomes $\mathbf{A}\mathbf{A}^{\mathrm{H}} = \mathbf{U}\mathbf{U}^{\mathrm{H}} = \mathbf{I}_K$ since $\mathbf{U}$ is unitary. It follows that $\mathbf{x} \sim \mathcal{N}_{\mathbb{C}}(\mathbf{0}, \mathbf{I}_K)$, which is the same distribution as $\tilde{\mathbf{x}}$ has. The conclusion is that a vector with uncorrelated complex Gaussian entries retains its distribution when multiplied by a unitary matrix.

In general, we can define the correlated multivariate complex Gaussian distribution

$$\mathbf{x} \sim \mathcal{N}_{\mathbb{C}}(\mathbf{0}, \mathbf{R}) \tag{2.84}$$

for an arbitrary positive definite covariance matrix $\mathbf{R}$. The special case considered above correspond to $\mathbf{R} = \mathbf{A}\mathbf{A}^{\mathrm{H}}$. The PDF is given by

$$f_{\mathbf{x}}(\mathbf{x}) = \frac{1}{\pi^M \det(\mathbf{R})} e^{-\mathbf{x}^{\mathrm{H}}\mathbf{R}^{-1}\mathbf{x}}. \tag{2.85}$$

Such correlated complex Gaussian vectors are circularly symmetric since $f_{\mathbf{x}}(\mathbf{x}) = f_{\mathbf{x}}(\mathbf{x}e^{\mathrm{j}\psi})$ for any constant phase-shift ψ.

An important property of complex Gaussian random vectors is that the joint PDF in (2.85) reduces to the one for independent entries in (2.80) if we insert the diagonal covariance matrix $\mathbf{R} = \sigma^2 \mathbf{I}_M$. Hence, it is sufficient to assume that all the entries of $\mathbf{x}$ are uncorrelated (i.e., the off-diagonal entries of $\mathbf{R}$ are zero) to get statistical independence as a side-effect. This property follows from the shape of the multivariate complex Gaussian distribution and does generally not hold for other random distributions. We will use this property repeatedly in the book.

Lemma 2.7. If two random variables are jointly complex Gaussian distributed and uncorrelated, the variables are also statistically independent.

When exposed to a correlated complex Gaussian random vector $\mathbf{x}$, removing the correlation through signal processing can sometimes be helpful. Since the covariance matrix $\mathbf{R}$ in (2.84) is positive definite, its square root $\mathbf{R}^{1/2}$ (computed as in Lemma 2.2) is invertible and its inverse will be denoted as $\mathbf{R}^{-1/2}$. Let us define the random variable $\mathbf{n} = \mathbf{R}^{-1/2}\mathbf{x}$. It is complex Gaussian distributed with zero mean and the covariance matrix

$$\begin{aligned} \mathrm{Cov}\{\mathbf{n}\} &= \mathbb{E}\{(\mathbf{n} - \mathbb{E}\{\mathbf{n}\})(\mathbf{n} - \mathbb{E}\{\mathbf{n}\})^{\mathrm{H}}\} = \mathbb{E}\{\mathbf{n}\mathbf{n}^{\mathrm{H}}\} \\ &= \mathbf{R}^{-1/2} \underbrace{\mathbb{E}\{\mathbf{x}\mathbf{x}^{\mathrm{H}}\}}_{=\mathbf{R}} \mathbf{R}^{-1/2} = \mathbf{I}_M. \end{aligned} \tag{2.86}$$

Hence, $\mathbf{n} = \mathbf{R}^{-1/2}\mathbf{x} \sim \mathcal{N}_{\mathbb{C}}(\mathbf{0}, \mathbf{I}_M)$ has uncorrelated entries, which are also statistically independent thanks to Lemma 2.7. This procedure of removing correlation from a random vector is known as *whitening*, particularly when dealing with Gaussian noise. A noise vector with correlated entries is called *colored* noise, and the whitening procedure transforms it into white noise, as defined in (2.79). The theory developed in this book will be based on the assumption of having white noise, but it can also be applied in the presence of colored noise by adding a whitening step at the receiver.

Example 2.11. What is the PDF of a multivariate real Gaussian distribution?

If $x_m \sim \mathcal{N}(\mu_m, \sigma^2)$ for $m = 1, \ldots, M$ are independent variables, then the PDF of $\mathbf{x} = [x_1, \ldots, x_M]^{\mathrm{T}}$ is $\frac{1}{(2\pi\sigma^2)^{M/2}} e^{-(\mathbf{x}-\boldsymbol{\mu})^{\mathrm{T}}(\mathbf{x}-\boldsymbol{\mu})/(2\sigma^2)}$. We obtain this expression by taking the product of M PDFs of the kind in (2.63) and defining $\boldsymbol{\mu} = [\mu_1, \ldots, \mu_M]^{\mathrm{T}}$. When the variables are correlated with the covariance matrix $\mathbf{R}$, the resulting PDF is

$$f_{\mathbf{x}}(\mathbf{x}) = \frac{1}{(2\pi)^{\frac{M}{2}} \sqrt{\det(\mathbf{R})}} e^{-\frac{(\mathbf{x}-\boldsymbol{\mu})^{\mathrm{T}}\mathbf{R}^{-1}(\mathbf{x}-\boldsymbol{\mu})}{2}}. \tag{2.87}$$

We denote such a real Gaussian distribution as $\mathbf{x} \sim \mathcal{N}(\boldsymbol{\mu}, \mathbf{R})$.

2.2.5 Rayleigh, Exponential, and χ^2 Distribution

The PDF of a complex random variable x determines how the magnitude $|x|$ and the argument $\arg(x)$ are distributed. These components are generally correlated, but if x is complex Gaussian, the circular symmetry implies that they are independent. In wireless communications, we are particularly interested in the magnitude since it can describe the amplitude of a signal. We denote the magnitude as $y = |x| \geq 0$ and the argument as $\psi = \arg(x) \in [-\pi, \pi)$, so that $x = ye^{\mathrm{j}\psi}$. Since the PDF of the complex Gaussian distribution in (2.67) is defined using the Cartesian form $x = \Re(x) + \mathrm{j}\Im(x)$, a change of variables to the polar form consists of two steps: replacing the old variables with the new variables, followed by the multiplication with the magnitude of the Jacobian determinant, $|J(y, \psi)|$. We can compute the latter term based on the definition of Jacobian matrices as

$$\begin{aligned} |J(y,\psi)| &= \left|\det\left(\begin{bmatrix} \frac{\partial \Re(x)}{\partial y} & \frac{\partial \Im(x)}{\partial y} \\ \frac{\partial \Re(x)}{\partial \psi} & \frac{\partial \Im(x)}{\partial \psi} \end{bmatrix}\right)\right| = \left|\det\left(\begin{bmatrix} \frac{\partial y\cos(\psi)}{\partial y} & \frac{\partial y\sin(\psi)}{\partial y} \\ \frac{\partial y\cos(\psi)}{\partial \psi} & \frac{\partial y\sin(\psi)}{\partial \psi} \end{bmatrix}\right)\right| \\ &= \left|\det\left(\begin{bmatrix} \cos(\psi) & \sin(\psi) \\ -y\sin(\psi) & y\cos(\psi) \end{bmatrix}\right)\right| = y\left(\cos^2(\psi) + \sin^2(\psi)\right) = y. \end{aligned} \tag{2.88}$$

Using this method, we can rewrite the PDF in (2.67) of the complex Gaussian distribution as a function of the magnitude and argument:

$$f_{\mathrm{y},\psi}(y,\psi) = \frac{y}{\pi\sigma^2}e^{-\frac{y^2}{\sigma^2}} \tag{2.89}$$

for $y \geq 0$ while it is zero for $y < 0$. Since the PDF does not depend on ψ, we can conclude that ψ is uniformly distributed between $-\pi$ and π (or any other interval of length 2π) and independent of y. We can compute the marginal distribution of the magnitude as

$$f_{\mathrm{y}}(y) = \int_{-\pi}^{\pi} f_{\mathrm{y},\psi}(y,\psi)\partial\psi = \frac{2y}{\sigma^2}e^{-\frac{y^2}{\sigma^2}} \quad \text{for } y \geq 0. \tag{2.90}$$

This PDF characterizes the variations in the magnitude of a complex Gaussian random variable. It matches with what is known as the *Rayleigh distribution*. Just as the complex Gaussian distribution is characterized by its variance σ^2, the Rayleigh distribution is characterized by a scale parameter. For the PDF in (2.90), the scale parameter can be identified to be $\sigma/\sqrt{2}$ and, thus, we can express the distribution of the magnitude as $y \sim \text{Rayleigh}(\sigma/\sqrt{2})$. The PDF with $\sigma = 1$ is illustrated in Figure 2.7. When a communication channel is complex Gaussian distributed, it is referred to as *Rayleigh fading* since the magnitude is Rayleigh distributed. We will return to this later in the book.

When analyzing the SNR of a communication system, we are not interested in the amplitude $y = |x|$ but its square $y^2 = |x|^2$ (the SNR is a ratio between the signal power and noise power). Let us denote this random variable as $z = y^2$. We can obtain the PDF of z by following the same two steps as above: replace the y in (2.90) with $\sqrt{z}$ and then multiply by the magnitude of the Jacobian determinant $|J(z)|$, which is $|\partial y/\partial z| = 1/(2\sqrt{z})$ in this case. Using this method, we obtain

$$f_{\mathrm{z}}(z) = \frac{1}{\sigma^2} e^{-\frac{z}{\sigma^2}} \quad \text{for } z \geq 0, \tag{2.91}$$

while it is zero for $z < 0$. This PDF characterizes the variations in the squared magnitude of a complex Gaussian random variable. It matches what is known as the *exponential distribution*. This distribution is generally characterized by a so-called rate parameter, which in this case equals $1/\sigma^2$. Hence, we can express the distribution of the squared magnitude as $z \sim \mathrm{Exp}(1/\sigma^2)$. The PDF with $\sigma^2 = 1$ is illustrated in Figure 2.7.

A useful property of the exponential distribution is that

$$\mathbb{E}\{z^n\} = n!(\sigma^2)^n \tag{2.92}$$

for any positive integer n, where $n!$ denotes the factorial.

Example 2.12. Suppose $x \sim \mathcal{N}_{\mathbb{C}}(0, \sigma^2)$. What are the mean, quadratic mean, and variance of $|x|^2$?

Since $z = |x|^2 \sim \mathrm{Exp}(1/\sigma^2)$, we can utilize the property in (2.92) to compute the mean, quadratic mean, and variance of $|x|^2$ as follows:

$$\mathbb{E}\left\{|x|^2\right\} = \mathbb{E}\{z\} = \sigma^2, \tag{2.93}$$

$$\mathbb{E}\left\{|x|^4\right\} = \mathbb{E}\left\{z^2\right\} = 2\sigma^4, \tag{2.94}$$

$$\mathbb{V}\mathrm{ar}\left\{|x|^2\right\} = \mathbb{E}\left\{|x|^4\right\} - \left(\mathbb{E}\left\{|x|^2\right\}\right)^2 = \sigma^4. \tag{2.95}$$

We can also utilize the property in (2.92) when computing mean values that involve an M-dimensional complex Gaussian random vector with independent entries: $\mathbf{x} = [x_1, \ldots, x_M]^{\mathrm{T}} \sim \mathcal{N}_{\mathbb{C}}(\mathbf{0}, \sigma^2 \mathbf{I}_M)$. Since $z_m = |x_m|^2 \sim \mathrm{Exp}(1/\sigma^2)$ for $m = 1, \ldots, M$, we can compute mean, quadratic mean, and variance of the squared norm $\|\mathbf{x}\|^2$ as follows:

$$\mathbb{E}\left\{\|\mathbf{x}\|^2\right\} = \sum_{m=1}^{M} \mathbb{E}\{z_m\} = M\sigma^2, \tag{2.96}$$

$$\begin{aligned}\mathbb{E}\left\{\|\mathbf{x}\|^4\right\} &= \mathbb{E}\left\{\left(\sum_{m=1}^{M} z_m\right)^2\right\} = \sum_{m=1}^{M} \mathbb{E}\left\{z_m^2\right\} + \sum_{m=1}^{M} \sum_{\substack{n=1 \\ n \neq m}}^{M} \mathbb{E}\{z_m\}\mathbb{E}\{z_n\} \\ &= 2M\sigma^4 + M(M-1)\sigma^2\sigma^2 = (M^2 + M)\sigma^4, \end{aligned} \tag{2.97}$$

$$\mathbb{V}\mathrm{ar}\left\{\|\mathbf{x}\|^2\right\} = \mathbb{E}\left\{\|\mathbf{x}\|^4\right\} - \left(\mathbb{E}\left\{\|\mathbf{x}\|^2\right\}\right)^2 = M\sigma^4. \tag{2.98}$$

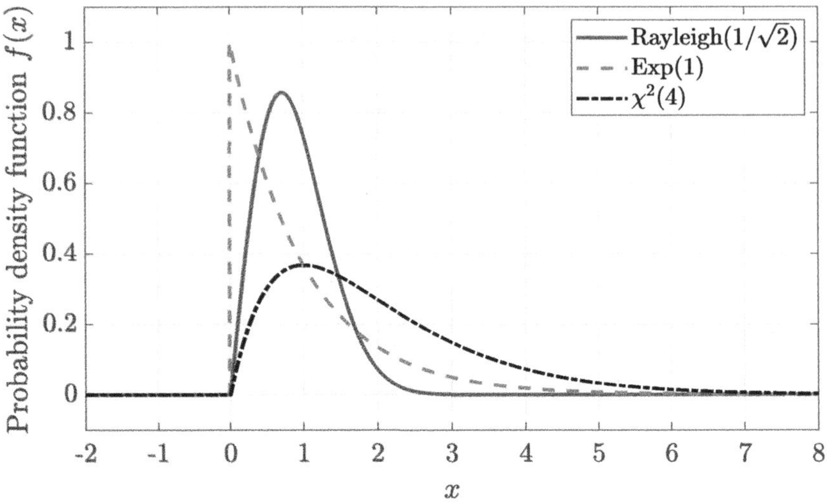

Figure 2.7: Examples of the PDFs of the Rayleigh distribution, exponential distribution, and χ^2-distribution.

These results were obtained by utilizing the fact that $\|\mathbf{x}\|^2 = \sum_{m=1}^{M} z_m$ is the sum of M independent random variables with identical exponential distribution. By utilizing the fact that the PDF of a sum of independent random variables is the convolution of the marginal PDFs, one can show that the squared norm has the PDF

$$f_{\|\mathbf{x}\|^2}(x) = \frac{x^{M-1}e^{-\frac{x}{\sigma^2}}}{(\sigma^2)^M(M-1)!} \quad \text{for } x \geq 0, \tag{2.99}$$

while it is zero for $x < 0$. This distribution is often referred to as the χ^2-distribution in the communication literature and denoted as $\chi^2(2M)$, where $2M$ is called the degrees of freedom since $\|\mathbf{x}\|^2$ is the sum of $2M$ squared real Gaussian variables. However, formally speaking, it is only in the special case of $\sigma^2 = 2$ that one obtains that random distribution. Hence, we will refer to (2.99) as the *scaled* χ^2*-distribution* in this book. The mean $M\sigma^2$ of $\|\mathbf{x}\|^2$ was computed in (2.96), while the variance $M\sigma^4$ was computed in (2.98). If we set $M = 1$, then the $\chi^2(2M)$-distribution reduces to the exponential distribution. The PDF with $M = 2$ and $\sigma^2 = 1$ is illustrated in Figure 2.7.

2.2.6 Cumulative Distribution Function

It is common to compare the realization of a real-valued random variable with a threshold when analyzing the performance of a communication system. Suppose the random variable is x and the threshold is a, then the probability $\Pr\{x \leq a\}$ of x taking realizations smaller than or equal to a is important. To characterize how its value depends on the threshold, we can define the

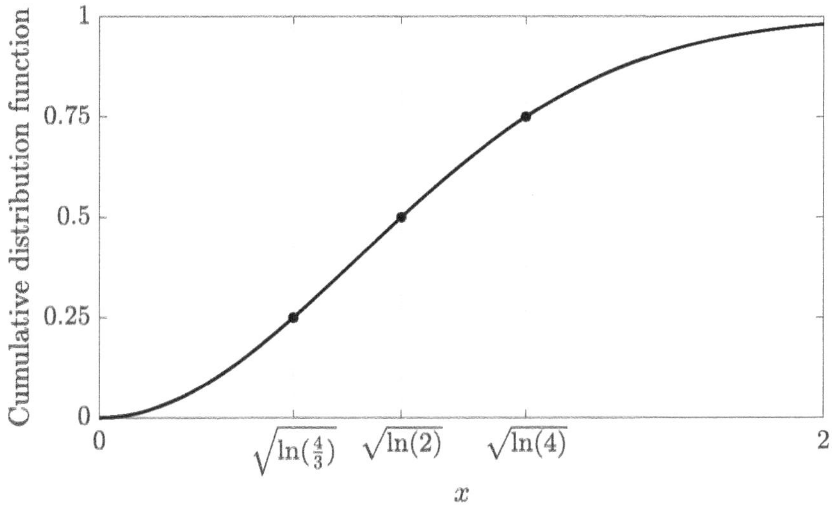

Figure 2.8: The CDF of the Rayleigh distribution for $\sigma^2 = 1$, where the 25% percentile, median, and 75% percentile points are marked.

cumulative distribution function (CDF) $F_{\mathsf{x}}(a)$ as

$$F_{\mathsf{x}}(a) = \Pr\{x \leq a\} = \int_{-\infty}^{a} f_{\mathsf{x}}(x)\partial x, \tag{2.100}$$

which is computed by integrating the PDF from its lower limit (generally from $-\infty$, but we can start from 0 for positive random variables) to a. The CDF is a monotonically increasing function of a since we are integrating the non-negative PDF $f_{\mathsf{x}}(x)$ over an increasing interval. Moreover, it only takes values between 0 and 1, which equal the probability of the event $\Pr\{x \leq a\}$. The CDF provides a full characterization of the random distribution, just as the PDF does; for example, the PDF can be retained from the CDF by computing the first-order derivative:

$$\frac{\partial}{\partial x} F_{\mathsf{x}}(x) = f_{\mathsf{x}}(x). \tag{2.101}$$

The value of a for which $F_{\mathsf{x}}(a) = 0.5$ is known as the *median* of the distribution because it is equally likely to obtain a realization above and below it. If the CDF is strictly increasing and continuous, the inverse CDF $F_{\mathsf{x}}^{-1}(y)$ exists and is called the *percentile function*. We can then compute the median as $F_{\mathsf{x}}^{-1}(0.5)$. The point $F_{\mathsf{x}}^{-1}(0.25)$ is called the 25% percentile since 25% of all random realizations are below it, while the point $F_{\mathsf{x}}^{-1}(0.75)$ is called the 75% percentile since 75% of all random realizations are below it (and 25% are above it). The small and large percentiles are of interest when analyzing a random variable's worst-case and best-case realizations.

Figure 2.8 shows the CDF of the Rayleigh distribution for $\sigma^2 = 1$. The horizontal axis emphasizes the 25% percentile point $\sqrt{\ln(4/3)}$ where the CDF is 0.25, the median $\sqrt{\ln(2)}$ where the CDF is 0.5, and the 75% percentile point $\sqrt{\ln(4)}$ where the CDF is 0.75. Many different CDF curves can be drawn through these three points; thus, the entire CDF is required to obtain a complete statistical characterization of the Rayleigh distribution. The CDF and percentiles used in the figure are computed as follows.

Example 2.13. Consider the Rayleigh distribution $x \sim \text{Rayleigh}(\sigma/\sqrt{2})$ in (2.90). What CDF and percentile function does it have?

The PDF is $f_{\mathsf{x}}(x) = \frac{2x}{\sigma^2} e^{-\frac{x^2}{\sigma^2}}$ for $x \geq 0$, thus the CDF becomes

$$F_{\mathsf{x}}(a) = \int_0^a \frac{2x}{\sigma^2} e^{-\frac{x^2}{\sigma^2}} \partial x = \left[-e^{-\frac{x^2}{\sigma^2}} \right]_0^a = 1 - e^{-\frac{a^2}{\sigma^2}}. \tag{2.102}$$

The percentile function $F_{\mathsf{x}}^{-1}(y)$ can be obtained by inverting the CDF in (2.102) as

$$y = 1 - e^{-\frac{a^2}{\sigma^2}} \quad \Rightarrow \quad 1 - y = e^{-\frac{a^2}{\sigma^2}} \quad \Rightarrow \quad \ln(1-y) = -\frac{a^2}{\sigma^2}$$

$$\Rightarrow F_{\mathsf{x}}^{-1}(y) = a = \sigma \sqrt{\ln\left(\frac{1}{1-y}\right)}. \tag{2.103}$$

We can use this function to identify any percentile of the distribution; for example, the median is $F_{\mathsf{x}}^{-1}(0.5) = \sigma\sqrt{\ln(2)}$, the 25% percentile is $F_{\mathsf{x}}^{-1}(0.25) = \sigma\sqrt{\ln(4/3)}$, and the 75% percentile is $F_{\mathsf{x}}^{-1}(0.75) = \sigma\sqrt{\ln(4)}$. These values are indicated on the horizontal axis in Figure 2.8 for $\sigma^2 = 1$.

2.2.7 Random Process

A random continuous-time signal $x(t)$ is called a random process and is a generalization of a multivariate random variable. More precisely, if we take samples of a random process at the M time instances $t_1, \ldots, t_M$ and collect them in a vector

$$\begin{bmatrix} x(t_1) \\ \vdots \\ x(t_M) \end{bmatrix}, \tag{2.104}$$

then we obtain a multivariate random variable.

The random processes considered in this book are *wide-sense stationary*, which means that the random distribution is constant over time. Three specific properties are satisfied for such processes. Firstly, the mean value $\mu = \mathbb{E}\{x(t)\}$ does not depend on the time t. Secondly, the variance $\sigma^2 = \mathbb{E}\{|x(t) - \mu|^2\}$ also does not depend on the time t. The third property relates to how the random process is correlated in time, measured by the *autocorrelation function*. The correlation between the samples at time t_1 and t_2 should only depend on the time lag $t_2 - t_1$ between the samples and not on their individual values. Hence, the autocorrelation function of a wide-sense stationary process is denoted as

$$r(t_2 - t_1) = \mathbb{E}\{x(t_1)x^*(t_2)\}. \tag{2.105}$$

A *white* random process changes so rapidly with the time that $x(t_1)$ and $x(t_2)$ are only correlated when $t_1 = t_2$. This is represented by the autocorrelation function

$$r(t_2 - t_1) = c\delta(t_2 - t_1), \tag{2.106}$$

where $c = |\mu|^2 + \sigma^2$ is called the *power spectral density* and $\delta(t)$ is the Dirac delta function.

A *complex Gaussian* random process has the property that the vector in (2.104) becomes a multivariate complex Gaussian distribution, irrespective of the time instances at which the samples are taken. The noise in wireless communications is often modeled as a white complex Gaussian random process.

2.3 Signal Modeling

Wireless communication systems transfer data by utilizing electromagnetic *signals*. These signals propagate from the transmitter to the receiver over an analog wireless channel that acts as a *system* that filters the signal. This section provides the fundamental connection between the physical continuous-time signal models and the simple discrete-time models used in later book sections. We will use standard results from signals-and-systems theory to establish the connection.

Suppose we are allowed to communicate using a real-valued passband signal with bandwidth B centered around a carrier frequency f_c. For example, a typical scenario in the first 5G deployments is $f_c = 3\,\text{GHz}$ and $B = 100\,\text{MHz}$. The passband assumption implies that $B < 2f_c$ so that the signal does not contain the near-zero frequency range. In practice, we typically have $B \ll f_c$, as in the given example. Let the transmitted signal be denoted as $z_p(t)$, where $t \in \mathbb{R}$ is the continuous time variable and the subscript p indicates it is a passband signal. The amplitude spectrum of such a signal is sketched in Figure 2.9(a). The signal $z_p(t)$ is real-valued; thus, the spectrum is symmetric for positive and negative frequencies.

Wireless channels generally have time-varying properties, for example, due to the movement of the transmitter, receiver, or objects in the propagation environment. However, we can divide the transmission into blocks such that the channel is (approximately) time-invariant within each block. Following that approach, we assume that the wireless channel can be represented by a *linear time-invariant (LTI)* system. A key property of such systems is that the filtering is entirely determined by the real-valued impulse response $g_p(t)$.

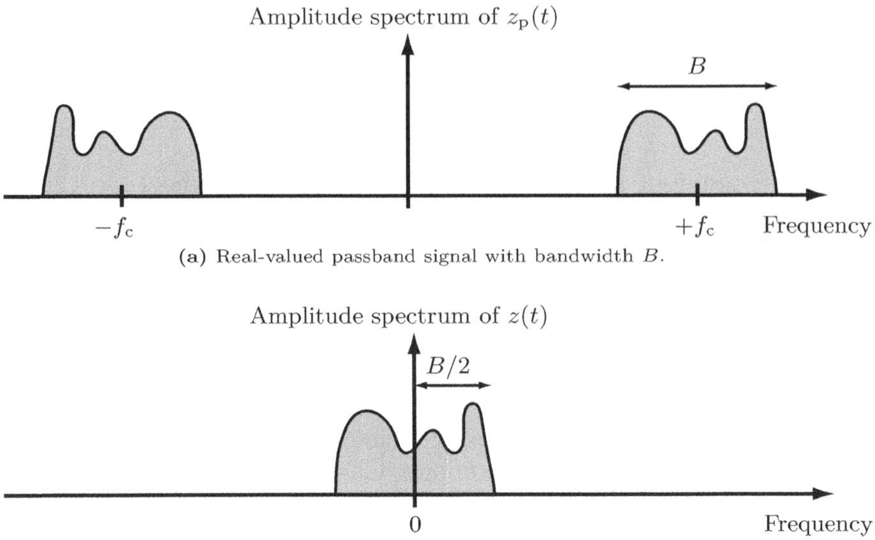

(a) Real-valued passband signal with bandwidth B.

(b) Complex-valued baseband signal with bandwidth $B/2$.

Figure 2.9: Sketch of a real-valued passband signal $z_p(t)$ with center frequency f_c and bandwidth B that can be communicated over a wireless channel, and the equivalent complex-valued baseband signal $z(t)$ that can be communicated over the complex baseband representation of the channel. The mathematical relation between the two signals is given in (2.111).

In particular, the output signal $v_p(t)$ is the convolution between the input signal and impulse response:

$$v_p(t) = (g_p * z_p)(t) = \int_{-\infty}^{\infty} g_p(u) z_p(t-u) \partial u. \tag{2.107}$$

The impulse response must satisfy the technical condition $\int_{-\infty}^{\infty} |g_p(t)| \partial t < \infty$ for (2.107) to hold, but this is always the case in wireless communications since otherwise, one could receive more signal energy than was transmitted.

The input-output relation in (2.107) is illustrated in Figure 2.10(a). We will later add the transmitter and receiver hardware to this model, including the additive noise, but we will first reformulate the basic relation.

2.3.1 Complex Baseband Representation

To avoid making the communication system design dependent on a particular value of f_c, the signal processing algorithms used in wireless communications are developed for an equivalent baseband system where the signals are centered around the zero frequency. If we take the spectrum of the passband signal in Figure 2.9(a) and downshift it to the baseband, we obtain the equivalent signal $z(t)$ whose amplitude spectrum is illustrated in Figure 2.9(b). This is called the *complex baseband representation* of the signal in Figure 2.9(a). If the hardware is designed to generate baseband signals of this type, we can

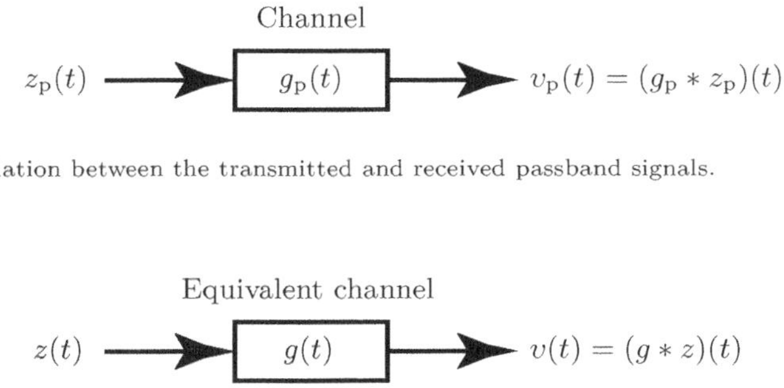

(a) Relation between the transmitted and received passband signals.

(b) Equivalent relation using complex-baseband signals.

Figure 2.10: Block diagrams of the input-output relations when transmitting a signal over a wireless channel. The practical system transmits passband signals but can be equivalently represented in the complex baseband.

modulate the signals up to different carrier frequencies at different times (e.g., a mobile phone supports many bands so that it can be used worldwide).

We can establish a mathematical connection between $z_{\mathrm{p}}(t)$ and $z(t)$ in the frequency domain by utilizing the Fourier transform $\mathcal{F}\{\cdot\}$. The frequency-domain representation of an arbitrary continuous-time signal $a(t)$ is defined as

$$A(f) = \mathcal{F}\{a(t)\} = \int_{-\infty}^{\infty} a(t)e^{-\mathrm{j}2\pi f t}\partial t. \tag{2.108}$$

The Fourier transform is generally complex-valued, but it is conjugate symmetric if the signal $a(t)$ is real-valued: $A^*(-f) = A(f)$. This is proved as

$$A^*(-f) = \left(\int_{-\infty}^{\infty} a(t)e^{-\mathrm{j}2\pi(-f)t}\partial t\right)^* = \int_{-\infty}^{\infty} a^*(t)e^{-\mathrm{j}2\pi f t}\partial t = A(f), \tag{2.109}$$

where the last equality follows from that $a(t) = a^*(t)$ for real-valued signals.

The frequency-domain representations of the passband signal and baseband signal respectively become $Z_{\mathrm{p}}(f) = \mathcal{F}\{z_{\mathrm{p}}(t)\}$ and $Z(f) = \mathcal{F}\{z(t)\}$ when using the Fourier transform. We can then express the relation shown in Figure 2.9 as

$$Z_{\mathrm{p}}(f) = \frac{Z(f - f_{\mathrm{c}}) + Z^*(-f - f_{\mathrm{c}})}{\sqrt{2}}. \tag{2.110}$$

The scaling factor $1/\sqrt{2}$ ensures that the passband and baseband signals have the same energy; that is, $\int_{-\infty}^{\infty} |Z_{\mathrm{p}}(f)|^2 \partial f = \int_{-\infty}^{\infty} |Z(f)|^2 \partial f$. By taking the inverse Fourier transform of both sides of (2.110), it follows that the time-domain signals $z_{\mathrm{p}}(t)$ and $z(t)$ are related as

$$\begin{aligned} z_{\mathrm{p}}(t) &= \frac{z(t)e^{\mathrm{j}2\pi f_{\mathrm{c}} t} + z^*(t)e^{-\mathrm{j}2\pi f_{\mathrm{c}} t}}{\sqrt{2}} \\ &= \sqrt{2}\Re\left(z(t)e^{\mathrm{j}2\pi f_{\mathrm{c}} t}\right). \end{aligned} \tag{2.111}$$

We notice that the amplitude spectrum of $z(t)$ in Figure 2.9(b) is not symmetric for positive and negative frequencies, which implies that it is a complex-valued signal. Any real-valued passband signal $z_\mathrm{p}(t)$ with bandwidth B can be equivalently represented by a complex-valued signal $z(t)$ with bandwidth $B/2$ according to (2.111). The bandwidth is halved, but there are instead both real and imaginary signal dimensions. The signal $z(t)$ has the same total energy as $z_\mathrm{p}(t)$, meaning that $\int_{-\infty}^{\infty} |z(t)|^2 \partial t = \int_{-\infty}^{\infty} |z_\mathrm{p}(t)|^2 \partial t$, but the energy is moved to different frequencies.[1]

Next, we would like to find a complex baseband representation of the entire output-input relation in (2.107), so we can abstract away the carrier frequency and only analyze the baseband. To this end, we let $G_\mathrm{p}(f) = \mathcal{F}\{g_\mathrm{p}(t)\}$ denote the frequency response of the system, which determines how the channel filters different frequencies of the input signal. By taking the Fourier transform of both sides of (2.107) and utilizing (2.110), we obtain

$$\begin{aligned} \Upsilon_\mathrm{p}(f) = \mathcal{F}\{v_\mathrm{p}(t)\} &= G_\mathrm{p}(f) Z_\mathrm{p}(f) \\ &= G_\mathrm{p}(f) \frac{Z(f-f_\mathrm{c}) + Z^*(-f-f_\mathrm{c})}{\sqrt{2}} \\ &= \frac{G_\mathrm{p}(f) Z(f-f_\mathrm{c}) + G_\mathrm{p}^*(-f) Z^*(-f-f_\mathrm{c})}{\sqrt{2}}. \end{aligned} \tag{2.112}$$

The last equality in (2.112) follows the fact that $G_\mathrm{p}(f) = G_\mathrm{p}^*(-f)$ for real-valued systems. Since (2.110) and (2.111) provide a general connection between a passband signal and its equivalent complex-baseband signal, we can define the received signal $v(t)$ in the complex baseband and relate it to the received passband signal as

$$v_\mathrm{p}(t) = \Re\left(\sqrt{2} v(t) e^{\mathrm{j}2\pi f_\mathrm{c} t}\right), \tag{2.113}$$

$$\Upsilon_\mathrm{p}(f) = \mathcal{F}\{v_\mathrm{p}(t)\} = \frac{\Upsilon(f-f_\mathrm{c}) + \Upsilon^*(-f-f_\mathrm{c})}{\sqrt{2}}, \tag{2.114}$$

where $\Upsilon(f) = \mathcal{F}\{v(t)\}$. By comparing (2.112) with (2.114), we can identify the Fourier transform of the received baseband signal as

$$\Upsilon(f-f_\mathrm{c}) = G_\mathrm{p}(f) Z(f-f_\mathrm{c}) \quad \Rightarrow \quad \Upsilon(f) = G_\mathrm{p}(f+f_\mathrm{c}) Z(f). \tag{2.115}$$

Taking the inverse Fourier transform of (2.115) yields

$$v(t) = (g * z)(t) = \int_{-\infty}^{\infty} g(u) z(t-u) \partial u, \tag{2.116}$$

where the complex baseband representation of the system has the frequency response $G(f) = G_\mathrm{p}(f+f_\mathrm{c})$ and impulse response

$$g(t) = g_\mathrm{p}(t) e^{-\mathrm{j}2\pi f_\mathrm{c} t}. \tag{2.117}$$

[1]If the total signal energy is infinite, we can compare the signal powers and conclude that these are equal. The power of a signal $a(t)$ is computed as $\lim_{T\to\infty} \frac{1}{2T} \int_{-T}^{T} |a(t)|^2 \partial t$.

We identify (2.116) as an equivalent way to describe a continuous-time communication channel in the complex baseband. The input-output relation is illustrated in Figure 2.10(b). Note that the complex-baseband terminology only refers to the signals: we have taken the passband signal $z_{\mathrm{p}}(t)$ and downshifted it to the complex-baseband signal $z(t)$. In contrast, the impulse responses in wireless communications are neither passband nor baseband filters. In fact, the wireless medium supports communication at any frequency and bandwidth, and causes varying attenuation and delays to signals in different bands. However, by sending signals confined to a specific frequency range $[f_{\mathrm{c}} - B/2, f_{\mathrm{c}} + B/2]$, we are only using the corresponding part of the wireless medium. In contrast, other systems can use different parts simultaneously. The only difference between $g(t)$ in (2.117) and the original impulse response $g_{\mathrm{p}}(t)$ is that it has been downshifted along the frequency axis so that the channel filters the signal in an equivalent manner.

Without loss of generality, we will consider the complex baseband in the remainder of this book, except at a few places where we model the impulse response $g_{\mathrm{p}}(t)$ of a particular wireless channel and then use (2.117) to obtain the equivalent impulse response in the complex baseband.

2.3.2 From Continuous Time to Discrete Time

Digital data is described by a sequence of bits. In digital communications, these bits are further represented by a discrete data sequence $\{x[l]\}$ of symbols selected based on the bits, where the integer l is the discrete time index. The symbols are selected from the complex set $\mathbb{C}$, such that $x[l] \in \mathbb{C}$. More precisely, a *modulation and channel coding* scheme is utilized to decide how many bits each symbol represents and how much redundancy is introduced to enable error correction in the receiver. We need to create a continuous-time signal $z(t)$ that contains the data symbols $\{x[l]\}$ and can be transmitted as an analog electromagnetic wave over the wireless channel. This is achieved by *pulse-amplitude modulation* (PAM). We will not explain all the underlying theory but focus on the properties needed to derive the discrete-time model we will use in the remainder of the book.

The essence of PAM is that each of the symbols $\{x[l]\}$ is multiplied by a continuous-time pulse and then transmitted one after the other. We consider PAM with the ideal sinc-pulse[2]

$$p(t) = \sqrt{B}\mathrm{sinc}(Bt) = \sqrt{B}\frac{\sin(\pi Bt)}{\pi Bt}, \tag{2.118}$$

which has the Fourier transform

$$P(f) = \mathcal{F}\{p(t)\} = \begin{cases} 1/\sqrt{B}, & \text{if } |f| \leq B/2, \\ 0, & \text{if } |f| > B/2. \end{cases} \tag{2.119}$$

[2]In the communications and signal processing literature, the sinc function is defined as $\mathrm{sinc}(t) = \sin(\pi t)/(\pi t)$ for $t \neq 0$ and $\mathrm{sinc}(0) = 1$. Other definitions exist in other contexts.

This baseband pulse has bandwidth $B/2$, can be used as an ideal lowpass filter in the frequency domain, and has unit energy: $\int_{-\infty}^{\infty} |P(f)|^2 \partial f = 1$. An illustration of these functions is provided in Figure 2.11. The sinc-function $\mathrm{sinc}(t)$ oscillates in the time domain with a linearly reducing amplitude and zero-crossings when t is a non-zero integer. Hence, $\sqrt{B}\mathrm{sinc}(Bt)$ has zero-crossing when t is a non-zero integer divided by B. We will exploit this feature to transmit a new data symbol $x[l]$ every $1/B$ seconds while keeping them separable at the receiver. Any pulse function with these zero-crossings is said to satisfy the *Nyquist criterion* and could be used instead of the sinc-function, but one can prove that the feasible alternatives have a strictly larger bandwidth than $B/2$. The bandwidth of the transmitted signal will match that of the pulse; thus, we will consider PAM using the most bandwidth-efficient pulse in this book. If we increase B, then $\sqrt{B}\mathrm{sinc}(Bt)$ will be compressed in the time domain (i.e., having more zero-crossings per second so we can send more data symbols), while it will expand in the frequency domain.

When using PAM, the continuous-time complex-baseband signal is

$$z(t) = \sum_{k=-\infty}^{\infty} x[k]\, p\left(t - \frac{k}{B}\right), \tag{2.120}$$

where we notice that a new symbol is transmitted every $1/B$ seconds and multiplied by a time-delayed version of $p(t)$. It is common to refer to $1/B$ as the *symbol time* and B as the *symbol rate* (in addition to being the bandwidth). Notably, B complex-valued symbols are transmitted per second, and more bandwidth leads to a shorter time between the symbols. The PAM procedure is tightly connected to the Nyquist-Shannon sampling theorem [38, Th. 1], which can be stated for complex signals as follows [39, Sec. 2.8].

Lemma 2.8. If a complex-valued continuous-time signal $z(t)$ only contains frequencies in an interval smaller than B Hz, it is entirely determined by a series of samples spaced $1/B$ seconds apart.

Two commonly considered frequency intervals that satisfy this condition are $-B/2 < f \leq B/2$ and $-B/2 \leq f < B/2$, which can be written in short form as $(-B/2, B/2]$ and $[-B/2, B/2)$, respectively. The interval shrinks to $(-B/2, B/2)$ for real-valued signals since such signals always contain the same positive and negative frequencies. The intuition behind the sampling theorem is that the largest frequency (in magnitude) determines how rapidly the signal can change. If the largest frequency is $B/2$ or $-B/2$ (but not both), then the fastest signal components have a period of $2/B$. We can uniquely capture all signal variations if we sample the signal twice per period (i.e., at a sampling rate of B Hz). This specific sampling rate is known as the *Nyquist rate* and gives rise to B samples per second. It is also called the *critical sampling rate* to signify that it is fully acceptable to sample the signal more densely, but it

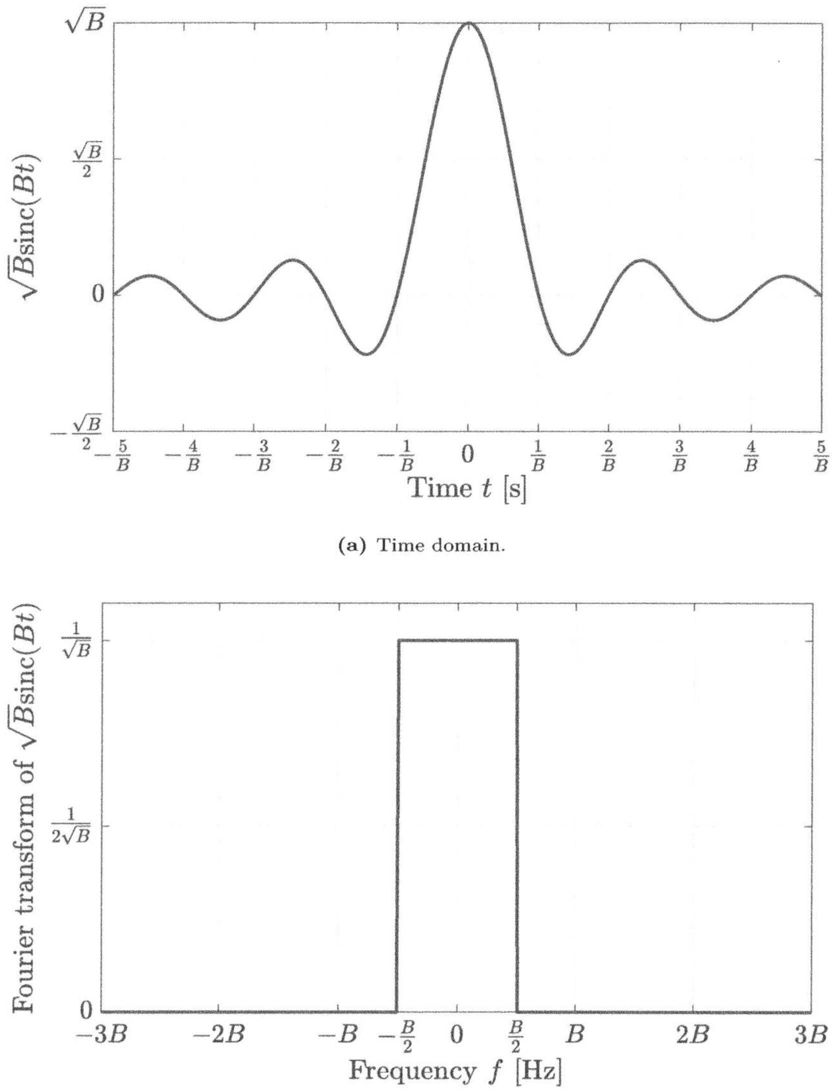

(a) Time domain.

(b) Frequency domain.

Figure 2.11: The unit-energy sinc function $\sqrt{B}\mathrm{sinc}(Bt)$ is shown in the time domain in (a), while the Fourier transform is shown in (b).

is critically important not to sample more sparsely in time because that will create ambiguity; that is, multiple signals can give rise to the same samples, which is known as *aliasing*. We are transmitting data at the Nyquist rate in digital communications, and it is the corresponding signal samples that we call "symbols" and select to represent information bits.[3] Since we are dealing with a complex-valued baseband signal, the B samples are also complex-valued.

Figure 2.12(a) shows the pulses that are utilized for transmitting three subsequent symbols in PAM: $p(t) = \sqrt{B}\text{sinc}(Bt)$, $p(t-1/B)$, and $p(t-2/B)$. More precisely, $p(t)$ is multiplied by $x[0]$, $p(t-1/B)$ is multiplied by $x[1]$, and $p(t-2/B)$ is multiplied by $x[2]$, and then summed up to create $z(t)$. The symbol values become the amplitudes of the respective pulses, which explains why PAM stands for pulse-amplitude modulation. Figure 2.12(b) exemplifies the resulting PAM signal $z(t)$ in (2.120) with $x[0] = 1$, $x[1] = 0.5$, and $x[2] = -0.5$ (and $x[k] = 0$ for all other k). We notice that the duration of each pulse is much larger than the symbol time; thus, each symbol affects the shape of $z(t)$ in a relatively broad time interval. This is an unavoidable side-effect of using pulses with as little bandwidth as possible. Nevertheless, we have $z(k/B) = p(0)x[k] = \sqrt{B}x[k]$ since the pulses are designed to have zero-crossings at all non-zero integers divided by B. This can be observed in Figure 2.12(b) where $z(t)$ intersects the peak values of the respective pulses.

We have now designed a transmitter that maps the discrete-time symbol sequence $\{x[l]\}$ to a continuous-time signal $z(t)$ that can be transmitted over the complex-baseband system. The transmitter operation is illustrated in Figure 2.13, where it is attached to the channel from Figure 2.10(b).

Next, we will design a receiver that can extract the transmitted discrete-time signals by taking samples of the received signal. The main complication is that thermal noise is added to $v(t)$ in the receiver hardware due to the random motion of free electrons caused by thermal agitation. We model the noise by a white circularly symmetric complex Gaussian random process $w(t)$ with constant power spectral density N_0 W/Hz for all (relevant) frequencies.[4] The Gaussian distribution can be motivated by the central limit theorem in Lemma 2.6 since the random motion of many electrons gives rise to approximately Gaussian randomness. By adding the noise to the channel output $v(t)$ in (2.116), we obtain

$$\begin{aligned} \mu(t) &= v(t) + w(t) = (g * z)(t) + w(t) \\ &= \sum_{k=-\infty}^{\infty} x[k]\,(g*p)\left(t - \frac{k}{B}\right) + w(t), \end{aligned} \tag{2.121}$$

[3]The sampling rate must be strictly larger than the Nyquist rate if a signal that contains the frequencies $\pm B/2$ should be identifiable after sampling. This can be seen from the fact that Nyquist sampling of a sine signal results in all samples being zero because they are taken every time the signal crosses zero. Practical communication signals are never perfectly bandlimited; thus, oversampling is often utilized to avoid aliasing and enable digital filtering that deals with the out-of-band signal components. These implementation details are beyond the scope of this book, where we consider ideal pulses and sampling at the Nyquist rate for conceptual simplicity.

[4]A practical signal cannot have a constant power spectral density for *all* frequencies because then it will have infinite power. Hence, we assume that the power spectral density is constant for all relevant frequencies to consider in wireless communications but can drop to zero for other frequencies to keep the power finite (this happens in practice for extremely large frequencies).

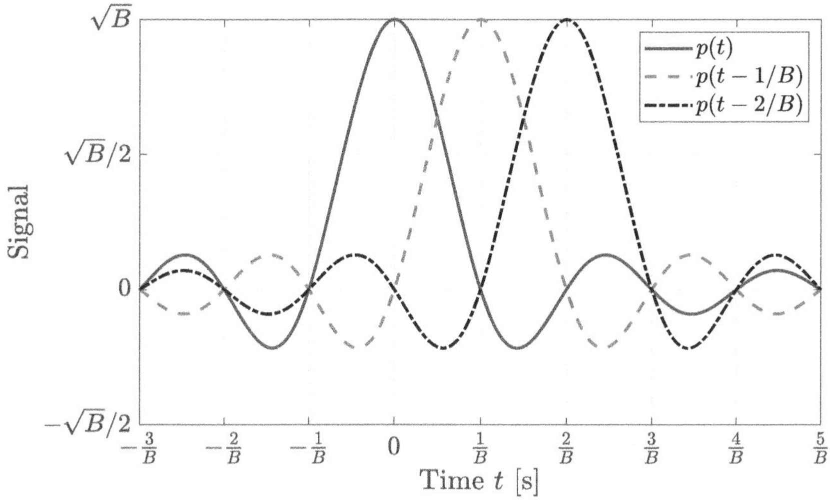

(a) Three subsequent pulses in PAM.

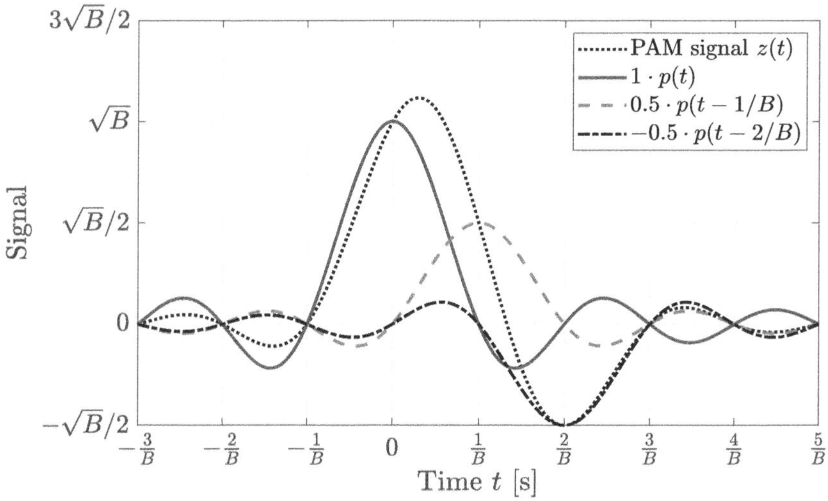

(b) Example of PAM for $x[0] = 1$, $x[1] = 0.5$, and $x[2] = -0.5$.

Figure 2.12: The PAM signal $z(t)$ defined in (2.120) uses time-shifted pulses, as illustrated in (a) for $p(t) = \sqrt{B}\text{sinc}(Bt)$. These pulses are multiplied by different symbol values and summed up to create $z(t)$, as shown in (b).

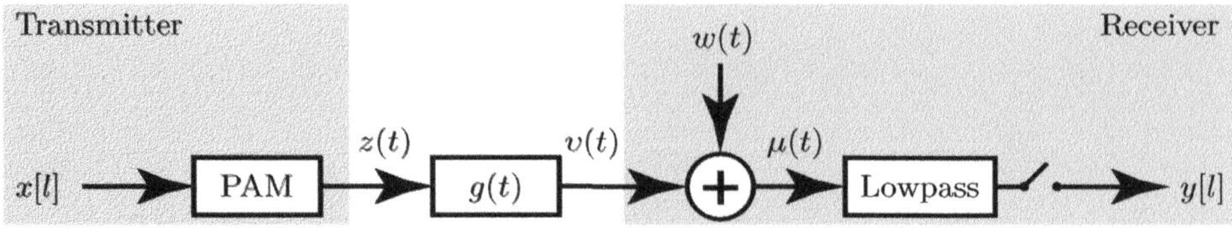

Figure 2.13: The transmitter of a communication system generates a continuous-time signal $z(t)$ from the discrete-time symbol sequence $\{x[l]\}$, using PAM. The receiver is undoing this operation by lowpass filtering (to suppress noise) and sampling. The channel in the middle is the same as in Figure 2.10(b).

where the last equality follows from (2.120). The additive noise is spread over all frequencies, while the desired signal $v(t)$ is bandlimited to $|f| \leq B/2$ by design. Hence, we can remove the out-of-band noise by lowpass filtering $\mu(t)$ without affecting the desired signal.[5] The sinc-pulse $p(t)$ defined in (2.118) and (2.119) is an ideal lowpass filter that can be used for this purpose. We will filter $\mu(t)$ by $p(t)$ and take samples of the output at the same rate as the symbols are transmitted; that is, one sample every $1/B$ seconds. We denote the time instances of the samples as $t = l/B$, where l is the integer sample index, and thereby obtain the sampled received signal

$$\begin{aligned} y[l] &= (p * \mu)(t)\Big|_{t=l/B} \\ &= \sum_{k=-\infty}^{\infty} x[k]\,(p * g * p)\left(t - \frac{k}{B}\right)\Big|_{t=l/B} + (p * w)(t)\Big|_{t=l/B} \\ &= \sum_{k=-\infty}^{\infty} x[k](p * g * p)\left(\frac{l-k}{B}\right) + n[l], \end{aligned} \tag{2.122}$$

where the discrete-time noise $n[l]$ can be shown (see Exercise 2.5) to be complex Gaussian distributed and independent for different l:

$$n[l] = (p * w)\,(t)\Big|_{t=l/B} \sim \mathcal{N}_{\mathbb{C}}(0, N_0). \tag{2.123}$$

We have now derived the discrete-time system model (2.122) that determines how the sampled received signal $y[l]$ depends on the input symbol sequence $\{x[k]\}$. Hence, we can abstract away the notationally complicated continuous-time description of the communication system and only consider discrete-time models in the remainder of this book.

2.3.3 Basic Wireless Channel Modeling

Wireless channels have a particular structure that we can utilize to simplify the system model: the received signal is a summation of several attenuated and

[5]This operation is also necessary in practice to filter out interference from other wireless systems operating in neighboring frequency bands.

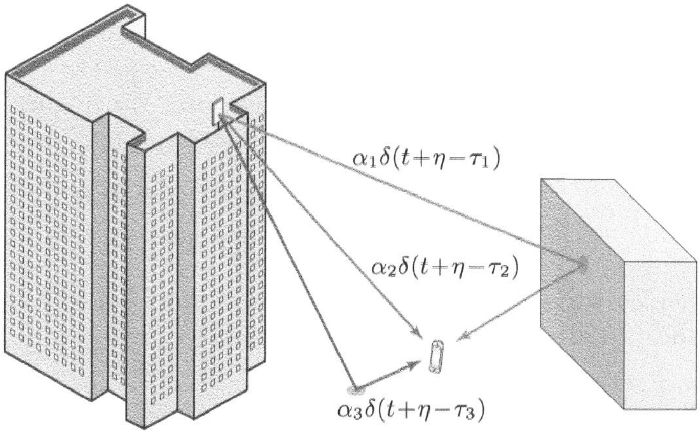

Figure 2.14: The passband channel model in (2.124) consists of L components with different attenuation α_i and delay τ_i. This figure illustrates how these components can be connected to different propagation paths.

delayed copies of the transmitted signal (i.e., a superposition of echos). Suppose the received signal consists of L copies, each having an attenuation $\alpha_i \in [0,1]$ and a delay $\tau_i \geq 0$ seconds, for $i = 1, \ldots, L$. The receiver synchronizes its clock to the transmitter by delaying it by $\eta \geq 0$ seconds to compensate for the propagation delays. The receiver will then observe a superposition of L signal copies that are delayed by $\tau_i - \eta \in \mathbb{R}$, for $i = 1, \ldots, L$. We can write the impulse response of the channel in the passband as

$$g_{\mathrm{p}}(t) = \sum_{i=1}^{L} \alpha_i \delta(t + \eta - \tau_i) \tag{2.124}$$

and it then follows from (2.117) that the equivalent impulse response in the complex baseband is

$$g(t) = \sum_{i=1}^{L} \alpha_i e^{-\mathrm{j}2\pi f_{\mathrm{c}} t} \delta(t + \eta - \tau_i). \tag{2.125}$$

Figure 2.14 illustrates how the L copies can be connected to different propagation paths in the environment. The delay of a path is closely related to the length of the corresponding path, while the attenuation is determined by the distance that the signal has traveled (as in free-space propagation) and which objects the signal has interacted with along the way. Note that the impulse response in the complex baseband contains additional phase-shifts that depend on the carrier frequency.

The channel $g(t)$ appears in (2.122) as the convolution $(p*g*p)(t)$ sampled at time $t = \frac{l-k}{B}$. For the model in (2.125), this convolution term becomes

$$(p * g * p)(t) = \sum_{i=1}^{L} \alpha_i e^{-\mathrm{j}2\pi f_{\mathrm{c}}(\tau_i - \eta)} (p * p)(t + \eta - \tau_i) \tag{2.126}$$

by utilizing the fact[6] that the convolution between an arbitrary function $f(t)$ and the delayed Dirac delta function $e^{-\mathrm{j}2\pi f_c t}\delta(t-\tau)$ is equal to $e^{-\mathrm{j}2\pi f_c \tau} f(t-\tau)$, where $\tau = \tau_i - \eta$ and $f(t) = (p * p)(t)$ in this case. We further notice that $(p * p)(t) = \mathrm{sinc}(Bt)$ since the Fourier transform of $(p * p)(t)$ is

$$\mathcal{F}\{(p * p)(t)\} = P(f)P(f) = \begin{cases} \frac{1}{B}, & \text{if } |f| \leq B/2, \\ 0, & \text{if } |f| > B/2, \end{cases} \tag{2.127}$$

which coincides with the Fourier transform of $\mathrm{sinc}(Bt)$. By utilizing this property and (2.126), we can simplify (2.122) as

$$y[l] = \sum_{k=-\infty}^{\infty} x[k] \sum_{i=1}^{L} \alpha_i e^{-\mathrm{j}2\pi f_c(\tau_i - \eta)} \mathrm{sinc}((l-k) + B(\eta - \tau_i)) + n[l]. \tag{2.128}$$

2.3.4 Discrete Memoryless Channel Model

The received signal $y[l]$ in (2.128) at time l depends on multiple transmitted symbols, as can be seen by the summation over k. Since the symbols were transmitted one after the other, the channel has created the *intersymbol interference*. This happens when the L paths in our channel model have widely different lengths/delays so that a symbol that reaches the receiver over a short path arrives at the same time as a previous symbol arrives over a longer path. Another way to view it is that the received signal $y[l]$ is not only containing the latest transmitted symbol $x[l]$ but also has a *memory* of previously transmitted symbols (and potentially future symbols due to the non-causal sinc-pulse). The memory effect is undesired and can be combatted in various ways. We will identify a condition for when the memory vanishes.

If all the channel components have roughly the same delay, we can synchronize the receiver by selecting η such that $B(\eta - \tau_i) \approx 0$ for all i. To alleviate the channel memory, we want the following approximation to hold:

$$\mathrm{sinc}\big((l-k) + B(\eta - \tau_i)\big) \approx \mathrm{sinc}(l-k) = \begin{cases} 1, & \text{if } l = k, \\ 0, & \text{if } l \neq k. \end{cases} \tag{2.129}$$

Since we can always make this approximation tight by selecting a sufficiently small bandwidth B, this is known as the *narrowband signal assumption*. This result follows from two assumptions that we have made. First, $p(t)$ was selected to be the pulse in the PAM since it satisfies the Nyquist criterion; that is, $(p * p)(l/B)$ is zero for all integers l except $l = 0$. Second, the narrowband assumption implies that the channel will not tamper with the Nyquist criterion. We stress that the narrowband assumption is valid for large bandwidths in environments with tiny path delay differences (or only one path).

[6]The convolution is computed as $\int_{-\infty}^{\infty} e^{-\mathrm{j}2\pi f_c u}\delta(u-\tau) f(t-u)\partial u = e^{-\mathrm{j}2\pi f_c \tau} f(t-\tau)$ by using the sifting property of the Dirac delta function.

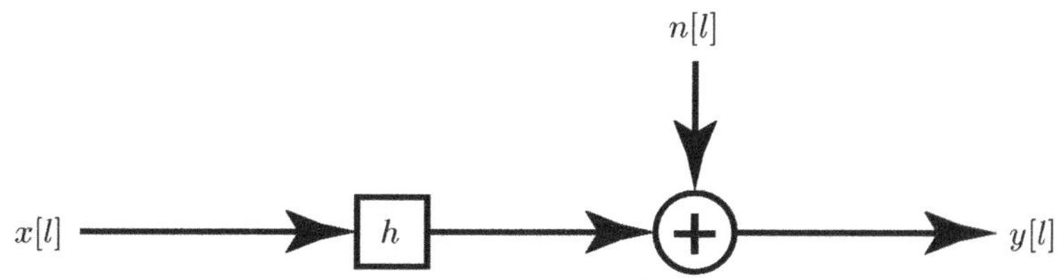

Figure 2.15: A discrete memoryless SISO channel with input $x[l]$ and output $y[l] = h \cdot x[l] + n[l]$, where l is a discrete time index, h is the channel response, and $n[l]$ is the independent complex Gaussian receiver noise.

By inserting (2.129) into (2.128), the system model simplifies to

$$\begin{aligned} y[l] &= \sum_{k=-\infty}^{\infty} x[k] \sum_{i=1}^{L} \alpha_i e^{-\mathrm{j}2\pi f_c(\tau_i - \eta)} \mathrm{sinc}(l-k) + n[l] \\ &= h \cdot x[l] + n[l], \end{aligned} \tag{2.130}$$

where l is a discrete-time index, and the channel is now represented by

$$h = \sum_{i=1}^{L} \alpha_i e^{-\mathrm{j}2\pi f_c(\tau_i - \eta)}. \tag{2.131}$$

From now on, we will refer to $h \in \mathbb{C}$ as the channel response and note that $\beta = |h|^2$ is the channel gain described in Chapter 1. In some parts of this book, we will utilize h as an arbitrary channel response, while there are other parts where we will utilize and generalize the specific structure in (2.131).

Interestingly, we can represent the entire continuous-time communication system in Figure 2.13 by the simple equation (2.130). This is called the *symbol-sampled discrete-time representation* of the channel and will be used in the remainder of this book without loss of generality. A block diagram for this channel is given in Figure 2.15, where we also stress that this is a single-input single-output (SISO) channel with one input to the channel and one output.

The type of channel in (2.130) is also known as a *discrete memoryless channel* since the received signal $y[l]$ only depends on one transmitted signal $x[l]$ and one independent noise realization $n[l]$; there is no memory of previous time instances or impact from later time instances. For this reason, we can just as well drop the time index l and get the system model

$$y = h \cdot x + n. \tag{2.132}$$

When designing the input signal x, we often treat it as a random variable. We will let q denote the average signal energy per symbol (which is a measure of signal power), which implies $\mathbb{E}\{|x|^2\} = q$. The system in (2.132) is also known as an *additive white Gaussian noise (AWGN)* channel.

2.4 Performance Metrics

This section explains how much information can be transmitted reliably over the discrete memoryless channel in (2.132). We consider the transmission of a finite-sized data packet that represents a particular piece of information (e.g., an image, a text document, a piece of a video, or a control command). In this book, we use the words *information* and *data* interchangeably because we consider transferring bits from a transmitter to a receiver, while abstracting away what those bits might represent. However, we stress that data generally refers to the raw sequence of bits considered within a communication system, while information is the application-level interpretation of these bits.

A data packet is characterized by the following:

- How many symbols the packet contains, which is the number of times we will transmit over the channel in (2.132);
- How many data bits each of these symbols represent, determined by the modulation and coding scheme;
- How large the probability of incorrect decoding is at the receiver.

When transmitting a packet containing a small number of symbols, the probability of incorrect decoding is a major concern. Hence, a common performance metric is the *symbol error probability* (also called the symbol error rate), which is the probability that an arbitrary symbol $x[l]$ is decoded incorrectly. This metric has many variations, such as the bit error probability and packet error probability. The values of these error probabilities depend on the choice of the modulation and coding scheme, and the SNR of the channel. In each case, one can derive exact or approximate error probability expressions, which often contain the Gaussian Q-function due to the Gaussian noise.

Letting each symbol describe many data bits is desirable, but the error probability increases when more bits are represented. Hence, there is a tradeoff between low error probability and many data bits per symbol. This tradeoff is non-trivial when transmitting packets with a small number of symbols. It typically boils down to selecting a non-zero target error probability based on experiments (e.g., 0.01) and then selecting the "best" modulation and coding scheme that satisfies that target from a predefined list of schemes. When an error occurs, we need to retransmit the packet. This tradeoff is illustrated in Figure 2.16(a), where there are few errors when transmitting a few bit/symbol and many errors when transmitting many bit/symbol. The gradual color change shows how the error probability increases gradually.

In contrast, when transmitting a packet with many symbols, the error probability can be made negligible by selecting the proper modulation and coding scheme, which renders the error metric superfluous. The "right" scheme should operate close to, but below, the *channel capacity*. Claude Shannon

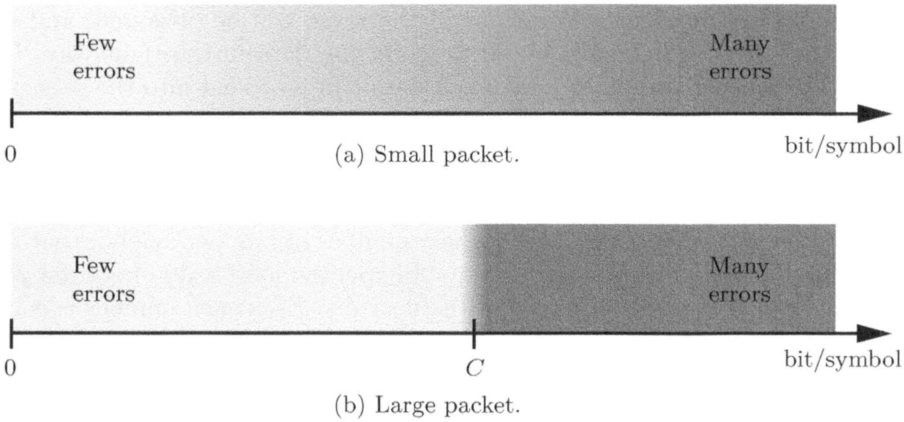

Figure 2.16: The packet error probability increases the more bits are transmitted per symbol. When the packet is small, then there is a gradual transition between a few and many errors, as shown in (a). However, when the packet is large, the transition is concentrated in a small interval around the channel capacity C.

defined the capacity in the seminal paper [40] from 1948, therefore, it is also known as the *Shannon capacity*. Figure 2.16(b) illustrates the essence of this result, namely that the transition between having few and many errors in the transmission happens in a small interval around a value C bit/symbol called the capacity when the packet is large. It can be formally defined as follows.

Definition 2.6. The channel capacity of a given channel is the highest number of bits per symbol that can be communicated with arbitrarily low error probability as the number of symbols in the packet approaches infinity.

The interpretation of the channel capacity is that we can communicate without error when sending long sequences of symbols, if we carefully select how many bits each symbol represents. This implies that the gradual colored transition interval shown in Figure 2.16(b) vanishes asymptotically so that we get a sudden shift between no errors when operating below the capacity C and many errors when operating above the capacity. In this context, "long" means (at least) 10000 symbols [41], which takes 1 ms to transmit when using $B = 10\,\text{MHz}$. This is relatively short in practice; thus, many wireless communication systems operate very close to the capacity. Since one of the core motivating factors of multiple antenna communications is to transmit a large amount of data in a way that is faster and/or requires less power than in single-antenna communications, it is natural to adopt the channel capacity as the performance metric in this book. That said, methods to achieve high capacity with multiple antennas coincide, to a large extent, with methods that provide low error probabilities when transmitting small data packets.

The unknown randomness of the noise must be combatted to achieve reliable (error-free) communications. When sending long sequences of symbols, many independent realizations of the noise will be observed, and the uncertainty can be averaged out if we code the data in the right way. The channel capacity determines how much data can be coded into the sequence of symbols while enabling the noise effect to average out.

2.4.1 Basic Capacity Results

Since the capacity represents the highest number of bits per symbol that can be communicated without errors, any "bit per symbol" value between zero and the capacity can also be used without errors. Each such number is called an *achievable data rate*, an achievable rate, or a rate.

Definition 2.7. An achievable data rate is a positive number below the channel capacity. It is possible to communicate at this rate with arbitrarily low error probability as the number of symbols in the packet approaches infinity.

Although the capacity is of primary interest, there are situations where the capacity is unknown. Therefore, it is crucial to find achievable data rates that can be used to communicate without error.

The channel capacity can be rigorously defined for any communication channel, but we refer to [40] and [42] for the general details. This book only considers the general concept of discrete memoryless channels. For such a channel that takes the data symbol x as input and produces y as output, the channel capacity takes the following form as proved in [38], [40], [42].

Theorem 2.1. Consider a discrete memoryless channel with input $x \in \mathbb{C}$ and output $y \in \mathbb{C}$, which are two random variables specified by the conditional PDF $f_{\mathsf{y}|\mathsf{x}}(y|x)$. The channel capacity is

$$C = \max_{f_{\mathsf{x}}(x)} \left(\mathcal{H}(y) - \mathcal{H}(y|x)\right), \tag{2.133}$$

where the maximum is taken with respect to all distributions $f_{\mathsf{x}}(x)$ of the input that are considered feasible. The differential entropy $\mathcal{H}(y)$ is defined as

$$\mathcal{H}(y) = -\int_{y\in\mathbb{C}} \log_2\left(f_{\mathsf{y}}(y)\right) f_{\mathsf{y}}(y)\partial y \tag{2.134}$$

using the marginal distribution $f_{\mathsf{y}}(y) = \int_{x\in\mathbb{C}} f_{\mathsf{y}|\mathsf{x}}(y|x) f_{\mathsf{x}}(x)\partial x$ of y and the conditional differential entropy $\mathcal{H}(y|x)$ is defined as

$$\mathcal{H}(y|x) = -\int_{y\in\mathbb{C}}\int_{x\in\mathbb{C}} \log_2\left(f_{\mathsf{y}|\mathsf{x}}(y|x)\right) f_{\mathsf{y}|\mathsf{x}}(y|x) f_{\mathsf{x}}(x)\partial x\partial y. \tag{2.135}$$

We note that all the integrals in Theorem 2.1 are computed over the entire complex plane, which is the same as considering a double integral where both

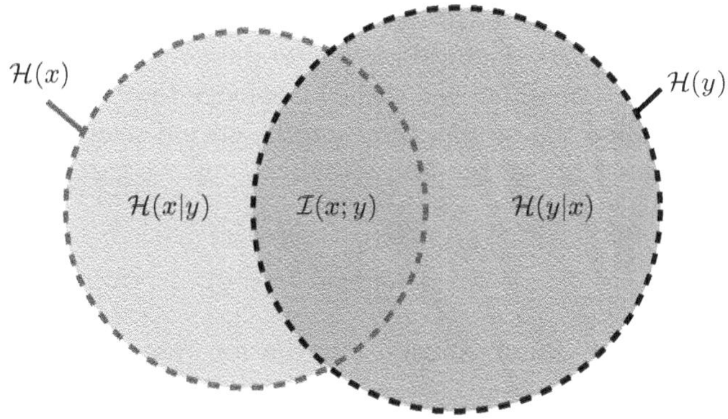

Figure 2.17: The left circle (green) represents the random variable x and the right circle (red) represents the random variable y. The circle areas match the respective differential entropies $\mathcal{H}(x)$ and $\mathcal{H}(y)$, while the intersection is the mutual information $\mathcal{I}(x;y)$ that can be computed in the two ways described in (2.137). The capacity is the maximum mutual information; that is, the information that is contained in both the transmitted signal x and the received signal y.

the real and imaginary parts are integrated from $-\infty$ to $+\infty$. The capacity in (2.133) is given by the difference between two terms: $\mathcal{H}(y) - \mathcal{H}(y|x)$. The differential entropy $\mathcal{H}(y)$ measures our surprisal when observing a realization of the random variable y at the receiver, which also measures the amount of unknown information that the variable conveys. The differential entropy can take any value from $-\infty$ to $+\infty$, where a larger value implies a larger surprisal. Similarly, $\mathcal{H}(y|x)$ measures the amount of additional information we obtain by observing y if we already know x. It holds that $\mathcal{H}(y) \geq \mathcal{H}(y|x)$ since observing x cannot increase our surprisal when we later observe y, but it can usually reduce the surprisal substantially. Hence, $\mathcal{H}(y) - \mathcal{H}(y|x) \geq 0$ and the channel capacity must be greater than or equal to zero.

More generally, the differential entropy of a sequence $x_1, \ldots, x_L$ of random variables can be expressed using the following chain rule:

$$\mathcal{H}(x_1, \ldots, x_L) = \sum_{l=1}^{L} \mathcal{H}(x_l|x_1, \ldots, x_{l-1}). \tag{2.136}$$

Since the conditioning cannot increase the surprisal, the lth term in the sum can be upper bounded by $\mathcal{H}(x_l)$. It follows that $\mathcal{H}(x_1, \ldots, x_L) \leq \sum_{l=1}^{L} \mathcal{H}(x_l)$, where equality is achieved if and only if the random variables are independent.

Figure 2.17 shows a Venn diagram where the circles represent the random variables x and y, and their areas equal the respective differential entropies $\mathcal{H}(x)$ and $\mathcal{H}(y)$. The intersection between the circles determines the ability to extract information about the transmitted signal x from observing the received signal y. The area of the intersection is $\mathcal{H}(y) - \mathcal{H}(y|x)$. If we select the input distribution $f_{\mathsf{x}}(x)$ to maximize this area, then it equals the channel capacity C in Theorem 2.1. There is an important statistical symmetry in this figure, which implies that the intersection area can also be expressed as $\mathcal{H}(x) - \mathcal{H}(x|y)$. This expression has an intuitive interpretation: The entropy/uncertainty about the transmitted signal x minus the entropy/uncertainty remaining after observing the received signal y. The difference is the knowledge that we learned from our observation. It is called the *mutual information* since it measures the common information contained in the random variables, and we can denote it as

$$\begin{aligned}\mathcal{I}(x;y) &= \mathcal{H}(y) - \mathcal{H}(y|x) \\ &= \mathcal{H}(x) - \mathcal{H}(x|y). \end{aligned} \tag{2.137}$$

The capacity is the maximum mutual information that can be achieved.

Example 2.14. What is the channel capacity if x and y are independent?

In this case, the conditional PDF that determines the capacity reduces to $f_{\mathsf{y}|\mathsf{x}}(y|x) = f_{\mathsf{y}}(y)$, which is the marginal PDF of the output y. The conditional differential entropy in (2.135) can now be computed as

$$\begin{aligned}\mathcal{H}(y|x) &= -\int_{y\in\mathbb{C}}\int_{x\in\mathbb{C}} \log_2\left(f_{\mathsf{y}}(y)\right) f_{\mathsf{y}}(y) f_{\mathsf{x}}(x) \partial x \partial y \\ &= -\int_{y\in\mathbb{C}} \log_2\left(f_{\mathsf{y}}(y)\right) f_{\mathsf{y}}(y) \partial y \underbrace{\int_{x\in\mathbb{C}} f_{\mathsf{x}}(x)\partial x}_{=1} = \mathcal{H}(y). \end{aligned} \tag{2.138}$$

The capacity in (2.133) becomes zero in this case since $\mathcal{H}(y) = \mathcal{H}(y|x)$, so there is no intersection between the circles in the Venn diagram in Figure 2.17. Consequently, the ability to transfer information lies in the correlation between the random variables at the input and output of the channel.

To compute the capacity in (2.133), we need to identify the PDF $f_{\mathsf{x}}(x)$ of the input x that maximizes $\mathcal{H}(y) - \mathcal{H}(y|x)$. This is the same as finding an optimal modulation and coding scheme. Theorem 2.1 says we can only select distributions that are "considered feasible", so we must specify some requirements on x. It is common to consider all distributions for which the symbol power $\mathbb{E}\{|x|^2\}$ is upper limited by a constant representing the maximum power. To find the optimal PDF, we need the following key result that says which distribution maximizes our surprisal [42], [1, Lemma B.20].

Lemma 2.9. For any continuous random variable $z \in \mathbb{C}$ with $\mathbb{E}\{|z|^2\} = p$, the differential entropy of z is upper bounded as

$$\mathcal{H}(z) \leq \log_2(e\pi p), \tag{2.139}$$

where $e \approx 2.71828$ is Euler's number. Equality is achieved in (2.139) if and only if $z \sim \mathcal{N}_{\mathbb{C}}(0, p)$; that is, the complex Gaussian distribution has the largest possible differential entropy.

For the kind of discrete memoryless channel in (2.132) and Figure 2.15, we have $y = hx + n$, where h is a deterministic scalar, $n \sim \mathcal{N}_{\mathbb{C}}(0, N_0)$, and the signal x has the symbol power $\mathbb{E}\{|x|^2\} = q$. Hence, the feasible input distributions are all $f_{\mathsf{x}}(x)$ that satisfy $\mathbb{E}\{|x|^2\} = q$. The choice of input distribution only affects $\mathcal{H}(y)$ because x is known in $\mathcal{H}(y|x)$, thus we want to select the distribution of x to maximize $\mathcal{H}(y)$. Since the signal and noise are independent, we obtain

$$\mathbb{E}\{|y|^2\} = \mathbb{E}\{|x|^2\}|h|^2 + \mathbb{E}\{|n|^2\} = q|h|^2 + N_0. \tag{2.140}$$

We can utilize the result in (2.139) to conclude that

$$\mathcal{H}(y) \leq \log_2\left(e\pi(q|h|^2 + N_0)\right) \tag{2.141}$$

with equality if and only if $y \sim \mathcal{N}_{\mathbb{C}}(0, q|h|^2 + N_0)$. This maximum entropy is achieved when $x \sim \mathcal{N}_{\mathbb{C}}(0, q)$; thus, we have found the input distribution corresponding to the maximum in the capacity expression in (2.133). This is called the *capacity-achieving* input distribution.

To obtain a closed-form capacity expression, it remains to compute $\mathcal{H}(y|x)$. When x is known, the only randomness that remains in $y = hx + n$ is that of the noise $n \sim \mathcal{N}_{\mathbb{C}}(0, N_0)$ since h is deterministic, thus

$$\mathcal{H}(y|x) = \mathcal{H}(n) = \log_2\left(e\pi N_0\right), \tag{2.142}$$

where the last equality follows from Lemma 2.9. As a final step, we notice that

$$C = \log_2\left(e\pi(q|h|^2 + N_0)\right) - \log_2\left(e\pi N_0\right) = \log_2\left(1 + \frac{q|h|^2}{N_0}\right). \tag{2.143}$$

We can summarize the capacity of an AWGN channel as follows.

Corollary 2.1. Consider the discrete memoryless channel in Figure 2.15 with input $x \in \mathbb{C}$ and output $y \in \mathbb{C}$ given by

$$y = h \cdot x + n, \tag{2.144}$$

where $n \sim \mathcal{N}_{\mathbb{C}}(0, N_0)$ is independent noise. Suppose the input distribution is feasible whenever the symbol power satisfies $\mathbb{E}\{|x|^2\} \leq q$ and $h \in \mathbb{C}$ is a constant known at the output. The channel capacity is

$$C = \log_2\left(1 + \frac{q|h|^2}{N_0}\right) \quad \text{bit/symbol} \tag{2.145}$$

and is achieved when the input is distributed as $x \sim \mathcal{N}_{\mathbb{C}}(0, q)$.

The channel capacity in (2.145) is expressed in *bit per symbol*, but many equivalent units appear in the communication literature: *bit per sample*, *bit per channel use*, *bit/s/Hz*, and *bit per complex degree of freedom*.

The complex Gaussian input distribution creates a continuous signal constellation, where the transmitted signal x can take any value in $\mathbb{C}$. We will transmit packets containing N symbols to showcase how a capacity-achieving system operates. For a capacity of C bit/symbol, we can convey NC data bits per packet. Hence, 2^{NC} different potential data sequences can be communicated. We then need to create a *codebook* containing 2^{NC} different symbol sequences, and each is called a *codeword* and represents one of the 2^{NC} data sequences. When we want to transfer a packet containing specific data, we transmit the corresponding codeword from the codebook. The receiver's task is determining which of the 2^{NC} codewords was most likely to have been transmitted. With the capacity-achieving complex Gaussian input distribution, each codeword is generated by taking N independent and identically distributed (i.i.d.) realizations from $\mathcal{N}_{\mathbb{C}}(0, q)$. This is called a *Gaussian codebook*. The codebook generation is done once and for all when designing the communication system. The codewords must be stored in the transmitter to enable encoding (i.e., transmitting the correct codeword) and in the receiver to facilitate decoding (i.e., identifying which codeword was sent). More precise details can be found in [42, Ch. 10]. Since the channel capacity is achieved as the packet length $N \to \infty$, this communication method is impractical since the complexity of finding the correct codeword and the storage requirements for the codewords grow exponentially with N.

In practice, the capacity-achieving system operation is approximated by imposing a structure that alleviates the need for storing the codewords and simplifies the encoding/decoding. It is common to utilize a discrete signal constellation where each symbol x can only take values on a square grid containing $2^{\tilde{C}}$ points, where $\tilde{C}$ is the closest even integer above C. This is called *quadrature amplitude modulation (QAM)*. To not attempt transferring more data than the capacity allows, only a subset of 2^{NC} symbol sequences among the $2^{N\tilde{C}}$ possible sequences is utilized, where the ratio $C/\tilde{C}$ is called the *coding rate*. The subset is selected by a channel coding scheme designed to minimize the risk of mixing up the selected sequences at the receiver side (i.e., minimizing the probability of decoding error).

To give a concrete example, the 5G NR standard utilizes the modulation formats 4-QAM, 16-QAM, 64-QAM, and 256-QAM along with the low-density parity-check (LDPC) coding scheme, where the coding is designed to operate close to the capacity while enabling efficient encoding and decoding.[7] Figure 2.18 exemplifies 28 predefined combinations of *modulation and coding schemes (MCSs)* from [43, Table 5.1.3.1-2], where the first column is an index that the transmitter and receiver can use when agreeing upon which combination to utilize. The second column describes the modulation format, the third column is the coding rate, and the fourth column is the number of bits per symbol. If the channel capacity would be 4 bit/symbol, then we should

[7] Polar codes are also used in 5G NR but for transmission of small blocks.

Index	Modulation format	Coding rate	bit/symbol
0	4-QAM	0.12	0.24
1	4-QAM	0.19	0.38
2	4-QAM	0.30	0.60
3	4-QAM	0.44	0.88
4	4-QAM	0.59	1.18
5	16-QAM	0.37	1.48
6	16-QAM	0.42	1.70
7	16-QAM	0.48	1.91
8	16-QAM	0.54	2.16
9	16-QAM	0.60	2.41
10	16-QAM	0.64	2.57
11	64-QAM	0.46	2.73
12	64-QAM	0.50	3.03
13	64-QAM	0.55	3.32
14	64-QAM	0.60	3.61
15	64-QAM	0.65	3.90
16	64-QAM	0.70	4.21
17	64-QAM	0.75	4.52
18	64-QAM	0.80	4.82
19	64-QAM	0.86	5.12
20	256-QAM	0.67	5.33
21	256-QAM	0.69	5.55
22	256-QAM	0.74	5.89
23	256-QAM	0.78	6.23
24	256-QAM	0.82	6.57
25	256-QAM	0.86	6.91
26	256-QAM	0.90	7.16
27	256-QAM	0.93	7.41

Figure 2.18: The list of 28 MCS combinations utilized in the 5G NR standard. The list is adapted from [43, Table 5.1.3.1-2].

search the table for the closest but smaller number, which in this case is index 15 that provides 3.9 bit/symbol. Hence, 64-QAM should be used to transmit $\log_2(64) = 6$ codeword bits per symbol, whereof a fraction 0.65 contains data bits, resulting in $6 \cdot 0.65 = 3.9$ bit/symbol. We will not consider any of these specific details in the remainder of this book but utilize the channel capacity as the performance metric while keeping in mind that there are practical ways to communicate at data rates close to the capacity.

We can rewrite the capacity expression in (2.145) taking the following three facts into account:

1. B symbols are transmitted per second;
2. The channel gain is $\beta = |h|^2$;
3. The symbol power q is measured in energy per symbol. It can be expressed as $q = P/B$, where P is the transmit power in Watt and B is the number of symbols per second.

The first fact means we can multiply (2.145) with B to change the unit from bit/symbol to bit/s. This is why the unit bit/symbol is also equivalent to the unit bit/s/Hz. The latter two facts can be used to make changes of variables, leading to

$$C = B \log_2 \left(1 + \frac{P\beta}{BN_0} \right) \quad \text{bit/s}. \tag{2.146}$$

We notice that the channel capacity is given by the bandwidth multiplied by the base-two logarithm of one plus

$$\text{SNR} = \frac{P\beta}{BN_0} \tag{2.147}$$

that was previously stated in (1.13). Hence, the channel capacity is tightly connected to the SNR, just as many other communication performance metrics.

2.5 Estimation Theory

The goal of estimation is to compute a good approximate value of an unknown parameter based on measurements. The estimation procedure is particularly challenging when the measurements are limited and noisy. There are two main subfields of estimation theory [44]. In *classical estimation*, the unknown variable is deterministic and, thus, has the same constant value forever. In *Bayesian estimation*, the unknown variable is instead a realization of a random variable with a known statistical distribution (also known as *the prior*).

In wireless communications, the transmission of very large data packets is implicitly assumed whenever the channel capacity is used as the performance metric. Hence, unknown variables that are constant throughout the transmission are relatively easy to estimate; for example, a negligibly small preamble can be attached to the packet to obtain the necessary measurements. In contrast, unknown variables that take different values during the transmission must be estimated using few measurements because there is insufficient time to make extensive measurements. This can be modeled as if the unknown variable takes different realizations from the same random variable at different times. For this reason, Bayesian estimation will mostly be considered in this book. It is generally assumed that the statistics are known, but Section 2.6 describes how they can be estimated in practice.

The general principle is that we want to compute an estimate of a realization h of a random variable. The available information is an observation y connected statistically with the unknown variable. More precisely, we have measured the current value of y and know the conditional PDF $f_{\mathsf{h}|\mathsf{y}}(h|y)$ of h given the value of y. There is a rich theory for Bayesian estimation of both real and complex variables and different ways of measuring what is a *good* approximate value [44]. We will only consider the *mean-squared error (MSE)* as the performance metric for the estimation.

Definition 2.8. Consider a random variable $h \in \mathbb{C}$ and let $\hat{h}(y)$ denote an arbitrary estimator of h based on the observation $y \in \mathbb{C}$. The estimation error is $h - \hat{h}(y)$ and the MSE is defined as

$$\mathrm{MSE}_h = \mathbb{E}\left\{|h - \hat{h}(y)|^2\right\}, \tag{2.148}$$

by taking the average squared estimation error.

Lemma 2.10. The estimator that minimizes the MSE in (2.148) is called the *minimum mean-squared error (MMSE)* estimator. It can be computed as

$$\hat{h}_{\mathrm{MMSE}}(y) = \mathbb{E}\{h|y\} = \int_{h \in \mathbb{C}} h f_{\mathsf{h}|\mathsf{y}}(h|y) \partial h \tag{2.149}$$

where $f_{\mathsf{h}|\mathsf{y}}(h|y)$ is the conditional PDF of h given the observation y.

The MMSE estimator is the conditional mean of h given y. By definition, it minimizes the variance of the estimation error. Since the estimator depends on the conditional PDF $f_{\mathsf{h}|\mathsf{y}}(h|y)$, it will be different depending on how h is distributed. The integral in (2.149) cannot be computed analytically in general, so it must be evaluated numerically. The Gaussian case is an exception.

2.5.1 MMSE Estimation of Complex Gaussian Variables

We are particularly interested in the memoryless channel model in (2.132), which we restate as

$$y = h \cdot x + n, \quad n \sim \mathcal{N}_{\mathbb{C}}(0, N_0). \tag{2.150}$$

Suppose the channel response h is unknown and should be estimated. We can then select the transmitted signal x as a deterministic number known at both the transmitter and receiver, so that only h is unknown in the product $h \cdot x$ in (2.150). We also know the distribution of the additive complex Gaussian noise n, but the realization is unknown. The goal is to compute the MMSE estimate of h based on the observation y obtained at the receiver.

We consider the case when the channel is complex Gaussian distributed:

$$h \sim \mathcal{N}_{\mathbb{C}}(0, \beta). \tag{2.151}$$

To compute the MMSE estimator in (2.149), we must first determine the conditional PDF $f_{\mathsf{h}|\mathsf{y}}(h|y)$. This problem resembles the one considered in Section 2.2.3. If we divide all terms in (2.150) by x, we obtain

$$\underbrace{\frac{1}{x}y}_{=z} = \underbrace{h}_{=v} + \underbrace{\frac{1}{x}n}_{=w}, \tag{2.152}$$

which is of the same form as (2.69) but with $\sigma_v^2 = \beta$ and $\sigma_w^2 = N_0/|x|^2$. Hence, we can utilize (2.74) to obtain

$$f_{\mathsf{h}|\mathsf{y}}(h|y) = \frac{\beta + \frac{N_0}{|x|^2}}{\pi \beta \frac{N_0}{|x|^2}} e^{-\frac{\beta + \frac{N_0}{|x|^2}}{\beta \frac{N_0}{|x|^2}} \left| h - \frac{\beta}{\beta + \frac{N_0}{|x|^2}} \frac{y}{x} \right|^2} = \frac{\beta |x|^2 + N_0}{\pi \beta N_0} e^{-\frac{\beta |x|^2 + N_0}{\beta N_0} \left| h - \frac{\beta x^*}{\beta |x|^2 + N_0} y \right|^2}. \tag{2.153}$$

The MMSE estimate is the mean value of this conditional PDF. By comparing (2.153) with the PDF of a complex Gaussian distribution, we notice that

$$h - \frac{\beta x^*}{\beta |x|^2 + N_0} y \sim \mathcal{N}_{\mathbb{C}}\left(0, \frac{\beta N_0}{\beta |x|^2 + N_0}\right) \tag{2.154}$$

when y is known. Hence, the conditional mean value is $\mathbb{E}\{h|y\} = \frac{\beta x^*}{\beta |x|^2 + N_0} y$. The variance $\frac{\beta N_0}{\beta |x|^2 + N_0}$ in (2.154) is the MSE of the estimate.

Lemma 2.11. Consider the estimation of $h \sim \mathcal{N}_{\mathbb{C}}(0, \beta)$ from the observation $y = h \cdot x + n$, when the signal $x \in \mathbb{C}$ is known and $n \sim \mathcal{N}_{\mathbb{C}}(0, N_0)$ is independent noise. The MMSE estimator of h is

$$\hat{h}_{\mathrm{MMSE}}(y) = \frac{\beta x^*}{\beta |x|^2 + N_0} y. \tag{2.155}$$

The corresponding minimum MSE is

$$\mathrm{MSE}_h = \frac{\beta N_0}{\beta |x|^2 + N_0}. \tag{2.156}$$

Among all possible estimators that utilize the observation y and the channel statistics, the MMSE estimator minimizes the MSE. The MMSE estimate in (2.155) will be expressed as $\hat{h}$ later in this book without explicitly specifying what observation it is based on and which type of estimate it is.

We notice that this MMSE estimate is a linear function ay of the observation y, which is scaled by the factor $a = \frac{\beta x^*}{\beta |x|^2 + N_0}$ to obtain the estimate that is closest to the true value of h in the MSE sense. For this reason, the estimator in Lemma 2.11 is sometimes referred to as the *linear MMSE (LMMSE) estimator*; that is, the estimator that obtains the lowest MSE among all linear estimators. While it is formally correct to use that terminology, the naming devalues its properties by giving the wrong impression that there might exist better estimators that are non-linear functions of y. Hence, in the remainder of this book, we will call (2.155) the MMSE estimator.

A useful benefit of the expression in (2.155) is that we can directly generate random realizations of $\hat{h}$ without first generating realizations of y, h, and n. Since $y \sim \mathcal{N}_{\mathbb{C}}(0, \beta|x|^2 + N_0)$, it follows that

$$\begin{aligned}\hat{h} &\sim \mathcal{N}_{\mathbb{C}}\left(0, \left|\frac{\beta x^*}{\beta|x|^2 + N_0}\right|^2 (\beta|x|^2 + N_0)\right) = \mathcal{N}_{\mathbb{C}}\left(0, \frac{\beta^2|x|^2}{\beta|x|^2 + N_0}\right) \\ &= \mathcal{N}_{\mathbb{C}}\left(0, \beta - \mathrm{MSE}_h\right).\end{aligned} \tag{2.157}$$

Moreover, the estimation error $\tilde{h} = h - \hat{h}$ is distributed as

$$\tilde{h} \sim \mathcal{N}_{\mathbb{C}}(0, \mathrm{MSE}_h) \tag{2.158}$$

with the MSE in (2.156) being the variance since

$$\mathbb{V}\mathrm{ar}\left\{\tilde{h}\right\} = \mathbb{E}\left\{|\tilde{h}|^2\right\} = \mathbb{E}\left\{\left|h - \hat{h}\right|^2\right\} = \mathrm{MSE}_h. \tag{2.159}$$

The estimate and estimation error are statistically independent, which can be seen from the fact that they are complex Gaussian distributed and uncorrelated. Their variances add up to that of the original unknown variable h: $\mathbb{V}\mathrm{ar}\{\hat{h}\} + \mathbb{V}\mathrm{ar}\{\tilde{h}\} = \beta - \mathrm{MSE}_h + \mathrm{MSE}_h = \beta$. This showcases how the MMSE estimator extracts all useful information from the observation y so that the error term only contains information that was not observed. Consequently, the estimation error is also statistically independent of the observed signal y.

Intuitively, the estimation quality should be better when the factor hx in (2.150) is much larger than the noise term when comparing their magnitudes. If we let $|x| \to \infty$, it follows that the MSE in (2.156) goes to zero and that the estimate's variance in (2.157) approaches β. This means we can estimate the channel without error when the SNR is large.

The MSE in (2.156) is an increasing function of β, so we should expect larger estimation errors when estimating a variable with a large variance compared to a small variance. However, it is the relative size of the estimation error that matters in many contexts, and it is quantified by the *normalized MSE (NMSE)* that is computed as

$$\mathrm{NMSE}_h = \frac{\mathbb{E}\left\{|h - \hat{h}|^2\right\}}{\mathbb{E}\left\{|h|^2\right\}} = \frac{\mathrm{MSE}_h}{\beta} = \frac{N_0}{\beta|x|^2 + N_0}. \tag{2.160}$$

The NMSE is a decreasing function of β, so it is easier to estimate a variable with a large variance than a small one as it stands out more from the noise.

We have now described how to estimate the channel coefficient h. The next example shows how to estimate signals using the MMSE estimator.

Example 2.15. Suppose we want to estimate the data signal $x \sim \mathcal{N}_{\mathbb{C}}(0, q)$ from the received signal

$$y = h \cdot x + n, \tag{2.161}$$

where $h \in \mathbb{C}$ is a known constant channel and $n \sim \mathcal{N}_{\mathbb{C}}(0, N_0)$ is independent noise. What is the MSE if the MMSE estimator is used? Use the MSE expression to compute the mutual information $\mathcal{I}(x; y)$.

The MMSE estimation problem is the same as in Lemma 2.11, except that x and h have interchanged the roles of being known and unknown. We can denote the MMSE estimate as $\hat{x}$. By making the variable substitutions $\beta \to q$ and $x \to h$ in (2.156), the MSE when estimating x becomes

$$\mathrm{MSE}_x = \frac{qN_0}{q|h|^2 + N_0}. \tag{2.162}$$

The error is independent of $\hat{x}$ and distributed as $\tilde{x} = x - \hat{x} \sim \mathcal{N}_{\mathbb{C}}(0, \mathrm{MSE}_x)$.

The mutual information in (2.137) is equal to $\mathcal{H}(x) - \mathcal{H}(x|y)$ and Lemma 2.9 states that $\mathcal{H}(x) = \log_2(e\pi q)$ since the signal is complex Gaussian distributed with variance q. It further holds that

$$\mathcal{H}(x|y) = \mathcal{H}(x - \hat{x}|y) = \mathcal{H}(\tilde{x}|y) = \log_2(e\pi \mathrm{MSE}_x), \tag{2.163}$$

where the first equality follows from subtracting the MMSE estimate from x, which can be done without changing the entropy since y is known. The last equality follows from noticing that the estimation error is independent of y and complex Gaussian distributed with variance MSE_x. The mutual information can finally be computed as

$$\begin{aligned}\mathcal{H}(x) - \mathcal{H}(x|y) &= \log_2(e\pi q) - \log_2(e\pi \mathrm{MSE}_x) \\ &= \log_2\left(\frac{q}{\mathrm{MSE}_x}\right) = \log_2\left(1 + \frac{q|h|^2}{N_0}\right).\end{aligned} \tag{2.164}$$

This is an alternative way of computing the capacity in (2.145).

2.5.2 LMMSE Estimation of Arbitrarily Distributed Variables

We will now consider LMMSE estimation when the received signal is $y = h \cdot x + n$ as before, but the unknown variable h and the noise n might not be Gaussian distributed. An LMMSE estimator has the form $\hat{h} = ay$, where a is selected to minimize the MSE. The MSE is a function of a and can be minimized by equating the first-order derivative to zero:[8]

$$0 = \frac{\partial}{\partial a^*}\mathbb{E}\left\{|\tilde{h}|^2\right\} = \frac{\partial}{\partial a^*}\mathbb{E}\{(h-ay)(h-ay)^*\} = -\mathbb{E}\{\tilde{h}y^*\}. \tag{2.165}$$

This sufficient and necessary condition for selecting a is called the *orthogonality principle*: $\mathbb{E}\{\tilde{h}y^*\} = 0$. The interpretation is that the scaling factor a must be designed so that the error term $\tilde{h} = h - \hat{h}$ is uncorrelated with the received signal y; that is, there is no useful information left that can be extracted using linear methods. It follows from the orthogonality principle that $\mathbb{E}\{\tilde{h}\hat{h}^*\} = \mathbb{E}\{\tilde{h}y^*\}a^* = 0$, which implies that the estimate and estimation error are uncorrelated random variables. In the special case where the estimate and estimation error are complex Gaussian distributed (which happens when h and n are Gaussian, as in the last section), it follows from Lemma 2.7 that the uncorrelated variables $\hat{h}$ and $\tilde{h}$ are also independent random variables. In the general non-Gaussian case, the estimate and error are only uncorrelated.

The orthogonality principle can be used to find the LMMSE estimator, which we will show through an example.

Example 2.16. Use the orthogonality principle to derive the LMMSE estimator of h given the received signal $y = h \cdot x + n$. Assume that $\mathbb{E}\{h\} = \mathbb{E}\{n\} = \mathbb{E}\{hn^*\} = 0$, $\mathbb{E}\{|h|^2\} = \beta$, and $\mathbb{E}\{|n|^2\} = N_0$.

An arbitrary linear estimator has the form $\hat{h} = ay$. We need to find the value of a that satisfies the orthogonality principle $\mathbb{E}\{\tilde{h}y^*\} = 0$:

$$0 = \mathbb{E}\left\{\tilde{h}y^*\right\} = \mathbb{E}\left\{(h-ay)\,y^*\right\} = \mathbb{E}\left\{hy^*\right\} - a\mathbb{E}\left\{|y|^2\right\}. \tag{2.166}$$

By solving for a in (2.166), we obtain

$$a = \frac{\mathbb{E}\{hy^*\}}{\mathbb{E}\left\{|y|^2\right\}} = \frac{\mathbb{E}\left\{h\,(hx+n)^*\right\}}{\mathbb{E}\left\{|hx+n|^2\right\}} = \frac{\mathbb{E}\left\{|h|^2\right\}x^* + \mathbb{E}\{hn^*\}}{\mathbb{E}\left\{|h|^2\right\}|x|^2 + \mathbb{E}\left\{|n|^2\right\}} = \frac{\beta x^*}{\beta|x|^2 + N_0} \tag{2.167}$$

by utilizing that h and n are uncorrelated. In summary, the LMMSE estimator is $\hat{h} = ay$ with a given in (2.167). It coincides with the MMSE estimator in (2.155) for complex Gaussian variables with the specified variances.

[8] Since a is a complex-valued parameter, we compute the Wirtinger derivative $\frac{\partial}{\partial a^*} = \frac{1}{2}\left(\frac{\partial}{\partial \Re(a)} + \mathrm{j}\frac{\partial}{\partial \Im(a)}\right)$, which includes the derivatives with respect to $\Re(a)$ and $\Im(a)$.

The derivation of the LMMSE estimator only used the mean, variance, and covariance of h and n. This implies that the LMMSE estimator is the same irrespectively of the exact distribution of h and n, as long as the mean and (co)variance are as specified. On the other hand, the general MMSE estimator utilizes the complete statistical distributions and will not only change in the non-Gaussian case but likely be harder to derive analytically. The equivalence between the MMSE and LMMSE estimators only holds in the Gaussian case, so the MMSE estimator must give a strictly smaller MSE in non-Gaussian cases. This implies that estimating Gaussian variables that are observed in Gaussian noise is the hardest situation, which is aligned with the fact that the Gaussian distribution maximizes the differential entropy.

We have established the following result regarding the LMMSE estimator when h is not necessarily Gaussian distributed.

Lemma 2.12. Consider the estimation of h from the observation $y = h \cdot x + n$, when the signal $x \in \mathbb{C}$ is known and n is noise with zero mean and variance N_0. Suppose the variable h has zero mean, variance β, and is uncorrelated with the noise (i.e., $\mathbb{E}\{hn^*\} = 0$). The LMMSE estimator of h is

$$\hat{h}_{\text{LMMSE}}(y) = \frac{\beta x^*}{\beta |x|^2 + N_0} y. \tag{2.168}$$

The corresponding minimum MSE is

$$\text{MSE}_h = \frac{\beta N_0}{\beta |x|^2 + N_0}. \tag{2.169}$$

2.6 Monte Carlo Methods for Statistical Inference

The previous section described how to estimate the realization of a random variable from noisy observations. An underlying assumption was that the statistics are known, but, in practice, we must also have a mechanism to acquire the statistics. In this section, we will describe how the statistical properties of functions of random variables can be inferred. The statistics might determine the performance of a communication system or an estimator. There are many categories of methods that can be utilized for this purpose. We will consider *Monte Carlo methods* that use random samples of the underlying variables and process them to infer the unknown deterministic quantities. We will estimate the mean value of a function of random variables, estimate the error probability of a system that performs a task either resulting in success or error, and estimate the CDF of a random variable. Particular attention will be given to quantifying the estimation precision, which is essential when drawing conclusions based on the outcome of statistical inference.

2.6.1 Estimating the Mean Value

Consider a real-valued random variable x with the PDF $f_{\mathsf{x}}(x)$ and mean value denoted as μ. We recall from (2.56) that the mean value is defined as

$$\mu = \mathbb{E}\{x\} = \int_{-\infty}^{\infty} x f_{\mathsf{x}}(x) \partial x. \tag{2.170}$$

There are many situations where this integral cannot be computed analytically, and then we have to resort to numerical methods for computing an approximate value of μ. One example is the Monte Carlo method that takes L independent samples $x_1, \ldots, x_L$ from the random distribution and uses them to estimate μ. Two properties are essential when designing the estimator: accuracy and precision. An estimator $\hat{\mu}_L$ is accurate if its mean is equal to the value to be estimated (i.e., $\mathbb{E}\{\hat{\mu}_L\} = \mathbb{E}\{x\} = \mu$) and it is precise if its variance $\mathbb{V}\mathrm{ar}\{\hat{\mu}_L\}$ is small. The sample average is an accurate (also known as unbiased) estimator of $\mathbb{E}\{x\}$ and is computed as

$$\hat{\mu}_L = \frac{1}{L} \sum_{i=1}^{L} x_i, \tag{2.171}$$

where the subscript denotes the number of samples. We only need a way to generate independent samples to compute this estimate, while the PDF can be unknown. The motivation behind using the sample average in (2.171) is the law of large numbers in Lemma 2.4, which says that the sample average approaches the statistical mean when the number of samples L goes to infinity:

$$\hat{\mu}_L \to \mathbb{E}\{x\} \quad \text{as } L \to \infty. \tag{2.172}$$

The only required condition for the convergence is that the variance $\mathbb{V}\mathrm{ar}\{x\}$ of the random variable must be finite. To see the reason for that, we can compute the variance of the sample average as

$$\mathbb{V}\mathrm{ar}\left\{\hat{\mu}_L\right\} = \frac{1}{L^2} \mathbb{V}\mathrm{ar}\left\{\sum_{i=1}^{L} x_i\right\} = \frac{1}{L^2} \sum_{i=1}^{L} \mathbb{V}\mathrm{ar}\left\{x_i\right\} = \frac{\mathbb{V}\mathrm{ar}\{x\}}{L}, \tag{2.173}$$

where the second equality utilizes the fact that the samples are independent. The variance in (2.173) reduces proportionally to $1/L$ when the number of samples increases, starting from the original variance value. Hence, as long as the original value is finite, the variance of the sample average goes to zero as $L \to \infty$. Furthermore, the standard deviation is the square root of the variance and becomes $\sqrt{\mathbb{V}\mathrm{ar}\{x\}/L}$, which goes to zero proportionally to $1/\sqrt{L}$ when increasing the number of samples.

Depending on the application, the number of samples, L, should be selected to achieve an estimate with the desired precision. Since the Monte Carlo method uses random samples, we can only guarantee the precision in a probabilistic sense; that is, we can make sure that the estimation error $|\hat{\mu}_L - \mu|$ is smaller than some specified error tolerance $\delta > 0$ with the (high) probability $1-\epsilon$, where $\epsilon > 0$ is the (small) probability that the requirement is unsatisfied. In other words, we want to find L such that

$$\begin{aligned}\Pr\{|\hat{\mu}_L - \mu| \leq \delta\} &= \Pr\{\mu - \delta \leq \hat{\mu}_L \leq \mu + \delta\} \\ &= \Pr\underbrace{\{\hat{\mu}_L - \delta \leq \mu \leq \hat{\mu}_L + \delta\}}_{\text{Confidence interval}} \geq 1 - \epsilon, \end{aligned} \tag{2.174}$$

where the third expression is known as a *confidence interval* with confidence level $1-\epsilon$. It says that a fraction $1-\epsilon$ of all realizations of the estimator $\hat{\mu}_L$ are so close to the true value μ that it lies between $\hat{\mu}_L - \delta$ and $\hat{\mu}_L + \delta$. It is common to begin by specifying ϵ to reach a desired confidence level and then either determine how large δ becomes in a given experimental setup (i.e., for a given L) or design the experiment (i.e., select L) to reach a desired value of δ.

We can utilize Chebyshev's inequality from Lemma 2.5 to derive an upper bound on how many samples are needed to satisfy (2.174) for given ϵ and δ. However, the result will be overly conservative since it considers the worst-case random distribution. Since we consider the summation of L independent and identically distributed realizations, the central limit theorem implies that $\hat{\mu}_L$ is approximately Gaussian distributed, as previously stated in (2.65). Hence, we can utilize that distribution when characterizing the required number of samples. Recall from (2.66) that 95% of all realizations are within two standard deviations from the mean value. If we set $\epsilon = 0.05$ and want to guarantee an estimation error smaller than δ, then we need

$$2\sqrt{\frac{\mathbb{V}\text{ar}\{x\}}{L}} \leq \delta \quad \Rightarrow \quad L \geq \frac{4\mathbb{V}\text{ar}\{x\}}{\delta^2}. \tag{2.175}$$

For example, if $\mathbb{V}\text{ar}\{x\} = 1$ and we want a precision of $\delta = 0.1$, then at least $L = 400$ samples are required to satisfy that requirement with 95% probability. The variance might also be unknown, in which case an approximation of it can be utilized when determining the number of samples.

Figure 2.19 exemplifies how the Monte Carlo method can be utilized to estimate the mean value $\mu = 1$ of $x \sim \text{Exp}(1)$, which has an exponential distribution. The number of samples, L, is shown on the horizontal axis, and the vertical axis shows potential estimates of μ. Figure 2.19(a) and (b) show how the value of $\hat{\mu}_L$ progresses in two different experiments where we add more and more samples to the estimator. The shaded area between the dashed lines shows the (approximate) confidence interval around $\hat{\mu}_L$ where μ exists with 95% probability. It is computed using the Gaussian approximation. The width of this interval reduces as $1/\sqrt{L}$ when L increases because the width is proportional to the standard deviation. In both experiments, the estimator fluctuates, but the general trend is that more samples lead to a better estimate

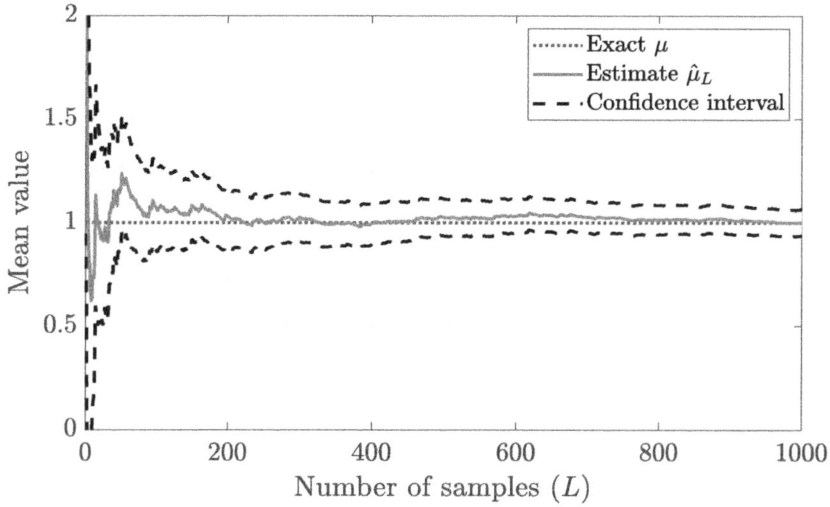

(a) Experiment 1.

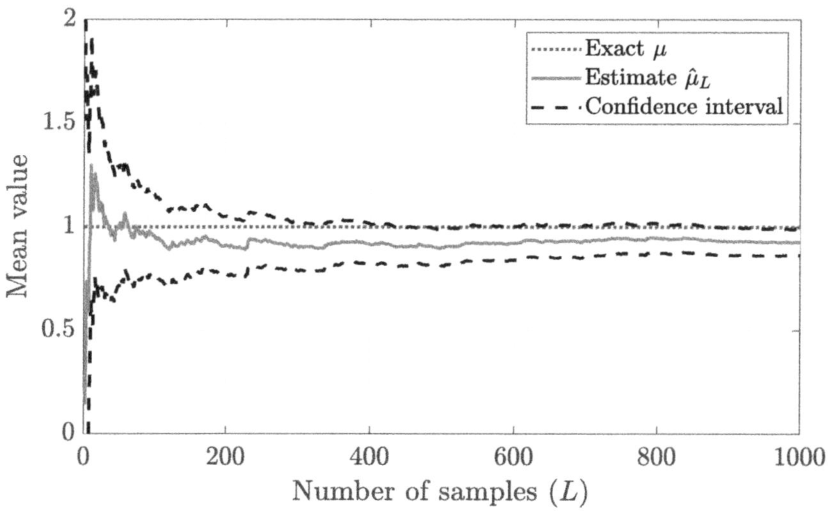

(b) Experiment 2.

Figure 2.19: Example of estimation of the mean $\mu = 1$ of a random variable with exponential distribution using the Monte Carlo method. The value of $\hat{\mu}_L$ is shown as a function of L in two different experiments. The 95% confidence interval is indicated, as well as the true value.

of μ. Nevertheless, Experiment 2 shows that even with $L = 1000$ samples, the exact μ might be outside the confidence interval.

Instead of resorting to taking random samples, as in the Monte Carlo method, one can approximate the integral in (2.170) in a deterministic manner by approximating the integrand $x f_{\mathsf{x}}(x)$ as a piecewise constant function. This is called a Riemann sum and the approximation error can then be bounded in a non-probabilistic manner, but it only works if the PDF $f_{\mathsf{x}}(x)$ is known. In contrast, the Monte Carlo method is convenient in practical situations where the PDF is unknown. For example, suppose the random variable x is obtained as a function of some multi-variate random variable $\mathbf{y}$; that is, $x = a(\mathbf{y})$ where $a(\cdot)$ can be any deterministic function. In this case, the PDF of x might be hard to characterize, even if the PDF of $\mathbf{y}$ is known. The Monte Carlo method can even be utilized when $\mathbf{y}$ has an unknown PDF, as long as samples from this random variable can be obtained from measurements. In wireless communications, $\mathbf{y}$ might be the randomness occurring in the propagation environment, while $a(\cdot)$ could be a complicated function that determines the communication performance.

Under these circumstances, we can still obtain an approximation of the mean value by following the following procedure:

1. Determine the required number of samples L;
2. Draw L independent samples $\mathbf{y}_1, \ldots, \mathbf{y}_L$ of the random variable $\mathbf{y}$;
3. Compute the L corresponding samples of the random variable x, denoted as $x_i = a(\mathbf{y}_i)$ for $i = 1, \ldots, L$;
4. Compute the sample average $\hat{\mu}_L = \frac{1}{L}\sum_{i=1}^{L} x_i$ to estimate $\mu = \mathbb{E}\{x\}$.

The samples must be generated independently and from the same distribution. Otherwise, the sample average might not converge to the correct number or not converge at all as L increases. These conditions put constraints on the methodology used when gathering the samples. One should, for example, be careful when merging measurements taken at different points in time, with different equipment, or at different locations. Computer simulations are robust against some of these effects but can nevertheless be affected by correlation in the (pseudo)random number generator (e.g., if multiple computers generate samples using the same random seed), limited arithmetic precision, other processes running in the same hardware, etc.

If the same L samples are utilized to estimate multiple quantities, then their respective estimation errors will be correlated, leading to undiscoverable systematic errors. As an example, suppose we want to use the Monte Carlo method to compute the MSE $\frac{\beta N_0}{\beta|x|^2+N_0}$ in (2.156) of the MMSE estimator for a range of different signal strengths $|x|^2$. This might be the only way of determining the MSE in situations where it cannot be computed analytically.

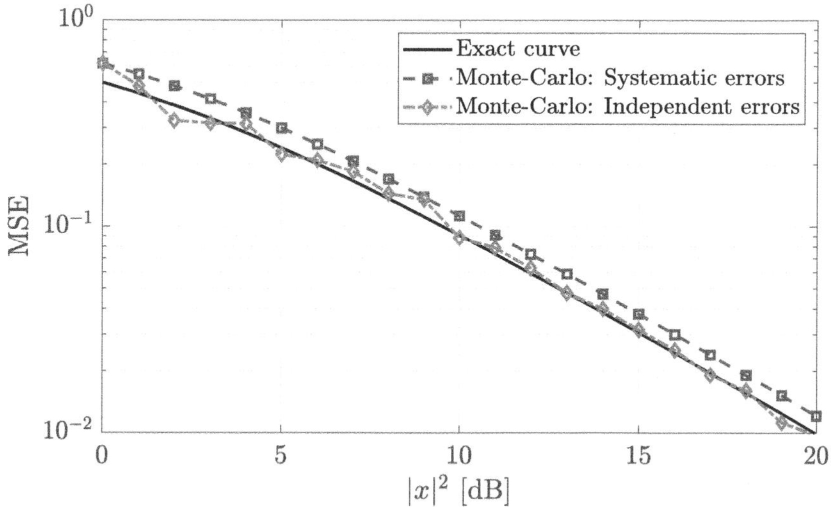

Figure 2.20: Example of estimation of MSE in (2.156) with $\beta = N_0 = 1$ using the Monte Carlo method with $L = 100$ samples. Independent samples must be utilized when estimating different points on the curve, otherwise, systematic errors occur.

According to (2.159), the MSE is equal to $\mathbb{E}\{|h - \hat{h}|^2\}$, where h is the desired variable and $\hat{h}$ is its MMSE estimate. To compute this mean using the Monte Carlo method, we should generate L independent realizations of $|h - \hat{h}|^2$ and compute the sample mean. For any non-zero value of $|x|^2$, we can generate L independent realizations of h and the noise n, then compute the observation $y = hx + n$, and finally compute $|h - \hat{h}|^2$ using (2.155).

Figure 2.20 shows the exact MSE and estimated MSE for $\beta = N_0 = 1$ and varying signal strength $|x|^2$. Since there are many points on the estimated curves, we can implement the Monte Carlo method in different ways: a) we can generate $L = 100$ independent samples of h and n and then utilize this set to estimate every point on the curve (by varying $|x|^2$ when computing $|h - \hat{h}|^2$); b) each point on the curve is estimated using L *new* independent realizations of h and n. From a programming perspective, the difference is whether the L samples are generated before the for-loop that goes through each value of $|x|^2$ or if L new samples are generated in each iteration of the loop. The blue curve is generated in the former way, where the same realizations are utilized to estimate every point. This results in a smooth curve that gives the impression of being highly accurate, but this is deceiving, as seen from the gap to the exact curve. The fact that the same randomness is used when estimating every point leads to such unnoticeable systematic errors because the estimation errors are correlated. The latter approach is recommended: generate L new independent samples for every value of $|x|^2$, which was done when generating the red curve. This curve is not smooth, showcasing the limited precision obtained when only using $L = 100$ samples in the Monte

Carlo method. In summary, to obtain an estimate of the curve that is both precise (i.e., smooth) and accurate (i.e., without systematic errors), we must use an even larger number of samples generated independently for every point on the curve.

2.6.2 Estimating the Error Probability

Another common problem in communications is computing the error probability or its converse, the success probability. For instance, we might design a communication protocol to convey messages over a random channel and want to determine the probability that a message is received in error. The more complicated the protocol and communication channel are, the smaller the chance that we can compute the error probability analytically. However, we can use Monte Carlo methods to estimate the error probability. Since there are only two possible outcomes—success or error—the randomness can be modeled by a Bernoulli distribution, which is a random variable x with two outcomes: the value 1 with probability p and the value 0 with probability $1-p$. The mean is $\mathbb{E}\{x\} = p$ and the variance is $\mathbb{V}\text{ar}\{x\} = p(1-p)$.

Suppose we associate the outcome 1 of the Bernoulli distribution with an error, then our goal is to obtain an estimate $\hat{p}$ of the mean p, representing the error probability. Hence, we can follow the same procedure as in the previous section: Generate L independent samples $x_1, \ldots, x_L$ of the Bernoulli distribution and then use the sample average $\frac{1}{L}\sum_{i=1}^{L} x_i$ as the estimate of p. Each sample can be obtained from one independent trial of the communication protocol by determining whether an error occurred or not. This is a feasible approach, but the main practical hurdle is determining the number of samples that need to be taken. The error probabilities in communication systems can range between 0.1 and 10^{-9}, which require very different error tolerances and numbers of samples when being estimated.

Suppose we select the error tolerance proportionally to p as $\delta = \alpha p$, where $\alpha \in [0, 1]$ is the relative error tolerance. The goal is then to find an estimate $\hat{p}_L$ that falls into the interval $[(1-\alpha)p, (1+\alpha)p]$ with high certainty. By substituting this value of δ into (2.175), we need

$$L \geq \frac{4\mathbb{V}\text{ar}\{x\}}{\delta^2} = \frac{4p(1-p)}{\alpha^2 p^2} = \frac{4(1-p)}{\alpha^2 p} \tag{2.176}$$

samples to satisfy the tolerance with 95% certainty. This value depends on p, so we need a good sense of the (worst-case) error probability when selecting L, which severely limits its applicability. However, one important observation can be made from (2.176): if p is much smaller than one, then $(1-p)/p \approx 1/p$ and the required number of samples is inversely proportional to p. Hence, the more unlikely an error is to occur, the more samples are needed to obtain an accurate estimate, which is rather intuitive. A classical rule-of-thumb is that $L \geq 10/p$ samples are needed to obtain a rough estimate of p [45], which implies that we need $L = 1000$ if $p = 10^{-2}$ and $L = 10^6$ if $p = 10^{-5}$. Using at least $L \geq 100/p$ samples is recommended to get a precise estimate.

There is an alternative estimation approach that is particularly well suited for estimating error probabilities without requiring prior knowledge when determining the sample size [46]: We generate independent samples repeatedly until we have gathered L_{error} errors, where $L_{\text{error}} \geq 2$ is a predefined constant. The number of successful samples L_{success} that are observed before we reach L_{error} errors is a random variable that has the negative binomial distribution. Based on a random realization of L_{success}, we can estimate p as

$$\hat{p} = \frac{L_{\text{error}} - 1}{L_{\text{success}} + L_{\text{error}} - 1}. \tag{2.177}$$

This estimator is unbiased (i.e., $\mathbb{E}\{\hat{p}\} = p$) and is also the one minimizing the error variance [47]. The standard deviation of this estimator is approximately $p/\sqrt{L_{\text{error}} - 2}$ when p is small, thus it is proportional to p and reduces roughly as $1/\sqrt{L_{\text{error}}}$. Suppose p is relatively large, in the sense that $1 - p$ cannot be approximated as 1. Then the standard deviation is larger because we gather errors too quickly to reach a sufficient total number $L_{\text{success}} + L_{\text{error}}$ of measurements to get an accurate estimate.

A classical rule-of-thumb is to make measurements until we have observed $L_{\text{error}} = 10$ errors [46], which gives a rough estimate of p with a standard deviation of roughly $p/\sqrt{8} \approx 0.35p$ when p is small. To get a precise estimate with a smaller standard deviation, observing at least $L_{\text{error}} = 100$ errors is recommended. In those cases, the -1 terms in (2.177) can be neglected.

Figure 2.21 exemplifies the error probability p as a function of the SNR. The true relation is $p = 1 - e^{-\frac{1}{\text{SNR}}}$ in this example, which is a formula that will be derived in Chapter 5. In addition to showing the exact curve, Figure 2.21 also shows estimated curves obtained using the two approaches described above. The blue curve uses $L = 10000$ samples and provides excellent estimates for $p \geq 10^{-3}$ and decent estimates for $10^{-4} \leq p \leq 10^{-3}$, as predicted by the first rule-of-thumb. The curve then vanishes since there are too few samples to measure any error events; whenever less than ten errors have been observed, we should discard the result as unreliable (recall the second rule-of-thumb). The red curve uses the alternative approach of running the simulation until $L_{\text{error}} = 100$ has been observed. This curve provides accurate estimates of p for all the considered SNR values.

In summary, to avoid selecting L in advance, we can estimate the error probability p by counting the number of successes that occurred before we reached a predefined number of errors. The number of samples to gather is then determined dynamically and increases linearly with the true value of p. This approach is particularly useful when a complicated communication protocol is used so the error probability cannot be determined analytically.

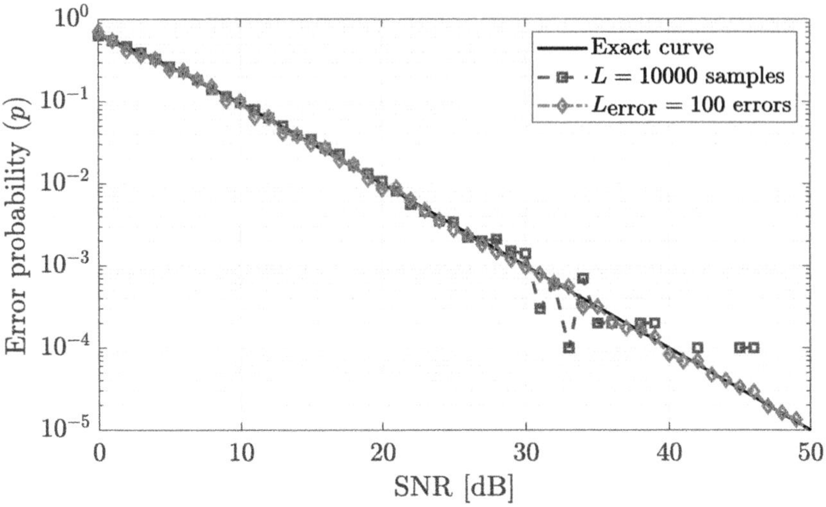

Figure 2.21: Example of estimation of the error probability curve $p = 1 - e^{-\frac{1}{\text{SNR}}}$ using the Monte Carlo method, by either using 10000 random samples or running the simulation until 100 errors have been observed.

2.6.3 Empirical Cumulative Distribution Function

In addition to estimating the mean value of a random variable from observations, we can estimate its entire distribution. In this section, we will estimate the CDF, defined in (2.100), which fully characterizes the random distribution. Suppose we obtain L independent samples $x_1, \ldots, x_L$ from a random distribution with the CDF $F_{\mathsf{x}}(a)$. For a given value a, the CDF represents the probability of obtaining a realization below or equal to the threshold a. Hence, we can estimate $F_{\mathsf{x}}(a)$ by counting the fraction of the L samples that is lower than or equal to a. This estimator can be defined as

$$\hat{F}_{X,L}(a) = \frac{1}{L}\sum_{i=1}^{L} \mathbb{I}_{x_i \leq a}, \tag{2.178}$$

by utilizing the indicator function

$$\mathbb{I}_{x \leq a} = \begin{cases} 1, & \text{if } x \leq a, \\ 0, & \text{if } x > a. \end{cases} \tag{2.179}$$

We can treat $\hat{F}_{X,L}(a)$ as an estimate of the entire CDF and call it the *empirical cumulative distribution function (eCDF)*. The true CDF might be a continuous function, but the eCDF is always a piecewise constant function. It will look like a staircase with L steps, each having a vertical height of $1/L$ but varying horizontal widths that determine the shape of the estimated curve.

The eCDF converges to the true CDF as L goes to infinity, and the convergence can be proved in various ways. For example, we can prove pointwise convergence by comparing $F_{\mathsf{x}}(a)$ to its estimate $\hat{F}_{X,L}(a)$ for any given point a. For a random x, the indicator function $\mathbb{I}_{x \leq a}$ will output a random variable with a Bernoulli distribution that gives 1 with probability $F_{\mathsf{x}}(a)$ and 0 with probability $1 - F_{\mathsf{x}}(a)$. As discussed in the previous section, such a random variable has the mean $F_{\mathsf{x}}(a)$ and variance $F_{\mathsf{x}}(a)(1 - F_{\mathsf{x}}(a))$. Hence, $\hat{F}_{X,L}(a)$ is the sample average of L independent Bernoulli variables having that mean and variance. The mean of $\hat{F}_{X,L}(a)$ is the true CDF value $F_{\mathsf{x}}(a)$ and the variance can be determined using (2.173) as

$$\mathbb{V}\mathrm{ar}\left\{\hat{F}_{X,L}(a)\right\} = \frac{1}{L^2}\sum_{i=1}^{L} \mathbb{V}\mathrm{ar}\left\{\mathbb{I}_{x_i \leq a}\right\} = \frac{F_{\mathsf{x}}(a)(1 - F_{\mathsf{x}}(a))}{L}. \tag{2.180}$$

The variance goes to zero as $L \to \infty$, which is the property used by the law of large numbers to establish asymptotic convergence to the mean. When L is large but finite, the central limit theorem implies that $\hat{F}_{X,L}(a)$ is approximately Gaussian distributed with mean $F_{\mathsf{x}}(a)$ and variance $F_{\mathsf{x}}(a)(1 - F_{\mathsf{x}}(a))/L$. We recall from Section 2.2.1 that 95% of all realizations of a Gaussian random variable are within two standard deviations from the mean.

The precision of the eCDF varies over the curve, reflected by the fact that the standard deviation $\sqrt{F_{\mathsf{x}}(a)(1 - F_{\mathsf{x}}(a))/L}$ depends on $F_{\mathsf{x}}(a)$. The largest value appears at the median where $F_{\mathsf{x}}(a) = 0.5$. However, it might be more important to consider the relative deviation from the true CDF value. If we divide the standard deviation by $F_{\mathsf{x}}(a)$, we obtain $\sqrt{(1 - F_{\mathsf{x}}(a))/(LF_{\mathsf{x}}(a))}$ and it is maximized as $F_{\mathsf{x}}(a) \to 0$. This reveals that it is hardest to precisely approximate the lower-left tail of the curve because very few samples appear in that tail, and small deviations are large in the relative sense. When selecting the number of samples L in a practical experiment, one can either target a desired precision in the crucial parts of the CDF curve (e.g., center or tails) or run the simulation until a visually smooth eCDF curve is obtained.

Figure 2.22 considers the estimation of the CDF of $x \sim \text{Rayleigh}(1/\sqrt{2})$. The analytical CDF expression $F_{\mathsf{x}}(x) = 1 - e^{-x^2}$ of this Rayleigh distribution was provided in (2.102). The red curve shows the eCDF obtained using $L = 100$ independent samples of the random variable. The eCDF has the same general shape as the true CDF but fluctuates between being well aligned with it and deviating. The estimation errors are correlated along the curve since the same L samples are utilized to estimate all the points on the curve, but this property is unavoidable when computing an eCDF. The 95% confidence interval around the eCDF (obtained using the Gaussian approximation) is also shown in the figure. This interval is relatively wide, which shows that more than 100 samples are needed to obtain a precise eCDF. The staircase shape of the eCDF is particularly evident in the lower tail, where there are too few samples to estimate the precise shape of the CDF.

The precision is essential when comparing different random variables based on estimates of their respective distributions. For example, we might obtain

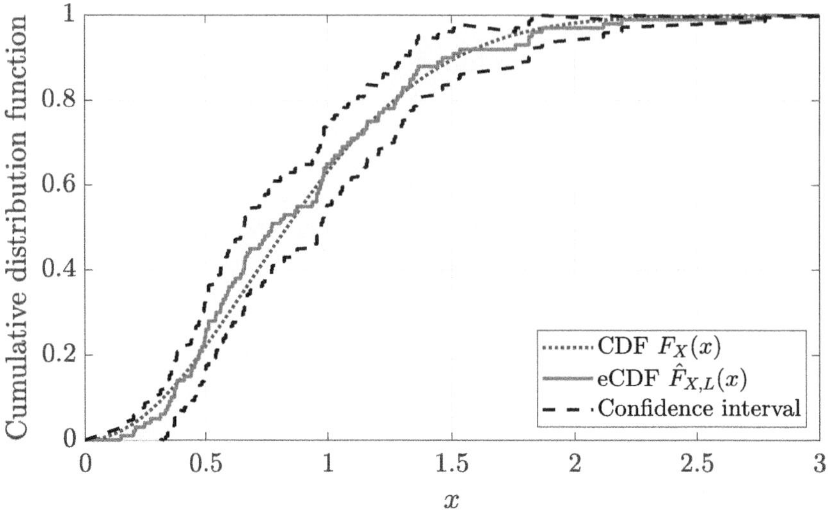

Figure 2.22: The CDF $F_x(x) = 1 - e^{-x^2}$ of a Rayleigh distributed random variable is compared with the eCDF obtained using $L = 100$ samples from the distribution. The approximate 95% confidence interval is indicated as a reference.

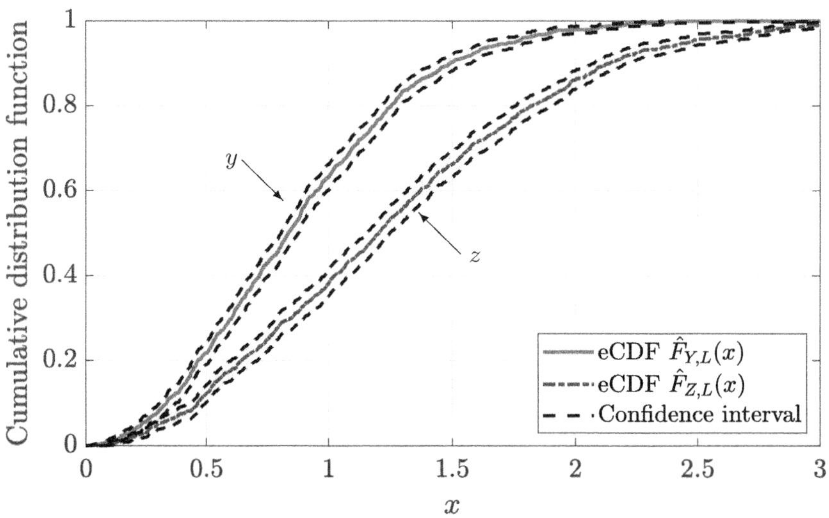

Figure 2.23: The eCDFs and confidence intervals of $y \sim \text{Rayleigh}(1/\sqrt{2})$ and $z \sim \text{Rayleigh}(1)$, based on $L = 1000$ samples from each distribution.

measurements of the performance variations in two different communication systems and plot their respective eCDFs to determine which system is preferable. For the sake of argument, Figure 2.23 shows the eCDFs obtained by $L = 1000$ samples from $y \sim \text{Rayleigh}(1/\sqrt{2})$ and $z \sim \text{Rayleigh}(1)$, respectively. The two eCDFs are different, but most importantly, the 95% confidence intervals (also shown in the figure) are different and mostly non-overlapping. Whenever that happens, we can make meaningful comparisons of the eCDFs.

Since $\hat{F}_{Y,L}(x) \geq \hat{F}_{Z,L}(x)$ for most values of x (i.e., the y-curve is above the z-curve), we can conclude that the system represented by y is likely to provide smaller performance values. For example, y is smaller than 1 with probability $\hat{F}_{Y,L}(1) \approx 0.6$, while z is smaller than 1 with probability $\hat{F}_{Z,L}(1) \approx 0.4$. If it is preferable to have a large value, the system represented by z should be selected. The only uncertainty occurs in the lower left tail, where the confidence intervals partially overlap. When this happens, we can only conclude that their performance is so similar that we cannot tell the systems apart with statistical significance. This issue can be mitigated by increasing L to improve the precision (i.e., reduce the standard deviation).

2.7 Detection Theory

Detection theory provides a structured way to determine which event occurred among a finite number of possibilities based on probabilistic observations. It is commonly used in several areas, particularly radar signal processing and communications [48]. The task of the detector is to determine which event has happened by processing the observed signal and exploiting prior information regarding the received signal's characteristics and statistics. The events are mutually exclusive, and each is called a *hypothesis* under testing. Due to this terminology, detection theory is also known as *hypothesis testing* [48].

To exemplify the basics, we consider a fire-alarm sensor that measures the smoke density in its surroundings. If there is smoke, it sends a wireless message representing "1". If there is no smoke, the sensor does not transmit anything, representing the message "0". A wireless receiver monitors the transmission and wants to detect the message. Regardless of what message is sent, noise is added to the received signal. Hence, the receiver should use the received signal to determine if there is a non-zero signal or only noise. There are two events in this example: i) there is no smoke, and ii) there is smoke. Since there are two possibilities, we call this a *binary hypothesis test*.

In binary hypothesis testing, it is common to let the *null hypothesis* represent the case when the event of interest does not happen. It is denoted as $\mathcal{H}_0$. The opposite hypothesis is denoted as $\mathcal{H}_1$ and called the *alternative hypothesis*. Mathematically, we can express the corresponding detection problem as

$$\mathcal{H}_0 \quad : \quad y = n, \tag{2.181}$$

$$\mathcal{H}_1 \quad : \quad y = 1 + n, \tag{2.182}$$

where the detector determines if "1" is transmitted or not by observing y and exploiting any other prior information, such as the statistical models of (2.181) and (2.182). In this section, we will assume that the additive noise is distributed as $n \sim \mathcal{N}(0, \sigma^2)$. The goal of detection theory is to select a detection performance metric and then develop the detection rule (i.e., selection rule between $\mathcal{H}_0$ and $\mathcal{H}_1$) that optimizes that metric. In the example

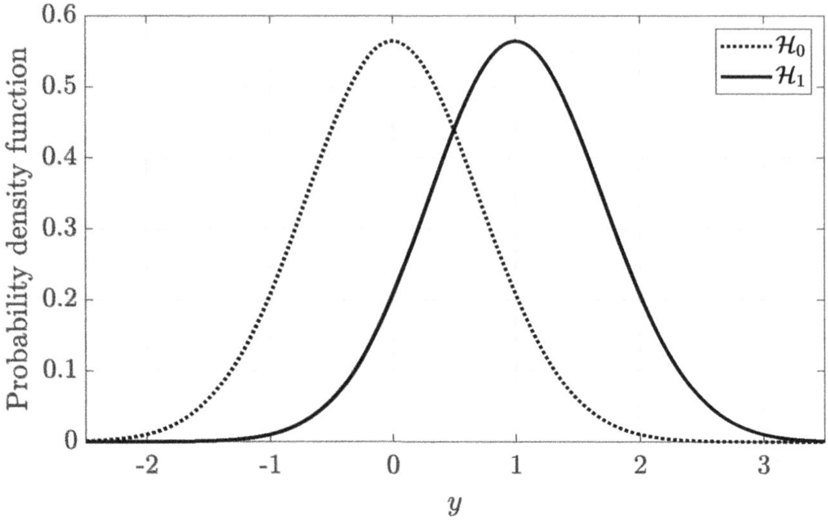

Figure 2.24: The PDF of the received signal y under two hypotheses $\mathcal{H}_0$ and $\mathcal{H}_1$. Under the null hypothesis $\mathcal{H}_0$, the signal is distributed as $y \sim \mathcal{N}(0, 0.5)$ whereas it holds that $y \sim \mathcal{N}(1, 0.5)$ under the alternative hypothesis $\mathcal{H}_1$.

above, the metric is the probability of making an incorrect detection, and the goal is to minimize it. If the a priori probabilities of transmitting 1 or nothing are defined and known, they can be used to minimize the error.

Figure 2.24 shows the PDFs of the received signal y under the null hypothesis $\mathcal{H}_0$ and the alternative hypothesis $\mathcal{H}_1$. Under $\mathcal{H}_0$ it follows that $y = n \sim \mathcal{N}(0, \sigma^2)$, whereas under $\mathcal{H}_1$ we have $y = 1 + n \sim \mathcal{N}(1, \sigma^2)$. The figure shows the case when $\sigma^2 = 0.5$. Suppose we use a detector of the form

$$\hat{\mathcal{H}} = \begin{cases} \mathcal{H}_1, & \text{if } y \geq \gamma, \\ \mathcal{H}_0, & \text{if } y < \gamma, \end{cases} \tag{2.183}$$

where there is a threshold γ that determines when to select each hypothesis. The two PDFs in Figure 2.24 intersect at $y = 1/2$, which will also happen for other values of σ^2. Hence, if we select the threshold as $\gamma = 1/2$, the detection rule in (2.183) will select the hypothesis most likely to have generated the received observation y. This threshold divides the decision region symmetrically into two parts, as illustrated by the red dashed line in Figure 2.25. This threshold maximizes the probability of making a correct detection if the two events are equally likely, which is seemingly a good performance metric. However, it is not the only metric of practical importance. Three other important metrics are:

- The detection probability, P_{D}, which is the correct detection probability when the event of interest happens, i.e., under hypothesis $\mathcal{H}_1$;
- The false alarm probability, P_{FA}, which is the wrong detection probability

when the event of interest did not happen, i.e., under hypothesis $\mathcal{H}_0$;

- The missing probability, $P_{\mathrm{M}} = 1 - P_{\mathrm{D}}$, which is the wrong detection probability when the event happens, i.e., under hypothesis $\mathcal{H}_1$.

In Figure 2.25(a), the yellow and purple shaded regions represent the detection probability, P_{D}, and $1 - P_{\mathrm{FA}}$, respectively (i.e., the areas under the curves equal the probabilities). When hypothesis $\mathcal{H}_1$ is true, we detect the event correctly when the received signal is greater than the threshold $\gamma = 1/2$, which happens with the probability P_{D}. On the other hand, when hypothesis $\mathcal{H}_0$ is true, we detect the event correctly when the received signal is below the threshold, which happens with the probability $1 - P_{\mathrm{FA}}$. Figure 2.25(b) shows the probabilities of false detection. When $\mathcal{H}_1$ is true, but the noise takes a big negative realization so that the received signal is below the threshold, we miss the event, and the resulting probability is $P_{\mathrm{M}} = 1 - P_{\mathrm{D}}$. When $\mathcal{H}_0$ is true, but the noise takes a big positive realization so that the received signal is above the threshold, a false alarm occurs. The associated probability is P_{FA}.

It is good to have high values of P_{D} (corresponding to low values of P_{M}) and low values of P_{FA}, but there is unfortunately always a tradeoff between these metrics. To illustrate this, we increase the threshold value to $\gamma = 1$ in Figure 2.26. As shown in Figure 2.26(a), the correct detection probability when there is no transmitted signal (i.e., $\mathcal{H}_0$ is true) increases compared to the last figure. Similarly, the false alarm probability decreases, as shown in Figure 2.26(b). However, this improvement is associated with a decrease in P_{D} since a larger threshold makes it less likely to make the correct detection decision when there a signal is transmitted (i.e., $\mathcal{H}_1$ is true). Moreover, the missing probability $P_{\mathrm{M}} = 1 - P_{\mathrm{D}}$ increases when P_{D} decreases.

The fact that there are multiple conflicting design objectives implies that we need to actively design the decision rule for every detection application, even if the underlying mathematical models are the same. For example, a fire-alarm sensor might be designed to have a very high detection probability, P_{D}, since missing the event of interest can be dangerous. On the other hand, a radar surveillance system might be designed to have a very low false alarm probability, so it only identifies large objects.

The hypothesis testing we have considered so far assumed that the PDF of the received signal is fully known for all the hypotheses, which is known as *simple hypothesis testing*. For example, in the previous example, we know that $y \sim \mathcal{N}(0, 0.5)$ when $\mathcal{H}_0$ is true, whereas $y \sim \mathcal{N}(1, 0.5)$ when $\mathcal{H}_1$ is true. We will focus on simple hypothesis testing in this book. Another class of problems is *composite hypothesis tests* in which there are unknown deterministic parameters or random variables with unknown distributions. An example of this is the detection problem in (2.181)–(2.182) when the noise variance σ^2 is unknown; the PDF of y is unknown for all the hypotheses because we only know the Gaussian shape but not the associated variance.

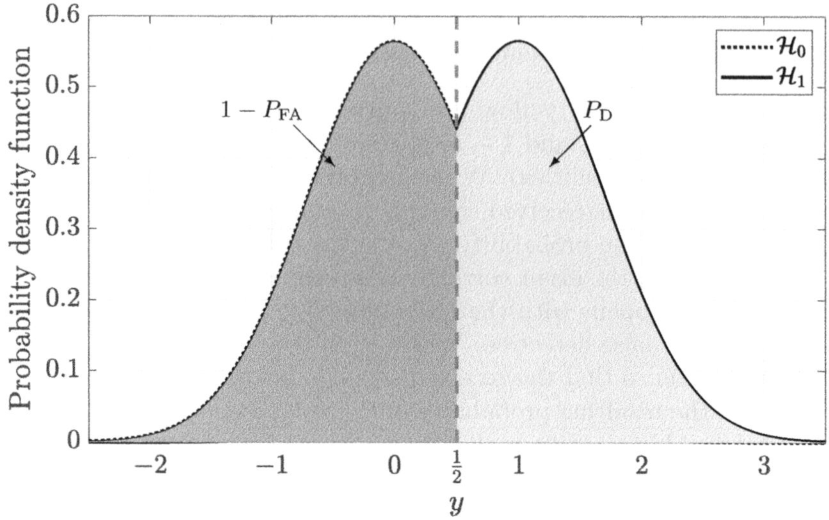

(a) The correct detection probability under $\mathcal{H}_0$ is $1 - P_{\mathrm{FA}}$ (the area of the purple region), while it is P_{D} under $\mathcal{H}_1$ (the area of the yellow region).

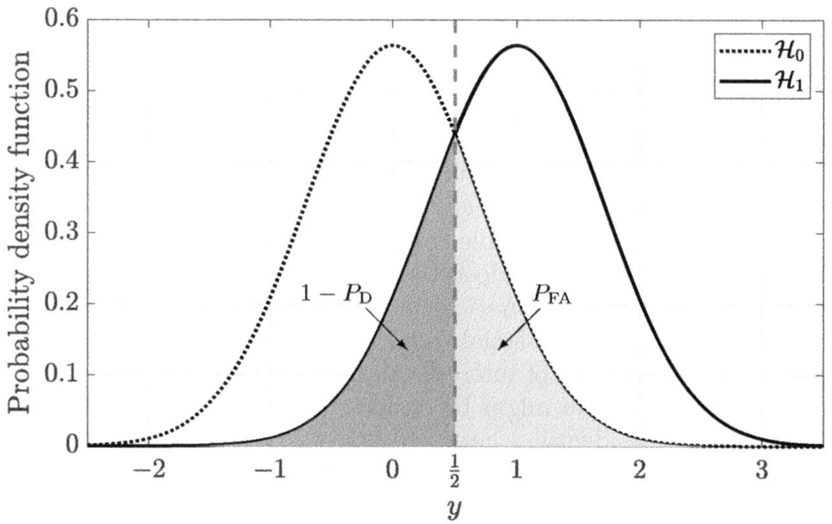

(b) The wrong detection probability under $\mathcal{H}_0$ is P_{FA} (the area of the yellow region), while it is $1 - P_{\mathrm{D}}$ under $\mathcal{H}_1$ (the area of the purple region).

Figure 2.25: The probabilities of correct and incorrect detection under the hypotheses $\mathcal{H}_0$ and $\mathcal{H}_1$ when the detection threshold is $1/2$. The dashed red line shows the corresponding detection boundary. The areas of the shaded regions represent the respective probabilities.

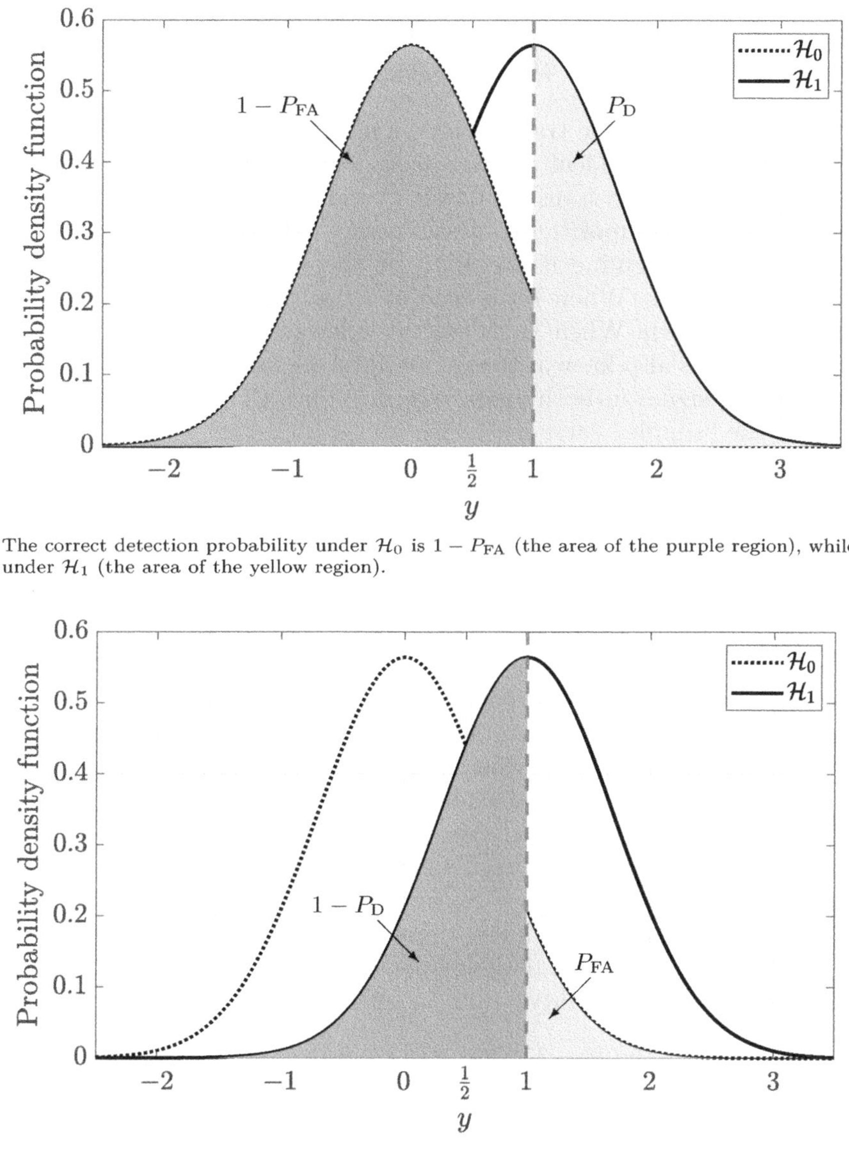

(a) The correct detection probability under $\mathcal{H}_0$ is $1 - P_{\text{FA}}$ (the area of the purple region), while it is P_{D} under $\mathcal{H}_1$ (the area of the yellow region).

(b) The wrong detection probability under $\mathcal{H}_0$ is P_{FA} (the area of the yellow region), while it is $1 - P_{\text{D}}$ under $\mathcal{H}_1$ (the area of the purple region).

Figure 2.26: The probabilities of correct and wrong detection under the hypotheses $\mathcal{H}_0$ and $\mathcal{H}_1$ when the detection threshold is 1. The dashed red line shows the corresponding detection boundary. The areas of the shaded regions represent the respective probabilities.

Example 2.17. Consider the binary hypothesis test

$$\mathcal{H}_0 \quad : \quad y = n, \tag{2.184}$$
$$\mathcal{H}_1 \quad : \quad y = x + n, \tag{2.185}$$

where $x \sim \mathcal{N}_{\mathbb{C}}(0, P)$ is the transmitted signal under the hypothesis $\mathcal{H}_1$ and $n \sim \mathcal{N}_{\mathbb{C}}(0, \sigma^2)$ is independent receiver noise. The random signal x is unknown at the detector, but the transmit power P and noise variance σ^2 are known. Is this a simple or composite hypothesis test?

We need to determine if the PDF of the received signal y is known under all hypotheses. When $\mathcal{H}_0$ is true, it follows that $y \sim \mathcal{N}_{\mathbb{C}}(0, \sigma^2)$ so the distribution is known. When $\mathcal{H}_1$ is true, it follows that $y \sim \mathcal{N}_{\mathbb{C}}(0, P + \sigma^2)$ so this distribution is also known. Hence, we have the full knowledge of the PDF of the received signal in both cases, which implies that this hypothesis test belongs to the "simple" category.

In the following sections, we will consider two approaches to simple hypothesis testing. The fundamental difference is whether the occurrences of the different events are modeled statistically or not.

2.7.1 Bayesian Detection

In the *Bayesian detection* approach, we assume that the occurrence of each hypothesis can be modeled statistically and has a specific probability. This approach is particularly useful when the underlying events happen repeatedly so that statistics can be inferred as described in Section 2.6, and the detector will be applied many times so that its average performance is essential. Consider a binary hypothesis test where $\Pr\{\mathcal{H}_0\}$ and $\Pr\{\mathcal{H}_1\}$ denote the probabilities that the hypotheses $\mathcal{H}_0$ and $\mathcal{H}_1$ take place, respectively. In the detection problems where we know these probabilities (e.g., communication tasks where the messages are designed to be equally likely), it is of interest to minimize the error probability, which is defined as

$$P_{\mathrm{e}} = \Pr\{\mathcal{H}_0\} \underbrace{\Pr\{\hat{\mathcal{H}} = \mathcal{H}_1 | \mathcal{H}_0\}}_{=P_{\mathrm{FA}}} + \Pr\{\mathcal{H}_1\} \underbrace{\Pr\{\hat{\mathcal{H}} = \mathcal{H}_0 | \mathcal{H}_1\}}_{=P_{\mathrm{M}}=1-P_{\mathrm{D}}}, \tag{2.186}$$

where the conditional probability $\Pr\{\hat{\mathcal{H}} = \mathcal{H}_1 | \mathcal{H}_0\}$ is the probability of detecting the hypothesis $\mathcal{H}_1$ when $\mathcal{H}_0$ is true, which we previously called the false alarm probability, P_{FA}. Similarly, $\Pr\{\hat{\mathcal{H}} = \mathcal{H}_0 | \mathcal{H}_1\}$ is the conditional probability of selecting the hypothesis $\mathcal{H}_0$ when $\mathcal{H}_1$ is true, which we previously called the missing probability, $P_{\mathrm{M}} = 1 - P_{\mathrm{D}}$. The detector that minimizes the error probability, P_{e} is as follows [48, Ch. 3].

Lemma 2.13. The detector that minimizes the error probability in (2.186) selects the hypothesis $\mathcal{H}_1$ if

$$\frac{f_{\mathsf{y}|\mathcal{H}_1}(y|\mathcal{H}_1)}{f_{\mathsf{y}|\mathcal{H}_0}(y|\mathcal{H}_0)} \geq \frac{\Pr\{\mathcal{H}_0\}}{\Pr\{\mathcal{H}_1\}} = \gamma \tag{2.187}$$

where $f_{\mathsf{y}|\mathcal{H}_1}(y|\mathcal{H}_1)$ and $f_{\mathsf{y}|\mathcal{H}_0}(y|\mathcal{H}_0)$ denote the conditional PDFs of the received signal y when $\mathcal{H}_1$ and $\mathcal{H}_0$ are true, respectively.

The ratio of the conditional PDFs on the left-hand side of (2.187) is called the *likelihood ratio*. The detector that minimizes P_{e} compares it to the threshold γ, which is the ratio of the a priori probabilities of the hypotheses. The threshold is 1 when the hypotheses are equally likely. On the other hand, when hypothesis $\mathcal{H}_1$ is more likely, then γ is smaller to decrease the missing probability, P_{M}, since its contribution to (2.186) is more dominant compared to the false alarm probability. When hypothesis $\mathcal{H}_0$ is more likely, the optimal γ is greater than 1 to force P_{FA} to become smaller.

Example 2.18. Consider the binary hypothesis test in (2.181). For a given value of $\gamma = \Pr\{\mathcal{H}_0\}/\Pr\{\mathcal{H}_1\}$, derive the Bayesian detector that minimizes the error probability. What are P_{D} and P_{FA} for this detector?

The received signal y is distributed as $y \sim \mathcal{N}(1, \sigma^2)$ when $\mathcal{H}_1$ is true. On the other hand, it is distributed as $y \sim \mathcal{N}(0, \sigma^2)$ when $\mathcal{H}_0$ is true. Inserting the respective Gaussian distributions from (2.63) into the likelihood ratio in (2.187), we obtain the minimum P_{e} detector as

$$\frac{\frac{1}{\sqrt{2\pi\sigma^2}}e^{-\frac{(y-1)^2}{2\sigma^2}}}{\frac{1}{\sqrt{2\pi\sigma^2}}e^{-\frac{y^2}{2\sigma^2}}} \geq \gamma \Rightarrow \ln\left(\frac{e^{-\frac{(y-1)^2}{2\sigma^2}}}{e^{-\frac{y^2}{2\sigma^2}}}\right) \geq \ln(\gamma)$$

$$\Rightarrow -\frac{(y-1)^2}{2\sigma^2} + \frac{y^2}{2\sigma^2} \geq \ln(\gamma) \Rightarrow y \geq \sigma^2 \ln(\gamma) + \frac{1}{2}, \tag{2.188}$$

where we used the fact that $\ln(\gamma)$ is a monotonically increasing function of $\gamma \geq 0$, so it can be applied to both sides of the inequality without changing the inequality sign. By using the notation $\gamma' = \sigma^2 \ln(\gamma) + \frac{1}{2}$, the detection probability is associated with the event $y \geq \gamma'$ and computed as

$$P_{\mathrm{D}} = \int_{\gamma'}^{\infty} f_{\mathsf{y}|\mathcal{H}_1}(y|\mathcal{H}_1)\,\partial y = \frac{1}{\sqrt{2\pi\sigma^2}}e^{-\frac{(y-1)^2}{2\sigma^2}}\,\partial y. \tag{2.189}$$

Similarly, the false alarm probability is associated with the event $y \geq \gamma'$ and is computed using $f_{\mathsf{y}|\mathcal{H}_0}(y|\mathcal{H}_0)$ as

$$P_{\mathrm{FA}} = \int_{\gamma'}^{\infty} \frac{1}{\sqrt{2\pi\sigma^2}}e^{-\frac{y^2}{2\sigma^2}}\,\partial y. \tag{2.190}$$

In Figure 2.27, we show the missing probability ($P_{\mathrm{M}} = 1 - P_{\mathrm{D}}$), the false alarm probability (P_{FA}), and the error probability (P_{e}) for the considered binary hypothesis test, as a function of the threshold γ. The curves are generated using the formulas derived in Example 2.18. In Figure 2.27(a), we consider equally likely hypotheses (i.e., $\Pr\{\mathcal{H}_0\} = \Pr\{\mathcal{H}_1\} = \frac{1}{2}$). The P_{M}-curve increases with γ, while the P_{FA}-curve decreases. The error probability is a weighted sum of these metrics, so it goes down and then up again when γ increases. The threshold that minimizes P_{e} is $\gamma = \Pr\{\mathcal{H}_0\}/\Pr\{\mathcal{H}_1\} = 1$, which is denoted by a cross in the figure. We notice that the optimal threshold occurs where $P_{\mathrm{M}} = P_{\mathrm{FA}}$, which can be proved analytically. As the threshold increases beyond 1, P_{FA} decreases but P_{M} increases faster, which leads to an increased error probability P_{e}. If γ instead becomes smaller than 1, then P_{M} decreases but P_{FA} increases faster, leading to an increased error probability.

In Figure 2.27(b), we set $\Pr\{\mathcal{H}_1\} = \frac{1}{3}$ and $\Pr\{\mathcal{H}_0\} = \frac{2}{3}$, which leads to the optimal threshold $\gamma = \Pr\{\mathcal{H}_0\}/\Pr\{\mathcal{H}_1\} = 2$. The figure confirms that the minimum error probability is obtained when $\gamma = 2$. P_{FA} is less than P_{M} at this point, which is expected since the contribution of P_{FA} to the error probability in (2.186) is more dominant since it is multiplied by $\Pr\{\mathcal{H}_0\}$, which is larger than $\Pr\{\mathcal{H}_1\}$ that is multiplied by P_{M}.

2.7.2 Neyman-Pearson Detection

There are situations when prior information about the hypothesis probabilities is unavailable, either because the statistics are hard to obtain or because the events only occur once, so statistical modeling is not viable. We can then follow the *Neyman-Pearson detection* approach where a priori probabilities of the hypotheses are not considered. This approach is common in radar applications; for example, in target detection, it is hard to set a probability for the existence of a target. Instead, a desired value of $P_{\mathrm{FA}} = \alpha$ is set, and the detector is designed to maximize P_{D} under the condition that $P_{\mathrm{FA}} = \alpha$. The detector that maximizes the detection probability in such a constrained detection problem is as follows [48, Ch. 3].

Lemma 2.14. The detector that maximizes the detection probability, P_{D}, under the constraint that $P_{\mathrm{FA}} = \alpha$ selects the hypothesis $\mathcal{H}_1$ if

$$\frac{f_{\mathsf{y}|\mathcal{H}_1}(y|\mathcal{H}_1)}{f_{\mathsf{y}|\mathcal{H}_0}(y|\mathcal{H}_0)} \geq \gamma, \tag{2.191}$$

where the threshold γ is selected to satisfy

$$P_{\mathrm{FA}} = \int_{\frac{f_{\mathsf{y}|\mathcal{H}_1}(y|\mathcal{H}_1)}{f_{\mathsf{y}|\mathcal{H}_0}(y|\mathcal{H}_0)} \geq \gamma} f_{\mathsf{y}|\mathcal{H}_0}(y|\mathcal{H}_0)\, \partial y = \alpha. \tag{2.192}$$

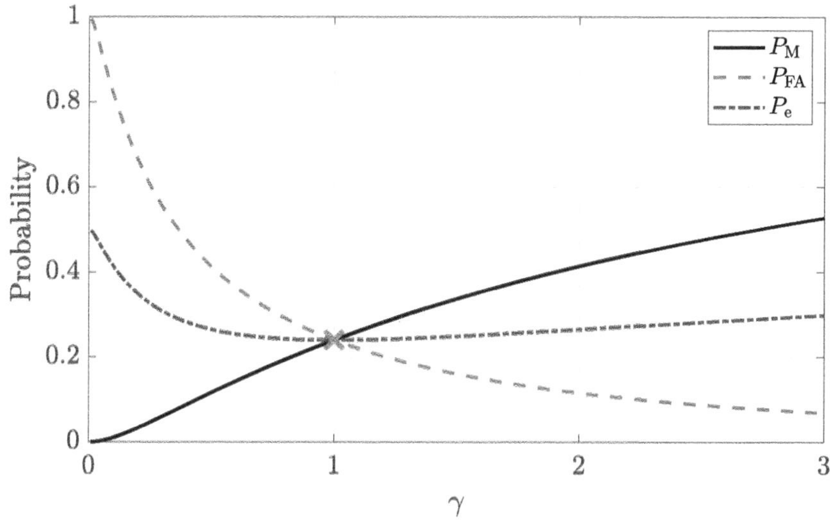

(a) $\Pr\{\mathcal{H}_0\} = \Pr\{\mathcal{H}_1\} = \frac{1}{2}$ and the optimal threshold is $\gamma = 1$.

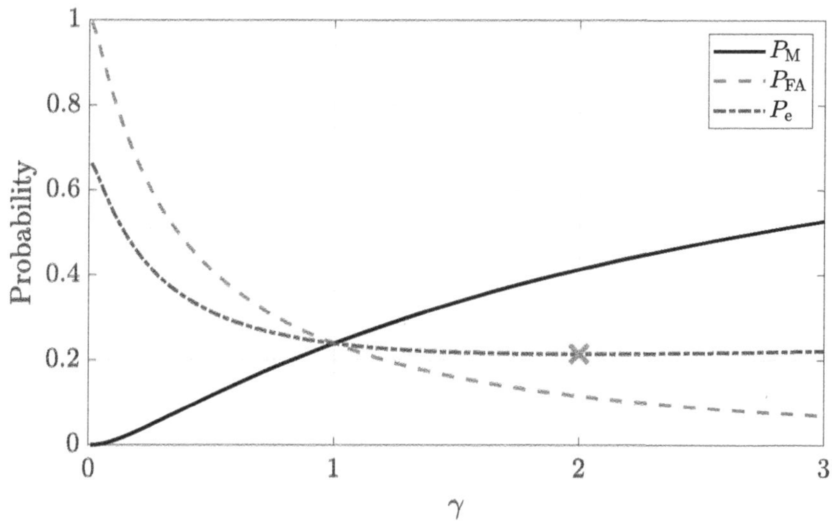

(b) $\Pr\{\mathcal{H}_0\} = 2\Pr\{\mathcal{H}_1\} = \frac{2}{3}$ and the optimal threshold is $\gamma = 2$.

Figure 2.27: The missing probability ($P_{\rm M}$), the false alarm probability ($P_{\rm FA}$), and the error probability ($P_{\rm e}$) as a function of the threshold γ for the binary hypothesis test in Example 2.18 with $\sigma^2 = 0.5$. The cross shows the threshold from Lemma 2.13 that minimizes the error probability: $\gamma = \Pr\{\mathcal{H}_0\}/\Pr\{\mathcal{H}_1\}$.

Example 2.19. Consider the binary hypothesis test in (2.181). Derive the Neyman-Pearson detector that satisfies $P_{\text{FA}} = \alpha$. What is P_{D} for this detector?

The condition in (2.191) appeared already in the Bayesian detector but for a predefined value of γ. We now need to find the value that gives equality in (2.192). If we rewrite this condition in the way previously done in (2.188), our goal becomes to find the value of γ' that results in $P_{\text{FA}} = \alpha$. This value is found by solving the equation

$$P_{\text{FA}} = \alpha = \int_{\gamma'}^{\infty} \frac{1}{\sqrt{2\pi\sigma^2}} e^{-\frac{y^2}{2\sigma^2}} \partial y = 1 - F_{\mathsf{y}}\left(\frac{\gamma'}{\sigma}\right), \tag{2.193}$$

where $F_{\mathsf{y}}(y)$ denotes the CDF of the standard Gaussian distribution with zero mean and variance 1. Since the CDF of a continuous random variable is an invertible function, we can solve for γ'/σ and obtain $\gamma' = \sigma F_{\mathsf{y}}^{-1}(1-\alpha)$. In conclusion, the Neyman-Pearson detector selects the hypothesis $\mathcal{H}_1$ if $y \geq \sigma F_{\mathsf{y}}^{-1}(1-\alpha)$ and selects $\mathcal{H}_0$ otherwise. If we insert that value into (2.189), we obtain the detection probability

$$P_{\text{D}} = \int_{\sigma F_{\mathsf{y}}^{-1}(1-\alpha)}^{\infty} \frac{1}{\sqrt{2\pi\sigma^2}} e^{-\frac{(y-1)^2}{2\sigma^2}} \partial y = 1 - F_{\mathsf{y}}\left(F_{\mathsf{y}}^{-1}(1-\alpha) - \sigma^{-1}\right), \tag{2.194}$$

where we made a change of integration variable from y to $(y-1)/\sigma$ when obtaining the final result.

We can use the Neyman-Pearson detector to handle the binary hypothesis test in (2.181), using the formulas derived in Example 2.19. Figure 2.28 shows how the detection probability, P_{D}, varies with the SNR. Three different false alarm probabilities are considered: $\alpha = 10^{-1}$, $\alpha = 10^{-3}$, and $\alpha = 10^{-5}$. Since the signal of interest is 1 under $\mathcal{H}_1$, the SNR is defined as $\text{SNR} = 1/\sigma^2$. The detection probability improves as the SNR increases for any given value of P_{FA}. We notice that P_{D} is higher when the false alarm probability is set to a higher value. This is expected since the challenge in detection is to handle uncertain cases. If we select $\mathcal{H}_1$ for most of these cases, we get a high value of P_{D} but also many false alarms. When the desired value of P_{FA} is smaller, a higher SNR is needed to achieve the same P_{D}.

2.8 Frequency Domain and Discrete Fourier Transform

Wireless signals can be equivalently represented in the time domain and frequency domain. The Fourier transform was used earlier in this chapter to obtain the frequency-domain representation of continuous-time signals. In this section, we will describe the mathematical transformation between these domains for discrete signals. In particular, we will define the *discrete Fourier transform (DFT)* and describe how it can be utilized to analyze the frequency

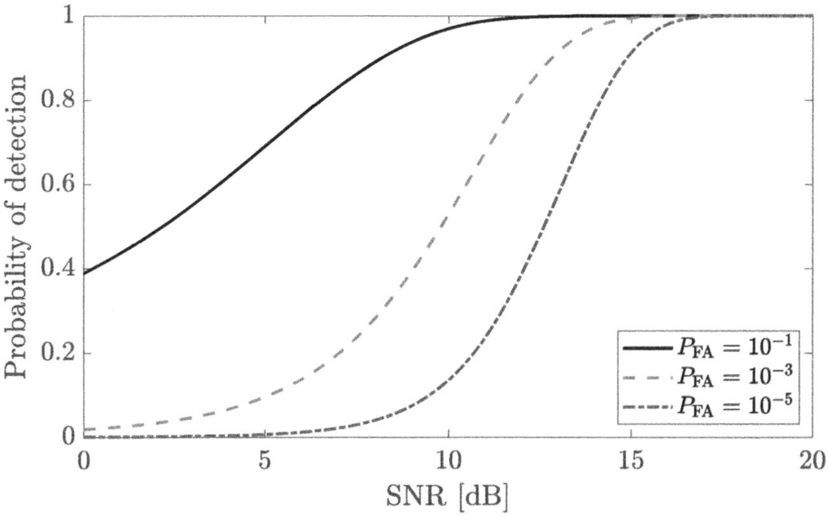

Figure 2.28: The detection probability, P_D, versus the SNR $= 1/\sigma^2$, for three different values of P_{FA}. The Neyman-Pearson detector is used for the binary hypothesis test in Example 2.19.

content of a sampled time-domain signal of finite length. In communication systems that operate over a large bandwidth, it is common to insert the data content into the frequency-domain representation of the signal instead of the time-domain representation. The reason can be to efficiently handle channels that change substantially over the signal bandwidth. We will provide key results regarding the DFT and *inverse DFT (IDFT)* that will be utilized in later chapters.

Consider an S-length sequence $\chi[0], \dots, \chi[S-1]$ with samples of a time-domain signal. The DFT of this sequence is a sequence $\bar{\chi}[0], \dots, \bar{\chi}[S-1]$ of equal length that describes the frequency-domain content and is given by

$$\bar{\chi}[\nu] = \mathcal{F}_{\mathrm{d}}\{\chi[s]\} = \frac{1}{\sqrt{S}} \sum_{s=0}^{S-1} \chi[s] e^{-\mathrm{j}2\pi s\nu/S} \quad \text{for } \nu = 0, \dots, S-1. \tag{2.195}$$

The constant $1/\sqrt{S}$ in (2.195) ensures that the energy is the same in both the time-domain sequence and the corresponding frequency-domain sequence:

$$\sum_{s=0}^{S-1} |\chi[s]|^2 = \sum_{\nu=0}^{S-1} |\bar{\chi}[\nu]|^2, \tag{2.196}$$

which is known as *Parseval's relation.* Many other textbooks omit this scaling factor, which results in an energy mismatch that must be compensated for when taking the IDFT. However, the scaling factor is vital in communications since the signal energy is constrained, and we want to be able to measure it over both time and frequency. The IDFT of the sequence $\bar{\chi}[0], \dots, \bar{\chi}[S-1]$ is computed as

$$\chi[s] = \mathcal{F}_{\mathrm{d}}^{-1}\{\bar{\chi}[\nu]\} = \frac{1}{\sqrt{S}} \sum_{\nu=0}^{S-1} \bar{\chi}[\nu] e^{\mathrm{j}2\pi s\nu/S} \quad \text{for } s = 0, \ldots, S-1 \tag{2.197}$$

and returns the original time-domain sequence.

The DFT is a linear transform, which can be seen by defining the $S \times S$ *DFT matrix*

$$\mathbf{F}_S = \frac{1}{\sqrt{S}} \begin{bmatrix} 1 & 1 & 1 & \ldots & 1 \\ 1 & \upsilon_S & \upsilon_S^2 & \ldots & \upsilon_S^{S-1} \\ \vdots & \vdots & \vdots & \vdots & \vdots \\ 1 & \upsilon_S^{S-1} & \upsilon_S^{2(S-1)} & \ldots & \upsilon_S^{(S-1)(S-1)} \end{bmatrix}, \tag{2.198}$$

where $\upsilon_S = e^{-\mathrm{j}2\pi/S}$. We can use $\mathbf{F}_S$ to write the DFT in (2.195) in vector/matrix form as

$$\underbrace{\begin{bmatrix} \bar{\chi}[0] \\ \vdots \\ \bar{\chi}[S-1] \end{bmatrix}}_{=\bar{\boldsymbol{\chi}}} = \mathbf{F}_S \underbrace{\begin{bmatrix} \chi[0] \\ \vdots \\ \chi[S-1] \end{bmatrix}}_{=\boldsymbol{\chi}}, \tag{2.199}$$

or more concisely as $\bar{\boldsymbol{\chi}} = \mathbf{F}_S \boldsymbol{\chi}$. The DFT matrix is unitary (i.e., $\mathbf{F}_S^{\mathrm{H}} \mathbf{F}_S = \mathbf{F}_S \mathbf{F}_S^{\mathrm{H}} = \mathbf{I}_S$), thus the IDFT can be obtained from (2.199) by multiplying with the *IDFT matrix* $\mathbf{F}_S^{\mathrm{H}}$ from the left-hand side:

$$\boldsymbol{\chi} = \mathbf{F}_S^{\mathrm{H}} \bar{\boldsymbol{\chi}}. \tag{2.200}$$

The columns of $\mathbf{F}_S^{\mathrm{H}}$ are an orthonormal basis in $\mathbb{C}^S$ since the DFT matrix is unitary. Any signal vector $\boldsymbol{\chi}$ is spanned by this basis, and the basis vectors can be shown to represent a set of specific signal frequencies.

2.8.1 Interpretation of Signal Frequencies

Any S-length signal can be represented by a vector $\boldsymbol{\chi} = [\chi[0], \ldots, \chi[S-1]]^{\mathrm{T}} \in \mathbb{C}^S$. The IDFT formula in (2.200) shows that this vector can also be represented as a linear combination of the columns of $\mathbf{F}_S^{\mathrm{H}}$ with the coefficients given by the DFT vector $\bar{\boldsymbol{\chi}}$. The columns of $\mathbf{F}_S^{\mathrm{H}}$ take the role of an orthonormal basis in $\mathbb{C}^S$ and are not selected arbitrarily but to represent different signal frequencies. If we count the columns of $\mathbf{F}_S^{\mathrm{H}}$ from 0 to $S-1$, then column $\nu \in \{0, \ldots, S-1\}$ is

$$\frac{1}{\sqrt{S}} \begin{bmatrix} 1 \\ (\upsilon_S^{\nu})^* \\ (\upsilon_S^{2\nu})^* \\ \vdots \\ \left(\upsilon_S^{(S-1)\nu}\right)^* \end{bmatrix} = \frac{1}{\sqrt{S}} \begin{bmatrix} e^{\mathrm{j}\frac{2\pi\nu}{S} \cdot 0} \\ e^{\mathrm{j}\frac{2\pi\nu}{S} \cdot 1} \\ e^{\mathrm{j}\frac{2\pi\nu}{S} \cdot 2} \\ \vdots \\ e^{\mathrm{j}\frac{2\pi\nu}{S} \cdot (S-1)} \end{bmatrix}. \tag{2.201}$$

This basis vector contains S equal-spaced samples of the complex exponential $e^{\mathrm{j}\frac{2\pi\nu}{S}\cdot l}$, with the *normalized frequency* ν/S and the integer sample times $l = 0, 1, \ldots, S-1$. The frequency is normalized in the sense that the time between the samples is unspecified. Suppose the discrete samples are obtained from a complex exponential $e^{\mathrm{j}2\pi f t}$ with the frequency f Hz and time variable $t \in \mathbb{R}$. In that case, we need to know the sampling rate (samples/second) to connect the normalized frequency to the original frequency.

We considered the normalized frequencies $\nu/S \in \{0, \ldots, (S-1)/S\}$ between 0 and 1 when computing the IDFT in (2.197), but we can as well consider another interval of length 1. The reason is that the complex exponential $e^{\mathrm{j}\frac{2\pi\nu}{S}\cdot l}$ is a periodic function of ν. In particular, we obtain the same column in (2.201) with ν/S and $\nu/S + n$ for any integer n because

$$e^{\mathrm{j}2\pi\left(\frac{\nu}{S}+n\right)\cdot l} = e^{\mathrm{j}\frac{2\pi\nu}{S}\cdot l} \underbrace{e^{\mathrm{j}2\pi n l}}_{=1} = e^{\mathrm{j}\frac{2\pi\nu}{S}\cdot l} \tag{2.202}$$

when l is an integer. Since positive and negative frequencies often come in pairs in practical signals (e.g., in the complex baseband representation), it is common to consider a symmetric frequency interval such as $f \in [-B/2, B/2)$, where the upper limit is excluded so that the Nyquist-Shannon sampling theorem stated in Lemma 2.8 is satisfied. Hence, utilizing the normalized frequency interval $\bar{f} \in [-1/2, 1/2)$ that is also symmetric around zero can be convenient. There is then a simple bijective mapping where the sampling of a signal with the original frequency f results in the normalized frequency

$$\bar{f} = \frac{f}{B} \tag{2.203}$$

when the sampling rate is B sample/second. Half of the normalized frequencies in $[-1/2, 1/2)$ are negative, and the concept of negative frequencies might seem illogical but is fundamentally important. The complex exponentials with the positive normalized frequency ν/S and with the negative counterpart $-\nu/S$ only differ by a complex conjugate:

$$e^{-\mathrm{j}\frac{2\pi\nu}{S}\cdot l} = \left(e^{\mathrm{j}\frac{2\pi\nu}{S}\cdot l}\right)^{*}. \tag{2.204}$$

Hence, the real parts are equal, while the imaginary parts have opposite signs. Euler's formula in (2.3) can be utilized to create any discrete-time sinusoidal signal with the normalized frequency ν/S as a linear combination of $e^{\mathrm{j}\frac{2\pi\nu}{S}\cdot l}$ and $e^{-\mathrm{j}\frac{2\pi\nu}{S}\cdot l}$; for example, we can create the cosine and sine signals as

$$\cos\left(\frac{2\pi\nu}{S}\cdot l\right) = \frac{1}{2}e^{\mathrm{j}\frac{2\pi\nu}{S}\cdot l} + \frac{1}{2}e^{-\mathrm{j}\frac{2\pi\nu}{S}\cdot l}, \tag{2.205}$$

$$\sin\left(\frac{2\pi\nu}{S}\cdot l\right) = \frac{1}{2\mathrm{j}}e^{\mathrm{j}\frac{2\pi\nu}{S}\cdot l} - \frac{1}{2\mathrm{j}}e^{-\mathrm{j}\frac{2\pi\nu}{S}\cdot l}. \tag{2.206}$$

This is why we need pairs of positive and negative frequencies to synthesize arbitrary signals using the IDFT.

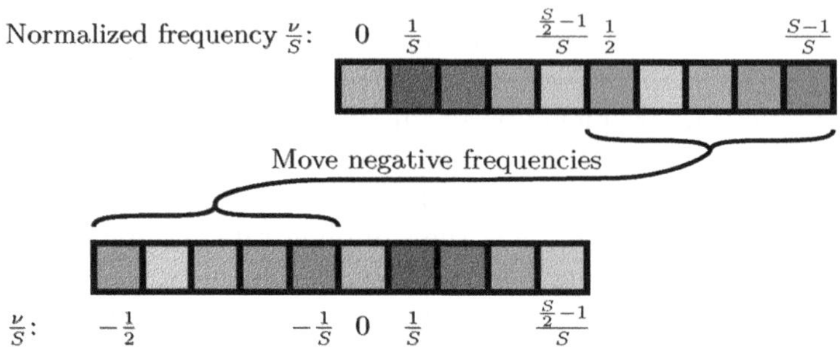

Figure 2.29: Illustration of how the positive range $\nu/S \in \{0, \ldots, (S-1)/S\}$ of normalized frequencies can be turned into the symmetric range in (2.207) with both positive and negative frequencies through a cyclic shift. $S = 10$ samples are considered in this example.

Example 2.20. Which are the S normalized frequencies $\bar{f} \in [-1/2, 1/2)$ that the IDFT utilizes?

The columns of the IDFT matrix are generated by the normalized frequencies $\nu/S \in \{0, \ldots, (S-1)/S\}$. The lower half is in the intended range $[0, 1/2)$, while the upper half is in the interval $[1/2, 1)$ that is larger than $1/2$. We can use the periodicity property from (2.202) to subtract 1 from these normalized frequencies and obtain an equivalent representation in the range $[-1/2, 0)$. The IDFT is therefore synthesizing signals using the following S normalized frequencies $\bar{f}$ between $-1/2$ and $1/2$:

$$\begin{aligned}\bar{f} &\in \left\{ \frac{\lceil \frac{S}{2} \rceil}{S} - 1, \ldots, -\frac{1}{S}, 0, \frac{1}{S}, \ldots, \frac{\lceil \frac{S}{2} \rceil - 1}{S} \right\} \\ &= \begin{cases} -\frac{1}{2}, \ldots, \frac{1}{2} - \frac{1}{S} & \text{if } S \text{ is even}, \\ -\frac{1}{2} + \frac{1}{2S}, \ldots, \frac{1}{2} - \frac{1}{2S} & \text{if } S \text{ is odd}, \end{cases}\end{aligned} \tag{2.207}$$

where the operator $\lceil \cdot \rceil$ returns the closest integer larger than or equal to its argument. The first and last frequencies differ for even and odd values of S.

Figure 2.29 shows how to switch from the range $\nu/S \in \{0, \ldots, (S-1)/S\}$ of positive normalized frequencies to the symmetric range in (2.207) with both positive and negative frequencies. This is achieved through a cyclic shift where the upper half is moved to the beginning. The figure shows the case of $S = 10$, which is an even number, so $1/2$ is one of the original normalized frequencies (this will not happen if S is odd). This frequency is equivalent to $1/2 - 1 = -1/2$, so we can put it in either the beginning or the end of the symmetric range. We follow the convention of starting with $-1/2$ so that the cyclic shift divides the range into two equal halves and shifts their order.

Figure 2.30(a) shows the real and imaginary parts of $e^{j\frac{2\pi\nu}{S} \cdot l}$ for $\nu = 5$ and $S = 7$. The curves are drawn as a function of a continuous variable l, but the samples obtained at the integer times $l = 0, \ldots, 6$ are marked with circles.

When the DFT is applied to these 7 time-domain samples, we obtain the vector $\bar{\chi} = [0, \ldots, 0, \sqrt{7}, 0]^{\mathrm{T}}$ where only the sixth entry corresponding to $\nu = 5$ is non-zero. The normalized frequency is $\nu/S = 5/7$, which is not within the interval $[-1/2, 1/2)$ but can be identically represented within that range by $\bar{f} = 5/7 - 1 = -2/7$. Figure 2.30(b) shows the DFT representation of the signal using the set of normalized frequencies from (2.207) that are within the desired interval $[-1/2, 1/2)$.

Figure 2.31 illustrates how the IDFT formula $\chi = \mathbf{F}_S^{\mathrm{H}} \bar{\chi}$ in (2.200) synthesizes the time-domain signal by showing how each column of $\mathbf{F}_S^{\mathrm{H}}$ contains samples of a complex exponential with a different frequency. The time axis points downwards, with positive values to the left of the vertical lines. We consider $S = 7$ as in the last figure. The curves in the first four columns are obtained for the positive normalized frequencies 0, 1/7, 2/7, and 3/7. The curves in the last three columns are obtained using the negative frequencies $-3/7$, $-2/7$, and $-1/7$, which are equivalent to the normalized frequencies 4/7, 5/7, and 6/7 that are outside the range $[-1/2, 1/2)$. The color coding identifies the columns that oscillate at the same frequency except for a different sign, leading to the same real parts but inverted imaginary parts.

The considered time-domain signal χ and its DFT $\bar{\chi}$ are sequences of the same finite length S, but the DFT and IDFT definitions can be easily extended into infinite sequences. The IDFT formula in (2.197) can be evaluated for any integer s, but the sequence is S-periodic since $\chi[s \pm S] = e^{\pm \mathrm{j}2\pi S\nu/S} \chi[s] = \chi[s]$ follows by the fact that $e^{\pm \mathrm{j}2\pi S\nu/S} = 1$. This can be pictured by considering Figure 2.31 and adding additional rows to the matrix by extending the oscillating curves up and down. No additional frequencies would be added to the signal when doing that. Similarly, for any integer ν, the DFT in (2.195) satisfies $\bar{\chi}[\nu \pm S] = e^{\mp \mathrm{j}2\pi sS/S} \bar{\chi}[\nu] = \bar{\chi}[\nu]$ since $e^{\mp \mathrm{j}2\pi sS/S} = 1$. This frequency-domain periodicity is the property we utilized when shifting the interval of normalized frequencies from $[0, 1)$ to $[-1/2, 1/2)$.

In summary, any S-length signal vector χ can be expressed as a linear combination of (samples from) complex exponentials having the S normalized frequencies stated in (2.207). This is why the DFT gives a frequency-domain representation, and the coefficients of the linear combination are stored in the DFT vector $\bar{\chi}$.

2.8.2 Finite Impulse Response Filters

The discrete-time representation of a communication system might contain the filtering of a signal sequence by a *finite impulse response (FIR)* filter, which might represent the communication channel in a discrete time. A causal discrete-time FIR filter of order T provides the output signal

$$y[k] = h[0]\chi[k] + h[1]\chi[k-1] + \ldots + h[T]\chi[k-T], \tag{2.208}$$

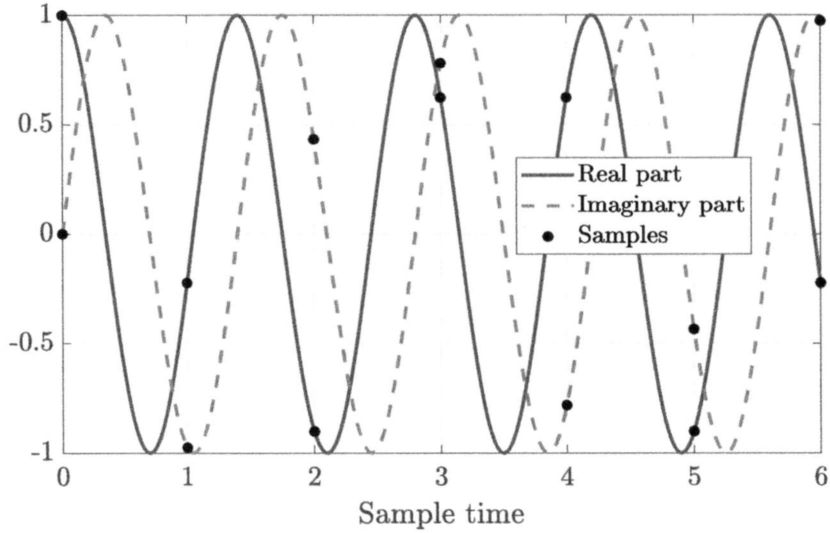

(a) The time-domain representation of $e^{\mathrm{j}\frac{2\pi\nu}{S}\cdot l}$ with samples taken at $l = 0, 1, \ldots, 6$.

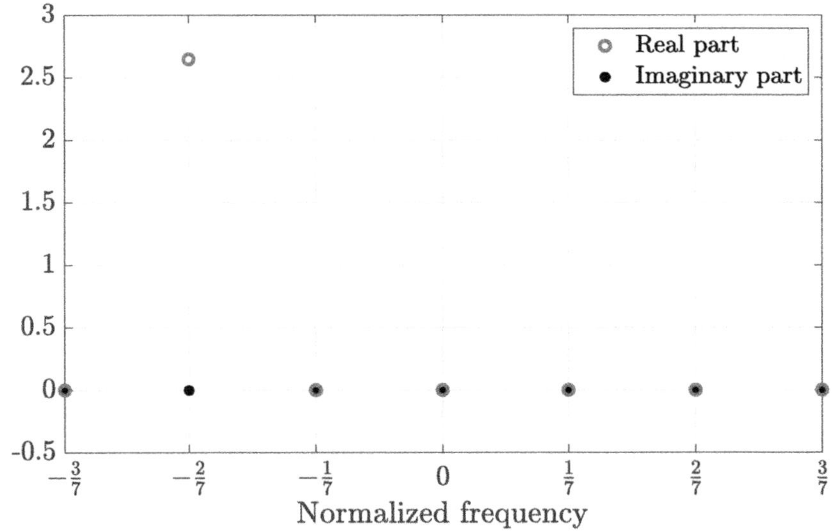

(b) The DFT representation of $e^{\mathrm{j}\frac{2\pi\nu}{S}\cdot l}$ using normalized frequencies.

Figure 2.30: The signal $e^{\mathrm{j}\frac{2\pi\nu}{S}\cdot l}$ with $\nu = 5$ and $S = 7$ is sampled at the integer times $l = 0, 1, \ldots, 6$. The time-domain representation is shown in (a), and the frequency-domain representation is shown in (b) using the set of normalized frequencies from (2.207).

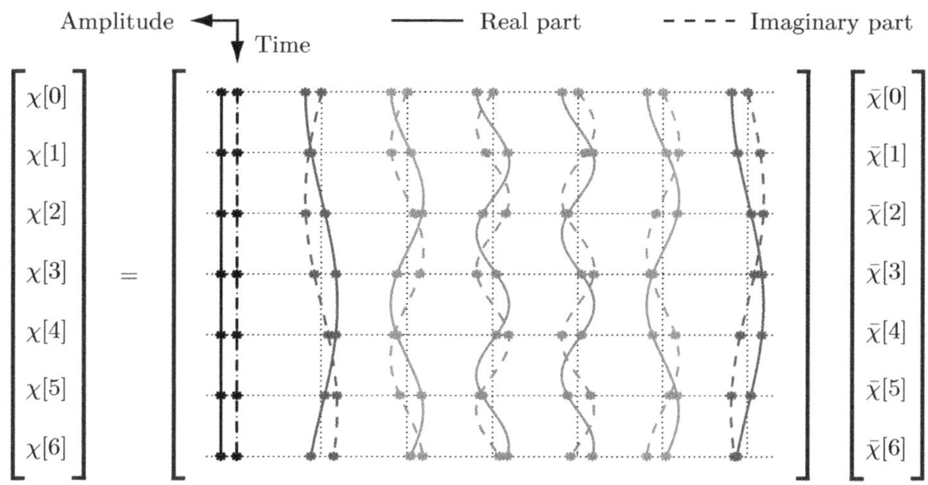

Figure 2.31: The IDFT formula in (2.200) is illustrated with a connection to the complex exponentials that are sampled to obtain the entries of $\mathbf{F}_S^{\mathrm{H}}$. The solid lines show the real parts, the dashed lines show the imaginary parts, the dotted vertical lines show the time axis, and the stars show the sampling points.

where $\chi[k]$ is the input signal and $h[0], \ldots, h[T]$ is the $(T+1)$-length discrete impulse response that characterizes the filter. Figure 2.32 illustrates the filtering operation in (2.208). The individual terms $h[k]$ are often called *taps*, and the entire filter can be referred to as a *tapped delay line* since the output contains delayed copies of the input multiplied by different taps.

When the signal sequence $\chi[0], \ldots, \chi[S-1]$ is sent as input to an FIR filter of order $T < S$, the output (2.208) can be expressed as a *linear convolution* (denoted by $*$) between the input sequence and the impulse response:

$$y[k] = (h * \chi)[k] = \sum_{\ell=0}^{T} h[\ell]\chi[k-\ell] \quad \text{for } k = 0, \ldots, S-1. \tag{2.209}$$

This equation also depends on the T signal values $\chi[-T], \ldots, \chi[-1]$ sent before the actual transmission began. This is a major issue if we want to identify all input signal values from the output sequence $y[0], \ldots, y[S-1]$ because there are $S+T$ parameters to identify but only S observations. Hence, controlling the content of the extra T signal values is desirable to avoid transient effects where unknown signals are mixed with the intended ones to create an ill-posed signal identification problem. A simple solution is to actively send a *prefix* containing $\chi[-T], \ldots, \chi[-1]$ into the FIR filter before the actual intended transmission of $\chi[0], \ldots, \chi[S-1]$ begins. The prefix can be designed in different ways under the constraint that it is not introducing any additional unknown signal values: we can only handle S unknowns when having S observations. One option is to use a silent prefix represented by

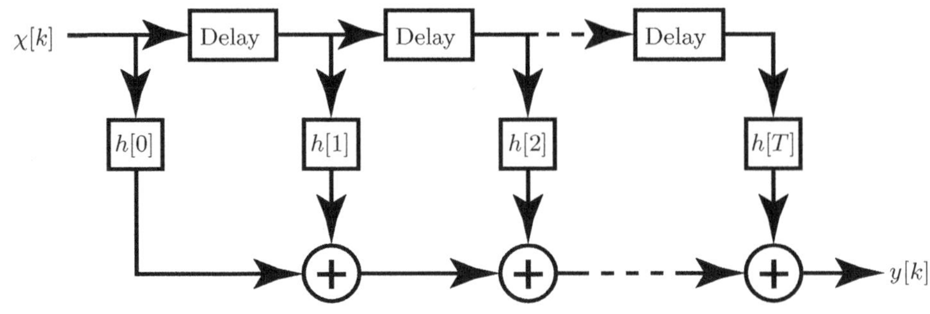

Figure 2.32: A block diagram of a discrete-time FIR filter of order T, which takes $\chi[k]$ as input and provides $y[k]$ as output.

$\chi[-1] = \ldots = \chi[-T] = 0$, so the corresponding terms vanish from (2.209). This prefix has the benefit of not increasing the total signal energy but has the drawback that we must design an inverse filter based on the channel taps to recover the input signal sequence. As will soon become apparent, a more convenient option is to add a *cyclic prefix* where we use values from the end of the sequence: $\chi[-1] = \chi[S-1]$, $\chi[-2] = \chi[S-2]$, and so on until $\chi[-T] = \chi[S-T]$. This option has the important consequence that the input signal sequence will appear to be periodic, in the sense that the received signal in (2.209) can be expressed as

$$\begin{aligned} y[k] &= \sum_{\ell=0}^{T} h[\ell]\chi[k-\ell] \\ &= \sum_{\ell=0}^{T} h[\ell]\chi[(k-\ell)_{\mathrm{mod}\, S}] = (h \circledast \chi)[k] \quad \text{for } k = 0, \ldots, S-1, \end{aligned} \tag{2.210}$$

where "mod S" is the modulo operation that adds S to $k-\ell$ whenever needed to get a value between 0 and $S-1$. Even if the FIR filter performs a linear convolution, the addition of the cyclic prefix makes the output signal mathematically equivalent to a *cyclic convolution* between $h[0], \ldots, h[T]$ and an infinite S-periodic extension of $\chi[0], \ldots, \chi[S-1]$. Recall that S-length sequences behave as S-periodic sequences when analyzed using the DFT, so this is the property that we want to maintain by adding the cyclic prefix. It is called cyclic (or circular) convolution since the modulo operation provides indices from the end of the signal sequence when $k-\ell$ is negative; for example, $(-1)_{\mathrm{mod}\, S} = S-1$, $(-2)_{\mathrm{mod}\, S} = S-2$, etc.

The DFT of the output $y[0], \ldots, y[S-1]$ can be expressed as

$$\begin{aligned} \bar{y}[\nu] &= \frac{1}{\sqrt{S}} \sum_{s=0}^{S-1} y[s] e^{-\mathrm{j}2\pi s\nu/S} = \frac{1}{\sqrt{S}} \sum_{s=0}^{S-1} \sum_{\ell=0}^{T} h[\ell]\chi[(s-\ell)_{\mathrm{mod}\, S}] e^{-\mathrm{j}2\pi s\nu/S} \\ &= \frac{1}{\sqrt{S}} \sum_{\ell=0}^{T} \sum_{i=-\ell}^{S-1-\ell} h[\ell]\chi[(i)_{\mathrm{mod}\, S}] e^{-\mathrm{j}2\pi(i+\ell)\nu/S} \end{aligned} \tag{2.211}$$

by changing the summation index from s to $i = s - \ell$. We can further rewrite the expression by adding S to all negative values of i and exploiting the cyclic signal structure, which results in

$$\begin{aligned}\bar{y}[\nu] &= \sum_{\ell=0}^{T} h[\ell]\frac{1}{\sqrt{S}}\left(\underbrace{\sum_{i=-\ell}^{-1} \chi[(i)_{\text{mod } S}]e^{-\mathrm{j}2\pi i\nu/S}}_{=\sum_{i=S-\ell}^{S-1}\chi[i]e^{-\mathrm{j}2\pi(i-S)\nu/S}} + \sum_{i=0}^{S-1-\ell} \chi[i]e^{-\mathrm{j}2\pi i\nu/S}\right)e^{-\mathrm{j}2\pi\ell\nu/S} \\ &= \underbrace{\sum_{\ell=0}^{T} h[\ell]e^{-\mathrm{j}2\pi\ell\nu/S}}_{\bar{h}[\nu]} \underbrace{\frac{1}{\sqrt{S}}\sum_{i=0}^{S-1}\chi[i]e^{-\mathrm{j}2\pi i\nu/S}}_{=\bar{\chi}[\nu]}, \end{aligned} \tag{2.212}$$

where the equality follows by using that $e^{\mathrm{j}2\pi S\nu/S} = 1$ since ν is an integer. The final expression in (2.212) shows that $\bar{y}[\nu]$ is the product between the DFT of the input signal and frequency response of the FIR filter, defined as

$$\bar{h}[\nu] = \sum_{\ell=0}^{T} h[\ell]e^{-\mathrm{j}2\pi\ell\nu/S} \quad \text{for } \nu = 0,\dots,S-1. \tag{2.213}$$

The frequency response is defined similarly to the DFT of a signal, except for the lack of a $1/\sqrt{S}$ scaling factor.[9] The property we derived above is known as the *cyclic convolution theorem*.

Lemma 2.15. Let $y[k] = (h \circledast \chi)[k]$ denote the S-length sequence obtained by cyclic convolution between the sequence $\chi[0],\dots,\chi[S-1]$ and the FIR filter $h[0],\dots,h[T]$ with order $T < S$. The DFT of $y[k]$ is given by

$$\bar{y}[\nu] = \bar{h}[\nu]\bar{\chi}[\nu] \quad \text{for } \nu = 0,\dots,S-1, \tag{2.214}$$

where $\bar{\chi}[\nu]$ is the DFT in (2.195) and $\bar{h}[\nu]$ is the frequency response in (2.213).

This lemma states that the DFT of a cyclic convolution between a signal sequence and the impulse response of a filter is the product of the respective frequency-domain representations. This is the discrete counterpart of the (perhaps) more widely used property that the continuous Fourier transform of the convolution between two functions is the product of the Fourier transforms of the respective functions. The practical consequence of this lemma is that we identify the DFT of the input signal sequence by computing the DFT of the output signal sequence and then simply dividing $\bar{y}[\nu]$ in (2.214) by $\bar{h}[\nu]$.

[9]Many textbooks omit the $1/\sqrt{S}$ factor when defining the DFT to achieve symmetry between how signals and impulse response are transformed to the frequency domain. As mentioned earlier, the drawback of that convention is that the signal energy will differ between the time and frequency domains, which we circumvent by using (2.195) for the DFT of a signal and (2.213) for the frequency response of an FIR filter.

The DFT operation is the same irrespective of the channel taps, which makes it convenient to implement in hardware. We will return to this in Section 7.1.1 when considering orthogonal frequency-division multiplexing (OFDM).

We can also establish a matrix-vector representation of the FIR filter. If we begin by considering the cyclic convolution in (2.210) and assume $T = 3$ (for brevity), we can connect the S outputs with the S inputs as

$$\underbrace{\begin{bmatrix} y[0] \\ \vdots \\ y[S-1] \end{bmatrix}}_{=\mathbf{y}} = \underbrace{\begin{bmatrix} h[0] & 0 & \dots & \dots & \dots & 0 & h[3] & h[2] & h[1] \\ h[1] & h[0] & 0 & \dots & \dots & \dots & 0 & h[3] & h[2] \\ h[2] & h[1] & h[0] & 0 & \dots & \dots & \dots & 0 & h[3] \\ h[3] & h[2] & h[1] & h[0] & 0 & \dots & \dots & \dots & 0 \\ 0 & \ddots & \ddots & \ddots & \ddots & \ddots & \ddots & \ddots & \vdots \\ \vdots & \ddots & h[3] & h[2] & h[1] & h[0] & 0 & \ddots & \vdots \\ \vdots & \dots & 0 & h[3] & h[2] & h[1] & h[0] & 0 & \vdots \\ \vdots & \dots & \dots & 0 & h[3] & h[2] & h[1] & h[0] & 0 \\ 0 & \dots & \dots & \dots & 0 & h[3] & h[2] & h[1] & h[0] \end{bmatrix}}_{=\mathbf{C}_h} \underbrace{\begin{bmatrix} \chi[0] \\ \vdots \\ \chi[S-1] \end{bmatrix}}_{=\boldsymbol{\chi}}, \tag{2.215}$$

which can be written in short form as $\mathbf{y} = \mathbf{C}_h \boldsymbol{\chi}$. The filtering is carried out by the $S \times S$ matrix called $\mathbf{C}_h$, where each row contains all the channel taps but shifted cyclically one entry to the right for each row. This kind of matrix is known as a *circulant matrix* and can be created for any value of $T < S$. Any such matrix can be viewed as the matrix representation of the cyclic convolution that an FIR filter carries out when the input has a cyclic prefix.

Another matrix-vector representation can be established by considering the frequency-domain expression in (2.214), which we can write as

$$\underbrace{\begin{bmatrix} \bar{y}[0] \\ \vdots \\ \bar{y}[S-1] \end{bmatrix}}_{=\bar{\mathbf{y}}} = \underbrace{\begin{bmatrix} \bar{h}[0] & 0 & \dots & 0 \\ 0 & \bar{h}[1] & \ddots & \vdots \\ \vdots & \ddots & \ddots & 0 \\ 0 & \dots & 0 & \bar{h}[S-1] \end{bmatrix}}_{=\mathbf{D}_{\bar{h}}} \underbrace{\begin{bmatrix} \bar{\chi}[0] \\ \vdots \\ \bar{\chi}[S-1] \end{bmatrix}}_{=\bar{\boldsymbol{\chi}}} \tag{2.216}$$

or in short form as $\bar{\mathbf{y}} = \mathbf{D}_{\bar{h}} \bar{\boldsymbol{\chi}}$. We notice that $\mathbf{D}_{\bar{h}}$ is a diagonal matrix containing the frequency response of the FIR filter. We can connect the time-domain representation in (2.215) and the frequency-domain representation in (2.216) using the DFT matrix $\mathbf{F}_S$. We known from (2.199) that $\bar{\boldsymbol{\chi}} = \mathbf{F}_S \boldsymbol{\chi}$, which also implies that $\bar{\mathbf{y}} = \mathbf{F}_S \mathbf{y}$. By substituting these expressions into (2.216), we obtain

$$\mathbf{F}_S \mathbf{y} = \mathbf{D}_{\bar{h}} \mathbf{F}_S \boldsymbol{\chi} \quad \Rightarrow \quad \mathbf{y} = \mathbf{F}_S^{\mathrm{H}} \mathbf{D}_{\bar{h}} \mathbf{F}_S \boldsymbol{\chi}. \tag{2.217}$$

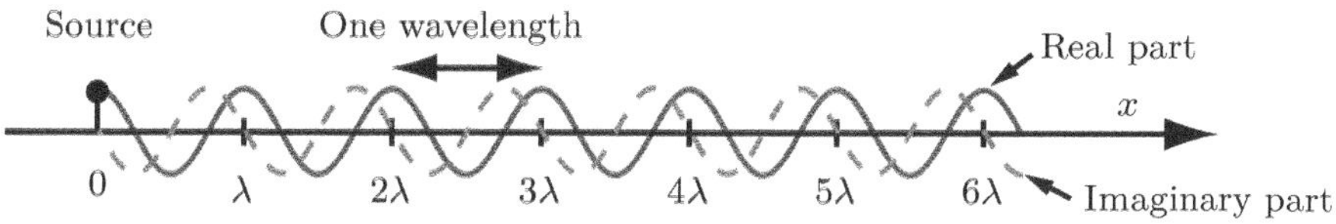

Figure 2.33: The complex exponential signal in (2.219) travels along the x-axis and the signal at time $t = 0$ is shown. The wavelength λ is the spatial interval between two peaks. The spatial frequency $1/\lambda$ is the number of wavelengths that fit into one meter.

By comparing (2.217) with (2.215), we notice that the circulant matrix $\mathbf{C}_h$ can be alternatively expressed as

$$\mathbf{C}_h = \mathbf{F}_S^{\mathrm{H}} \mathbf{D}_{\bar{h}} \mathbf{F}_S. \tag{2.218}$$

Since $\mathbf{D}_{\bar{h}}$ is a diagonal matrix and $\mathbf{F}_S$ is a unitary matrix, we recognize (2.218) as the eigendecomposition of $\mathbf{C}_h$; it has the same structure as in Lemma 2.1, except that the eigenvalues can be complex in this case since $\mathbf{C}_h$ is not Hermitian. The eigenvalues are the entries $\bar{h}[0], \ldots, \bar{h}[S-1]$ of the frequency response of the filter, while the eigenvectors are the columns of the IDFT matrix $\mathbf{F}_S^{\mathrm{H}}$. Since this result holds for any circular convolution, we can conclude that the DFT matrix diagonalizes any circulant matrix.

2.8.3 Temporal and Spatial Frequencies

The DFT was introduced in this section to study the *temporal frequencies* contained in a time-varying signal, but there is another related concept: *spatial frequencies*. When an electromagnetic signal propagates through free space, it can be observed simultaneously at many spatial locations, but it will be delayed differently depending on how far it has traveled from the signal source. Suppose the complex exponential signal $e^{\mathrm{j}2\pi f_{\mathrm{c}} t} = \cos(2\pi f_{\mathrm{c}} t) + \mathrm{j}\sin(2\pi f_{\mathrm{c}} t)$ with the temporal frequency f_{c} is emitted from a source located in the origin, as illustrated in Figure 2.33. The signal observed at the spatial location $x \geq 0$ along the positive x-axis at the time t is

$$e^{\mathrm{j}2\pi f_{\mathrm{c}}\left(t - \frac{x}{c}\right)} = e^{\mathrm{j}2\pi f_{\mathrm{c}} t} e^{-\mathrm{j}\frac{2\pi x}{\lambda}}, \tag{2.219}$$

where x/c is the propagation delay, c is the speed of light, and the wavelength at the carrier frequency is denoted by $\lambda = c/f_{\mathrm{c}}$. For a given communication system, the carrier frequency is predetermined, while the wavelength might change depending on the speed of light, which is reduced in some propagation media compared to its maximum value 299 792 458 m/s obtained in free space (i.e., vacuum). We will treat c as equal to the maximum in this book since the waves reach the receiver through the air. The factor $e^{\mathrm{j}2\pi f_{\mathrm{c}} t}$ in (2.219) determines the temporal signal variations while the factor $e^{-\mathrm{j}\frac{2\pi x}{\lambda}}$ determines the spatial variations. At time $t = 0$, the signal observed along the x-axis is

$$e^{-\mathrm{j}\frac{2\pi x}{\lambda}} = \cos\left(\frac{2\pi}{\lambda} x\right) - \mathrm{j}\sin\left(\frac{2\pi}{\lambda} x\right), \tag{2.220}$$

which is a periodic function that repeats itself every λ meters, thus the spatial frequency is $1/\lambda$, representing the number of periods per meter. The spatial frequency is also called the *wavenumber*, but we will use the spatial frequency terminology in this book to highlight that signals obtained in the time and space domains can be studied using the same methods (e.g., the sampling theorem and filtering). Spatial frequencies can be positive and negative, but the convention is that there is a minus sign in the complex exponential as in (2.220) when the spatial frequency is positive. In this way, the positive temporal frequency f_c gives rise to the positive spatial frequency $1/\lambda$. In this example, the spatial frequency is the same at any time t since the wave is shifted to the right as it travels along the line. This follows from the fact that the time variable t and spatial variable x affect different factors in (2.219).

The temporal frequency f_c and the spatial frequency $1/\lambda$ are closely related in wireless signaling (they only differ by a factor c), but there is a distinct conceptual difference. One way to separate the concepts is to consider a video recording of wave propagation (e.g., ocean waves). A video contains a sequence of frames (pictures shown at different times), and each frame consists of colored pixels at different screen locations. The temporal frequency describes how the wave observed at a particular pixel evolves with time. In contrast, the spatial frequency describes how the waves at a particular time instance oscillate between the pixels in the current frame. The fundamental relation between temporal and spatial frequency breaks down when static objects are introduced in the propagation environment. In that case, the temporal frequency remains the same, but the waves change directions when interacting with the objects, changing the spatial frequency observed along the given line. The connection also breaks down when observing the wave along a line that is not parallel to the direction the wave travels.

Figure 2.34 shows how the sinusoid $\cos(2\pi f_c t)$ propagates radially in two dimensions from a transmitter located in the origin, where the coloring describes its value. The signal observed at the point (x, y) at time t is

$$\cos\left(2\pi f_c\left(t-\frac{\sqrt{x^2+y^2}}{c}\right)\right) = \cos\left(2\pi f_c t - \frac{2\pi\sqrt{x^2+y^2}}{\lambda}\right) = \frac{1}{2}e^{j2\pi f_c t}e^{-j2\pi\frac{\sqrt{x^2+y^2}}{\lambda}} + \frac{1}{2}e^{-j2\pi f_c t}e^{j2\pi\frac{\sqrt{x^2+y^2}}{\lambda}}, \tag{2.221}$$

which is obtained similarly to (2.219) but with the propagation distance computed as $\sqrt{x^2+y^2}$. We also used Euler's formula as in (2.8) to express the cosine as two complex exponentials, which reveals that the considered signal contains the spatial frequencies $1/\lambda$ and $-1/\lambda$. The figure shows this signal at time $t = 0$, and we observe that the pattern is invariant to radial rotations since the signal propagates equally in all angular directions. The distance between two adjacent peaks in any radial direction equals the wavelength λ.

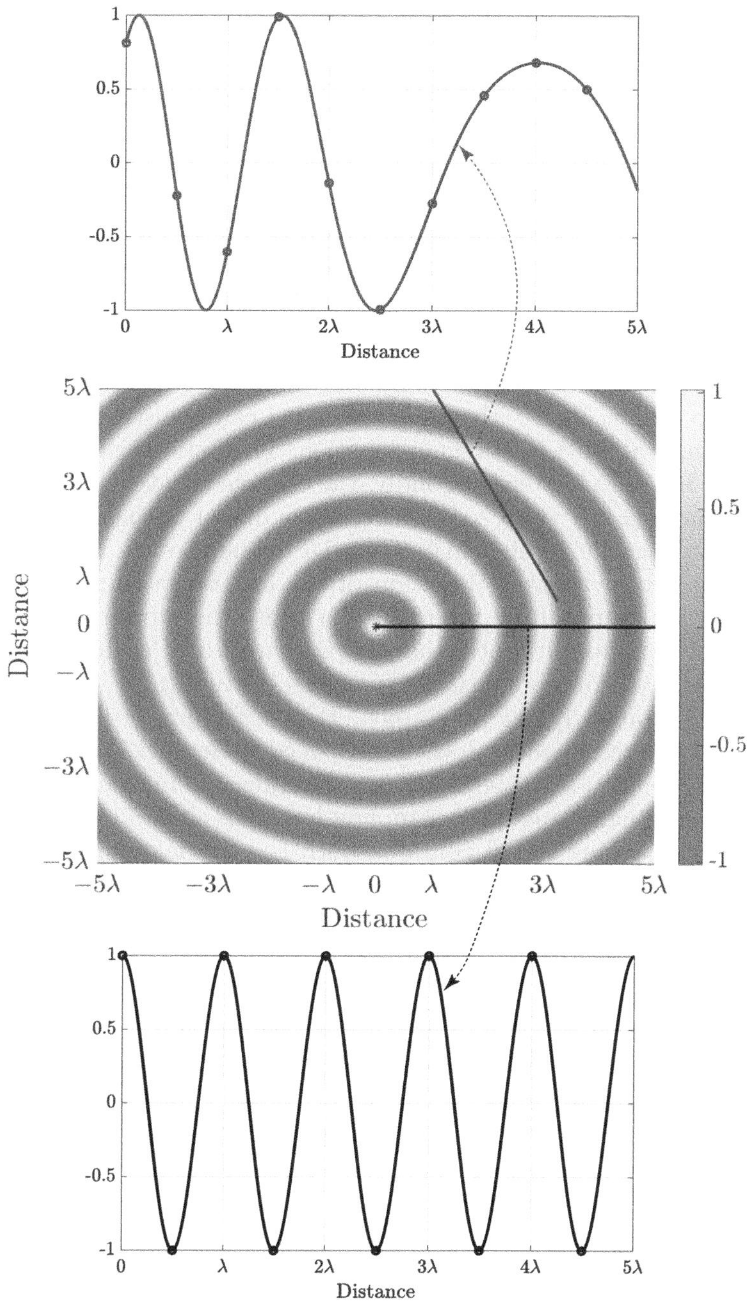

Figure 2.34: A sinusoidal wave propagates radially from a transmitter located in the origin. The middle figure shows the signal at different locations at $t = 0$. The upper and lower figures show how the signals observed along two lines contain different spatial frequencies.

At the bottom of the figure, the waveform observed along the black line is shown. This line covers the positive x-axis, a radial direction from the origin. At time $t = 0$, we observe a sinusoid $\cos(\frac{2\pi x}{\lambda})$ with the wavelength λ and the spatial frequencies $\pm 1/\lambda$. The signal observed along the blue line is shown at the top of the figure. This signal appears aperiodic and contains a broader range of spatial frequencies. The reason is that the wave propagation is not aligned with the direction of the line. The distance between the adjacent peaks varies but is larger than λ, which indicates that the observed signal only contains spatial frequencies in the range $[-1/\lambda, 1/\lambda]$.

There are two main messages from this example. Firstly, the spatial frequencies of the signal observed along a given line segment depend on the location of the source. Hence, the observed signal can be used to identify the source location or at least its angular direction. This estimation problem will be considered in later chapters. Secondly, the observed signal has the original spatial frequencies $\pm 1/\lambda$ when considering a line drawn in the same direction as the wave propagation, while smaller spatial frequencies (in the magnitude sense) are observed when the direction of the line is not aligned with the wave propagation. Suppose we insert $B = 2/\lambda$ into the sampling theorem in Lemma 2.8. In that case, it states that we can capture all useful information from any signal containing spatial frequencies in the range $[-1/\lambda, 1/\lambda)$ by taking samples spaced $1/B = \lambda/2$ apart. Hence, for any of the considered lines, measuring the signal at locations spaced apart by $\lambda/2$ is sufficient. This principle will guide us later when designing antenna arrays. Strictly speaking, the spacing between the sampling locations should be smaller than $\lambda/2$, because the cosine signal contains both the spatial frequencies $-1/\lambda$ and $1/\lambda$. Aliasing might appear when sampling precisely at the Nyquist rate, which we will discuss further in Chapter 4. We will also show that an antenna array's ability to distinguish between signals arriving from different directions is determined by its ability to separate the spatial frequencies of these signals.

The DFT can be applied to samples obtained at the same time but at different spatial locations. It will then reveal the spatial frequencies present in the spatial signal samples. An example of this is shown in Figure 2.35, where we take samples from the upper and lower curves in Figure 2.34. The $S = 10$ sample points per curve are indicated by circles in that previous figure and are spaced apart by $\lambda/2$, as suggested by the sampling theorem. Since we are taking spatial samples, the DFT computes the normalized spatial frequencies. Figure 2.35(a) shows the DFT of the blue upper curve in Figure 2.34, which contains a wide range of spatial frequencies since the waveform is sampled in a dimension that is not aligned with the direction of the propagating waveform. Since the original signal is real-valued, there is a symmetry between the positive and negative spatial frequencies. Figure 2.35(b) shows the DFT of the black signal, which only contains the normalized frequency $-1/2 = 1/2$. A single point in the DFT represents both frequencies due to the aliasing that

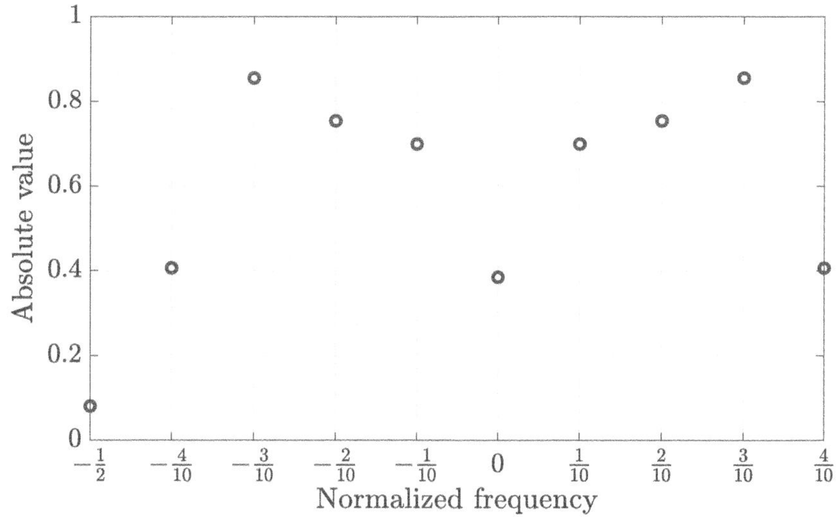

(a) DFT of the upper curve in Figure 2.34.

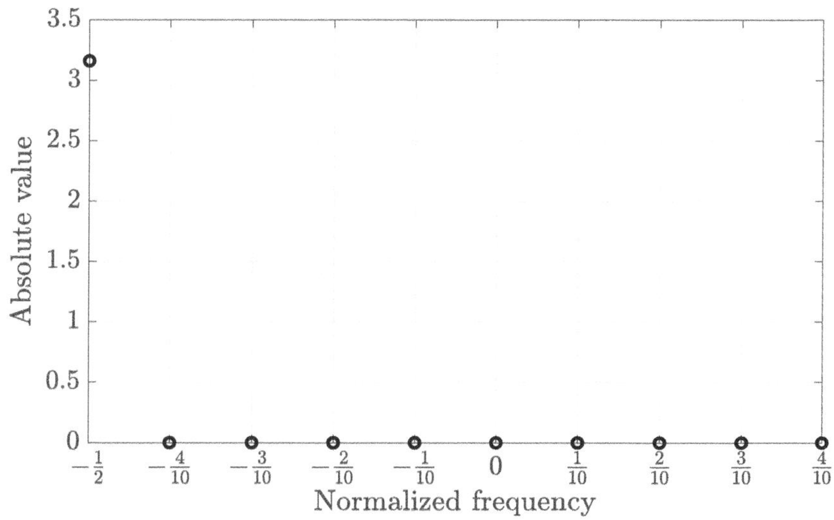

(b) DFT of the lower curve in Figure 2.34.

Figure 2.35: The DFTs of the spatially sampled waveforms from the upper and lower curves of Figure 2.34. In this case, the DFT describes spatial frequencies, and the figures show the magnitudes (i.e., absolute values) since the DFTs can be complex-valued.

can appear when sampling precisely at the Nyquist rate. However, from the preceding discussion, we know that the signal only contains spatial frequencies smaller or equal to $1/\lambda$; the fastest changes always occur in the direction the wave propagates. When combined with the prior knowledge that the signal is real-valued, it is possible to reconstruct the original signal in this special case. Since the spatial sampling rate is $2/\lambda$ samples per meter, the true spatial frequencies are $\pm\frac{1}{2}\frac{2}{\lambda} = \pm\frac{1}{\lambda}$, as anticipated from the previous discussion.

2.9 Exercises

Exercise 2.1. Consider two orthogonal vectors $\mathbf{x}_1 \in \mathbb{C}^M$ and $\mathbf{x}_2 \in \mathbb{C}^M$.

(a) What is the projection $\mathbf{y}_{\text{proj},\mathbf{x}_1,\mathbf{x}_2}$ of another vector $\mathbf{y} \in \mathbb{C}^M$ onto the space spanned by $\mathbf{x}_1$ and $\mathbf{x}_2$? Hint: Express the projection as $\mathbf{y}_{\text{proj},\mathbf{x}_1,\mathbf{x}_2} = \alpha_1\mathbf{x}_1 + \alpha_2\mathbf{x}_2$ and find the coefficients $\alpha_1, \alpha_2 \in \mathbb{C}$ that make the residual vector $\mathbf{y} - \mathbf{y}_{\text{proj},\mathbf{x}_1,\mathbf{x}_2}$ orthogonal to $\mathbf{x}_1$ and $\mathbf{x}_2$.

(b) Generalize the result from (a) to the case where we project $\mathbf{y}$ onto the space that is spanned by the $L < M$ orthogonal vectors $\mathbf{x}_1, \ldots, \mathbf{x}_L \in \mathbb{C}^M$. Show that we can write the projection as $\mathbf{y}_{\text{proj},\mathbf{x}_1,\ldots,\mathbf{x}_L} = \mathbf{P}\mathbf{y}$ and obtain an expression for the *projection matrix* $\mathbf{P}$.

Exercise 2.2. Let $\mathbf{x} \sim \mathcal{N}_{\mathbb{C}}(\mathbf{0}, \mathbf{I}_M)$ and $\mathbf{y} \sim \mathcal{N}_{\mathbb{C}}(\mathbf{0}, \mathbf{R})$ be M-dimensional complex Gaussian random vectors. Moreover, let $\mathbf{z} = [z_1, \ldots, z_M]^{\mathrm{T}}$ be a random vector with independent and identically distributed entries $z_m \sim \text{Exp}(1/3)$ for $m = 1, \ldots, M$.

(a) Compute $\mathbb{E}\{|\mathbf{v}^{\mathrm{H}}\mathbf{y}|^2\}$ for a given deterministic vector $\mathbf{v} = [v_1, \ldots, v_M]^{\mathrm{T}} \in \mathbb{C}^M$.

(b) Compute $\mathbb{E}\{|\mathbf{v}^{\mathrm{H}}\mathbf{z}|^2\}$ for a given deterministic vector $\mathbf{v} = [v_1, \ldots, v_M]^{\mathrm{T}} \in \mathbb{C}^M$.

(c) Compute $\mathbb{V}\text{ar}\{\|\mathbf{A}\mathbf{x}\|^2\}$ where $\mathbf{A} \in \mathbb{C}^{K \times M}$ is a deterministic matrix with $K \geq M$. Each column of $\mathbf{A}$ has a norm equal to 2 and is orthogonal to the other columns.

Exercise 2.3. When using PAM, the continuous-time complex-baseband signal can be expressed as in (2.120), which we repeat here as

$$z(t) = \sum_{k=-\infty}^{\infty} x[k]\, p\left(t - \frac{k}{B}\right). \tag{2.222}$$

The Nyquist criterion says that $z(n/B) = Ax[n]$, where $A \neq 0$ is an arbitrary constant. It can be equivalently expressed by multiplying $z(t)$ by the impulse train $\sum_{r=-\infty}^{\infty} \delta(t - r/B)$ and equating it to the impulse train weighted by the desired symbols:

$$z(t) \sum_{r=-\infty}^{\infty} \delta\left(t - \frac{r}{B}\right) = A \sum_{r=-\infty}^{\infty} x[r]\delta\left(t - \frac{r}{B}\right). \tag{2.223}$$

(a) By taking the Fourier transform of both sides of (2.223), derive the condition that the Fourier transform of the pulse must satisfy

$$B \sum_{r=-\infty}^{\infty} P(f - rB) = A \tag{2.224}$$

for the Nyquist criterion to hold. Hint: Use the fact that the Fourier transform of the impulse train is given by $\mathcal{F}\{\sum_{r=-\infty}^{\infty} \delta(t - r/B)\} = B \sum_{r=-\infty}^{\infty} \delta(f - rB)$.

(b) Verify that the sinc pulse is the most bandwidth-efficient pulse that satisfies the Nyquist criterion using the condition in (2.224).

(c) Determine whether the Nyquist criterion holds or not for the so-called *raised-cosine pulse* (with roll-off factor 0.5) that has the Fourier transform

$$P(f) = \begin{cases} 1 & \text{if } |f| \leq \frac{B}{4}, \\ \frac{1}{2}\left(1 + \sin\left(\frac{2\pi|f|}{B}\right)\right) & \text{if } \frac{B}{4} < |f| \leq \frac{3B}{4}, \\ 0 & \text{if } |f| > \frac{3B}{4}. \end{cases} \tag{2.225}$$

Exercise 2.4. Consider the LTI system in Figure 2.10(a) with the impulse response

$$g_{\mathrm{p}}(t) = \mathrm{rect}\left(\frac{t - T/2}{T}\right) = \begin{cases} 1, & \text{if } 0 \le t \le T, \\ 0, & \text{otherwise.} \end{cases} \tag{2.226}$$

The input signal $z_{\mathrm{p}}(t)$ is arbitrary and the complex-baseband equivalent input signal is denoted as $z(t)$.

(a) Find the complex-baseband representation of the output signal $v_{\mathrm{p}}(t)$ in terms of $z(t)$ and the carrier frequency f_{c} by first filtering the signal in the passband and then downshifting $v_{\mathrm{p}}(t)$ to the complex baseband.

(b) Compare the result obtained in (a) with the one obtained by first transforming the input signal $z_{\mathrm{p}}(t)$ to the complex baseband and then filtering it with the equivalent complex-baseband filter from (2.117).

Exercise 2.5. Consider the noise samples $n[l]$ in (2.123), where $w(t)$ is a white circularly symmetric complex Gaussian random process with the constant power-spectral density N_0 and $p(t)$ is the sinc-pulse defined in (2.118).

(a) Prove that the variance of $n[l]$ is N_0.

(b) Prove that the noise samples $n[l]$ and $n[m]$ obtained from (2.123) for $l \neq m$ are independent. Hint: Use the identity

$$\int_{-\infty}^{\infty} \mathrm{sinc}(l - t)\mathrm{sinc}(m - t)\partial t = 0, \tag{2.227}$$

which holds for any integers l and m such that $l \neq m$.

Exercise 2.6. Consider the linear observation model

$$\mathbf{z} = \mathbf{A}\mathbf{v} + \mathbf{n}, \tag{2.228}$$

where $\mathbf{v} \in \mathbb{C}^K$ and $\mathbf{n} \in \mathbb{C}^M$ are independent random vectors. Their entries are independent and identically distributed with zero mean and unit variance. The matrix $\mathbf{A} \in \mathbb{C}^{M\times K}$ is deterministic. Hence, the covariance matrices of $\mathbf{v}$ and $\mathbf{z}$ are $\mathbb{E}\{\mathbf{v}\mathbf{v}^{\mathrm{H}}\} = \mathbf{I}_K$ and $\mathbb{E}\{\mathbf{z}\mathbf{z}^{\mathrm{H}}\} = \mathbf{A}\mathbf{A}^{\mathrm{H}} + \mathbf{I}_M$, respectively. The LMMSE estimate of $\mathbf{v}$ based on the observation $\mathbf{z}$ is

$$\hat{\mathbf{v}} = \mathbf{A}^{\mathrm{H}}\left(\mathbf{A}\mathbf{A}^{\mathrm{H}} + \mathbf{I}_M\right)^{-1}\mathbf{z}. \tag{2.229}$$

(a) Verify the orthogonality principle $\mathbb{E}\{\tilde{\mathbf{v}}\mathbf{z}^{\mathrm{H}}\} = \mathbb{E}\{(\mathbf{v} - \hat{\mathbf{v}})\mathbf{z}^{\mathrm{H}}\} = \mathbf{0}$ for the given LMMSE estimator.

(b) Suppose $\mathbf{v} \sim \mathcal{N}_{\mathbb{C}}(\mathbf{0}, \mathbf{I}_K)$ and $\mathbf{n} \sim \mathcal{N}_{\mathbb{C}}(\mathbf{0}, \mathbf{I}_M)$. Show that the LMMSE estimator in (2.229) is also the MMSE estimator by verifying that $\mathbf{A}^{\mathrm{H}}\left(\mathbf{A}\mathbf{A}^{\mathrm{H}} + \mathbf{I}_M\right)^{-1}\mathbf{z}$ is the mean of the conditional PDF $f_{\mathbf{v}|\mathbf{z}}(\mathbf{v}|\mathbf{z})$. Hint: Use the matrix identity $\det\left(\mathbf{A}\mathbf{A}^{\mathrm{H}} + \mathbf{I}_M\right) = \det\left(\mathbf{A}^{\mathrm{H}}\mathbf{A} + \mathbf{I}_K\right) = \frac{1}{\det\left(\left(\mathbf{A}^{\mathrm{H}}\mathbf{A}+\mathbf{I}_K\right)^{-1}\right)}$ and the identity in (2.50).

(c) Find the MMSE estimate of $\mathbf{v} \sim \mathcal{N}_{\mathbb{C}}(\mathbf{0}, \mathbf{I}_M)$ based on the alternative observation

$$\mathbf{y} = \mathbf{v} + \mathbf{c}, \tag{2.230}$$

where $\mathbf{c} \sim \mathcal{N}_{\mathbb{C}}(\mathbf{0}, \mathbf{C})$ is the independent noise with an invertible covariance matrix $\mathbf{C}$. Hint: Use whitening and then (2.229).

Exercise 2.7. Consider a narrowband channel with L paths. The channel response is modeled according to (2.131) as

$$h = \sum_{i=1}^{L} \alpha_i e^{-\mathrm{j}2\pi f_{\mathrm{c}}(\tau_i - \eta)}. \tag{2.231}$$

(a) Is $|h|$ dependent of the value of η?

(b) Suppose there are $L = 2$ paths and $\alpha_1 = \alpha_2 = 1$. For which values of τ_1 and τ_2 is $|h|$ maximized? For which values is $|h|$ minimized?

(c) Define $\psi_i = 2\pi f_{\mathrm{c}}(\tau_i - \eta)$ and assume that it is a uniformly distributed random variable between $-\pi$ and π. Compute $\mathbb{E}\{|h|^2\}$ assuming that $\alpha_1, \ldots, \alpha_L$ are deterministic, while $\psi_1, \ldots, \psi_L$ are mutually independent. Hint: Use that $\mathbb{E}\{e^{-\mathrm{j}\psi_i}\} = 0$.

(d) Redo (c) under the assumption that $\alpha_1, \ldots, \alpha_L$ are also independent random variables, uniformly distributed between 0 and 1.

Exercise 2.8. Consider the complex-valued AWGN channel $y = x + n$ with B samples per second. Its capacity is $B \log_2(1 + P/(BN_0))$, which follows from (2.146) with $\beta = 1$. Decompose the channel into two real-valued AWGN channels.

(a) Are the two real-valued AWGN channels independent?

(b) How many samples per second do we have for each of the two channels?

(c) Suppose we transmit with a power of $P > 0$ Watt and place all the power in only one of the two real-valued AWGN channels. What is the capacity expressed in bits per second?

(d) Is the result in (c) higher or lower than the capacity of the complex-valued AWGN channel?

Exercise 2.9. A friend claims we can double the capacity (in bit/s) by doubling the bandwidth. Is this correct? If yes, use the capacity formula to prove it. If no, explain what else needs to be done to achieve twice the capacity.

Exercise 2.10. The received signal power reduces with the propagation distance d. This can be modeled as $\Upsilon \left(\frac{1\,\mathrm{m}}{d}\right)^{\alpha} P$ using the parametric channel gain model in (1.9), where P is the transmit power, $\alpha > 1$ is the pathloss exponent, and $\Upsilon > 0$ is a constant propagation loss.

(a) Suppose the channel is modeled as in (2.144). How can we select h to get the right received signal power? What is the resulting capacity expression?

(b) Consider $B = 10\,\mathrm{MHz}$, $N_0 = -174\,\mathrm{dBm}$, $P = 30\,\mathrm{dBm}$, $\Upsilon = -37\,\mathrm{dB}$, and $\alpha = 3.7$. What is the SNR for a user at the distance $d = 200\,\mathrm{m}$? What is the capacity (in bit/s)?

(c) What will the capacity be for a user at the distance $4d$? How can the transmit power be scaled to achieve the same capacity as in (b)?

(d) What will the capacity be for a user at the distance $d/2$? How can the transmit power be scaled to achieve the same capacity as in (b)?

Exercise 2.11. The capacity of the discrete memoryless channel $y = h \cdot x + n$ is achieved by the input signal $x \sim \mathcal{N}_{\mathbb{C}}(0, q)$, as proved in Corollary 2.1. Suppose we instead send two independent signals over the channel: $x_1 \sim \mathcal{N}_{\mathbb{C}}(0, q_1)$ and $x_2 \sim \mathcal{N}_{\mathbb{C}}(0, q_2)$. The resulting received signal is

$$y = h \cdot (x_1 + x_2) + n, \tag{2.232}$$

where $n \sim \mathcal{N}_{\mathbb{C}}(0, N_0)$ is independent complex Gaussian noise. What is the corresponding channel capacity, which is achieved by selecting q_1, q_2 to maximize the mutual information $\mathcal{H}(y) - \mathcal{H}(y|x_1, x_2)$ under the constraint $q_1 + q_2 \leq q$?

Exercise 2.12. Consider a random variable x with zero mean and variance σ^2. We want to estimate σ^2 from the L independent random realizations of x, which are denoted $x_1, \ldots, x_L$. The following estimator is utilized:

$$\hat{\sigma}_L^2 = \frac{\sum_{i=1}^{L} |x_i|^2}{K}, \tag{2.233}$$

where K is a pre-determined scalar.

(a) For which value of K is the considered estimator unbiased? Is the answer dependent on the specific distribution of x?

(b) For which value of K will the considered estimator achieve the minimum MSE? Is the answer dependent on more than the mean and variance of x? What is the MSE-minimizing value of K if $x \sim \mathcal{N}_{\mathbb{C}}(0, \sigma^2)$?

Exercise 2.13. Consider the binary hypothesis test

$$\mathcal{H}_0 \quad : \quad y[l] = n[l], \quad l = 1, \ldots, L, \tag{2.234}$$

$$\mathcal{H}_1 \quad : \quad y[l] = 1 + n[l], \quad l = 1, \ldots, L, \tag{2.235}$$

where the detector decides whether "1" is transmitted or not by observing multiple received signals $y[l]$. Unlike the hypothesis test in (2.181), L consecutive received signals are considered. The real-valued noise samples $n[l]$ are independent and identically distributed as $n[l] \sim \mathcal{N}(0, \sigma^2)$, for $l = 1, \ldots, L$.

(a) For a given value of $\gamma = \Pr\{\mathcal{H}_1\}/\Pr\{\mathcal{H}_0\}$, derive the Bayesian detector that minimizes the error probability. What are P_{D} and P_{FA} for this detector? Hint: The answers are integral expressions.

(b) For a given value of $P_{\mathrm{FA}} = \alpha$, derive the Neyman-Pearson detector that maximizes the detection probability, P_{D}. What is P_{D} for this detector?

Exercise 2.14. Consider the continuous-time signal $x(u) = 2\cos(200\pi u) + 3\sin(600\pi u)$, which is sampled to obtain the $S = 7$-length sequence $\chi[s] = x(s/B)$, for $s = 0, \ldots, 6$. What is the DFT of the sequence $\chi[s]$ when the sampling rate is $B = 700$ samples/s?

Exercise 2.15. Prove Parseval's relation in (2.196) using the unitary property of the DFT matrix $\mathbf{F}_S$.

Exercise 2.16. Suppose the S-length sequence $\mathbf{a} = [a[0], \ldots, a[S-1]]^{\mathrm{T}}$ has the DFT $\bar{\mathbf{a}} = [\bar{a}[0], \ldots, \bar{a}[S-1]]^{\mathrm{T}}$. Consider the $S \times S$ circulant matrix defined similar to (2.215) as

$$\mathbf{C}_{\bar{a}} = \begin{bmatrix} \bar{a}[0] & \bar{a}[S-1] & \bar{a}[S-2] & \ldots & \ldots & \bar{a}[1] \\ \bar{a}[1] & \bar{a}[0] & \bar{a}[S-1] & \ldots & \bar{a}[3] & \bar{a}[2] \\ \vdots & \ddots & \ddots & \ddots & \ddots & \vdots \\ \bar{a}[S-1] & \ldots & \ldots & \ldots & \ldots & \bar{a}[0] \end{bmatrix}, \tag{2.236}$$

but for the DFT sequence. We further define a diagonal matrix containing the time-domain sequence $\mathbf{a}$:

$$\mathbf{D}_a = \begin{bmatrix} a[0] & 0 & \ldots & 0 \\ 0 & a[1] & \ddots & \vdots \\ \vdots & \ddots & \ddots & 0 \\ 0 & \ldots & 0 & a[S-1] \end{bmatrix}. \tag{2.237}$$

(a) Derive a decomposition of $\mathbf{C}_{\bar{a}}$ in terms of $\mathbf{D}_a$ and the DFT matrix $\mathbf{F}_S$ similar to (2.218). Hint: Switch the roles of the sequences in time and frequency. From the structure of the DFT matrix, it holds that $\mathbf{F}_S^{\mathrm{T}} = \mathbf{F}_S$ and $\mathbf{F}_S^{\mathrm{H}} = \mathbf{F}_S^*$.

(b) Consider another S-length sequence $\mathbf{b} = [b[0], \ldots, b[S-1]]^{\mathrm{T}}$ which has the DFT $\bar{\mathbf{b}} = [\bar{b}[0], \ldots, \bar{b}[S-1]]^{\mathrm{T}}$. Prove that the DFT of the sequence $a[k]b[k]$ (for $k = 0, \ldots, S-1$) is given by $(\bar{a} \circledast \bar{b})[\nu]/\sqrt{S}$ (for $\nu = 0, \ldots, S-1$) by using the obtained decomposition of $\mathbf{C}_{\bar{a}}$ and the properties of $\mathbf{F}_S$. Hint: The kth entry of the S-length vector $\mathbf{D}_a\mathbf{b}$ is $a[k-1]b[k-1]$.

(c) For the given sequences $a[k] = e^{\mathrm{j}\frac{2\pi k}{10}}$, $b[k] = e^{\mathrm{j}\frac{6\pi k}{10}}$, for $k = 0, \ldots, 9$, verify that the DFT of the sequence $a[k]b[k]$ is given by $(\bar{a} \circledast \bar{b})[\nu]/\sqrt{S}$.

Exercise 2.17. The signal $x(t) = \cos(2\pi f_1 t)$ is modulated to the carrier frequency f_{c} by computing $x_{\mathrm{p}}(t) = x(t)\cos(2\pi f_{\mathrm{c}} t)$, where $f_{\mathrm{c}} > f_1$.

(a) Which positive and negative (temporal) frequencies does $x_{\mathrm{p}}(t)$ contain?

(b) The signal $x_{\mathrm{p}}(t)$ is radiated from an antenna located in the origin and propagates at the speed of light c. Which spatial frequencies can be observed along the y-axis?

(c) What happens to the temporal and spatial frequencies if the signal propagates through a medium where the propagation speed v is smaller than c (i.e., the speed of light in free space)?

Chapter 3

Capacity of Point-to-Point MIMO Channels

In this chapter, we will characterize the channel capacity in memoryless *point-to-point* scenarios where one transmitter communicates with one receiver without impacting other systems. We will distinguish between four cases:

1. *Single-input single-output (SISO)* channel: The transmitter and receiver have one antenna each.

2. *Single-input multiple-output (SIMO)* channel: The transmitter has one antenna and the receiver has multiple antennas.

3. *Multiple-input single-output (MISO)* channel: The transmitter has multiple antennas and the receiver has one antenna.

4. *Multiple-input multiple-output (MIMO)* channel: Both the transmitter and receiver have multiple antennas.

These cases are illustrated in Figure 3.1. The capacity of the SISO channel was derived and discussed in Section 2.4.1. This chapter will generalize the theory to capture the other three cases, one after the other. The results will be utilized in the remainder of the book to study specific communication scenarios and channel conditions.

3.1 Impact of Power and Bandwidth on the Capacity

Before introducing multiple antennas, we will return to the channel capacity for SISO channels in (2.146) and shed some light on how it depends on the transmit power P and the bandwidth B. The purpose is to understand how the capacity can be improved. For notational convenience, we now explicitly write the capacity in (2.146) as a function $C(P, B)$ of these variables:

$$C(P, B) = B \log_2 \left(1 + \frac{P\beta}{BN_0} \right) \quad \text{bit/s}. \tag{3.1}$$

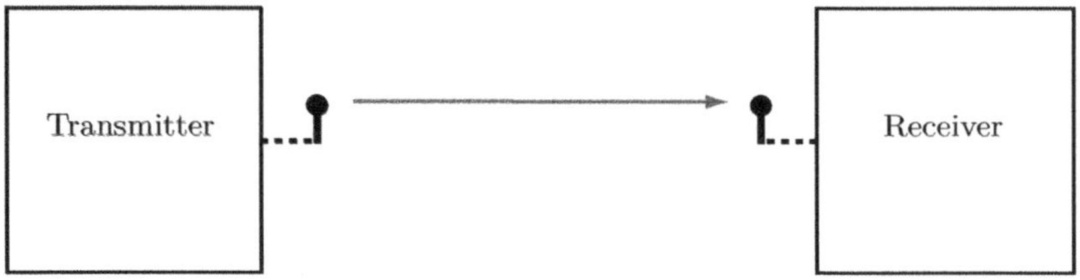

(a) Point-to-point SISO channel.

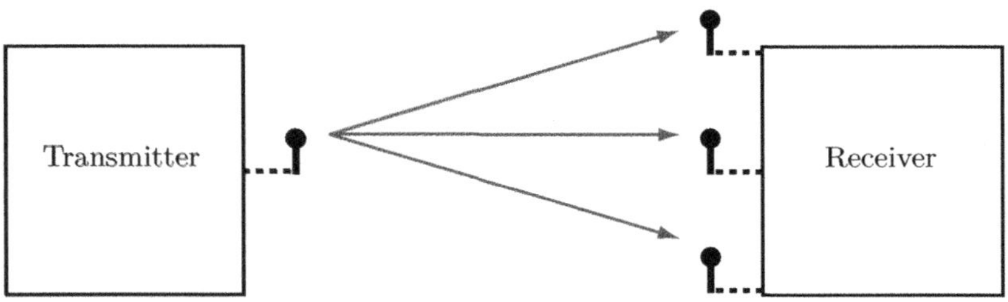

(b) Point-to-point SIMO channel.

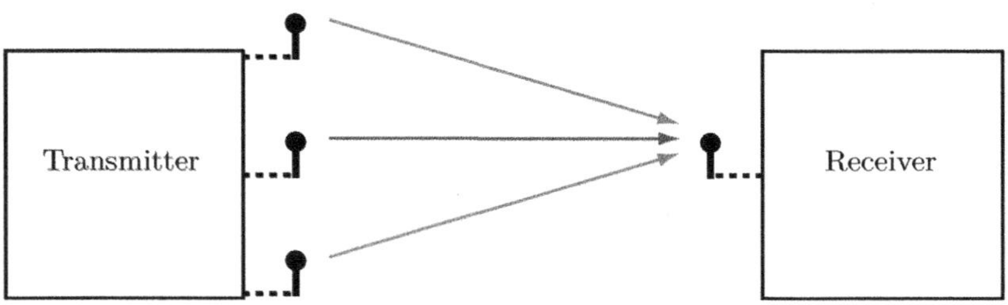

(c) Point-to-point MISO channel.

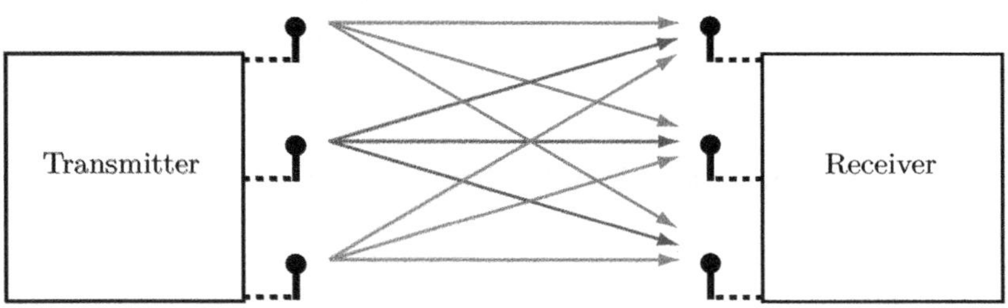

(d) Point-to-point MIMO channel.

Figure 3.1: The four kinds of point-to-point communication channels where the transmitter and receiver have either one or multiple antennas.

Since the capacity involves a logarithm, it is useful to notice that

$$\log_2(1+z) \approx \log_2(e)z \quad \text{if } z \approx 0, \tag{3.2}$$

$$\log_2(1+z) \approx \log_2(z) \quad \text{if } z \gg 0, \tag{3.3}$$

where $e \approx 2.71828$ is Euler's number. The expression in (3.2) is the first-order Taylor approximation of $\log_2(1+z)$ around $z = 0$. Since $\log_2(1+z)$ with the SNR $z = \frac{P\beta}{BN_0}$ appears in the capacity expression (3.1), (3.2) and (3.3) will help us to understand the capacity behavior at low and high SNR, respectively. The notions of low/high SNRs can be interpreted as follows.

Example 3.1. For which ranges of $z \geq 0$ will the approximations of $\log_2(1+z)$ in (3.2) and (3.3) lead to absolute errors that are smaller than 0.1?

The low SNR approximation $\log_2(1+z) \approx \log_2(e)z$ in (3.2) is based on a first-order Taylor approximation, and it can be written in an exact form as

$$\log_2(1+z) = \log_2(e)z - \log_2(e)\frac{z^2}{2(1+a)^2} \quad \text{for some } 0 \leq a \leq z, \tag{3.4}$$

where the second term is known as the Lagrange error bound. The absolute approximation error can be upper bounded using (3.4) as

$$\left|\log_2(1+z) - \log_2(e)z\right| = \log_2(e)\frac{z^2}{2(1+a)^2} \leq \frac{\log_2(e)}{2}z^2, \tag{3.5}$$

where the last step follows from setting $a = 0$ to get the largest possible error. Based on this upper bound, the absolute error is smaller than 0.1 when

$$\frac{\log_2(e)}{2}z^2 \leq 0.1 \quad \Rightarrow \quad z \leq \sqrt{\frac{0.2}{\log_2(e)}} \approx 0.37 \approx -4.3\,\text{dB}. \tag{3.6}$$

We can find the exact solution by solving $\log_2(e)z - \log_2(1+z) \leq 0.1$ numerically, which results in the somewhat larger range $z \lessapprox 0.42 \approx -3.8\,\text{dB}$.

For the high SNR approximation $\log_2(1+z) \approx \log_2(z)$ in (3.3), the absolute error is $\log_2(1+z) - \log_2(z) = \log_2(1+1/z)$. To guarantee $\log_2(1+1/z) \leq 0.1$, we should have

$$z \geq \frac{1}{2^{0.1}-1} \approx 13.93 \approx 11.4\,\text{dB}. \tag{3.7}$$

When varying the transmit power P, we notice that $C(P, B)$ is a monotonically increasing function of P. It starts at $C(0, B) = 0$ and then grows linearly with P when the SNR $\frac{P\beta}{BN_0}$ is small. We can utilize (3.2) to obtain

$$C(P, B) \approx B\log_2(e)\frac{P\beta}{BN_0} = \log_2(e)\frac{P\beta}{N_0} \tag{3.8}$$

at low SNR, which is independent of the bandwidth.

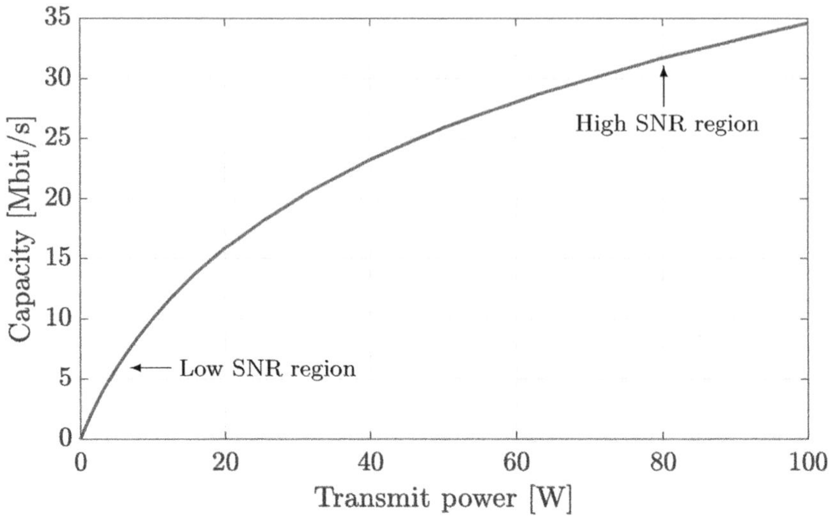

Figure 3.2: The capacity behavior in a single-antenna system when changing the transmit power P, for $B = 10\,\text{MHz}$ and $\beta/N_0 = 10^6\,\text{Hz/W}$.

When the SNR is large, the capacity only grows logarithmically with an increasing P due to (3.3). There is no upper limit on how large capacity we can achieve by increasing P, but the capacity growth is slow when we have reached the logarithmic growth rate at high SNR. Figure 3.2 illustrates these behaviors by showing $C(P, B)$ as a function of P for $B = 10\,\text{MHz}$ and $\beta/N_0 = 10^6\,\text{Hz/W}$. The capacity grows linearly with P in the *low SNR region*, while the logarithmic behavior appears in the *high SNR region*.

Example 3.2. Consider the capacity in (3.1) in a scenario where P and B have been selected such that $P\beta/(BN_0) = 1$. Suppose we change the transmit power from P to cP for some scalar $c > 0$. Which values of c will double and quadruple the capacity (compared to $c = 1$)?

The capacity in (3.1) becomes $C = B\log_2(1+1) = B$ under the assumption that $P\beta/(BN_0) = 1$ (i.e., when $c = 1$). Our first target is to double the capacity to $2B$ by increasing the transmit power to cP. This means that

$$B\log_2\left(1 + \frac{cP\beta}{BN_0}\right) = 2B \Leftrightarrow \log_2(1+c) = 2 \Leftrightarrow c = 2^2 - 1 = 3. \tag{3.9}$$

Hence, we need to triple the transmit power to double the capacity.

Next, we want to find the value of c that gives the capacity $4B$:

$$B\log_2\left(1 + \frac{cP\beta}{BN_0}\right) = 4B \Leftrightarrow \log_2(1+c) = 4 \Leftrightarrow c = 2^4 - 1 = 15. \tag{3.10}$$

Hence, we must transmit 15 times more power to quadruple the capacity.

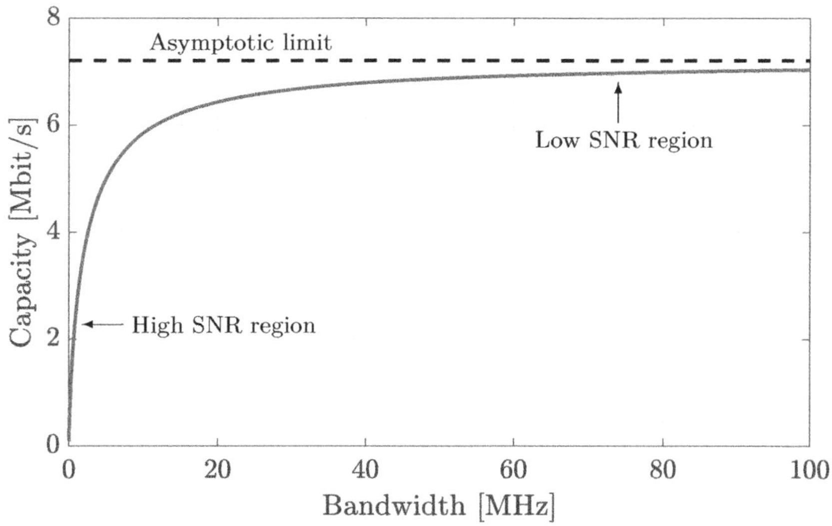

Figure 3.3: The capacity behavior in a single-antenna system when changing the bandwidth B, for $P\beta/N_0 = 5 \cdot 10^6$ Hz.

When varying the bandwidth B, we notice that $C(P, B)$ is a monotonically increasing function also of this variable, which can be shown by taking the first derivative and proving that it is positive. The capacity starts at $C(P, 0) = 0$, which can be shown by taking the limit $B \to 0$. This represents a *high SNR region* where the SNR $\frac{P\beta}{BN_0} \to \infty$, but the performance is anyway low due to the small bandwidth. This also implies that the capacity grows almost linearly when increasing B in the high SNR region since the factor in front of the logarithm in (3.1) grows linearly. However, the logarithm is almost unaffected by a small change in B at high SNR. If we instead consider the case when B is large, we can utilize that we operate in the low SNR region where $\frac{P\beta}{BN_0}$ is small, thus

$$C(P, B) \approx B \log_2(e) \frac{P\beta}{BN_0} = \log_2(e) \frac{P\beta}{N_0}. \tag{3.11}$$

One can prove that $C(P, B) \to \log_2(e)\frac{P\beta}{N_0}$ as $B \to \infty$, so there is an upper limit on how high capacity we can achieve when having a huge bandwidth. The reason is that the fixed transmit power P needs to be divided over the bandwidth, leading to a gradually lower SNR when using more bandwidth. This is directly seen from the signal energy per symbol $q = P/B$ used in Corollary 2.1. Figure 3.3 illustrates these behaviors by showing $C(P, B)$ as a function of B for $P\beta/N_0 = 5 \cdot 10^6$ Hz. The capacity grows linearly with B in the high SNR region but converges to an upper limit in the low SNR region.

With these behaviors in mind, we can conclude how to improve the channel capacity most efficiently in different cases. If we have a system that operates in the high SNR region, the capacity grows linearly with the bandwidth B but relatively slowly with the power P. Since changes in the bandwidth greatly impact the capacity, the high SNR region is called the *bandwidth-limited region*. In contrast, if we have a system that operates in the low SNR region, the capacity grows linearly with the power, while the bandwidth has little impact. Since changes in the power strongly impact the capacity, the low SNR region is called the *power-limited region*. Alternatively, we can increase both P and B while keeping their ratio P/B fixed. In that case, the SNR $\frac{P\beta}{BN_0}$ is constant, and the capacity (3.1) will always be linearly increasing, irrespective of the SNR value. The intuition is that we get more symbols per second, and each can carry the same amount of information since we keep the energy per symbol constant by increasing the transmit power at the same pace as we increase the bandwidth (i.e., the number of symbols per second). For example, if we need to double the capacity of a system, we can achieve that using twice the power and twice the bandwidth. If the original system operates in the power-limited region, we can achieve almost the same capacity gain by only doubling the power. On the other hand, if the original system operates in the bandwidth-limited region, we can achieve almost the same capacity gain by only doubling the bandwidth. However, in general, we need to increase both the power and bandwidth to achieve a significant capacity gain.

3.2 Capacity of SIMO Channels

We now know how the channel capacity is affected by power and bandwidth. To maximize the capacity, the communication systems should be designed to use the maximum available transmit power and bandwidth. This is rather obvious and has been the standard practice for decades. The purpose of multiple antenna communications is to design systems to further enhance the capacity without requiring more transmit power and bandwidth resources.

It is vital to notice that it is not the transmit power P that determines the channel capacity but the received power $P\beta$. If we want to achieve a higher received power, we can increase P. Alternatively, we can use multiple receive antennas to capture a larger share of the transmitted power, thereby increasing β. This case will be considered in this section, where the goal is to characterize the channel capacity when having multiple receive antennas.

A channel with one transmit antenna and multiple receive antennas is called a SIMO channel; see Figure 3.1(b). We denote the number of receive antennas as M. The channel to each receive antenna can be modeled as before, using the discrete memoryless channel model in (2.130). However, the channel responses will generally differ for every antenna, and the additive noise is statistically independent since it is created by randomness in the receiver hardware connected to the respective receive antennas. Hence, the received signal at the mth receive antenna is given by

$$y_m[l] = h_m x[l] + n_m[l], \quad \text{for } m = 1, \dots, M, \tag{3.12}$$

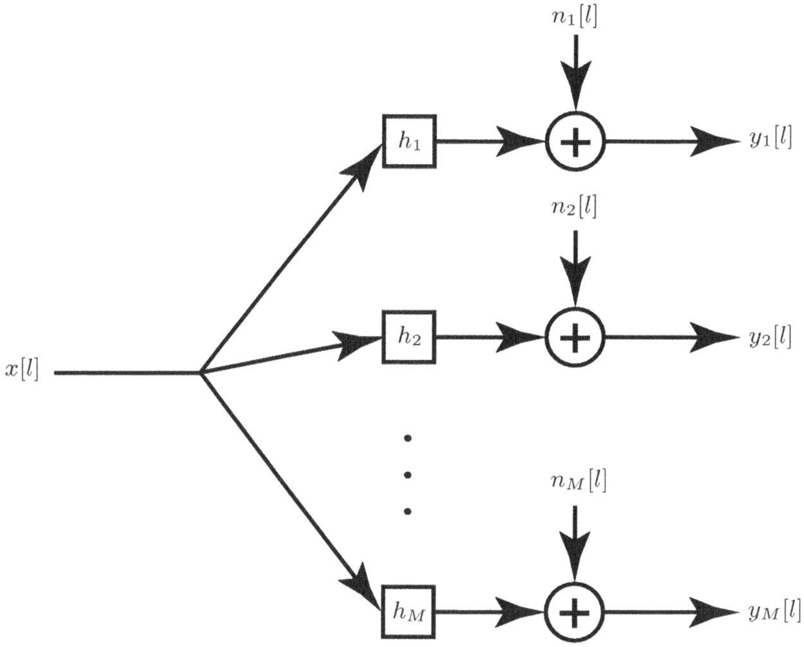

Figure 3.4: A discrete memoryless SIMO channel with the input $x[l]$ and M outputs $y_m[l] = h_m x[l] + n_m[l]$, for $m = 1, \dots, M$, where l is a discrete time index, h_m is the channel response to the mth receive antenna, and $n_m[l]$ is the independent Gaussian receiver noise at that antenna.

where $x[l]$ is the transmitted signal, l is the discrete time index, h_m is the channel response, and $n_m[l] \sim \mathcal{N}_{\mathbb{C}}(0, N_0)$ is the independent receiver noise. Note that the transmitted signal is the same for all m, while all other variables have an antenna index. A block diagram of this discrete memoryless SIMO channel is shown in Figure 3.4. Since this is a memoryless channel, we can just as well neglect the time index l and write the channel in (3.12) as

$$y_m = h_m \cdot x + n_m, \quad \text{for } m = 1, \dots, M. \tag{3.13}$$

Instead of representing the transmission over the SIMO channel using the M equations in (3.13), it is convenient to represent the entire system model in vector form as

$$\mathbf{y} = \mathbf{h}x + \mathbf{n} \tag{3.14}$$

by defining the M-dimensional received signal vector $\mathbf{y}$, the channel vector $\mathbf{h}$, and the noise vector $\mathbf{n}$:

$$\mathbf{y} = \begin{bmatrix} y_1 \\ \vdots \\ y_M \end{bmatrix}, \quad \mathbf{h} = \begin{bmatrix} h_1 \\ \vdots \\ h_M \end{bmatrix}, \quad \mathbf{n} = \begin{bmatrix} n_1 \\ \vdots \\ n_M \end{bmatrix}. \tag{3.15}$$

The geometric relation between these vectors is illustrated in Figure 3.5. The received signal vector $\mathbf{y}$ is the summation of two vectors: $\mathbf{h}x$ and $\mathbf{n}$. The

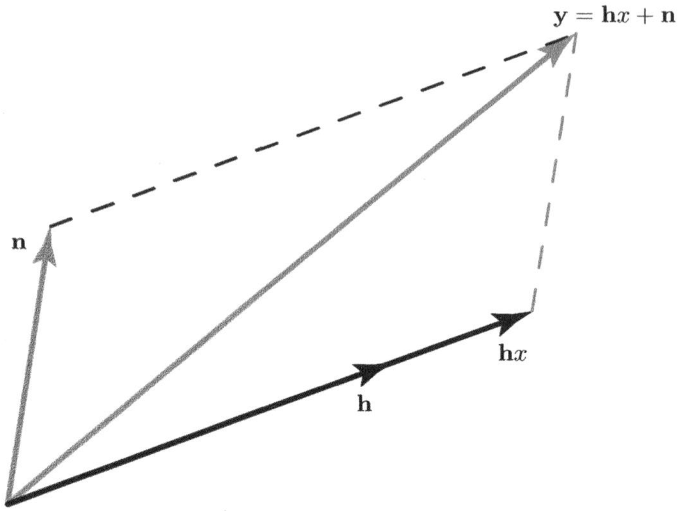

Figure 3.5: The received signal vector $\mathbf{y}$ is the summation of the noise vector $\mathbf{n}$ and the channel vector $\mathbf{h}$ that is multiplied by the data signal x.

former is a vector that points in the same direction as the channel vector $\mathbf{h}$ but is scaled by the unknown data signal x. The latter is the noise vector with independent entries distributed as $\mathcal{N}_{\mathbb{C}}(0, N_0)$. We can express the entire distribution as $\mathbf{n} \sim \mathcal{N}_{\mathbb{C}}(\mathbf{0}, N_0\mathbf{I}_M)$ using the multivariate notation introduced in Section 2.2.4, where $\mathrm{Cov}\{\mathbf{n}\} = N_0\mathbf{I}_M$ is the covariance matrix. The direction $\mathbf{n}/\|\mathbf{n}\|$ of the noise vector is uniformly distributed over all possible directions. The word "direction" refers to the geometry in the M-dimensional vector space $\mathbb{C}^M$ where these vectors reside. There is no simple connection to physical directions in our three-dimensional world, but we will return to the physical modeling of channels in Chapter 4.

The receiver wants to detect the data signal x based on the received signal $\mathbf{y}$. Since the received signal is a vector and the data signal is a scalar, the detection algorithm must somehow include a projection of $\mathbf{y}$ onto a scalar that we call $\hat{x}$. The projection should make $\hat{x}$ as similar to x as possible, and there should be no information loss in the projection. In general, a vector projection is carried out by selecting a unit-length vector $\mathbf{w}$ and computing the inner product $\hat{x} = \mathbf{w}^{\mathrm{H}}\mathbf{y}$. This scalar represents how far in the direction $\mathbf{w}$ that $\mathbf{y}$ points; that is, $\hat{x}$ is the (orthogonal) projection of $\mathbf{y}$ onto $\mathbf{w}$. The vector $\mathbf{w}$ is called the *receive combining vector* when dealing with SIMO channels, and it can also be called the *detection vector* or *receive beamforming vector*.

We want to find the capacity of the SIMO channel in (3.14). As a first step, we will compute an achievable data rate for an arbitrary $\mathbf{w}$ and then identify the best projection, which is the one that gives the channel capacity. We notice that

$$\hat{x} = \mathbf{w}^{\mathrm{H}}\mathbf{y} = \mathbf{w}^{\mathrm{H}}\mathbf{h}x + \mathbf{w}^{\mathrm{H}}\mathbf{n}, \tag{3.16}$$

where $\mathbf{w}^{\mathrm{H}}\mathbf{h}$ is a scalar and $\mathbf{w}^{\mathrm{H}}\mathbf{n} \sim \mathcal{N}_{\mathbb{C}}(0, N_0)$ is the component of the noise that points in the direction of $\mathbf{w}$.[1] Hence, (3.16) is effectively a memoryless SISO channel of the kind in (2.130) with $y = \hat{x}$ and $h = \mathbf{w}^{\mathrm{H}}\mathbf{h}$. It then follows from Corollary 2.1 that an achievable data rate is

$$\log_2\left(1 + \frac{q|\mathbf{w}^{\mathrm{H}}\mathbf{h}|^2}{N_0}\right) \quad \text{bit/symbol,} \tag{3.17}$$

where $q = \mathbb{E}\{|x|^2\} = P/B$ denotes the energy per symbol, which we will refer to as the *symbol power* in the remainder of this book. This variable is proportional to the transmit power P, so when we later optimize the symbol powers of multiple data streams, this is identical to optimizing the transmit powers (measured in Watt, i.e., energy per second).

The value in (3.17) depends on how we select the unit-length vector $\mathbf{w}$. Recall that the Cauchy-Schwarz inequality in (2.18) states that

$$|\mathbf{w}^{\mathrm{H}}\mathbf{h}|^2 \leq \underbrace{\|\mathbf{w}\|^2}_{=1} \|\mathbf{h}\|^2 = \|\mathbf{h}\|^2 \tag{3.18}$$

with equality if and only if $\mathbf{w}$ and $\mathbf{h}$ are parallel. Hence, we can maximize the SNR $\frac{q|\mathbf{w}^{\mathrm{H}}\mathbf{h}|^2}{N_0}$ in (3.17) by selecting the unit-length vector

$$\mathbf{w} = \frac{\mathbf{h}}{\|\mathbf{h}\|} \tag{3.19}$$

that is parallel to $\mathbf{h}$. By inserting (3.19) into (3.17), we obtain the achievable data rate

$$\log_2\left(1 + \frac{q\|\mathbf{h}\|^2}{N_0}\right) \quad \text{bit/symbol.} \tag{3.20}$$

The receive combining vector in (3.19) is called *maximum-ratio combining (MRC)* since it maximizes the SNR. It has also been called the *matched filter* since the combining vector is effectively matched to the channel. Recall from Figure 2.4 that the inner product with a unit-length vector can be interpreted as an orthogonal projection onto that vector. In this case, we take the received signal vector $\mathbf{y}$ and project it onto a unit-length version of the channel vector $\mathbf{h}$, as illustrated in Figure 3.6. Since the received signal contains the data signal with the form $\mathbf{h}x$, the projection will not remove any part of the data signal. The projection will, however, remove the parts of the noise vector $\mathbf{n}$ that point in other directions than $\mathbf{h}$. This noise suppression approach is conceptually similar to the lowpass filtering in Figure 2.13, where the receiver removes the noise in the part of the frequency domain where there is no signal. In the case of MRC, we instead remove noise from the part of the spatial domain where there is no signal.

[1] Since $\mathbf{w}^{\mathrm{H}}\mathbf{n}$ is the weighted sum of independent complex Gaussian distributed random variables, it is also complex Gaussian distributed. Since the mean is zero, the variance is computed as $\mathbb{V}\mathrm{ar}\{\mathbf{w}^{\mathrm{H}}\mathbf{n}\} = \mathbb{E}\{|\mathbf{w}^{\mathrm{H}}\mathbf{n}|^2\} = \mathbf{w}^{\mathrm{H}}\mathbb{E}\{\mathbf{n}\mathbf{n}^{\mathrm{H}}\}\mathbf{w} = N_0\mathbf{w}^{\mathrm{H}}\mathbf{I}_M\mathbf{w} = N_0\|\mathbf{w}\|^2 = N_0$.

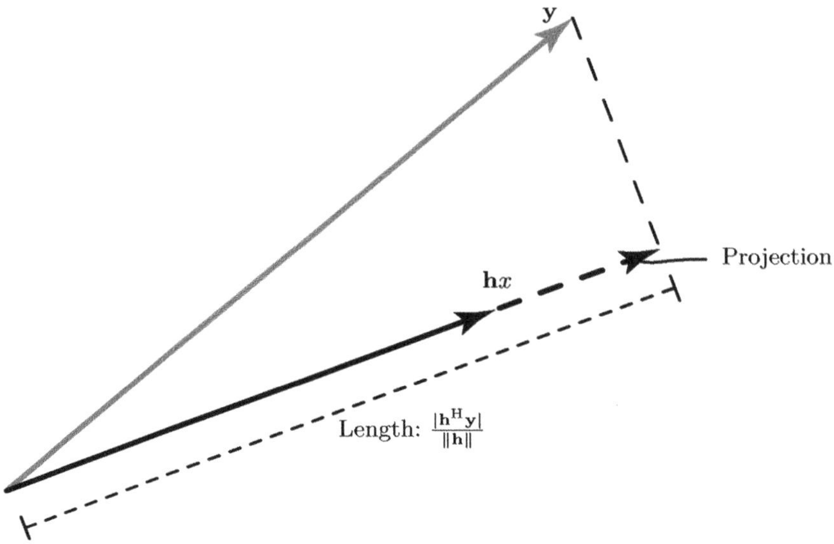

Figure 3.6: To achieve the SIMO capacity, we should use MRC to project the received signal $\mathbf{y}$ orthogonally onto the channel vector $\mathbf{h}$. The data-bearing vector $\mathbf{h}x$ is unaffected by this projection, but the parts of the noise that gave $\mathbf{y}$ another direction will be removed.

In estimation theory, $\frac{\mathbf{h}^{\mathrm{H}}\mathbf{y}}{\|\mathbf{h}\|}$ is called the sufficient statistics for estimating x since the projection removes only parts of the independent noise. Since MRC is the optimal projection, the achievable data rate in (3.20) is the channel capacity of the SIMO channel.

Corollary 3.1. Consider the discrete memoryless point-to-point SIMO channel in Figure 3.4 with the input $x \in \mathbb{C}$ and output $\mathbf{y} \in \mathbb{C}^M$ given by

$$\mathbf{y} = \mathbf{h}x + \mathbf{n}, \tag{3.21}$$

where $\mathbf{n} \sim \mathcal{N}_{\mathbb{C}}(\mathbf{0}, N_0\mathbf{I}_M)$ is independent noise. Suppose the input distribution is feasible whenever the symbol power satisfies $\mathbb{E}\{|x|^2\} \leq q$ and $\mathbf{h} \in \mathbb{C}^M$ is a constant vector known at the output. The channel capacity is

$$C = \log_2\left(1 + \frac{q\|\mathbf{h}\|^2}{N_0}\right) \quad \text{bit/symbol} \tag{3.22}$$

and is achieved when the input is distributed as $x \sim \mathcal{N}_{\mathbb{C}}(0, q)$.

When comparing the SIMO capacity expression in (3.22) with the SISO capacity in (2.145), we notice that the only difference is the channel gain. It is $|h|^2$ in the SISO case and has now been replaced by $\|\mathbf{h}\|^2 = \sum_{m=1}^{M} |h_m|^2$, which is the summation of the individual channel gains to all the M receive antennas. Hence, using multiple receive antennas leads to a beamforming gain compared to having a single antenna. For example, if $h_m = h$ for $m = 1, \ldots, M$, then $\|\mathbf{h}\|^2 = M|h|^2$ and the SNR will grow proportionally to the number of antennas. This is the beamforming gain introduced in Section 1.2.1, and it can be either used to get a larger SNR, or we can reduce q by a factor $1/\|\mathbf{h}\|^2$ to get the same SNR as in the single-antenna case using less transmit power.

As explained in Section 2.4.1, we can express the symbol power as $q = P/B$ and multiply the capacity expression in (3.22) by B to change the unit to bit/s. This leads to the alternative SIMO channel capacity expression

$$C = B \log_2 \left(1 + \frac{P\|\mathbf{h}\|^2}{BN_0} \right) \quad \text{bit/s.} \tag{3.23}$$

Example 3.3. Consider a SIMO system with M antennas and $\mathbf{h} = \sqrt{\beta}[1 \ \ldots \ 1]^{\mathrm{T}}$. What is the capacity C_{SIMO}? Determine the relative capacity gain $C_{\text{SIMO}}/C_{\text{SISO}}$ compared with the capacity C_{SISO} of the corresponding SISO system.

The capacity of this SIMO system is computed using (3.23) as

$$C_{\text{SIMO}} = B \log_2 \left(1 + \frac{P\|\mathbf{h}\|^2}{BN_0} \right) = B \log_2 \left(1 + \frac{PM\beta}{BN_0} \right) \quad \text{bit/s} \tag{3.24}$$

since $\|\mathbf{h}\|^2 = \sum_{m=1}^{M} |h_m|^2 = M\beta$ in this case. The corresponding SISO system with $M = 1$ has the capacity

$$C_{\text{SISO}} = B \log_2 \left(1 + \frac{P\beta}{BN_0} \right) \quad \text{bit/s.} \tag{3.25}$$

The SIMO system provides an M times larger SNR than the SISO system. Using the low-SNR approximation in (3.2), the relative capacity gain becomes

$$\frac{C_{\text{SIMO}}}{C_{\text{SISO}}} \approx \frac{B \log_2(e) \frac{PM\beta}{BN_0}}{B \log_2(e) \frac{P\beta}{BN_0}} = M, \tag{3.26}$$

which grows linearly with the number of antennas and equals the beamforming gain. Using (3.3), the relative capacity gain at high SNR is approximated as

$$\frac{C_{\text{SIMO}}}{C_{\text{SISO}}} \approx \frac{B \log_2 \left(\frac{PM\beta}{BN_0} \right)}{B \log_2 \left(\frac{P\beta}{BN_0} \right)} = 1 + \frac{\log_2(M)}{\log_2 \left(\frac{P\beta}{BN_0} \right)}. \tag{3.27}$$

The relative capacity gain only grows logarithmically at high SNR. The absolute difference becomes $C_{\text{SIMO}} - C_{\text{SISO}} \approx B \log_2(M)$ at high SNR.

In summary, since the beamforming gain increases the received power, the most significant relative capacity gain is achieved in the power-limited region where the SNR is low, while the gain is small in the bandwidth-limited region.

3.2.1 Alternative Combining Vectors

The derivation of MRC relied on the assumption that $\mathbf{w}$ is a unit-length vector, but this condition can be relaxed without changing the final result. For the matter of argument, suppose we select $\mathbf{w} = c\mathbf{h}$ for some arbitrary scaling factor $c \neq 0$. Substituting this vector into (3.16) yields

$$\hat{x} = \mathbf{w}^{\mathrm{H}}\mathbf{y} = c^* \underbrace{\mathbf{h}^{\mathrm{H}}\mathbf{h}}_{=\|\mathbf{h}\|^2} x + \underbrace{c^*\mathbf{h}^{\mathrm{H}}\mathbf{n}}_{\sim\mathcal{N}_{\mathbb{C}}(0,|c|^2\|\mathbf{h}\|^2 N_0)}, \tag{3.28}$$

which is a SISO channel with $h = c^*\|\mathbf{h}\|^2$ and noise with the variance $|c|^2\|\mathbf{h}\|^2 N_0$. It follows from Corollary 2.1 that an achievable data rate is

$$\log_2\left(1 + \frac{q|c|^2\|\mathbf{h}\|^4}{|c|^2\|\mathbf{h}\|^2 N_0}\right) = \log_2\left(1 + \frac{q\|\mathbf{h}\|^2}{N_0}\right) \quad \text{bit/symbol}, \tag{3.29}$$

which equals the capacity in (3.22). Hence, any combining vector parallel to $\mathbf{h}$ can be utilized to achieve the capacity.

In practical implementations, it might be desirable to identify the value of c that minimizes the MSE between the transmitted signal x and its estimate $\hat{x} = c^*\|\mathbf{h}\|^2 x + c^*\mathbf{h}^{\mathrm{H}}\mathbf{n}$ in (3.28):

$$\begin{aligned}
\mathbb{E}\{|x-\hat{x}|^2\} &= \mathbb{E}\left\{\left|x(1-c^*\|\mathbf{h}\|^2) - c^*\mathbf{h}^{\mathrm{H}}\mathbf{n}\right|^2\right\} \\
&\overset{(a)}{=} \mathbb{E}\{|x|^2\}\left|1-c^*\|\mathbf{h}\|^2\right|^2 + \mathbb{E}\{|c^*\mathbf{h}^{\mathrm{H}}\mathbf{n}|^2\} \\
&= q\left(1+|c|^2\|\mathbf{h}\|^4 - c\|\mathbf{h}\|^2 - c^*\|\mathbf{h}\|^2\right) + |c|^2\|\mathbf{h}\|^2 N_0 \\
&\overset{(b)}{=} \left|c - \frac{q}{q\|\mathbf{h}\|^2+N_0}\right|^2 \left(q\|\mathbf{h}\|^4 + N_0\|\mathbf{h}\|^2\right) + \frac{qN_0}{q\|\mathbf{h}\|^2+N_0},
\end{aligned} \tag{3.30}$$

where (a) follows from utilizing the independence between the signal x and the noise $\mathbf{n}$ (which both have zero mean), while (b) follows from completing the squares with respect to the variable c. Since the first term in (3.30) is quadratic, it cannot be negative. Hence, the MSE is minimized by selecting c to make the first term equal to zero, which is achieved by $c = \frac{q}{q\|\mathbf{h}\|^2+N_0}$. This results in the alternative MRC vector

$$\mathbf{w} = \frac{q}{q\|\mathbf{h}\|^2+N_0}\mathbf{h} \tag{3.31}$$

that will simultaneously achieve the capacity and minimize the MSE between the transmitted data symbol and the receiver's estimate $\hat{x}$. This is a suitable scaling factor since many decoding algorithms use Euclidean distances between constellation points and received signals when determining the likelihood of different symbols being transmitted, which is aligned with the MSE being the average squared Euclidean distance. The capacity can be achieved using MRC with any scaling factor $c \neq 0$ because the capacity expression implicitly assumes an optimal receiver, which can compensate for any scaling factor. In general, any receiver processing that is invertible has no impact on capacity.

Example 3.4. Suppose the received signal is $\mathbf{y} = \mathbf{h}x + \mathbf{n}$ as earlier in this section, but the noise is colored in the sense that $\mathbf{n} \sim \mathcal{N}_{\mathbb{C}}(\mathbf{0}, \mathbf{C})$. What is the LMMSE estimate of x given the received signal?

The LMMSE estimator concept was described in Section 2.5.2. It obtains an estimate of x through a linear operation: $\hat{x} = \mathbf{w}^{\mathrm{H}}\mathbf{y}$. Hence, we need to find the combining vector $\mathbf{w}$ that minimizes the MSE, which can be expressed as

$$\begin{aligned}
\mathbb{E}\left\{|x-\hat{x}|^2\right\} &= \mathbb{E}\left\{|x(1-\mathbf{w}^{\mathrm{H}}\mathbf{h}) - \mathbf{w}^{\mathrm{H}}\mathbf{n}|^2\right\} \\
&\overset{(a)}{=} \mathbb{E}\{|x|^2\}|1-\mathbf{w}^{\mathrm{H}}\mathbf{h}|^2 + \mathbb{E}\left\{|\mathbf{w}^{\mathrm{H}}\mathbf{n}|^2\right\} \\
&= q\left(1 + \mathbf{w}^{\mathrm{H}}\mathbf{h}\mathbf{h}^{\mathrm{H}}\mathbf{w} - \mathbf{w}^{\mathrm{H}}\mathbf{h} - \mathbf{h}^{\mathrm{H}}\mathbf{w}\right) + \mathbf{w}^{\mathrm{H}}\mathbf{C}\mathbf{w} \\
&= q + \mathbf{w}^{\mathrm{H}}\underbrace{\left(q\mathbf{h}\mathbf{h}^{\mathrm{H}} + \mathbf{C}\right)}_{=\mathbf{B}}\mathbf{w} - \mathbf{w}^{\mathrm{H}}\underbrace{q\mathbf{h}}_{=\mathbf{a}} - \underbrace{q\mathbf{h}^{\mathrm{H}}}_{=\mathbf{a}^{\mathrm{H}}}\mathbf{w},
\end{aligned} \tag{3.32}$$

where (a) follows from utilizing that the signal and noise are independent. By using the notation $\mathbf{a}$ and $\mathbf{B}$ introduced in (3.32), we can rewrite the MSE as

$$\begin{aligned}
\mathbb{E}\left\{|x-\hat{x}|^2\right\} &= q + \mathbf{w}^{\mathrm{H}}\mathbf{B}\mathbf{w} - \mathbf{w}^{\mathrm{H}}\mathbf{a} - \mathbf{a}^{\mathrm{H}}\mathbf{w} \\
&= q - \mathbf{a}^{\mathrm{H}}\mathbf{B}^{-1}\mathbf{a} + (\mathbf{w}-\mathbf{B}^{-1}\mathbf{a})^{\mathrm{H}}\mathbf{B}(\mathbf{w}-\mathbf{B}^{-1}\mathbf{a})
\end{aligned} \tag{3.33}$$

by completing the squares with respect to the vector $\mathbf{w}$. The last term is then a quadratic form that attains its minimum value of zero if $\mathbf{w} = \mathbf{B}^{-1}\mathbf{a}$. We can utilize the matrix identity in (2.49) to rewrite the expression as

$$\mathbf{w} = \mathbf{B}^{-1}\mathbf{a} = q\left(q\mathbf{h}\mathbf{h}^{\mathrm{H}} + \mathbf{C}\right)^{-1}\mathbf{h} = \frac{q}{q\mathbf{h}^{\mathrm{H}}\mathbf{C}^{-1}\mathbf{h}+1}\mathbf{C}^{-1}\mathbf{h}. \tag{3.34}$$

This vector is called *LMMSE combining* since it minimizes the MSE. It can also be proved to be the capacity-achieving combining scheme for the considered channel. LMMSE combining reduces to the MRC vector in (3.31) in the special case of $\mathbf{C} = N_0\mathbf{I}_M$. The LMMSE combining terminology is usually only used when it differs from conventional MRC; that is, when there is colored noise or interference, which we will come across later in the book. Otherwise, it is referred to as MRC, as earlier in this section.

3.3 Capacity of MISO Channels

We will now consider the opposite scenario of a channel with multiple transmit antennas and a single receive antenna, known as a MISO channel; see Figure 3.1(c). To emphasize the similarities with the SIMO case considered in the previous section, we consider the case when the transmitter and receiver from the SIMO channel have exchanged their roles. Hence, we assume there are M transmit antennas, and the channel response from the transmit antenna m to the receive antenna is denoted by h_m.

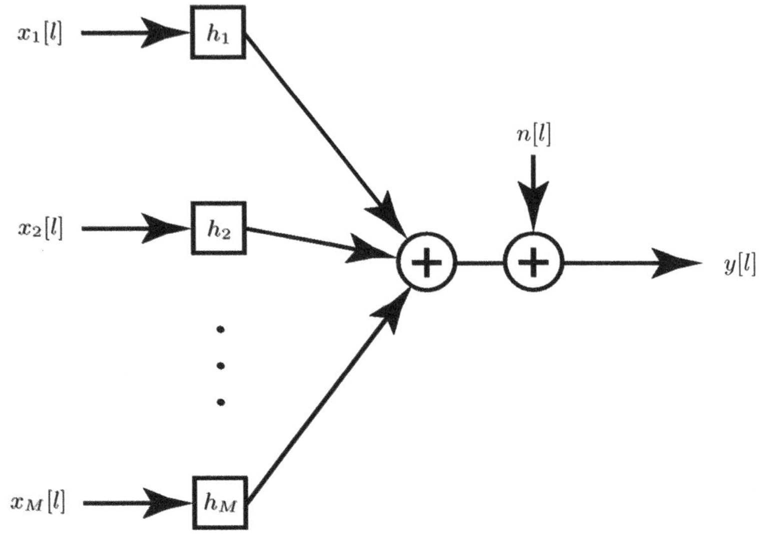

Figure 3.7: A discrete memoryless MISO channel with the inputs $x_m[l]$ for $m = 1, \ldots, M$ and output $y[l] = \sum_{m=1}^{M} h_m x_m[l] + n[l]$, where l is a discrete time index, h_m is the channel response from transmit antenna m, and $n[l]$ is the independent complex Gaussian receiver noise.

The channel from each transmit antenna to the receive antenna can be described by the discrete memoryless channel model in (2.130), but when we put it all together, we get the received signal

$$y[l] = \sum_{m=1}^{M} h_m x_m[l] + n[l], \tag{3.35}$$

where l is the discrete time index, $x_m[l]$ is the transmitted signal from antenna m, h_m is the channel response from transmit antenna m, and $n[l] \sim \mathcal{N}_{\mathbb{C}}(0, N_0)$ is the receiver noise. A block diagram of this discrete memoryless MISO channel is shown in Figure 3.7. Notice that there is only a single noise term and that the signal contributions $h_m x_m[l]$ from the different antennas are added together (superimposed) by the wireless channel. This makes the setup analytically different from the SIMO case. Since (3.35) is a memoryless channel, we can just as well neglect the time index and write the channel as

$$y = \sum_{m=1}^{M} h_m x_m + n. \tag{3.36}$$

To derive the channel capacity, it will be helpful to use the vector notation

$$\mathbf{x} = \begin{bmatrix} x_1 \\ \vdots \\ x_M \end{bmatrix}, \quad \mathbf{h} = \begin{bmatrix} h_1 \\ \vdots \\ h_M \end{bmatrix}, \tag{3.37}$$

where $\mathbf{x}$ is the signal vector and $\mathbf{h}$ is the channel vector. With this notation, we can rewrite the system model in (3.36) as

$$y = \mathbf{h}^{\mathrm{T}}\mathbf{x} + n. \tag{3.38}$$

Two different types of transposes were defined in Section 2.1.1 to be used when dealing with complex vectors and matrices: the conventional *transpose* ^{T} that flips a matrix over its diagonal and the *conjugate transpose* ^{H} that both flips the matrix and replaces each entry with its complex conjugate. The conjugate transpose is probably the most common when dealing with complex vectors/matrices due to its connection to the inner product and norm. Nevertheless, it is a conventional transpose on $\mathbf{h}$ in (3.38) because the physical channels do not give rise to any complex conjugation.[2] Recall from (2.17) that the inner product between two arbitrary complex-valued vectors $\mathbf{a}$ and $\mathbf{b}$ of the same dimension is computed as $\mathbf{a}^{\mathrm{H}}\mathbf{b}$ using the conjugate transpose. Hence, the term $\mathbf{h}^{\mathrm{T}}\mathbf{x}$ in (3.38) is an inner product between $\mathbf{h}^*$ and $\mathbf{x}$.

The M-dimensional signal vector $\mathbf{x}$ should be selected to send data to the receiver. Since the receiver only observes the scalar y, it can only estimate one scalar data-bearing signal based on its observation.[3] Hence, we can, without loss of optimality, select the signal vector as

$$\mathbf{x} = \mathbf{p}\bar{x}, \tag{3.39}$$

where $\mathbf{p}$ is an M-dimensional unit-length vector and $\bar{x}$ is the data signal having the symbol power $\mathbb{E}\{|\bar{x}|^2\} = q$. The vector $\mathbf{p}$ is called the *precoding vector* or *transmit beamforming vector*, and the unit-length requirement means that the total symbol power of the transmitted signal is

$$\mathbb{E}\{\|\mathbf{x}\|^2\} = \mathbb{E}\{\underbrace{\|\mathbf{p}\|^2}_{=1} |\bar{x}|^2\} = \mathbb{E}\{|\bar{x}|^2\} = q, \tag{3.40}$$

independently of how many antennas are used. This effectively means that the more transmit antennas are used, the less power is transmitted from each one of the antennas. By substituting (3.39) into (3.38), we obtain

$$y = \mathbf{h}^{\mathrm{T}}\mathbf{p}\bar{x} + n, \tag{3.41}$$

where $\mathbf{h}^{\mathrm{T}}\mathbf{p}$ is a scalar. This scalar is the inner product between the conjugate $\mathbf{h}^*$ of the channel and the precoding vector $\mathbf{p}$. Hence, (3.41) is effectively

---

[2]Many other textbooks on multiple antenna communications, however, write (3.38) as $y = \mathbf{h}^{\mathrm{H}}\mathbf{x} + n$ since the use of a conjugate transpose makes the analysis/notation slightly simpler. The downside with that approach is that the obtained algorithms cannot be directly applied to a practical system, but we must first compensate for the conjugation.

[3]It is theoretically possible to send more than one data-bearing signal to a single-antenna receiver, but it can be proved that this will not increase the capacity of the system since the channel will add these signals together when taking the inner product $\mathbf{h}^{\mathrm{T}}\mathbf{x}$.

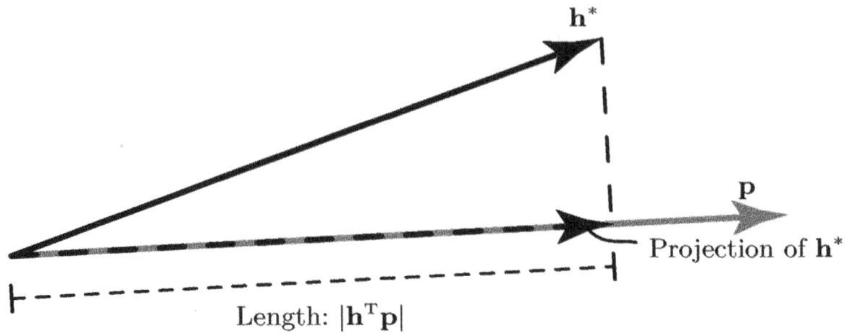

Figure 3.8: A MISO channel projects the channel vector $\mathbf{h}^*$ onto the unit-length precoding vector $\mathbf{p}$, and it is $|\mathbf{h}^{\mathrm{T}}\mathbf{p}|$ that determines the SNR. Hence, if the precoding vector $\mathbf{p}$ is not parallel to the channel vector $\mathbf{h}^*$, the SNR will be the same as if a shorter channel vector $|\mathbf{h}^{\mathrm{T}}\mathbf{p}|\mathbf{p}$ parallel to $\mathbf{p}$ was used.

a memoryless SISO channel of the kind in (2.130) with $h = \mathbf{h}^{\mathrm{T}}\mathbf{p}$ and noise variance N_0. It then follows from Corollary 2.1 that an achievable data rate is

$$\log_2\left(1 + \frac{q|\mathbf{h}^{\mathrm{T}}\mathbf{p}|^2}{N_0}\right) \quad \text{bit/symbol.} \tag{3.42}$$

To obtain the channel capacity, it remains to identify the precoding vector that maximizes (3.42), which corresponds to maximizing $|\mathbf{h}^{\mathrm{T}}\mathbf{p}|^2$. As in the last section, we can utilize the Cauchy-Schwarz inequality from (2.18), which states that

$$|\mathbf{h}^{\mathrm{T}}\mathbf{p}|^2 \leq \|\mathbf{h}^*\|^2 \underbrace{\|\mathbf{p}\|^2}_{=1} = \|\mathbf{h}\|^2 \tag{3.43}$$

with equality if and only if $\mathbf{h}^*$ and $\mathbf{p}$ are parallel. Note that $\|\mathbf{h}^*\|^2 = \|\mathbf{h}\|^2$ since the conjugate only changes the phase of the entries, not their magnitudes. Hence, we can maximize the SNR $\frac{q|\mathbf{h}^{\mathrm{T}}\mathbf{p}|^2}{N_0}$ in (3.42) by selecting the precoding vector as

$$\mathbf{p} = \frac{\mathbf{h}^*}{\|\mathbf{h}\|}, \tag{3.44}$$

which is a unit-length vector parallel to $\mathbf{h}^*$. This precoding gives the achievable data rate

$$\log_2\left(1 + \frac{q\|\mathbf{h}\|^2}{N_0}\right) \quad \text{bit/symbol.} \tag{3.45}$$

The precoding vector in (3.44) is called *maximum-ratio transmission (MRT)* since it maximizes the SNR. It has also been called *conjugate beamforming* since the precoding vector is selected based on the complex conjugate of the channel vector. This selection of the precoding vector is intuitive if we look at it geometrically as in Figure 3.8: $|\mathbf{h}^{\mathrm{T}}\mathbf{p}|$ is the length of the effective channel vector that is obtained when orthogonally projecting $\mathbf{h}^*$ onto $\mathbf{p}$. This vector has only the same length as $\mathbf{h}^*$ (i.e., same norm) when $\mathbf{h}^*$ and $\mathbf{p}$ are parallel, which is the case with MRT. For the matter of argument, suppose we select another precoding vector $\mathbf{p}$ that is not parallel to $\mathbf{h}^*$. The component of this precoding vector that is orthogonal to the conjugate of the channel (in the vector space $\mathbb{C}^M$) will vanish when taking the inner product $\mathbf{h}^{\mathrm{T}}\mathbf{p}$ and the corresponding transmit power is lost. In conclusion, MRT is the optimal precoding and the channel capacity is the achievable data rate in (3.45).

Corollary 3.2. Consider the discrete memoryless point-to-point MISO channel in Figure 3.7 with the input $\mathbf{x} \in \mathbb{C}^M$ and output $y \in \mathbb{C}$ given by

$$y = \mathbf{h}^{\mathrm{T}}\mathbf{x} + n, \tag{3.46}$$

where $n \sim \mathcal{N}_{\mathbb{C}}(0, N_0)$ is independent noise. Suppose the input distribution is feasible whenever the symbol power satisfies $\mathbb{E}\{\|\mathbf{x}\|^2\} \leq q$ and $\mathbf{h} \in \mathbb{C}^M$ is a constant vector known at the output. The channel capacity is

$$C = \log_2\left(1 + \frac{q\|\mathbf{h}\|^2}{N_0}\right) \quad \text{bit/symbol} \tag{3.47}$$

and is achieved when the input is $\mathbf{x} = \frac{\mathbf{h}^*}{\|\mathbf{h}\|}\bar{x}$ with $\bar{x} \sim \mathcal{N}_{\mathbb{C}}(0, q)$.

Comparing the MISO channel capacity in (3.47) with the capacity expression in (3.22) of the corresponding SIMO channel, we notice that these are identical. Hence, the benefit of transmitting from M antennas is that the channel gain $\|\mathbf{h}\|^2 = \sum_{m=1}^{M} |h_m|^2$ becomes the sum of the channel gains of the individual antennas. If $h_m = h$ for $m = 1, \ldots, M$, then $\|\mathbf{h}\|^2 = M|h|^2$ and the SNR is precisely proportional to the number of antennas. This gain is achieved by directing the transmission towards the receiver, as illustrated in Figure 1.17 and Figure 1.19. Another similarity is that the capacity-achieving combining and precoding vectors, called MRC and MRT, respectively, are equal except for a complex conjugate:

$$\mathbf{w} = \mathbf{p}^*. \tag{3.48}$$

In fact, MRT and MRC process the channel vector identically, so the conjugate in (3.48) is merely due to notational differences: the combining vector is applied as $\mathbf{w}^{\mathrm{H}}\mathbf{h}$ with a conjugate transpose, while the precoding vector is applied as $\mathbf{h}^{\mathrm{T}}\mathbf{p}$ without a conjugate so it needs to be placed in $\mathbf{p}$ beforehand.

Even if the channel capacities are equal, there are essential differences between the SIMO and MISO channels. When transmitting from M antennas to a single-antenna receiver, the transmit power is directed towards that receiver, as illustrated in Figure 1.17 and Figure 1.19. MRT basically selects the time delays of the different signals to achieve constructive interference at the point of the receiver; thus, the radiated signal resembles that of a directive transmit antenna but with the critical difference that the directivity is adapted to the channel. The precoding and directivity of the transmission will change when the channel changes, which cannot happen when using a directive antenna. In contrast, when a single-antenna device transmits to a receiver equipped with M antennas, the emitted signal propagates isotropically as illustrated in Figure 1.16 (or according to some other fixed antenna gain function, such as the one in Figure 1.10). Each receive antenna observes one component of the signal in additive noise with variance N_0. MRC combines the signal components constructively, while the noise components are neither constructively nor destructively combined, so the resulting noise term $\mathbf{w}^{\mathrm{H}}\mathbf{n}$ still has variance N_0. The combining creates a spatially directive reception resembling that of a directive receive antenna but with the vital difference that the directivity is adapted to the direction of the arriving signal.

Example 3.5. Is the MRT vector $\mathbf{p} = \frac{\mathbf{h}^*}{\|\mathbf{h}\|}$ unique, or are there capacity-achieving alternatives similar to the alternative MRC vectors in Section 3.2.1?

The precoding vector is selected under the constraint that $\|\mathbf{p}\| = 1$, which ensures that the symbol power equals the power of the signal $\bar{x}$. This is a crucial difference from the selection of combining vectors, which can be scaled arbitrarily since the scaling factor affects the signal and noise identically. However, there is still some flexibility in the MRT vector. The derivation in (3.43) is based on the Cauchy-Schwartz inequality where the maximum value is achieved when $\mathbf{h}^*$ and $\mathbf{p}$ are parallel. All the unit-norm vectors that satisfy this condition are MRT vectors and can be expressed as $\mathbf{p} = e^{\mathrm{j}\phi}\frac{\mathbf{h}^*}{\|\mathbf{h}\|}$, where the common phase-shift $\phi \in [-\pi, \pi)$ can be selected arbitrarily.

MRT effectively turns a MISO channel into a SISO channel with an improved SNR, and the same applies when using MRC in SIMO channels. Hence, in practice, the data encoding and decoding can be carried out like in SISO systems. For example, Figure 2.18 gave an example of 28 data rates that can be achieved by selecting different MCS combinations in 5G NR. When the capacity has been computed using the expressions provided in this chapter, we can identify the closest smaller data rate in the table and use that MCS. The same table can be utilized irrespective of how many antennas are utilized or whether it is a SIMO or MISO channel. In fact, a base station can hide the fact that it is equipped with multiple antennas from the user devices, which has the positive side-effect that one can add beamforming functionalities into existing systems without changing the fundamental communication protocols.

As explained in Section 2.4.1, we can also express the symbol power as $q = P/B$ and multiply the capacity expression in (3.47) with B to change the unit to bit/s. This leads to the alternative but equivalent way to write the capacity of a MISO channel as

$$C = B\log_2\left(1 + \frac{P\|\mathbf{h}\|^2}{BN_0}\right) \quad \text{bit/s}. \tag{3.49}$$

Example 3.6. Suppose we would transmit the signal $\mathbf{x} = \mathbf{p}_1\bar{x}_1 + \mathbf{p}_2\bar{x}_2$, where $\mathbf{p}_1, \mathbf{p}_2$ are two unit-norm precoding vectors and $\bar{x}_1, \bar{x}_2 \sim \mathcal{N}_{\mathbb{C}}(0, q/2)$ are independent data signals containing half the power. How large data rate can we achieve over a MISO channel? Can we achieve the capacity?

The received signal in (3.38) now becomes

$$y = \mathbf{h}^{\mathrm{T}}(\mathbf{p}_1\bar{x}_1 + \mathbf{p}_2\bar{x}_2) + n = \mathbf{h}^{\mathrm{T}}\mathbf{p}_1\bar{x}_1 + \mathbf{h}^{\mathrm{T}}\mathbf{p}_2\bar{x}_2 + n. \tag{3.50}$$

We need to detect the signal $\bar{x}_1$ under the independent additive distortion $\mathbf{h}^{\mathrm{T}}\mathbf{p}_2\bar{x}_2 + n \sim \mathcal{N}_{\mathbb{C}}(0, \frac{q}{2}\|\mathbf{h}^{\mathrm{T}}\mathbf{p}_2\|^2 + N_0)$ with both interference from $\bar{x}_2$ and noise. Since $\bar{x}_2$ is unknown, it is indistinguishable from the noise, and we can achieve a data rate similar to (3.42) but by using the noise variance $\frac{q}{2}\|\mathbf{h}^{\mathrm{T}}\mathbf{p}_2\|^2 + N_0$:

$$R_1 = \log_2\left(1 + \frac{\frac{q}{2}\|\mathbf{h}^{\mathrm{T}}\mathbf{p}_1\|^2}{\frac{q}{2}\|\mathbf{h}^{\mathrm{T}}\mathbf{p}_2\|^2 + N_0}\right) \quad \text{bit/symbol.} \tag{3.51}$$

Now when we have decoded the data contained in $\bar{x}_1$, we know the term $\mathbf{h}^{\mathrm{T}}\mathbf{p}_1\bar{x}_1$ in (3.50) and can subtract it from the received signal: $y - \mathbf{h}^{\mathrm{T}}\mathbf{p}_1\bar{x}_1 = \mathbf{h}^{\mathrm{T}}\mathbf{p}_2\bar{x}_2 + n$. This residual received signal is of the kind in (3.41), and the data rate that we can achieve when extracting the data contained in $\bar{x}_2$ is

$$R_2 = \log_2\left(1 + \frac{\frac{q}{2}\|\mathbf{h}^{\mathrm{T}}\mathbf{p}_2\|^2}{N_0}\right) \quad \text{bit/symbol.} \tag{3.52}$$

The total data rate of this system is

$$\begin{aligned}
R_1 + R_2 &= \log_2\left(1 + \frac{\frac{q}{2}\|\mathbf{h}^{\mathrm{T}}\mathbf{p}_1\|^2}{\frac{q}{2}\|\mathbf{h}^{\mathrm{T}}\mathbf{p}_2\|^2 + N_0}\right) + \log_2\left(1 + \frac{\frac{q}{2}\|\mathbf{h}^{\mathrm{T}}\mathbf{p}_2\|^2}{N_0}\right) \\
&= \log_2\left(\frac{\frac{q}{2}\|\mathbf{h}^{\mathrm{T}}\mathbf{p}_1\|^2 + \frac{q}{2}\|\mathbf{h}^{\mathrm{T}}\mathbf{p}_2\|^2 + N_0}{\frac{q}{2}\|\mathbf{h}^{\mathrm{T}}\mathbf{p}_2\|^2 + N_0} \, \frac{\frac{q}{2}\|\mathbf{h}^{\mathrm{T}}\mathbf{p}_2\|^2 + N_0}{N_0}\right) \\
&= \log_2\left(1 + \frac{\frac{q}{2}\|\mathbf{h}^{\mathrm{T}}\mathbf{p}_1\|^2 + \frac{q}{2}\|\mathbf{h}^{\mathrm{T}}\mathbf{p}_2\|^2}{N_0}\right) \\
&\leq \log_2\left(1 + \frac{\frac{q}{2}\|\mathbf{h}\|^2 + \frac{q}{2}\|\mathbf{h}\|^2}{N_0}\right) = \log_2\left(1 + \frac{q\|\mathbf{h}\|^2}{N_0}\right),
\end{aligned} \tag{3.53}$$

where the upper bound is achieved by recalling that MRT with $\mathbf{p}_1 = \mathbf{p}_2 = \frac{\mathbf{h}^*}{\|\mathbf{h}\|}$ has the largest inner product with the conjugate of the channel vector. The rate expression in (3.53) coincides with the capacity in (3.47). Hence, we have identified an alternative way to achieve the capacity, but it is more complicated since we transmit two independent data signals and decode them sequentially. Hence, the solution in Corollary 3.2 is preferable.

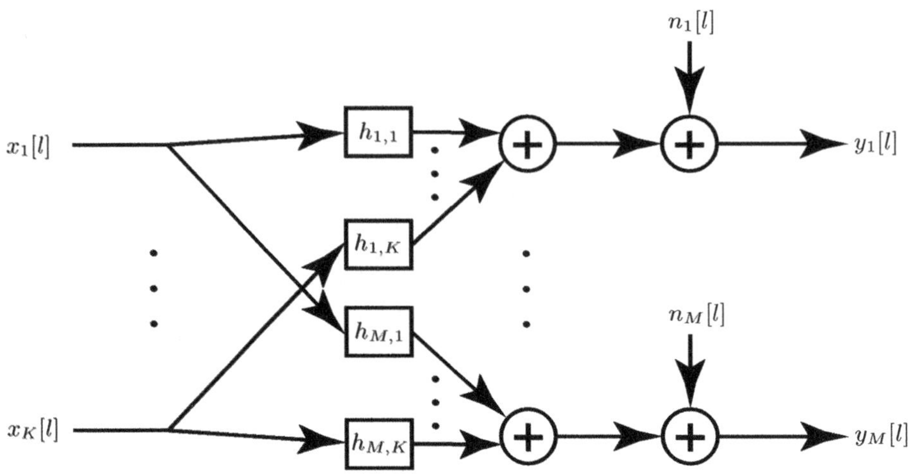

Figure 3.9: A discrete memoryless MIMO channel with the inputs $x_k[l]$ for $k = 1, \ldots, K$ and outputs $y_m[l] = \sum_{k=1}^{K} h_{m,k}x_k[l] + n_m[l]$ for $m = 1, \ldots, M$, where l is a discrete-time index, $h_{m,k}$ is the channel response from transmit antenna k to receive antenna m, and $n_m[l]$ is the independent complex Gaussian receiver noise at receive antenna m.

3.4 Capacity of MIMO Channels

We will conclude this chapter by considering the most general point-to-point scenario: the MIMO channel illustrated in Figure 3.1(d). We assume there are K transmit antennas and M receive antennas; thus, we need two indices to denote each channel response: $h_{m,k} \in \mathbb{C}$ is the channel response from transmit antenna k to receive antenna m, for $k = 1, \ldots, K$ and $m = 1, \ldots, M$. By modeling the channel between each transmit antenna and receive antenna using the discrete memoryless channel model in (2.130), the received signal at antenna m becomes

$$y_m[l] = \sum_{k=1}^{K} h_{m,k}x_k[l] + n_m[l], \quad \text{for } m = 1, \ldots, M, \tag{3.54}$$

where l is the discrete time index, $x_k[l]$ is the transmitted signal from antenna k, and $n_m[l] \sim \mathcal{N}_{\mathbb{C}}(0, N_0)$ is the receiver noise that is independent across antennas. A block diagram of the MIMO channel in (3.54) is shown in Figure 3.9. Note that the SISO, SIMO, and MISO channels are all special cases of the MIMO channel considered in this section.

To derive the MIMO channel capacity, we need to utilize all the M received signals $y_1[l], \ldots, y_M[l]$ for joint signal detection, which calls for a vector/matrix representation of (3.54). If we use the memoryless channel property to drop

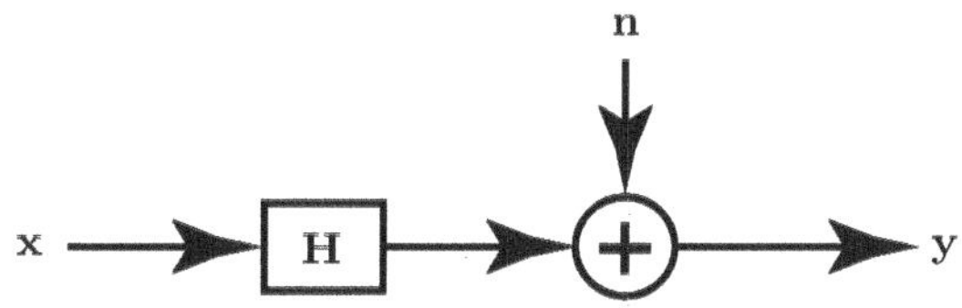

Figure 3.10: A discrete memoryless MIMO channel with vector input $\mathbf{x} \in \mathbb{C}^K$ and vector output $\mathbf{y} \in \mathbb{C}^M$. The channel is characterized by the $M \times K$ channel matrix $\mathbf{H}$ and the receiver noise vector $\mathbf{n} \in \mathbb{C}^M$, which contains M independent complex Gaussian variables. This block diagram is equivalent to the one in Figure 3.9 but uses the vector/matrix notation.

the time index l, the complete received signal at an arbitrary time instance is

$$\begin{bmatrix} y_1 \\ \vdots \\ y_M \end{bmatrix} = \begin{bmatrix} \sum_{k=1}^{K} h_{1,k} x_k \\ \vdots \\ \sum_{k=1}^{K} h_{M,k} x_k \end{bmatrix} + \begin{bmatrix} n_1 \\ \vdots \\ n_M \end{bmatrix}$$
$$= \begin{bmatrix} h_{1,1} & \dots & h_{1,K} \\ \vdots & \ddots & \vdots \\ h_{M,1} & \dots & h_{M,K} \end{bmatrix} \begin{bmatrix} x_1 \\ \vdots \\ x_K \end{bmatrix} + \begin{bmatrix} n_1 \\ \vdots \\ n_M \end{bmatrix}. \tag{3.55}$$

This system model can be written in a concise matrix form as

$$\mathbf{y} = \mathbf{H}\mathbf{x} + \mathbf{n} \tag{3.56}$$

by defining the $M \times K$ channel matrix

$$\mathbf{H} = \begin{bmatrix} h_{1,1} & \dots & h_{1,K} \\ \vdots & \ddots & \vdots \\ h_{M,1} & \dots & h_{M,K} \end{bmatrix} \tag{3.57}$$

and the vectors

$$\mathbf{y} = \begin{bmatrix} y_1 \\ \vdots \\ y_M \end{bmatrix}, \quad \mathbf{x} = \begin{bmatrix} x_1 \\ \vdots \\ x_K \end{bmatrix}, \quad \mathbf{n} = \begin{bmatrix} n_1 \\ \vdots \\ n_M \end{bmatrix}. \tag{3.58}$$

Note that the transmitted data signal vector $\mathbf{x}$ is K-dimensional since there are K transmit antennas, while the received signal vector $\mathbf{y}$ and the noise vector $\mathbf{n}$ are M-dimensional since there are M receive antennas. Since the noise terms are independent, the noise vector $\mathbf{n}$ has the distribution $\mathbf{n} \sim \mathcal{N}_{\mathbb{C}}(\mathbf{0}, N_0\mathbf{I}_M)$. Figure 3.10 shows a block diagram of (3.56) that is equivalent to Figure 3.9 but uses the matrix/vector notation, which makes it more concise.

The main goal of this section is to compute the channel's capacity from $\mathbf{x}$ to $\mathbf{y}$ under a constraint on the maximum symbol power. We let q denote the total symbol power of all antennas, which implies that $\mathbb{E}\{\|\mathbf{x}\|^2\} = q$ where the mean is computed since the data signal vector $\mathbf{x}$ is random. The matrix form in (3.56) invites to apply linear algebra results to determine how the transmitter and receiver should process their signals. We will use the following matrix factorization, called the *singular-value decomposition (SVD)* [49].

Lemma 3.1. Every complex $M \times K$ matrix $\mathbf{H}$ can be factorized as

$$\mathbf{H} = \mathbf{U}\boldsymbol{\Sigma}\mathbf{V}^{\mathrm{H}}, \tag{3.59}$$

where $\mathbf{U}$ is a unitary[a] $M \times M$ matrix containing the eigenvectors of $\mathbf{H}\mathbf{H}^{\mathrm{H}}$, $\mathbf{V}$ is a unitary $K \times K$ matrix containing the eigenvectors of $\mathbf{H}^{\mathrm{H}}\mathbf{H}$, and $\boldsymbol{\Sigma}$ is a rectangular $M \times K$ diagonal matrix[b] with the real numbers $s_1 \geq \ldots \geq s_{\min(M,K)} \geq 0$ on the diagonal.

[a]Unitary matrices are described in Definition 2.4.

[b]A rectangular diagonal matrix of size $M \times K$ can be viewed as a diagonal matrix of size $\min(M,K) \times \min(M,K)$ that has been appended with zeros to become an $M \times K$ matrix.

The SVD can factorize an arbitrary matrix using two specific unitary matrices, $\mathbf{U}$ and $\mathbf{V}$, whose columns are called the left and right *singular vectors*. The non-negative numbers $s_1, \ldots, s_{\min(M,K)}$ are assumed to be ordered in decreasing order and are called the *singular values* of $\mathbf{H}$.

Example 3.7. Compute $\mathbf{H}\mathbf{H}^{\mathrm{H}}$ and $\mathbf{H}^{\mathrm{H}}\mathbf{H}$ using the SVD of $\mathbf{H}$ from (3.59). How are eigenvalues of $\mathbf{H}\mathbf{H}^{\mathrm{H}}$ and $\mathbf{H}^{\mathrm{H}}\mathbf{H}$ related to the singular values of $\mathbf{H}$?

We can express $\mathbf{H}\mathbf{H}^{\mathrm{H}}$ using the SVD of $\mathbf{H}$ from (3.59) as

$$\mathbf{H}\mathbf{H}^{\mathrm{H}} = \mathbf{U}\boldsymbol{\Sigma}\mathbf{V}^{\mathrm{H}}(\mathbf{U}\boldsymbol{\Sigma}\mathbf{V}^{\mathrm{H}})^{\mathrm{H}} = \mathbf{U}\boldsymbol{\Sigma}\underbrace{\mathbf{V}^{\mathrm{H}}\mathbf{V}}_{=\mathbf{I}_K}\boldsymbol{\Sigma}^{\mathrm{H}}\mathbf{U}^{\mathrm{H}} = \mathbf{U}\boldsymbol{\Sigma}\boldsymbol{\Sigma}^{\mathrm{H}}\mathbf{U}^{\mathrm{H}}, \tag{3.60}$$

where we utilized that $\mathbf{V}$ is a unitary matrix. We notice that $\boldsymbol{\Sigma}\boldsymbol{\Sigma}^{\mathrm{H}}$ is a diagonal matrix, thus, $\mathbf{U}\boldsymbol{\Sigma}\boldsymbol{\Sigma}^{\mathrm{H}}\mathbf{U}^{\mathrm{H}}$ fits the eigendecomposition form in Lemma 2.1. Hence, $\mathbf{U}$ contains the orthonormal eigenvectors of $\mathbf{H}\mathbf{H}^{\mathrm{H}}$ and the $M \times M$ diagonal matrix $\boldsymbol{\Sigma}\boldsymbol{\Sigma}^{\mathrm{H}}$ contains the real-valued eigenvalues $s_1^2 \geq \ldots \geq s_{\min(M,K)}^2 \geq 0$, and an additional $M - \min(M,K)$ zero-valued eigenvalues if $M > \min(M,K)$.

Similarly, we can express $\mathbf{H}^{\mathrm{H}}\mathbf{H}$ using the SVD of $\mathbf{H}$ from (3.59) as

$$\mathbf{H}^{\mathrm{H}}\mathbf{H} = (\mathbf{U}\boldsymbol{\Sigma}\mathbf{V}^{\mathrm{H}})^{\mathrm{H}}\mathbf{U}\boldsymbol{\Sigma}\mathbf{V}^{\mathrm{H}} = \mathbf{V}\boldsymbol{\Sigma}^{\mathrm{H}}\underbrace{\mathbf{U}^{\mathrm{H}}\mathbf{U}}_{=\mathbf{I}_M}\boldsymbol{\Sigma}\mathbf{V}^{\mathrm{H}} = \mathbf{V}\boldsymbol{\Sigma}^{\mathrm{H}}\boldsymbol{\Sigma}\mathbf{V}^{\mathrm{H}}, \tag{3.61}$$

which we identify as the eigendecomposition of $\mathbf{H}^{\mathrm{H}}\mathbf{H}$. The unitary matrix $\mathbf{V}$ contains the orthonormal eigenvectors and the $K \times K$ diagonal matrix $\boldsymbol{\Sigma}^{\mathrm{H}}\boldsymbol{\Sigma}$ contains the real-valued eigenvalues, which are $s_1^2 \geq \ldots \geq s_{\min(M,K)}^2 \geq 0$ and the additional $K - \min(M,K)$ zero eigenvalues if $K - \min(M,K) > 0$.

The SVD can be viewed as a generalization of the conventional eigendecomposition and can be used to diagonalize any matrix. In contrast, only some square matrices can be diagonalized using the eigendecomposition. For Hermitian square matrices, the SVD coincides with the eigendecomposition in Lemma 2.1, in the sense that $\mathbf{U} = \mathbf{V}$ contains the eigenvectors and $\mathbf{\Sigma}$ contains the corresponding eigenvalues.

The last example demonstrates a way to derive the singular values of $\mathbf{H}$:

1. Compute either $\mathbf{H}\mathbf{H}^{\mathrm{H}}$ or $\mathbf{H}^{\mathrm{H}}\mathbf{H}$ (preferably the one resulting in the smallest matrix dimensions) and call it $\mathbf{A}$;
2. Compute the eigenvalues of $\mathbf{A}$ by finding the roots to its characteristic polynomial $\det(\mathbf{A} - \lambda\mathbf{I})$;
3. Obtain the singular values by taking the square root of the eigenvalues.

The SVD has the same structure for any matrix but with different values in $\mathbf{U}$, $\mathbf{\Sigma}$, and $\mathbf{V}$. To derive the MIMO channel capacity, we specifically utilize the SVD $\mathbf{H} = \mathbf{U}\mathbf{\Sigma}\mathbf{V}^{\mathrm{H}}$ to the channel matrix in (3.57). Suppose the transmitter creates its transmit signal as $\mathbf{x} = \mathbf{V}\bar{\mathbf{x}}$ for some $\bar{\mathbf{x}}$, while the receiver processes its received signal $\mathbf{y}$ by multiplying it with $\mathbf{U}^{\mathrm{H}}$ to obtain $\bar{\mathbf{y}} = \mathbf{U}^{\mathrm{H}}\mathbf{y}$. It then follows that

$$\begin{aligned}\bar{\mathbf{y}} &= \mathbf{U}^{\mathrm{H}}\mathbf{H}\mathbf{x} + \mathbf{U}^{\mathrm{H}}\mathbf{n}\\ &= \mathbf{U}^{\mathrm{H}}\mathbf{U}\mathbf{\Sigma}\mathbf{V}^{\mathrm{H}}\mathbf{V}\bar{\mathbf{x}} + \mathbf{U}^{\mathrm{H}}\mathbf{n}\\ &= \mathbf{\Sigma}\bar{\mathbf{x}} + \bar{\mathbf{n}}, \end{aligned} \tag{3.62}$$

where we defined $\bar{\mathbf{n}} = \mathbf{U}^{\mathrm{H}}\mathbf{n} \sim \mathcal{N}_{\mathbb{C}}(\mathbf{0}, N_0\mathbf{I}_M)$ and notice that this "rotated" noise vector has the same distribution as $\mathbf{n}$.[4] The last equality in (3.62) utilizes that $\mathbf{U}^{\mathrm{H}}\mathbf{U} = \mathbf{I}_M$ and $\mathbf{V}^{\mathrm{H}}\mathbf{V} = \mathbf{I}_K$ for unitary matrices. The proposed transmitter and receiver processing is non-destructive, meaning that we can get $\mathbf{y}$ back by computing $\mathbf{U}\bar{\mathbf{y}}$ (since $\mathbf{U}\mathbf{U}^{\mathrm{H}} = \mathbf{I}_M$). In contrast, any vector $\mathbf{x}$ can be expressed as $\mathbf{V}\bar{\mathbf{x}}$ by selecting $\bar{\mathbf{x}} = \mathbf{V}^{\mathrm{H}}\mathbf{x}$. Hence, there is no loss of information when going from (3.56) to (3.62), and the channel capacities must be identical. However, (3.62) will be more convenient to analyze since $\mathbf{\Sigma}$ is a (rectangular) diagonal matrix.

Let r denote the number of non-zero singular values of $\mathbf{\Sigma}$, which is equal to the rank of $\mathbf{\Sigma}$ (and $\mathbf{H}$). This means that $s_1 > 0, \ldots, s_r > 0$ while the remaining singular values are zero. It follows that $r \leq \min(M, K)$ since $\mathbf{H}$ has $\min(M, K)$ singular values. If $r < \min(M, K)$, it holds that $s_{r+1} = \ldots = s_{\min(M,K)} = 0$. By utilizing r and the fact that $\mathbf{\Sigma}$ is a rectangular diagonal matrix, we can write (3.62) in scalar form as

$$\bar{y}_k = \begin{cases} s_k \bar{x}_k + \bar{n}_k, & \text{if } k = 1, \ldots, r, \\ \bar{n}_k, & \text{if } k = r+1, \ldots, M, \end{cases} \tag{3.63}$$

[4]This can be proved by computing the covariance matrix of $\bar{\mathbf{n}}$ as $\mathrm{Cov}\{\bar{\mathbf{n}}\} = \mathbb{E}\{\bar{\mathbf{n}}\bar{\mathbf{n}}^{\mathrm{H}}\} = \mathbf{U}^{\mathrm{H}}\mathbb{E}\{\mathbf{n}\mathbf{n}^{\mathrm{H}}\}\mathbf{U} = N_0\mathbf{U}^{\mathrm{H}}\mathbf{I}_M\mathbf{U} = N_0\mathbf{I}_M$.

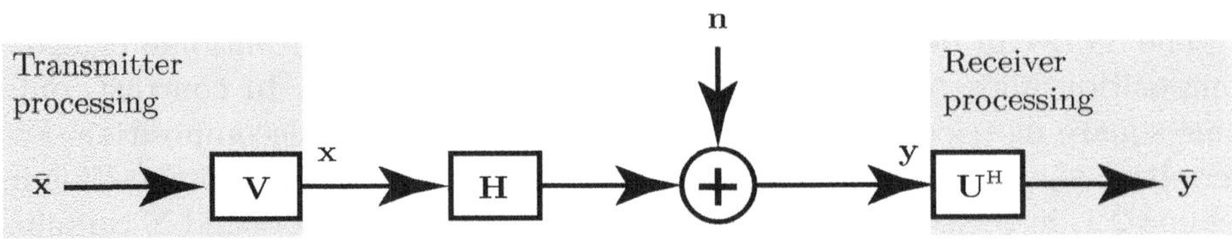

(a) The transmitter and receiver processing that diagonalizes the MIMO channel.

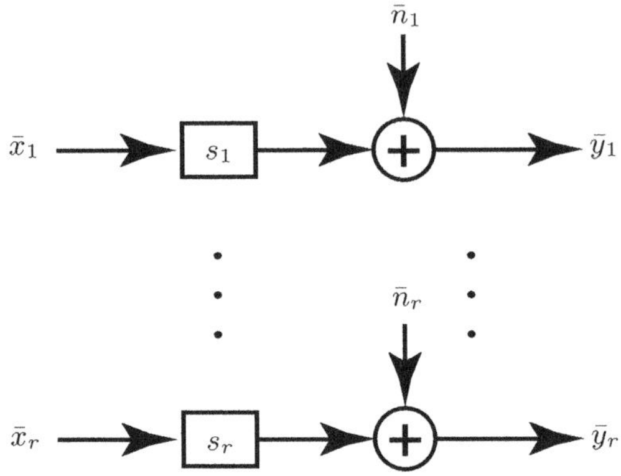

(b) An equivalent representation with r parallel SISO channels.

Figure 3.11: By utilizing the SVD $\mathbf{H} = \mathbf{U}\mathbf{\Sigma}\mathbf{V}^{\mathrm{H}}$ of the MIMO channel matrix, the transmitter and receiver can process the signals as shown in (a) to achieve r parallel SISO channels as shown in (b). The channel response in each parallel channel is a non-zero singular value of $\mathbf{H}$.

for $k = 1, \ldots, M$. Notice that the entries in (3.63) can be denoted in vector form as follows: $\bar{\mathbf{y}} = [\bar{y}_1, \ldots, \bar{y}_M]^{\mathrm{T}}$, $\bar{\mathbf{x}} = [\bar{x}_1, \ldots, \bar{x}_K]^{\mathrm{T}}$, and $\bar{\mathbf{n}} = [\bar{n}_1, \ldots, \bar{n}_M]^{\mathrm{T}}$.

Interestingly, each of the first r received signals $\bar{y}_k$ in (3.63) only depends on one channel response s_k obtained from the SVD, one transmitted signal parameter $\bar{x}_k$, and one independent noise variable $\bar{n}_k$. Hence, we can interpret the first row of (3.63) as being r parallel discrete memoryless SISO channels useful for communication. The processing that turns the MIMO channel into r parallel SISO channels is illustrated in Figure 3.11.

If $M > r$, there are $M - r$ additional received signals $\bar{y}_{r+1}, \ldots, \bar{y}_M$ in (3.63) that only contain the independent noise variables $\bar{n}_{r+1}, \ldots, \bar{n}_M$. This happens especially when $M > K$ since r cannot be larger than $\min(M, K) = K$; thus, the transmitter sends a K-dimensional signal, while the receiver obtains a higher-dimensional received signal where the extra dimensions contain no signal information. We might also have $r < \min(M, K)$ when the channel matrix is rank-deficient so that we have fewer than $\min(M, K)$ useful parallel channels between the transmitter and receiver. The $M - r$ received signals in (3.63) that only contain noise are not helpful for communication and are disregarded in the remainder of this chapter without loss of optimality.

Example 3.8. Consider a discrete memoryless MIMO channel with the channel matrix $\mathbf{H} \in \mathbb{C}^{6\times 4}$. The eigenvalues of the matrix $\mathbf{H}\mathbf{H}^{\mathrm{H}}$ are $\lambda_1 = 3$, $\lambda_2 = 2.1$, $\lambda_3 = 1.7$, and $\lambda_4 = \lambda_5 = \lambda_6 = 0$. What is the rank r of $\mathbf{H}$? What are the expressions of the r useful parallel SISO channels?

The singular values of $\mathbf{H}$ equals the square roots of the $\min(M, K) = 4$ largest eigenvalues of $\mathbf{H}\mathbf{H}^{\mathrm{H}}$. Hence, $s_1 = \sqrt{\lambda_1} = \sqrt{3}$, $s_2 = \sqrt{\lambda_2} = \sqrt{2.1}$, $s_3 = \sqrt{\lambda_3} = \sqrt{1.7}$, and $s_4 = \sqrt{\lambda_4} = 0$. The rank of $\mathbf{H}$ is $r = 3$ since there are three non-zero singular values. In this case, we have $r < \min(M, K)$.

By substituting the three non-zero singular values into (3.63), we obtain the following $r = 3$ parallel SISO channels that can be used for data transmission:

$$\bar{y}_1 = \sqrt{3}\bar{x}_1 + \bar{n}_1, \quad \bar{y}_2 = \sqrt{2.1}\bar{x}_2 + \bar{n}_2, \quad \bar{y}_3 = \sqrt{1.7}\bar{x}_3 + \bar{n}_3. \tag{3.64}$$

It remains to compute the joint capacity of the r parallel channels in (3.63). We know from Corollary 2.1 how to compute the channel capacity of one such channel, but we cannot directly use this result to deal with the parallel channels in (3.63) since there is one thing that couples them: the transmitter has a total symbol power q that it must divide between $\bar{x}_1, \dots, \bar{x}_K$, and we need to find the optimal way to do this.

As a first step, we let $q_1, \dots, q_K$ denote the symbol power of each of these signals, such that $\mathbb{E}\{|\bar{x}_k|^2\} = q_k$. These K power variables must be non-negative. It then follows that[5]

$$q = \mathbb{E}\{\|\mathbf{x}\|^2\} = \mathbb{E}\{\|\bar{\mathbf{x}}\|^2\} = \sum_{k=1}^{K} \mathbb{E}\{|\bar{x}_k|^2\} = \sum_{k=1}^{K} q_k. \tag{3.65}$$

For any given values of $q_1, \dots, q_K$, the maximum data rate is the sum of the capacities of the individual channels, each obtained using Corollary 2.1:[6]

$$\sum_{k=1}^{r} \log_2\left(1 + \frac{q_k s_k^2}{N_0}\right). \tag{3.66}$$

Since this expression only depends on the power variables $q_1, \dots, q_r$, the values that we assign to $q_{r+1}, \dots, q_K$ for the unused dimensions will not affect the data rate. Hence, we can set $q_{r+1} = \dots = q_K = 0$ so that all the available power can be used for the r parallel SISO channels between the transmitter and receiver. The channel capacity of the MIMO channel is obtained by maximizing (3.66) with respect to the allocation of power over $q_1, \dots, q_r$,

[5]Note that $\|\mathbf{x}\|^2 = \mathbf{x}^{\mathrm{H}}\mathbf{x} = \bar{\mathbf{x}}^{\mathrm{H}}\mathbf{V}^{\mathrm{H}}\mathbf{V}\bar{\mathbf{x}} = \bar{\mathbf{x}}^{\mathrm{H}}\bar{\mathbf{x}} = \|\bar{\mathbf{x}}\|^2$ since $\mathbf{V}$ is a unitary matrix.

[6]This step utilizes the fact that the transmitted signals $\bar{x}_1, \dots, \bar{x}_K$ are independent. Hence, the received signals $\bar{y}_1, \dots, \bar{y}_M$ are also independent, which is a property that follows from the fact that the noise terms $\bar{n}_1, \dots, \bar{n}_M$ are independent in (3.63), so there is no reason to introduce any statistical dependence between the parallel channels. More precisely, the differential entropy of $\bar{\mathbf{y}}$ is maximized when its entries are independent.

under the constraint that the total symbol power is q:

$$C = \max_{\substack{q_1 \geq 0, \ldots, q_r \geq 0: \\ \sum_{k=1}^{r} q_k = q}} \sum_{k=1}^{r} \log_2\left(1 + \frac{q_k s_k^2}{N_0}\right). \tag{3.67}$$

To obtain the MIMO channel capacity, it remains to derive the capacity-achieving values of the power variables. Some power variables might be zero at the optimal solution to (3.67). For the sake of argument, suppose we know that $N_+ \in \{1, \ldots, r\}$ of the variables are non-zero. We can then be sure that $q_1 > 0, \ldots, q_{N_+} > 0$ and $q_{N_+ +1} = \ldots = q_r = 0$, because $s_1, \ldots, s_{N_+}$ are the largest singular values.[7] In this case, we observe that

$$\begin{aligned}\sum_{k=1}^{r} \log_2\left(1 + \frac{q_k s_k^2}{N_0}\right) &= \sum_{k=1}^{N_+} \log_2\left(1 + \frac{q_k s_k^2}{N_0}\right) \\ &= \sum_{k=1}^{N_+} \log_2\left(\frac{s_k^2}{N_0}\right) + \sum_{k=1}^{N_+} \log_2\left(\frac{N_0}{s_k^2} + q_k\right),\end{aligned} \tag{3.68}$$

where only the second term depends on the power variables and is the one that should be maximized. This term can be upper bounded by utilizing the following classical inequality of arithmetic and geometric means.

Lemma 3.2. For any set of n real positive numbers $x_1, \ldots, x_n$ it holds that

$$\sqrt[n]{x_1 \cdot \ldots \cdot x_n} \leq \frac{1}{n} \sum_{k=1}^{n} x_k. \tag{3.69}$$

The equality in (3.69) holds if and only if $x_1 = \ldots = x_n$.

We now apply Lemma 3.2 to the second term in (3.68) to obtain

$$\begin{aligned}\sum_{k=1}^{N_+} \log_2\left(\frac{N_0}{s_k^2} + q_k\right) &= \log_2\left(\prod_{k=1}^{N_+}\left(\frac{N_0}{s_k^2} + q_k\right)\right) \\ &= N_+ \log_2\left(\sqrt[N_+]{\prod_{k=1}^{N_+}\left(\frac{N_0}{s_k^2} + q_k\right)}\right) \leq N_+ \log_2\left(\frac{1}{N_+}\sum_{k=1}^{N_+}\left(\frac{N_0}{s_k^2} + q_k\right)\right) \\ &= N_+ \log_2\left(\frac{1}{N_+}\left(q + \sum_{k=1}^{N_+} \frac{N_0}{s_k^2}\right)\right),\end{aligned} \tag{3.70}$$

[7]If this was not the case, we would have $q_k = 0$ for some $k \leq N_+$ and $q_i > 0$ for some $i > N_+$. Since $s_k \geq s_i$, we can switch the power between q_k and q_i, thereby getting a higher capacity. That is impossible if we start from the power allocation that maximizes (3.67) and, hence, we must only use the N_+ largest singular values at the solution.

where the last equality follows from the fact that $\sum_{k=1}^{N_+} q_k = q$ due to the constraint in (3.67). The upper bound in (3.70) is independent of the optimization variables. We can achieve this upper bound if the power variables are selected to achieve equality in the inequality of arithmetic and geometric means. From Lemma 3.2 we know that this happens if $\frac{N_0}{s_k^2} + q_k$ takes the same value for $k = 1, \ldots, N_+$. If we call this common value $\mu_{N_+} \geq 0$, it follows from $\frac{N_0}{s_k^2} + q_k = \mu_{N_+}$ that we should select the symbol powers to satisfy

$$q_k = \mu_{N_+} - \frac{N_0}{s_k^2} \quad \text{for } k = 1, \ldots, N_+. \tag{3.71}$$

Moreover, the common value must be

$$\begin{aligned} \mu_{N_+} &= \frac{1}{N_+}\left(q + \sum_{k=1}^{N_+} \frac{N_0}{s_k^2}\right) \\ &= \frac{q}{N_+} + \frac{1}{N_+}\sum_{k=1}^{N_+} \frac{N_0}{s_k^2} \end{aligned} \tag{3.72}$$

since this is the argument of the logarithm on the right-hand side of (3.70).

We have now determined how to compute the optimal symbol powers if we know that exactly N_+ power values will be non-zero. The remaining issue is that the value of N_+ is not known in advance. As we increase N_+, we maximize an expression in (3.68) with additional terms and power variables. This might give the impression that the data rate will increase with N_+, but we must recall that N_+ equals the number of channels we provide with non-zero power. Some SISO channels might have such small singular values that it is not helpful to allocate any power to them, even if we can. This can be observed from the optimized expression for q_k in (3.71), which becomes negative for $k = N_+$ if $s_{N_+}^2$ is so small that $N_0/s_{N_+}^2$ is larger than μ_{N_+}. We should reduce N_+ when that happens. On other hand, if we select N_+ too small, then $\mu_{N_+} - \frac{N_0}{s_k^2} > 0$ not only for $k \in \{1, \ldots, N_+\}$ but also for $k = N_+ + 1$. This indicates that we should increase N_+ to find the solution.

The final solution is to select the capacity-achieving symbol powers as

$$q_k = \max\left(\mu - \frac{N_0}{s_k^2}, 0\right), \quad k = 1, \ldots, r, \tag{3.73}$$

where we choose the value of $\mu \in \{\mu_1, \ldots, \mu_r\}$ that results in $\sum_{k=1}^{r} q_k = q$. This condition only applies when choosing the value $\mu = \mu_{N_+}$ that gives exactly N_+ non-zero powers, while all other options will assign too little or too much power. We have now proved the following MIMO capacity.

Theorem 3.1. Consider the discrete memoryless point-to-point MIMO channel in Figure 3.10 with the input $\mathbf{x} \in \mathbb{C}^K$ and output $\mathbf{y} \in \mathbb{C}^M$ given by

$$\mathbf{y} = \mathbf{H}\mathbf{x} + \mathbf{n}, \tag{3.74}$$

where $\mathbf{n} \sim \mathcal{N}_{\mathbb{C}}(\mathbf{0}, N_0\mathbf{I}_M)$ is independent noise. Suppose the input distribution is feasible whenever the symbol power satisfies $\mathbb{E}\{\|\mathbf{x}\|^2\} \leq q$. Let $\mathbf{H} \in \mathbb{C}^{M\times K}$ be a constant matrix known at the input and output with r non-zero singular values $s_1, \ldots, s_r$. The channel capacity is

$$C = \sum_{k=1}^{r} \log_2\left(1 + \frac{q_k^{\text{opt}} s_k^2}{N_0}\right) \quad \text{bit/symbol}, \tag{3.75}$$

where

$$q_k^{\text{opt}} = \max\left(\mu - \frac{N_0}{s_k^2}, 0\right), \quad k = 1, \ldots, r \tag{3.76}$$

and the variable μ is selected to make $\sum_{k=1}^{r} q_k^{\text{opt}} = q$.

The capacity is achieved by the input distribution $\mathbf{x} \sim \mathcal{N}_{\mathbb{C}}(\mathbf{0}, \mathbf{V}\mathbf{Q}^{\text{opt}}\mathbf{V}^{\text{H}})$, where $\mathbf{Q}^{\text{opt}} = \text{diag}(q_1^{\text{opt}}, \ldots, q_r^{\text{opt}}, 0, \ldots, 0)$ is a $K \times K$ diagonal matrix and $\mathbf{V}$ contains the ordered right singular vectors of $\mathbf{H}$.

We have now proved that the transmitter should select the data signal $\mathbf{x}$ to have a covariance matrix $\text{Cov}\{\mathbf{x}\} = \mathbf{V}\mathbf{Q}^{\text{opt}}\mathbf{V}^{\text{H}}$, where $\mathbf{V}$ contains the right singular vectors of the channel matrix $\mathbf{H}$, as defined in Lemma 3.1. This optimal choice diagonalizes the point-to-point MIMO channel into r parallel SISO channels with the channel gains s_k^2 for $k = 1, \ldots, r$. Recall that s_k is the kth singular value of the channel matrix $\mathbf{H}$. The singular values were defined in Lemma 3.1 to be in decreasing order, which implies that s_1 is the "strongest" channel and s_r is the "weakest" channel with non-zero gain. This fact is also reflected in how the transmitter allocates its transmit power over the parallel channels. Suppose we know the optimal value of μ in (3.76). If $\mu - \frac{N_0}{s_k^2} > 0$, then the transmitter allocates the power $q_k^{\text{opt}} = \mu - \frac{N_0}{s_k^2}$ to the kth parallel channel. Otherwise, it allocates no power to this channel: $q_k^{\text{opt}} = 0$. Since the singular values are in decreasing order, it follows that

$$\frac{N_0}{s_1^2} \leq \frac{N_0}{s_2^2} \leq \ldots \leq \frac{N_0}{s_r^2} \tag{3.77}$$

and, therefore,

$$\mu - \frac{N_0}{s_1^2} \geq \mu - \frac{N_0}{s_2^2} \geq \ldots \geq \mu - \frac{N_0}{s_r^2}. \tag{3.78}$$

Hence, a capacity-achieving transmitter allocates more power to a channel with a stronger gain than a weaker one. It might also put $q_k^{\text{opt}} = 0$ to some of the weakest channels, even if the channel gain is non-zero. Two properties govern

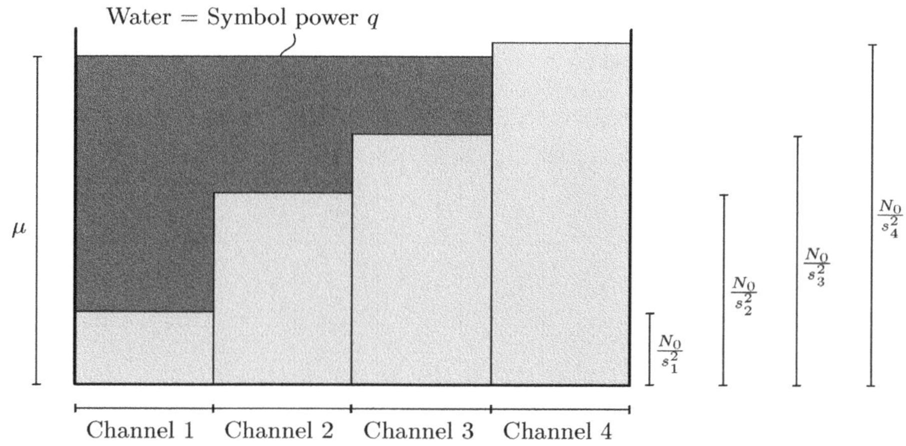

Figure 3.12: The optimal power allocation for a point-to-point MIMO channel can be described as filling a tank with a volume of water corresponding to the total symbol power q. The height of each segment of the bottom of the tank is inversely proportional to the channel gain.

the behavior. Logically, stronger channels should be allocated more power than weaker channels. However, the capacity expression in (3.75) contains the logarithmic function $\log_2(1 + q_k \frac{s_k^2}{N_0})$. We recall from Section 3.1 that it grows linearly with q_k as $q_k \frac{s_k^2}{N_0} \log_2(e)$ when the SNR is small, but then grows at a slower and slower pace; therefore, it eventually becomes preferable to also allocate power to weaker channels (with smaller s_k^2 values) because these can initially deliver a linear capacity growth, even if the slope is weaker.

This optimal power allocation solution is called *water-filling* since the implementation can be illustrated by filling a tank with an uneven bottom with water. This is illustrated in Figure 3.12 for the case of $r = 4$. The bottom is divided into four equal-sized segments representing each parallel channel. The segment related to channel k has a height of N_0/s_k^2, and the power allocated to this channel is the water that is above it. When we pour water into the tank, it will first be allocated to the strongest channel. We continue pouring water until the water volume is q. If $q > N_0/s_2^2 - N_0/s_1^2$, the water level will eventually reach a point where also the second channel is used. As we continue pouring water into the tank, the first and second channels will receive an equal share of the additional water until the point where also the third channel is activated. In the example shown in Figure 3.12, the total symbol power q is divided over the three strongest channels, while the fourth channel is not used, although it has a non-zero channel gain (i.e., the height of the bottom is finite).

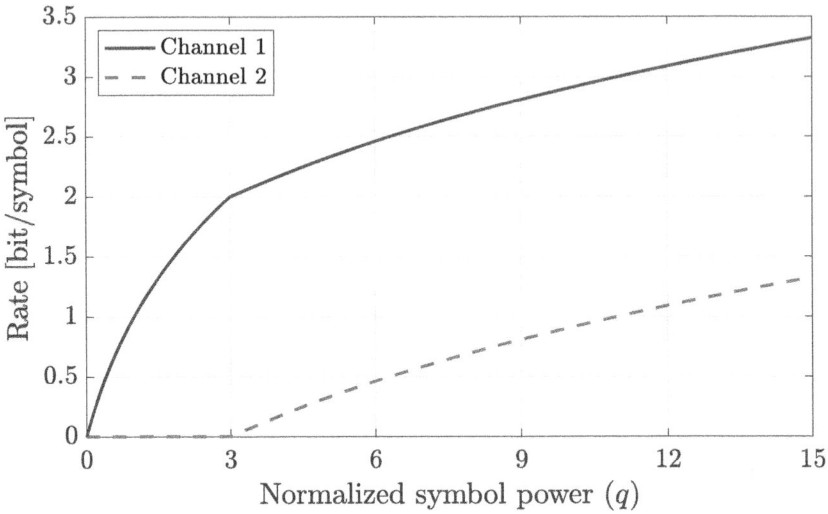

Figure 3.13: The rates achieved by the two parallel channels from Example 3.9 when optimal water-filling power allocation is used. When the weaker channel 2 begins to be used, it contributes equally much to the capacity growth as the stronger channel 1.

Example 3.9. Consider a MIMO channel with $r = 2$, $s_1^2/N_0 = 1$, and $s_2^2/N_0 = 1/4$. How is the transmit power allocated when using water-filling?

According to the water-filling expression in (3.76), we will select $q_2^{\text{opt}} > 0$ if $\mu - s_2^2/N_0 = \mu - 4 > 0$, which implies that the water level must be $\mu > 4$. By contrast, for $\mu \in [1, 4]$, we assign all power to the strongest channel, resulting in $q_1^{\text{opt}} = \mu - s_1^2/N_0 = \mu - 1 \in [0, 3]$. In the range $\mu > 4$ where both channels are used, they contribute equally to the capacity growth because

$$\log_2\left(1 + \frac{q_k^{\text{opt}} s_k^2}{N_0}\right) = \log_2\left(1 + \left(\mu - \frac{N_0}{s_k^2}\right)\frac{s_k^2}{N_0}\right) = \log_2(\mu) + \log_2\left(\frac{s_k^2}{N_0}\right) \tag{3.79}$$

increases with μ in the same way regardless of the index k.

Figure 3.13 illustrates this behavior as a function of the total symbol power q, which is normalized in the sense of being dimensionless in this example. We notice that the rate of channel 1 grows rapidly in the beginning. However, for $q > 3$, we allocate the additional power $q - 3$ equally among the two channels, and this results in rate curves for the two channels that grow equally fast.

The two extreme cases of the water-filling power allocation are illustrated in Figure 3.14. If the symbol power is low, only the strongest channel will be used, as shown in Figure 3.14(a). If the power is high, the total symbol power will be allocated over all the r parallel channels. The stronger channels are always allocated more power than the weaker channels, but the relative

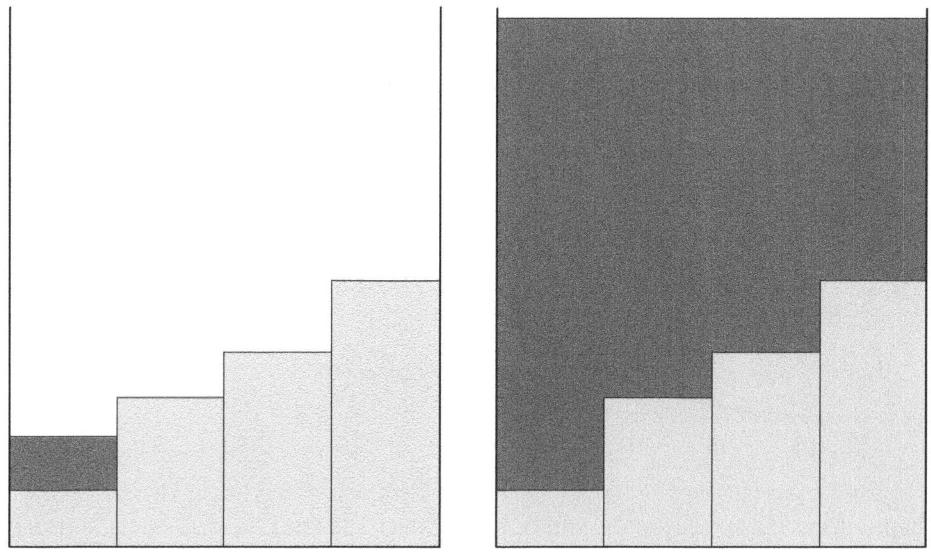

(a) Low power. (b) High power.

Figure 3.14: Illustration of the water-filling power allocation at low and high power: (a) Only the strongest channel is used when the power is low. (b) All channels are used when the power is high, and the power allocation becomes almost equal.

difference gradually disappears. In fact, we get an asymptotically equal power allocation of q/r per channel as $q \to \infty$. Notice that when we say "high" or "low" power in this context, it typically means that the SNR is high or low. As mentioned earlier, it is the fact that the logarithm grows slowly at higher SNRs that motivates the water-filling power allocation to use more than one channel when the strongest channel has reached a high SNR.

The variable r is called the *multiplexing gain* of the point-to-point MIMO channel since it represents the number of parallel data streams the channel supports with non-zero channel gain. This is an important performance indicator when the water-filling power allocation assigns non-zero power to all the r channels (e.g., at high SNR) because then the MIMO capacity is roughly r times larger than the capacity of a corresponding SISO channel.

To demonstrate how the multiplexing gain can greatly increase the capacity, we will compare a SISO channel with $|h|^2 = 1$ with a SIMO/MISO channel with $\|\mathbf{h}\|^2 = M$ and a MIMO channel with $M = K$ in which all entries of the channel matrix $\mathbf{H}$ also have unit magnitude. The singular values of this MIMO channel will satisfy $\sum_{k=1}^{M} s_k^2 = \mathrm{tr}(\mathbf{H}^{\mathrm{H}}\mathbf{H}) = MK = M^2$, but their individual values will vary depending on how we select the phases of the individual entries in $\mathbf{H}$. Let us consider an "ideal" MIMO channel where all singular

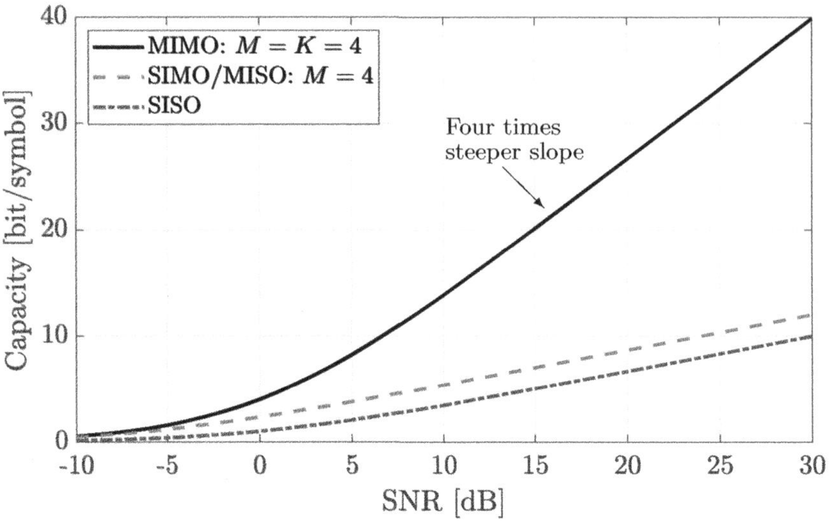

Figure 3.15: The capacity in the MIMO, SIMO/MISO, and SISO cases over an ideal channel where all entries have unit magnitude and all singular values of $\mathbf{H}$ are equal. The MIMO capacity is $M\log_2(1+\text{SNR})$, the SIMO/MISO capacity is $\log_2(1+M\text{SNR})$, and the SISO capacity is $\log_2(1+\text{SNR})$.

values are equal: $s_1 = \ldots = s_M = \sqrt{M}$.[8] The MIMO capacity in (3.75) then becomes

$$C = \sum_{k=1}^{r} \log_2\left(1 + \frac{q_k^{\text{opt}} s_k^2}{N_0}\right) = M\log_2\left(1+\frac{q}{N_0}\right) \tag{3.80}$$

since $r = M$, $s_k^2 = M$, and equal power allocation $q_k^{\text{opt}} = q/M$ is optimal. The value in (3.80) is exactly M times larger than the corresponding SISO capacity $\log_2(1+\frac{q}{N_0})$ in (2.145). Moreover, the SIMO/MISO capacity is $\log_2(1+M\frac{q}{N_0})$ in this example. The key difference from (3.80) is that the factor M appears inside the logarithm instead of in front of the logarithm. This makes a huge difference when the SNR is large; the multiplexing gain is greatly preferred over a beamforming gain since the capacity grows linearly with M instead of logarithmically. The multiplexing gain is also called the *pre-log factor* since it appears in front of the logarithm in the capacity expression.

We show the capacities in Figure 3.15 as a function of $\text{SNR} = \frac{q}{N_0}$ for $r = M = 4$. Note that the SNR is shown in the decibel scale. The lowest curve is the SISO case, which represents the baseline performance. The SIMO/MISO case gives a curve with the same shape as in the SISO case, but it is shifted to the left by 6 dB, due to the beamforming gain of $M = 4$. The MIMO case gives the same capacity as the SIMO/MISO case at low SNR (when the

[8]Equal singular values can be achieved by letting $\mathbf{H}$ be a unitary matrix that is scaled by a factor $\sqrt{M}$, which leads to an SVD with $\mathbf{\Sigma} = \sqrt{M}\mathbf{I}_M$. Two concrete examples are when $\mathbf{H}$ is a Hadamard matrix or a properly scaled discrete Fourier transform matrix. Section 4.4.3 describes a way to deploy practical antenna arrays to achieve equal singular values.

logarithm is approximately a linear function), but then it grows much faster with the SNR thanks to the multiplexing gain. More precisely, the slope of the curve is $r = 4$ times steeper; therefore, the performance gain of having a MIMO channel becomes larger the higher the SNR becomes.

Example 3.10. Consider a point-to-point MIMO channel where the channel matrix has the singular values: $s_1 = 1$, $s_2 = \frac{1}{2}$, $s_3 = \frac{1}{3}$, and $s_4 = \frac{1}{4}$. The optimal water-filling power allocation is used.
(a) If $q/N_0 = 2$, what is the optimal power allocation?
(b) For which values of q/N_0 are all singular values assigned non-zero power?
(c) If $q/N_0 = 434$, what is the optimal water-filling power allocation?

(a) We can notice from Figure 3.12 that only the strongest channel s_1 is utilized if $q \leq N_0/s_2^2 - N_0/s_1^2 = 2^2 N_0 - N_0 = 3N_0$. This is the case when $q = 2N_0$, thus the power allocation is $q_1 = q = 2N_0$ and $q_2 = q_3 = q_4 = 0$.

(b) All the parallel SISO channels are allocated non-zero power when the water height μ is above the height of the fourth segment in Figure 3.12. The breaking point occurs at $\mu = N_0/s_4^2 = 16N_0$, in which case the total power is

$$q = \sum_{k=1}^{4} \left(\mu - \frac{N_0}{s_k^2} \right) = 4\mu - N_0 - 4N_0 - 9N_0 - 16N_0 = 34N_0. \tag{3.81}$$

Hence, the four singular values are assigned non-zero power when $q/N_0 > 34$.

(c) All the channels are utilized since $q/N_0 = 434 > 34$. After filling the tank with the water volume $34N_0$, the remaining $434N_0 - 34N_0$ is divided equally among the four channels. An additional $100N_0$ of water is added to each segment, resulting in the new water height $\bar{\mu} = 116N_0$. The optimal power allocation is $q_1 = \bar{\mu} - N_0 = 115N_0$, $q_2 = \bar{\mu} - 4N_0 = 112N_0$, $q_3 = \bar{\mu} - 9N_0 = 107N_0$, and $q_4 = \bar{\mu} - 16N_0 = 100N_0$. This allocation is almost equal, which is expected when the transmit power is high.

In (3.80) and the last example, we assumed the MIMO channel matrix has the full rank $\min(M, K)$. The multiplexing gain r is generally upper bounded as $r \leq \min(M, K)$. Hence, there is no need to transmit more parallel data streams than the minimum of the number of transmit and receive antennas. This explains why only one data stream was sent over the SIMO and MISO channels we considered earlier in this chapter. In some cases, r is strictly smaller than $\min(M, K)$, so we have a lower multiplexing gain than in the ideal case. If the singular values are very different, we need a huge SNR before the water-filling power allocation uses all r channels. It is only then that the entire multiplexing gain is helpful in practice. For a given channel matrix and power level, the *effective multiplexing gain* N_+ (i.e., the number of non-zero power variables) is more indicative of the multiplexing behavior. Even if $N_+ = 1$, having multiple antennas on both sides of the channel is beneficial because the singular value s_1 grows the more antennas are used.

Example 3.11. Consider a point-to-point MIMO system with the total symbol power q, noise variance is N_0, and the channel matrix

$$\mathbf{H} = \begin{bmatrix} 1 & 1 \\ 1 & 1 \end{bmatrix}. \tag{3.82}$$

What is the channel capacity C_{MIMO}? Compare it with the MISO channel capacity C_{MISO} obtained when only one of the receive antennas is used.

We begin by computing the singular values of $\mathbf{H}$, which are the square roots of the eigenvalues of

$$\mathbf{H}\mathbf{H}^{\text{H}} = \begin{bmatrix} 1 & 1 \\ 1 & 1 \end{bmatrix} \begin{bmatrix} 1 & 1 \\ 1 & 1 \end{bmatrix} = \begin{bmatrix} 2 & 2 \\ 2 & 2 \end{bmatrix}. \tag{3.83}$$

The eigenvalues can be obtained by solving characteristic polynomial equation

$$0 = \det\left(\mathbf{H}\mathbf{H}^{\text{H}} - \lambda \mathbf{I}_2\right) = \det\left(\begin{bmatrix} 2-\lambda & 2 \\ 2 & 2-\lambda \end{bmatrix}\right) = (2-\lambda)^2 - 4, \tag{3.84}$$

from which we obtain $\lambda_1 = 4$ and $\lambda_2 = 0$. Hence, the singular values of $\mathbf{H}$ are $s_1 = \sqrt{4} = 2$ and $s_2 = 0$. The rank is $r = 1$, which is also the multiplexing gain. Since there is only one non-zero singular value, assigning all power to it is optimal: $q_1 = q$ and $q_2 = 0$. This results in the MIMO channel capacity

$$C_{\text{MIMO}} = \log_2\left(1 + \frac{4q}{N_0}\right) \quad \text{bit/symbol.} \tag{3.85}$$

When the receiver only uses a single antenna, we obtain a MISO channel with the channel vector $\mathbf{h} = [1, 1]^{\text{T}}$ being one of the columns of $\mathbf{H}$. The corresponding MISO channel capacity is obtained from (3.47) as

$$C_{\text{MISO}} = \log_2\left(1 + \frac{q}{N_0}\|\mathbf{h}\|^2\right) = \log_2\left(1 + \frac{2q}{N_0}\right) \quad \text{bit/symbol.} \tag{3.86}$$

The MIMO channel obtains a beamforming gain of $MK = 4$, while the MISO channel only achieves a beamforming gain of $K = 2$. Hence, the MIMO channel has a distinct benefit even when the multiplexing gain is $r = 1$.

We will now take a closer look at the water-filling power allocation. The variable μ represents the water level in Figure 3.12. Recall that this variable equals μ_{N_+} in (3.72) for some $N_+ \in \{1, \ldots, r\}$. For each potential value of N_+ we can verify if $\mu = \mu_{N_+}$ indeed gives N_+ non-zero powers in (3.76); nothing more and nothing less. This implies that we must have $\mu_{N_+} - \frac{N_0}{s_{N_+}^2} \geq 0$ and $\mu_{N_+} - \frac{N_0}{s_{N_+ + 1}^2} < 0$. Only one value of N_+ satisfies both conditions because the water level is always between two consecutive segments in Figure 3.12. Hence, the recipe for computing the optimal water level is as follows.

Corollary 3.3. The optimal water level is

$$\mu = \begin{cases} \mu_1, & \text{if } \mu_1 - \frac{N_0}{s_2^2} < 0, \\ \mu_{N_+}, & \text{if } \mu_{N_+} - \frac{N_0}{s_{N_++1}^2} < 0 \text{ and } \mu_{N_+} - \frac{N_0}{s_{N_+}^2} \geq 0, \\ & \quad \text{for } N_+ \in \{2, \ldots, r-1\}, \\ \mu_r, & \text{if } \mu_r - \frac{N_0}{s_r^2} \geq 0, \end{cases} \tag{3.87}$$

where μ_{N_+} is given in (3.72) for $N_+ \in \{1, \ldots, r\}$.

Since only one of the r possible values of μ in Corollary 3.3 has conditions that hold, one way to implement the water-filling power allocation is to start with computing μ_r and check if the condition in (3.87) holds. If not, we compute μ_{r-1} and check if its condition holds. We continue until we find one μ for which the conditions in (3.87) hold, and this is the optimal solution.

Example 3.12. Consider a point-to-point MIMO channel with the $r = 7$ non-zero singular values $s_1 = 1$, $s_2 = \frac{1}{\sqrt{3}}$, $s_3 = \frac{1}{\sqrt{5}}$, $s_4 = \frac{1}{\sqrt{6}}$, $s_5 = \frac{1}{\sqrt{7}}$, $s_6 = \frac{1}{\sqrt{10}}$, and $s_7 = \frac{1}{\sqrt{16}}$. What is the water-filling power allocation if $q/N_0 = 23$?

We must identify the optimal water level to find the capacity-achieving power allocation. Corollary 3.3 provides the options $\mu_1, \ldots, \mu_7$, along with their respective optimality conditions. We begin by computing μ_7 using (3.72):

$$\mu_7 = \frac{q}{7} + \frac{N_0}{7}(1 + 3 + 5 + 6 + 7 + 10 + 16) = \frac{71}{7} N_0. \tag{3.88}$$

We notice that $\mu_7 - \frac{N_0}{s_7^2} = \frac{71}{7} N_0 - 16N_0 \ngeq 0$, thus, the condition in (3.87) is not satisfied. We continue by computing μ_6 using (3.72), which results in

$$\mu_6 = \frac{q}{6} + \frac{N_0}{6}(1 + 3 + 5 + 6 + 7 + 10) = \frac{55}{6} N_0. \tag{3.89}$$

We notice that $\mu_6 - \frac{N_0}{s_7^2} = \frac{55}{6} N_0 - 16N_0 < 0$ but $\mu_6 - \frac{N_0}{s_6^2} = \frac{55}{6} N_0 - 10N_0 \ngeq 0$, so μ_6 is not satisfying its optimality conditions in (3.87). Next, we compute

$$\mu_5 = \frac{q}{5} + \frac{N_0}{5}(1 + 3 + 5 + 6 + 7) = 9N_0. \tag{3.90}$$

We note that $\mu_5 - \frac{N_0}{s_6^2} = 9N_0 - 10N_0 < 0$ and $\mu_5 - \frac{N_0}{s_5^2} = 9N_0 - 7N_0 \geq 0$, hence, the optimality conditions in Corollary 3.3 are satisfied. Since only one water level satisfies its conditions, there is no need to consider $\mu_1, \ldots, \mu_4$.

In conclusion, $N_+ = 5$, and $\mu_5 = 9N_0$ is the optimal water level. Substituting these values into (3.76), we obtain the water-filling power allocation $q_1^{\text{opt}} = 8N_0$, $q_2^{\text{opt}} = 6N_0$, $q_3^{\text{opt}} = 4N_0$, $q_4^{\text{opt}} = 3N_0$, $q_5^{\text{opt}} = 2N_0$, $q_6^{\text{opt}} = q_7^{\text{opt}} = 0$.

In practical systems, we cannot operate at arbitrary capacity values but only those achievable by predefined MCS combinations; for example, those listed in Table 2.18 for 5G NR. For stream $k \in \{1, \dots, r\}$, we should select an MCS delivering a number of bits per symbol that is close to but smaller than the capacity $\log_2(1 + \frac{q_k s_k^2}{N_0})$ of that stream. The r streams will generally use different MCSs. The water-filling solution is not optimal when considering this mapping between the continuous channel capacity and the discrete set of data rates supported by the available MCS combinations. In particular, one can sometimes modify the power allocation to push some streams to the next row in the table (i.e., achieve a larger data rate) without reducing the other ones. This principle is called mercury/water-filling and is described in [50].

3.4.1 Geometric Interpretation of MIMO Transmission

We will now provide a basic physical interpretation of how we achieve the MIMO capacity. Let us write the $K \times K$ matrix $\mathbf{V}$ from the SVD of the channel matrix as $\mathbf{V} = [\mathbf{v}_1, \dots, \mathbf{v}_K]$, where $\mathbf{v}_k$ is the kth column. To achieve the capacity, the transmitter sends the signal vector

$$\mathbf{x} = \mathbf{V}\bar{\mathbf{x}} = \sum_{k=1}^{K} \mathbf{v}_k \bar{x}_k, \tag{3.91}$$

which consists of K data signals $\bar{x}_1, \dots, \bar{x}_K$, each being multiplied by a column $\mathbf{v}_k$ from $\mathbf{V}$ that acts as a precoding vector. This is a generalization of the MISO setup in (3.39) where we only sent one data signal multiplied by one precoding vector. We call this type of transmission *spatial multiplexing* since we send (up to) K signals simultaneously, but with different spatial directivity determined by the precoding vectors. These vectors are mutually orthogonal since $\mathbf{V}^{\mathrm{H}}\mathbf{V} = \mathbf{I}_K$ but might be assigned different symbol powers since $\bar{x}_k \sim \mathcal{N}_{\mathbb{C}}(0, q_k)$ for $k = 1, \dots, K$. We call $\mathbf{V}$ the *precoding matrix*.

Similarly, let us write the $M \times M$ matrix $\mathbf{U}$ from the SVD as $\mathbf{U} = [\mathbf{u}_1, \dots, \mathbf{u}_M]$, where $\mathbf{u}_m$ is the mth column. When the receiver computes $\bar{\mathbf{y}} = \mathbf{U}^{\mathrm{H}}\mathbf{y}$, it obtains

$$\bar{\mathbf{y}} = \begin{bmatrix} \mathbf{u}_1^{\mathrm{H}}\mathbf{y} \\ \vdots \\ \mathbf{u}_M^{\mathrm{H}}\mathbf{y} \end{bmatrix}, \tag{3.92}$$

which can be interpreted as applying M different receive combining vectors, in the same way as we did with one combining vector in the SIMO case in (3.16). The receive combining vectors are mutually orthogonal since $\mathbf{U}^{\mathrm{H}}\mathbf{U} = \mathbf{I}_M$. Since the precoding vectors $\mathbf{v}_1, \dots, \mathbf{v}_K$ and combining vectors $\mathbf{u}_1, \dots, \mathbf{u}_M$ are selected based on the SVD of the channel matrix, it follows that

$$\mathbf{u}_m^{\mathrm{H}}\mathbf{y} = \mathbf{u}_m^{\mathrm{H}} \left(\mathbf{H} \sum_{k=1}^{K} \mathbf{v}_k \bar{x}_k + \mathbf{n} \right) = s_m \bar{x}_m + \bar{n}_m, \quad m = 1, \dots, r. \tag{3.93}$$

This is the first row in (3.63). The precoding and combining vectors $\mathbf{v}_m, \mathbf{u}_m$ for $m > r$ are not used since no data signal can reach the receiver when using those because the corresponding singular values are zero.

The vectors $\mathbf{u}_1, \ldots, \mathbf{u}_M$ are called the left singular vectors of $\mathbf{H}$, while $\mathbf{v}_1, \ldots, \mathbf{v}_K$ are called the right singular vectors. Using this notation, we can decompose the MIMO channel matrix as

$$\mathbf{H} = \mathbf{U}\boldsymbol{\Sigma}\mathbf{V}^{\mathrm{H}} = \sum_{k=1}^{r} s_k \mathbf{u}_k \mathbf{v}_k^{\mathrm{H}}, \tag{3.94}$$

in the same way as we did for the eigendecomposition in (2.40). The above decomposition can be verified by directly computing the matrix entries on the right-hand side. Hence, the channel matrix consists of r components, which might represent different propagation paths. Figure 3.16 provides a rough geometric interpretation for the case with $K = 3$ transmit antennas, $M = 3$ receive antennas, and $r = 3$. In this figure, each of the three components is represented by one physical propagation path, which either is the direct path between the transmitter and receiver, or a path where the transmitted signal bounces off a scattering object before reaching the receiver. The channel responses of the respective three paths are s_1, s_2, s_3, which are the singular values of the channel matrix. To achieve the MIMO capacity, the transmitter should precode its signals to transmit along the three beams indicated in Figure 3.16. The receiver "listens" to the corresponding signals by applying the corresponding receive combining vectors $\mathbf{u}_1, \mathbf{u}_2, \mathbf{u}_3$. The water-filling power allocation determines how much of the total power is assigned to each path.

This example exposes another critical difference between having multiple antennas at the transmitter/receiver versus having a single directive antenna at the transmitter/receiver. A directive antenna can only transmit/receive with one directivity, while multiple beams with different directivity (each adapted to the MIMO channel) are needed to achieve a multiplexing gain.

It is important to note that a direct mapping between precoding/combining vectors and physical propagation paths, as sketched in Figure 3.16, is not possible in general. It mainly happens when a few propagation paths (compared to the number of antennas) are spread out spatially. In all other cases, each component $s_k \mathbf{u}_k \mathbf{v}_k^{\mathrm{H}}$ in (3.94) represents some complicated linear combination of many different propagation paths, which happen to lead to an orthogonal transmission. Hence, when talking about the spatial directivity of an M-dimensional precoding/combining vector, this should not be interpreted as a distinct angular direction in our three-dimensional world but as the direction of a vector in an M-dimensional vector space. We will return to this matter in later chapters when we study MIMO channels in different deployment scenarios and identify ways to generate the channel matrices from the geometry of the propagation environment.

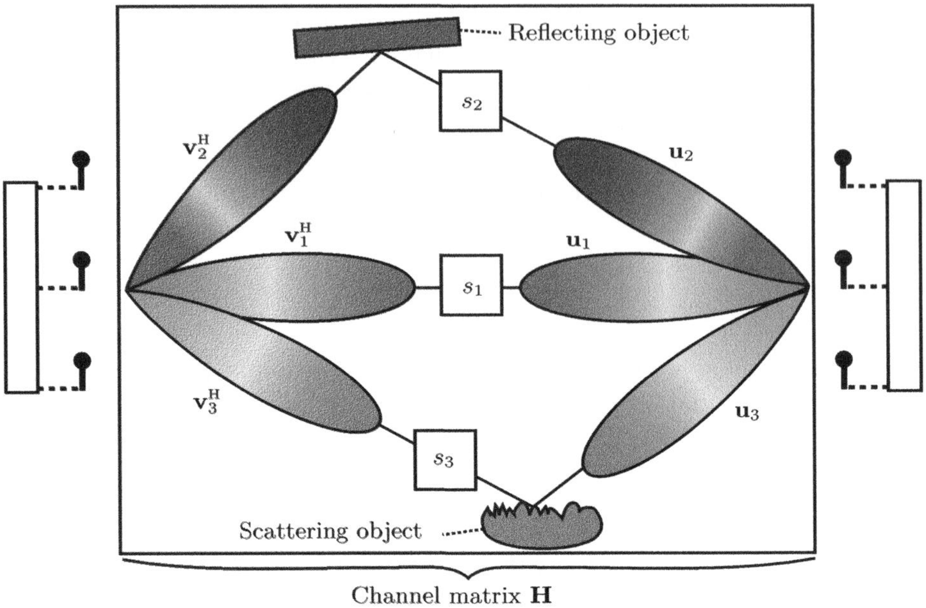

Figure 3.16: The SVD divides the channel matrix $\mathbf{H}$ into r paths of the form $s_k \mathbf{u}_k \mathbf{v}_k^{\mathrm{H}}$, where s_k^2 describes the channel gain of the kth path, $\mathbf{v}_k$ describes the spatial direction of the path seen from the transmitter, and $\mathbf{u}_k$ describes the spatial direction seen from the receiver.

Example 3.13. Consider a MIMO channel matrix that is decomposed as

$$\mathbf{H} = 3\mathbf{a}_1 \mathbf{b}_1^{\mathrm{H}} + \mathbf{a}_2 \mathbf{b}_2^{\mathrm{H}}, \tag{3.95}$$

where

$$\mathbf{a}_1 = \begin{bmatrix} 1 \\ 0 \end{bmatrix}, \quad \mathbf{a}_2 = \begin{bmatrix} 0 \\ \mathrm{j} \end{bmatrix}, \quad \mathbf{b}_1 = \begin{bmatrix} 2 \\ 1 \end{bmatrix}, \quad \mathbf{b}_2 = \begin{bmatrix} 1+\mathrm{j} \\ -2-2\mathrm{j} \end{bmatrix}. \tag{3.96}$$

What is the multiplexing gain of this channel? What are the channel gains of the SISO channels through which parallel data streams can be sent?

We first notice that $\mathbf{a}_1$ and $\mathbf{a}_2$ are orthogonal since $\mathbf{a}_1^{\mathrm{H}} \mathbf{a}_2 = 0$. Moreover, $\mathbf{b}_1$ and $\mathbf{b}_2$ are orthogonal since $\mathbf{b}_1^{\mathrm{H}} \mathbf{b}_2 = 0$. Hence, the given decomposition can be used to obtain the SVD as in (3.94). Recalling that the left and right singular vectors have unit norms, we can obtain them as

$$\mathbf{u}_1 = \frac{\mathbf{a}_1}{\|\mathbf{a}_1\|}, \quad \mathbf{u}_2 = \frac{\mathbf{a}_2}{\|\mathbf{a}_2\|}, \quad \mathbf{v}_1 = \frac{\mathbf{b}_1}{\|\mathbf{b}_1\|}, \quad \mathbf{v}_2 = \frac{\mathbf{b}_2}{\|\mathbf{b}_2\|}. \tag{3.97}$$

Accordingly, the singular values are computed as $s_1 = 3\|\mathbf{a}_1\|\|\mathbf{b}_1\| = 3\sqrt{5}$ and $s_2 = \|\mathbf{a}_2\|\|\mathbf{b}_2\| = \sqrt{10}$. The multiplexing gain is $r = 2$ since the rank of $\mathbf{H}$ is two. The channel gains of the parallel SISO channels are $s_1^2 = 45$ and $s_2^2 = 10$.

3.4.2 Duality and Alternative Capacity Expressions

The channel capacity of the MIMO channel is determined by the total symbol power q, the singular values of $\mathbf{H}$, and the noise variance N_0. Suppose we would transmit with power q in the opposite direction; that is, over the MIMO channel $\mathbf{H}^{\mathrm{T}}$ from M transmit antennas to K receive antennas. The SVD of this channel matrix is $\mathbf{H}^{\mathrm{T}} = (\mathbf{U}\boldsymbol{\Sigma}\mathbf{V}^{\mathrm{H}})^{\mathrm{T}} = \mathbf{V}^*\boldsymbol{\Sigma}^{\mathrm{T}}\mathbf{U}^{\mathrm{T}}$. Since $\boldsymbol{\Sigma}^{\mathrm{T}}$ has the same diagonal values as $\boldsymbol{\Sigma}$, the singular values of $\mathbf{H}$ and $\mathbf{H}^{\mathrm{T}}$ coincide and the water-filling power allocation will be identical if N_0 is also unchanged. We have obtained the following result.

Corollary 3.4. The capacity of the MIMO channel with channel matrix $\mathbf{H}$ is the same as the capacity of the MIMO channel with channel matrix $\mathbf{H}^{\mathrm{T}}$ if the transmit power to noise power ratio is the same.

This corollary establishes a strong connection between a *primal system* with channel matrix $\mathbf{H}$ and a *dual system* with channel matrix $\mathbf{H}^{\mathrm{T}}$. The fact that the capacity is the same in both directions of a communication channel is called *duality*. One instance of the duality is that SIMO and MISO channels have the same capacity, as we previously observed in this chapter. Duality might not be achieved in practice because different devices might have different transmit power (recall the comparison between uplink and downlink in Figure 1.7) and noise power due to different hardware characteristics.

Example 3.14. What is the capacity-achieving input distribution for the dual system with channel matrix $\mathbf{H}^{\mathrm{T}}$? Assume q and N_0 remain the same.

The optimal input distribution is $\mathbf{x} \sim \mathcal{N}_{\mathbb{C}}(\mathbf{0}, \mathbf{V}\mathbf{Q}^{\mathrm{opt}}\mathbf{V}^{\mathrm{H}})$ for the primal system, which depends on the right singular vectors $\mathbf{V}$ of $\mathbf{H} \in \mathbb{C}^{M\times K}$ and the $K \times K$ power allocation matrix $\mathbf{Q}^{\mathrm{opt}} = \mathrm{diag}(q_1^{\mathrm{opt}}, \ldots, q_r^{\mathrm{opt}}, 0, \ldots, 0)$ computed using Theorem 3.1. The SVD of the dual channel matrix is $\mathbf{H}^{\mathrm{T}} = \mathbf{V}^*\boldsymbol{\Sigma}^{\mathrm{T}}\mathbf{U}^{\mathrm{T}}$, which instead has the matrix $\mathbf{U}^* \in \mathbb{C}^{M\times M}$ containing its right singular vectors. Even if the water-filling power allocation is the same in the primal and dual systems, the number of transmit antennas might differ, so the power allocation matrix for the dual system must be defined differently: $\dot{\mathbf{Q}}^{\mathrm{opt}} = \mathrm{diag}(q_1^{\mathrm{opt}}, \ldots, q_r^{\mathrm{opt}}, 0, \ldots, 0)$ is an $M \times M$ diagonal matrix. The first r diagonal entries are the same as in $\mathbf{Q}^{\mathrm{opt}}$ but are proceeded by $M - r$ zero-valued entries.

In conclusion, the capacity-achieving input distribution for the dual system is $\dot{\mathbf{x}} \sim \mathcal{N}_{\mathbb{C}}(\mathbf{0}, \mathbf{U}^*\dot{\mathbf{Q}}^{\mathrm{opt}}\mathbf{U}^{\mathrm{T}})$.

There are several alternative ways to express the MIMO capacity. Recall that we can multiply the capacity expression in (3.75) with B to change the unit to bit/s. This leads to the alternative but equivalent way to write the capacity of a MIMO channel as

$$C = \sum_{k=1}^{r} B \log_2 \left(1 + \frac{q_k^{\text{opt}} s_k^2}{N_0}\right) \quad \text{bit/s.} \tag{3.98}$$

By substituting the expression for q_k^{opt} in (3.76) into (3.98), we obtain

$$C = \sum_{k=1}^{r} \max\left(B \log_2 \left(\frac{\mu s_k^2}{N_0}\right), 0\right) \quad \text{bit/s.} \tag{3.99}$$

The capacity expression in (3.75) can also be rewritten using the determinant in the following way:

$$\begin{aligned} C &= \sum_{k=1}^{r} \log_2 \left(1 + \frac{q_k^{\text{opt}} s_k^2}{N_0}\right) = \log_2 \left(\prod_{k=1}^{r} \left(1 + \frac{q_k^{\text{opt}} s_k^2}{N_0}\right)\right) \\ &= \log_2 \left(\det\left(\mathbf{I}_M + \frac{1}{N_0} \boldsymbol{\Sigma} \mathbf{Q}^{\text{opt}} \boldsymbol{\Sigma}^{\text{H}}\right)\right) \\ &= \log_2 \left(\det\left(\mathbf{I}_M + \frac{1}{N_0} \mathbf{H} \mathbf{V} \mathbf{Q}^{\text{opt}} \mathbf{V}^{\text{H}} \mathbf{H}^{\text{H}}\right)\right), \end{aligned} \tag{3.100}$$

where the last step follows from some matrix algebra that exploits the fact that $\mathbf{U}$ is a unitary matrix.[9] We notice that $\mathbf{V}\mathbf{Q}^{\text{opt}}\mathbf{V}^{\text{H}}$ in (3.100) is the covariance matrix of the transmitted signal in Theorem 3.1. It is a quadratic form containing the optimal precoding matrix $\mathbf{V}$ and the optimal diagonal power allocation matrix $\mathbf{Q}^{\text{opt}}$. Moreover, the matrix inside the determinant in (3.100) is the covariance matrix of the received signal $\mathbf{y}$, divided by the noise variance N_0, so we can also express the MIMO capacity as $C = \log_2(\det(\frac{1}{N_0}\text{Cov}\{\mathbf{y}\}))$. Hence, the capacity-achieving transmission over a MIMO channel is the one that maximizes the determinant of the received signal's covariance matrix.

3.4.3 Arbitrary Precoding and Successive Interference Cancellation

There are various reasons for not transmitting in a capacity-achieving way in practice, such as having imperfect channel knowledge at the transmitter or limited hardware capabilities. We will return to such issues in later chapters but cover the fundamental theory here. Recall that the capacity-achieving precoding creates many parallel SISO channels, as illustrated in Figure 3.11(b). This will not happen when suboptimal precoding is used to transmit multiple data streams; thus, the spatially multiplexed signals partially collide at the receiver and must be appropriately decoded to deal with mutual interference.

[9] For any $M \times M$ square matrices $\mathbf{A}$ and $\mathbf{B}$, it holds that $\det(\mathbf{AB}) = \det(\mathbf{A})\det(\mathbf{B})$. We can utilize this property to achieve the last equality in (3.100): $\det(\mathbf{I}_M + \frac{1}{N_0}\boldsymbol{\Sigma}\mathbf{Q}^{\text{opt}}\boldsymbol{\Sigma}^{\text{H}}) = \det(\mathbf{U}^{\text{H}}\mathbf{U} + \frac{1}{N_0}\mathbf{U}^{\text{H}}\mathbf{H}\mathbf{V}\mathbf{Q}^{\text{opt}}\mathbf{V}^{\text{H}}\mathbf{H}^{\text{H}}\mathbf{U}) = \det(\mathbf{U}^{\text{H}})\det(\mathbf{I}_M + \frac{1}{N_0}\mathbf{H}\mathbf{V}\mathbf{Q}^{\text{opt}}\mathbf{V}^{\text{H}}\mathbf{H}^{\text{H}})\det(\mathbf{U})$, where $\det(\mathbf{U}^{\text{H}}) = \det(\mathbf{U}) = 1$ since these are unitary matrices.

Suppose the transmitted signal is generated as

$$\mathbf{x} = \mathbf{P}\bar{\mathbf{x}} = \sum_{k=1}^{K} \mathbf{p}_k \bar{x}_k \tag{3.101}$$

using an arbitrary precoding matrix $\mathbf{P} = [\mathbf{p}_1, \ldots, \mathbf{p}_K] \in \mathbb{C}^{K \times K}$ with unit-norm columns $\mathbf{p}_k$ to send the K independent data signals from $\bar{\mathbf{x}} = [\bar{x}_1, \ldots, \bar{x}_K]^{\mathrm{T}}$ in different spatial directions. As the complex Gaussian input distribution is capacity-achieving, we assume that $\bar{\mathbf{x}} \sim \mathcal{N}_{\mathbb{C}}(\mathbf{0}, \mathbf{Q})$, where $\mathbf{Q} = \mathrm{diag}(q_1, \ldots, q_K)$ is an arbitrary diagonal power allocation matrix satisfying the power constraint $\sum_{k=1}^{K} q_k \leq q$. This is a feasible way to communicate over a MIMO channel, but it is suboptimal unless we select $\mathbf{P} = \mathbf{V}$ and $\mathbf{Q} = \mathbf{Q}^{\mathrm{opt}}$. Before deriving the achievable rate with an arbitrary $\mathbf{P}$, we begin with a helpful example.

Example 3.15. Suppose we use a fixed precoding vector $\mathbf{p} \in \mathbb{C}^K$ to transmit a data signal $\bar{x} \sim \mathcal{N}_{\mathbb{C}}(0, q)$ over a MIMO channel so that the received signal is

$$\mathbf{y} = \mathbf{H}\mathbf{p}\bar{x} + \mathbf{n}, \tag{3.102}$$

where the noise has the colored distribution $\mathbf{n} \sim \mathcal{N}_{\mathbb{C}}(\mathbf{0}, N_0\mathbf{C})$ for some invertible covariance matrix $\mathbf{C}^{M \times M}$. What is the capacity of this channel?

With a fixed precoding vector, the MIMO channel effectively becomes a SIMO channel with the channel vector $\mathbf{H}\mathbf{p}$. An unusual property is that the noise is colored since $\mathbf{C}$ is generally not an identity matrix. This can be dealt with using the whitening procedure in (2.86), by transforming (3.102) as

$$\mathbf{C}^{-1/2}\mathbf{y} = \mathbf{C}^{-1/2}\mathbf{H}\mathbf{p}\bar{x} + \underbrace{\mathbf{C}^{-1/2}\mathbf{n}}_{\sim \mathcal{N}_{\mathbb{C}}(\mathbf{0}, N_0 \mathbf{I}_M)}. \tag{3.103}$$

The whitening operation is reversible (i.e., it causes no information loss), so we can use (3.103) to compute the capacity. We now have a received signal with white noise as in Corollary 3.1 and with the effective SIMO channel vector $\mathbf{C}^{-1/2}\mathbf{H}\mathbf{p}$. The capacity with the fixed precoding becomes

$$\log_2\left(1 + \frac{q\|\mathbf{C}^{-1/2}\mathbf{H}\mathbf{p}\|^2}{N_0}\right) = \log_2\left(1 + \frac{q}{N_0}\mathbf{p}^{\mathrm{H}}\mathbf{H}^{\mathrm{H}}\mathbf{C}^{-1}\mathbf{H}\mathbf{p}\right). \tag{3.104}$$

This capacity is achieved when applying MRC to (3.103) based on the effective SIMO channel vector. Hence, we need to apply the combining vector

$$\mathbf{w} = \mathbf{C}^{-1/2}\mathbf{C}^{-1/2}\mathbf{H}\mathbf{p} = \mathbf{C}^{-1}\mathbf{H}\mathbf{p} \tag{3.105}$$

to the original received signal in (3.102) to first perform whitening and then MRC. This vector is equal to *LMMSE combining* vector in Example 3.4, except for a scaling factor, so we will use that terminology in this section.

The transmitted signal $\mathbf{x}$ in (3.102) has the (suboptimal) covariance matrix $\mathbf{PQP}^{\mathrm{H}}$ and the corresponding data rate is

$$\log_2\left(\det\left(\mathbf{I}_M + \frac{1}{N_0}\mathbf{HPQP}^{\mathrm{H}}\mathbf{H}^{\mathrm{H}}\right)\right) \quad \text{bit/symbol}, \tag{3.106}$$

which naturally is smaller than the capacity in (3.100). We will prove that (3.106) is an achievable rate by expanding the expression until we reach a familiar form, which reveals how the receiver can operate to achieve this rate. The rate expression from the last example will be useful in the derivation.

With arbitrary precoding, the received signal in (3.56) can be expressed as

$$\begin{aligned}\mathbf{y} &= \mathbf{Hx} + \mathbf{n} = \mathbf{HP\bar{x}} + \mathbf{n}\\ &= \sum_{k=1}^{K} \mathbf{Hp}_k \bar{x}_k + \mathbf{n}\end{aligned} \tag{3.107}$$

and its covariance matrix is $\sum_{k=1}^{K} q_k \mathbf{Hp}_k\mathbf{p}_k^{\mathrm{H}}\mathbf{H}^{\mathrm{H}} + N_0\mathbf{I}_M$. Each signal appears at the receiver in a unique direction $\mathbf{Hp}_k$ in the M-dimensional vector space, and the K directions might be linearly independent (if $K \leq M$), but generally not mutually orthogonal. Therefore, the signals are interfering with each other, which we can deal with by decoding them sequentially and successively removing the already decoded signals—known signals cease to be interference. For notational convenience, we define

$$\mathbf{y}_i = \mathbf{y} - \sum_{k=1}^{i-1} \mathbf{Hp}_k \bar{x}_k, \quad i = 1, \ldots, K+1 \tag{3.108}$$

as the residual received signal when the first $i-1$ data signals have been decoded and removed. This vector has the covariance matrix $N_0\mathbf{C}_i$, where

$$\mathbf{C}_i = \begin{cases} \mathbf{I}_M + \sum\limits_{k=i}^{K} \frac{q_k}{N_0}\mathbf{Hp}_k\mathbf{p}_k^{\mathrm{H}}\mathbf{H}^{\mathrm{H}}, & \text{if } i = 1, \ldots, K, \\ \mathbf{I}_M, & \text{if } i = K+1. \end{cases} \tag{3.109}$$

The data rate in (3.106) can be rewritten using this notation as

$$\begin{aligned}\log_2\left(\det\left(\mathbf{I}_M + \frac{1}{N_0}\mathbf{HPQP}^{\mathrm{H}}\mathbf{H}^{\mathrm{H}}\right)\right) &= \log_2\left(\det\left(\mathbf{I}_M + \sum_{k=1}^{K}\frac{q_k}{N_0}\mathbf{Hp}_k\mathbf{p}_k^{\mathrm{H}}\mathbf{H}^{\mathrm{H}}\right)\right)\\ &= \log_2\left(\det\left(\mathbf{C}_1\right)\right)\end{aligned} \tag{3.110}$$

by utilizing (3.109) and the fact that the signal covariance matrix can be expanded as $\mathbf{PQP}^{\mathrm{H}} = \sum_{k=1}^{K} q_k \mathbf{p}_k \mathbf{p}_k^{\mathrm{H}}$. We can further rewrite (3.110) as

$$\begin{aligned}
\log_2(\det(\mathbf{C}_1)) &= \log_2\left(\det\left(\mathbf{C}_2 + \frac{q_1}{N_0}\mathbf{H}\mathbf{p}_1\mathbf{p}_1^{\mathrm{H}}\mathbf{H}^{\mathrm{H}}\right)\right) \\
&= \log_2\left(\det\left(\mathbf{C}_2 + \frac{q_1}{N_0}\mathbf{H}\mathbf{p}_1\mathbf{p}_1^{\mathrm{H}}\mathbf{H}^{\mathrm{H}}\mathbf{C}_2^{-1}\mathbf{C}_2\right)\right) \\
&= \log_2\left(\det\left(\mathbf{I}_M + \frac{q_1}{N_0}\mathbf{H}\mathbf{p}_1\mathbf{p}_1^{\mathrm{H}}\mathbf{H}^{\mathrm{H}}\mathbf{C}_2^{-1}\right)\right) + \log_2(\det(\mathbf{C}_2)) \\
&= \log_2\left(1 + \frac{q_1}{N_0}\mathbf{p}_1^{\mathrm{H}}\mathbf{H}^{\mathrm{H}}\mathbf{C}_2^{-1}\mathbf{H}\mathbf{p}_1\right) + \log_2(\det(\mathbf{C}_2)), \quad (3.111)
\end{aligned}$$

where the last equality follows from Sylvester's determinant theorem in (2.53). The first term in (3.111) has the same structure as the capacity expression in Example 3.15; that is, it is the capacity when transmitting using the precoding vector $\mathbf{p}_1$ and having colored complex Gaussian noise with the covariance matrix $N_0\mathbf{C}_2$. This is precisely how the received signal in (3.107) is structured if we decompose it as

$$\mathbf{y} = \mathbf{H}\mathbf{p}_1\bar{x}_1 + \underbrace{\sum_{k=2}^{K} \mathbf{H}\mathbf{p}_k\bar{x}_k + \mathbf{n}}_{\sim\mathcal{N}_{\mathbb{C}}(\mathbf{0}, N_0\mathbf{C}_2)}. \quad (3.112)$$

The latter term is not conventional noise since it contains both interfering signals and receiver noise. However, from the decoding perspective, it is distributed as colored complex Gaussian noise, so it takes the role of an *effective noise* term. Hence, if we decode the data signal $\bar{x}_1$ while treating the remaining $K-1$ interfering signals as part of the noise, then we can achieve a data rate equal to the first term in (3.111) using LMMSE combining of the kind described in Example 3.15.

The second term $\log_2(\det(\mathbf{C}_2))$ in (3.111) can be expanded similarly. In fact, for any $i = 1, \ldots, K$, it holds that

$$\begin{aligned}
\log_2(\det(\mathbf{C}_i)) &= \log_2\left(\det\left(\mathbf{C}_{i+1} + \frac{q_i}{N_0}\mathbf{H}\mathbf{p}_i\mathbf{p}_i^{\mathrm{H}}\mathbf{H}^{\mathrm{H}}\right)\right) \\
&= \log_2\left(\det\left(\mathbf{I}_M + \frac{q_i}{N_0}\mathbf{H}\mathbf{p}_i\mathbf{p}_i^{\mathrm{H}}\mathbf{H}^{\mathrm{H}}\mathbf{C}_{i+1}^{-1}\right)\right) + \log_2(\det(\mathbf{C}_{i+1})) \\
&= \log_2\left(1 + \frac{q_i}{N_0}\mathbf{p}_i^{\mathrm{H}}\mathbf{H}^{\mathrm{H}}\mathbf{C}_{i+1}^{-1}\mathbf{H}\mathbf{p}_i\right) + \log_2(\det(\mathbf{C}_{i+1})), \quad (3.113)
\end{aligned}$$

where the first term is the capacity when transmitting using the precoding vector $\mathbf{p}_i$ and having colored complex Gaussian noise with the covariance matrix $N_0\mathbf{C}_{i+1}$. This is how the residual received signal in (3.108) is structured because it can be decomposed as

$$\mathbf{y}_i = \mathbf{H}\mathbf{p}_i\bar{x}_i + \underbrace{\sum_{k=i+1}^{K} \mathbf{H}\mathbf{p}_k\bar{x}_k + \mathbf{n}}_{\sim\mathcal{N}_{\mathbb{C}}(\mathbf{0},N_0\mathbf{C}_{i+1})}. \tag{3.114}$$

The first term in (3.113) is, therefore, a data rate we can achieve by removing the first $i-1$ data signals from $\mathbf{y}$ to obtain $\mathbf{y}_i$ and then decode $\bar{x}_i$ while treating the remaining $K-i$ interfering signals as part of the colored noise. The iterative expansion in (3.113) terminates when $i = K$ since $\log_2(\det(\mathbf{C}_{K+1})) = 0$.

In summary, the data rate in (3.106) can be expanded as

$$\log_2\left(\det\left(\mathbf{I}_M + \frac{1}{N_0}\mathbf{H}\mathbf{P}\mathbf{Q}\mathbf{P}^{\mathrm{H}}\mathbf{H}^{\mathrm{H}}\right)\right) = \sum_{i=1}^{K}\log_2\left(1 + \frac{q_i}{N_0}\mathbf{p}_i^{\mathrm{H}}\mathbf{H}^{\mathrm{H}}\mathbf{C}_{i+1}^{-1}\mathbf{H}\mathbf{p}_i\right) \tag{3.115}$$

and is achieved by decoding the signals sequentially while removing the previously decoded signals and treating the uncoded signals as noise. This procedure is known as *successive interference cancellation (SIC)* and is summarized in Figure 3.17. The signal $\bar{x}_1$ is first decoded using $\mathbf{y}_1 = \mathbf{y}$. Next, $\mathbf{y}_2$ is computed and $\bar{x}_2$ is decoded using it. This procedure continues successively until $\bar{x}_K$ has been decoded. The whole procedure is also known as LMMSE-SIC because each signal is decoded using LMMSE combining, as discussed in Example 3.15.

The signals were assumed to be decoded in increasing numerical order, which can be done without loss of generality because the precoding vectors are numbered arbitrarily. The expression on the left-hand side of (3.115) takes the same value regardless of how the precoding vectors are numbered; however, individual terms in the right-hand side expression will take different values depending on the numbering. Moreover, the right-hand side expression is explicitly achieved using LMMSE combining, as described in Example 3.15, but the choice of receiver processing is not visible in the left-hand side expression. The reason is that rate expressions implicitly assume an optimal receiver based on the available information.

The SIC procedure is information-theoretically optimal but has several practical issues. Recall from Definition 2.6 that the capacity determines the data rate we can communicate at while achieving an arbitrarily low error probability as the number of symbols in the packet approaches infinity. Hence, to decode the signal $\bar{x}_1$ actually means to decode an N-length codeword $\bar{x}_1[1], \ldots, \bar{x}_1[N]$ where $N \to \infty$ or at least is very large. Next, we need to recreate $\mathbf{H}\mathbf{p}_1\bar{x}_1[l]$ for the time instances $l = 1, \ldots, N$ in the packet and subtract it from the entire sequence of received signals $\mathbf{y}[1], \ldots, \mathbf{y}[N]$. This procedure requires extensive memory storage and causes delays proportional to K. Since N is finite in practice, there will also be a non-zero error probability for each stream, and when an error occurs, the wrong data signal will be subtracted from the received signals. This increases rather than reduces the amount of interference and is called *error propagation* because it will likely result in decoding errors for all the remaining uncoded streams.

Example 3.16. What is the achievable data rate if we decode the K signals separately without using SIC?

When we decode signal i, we can express the received signal in (3.107) as

$$\mathbf{y} = \mathbf{H}\mathbf{p}_i \bar{x}_i + \underbrace{\sum_{k=1, k\neq i}^{K} \mathbf{H}\mathbf{p}_k \bar{x}_k + \mathbf{n}}_{\sim \mathcal{N}_{\mathbb{C}}(\mathbf{0}, N_0 \mathbf{C}_{-i})}, \tag{3.116}$$

where the colored noise is based on the covariance matrix

$$\mathbf{C}_{-i} = \mathbf{I}_M + \sum_{k=1, k\neq i}^{K} \frac{q_k}{N_0} \mathbf{H}\mathbf{p}_k \mathbf{p}_k^{\mathrm{H}} \mathbf{H}^{\mathrm{H}}. \tag{3.117}$$

It follows from Example 3.15 that the achievable data rate when treating interference as colored noise is

$$\log_2 \left(1 + \frac{q_i}{N_0} \mathbf{p}_i^{\mathrm{H}} \mathbf{H}^{\mathrm{H}} \mathbf{C}_{-i}^{-1} \mathbf{H}\mathbf{p}_i \right), \tag{3.118}$$

which is achieved using the LMMSE combining $\mathbf{w}_i = \frac{1}{N_0} \mathbf{C}_{-i}^{-1} \mathbf{H}\mathbf{p}_i$. The achievable data rate of all the K data streams then becomes

$$\sum_{i=1}^{K} \log_2 \left(1 + \frac{q_i}{N_0} \mathbf{p}_i^{\mathrm{H}} \mathbf{H}^{\mathrm{H}} \mathbf{C}_{-i}^{-1} \mathbf{H}\mathbf{p}_i \right). \tag{3.119}$$

This value is smaller than (3.115) because of the lack of SIC (i.e., there is more interference). However, it is easier to implement the receiver processing in practice since the K data streams can be decoded in parallel. Moreover, unlike SIC, there is no risk of error propagation.

This example describes a setup where each data stream is decoded independently while treating the other streams as colored noise. This is called *linear processing* because the receiver only performs a linear algebra operation before the signal decoding: it multiplies the received signal with a receive combining vector of the LMMSE-kind in (3.105). A block diagram is shown in Figure 3.18, where we can notice that the K decoding branches are parallel and independent. By contrast, the LMMSE-SIC receiver processing in Figure 3.17 is non-linear because of the successive removal of interference, which connects the decoding of the different signals.

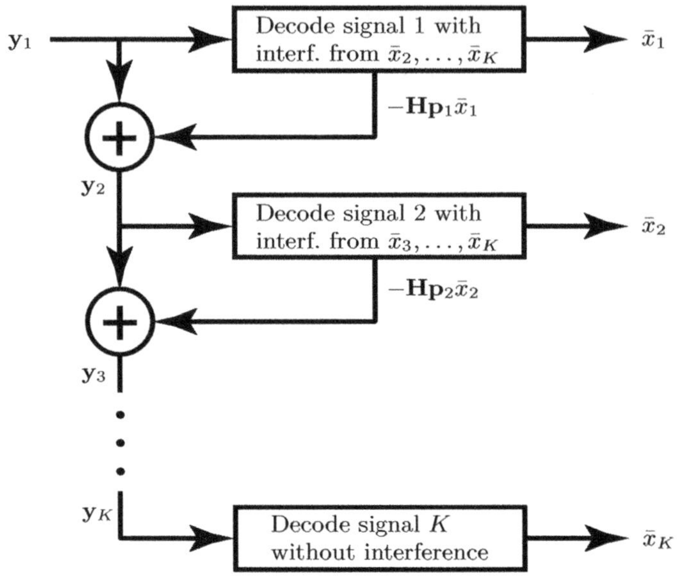

Figure 3.17: A block diagram of the LMMSE-SIC receiver processing. When the precoding is not dividing the MIMO channel into parallel SISO channels, the receiver can instead decode the data signals sequentially to deal with interference. Each decoded signal is subtracted from the received signal vector before the next signal is decoded, while the remaining interfering signals are treated as colored noise. This is called successive interference cancellation.

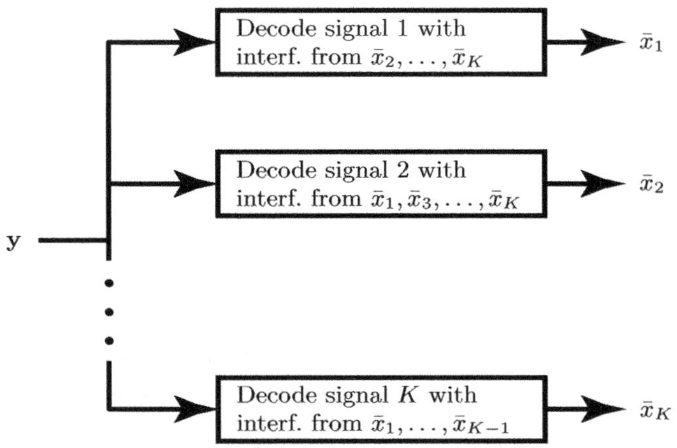

Figure 3.18: A block diagram of a linear MIMO receiver processing. Each data stream is decoded separately while treating the interference from the other signals as colored noise.

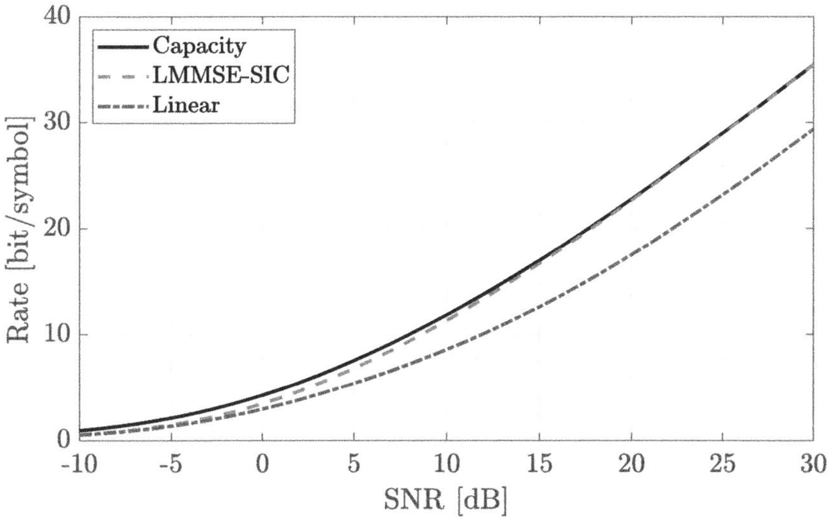

Figure 3.19: The capacity is compared with the data rates achieved with suboptimal precoding (i.e., sending an independent data stream per antenna) and two types of receiver processing: The LMMSE-SIC receiver in Figure 3.17 and the linear receiver in Figure 3.18.

Figure 3.19 compares the MIMO channel capacity with the data rates achieved with equal power allocation and the precoding $\mathbf{P} = \mathbf{I}_K$ that transmits one independent signal per antenna. In the latter case, we consider both the non-linear LMMSE-SIC receiver in Figure 3.17 and the simplified linear receiver in Figure 3.18. We consider a setup with $M = K = 4$ antennas. To obtain a slightly asymmetric channel matrix, we let the entries have unit magnitude but independent random phases between 0 and 2π. The figure shows the average rates (over different random phases) as a function of $\mathrm{SNR} = \frac{q}{N_0}$. The LMMSE-SIC curve is below the capacity due to the suboptimal precoding. However, it approaches the capacity at high SNR because water-filling converges to equal power allocation in this regime so that $\mathbf{V}\mathbf{Q}^{\mathrm{opt}}\mathbf{V}^{\mathrm{H}} = \frac{q}{K}\mathbf{V}\mathbf{V}^{\mathrm{H}} = \frac{q}{K}\mathbf{I}_K$. This is the same signal covariance matrix as when $\mathbf{P} = \mathbf{I}_K$ and equal power allocation are used. The linear receiver is affected by more interference than the LMMSE-SIC receiver, but they perform equally well at low SNRs where the interference is anyway negligible. There is a substantial performance loss at high SNRs, but the curve with the linear receiver has the same slope as the capacity curve, which showcases that the same multiplexing gain of $r = 4$ is achieved in all cases.

3.5 Exercises

Exercise 3.1. Consider the capacity $B \log_2 \left(1 + \frac{P\beta}{BN_0}\right)$ of a SISO channel.

(a) Show that the capacity goes to zero when $B \to 0$. What is the name of this operating regime?

(b) What happens to the capacity when $B \to \infty$? What is the name of this operating regime?

Exercise 3.2. Consider the capacity $C(P, B) = B \log_2 \left(1 + \frac{P\beta}{BN_0}\right)$ of a SISO channel.

(a) Compute the first-order derivative of the capacity with respect to P. At what value of P does the capacity grow the fastest? What happens with the capacity growth as $P \to \infty$?

(b) Compute the second-order derivative of the capacity with respect to P. Show that it is negative; that is, the capacity is a concave function of P.

(c) Compute the first-order derivative of the capacity, with respect to B. At what value of B does the capacity grow the fastest? What happens with the capacity growth as $B \to \infty$? Hint: Use the inequality $\frac{x}{1+x} < \ln(1 + x)$ for $x > 0$.

(d) Compute the second-order derivative of the capacity with respect to B. Show that it is negative; that is, the capacity is a concave function of B.

Exercise 3.3. Consider the capacity $C(P, B) = B \log_2 \left(1 + \frac{P\beta}{BN_0}\right)$ of a SISO channel.

(a) Suppose there is a reference setup where P and B have been selected such that $P\beta/(BN_0) = 7$. We want to change the bandwidth from B to cB for some scalar $c > 1$ to at least double the capacity while keeping all other variables constant. What will at least double the capacity (compared to $c = 1$): increasing the bandwidth to $2B$ or $6B$?

(b) Repeat (a) for the case when the reference setup has $P\beta/(BN_0) = 1$. Can we find a value of c that doubles the capacity (compared to $c = 1$)? Hint: Utilize the fact that $f(c) = \log_2 \left(1 + \frac{1}{c}\right) - \frac{2}{c} < 0$ for all $c > 1$.

(c) Use the asymptotic limit of the capacity as $B \to \infty$ (i.e., $\log_2(e)\frac{P\beta}{N_0}$) to derive the condition on the initial selection of $P\beta/(BN_0)$ so that we can double the capacity by increasing the bandwidth to cB for some $c > 0$. Use this relation to verify your answers to parts (a) and (b).

Exercise 3.4. Consider a system where the received signal power is $P_{\text{rx}} = 10^{-9}$ W and the bandwidth is $B = 100$ MHz. There is an AWGN channel between a transmitting single-antenna user device and a receiving single-antenna base station.

(a) Give an expression for the channel capacity, as a function of P_{rx}, B, and the noise power spectral density N_0.

(b) What is the channel capacity in bit/s using the numbers given above and $N_0 = 10^{-17}$ W/Hz?

(c) Suppose we would equip the base station with multiple antennas, and each antenna receives $P_{\text{rx}} = 10^{-9}$ W. How many antennas do we need to get 8 times higher capacity than in (b)?

Exercise 3.5. Consider the SIMO channel $\mathbf{y} = \mathbf{h}x + \mathbf{n}$, where the input signal x has the power limit $\mathbb{E}\{|x|^2\} \leq q$ and the noise vector $\mathbf{n}$ has independent and identically distributed $\mathcal{N}_{\mathbb{C}}(0, N_0)$-entries. The channel $\mathbf{h}$ is an M-length vector with only ones, where M denotes the number of receive antennas.

(a) What is the capacity of this channel? What kind of input distribution achieves the capacity?

(b) Suppose $q/N_0 = 1$. How many antennas do we need to achieve a capacity of 6 bit/symbol?

(c) Suppose we have $M = 10$ antennas. How large SNR q/N_0 do we need to achieve a capacity of 6 bit/symbol?

(d) Suppose all entries of $\mathbf{h}$ are equal to two, instead of one. What is the capacity of this channel?

(e) Suppose all entries of $\mathbf{h}$ are equal to -1, instead of $+1$. What is the capacity of this channel? Compare it with (a) and explain the intuition behind the result.

Exercise 3.6. Suppose we are designing an uplink communication system that should provide (at least) 400 Mbit/s at every point in its coverage area. The transmit power is 0.1 W, the bandwidth is 100 MHz, and the noise power spectral density is $N_0 = 10^{-17}$ W/Hz. The propagation distance is denoted by d and the gain of the channel is $|h|^2 = 10^{-8}(1\,\text{km}/d)^4$.

(a) Use the capacity formula for a SISO channel to determine for which range of distances, d, we can deliver the required data rate.

(b) We would like to extend the range to $d = 2\,\text{km}$, but we cannot increase the transmit power at the user devices. Instead, we will use multiple antennas at the receiving base station. Suppose the channel gain $|h_m|^2$ is the same for each receive antenna m and matches the SISO case. How many antennas are needed?

Exercise 3.7. Consider a SIMO channel where the single-antenna transmitter sends the signal $x \sim \mathcal{N}_{\mathbb{C}}(0, q)$ to a receiver with M antennas. The received signal is denoted as $\mathbf{y} = \mathbf{h}x + \mathbf{n}$, where $\mathbf{h}$ is constant and $\mathbf{n}$ is complex Gaussian noise. The receive combining vector $\mathbf{w}$ is applied to the received signal $\mathbf{y}$ to detect the signal x from $\mathbf{w}^{\mathrm{H}}\mathbf{y}$.

(a) Suppose the noise vector is colored $\mathbf{n}$, which means that the covariance matrix $\mathrm{Cov}\{\mathbf{n}\} = \mathbf{C}$ is not equal to a scaled identity matrix but invertible. Derive the receive combining vector $\mathbf{w}$ that maximizes the SNR. Hint: Define $\mathbf{a} = \mathbf{C}^{1/2}\mathbf{w}$ and optimize $\mathbf{a}$ instead.

(b) Consider a hypothetical system where $\mathbf{C}$ is a singular matrix and $\mathbf{h}$ is a non-zero vector in the nullspace of $\mathbf{C}$ (i.e., $\mathbf{Ch} = \mathbf{0}$). What is the largest SNR that we can achieve in such a system?

Exercise 3.8. Consider a MISO system with two transmit antennas where the received signal is $y = h_1 \cdot x_1 + h_2 \cdot x_2 + n$, and $n \sim \mathcal{N}_{\mathbb{C}}(0, N_0)$ is the independent receiver noise.

(a) Suppose the two transmit antennas send the independent signals $x_1 \sim \mathcal{N}_{\mathbb{C}}(0, q_1)$ and $x_2 \sim \mathcal{N}_{\mathbb{C}}(0, q_2)$, where the powers satisfy the constraint $q_1 + q_2 \leq q$. What is the resulting data rate? Which values of q_1 and q_2 will maximize that rate? Hint: The data rate expression in (3.106) can be utilized.

(b) Compare the data rate from (a) with the MISO channel capacity. Under which conditions on h_1 and h_2 are they equal?

Exercise 3.9. Consider a downlink channel where the user device has one receive antenna and the base station has three transmit antennas. The transmit power P, bandwidth $B = 100\,\text{MHz}$, and noise power spectral density N_0 are selected such that $P/(BN_0) = 2$. Suppose the channel vector is $\mathbf{h} = [3, 1, -4]^{\mathrm{T}}$.

(a) What is the capacity of this channel in bit/s? What does the capacity-achieving MRT vector become?

(b) Suppose the base station hardware has restricted capabilities so that each entry of the precoding vector $\mathbf{p}$ must have a magnitude equal to $1/\sqrt{3}$, such that $\|\mathbf{p}\| = 1$. However, we can choose any sign/phase of the entries in the complex-valued vector $\mathbf{p}$. How should we select the precoding vector to achieve the largest possible data rate (bit/s)?

(c) Compare the rate values from (a) and (b), and provide a high-level explanation of the difference.

Exercise 3.10. Consider the discrete memoryless point-to-point MIMO channel with the input $\mathbf{x} \in \mathbb{C}^K$ and output $\mathbf{y} \in \mathbb{C}^M$ given by $\mathbf{y} = \mathbf{H}\mathbf{x} + \mathbf{n}$. The receiver noise $\mathbf{n} \sim \mathcal{N}_{\mathbb{C}}(\mathbf{0}, \mathbf{C})$ is independent of $\mathbf{x}$ but has an arbitrary non-singular covariance matrix $\mathbf{C} \in \mathbb{C}^{M \times M}$. State the generalized version of Theorem 3.1 that supports such noise covariance matrices. Hint: Begin by whitening the noise.

Exercise 3.11. Consider a point-to-point MIMO system with $\frac{q}{N_0} = 1$ and the channel matrix

$$\mathbf{H} = \begin{bmatrix} 1 & 0 \\ 1-2\mathrm{j} & 0 \\ 0 & 3+4\mathrm{j} \\ 0 & -\sqrt{5} \end{bmatrix}. \tag{3.120}$$

(a) What is the channel capacity? What is the covariance matrix of the capacity-achieving input distribution?

(b) Consider the dual channel $\mathbf{H}^{\mathrm{T}}$ with $\frac{q}{N_0} = 1$. What is the channel capacity? What is the covariance matrix of the capacity-achieving input distribution?

Exercise 3.12. Consider a point-to-point MIMO system with $q/N_0 = 2$. Find the water-filling power allocation and capacity for each of the following channel matrices:

(a) $\mathbf{H} = \begin{bmatrix} 1 & 1 \\ -1 & 1 \end{bmatrix}$. (b) $\mathbf{H} = \begin{bmatrix} 1 & 0 \\ 0 & 1/2 \end{bmatrix}$. (c) $\mathbf{H} = \begin{bmatrix} 1 & 1 \\ 1 & 1 \end{bmatrix}$.

Exercise 3.13. Consider a point-to-point MIMO channel with the channel matrix

$$\mathbf{H} = \begin{bmatrix} \frac{1}{\sqrt{3}} & \frac{1}{\sqrt{3}} \\ \frac{1}{\sqrt{3}} & \frac{-1}{\sqrt{3}} \end{bmatrix}. \tag{3.121}$$

(a) For what value of $\frac{q}{N_0}$ is the capacity 2 bit/symbol if we use only the first antenna at the transmitter and the first antenna at the receiver?

(b) For what value of $\frac{q}{N_0}$ is the capacity 2 bit/symbol if we use only the first antenna of the transmitter but both antennas at the receiver?

(c) For what value of $\frac{q}{N_0}$ is the capacity 2 bit/symbol if we use the whole 2×2 MIMO channel? Compare the results in (a), (b), and (c).

Exercise 3.14. Consider the transmission over a point-to-point MIMO channel with $M = K = 2$. We will use the SNR notation $\varrho = q/N_0$.

(a) Suppose the channel matrix is

$$\mathbf{H} = \begin{bmatrix} e^{-\mathrm{j}\pi/3} & e^{-\mathrm{j}\pi/3} \\ 1 & 1 \end{bmatrix}. \tag{3.122}$$

Compute the capacity of this channel as a function of ϱ. Explain how the capacity is achieved and what kind of gain is achieved compared to the corresponding SISO channel, which has capacity $\log_2(1+\varrho)$.

(b) Suppose the channel matrix is

$$\mathbf{H} = \begin{bmatrix} 1 & -1 \\ 1 & 1 \end{bmatrix}. \tag{3.123}$$

Compute the capacity of this channel as a function of ϱ. Explain how the capacity is achieved and what kind of gain is achieved compared to the corresponding SISO channel.

(c) For which values of the SNR ϱ is the capacity in (b) larger than in (a)?

Exercise 3.15. Consider a MIMO channel with the channel matrix $\mathbf{H} \in \mathbb{C}^{M \times K}$. All the entries of $\mathbf{H}$ have unit magnitude.

(a) Assume that $M \geq K$ and all the columns of $\mathbf{H}$ are mutually orthogonal. What is the channel capacity for a given value of $\frac{q}{N_0}$?

(b) Compute the first-order derivative of the capacity expression in (a) with respect to K. Is it an increasing or decreasing function? Hint: Use the inequality $\frac{x}{1+x} < \ln(1+x)$ for $x > 0$.

(c) Compute the second-order derivative of the capacity expression with respect to K. At what value of K does the capacity grow the fastest?

(d) Suppose that $K = M$. How does the capacity depend on K (and M) in this case?

(e) How does the capacity depend on M and K when $\frac{q}{N_0}$ is close to zero?

Exercise 3.16. Consider an $M \times M$ MIMO channel matrix $\mathbf{H}$ with the singular values $s_1, \ldots, s_M$. The eigenvalues $\lambda_1, \ldots, \lambda_M$ of $\mathbf{H}\mathbf{H}^{\mathrm{H}}$ satisfies $\lambda_m = s_m^2$ for $m = 1, \ldots, M$. Suppose we are free to select the eigenvalues freely, under the constraint that they are positive and that $\sum_{k=1}^{M} \lambda_k = \lambda_{\mathrm{sum}}$.

(a) Which selection of eigenvalues maximizes the capacity at low SNRs? Hint: Use that the water-filling only assigns power to the largest eigenvalue at low SNRs.

(b) Which selection of eigenvalues maximizes the capacity at high SNRs? Hint: Use (3.3), Lemma 3.2, and that water-filling assigns power equally among the eigenvalues at high SNRs.

Exercise 3.17. Consider the $M \times M$ MIMO channel where the received signal is

$$\mathbf{y} = \mathbf{H}\mathbf{P}\bar{\mathbf{x}} + \mathbf{n}. \tag{3.124}$$

Suppose the precoding matrix is $\mathbf{P} = \mathbf{H}^{\mathrm{H}}(\mathbf{H}\mathbf{H}^{\mathrm{H}})^{-1}\mathbf{D}$, where $\mathbf{D} = \mathrm{diag}(d_1, \ldots, d_M)$ is a diagonal matrix.

(a) The columns of the precoding matrix are the precoding vectors $\mathbf{p}_1, \ldots, \mathbf{p}_M$. How can $\mathbf{D}$ be selected to ensure each precoding vector has a unit norm?

(b) Show that this precoding matrix creates M parallel SISO channels. What is the channel gain on each such channel?

(c) Suppose $\mathbf{H} = \begin{bmatrix} 2 & 1 \\ 1 & 2 \end{bmatrix}$. Compare the gains of the parallel channels achieved in (b) with the gains of the parallel channels obtained using the SVD. Which approach gives the largest sum of the channel gains?

Exercise 3.18. Consider the $M \times K$ MIMO channel with the received signal

$$\mathbf{y} = \mathbf{H}\mathbf{P}\bar{\mathbf{x}} + \mathbf{n} = \sum_{k=1}^{K} \mathbf{p}_k \bar{x}_k + \mathbf{n}_k. \tag{3.125}$$

An arbitrary precoding matrix $\mathbf{P} = [\mathbf{p}_1, \ldots, \mathbf{p}_K] \in \mathbb{C}^{K \times K}$ with unit-norm columns $\mathbf{p}_k$ is used to send the K independent data signals from $\bar{\mathbf{x}} = [\bar{x}_1, \ldots, \bar{x}_K]^{\mathrm{T}}$. We assume that $\mathbf{n} \sim \mathcal{N}_{\mathbb{C}}(\mathbf{0}, N_0\mathbf{I}_M)$ and $\bar{\mathbf{x}} \sim \mathcal{N}_{\mathbb{C}}(\mathbf{0}, \mathbf{Q})$ with a fixed power allocation matrix $\mathbf{Q} = \mathrm{diag}(q_1, \ldots, q_K)$. The resulting data rate in (3.106) is equal to the mutual information between $\bar{\mathbf{x}}$ and $\mathbf{y}$, i.e., $\mathcal{I}(\bar{\mathbf{x}}; \mathbf{y})$.

(a) The chain rule for mutual information is given as

$$\begin{aligned} \mathcal{I}(x_1, \ldots, x_n; y) &= \mathcal{I}(x_1; y) + \mathcal{I}(x_2; y|x_1) + \mathcal{I}(x_3; y|x_1, x_2) \\ &\quad + \ldots + \mathcal{I}(x_n; y|x_1, x_2, \ldots, x_{n-1}), \end{aligned} \tag{3.126}$$

where $\mathcal{I}(x_n; y|x_1, x_2, \ldots, x_{n-1})$ is the mutual information between x_n and y given the knowledge of $x_1, x_2, \ldots, x_{n-1}$. Express the data rate for the considered MIMO channel $\mathcal{I}(\bar{\mathbf{x}}; \mathbf{y})$ in terms of $\mathcal{I}(\bar{x}_i; \mathbf{y}|\bar{x}_1, \bar{x}_2, \ldots, \bar{x}_{i-1})$ using the chain rule for mutual information.

(b) Consider the LMMSE-SIC receiver processing illustrated in Figure 3.17. At stage i, the LMMSE receiver $\mathbf{w}_i = \mathbf{C}_i^{-1}\mathbf{H}\mathbf{p}_i$ is applied to the residual $\mathbf{y}_i = \mathbf{y} - \sum_{k=1}^{i-1} \mathbf{H}\mathbf{p}_k \bar{x}_k$ to decode $\bar{x}_i$. Since $\bar{\mathbf{x}}$ and $\mathbf{n}$ are Gaussian distributed, the LMMSE receiver is also the MMSE receiver. Using this, show that $\mathcal{I}(\bar{x}_i; \mathbf{y}|\bar{x}_1, \bar{x}_2, \ldots, \bar{x}_{i-1}) = \mathcal{I}(\bar{x}_i; \mathbf{w}_i^{\mathrm{H}}\mathbf{y}_i|\bar{x}_1, \bar{x}_2, \ldots, \bar{x}_{i-1})$, i.e., that the MMSE receiver at each stage is information lossless.

(c) Using (a) and (b), conclude that the LMMSE-SIC receiver processing is information-theoretically optimal.

Chapter 4

Line-of-Sight Point-to-Point MIMO Channels

The capacity of a point-to-point MIMO channel with an arbitrary channel matrix $\mathbf{H} \in \mathbb{C}^{M \times K}$ was derived in the last chapter. In this chapter, we will derive a model for $\mathbf{H}$ in free-space line-of-sight (LOS) propagation and use it to analyze the capacity behavior further using the previously derived expressions. There is only one path between each transmit antenna and each receive antenna in free-space LOS channels, namely the direct path obtained by drawing a straight line between the antennas. This is an exact model of space communications, where no objects create additional signal paths by reflecting or scattering the transmitted signal. It can also be a reasonably accurate model of LOS channels on Earth, where there are objects that create additional signal paths, but these generally have much smaller channel gains than the direct path. This is particularly the case for the high-band spectrum, where the reflected paths typically are weaker while the LOS path is not.

As in Chapter 3, we start with the special cases of SIMO and MISO channels, where only one side of the channel utilizes multiple antennas. The results will then be extended to the MIMO case.

4.1 Basic Properties of Antenna Arrays

Within the context of this book, an antenna array is a collection of antennas that operate jointly at the transmitter or receiver side of a communication system. We used K and M to respectively denote the number of transmit antennas and receive antennas in Section 3, which then became the dimensions of the channel matrix $\mathbf{H} \in \mathbb{C}^{M \times K}$. Two properties determine the channel matrix: the array geometries at the transmitter and receiver, and the propagation environment between them.

Figure 4.1 exemplifies three different array geometries where all the antennas are deployed on a two-dimensional plane. Each array is characterized by the convex enclosure containing all the antennas, called the *aperture*, and its antenna arrangement. Figure 4.1(a) shows an antenna array with an irregu-

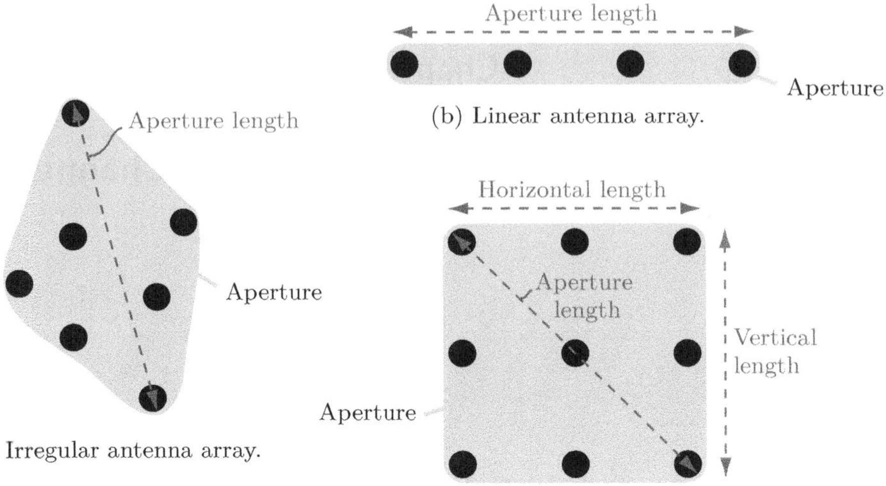

(a) Irregular antenna array.

(b) Linear antenna array.

(c) Planar antenna array.

Figure 4.1: An antenna array is characterized by its aperture (i.e., the convex enclosure of all antennas) and the antenna arrangement within the aperture. Three examples are given in this figure, where the filled circles represent the individual antennas. The largest dimension of the aperture is called the aperture length.

larly shaped aperture and a non-uniform antenna arrangement. Such arrays are seldom encountered in practice; in fact, the word *array* is often associated with a regular geometrical arrangement. Figure 4.1(b) shows a *linear array* with a uniform antenna spacing in one dimension, while Figure 4.1(c) shows a *planar array* with a uniform antenna spacing in two dimensions.

Definition 4.1. The *aperture length* D is the largest separation between any two antennas in an array. It is called the *normalized aperture length* when normalized by the wavelength λ and is then denoted as $D_\lambda = D/\lambda$.

The aperture length is indicated for each of the three examples in Figure 4.1. It is the distance between the first and last antenna in a linear array. In contrast, it is the distance between the antennas in opposite corners in a planar array. The aperture length will play an important role throughout this chapter. As indicated in Figure 4.1(c), a planar array's horizontal and vertical lengths will also play a role when analyzing such arrays.

4.2 Modeling of Line-of-Sight SIMO Channels

The SIMO capacity analysis in the previous chapter was based on the discrete-time complex-baseband channel model $y_m[l] = h_m x[l] + n_m[l]$ in (3.12), where $y_m[l]$ is the received signal at the mth antenna, h_m is the corresponding

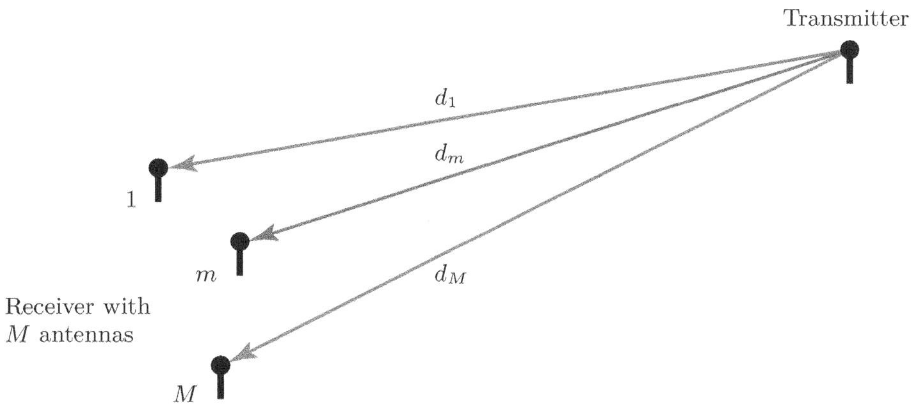

Figure 4.2: A free-space SIMO LOS channel where d_m is the distance between the transmitter and the mth receive antenna for $m = 1, \dots, M$. The array of receive antennas has an arbitrary geometry in this figure.

channel coefficient, $x[l]$ is the transmitted signal, and $n_m[l]$ is the noise. The analysis considered arbitrary values of $h_1, \dots, h_M$. The purpose of this section is to derive expressions for these coefficients in a scenario where an isotropic antenna transmits in free space to an array of M isotropic antennas, as illustrated in Figure 4.2.

We need to start the derivation from a continuous-time signal model since the physical channel affects this physical signal. The single-antenna transmitter sends the passband signal

$$z_p(t) = \sqrt{2}\Re\left(z(t)e^{\mathrm{j}2\pi f_c t}\right) \tag{4.1}$$

of the kind previously defined in (2.111), where $z(t)$ is the complex-baseband PAM signal in (2.120) and f_c is the carrier frequency.

We denote by d_m the physical distance (in meters) between the transmitter and the mth receive antenna, for $m = 1, \dots, M$. Using this notation, the received passband signal at the mth antenna is (before noise is added)

$$v_{p,m}(t) = \sqrt{\beta_m} z_p\left(t - \frac{d_m}{c}\right), \tag{4.2}$$

where d_m/c is the propagation time delay, c denotes the speed of light, and

$$\beta_m = \frac{\lambda^2}{(4\pi)^2} \frac{1}{d_m^2} \tag{4.3}$$

is the free-space channel gain computed as in (1.7).

From (4.2), we notice that the transmitted signal is attenuated by a factor $\sqrt{\beta_m}$ and delayed by $\frac{d_m}{c}$ seconds. This matches the channel model type introduced in Section 2.3.3 with $L = 1$ path. As mentioned in that section, the receiver must delay its clock by η seconds to compensate for the propagation delay before sampling the received signal. Following (2.125), the continuous-time channel impulse response to the mth receive antenna in the complex baseband then becomes

$$g_m(t) = \sqrt{\beta_m} e^{-j2\pi f_c t} \delta\left(t + \eta - \frac{d_m}{c}\right), \quad m = 1, \ldots, M. \tag{4.4}$$

Furthermore, it follows from (2.128) that the received signal at the mth antenna after sampling is

$$y_m[l] = \sum_{k=-\infty}^{\infty} x[k] \sqrt{\beta_m} e^{-j2\pi f_c \left(\frac{d_m}{c} - \eta\right)} \text{sinc}\left((l-k) + B\left(\eta - \frac{d_m}{c}\right)\right) + n_m[l]. \tag{4.5}$$

To avoid intersymbol interference, we would like to select the sampling delay η such that

$$\text{sinc}\left((l-k) + B\left(\eta - \frac{d_m}{c}\right)\right) \approx \text{sinc}(l-k) = \begin{cases} 1, & l = k, \\ 0, & l \neq k, \end{cases} \quad m = 1, \ldots, M. \tag{4.6}$$

Exact equality is achieved for $\eta = \frac{d_m}{c}$, but this value depends on the antenna index m. Since each antenna experiences a different propagation delay, we generally cannot find one value of η that achieves exact equality for all of them. Suppose we use the first antenna ($m = 1$) as the timing reference for the sampling by setting $\eta = \frac{d_1}{c}$. We then want $B\frac{(d_1 - d_m)}{c}$ to be close to zero for $m = 2, \ldots, M$. This means that the maximum difference in propagation delay $\max_{m \in \{2,\ldots,M\}} \frac{|d_m - d_1|}{c}$ with respect to the reference antenna should be much shorter than the symbol time $\frac{1}{B}$:

$$\max_{m \in \{2,\ldots,M\}} \frac{|d_m - d_1|}{c} \ll \frac{1}{B}. \tag{4.7}$$

The distances $d_1, \ldots, d_M$ depend on the transmitter's location compared to the array, but the maximum difference only depends on the array geometry.

The aperture length D was introduced in Definition 4.1 as the maximum separation between any two antennas in the array. The worst-case delay scenario can be constructed by identifying two receive antennas separated by D, making one of them antenna 1, and then placing the transmitter on the line that connects these receive antennas. The condition in (4.7) becomes

$$\frac{D}{c} \ll \frac{1}{B} \tag{4.8}$$

in this worst-case scenario and is satisfied for many array sizes and signal bandwidths. For example, if the aperture length is $D = 1\,\text{m}$, then $\frac{1\,\text{m}}{c} = \frac{1}{B}$

implies that $B = \frac{c}{1\,\mathrm{m}} = 300\,\mathrm{MHz}$ gives equality in (4.8). Many practical systems use much smaller bandwidths (e.g., 20 MHz) and often smaller arrays. Hence, we will use the approximation in (4.6) in this chapter.[1] General channel modeling without this approximation will be considered in Chapter 7.

Example 4.1. Suppose the condition in (4.8) is assumed to hold if $\frac{D}{c} \leq \frac{0.1}{B}$. What is the maximum allowed normalized aperture length if

(a) $f_\mathrm{c} = 3\,\mathrm{GHz}$, and $B = 20\,\mathrm{MHz}$;

(b) $f_\mathrm{c} = 30\,\mathrm{GHz}$, and $B = 300\,\mathrm{MHz}$;

(c) $f_\mathrm{c} = 30\,\mathrm{GHz}$, and $B = 1\,\mathrm{GHz}$?

The normalized aperture length is defined as $D_\lambda = D/\lambda$. Since $c = \lambda f_\mathrm{c}$, we need to satisfy the condition

$$\frac{D_\lambda}{f_\mathrm{c}} \leq \frac{0.1}{B},$$

which leads to the following maximum allowed normalized aperture lengths:

(a) $D_\lambda \leq \frac{0.1 f_\mathrm{c}}{B} = \frac{0.1 \cdot 3 \cdot 10^9}{20 \cdot 10^6} = 15$ wavelengths;

(b) $D_\lambda \leq \frac{0.1 \cdot 30 \cdot 10^9}{300 \cdot 10^6} = 10$ wavelengths;

(c) $D_\lambda \leq \frac{0.1 \cdot 30 \cdot 10^9}{10^9} = 3$ wavelengths.

By substituting (4.6) into (4.5), the system model simplifies to

$$y_m[l] = \sqrt{\beta_m} e^{-\mathrm{j}2\pi f_\mathrm{c} \frac{(d_m - d_1)}{c}} x[l] + n_m[l] = \underbrace{\sqrt{\beta_m} e^{-\mathrm{j}2\pi \frac{(d_m - d_1)}{\lambda}}}_{=h_m} x[l] + n_m[l], \quad (4.9)$$

where the second equality utilizes the fact that the wavelength at the carrier frequency is $\lambda = c/f_\mathrm{c}$. We can identify the value of h_m from (4.9):

$$h_m = \sqrt{\beta_m} e^{-\mathrm{j}2\pi \frac{(d_m - d_1)}{\lambda}}. \quad (4.10)$$

This channel response consists of a channel gain β_m and a complex exponential $e^{-\mathrm{j}2\pi \frac{(d_m - d_1)}{\lambda}}$ containing a phase-shift proportional to $(d_m - d_1)/\lambda$. This is not the absolute phase-shift of the propagation but the relative phase-shift

[1] In practice, it is preferable to select antenna 1 to *minimize* the maximum separation to all other antennas; for example, if the array has a square shape as in Figure 4.1(c), we should pick the antenna in the center as the timing reference, instead of an antenna in one of the corners as in the worst-case scenario. This will reduce the maximum delay from D/c to $D/(2c)$. However, to obtain expressions that resemble those in other textbooks, we will nevertheless use one of the corners as the reference antenna in this chapter.

compared to the reference antenna. We can collect all the channel responses in the channel vector

$$\mathbf{h} = \begin{bmatrix} h_1 \\ \vdots \\ h_M \end{bmatrix} = \begin{bmatrix} \sqrt{\beta_1} \\ \sqrt{\beta_2} e^{-\mathrm{j}2\pi \frac{(d_2 - d_1)}{\lambda}} \\ \vdots \\ \sqrt{\beta_M} e^{-\mathrm{j}2\pi \frac{(d_M - d_1)}{\lambda}} \end{bmatrix}, \tag{4.11}$$

where we recall that $\beta_m = \frac{\lambda^2}{(4\pi)^2} \frac{1}{d_m^2}$ for $m = 1, \ldots, M$.

The approximation that we utilized above results in *frequency flatness* since the impulse response in (4.4) has effectively been approximated as

$$g_m(t) = h_m \delta(t), \quad m = 1, \ldots, M, \tag{4.12}$$

which has a Fourier transform with the constant value h_m across all frequencies. We can utilize the derived expression in (4.11) when dealing with any practical receiver array of limited size. When the array has a regular geometrical structure, it can be utilized to simplify the expression. A particular example will be considered next.

4.2.1 Uniform Linear Array at the Receiver

One type of antenna array is particularly common to deploy and analyze: the *uniform linear array (ULA)*. In this array type, the M antennas are deployed with uniform spacing, and the centers are located on a straight line, as in Figure 4.1(b). We let Δ denote the spacing between the centers of any two adjacent antennas. The spacing between the centers of the two outermost antennas will then be $(M-1)\Delta$. The total length of the ULA, measured between the outer edges of the outermost antennas, depends on the physical width of the individual antennas, which depends on the hardware implementation. For convenience, we will denote the aperture length of the ULA as $M\Delta$, because this expression will appear in many expressions derived in this chapter. A setup with a receiving ULA is shown in Figure 4.3. We continue to use receive antenna 1 as the reference point and define the *angle-of-arrival* $\varphi \in [-\pi, \pi)$ of the impinging signal at this antenna, as shown in the figure. More precisely, we consider a two-dimensional plane (in the three-dimensional world) that contains the ULA and transmitter, and define angles in that plane.[2] Note that $\varphi = 0$ corresponds to a transmitter on a line perpendicular to the line where the ULA is deployed. This is called the *broadside* or *front-fire* direction of the array, which is a terminology borrowed from how the canons on a warship are lined up to fire toward the sides. Two

[2]This assumption can be made without loss of generality. A plane is defined by two linearly independent vectors that lie in the plane; thus, we can create the plane by selecting one vector pointing along the ULA and the other vector pointing from the reference antenna to the transmitter.

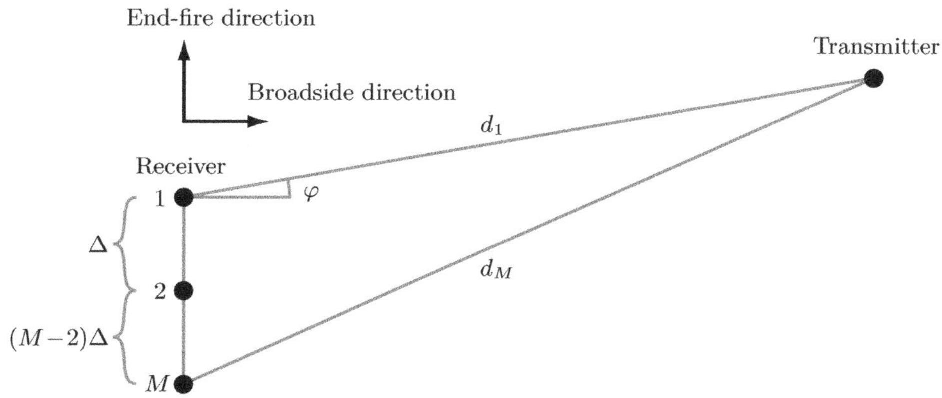

Figure 4.3: Illustration of communication from a single-antenna transmitter to a receiver equipped with a ULA. The antenna spacing is Δ, and the distance to receive antenna m is d_m for $m = 1, \ldots, M$. The angle-of-arrival φ is measured at the first antenna. The transmitter is in the broadside direction if $\varphi = 0$, while it is in the end-fire direction if $\varphi = \pm\pi/2$.

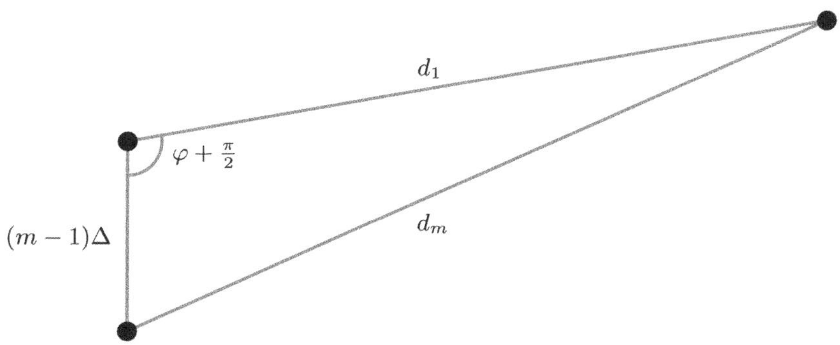

Figure 4.4: The distance d_m between the transmitter and mth receive antenna can be computed based on d_1, Δ, and φ using the law of cosines.

other important directions are $\varphi = \pm\pi/2$, where the transmitter is on the same line as the ULA. These are called the *end-fire* directions of the array.

We can now use the geometry to compute the distance d_m to the mth antenna as a function of d_1, φ, and Δ. Their relationship is illustrated in Figure 4.4, and we can utilize the law of cosines to establish the relationship

$$\begin{aligned} d_m^2 &= d_1^2 + (m-1)^2\Delta^2 - 2d_1(m-1)\Delta\cos\left(\varphi + \frac{\pi}{2}\right) \\ &= d_1^2 + (m-1)^2\Delta^2 + 2d_1(m-1)\Delta\sin(\varphi). \end{aligned} \tag{4.13}$$

This difference in propagation distance will affect both the channel gain β_m and the phase-shift $2\pi\frac{(d_m-d_1)}{\lambda}$ in (4.10). In many cases of practical interest, it holds that $d_1 \gg M\Delta$, which means that the distance between the transmitter and the first antenna is much larger than the aperture length of the ULA. Since the channel gain depends on the total distance, it then follows that

$$\beta_m = \frac{\lambda^2}{(4\pi)^2} \frac{1}{d_1^2 + (m-1)^2\Delta^2 + 2d_1(m-1)\Delta \sin(\varphi)} \approx \frac{\lambda^2}{(4\pi)^2} \frac{1}{d_1^2} = \beta_1 \tag{4.14}$$

for $m = 1, \dots, M$. This means the channel gain is approximately the same for all antennas in most free-space LOS scenarios. For simplicity, we will use the notation

$$\beta = \beta_1 = \frac{\lambda^2}{(4\pi)^2} \frac{1}{d^2} \tag{4.15}$$

without an antenna index to denote the common channel gain of all antennas, where $d = d_1$ denotes the distance to the reference antenna.

In contrast, the phase-shift $2\pi \frac{(d_m - d_1)}{\lambda}$ at the mth antenna depends on the relative distance $d_m - d_1$ between the mth and first antenna, and this variation cannot be neglected, even if the total distance is considerable. Recall from Section 1.1.2 that the transmit antenna emits a spherical wave, which can be approximated as a plane wave when the receive antenna is beyond the Fraunhofer distance defined in (1.18). The same argument can be applied when considering a receiving ULA, but the aperture length of the ULA should be considered instead of the width of a single receive antenna. More specifically, the impinging wave will have an approximately planar wavefront if

$$d_1 \geq \frac{2M^2\Delta^2}{\lambda}. \tag{4.16}$$

When this holds, we say that the ULA is in the far field of the transmitter. The far-field condition is often satisfied in practice; for example, if the ULA has an aperture length of 1 m and $\lambda = 0.1$ m (i.e., $f_c = 3$ GHz), then we need $d_1 \geq 20$ m to be in the far-field, which is typically the case (at least in the practical scenarios where such large arrays are being used). Moreover, the condition (4.16) generally implies $d_1 \gg M\Delta$, as assumed in (4.14) when approximating the channel gain, because $2M\Delta/\lambda \gg 1$ for most arrays.

The difference between spherical and planar wavefronts is illustrated in Figure 4.5, which shows snapshots of sinusoidal waves propagating in the xy-plane. The shape of the wavefront is seen by inspecting the points that attain the maximum value at the same time: these points lie on circular curves in Figure 4.5(a) and on straight lines in Figure 4.5(b). When this example is extended to wave propagation in three dimensions, the circular curves become spherical, while the straight lines become planes. The wave in Figure 4.5(b) propagates along the x-axis. The spatial frequency is zero along the wavefront (i.e., no phase difference along the y-axis). We recall from Section 2.8.3 that the spatial frequency is $\pm 1/\lambda$ in the direction the wave propagates.

Even if the impinging wavefronts are planar, the receiving ULA might not be deployed in a direction that matches the wavefronts. In particular, the

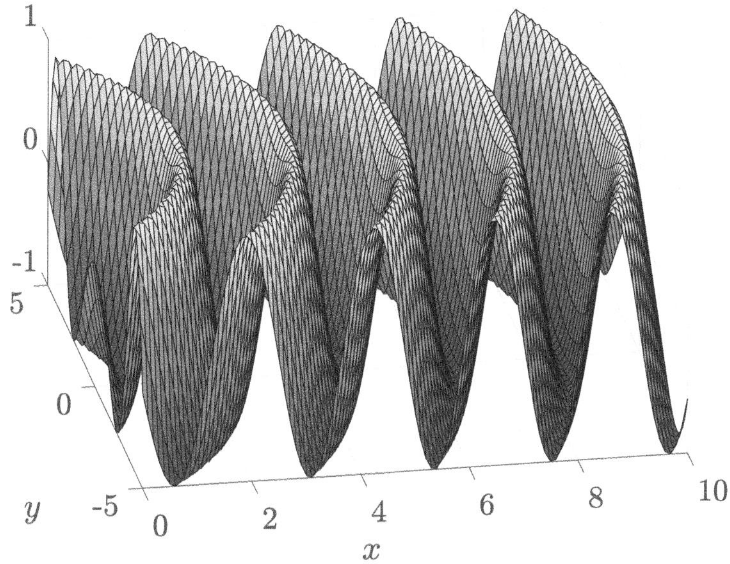

(a) Spherical wavefronts.

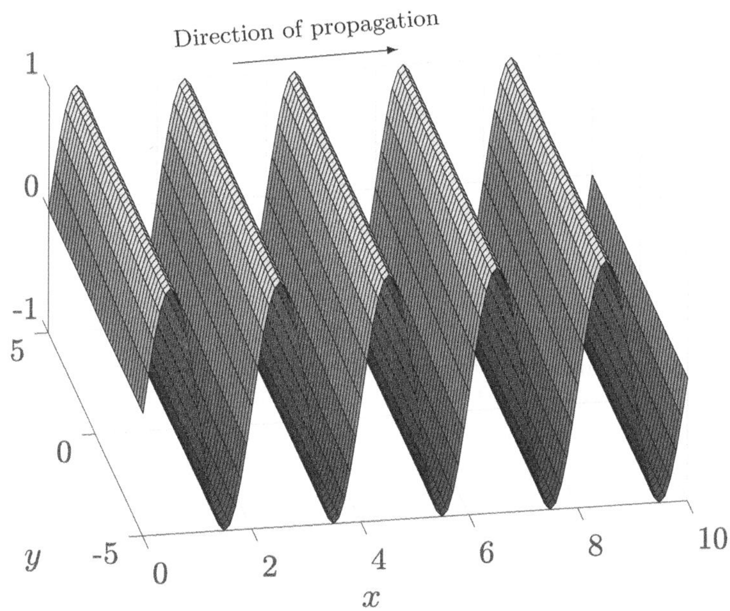

(b) Planar wavefronts.

Figure 4.5: Example of two sinusoidal waves propagating in the xy-plane. The vertical axis shows the value at a particular time instance. The shape of the wavefronts can be seen by drawing lines between the neighboring points that attain the same value simultaneously. The wavefronts are spherical in (a) and planar in (b), represented by circular and straight lines in this two-dimensional example.

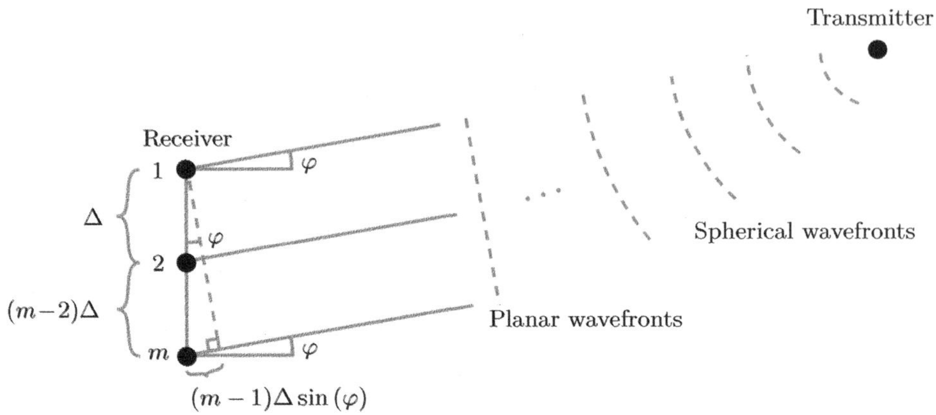

Figure 4.6: The isotropic transmitter emits spherical waves, which look like planar waves when the receiver is far from the transmitter. The angle-of-arrival φ is approximately the same for all antennas, and the difference in propagation distance between antenna 1 and antenna m is $(m-1)\Delta\sin(\varphi)$.

distances to the receive antennas will differ when the wave arrives from a non-broadside direction with angle $\varphi \notin \{0, \pm\pi\}$. As shown in Figure 4.6, the difference in propagation distance between the first and the mth antenna can be computed using trigonometry when having planar wavefronts:

$$d_m - d_1 = (m-1)\Delta\sin(\varphi). \tag{4.17}$$

This happens because the opposite angle is φ and the triangle's longest side is $(m-1)\Delta$. The phase difference between the considered antennas is $2\pi(m-1)\Delta\sin(\varphi)/\lambda$. As the distance between the antennas is $(m-1)\Delta$, the phase variations between the signals observed simultaneously at the different antennas in the ULA vary with a spatial frequency of $\sin(\varphi)/\lambda$ periods per meter. The information-bearing signal still oscillates with time at the temporal frequency f_c, but the relative phase difference between the antennas remains constant and is determined by the spatial frequency. Hence, the (spatial) channel vector $\mathbf{h}$ contains this spatial frequency, not the signal. One can view it as the spatial counterpart to how the (temporal) impulse response of the channel has a frequency response containing a collection of different frequencies. For the considered ULA, the channel contains the zero-valued spatial frequency when deployed parallel to the wavefronts (i.e., $\varphi \in \{0, \pm\pi\}$). Similarly, the channel contains the spatial frequency $\pm 1/\lambda$ when deployed along the direction of propagation, which is the case when $\varphi = \pm\pi/2$.

We have now derived far-field approximations of the channel gain and phase-shifts when using a ULA. By substituting (4.14) and (4.17) into the general expression for h_m in (4.10), we obtain

$$\sqrt{\beta_m}e^{-\mathrm{j}2\pi\frac{(d_m-d_1)}{\lambda}} \approx \sqrt{\beta}e^{-\mathrm{j}2\pi\frac{(m-1)\Delta\sin(\varphi)}{\lambda}}, \quad m = 1, \ldots, M. \tag{4.18}$$

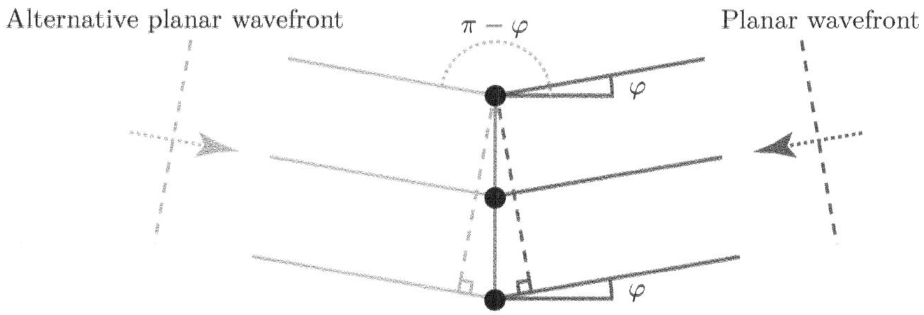

Figure 4.7: The channel vector in (4.19) depends on $\sin(\varphi)$, where φ is the angle-of-arrival of the impinging planar wavefront. The same channel vector is obtained if a planar wavefront impinges from the alternative angle $\pi - \varphi$, which leads to a mirror-like ambiguity.

This expression is unique to ULAs because it explicitly utilizes how the antennas are located with respect to each other. In summary, in the free-space SIMO channel with a ULA, the (approximate) channel vector is

$$\mathbf{h} = \begin{bmatrix} h_1 \\ \vdots \\ h_M \end{bmatrix} = \sqrt{\beta} \begin{bmatrix} 1 \\ e^{-\mathrm{j}2\pi \frac{\Delta \sin(\varphi)}{\lambda}} \\ e^{-\mathrm{j}2\pi \frac{2\Delta \sin(\varphi)}{\lambda}} \\ \vdots \\ e^{-\mathrm{j}2\pi \frac{(M-1)\Delta \sin(\varphi)}{\lambda}} \end{bmatrix}. \tag{4.19}$$

Two variables determine the transmitter's location: the channel gain β and the angle φ. These variables affect $\mathbf{h}$ differently. The norm $\|\mathbf{h}\| = \sqrt{M\beta}$ depends only on the channel gain, while the direction $\frac{\mathbf{h}}{\|\mathbf{h}\|}$ only depends on the angle φ. This is a characteristic feature of far-field propagation.

The angle-of-arrival is measured from the broadside direction, as indicated in Figure 4.6, and can take any value from $-\pi$ and π. The channel vector in (4.19) depends on this angle but only through the sine of it, which creates ambiguity because $\sin(\varphi)$ is not a bijective function for $\varphi \in [-\pi, \pi)$. More precisely, $\sin(\varphi) = \sin(\pi - \varphi)$ for any φ, which implies that every feasible channel vector can be obtained by two different angles-of-arrival. This happens for pairs of incident wavefronts that are each others' mirror reflections, as illustrated in Figure 4.7. This is the reception counterpart of the phenomenon previously illustrated in Figure 1.17 and Figure 1.19: when a ULA with isotropic antennas beamforms in one angular direction, it will also beamform in the mirror-reflected direction. When we continue analyzing ULAs in this chapter, we will mostly consider signals arriving from (or transmitted into) the half-space represented by $\varphi \in [-\pi/2, \pi/2]$. There are two main reasons for this. Firstly, we can illustrate the beamforming concepts more clearly since there will mainly be one beam direction. Secondly, many ULAs deployed in practice use directive antennas that only radiate signals into the directions given by $\varphi \in [-\pi/2, \pi/2]$. The cosine antenna in Figure 1.10 has this property and is suitable for deployment on a wall to cover the half-space in front of the wall. When considering the half-space represented by $\varphi \in [-\pi/2, \pi/2]$, we span the entire range of spatial frequencies from $-1/\lambda$ (when $\varphi = -\pi/2$) to $1/\lambda$ (when $\varphi = \pi/2$) that the channel vector can contain. In other words, we can distinguish between signals impinging from different angles at the same side of the array because they give rise to channel vectors containing different spatial frequencies. However, we cannot uniquely distinguish these signals from their respective mirror reflections.

Example 4.2. Consider a ULA with antenna spacing $\Delta = \lambda/2$ designed for the carrier frequency $f_c = 3$ GHz and bandwidth $B = 20$ MHz. The aperture length is 15λ (i.e., the maximum length from Example 4.1) and the transmitter is located at a distance $d_1 = 50$ m in the angular direction $\varphi = \pi/6$.

(a) What is the number of antennas, M, in the array?

(b) Compute the channel gains of the outermost antennas using the exact formula in (4.14) and comment on the differences.

(c) Compute and compare the two expressions in (4.17) for $m = M$.

The aperture length of the ULA is $M\Delta = M\lambda/2$ in this example.

(a) The length is said to satisfy $M\lambda/2 = 15\lambda$, which implies $M = 30$.

(b) Using $\sin(\pi/6) = 0.5$, $\lambda = 0.1$ m ($f_c = 3$ GHz), $d_1 = 50$ m, and $\Delta = \lambda/2 = 0.05$ m, the squared distance in (4.13) to the last antenna becomes

$$d_M^2 = 50^2 + 29^2 \cdot 0.05^2 + 2 \cdot 50 \cdot 29 \cdot 0.05 \cdot 0.5 \approx 2575\,\text{m}^2. \tag{4.20}$$

The channel gains can now be computed using (4.14) as

$$\beta_1 = \frac{0.1^2}{(4\pi)^2} \frac{1}{50^2} \approx 2.53 \cdot 10^{-8} \approx -76.0\,\text{dB}, \tag{4.21}$$

$$\beta_M \approx \frac{0.1^2}{(4\pi)^2} \frac{1}{2575} \approx 2.46 \cdot 10^{-8} \approx -76.1\,\text{dB}. \tag{4.22}$$

We notice that β_1 and β_M differ by as little as 0.1 dB; thus, the far-field approximation is highly accurate.

(c) The exact distance difference in the left-hand side of (4.17) is $d_M - d_1 \approx \sqrt{2575} - 50 \approx 0.74$ m. The far-field approximation in the right-hand side of (4.17) becomes $(M-1)\Delta\sin(\varphi) = 29 \cdot 0.05 \cdot 0.5 \approx 0.73$ m. The approximation error is around 0.01 m, which is roughly one-tenth of the wavelength; thus, the far-field approximation is highly accurate.

As explained in Section 2.8.3, separating adjacent antennas by half a wavelength is common to obtain spatial samples at twice the maximum spatial frequency $1/\lambda$ that the channel might contain. This corresponds to the antenna spacing $\Delta = \lambda/2$. If we substitute this value into (4.19), it simplifies to

$$\mathbf{h} = \begin{bmatrix} h_1 \\ \vdots \\ h_M \end{bmatrix} = \sqrt{\beta} \begin{bmatrix} 1 \\ e^{-\mathrm{j}\pi \sin(\varphi)} \\ e^{-\mathrm{j}\pi 2 \sin(\varphi)} \\ \vdots \\ e^{-\mathrm{j}\pi (M-1) \sin(\varphi)} \end{bmatrix}. \tag{4.23}$$

We will analyze the impact of other antenna spacings in Section 4.3.4.

4.2.2 SIMO Channel Capacity with ULA

The capacity of a SIMO channel was presented in (3.23) as

$$C = B \log_2 \left(1 + \frac{P \|\mathbf{h}\|^2}{B N_0} \right) \quad \text{bit/s.} \tag{4.24}$$

For a ULA with $\mathbf{h}$ given by (4.19), we have $\|\mathbf{h}\|^2 = M\beta$ which is independent of the antenna spacing. By substituting this value into (4.24), we obtain

$$C = B \log_2 \left(1 + \frac{P M \beta}{B N_0} \right) \quad \text{bit/s.} \tag{4.25}$$

If we compare this expression with the SISO capacity $B \log_2(1 + \frac{P\beta}{BN_0})$ from (2.146), we notice that the SNR is M times larger in the SIMO case. This is the beamforming gain obtained when receiving the same signal at M antennas and optimally combining the observations using MRC. In this case, the MRC vector in (3.19) becomes

$$\mathbf{w} = \frac{\mathbf{h}}{\|\mathbf{h}\|} = \frac{1}{\sqrt{M}} \begin{bmatrix} 1 \\ e^{-\mathrm{j}2\pi \frac{\Delta \sin(\varphi)}{\lambda}} \\ e^{-\mathrm{j}2\pi \frac{2\Delta \sin(\varphi)}{\lambda}} \\ \vdots \\ e^{-\mathrm{j}2\pi \frac{(M-1)\Delta \sin(\varphi)}{\lambda}} \end{bmatrix}. \tag{4.26}$$

Since the channel gain is the same for all receive antennas, all the elements in $\mathbf{w}$ have the same magnitude. In other words, all antennas contribute equally much to improve the SNR achieved over free-space LOS channels. MRC rotates the phases of the received signals so that $\mathbf{w}^{\mathrm{H}}\mathbf{h}$ becomes a sum of M positive terms, each equal to $\sqrt{\beta/M}$. Recall that the phase-shifts in the channel vector are caused by having different propagation delays to the different receive antennas. Since a conjugate transpose is applied to the MRC vector when multiplying it with the channel vector in (3.17), MRC compensates for these delay variations. The result is essentially the same as if the received signal had been sampled at slightly different times at the different antennas.

Example 4.3. How does the SIMO capacity in (4.25) depend on the wavelength λ if the number of antennas is fixed?

The capacity expression depends on the wavelength λ through $\beta = \frac{\lambda^2}{(4\pi)^2}\frac{1}{d^2}$, which was defined in (4.15). Hence, we can express (4.25) as

$$C = B\log_2\left(1 + \frac{PM\lambda^2}{BN_0(4\pi d)^2}\right). \tag{4.27}$$

If M is constant, then the capacity in (4.27) is an increasing function of λ, since the SNR is proportional to λ^2. This implies that the capacity is larger when using low-band spectrum than with high-band spectrum. The reason is that the ULA consists of M isotropic antennas with the wavelength-dependent area $\lambda^2/(4\pi)$ from (1.3). The strength of the electric field that impinges on the ULA is independent of the wavelength, but the array captures less power when the individual receive antennas shrink in size when λ is reduced.

4.2.3 Array Factor and Spatial Filtering

In LOS communications, MRC acts as a *spatial filter* that attenuates any component of the received signal that arrives from an angle that is (substantially) different from φ. This applies to noise as well as interfering signals. The channel vector in (4.19) can be expressed as $\mathbf{h} = \sqrt{\beta}\mathbf{a}(\varphi)$, where the vector

$$\mathbf{a}(\varphi) = \begin{bmatrix} 1 \\ e^{-\mathrm{j}2\pi\frac{\Delta\sin(\varphi)}{\lambda}} \\ e^{-\mathrm{j}2\pi\frac{2\Delta\sin(\varphi)}{\lambda}} \\ \vdots \\ e^{-\mathrm{j}2\pi\frac{(M-1)\Delta\sin(\varphi)}{\lambda}} \end{bmatrix} \in \mathbb{C}^M \tag{4.28}$$

is called the *array response vector* or steering vector. This vector depends on the angle-of-arrival φ through the function $\sin(\varphi)/\lambda$, which we recognize as the spatial frequency the channel vector contains. If two signals arrive from vastly different angular directions, their respective channel vectors contain vastly different spatial frequencies. Consequently, their respective array response vectors point in vastly different directions in the vector space $\mathbb{C}^M$.

To give a concrete example, suppose the desired signal arrives from $\varphi = 0$. The MRC vector in (4.26) then becomes $\mathbf{w} = \mathbf{a}(0)/\sqrt{M} = [1,\ldots,1]^{\mathrm{T}}/\sqrt{M}$. If an interfering signal $\sqrt{\beta_{\text{interf}}}e^{-\mathrm{j}\psi_{\text{interf}}}$ reaches the reference antenna from the angle φ_{interf}, then the signals reaching each of the antennas is computed as

$$\sqrt{\beta_{\text{interf}}}e^{-\mathrm{j}\psi_{\text{interf}}}\mathbf{a}(\varphi_{\text{interf}}), \tag{4.29}$$

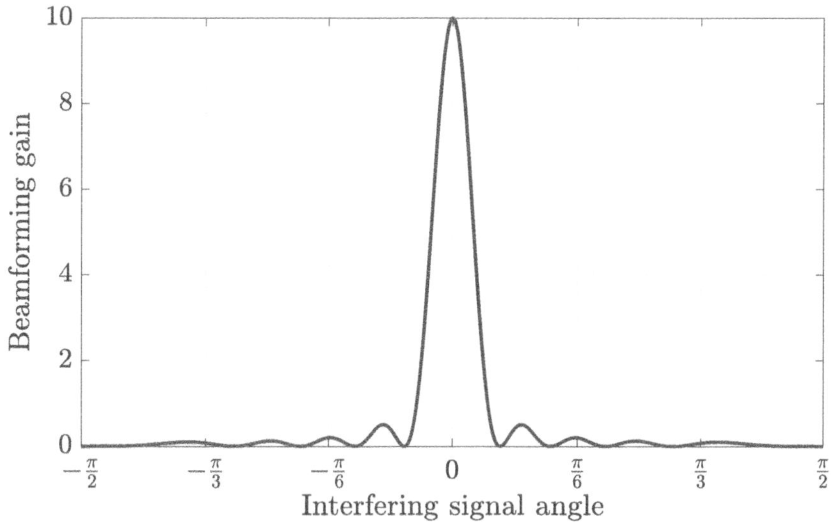

Figure 4.8: An MRC filter designed for an incoming signal from the angle 0 acts as a spatial filter that attenuates any interfering signal arriving from an angle much different than 0. There are $M = 10$ antennas in this example, so the maximum beamforming gain is 10.

since the array response vector determines the relative phase-shift compared to the reference antenna. If we now apply MRC to (4.29), the resulting scalar signal is

$$\mathbf{w}^{\mathrm{H}}\left(\sqrt{\beta_{\text{interf}}}e^{-\mathrm{j}\psi_{\text{interf}}}\mathbf{a}(\varphi_{\text{interf}})\right) = \underbrace{\sqrt{\beta_{\text{interf}}}e^{-\mathrm{j}\psi_{\text{interf}}}}_{\text{Interfering signal}}\underbrace{\mathbf{w}^{\mathrm{H}}\mathbf{a}(\varphi_{\text{interf}})}_{\text{Array factor}}. \tag{4.30}$$

This is the original interfering signal $\sqrt{\beta_{\text{interf}}}e^{-\mathrm{j}\psi_{\text{interf}}}$ at the reference antenna multiplied by the factor $\mathbf{w}^{\mathrm{H}}\mathbf{a}(\varphi_{\text{interf}})$, which is the inner product between the MRC vector and the array response vector for the direction that the interfering signal arrives from. This inner product is called the *array factor* and determines how the array as a whole amplifies/attenuates and phase-shifts the signal by its processing. When talking about spatial filtering, we are interested in the squared magnitude of the array factor:

$$|\mathbf{w}^{\mathrm{H}}\mathbf{a}(\varphi_{\text{interf}})|^2 = \frac{1}{M}|\mathbf{a}^{\mathrm{H}}(0)\mathbf{a}(\varphi_{\text{interf}})|^2 = \frac{1}{M}\left|\sum_{m=1}^{M} e^{-\mathrm{j}2\pi\frac{(m-1)\Delta\sin(\varphi_{\text{interf}})}{\lambda}}\right|^2, \tag{4.31}$$

which determines the relative signal strength compared to the case with a single-antenna receiver. The beamforming gain the filter applies to the interfering signal can attain any value between 0 and M. The precise value in (4.31) depends on the angle φ_{interf}, as it should for a spatial filter.

Figure 4.8 shows $|\mathbf{w}^{\mathrm{H}}\mathbf{a}(\varphi_{\text{interf}})|^2$ for the antenna spacing $\Delta = \lambda/2$, $M = 10$ receive antennas, and varying values of φ_{interf}. The MRC vector is designed for a signal arriving from the angle 0, which has a channel vector with the spatial frequency $\sin(0)/\lambda = 0$. There is an angular interval around 0 where MRC will amplify any arriving signal, with at most a factor $M = 10$. However, any interfering signal arriving outside that angular interval will be greatly attenuated. In this sense, MRC acts as a *spatial bandpass filter*: it only passes through signals from specific spatial directions (i.e., their channel vectors contain spatial frequencies within a specific range). The width of the spatial filter can be quantified analytically as a function of M and Δ, and is generally inversely proportional to the aperture length $M\Delta$. We will postpone the detailed analysis to Section 4.3.2 since this example only intends to introduce spatial filtering from a qualitative perspective.

The derivations in this section have assumed isotropic antennas but can also be applied when using directional antennas. The generalization is obtained by including the antenna gains in β, following the approach in Section 1.1.4, and will be provided later in Section 4.5.

4.2.4 Acquiring Channel State Information

The channel vector $\mathbf{h}$ completely characterizes a deterministic frequency-flat channel, irrespective of whether the general model in (4.11) is used or the ULA-specific model in (4.19). To achieve the SIMO channel capacity in (4.24), the receiver must know $\mathbf{h}$ so that it can first apply MRC and then decode the data symbols. Moreover, the transmitter must know the capacity value C to encode the data accordingly. Thus far, we have assumed this information to be available automatically, but an acquisition mechanism is required in practice. The vector $\mathbf{h}$ is often referred to as the *channel state*, while the available knowledge of it is called the *channel state information (CSI)*. It is sometimes necessary to distinguish between the CSI available at the transmitter and the receiver if these are substantially different. When communicating over deterministic channels, as in this chapter, it is common to assume that perfect CSI is available at both the transmitter and receiver; that is, the channel vector is known precisely. The goal of this section is to justify that statement.

Suppose we transmit a packet of L symbols $\{x[l] : l = 1, \dots, L\}$ over the discrete memoryless SIMO channel

$$\mathbf{y}[l] = \mathbf{h}x[l] + \mathbf{n}[l] \quad \text{for } l = 1, \dots, L, \tag{4.32}$$

where $\mathbf{n}[l] \sim \mathcal{N}_{\mathbb{C}}(\mathbf{0}, N_0\mathbf{I}_M)$ is the receiver noise. We consider the case when $\mathbf{h}$ is unknown at the receiver when initiating the transmission. Since the received signal in (4.32) contains products $\mathbf{h}x[l]$ between the channel and the transmitted symbols, it is hard to separate $\mathbf{h}$ and $x[l]$ if they are both unknown. To resolve this ambiguity, we can divide the packet into two parts:

1. A preamble part with L_{p} predefined symbols that enables estimation of $\mathbf{h}$ since $x[l]$ is known for $l = 1, \dots, L_{\mathrm{p}}$;
2. A payload part with $L - L_{\mathrm{p}}$ symbols where detection of the random data symbols $x[l]$ (for $l = L_{\mathrm{p}} + 1, \dots, L$) is possible since $\mathbf{h}$ is now known.

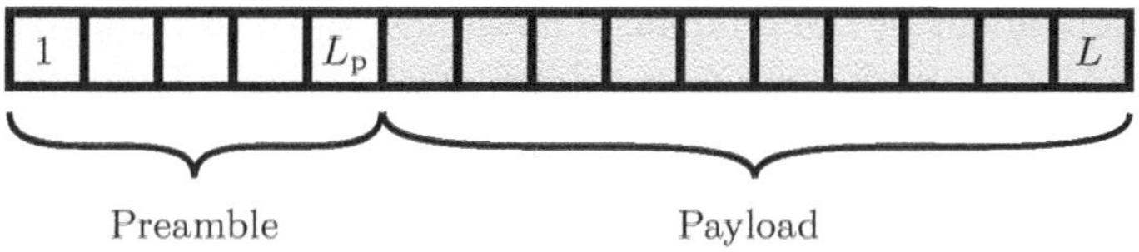

Figure 4.9: A data packet contains a preamble that can be used for channel estimation and a payload with data symbols.

Such a packet is illustrated in Figure 4.9. We have analyzed the second part when characterizing the capacity; thus, we will focus on the preamble in this section. The preamble is often called a *pilot* sequence since it tests the channel quality before the data transmission commences. To comply with the symbol power constraint $\mathbb{E}\{|x[l]|^2\} \leq q$, we can utilize the constant symbols

$$x[l] = \sqrt{q}, \quad l = 1, \dots, L_\mathrm{p}, \tag{4.33}$$

in the preamble. Suppose we compute the sample average of the received preamble signals (divided by $\sqrt{q}$):

$$\frac{1}{L_\mathrm{p}\sqrt{q}} \sum_{l=1}^{L_\mathrm{p}} \mathbf{y}[l] = \mathbf{h} \underbrace{\frac{1}{L_\mathrm{p}\sqrt{q}} \sum_{l=1}^{L_\mathrm{p}} \sqrt{q}}_{=1} + \frac{1}{L_\mathrm{p}\sqrt{q}} \sum_{l=1}^{L_\mathrm{p}} \mathbf{n}[l] = \mathbf{h} + \mathbf{n}' \tag{4.34}$$

where $\mathbf{n}' = \frac{1}{L_\mathrm{p}\sqrt{q}} \sum_{l=1}^{L_\mathrm{p}} \mathbf{n}[l] \sim \mathcal{N}_\mathbb{C}(\mathbf{0}, \frac{N_0}{L_\mathrm{p}q}\mathbf{I}_M)$ since we are computing a weighted sum of independent noise vectors, each having variance N_0. We can notice that (4.34) is equal to the channel vector $\mathbf{h}$ plus a noise vector $\mathbf{n}'$ whose entries have a variance $\frac{N_0}{L_\mathrm{p}q}$ that is inversely proportional to the length L_p of the preamble (called a *processing gain*). The noise variance will go to zero as $L_\mathrm{p} \to \infty$, and so will all random realizations of $\mathbf{n}'$ that are likely to occur.[3] Hence, $\frac{1}{L_\mathrm{p}\sqrt{q}} \sum_{l=1}^{L_\mathrm{p}} \mathbf{y}[l] \to \mathbf{h}$ so that the receiver has acquired a perfect estimate of $\mathbf{h}$.

This example shows that, by making the preamble sufficiently long, we can achieve any desired exactness of the CSI. When quantifying the estimation error for finite values of L_p, it is important to relate the magnitude of the error to the magnitude of the channel. This can be done by considering the NMSE metric from (2.160), which in this context becomes

$$\mathrm{NMSE} = \frac{\mathbb{E}\{\|\mathbf{n}'\|^2\}}{\mathbb{E}\{\|\mathbf{h}\|^2\}} = \frac{N_0}{L_\mathrm{p}q\beta}. \tag{4.35}$$

The NMSE is a decreasing function of the channel gain β, so we can generally use shorter preambles when the channel is strong. Moreover, the NMSE in

[3]The Gaussian distribution has unbounded support; thus, it can give rise to arbitrarily large realizations, but the probability of them occurring goes to zero as $L_\mathrm{p} \to \infty$. It is also important to remember that the Gaussian modeling of receiver noise is approximate since we cannot get arbitrarily large noise realizations in practice.

(4.35) is independent of the number of antennas; thus, it is equally easy/hard to estimate the channel vector to a ULA with $M = 100$ antennas as to a single-antenna receiver. This is because all the receive antennas listen to the same pilot transmission and perform their channel estimation simultaneously.

If we switch focus to the payload part, we recall from Definition 2.6 that the channel capacity is achieved "as the number of symbols in the packet approaches infinity". This implies that we need $L - L_{\mathrm{p}} \to \infty$. However, since only a fraction

$$\frac{L - L_{\mathrm{p}}}{L} = 1 - \frac{L_{\mathrm{p}}}{L} \tag{4.36}$$

of the packet in Figure 4.9 is used for data symbols, the capacity is obtained by multiplying the fraction in (4.36) with the capacity value computed earlier in this chapter. Interestingly, it is possible to operate the system such that this fraction becomes arbitrarily close to 1. For example, if we let $L_{\mathrm{p}} = \sqrt{L}$, we will get perfect CSI as $L \to \infty$ since this implies $L_{\mathrm{p}} \to \infty$. However, the fraction in (4.36) becomes $1 - L_{\mathrm{p}}/L = 1 - 1/\sqrt{L} \to 1$, so asymptotically there is no loss in capacity from having the preamble. In other words, we can safely assume that the receiver has perfect CSI when evaluating the capacity of deterministic channels because we can simultaneously make the preamble large enough to acquire perfect CSI and negligibly small compared to the packet's total length to avoid a capacity loss.

When the preamble has been transmitted, the receiver can compute the channel capacity and feed back this information to the transmitter so that it can encode the data accordingly. In practical systems, there is usually a predefined table of data rates that the system supports using different MCS, such as the one for 5G NR exemplified in Table 2.18. It is then sufficient to feed back a few bits to indicate which table entry to utilize (e.g., 5 bits when there are 28 rows, as in the table).

Example 4.4. Suppose we want to transmit 10 kbit of data over an LOS SIMO channel. We have $M = 8$ antennas and the SNR $\frac{q\beta}{N_0} = 10\,\mathrm{dB}$. How long preamble is needed to achieve an NMSE of 0.01? How many data symbols are needed if we communicate at the capacity?

We need the NMSE in (4.35) to become 0.01, which for the given SNR value means that $\frac{1}{10L_{\mathrm{p}}} = 0.01$. A preamble of $L_{\mathrm{p}} = 10$ symbols satisfies this requirement.

The SIMO capacity in (4.25) can be expressed as $\log_2(1 + \frac{q\beta M}{N_0}) = \log_2(81) = 6.34\,\mathrm{bit/symbol}$ since $q = P/B$. We therefore need $\frac{10 \cdot 10^3}{6.34} \approx 1577$ symbols to transmit 10 kbit. The packet's total length will be $L \approx 1587$, where 99.4% is used for data and 0.6% for the preamble.

4.2.5 Maximum Likelihood Channel Estimation

The previous section described a protocol for acquiring CSI by transmitting a preamble of length L_{p} and then computing the average of the received signals according to (4.34). The result is a *consistent estimate* of the channel vector $\mathbf{h}$, meaning that we obtain an exact estimate as $L_{\mathrm{p}} \to \infty$, but it is not the most accurate estimator for a given finite value of L_{p}. The array geometry and propagation scenario provide valuable information that can be utilized for improved estimation. For example, when considering a ULA in a free-space LOS scenario, we know from (4.19) that only channel vectors with a particular structure can appear:

$$\mathbf{h} = \sqrt{\beta} \underbrace{\begin{bmatrix} 1 \\ e^{-\mathrm{j}2\pi \frac{\Delta \sin(\varphi)}{\lambda}} \\ e^{-\mathrm{j}2\pi \frac{2\Delta \sin(\varphi)}{\lambda}} \\ \vdots \\ e^{-\mathrm{j}2\pi \frac{(M-1)\Delta \sin(\varphi)}{\lambda}} \end{bmatrix}}_{=\mathbf{a}(\varphi)}. \tag{4.37}$$

These feasible channel vectors are parametrized by the channel gain $\beta \in [0,1]$ and array response vector $\mathbf{a}(\varphi) \in \mathbb{C}^M$ from (4.28), which only depends on the angle-of-arrival $\varphi \in [-\pi/2, \pi/2]$ because the impinging wavefront is planar. The fact that the M-dimensional complex-valued channel vector $\mathbf{h}$ is entirely determined by two real-valued variables indicates that only a tiny subset of all vectors in $\mathbb{C}^M$ can appear as channel vectors in LOS communications. We must select a suitable design criterion when designing a *parametric estimator* that utilizes this structural knowledge. We will consider the *maximum likelihood (ML)* criterion that identifies the feasible channel vector most likely to have provided the received signals during the preamble transmission.

The channel vector $\mathbf{h} = \sqrt{\beta}\mathbf{a}(\varphi)$ is deterministic but unknown. The PDF of the received signal $\mathbf{y}[l] = \mathbf{h}\sqrt{q} + \mathbf{n}[l]$ in (4.32) can be expressed as

$$f_{\mathbf{y}[l]}(\mathbf{y}[l]) = \frac{1}{(\pi N_0)^M} e^{-\frac{\|\mathbf{y}[l] - \mathbf{h}\sqrt{q}\|^2}{N_0}} \tag{4.38}$$

because $\mathbf{y}[l] - \mathbf{h}\sqrt{q} \sim \mathcal{N}_{\mathbb{C}}(\mathbf{0}, N_0 \mathbf{I}_M)$ whose PDF is given in (2.80). Since the noise realizations in $\mathbf{y}[1], \ldots, \mathbf{y}[L_{\mathrm{p}}]$ are independent, the joint PDF is

$$\begin{aligned} \prod_{l=1}^{L_{\mathrm{p}}} f_{\mathbf{y}[l]}(\mathbf{y}[l]) &= \frac{1}{(\pi N_0)^{M L_{\mathrm{p}}}} e^{-\sum_{l=1}^{L_{\mathrm{p}}} \frac{\|\mathbf{y}[l] - \mathbf{h}\sqrt{q}\|^2}{N_0}} \\ &= \frac{1}{(\pi N_0)^{M L_{\mathrm{p}}}} e^{-\sum_{l=1}^{L_{\mathrm{p}}} \frac{\|\mathbf{y}[l]\|^2 + \|\mathbf{h}\|^2 q - 2\sqrt{q}\Re\left(\mathbf{h}^{\mathrm{H}}\mathbf{y}[l]\right)}{N_0}} \\ &= \frac{1}{(\pi N_0)^{M L_{\mathrm{p}}}} e^{\frac{2\sqrt{q\beta}}{N_0}\Re\left(\mathbf{a}^{\mathrm{H}}(\varphi)\sum_{l=1}^{L_{\mathrm{p}}} \mathbf{y}[l]\right) - \sum_{l=1}^{L_{\mathrm{p}}} \frac{\|\mathbf{y}[l]\|^2}{N_0} - \frac{L_{\mathrm{p}} M \beta q}{N_0}}, \end{aligned} \tag{4.39}$$

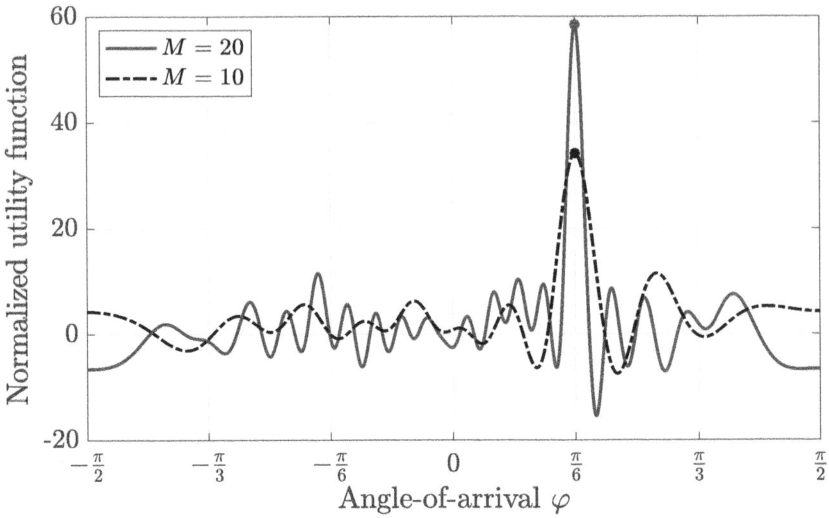

Figure 4.10: The utility function in (4.40) depends on the potential angles-of-arrival φ, as shown in the figure for one random noise realization. The utility is maximized to obtain the ML estimate $\hat{\varphi}$ in (4.40). The true angle-of-arrival is $\pi/6$ and the ML estimates are marked by stars. The peak values become more distinct as the number of antennas increases.

where the last equality utilizes the fact that $\|\mathbf{h}\|^2 = M\beta$. The ML estimates of β and φ are the values that jointly maximize (4.39), which is equivalent to maximizing the argument of the exponential function. If we begin by considering the angle-of-arrival estimation, the angle only appears in the term $\Re(\mathbf{a}^{\mathrm{H}}(\varphi)\sum_{l=1}^{L_{\mathrm{p}}}\mathbf{y}[l])$; thus, the ML estimate is obtained as

$$\hat{\varphi} = \underset{\varphi \in \left[-\frac{\pi}{2}, \frac{\pi}{2}\right]}{\arg\max} \; \Re\left(\mathbf{a}^{\mathrm{H}}(\varphi)\sum_{l=1}^{L_{\mathrm{p}}}\mathbf{y}[l]\right). \tag{4.40}$$

We should look for the array response vector $\mathbf{a}(\varphi)$ that has the largest real part of the inner product with the sample average of the received signals (except for some missing scaling factors that will not affect the solution). This corresponds to comparing the average received signal with the plausible signal vectors obtained with different spatial frequencies to determine the best match. The maximum can be found by doing a one-dimensional search over the range of possible angles.

Figure 4.10 shows the utility function in (4.40), normalized by $\sqrt{L_{\mathrm{p}}N_0}$ so that each entry of $\sum_{l=1}^{L_{\mathrm{p}}}\mathbf{n}[l]$ has unit variance, for different potential values of φ. We consider a scenario with $\Delta = \lambda/2$ and $\mathrm{SNR} = q\beta L_{\mathrm{p}}/N_0 = 10\,\mathrm{dB}$. The true angle-of-arrival is $\pi/6$ and the number of antennas is either $M = 10$ or $M = 20$. The utility function oscillates, but there are distinct maximum peaks in both cases, and the estimate $\hat{\varphi}$ in (4.40) is obtained at the peak values (marked by stars). The ML estimator exploits the ULA's spatial filtering feature to identify the angle of the arriving signal. As the number of antennas increases, the peak becomes taller and narrower, which implies that the estimation accuracy improves with M. An equivalent interpretation is that we estimate the spatial frequency of the channel as $\sin(\hat{\varphi})/\lambda$, and we can distinguish between smaller variations when having more antennas. This is a vital benefit of the parametric ML estimator compared to the non-parametric sample-average estimator in (4.34), which achieves an NMSE independent of M. The reason is that the former only needs to estimate the two parameters β and φ, while the latter needs to estimate M parameters.

Example 4.5. Consider the ML channel estimation method described in this section and assume the angle-of-arrival is estimated perfectly: $\hat{\varphi} = \varphi$. Let $\alpha = \sqrt{\beta}$ denote the square root of the channel gain. What is the ML estimate of α? What are the mean and variance of the estimation error?

The ML estimate of α is obtained by modifying (4.43) as

$$\hat{\alpha} = \underset{\alpha \in [0,1]}{\arg\max} \ \frac{2\sqrt{q}\alpha}{N_0} \Re\left(\mathbf{a}^{\mathrm{H}}(\varphi) \sum_{l=1}^{L_{\mathrm{p}}} \mathbf{y}[l] \right) - \frac{L_{\mathrm{p}} M \alpha^2 q}{N_0} = \frac{\Re\left(\mathbf{a}^{\mathrm{H}}(\varphi) \sum_{l=1}^{L_{\mathrm{p}}} \mathbf{y}[l] \right)}{L_{\mathrm{p}} M \sqrt{q}},$$

where the solution is obtained by taking the first-order derivative with respect to α, equating it to zero, and solving the equation. We notice that $\hat{\alpha}$ is the square root of the ML estimate of β in (4.43) (when $\hat{\varphi} = \varphi$).

Recalling $\mathbf{h} = \sqrt{\beta}\mathbf{a}(\varphi)$ and the received signals from (4.32), we write $\hat{\alpha}$ as

$$\begin{aligned} \hat{\alpha} &= \frac{\Re\left(\mathbf{a}^{\mathrm{H}}(\varphi) \sum_{l=1}^{L_{\mathrm{p}}} \left(\sqrt{\beta}\mathbf{a}(\varphi)\sqrt{q} + \mathbf{n}[l] \right) \right)}{L_{\mathrm{p}} M \sqrt{q}} \\ &= \underbrace{\frac{\sqrt{\beta}\sqrt{q} \sum_{l=1}^{L_{\mathrm{p}}} \mathbf{a}^{\mathrm{H}}(\varphi)\mathbf{a}(\varphi)}{L_{\mathrm{p}} M \sqrt{q}}}_{=\sqrt{\beta}} + \underbrace{\frac{1}{L_{\mathrm{p}} M \sqrt{q}} \sum_{l=1}^{L_{\mathrm{p}}} \Re\left(\mathbf{a}^{\mathrm{H}}(\varphi)\mathbf{n}[l] \right)}_{=n_\alpha}, \end{aligned} \tag{4.41}$$

where n_α denotes the estimation error and we used $\mathbf{a}^{\mathrm{H}}(\varphi)\mathbf{a}(\varphi) = M$. Since $\mathbb{E}\{\mathbf{a}^{\mathrm{H}}(\varphi)\mathbf{n}[l]\mathbf{n}^{\mathrm{H}}[l]\mathbf{a}(\varphi)\} = \mathbf{a}^{\mathrm{H}}(\varphi)\mathbb{E}\{\mathbf{n}[l]\mathbf{n}^{\mathrm{H}}[l]\}\mathbf{a}(\varphi) = \mathbf{a}^{\mathrm{H}}(\varphi)\mathbf{a}(\varphi)N_0 = MN_0$, it follows that $\mathbf{a}^{\mathrm{H}}(\varphi)\mathbf{n}[l] \sim \mathcal{N}_{\mathbb{C}}(0, MN_0)$ and $\Re(\mathbf{a}^{\mathrm{H}}(\varphi)\mathbf{n}[l]) \sim \mathcal{N}(0, MN_0/2)$. The mean of the estimation error is $\mathbb{E}\{n_\alpha\} = 0$ and the variance is

$$\mathbb{E}\{n_\alpha^2\} = \frac{N_0}{2L_{\mathrm{p}} M q} \tag{4.42}$$

because n_α is the summation of many independent random variables. The error variance in (4.42) decreases with L_{p}, M, and q/N_0. Hence, increasing the number of antennas improves the estimation quality of the channel gain when using the parametric ML estimator.

Even if the angle-of-arrival is estimated imperfectly, we can still use it to estimate the channel gain. Unlike the last example, we will directly estimate the channel gain instead of estimating its square root. Specifically, we can substitute $\hat{\varphi}$ back into (4.39) and look for the value of β that maximizes the PDF. Since only two terms in the exponent contain β, the ML estimate is obtained as

$$\hat{\beta} = \underset{\beta \in [0,1]}{\arg\max} \quad \frac{2\sqrt{q\beta}}{N_0} \Re\left(\mathbf{a}^{\mathrm{H}}(\hat{\varphi}) \sum_{l=1}^{L_{\mathrm{p}}} \mathbf{y}[l]\right) - \frac{L_{\mathrm{p}} M \beta q}{N_0}$$
$$= \frac{\left(\Re\left(\mathbf{a}^{\mathrm{H}}(\hat{\varphi}) \sum_{l=1}^{L_{\mathrm{p}}} \mathbf{y}[l]\right)\right)^2}{L_{\mathrm{p}}^2 M^2 q}. \tag{4.43}$$

The solution is obtained by taking the first-order derivative of the exponent with respect to β, equating it to zero, and solving for β.[4]

In summary, the ML estimate of the channel vector is $\sqrt{\hat{\beta}}\mathbf{a}(\hat{\varphi})$ and computed using (4.40) and (4.43). There are many other channel estimation methods for LOS channels, including those that can simultaneously identify multiple signals arriving from different angles. This is a common problem in radar applications. We refer to [51] for a classic overview of such algorithms. Section 8.1 describes a few of these algorithms.

4.3 Modeling of Line-of-Sight MISO Channels

A MISO channel can be obtained from a SIMO channel by switching the transmitter and receiver roles, as discussed in Section 3.3. Figure 4.11 shows a general free-space MISO LOS setup of the same kind as in Figure 4.2, but with the opposite transmitter/receiver roles. The distances remain the same, with d_m denoting the distance between the transmit antenna m and the receiver. Hence, the SIMO and MISO channels are reciprocal so that the channel vector $\mathbf{h} = [h_1, \dots, h_M]^{\mathrm{T}}$ is the same in both cases. There is no need to repeat any derivations, but we will summarize the results from the last sections. With arbitrary antenna locations and the assumptions leading to frequency flatness, h_m can be computed using (4.10) when antenna 1 is the reference antenna. The channel vector becomes

$$\mathbf{h} = \begin{bmatrix} h_1 \\ \vdots \\ h_M \end{bmatrix} = \begin{bmatrix} \sqrt{\beta_1} \\ \sqrt{\beta_2} e^{-\mathrm{j}2\pi \frac{(d_2 - d_1)}{\lambda}} \\ \sqrt{\beta_3} e^{-\mathrm{j}2\pi \frac{(d_3 - d_1)}{\lambda}} \\ \vdots \\ \sqrt{\beta_M} e^{-\mathrm{j}2\pi \frac{(d_M - d_1)}{\lambda}} \end{bmatrix}. \tag{4.44}$$

[4] We implicitly assumed that the estimate $\hat{\beta}$ in (4.43) is not larger than 1. This is a meaningful assumption since the magnitude of the sample average of the received signals is sufficiently small in practice so that $\hat{\beta} \leq 1$.

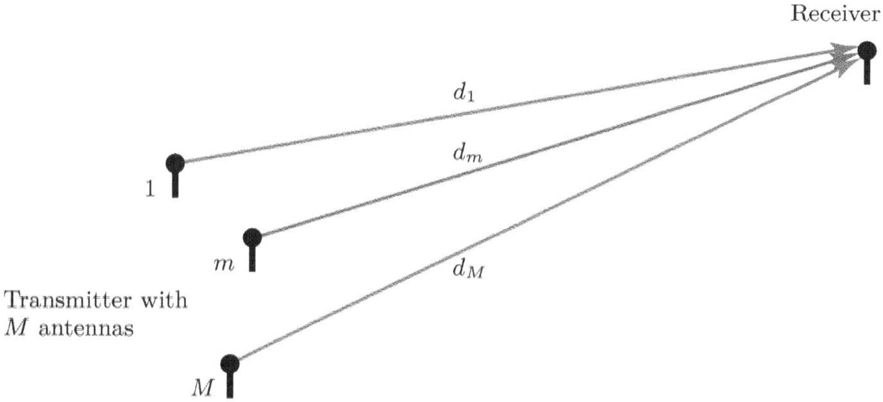

Figure 4.11: A free-space MISO LOS channel where d_m is the distance between the transmit antenna m and the receive antenna for $m = 1, \ldots, M$.

If we restrict ourselves to a ULA at the transmitter with the antenna spacing Δ and φ being the *angle-of-departure* from the first transmit antenna to the receiver, then we obtain the same geometry as in Figure 4.3, except that the transmitter and receiver roles are interchanged. Assuming $d_1 \geq 2M^2\Delta^2/\lambda$, so that the receiver is in the far-field of the ULA, (4.44) simplifies (approximately) to

$$\mathbf{h} = \begin{bmatrix} h_1 \\ \vdots \\ h_M \end{bmatrix} = \sqrt{\beta} \begin{bmatrix} 1 \\ e^{-\mathrm{j}2\pi\frac{\Delta\sin(\varphi)}{\lambda}} \\ e^{-\mathrm{j}2\pi\frac{2\Delta\sin(\varphi)}{\lambda}} \\ \vdots \\ e^{-\mathrm{j}2\pi\frac{(M-1)\Delta\sin(\varphi)}{\lambda}} \end{bmatrix}, \tag{4.45}$$

where β is the common channel gain of all the antennas. The expression depends on the angle-of-departure via the spatial frequency $\sin(\varphi)/\lambda$. This is the same spatial frequency as when the array receives a signal from the angle φ. The expression can be further simplified by setting $\Delta = \lambda/2$ to obtain the typical expression in (4.23) for a half-wavelength-spaced ULA.

4.3.1 MISO Channel Capacity with ULA

The capacity of a MISO channel was presented in (3.49) as

$$C = B\log_2\left(1 + \frac{P\|\mathbf{h}\|^2}{BN_0}\right) \quad \text{bit/s.} \tag{4.46}$$

For the ULA with $\mathbf{h}$ given by (4.45), we have $\|\mathbf{h}\|^2 = M\beta$ independently of the antenna spacing and angle. If we substitute this into (4.46), we obtain

$$C = B\log_2\left(1 + \frac{PM\beta}{BN_0}\right) \quad \text{bit/s,} \tag{4.47}$$

which is the same as for the corresponding SIMO channel. The SNR is M times larger than the corresponding SISO case where only one transmit antenna is used. This beamforming gain is achieved by using the M antennas of the ULA to focus the transmitted signal on the receiver using MRT. In this case, the MRT vector in (3.44) becomes

$$\mathbf{p} = \frac{\mathbf{h}^*}{\|\mathbf{h}\|} = \frac{1}{\sqrt{M}} \begin{bmatrix} 1 \\ e^{\mathrm{j}2\pi \frac{\Delta \sin(\varphi)}{\lambda}} \\ e^{\mathrm{j}2\pi \frac{2\Delta \sin(\varphi)}{\lambda}} \\ \vdots \\ e^{\mathrm{j}2\pi \frac{(M-1)\Delta \sin(\varphi)}{\lambda}} \end{bmatrix}. \tag{4.48}$$

All the elements in $\mathbf{p}$ have the same magnitude $1/\sqrt{M}$ since the channel gain is the same for all transmit antennas, a unique feature of LOS channels. Hence, MRT consists of dividing the transmit power equally between the M antennas and phase-shifting the signals before transmission to make sure that the M signal components combine coherently at the receiver; that is, $\mathbf{h}^{\mathrm{T}}\mathbf{p}$ becomes a sum of M positive terms, each being equal to $\sqrt{\beta/M}$.

The phase-shifts in MRT actually describe different time delays; antennas with slightly longer distances to the receiver will transmit their signals slightly earlier so that all M signals are received synchronously. This principle is illustrated in Figure 4.12, where two of the antennas must transmit earlier to compensate for their longer distances to the receiver. This corresponds to a virtual rotation of the ULA by φ to mimic the situation where the receiver is in the broadside direction. The equivalence between phase-shifts and time delays appears under frequency flatness. ULAs can also be used for a channel that does not satisfy the frequency flatness condition $\max_m \frac{|d_m - d_1|}{c} \ll \frac{1}{B}$ (e.g., due to a huge bandwidth B or vast distance between the outermost antennas), but in this case, we must select the precoding vector differently to match the corresponding channel vector.

An equivalent description is that the channel vector contains the spatial frequency $\sin(\varphi)/\lambda$, and the MRT vector must be matched to that frequency to ensure that the M signal components combine coherently. This interpretation will be instrumental in understanding beamforming from ULAs.

To achieve the MISO capacity, the transmitter needs to know the channel $\mathbf{h}$ so that it can compute the MRT vector in (4.48) and the capacity value in (4.47) that determines how to encode the data symbols. The receiver needs comparably less CSI to decode the transmitted data: it needs to know the factor $\mathbf{h}^{\mathrm{T}}\mathbf{p} = \|\mathbf{h}\|$ that the data signal is multiplied by in the received signal $y = \mathbf{h}^{\mathrm{T}}\mathbf{p}\bar{x} + n$ in (3.41) and the capacity value. The CSI can be acquired by transmitting a preamble, similar to what was described in Section 4.2.4. One option is that the multi-antenna transmitter sends multiple preambles in a sequence (one from each of the M antennas) and lets the receiver feed back

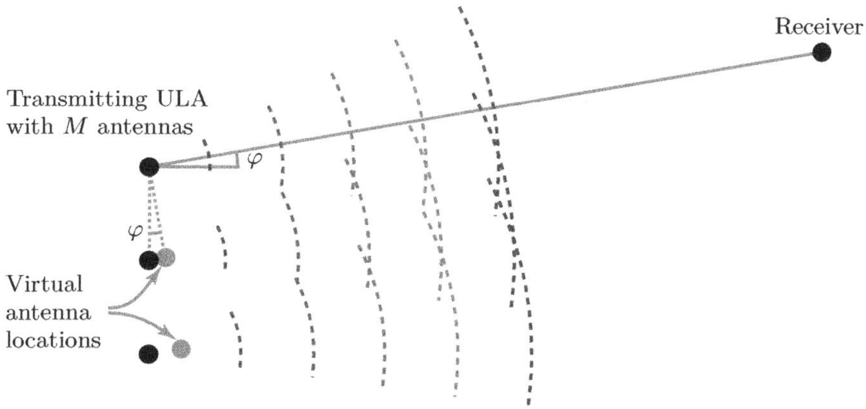

Figure 4.12: When a ULA transmits to a far-field receiver located in a non-broadside direction φ, all antennas except the reference antenna will phase-shift their signals to ensure they are received synchronously at the intended receiver. This is equivalent to a virtual rotation of the ULA by the angle φ to synthesize that the receiver is in its broadside direction. The coloring in this figure represents wave components that are supposed to be received synchronously.

the channel estimates. Since this requires the receiver to transmit something, an alternative is to let the single-antenna receiver transmit the preamble, in which case we can precisely follow the approach for SIMO channels in Section 4.2.4. The fact that the SIMO and MISO channels are reciprocal enables us to send the preamble in any direction. It is typically more efficient to send preambles in the SIMO direction since one can estimate the entire channel vector from a single preamble, while the MISO direction requires one preamble per transmit antenna.[5] Moreover, it is only the multi-antenna device that needs to know the complete channel vector, so it is convenient if it is the one that computes the estimate.

4.3.2 Beamwidth of the Transmitted Signal

When using MRT in free-space LOS communications, the transmitted signal from a ULA takes the shape of a directional beam when measured in the far-field. This was illustrated already in Section 1.2.1. Figure 1.17 shows beamforming in the direction $\varphi = 0$ from a ULA with $M = 10$ antennas and $\Delta = \lambda/2$, while the corresponding case of $\varphi = \pi/2$ is illustrated in Figure 1.19. The equivalent to MRT when using the ULA for reception was also exemplified in Figure 4.8, where we noticed that MRC acts as a spatial filter that only amplifies signals arriving from the preferred angular directions.

When the transmitted signal is directed in the angular direction φ, a receiver located in precisely that direction will obtain a beamforming gain of M.

[5]If the parametric ML estimator is used and the SNR is high, it is sufficient to transmit two preambles in the MISO direction to estimate the two unknown parameters: channel gain and angle. This is, nevertheless, more preambles than in the SIMO direction.

Receivers in other nearby angular directions will also achieve a beamforming gain, but it is smaller than M. The angular interval where a beamforming gain is observed is called the *beamwidth*. The purpose of this section is to quantify the beamwidth for a ULA with $\Delta = \lambda/2$.

We begin by defining the array response vector of dimension M as

$$\mathbf{a}_M(\varphi) = \begin{bmatrix} 1 \\ e^{-\mathrm{j}\pi \sin(\varphi)} \\ e^{-\mathrm{j}\pi 2 \sin(\varphi)} \\ \vdots \\ e^{-\mathrm{j}\pi (M-1) \sin(\varphi)} \end{bmatrix}. \tag{4.49}$$

This is a special case of (4.28), where we considered an arbitrary antenna spacing. The array response vector is equal to $\mathbf{h}/\sqrt{\beta}$ where $\mathbf{h}$ is taken from (4.23); thus, it describes the normalized channel to any receiver located in the far-field in the angular direction φ. The normalization removes β from the channel expression and thereby eliminates the dependence on the propagation distance, which has no impact on the angular properties of the beam.

Suppose we transmit a signal in the direction $\varphi_{\text{beam}} \in [-\pi/2, \pi/2]$ using the MRT vector $\mathbf{p} = \mathbf{a}_M^*(\varphi_{\text{beam}})/\|\mathbf{a}_M(\varphi_{\text{beam}})\|$, then the array factor observed by a receiver located in another direction $\varphi \in [-\pi/2, \pi/2]$ is $\mathbf{a}_M^{\mathrm{T}}(\varphi)\mathbf{p}$. This represents the complex scaling factor the signal will experience compared to the single-antenna case. The beamforming gain is the squared magnitude of the array factor:

$$\begin{aligned} \left| \mathbf{a}_M^{\mathrm{T}}(\varphi) \frac{\mathbf{a}_M^*(\varphi_{\text{beam}})}{\|\mathbf{a}_M(\varphi_{\text{beam}})\|} \right|^2 &= \frac{\left|\mathbf{a}_M^{\mathrm{T}}(\varphi)\mathbf{a}_M^*(\varphi_{\text{beam}})\right|^2}{M} \\ &= \frac{1}{M} \left| \sum_{m=1}^{M} e^{-\mathrm{j}\pi(m-1)(\sin(\varphi)-\sin(\varphi_{\text{beam}}))} \right|^2, \end{aligned} \tag{4.50}$$

where we utilized the fact that $\|\mathbf{a}_M(\varphi_{\text{beam}})\|^2 = M$. The beamforming gain in (4.50) is $\frac{1}{M}|\sum_{m=1}^{M} 1|^2 = M$ for a user in direction $\varphi = \varphi_{\text{beam}}$, as expected. To compute the beamforming gain achieved/observed in other angular directions, we make use of the summation formula for geometric series:

$$\sum_{m=1}^{M} x^{m-1} = \begin{cases} \frac{1-x^M}{1-x}, & \text{if } x \neq 1, \\ M, & \text{if } x = 1. \end{cases} \tag{4.51}$$

The summation in (4.50) is a geometric series with $x = e^{-\mathrm{j}\pi(\sin(\varphi)-\sin(\varphi_{\text{beam}}))}$. The case $x = 1$ occurs when the two angles are equal, $\varphi = \varphi_{\text{beam}}$, and then the beamforming gain in (4.50) becomes M.[6] For any other $\varphi \in [-\pi/2, \pi/2]$,

[6] If we extend the range to $\varphi \in [-\pi, \pi)$, we will also obtain $x = 1$ for $\varphi = \pi - \varphi_{\text{beam}}$. This demonstrates that a ULA with isotropic antennas cannot beamform in one direction without also sending a beam in the mirror-reflection direction that was illustrated in Figure 4.7.

we have $\sin(\varphi) \neq \sin(\varphi_{\text{beam}})$ (leading to $x \neq 1$) and the beamforming gain in (4.50) can be rewritten as

$$\frac{1}{M}\left|\sum_{m=1}^{M} e^{-\mathrm{j}\pi(m-1)(\sin(\varphi)-\sin(\varphi_{\text{beam}}))}\right|^2 = \frac{1}{M}\left|\frac{1-e^{-\mathrm{j}\pi M(\sin(\varphi)-\sin(\varphi_{\text{beam}}))}}{1-e^{-\mathrm{j}\pi(\sin(\varphi)-\sin(\varphi_{\text{beam}}))}}\right|^2$$
$$= \frac{1}{M}\frac{\sin^2\left(M\frac{\pi(\sin(\varphi)-\sin(\varphi_{\text{beam}}))}{2}\right)}{\sin^2\left(\frac{\pi(\sin(\varphi)-\sin(\varphi_{\text{beam}}))}{2}\right)}, \tag{4.52}$$

where the second equality follows from Euler's formula:

$$\sin(x) = \frac{e^{\mathrm{j}x}-e^{-\mathrm{j}x}}{2\mathrm{j}} = e^{\mathrm{j}x}\frac{1-e^{-2\mathrm{j}x}}{2\mathrm{j}}. \tag{4.53}$$

This formula is applied in both the numerator and the denominator. In particular, we utilize that $|1-e^{-2\mathrm{j}x}|^2 = 4|\sin(x)|^2 = 4\sin^2(x)$.

The ratio in (4.52) can be recognized as a squared *Dirichlet kernel/function*, but this terminology from Fourier analysis does not make it easier to grasp its behavior. However, it can be well approximated for small angle differences by a squared sinc-function. By exploiting the fact that $\sin^2(x) \approx x^2$ for argument values close to zero, we obtain[7]

$$\frac{1}{M}\frac{\sin^2\left(M\frac{\pi(\sin(\varphi)-\sin(\varphi_{\text{beam}}))}{2}\right)}{\sin^2\left(\frac{\pi(\sin(\varphi)-\sin(\varphi_{\text{beam}}))}{2}\right)} \approx \frac{1}{M}\frac{\sin^2\left(M\frac{\pi(\sin(\varphi)-\sin(\varphi_{\text{beam}}))}{2}\right)}{\left(\frac{\pi(\sin(\varphi)-\sin(\varphi_{\text{beam}}))}{2}\right)^2}$$
$$= M\mathrm{sinc}^2\left(\frac{M\left(\sin(\varphi)-\sin(\varphi_{\text{beam}})\right)}{2}\right). \tag{4.54}$$

This approximation is tight when the beam angle φ_{beam} and the observation angle φ are similar, in the sense that $\sin(\varphi) \approx \sin(\varphi_{\text{beam}})$. The argument of the sinc-function in (4.54) is the aperture length $M\Delta = M\lambda/2$ of the ULA multiplied by $(\sin(\varphi) - \sin(\varphi_{\text{beam}}))/\lambda$, which is the difference between the spatial frequencies of the channel vector and the MRT vector being used. The sinc-function attains its largest value when the argument is zero, which happens for $\sin(\varphi) = \sin(\varphi_{\text{beam}})$. The general trend is that the function in (4.54)x reduces as the argument attains larger positive or negative values, but it also oscillates and has zero-crossings for integer-valued arguments. This indicates that the beamforming gain is largest in the intended angular direction $\varphi = \varphi_{\text{beam}}$ and then reduces in an oscillating manner.

[7]Another way to obtain this approximation is to interpret the summation in (4.50) as the left Riemann sum of the function $e^{-\mathrm{j}\pi m(\sin(\varphi)-\sin(\varphi_{\text{beam}}))}$ with unit-sized partitions. By replacing it with the corresponding Riemann integral $\int_0^M e^{-\mathrm{j}\pi m(\sin(\varphi)-\sin(\varphi_{\text{beam}}))}\partial m$ and computing its value, we obtain the final expression in (4.54) after some algebra.

Example 4.6. Consider a ULA with M antennas and $\Delta = \lambda/2$. Suppose a signal is transmitted in the direction $\varphi_{\text{beam}} = 0$. Use the exact formula in (4.52) and sinc-approximation in (4.54) to determine the beamforming gain

(a) in the direction $\varphi = \pi/6$ when $M = 10$;

(b) in the direction $\varphi = \pi/60$ when $M = 10$;

(c) in the direction $\varphi = \pi/60$ when $M = 100$.

By inserting the corresponding values into the expressions in (4.52) and (4.54), we obtain the exact and approximate beamforming gains as

(a) exact: $\frac{1}{10}\frac{\sin^2\left(10\frac{\pi\sin(\pi/6)}{2}\right)}{\sin^2\left(\frac{\pi\sin(\pi/6)}{2}\right)} = 0.2,$

approximate: $10 \cdot \mathrm{sinc}^2\left(\frac{10\sin(\pi/6)}{2}\right) \approx 0.16.$

(b) exact: $\frac{1}{10}\frac{\sin^2\left(10\frac{\pi\sin(\pi/60)}{2}\right)}{\sin^2\left(\frac{\pi\sin(\pi/60)}{2}\right)} \approx 7.96,$

approximate: $10 \cdot \mathrm{sinc}^2\left(\frac{10\sin(\pi/60)}{2}\right) \approx 7.94.$

(c) exact: $\frac{1}{100}\frac{\sin^2\left(100\frac{\pi\sin(\pi/60)}{2}\right)}{\sin^2\left(\frac{\pi\sin(\pi/60)}{2}\right)} \approx 1.29,$

approximate: $100 \cdot \mathrm{sinc}^2\left(\frac{100\sin(\pi/60)}{2}\right) \approx 1.29.$

We notice a large beamforming gain of 7.96 in the direction $\varphi = \pi/60$ with $M = 10$ antennas, while it reduces to 1.29 for $M = 100$. This is remarkable since the maximum beamforming gain equals the number of antennas and simultaneously increases from 10 to 100. We notice that the approximate beamforming gains are very similar to the exact gains when φ is close to φ_{beam}, but can otherwise slightly underestimate the gain.

To gain further insights, we will analyze the special case when we transmit in the broadside direction perpendicularly to the array: $\varphi_{\text{beam}} = 0$. It then follows from (4.50) and (4.52) that

$$\left|\mathbf{a}_M^{\mathrm{T}}(\varphi)\frac{\mathbf{a}_M^*(0)}{\|\mathbf{a}_M(0)\|}\right|^2 = \frac{1}{M}\frac{\sin^2\left(M\frac{\pi\sin(\varphi)}{2}\right)}{\sin^2\left(\frac{\pi\sin(\varphi)}{2}\right)}. \tag{4.55}$$

The solid line in Figure 4.13(a) shows the beamforming gain in (4.55) that is observed for angles φ between $-\pi/2$ and $\pi/2$ (i.e., from $-90°$ to $90°$). A plot like this is called the *beam pattern*. We consider $M = 10$ antennas, and the vertical axis is shown in the decibel scale since the beamforming gain variations are substantial. The maximum beamforming gain is 10 dB and is achieved for $\varphi = 0 = \varphi_{\text{beam}}$. This is expected since the ULA focuses its

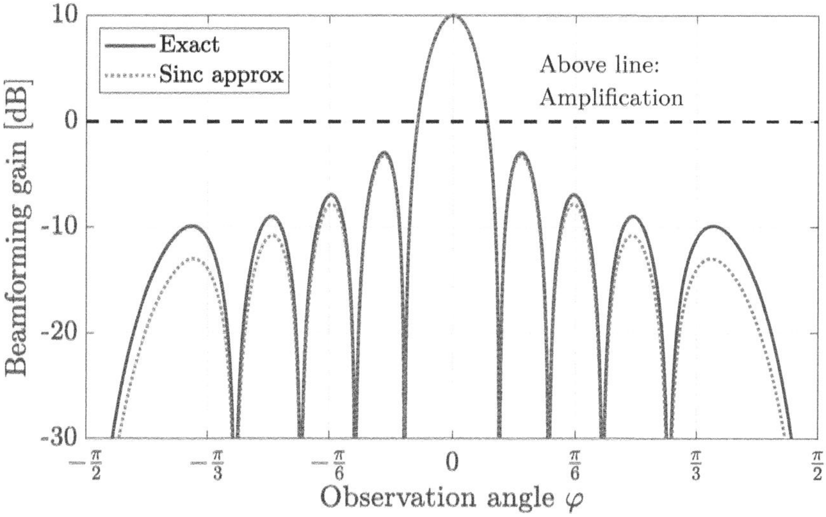

(a) Beamforming gain shown using a rectangular plot.

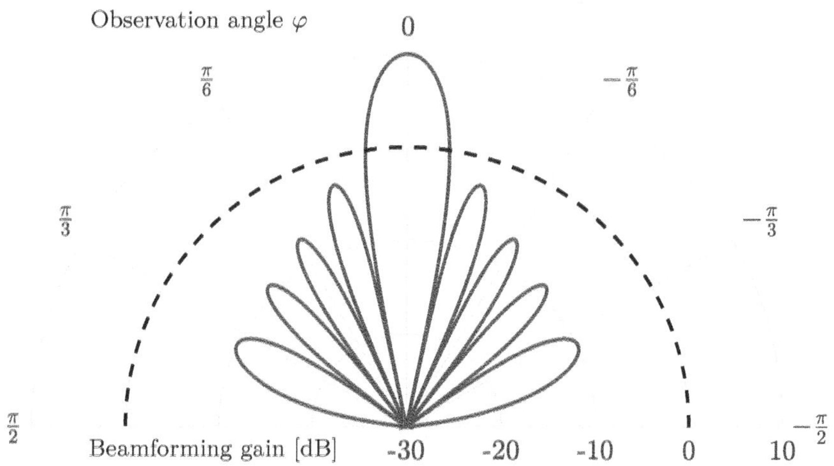

(b) Beamforming gain shown using a polar plot.

Figure 4.13: The beamforming gain that is observed in different directions φ when a ULA with $M = 10$ antennas transmits in the zero-angle direction: $\varphi_{\text{beam}} = 0$. The beamforming gain is computed using (4.52). The angles are measured in radians, but the scale is easy to convert to degrees since $\pi/6$ is $30°$, $\pi/3$ is $60°$, and $\pi/2$ is $90°$.

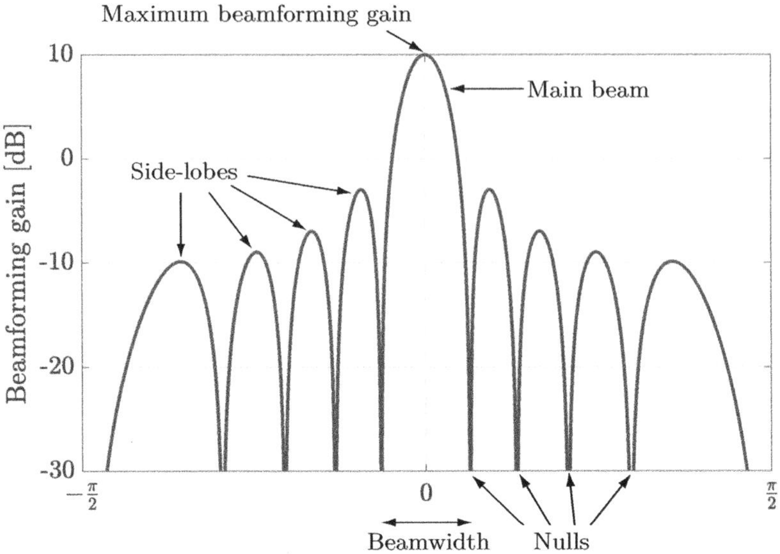

Figure 4.14: A typical beam pattern and summary of the related terminology.

signal in that direction and $10\log_{10}(M) = 10\,\mathrm{dB}$. The beamforming gain gradually reduces when φ is changed, and after a while, it drops below $0\,\mathrm{dB}$. The dashed horizontal line in Figure 4.13(a) corresponds to $0\,\mathrm{dB}$. When the beamforming gain is below this line, the signal is not amplified by the ULA but attenuated compared to the transmission from a single isotropic antenna. This demonstrates that beamforming does not create signal power but merely redistributes power between different angular directions. The beamforming gain oscillates in the left and right parts of the beam pattern, but the general trend is that it decreases as $|\varphi|$ increases. The dotted curve is computed using the sinc-approximation from (4.54). This approximation has the correct zero-crossings but underestimates the maximum gains of the oscillations when φ is much different from φ_{beam}.

Figure 4.13(b) shows the same beam pattern using a polar plot. This type of plot gives a better visualization of the beam directivities since each beam points in its actual angular direction seen from the origin. To achieve this, the angles are presented in the opposite order compared to Figure 4.13(a). The same nine beams can be observed in both cases: A strong *main beam* is in the middle (around $\varphi = 0$) and four *side-beams* on each side. We will refer to the latter as *side-lobes* to reserve the word "beam" for the intended direction. The beamforming gain is precisely zero in between these beams, and those points are called *nulls*. The null locations and the angular widths are easier to measure and compare using the rectangular plot in Figure 4.13(a) because the strength of a beam does not affect how wide it appears in the plot. Hence, we will mainly consider rectangular plots in this book. We summarize this terminology in Figure 4.14.

We want to characterize the width of the main beam because it is within this interval that the beamforming gain is large. The beamwidth can be defined in several different ways. The *half-power beamwidth* is the width of the angular interval in which the beamforming gain is between M and $M/2$. This definition quantifies the angular interval where the beamforming gain is close to the maximum gain. It is also known as the 3 dB-beamwidth since a loss of 1/2 in linear scale can also be expressed as $10\log_{10}(1/2) \approx -3\,\text{dB}$.

Example 4.7. What is the half-power beamwidth when a ULA transmits in the direction $\varphi_{\text{beam}} = 0$? Utilize the sinc-approximation and the fact that $\text{sinc}^2(0.443) \approx \frac{1}{2}$.

Under the sinc-approximation, the lower and upper limits of the half-power beamwidth are obtained by equating the beamforming gain in (4.54) to $M/2$:

$$M\text{sinc}^2\left(\frac{M\sin(\varphi)}{2}\right) = \frac{M}{2} \quad \Rightarrow \quad \text{sinc}^2\left(\frac{M\sin(\varphi)}{2}\right) = \frac{1}{2}. \tag{4.56}$$

Using the facts that $\text{sinc}(0.443) \approx \frac{1}{\sqrt{2}}$ and $\text{sinc}(-0.443) \approx \frac{1}{\sqrt{2}}$, we obtain the lower and upper limits as

$$\frac{M\sin(\varphi)}{2} \approx \pm 0.443 \Leftrightarrow \varphi \approx \pm\arcsin\left(\frac{0.886}{M}\right). \tag{4.57}$$

The half-power beamwidth is the difference between these limits and becomes approximately

$$2\arcsin\left(\frac{0.886}{M}\right), \tag{4.58}$$

which decreases when M increases. When M is large, we can utilize the Taylor approximation $\arcsin(x) \approx x$ that holds for $x \approx 0$ to simplify the half-power beamwidth to $2 \cdot 0.886/M = 1.772/M$. The considered beam transmission is matched to a channel vector containing the spatial frequency $\sin(\varphi_{\text{beam}})/\lambda = 0$ but will provide substantial beamforming gains over channels with spatial frequencies in the interval $[-\frac{0.886}{M\lambda}, \frac{0.886}{M\lambda}]$.

Figure 4.15 shows the main beam from Figure 4.13. The half-power beamwidth is indicated, as well as two alternative beamwidth definitions. Another option is determining the angular interval within the main beam, where the beamforming gain is above 0 dB. We call this the *amplification beamwidth*. This definition quantifies the angular interval where the received power is larger than when transmitting from an isotropic antenna.

One can also measure the total width of the main beam, which we call the *first-null beamwidth*. The benefit of this definition is that it is relatively easy to compute an exact analytical expression, while the drawback is that the beamforming gain is tiny at the edges of the main beam. The expression in

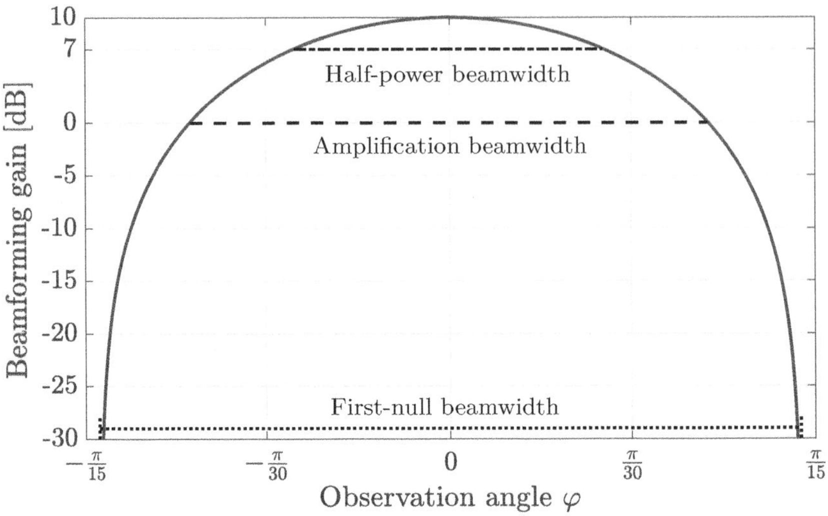

Figure 4.15: There are three possible beamwidth definitions, which have different benefits and drawbacks. This figure illustrates what is measured with these definitions in the setup considered in Figure 4.13.

(4.55) contains the ratio between two sine functions that depend on φ. The lower and upper limits of the main beam occur when the numerator is zero while the denominator is not. This happens for angles φ such that

$$M\frac{\pi \sin(\varphi)}{2} = n\pi \tag{4.59}$$

for some integer $n \neq 0$. The range of solutions to (4.59) is limited by the fact that $\sin(\varphi) \in [-1, 1]$. This implies that there are nulls at the M angles

$$\varphi = \arcsin\left(\frac{2n}{M}\right) \tag{4.60}$$

for $n = \pm 1, \pm 2, \ldots, \pm\lfloor \frac{M}{2} \rfloor$, where $\lfloor \cdot \rfloor$ rounds the argument to the closest smaller or equal integer. The nulls that specify the left and right limits of the main beam are given by $n = \pm 1$, for which (4.60) reduces to

$$\varphi = \pm \arcsin\left(\frac{2}{M}\right). \tag{4.61}$$

Hence, the first-null beamwidth is $2\arcsin(\frac{2}{M})$. If $M \geq 5$, we can utilize the Taylor approximation $\arcsin(x) \approx x$, which is very tight for $x \in [0, 0.4]$, to conclude that the lower and upper limits of the main beam are

$$\varphi \approx \pm\frac{2}{M}. \tag{4.62}$$

Hence, the width of the main beam is approximately $4/M$ radians, which can also be expressed as $(180/\pi) \cdot (4/M) = 720/(M\pi)$ degrees. This expression for the first-null beamwidth is more than twice as large as the approximate half-power beamwidth $1.772/M$ that was derived in Example 4.7 (as can also be seen in Figure 4.15). However, the two beamwidth definitions share the following general behavior: the beamwidth is inversely proportional to the number of antennas M. The more antennas are used, the narrower the beams will be, which has two benefits: the receiver obtains a stronger signal, and less interference is transmitted in other non-intended directions. This is yet another reason for using many antennas in wireless communications.

We can also measure the purity of the beamforming directivity by comparing the gain of the main beam with the peak gains of the largest side-lobes. As noticed earlier, the main beam has a beamforming gain of M. The sinc-approximation in (4.54) implies that the peak of the first lobe has a beamforming gain of $M\text{sinc}^2(3/2) = M(\frac{2}{3\pi})^2$ since a sinc-function has peak values roughly at $0, \pm\frac{3}{2}, \pm\frac{5}{2}, \ldots$. Hence, the main beam is roughly $(\frac{3\pi}{2})^2 \approx 13.5\,\text{dB}$ stronger than the largest side-lobe. This ratio is independent of the number of antennas; thus, we can shrink the beamwidth by adding extra antennas, but it will not reduce the relative strength of the side-lobes. Following the same approach, we can conclude that the main beam always has a gain that is roughly $(\frac{5\pi}{2})^2 \approx 17.9\,\text{dB}$ stronger than the gain of the second largest side-lobe.

In addition to communication applications, ULAs have been considered for radar applications for many years. The goal is then to detect the angle of a target (e.g., a vehicle) using methods similar to the angle-of-arrival estimation described in Section 4.2.5. In these cases, it is not only the SNR that matters, but the beamwidth determines the *spatial resolution* of the array, also known as the *angular resolution*. For example, Figure 4.10 showed how the utility function in ML estimation looks like a beam around the correct angle. The width matches the beamwidth; thus, more antennas lead to a smaller width, resulting in better estimation accuracy. In radar applications, detecting two targets with an angle difference smaller than the beamwidth is hard because they appear as a single target with a somewhat larger size. Similarly, if we want to transmit communication signals to two LOS receivers simultaneously, the mutual interference is small if their angle separation is larger than the beamwidth. We will consider localization and sensing in Chapter 8.

Figure 4.16 considers the same setup as in the previous figure, except that we are now comparing $M = 10$ and $M = 20$ to show how the beamwidth shrinks as we increase the number of antennas (irrespective of which beamwidth definition we consider). When having $M = 20$ antennas, we get roughly half the beamwidth compared to the case of $M = 10$. The width of the side-lobes shrinks similarly, which also means there are more side-lobes. Since we assumed an antenna spacing of $\Delta = \lambda/2$ in this section, increasing the number of antennas is equivalent to making the ULA wider. If we generalized the results to consider other antenna spacings, we would observe that the aperture length of the ULA determines the beamwidth and not the number of antennas. We will return to this in Section 4.3.4.

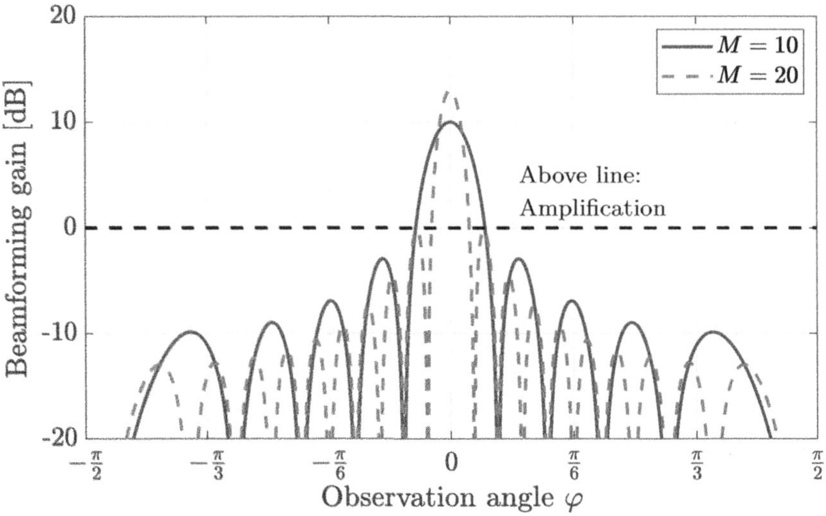

Figure 4.16: Comparison of the beamforming gains with $M = 10$ and $M = 20$ in the same setup as in Figure 4.13.

The beamwidth also depends on which angular direction we point the beam in. If we consider an arbitrary beam direction $\varphi_{\text{beam}} \in [-\pi/2, \pi/2]$, then the nulls appear when the sine function in the numerator of (4.52) is zero while the denominator is non-zero. This happens for angles φ such that

$$M \frac{\pi (\sin(\varphi) - \sin(\varphi_{\text{beam}}))}{2} = n\pi \tag{4.63}$$

for non-zero integers n satisfying $-\frac{M}{2}(1+\sin(\varphi_{\text{beam}})) \leq n \leq \frac{M}{2}(1-\sin(\varphi_{\text{beam}}))$. The limits of the main beam are usually obtained by $n = \pm 1$, which results in

$$\varphi = +\arcsin\left(\frac{2}{M} + \sin(\varphi_{\text{beam}})\right), \tag{4.64}$$

$$\varphi = -\arcsin\left(\frac{2}{M} - \sin(\varphi_{\text{beam}})\right). \tag{4.65}$$

The first-null beamwidth is the difference between (4.64) and (4.65):

$$\arcsin\left(\frac{2}{M} + \sin(\varphi_{\text{beam}})\right) + \arcsin\left(\frac{2}{M} - \sin(\varphi_{\text{beam}})\right). \tag{4.66}$$

This is an increasing function of $|\sin(\varphi_{\text{beam}})|$, as can be proved by showing that its first-order derivative is positive for $\sin(\varphi_{\text{beam}}) \geq 0$ and noting that it is a symmetric function of $\sin(\varphi_{\text{beam}})$. Hence, the beamwidth gradually increases as the beam direction is changed from the broadside direction $\varphi_{\text{beam}} = 0$ to the end-fire direction $\varphi_{\text{beam}} = \pm\pi/2$. In other words, the angular resolution is worse in the vicinity of the end-fire direction because the spatial frequency $\sin(\varphi_{\text{beam}})/\lambda$ varies slowly with the beam angle in these situations.

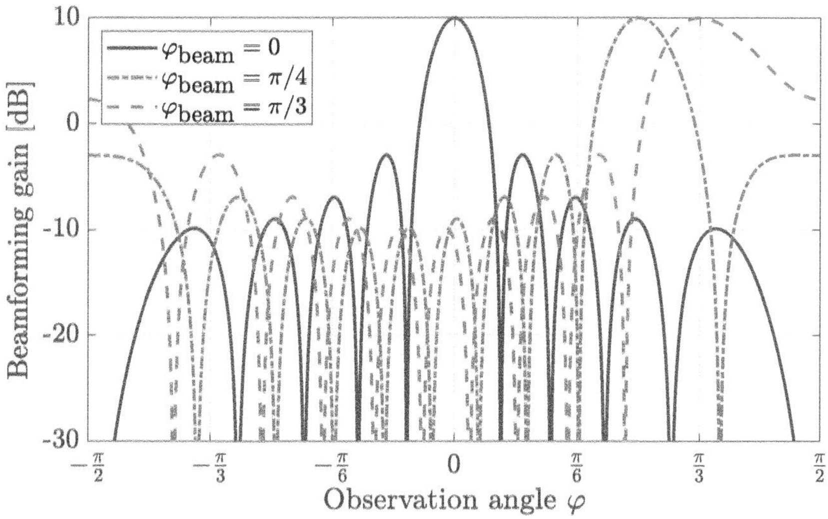

Figure 4.17: The beamforming gain that is observed in different directions φ when a ULA with $M = 10$ antennas transmits a beam in the directions $\varphi_{\text{beam}} = \pi/3$, $\varphi_{\text{beam}} = \pi/4$, or $\varphi_{\text{beam}} = 0$.

When the beam direction is close to $\pi/2$ or $-\pi/2$, it happens that the main beam is divided within the interval $[-\pi/2, \pi/2]$ so that one part appears close to $-\pi/2$ and the other part is close to $+\pi/2$. In this case, the interval $-\frac{M}{2}(1 + \sin(\varphi_{\text{beam}})) \leq n \leq \frac{M}{2}(1 - \sin(\varphi_{\text{beam}}))$ either contains only positive integers or only negative integers. The smallest and largest n then give the nulls of the main beam in the interval.

Figure 4.17 shows the beamforming gains that can be observed in different directions when the beam is transmitted in directions $\varphi_{\text{beam}} = \pi/3$ radians (60°), $\varphi_{\text{beam}} = \pi/4$ radians (45°), or $\varphi_{\text{beam}} = 0$ (the broadside direction as in the previous figures). As expected, the beamwidth is smallest when transmitting in the broadside direction, while it grows when we increase $|\varphi_{\text{beam}}|$ towards any of the end-fire directions $\pm\pi/2$. The main beam is divided into two pieces when $\varphi_{\text{beam}} = \pi/3$, of which the majority appears in the right part of the figure and a small piece appears to the left. The maximum beamforming gain is equal to M in all three cases, but the shape of the signal leakage in other directions is different.

The wider beamwidths obtained with $\varphi_{\text{beam}} = \pi/3$ and $\varphi_{\text{beam}} = \pi/4$ might give the impression that there is more signal power in these cases (e.g., the areas under the curves are larger). However, the precise interpretation is that a larger fraction of the signal power is radiated into the horizontal plane than with broadside beamforming. To demonstrate this property, Figure 4.18 shows the beam patterns for $\varphi_{\text{beam}} \in \{0, \pi/4\}$ in all three dimensions. The ULA is deployed along the y-axis and the xy-plane is the horizontal plane; thus, it is the beam patterns along the dotted curves shown in Figure 4.17. Note that

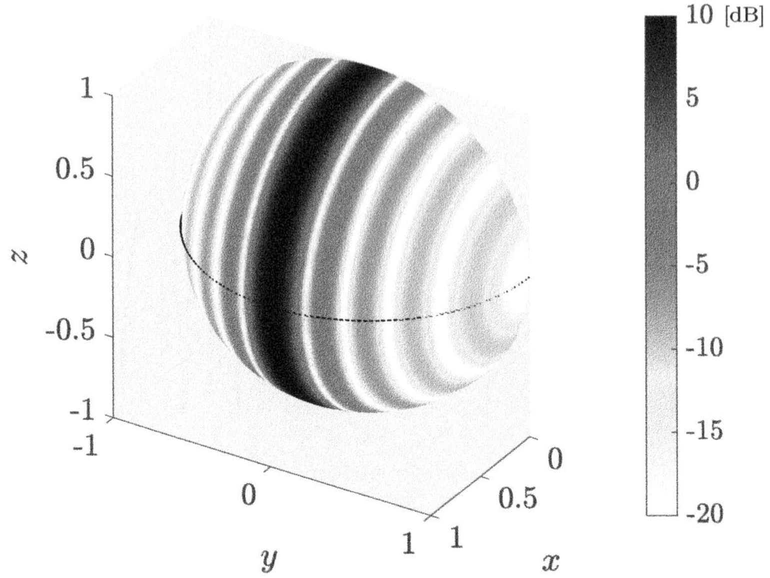

(a) Beamforming in the broadside azimuth direction $\varphi_{\text{beam}} = 0$.

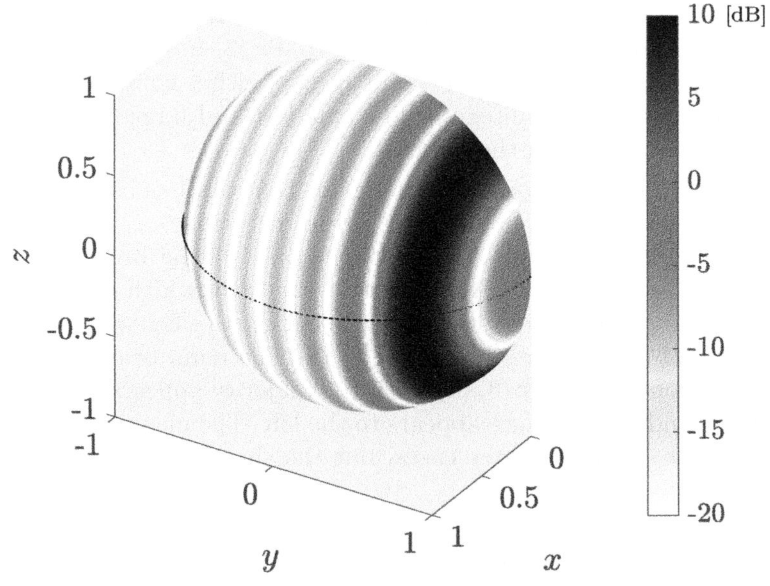

(b) Beamforming in the azimuth direction $\varphi_{\text{beam}} = \pi/4$.

Figure 4.18: The beamforming gain observed in different 3D directions when a ULA with $M = 10$ antennas is deployed along the y-axis. Beamforming in two different azimuth directions is considered, and the dotted curves show the gain variations in the horizontal plane. These are the same beam patterns as shown in Figure 4.17.

the beam patterns are invariant if we rotate them around the y-axis; thus, the beam points in the desired direction in the azimuth plane and many other directions with non-zero elevation angles. The beamwidth in the horizontal plane is indeed smaller with $\varphi_{\text{beam}} = 0$ than with $\varphi_{\text{beam}} = \pi/4$, but this is compensated for since the beam covers more area in the elevation dimension in the former case. The average beamforming gain over the sphere is 1 in both cases. Since the beamwidth in the elevation dimension is broad, covering all elevation angles, the beams created by a ULA have roughly the same shape as orange slices.

It is a common practice to deploy ULAs so that most of the intended receivers are located close to the broadside direction, where the beams are sharper (smaller beamwidth). This feature is essential in radar applications where one can detect targets in different angular directions if the main beams leading to those targets are non-overlapping. Since the total signal power is constant irrespective of the value of φ_{beam}, having a sharp main beam in the desired plane leads to the main beam extending into other dimensions and/or the side-lobes becoming larger. In wireless communications, all the signal power that does not reach the desired receiver can cause interference to other receivers, depending on where those receivers are. Which type of angular beam pattern causes the least interference varies depending on the deployment scenario and distribution of users over the propagation environment. If the users are closely located, the beamwidth should be small so that the non-intended user is not within the main beam. However, if the users have well-separated angles, the beamwidth can be broad because the side-lobes anyway cause all interference.

Example 4.8. An M-antenna ULA with $\Delta = \lambda/2$ transmits with MRT to a receiver in the direction $\varphi_{\text{beam}} = \pi/3$ ($60°$). An unintended receiver is in the direction $\varphi = 13\pi/36$ ($65°$). How many antennas are needed to ensure the unintended receiver is outside the half-power beamwidth?

We must find how many antennas are needed to achieve a beamforming gain smaller than $M/2$ at the angle φ of the unintended receiver. By using the sinc-approximation in (4.54), we can express this condition as

$$M\text{sinc}^2\left(\frac{M\left(\sin(\varphi) - \sin(\varphi_{\text{beam}})\right)}{2}\right) < \frac{M}{2} \Rightarrow \text{sinc}^2\left(0.02014 \cdot M\right) < \frac{1}{2} \quad (4.67)$$

since $(\sin(\varphi) - \sin(\varphi_{\text{beam}}))/2 \approx 0.02014$. We recall from Example 4.7 that $\text{sinc}^2(0.443) \approx \frac{1}{2}$ and notice that $\text{sinc}(x)$ is a decreasing function for $x \in [0, 0.443]$. This implies that (4.67) can be rewritten as

$$0.02014 \cdot M > 0.443 \quad \Rightarrow \quad M > 21.996. \quad (4.68)$$

Since the number of antennas must be an integer, we need at least 22 antennas to ensure that the unintended receiver is outside the half-power beamwidth.

The beamwidth concept is most easily explained when transmitting from a ULA but also exists for arrays with other geometries. In the case of a SIMO channel, the counterpart for signal reception using a ULA is the spatial filtering illustrated in Figure 4.8. If MRC is used to coherently combine the signals from a transmitter with angle-of-arrival $\varphi = 0$, then the beamwidth defines the angular interval around $\varphi = 0$ for which other incoming (interfering) signals will also be partially coherently combined.

4.3.3 Grid of Orthogonal Beams

The beamwidth demonstrates how beamforming focuses the emitted power into a limited angular interval. A receiver in other directions will be reached by comparably less power and possibly zero power if it is located in a null direction. This is a desired feature when transmitting data to a known receiver in a known direction, but it is problematic when the goal is to broadcast signals to unknown receivers in unknown directions. For example, a cellular base station must occasionally announce its existence by broadcasting common messages over its entire coverage area to tell prospective user devices how to connect to the base station, thereby becoming one of the receivers with a known direction. In 5G, the broadcasting starts with the primary synchronization signal (PSS). Generally speaking, we need a procedure to reach prospective users with a relatively high beamforming gain without knowing their locations. To this end, we can preselect a collection of beams and transmit the same common message through each of them. This procedure is called *beam sweeping*. We want to select the beam directions to ensure that any prospective user is located within the main beam of at least one of the beams in the collection.

Consider the MISO channel to an unknown receiver, represented by an unknown M-dimensional vector $\mathbf{h}$ in the vector space $\mathbb{C}^M$. Any non-zero vector can be written as a linear combination of M orthonormal basis vectors $\mathbf{b}_1, \ldots, \mathbf{b}_M \in \mathbb{C}^M$:

$$\mathbf{h} = c_1\mathbf{b}_1 + \ldots + c_M\mathbf{b}_M, \tag{4.69}$$

where at least one of the scalar coefficients $c_1, \ldots, c_M$ is non-zero. If we use the conjugate of the basis vector $\mathbf{b}_i$ as the precoding vector, the SNR at the unknown receiver will be proportional to

$$|\mathbf{h}^{\mathrm{T}}\mathbf{b}_i^*|^2 = |(c_1\mathbf{b}_1^{\mathrm{T}} + \ldots + c_M\mathbf{b}_M^{\mathrm{T}})\,\mathbf{b}_i^*|^2 = |c_i|^2, \tag{4.70}$$

where we utilized that orthonormal vectors satisfy $\mathbf{b}_i^{\mathrm{T}}\mathbf{b}_i^* = 1$ and $\mathbf{b}_m^{\mathrm{T}}\mathbf{b}_i^* = 0$ for $m \neq i$. It is plausible that only one of the coefficients in (4.69) is non-zero for a particular receiver; thus, we will have to transmit beams using all M basis vectors to ensure that the SNR is non-zero for at least one beam. The conclusion is that we need a collection of M beams to reach all users and that those beams should constitute an orthonormal basis. However, many different orthonormal bases can be created. One option is to utilize the columns of the identity matrix $\mathbf{I}_M$ as the basis vectors, which effectively means that we will transmit from one antenna at a time. This is undesirable because there will never be any beamforming gains, but each antenna will spread a relatively weak signal over the coverage area. Instead, we want to divide the coverage area into M subareas (i.e., angular sectors) and let each basis vector beamform towards one of these subareas. We will describe how to do that in the situation considered in the previous sections: the transmitter is equipped with a ULA with $\Delta = \lambda/2$ as the antenna spacing, and there are far-field free-space LOS channels to all the prospective user locations.

If we transmit a beam in the broadside direction $\varphi_{\text{beam}} = 0$, then we recall from (4.60) that there are nulls at the angles

$$\varphi = \arcsin\left(\frac{2n}{M}\right) \tag{4.71}$$

for $n = \pm 1, \pm 2, \dots, \pm\lfloor \frac{M}{2} \rfloor$. If we substitute these angles into (4.49), we can find the array response vectors that are orthogonal to the one obtained by $\varphi_{\text{beam}} = 0$. If we normalize these vectors to have unit length, as we normally do when selecting precoding vectors, we obtain

$$\frac{1}{\sqrt{M}}\mathbf{a}_M\left(\arcsin\left(\frac{2n}{M}\right)\right) = \frac{1}{\sqrt{M}}\begin{bmatrix} 1 \\ e^{-\mathrm{j}\pi n \frac{2}{M}} \\ e^{-\mathrm{j}\pi 2n \frac{2}{M}} \\ \vdots \\ e^{-\mathrm{j}\pi (M-1)n \frac{2}{M}} \end{bmatrix} = \frac{1}{\sqrt{M}}\begin{bmatrix} 1 \\ \upsilon_M^{n} \\ \upsilon_M^{2n} \\ \vdots \\ \upsilon_M^{(M-1)n} \end{bmatrix}, \tag{4.72}$$

where the last expression uses the notation $\upsilon_M = e^{-\mathrm{j}2\pi/M}$ that was first introduced in (2.198) when defining the DFT matrix. Several observations can be made by inspecting (4.72) and comparing it to the $M \times M$ DFT matrix $\mathbf{F}_M$. Firstly, the vector with index n contains samples of the complex exponential $\upsilon_M^n = e^{-\mathrm{j}2\pi\Delta\frac{2n}{M\lambda}}$ with the spatial frequency $\frac{2n}{M\lambda}$. The distance between the samples is the antenna spacing $\Delta = \lambda/2$. Secondly, the vectors in (4.72) for the positive values $n = 1, \dots, \lfloor \frac{M}{2} \rfloor$ are columns of the DFT matrix. Thirdly, the vectors in (4.72) for the negative values $n = -1, -2, \dots, -\lfloor \frac{M}{2} \rfloor$ can be rewritten to only contain positive exponents of υ_M. The key is to utilize the property $\upsilon_M^M = 1$ to rewrite expression as $\upsilon_M^n = \upsilon_M^{M+n}$ for these negative values of n. Hence, all vectors with negative values of n are identical to the last $\lfloor \frac{M}{2} \rfloor$ columns of the DFT matrix. If M is even, then $n = M/2$ and $n = -M/2$ result in the same precoding vector. The convention is to only consider $n = -M/2$ in this case. Finally, the precoding vector for broadside transmission contains only ones, just like the first column of the DFT matrix.

The conclusion is that the columns of the M-dimensional DFT matrix contain a collection of M beams, each associated with beamforming in a distinct angular direction. This is called a *grid of beams* since it is obtained by sampling the range of azimuth angles φ to obtain M grid points. The points are not equally spaced in the angular domain $[-\pi/2, \pi/2]$ but in the spatial frequency domain $[-1/\lambda, 1/\lambda]$ to take the angle-dependent beamwidths into account; we need more beams close to the broadside direction and fewer beams close to the end-fire directions. We identified this particular grid of beams by starting from the first column of the DFT matrix (i.e., beamforming in the broadside direction) and noticing that the other columns of the DFT matrix are orthogonal to it and, thus, can be used to beamform in its null directions. However, we also know that the DFT matrix is unitary, which implies that all columns are mutually orthogonal and constitute an orthonormal basis in $\mathbb{C}^M$. This implies that each column will form a main beam centered at a null of all the other beams. It is, therefore, appropriate to call this a *grid of orthogonal beams* to distinguish it from other prospective collections of beams. It is also known as the *DFT beams* for apparent reasons.

Figure 4.19(a) shows the ten orthogonal beams obtained from the DFT matrix when having $M = 10$ antennas. The seven beams in the middle have similar beamwidths and angular separation, while the outermost beams are substantially broader and, thus, more spread out. This aligns with the previous beamwidth discussion and demonstrates how a ULA has a worse angular resolution close to the end-fire directions. In fact, the outermost beam is split into two parts, of which one-half points to the left and the other half points to the right. Figure 4.19(b) shows the same ten beams but considering the spatial frequency $\sin(\varphi)/\lambda$ on the horizontal axis, in which case all the beams become equally wide. The spatial resolution of a ULA is fundamentally a *spatial frequency resolution*, as we have hinted earlier in this chapter. The maximum beamforming gain is 10 dB for all the beams because $M = 10$. The neighboring beams intersect at the point where the beamforming gain has reduced by almost 4 dB to around 6 dB. Hence, if the considered multi-antenna base station broadcasts a common message by repeating it ten times using these different beams, any LOS user is guaranteed a beamforming gain of at least 6 dB, although the beams were not adapted to any user location.

The loss in beamforming gain remains around 4 dB at the intersection point between two adjacent DFT beams, even if we increase the number of antennas. To prove this, we consider the adjacent beams in the directions $\arcsin(\frac{2n}{M})$ and $\arcsin(\frac{2n+2}{M})$ for some feasible n. The intersection point between these beams is at the angle $\arcsin(\frac{2n+1}{M})$.[8] If we substitute the first beam and the intersection point into (4.52), we obtain the beamforming gain

$$\frac{1}{M}\frac{\sin^2\left(M\frac{\pi\left(\frac{2n+1}{M}-\frac{2n}{M}\right)}{2}\right)}{\sin^2\left(\frac{\pi\left(\frac{2n+1}{M}-\frac{2n}{M}\right)}{2}\right)} = \frac{1}{M}\frac{\sin^2\left(\frac{\pi}{2}\right)}{\sin^2\left(\frac{\pi}{2M}\right)} = \frac{1}{M\sin^2\left(\frac{\pi}{2M}\right)} \geq M\left(\frac{2}{\pi}\right)^2, \tag{4.73}$$

[8]The intersection point is found by comparing (4.52) for the two beam directions and identifying the value of φ where the expressions are equal.

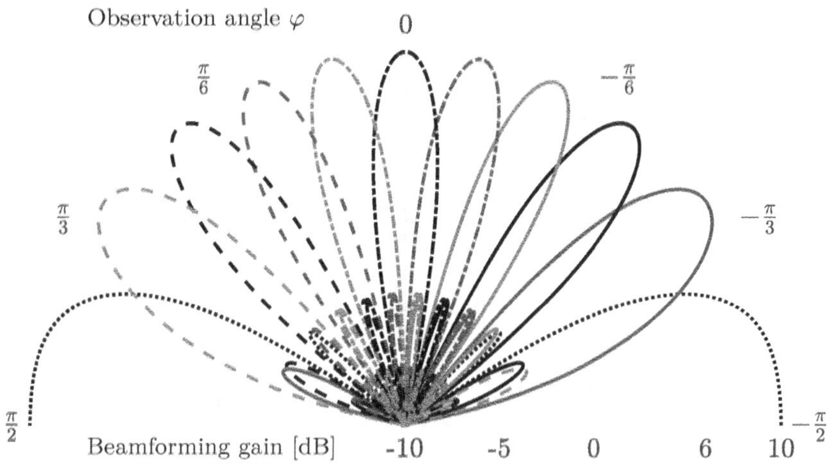

(a) Grid of orthogonal beams shown using a polar plot as a function of the azimuth observation angle φ.

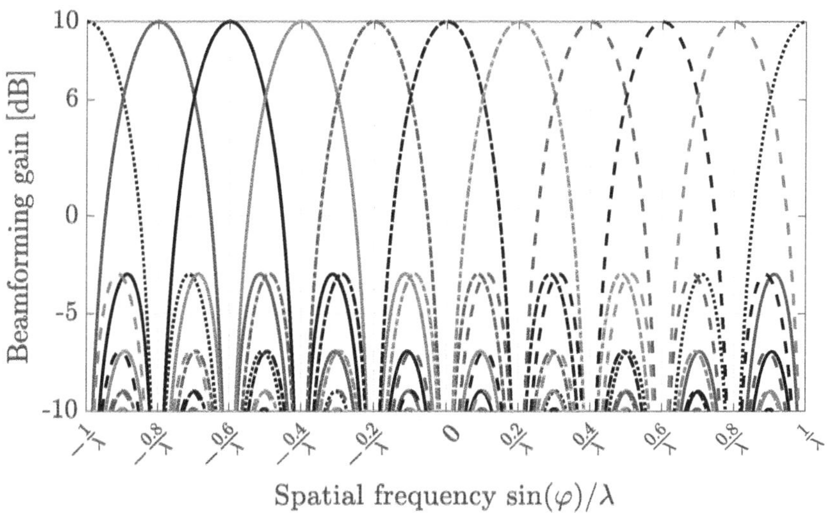

(b) Grid of orthogonal beams as a function of the spatial frequency.

Figure 4.19: The beam patterns of the grid of ten orthogonal beams obtained from the columns of the DFT matrix when using a ULA with $M = 10$ antennas. The angles in (a) are measured in radians, but the scale can easily be converted into degrees since $\pi/6$ is 30°, $\pi/3$ is 60°, and $\pi/2$ is 90°. An equivalent representation as a function of the spatial frequency is given in (b) to showcase that all beams are equally wide in that domain.

where the last step follows from that $\sin^2(x) \leq x^2$. The lower bound is approached when M is large, as seen from making a first-order Taylor approximation of the sine function. This reveals that the reduction in beamforming gain at the intersection point between two DFT beams is at most $20\log_{10}(2/\pi) \approx -3.9\,\text{dB}$ and reaches this value when M grows large. In summary, when a ULA with M antennas transmits a grid of orthogonal beams, all LOS users are guaranteed to find one beam where it achieves a beamforming gain of at least $M(2/\pi)^2$.

4.3.4 Impact of Aperture Length and Antenna Spacing

The beamwidth analysis in the previous sections shows how the directivity of the radiation pattern can be controlled under the assumption of a ULA with antenna spacing $\Delta = \lambda/2$. If the antenna array has a different geometry, the radiation patterns that beamforming creates will be different—a more irregular geometry will result in a more irregular pattern. In this section, we will still consider ULAs but determine the impact of the antenna spacing. Recall from (4.19) that the LOS channel vector with an arbitrary antenna spacing Δ and wavelength λ can be expressed as $\mathbf{h} = \sqrt{\beta}\mathbf{a}(\varphi)$, where the array response vector for the angle-of-departure φ is

$$\mathbf{a}(\varphi) = \begin{bmatrix} 1 \\ e^{-\mathrm{j}2\pi\frac{\Delta\sin(\varphi)}{\lambda}} \\ e^{-\mathrm{j}2\pi\frac{2\Delta\sin(\varphi)}{\lambda}} \\ \vdots \\ e^{-\mathrm{j}2\pi\frac{(M-1)\Delta\sin(\varphi)}{\lambda}} \end{bmatrix}. \tag{4.74}$$

If we transmit a signal in the direction $\varphi_{\text{beam}} \in [-\pi/2, \pi/2]$ using the MRT vector $\mathbf{p} = \mathbf{a}^*(\varphi_{\text{beam}})/\|\mathbf{a}(\varphi_{\text{beam}})\|$, then we can follow the approach in (4.50)–(4.52) to determine the beamforming gain that is observed by a receiver located in any direction $\varphi \in [-\pi/2, \pi/2]$:

$$\left|\mathbf{a}^{\mathrm{T}}(\varphi)\frac{\mathbf{a}^*(\varphi_{\text{beam}})}{\|\mathbf{a}(\varphi_{\text{beam}})\|}\right|^2 = \frac{1}{M}\left|\sum_{m=1}^{M} e^{-\mathrm{j}\frac{2\pi\Delta(m-1)}{\lambda}(\sin(\varphi)-\sin(\varphi_{\text{beam}}))}\right|^2$$
$$= \begin{cases} M, & \text{if } \frac{\Delta}{\lambda}(\sin(\varphi)-\sin(\varphi_{\text{beam}})) \text{ is an integer}, \\ \frac{1}{M}\frac{\sin^2\left(M\frac{\pi\Delta(\sin(\varphi)-\sin(\varphi_{\text{beam}}))}{\lambda}\right)}{\sin^2\left(\frac{\pi\Delta(\sin(\varphi)-\sin(\varphi_{\text{beam}}))}{\lambda}\right)}, & \text{otherwise}. \end{cases} \tag{4.75}$$

The last equality follows from the summation formula for geometric series in (4.51) and Euler's formula. The first row in (4.75) shows that the maximum beamforming gain is M and is achieved for $\varphi = \varphi_{\text{beam}}$ (the intended beamforming direction) because in that case, we have $\frac{\Delta}{\lambda}(\sin(\varphi) - \sin(\varphi_{\text{beam}})) = 0$.

There might be additional values of φ for which $\frac{\Delta}{\lambda}\left(\sin(\varphi) - \sin(\varphi_{\text{beam}})\right)$ becomes and integer. Since the maximum gain is M, it depends on the number of antennas but not the antenna spacing. The reason is that MRT ensures that all antennas' signal components superimpose constructively in the desired angular direction, irrespective of the array geometry. The second expression in (4.75) characterizes the beamforming gain in other directions, and it depends on the normalized antenna spacing relative to the wavelength, which we will denote in this section as

$$\Delta_\lambda = \frac{\Delta}{\lambda}. \tag{4.76}$$

The numerator of the second expression in (4.75) also contains the product between the number of antennas and the normalized antenna spacing, which we will further denote as

$$D_\lambda = \frac{M\Delta}{\lambda} = M\Delta_\lambda. \tag{4.77}$$

This is the normalized aperture length, according to Definition 4.1, which is the physical aperture length $M\Delta$ of the ULA normalized by the wavelength. To analyze how the beamforming gain depends on Δ_λ, D_λ, and the observation angle φ, we first introduce the variable

$$\Phi = \sin(\varphi_{\text{beam}}) - \sin(\varphi). \tag{4.78}$$

For a given value of φ_{beam}, only the range $\Phi \in [\sin(\varphi_{\text{beam}}) - 1, \sin(\varphi_{\text{beam}}) + 1]$ can be achieved since $\sin(\varphi)$ takes values between -1 and 1. When considering all possible beamforming directions, we should consider values of Φ from -2 to 2. We notice that Φ/λ is the difference in spatial frequency between beam direction and observation direction.

The beamforming gain in (4.75) can be expressed as a function of Φ as

$$A(\Phi) = \frac{1}{M} \frac{\sin^2\left(M\frac{\pi\Delta\Phi}{\lambda}\right)}{\sin^2\left(\frac{\pi\Delta\Phi}{\lambda}\right)} = \frac{1}{M}\frac{\sin^2(\pi D_\lambda \Phi)}{\sin^2(\pi\Delta_\lambda\Phi)}. \tag{4.79}$$

The squared sine-function is a periodic function that repeats when the argument changes by $\pm\pi$. This implies that the numerator in (4.79) is a periodic function of Φ with period $1/D_\lambda$ and the denominator is periodic with a period of $1/\Delta_\lambda$. The numerator varies M times faster than the denominator because $D_\lambda/\Delta_\lambda = M$; thus, $A(\Phi)$ has a period of $1/\Delta_\lambda$. We also have that

$$A\left(\frac{m}{D_\lambda}\right) = 0, \quad m = \pm1, \ldots, \pm(M-1), \tag{4.80}$$

since the numerator is zero while the denominator is non-zero at these points. These values correspond to the nulls in the beam pattern. In particular, the main beam around $\Phi = 0$ has its nulls at $\pm 1/D_\lambda$; thus, the first-null beamwidth only depends on the normalized aperture length. The larger the aperture, the smaller the beamwidth, irrespective of whether the aperture is achieved using many antennas with small spacing or few antennas with large spacing. At the points $A(0)$ and $A(\pm 1/\Delta_\lambda)$, where the beam pattern repeats itself, the numerator and denominator in (4.79) are both zero, which makes the function seemingly undefined. However, the limit value is M, as in the first row of (4.75) that represents the maximum beamforming gain. For brevity, we will not write that out explicitly when analyzing $A(\Phi)$, but remember that it is indeed a well-defined continuous function of Φ.

Figure 4.20 shows $A(\Phi)$ in dB-scale for a ULA with $M = 10$ antennas and an antenna spacing of $\Delta_\lambda = 1/2$ wavelengths, which results in a normalized aperture length of $D_\lambda = 5$ wavelengths. The purpose of this figure is to illustrate how the beam patterns for different values of φ_{beam} are obtained from $A(\Phi)$ for $\Phi \in [\sin(\varphi_{\text{beam}}) - 1, \sin(\varphi_{\text{beam}}) + 1]$. In the upper part of the figure, the red dash-dotted curve is obtained by beamforming directed in the broadside direction $\varphi_{\text{beam}} = 0$, while the dotted green curve is obtained by beamforming in the end-fire direction $\varphi_{\text{beam}} = \pi/2$. These curves are drawn as a function of φ but are each obtained by taking the indicated intervals of $A(\Phi)$ in the lower part of the figure and "stretching" them out over all angles. The mapping of the horizontal axes is non-linear since $\Phi = \sin(\varphi_{\text{beam}}) - \sin(\varphi)$ contains the sine-function, but the shape along the vertical axis is unchanged.

The first-null beamwidth is $2/D_\lambda$ when considering $A(\Phi)$. In case of $\varphi_{\text{beam}} = 0$, we have $\varphi = -\arcsin(\Phi)$ and, thus, the beamwidth becomes

$$2 \arcsin\left(\frac{1}{D_\lambda}\right) \approx \frac{2}{D_\lambda} \quad \text{radians} \tag{4.81}$$

when expressed in terms of the observation angle φ. The approximation in (4.81) is tight for $D_\lambda \geq 2.5$ since $\arcsin(x) \approx x$ holds very well for $x \in [0, 0.4]$. Hence, for arrays with aperture lengths beyond a few wavelengths, the beamwidth becomes inversely proportional to the aperture length (irrespective of the antenna spacing). The beamwidth widens as φ_{beam} increases and reaches its maximum in the end-fire direction. However, the heights of the main beam and side-lobes are the same in all these cases.

Since $A(\Phi)$ has a period of $1/\Delta_\lambda = 2$, the main beam at $\Phi = 0$ repeats itself at $\Phi = \pm 2$. This explains why the main beam is divided into two pieces on the green curve that utilizes the range $\Phi \in [0, 2]$. This was previously observed in Figure 4.19(a), where beamforming in one end-fire direction also resulted in a beam pointing in the opposite end-fire direction.

Using the established connection between $A(\Phi)$ and the beam patterns, we will further study how the antenna spacing affects $A(\Phi)$. Figure 4.21 shows $A(\Phi)$ for a ULA with a normalized aperture length of $D_\lambda = M\Delta_\lambda = 5$ wavelengths, but three different configurations:

1. $M = 20$ with $\Delta_\lambda = 1/4$ wavelengths;

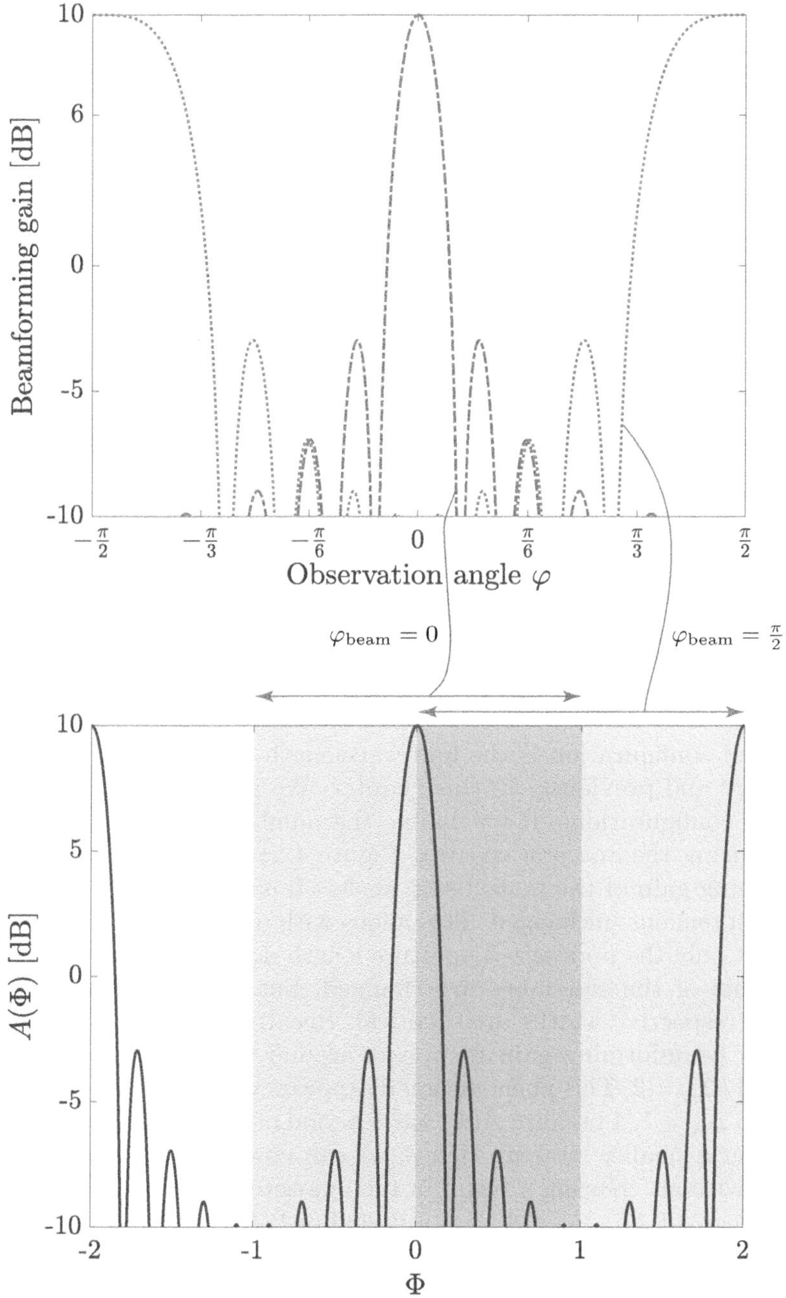

Figure 4.20: The function $A(\Phi)$ in (4.79) is shown in the bottom figure for $M = 10$ antennas and the antenna spacing $\Delta_\lambda = 1/2$ wavelengths. Depending on the beamforming direction φ_{beam}, we take a certain interval $\Phi \in [\sin(\varphi_{\text{beam}}) - 1, \sin(\varphi_{\text{beam}}) + 1]$ from $A(\Phi)$ and use it to generate the resulting beam pattern at the top. The horizontal axis is stretched since $\Phi = \sin(\varphi_{\text{beam}}) - \sin(\varphi)$.This is illustrated for $\varphi_{\text{beam}} = 0$ and $\varphi_{\text{beam}} = \pi/2$.

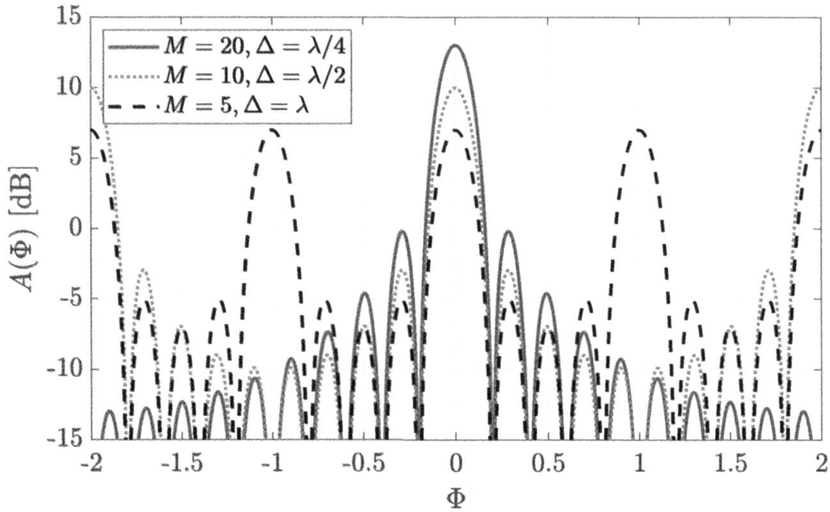

Figure 4.21: The function $A(\Phi)$ in (4.79) is shown for three different ULAs with a normalized aperture length of $D_\lambda = 5$ wavelengths. The first-null beamwidth and locations of other nulls are the same in all three cases, but the sizes of the side-lobes vary.

2. $M = 10$ with $\Delta_\lambda = 1/2$ wavelengths;
3. $M = 5$ with $\Delta_\lambda = 1$ wavelengths.

The second configuration is the half-wavelength-spacing case considered in Figure 4.20 and previously in this chapter. We will compare it to the first and third configurations. If we double the number of antennas to $M = 20$ while reducing the antenna spacing, Figure 4.21 shows that the maximum beamforming gain of the main beam at $\Phi = 0$ is doubled, but the first-null beamwidth remains unchanged. This aligns with our analytical observation in (4.80) that only the normalized aperture length determines the null locations. The heights of the side-lobes are changed, but the number of side-lobes and their respective widths are identical. Recall from Figure 4.20 that the maximum beamforming gain reappears around $\Phi = \pm 2$ since $A(\Phi)$ has a period of $1/\Delta_\lambda = 2$. This phenomenon disappears when the antenna spacing is reduced to $\Delta_\lambda = 1/4$ because $A(\Phi)$ has a period of $1/\Delta_\lambda = 4$ in that case, and we consider a smaller interval. Hence, we can now beamform in one end-fire direction without creating a beam in the opposite end-fire direction.

In the case of $M = 5$, $A(\Phi)$ has a period of $1/\Delta_\lambda = 1$; thus, the side-lobes at $\Phi = \pm 1$ are equally strong as the main beam. These are called *grating lobes* and show that the array cannot distinguish between specific angular directions when the antenna spacing is λ. Apart from this phenomenon, the first-null beamwidth and the locations of the other side-lobes remain the same since these are only determined by the normalized aperture length of the ULA.

Recall that the classical sampling theorem in Lemma 2.8 states that a signal must be sampled at least twice per period (of its largest frequency) to be uniquely distinguishable. We normally apply this theorem by letting the same device take samples at regular time instances. However, since a wireless signal propagates over the wireless medium, we can also take samples at the same time but at different spatial locations. The latter is what an antenna array does during reception. The array response vector $\mathbf{a}(\varphi)$ in (4.74) contains the entries

$$e^{-\mathrm{j}2\pi\frac{\sin(\varphi)}{\lambda}\Delta m}, \quad m = 0, \ldots, M-1, \tag{4.82}$$

which are obtained simultaneously by a ULA with the spatial antenna spacing Δ. The same entries could alternatively be obtained as samples taken once every Δ seconds from a complex exponential with the frequency $\sin(\varphi)/\lambda$ Hz. When the wavelength is λ, the largest observable spatial frequencies (in the magnitude sense) are $\pm 1/\lambda$. The ULA will observe a complex exponential with those spatial frequencies when the signal impinges from the end-fire directions $\varphi = \pm\pi/2$. Since the period is λ in this case, the sampling theorem dictates that complex exponentials with spatial frequencies in $[-1/\lambda, 1/\lambda)$ can only be uniquely distinguished from their samples if $\Delta \leq \lambda/2$ (i.e., sampling twice per period). In analogy to sampling at the Nyquist rate, a ULA with $\Delta_\lambda = 1/2$ is called a *critically spaced* array. A *sparsely spaced* array with $\Delta_\lambda > 1/2$ performs spatial undersampling and gives rise to grating lobes, which is a kind of spatial aliasing where some widely different directions are indistinguishable. A *densely spaced* array with $\Delta_\lambda < 1/2$ performs spatial oversampling, which cannot increase the spatial resolution, just as oversampling of a time-domain signal does not resolve any ambiguities since those disappear at the Nyquist rate. However, oversampling increases the maximum beamforming gain proportionally to M for a given aperture length.

A subtle but important point is that a critically spaced array cannot distinguish between $-1/\lambda$ and $1/\lambda$, which is why the same DFT beam covers both end-fire directions in Figure 4.19. This is because the sampling theorem requires the spatial bandwidth to be strictly smaller than $2/\lambda$ (i.e., equality is not permitted) when sampling at the spatial Nyquist rate of $1/\Delta = 2/\lambda$ (i.e., $\Delta = \lambda/2$). This issue can be disregarded when the ULA uses directive antennas that cannot receive anything from the end-fire directions, as is the case for the cosine antenna in (1.34),

We will now let the number of antennas be fixed but vary the antenna spacing. Figure 4.22(a) shows $A(\Phi)$ with $M = 10$ and either $\Delta_\lambda = 1/2$ (critically spaced) or $\Delta_\lambda = 1/4$ (densely spaced). The aperture length is smaller in the latter case since the number of antennas is fixed. The widths of the main beam and side-lobes increase when the antenna spacing is reduced; recall that the distance between the nulls in (4.80) increases when the normalized aperture length shrinks. The opposite result is seen in Figure 4.22(b), where we compare $\Delta_\lambda = 1/2$ (critically spaced) and $\Delta_\lambda = 1$ (sparsely spaced). A larger antenna spacing results in a narrower main beam but also gives rise to

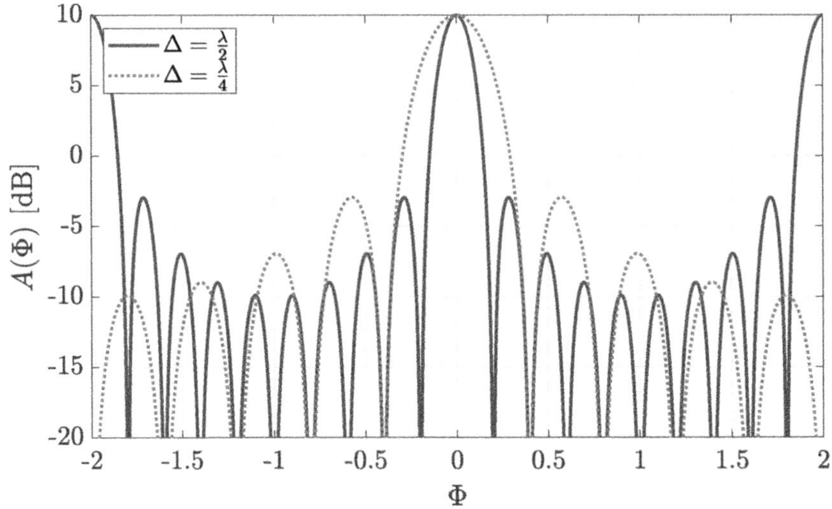

(a) Comparison between critically spaced and densely spaced arrays.

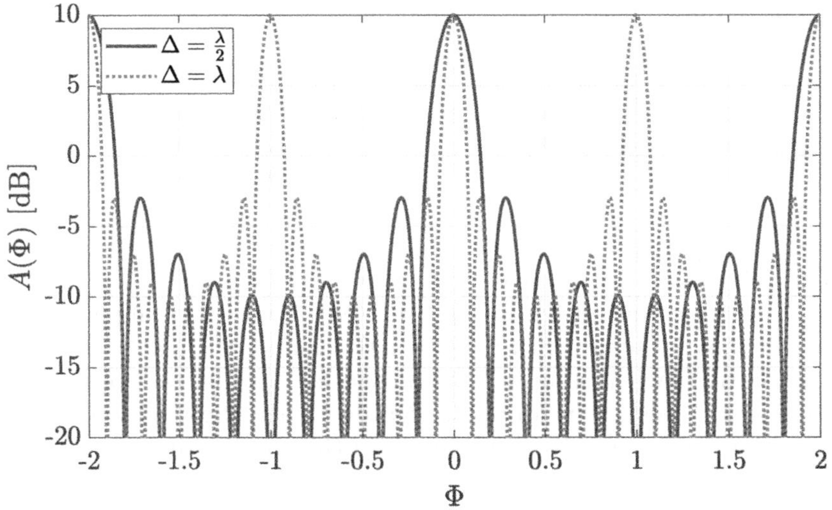

(b) Comparison between critically spaced and sparsely spaced arrays.

Figure 4.22: The function $A(\Phi)$ in (4.79) is shown for ULAs with $M = 10$ antennas, but either critical spacing ($\Delta = \lambda/2$), dense spacing ($\Delta = \lambda/4$), or sparse spacing ($\Delta = \lambda$). A larger antenna spacing leads to smaller beamwidth but will also give rise to grating lobes when the spacing is larger than $\lambda/2$.

grating lobes. The total width of the main beam and grating lobes are always the same. Grating lobes are undesirable if we want to extract information from the channel, such as determining the angle to the receiver, because we cannot distinguish whether the received signal power is strong because the main beam points to the receiver or one of the grating lobes. This ambiguity is primarily a concern in radar and not in communications, where we can even benefit from the fact that the main beam is narrower when there are grating lobes—it gives a higher spatial resolution around the intended beamforming direction, which might improve the ability of spatial multiplexing.

Example 4.9. Consider a ULA deployed vertically in a mast to serve user devices on the ground. The potential users are located in the angular interval $[0, \pi/2]$, so no grating lobes are allowed in this interval when beamforming towards the users. How should the antenna spacing be selected for a given number of antennas, M, to minimize the beamwidth?

The beamwidth is inversely proportional to the normalized aperture length $D_\lambda = M\Delta_\lambda$. Since M is fixed, the beamwidth can be minimized by selecting the largest permitted antenna spacing Δ_λ. The function $A(\Phi)$ in (4.79) is periodic with period $1/\Delta_\lambda$, so we need $|\Phi| \leq 1/\Delta_\lambda$ for all the values of $\Phi = \sin(\varphi_{\textrm{beam}}) - \sin(\varphi)$ that appear in this deployment scenario. We might send a beam to a user device in any direction $\varphi_{\textrm{beam}} \in [0, \pi/2]$ and have prospective receivers in any direction $\varphi \in [0, \pi/2]$. Hence, we require that

$$\max_{\varphi_{\textrm{beam}}, \varphi \in [0,\pi/2]} |\sin(\varphi_{\textrm{beam}}) - \sin(\varphi)| \leq \frac{1}{\Delta_\lambda}. \tag{4.83}$$

Since the sine function takes values between 0 and 1 in the given interval, the maximum difference is 1. As a result, we need to guarantee that

$$1 \leq \frac{1}{\Delta_\lambda} \quad \Rightarrow \quad \Delta_\lambda \leq 1 \text{ wavelength}. \tag{4.84}$$

In conclusion, we achieve the smallest beamwidth (highest spatial resolution) with an antenna spacing of one wavelength. This sparsely spaced array achieves beamwidths roughly half as wide as with the corresponding critically spaced array. The price to pay is that grating lobes are sent into the sky, but this will not cause interference to the users on the ground.

The conclusion is that $\Delta = \lambda/2$ is often the preferred antenna spacing because, for a given number of antennas, it gives the smallest beamwidth achievable without grating lobes (i.e., spatial aliasing), except in the end-fire directions. This is why we considered this spacing earlier in the chapter and will continue doing so in the remainder of this book. However, in situations where grating lobes are acceptable in certain angular intervals, increasing the antenna spacing to reduce the beamwidth in other intervals can be desirable.

4.4 Modeling of Line-of-Sight MIMO Channels

We will now reuse the analysis from the SIMO and MISO cases to characterize the point-to-point MIMO channel matrix $\mathbf{H}$. We assume there are K transmit antennas and M receive antennas. We let $d_{m,k}$ denote the distance between the transmit antenna k and receive antenna m. A detailed derivation of the MIMO channel can be obtained by following the same approach as in Section 4.2.1, but we will only provide the main results. The transmitter and receiver are time-synchronized, meaning that the receiver samples the received signal $\eta = d/c$ seconds after the transmission, where d is a reference distance between the transmitter and receiver. The channel response $h_{m,k}$ between transmit antenna k and receive antenna m is then obtained (similar to (4.10)) as

$$h_{m,k} = \sqrt{\beta_{m,k}} e^{-\mathrm{j}2\pi \frac{(d_{m,k}-d)}{\lambda}}, \tag{4.85}$$

where the phase-shift is $2\pi \frac{(d_{m,k}-d)}{\lambda}$ and the channel gain is

$$\beta_{m,k} = \frac{\lambda^2}{(4\pi)^2} \frac{1}{d_{m,k}^2}. \tag{4.86}$$

By gathering all the channel responses in an $M \times K$ channel matrix, we obtain

$$\mathbf{H} = \begin{bmatrix} h_{1,1} & \dots & h_{1,K} \\ \vdots & \ddots & \vdots \\ h_{M,1} & \dots & h_{M,K} \end{bmatrix} = \begin{bmatrix} \sqrt{\beta_{1,1}} e^{-\mathrm{j}2\pi \frac{(d_{1,1}-d)}{\lambda}} & \dots & \sqrt{\beta_{1,K}} e^{-\mathrm{j}2\pi \frac{(d_{1,K}-d)}{\lambda}} \\ \vdots & \ddots & \vdots \\ \sqrt{\beta_{M,1}} e^{-\mathrm{j}2\pi \frac{(d_{M,1}-d)}{\lambda}} & \dots & \sqrt{\beta_{M,K}} e^{-\mathrm{j}2\pi \frac{(d_{M,K}-d)}{\lambda}} \end{bmatrix}. \tag{4.87}$$

This channel matrix applies to any MIMO LOS setup, regardless of the antenna array geometries or distances. The channel capacity can be computed using Theorem 3.1. In the following sections, we will analyze three specific cases to shed light on the interplay between array deployment and capacity.

4.4.1 MIMO Channel Capacity with ULAs and Planar Wavefronts

We assume that the transmitter and receiver are equipped with ULAs with the same antenna spacing Δ to gain further insights into the channel matrix properties. Moreover, when synchronizing the transmitter and receiver, we use the distance $d = d_{1,1}$ between the first antennas in each array as the reference distance. The transmitter and receiver are assumed to be located in the same two-dimensional plane (e.g., at the same height above the ground).[9] We will use the same approximations as in the SIMO and MISO cases: Frequency

[9]This is a limiting assumption in the MIMO case, which for instance does not cover the case when one ULA is deployed horizontally and the other ULA is deployed vertically. The general case requires other angles to be defined and adjustments to be made in the channel model, but the main conclusions drawn in this section will not change.

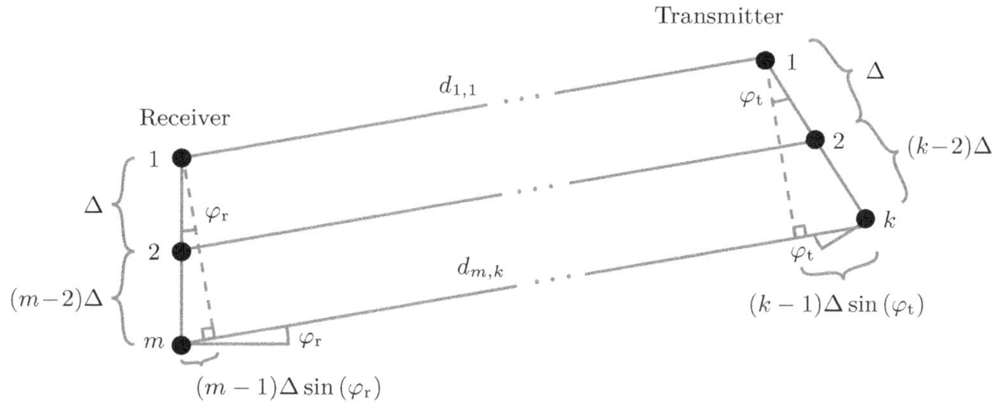

Figure 4.23: Illustration of a MIMO communication setup where the transmitter is equipped with a ULA with k antennas and the receiver is equipped with a ULA with m antennas. The antenna spacing is Δ in each array and the distance between transmit antenna k and receiver antenna m is denoted by $d_{m,k}$. The figure shows a far-field scenario where the angle-of-departure is φ_t for all the transmit antennas, while the angle-of-arrival is φ_r for all the receive antennas.

flatness and that each antenna is in the far-field of the other array. The latter means that $d \geq 2M^2\Delta^2/\lambda$ and $d \geq 2K^2\Delta^2/\lambda$ according to (4.16). Under these approximations, valid in many practical scenarios, there is a common angle-of-departure φ_t for all transmit antennas and a common angle-of-arrival φ_r among all the receive antennas. We illustrate this setup in Figure 4.23. As shown in the figure, the distance $d_{m,k}$ can be (approximately) computed as

$$d_{m,k} = d + (k-1)\Delta\sin(\varphi_t) + (m-1)\Delta\sin(\varphi_r), \tag{4.88}$$

which is the reference distance d plus two additional terms describing the phase differences among the transmit and receive antennas, respectively. These terms are computed trigonometrically, as shown in the figure. The term $(k-1)\Delta\sin(\varphi_t)$ represents the extra propagation distance at the transmitter side, while $(m-1)\Delta\sin(\varphi_r)$ represent the extra propagation distance at the receiver side. Their values can be either positive or negative, depending on the angles, and give rise to different phase-shifts between every pair of antennas.

The far-field assumption also implies that there is a common channel gain

$$\beta = \frac{\lambda^2}{(4\pi)^2}\frac{1}{d^2} \tag{4.89}$$

between any pair of transmit and receive antennas because d is much larger than the latter two terms in (4.88). Hence, $\beta_{m,k} \approx \beta$ for all m and k. Under these far-field conditions, the $M \times K$ channel matrix in (4.87) can be simplified as

$$\mathbf{H} = \begin{bmatrix} h_{1,1} & \dots & h_{1,K} \\ \vdots & \ddots & \vdots \\ h_{M,1} & \dots & h_{M,K} \end{bmatrix}$$

$$= \sqrt{\beta} \begin{bmatrix} 1 & \dots & e^{-\mathrm{j}2\pi \frac{(K-1)\Delta \sin(\varphi_{\mathrm{t}})}{\lambda}} \\ \vdots & \ddots & \vdots \\ e^{-\mathrm{j}2\pi \frac{(M-1)\Delta \sin(\varphi_{\mathrm{r}})}{\lambda}} & \dots & e^{-\mathrm{j}2\pi \frac{(M-1)\Delta \sin(\varphi_{\mathrm{r}})}{\lambda}} e^{-\mathrm{j}2\pi \frac{(K-1)\Delta \sin(\varphi_{\mathrm{t}})}{\lambda}} \end{bmatrix}$$

$$= \sqrt{\beta} \begin{bmatrix} 1 \\ \vdots \\ e^{-\mathrm{j}2\pi \frac{(M-1)\Delta \sin(\varphi_{\mathrm{r}})}{\lambda}} \end{bmatrix} \begin{bmatrix} 1 & \dots & e^{-\mathrm{j}2\pi \frac{(K-1)\Delta \sin(\varphi_{\mathrm{t}})}{\lambda}} \end{bmatrix}. \tag{4.90}$$

Interestingly, (4.90) shows that the matrix $\mathbf{H}$ can be written as an outer product of two vectors when considering free-space LOS channels under the far-field assumption. The two vectors are the channel vectors that one would get with a SIMO channel ($K = 1$) and a MISO channel ($M = 1$), except that the channel gain β only appears once in the expression. The channel matrix in (4.90) is derived for arbitrary antenna spacings, but it is common to consider $\Delta = \lambda/2$. In that special case, we can utilize the array response vector defined in (4.49) to write (4.90) as

$$\mathbf{H} = \sqrt{\beta}\mathbf{a}_M(\varphi_{\mathrm{r}})\mathbf{a}_K^{\mathrm{T}}(\varphi_{\mathrm{t}}). \tag{4.91}$$

We will now compute the capacity of MIMO channels that can be described using the channel matrix in (4.90). We recall that the MIMO channel capacity in Theorem 3.1 depends on the non-zero singular values of $\mathbf{H}$. Since the channel matrix in (4.90) is the outer product of two vectors, it is a matrix with rank one. We can then write its SVD as

$$\mathbf{H} = s_1 \mathbf{u}_1 \mathbf{v}_1^{\mathrm{H}}, \tag{4.92}$$

where

$$s_1 = \sqrt{\beta M K} \tag{4.93}$$

is the only non-zero singular value and the unit-length left and right singular vectors are given by the following normalized array response vectors:

$$\mathbf{u}_1 = \frac{1}{\sqrt{M}}\mathbf{a}_M(\varphi_{\mathrm{r}}) = \frac{1}{\sqrt{M}} \begin{bmatrix} 1 \\ \vdots \\ e^{-\mathrm{j}2\pi \frac{(M-1)\Delta \sin(\varphi_{\mathrm{r}})}{\lambda}} \end{bmatrix}, \tag{4.94}$$

$$\mathbf{v}_1 = \frac{1}{\sqrt{K}}\mathbf{a}_K(\varphi_{\mathrm{t}}) = \frac{1}{\sqrt{K}} \begin{bmatrix} 1 \\ \vdots \\ e^{+\mathrm{j}2\pi \frac{(K-1)\Delta \sin(\varphi_{\mathrm{t}})}{\lambda}} \end{bmatrix}. \tag{4.95}$$

If we substitute s_1 and $r = 1$ into (3.75), the MIMO channel capacity becomes

$$C = \log_2\left(1 + \frac{q_1^{\text{opt}} s_1^2}{N_0}\right) = \log_2\left(1 + \frac{q\beta MK}{N_0}\right), \tag{4.96}$$

where we utilized that $q_1^{\text{opt}} = q$ when there is only one non-zero singular value.

Example 4.10. An early 5G demonstration reached a data rate of 4.3 Gbit/s over a point-to-point LOS channel using $B = 800$ MHz of mmWave spectrum. This example will consider how this value might have been achieved. Suppose the wavelength is $\lambda = 10$ mm, the transmit power is $P = 10$ W, and the noise power spectral density is $N_0 = 10^{-17}$ W/Hz.

(a) If $M = K = 1$ isotropic antennas were used, how large was the propagation distance?

(b) If $M = 64$ and $K = 8$ isotropic antennas were used, how large was the propagation distance?

The capacity of the system is $B\log_2(1 + \text{SNR}) = 4.3 \cdot 10^9$ bit/s, which requires an SNR value of $2^{4.3\cdot 10^9/(8\cdot 10^8)} - 1 \approx 40.5$ for a bandwidth of $8 \cdot 10^8$ Hz.

(a) The SNR in the SISO case is $\text{SNR} = P\frac{\lambda^2}{(4\pi)^2 BN_0}\frac{1}{d^2}$. If we equate it to 40.5 and solve for the distance d, we obtain

$$d = \sqrt{P\frac{\lambda^2}{(4\pi)^2 BN_0 \text{SNR}}} \approx \sqrt{10 \cdot \frac{0.01^2}{(4\pi)^2 \cdot 8 \cdot 10^8 \cdot 10^{-17} \cdot 40.5}} \approx 4.4\,\text{m}. \tag{4.97}$$

(b) The SNR over an 64×8 LOS MIMO channel can be extracted from (4.96) as $\text{SNR} = P\frac{\lambda^2}{(4\pi)^2 BN_0}\frac{MK}{d^2}$, where $MK = 512$. The SNR should still be 40.5, but since the numerator of the SNR has increased by a factor of 512, the squared propagation distance d^2 can increase by the same factor. Hence, the distance is increased to $d \approx 4.4\sqrt{512} \approx 100$ m.

When the point-to-point MIMO channel capacity was discussed in Section 3.4, a major distinguishing factor from the SIMO and MISO capacities was the existence of a multiplexing gain; that is, the ability to transmit $r > 1$ parallel data streams, so that channel capacity grows proportional to r (particularly at high SNR). Since the multiplexing gain is multiplied in front of the logarithm in the capacity expression, it can improve the capacity much more than the beamforming gain, which appears inside the logarithm. Unfortunately, the spatial multiplexing gain cannot be harnessed in the considered LOS setup since $r = 1$. Only a beamforming gain of MK appears in (4.96), in the sense that the SNR is MK times larger than for the corresponding SISO system. The main reason is the far-field situation: the angle-of-departure φ_t is (approximately) the same from all the transmit antennas towards all the receive antennas. Hence, when the transmitter forms a beam towards the center of the receiver array, all the receive antennas are at the center of the main beam. Recall from (3.62) that the capacity is achieved using the left and right singular vectors to turn the channel into r parallel channels. In this case, we only get one such channel, achieved by transmitting the signal using the precoding vector $\mathbf{v}_1$ and processing the received signal using the combining vector $\mathbf{u}_1$. As illustrated in Figure 4.24, this is the same thing as performing MRT at the transmitter based on the MISO channel vector $\mathbf{v}_1^*$, followed by MRC at the receiver based on the SIMO channel vector $\mathbf{u}_1$. Hence, the transmitter and receiver can compute their precoding/combining independently without knowing how many antennas the other device has. To get a rank $r > 1$, the transmitter must be able to transmit multiple beams that are distinguishable at the receiver. This can happen when there are scattering objects that the signal can bounce off, as previously illustrated in Figure 3.16, but not in the considered setup.

Despite the lack of a multiplexing gain, the beamforming gain is larger in MIMO channels compared to SIMO and MISO channels having the same total number of antennas. The following example demonstrates the benefit of having multiple antennas on both sides of a communication system.

Example 4.11. If the total number of transmit and receiver antennas must satisfy $M + K = c$, for some integer c, how should we distribute them between the transmitter and receiver to maximize capacity?

The SNR is proportional to MK in the MIMO capacity expression in (4.96). Since we have the condition $M + K = c$, we can rewrite this beamforming gain as $MK = M(c - M)$. The first-order derivative with respect to M is $c - 2M$ and by equating it to zero, we find that the beamforming gain is maximized if $M = c/2$. Hence, we maximize the capacity by dividing the antennas equally between the transmitter and receiver.

Suppose we have $M + K = 10$ antennas that can be deployed on the transmitter or the receiver. If we create a SIMO system with $M = 9$ and $K = 1$, the beamforming gain is $MK = 9$. However, if we create a MIMO system with $M = 5$ and $K = 5$, the beamforming gain is $MK = 25$. Figure 4.25 shows how the corresponding channel capacities depend on the SNR. Since the SNR of a particular data signal depends on both the number of antennas and whether they are used for beamforming or multiplexing, we define the reference SNR as

$$\text{SNR} = \frac{q\beta}{N_0}. \tag{4.98}$$

This is the SNR that a SISO system achieves under the same propagation

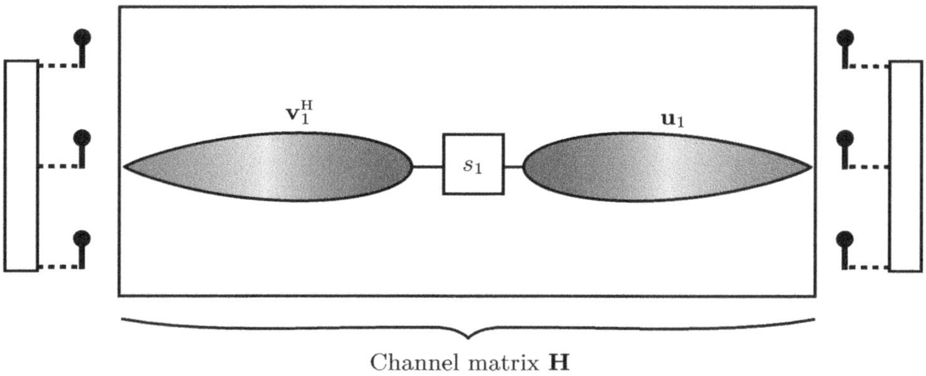

Figure 4.24: An LOS channel between two ULAs only features one propagation path between the transmitter and the receiver. The multiplexing gain is $r = 1$ and the SVD of the channel matrix can be expressed as $\mathbf{H} = s_1 \mathbf{u}_1 \mathbf{v}_1^{\mathrm{H}}$.

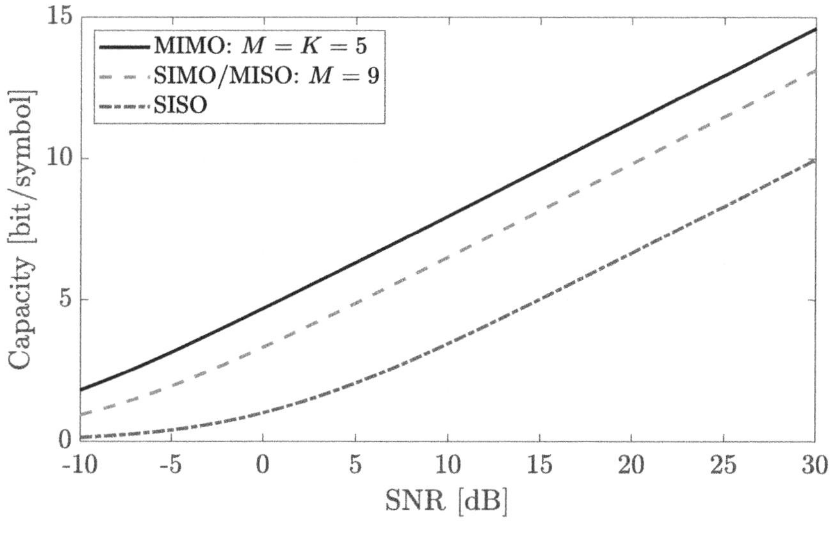

Figure 4.25: The capacity in the MIMO, SIMO/MISO, and SISO cases over far-field LOS channels. The MIMO capacity is $\log_2(1 + 25\,\mathrm{SNR})$ and the SIMO/MISO capacity is $\log_2(1 + 9\,\mathrm{SNR})$, but the total number of antennas is 10 in both scenarios. The SISO capacity $\log_2(1+\mathrm{SNR})$ is also shown and its SNR is used as the reference SNR.

conditions and is used on the horizontal axis in Figure 4.25. All the curves have the same slope since the multiplexing gain is $r = 1$ in all the considered cases, but the beamforming gain shifts the curves toward the left so that a higher capacity is achieved in the MIMO setup for any given SNR. In conclusion, it is beneficial to deploy a MIMO system even in a far-field LOS scenario with ULAs, although it is disappointing that there is no multiplexing gain.

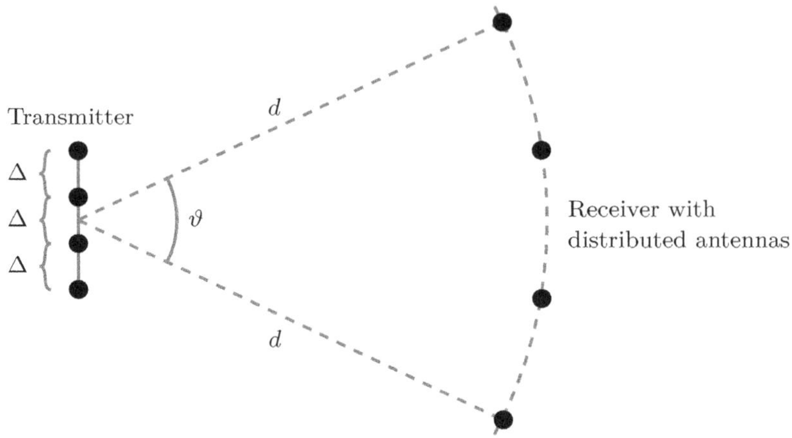

Figure 4.26: Illustration of a MIMO communication setup where the transmitter has a ULA with antenna spacing Δ. The receiver is equipped with a distributed array deployed along the arc of a circle with the radius d. The antennas are uniformly spaced over a circle sector with the central angle ϑ.

4.4.2 MIMO Channel Capacity with Distributed Antennas

The rank deficiency of the MIMO channel matrix that we observed in the last section is inherent in the *point-to-point* terminology. If we transmit from an array at one location to an array at another location, and the propagation distance is large, then the receiver will be at the center of the main beam and can only identify the angular direction of the incoming plane wave; it cannot distinguish between the individual transmit antennas. Similarly, if one watches a brick wall from a distance, one can identify the location of the building but not distinguish individual bricks.

In this section, we will demonstrate that it is the plane-wave/far-field assumption implies the rank-one channel matrix. If we spread out the antennas in one of the arrays to the point where the spherical wavefronts become noticeable, the rank of the channel matrix will increase. A potential such setup is illustrated in Figure 4.26, where the transmitter is equipped with a ULA while the receiver is equipped with a distributed array of antennas. For simplicity, we assume all the receive antennas are at the same distance d from the transmitter's center; thus, the antennas are deployed on the arc of a circle with radius d. The figure illustrates that the angular difference between the outermost receive antennas is called ϑ and the antennas are uniformly spaced on the arc. We can obtain the exact channel matrix based on these geometrical assumptions using (4.87), without making a plane-wave approximation, and then compute the channel capacity using Theorem 3.1.

Figure 4.27 exemplifies the capacity in a setup with $M = K = 4$ antennas and $d = 2000\lambda$ (e.g., 200 m if $\lambda = 0.1$ m). The transmitter has a ULA with $\Delta = \lambda/2$ spacing. The receiver either has an identical compact ULA or a distributed array of the kind illustrated in Figure 4.26 with either $\vartheta = 20°$ or

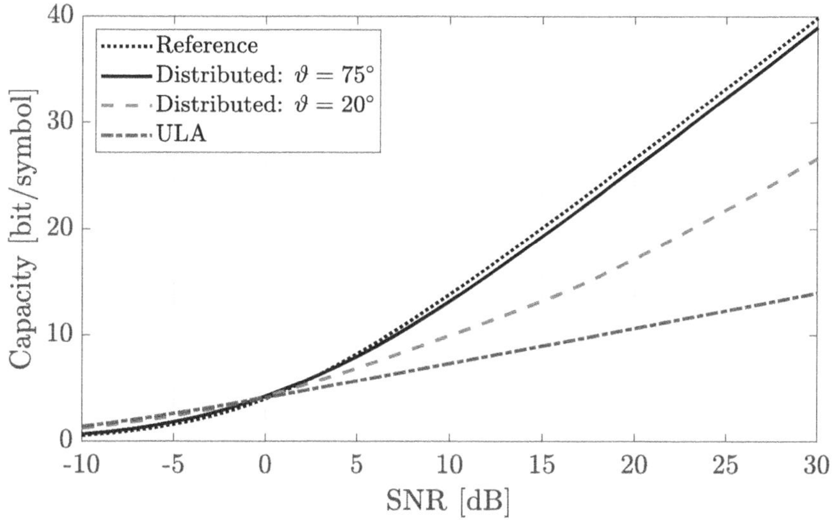

Figure 4.27: Capacity of LOS MIMO channels with $M = K = 4$ where the transmitter is equipped with a ULA, while the receiver is either equipped with a half-wavelength-spaced ULA or a distributed array of the kind illustrated in Figure 4.26. The ideal MIMO capacity $4\log_2(1 + \text{SNR})$ is shown as a reference.

$\vartheta = 75°$. The figure shows how the channel capacity depends on the reference SNR from (4.98), measured as in a SISO system. Recall that the multiplexing gain determines the high-SNR slope of a capacity curve. We notice that the slope differs between the curves; thus, the multiplexing gains differ. While a receiver equipped with a compact ULA only achieves a multiplexing gain of $r = 1$, a receiver with a distributed array can achieve a larger multiplexing gain and, thereby, a steeper slope. The benefit of distributing the antennas comes gradually as ϑ increases. The full multiplexing gain $r = 4$ is achieved for $\vartheta = 75°$, but not for $\vartheta = 20°$.[10] The reference curve $4\log_2(1 + \text{SNR})$ is included to represent the ideal case when all the singular values of $\mathbf{H}$ are equal. The setup with $\vartheta = 75°$ has the same slope but is slightly shifted to the right since the singular values are unequal.

We can achieve a larger multiplexing gain when having a distributed array because a single beam is too narrow to cover the entire receiver array. The half-power beamwidth can be computed for the simulation example using (4.58) and becomes $2\arcsin(0.886/4) \approx 0.45 \approx 26°$. The angle difference between the adjacent receive antennas is $25°$ when $\vartheta = 75°$; hence, if we point one beam towards each receive antenna, as illustrated in Figure 4.28, they will barely overlap. This explains why we can nearly reach the ideal MIMO

[10]Strictly speaking, the rank of $\mathbf{H}$ is 4 in all the considered setups because all the singular values are non-zero. However, it is not until the water-filling power allocation uses all the parallel channels that the effective multiplexing gain becomes 4 and the slope increases to its maximum. It is only for $\vartheta = 50°$ that all the parallel channels are utilized within the considered SNR range, while the other setups have several singular values that are negligibly small.

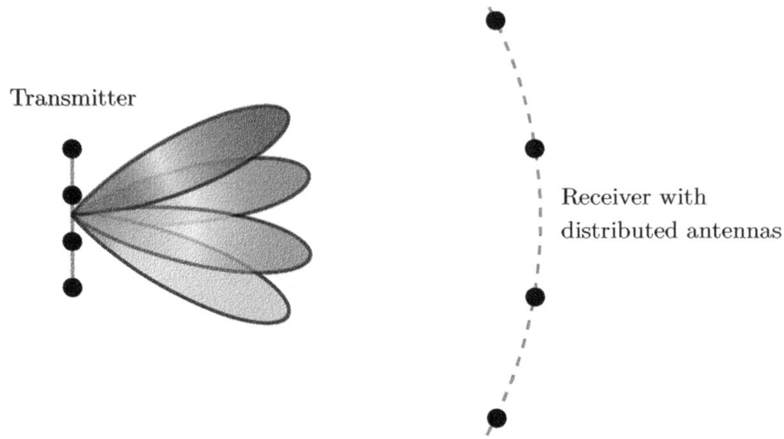

Figure 4.28: Illustration of how a transmitter with four antennas can transmit a superposition of four beams, each carrying different data. Each beam has a different direction and focuses on a different part of the receiver's distributed array. This is how the MIMO channel capacity is achieved when the receiver is so large that each main beam only reaches one antenna.

capacity in this setup. All the signals are sent from all the transmit antennas to achieve narrow beams and reach all the receive antennas, but with varying amplitudes and phases, so we can use the SVD to create four parallel channels that are almost equally strong. In practice, this property can be utilized by deploying base stations at different locations and serving each user device using all of them to create a MIMO channel with a high rank. Such systems are called *Cell-free MIMO* [2] or *coordinated multipoint* [52].

If we shift focus to the low-SNR regime, we can notice from Figure 4.27 that the ULA outperforms the two distributed arrays in this case (but the margin is smaller for $\vartheta = 20°$). The reason is that the water-filling power allocation will only utilize the subchannel with the largest singular value in this case; thus, having a rank-one channel is preferable at low SNR because then the maximum beamforming gain of MK can be achieved.

Example 4.12. Consider a MIMO system with $M = K = 2$ antennas. For which SNR values will we achieve a higher capacity with a full-rank channel matrix with two identical singular values than a rank-one channel matrix?

A rank-one channel has the capacity $\log_2(1 + MK\text{SNR}) = \log_2(1 + 4\text{SNR})$ from (4.96), where SNR denotes the SNR of a corresponding SISO channel. If the singular values are equal, the capacity becomes $2\log_2(1 + \text{SNR})$ where there is a multiplexing gain, but the power allocation cancels the beamforming gain. The full-rank channel matrix achieves a higher capacity if

$$\begin{aligned} 2\log_2(1+\text{SNR}) \geq \log_2(1+4\text{SNR}) \quad &\Rightarrow \quad (1+\text{SNR})^2 \geq 1+4\text{SNR} \\ &\Rightarrow \quad \text{SNR} \geq 2. \end{aligned} \tag{4.99}$$

A similar computation for the case $M = K = 4$ that was considered in Figure 4.27 would result in $(1 + \text{SNR})^4 \geq 1 + 16\text{SNR}$, which implies $\text{SNR} \gtrapprox 1.06 = 0.25\,\text{dB}$ (which is an approximate number). We can observe that this is the intersection point between the ULA curve and reference curve in Figure 4.27. The intersection point will gradually reduce if we continue increasing the number of antennas.

4.4.3 MIMO Channel Capacity in the Radiative Near-Field

The last section demonstrated that the rank of an LOS MIMO channel matrix becomes larger than 1 when the receiving array has a size larger than the transmitted signal's beamwidth. This might happen even if the antennas are gathered in a compact aperture but not under the far-field conditions considered earlier. In this section, we instead consider the radiative near-field, where the spherical curvature of the signals helps to increase the channel rank.

The half-power beamwidth of a ULA transmitting in the broadside direction was shown in (4.58) to be $2\arcsin(0.886/M)$ when the antenna spacing is $\Delta = \lambda/2$. The beamwidth with an arbitrary antenna spacing Δ can be expressed as

$$\psi = 2\arcsin\left(\frac{0.886}{M\Delta\frac{2}{\lambda}}\right) = 2\arcsin\left(\frac{0.443\lambda}{D_{\mathrm{t}}}\right) \approx \frac{0.886\lambda}{D_{\mathrm{t}}} \quad \text{radians}, \tag{4.100}$$

where $D_{\mathrm{t}} = M\Delta$ denotes the aperture length of the transmitting ULA. The approximation in (4.100) is based on that $\arcsin(x) \approx x$ for $x \in [0, 0.4]$ and is therefore tight when the aperture length is at least a few wavelengths. The expression in (4.100) shows that the beamwidth is narrow when either the wavelength λ is small or the aperture length D_{t} is large. Since the beam has a constant angular width, the physical width measured in meters grows linearly with the propagation distance. At a distance d from the transmitter, the physical beamwidth becomes

$$2d\tan\left(\frac{\psi}{2}\right) \approx 2d\frac{\psi}{2} = d\psi \approx \frac{0.886\lambda d}{D_{\mathrm{t}}} \quad \text{meters}, \tag{4.101}$$

where the approximations once again follow from the assumption that the angles are small. Figure 4.29 illustrates this relationship between the aperture length, half-power angular beamwidth, and physical beamwidth at a distance d. If the receiver has a smaller aperture length than the physical beamwidth when the beam is focused on the outermost antennas, as shown in the figure, we can expect the channel matrix to have rank 1. The rank is higher when the receiver's aperture length D_{r} is larger than half the physical beamwidth $\frac{\psi}{2}d$ because then we can focus different beams on the two outermost antennas (top and bottom) and have limited overlap. In practice, the aperture lengths of the transmitter and receiver are limited by the physical sizes of the respective devices. However,

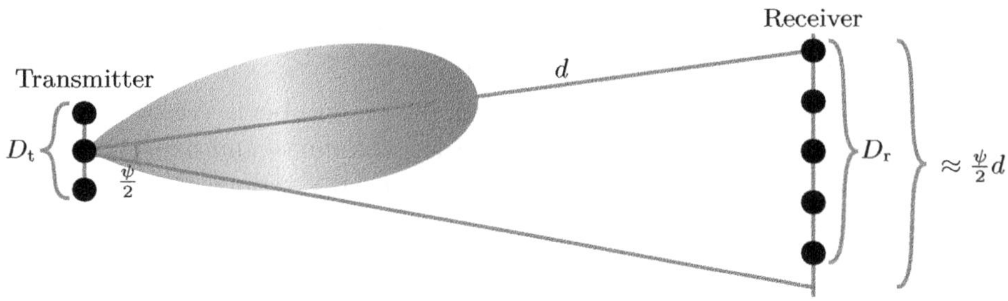

Figure 4.29: The half-power beamwidth $\psi \approx 0.886\lambda/D_t$ measures the angular width of the transmitted beam containing most of the power. The physical beamwidth (in meters) experienced at a distance d from the transmitter is approximately ψd. If the beam is focused on the outermost antenna, half the physical beamwidth should be compared with the receiver's aperture length D_r. If it is smaller than $\psi d/2$, as exemplified here, the channel matrix will have rank 1.

we can achieve a high-rank channel by reducing the wavelength (i.e., increasing the carrier frequency), which is a distinct benefit of using the high-band spectrum in LOS scenarios.

Example 4.13. Suppose the transmitter and receiver are equipped with ULAs with the aperture length $D_t = D_r = 1\,\text{m}$. For which propagation distances d is half the physical beamwidth in (4.101) smaller than the receiver's aperture? Consider the wavelengths $\lambda = 1\,\text{dm}$ (3 GHz) and $\lambda = 1\,\text{cm}$ (30 GHz).

Half the physical beamwidth in (4.101) is smaller than D_r if

$$\frac{1}{2}\frac{0.886\lambda d}{D_t} \leq D_r \quad \Rightarrow \quad d \leq \frac{2D_t D_r}{0.886\lambda}. \tag{4.102}$$

The upper bound is similar to the Fraunhofer distance $2D^2/\lambda$ if $D = D_t = D_r$. Hence, the far-field plane wave approximation is not applicable when half the transmitter's beamwidth is smaller than the receiver's aperture length. We obtain the range $d \leq 22.6\,\text{m}$ if $\lambda = 1\,\text{dm}$ and $d \leq 226\,\text{m}$ if $\lambda = 1\,\text{cm}$. Hence, for practically sized arrays, we can utilize spherical wavefronts to achieve a high-rank LOS channel if the distance and/or wavelength is small.

Figure 4.30 shows the capacity achieved when having ULAs with $M = K = 100$ antennas at the transmitter and receiver. The wavelength is $\lambda = 1\,\text{cm}$ (30 GHz) and the antenna spacing is $\Delta = \lambda$, which results in an aperture length of $D_t = D_r = 1\,\text{m}$ as in the last example. The arrays are deployed parallel to each other at a distance d that is varied on the horizontal axis. The SNR is defined as in (4.98) and fixed at 20 dB.[11] The capacity is substantially higher for all the considered distances than the reference curve $\log_2(1 + MK\,\text{SNR})$, which is achieved with a rank-1 channel matrix. However, the capacity curve converges to this value as the distance grows large because then the far-field

[11]If the transmit power is fixed, the SNR is also distance-dependent. However, this example focuses on showing how the singular values of the channel matrix depend on the propagation distance in the radiative near-field, so we keep the SNR fixed to highlight this effect.

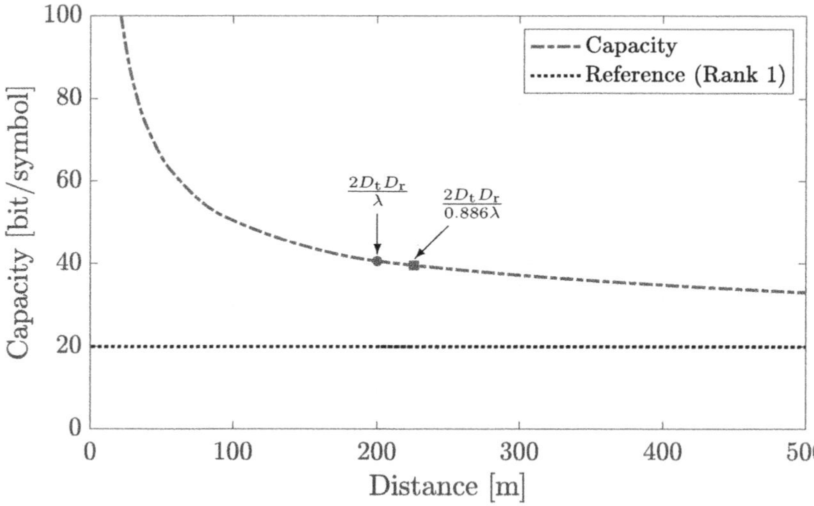

Figure 4.30: The capacity of LOS MIMO channels with $M = K = 100$ where the transmitter and receiver are equipped with ULAs. The channel matrix gets more non-zero singular values with similar strengths at shorter propagation distances, which results in a higher capacity thanks to the larger effective multiplexing gain. The rank-1 MIMO capacity $\log_2(1 + MK\,\text{SNR})$ is shown as a reference curve and SNR $= 20\,$dB.

approximation becomes valid. We get a higher capacity in the considered range because $\mathbf{H}$ has multiple non-zero singular values, and the number grows rapidly at short distances. When $d = \frac{2D_t D_r}{0.886\lambda} \approx 226\,$m, which was derived in (4.102) by comparing the transmitter's beamwidth with the receiver's aperture, the capacity is roughly twice as large as in the far-field. The capacity changes slowly with the distance, so we could alternatively use $\frac{2D_t D_r}{\lambda} = 200\,$m as the approximate maximum distance for near-field spatial multiplexing because it looks similar to the Fraunhofer distance. Both points are indicated in the figure. In conclusion, for fixed aperture sizes at the transmitter and receiver, the LOS channel matrix gets more non-zero singular values when the propagation distance d shrinks because the physical beamwidth becomes smaller than the receiver array.

When communicating between two locations separated by a distance of d, we can optimize the antenna deployment to achieve the maximum channel rank and equal singular values, which gives the ideal MIMO capacity at high SNRs. For notational convenience and inspired by [53], we will consider two ULAs with M antennas and matching antenna separation Δ. The ULAs are parallel and located in each other's broadside directions. The antennas with the same index in the two ULAs are separated by the distance d, as illustrated in Figure 4.31. It then follows from the Pythagorean theorem that the distance between antenna k at the transmitter and antenna m at the receiver is

$$d_{m,k} = \sqrt{d^2 + (m-k)^2\Delta^2} = d\sqrt{1 + \frac{(m-k)^2\Delta^2}{d^2}} \approx d\left(1 + \frac{(m-k)^2\Delta^2}{2d^2}\right), \tag{4.103}$$

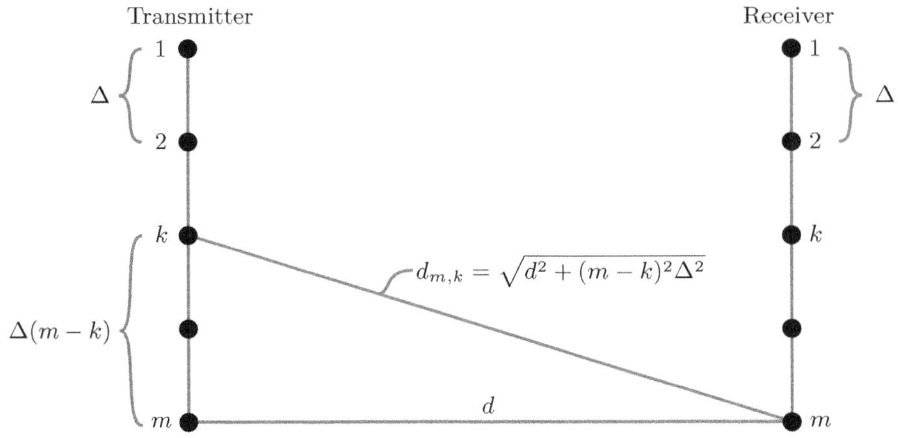

Figure 4.31: Illustration of a MIMO communication setup with two parallel ULAs with the same number of antennas and antenna spacing Δ. The distance between transmit antenna k and receive antenna m is $\sqrt{d^2+(m-k)^2\Delta^2}$ and can be Fresnel approximated as in (4.103).

where we used the first-order Taylor approximation $\sqrt{1+x^2} \approx 1+\frac{x^2}{2}$ that is tight for $0 \leq x \leq 0.25$. Therefore, the approximate expression is tight when the distance d exceeds the aperture length. Despite the approximation in (4.103), the derivations that will follow in this section are more accurate than the far-field approximation in (4.88), which corresponds to making the approximation $\sqrt{d^2+(m-k)^2\Delta^2} \approx d$. More precisely, it is the difference between using a first-order and zeroth-order Taylor approximation of the distance between antennas. The simplification in (4.103) is called the *Fresnel approximation* [54] and models the waves as parabolic. We say that we operate in the radiative near-field region (also known as the *Fresnel zone*) when the Fresnel approximation is tight, but the far-field approximation is not.

If we return to the general MIMO channel matrix expression in (4.87) and make use of the Fresnel approximation in (4.103) and that $K = M$, we obtain

$$\mathbf{H} = \sqrt{\beta}\begin{bmatrix} 1 & e^{-\mathrm{j}\pi\frac{1^2\Delta^2}{\lambda d}} & \dots & e^{-\mathrm{j}\pi\frac{(M-1)^2\Delta^2}{\lambda d}} \\ e^{-\mathrm{j}\pi\frac{1^2\Delta^2}{\lambda d}} & 1 & \dots & e^{-\mathrm{j}\pi\frac{(M-2)^2\Delta^2}{\lambda d}} \\ \vdots & \vdots & \ddots & \vdots \\ e^{-\mathrm{j}\pi\frac{(M-1)^2\Delta^2}{\lambda d}} & e^{-\mathrm{j}\pi\frac{(M-2)^2\Delta^2}{\lambda d}} & \dots & 1 \end{bmatrix}, \tag{4.104}$$

where $\beta = \frac{\lambda^2}{(4\pi)^2}\frac{1}{d^2}$ is an accurate approximation of the channel gain in these cases. We recall from Example 3.7 that the singular values of $\mathbf{H}$ are also the square roots of the eigenvalues of $\mathbf{H}^{\mathrm{H}}\mathbf{H}$. If the columns of $\mathbf{H}$ are mutually orthogonal, then $\mathbf{H}^{\mathrm{H}}\mathbf{H} = M\beta\mathbf{I}_M$ and all the singular values will be $\sqrt{M\beta}$. Orthogonality between the columns can be achieved by fine-tuning the antenna spacing Δ. The magnitude of the inner product between the kth and lth column (for $l \neq k$) can be computed as

$$\beta \left| \begin{bmatrix} e^{-\mathrm{j}\pi \frac{(k-1)^2\Delta^2}{\lambda d}} \\ e^{-\mathrm{j}\pi \frac{(k-2)^2\Delta^2}{\lambda d}} \\ \vdots \\ e^{-\mathrm{j}\pi \frac{(k-M)^2\Delta^2}{\lambda d}} \end{bmatrix}^{\mathrm{H}} \begin{bmatrix} e^{-\mathrm{j}\pi \frac{(l-1)^2\Delta^2}{\lambda d}} \\ e^{-\mathrm{j}\pi \frac{(l-2)^2\Delta^2}{\lambda d}} \\ \vdots \\ e^{-\mathrm{j}\pi \frac{(l-M)^2\Delta^2}{\lambda d}} \end{bmatrix} \right| = \beta \left| \sum_{m=1}^{M} e^{\mathrm{j}\pi \frac{(k-m)^2\Delta^2}{\lambda d}} e^{-\mathrm{j}\pi \frac{(l-m)^2\Delta^2}{\lambda d}} \right|$$

$$= \beta \left| \sum_{m=1}^{M} e^{\mathrm{j}\pi \frac{2(m-1)(l-k)\Delta^2}{\lambda d}} \right| = \beta \left| \frac{1 - e^{\mathrm{j}\pi \frac{2M(l-k)\Delta^2}{\lambda d}}}{1 - e^{\mathrm{j}\pi \frac{2(l-k)\Delta^2}{\lambda d}}} \right|^2 = \beta \left| \frac{\sin\left(\pi \frac{M(l-k)\Delta^2}{\lambda d}\right)}{\sin\left(\pi \frac{(l-k)\Delta^2}{\lambda d}\right)} \right|. \tag{4.105}$$

The second equality follows from multiplying with $e^{-\mathrm{j}\pi \frac{\Delta^2}{\lambda d}(k^2 - l^2 + 2(l-k))}$ inside the magnitude to remove terms that are independent of m. This can be done since this term has unit magnitude. We used the formula for geometric series similarly as in (4.52) to obtain the final expression in (4.105).

If we select the antenna spacing so that $\frac{M\Delta^2}{\lambda d} = 1$, then the numerator in (4.105) is zero while the denominator is non-zero, because $|l - k| \leq M - 1$ for $l, k \in \{1, \ldots, M\}$. Hence, if the antenna spacing in the two antenna arrays is

$$\Delta = \sqrt{\frac{\lambda d}{M}}, \tag{4.106}$$

the columns of the channel matrix in (4.104) are mutually orthogonal. The aperture lengths of the ULAs are $D = M\Delta = \sqrt{M\lambda d}$, which shrinks when the carrier frequency $f_{\mathrm{c}} = c/\lambda$ is increased. The length grows with the number of antennas but slower than linear since the antenna spacing reduces with M.

Since there are M singular values that equal $\sqrt{M\beta}$, the water-filling power allocation will assign the power equally between them. Hence, for the considered setup with two parallel M-antenna ULAs and the optimized antenna spacing in (4.106), the MIMO capacity in (3.75) becomes

$$C = M \log_2\left(1 + \frac{q}{M}\frac{M\beta}{N_0}\right) = M \log_2\left(1 + \frac{q\beta}{N_0}\right) \quad \text{bit/symbol}. \tag{4.107}$$

This is the ideal capacity from a multiplexing perspective.

Example 4.14. Suppose we want to design an LOS MIMO system with $M = K = 32$ antennas where the propagation distance is $d = 50\,\mathrm{m}$. Which antenna spacing is needed to achieve M identical singular values if $\lambda = 1\,\mathrm{cm}$ (30 GHz)? What is the resulting aperture length?

The antenna spacing in (4.106) becomes $\Delta = \sqrt{\frac{\lambda d}{M}} = 0.125\,\mathrm{m}$, which is 12.5 wavelengths. This unusually large separation will enable the receiver to detect the spherical wavefronts. The aperture length becomes $M\Delta = 4\,\mathrm{m}$.

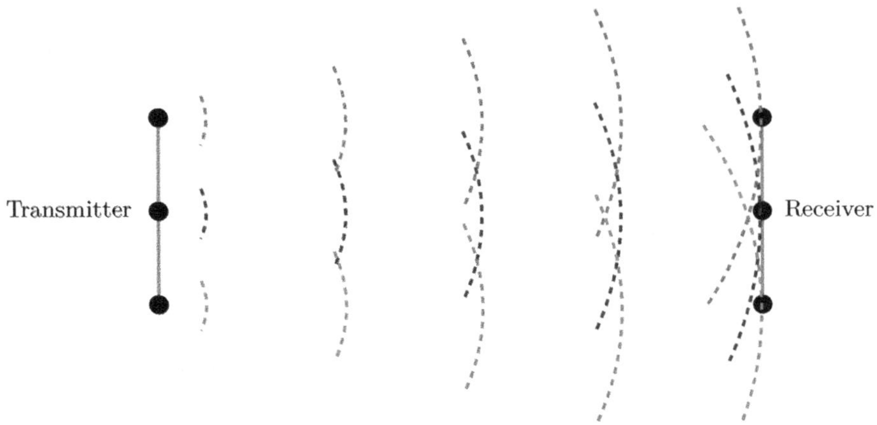

Figure 4.32: When the antenna spacing of the ULAs in Figure 4.31 is optimized according to (4.106), then the channel matrix has a full rank, all the singular values are equal, and the capacity is achieved by transmitting independent signals from each antenna (as illustrated by the coloring). The receiver will use the phase-shift variations created by the spherical wavefronts to separate the transmitted signals.

The antenna spacing was fine-tuned in (4.106) to achieve orthogonal columns in $\mathbf{H}$. The singular value decomposition is

$$\mathbf{H} = \underbrace{\frac{1}{\sqrt{M}} \begin{bmatrix} e^{-\jmath\pi \frac{0^2}{M}} & \dots & e^{-\jmath\pi \frac{(M-1)^2}{M}} \\ e^{-\jmath\pi \frac{1^2}{M}} & \dots & e^{-\jmath\pi \frac{(M-2)^2}{M}} \\ \vdots & \ddots & \vdots \\ e^{-\jmath\pi \frac{(M-1)^2}{M}} & \dots & e^{-\jmath\pi \frac{0^2}{M}} \end{bmatrix}}_{=\mathbf{U}} \underbrace{\begin{bmatrix} \sqrt{\beta M} & 0 & \dots \\ 0 & \ddots & 0 \\ \vdots & 0 & \sqrt{\beta M} \end{bmatrix}}_{=\mathbf{\Sigma}} \underbrace{\mathbf{I}_M}_{=\mathbf{V}^{\mathrm{H}}}, \tag{4.108}$$

where the left singular vectors in $\mathbf{U}$ are the normalized columns of $\mathbf{H}$ in (4.104) when using the antenna spacing in (4.106). The matrix $\mathbf{V}$ with the right singular vectors is an identity matrix and is used for precoding. The transmitted signal becomes $\mathbf{x} = \mathbf{V}\bar{\mathbf{x}} = \bar{\mathbf{x}}$, which implies that each of the independent signals in $\bar{\mathbf{x}}$ is transmitted from only one of the antennas. The receiver then utilizes the fact that the spherical wavefronts create varying phase-shifts over the receive antennas to separate the signals while achieving identical signal strengths for all of them. This mode of operation is illustrated in Figure 4.32. This structure is unique to the optimized antenna spacing, which is fine-tuned so that when the receiver focuses a beam on a transmit antenna, the other antennas are exactly at the nulls of the beam pattern. If we use slightly different spacings, the channel matrix will have slight variations in the singular values, and the precoding will not reduce to transmitting independent signals from the antennas.

We analyzed parallel ULAs in this section, in which case all the phase-shifts in $\mathbf{H}$ are caused by the spherical wavefronts. When the arrays are rotated differently, there will be further phase-shifts since plane waves will also give rise to that, as shown in (4.90) for far-field communication. The unique feature of communicating in the radiative near-field is that the MIMO channel matrix gets a higher rank than 1 because the phase-shifts caused by the spherical wavefronts vary non-linearly with the antenna index. The traditional Fraunhofer distance $2D^2/\lambda$ is unable to determine the upper limit of the near-field region of MIMO channels because it was derived for a MISO/SIMO channel with an aperture length of D on one side and a single isotropic antenna on the other side. In MIMO scenarios, we must take both the transmitter's aperture length D_{t} into account since it determines the narrowness of the beam and consider the receiver's aperture length D_{r} because it determines the ability to observe spherical wavefronts. One way to characterize the radiative near-field is that the propagation distance d must satisfy

$$d \leq \frac{2D_{\mathrm{t}}D_{\mathrm{r}}}{\lambda}, \tag{4.109}$$

where the upper bound is called the *near-field multiplexing distance*. We stress that this is a rule-of-thumb for when we can at least double the capacity compared to the far-field, but the capacity value varies slowly around this value, so alternative upper limits can be defined.

4.5 Three-Dimensional Far-Field Channel Modeling

The MIMO channel matrix expression in (4.87) depends on the exact propagation distances between every pair of transmit and receive antennas; thus, it can be utilized to model any free-space LOS channel. In contrast, all the simplified expressions for ULAs that have been derived so far are limited in their generality by the choice of array geometry and the assumption that all antennas are located in the same two-dimensional plane. When analyzing SIMO and MISO channels with ULAs, we can always define the coordinate system such that all antennas are located in the same plane. By contrast, this is not always possible in the MIMO case; for example, one array might be deployed horizontally and the other array vertically. In this section, we will consider the general case where the transmitter and receiver can have arbitrary array geometries. The only limiting assumption is that the receiver is in the far-field of the transmitter.

We begin by considering the SIMO setup illustrated in Figure 4.33, where a single-antenna transmitter sends a signal toward two receive antennas. The receive antennas are in the far-field of the transmitter; thus, the impinging wavefront can be approximated as planar. The location of each antenna is represented by a three-dimensional vector, representing a point in the three-dimensional world. Suppose we define the coordinate system such that one receive antenna is located in the origin, denoted by $\mathbf{0} = [0, 0, 0]^{\mathrm{T}}$. The

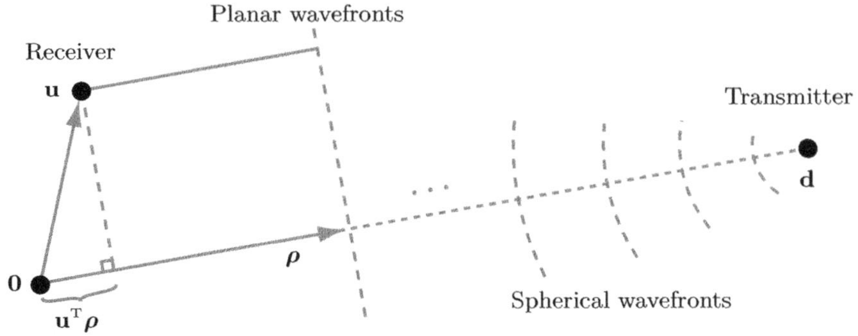

Figure 4.33: Illustration of a setup with two receive antennas, one located at the origin $\mathbf{0}$ and one at location $\mathbf{u}$, receiving a planar wavefront emitted by a transmitter at location $\mathbf{d}$. The unit-length vector $\boldsymbol{\rho}$ points out the direction leading towards the transmitter. The difference in propagation distance between the two receive antennas is $\mathbf{u}^{\mathrm{T}}\boldsymbol{\rho}$.

transmitter location is $\mathbf{d} \in \mathbb{R}^3$, while the location of the other receive antenna is $\mathbf{u} \in \mathbb{R}^3$. The impinging wave will generally reach the receive antennas at slightly different times, determined by the difference in propagation distances, leading to phase differences when the signals are sampled simultaneously. To determine this phase difference, we define the unit length vector

$$\boldsymbol{\rho} = \frac{\mathbf{d}}{\|\mathbf{d}\|} \tag{4.110}$$

that points from the origin towards the transmitter. Since the planar wavefront propagates perpendicular to $\boldsymbol{\rho}$, we can determine the path difference between the two receive antennas by projecting the location $\mathbf{u}$ of the second receive antenna onto $\boldsymbol{\rho}$. The orthogonal projection is given by $\mathbf{u}^{\mathrm{T}}\boldsymbol{\rho} \in \mathbb{R}$ and represents how much *shorter* the distance is to the second receive antenna, compared to the distance to the antenna in the origin. A negative value implies a longer distance to the second antenna.

Suppose the impinging signal has wavelength λ. In that case, $\mathbf{u}^{\mathrm{T}}\boldsymbol{\rho}/\lambda$ represents how many wavelengths shorter the propagation distance is to the second antenna, while the corresponding phase-shift is

$$-\frac{2\pi}{\lambda}\mathbf{u}^{\mathrm{T}}\boldsymbol{\rho}. \tag{4.111}$$

The channel response will then be $\sqrt{\beta}e^{\mathrm{j}\frac{2\pi}{\lambda}\mathbf{u}^{\mathrm{T}}\boldsymbol{\rho}}$, where $\beta \in [0,1]$ is the channel gain and the minus sign disappeared since phase-shifts appear with a minus in channel models. Recall that the unit-length vector $\boldsymbol{\rho}$ specifies the direction-of-arrival of the planar wavefront using Cartesian coordinates; however, it can be more instructive to describe it using angles. To this end, we will make use of the spherical coordinate system, defined in Figure 1.9, and parametrize the directional vector $\boldsymbol{\rho}$ in terms of the azimuth angle $\varphi \in [-\pi, \pi)$ and

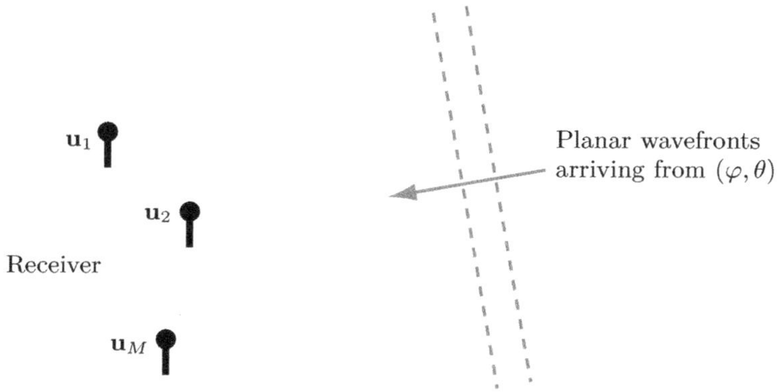

Figure 4.34: Illustration of a setup where a planar wave impinges on an array with M receive antennas from the azimuth angle φ and elevation angle θ.

the elevation angle $\theta \in [-\pi/2, \pi/2]$. The one-to-one mapping between these coordinate systems was stated in (1.22) and implies that

$$\boldsymbol{\rho} = \begin{bmatrix} \cos(\varphi)\cos(\theta) \\ \sin(\varphi)\cos(\theta) \\ \sin(\theta) \end{bmatrix}. \tag{4.112}$$

By substituting (4.112) into (4.111), we can represent the phase-shift using the azimuth and elevation angles.

We will now consider a SIMO channel with M receive antennas where the location of antenna m is denoted by $\mathbf{u}_m \in \mathbb{R}^3$, as illustrated in Figure 4.34. A planar wave is impinging from the angular direction (φ, θ), measured from the origin wherever it might be. None of the antennas need to be located at the origin, but we will still utilize it as the reference point when computing the phase-shifts. More precisely, the sampling delay is selected to obtain a zero-valued phase-shift at the origin. The phase-shift at the mth antenna will then be $-2\pi\mathbf{u}_m^{\mathrm{T}}\boldsymbol{\rho}/\lambda$, where $\boldsymbol{\rho}$ is computed using (4.112). We can define the array response vector

$$\mathbf{a}(\varphi, \theta) = \begin{bmatrix} e^{\mathrm{j}\frac{2\pi}{\lambda}\mathbf{u}_1^{\mathrm{T}}\boldsymbol{\rho}} \\ e^{\mathrm{j}\frac{2\pi}{\lambda}\mathbf{u}_2^{\mathrm{T}}\boldsymbol{\rho}} \\ \vdots \\ e^{\mathrm{j}\frac{2\pi}{\lambda}\mathbf{u}_M^{\mathrm{T}}\boldsymbol{\rho}} \end{bmatrix} \tag{4.113}$$

as the normalized channel vector (i.e., without the channel gain) for the case when the impinging signal has the angles-of-arrival (φ, θ). If all the antennas are isotropic and $\beta = \lambda^2/(4\pi d)^2$ denotes the channel gain at a propagation distance of d, then the SIMO channel vector can be expressed using (4.113) as

$$\mathbf{h} = \sqrt{\beta}\mathbf{a}(\varphi, \theta). \tag{4.114}$$

This channel vector can also be utilized for the MISO channel obtained by reversing the roles of the transmitter and receiver.

Example 4.15. Consider two receive antennas deployed on the y-axis at the locations $\mathbf{u}_1 = [0, \lambda/4, 0]^{\mathrm{T}}$ and $\mathbf{u}_2 = [0, -\lambda/4, 0]^{\mathrm{T}}$. What is the array response vector $\mathbf{a}(\varphi, \theta)$? How does $\mathbf{a}(\varphi, \theta)$ depend on the azimuth angle φ when the elevation angle is $\theta = 0$ or $\theta = \pi/2$?

The antenna separation is $\lambda/2$. The array response vector for an arbitrary angle (φ, θ) is obtained using (4.113) as

$$\mathbf{a}(\varphi, \theta) = \begin{bmatrix} e^{\mathrm{j}\frac{2\pi}{\lambda}\mathbf{u}_1^{\mathrm{T}}\boldsymbol{\rho}} \\ e^{\mathrm{j}\frac{2\pi}{\lambda}\mathbf{u}_2^{\mathrm{T}}\boldsymbol{\rho}} \end{bmatrix} = \begin{bmatrix} e^{\mathrm{j}\frac{\pi}{2}\sin(\varphi)\cos(\theta)} \\ e^{-\mathrm{j}\frac{\pi}{2}\sin(\varphi)\cos(\theta)} \end{bmatrix}. \tag{4.115}$$

When $\theta = 0$, (4.115) simplifies to

$$\mathbf{a}(\varphi, 0) = \begin{bmatrix} e^{\mathrm{j}\frac{\pi}{2}\sin(\varphi)} \\ e^{-\mathrm{j}\frac{\pi}{2}\sin(\varphi)} \end{bmatrix}, \tag{4.116}$$

where there is a phase-shift difference of $\pi\sin(\varphi)$ between the antennas. The same phase difference between the adjacent antennas was obtained in (4.23) for a ULA with $\Delta = \lambda/2$ that was also deployed along the y-axis.

When $\theta = \pi/2$, the transmitter is at a point along the z-axis and is unaffected by φ since $\boldsymbol{\rho} = [0, 0, 1]^{\mathrm{T}}$. Hence, the impinging wave is always from the broadside direction, and the corresponding array response vector is

$$\mathbf{a}(\varphi, \pi/2) = \begin{bmatrix} 1 \\ 1 \end{bmatrix}. \tag{4.117}$$

4.5.1 Array Response Vector with a ULA in Three Dimensions

We will now particularize the array response vector in (4.113) for a ULA with M antennas where the spacing is Δ. We assume that the ULA is deployed along the y-axis, with the first antenna located in the origin and the remaining antennas located along the negative side of the axis. This assumption can be made without losing generality since we can rotate the coordinate system as we like. Under these assumptions, the location of receive antenna m becomes

$$\mathbf{u}_m = \begin{bmatrix} 0 \\ -(m-1)\Delta \\ 0 \end{bmatrix}. \tag{4.118}$$

This setup is illustrated in Figure 4.35. The inner product between the location vector in (4.118) and the direction-of-arrival vector in (4.112) becomes

$$\mathbf{u}_m^{\mathrm{T}}\boldsymbol{\rho} = \begin{bmatrix} 0 \\ -(m-1)\Delta \\ 0 \end{bmatrix}^{\mathrm{T}} \begin{bmatrix} \cos(\varphi)\cos(\theta) \\ \sin(\varphi)\cos(\theta) \\ \sin(\theta) \end{bmatrix} = -(m-1)\Delta\sin(\varphi)\cos(\theta). \tag{4.119}$$

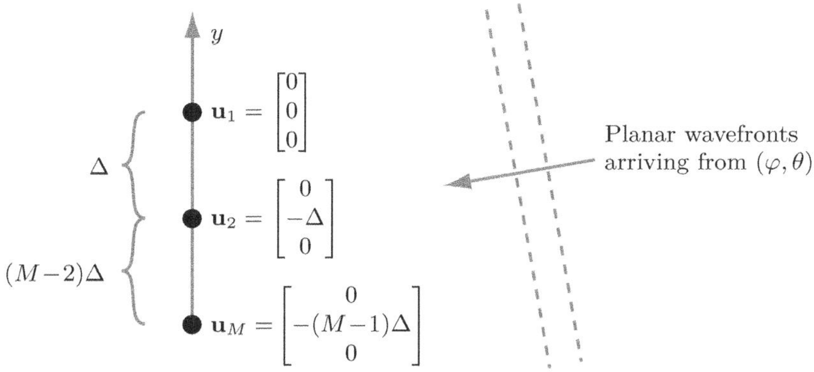

Figure 4.35: Illustration of a setup where a planar wave impinges on a ULA with M receive antennas from the azimuth angle φ and elevation angle θ.

If we substitute (4.119) into (4.113), we obtain the array response vector for a ULA as

$$\mathbf{a}_M(\varphi, \theta) = \begin{bmatrix} 1 \\ e^{-\mathrm{j}2\pi \frac{\Delta \sin(\varphi)\cos(\theta)}{\lambda}} \\ \vdots \\ e^{-\mathrm{j}2\pi \frac{(M-1)\Delta \sin(\varphi)\cos(\theta)}{\lambda}} \end{bmatrix} \tag{4.120}$$

where the subscript denotes the number of antennas. This is a generalization of the previous array response vector in (4.37) since the impinging wave can arrive from any angular direction (not limited to the horizontal xy-plane).

Example 4.16. Consider a ULA with M antennas deployed along the z-axis. The antenna spacing is Δ and the mth element is located at $\mathbf{u}_m = [0, 0, -(m-1)\Delta]^{\mathrm{T}}$. What is the array response vector?

This ULA is deployed vertically. The inner product between the location vector $\mathbf{u}_m$ and the direction-of-arrival vector in (4.112) becomes

$$\mathbf{u}_m^{\mathrm{T}} \boldsymbol{\rho} = \begin{bmatrix} 0 \\ 0 \\ -(m-1)\Delta \end{bmatrix}^{\mathrm{T}} \begin{bmatrix} \cos(\varphi)\cos(\theta) \\ \sin(\varphi)\cos(\theta) \\ \sin(\theta) \end{bmatrix} = -(m-1)\Delta \sin(\theta). \tag{4.121}$$

The array response vector of this vertical ULA is obtained by substituting (4.121) into (4.113), which yields

$$\mathbf{a}(\varphi, \theta) = \left[1, e^{-\mathrm{j}2\pi \frac{\Delta \sin(\theta)}{\lambda}}, \dots, e^{-\mathrm{j}2\pi \frac{(M-1)\Delta \sin(\theta)}{\lambda}}\right]^{\mathrm{T}}. \tag{4.122}$$

This expression differs from the one in (4.120) for a horizontal ULA, due to the different deployment directions compared to the assumed spherical coordinate system. However, (4.122) matches with $\mathbf{a}_M(\theta, 0)$ from (4.120) when the elevation angle is zero while the azimuth angle is replaced by θ.

There are many ways to express the array response vector of a ULA, depending on how the coordinate system is rotated compared to it. Two examples are given in (4.120) and (4.122). While the transmitter's location relative to the receiver matters when determining the beamwidth, the communication performance is the same irrespective of how the coordinate system is rotated. A three-dimensional array response model is necessary when there are multiple impinging wavefronts via different propagation paths, so we cannot rotate the coordinate system to place everything in a two-dimensional plane.

4.5.2 Array Response Vector with a Uniform Planar Array

Many practical antenna arrays are planar, which means that the antennas are deployed in two dimensions: one horizontally and one vertically. There are three main benefits of this. Firstly, if an array is designed with a maximally allowed aperture length, we can fit more antennas by distributing them over two dimensions. This is because the length is then measured diagonally, as previously illustrated in Figure 4.1(c). This allows for a larger beamforming gain in a size-constrained deployment. Secondly, a horizontal ULA can have a small beamwidth in the horizontal plane while it spreads the power equally over all elevation angles (see Figure 4.18). Since the prospective users are typically below the base station (i.e., the elevation angles of interest are $\theta \in [-\pi/2, 0]$), half of the power is lost by radiating it into the sky. A planar array can have a small beamwidth also in the vertical plane. Thirdly, a planar array is capable of *3D beamforming*, where it points different beams towards objects/users located in similar azimuth angles but different elevation angles.

We will analyze the canonical form of a planar array: the *uniform planar array (UPA)* where the antennas are deployed on an evenly spaced grid in two dimensions, as illustrated in Figure 4.36. Each row has M_{H} antennas with the spacing Δ between the adjacent antennas. Similarly, each column has M_{V} antennas with the spacing Δ between adjacent antennas.[12] Since there are M_{V} rows and M_{H} columns, the total number of antennas is $M = M_{\mathrm{H}} M_{\mathrm{V}}$. The horizontal spacing between the centers of the two outermost antennas in each row is $(M_{\mathrm{H}} - 1)\Delta$. Similarly, the vertical spacing between the centers of the two outermost antennas in each column is $(M_{\mathrm{V}} - 1)\Delta$. Since each antenna also has a physical size, we will denote the horizontal length as $M_{\mathrm{H}}\Delta$ and the vertical length as $M_{\mathrm{V}}\Delta$ (in line with what we did when analyzing ULAs).

The aperture length D is measured along the diagonal of the UPA. It follows from the Pythagorean theorem that

$$D = \sqrt{(M_{\mathrm{H}}\Delta)^2 + (M_{\mathrm{V}}\Delta)^2} = \sqrt{M_{\mathrm{H}}^2 + M_{\mathrm{V}}^2}\,\Delta. \tag{4.123}$$

[12]We have selected equal horizontal and vertical antenna spacings for notational convenience, but this assumption can be generalized. It makes sense for devices to have the same spacings in both dimensions because they can be rotated freely by the user. However, fixed base stations commonly use a larger vertical than horizontal spacing to achieve a narrower vertical beamwidth.

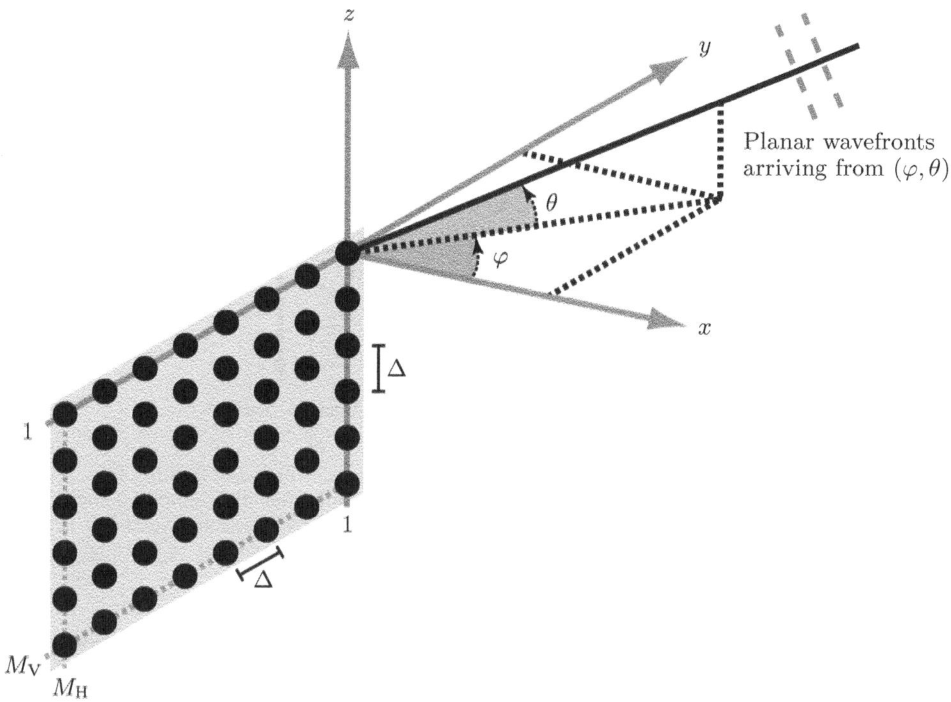

Figure 4.36: Illustration of a setup where a planar wave impinges on a UPA from the azimuth angle φ and elevation angle θ. The UPA has M_H antennas per row and M_V antennas per column on a grid in the yz-plane. The horizontal and vertical spacings are Δ.

Example 4.17. Consider an array with $M = 100$ antennas with $\Delta = \lambda/2$ spacing that is designed for $\lambda = 0.1\,\mathrm{m}$ (i.e., 3 GHz). Compare the aperture lengths obtained if the array is a ULA or a UPA with 10×10 antennas.

With the ULA configuration, the aperture length is $D = M\Delta = 100 \cdot 0.05 = 5\,\mathrm{m}$. With the square-shaped UPA configuration, the aperture length in (4.123) becomes $D = \sqrt{10^2 + 10^2} \cdot 0.05 = \sqrt{2} \cdot 0.5 \approx 0.7\,\mathrm{m}$. The horizontal and vertical lengths of the UPA are $10 \cdot 0.05 = 0.5\,\mathrm{m}$, which is ten times smaller than with the ULA. In conclusion, the UPA configuration enables the given number of antennas to be deployed in a physically smaller form factor. This feature enables large numbers of antennas in practical deployments.

We will now particularize the general array response vector expression in (4.113) for a UPA with the antenna spacing Δ. We assume that the UPA is deployed along the yz-plane with the first antenna located in the origin and the remaining antennas located along the negative side of the y-axis and z-axis, as illustrated in Figure 4.36. This assumption can be made without loss of generality since we can define/rotate the coordinate system as we like. There are M_V rows, each extending horizontally along the negative y-axis and

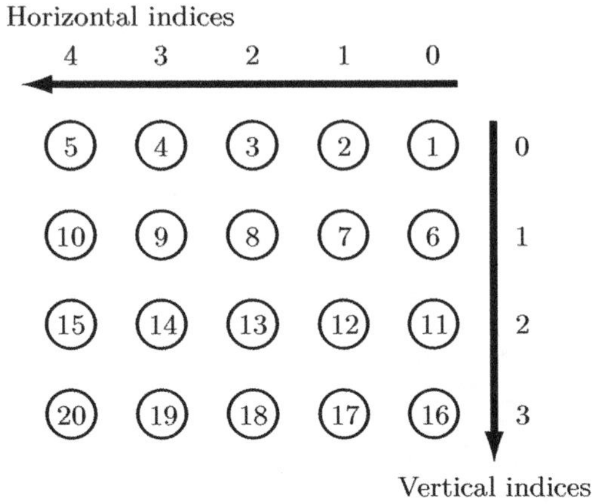

Figure 4.37: Illustration of a UPA with $M = 20$ antennas, which are divided into $M_{\mathrm{H}} = 5$ antennas per row and $M_{\mathrm{V}} = 4$ antennas per column. The antennas are numbered row-by-row from 1 to 20, as shown in the figure. When characterizing the array response vector, it is useful to characterize the horizontal index as in (4.124) and the vertical index as in (4.125).

containing M_{H} antennas. Similarly, there are M_{H} columns, each extending vertically along the negative z-axis and containing M_{V} antennas.

The first antenna is located in the origin. Suppose the antennas are then consecutively indexed row-by-row by $m \in \{1, \ldots, M\}$, where $M = M_{\mathrm{H}}M_{\mathrm{V}}$ is the total number of antennas. The horizontal index of the first M_{H} antennas will be $0, 1, \ldots, M_{\mathrm{H}} - 1$. These indices are repeated on each row; thus, the antennas $(n-1)M_{\mathrm{H}} + 1, \ldots, nM_{\mathrm{H}}$ also have the horizontal indices $0, 1, \ldots, M_{\mathrm{H}} - 1$, for $n = 2, \ldots, M_{\mathrm{V}}$. Using this row-by-row indexing, the horizontal index of antenna m can be computed as

$$i(m) = (m-1) - M_{\mathrm{H}} \left\lfloor \frac{m-1}{M_{\mathrm{H}}} \right\rfloor \in \{0, 1, \ldots, M_{\mathrm{H}} - 1\}, \tag{4.124}$$

where $\lfloor \cdot \rfloor$ rounds the argument to the closest smaller or equal integer. The computation in (4.124) gives the remainder when dividing $m-1$ by M_{H} and is known as the modulo operation.

Next, we will define the vertical index, which will be 0 for the M_{H} antennas on the first row. The vertical index will then be $n-1$ for the antennas $(n-1)M_{\mathrm{H}} + 1, \ldots, nM_{\mathrm{H}}$, for $n = 2, \ldots, M_{\mathrm{V}}$. Hence, the vertical index of antenna m can be obtained as

$$j(m) = \left\lfloor \frac{m-1}{M_{\mathrm{H}}} \right\rfloor \in \{0, 1, \ldots, M_{\mathrm{V}} - 1\}, \tag{4.125}$$

which returns the integer-valued quotient when dividing $m-1$ by M_{H}. The mapping between antenna numbers and horizontal/vertical indices is illustrated in Figure 4.37 for a UPA with $M_{\mathrm{H}} = 5$ and $M_{\mathrm{V}} = 4$.

Under these indexing assumptions, the location of antenna m becomes

$$\mathbf{u}_m = \begin{bmatrix} 0 \\ -i(m)\Delta \\ -j(m)\Delta \end{bmatrix}. \tag{4.126}$$

The inner product between the location vector in (4.126) and the direction-of-arrival vector in (4.112) becomes

$$\mathbf{u}_m^{\mathrm{T}} \boldsymbol{\rho} = \begin{bmatrix} 0 \\ -i(m)\Delta \\ -j(m)\Delta \end{bmatrix}^{\mathrm{T}} \begin{bmatrix} \cos(\varphi)\cos(\theta) \\ \sin(\varphi)\cos(\theta) \\ \sin(\theta) \end{bmatrix} = -i(m)\Delta\sin(\varphi)\cos(\theta) - j(m)\Delta\sin(\theta). \tag{4.127}$$

If we substitute (4.127) into (4.113), we obtain the array response vector for the UPA as

$$\mathbf{a}_{M_{\mathrm{H}},M_{\mathrm{V}}}(\varphi,\theta) = \begin{bmatrix} 1 \\ e^{-\mathrm{j}\frac{2\pi}{\lambda}(i(2)\Delta\sin(\varphi)\cos(\theta)+j(2)\Delta\sin(\theta))} \\ \vdots \\ e^{-\mathrm{j}\frac{2\pi}{\lambda}(i(M)\Delta\sin(\varphi)\cos(\theta)+j(M)\Delta\sin(\theta))} \end{bmatrix}$$

$$= \begin{bmatrix} 1 \cdot \begin{bmatrix} 1 \\ e^{-\mathrm{j}2\pi\frac{\Delta\sin(\varphi)\cos(\theta)}{\lambda}} \\ \vdots \\ e^{-\mathrm{j}2\pi\frac{(M_{\mathrm{H}}-1)\Delta\sin(\varphi)\cos(\theta)}{\lambda}} \end{bmatrix} \\ e^{-\mathrm{j}2\pi\frac{\Delta\sin(\theta)}{\lambda}} \cdot \begin{bmatrix} 1 \\ e^{-\mathrm{j}2\pi\frac{\Delta\sin(\varphi)\cos(\theta)}{\lambda}} \\ \vdots \\ e^{-\mathrm{j}2\pi\frac{(M_{\mathrm{H}}-1)\Delta\sin(\varphi)\cos(\theta)}{\lambda}} \end{bmatrix} \\ \vdots \\ e^{-\mathrm{j}2\pi\frac{(M_{\mathrm{V}}-1)\Delta\sin(\theta)}{\lambda}} \cdot \begin{bmatrix} 1 \\ e^{-\mathrm{j}2\pi\frac{\Delta\sin(\varphi)\cos(\theta)}{\lambda}} \\ \vdots \\ e^{-\mathrm{j}2\pi\frac{(M_{\mathrm{H}}-1)\Delta\sin(\varphi)\cos(\theta)}{\lambda}} \end{bmatrix} \end{bmatrix}$$

$$= \begin{bmatrix} 1 \cdot \mathbf{a}_{M_{\mathrm{H}}}(\varphi,\theta) \\ e^{-\mathrm{j}2\pi\frac{\Delta\sin(\theta)}{\lambda}} \cdot \mathbf{a}_{M_{\mathrm{H}}}(\varphi,\theta) \\ \vdots \\ e^{-\mathrm{j}2\pi\frac{(M_{\mathrm{V}}-1)\Delta\sin(\theta)}{\lambda}} \cdot \mathbf{a}_{M_{\mathrm{H}}}(\varphi,\theta) \end{bmatrix} = \mathbf{a}_{M_{\mathrm{V}}}(\theta,0) \otimes \mathbf{a}_{M_{\mathrm{H}}}(\varphi,\theta). \tag{4.128}$$

The subscript of $\mathbf{a}_{M_{\mathrm{H}},M_{\mathrm{V}}}(\varphi,\theta)$ denotes the number of antennas along the horizontal and vertical axes, respectively. The derivation utilizes the fact that the horizontal index changes between each entry while the vertical index only changes after M_{H} entries. On the last row in (4.128), we first recognize the array response vector $\mathbf{a}_{M_{\mathrm{H}}}(\varphi,\theta)$ in (4.120) for a horizontally deployed ULA with M_{H} antennas with separation Δ. Next, we identify a Kronecker product between this vector and the array response vector $\mathbf{a}_{M_{\mathrm{V}}}(\theta,0)$ of a vertically deployed ULA with M_{V} antennas with separation Δ. This setup was previously characterized in Example 4.16.

In summary, the array response vector of a UPA is a concatenation of the array response vectors of two ULAs computed through a Kronecker product. Each vector entry contains the phase-shift the corresponding antenna experiences compared to the first antenna in the UPA, located in the origin. The phase-shift depends on the azimuth angle φ, elevation angle θ, antenna spacing Δ, and the number of horizontal and vertical antennas. Note that if we set $M_{\mathrm{V}}=1$ and $M=M_{\mathrm{H}}$, the array response vector in (4.128) reduces to the previous result in (4.120) for an M-dimensional horizontal ULA.

Example 4.18. What is the array response vector of a UPA with $M_{\mathrm{H}}=2$, $M_{\mathrm{V}}=3$, and the antenna spacing $\Delta=\lambda/2$?

According to (4.128), we can compute the array response vector as

$$\mathbf{a}_{2,3}(\varphi,\theta)=\mathbf{a}_3(\theta,0)\otimes\mathbf{a}_2(\varphi,\theta), \tag{4.129}$$

which is the Kronecker product between the array response vectors of two ULAs. We can compute those vectors using (4.120) as

$$\mathbf{a}_3(\varphi,\theta)=\begin{bmatrix}1\\ e^{-\mathrm{j}\pi\sin(\theta)}\\ e^{-\mathrm{j}2\pi\sin(\theta)}\end{bmatrix},\quad \mathbf{a}_2(\varphi,\theta)=\begin{bmatrix}1\\ e^{-\mathrm{j}\pi\sin(\varphi)\cos(\theta)}\end{bmatrix}. \tag{4.130}$$

We can now use the definition (2.54) of a Kronecker product to compute the UPA's array response vector as

$$\mathbf{a}_{2,3}(\varphi,\theta)=\begin{bmatrix}1\cdot\begin{bmatrix}1\\ e^{-\mathrm{j}\pi\sin(\varphi)\cos(\theta)}\end{bmatrix}\\ e^{-\mathrm{j}\pi\sin(\theta)}\cdot\begin{bmatrix}1\\ e^{-\mathrm{j}\pi\sin(\varphi)\cos(\theta)}\end{bmatrix}\\ e^{-\mathrm{j}2\pi\sin(\theta)}\cdot\begin{bmatrix}1\\ e^{-\mathrm{j}\pi\sin(\varphi)\cos(\theta)}\end{bmatrix}\end{bmatrix}=\begin{bmatrix}1\\ e^{-\mathrm{j}\pi\sin(\varphi)\cos(\theta)}\\ e^{-\mathrm{j}\pi\sin(\theta)}\\ e^{-\mathrm{j}\pi\sin(\theta)}e^{-\mathrm{j}\pi\sin(\varphi)\cos(\theta)}\\ e^{-\mathrm{j}2\pi\sin(\theta)}\\ e^{-\mathrm{j}2\pi\sin(\theta)}e^{-\mathrm{j}\pi\sin(\varphi)\cos(\theta)}\end{bmatrix}. \tag{4.131}$$

4.5.3 Horizontal and Vertical Beamwidths with UPAs

The beamwidth has been characterized previously in this chapter for a ULA, but only considering the horizontal plane where $\theta=0$. In this section, we will extend the analysis from Section 4.3.4 to the case of a UPA and derive the beamwidths of 3D beamforming. The channel vector for a UPA with an arbitrary antenna spacing Δ and wavelength λ is $\mathbf{h} = \sqrt{\beta}\mathbf{a}_{M_{\mathrm{H}},M_{\mathrm{V}}}(\varphi,\theta)$, where β is the channel gain and the array response vector is given by (4.128).

Suppose we transmit a signal in the direction $(\varphi_{\mathrm{beam}},\theta_{\mathrm{beam}})$, where $\varphi_{\mathrm{beam}} \in [-\pi/2,\pi/2]$ is the azimuth angle and $\theta_{\mathrm{beam}} \in [-\pi/2,\pi/2]$ is the elevation angle, using MRT with $\mathbf{p} = \mathbf{a}^*_{M_{\mathrm{H}},M_{\mathrm{V}}}(\varphi_{\mathrm{beam}},\theta_{\mathrm{beam}})/\|\mathbf{a}_{M_{\mathrm{H}},M_{\mathrm{V}}}(\varphi_{\mathrm{beam}},\theta_{\mathrm{beam}})\|$. We can then follow the approach in (4.75) to determine the beamforming gain that is observed by a receiver located in any another direction $\varphi \in [-\pi/2,\pi/2]$, $\theta \in [-\pi/2,\pi/2]$:

$$
\begin{aligned}
&\left|\mathbf{a}^{\mathrm{T}}_{M_{\mathrm{H}},M_{\mathrm{V}}}(\varphi,\theta)\frac{\mathbf{a}^*_{M_{\mathrm{H}},M_{\mathrm{V}}}(\varphi_{\mathrm{beam}},\theta_{\mathrm{beam}})}{\|\mathbf{a}_{M_{\mathrm{H}},M_{\mathrm{V}}}(\varphi_{\mathrm{beam}},\theta_{\mathrm{beam}})\|}\right|^2 \\
&= \frac{1}{M_{\mathrm{H}}M_{\mathrm{V}}}\left|(\mathbf{a}_{M_{\mathrm{V}}}(\theta,0)\otimes\mathbf{a}_{M_{\mathrm{H}}}(\varphi,\theta))^{\mathrm{T}}(\mathbf{a}_{M_{\mathrm{V}}}(\theta_{\mathrm{beam}},0)\otimes\mathbf{a}_{M_{\mathrm{H}}}(\varphi_{\mathrm{beam}},\theta_{\mathrm{beam}}))^*\right|^2 \\
&= \frac{1}{M_{\mathrm{V}}}\left|\mathbf{a}^{\mathrm{T}}_{M_{\mathrm{V}}}(\theta,0)\mathbf{a}^*_{M_{\mathrm{V}}}(\theta_{\mathrm{beam}},0)\right|^2\frac{1}{M_{\mathrm{H}}}\left|\mathbf{a}^{\mathrm{T}}_{M_{\mathrm{H}}}(\varphi,\theta)\mathbf{a}^*_{M_{\mathrm{H}}}(\varphi_{\mathrm{beam}},\theta_{\mathrm{beam}})\right|^2 \\
&= \underbrace{\frac{1}{M_{\mathrm{H}}}\left|\sum_{m=1}^{M_{\mathrm{H}}} e^{-\mathrm{j}\frac{2\pi\Delta(m-1)}{\lambda}\underbrace{(\sin(\varphi)\cos(\theta)-\sin(\varphi_{\mathrm{beam}})\cos(\theta_{\mathrm{beam}}))}_{=\Phi}}\right|^2}_{=A_{M_{\mathrm{H}}}(\Phi)} \\
&\quad\times\underbrace{\frac{1}{M_{\mathrm{V}}}\left|\sum_{n=1}^{M_{\mathrm{V}}} e^{-\mathrm{j}\frac{2\pi\Delta(n-1)}{\lambda}\underbrace{(\sin(\theta)-\sin(\theta_{\mathrm{beam}}))}_{=\Omega}}\right|^2}_{=A_{M_{\mathrm{V}}}(\Omega)},
\end{aligned}
\tag{4.132}
$$

where the first equality utilizes the expression in (4.128) and the fact that $\|\mathbf{a}_{M_{\mathrm{H}},M_{\mathrm{V}}}(\varphi_{\mathrm{beam}},\theta_{\mathrm{beam}})\|^2 = M_{\mathrm{H}}M_{\mathrm{V}}$. The second equality utilizes the Kronecker product property $(\mathbf{a}\otimes\mathbf{b})^{\mathrm{T}}(\mathbf{c}\otimes\mathbf{d})^* = (\mathbf{a}^{\mathrm{T}}\mathbf{c}^*\otimes\mathbf{b}^{\mathrm{T}}\mathbf{d}^*)$ which holds for any vectors $\mathbf{a},\mathbf{b},\mathbf{c},\mathbf{d}$ with matching dimensions.

We notice that the beamforming gain expression in (4.132) is decomposed as the product of the two terms that we denote as $A_{M_{\mathrm{H}}}(\Phi)$ and $A_{M_{\mathrm{V}}}(\Omega)$, which depend on the angles through the variables

$$\Phi = \sin(\varphi)\cos(\theta) - \sin(\varphi_{\mathrm{beam}})\cos(\theta_{\mathrm{beam}}), \tag{4.133}$$

$$\Omega = \sin(\theta) - \sin(\theta_{\mathrm{beam}}). \tag{4.134}$$

Each of these functions has the same structure as the beamforming gain function considered in Section 4.3.4. More precisely, we can use the summation formula for geometric series in (4.51) and Euler's formula to compute the horizontal function $A_{M_{\mathrm{H}}}(\Phi)$ as

$$A_{M_{\mathrm{H}}}(\Phi) = \frac{1}{M_{\mathrm{H}}} \frac{\sin^2\left(M_{\mathrm{H}} \frac{\pi \Delta \Phi}{\lambda}\right)}{\sin^2\left(\frac{\pi \Delta \Phi}{\lambda}\right)} = \frac{1}{M_{\mathrm{H}}} \frac{\sin^2(\pi L_{\mathrm{H},\lambda} \Phi)}{\sin^2(\pi \Delta_\lambda \Phi)}, \tag{4.135}$$

which equals $A(\Phi)$ in (4.79) when $M = M_{\mathrm{H}}$. The last equality in (4.135) follows from using the notation $\Delta_\lambda = \Delta/\lambda$ for the normalized antenna spacing and by defining the *normalized horizontal length* of the UPA as

$$L_{\mathrm{H},\lambda} = \frac{M_{\mathrm{H}} \Delta}{\lambda} = M_{\mathrm{H}} \Delta_\lambda. \tag{4.136}$$

Whenever the numerator and denominator in (4.135) are zero simultaneously (e.g., if $\Phi = 0$), it follows from the geometric series formula that the function value is M_{H}. We will not write this out to keep the expression compact.

Similarly, the vertical function $A_{M_{\mathrm{V}}}(\Omega)$ in (4.132) can be expressed as

$$A_{M_{\mathrm{V}}}(\Omega) = \frac{1}{M_{\mathrm{V}}} \frac{\sin^2\left(M_{\mathrm{V}} \frac{\pi \Delta \Omega}{\lambda}\right)}{\sin^2\left(\frac{\pi \Delta \Omega}{\lambda}\right)} = \frac{1}{M_{\mathrm{V}}} \frac{\sin^2(\pi L_{\mathrm{V},\lambda} \Omega)}{\sin^2(\pi \Delta_\lambda \Omega)}, \tag{4.137}$$

which is equal to $A(\Phi)$ in (4.79) when $\Phi = \Omega$ and $M = M_{\mathrm{V}}$. The last equality in (4.137) follows from defining the *normalized vertical length* of the UPA as

$$L_{\mathrm{V},\lambda} = \frac{M_{\mathrm{V}} \Delta}{\lambda} = M_{\mathrm{V}} \Delta_\lambda. \tag{4.138}$$

In summary, the beamforming gain in (4.132) that is obtained in the angular direction (φ, θ) can be expressed as

$$A_{M_{\mathrm{H}}}(\Phi) A_{M_{\mathrm{V}}}(\Omega) = \frac{1}{M} \frac{\sin^2(\pi L_{\mathrm{H},\lambda} \Phi)}{\sin^2(\pi \Delta_\lambda \Phi)} \frac{\sin^2(\pi L_{\mathrm{V},\lambda} \Omega)}{\sin^2(\pi \Delta_\lambda \Omega)}. \tag{4.139}$$

The maximum beamforming gain is M and is achieved for $\varphi = \varphi_{\mathrm{beam}}$ and $\theta = \theta_{\mathrm{beam}}$ (the intended beamforming direction) because in that case, we have $\Phi = \Omega = 0$.[13] We further notice that the maximum gain, M, only depends on the number of antennas and not on the antenna spacing, which aligns with the analysis of ULAs earlier in this chapter.

The beamforming gain $A_{M_{\mathrm{H}}}(\Phi) A_{M_{\mathrm{V}}}(\Omega)$ is generally smaller than the number of antennas M. Depending on the angles (φ, θ) and $(\varphi_{\mathrm{beam}}, \theta_{\mathrm{beam}})$, the input variables can take values in the ranges $\Phi \in [-2, 2]$ and $\Omega \in [-2, 2]$. For given values of Φ and Ω, the function value is determined by M, the normalized antenna spacing Δ_λ, the normalized horizontal length $L_{\mathrm{H},\lambda}$, and the normalized vertical length $L_{\mathrm{V},\lambda}$. It is always the relative distances compared to the wavelength that matters in this context.

[13] It follows from (4.132) that $A_{M_{\mathrm{H}}}(0) = M_{\mathrm{H}}$ and $A_{M_{\mathrm{V}}}(0) = M_{\mathrm{V}}$.

Example 4.19. What is the area of a UPA? How does the SIMO capacity in (4.25) depend on the UPA's area if $\Delta = \lambda/2$?

The horizontal length of the UPA is $M_{\mathrm{H}}\Delta$ and the vertical length is $M_{\mathrm{V}}\Delta$; thus, the area is $\mathsf{Area} = M_{\mathrm{H}}\Delta \cdot M_{\mathrm{V}}\Delta = M\Delta^2$. By utilizing the fact that $\beta = \frac{\lambda^2}{(4\pi)^2}\frac{1}{d^2}$, we can express the SIMO capacity in (4.25) as

$$C = B\log_2\left(1 + \frac{PM\lambda^2}{BN_0(4\pi d)^2}\right) = B\log_2\left(1 + \frac{P}{BN_0(2\pi d)^2}\mathsf{Area}\right). \tag{4.140}$$

The SNR in this capacity expression is proportional to the array area but independent of the wavelength. The reason is that an isotropic transmit antenna radiates identically irrespective of the wavelength, while it is the area of the receiver array that determines what fraction of the signal power it captures. However, $\mathsf{Area} = M\lambda^2/4$, so if the wavelength reduces, the number of antennas must grow as $1/\lambda^2$ to maintain the array area.

By following the same steps as in Section 4.3.4, one can show that the numerator of $A_{M_{\mathrm{H}}}(\Phi)$ is a periodic function of Φ with period $1/L_{\mathrm{H},\lambda} = 1/(M_{\mathrm{H}}\Delta_\lambda)$ and the denominator is periodic with period $1/\Delta_\lambda$. Similarly, the numerator of $A_{M_{\mathrm{V}}}(\Omega)$ is a periodic function of Ω with period $1/L_{\mathrm{V},\lambda} = 1/(M_{\mathrm{V}}\Delta_\lambda)$ and the denominator is periodic with period $1/\Delta_\lambda$. Hence, both $A_{M_{\mathrm{H}}}(\Phi)$ and $A_{M_{\mathrm{V}}}(\Omega)$ have a period of $1/\Delta_\lambda$. We also have that

$$A_{M_{\mathrm{H}}}\left(\frac{m}{L_{\mathrm{H},\lambda}}\right) = 0, \quad m = \pm 1, \ldots, \pm(M_{\mathrm{H}} - 1), \tag{4.141}$$

$$A_{M_{\mathrm{V}}}\left(\frac{n}{L_{\mathrm{V},\lambda}}\right) = 0, \quad n = \pm 1, \ldots \pm (M_{\mathrm{V}} - 1). \tag{4.142}$$

These points correspond to the nulls of the beamforming gain pattern and thereby characterize the beamwidths. It is sufficient that one of these functions is zero to obtain a null. This implies that for a given value of θ, we can find a value of φ that results in a null (and vice versa). One can measure both the horizontal and vertical beamwidths, which are generally different.

We will consider a few special cases to shed light on the angular locations of the nulls. We begin by considering the horizontal plane where $\theta = \theta_{\mathrm{beam}} = 0$. We then have $\Omega = \sin(\theta) - \sin(\theta_{\mathrm{beam}}) = 0$ and $\Phi = \sin(\varphi)\cos(\theta) - \sin(\varphi_{\mathrm{beam}})\cos(\theta_{\mathrm{beam}}) = \sin(\varphi) - \sin(\varphi_{\mathrm{beam}})$. The latter coincides with (4.78) that was derived for a ULA; thus, the horizontal beamwidth of a UPA is the same as for a ULA with the same number of antennas horizontally.

Figure 4.38 shows the beamforming gain in (4.139) when transmitting in the broadside direction where $\varphi_{\mathrm{beam}} = 0$ and $\theta_{\mathrm{beam}} = 0$, which results in $A_{M_{\mathrm{H}}}(\Phi)A_{M_{\mathrm{V}}}(\Omega) = M_{\mathrm{V}}A_{M_{\mathrm{H}}}(\sin(\varphi))$. The antenna spacing is $\Delta_\lambda = 1/2$ wavelengths. We compare two arrays: i) a UPA with $M_{\mathrm{H}} = 10$ horizontal antennas and $M_{\mathrm{V}} = 4$ vertical antennas; ii) a ULA with $M = M_{\mathrm{H}} = 10$

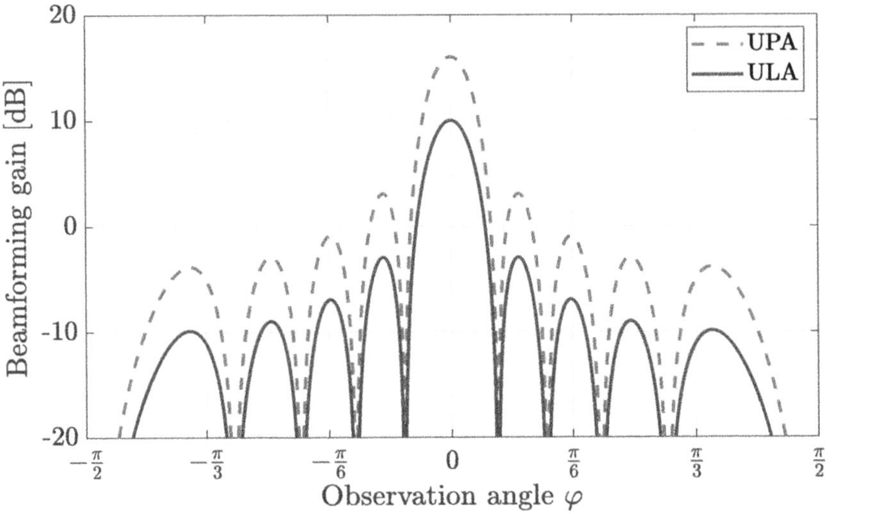

Figure 4.38: Comparison of the beamforming gain in (4.139) with a UPA ($M_{\mathrm{H}} = 10$, $M_{\mathrm{V}} = 4$) and a ULA ($M = M_{\mathrm{H}} = 10$, $M_{\mathrm{V}} = 1$) in the horizontal plane where $\theta = 0$. The arrays transmit in the broadside direction $\varphi_{\mathrm{beam}} = 0$ and $\theta_{\mathrm{beam}} = 0$.

horizontal antennas (and $M_{\mathrm{V}} = 1$). The beamforming gain is shown as a function of the observation azimuth angle φ in the horizontal plane (where $\theta = 0$). The figure validates that the beamforming gain in this plane is only determined by the horizontal lengths of the arrays, which are identical for the UPA and ULA. Hence, the null locations and the shape of the lobes are the same in both cases. However, since the UPA has $M = 40$ antennas while the ULA has $M = 10$, there is a $40/10 \approx 6\,\mathrm{dB}$ vertical difference between the beamforming gain patterns. The maximum beamforming gain for the UPA is $40 \approx 16\,\mathrm{dB}$, whereas the maximum beamforming gain for the ULA is $10\,\mathrm{dB}$.

If we keep transmitting in the broadside direction but consider the plane where $\theta = \pi/4$, then we have $\Omega = \sin(\theta) - \sin(\theta_{\mathrm{beam}}) = 1/\sqrt{2}$ and $\Phi = \sin(\varphi)\cos(\theta) - \sin(\varphi_{\mathrm{beam}})\cos(\theta_{\mathrm{beam}}) = \sin(\varphi)/\sqrt{2}$. The beamforming gain for different azimuth angles will then be determined by $A_{M_{\mathrm{H}}}(\Phi)A_{M_{\mathrm{V}}}(\Omega) = A_{M_{\mathrm{H}}}(\sin(\varphi)/\sqrt{2})A_{M_{\mathrm{V}}}(1/\sqrt{2})$. The division by $\sqrt{2}$ in the first factor leads to a widening of the beamwidth compared to having $\theta = 0$, while the second factor leads to a loss in beamforming gain since $A_{M_{\mathrm{V}}}(1/\sqrt{2}) < 1$ for $M_{\mathrm{V}} > 1$. Figure 4.39 compares the beamforming gains observed at different azimuth angles when the elevation angle is either $\theta = 0$ or $\theta = \pi/4$. As expected, the nulls move outwards when θ increases so that the beamwidth increases, but the beamforming gain is reduced.

The vertical beamwidth only depends on the number of vertical antennas. To show this, we continue transmitting in the broadside direction (i.e., $\varphi_{\mathrm{beam}} = \theta_{\mathrm{beam}} = 0$) and consider the vertical plane where $\varphi = 0$. We then have $\Phi = \sin(\varphi)\cos(\theta) - \sin(\varphi_{\mathrm{beam}})\cos(\theta_{\mathrm{beam}}) = 0$ and $A_{M_{\mathrm{H}}}(0) = M_{\mathrm{H}}$. On the

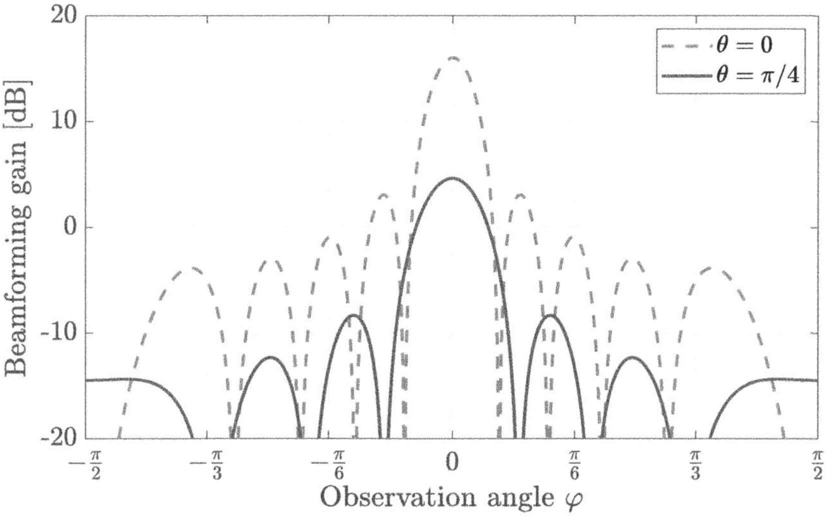

Figure 4.39: The beamforming gain that is observed in different azimuth directions φ when a UPA with $M_{\mathrm{H}} = 10$ and $M_{\mathrm{V}} = 4$ transmits a beam in the broadside direction $\varphi_{\mathrm{beam}} = 0$ and $\theta_{\mathrm{beam}} = 0$. The pattern depends on the elevation angle θ.

other hand, $\Omega = \sin(\theta) - \sin(\theta_{\mathrm{beam}}) = \sin(\theta)$. From (4.142), the $2\lfloor L_{\mathrm{V},\lambda} \rfloor$ null directions are obtained as[14]

$$\theta = \arcsin\left(\frac{n}{L_{\mathrm{V},\lambda}}\right), \quad n = \pm 1, \ldots, \pm \lfloor L_{\mathrm{V},\lambda} \rfloor. \tag{4.143}$$

A ULA with $M_{\mathrm{V}} = 1$ and normalized vertical length $L_{\mathrm{V},\lambda} = 0.5$ has no nulls in the vertical plane, so its beams have no vertical directivity. However, if we extend the ULA to a UPA by adding antennas in the vertical plane, we achieve a directive beam in both the horizontal and vertical planes. Figure 4.40 shows the beamforming gain in this vertical plane for the same UPA and ULA as in Figure 4.38. The ULA achieves a constant gain for all elevation angles. However, the UPA with $M_{\mathrm{V}} = 4$ has a normalized vertical length of $L_{\mathrm{V},\lambda} = 2$ so there are $2L_{\mathrm{V},\lambda} = 4$ nulls in the vertical plane. The null directions are $\theta = \pm \arcsin(1/2) = \pm\pi/6$ and $\theta = \pm \arcsin(1) = \pm\pi/2$.

Figure 4.41 shows the beamforming gain pattern in the 3D half-space $x \geq 0$ when the antenna array is deployed in the yz-plane and beamforms in the broadside direction. Figure 4.41(a) considers the UPA with $M_{\mathrm{H}} = 10$ and $M_{\mathrm{V}} = 4$, while Figure 4.41(b) considers the ULA with $M_{\mathrm{H}} = 10$ and $M_{\mathrm{V}} = 1$. The dotted black curve shows the angles representing the horizontal plane previously considered in Figure 4.38, while the dashed blue curve shows the vertical plane previously considered in Figure 4.40. The white areas

[14] All the nulls of the function $A_{M_{\mathrm{V}}}(\Omega)$ are given by (4.142). We should find the values of θ that give $\Omega = \sin(\theta) = n/L_{\mathrm{V},\lambda}$ for $n = \pm 1, \ldots, \pm(M_{\mathrm{V}} - 1)$. This equation can only be solved when $|n| \leq L_{\mathrm{V},\lambda}$, and the feasible solutions are given in (4.143).

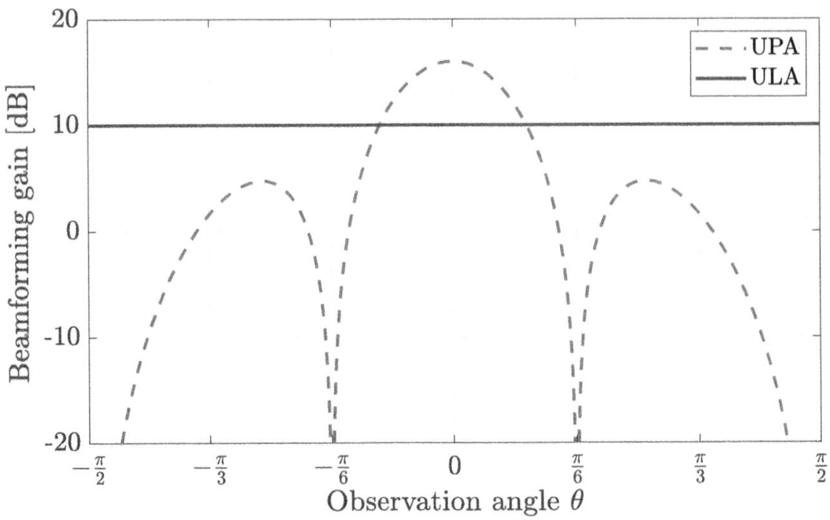

Figure 4.40: Comparison of the beamforming gain in (4.139) with a UPA ($M_{\mathrm{H}} = 10$, $M_{\mathrm{V}} = 4$) and a ULA ($M = M_{\mathrm{H}} = 10$, $M_{\mathrm{V}} = 1$) in the vertical plane where $\varphi = 0$. The arrays transmit in the broadside direction $\varphi_{\mathrm{beam}} = 0$ and $\theta_{\mathrm{beam}} = 0$.

represent directions that are close to the nulls. We notice that the UPA achieves directivity in both the azimuth and elevation directions, while the ULA has an azimuth directivity but spreads its signal equally for all elevation angles. The 3D directivity achieved by a UPA makes the transmission more confined to the directions close to the intended receiver.

Example 4.20. Consider a UPA with $M_{\mathrm{H}} = 10$, $M_{\mathrm{V}} = 4$, and $\Delta_{\lambda} = 1/2$. If it beamforms in the broadside direction as in Figure 4.41(a), what is the first-null beamwidth in the horizontal and vertical planes?

In the horizontal plane where $\theta = 0$, Φ in (4.133) becomes

$$\Phi = \sin(\varphi)\cos(0) - \sin(0)\cos(0) = \sin(\varphi) \tag{4.144}$$

since $\varphi_{\mathrm{beam}} = \theta_{\mathrm{beam}} = 0$. The nulls in the horizontal plane occur when $A_{M_{\mathrm{H}}}(\Phi) = 0$. Since $L_{\mathrm{H},\lambda} = M_{\mathrm{H}}\Delta_{\lambda} = 5$ wavelengths, it follows from (4.141) that the first nulls are at the azimuth angles $\varphi = \pm \arcsin(1/5)$. Hence, the first-null beamwidth is $2\arcsin(1/5) \approx 0.4$ (23°) in the horizontal plane.

In the vertical plane where $\varphi = 0$, Ω in (4.134) becomes

$$\Omega = \sin(\theta) - \sin(0) = \sin(\theta). \tag{4.145}$$

The nulls in the vertical plane occur when $A_{M_{\mathrm{V}}}(\Omega) = 0$. Since $L_{\mathrm{V},\lambda} = M_{\mathrm{V}}\Delta_{\lambda} = 2$ wavelengths, it follows from (4.142) that the first nulls are at the elevation angles $\theta = \pm \arcsin(1/2) = \pm\pi/6$. Hence, the first-null beamwidth is $2\pi/6 = \pi/3$ (60°) in the vertical plane.

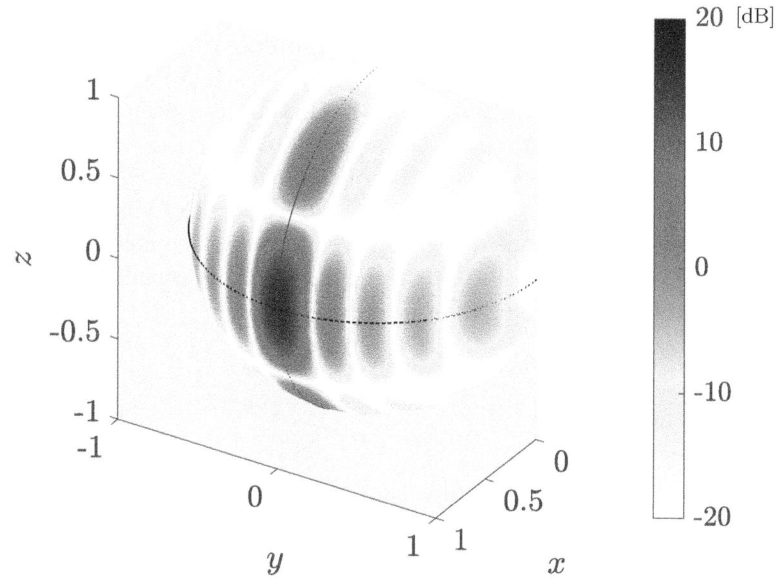

(a) Beamforming with a UPA having $M_{\mathrm{H}} = 10$ horizontal antennas and $M_{\mathrm{V}} = 4$ vertical antennas.

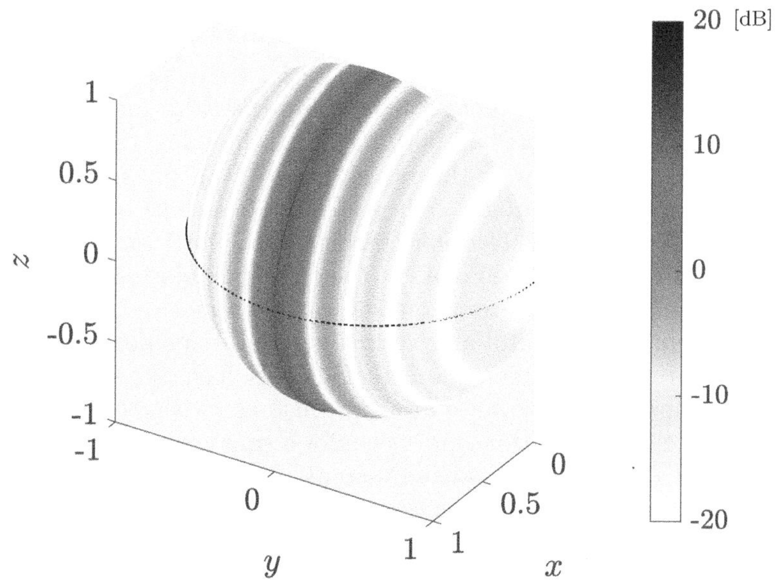

(b) Beamforming with a ULA having $M_{\mathrm{H}} = 10$ horizontal antennas.

Figure 4.41: The beamforming gain that is observed in different 3D directions for the UPA and ULA setups that were considered in Figure 4.38 and Figure 4.40. The dotted black curves show the gain variations in the horizontal plane where $\theta = 0$ and the dashed blue curves show the gain variations in the vertical plane where $\varphi = 0$. These are the same gain patterns as shown in the previous figures.

When a plane wave impinges on the UPA from the angle (φ, θ), the antennas in the array take simultaneous samples of the waveform. Suppose the information-bearing signal has the wavelength λ. The angle-of-arrival determines the phase-shift differences between the antennas and, thereby, what spatial frequencies in the range $[-1/\lambda, 1/\lambda]$ are present in the channel vector. Since the UPA extends in two dimensions, the channel vector can resolve spatial frequencies horizontally and vertically, which are generally different. Recall that we consider a UPA deployed in the negative parts of the yz-plane with one antenna in the origin. At any given time, the phase-shift seen along the horizontal negative y-axis is

$$\frac{2\pi}{\lambda}\begin{bmatrix} 0 \\ -y \\ 0 \end{bmatrix}^{\mathrm{T}} \begin{bmatrix} \cos(\varphi)\cos(\theta) \\ \sin(\varphi)\cos(\theta) \\ \sin(\theta) \end{bmatrix} = -\frac{2\pi}{\lambda} y \sin(\varphi)\cos(\theta), \tag{4.146}$$

relative to the origin. This implies that the channel contains the horizontal spatial frequency $\sin(\varphi)\cos(\theta)/\lambda$ periods per meter. Similarly, the phase-shift seen along the negative vertical z-axis is

$$\frac{2\pi}{\lambda}\begin{bmatrix} 0 \\ 0 \\ -z \end{bmatrix}^{\mathrm{T}} \begin{bmatrix} \cos(\varphi)\cos(\theta) \\ \sin(\varphi)\cos(\theta) \\ \sin(\theta) \end{bmatrix} = -\frac{2\pi}{\lambda} z \sin(\theta), \tag{4.147}$$

thus the channel contains the vertical spatial frequency $\sin(\theta)/\lambda$ periods per meter. The vertical frequency depends on the elevation angle, while the horizontal frequency depends on both the azimuth and elevation angles. Each frequency can take values in the range $[-1/\lambda, 1/\lambda]$, but since both values depend on the elevation angle, only some combinations of frequencies can occur. Figure 4.42 shows the feasible combinations, which are contained within a circle with a radius of $1/\lambda$. Points at the outer boundary are achieved when $\varphi = \pm\frac{\pi}{2}$ while θ is varied throughout its feasible range.

The concept of horizontal and vertical spatial frequencies is further illustrated in Figure 4.43 for different angle-of-arrivals. The coloring shows the real part of the impinging plane wave at a time instance when the phase is zero at the origin. The wave variations are shown for a square area with width 4λ and height 4λ, but the antenna array only samples it at 81 discrete points; that is, the UPA has $M_{\mathrm{H}} = M_{\mathrm{V}} = 9$ antennas per dimension with the spacing $\Delta = \lambda/2$. Figure 4.43(a) considers a plane wave arriving from the broadside direction $\varphi = \theta = 0$. In this case, the horizontal and vertical spatial frequencies are zero because the UPA is deployed perpendicularly to the direction the wave travels. No phase variations exist between the antennas; therefore, the entire array surface has the same color. Figure 4.43(b) considers a plane wave arriving from the direction $\varphi = \pi/6$, $\theta = 0$, which represents a rotation in the horizontal plane. The horizontal spatial frequency is $\sin(\pi/6)\cos(0)/\lambda = 1/(2\lambda)$, which explains why the waveform repeats itself at points that are separated by

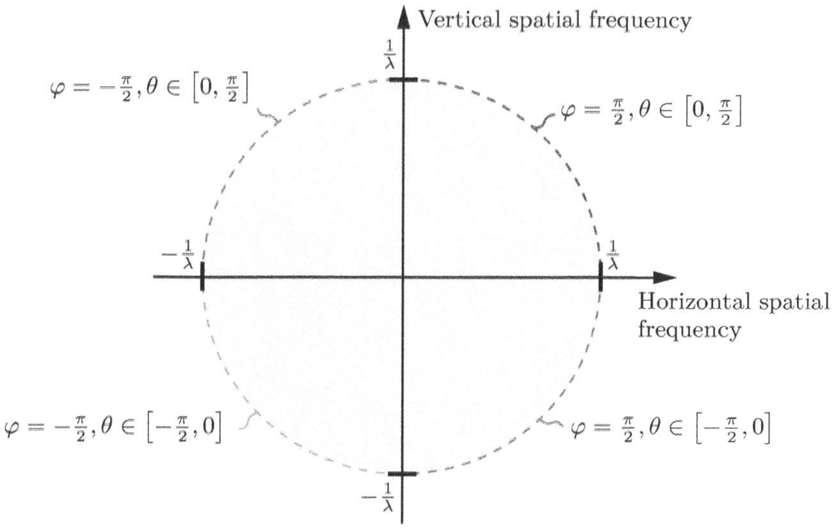

Figure 4.42: The combinations of horizontal and vertical spatial frequencies that the channel to/from a UPA can contain are all contained in a circle with the radius $1/\lambda$. The horizontal frequency is $\sin(\varphi)\cos(\theta)/\lambda$ and the vertical frequency is $\sin(\theta)/\lambda$, where φ is the azimuth angle and θ is the elevation angle.

2λ horizontally. The vertical spatial frequency is $\sin(0)/\lambda = 0$, as seen from the fact that there are no vertical variations in the waveform. Figure 4.43(c) considers the case of $\varphi = 0$, $\theta = \pi/4$, where the plane wave is rotated in the vertical plane compared to the UPA. The horizontal frequency is zero, while the vertical frequency becomes $\sin(\pi/4)/\lambda = 1/(\sqrt{2}\lambda)$, so the wave repeats itself at points that have a vertical separation of $\sqrt{2}\lambda$. Finally, Figure 4.43(d) considers a plane wave arriving from the direction $\varphi = \pi/4$, $\theta = \pi/4$, for which both the horizontal and vertical frequencies are non-zero. The horizontal spatial frequency is $\sin(\pi/4)\cos(\pi/4)/\lambda = 1/(2\lambda)$, which is the same as in Figure 4.43(b). The vertical spatial frequency is the same as in Figure 4.43(c); thus, these frequencies are simultaneously achievable (i.e., they are within the circle shown in Figure 4.42).

Example 4.21. What fraction of all horizontal/vertical spatial frequency combinations are practically achievable?

The horizontal spatial frequency can take any value in the range $[-1/\lambda, 1/\lambda]$, which is a range of length $2/\lambda$, and the same holds for the vertical spatial frequency. This corresponds to a total area of $4/\lambda^2$ of possible combinations. However, the practically achievable combinations are contained in the circle in Figure 4.42, which has the area $\pi(1/\lambda)^2$. Hence, a fraction $\pi(1/\lambda)^2/(4/\lambda^2) = \pi/4 \approx 0.79$ of all combinations are practically achievable.

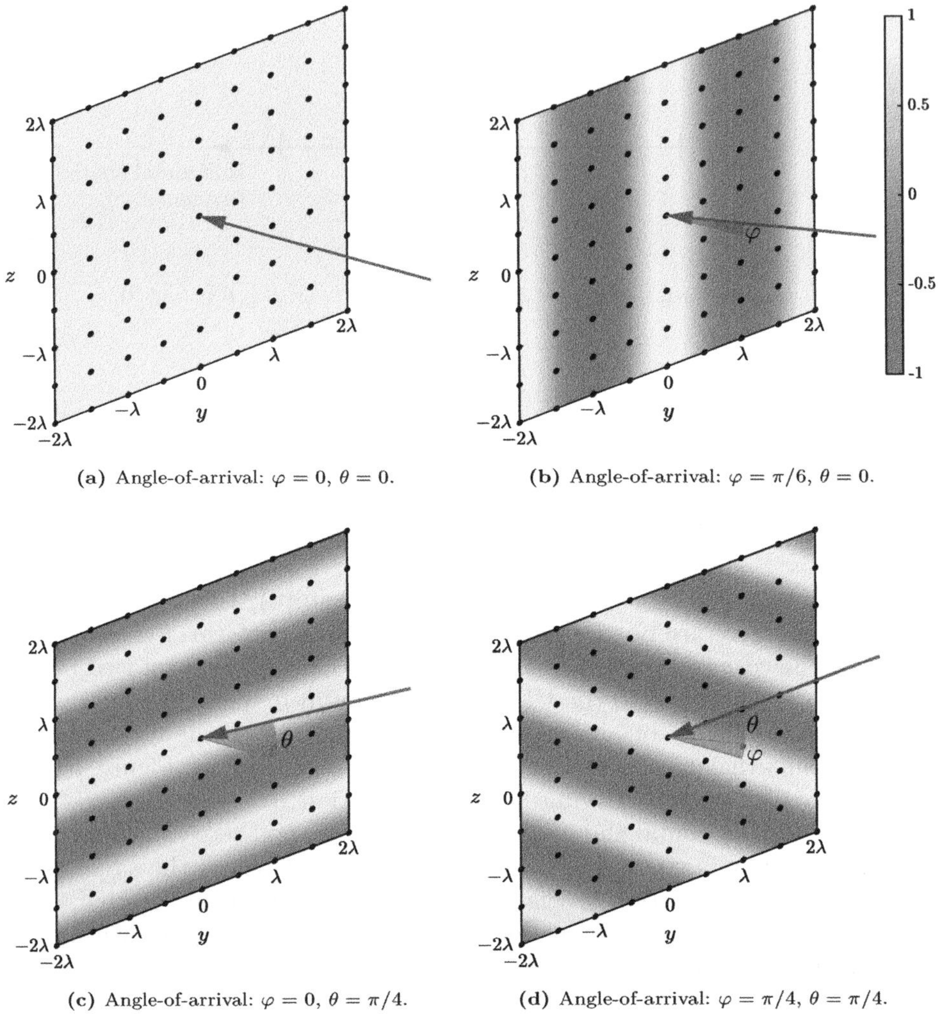

(a) Angle-of-arrival: $\varphi = 0$, $\theta = 0$.

(b) Angle-of-arrival: $\varphi = \pi/6$, $\theta = 0$.

(c) Angle-of-arrival: $\varphi = 0$, $\theta = \pi/4$.

(d) Angle-of-arrival: $\varphi = \pi/4$, $\theta = \pi/4$.

Figure 4.43: When a plane wave with wavelength λ impinges on a UPA, the angle-of-arrival (φ, θ) determines the wave variations simultaneously observable over the array's surface. The real part of the wave is shown for four different angle-of-arrivals using colors to represent the value. The horizontal and vertical spatial frequencies differ depending on the angle-of-arrival, as seen from the color patterns. The UPA consists of 81 antennas ($M_{\mathrm{H}} = M_{\mathrm{V}} = 9$) with the spacing $\Delta = \lambda/2$.

4.5.4 Effective Array Response with Directive Antennas

Another benefit of specifying the array response vectors in terms of both the azimuth and elevation angles is that we can readily extend the model to support arrays of directive antennas. Recall from Section 1.1.3 that the directivity of an antenna is determined by the antenna gain function $G(\varphi, \theta)$, which specifies the angular variations in the antenna gain compared to an isotropic antenna. We will analyze a MISO channel where the M transmit antennas have the same antenna gain function $G_{\mathrm{t}}(\varphi, \theta)$, while the receive antenna is isotropic. We can then define the *effective array response* as

$$\sqrt{G_{\mathrm{t}}(\varphi, \theta)}\mathbf{a}_M(\varphi, \theta). \tag{4.148}$$

and let $\beta_{\mathrm{iso}} = \lambda^2/(4\pi d)^2$ denote the channel gain when both the transmitter and receiver have isotropic antennas. If the transmitter has a ULA with antenna spacing Δ, then the channel vector becomes

$$\mathbf{h} = \sqrt{\beta_{\mathrm{iso}}}\sqrt{G_{\mathrm{t}}(\varphi, \theta)}\mathbf{a}_M(\varphi, \theta) = \sqrt{\beta_{\mathrm{iso}}}\sqrt{G_{\mathrm{t}}(\varphi, \theta)} \begin{bmatrix} 1 \\ e^{-\mathrm{j}2\pi \frac{\Delta \sin(\varphi)\cos(\theta)}{\lambda}} \\ \vdots \\ e^{-\mathrm{j}2\pi \frac{(M-1)\Delta \sin(\varphi)\cos(\theta)}{\lambda}} \end{bmatrix}. \tag{4.149}$$

The capacity of this MISO channel can be computed using (4.46) as

$$C = B\log_2\left(1 + \frac{P\|\mathbf{h}\|^2}{BN_0}\right) = B\log_2\left(1 + \frac{PG_{\mathrm{t}}(\varphi, \theta)M\beta_{\mathrm{iso}}}{BN_0}\right). \tag{4.150}$$

The only difference from the capacity expression in (4.47) for ULAs with isotropic antennas is that the SNR is multiplied by the antenna gain $G_{\mathrm{t}}(\varphi, \theta)$. Hence, if $G_{\mathrm{t}}(\varphi, \theta) \neq 0$, a beamforming gain of M can be achieved using an array of directive antennas. This is the same beamforming gain as with isotropic antennas; thus, the SNR grows proportionally to the number of antennas, but the proportionality constant depends on the antenna gain in the angular direction leading to the receiver. By replacing $G_{\mathrm{t}}(\varphi, \theta)$ with $G_{\mathrm{r}}(\varphi, \theta)$, the capacity expression in (4.150) applies to a SIMO channel where the receiver is equipped with M antennas with the gain function $G_{\mathrm{r}}(\varphi, \theta)$ and the transmitter is equipped with an isotropic antenna. Note that the MISO capacity is achieved using the MRT vector $\mathbf{a}_M^*(\varphi, \theta)/\sqrt{M}$, while the SIMO capacity is achieved using the MRC vector $\mathbf{a}_M(\varphi, \theta)/\sqrt{M}$. These are the same vectors as with isotropic antennas because the antenna gain only changes the scaling of the channel vector, not its direction in the vector space.

Recall that the primary purpose of beamforming is to control the directivity of the transmission. A directive antenna has a fixed directivity, while an array of isotropic antennas can form beams in any direction and always achieve the same maximum gain. In practice, arrays of *weakly* directive antennas are often

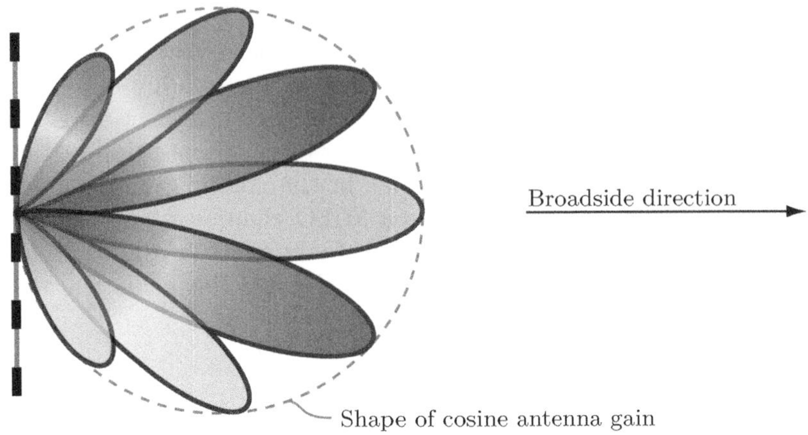

Figure 4.44: A ULA with cosine antennas can form beams in most directions, but the strength of the beams will depend on the beam direction. The maximum gain is achieved in the broadside direction and then tapers off for other angles, determined by the shape of the antenna gain function.

utilized when not all angular directions are important. For example, an array might be deployed to serve user devices in a 120° sector of the horizontal plane; that is, $\varphi \in [-\pi/3, \pi/3]$. We can then utilize an array of the cosine antenna from Section 1.1.3, which has the gain function

$$G(\varphi, \theta) = \begin{cases} 4\cos(\varphi)\cos(\theta), & \text{if } \varphi \in [-\pi/2, \pi/2], \theta \in [-\pi/2, \pi/2], \\ 0, & \text{elsewhere.} \end{cases} \tag{4.151}$$

It provides the maximum antenna gain of 4 (i.e., 6 dBi) when transmitting to receivers located in the direction $(\varphi, \theta) = (0, 0)$ and an antenna gain of 2 (i.e., 3 dBi) when transmitting to receivers located in the directions $(\varphi, \theta) = (\pm\pi/3, 0)$ at the edges of a 120° sector. Since these gains are larger than one, all users located in the intended sector will benefit from having this directive antenna (compared to having an isotropic antenna) but to a varying extent. Note that the antenna gain also varies with the elevation angle, but every user located in the interval $\varphi \in [-\pi/3, \pi/3]$, $\theta \in [-\pi/3, \pi/3]$ will obtain an antenna gain larger than one; thus, preferring the cosine antenna over an isotropic transmit antenna.

The combination of antenna directivity and beamforming makes the radiated signal even more directive than when using isotropic antennas, but the joint gain also becomes dependent on the beam direction. Figure 4.44 illustrates this property by showing a collection of beams transmitted in different angular directions from a ULA equipped with cosine antennas. The beam radiated in the broadside direction is substantially larger/stronger than the beams radiated towards angles closer to the end-fire directions. The overlaid

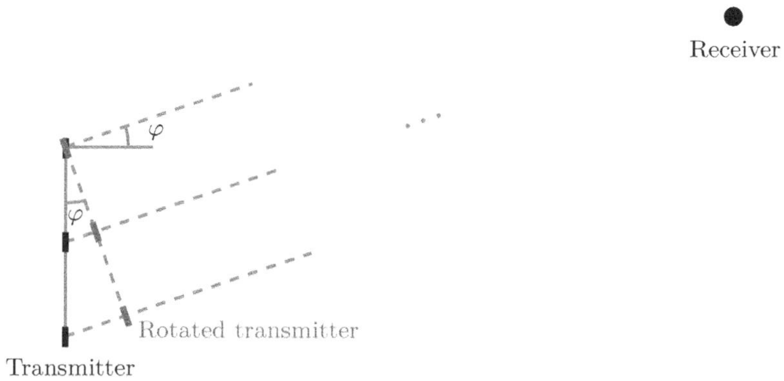

Figure 4.45: If the receiver is not located in the broadside direction of the ULA, electrical beamforming can be utilized to phase-shift the signals to form a beam in the angular direction φ leading toward the receiver. Alternatively, mechanical beamforming can be utilized where the transmitter is physically rotated so the receiver is in the new broadside direction.

shape of the cosine antenna gain demonstrates how it dictates the strength that beams can get in different directions.

We will now take a closer look at these properties. Figure 4.45 illustrates a setup where the transmitter is equipped with a ULA of cosine antennas with their maximum gain in the azimuth direction $\varphi = 0$. The antenna spacing is $\Delta = \lambda/2$. The receiver is located in another angular direction $\varphi \neq 0$. We will compare two ways of handling this situation. The first solution is to apply MRT, as described earlier in this chapter. We can refer to this as *electrical beamforming* since we are phase-shifting the radiated signals to form a beam in the desired direction. Another potential solution is to physically rotate the transmitter by the angle φ so that the maximum gain is achieved in the direction towards the receiver. We refer to this as *mechanical beamforming*, and we can then transmit the same signal from every antenna. These solutions have different pros and cons, which we will explain by a numerical example.

Figure 4.46 shows the joint beamforming and antenna gain achieved in different angular directions using a ULA with $M = 10$ cosine antennas. The receiver is in the direction $(\varphi, \theta) = (\pi/4, 0)$. Mechanical beamforming achieves a beamforming gain of 10 dB and the maximum antenna gain of 6 dBi, resulting in a joint gain of 16 dBi. Electrical beamforming also achieves a beamforming gain of 10 dB but the antenna gain is only $4\cos(\pi/4)\cos(0) = 2\sqrt{2} \approx 4.5\,\text{dB}$; thus, the joint gain is 14.5 dBi. If the transmit power is the same in both cases, the receiver will achieve a 1.5 dB lower SNR when using electrical beamforming. Nevertheless, it is beneficial to utilize directive antennas in this setup because the gain is 4.5 dB larger than if isotropic antennas would have been utilized, as was the case in Figure 4.17. Another difference between mechanical and electrical beamforming is that the beamwidth becomes broader in the latter case, as shown in Figure 4.46. This can lead to more interference towards undesired receivers located in roughly the same direction as the desired receiver.

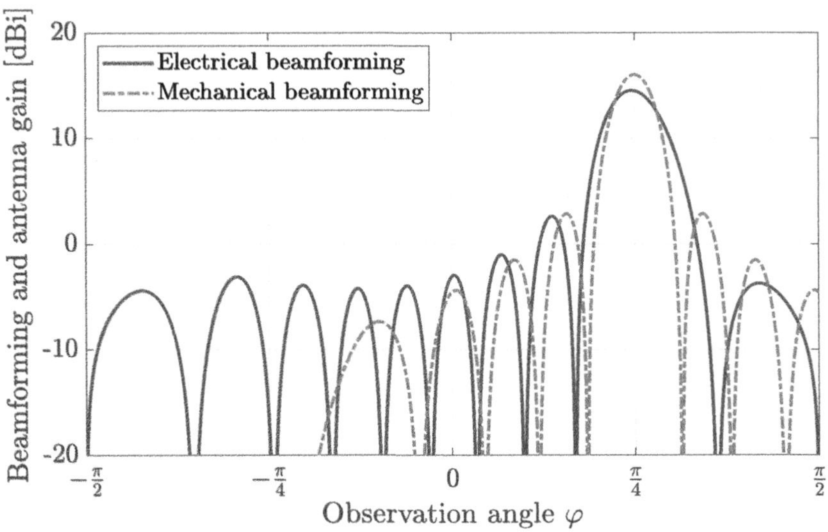

Figure 4.46: The joint beamforming and antenna gain that is observed in different azimuth directions φ when a ULA with $M = 10$ cosine antennas transmits a beam in the direction $\varphi_{\text{beam}} = \pi/4$. We compare electrical beamforming (i.e., phase-shifting the transmitted signals using MRT) with mechanical beamforming where the ULA is rotated by $\pi/4$ to have the broadcast towards the receiver, as shown in Figure 4.45.

However, the wider beam can also be a benefit if the receiver's angle is only approximately known.

Although mechanical beamforming might seem like a viable competing technology, it is seldom used in mobile communications since changes in the mechanical rotation at the millisecond level are associated with many practical implementation issues. The benefit also vanishes in MIMO scenarios where multiple beams are to be transmitted simultaneously in different directions to achieve multiplexing gains. The flexibility of electrical beamforming generally makes it a superior technology; however, a careful selection of the antennas and mechanical rotation is necessary when deploying an array to ensure that the antenna gain is sufficiently large within the intended coverage area. For example, in a cellular network where the base stations are deployed tens of meters above the ground, it is common to mechanically downtilt the base station array by around ten degrees in the elevation angle domain, to focus the antenna gain on the places where the prospective users are instead of towards the horizon. Electrical downtilt in the form of beamforming is then utilized to adapt the transmission to the current user location. Base station arrays are also rotated horizontally at the deployment stage to point toward the center of their intended coverage area, while electrical beamforming is used to point beams in the azimuth direction where the user currently resides.

Example 4.22. Consider a UPA with $M_\mathrm{H} = 10$, $M_\mathrm{V} = 4$, $\Delta_\lambda = 1/2$, and cosine antennas. What is the joint beamforming and antenna gain if the UPA is mechanically rotated to transmit to a receiver in the broadside direction? What is the joint beamforming and antenna gain if the UPA is electrically rotated to transmit to a receiver in the direction $\varphi = 0$ and $\theta = -\pi/4$? How does the first-null horizontal beamwidth differ between these setups?

The antenna gain function in (4.151) is $G(0, 0) = 4$ in the broadside direction, while the beamforming gain in (4.139) becomes $M = A_{10}(0)A_4(0) = 40$. Hence, the joint beamforming and antenna gain is $4 \cdot 40 = 160 \approx 22\,\mathrm{dBi}$.

With the electrical downtilt, the antenna gain becomes $G(0, -\pi/4) = 2\sqrt{2}$ while the beamforming gain remains $M = A_{10}(0)A_4(0) = 40$. Hence, the joint beamforming and antenna gain is $2\sqrt{2} \cdot 40 = 80\sqrt{2} \approx 20.5\,\mathrm{dBi}$.

Example 4.20 showed that $2\arcsin(1/5) \approx 0.4$ is the first-null horizontal beamwidth when transmitting in the broadside direction. In contrast, with the electrical downtilt, Φ in (4.133) becomes

$$\Phi = \sin(\varphi)\cos(-\pi/4) - \sin(0)\cos(-\pi/4) = \sin(\varphi)/\sqrt{2}. \tag{4.152}$$

According to (4.143), the first nulls in the horizontal plane occur when $\Phi = \pm 1/L_{\mathrm{H},\lambda}$. The normalized horizontal length of the considered UPA is $L_{\mathrm{H},\lambda} = M_\mathrm{H}\Delta_\lambda = 10 \cdot 0.5 = 5$ wavelengths. By solving for the azimuth angle φ, we obtain that the first nulls are at $\varphi = \pm\arcsin(\sqrt{2}/5)$ and the first-null horizontal beamwidth is therefore $2\arcsin(\sqrt{2}/5) \approx 0.57$.

In summary, the electrical downtilt results in a $1.5\,\mathrm{dBi}$ loss in antenna gain and an increased beamwidth by 42% since $\arcsin(\sqrt{2}/5)/\arcsin(1/5) \approx 1.42$. The benefit is that there is no need to mechanically rotate the array based on the receiver's location in the coverage area.

4.5.5 Effective Isotropic Radiated Power

The maximum radiated power in a wireless communication system is determined by regulations, which can differ between countries and frequency bands. The maximum power can be quantified in terms of the *total radiated power*, denoted by P in this book, without considering how this power is distributed over different angular directions. In regulations, it is common to consider the *effective isotropic radiated power (EIRP)*, which also includes the antenna and beamforming gains. Notice that the SNR in (4.150) for a MISO system can be factorized as

$$\frac{P\|\mathbf{h}\|^2}{BN_0} = \underbrace{PG_\mathrm{t}(\varphi,\theta)M}_{\mathrm{EIRP}}\frac{\beta_\mathrm{iso}}{BN_0}, \tag{4.153}$$

which is effectively the same as for a SISO channel with a single isotropic antenna that radiates $PG_\mathrm{t}(\varphi,\theta)M$. In terms of the received signal strength, the receiver cannot tell whether the signal was radiated isotropically or with a strong directivity. Hence, if the goal of the regulation is to limit the worst-case radiation intensity (e.g., to comply with health guidelines and limit out-of-band emissions), then the EIRP must be regulated. In particular, the maximum EIRP over all angles can be computed as

$$\mathrm{EIRP}_{\max} = \max_{\varphi,\theta} PG_{\mathrm{t}}(\varphi,\theta)M. \tag{4.154}$$

The maximum EIRP is proportional to total radiated power P, the maximum antenna gain $\max_{\varphi,\theta} G_{\mathrm{t}}(\varphi,\theta)$, and the beamforming gain M.

Example 4.23. What is the EIRP if the total radiated power is 1 W, the antenna gain is 4, and the beamforming gain is 10?

The EIRP is the product of these parameters: $1 \cdot 4 \cdot 10 = 40$ W, which is often expressed in decibel scale as 46 dBm. Thanks to the directive transmission, the receiver will experience a received signal equivalent to the transmission of 40 W from an isotropic antenna, although the transmitter only radiates 1 W.

The EIRP limits can vary significantly between different frequency bands and geographical regions. Within the European Union, the guidelines for the 3.5 GHz band (utilized for 5G NR) is to have a maximum EIRP of 68 dBm per 5 MHz of spectrum for base stations, while EIRP limit is only 25 dBm per user device [55]. There is a large power imbalance between the downlink and uplink transmissions, as previously discussed in relation to Figure 1.7. However, the EIRP numbers do not tell the whole story since the antenna/beamforming gains at the receiver side are omitted. For example, the downlink EIRP of 68 dBm might be reached using a total radiated power of 47 dBm (i.e., 50 W) and a joint antenna and beamforming gain of 21 dBi. The same antenna/beamforming gain can also be utilized when receiving the uplink transmission. Hence, the difference in total radiated power determines the SNR imbalance between uplink and downlink. The uplink EIRP limit of 25 dBm might be reached using a total radiated power of 23 dBm (i.e., 200 mW), an antenna gain of 2 dBi, and no beamforming gain. Importantly, base stations are often allowed to increase their power proportionally to the bandwidth. This is not the case for user devices, so the power imbalance between uplink and downlink becomes more severe as the bandwidth increases.

4.5.6 MIMO Channels with Arbitrary ULAs

A far-field MIMO channel model was provided in Section 4.4.1 for the case when the transmitter and receiver are equipped with ULAs of isotropic antennas located in the same two-dimensional plane. We will now utilize the array responses derived in Section 4.5.1 to generalize the expression

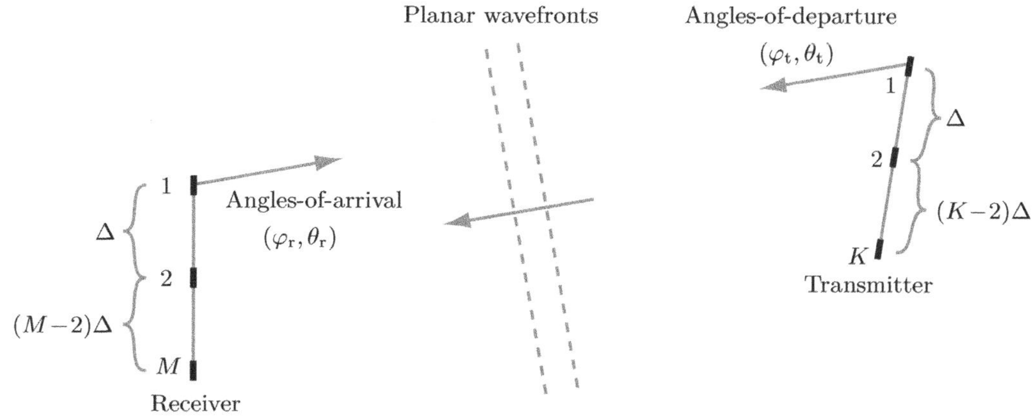

Figure 4.47: Illustration of a free-space MIMO LOS communication setup where the transmitter is equipped with a ULA with K antennas and the receiver with a ULA with M antennas. The antenna spacing is Δ in each array. From the transmitter's perspective, the angles-of-departure leading to the receiver are (φ_t, θ_t). From the receiver's perspective, the angles-of-arrival of the signal radiated by the transmitter are (φ_r, θ_r).

to support directive antennas, and ULAs arbitrarily rotated in the three-dimensional world. Figure 4.47 illustrates such a setup, where the receiver is in the far-field of the transmitter and the channel gain between any pair of transmit and receive antennas under the assumption of isotropic antennas is $\beta_{\text{iso}} = \lambda^2/(4\pi d)^2$. The transmitter is equipped with a ULA with K antennas and the antenna spacing is Δ. These antennas have the gain function $G_t(\varphi, \theta)$ and angles-of-departure leading towards the receiver is denoted by (φ_t, θ_t). The receiver has a ULA with M antennas and the same antenna spacing Δ. These antennas have the gain function $G_r(\varphi, \theta)$ and the radiated signal impinges as a planar wavefront with the angles-of-arrival denoted by (φ_r, θ_r).

We can utilize (4.149) to conclude that the channel from the first transmit antenna to the M receive antennas is $\sqrt{\beta_{\text{iso}}}\sqrt{G_t(\varphi_t, \theta_t)}\sqrt{G_r(\varphi_r, \theta_r)}\mathbf{a}_M(\varphi_r, \theta_r)$, by also taking the antenna gain $G_t(\varphi_t, \theta_t)$ of the transmitter into account. Similarly, from the transmitter's perspective, the channel from the K transmit antennas to the first receive antenna is $\sqrt{\beta_{\text{iso}}}\sqrt{G_t(\varphi_t, \theta_t)}\sqrt{G_r(\varphi_r, \theta_r)}\mathbf{a}_K(\varphi_t, \theta_t)$. By combining these results, we conclude that the complete channel matrix is

$$\mathbf{H} = \sqrt{\beta_{\text{iso}}}\sqrt{G_t(\varphi_t, \theta_t)}\sqrt{G_r(\varphi_r, \theta_r)}\mathbf{a}_M(\varphi_r, \theta_r)\mathbf{a}_K^{\mathrm{T}}(\varphi_t, \theta_t). \tag{4.155}$$

This is a rank-one matrix, which aligns with previous observations in this chapter. The non-zero singular value is

$$s_1 = \sqrt{\beta_{\text{iso}} MK G_t(\varphi_t, \theta_t) G_r(\varphi_r, \theta_r)}, \tag{4.156}$$

which now depends on the antenna gains at both the transmitter and receiver.

The MIMO channel capacity becomes

$$C = \log_2\left(1 + G_\mathrm{t}(\varphi_\mathrm{t},\theta_\mathrm{t})G_\mathrm{r}(\varphi_\mathrm{r},\theta_\mathrm{r})\frac{q\beta_\mathrm{iso}MK}{N_0}\right), \tag{4.157}$$

which coincides with (4.96) when $G_\mathrm{t}(\varphi_\mathrm{t},\theta_\mathrm{t})G_\mathrm{r}(\varphi_\mathrm{r},\theta_\mathrm{r}) = 1$. Depending on how the antennas are directed, the value of the capacity can be either higher, lower, or the same as with isotropic antennas. The same beamforming gain of MK is obtained irrespective of the choice of antennas (as long as the gain is non-zero). The capacity is achieved by using the MRT vector $\mathbf{a}_K^*(\varphi_\mathrm{t},\theta_\mathrm{t})/\sqrt{K}$ for precoding and the MRC vector $\mathbf{a}_M(\varphi_\mathrm{r},\theta_\mathrm{r})/\sqrt{M}$ for combining.

Example 4.24. Consider a free-space MIMO channel between a base station and a user device, both equipped with UPAs. The Cartesian coordinates of the center points of the base station and the user device are $(0,0,0)$ and $(300,300,300\sqrt{2})$ in meters, respectively. The number of antennas at the base station (receiver) and user device (transmitter) are $M = 32$ and $K = 4$, respectively. All antennas have the cosine gain function given in (4.151). The two UPAs are deployed along the yz-plane, and their broadside directions face each other (i.e., the base station has zero gain for $x < 0$ and the device has zero gain for $x > 300$). What is the capacity of the considered MIMO channel for $q = 10^{-8}$ W/Hz, $N_0 = 10^{-17}$ W/Hz, and $\lambda = 0.1$ m?

We can determine the capacity of the considered free-space MIMO channel using (4.157) and $\beta_\mathrm{iso} = \lambda^2/(4\pi d)^2$ with $M = 32$, $K = 4$, $q = 10^{-8}$ W/Hz, $N_0 = 10^{-17}$ W/Hz, and $\lambda = 0.1$ m. It becomes

$$C = \log_2\left(1 + G(\varphi_\mathrm{t},\theta_\mathrm{t})G(\varphi_\mathrm{r},\theta_\mathrm{r})\frac{10^{-8}\cdot 0.1^2\cdot 32\cdot 4}{10^{-17}\cdot(4\pi)^2 d^2}\right), \tag{4.158}$$

where $G(\varphi,\theta)$ is the antenna gain function in (4.151). The distance between the transmitter and receiver is computed as

$$d = \sqrt{(300-0)^2 + (300-0)^2 + (300\sqrt{2}-0)^2} = 600\,\mathrm{m}. \tag{4.159}$$

Let us first determine the angles-of-arrival $(\varphi_\mathrm{r},\theta_\mathrm{r})$ from the user device to the base station. From the given geometry of the UPAs, $\varphi_\mathrm{r} = \pi/4$ and $\theta_\mathrm{r} = \pi/4$. The angles-of-departure $(\varphi_\mathrm{t},\theta_\mathrm{t})$ from the user device to the base station are computed as $\varphi_\mathrm{t} = \pi/4$ and $\theta_\mathrm{t} = -\pi/4$. Hence, the cosine antenna gains are obtained as $G(\varphi_\mathrm{t},\theta_\mathrm{t}) = G(\varphi_\mathrm{r},\theta_\mathrm{r}) = \frac{4}{(\sqrt{2})^2} = 2$. Inserting these values into the capacity expression, we obtain

$$C = \log_2\left(1 + 2\cdot 2\frac{10^{-8}\cdot 0.1^2\cdot 32\cdot 4}{10^{-17}\cdot(4\pi)^2\cdot 600^2}\right) \approx 6.5\ \text{bit/symbol}. \tag{4.160}$$

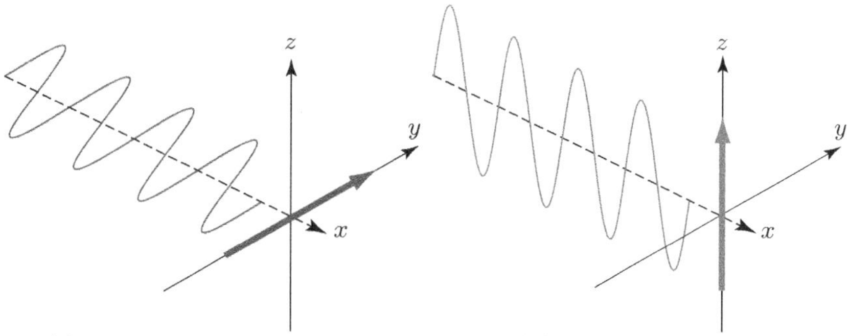

(a) Horizontally polarized wave. (b) Vertically polarized wave.

Figure 4.48: Electromagnetic waves can have different polarization, as represented by the direction in which the electric field oscillates. The figure shows two waves with orthogonal polarizations oscillating horizontally or vertically. The thick arrows show the dimensions of the oscillations and in which direction the amplitude is positive.

4.6 Polarization of Electromagnetic Waves

An electromagnetic wave travels in one direction, but the electric field oscillates (like a sinusoid) in a perpendicular direction. When a plane wave propagates along one dimension of our three-dimensional world, there are two possible perpendicular dimensions in which the electric field could oscillate, or it can be a linear combination of them. The direction of the oscillations is called the *polarization* of the wave. Figure 4.48 shows an example of the wave propagating along the x-axis. When the electric field oscillates in the horizontal plane (i.e., along the y-axis), we have a *horizontally polarized* wave. When the electric field oscillates in the vertical plane (i.e., along the z-axis), we have a *vertically polarized* wave. This is an example of a pair of orthogonally polarized waves since the electric fields exist in entirely different dimensions. One can find other pairs of orthogonal waves by rotating both waves with the same angle in the yz-plane. However, one cannot find more than two orthogonal waves since there are only two dimensions.[15]

Each antenna has predetermined polarization properties, in the sense that it radiates waves with a given polarization and responds to impinging waves with the same polarization. For example, a horizontally polarized antenna radiates and responds to waves of the horizontally polarized kind shown in Figure 4.48(a) and might have a physical shape similar to the thick arrow

[15]We have only exemplified linearly polarized waves for which the electric field oscillates in a single dimension. One can also create circularly/elliptically polarized waves for which the electric field rotates in the plane perpendicular to the direction of travel, which means that the direction of the oscillations is time-varying. For example, if the wave travels along the x-axis, then the wave's electric field could rotate in the yz-plane. In this case, a clockwise and a counter-clockwise rotation lead to orthogonal polarization. Electromagnetic fields are also characterized by their magnetic fields, which are orthogonal to the electric fields and the direction of travel, and thus oscillate as the orthogonal wave's electric field.

illustrated along the y-axis. Similarly, a vertically polarized antenna radiates and responds to waves of the vertically polarized kind shown in Figure 4.48(b) and might have a physical shape similar to the thick arrow illustrated along the z-axis. Note that the physical orientation of an antenna is essential when interpreting these things. A horizontally polarized antenna can be rotated by 90° to become a vertically polarized antenna and vice versa. If a horizontally polarized antenna is rotated by 180°, it remains horizontally polarized, but the notions of up and down are switched, corresponding to changing the signal's sign. The thick arrows in Figure 4.48 show the direction where the signal attains positive values in this coordinated system.

The analysis in this chapter has implicitly assumed that all antennas have matching polarization. However, there are three main reasons for generalizing the analysis. Firstly, we cannot control the device's orientation in mobile communications since the user must be allowed to hold and rotate it arbitrarily. For instance, a mobile phone antenna might be vertically polarized when held against the ear but horizontally polarized when put on a table. This calls for the use of multiple antennas with different polarization at the base station so that it can always generate waves with the device's currently preferred polarization.[16] Secondly, two antennas with opposite polarization can be co-located, which enables doubling the number of antennas that fit in a given physical aperture area. Thirdly, polarization creates an extra dimension that can be used for spatial multiplexing over LOS channels, which was recognized already in the 1980s [58] (i.e., before spatial multiplexing through beamforming was discovered). The latter property will be the focus of this section.

4.6.1 Channel Capacity with Dual-Polarized Antennas

Suppose the transmitter has two antennas with orthogonal polarization. Since the antennas have different orientations, they can be centered around the same point to create what is called a *dual-polarized antenna*: two antennas at one location but with orthogonal polarizations. In this section, we consider the setup in Figure 4.49, where the receiver is equipped with the same antenna configuration, including identical rotations. Just as any other MIMO system with $M = K = 2$, the considered setup can be described by the MIMO system model in (3.56):

$$\mathbf{y} = \mathbf{H}\mathbf{x} + \mathbf{n}, \tag{4.161}$$

[16]Any direction of the electric field can be obtained as a superposition of two orthogonal electric fields, for example, generated using horizontal and vertical polarizations. Practical base stations often utilize dual-polarized antennas with ±45° slanted polarizations, a pair of orthogonal polarizations between the horizontal and vertical directions. The reason is that many propagation environments provide better conditions for either the horizontally or vertically polarized wave component; for example, the vertical polarization often leads to stronger signals in mobile communications since the waves mostly travel horizontally between the base station and user device, and are reflected off vertical objects (e.g., buildings) that are better at reflecting vertically polarized waves [56], [57]. In any case, we can achieve a balanced power per antenna by dividing these dimensions equally between the antennas by using slanted polarization.

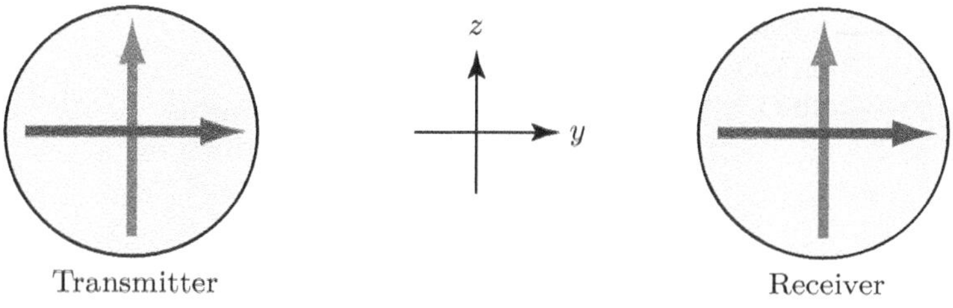

Figure 4.49: Illustration of a setup where the transmitter and receiver are equipped with dual-polarized antennas with identical rotations in the yz-plane.

where the elements of $\mathbf{y} \in \mathbb{C}^2$ correspond to the two receive antennas with orthogonal polarizations and the elements of $\mathbf{x} \in \mathbb{C}^2$ correspond to the two transmit antennas with orthogonal polarizations. The assumption of dual-polarized antennas only affects the modeling of the channel matrix $\mathbf{H} \in \mathbb{C}^{2\times 2}$.

We consider an LOS channel where d is the distance between the dual-polarized transmit antenna and the dual-polarized receive antenna; thus, the channel gain is $\beta = \frac{\lambda^2}{(4\pi)^2}\frac{1}{d^2}$. If we order the antennas so that transmit antenna m and receive antenna m have matching polarization, for $m = 1, 2$, then we obtain the channel matrix

$$\mathbf{H} = \sqrt{\beta}\begin{bmatrix} 1 & 0 \\ 0 & 1 \end{bmatrix} = \sqrt{\beta}\mathbf{I}_2. \tag{4.162}$$

The diagonal elements are $\sqrt{\beta}$ just as for a single-antenna LOS channel where the antennas have equal polarization. In contrast, the off-diagonal elements are zero because the corresponding antennas have orthogonal polarizations. Both singular values equal $\sqrt{\beta}$ since the channel matrix is a scaled identity matrix. This is an ideal type of MIMO channel for spatial multiplexing because we can transmit two parallel data streams that experience equally strong singular values. Hence, the water-filling power allocation will result in equal power allocation: $q_1 = q_2 = q/2$. The channel capacity in (3.75) becomes

$$C = 2\log_2\left(1 + \frac{q\beta}{2N_0}\right) \quad \text{bit/symbol}. \tag{4.163}$$

The transmit precoding and receive combining that achieves the capacity is trivial: send the mth stream from the mth transmit antenna and receive it using only the mth receive antenna. Since the orthogonality between the data streams is achieved by the different polarization rather than using different spatial beams, it is more appropriate to call this *polarization multiplexing* than spatial multiplexing. However, the underlying MIMO capacity theory is the same; it is just the physical interpretation that differs.

It is instructive to compare the capacity of the dual-polarized 2×2 MIMO channel in (4.163) with the capacity in (4.96) for a far-field MIMO setup with

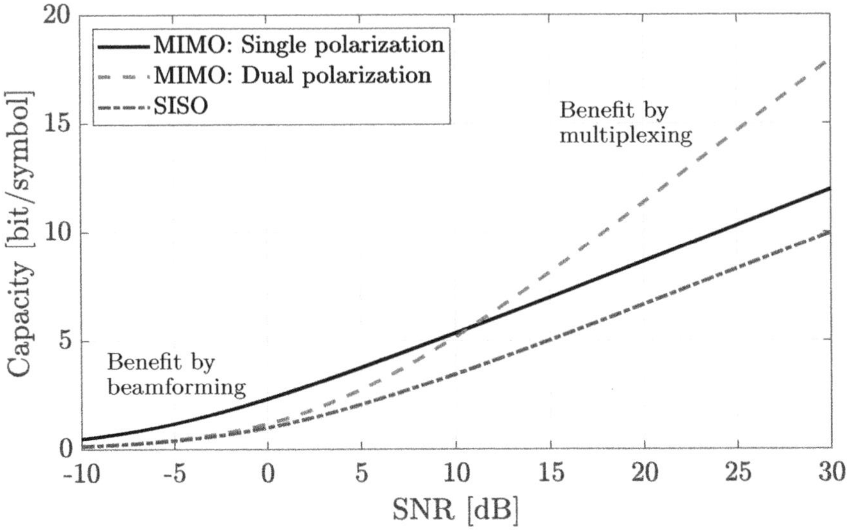

Figure 4.50: Comparison of the capacities of 2×2 MIMO LOS channels with either two single-polarized antennas or one dual-polarized antenna on each side. The SISO capacity is shown as a reference.

ULAs where all the antennas have the same polarization. For $M = K = 2$, (4.96) becomes $\log_2(1 + \frac{4q\beta}{N_0})$, where there is no multiplexing gain but a beamforming gain of $MK = 4$. Figure 4.50 shows the capacities as a function of $\mathrm{SNR} = q\beta/N_0$. This is the SNR achieved in a SISO system and its capacity $\log_2(1 + \mathrm{SNR})$ is shown as a reference. The single-polarized setup achieves the highest capacity at low SNR thanks to the beamforming gains obtained at both the transmitter and receiver. The dual-polarized setup performs identically to the SISO setup at low SNR because each antenna transmits isotropically, and each receive antenna only captures power from one transmit antenna. We can show this mathematically using the low SNR approximation in (3.2):

$$2\log_2\left(1 + \frac{\mathrm{SNR}}{2}\right) \approx 2\log_2(e)\frac{\mathrm{SNR}}{2} = \log_2(e)\mathrm{SNR} \approx \log_2(1 + \mathrm{SNR}). \quad (4.164)$$

However, the dual-polarized setup can use the multiplexing gain to achieve a significantly higher capacity at high SNR. Since the single-polarized MIMO channel in (4.90) has rank 1, while the dual-polarized MIMO channel in (4.162) has rank 2, the capacity curve has a steeper slope in the latter case and eventually provides the largest capacity. The SNR range where the dual-polarized setup provides the highest capacity can be identified as follows:

$$\begin{aligned} &2\log_2\left(1 + \frac{\mathrm{SNR}}{2}\right) \geq \log_2(1 + 4\mathrm{SNR}) \\ &\Rightarrow \left(1 + \frac{\mathrm{SNR}}{2}\right)^2 \geq 1 + 4\mathrm{SNR} \quad \Rightarrow \quad \mathrm{SNR} \geq 12. \end{aligned} \quad (4.165)$$

The intersection point is SNR $= 12 \approx 10.8\,\mathrm{dB}$, which can be observed in Figure 4.50. This SNR value is six times higher than in (4.99), where single-polarized MIMO channels with rank-one and rank-two were compared. The reason for the difference is that half the power is lost over the dual-polarized channel. In conclusion, dual-polarized antennas reduce the total received power but create an extra dimension that increases the high-SNR capacity.

Example 4.25. Suppose the dual-polarized transmit antenna is rotated by 45°, compared to the dual-polarized receive antenna, as illustrated in Figure 4.51. How will the channel matrix in (4.162) and channel capacity change?

The first antenna (blue) and the second antenna (red) of the receiver in Figure 4.51 have polarizations along the y-axis and z-axis, respectively. By contrast, the transmitter has an antenna configuration rotated counter-clockwise by 45° in the yz-plane. Each arrow points in the direction that the wave takes positive values. Hence, the first receive antenna obtains the summation of the signal components along the y-axis, which is $\frac{x_1 - x_2}{\sqrt{2}}$. The minus sign appears because the red arrow points toward the negative y-axis and the term $1/\sqrt{2}$ describes that only half the power is radiated in the y-dimension. Similarly, the second antenna of the receiver receives a summation of the components of the signals along the z-axis, which is $\frac{x_1 + x_2}{\sqrt{2}}$. Hence, the MIMO channel matrix is

$$\mathbf{H} = \sqrt{\beta} \begin{bmatrix} \frac{1}{\sqrt{2}} & -\frac{1}{\sqrt{2}} \\ \frac{1}{\sqrt{2}} & \frac{1}{\sqrt{2}} \end{bmatrix} = \underbrace{\begin{bmatrix} \frac{1}{\sqrt{2}} & -\frac{1}{\sqrt{2}} \\ \frac{1}{\sqrt{2}} & \frac{1}{\sqrt{2}} \end{bmatrix}}_{=\mathbf{U}} \underbrace{\sqrt{\beta}\mathbf{I}_2}_{=\boldsymbol{\Sigma}} \underbrace{\mathbf{I}_2}_{=\mathbf{V}^{\mathrm{H}}}, \tag{4.166}$$

where the second equality provides the SVD. The singular values are $s_1 = s_2 = \sqrt{\beta}$, just as with identically rotated antennas in (4.162). Therefore, the channel capacity is the same as in (4.163) but is achieved differently. Using Theorem 3.1, we can conclude that the capacity is achieved when the transmitter sends two independent data streams $x_1 \sim \mathcal{N}_{\mathbb{C}}(0, q/2)$ and $x_2 \sim \mathcal{N}_{\mathbb{C}}(0, q/2)$ and the receiver applies the receive combining $\bar{\mathbf{y}} = \mathbf{U}^{\mathrm{H}}\mathbf{y}$ to separate them, which corresponds to compensating for the polarization mismatch by computing $\bar{y}_1 = \frac{y_1 + y_2}{\sqrt{2}}$ and $\bar{y}_2 = \frac{y_1 - y_2}{\sqrt{2}}$.

Alternatively, the transmitter can generate two independent data streams $\bar{x}_1 \sim \mathcal{N}_{\mathbb{C}}(0, q/2)$ and $\bar{x}_2 \sim \mathcal{N}_{\mathbb{C}}(0, q/2)$ and apply the transmit precoding $\mathbf{x} = \mathbf{U}^{\mathrm{H}}\bar{\mathbf{x}}$ beacuse $\mathbf{H}\mathbf{x} = \sqrt{\beta}\mathbf{U}\mathbf{U}^{\mathrm{H}}\bar{\mathbf{x}} = \sqrt{\beta}\bar{\mathbf{x}}$. This corresponds to using the two rotated antennas to transmit signals with horizontal and vertical polarization.

In summary, it is sufficient to study the case with identical antenna rotations when characterizing the capacity with dual polarization. To achieve the capacity in practice, the exact channel matrix must be known so that the transmission can compensate for potential antenna rotations.

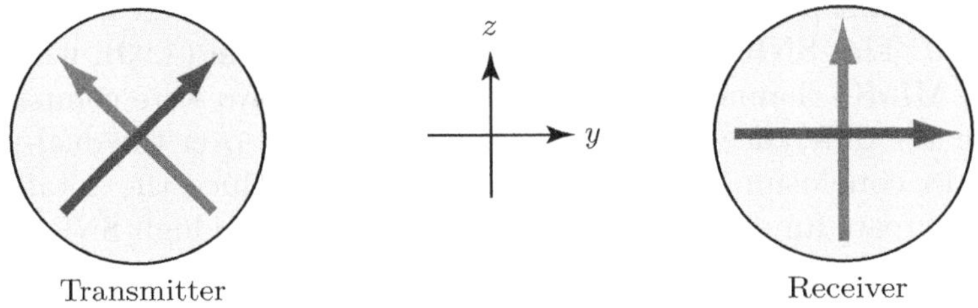

Figure 4.51: Illustration of a setup where the transmitter and receiver are equipped with dual-polarized antennas that differ in rotation by 45° in the yz-plane.

4.6.2 Impact of Finite Cross-Polar Discrimination

Even if antennas are designed to have orthogonal polarization, there is usually cross-talk between the polarizations; for example, caused by imperfect polarization discrimination in the individual antennas and imperfect isolation between the co-located antennas [59], [60]. These effects can be analyzed separately and in great detail, but that is beyond the scope of this chapter. To study their collective impact on the channel capacity, we measure the purity of a dual-polarized antenna by the *cross-polar discrimination (XPD)* factor, which is defined as the ratio between the power radiated into the intended polarization direction and the power transmitted into the orthogonal polarization direction. We will not distinguish whether this issue is created since the intended antenna partially radiates into the unintended polarization direction or if the signal leaks into the co-located opposite-polarized antenna and is then radiated into the unintended polarization direction. The XPD correspondingly affects the reception, so there is symmetry in the system.

We will now consider polarized antennas that transmit a fraction $(1-\gamma)$ of the total power into the intended polarization (for any of the reasons above) and a fraction γ into the opposite polarization. The parameter $\gamma \in [0,1]$ characterizes the impurity of the antenna (a smaller value is better). Note that $(1-\gamma)+\gamma = 1$, which implies that the total power is divided between the two orthogonal polarizations without losses. The XPD of such an antenna is

$$\text{XPD} = \frac{1-\gamma}{\gamma} \quad \rightarrow \quad \gamma = \frac{1}{1+\text{XPD}}. \tag{4.167}$$

A larger value of γ corresponds to a smaller XPD and vice versa.

Suppose we transmit a signal with power P to a polarized receive antenna of the same kind. A signal component with power $(1-\gamma)P$ is radiated with the intended polarization. If the channel gain is $\beta \in [0,1]$, a fraction β of the transmitted power reaches the receive antenna and a fraction $(1-\gamma)$ is properly received. Hence, the received signal component has power $(1-\gamma)^2 P\beta$. Moreover, a signal component with power γP will be radiated using the opposite polarization, and the receive antenna will then capture a fraction $\gamma\beta$

of it. Hence, the total received power is

$$(1-\gamma)^2 P\beta + \gamma^2 P\beta = (1 - 2(1-\gamma)\gamma)P\beta. \tag{4.168}$$

Suppose the receive antenna instead has the opposite polarization direction but the same XPD. It will then receive two signal components: one that leaks into the wrong polarization at the transmitter and one that leaks into the wrong polarization at the receiver. Due to the assumed XPD symmetry, each of them has power $(1-\gamma)\gamma P\beta$; thus, the total received power is

$$(1-\gamma)\gamma P\beta + (1-\gamma)\gamma P\beta = 2(1-\gamma)\gamma P\beta. \tag{4.169}$$

Note that the sum of (4.168) and (4.169) is $P\beta$; thus, a dual-polarized receive antenna can capture all the signal power that reaches the receiver, irrespective of the XPD. This observation can be extended to the case when the transmitter is a user device with an arbitrary orientation. The combined effect of the orientation and XPD will determine how the received power is distributed over the two polarizations of the dual-polarized receive antenna. However, the total power of the impinging wave over the antenna's effective area will always be captured. This is a somewhat intuitive result but requires much more notation to formalize mathematically; thus, it has been omitted here.

We will now study the impact of the XPD on the MIMO channel capacity. The discussion above is related to the power of signals, while the channel matrix describes how the amplitude and phase change. Hence, based on (4.168) and (4.169), we can write the channel matrix as

$$\mathbf{H} = \sqrt{\beta}\begin{bmatrix} \sqrt{1-2(1-\gamma)\gamma} & \sqrt{2(1-\gamma)\gamma} \\ \sqrt{2(1-\gamma)\gamma} & \sqrt{1-2(1-\gamma)\gamma} \end{bmatrix} = \sqrt{\beta}\begin{bmatrix} \sqrt{1-\kappa} & \sqrt{\kappa} \\ \sqrt{\kappa} & \sqrt{1-\kappa} \end{bmatrix}, \tag{4.170}$$

where β still denotes the channel gain and we have defined

$$\kappa = 2(1-\gamma)\gamma = \frac{2\,\mathrm{XPD}}{(1+\mathrm{XPD})^2} \tag{4.171}$$

as the total fraction of power that leaks from a transmitted signal with one polarization to a received signal with the opposite polarization. The derived model is equivalent to the ones presented in [59], [60]. Note that $\kappa \in [0, 0.5]$, where the largest value is achieved for $\gamma = 1/2$ and $\mathrm{XPD} = 1$. The channel matrix in (4.170) reduces to (4.162) in the special case of $\kappa = 0$ when the antenna polarizations are pure. The SVD of the channel matrix in (4.170) is[17]

$$\begin{aligned}\mathbf{H} &= \sqrt{\beta}\begin{bmatrix} \sqrt{1-\kappa} & \sqrt{\kappa} \\ \sqrt{\kappa} & \sqrt{1-\kappa} \end{bmatrix} \\ &= \begin{bmatrix} \frac{1}{\sqrt{2}} & \frac{1}{\sqrt{2}} \\ \frac{1}{\sqrt{2}} & -\frac{1}{\sqrt{2}} \end{bmatrix} \begin{bmatrix} \sqrt{\beta}(\sqrt{1-\kappa}+\sqrt{\kappa}) & 0 \\ 0 & \sqrt{\beta}(\sqrt{1-\kappa}-\sqrt{\kappa}) \end{bmatrix} \begin{bmatrix} \frac{1}{\sqrt{2}} & \frac{1}{\sqrt{2}} \\ \frac{1}{\sqrt{2}} & -\frac{1}{\sqrt{2}} \end{bmatrix}.\end{aligned} \tag{4.172}$$

[17]The SVD of $\mathbf{H}$ coincides with its eigendecomposition since $\mathbf{H}$ is a positive semi-definite Hermitian symmetric matrix.

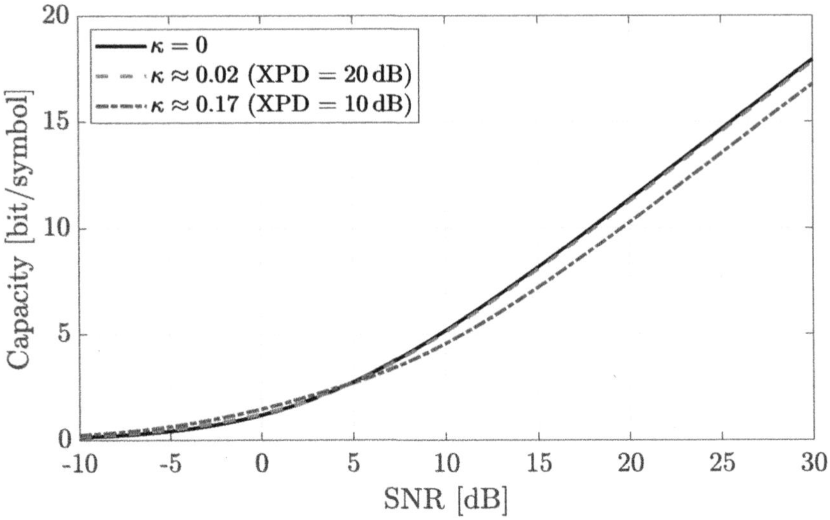

Figure 4.52: The capacity of a 2 × 2 MIMO channel with dual-polarized antennas is affected by the XPD, but the effect is negligible for large XPD values.

This implies that the MIMO channel can be divided into two parallel channels with $s_1 = \sqrt{\beta}(\sqrt{1-\kappa} + \sqrt{\kappa})$ and $s_2 = \sqrt{\beta}(\sqrt{1-\kappa} - \sqrt{\kappa})$ as the singular values. The channel capacity is achieved by transmitting each signal over both polarizations, using the precoding vectors $[1/\sqrt{2}\ \ 1/\sqrt{2}]^{\mathrm{T}}$ and $[1/\sqrt{2}\ \ -1/\sqrt{2}]^{\mathrm{T}}$. The same vectors are also utilized for receive combining.

The capacity can be computed using (3.75) as

$$C = \log_2\left(1 + \frac{q_1^{\mathrm{opt}} s_1^2}{N_0}\right) + \log_2\left(1 + \frac{q_2^{\mathrm{opt}} s_2^2}{N_0}\right), \tag{4.173}$$

where $q_k^{\mathrm{opt}} = \max(\mu - \frac{N_0}{s_k^2}, 0)$ is the transmit power obtained from the water-filling power allocation. It follows from Corollary 3.3 that

$$q_1^{\mathrm{opt}} = \begin{cases} q, & \text{if } q < \frac{N_0}{s_2^2} - \frac{N_0}{s_1^2}, \\ \frac{q}{2} + \frac{N_0}{2s_2^2} - \frac{N_0}{2s_1^2}, & \text{otherwise}, \end{cases} \tag{4.174}$$

$$q_2^{\mathrm{opt}} = \begin{cases} 0, & \text{if } q < \frac{N_0}{s_2^2} - \frac{N_0}{s_1^2}, \\ \frac{q}{2} + \frac{N_0}{2s_1^2} - \frac{N_0}{2s_2^2}, & \text{otherwise}, \end{cases} \tag{4.175}$$

where both channels are only utilized if the transmit power is above a threshold.

Figure 4.52 illustrates the impact of XPD on the MIMO capacity. The ideal case of $\kappa = 0$ is compared with XPD = 20 dB ($\kappa \approx 0.02$) and XPD = 10 dB ($\kappa \approx 0.17$), where the transformation from XPD to κ is achieved using (4.171). When the XPD is large, the relation becomes $\kappa \approx 2/\mathrm{XPD}$. The figure shows that the polarization impurity caused by having a low XPD (such as 10 dB) results in a capacity reduction at high SNR. The multiplexing gain is the same, as seen from the identical slopes of the curves, but the curve is shifted to the right, indicating a power loss from having singular values of different sizes. In contrast, the polarization impurity results in a minor capacity improvement at low SNR, where only the largest singular value s_1 is utilized, and s_1 is an increasing function of κ (in the range $[0, 0.5]$ of possible parameter values). When the XPD reaches 20 dB (or more), it has a negligible impact on the capacity. Many practical antennas operate in that regime.

4.6.3 MIMO Channel Capacity with Dual-Polarized ULAs

We will now consider a MIMO setup with arrays of dual-polarized antennas at both the transmitter and receiver. To keep the notation simple, we assume that the transmitter and receiver are equipped with ULAs located in the same two-dimensional plane (e.g., at the same height above the ground) and the antenna spacing is $\Delta = \lambda/2$. The transmitter has $K/2$ dual-polarized antennas, where K is an even number representing the total number of transmit antennas (counting both polarizations). The receiver has $M/2$ dual-polarized antennas, where M is an even number representing the total number of receive antennas. The XPD is characterized by the parameter $\kappa \in [0, 0.5]$ defined in (4.171).

We order the antennas according to their polarization so that transmit antennas $1, \ldots, K/2$ and receive antennas $1, \ldots, M/2$ become a $\frac{M}{2} \times \frac{K}{2}$ MIMO channel of the kind studied in Section 4.4. The same applies for transmit antennas $K/2+1, \ldots, K$ and receive antennas $M/2+1, \ldots, M$. Under the same frequency-flatness, far-field assumptions, and angle definitions as in Section 4.4.1, the channel matrix $\mathbf{H} \in \mathbb{C}^{M \times K}$ can be expressed using (4.91) as

$$\begin{aligned}\mathbf{H} &= \sqrt{\beta} \begin{bmatrix} \sqrt{1-\kappa}\mathbf{a}_{M/2}(\varphi_{\mathrm{r}})\mathbf{a}_{K/2}^{\mathrm{T}}(\varphi_{\mathrm{t}}) & \sqrt{\kappa}\mathbf{a}_{M/2}(\varphi_{\mathrm{r}})\mathbf{a}_{K/2}^{\mathrm{T}}(\varphi_{\mathrm{t}}) \\ \sqrt{\kappa}\mathbf{a}_{M/2}(\varphi_{\mathrm{r}})\mathbf{a}_{K/2}^{\mathrm{T}}(\varphi_{\mathrm{t}}) & \sqrt{1-\kappa}\mathbf{a}_{M/2}(\varphi_{\mathrm{r}})\mathbf{a}_{K/2}^{\mathrm{T}}(\varphi_{\mathrm{t}}) \end{bmatrix} \\ &= \sqrt{\beta} \begin{bmatrix} \sqrt{1-\kappa} & \sqrt{\kappa} \\ \sqrt{\kappa} & \sqrt{1-\kappa} \end{bmatrix} \otimes \left(\mathbf{a}_{M/2}(\varphi_{\mathrm{r}})\mathbf{a}_{K/2}^{\mathrm{T}}(\varphi_{\mathrm{t}}) \right), \end{aligned} \tag{4.176}$$

where $\mathbf{a}_{M/2}(\varphi)$ is the array response vector defined in (4.49). We notice that $\mathbf{H}$ is the Kronecker product between the channel matrix in (4.170) for two dual-polarized antennas and an $\frac{M}{2} \times \frac{K}{2}$ MIMO channel matrix $\mathbf{a}_{M/2}(\varphi_{\mathrm{r}})\mathbf{a}_{K/2}^{\mathrm{T}}(\varphi_{\mathrm{t}})$ with single-polarized antennas. Similarly to (4.172), the channel matrix in (4.176) can be factorized as

$$\mathbf{H} = \begin{bmatrix} \frac{\mathbf{a}_{M/2}(\varphi_{\mathrm{r}})}{\sqrt{M}} & \frac{\mathbf{a}_{M/2}(\varphi_{\mathrm{r}})}{\sqrt{M}} \\ \frac{\mathbf{a}_{M/2}(\varphi_{\mathrm{r}})}{\sqrt{M}} & -\frac{\mathbf{a}_{M/2}(\varphi_{\mathrm{r}})}{\sqrt{M}} \end{bmatrix} \begin{bmatrix} s_1 & 0 \\ 0 & s_2 \end{bmatrix} \begin{bmatrix} \frac{\mathbf{a}_{K/2}^{\mathrm{T}}(\varphi_{\mathrm{t}})}{\sqrt{K}} & \frac{\mathbf{a}_{K/2}^{\mathrm{T}}(\varphi_{\mathrm{t}})}{\sqrt{K}} \\ \frac{\mathbf{a}_{K/2}^{\mathrm{T}}(\varphi_{\mathrm{t}})}{\sqrt{K}} & -\frac{\mathbf{a}_{K/2}^{\mathrm{T}}(\varphi_{\mathrm{t}})}{\sqrt{K}} \end{bmatrix}. \tag{4.177}$$

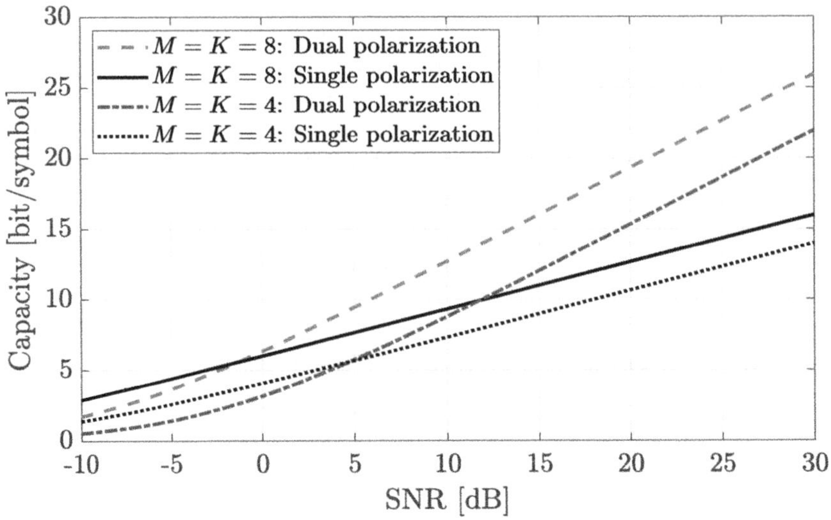

Figure 4.53: Comparison of the capacities of MIMO channels with either 4 or 8 single-polarized antennas on each side, or 2 or 4 dual-polarized antennas on each side (which is also 4 or 8 antennas).

This is the SVD where the two non-zero singular values are

$$s_1 = \frac{\sqrt{\beta MK}}{2}(\sqrt{1-\kappa} + \sqrt{\kappa}), \tag{4.178}$$

$$s_2 = \frac{\sqrt{\beta MK}}{2}(\sqrt{1-\kappa} - \sqrt{\kappa}). \tag{4.179}$$

Comparing with the case $M = K = 2$ in (4.172), we notice that having multiple dual-polarized antennas at the transmitter and receiver results in a beamforming gain of $MK/4$. This is the product of the number of transmit and receive antennas of each polarization. However, the multiplexing gain remains limited to $r = 2$ and is created by the different antenna polarizations. In other words, the varying polarization does not help to resolve the issue that far-field MIMO LOS channels have low rank since there is only one path between the transmitter and receiver. The capacity is achieved by using MRT with the precoding vector $\mathbf{a}^*_{K/2}(\varphi_{\mathrm{t}})/\sqrt{K}$ for each polarization at the transmitter side and MRC with the combining vector $\mathbf{a}_{M/2}(\varphi_{\mathrm{r}})/\sqrt{M}$ for each polarization at the receiver side. The polarization multiplexing is achieved by using the same or different signs in front of these vectors.

Figure 4.53 compares the capacities achieved with single-polarized and dual-polarized antennas with $\kappa = 0$. There are curves for $M = K = 8$ and $M = K = 4$. We previously observed in Figure 4.50 that the single-polarized setup has a benefit at low SNR because a higher beamforming gain is achievable when all antennas have matching polarization. This property remains when the number of antennas increases, but it only occurs at lower SNRs. When the SNR is greater than 0 dB in Figure 4.53, the improved multiplexing gain that dual-polarized antennas offer almost always compensates for the reduced beamforming gain. Hence, antenna arrays and dual polarization are a good combination when designing point-to-point MIMO LOS systems.

Example 4.26. Consider the MIMO channel with dual-polarized ULAs, whose channel matrix is given in (4.176). Compute the channel capacity with $\kappa = 0$ (best-case XPD) and $\kappa = 0.5$ (worst-case XPD). For which values of $\mathrm{SNR} = \frac{q\beta}{N_0}$ is $\kappa = 0$ giving the largest capacity?

The two non-zero singular values are given in (4.178) and (4.179). If $\kappa = 0$, we get $s_1 = s_2 = \sqrt{\beta M K}/2$ and then the water-filling power allocation gives $q_1 = q_2 = q/2$. The resulting channel capacity is

$$C_{\text{dual},\kappa=0} = 2\log_2\left(1 + \frac{q\beta MK}{8N_0}\right) = 2\log_2\left(1 + \frac{MK}{8}\mathrm{SNR}\right). \tag{4.180}$$

If $\kappa = 0.5$, we instead have $s_1 = \sqrt{\beta MK/2}$ and $s_2 = 0$. Hence, only a single subchannel is activated ($q_1 = q$), which leads to the channel capacity

$$C_{\text{dual},\kappa=0.5} = \log_2\left(1 + \frac{q\beta MK}{2N_0}\right) = \log_2\left(1 + \frac{MK}{2}\mathrm{SNR}\right). \tag{4.181}$$

This worst-case XPD scenario gives a smaller multiplexing gain but a larger beamforming gain. Yet, the beamforming gain is only half that achieved by the single-polarized MIMO channel in (4.96), so dual-polarized antennas are not desirable for pure beamforming.

We can identify the SNR range where $C_{\text{dual},\kappa=0} \geq C_{\text{dual},\kappa=0.5}$ as follows:

$$\begin{aligned} &2\log_2\left(1 + \frac{MK}{8}\mathrm{SNR}\right) \geq \log_2\left(1 + \frac{MK}{2}\mathrm{SNR}\right) \\ \Rightarrow \quad &\left(1 + \frac{MK}{8}\mathrm{SNR}\right)^2 \geq 1 + \frac{MK}{2}\mathrm{SNR} \quad \Rightarrow \quad \mathrm{SNR} \geq \frac{16}{MK}. \end{aligned} \tag{4.182}$$

The more antennas are used, the larger the SNR range where the setup with $\kappa = 0$ outperforms the setup with $\kappa = 0.5$.

4.7 Exercises

Exercise 4.1. Consider a ULA with M antennas that receive a signal from a single-antenna transmitter located at a distance d_1 in the angular direction φ. A general relationship between the distances d_1 and d_M is given in (4.13) for spherical waves, while the corresponding expression for plane waves is provided in (4.17). Show that we can obtain (4.17) as an approximation of (4.13) when $d_1 \gg M\Delta$. Hint: Use the Taylor approximation $\sqrt{1+x^2} \approx 1 + \frac{x^2}{2}$ that is tight for $0 \leq x \leq 0.25$, which was previously considered in Section 1.1.2.

Exercise 4.2. Consider a SIMO system with isotropic antennas operating over a far-field LOS channel. The transmit power is P, the bandwidth is B, and there are M receive antennas.

(a) State the channel capacity expression as a function of M, P, B, the wavelength λ, the propagation distance d, and the noise power spectral density N_0.

(b) Suppose $M = 1$, $P = 1\,\mathrm{W}$, $B = 10\,\mathrm{MHz}$, $\lambda = 10\,\mathrm{cm}$, and $N_0 = 10^{-17}$ W/Hz. At what distance d is the channel capacity 10 Mbit/s?

(c) We want to increase the number of antennas to achieve a channel capacity of 100 Mbit/s at the same distance as in (b). How many antennas are needed if the other parameters are unchanged? How large is the total effective area of the antennas in the receiver array?

(d) We now reduce the wavelength to $\lambda = 1\,\mathrm{cm}$. How many antennas are needed to achieve a channel capacity of 100 Mbit/s at the same distance as in (b)? How large is the total effective area of the antennas in the receiver array?

Exercise 4.3. Consider a ULA with $M = 10$ antennas and half-wavelength antenna spacing. Suppose it beamforms in the broadside direction in a free-space LOS scenario.

(a) What is the beamforming gain obtained in the direction $\varphi = 0$?

(b) What is the beamforming gain obtained in the direction $\varphi = \pi/6$?

Exercise 4.4. Reproduce the exact curve in Figure 4.13 but for a ULA with cosine antennas. Use the simulation results to discuss what happens to the half-power beamwidth, the amplification beamwidth, and the first-null beamwidth compared to having isotropic antennas. Which of these becomes smaller, wider, or unchanged in this example?

Exercise 4.5. Consider a MISO channel with a ULA having M antennas and $\Delta = \lambda/2$. The ULA transmits a signal in the end-fire direction $\varphi_{\text{beam}} = \pi/2$.

(a) If the receiver is located in another angular direction φ, what is the beamforming gain?

(b) Compute an approximate expression for the first-null beamwidth using the Taylor approximation $\arcsin(x) \approx x$, which is very tight if $M \geq 5$. Hint: Consider angular directions close to $\pm\pi/2$ by setting $\varphi = \pm\pi/2 + x$ and looking for small x. Use that $\sin(\pm\pi/2 + x) = \pm(1 - 2\sin^2(x/2))$.

(c) Compare the result with the beamwidth $4/M$ in (4.62) for broadside beamforming. Is the beamwidth smaller when transmitting in the broadside or end-fire direction? Does the beamwidth for end-fire beamforming decrease when M increases?

Exercise 4.6. A base station with $M = 4$ antennas transmits to a single-antenna device located at an angle φ. Suppose the channel vector is

$$\mathbf{h} = \sqrt{\beta} \begin{bmatrix} 1 \\ e^{-j\pi \sin(\varphi)} \\ e^{-j2\pi \sin(\varphi)} \\ e^{-j3\pi \sin(\varphi)} \end{bmatrix}. \tag{4.183}$$

(a) What is the capacity of this channel when $P\beta/(BN_0) = 10$? Explain how the capacity value depends on φ.

(b) Suppose the base station believes the user is located at $\varphi_{\text{beam}} = 0°$ and transmits using MRT. If the true angle of the user is $\varphi = 60°$, what is the achievable data rate? Compare the result with (a).

(c) Repeat (b) but with $\varphi = 30°$. Explain the result.

Exercise 4.7. It is possible to create other grids of orthogonal angular beams than the DFT beams defined in Section 4.3.3. Suppose we construct M beams using the angles

$$\varphi = \arcsin\left(\frac{2n + a}{M}\right) \tag{4.184}$$

for some $0 < a < 1$ and for the integers n satisfying $-\frac{M}{2} - \frac{a}{2} \leq n \leq \frac{M}{2} - \frac{a}{2}$ (there are M such integers). Show that these beams are mutually orthogonal when using a ULA with $\Delta = \lambda/2$.

Exercise 4.8. An M-antenna ULA receives the signal from $\varphi = -\pi/6$ and uses MRC. An interfering signal arrives from the angle $\varphi_{\text{interf}} = -\pi/9$.

(a) Obtain the sinc approximation of the beamforming gain in (4.31) for $\varphi = -\pi/6$ and $\varphi_{\text{interf}} = -\pi/9$ by using that $\sin(x^2) \approx x^2$ for arguments close to zero.

(b) Use the obtained sinc-expression from (a) to determine how many antennas are needed to ensure that the interfering transmitter is outside the half-power beamwidth if $\Delta = \lambda/2$.

(c) Repeat (b) for $\Delta = \lambda$.

Exercise 4.9. Consider a ULA with M antennas deployed to transmit beams toward user devices located in angular directions between 30° and 60°. When doing so, grating lobes are allowed if they do not appear in the angular interval $\varphi \in [10°, 80°]$. How should the antenna spacing be selected to achieve the smallest beamwidth?

Exercise 4.10. Consider a MIMO LOS channel where the transmitter is equipped with a ULA with $K = 4$ antennas and antenna spacing $\Delta = \lambda/2$. The receiver is equipped with $M = 4$ distributed antennas deployed along the arc of a circle with radius d (similar to Figure 4.26). The antennas are located in the angular directions $\varphi_1 = 0$, $\varphi_2 = \pi/6$, $\varphi_3 = \pi/2$, and $\varphi_4 = -\pi/6$ as seen from the transmitter. Compute the channel capacity in terms of q, β, and N_0. Hint: Express the channel matrix using array response vectors and show that these are mutually orthogonal.

Exercise 4.11. Consider an array with three isotropic antennas deployed at the corners of an equilateral triangle with the side length Δ. The antennas are placed in the yz-plane. Compute an expression for the array response vector $\mathbf{a}(\varphi, \theta)$.

Exercise 4.12. Consider a UPA transmitter with $M_{\mathrm{H}} = 10$ horizontal antennas along the y-axis and $M_{\mathrm{V}} = 4$ vertical antennas along the z-axis. Each antenna has the cosine gain function given in (4.151). The single-antenna receiver is in the direction $(\varphi, \theta) = (\pi/6, \pi/6)$.

(a) What is the joint antenna and beamforming gain achieved by mechanical beamforming?

(b) What is the joint antenna and beamforming gain achieved by electrical beamforming?

(c) Will the results in (a) and (b) change if the UPA instead has $M_{\mathrm{H}} = 5$ horizontal antennas and $M_{\mathrm{V}} = 8$ vertical antennas?

Exercise 4.13. Suppose we are building a MISO system that will operate under a maximum EIRP limit of 68 dBm. We use cosine antennas and let P denote the total transmit power. The power consumption of the system is measured as

$$\frac{P}{0.25} + M \cdot 1 + P_{\text{circuit}} \quad \text{W}, \tag{4.185}$$

where the first term models signal transmission with a power amplifier efficiency of 25%, the second term models that the transceiver hardware connected to each antenna consumes 1 W, and the fixed term $P_{\text{circuit}} \geq 0$ models the remaining power consumption. Which combination of P and M will reach the EIRP limit while minimizing the power consumption in (4.185)?

Exercise 4.14. Consider a point-to-point LOS channel. Suppose the wavelength is $\lambda = 0.1$ m, the transmit power is $P = 10$ W, the bandwidth is $B = 100$ MHz, and the noise power spectral density is $N_0 = 10^{-17}$ W/Hz.

(a) If $M = K = 1$ isotropic antennas are used, what is the capacity when the propagation distance is $d = 100$ m?

(b) If $M = 1$ isotropic antenna is used at the receiver, how many isotropic antennas, K, are needed at the transmitter to reach the same data rate in (a) at the propagation distance $d = 400$ m?

(c) Assuming the transmitter has K isotropic antennas, where K is obtained from (b), how many isotropic antennas are needed at the receiver, M, to reach the same data rate in (a) at the propagation distance $d = 800$ m?

(d) Is it possible to reach the same data rate as in (a) at the propagation distance $d = 800$ m by using a smaller total number of antennas $M + K$ than in (c)? What are M and K in that case?

Exercise 4.15. A UPA transmits a signal in the direction $(\varphi_{\text{beam}}, \theta_{\text{beam}})$, where $\varphi_{\text{beam}} \in [-\pi/2, \pi/2]$ is the azimuth angle and $\theta_{\text{beam}} \in [-\pi/2, \pi/2]$ is the elevation angle, using MRT.

(a) Compute the first-null beamwidths in the horizontal plane (i.e., $\theta = \theta_{\text{beam}}$) and vertical plane (i.e., $\varphi = \varphi_{\text{beam}}$).

(b) Suppose $M_{\mathrm{H}} = 10$, $M_{\mathrm{V}} = 4$, and $\Delta_\lambda = 1/2$. If the UPA beamforms in the direction $\varphi_{\text{beam}} = 0$, $\theta_{\text{beam}} = \pi/10$, what is the first-null beamwidth in the horizontal plane $(\theta = \pi/10)$ and vertical plane $(\varphi = 0)$? Compare the beamwidths with those achieved for $\varphi_{\text{beam}} = 0, \theta_{\text{beam}} = 0$ in Example 4.20.

Exercise 4.16. Consider a MIMO setup with two parallel M-antenna ULAs that are separated by a distance d. The transmitter and receiver have the antenna spacings Δ and $\lambda/2$, respectively. If the first antenna in each array are aligned, the distance between transmit antenna k and receive antenna m becomes

$$d_{m,k} = \sqrt{d^2 + \left(m\frac{\lambda}{2} - k\Delta\right)^2} = d\sqrt{1 + \frac{\left(m\frac{\lambda}{2} - k\Delta\right)^2}{d^2}} \approx d\left(1 + \frac{\left(m\frac{\lambda}{2} - k\Delta\right)^2}{2d^2}\right). \tag{4.186}$$

Use this approximation and that the channel gain is the same between every transmit and receive antenna pair.

(a) Find an antenna spacing Δ that makes all the singular values of the channel matrix equal. Hint: Follow the approach in Section 4.4.3 but with different antenna spacings at the transmitter and receiver.

(b) Consider the nulls of the beam pattern in (4.60). What is the physical distance between the nulls at the distance d from the transmitter? Compare it to Δ in (a).

Exercise 4.17. Consider a free-space point-to-point MIMO channel. The antennas at the transmitter and receiver are $K = 8$ and $M = 16$, respectively. The angles-of-arrival and angles-of-departure are the same and given as $\varphi_{\mathrm{r}} = \varphi_{\mathrm{t}} = \pi/3$ and $\theta_{\mathrm{r}} = \theta_{\mathrm{t}} = 0$. Suppose that $q = 10^{-8}$ W/Hz, $N_0 = 10^{-17}$ W/Hz, and $\lambda = 0.1$ m.

(a) At what propagation distance d is the channel capacity $C = 7$ bit/symbol if single-polarized isotropic antennas are used?

(b) Suppose single-polarized cosine antennas with the antenna gain function in (4.151) are used at both the transmitter and receiver. At what distance d is the channel capacity the same as in (a)?

(c) At what distance d is the same channel capacity as in (a) achieved when the transmitter and receiver use dual-polarized cosine antennas with the antenna gain function in (4.151) and $\kappa = 0$ (i.e., best-case XPD)?

Exercise 4.18. A packet of symbols is transmitted over the SIMO channel in (4.32). Suppose the transmitter sends the constant $\sqrt{q}$ symbol L_{p} times, as explained in Section 4.2.5, so that the ULA receiver can estimate the deterministic but unknown channel $\mathbf{h} = \sqrt{\beta}\mathbf{a}(\varphi)$. Due to hardware impairments, a deterministic but unknown phase-shift is introduced on the transmitted symbols. Hence, the received signals are $\mathbf{y}[l] = \mathbf{h}\sqrt{q}e^{-\mathrm{j}\phi} + \mathbf{n}[l]$, for $l = 1, \ldots, L_{\mathrm{p}}$, where $\phi \in [-\pi, \pi)$ represents the phase-shift. The power q of the transmitted symbols is known at the receiver. Find the ML estimates of φ, β, and ϕ.

Exercise 4.19. A packet of symbols is transmitted over the SIMO channel in (4.32). Suppose the transmitter sends the constant $\sqrt{q}$ symbol L_{p} times so that the receiver can estimate the deterministic but unknown channel $\mathbf{h} = \sqrt{\beta}\mathbf{a}(\varphi, \theta)$ using the received signals $\mathbf{y}[l] = \mathbf{h}\sqrt{q} + \mathbf{n}[l]$, for $l = 1, \ldots, L_{\mathrm{p}}$.

(a) Suppose the receiver has a ULA with $M = 2$ and $\Delta = \lambda/2$. The array response vector is obtained from (4.120) as $\mathbf{a}(\varphi, \theta) = \left[1,\ e^{-\mathrm{j}\pi \sin(\varphi)\cos(\theta)}\right]^{\mathrm{T}}$. Can the receiver uniquely find the ML estimates of φ and θ?

(b) Can we find an $M > 2$ so the receiver can uniquely find the ML estimates of φ and θ?

Chapter 5

Non-Line-of-Sight Point-to-Point MIMO Channels

The previous chapter considered free-space LOS channels, where the transmitted signal only reaches the receiver through a direct, unobstructed path. This chapter considers an entirely different setup: There is no LOS path (some object blocks it), but many reflected paths. We will first show that these multipath channels behave randomly and, thus, can be modeled statistically. This is known as a *fading channel* since the current SNR depends on the current random realization of the channel. We will then extend the statistical model to MIMO channels and obtain what is known as *independent and identically distributed (i.i.d.) Rayleigh fading*. Depending on how quickly the channels vary over time, we will consider different ways of extending the capacity concept to handle fading channels. The benefit of using multiple antennas to combat fading variations will be demonstrated and the spatial fading correlation created by the propagation environment's geometry will be studied.

5.1 Basics of Multipath Propagation and Rayleigh Fading

We begin by considering a non-LOS (NLOS) SISO channel where there are L different paths that the signal can travel between the transmitter and receiver. This is called a *multipath propagation* channel. The paths are created when the electromagnetic wave interacts with various objects in the propagation environment, as illustrated in Figure 5.1. In this figure, the LOS path is blocked, but one can draw an unobstructed line between the transmitter and each object, and between each such object and the receiver.

The interaction between the wave and the object depends on the shape of the object. Figure 5.2 showcases four main categories of interactions. *Specular reflection* refers to the case when the signal wave bounces off the surface in a mirror-like way; that is, the incident and outgoing angles are the same but on the opposite side of the normal to the object. This type of reflection occurs when the object is large and smooth (as compared to the wavelength).

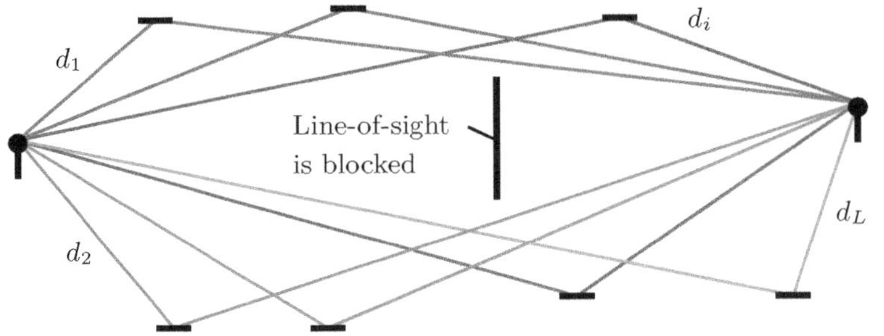

Figure 5.1: An NLOS SISO channel with L propagation paths, where d_i denotes the total length of the ith path. Each path is generated through interaction with an object in the environment. Figure 5.2 illustrates different types of interactions.

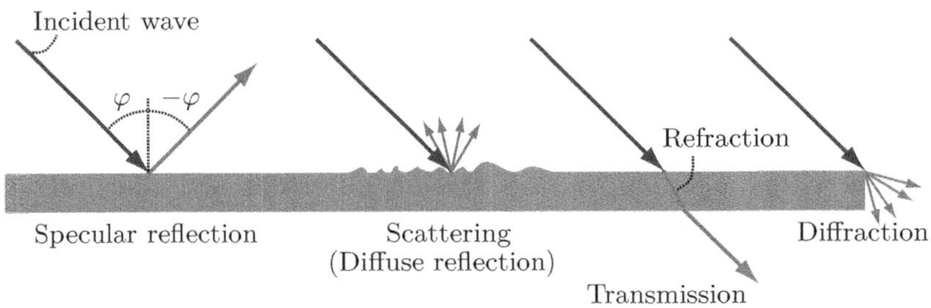

Figure 5.2: An electromagnetic wave can interact with an object in the propagation environment in various ways. The interaction might change both the direction of the signal and its shape, which can become more diffuse.

Scattering is a phenomenon that occurs when the signal wave impinges on a rough surface. When the signal bounces off the surface, it will be spread out in many different directions, also known as *diffuse reflection*. Compared to specular reflection, the benefit of scattering is the greater chance that one of the outgoing wave components propagates toward the receiver. The downside is that each component carries relatively little signal energy since the total energy of the impinging wave is spread between many directions. *Transmission* refers to when the signal passes through the object, often resulting in a slight shift in the propagation direction due to refraction inside the object. The object's material determines what fractions of the signal energy are reflected, absorbed, and transmitted to the other side. Finally, diffraction refers to the phenomenon that electromagnetic waves can bend around sharp corners, resulting in the signal spreading diffusely on the other side. Diffraction also happens when the signal passes through holes in an object with a smaller size than the wavelength.

In the context of communications, the important thing is the existence of multiple propagation paths, while the type of interaction with the objects is secondary. We considered a multipath channel with L paths already in Section 2.3.3 when deriving the memoryless channel model we used in previous chapters. We will now recall some main results and introduce new notation to study the channel properties further. We denote the total length of the ith propagation path by d_i meters. Hence, the signal that is received through the ith path will be time-delayed by $\tau_i = d_i/c = d_i/(f_c\lambda)$ seconds, where f_c is the carrier frequency, λ is the wavelength, and c is the speed of light. We let $d = \frac{1}{L}\sum_{i=1}^{L} d_i$ denote the average path length. The average path length in wireless communications is usually much larger than the variations $|d_i - d|$ around the average. This is because mainly objects surrounding the transmitter or receiver will create propagation paths that reach the receiver. Hence, the receiver can sample the received signal using the delay $\eta = d/c = d/(f_c\lambda)$, and we will obtain a memoryless channel if $B(d - d_i)/c \approx 0$ for $i = 1, \ldots, L$. Recall from Section 2.3.4 that this is known as the narrowband assumption since it is always satisfied when the bandwidth is sufficiently small. For example, suppose we interpret $B(d - d_i)/c \approx 0$ as requiring that $B|d - d_i|/c \leq 0.1$ for all paths.[1] If the maximum deviation from the average path length d is $\max_i |d - d_i| = 30\,\text{m}$, then we will obtain a narrowband channel for $B \leq 1\,\text{MHz}$. If $\max_i |d - d_i| = 3\,\text{m}$, then $B \leq 10\,\text{MHz}$ will result in a narrowband channel.

We let $\alpha_i \in [0, 1]$ denote attenuation of the ith path, while α_i^2 is the gain. We stress that α_i^2 should not be computed using the LOS formula in (1.7) since it only applies to direct free-space LOS paths. A path generated through specular reflection might have a gain that resembles that formula, but only when the reflecting surface is huge compared to the wavelength.[2] Diffusely reflected/scattered paths generally have a much lower channel gain due to the additional spatial dispersion created by the scattering and the material's absorption losses. We will not assume a specific model in this chapter.

By utilizing the narrowband assumption, we previously showed in (2.131) that the channel response $h \in \mathbb{C}$ when having L paths can be written as

$$h = \sum_{i=1}^{L} \alpha_i e^{-\mathrm{j}2\pi f_c(\tau_i - \eta)} = \sum_{i=1}^{L} \alpha_i e^{-\mathrm{j}2\pi \frac{(d_i - d)}{\lambda}}, \tag{5.1}$$

where the second equality follows from $\tau_i = d_i/(f_c\lambda)$ and our assumption of $\eta = d/(f_c\lambda)$. The value of this channel response depends on the distances and attenuations of the L individual paths. Since the attenuations α_i are multiplied by the phase-shift terms $e^{-\mathrm{j}2\pi\frac{(d_i-d)}{\lambda}}$, it is hard to tell whether the

[1] The upper bound depends on which pulse function $p(t)$ is utilized in the PAM because this determines for which time-shifts the intersymbol interference can be neglected. In this case, we selected 0.1 based on the fact that $\text{sinc}(\pm 0.1) \approx \text{sinc}(0) = 1$ and $\text{sinc}(l \pm 0.1) \approx \text{sinc}(l) = 0$ for $l = \pm 1, \pm 2, \ldots$ If we would use a pulse that varies more slowly than the sinc function, then we might expand the delay spread that we can manage without having intersymbol interference.

[2] Many objects that behave as specularly reflecting mirrors for visible light are too small to behave in that way in wireless communications because the wavelength might be a 100 000 times larger (compare 4 GHz communication with visible light that starts at 400 THz).

terms in the sum will reinforce or cancel each other. This depends on whether the phases happen to be aligned or not. When the transmitter or receiver moves, $d_1, \ldots, d_L$ will change. Even if the movement is only over a distance proportional to the wavelength λ, this might substantially change all the phase-shifts in (5.1) and thereby change if the terms reinforce or cancel each other. This phenomenon is called *multipath fading* and motivates why small movements can give rise to seemingly random changes in the channel response.

Example 5.1. Consider an environment with $L = 2$ objects creating propagation paths with identical phases. What is the shortest distance the receiver can move to risk that the paths have phases that differ by π instead?

Suppose for simplicity that $e^{-j2\pi\frac{(d_1-d)}{\lambda}} = e^{-j2\pi\frac{(d_2-d)}{\lambda}} = 1$ at the initial point. When the receiver moves, the distances d_1 and d_2 will change, and the phases will rotate. If the receiver moves a distance $\delta > 0$, the path lengths d_1, d_2 can at most increase or decrease by δ, depending on the direction of motion compared to the direction that the signal components arrive from. The largest phase difference between the two paths occurs when the receiver moves right towards object 2 so that d_2 shrinks to $d_2 - \delta$, while simultaneously moving away from object 1 so that d_1 increases to $d_1 + \delta$ (or the other way around). The respective phase-shifts will then become

$$e^{-j2\pi\frac{(d_1+\delta-d)}{\lambda}} = e^{-j2\pi\frac{\delta}{\lambda}}, \tag{5.2}$$

$$e^{-j2\pi\frac{(d_2-\delta-d)}{\lambda}} = e^{+j2\pi\frac{\delta}{\lambda}}. \tag{5.3}$$

The difference between these phases is $4\pi\frac{\delta}{\lambda}$, which becomes π if $\delta = \lambda/4$. Hence, whenever the receiver moves a quarter of the wavelength, there is a risk that the multipath propagation will change so radically that the paths are canceling out instead of reinforcing one another.

5.1.1 Rich Multipath Propagation: Rayleigh Fading

When the number of paths (L) is huge, we have a scenario known as *rich multipath propagation*. This is often a valid assumption in NLOS communications due to the many paths created by scattering. We will derive a statistical distribution for the channel response h in this case. Since there is a large set of path attenuations α_i and distances d_i, it makes sense to model their values statistically. We assume that $\alpha_1, \ldots, \alpha_L$ are independent realizations of a random variable $\mathcal{A}$, which describes how the channel attenuations vary between different objects in the environment.

We further use $\psi_i = 2\pi\frac{(d_i-d)}{\lambda} + 2\pi k_i$ to denote the phase-shift of the ith path in (5.1), where the integer k_i is selected so that $\psi_i \in [-\pi, \pi)$, for $i = 1, \ldots, L$. We can wrap any phase-shift into the interval $[-\pi, \pi)$ without loss of generality since $e^{-j\psi_i}$ is a periodic function with period 2π. When there are many paths, it is likely that the path difference $|d_i - d|$ ranges from zero up to many wavelengths, which implies that ψ_i will likely be uniformly distributed between $-\pi$ and π. Hence, we assume $\psi_1, \ldots, \psi_L$ are independent realizations of a random variable with a continuous uniform distribution between $-\pi$ and π. This is denoted as $\psi_i \sim U[-\pi, \pi)$ and the corresponding PDF is

$$f_{\psi}(\psi) = \begin{cases} \frac{1}{2\pi}, & \text{if } -\pi \leq \psi < \pi, \\ 0, & \text{otherwise.} \end{cases} \tag{5.4}$$

The purpose of this statistical modeling is to emphasize that (5.1) can be viewed as the summation of L independent realizations drawn from the same random distribution. We can separate the sum into two parts:

$$h = \sum_{i=1}^{L} \alpha_i e^{-\mathrm{j}\psi_i} = \underbrace{\sum_{i=1}^{L} \alpha_i \cos(\psi_i)}_{\text{Real part}} - \mathrm{j} \underbrace{\sum_{i=1}^{L} \alpha_i \sin(\psi_i)}_{\text{Imaginary part}}. \tag{5.5}$$

The real and imaginary parts have zero means because the integral over a period of a cosine/sine function is zero. These parts are uncorrelated since

$$\mathbb{E}\left\{\sum_{i=1}^{L} \alpha_i \cos(\psi_i) \sum_{j=1}^{L} \alpha_j \sin(\psi_j)\right\} = \sum_{i=1}^{L} \mathbb{E}\left\{\alpha_i^2\right\} \underbrace{\mathbb{E}\{\cos(\psi_i)\sin(\psi_i)\}}_{=0} = 0 \tag{5.6}$$

when $\psi_i \sim U[-\pi, \pi)$.[3] One can also show that the real and imaginary parts in (5.5) have the same variance since $\mathbb{E}\{\cos^2(\psi_i)\} = \mathbb{E}\{\sin^2(\psi_i)\} = 1/2$.[4]

When L is large, we can utilize the central limit theorem, stated in Lemma 2.6, to obtain an approximate random distribution of h. This result manifests that the sum of many independent and identically distributed real-valued random variables becomes approximately Gaussian distributed. We can apply this theorem to the real and imaginary parts of h in (5.5) to motivate that both are approximately Gaussian distributed. Since we have shown that the real and imaginary parts are also uncorrelated, it follows from the Gaussian distribution that they are also approximately independent. Hence, the channel response in a rich multipath environment is approximately complex Gaussian distributed:

$$h \sim \mathcal{N}_{\mathbb{C}}(0, \beta). \tag{5.7}$$

We let β denote the average channel gain $\mathbb{E}\{|h|^2\} = \beta$ of h to obtain a notation where the average SNR is denoted in the same way as in LOS

[3]There is a trigonometric identity saying that $\cos(\psi_i)\sin(\psi_i) = \sin(2\psi_i)/2$. For $\psi_i \sim U[-\pi, \pi)$, it follows that $\mathbb{E}\{\cos(\psi_i)\sin(\psi_i)\} = \mathbb{E}\{\sin(2\psi_i)\}/2$ is an integral over two periods of the sine function; thus, it is equal to zero.

[4]There is another trigonometric identity saying that $\cos^2(\psi_i) = 1/2 + \cos(2\psi_i)/2$. For $\psi_i \sim U[-\pi, \pi)$, it follows that $\mathbb{E}\{\cos^2(\psi_i)\} = 1/2$. Similarly, using the trigonometric identity $\sin^2(\psi_i) = 1/2 - \cos(2\psi_i)/2$, it follows that $\mathbb{E}\{\sin^2(\psi_i)\} = 1/2$.

communications. Note that β is also the variance of h since the mean value is zero. As explained in Section 2.2.2, the full name of the distribution in (5.7) is the circularly symmetric complex Gaussian distribution, where the circular symmetry means that h and $he^{-\mathrm{j}\psi}$ have the same distribution for any phase-shift ψ. This property can be observed in Figure 2.6, where the PDF remains the same if it is rotated around the origin.

The type of channel distribution in (5.7) is commonly known as *Rayleigh fading*. The reason is that the channel magnitude $|h|$ is Rayleigh distributed, as previously described in Section 2.2.5. Specifically, $|h| \sim \text{Rayleigh}(\sqrt{\beta/2})$, resulting in the PDF

$$f_{|\mathsf{h}|}(x) = \frac{2x}{\beta} e^{-\frac{x^2}{\beta}}, \quad \text{for } x \geq 0. \tag{5.8}$$

This PDF is illustrated in Figure 5.3(a) for $\beta = 1$. We observe that $|h|$ has most of its probability mass between 0 and 3. The mean value can be shown to be $\sqrt{\pi}/2 \approx 0.9$. The PDF does not look particularly strange, but an important characteristic is emphasized in Figure 5.3(b), where we show the same PDF using a logarithmic scale on the horizontal axis. We can then notice that most channel realizations will give $|h| \approx 1$, but there is also a substantial risk of getting a value closer to zero. For example, $|h| < 0.5$ happens in 22% of the realizations and $|h| < 10^{-1}$ happens in 1% of the realizations. When the magnitude of the channel is this small, we say that it is in *deep fade*.

Relatively few propagation paths are sufficient to approximately observe Rayleigh fading, especially if the paths have roughly the same channel gains α_i^2. This will be the case when the scattering is close to the transmitter and/or receiver, so the path lengths $d_1, \ldots, d_L$ are roughly the same. The convergence to Rayleigh fading is illustrated in Figure 5.4 by showing the CDF

$$F_{|\mathsf{h}|}(x) = \int_{-\infty}^{x} f_{|\mathsf{h}|}(u) \partial u \tag{5.9}$$

of $|h|$ with $L = 2$, $L = 5$, and Rayleigh fading that is obtained as $L \to \infty$. Recall that the CDF of the Rayleigh distribution was given in (2.102). We have assumed $\alpha_i = 1/\sqrt{L}$ and $\psi_i \sim U[-\pi, \pi)$. The curves with Rayleigh fading and $L = 5$ are nearly the same, while $L = 2$ gives a different shape. Hence, it is sufficient with five propagation paths with equal attenuation and random phases to obtain a channel that can be modeled by Rayleigh fading.

Random channels are generally called fading channels since the SNR $\frac{q|h|^2}{N_0}$ varies depending on the realization of h. The cause of the variations is the summation of the many complex exponentials in (5.1), which cancel out each other by having very different phases when the channel is in a deep fade. This is essentially an extension to Example 5.1, which showcased how two paths are canceled when the phase-shifts differ by π. As we will see later in this chapter, the fading variations are problematic in wireless communications and require us to define the channel capacity differently.

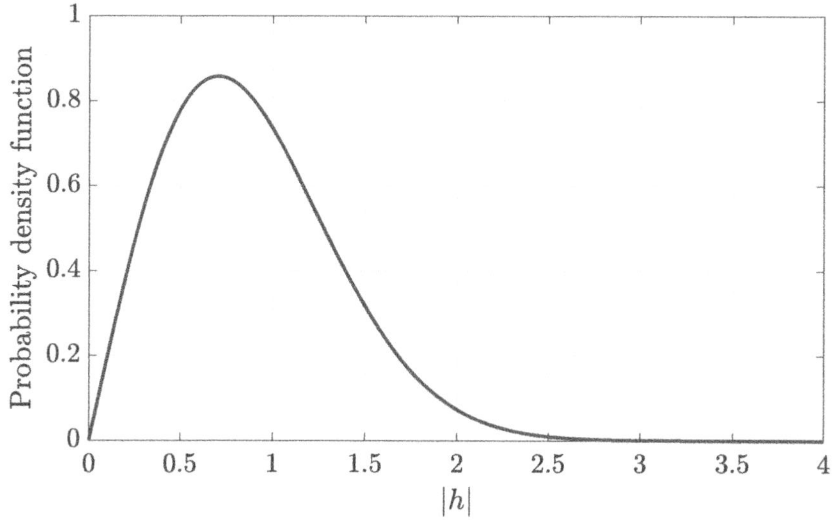

(a) Probability density function using a linear scale on the horizontal axis.

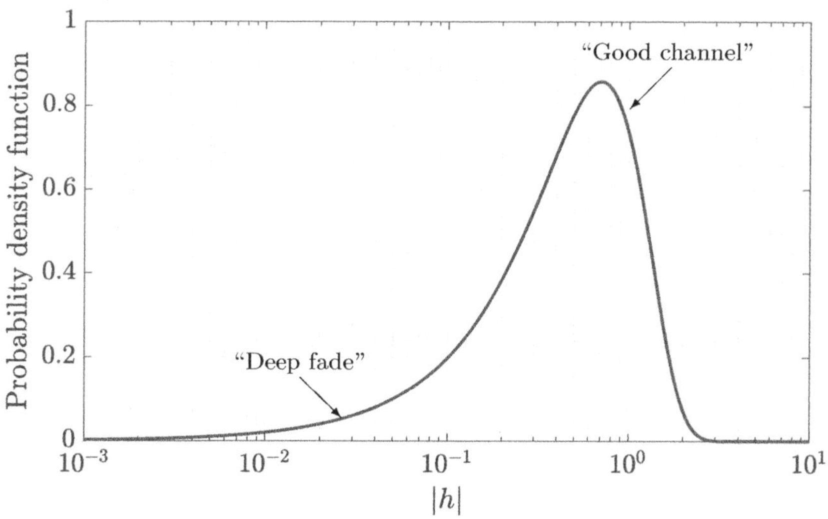

(b) Probability density function using a log-scale on the horizontal axis.

Figure 5.3: The probability density function $2xe^{-x^2}$ of $x = |h|$, when $h \sim \mathcal{N}_{\mathbb{C}}(0, 1)$. This channel distribution is known as Rayleigh fading and is characterized by occasional deep fades where $|h|$ is much smaller than the average value.

Although the distribution of the magnitude $|h|$ has given rise to the term "Rayleigh fading," mostly the distribution of the squared magnitude $|h|^2$ is useful when analyzing the communication over Rayleigh fading channels. We will utilize it later in this chapter. It was shown in Section 2.2.5 that it has an exponential distribution: $|h|^2 \sim \text{Exp}(1/\beta)$ with the PDF

$$f_{|\mathsf{h}|^2}(x) = \frac{1}{\beta} e^{-\frac{x}{\beta}}, \quad \text{for } x \geq 0. \tag{5.10}$$

Example 5.2. The derivation of Rayleigh fading distribution assumes that there are L independent and identically distributed propagation paths. What happens to the distribution if there is also an LOS path?

The distinguishing property of the LOS path is that its gain is much stronger than that of the NLOS paths, so it cannot be included when applying the central limit theorem. If we denote the LOS channel gain as α_0^2 and phase-shift as $\psi_0 \sim U[-\pi, \pi)$, then the channel response can be expressed as

$$h = \alpha_0 e^{-\mathrm{j}\psi_0} + \sum_{i=1}^{L} \alpha_i e^{-\mathrm{j}\psi_i} \tag{5.11}$$

$$\to \alpha_0 e^{-\mathrm{j}\psi_0} + h_{\text{NLOS}} \quad \text{as } L \to \infty, \tag{5.12}$$

where $h_{\text{NLOS}} \sim \mathcal{N}_{\mathbb{C}}(0, \beta_{\text{NLOS}})$ is Rayleigh fading created by the NLOS paths. This channel model is called *Rician fading* since $|h| \sim \text{Rice}(\alpha_0, \sqrt{\beta_{\text{NLOS}}/2})$ has a Rician distribution.[a] When using this alternative model, it is common to let $\beta = \mathbb{E}\{|h|^2\} = \alpha_0^2 + \beta_{\text{NLOS}}$ denote the average gain of the entire channel and define the so-called κ-factor determining how the gain is divided between the LOS and NLOS paths:

$$\kappa = \frac{\alpha_0^2}{\beta_{\text{NLOS}}}. \tag{5.13}$$

Using this notation, we can generate random channel realizations as

$$h = \sqrt{\frac{\kappa}{\kappa+1}} \sqrt{\beta} e^{-\mathrm{j} \cdot U[-\pi,\pi)} + \sqrt{\frac{1}{\kappa+1}} \mathcal{N}_{\mathbb{C}}(0, \beta), \tag{5.14}$$

where the phase of the LOS path and the Rayleigh fading created by the NLOS paths are the two sources of randomness. Using this notation, we also have that $|h| \sim \text{Rice}(\sqrt{\beta\kappa/(\kappa+1)}, \sqrt{\beta/(2(\kappa+1))})$.

[a] The Rician distribution $x \sim \text{Rice}(\nu, \sigma)$ has the PDF $f(x) = \frac{x}{\sigma^2} e^{-\frac{x^2+\nu^2}{2\sigma^2}} I_0\left(\frac{x\nu}{\sigma^2}\right)$, where $I_0(z) = \sum_{n=0}^{\infty} \frac{(z/2)^{2n}}{(n!)^2}$ is the zeroth-order modified Bessel function of the first kind.

Under the Rician fading model, the PDF of $|h|$ can have a shape that differs substantially from Rayleigh fading. Figure 5.5 shows the PDF with

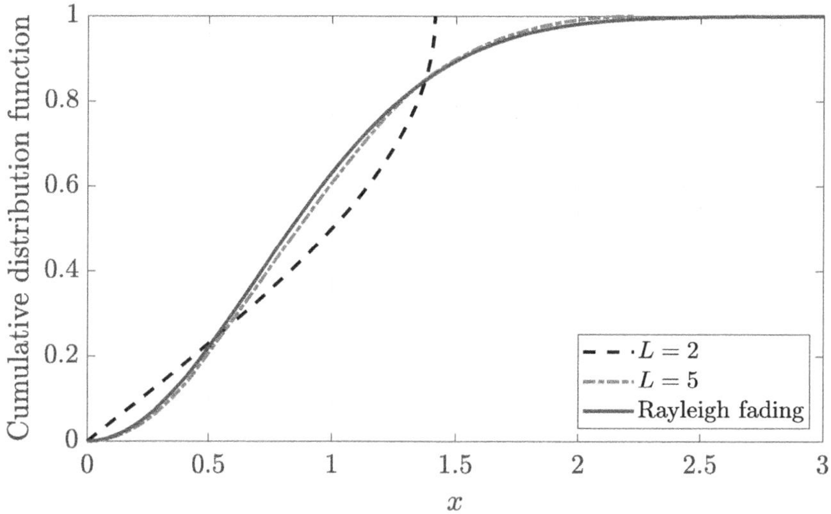

Figure 5.4: The CDF $\Pr\{|h| \leq x\}$ of the channel magnitude $|h|$ with Rayleigh fading and with $L = 2$ or $L = 5$ paths with constant gain and uniformly distributed phases. It is sufficient to have five paths to approximately observe Rayleigh fading.

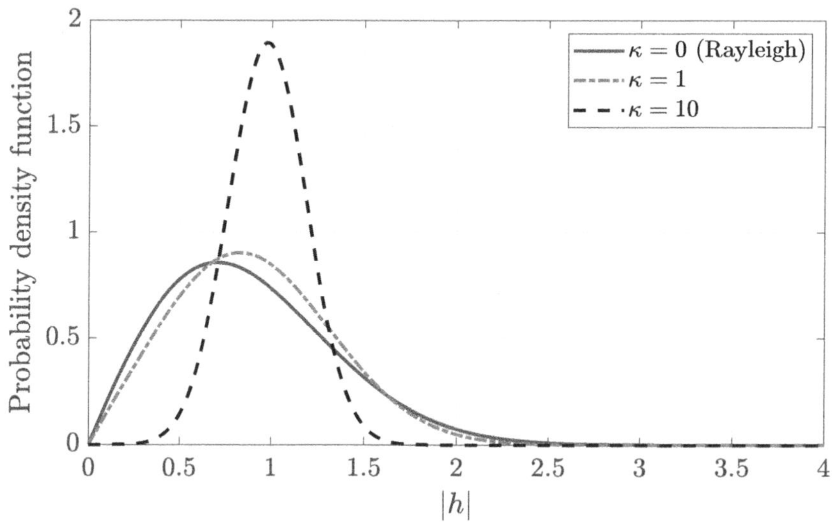

Figure 5.5: The probability density function of $|h|$ for Rician fading with $\beta = 1$ and different values of the κ-factor.

$\beta = 1$ and three values of the κ-factor. Rayleigh fading is given by $\kappa = 0$ while $\kappa = 1$ represents the scenario when the strength of the LOS path is identical to the average combined strength of all the NLOS paths. The existence of the LOS path shifts the probability mass slightly towards $\sqrt{\beta} = 1$, but the difference from Rayleigh fading is not so large, and deep fades still occur. In contrast, $\kappa = 10$ results in a PDF more confined around 1, so small and large realizations are much less likely than under Rayleigh fading.

The remainder of this chapter considers Rayleigh fading since this is the more problematic scenario. The corresponding PDF expression is also tractable for performance analysis and developing methods that counteract fading.

5.1.2 Independent Rayleigh Fading in SIMO and MISO Channels

The Rayleigh fading channel model will now be extended to systems with multiple antennas. Building on the results in the previous section, we can expect that the channel between one transmit antenna and one receive antenna can be modeled by the complex Gaussian distribution in (5.7) under rich multipath conditions. Hence, every entry $h_{m,k}$ of the MIMO channel matrix $\mathbf{H}$ can be modeled this way. The remaining question is how these channel coefficients are related to each other. Will $h_{1,1}$ and $h_{m,k}$ be statistically independent or correlated? Will they have different variances? These questions will be answered in this section for the considered frequency-flat channel.

We begin by considering a SIMO channel where a ULA with M antennas and antenna spacing Δ receives a signal from a single-antenna transmitter. We assume the ULA receives signal components via L objects in different angular directions in the three-dimensional world. We will use the spherical coordinate system in Figure 1.9 and assume the ULA is located along the z-axis, as in Example 4.16. The reason is that the corresponding array response in (4.122) is independent of the azimuth angle, which will simplify the presentation in this section. We let $\alpha_i \in [0,1]$ denote the attenuation of the ith path, ψ_i is the non-zero phase-shift[5] at the reference antenna, and θ_i is the angle-of-arrival in the elevation domain. The channel response $\mathbf{h} \in \mathbb{C}^M$ can then be written as

$$\mathbf{h} = \sum_{i=1}^{L} \alpha_i e^{-\mathrm{j}\psi_i} \begin{bmatrix} 1 \\ e^{-\mathrm{j}2\pi \frac{\Delta \sin(\theta_i)}{\lambda}} \\ e^{-\mathrm{j}2\pi \frac{2\Delta \sin(\theta_i)}{\lambda}} \\ \vdots \\ e^{-\mathrm{j}2\pi \frac{(M-1)\Delta \sin(\theta_i)}{\lambda}} \end{bmatrix}, \tag{5.15}$$

where the summation resembles the SISO channel in (5.5) but the ith term is multiplied by the array response vector from (4.122) for a signal arriving from θ_i. This setup is illustrated in Figure 5.6. We assume $\psi_1, \ldots, \psi_L$ are independent realizations of a uniform distribution between 0 and 2π. We further assume $\alpha_1, \ldots, \alpha_L$ are independent realizations of a random variable and denote the average channel gain as

$$\beta = \sum_{i=1}^{L} \mathbb{E}\left\{\alpha_i^2\right\}. \tag{5.16}$$

[5]In free-space LOS communications, there is only one path so we can synchronize the receiver such that there is no phase-shift at the reference antenna. This cannot be done when there are multiple paths, which is why ψ_i is needed in this case.

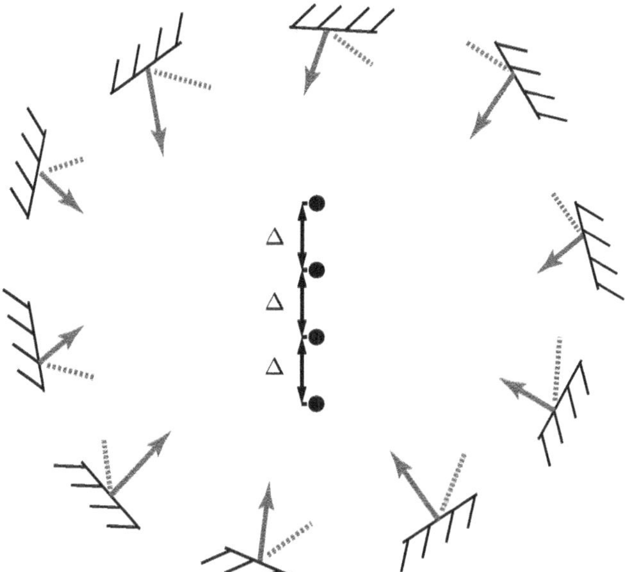

Figure 5.6: A ULA in an isotropic rich multipath environment where the multipath components are received from random elevation angle directions with uniform distribution.

In an *isotropic rich multipath environment*, the number of multipath components L is large, and their locations are uniformly/isotropically distributed over all directions. Recall that $\varphi \in [-\pi, \pi)$ denotes the azimuth angle, while $\theta \in [-\pi/2, \pi/2]$ denotes the elevation angle in the spherical coordinate system. The PDF of a uniform distribution over a unit sphere is given by

$$f_{\varphi,\theta}(\varphi, \theta) = \frac{\cos(\theta)}{4\pi}, \quad -\pi \leq \varphi < \pi, \quad -\frac{\pi}{2} \leq \theta \leq \frac{\pi}{2} \tag{5.17}$$

where 4π is the surface area of the unit sphere and $\cos(\theta)\partial\theta\partial\varphi$ is the area of a surface element in direction (φ, θ) that appears when integrating over a sphere using spherical coordinates as in (1.27). The channel in (5.15) does not depend on the azimuth angle φ; there is a rotational invariance when using a ULA that can be observed in Figure 1.18 and Figure 1.20. Hence, we only need the marginal PDF

$$f_{\theta}(\theta) = \int_{-\pi}^{\pi} f_{\varphi,\theta}(\varphi, \theta)\partial\varphi = \frac{\cos(\theta)}{2}, \quad -\frac{\pi}{2} \leq \theta \leq \frac{\pi}{2}, \tag{5.18}$$

when characterizing the statistical channel properties in this section. The entry $h_m = \sum_{i=1}^{L} \alpha_i e^{-\mathrm{j}\psi_i} e^{-\mathrm{j}2\pi\frac{(m-1)\Delta\sin(\theta_i)}{\lambda}}$ in $\mathbf{h} = [h_1, \ldots, h_M]^{\mathrm{T}}$ has the mean

$$\mathbb{E}\{h_m\} = \sum_{i=1}^{L} \mathbb{E}\{\alpha_i\} \underbrace{\mathbb{E}\{e^{-\mathrm{j}\psi_i}\}}_{=0} \mathbb{E}\left\{e^{-\mathrm{j}2\pi\frac{(m-1)\Delta\sin(\theta_i)}{\lambda}}\right\} = 0, \tag{5.19}$$

where $\mathbb{E}\{e^{-\mathrm{j}\psi_i}\} = 0$ follows from that the angles are uniformly distributed between 0 and 2π. Furthermore, the variance is

$$\begin{aligned}\mathbb{V}\mathrm{ar}\{h_m\} = \mathbb{E}\left\{|h_m|^2\right\} &= \mathbb{E}\left\{\left|\sum_{i=1}^{L} \alpha_i e^{-\mathrm{j}\psi_i} e^{-\mathrm{j}2\pi \frac{(m-1)\Delta \sin(\theta_i)}{\lambda}}\right|^2\right\} \\ &= \sum_{i=1}^{L} \mathbb{E}\left\{\alpha_i^2\right\} = \beta. \end{aligned} \tag{5.20}$$

If L is large, we can model this channel coefficient as

$$h_m \sim \mathcal{N}_{\mathbb{C}}(0, \beta), \tag{5.21}$$

according to the central limit theorem in Lemma 2.6 (following the same procedure as in the last section). All the channel coefficients in $\mathbf{h}$ have the same marginal distribution, including mean and variance. The multipath environment creates a random process that determines the fading realizations at all spatial locations, and the channel coefficients are samples taken from that process at antenna locations. Hence, the coefficients are also jointly complex Gaussian distributed. It remains to determine if the coefficients are correlated. To this end, we consider two different channel coefficients h_m and h_n, for which $m \neq n$, and compute the correlation

$$\begin{aligned}\mathbb{E}\{h_m h_n^*\} &= \mathbb{E}\left\{\sum_{i=1}^{L} \alpha_i e^{-\mathrm{j}\psi_i} e^{-\mathrm{j}2\pi \frac{(m-1)\Delta \sin(\theta_i)}{\lambda}} \sum_{j=1}^{L} \alpha_j e^{\mathrm{j}\psi_j} e^{\mathrm{j}2\pi \frac{(n-1)\Delta \sin(\theta_j)}{\lambda}}\right\} \\ &= \sum_{i=1}^{L} \mathbb{E}\left\{\alpha_i^2\right\} \mathbb{E}\left\{e^{\mathrm{j}2\pi \frac{(n-m)\Delta \sin(\theta_i)}{\lambda}}\right\}, \end{aligned} \tag{5.22}$$

where the last equality follows from that $\mathbb{E}\{e^{-\mathrm{j}\psi_i} e^{\mathrm{j}\psi_j}\} = \mathbb{E}\{e^{-\mathrm{j}\psi_i}\}\mathbb{E}\{e^{\mathrm{j}\psi_j}\} = 0$ for $i \neq j$ because the angles are independent and uniformly distributed between 0 and 2π. We can use the PDF in (5.18) to compute the last mean value in (5.22):

$$\begin{aligned}\mathbb{E}\left\{e^{\mathrm{j}2\pi \frac{(n-m)\Delta \sin(\theta_i)}{\lambda}}\right\} &= \int_{-\pi/2}^{\pi/2} e^{\mathrm{j}2\pi \frac{(n-m)\Delta \sin(\theta_i)}{\lambda}} \frac{\cos(\theta_i)}{2} \partial \theta_i \\ &= \frac{\lambda}{2\pi(n-m)\Delta} \frac{e^{\mathrm{j}2\pi \frac{(n-m)\Delta}{\lambda}} - e^{-\mathrm{j}2\pi \frac{(n-m)\Delta}{\lambda}}}{\mathrm{j}2} \\ &= \frac{\lambda \sin\left(2\pi \frac{(n-m)\Delta}{\lambda}\right)}{2\pi(n-m)\Delta} = \mathrm{sinc}\left(\frac{2(n-m)\Delta}{\lambda}\right). \end{aligned} \tag{5.23}$$

By using this expression and (5.16), the correlation in (5.22) becomes

$$\mathbb{E}\{h_m h_n^*\} = \beta \, \mathrm{sinc}\left(\frac{2(n-m)\Delta}{\lambda}\right). \tag{5.24}$$

This value is generally non-zero, meaning the channel coefficients are generally statistically correlated. This is known as *spatial correlation* since we measure the correlation between the channel coefficients observed at different spatial locations. However, we can identify specific antenna spacings that give uncorrelated channels. Since $(n - m)$ is an integer and the sinc function is zero for integer arguments (except for zero), the expression in (5.24) is zero if $2\Delta/\lambda$ is an integer. In particular, this happens for the antenna spacing $\Delta = \lambda/2$, which is yet another reason why the half-wavelength spacing is popular when considering ULAs. Intuitively, having uncorrelated channel coefficients in the array is preferable because every receive antenna provides unique information. We will see later that it is an important property to combat the adverse effects of fading. Note that if h_m and h_n are uncorrelated, they are also independent since they are jointly complex Gaussian distributed.

Spatial correlation is sometimes referred to as *antenna correlation*, but this is a misnomer since the antennas are physical objects, not random variables. It is channel coefficients observed at different antennas that can be correlated, both in time and space. It is the spatial correlation we focus on in this chapter.

Example 5.3. Consider two antennas located at the Cartesian coordinates (x_1, y_1, z_1) and (x_2, y_2, z_2), respectively. What is the spatial correlation between their channel coefficients in an isotropic rich multipath environment?

There is a distance $\delta = \sqrt{(x_1 - x_2)^2 + (y_1 - y_2)^2 + (z_1 - z_2)^2}$ between the antennas. Since the multipath components are uniformly distributed over all directions, we can shift and rotate the Cartesian coordinate system without changing the statistical distribution. In particular, we can place the origin at the first antenna and point the z-axis in the direction of the second antenna. We then have a ULA along the z-axis with $M = 2$ antennas and antenna spacing δ. It then follows from (5.24) that the correlation between the channel coefficients h_1 and h_2 at the two antennas is

$$\mathbb{E}\{h_1 h_2^*\} = \beta \operatorname{sinc}\left(\frac{2\delta}{\lambda}\right), \tag{5.25}$$

where β denotes the common channel gain. The conclusion is that spatial correlation in an isotropic rich multipath environment only depends on the distance between the antennas, not their exact locations.

The derivation of the spatial correlation in this section is based on the assumption of having a ULA deployed along the z-axis. Example 5.3 shows that we can rotate the coordinate system arbitrarily and get the same result. Only the antenna spacing matters when determining the correlation in an isotropic rich multipath environment. For example, a ULA deployed in the horizontal plane will also give rise to Rayleigh fading with the spatial correlation given by (5.24). The correlation is zero if the antennas are half-wavelength-spaced (or an integer times that), but not otherwise.

There is a connection between the preferable spacing between antennas and the classical sampling theorem in Lemma 2.8. The sampling theorem of complex time-domain signals says we can reconstruct a signal with bandwidth B (counting both positive and negative frequencies) from samples spaced apart by $1/B$ in time. As the signal bandwidth is typically distributed between $-B/2$ and $B/2$, the samples are taken twice per period of the largest signal frequency $\pm B/2$. What we have observed in this section is that we should sample a wireless signal in space using an antenna spacing of $\lambda/2$ apart. We recall from Section 2.8.3 that a signal with wavelength λ has the spatial frequencies $\pm 1/\lambda$ in the direction of propagation. In contrast, spatial frequencies in the range $(-1/\lambda, 1/\lambda)$ can be observed in other directions. In an isotropic rich multipath environment where signal components impinge on the ULA from all possible angular directions, the channel will contain all spatial frequencies from $-1/\lambda$ to $1/\lambda$ (not only one frequency as in Chapter 4). We can say that the *spatial bandwidth* is $2/\lambda$, and the sampling theorem then recommends taking samples that are spatially separated by $\lambda/2$ (i.e., twice per period). This is why the ULA should have that antenna spacing.

The conclusion from the analysis above is the following. If a ULA with M antennas and $\Delta = \lambda/2$ spacing receives a signal in a rich multipath environment (with scatterers being equally distributed over all angular directions), then the SIMO channel $\mathbf{h} = [h_1, \ldots, h_M]^{\mathrm{T}}$ contains independent entries that are equally distributed according to (5.21). We can write this distribution in vector form as

$$\mathbf{h} \sim \mathcal{N}_{\mathbb{C}}(\mathbf{0}, \beta \mathbf{I}_M). \tag{5.26}$$

This channel model is known as *i.i.d. Rayleigh fading* and can be utilized for SIMO and MISO channels.

5.1.3 Independent Rayleigh Fading in MIMO Channels

We can extend this result to MIMO channels where both the transmitter and receiver are equipped with ULAs with $\Delta = \lambda/2$ as antenna spacing. If each array is surrounded by many isotropically distributed scatterers (according to the conditions above), then every entry $h_{m,k}$ of the channel matrix $\mathbf{H} \in \mathbb{C}^{M \times K}$ will be independent and identically distributed as

$$h_{m,k} \sim \mathcal{N}_{\mathbb{C}}(0, \beta). \tag{5.27}$$

If the antennas are arranged in other ways, the entries of $\mathbf{H}$ will generally be correlated. For example, MIMO systems with UPAs always feature spatial correlation because one cannot achieve $\lambda/2$-spacing along the many diagonals in the arrays. We will focus on i.i.d. fading in this chapter since this is practically achievable and analytically tractable.

Example 5.4. What is the rank of $\mathbf{H}$ in i.i.d. Rayleigh fading?

The rank of a matrix is equal to the maximal number of linearly independent columns. The matrix $\mathbf{H} \in \mathbb{C}^{M \times K}$ has K column vectors from $\mathbb{C}^M$. We begin by considering the case when $K \leq M$, so there are fewer columns than rows. The probability that a collection of $K \leq M$ randomly generated columns will happen to be linearly dependent is zero when the randomness is independent and originates from a continuous distribution. The formal proof builds on generating random realizations for all entries of $\mathbf{H}$ except the last one: $h_{M,K}$. For the last column to be a linear combination of the previous ones, there will only be one or a few discrete values that $h_{M,K}$ can take (or there is no value at all). The probability of obtaining one of a few specific values from a continuous distribution is zero.

If $K > M$, then the first M columns will be linearly independent with probability one, and the same holds for any subset of M columns from $\mathbf{H}$. Hence, we will get a realization of $\mathbf{H}$ that has the maximum rank $\min(M, K)$ with probability one. This also means that a MIMO channel with i.i.d. Rayleigh fading can achieve the maximum multiplexing gain $r = \min(M, K)$.

5.2 Slow and Fast Fading Versus the Channel Coherence Time

When the channel capacity was analyzed in Section 3, it was assumed that the channels are fixed throughout the transmission and known at both the transmitter and receiver. These are reasonable assumptions for LOS channels but not necessarily for fading channels. Recall from Definition 2.7 that the capacity describes the number of bits per second that can be "communicated with arbitrarily low error probability as the number of symbols in the packet approaches infinity". The second part of this sentence is crucial in the context of fading channels: When we send a long packet, how many random realizations of the fading channels will we observe in the meantime?

The answer to this question depends on many factors, such as the packet size, the geometry of the propagation environment, and the mobility of the transmitter, receiver, and objects that interact with the waves. We will quantify the time a channel coefficient is approximately constant to shed light on this.

The worst-case scenario for channel variations was identified in Example 5.1. A practical situation where the same thing occurs is illustrated in Figure 5.7. The transmitter can reach the receiver via two reflecting objects, although the LOS path is blocked. Initially, the two propagation paths are of equal length d (i.e., $d_1 = d_2 = d$) and have the same attenuation α. The receiver then moves towards object 2 at a speed of $v\,\mathrm{m/s}$ without changing the attenuation (for simplicity). Hence, the two propagation distances can be expressed as functions of the time t as

$$d_1(t) = d + vt, \qquad d_2(t) = d - vt. \tag{5.28}$$

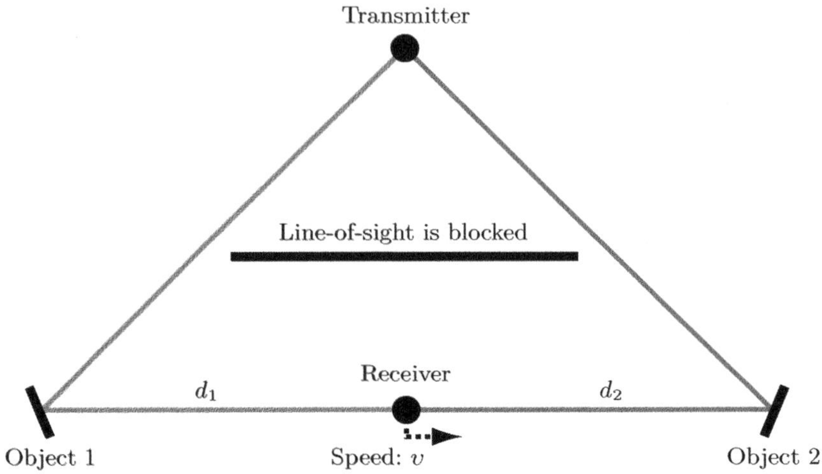

Figure 5.7: An NLOS SISO channel with two propagation paths that are initially of equal length, but the receiver then moves towards object 2 at a speed of v m/s.

The SISO channel coefficient in (5.1) also becomes a function of time:

$$h(t) = \alpha \left(e^{-j2\pi \frac{(d_1(t)-d)}{\lambda}} + e^{-j2\pi \frac{(d_2(t)-d)}{\lambda}} \right) = \alpha \left(e^{-j2\pi \frac{vt}{\lambda}} + e^{+j2\pi \frac{vt}{\lambda}} \right)$$
$$= 2\alpha \cos \left(2\pi \frac{vt}{\lambda} \right). \tag{5.29}$$

We can notice several things from this expression. Firstly, the two paths have aligned phases at $t = 0$ and cancel out when $vt = \lambda/4$ which happens at the time $t = T_c$, where

$$T_c = \frac{\lambda}{4v}. \tag{5.30}$$

This is called the *channel coherence time* since it represents the shortest time to move from constructive superposition to a deep fade. The expression in (5.30) is often used to approximate the time a channel response remains approximately constant. The coherence time is proportional to the wavelength and inversely proportional to the speed of motion. One can rightfully criticize whether the proportionality constant in (5.30) should be 1/4 because the channel in (5.29) will change drastically from $h(0) = 2\alpha$ to $h(T_c) = 0$ in that time period. On the other hand, we considered a worst-case scenario that is unlikely to happen in practice, which is why it is a common rule-of-thumb [26, Sec. 2.1.4].[6]

The second observation is that the phase-shifts in (5.29) due to mobility are $e^{\pm j2\pi \frac{vt}{\lambda}} = e^{\pm j2\pi f_c \frac{vt}{c}}$, which corresponds to shifting the instantaneous frequency of the received signal by $\pm f_c \frac{v}{c}$. This is known as *Doppler shifts*.

[6] One can find alternative definitions of the coherence time in other textbooks; for example, based on maintaining temporal correlation above 0.5 [61] or based on the sampling theorem [3].

Example 5.5. Consider transmitting a data packet with a time duration of 100 ms. The communication takes place in the 3 GHz band (i.e., $\lambda = 0.1$ m). What is the coherence time if $v = 0.1$ m/s or $v = 25$ m/s?

The two speeds correspond to slow indoor mobility and driving on a highway, respectively. The coherence time in (5.30) becomes

$$T_{\mathrm{c}} = \frac{\lambda}{4v} = \frac{0.1}{4 \cdot 0.1} = 250\,\text{ms} \qquad \text{if } v = 0.1\,\text{m/s}, \tag{5.31}$$

$$T_{\mathrm{c}} = \frac{\lambda}{4v} = \frac{0.1}{4 \cdot 25} = 1\,\text{ms} \qquad \text{if } v = 25\,\text{m/s}. \tag{5.32}$$

In the former case, the time duration of the packet is substantially smaller than the coherence time; thus, the channel will be approximately constant throughout the transmission. In the latter case, the coherence time is 100 times shorter than the duration of the data packet; thus, the communication will be subject to roughly 100 different channel realizations.

An additional perspective on the coherence time concept can be obtained by revisiting the rich multipath environment from Section 5.1.2. The channel response at a given location is then a realization of a complex Gaussian random variable: $\mathcal{N}_{\mathbb{C}}(0, \beta)$. Suppose the receiver starts at an arbitrary location at time 0 and then moves along a straight line at the speed v m/s. At the time t, it will be at a location $\delta = vt$ meters away from the initial location. Suppose we let $h(0)$ and $h(t)$ denote the channel realizations at these locations. In that case, we basically have a ULA with antennas separated by δ, except that the receive antenna is not simultaneously at both locations. It then follows from (5.25) that the *temporal* channel correlation is

$$\mathbb{E}\{h(0)h^*(t)\} = \beta\,\mathrm{sinc}\left(\frac{2\delta}{\lambda}\right) = \beta\,\mathrm{sinc}\left(\frac{2vt}{\lambda}\right). \tag{5.33}$$

If we continue using $T_{\mathrm{c}} = \frac{\lambda}{4v}$ from (5.30) as the channel coherence time definition, the correlation in (5.33) becomes $\beta\mathrm{sinc}(2vT_{\mathrm{c}}/\lambda) = \beta\mathrm{sinc}(1/2) \approx 0.64\beta$. One way to interpret this correlation value is that

$$h(T_{\mathrm{c}}) \approx 0.64\,h(0) + \sqrt{1 - 0.64^2}\,\mathcal{N}_{\mathbb{C}}(0, \beta), \tag{5.34}$$

which is a linear combination of the old channel $h(0)$ and a new independent realization of the complex Gaussian distribution. The coefficient ensures that $\mathbb{E}\{|h(T_{\mathrm{c}})|^2\} = \beta$. We can expect the random channel fluctuations to be small within the coherence time. Beyond that time interval, the temporal correlation reduces more rapidly and becomes $\beta\mathrm{sinc}(1) = 0$ when $t = \frac{\lambda}{2v}$.

Figure 5.8 shows random realizations of $|h(t)|$, as a function of the time t, that are generated based on the temporal correlation model in (5.33). We consider $\beta = 1$, $\lambda = 0.1$ m, and the same two speeds of motion as in Example 5.5: $v = 0.1$ m/s or $v = 25$ m/s. We notice that channel magnitude

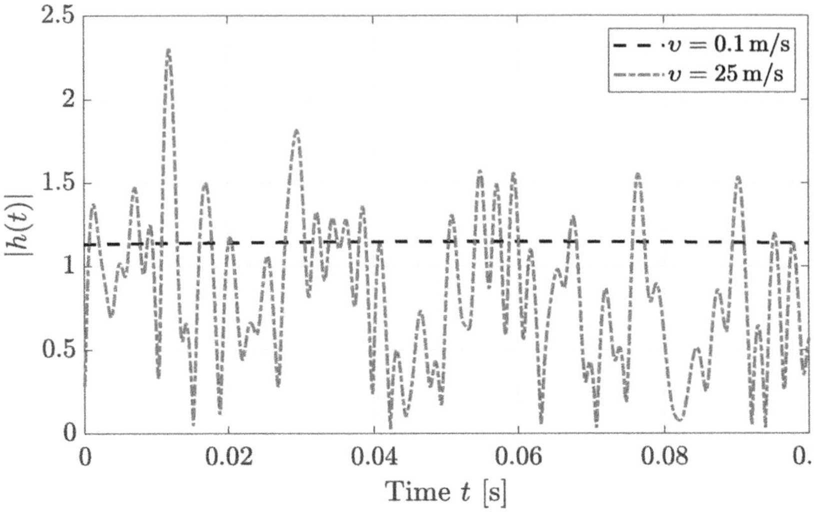

Figure 5.8: A sequence of random realizations of $|h(t)|$ that was generated using the temporal correlation model in (5.33). The speed v determines how quickly the channel changes over time.

is almost constant over the considered 100 ms time interval when the speed is low, while there are very rapid variations when the speed is high.

In conclusion, depending on how quickly things are moving in the propagation environment, a fading channel takes one random realization throughout the time interval required to send a data packet, or the channel magnitude oscillates rapidly. Even in the latter case, there is a channel coherence time within which the channel is approximately constant. Hence, we can treat a fading channel as being piecewise constant over short blocks of time and jumps between different random fading realizations across these blocks. Figure 5.9 illustrates how a continuously time-varying channel can be approximated to be piecewise constant in time intervals that match the coherence time of the channel. If we further assume that the random realizations are independent across these blocks but originate from the same distribution, we obtain what is known as the *block fading model*.

5.2.1 Definitions of Slow and Fast Fading

The relation between the channel coherence time and the packet length determines how many fading realizations will be observed during communication. When studying the impact of fading on the channel capacity, two canonical setups (or extreme cases) are normally considered:

1. *Slow fading*: The channel takes only one random realization throughout the entire transmission.

2. *Fast fading*: The channel takes a new independent random realization at every time instance.

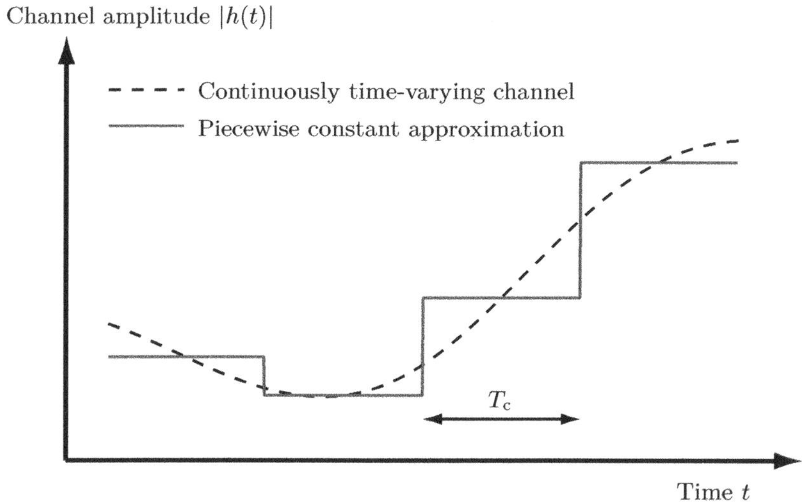

Figure 5.9: The block fading model approximates a continuously time-varying channel as being piecewise constant over time. The channel response changes every T_c second, based on the channel coherence time, and takes independent and identically distributed realizations.

We will study these cases separately in the remainder of this chapter. In both cases, the receiver is assumed to know the channel realization, while the transmitter does not. The motivation for this assumption is that the receiver can learn the channel realization after/during the transmission by analyzing the received signal. By contrast, the transmitter must decide how to transmit in advance, and then the random fading realization is generated.

One can also relate the slow and fast fading concepts to the latency requirements of the communication link; that is, the time delay from a bit is transmitted until it must be decoded at the receiver. For a given channel coherence time, we can choose between transmitting a relatively short data packet only exposed to one fading realization (i.e., slow fading) or a very long data packet exposed to many fading realizations (i.e., fast fading). Since the receiver cannot finish the data decoding until the entire packet has been received, the former option will result in lower latency, while the latter option will result in higher latency. On the other hand, we will observe later in this chapter that the performance loss due to channel fading is lower under fast-fading conditions, so it is the preferred operating regime whenever latency is of little concern.

5.3 Capacity Concept with Slow Fading

In the slow-fading scenario, the channel responses are constant throughout the communication, but their values are generated as realizations of random variables. We consider the transmission of a packet containing sufficiently much data to use the channel capacity as the performance metric. We further assume that the receiver knows the realization of the channel response, which we refer to as having *perfect CSI*. The transmitter can enable channel estimation at the receiver by transmitting a known preamble, following the procedure described in Section 4.2.4. Regarding channel knowledge at the transmitter, there are two possible modes of operation.

In a *closed-loop* system, the receiver can feed back its channel estimate to the transmitter, which will then also have perfect CSI. The capacity of such a channel can be computed as described in Chapter 3, with the only addition that the channel coefficients are now drawn randomly from a specific distribution (e.g., i.i.d. Rayleigh fading).

This section considers *open-loop* systems, where the transmitter is unaware of the current channel realization but knows the statistical distribution. This situation especially appears in systems where a reverse feedback link does not exist (e.g., when broadcasting data to many unknown user devices) or when the feedback functionality is too slow to provide the transmitter with CSI (e.g., when the latency requirements are strict). The capacity results from Chapter 3 cannot be applied under these circumstances; thus, a new capacity concept will be developed in this section.

To this end, we begin by returning to the memoryless SISO channel that was initially described in (2.130):

$$y[l] = h \cdot x[l] + n[l]. \tag{5.35}$$

We will mainly consider a Rayleigh fading channel where the channel response h is distributed as

$$h \sim \mathcal{N}_{\mathbb{C}}(0, \beta) \tag{5.36}$$

and takes only one realization throughout the communication. This might happen in practice when the transmitter and/or receiver are at random but fixed locations in a rich multipath environment.

For a given realization h, we know from Corollary 2.1 that the (conditional) capacity is

$$C_h = \log_2\left(1 + \frac{q|h|^2}{N_0}\right) \quad \text{bit/symbol}, \tag{5.37}$$

where the subscript h indicates that we have conditioned on the realization h. The receiver can decode a signal transmitted using any data rate $R \leq C_h$ since it has perfect CSI. The critical challenge in slow fading is that the transmitter does not know the realization h, but only the statistics (i.e., Rayleigh fading with variance β). Hence, the transmitter needs to select a data rate R bit/symbol, encode its data at that rate, and then hope that the communication will be successful so that the receiver can decode the data. The randomness can give rise to two different events:

- If $R \leq C_h$, the transmitter has selected a data rate below the capacity. The communication will then be successful in the sense of achieving an arbitrarily low packet error probability.

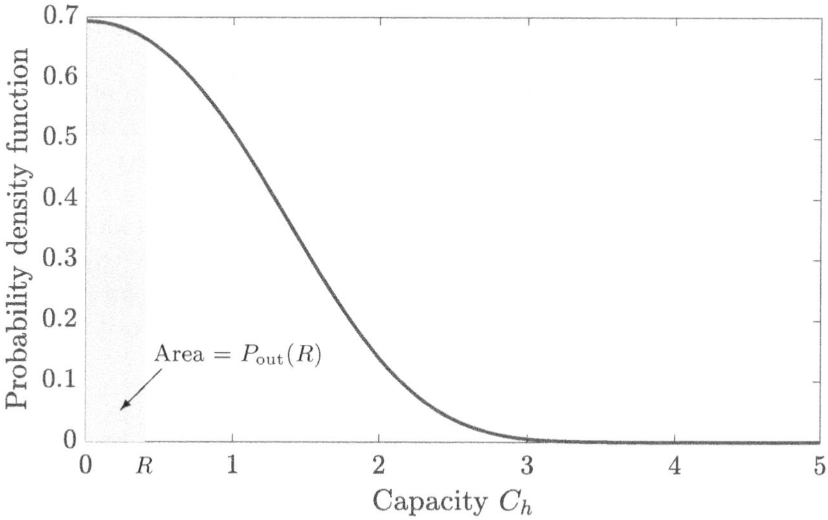

Figure 5.10: In a slow-fading scenario, the fading realization h determines the supported channel capacity C_h. The PDF of C_h is shown in this figure for $h \sim \mathcal{N}_{\mathbb{C}}(0,1)$ and $q/N_0 = 1$. For a given R, the outage probability $P_{\text{out}}(R)$ is the area under the curve for which $C_h < R$.

- If $R > C_h$, the transmitter has selected a data rate above the capacity. The communication will then be unsuccessful in the sense of having a very high packet error probability.

When the latter happens, the system is said to be in an *outage*. For a given rate R, we can define the *outage probability*:

$$P_{\text{out}}(R) = \Pr\{R > C_h\} = \Pr\left\{R > \log_2\left(1 + \frac{q|h|^2}{N_0}\right)\right\}. \tag{5.38}$$

The outage probability is a strictly increasing function of R and $P_{\text{out}}(0) = 0$. Since the only rate guaranteed to provide zero packet error for any channel realization is $R = 0$, the channel capacity is strictly speaking equal to zero.

We can nevertheless communicate over the channel, but the selection of R becomes a gamble. We can communicate relatively reliably by selecting a low R (resulting in a low outage probability), but then we will get little data through the channel. Alternatively, we can communicate unreliably by selecting a high R (resulting in a high outage probability). In this case, we can get a lot of data through the channel, but only on those few occasions when there is no outage. Figure 5.10 illustrates this situation by showing the PDF of the capacity C_h for $h \sim \mathcal{N}_{\mathbb{C}}(0,1)$ and $q/N_0 = 1$. The larger R is, the more probability mass will be under the curve between $C_h = 0$ and $C_h = R$. The outage probability equals this probability mass.

We can compare the variations in Figure 5.10 with a non-fading/LOS channel having the same average SNR: $\mathbb{E}\{\frac{q|h|^2}{N_0}\} = \frac{q}{N_0} = 1$. Such a channel would have a capacity of $C = \log_2(1+1) = 1$ bit/symbol. The figure shows that a fading channel can provide both larger and smaller values of C_h, which might give the impression that fading can be both positive and negative. Unfortunately, the adverse effect dominates in slow-fading scenarios since the transmitter does not know the value of C_h, so it must be very conservative when selecting R to avoid getting a large outage probability.

The outage probability expression in (5.38) can be utilized along with any fading distribution. We will now compute the probability by exploiting the assumption that $h \sim \mathcal{N}_{\mathbb{C}}(0, \beta)$, which implies that $|h|^2$ has an exponential distribution with its PDF $f_{|\mathsf{h}|^2}(x)$ given in (5.10). In particular, it follows that

$$\Pr\left\{|h|^2 < x\right\} = \int_0^x f_{|\mathsf{h}|^2}(t)\partial t = 1 - e^{-\frac{x}{\beta}}. \tag{5.39}$$

By rearranging the expression in (5.38), we can obtain

$$\begin{aligned} P_{\text{out}}(R) &= \Pr\left\{R > \log_2\left(1 + \frac{q|h|^2}{N_0}\right)\right\} \\ &= \Pr\left\{2^R > 1 + \frac{q|h|^2}{N_0}\right\} \\ &= \Pr\left\{|h|^2 < \frac{N_0(2^R - 1)}{q}\right\} = 1 - e^{-\frac{N_0\left(2^R-1\right)}{q\beta}}. \end{aligned} \tag{5.40}$$

If we denote the average SNR (similar to the case of non-fading channels) as

$$\text{SNR} = \mathbb{E}\left\{\frac{q|h|^2}{N_0}\right\} = \frac{q\beta}{N_0}, \tag{5.41}$$

then the outage probability in (5.40) can be expressed as

$$P_{\text{out}}(R) = 1 - e^{-\frac{2^R-1}{\text{SNR}}}. \tag{5.42}$$

This is a decreasing function of the SNR, which is logical since a higher SNR should make it easier for the channel to support a given rate R. We want to operate communication systems at relatively high SNRs to achieve high data rates. Hence, analyzing the scaling behavior of the outage probability in the high-SNR regime is essential. We can utilize the first-order Taylor approximation $e^{-x} \approx 1 - x$ for $x \approx 0$ to observe that

$$P_{\text{out}}(R) \approx 1 - \left(1 - \frac{2^R - 1}{\text{SNR}}\right) = \frac{2^R - 1}{\text{SNR}} \tag{5.43}$$

when the SNR is high. Hence, the outage probability is proportional to SNR^{-1} in the high-SNR regime and will go to zero as $\text{SNR} \to \infty$. We will show later that we can improve this high-SNR behavior by utilizing multiple antennas, but we will first provide an alternative way of formulating the outage situation.

Example 5.6. Suppose the channel response h has a fading distribution such that $|h|^2$ is uniformly distributed between 0 and 2β. What is the average SNR? How does the outage probability depend on the SNR?

The mean value of a uniform distribution with support in $[0, 2\beta]$ equals the interval's midpoint: β. Hence, the average SNR is

$$\mathrm{SNR} = \mathbb{E}\left\{\frac{q|h|^2}{N_0}\right\} = \frac{q\beta}{N_0}, \tag{5.44}$$

which is the same as in (5.41). The assumed channel distribution implies that

$$\Pr\left\{|h|^2 < x\right\} = \begin{cases} 0 & x < 0, \\ \frac{x}{2\beta} & x \in [0, 2\beta], \\ 1 & x > 2\beta. \end{cases} \tag{5.45}$$

By utilizing this property, we can calculate the outage probability in (5.38) as

$$\begin{aligned} P_{\mathrm{out}}(R) &= \Pr\left\{R > \log_2\left(1 + \frac{q|h|^2}{N_0}\right)\right\} = \Pr\left\{|h|^2 < \frac{N_0(2^R - 1)}{q}\right\} \\ &= \begin{cases} \frac{2^R-1}{2\,\mathrm{SNR}} & R \in [0, \log_2(1 + 2\,\mathrm{SNR})], \\ 1 & R > \log_2(1 + 2\,\mathrm{SNR}). \end{cases} \end{aligned} \tag{5.46}$$

This expression is proportional to SNR^{-1}, just as the outage probability in (5.43) with Rayleigh fading. However, the proportionality constant is only half as large, so there is a smaller risk of an outage when having a uniform fading distribution than with Rayleigh fading.

5.3.1 ϵ-Outage Capacity

Instead of specifying the desired rate $R \geq 0$ and computing the resulting outage probability $P_{\mathrm{out}}(R)$, we can specify a desired outage probability $\epsilon > 0$ and compute the resulting maximum rate that can be supported over the channel. That rate is called the *ϵ-outage capacity* and will be denoted as C_ϵ. It represents a capacity that can be achieved with probability $1 - \epsilon$.

In the considered Rayleigh fading SISO setup, we can set $\epsilon = P_{\mathrm{out}}(R)$ and solve for R to find C_ϵ. By utilizing (5.42), we obtain

$$\begin{aligned} \epsilon = 1 - e^{-\frac{2^R-1}{\mathrm{SNR}}} \quad &\Leftrightarrow \quad 1 - \epsilon = e^{-\frac{2^R-1}{\mathrm{SNR}}} \\ &\Leftrightarrow \quad \frac{2^R - 1}{\mathrm{SNR}} = -\ln(1 - \epsilon) \\ &\Leftrightarrow \quad 2^R - 1 = \mathrm{SNR}\ln\left((1-\epsilon)^{-1}\right) \\ &\Leftrightarrow \quad R = \log_2\left(1 + \mathrm{SNR}\ln\left((1-\epsilon)^{-1}\right)\right). \end{aligned} \tag{5.47}$$

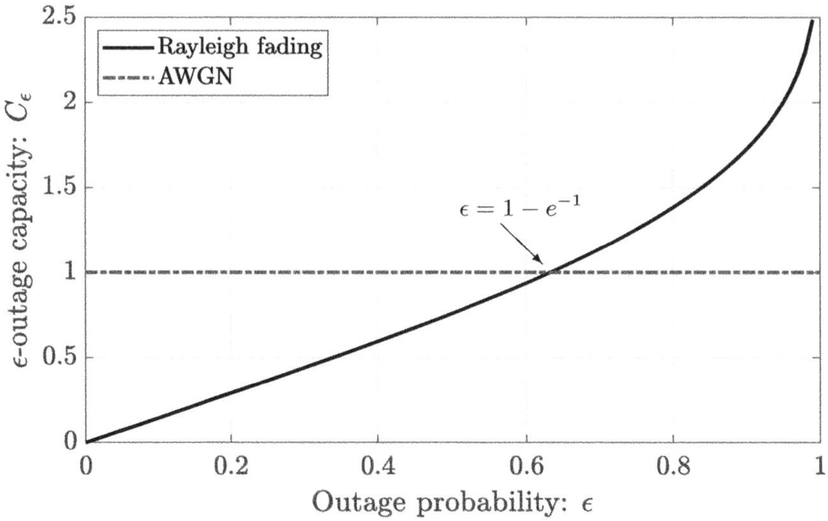

Figure 5.11: The ϵ-outage capacity C_ϵ of a Rayleigh fading channel is compared with the capacity of an AWGN channel when the (average) SNR is $\mathrm{SNR} = 0\,\mathrm{dB}$. The ϵ-outage capacity is much smaller than the AWGN capacity in the practically interesting range of outage capacities (e.g., $\epsilon \leq 0.1$) but is higher than the AWGN capacity for $\epsilon > 1 - e^{-1}$.

Hence, the ϵ-outage capacity is

$$C_\epsilon = \log_2\left(1 + \mathrm{SNR}\ln\left((1-\epsilon)^{-1}\right)\right) \tag{5.48}$$

and depends on the SNR and ϵ. By computing the first-order derivative of C_ϵ with respect to SNR and ϵ, one can respectively show that C_ϵ is an increasing function of the SNR and also an increasing function of the outage probability. Naturally, a higher SNR allows us to send more data. The reason that C_ϵ increases with ϵ is that we can then select a rate that is supported by the channel only when we get "good" fading realizations.

It is instructive to compare C_ϵ with the capacity $C = \log_2(1 + \mathrm{SNR})$ of a non-fading AWGN channel having the same (average) SNR. The only difference is the additional term $\ln((1-\epsilon)^{-1})$ that the SNR is multiplied by in (5.48). Interestingly, this term can be both smaller and larger than one. More precisely, it is smaller than one if $\epsilon < 1 - e^{-1} \approx 0.63$, because this is the probability that $|h|^2$ is smaller than its average value $\mathbb{E}\{|h|^2\} = \beta$. In these cases, the ϵ-outage capacity is smaller than the AWGN capacity. The opposite is true for $\epsilon > 1 - e^{-1}$ because then the communication is only successful when the channel realization is stronger than its average.

Figure 5.11 compares C_ϵ and $C = \log_2(1 + \mathrm{SNR})$ for $\mathrm{SNR} = 0\,\mathrm{dB}$. The ϵ-outage capacity is much smaller than the AWGN capacity for the vast majority of ϵ-values. Typical desired values of the outage probability are $\epsilon \leq 0.1$, for which the ϵ-outage capacity is less than 15% of the AWGN capacity. Hence, fading is generally considered a detrimental property of wireless channels.

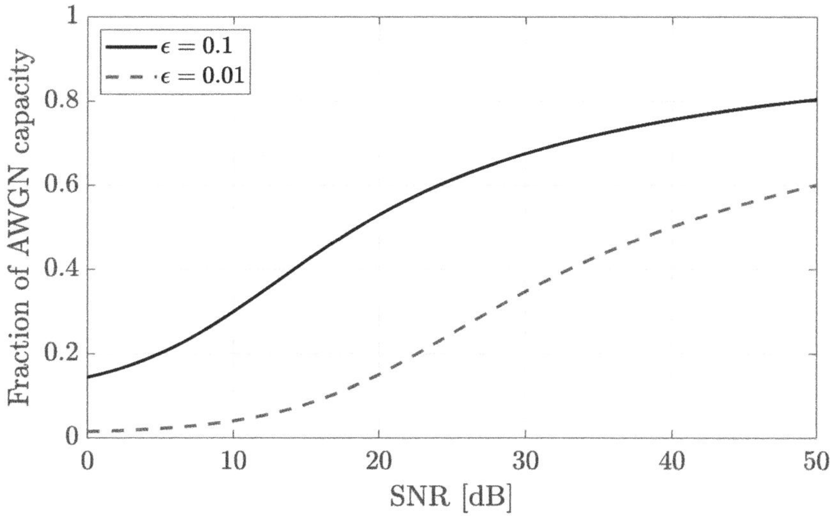

Figure 5.12: The ratio between the ϵ-outage capacity C_ϵ in (5.48) and the AWGN capacity $C = \log_2(1 + \text{SNR})$ for different SNRs. The fading channel achieves a higher fraction of the AWGN capacity at high SNR, but the convergence in (5.49) is not visible.

However, the figure also shows that for large values of ϵ, the ϵ-outage capacity is larger than the AWGN channel capacity. If reliability is unimportant, the fading can occasionally be exploited to achieve high rates. However, the transmitter needs to know when the channel has good realizations, which is inconsistent with the considered setup.

The last figure considered a rather low SNR value. The difference between C_ϵ and C depends on the SNR. The fraction of the AWGN capacity that is achieved with a fading channel converges as

$$\frac{C_\epsilon}{C} = \frac{\log_2\left(1 + \text{SNR} \ln\left((1-\epsilon)^{-1}\right)\right)}{\log_2(1 + \text{SNR})} \to 1 \tag{5.49}$$

when $\text{SNR} \to \infty$, where the limit can be established using L'Hospital's rule. Hence, the relative difference vanishes asymptotically at high SNR.

Figure 5.12 shows the fraction $\frac{C_\epsilon}{C}$ from (5.49) for a range of practical SNR values. Two practical outage probability values are considered: $\epsilon = 0.1$ and $\epsilon = 0.01$. The figure shows that fading channels operate closer to the AWGN capacity at higher SNRs; however, the convergence to the upper limit in (5.49) is not apparent in the considered SNR range. An SNR of hundreds of dB is necessary to approach the upper limit for these values of ϵ. The conclusions are that channel fading has a detrimental impact on the capacity at practical SNRs and that the asymptotic result in (5.49) is not practically useful.

5.3.2 Receive Diversity in SIMO Systems

The issue with slow fading channels is the substantial risk that the channel coefficient is in a deep fade; thus, we need to select a low rate value of R to keep the outage probability reasonably low. This problem can be mitigated by using multiple antennas that have been deployed to observe different fading realizations. In this section, we consider a SIMO system with i.i.d. Rayleigh fading. We will demonstrate that having multiple independent channel coefficients under slow fading is beneficial since there is a good chance that at least one of the M antennas experiences a decent channel realization.

The SIMO channel $\mathbf{h} \sim \mathcal{N}_{\mathbb{C}}(\mathbf{0}, \beta \mathbf{I}_M)$ is considered in this section. For a given channel realization, we can utilize (3.22) to obtain the (conditional) capacity value

$$C_{\mathbf{h}} = \log_2\left(1 + \frac{q\|\mathbf{h}\|^2}{N_0}\right) \tag{5.50}$$

for a given realization of $\mathbf{h}$. This expression does not depend on the individual entries of $\mathbf{h}$ but only on the squared norm $\|\mathbf{h}\|^2$. The norm is only small when all the entries of $\mathbf{h}$ are simultaneously small. Under i.i.d. Rayleigh fading, this variable has the scaled $\chi^2(2M)$-distribution introduced in Section 2.2.5. The PDF of $\|\mathbf{h}\|^2$ was stated in (2.99) as

$$f_{\|\mathbf{h}\|^2}(x) = \frac{x^{M-1} e^{-\frac{x}{\beta}}}{\beta^M (M-1)!}, \quad \text{for } x \geq 0. \tag{5.51}$$

We can define the outage probability when the transmitter uses the rate R as

$$\begin{aligned} P_{\text{out}}(R) &= \Pr\{R > C_{\mathbf{h}}\} = \Pr\left\{R > \log_2\left(1 + \frac{q\|\mathbf{h}\|^2}{N_0}\right)\right\} \\ &= \Pr\left\{\|\mathbf{h}\|^2 < \frac{N_0\left(2^R - 1\right)}{q}\right\}. \end{aligned} \tag{5.52}$$

The exact outage probability can then be computed using (5.51) as

$$\begin{aligned} P_{\text{out}}(R) &= \int_0^{\frac{N_0\left(2^R-1\right)}{q}} \frac{x^{M-1} e^{-\frac{x}{\beta}}}{\beta^M (M-1)!} \partial x \\ &= -\frac{\left(\frac{N_0\left(2^R-1\right)}{q}\right)^{M-1} e^{-\frac{N_0\left(2^R-1\right)}{q\beta}}}{\beta^{M-1}(M-1)!} + \int_0^{\frac{N_0\left(2^R-1\right)}{q}} \frac{x^{M-2} e^{-\frac{x}{\beta}}}{\beta^{M-1}(M-2)!} \partial x \\ &= 1 - e^{-\frac{N_0\left(2^R-1\right)}{q\beta}} \sum_{m=0}^{M-1} \frac{\left(\frac{N_0\left(2^R-1\right)}{q\beta}\right)^m}{m!} = 1 - e^{-\frac{2^R-1}{\text{SNR}}} \sum_{m=0}^{M-1} \frac{\left(\frac{2^R-1}{\text{SNR}}\right)^m}{m!} \end{aligned} \tag{5.53}$$

by integrating by parts repeatedly and then using the SNR definition $\text{SNR} = \frac{q\beta}{N_0}$ from (5.41). This expression is complicated to analyze since there are many terms. We are primarily interested in the behavior at high SNRs since this is where we want the system to operate and where outages can be avoided through a good system design. In those cases, the outage probability is determined by the behavior of $f_{\|\mathbf{h}\|^2}(x)$ for $x \approx 0$. By utilizing the fact that $e^{-x/\beta} \leq 1$ for $x \geq 0$, we obtain the inequality

$$f_{\|\mathbf{h}\|^2}(x) \leq \frac{x^{M-1}}{\beta^M (M-1)!}. \tag{5.54}$$

We can expect to achieve equality approximately in (5.54) when $x \approx 0$. Hence, a tight upper bound on the outage probability in (5.52) can be computed as

$$\begin{aligned} P_{\text{out}}(R) &= \int_0^{\frac{N_0(2^R-1)}{q}} f_{\|\mathbf{h}\|^2}(x) \partial x \\ &\leq \int_0^{\frac{N_0(2^R-1)}{q}} \frac{x^{M-1}}{\beta^M (M-1)!} \partial x = \left(\frac{N_0 \left(2^R - 1\right)}{q\beta} \right)^M \frac{1}{M!}. \end{aligned} \tag{5.55}$$

By using the SNR definition in (5.41), we can write (5.55) as

$$P_{\text{out}}(R) \leq \left(\frac{2^R - 1}{\text{SNR}} \right)^M \frac{1}{M!}, \tag{5.56}$$

where the upper bound is proportional to SNR^{-M} and is approximately achieved when the SNR is high. Hence, the outage probability reduces with the SNR much more rapidly when multiple antennas exist. This is known as a *spatial diversity gain* and M is the *diversity order*. The more antennas are used, the less probable it is that all the antennas are simultaneously in deep fades; each independent channel coefficient contributes $+1$ to the diversity order. Moreover, the higher the SNR is, the deeper the fade must be to get an outage for a given value of R, and this becomes even less probable when there are multiple antennas.

Figure 5.13 illustrates the diversity gain in a setup with i.i.d. Rayleigh fading and $R = 1$ bit/symbol. Figure 5.13(a) shows the outage probability for different SNRs with $M = 1$, $M = 2$, and $M = 4$ antennas. The outage probabilities are roughly the same at low SNRs, while the curves behave very differently at higher SNRs. We know that the outage probability is proportional to SNR^{-M} at high SNR. This means that for every 10 dB that the SNR increases, the outage probability is reduced by a factor of $1/10^M$. Since logarithmic scales are used on both axes in Figure 5.13(a), this results in lines with the slope $-M$. As the SNR increases, a steeper slope leads to a rapidly lower outage probability. If we want to achieve $P_{\text{out}} = 10^{-3}$, then we need SNR $= 30$ dB with $M = 1$, SNR $= 13$ dB with $M = 2$, and only SNR $= 4$ dB with $M = 4$.

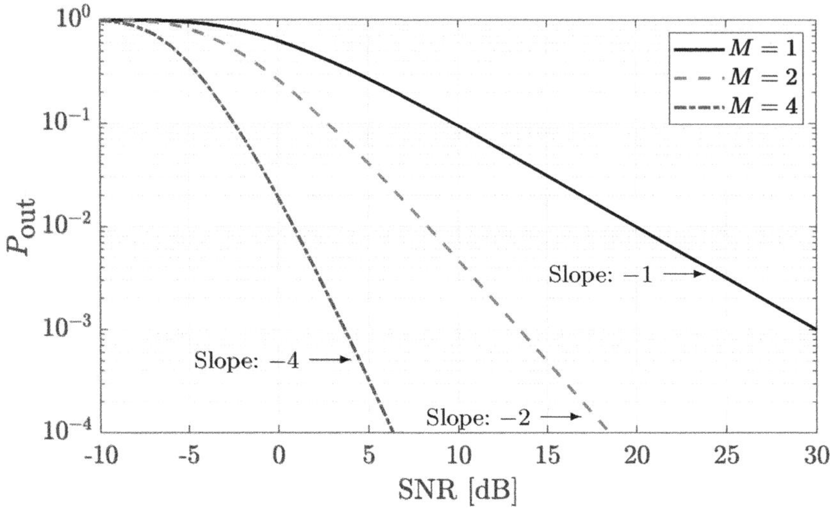

(a) The exact outage probabilities with $M = 1$, $M = 2$, and $M = 4$.

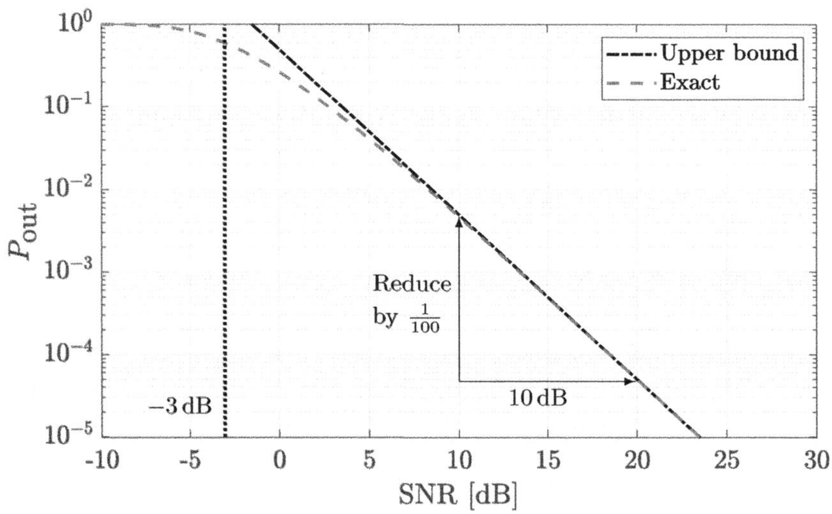

(b) The exact outage probability and the upper bound in (5.56) for $M = 2$.

Figure 5.13: The outage probability is proportional to SNR^{-M} at high SNRs when communicating over a SIMO channel with i.i.d. Rayleigh fading. In this case, we consider $R = 1$ bit/symbol.

The origin of the performance gain lies in the behavior of $\|\mathbf{h}\|^2$, and there are two contributing phenomena. Firstly, the average gain

$$\mathbb{E}\left\{\|\mathbf{h}\|^2\right\} = \sum_{m=1}^{M} \mathbb{E}\left\{|h_m|^2\right\} = M\beta \tag{5.57}$$

is proportional to M (see Section 2.2.5 for further details), which is how the beamforming gain is manifested in NLOS channels. The gain is the same as for LOS channels, except that it now varies around the mean value M depending on the channel realizations instead of always having that exact value. Secondly, the variations in $\|\mathbf{h}\|^2$ around its mean value reduce in relative terms (i.e., normalized by the average gain). The variance can be computed as

$$\begin{aligned}\mathbb{V}\mathrm{ar}\left\{\frac{\|\mathbf{h}\|^2}{\mathbb{E}\{\|\mathbf{h}\|^2\}}\right\} &= \mathbb{E}\left\{\left|\frac{\|\mathbf{h}\|^2}{\mathbb{E}\{\|\mathbf{h}\|^2\}}\right|^2\right\} - \left|\mathbb{E}\left\{\frac{\|\mathbf{h}\|^2}{\mathbb{E}\left\{\|\mathbf{h}\|^2\right\}}\right\}\right|^2 \\ &= \frac{\mathbb{E}\left\{\|\mathbf{h}\|^4\right\}}{M^2\beta^2} - \frac{\left|\mathbb{E}\left\{\|\mathbf{h}\|^2\right\}\right|^2}{\left|\mathbb{E}\left\{\|\mathbf{h}\|^2\right\}\right|^2} = \frac{(M^2+M)\beta^2}{M^2\beta^2} - 1 = \frac{1}{M},\end{aligned} \tag{5.58}$$

by using (5.57) and the following result from (2.97): $\mathbb{E}\left\{\|\mathbf{h}\|^4\right\} = (M^2+M)\beta^2$. We notice that the variance in (5.58) reduces with M. This is the statistical property that gives rise to the diversity gain. In general, the beamforming gain shifts the outage probability curves to the left in Figure 5.13(a) as we increase M, while the diversity gain makes the curves steeper.

We will now continue describing the simulation example. Figure 5.13(b) compares the exact outage probability $P_{\text{out}}(1)$ in (5.53) with $M = 2$ and the upper bound in the right-hand side of (5.56), for different SNR values. As previously claimed, the upper bound overlaps with the exact curve when the SNR is large. At high SNRs, we can thus increase the SNR by 10 dB and expect the outage probability to reduce by a factor $1/10^M = 1/100$ since $M = 2$. This is indicated in the figure.

Recall that we want to achieve $R = 1$ bit/symbol in this example. It is instructive to compare the fading channel with an LOS channel with $M = 2$ antennas and an SNR of -3 dB because it has a matching capacity of 1 bit/symbol. There are no outage issues for such a non-fading channel: we need the SNR to be at least -3 dB, and then we are guaranteed to achieve a data rate of 1 bit/symbol. The dotted vertical line in Figure 5.13(b) indicates this SNR level. If the i.i.d. Rayleigh fading channel has the same SNR, the outage probability is approximately 0.6 (this is where the curves intersect), which is too high to get reliable communication. We must increase the SNR to achieve a reasonably low outage probability. This is the price to pay for reliability over fading channels. The price reduces when we add more receive antennas, thanks to the diversity gain, but there is always a need for operating at somewhat higher SNRs than in the corresponding non-fading channel.

Example 5.7. What is the ϵ-outage capacity of the SIMO channel?

The exact ϵ-outage capacity C_ϵ can be obtained by solving the equation $P_{\text{out}}(R) = \epsilon$ for R. The outage probability in (5.53) can be expressed as

$$P_{\text{out}}(R) = F_{\|\mathbf{h}\|^2}\left(\frac{N_0\left(2^R - 1\right)}{q}\right), \tag{5.59}$$

where $F_{\|\mathbf{h}\|^2}(x) = 1 - e^{-\frac{x}{\beta}} \sum_{m=0}^{M-1} \frac{\left(\frac{x}{\beta}\right)^m}{m!}$ is the CDF of $\|\mathbf{h}\|^2$ with i.i.d. Rayleigh fading. By inverting the CDF, we can compute the outage capacity as

$$\begin{aligned} \epsilon = F_{\|\mathbf{h}\|^2}\left(\frac{N_0\left(2^R - 1\right)}{q}\right) \quad &\Leftrightarrow \quad F^{-1}_{\|\mathbf{h}\|^2}(\epsilon) = \frac{N_0\left(2^R - 1\right)}{q} \\ &\Leftrightarrow \quad \log_2\left(1 + \frac{q}{N_0} F^{-1}_{\|\mathbf{h}\|^2}(\epsilon)\right) = R. \end{aligned} \tag{5.60}$$

Hence, the ϵ-outage capacity becomes

$$C_\epsilon = \log_2\left(1 + \frac{q}{N_0} F^{-1}_{\|\mathbf{h}\|^2}(\epsilon)\right). \tag{5.61}$$

Unfortunately, there is no simple expression for the inverse CDF, but the inverse exists since the CDF is a strictly increasing function. We can use the expression in (5.61) for any fading distribution, not only i.i.d. Rayleigh fading.

5.3.3 Transmit Diversity in MISO Systems

We now turn our attention to a MISO system. We know from Section 3.3 that the capacity of SIMO and MISO channels are the same, but that result was obtained assuming that both the transmitter and receiver know the channel vector $\mathbf{h}$. Only the receiver knows the channel in the slow-fading scenario we consider in this section. Hence, the receiver could apply the optimal MRC vector $\mathbf{w} = \frac{\mathbf{h}}{\|\mathbf{h}\|}$ to the received signal in the SIMO system in the previous section. By contrast, the transmitter cannot apply the optimal MRT vector $\mathbf{p} = \frac{\mathbf{h}^*}{\|\mathbf{h}\|}$ in the corresponding MISO system since it does not know $\mathbf{h}$. However, a way to achieve a diversity gain in MISO systems is to use a *space-time block code (STBC)*. We will provide a few basic examples to introduce the main characteristics while we refer to [62] for a textbook dedicated to the topic.

The received signal of a MISO system with linear precoding was given in (3.41) as $y = \mathbf{h}^{\mathrm{T}}\mathbf{p}\bar{x} + n$. If the transmitter selects a fixed unit-norm precoding vector $\mathbf{p}$ that is independent of the channel $\mathbf{h}$, then we obtain

$$\mathbf{h}^{\mathrm{T}}\mathbf{p} \sim \mathcal{N}_{\mathbb{C}}(0, \beta) \tag{5.62}$$

under i.i.d. Rayleigh fading with $\mathbf{h} \sim \mathcal{N}_{\mathbb{C}}(\mathbf{0}, \beta\mathbf{I}_M)$. This follows from the fact that the weighted sum of independent Gaussian random variables is also Gaussian distributed, and from that $\mathbb{E}\{|\mathbf{h}^{\mathrm{T}}\mathbf{p}|^2\} = \mathbf{p}^{\mathrm{H}}\mathbb{E}\{\mathbf{h}^*\mathbf{h}^{\mathrm{T}}\}\mathbf{p} = \beta\mathbf{p}^{\mathrm{H}}\mathbf{I}_M\mathbf{p} = \beta$. The effective SISO channel $\mathbf{h}^{\mathrm{T}}\mathbf{p}$ that we obtained has the same fading distribution as the SISO channel we analyzed earlier in this chapter; thus, there is no additional diversity. To achieve transmit diversity, we need a more intricate transmission scheme than precoding in a fixed direction.

The technique for achieving the maximum transmit diversity with $M = 2$ is known as the *Alamouti code* because it was first proposed by Alamouti in [36]. The main idea is to transmit the same set of two data symbols two times using different precoding. The precoding vectors are not selected based on the channel $\mathbf{h} = [h_1, h_2]^{\mathrm{T}}$ but in a clever way that works for any realization of the channel and yet enables the receiver to separate the data symbols.

We consider two consecutive transmissions over the MISO channel in (3.35) with time indices $l = 1$ and $l = 2$:

$$y[1] = \sum_{m=1}^{M} h_m x_m[1] + n[1], \tag{5.63}$$

$$y[2] = \sum_{m=1}^{M} h_m x_m[2] + n[2], \tag{5.64}$$

where $y[l]$ is the received signal at time l, $x_m[l]$ is the transmitted signal from the mth antenna, and $n[l]$ is the noise. We can write this entire system in matrix form as

$$\underbrace{\begin{bmatrix} y[1] \\ y[2] \end{bmatrix}}_{=\mathbf{y}} = \underbrace{\begin{bmatrix} x_1[1] & x_2[1] \\ x_1[2] & x_2[2] \end{bmatrix}}_{=\mathbf{X}} \underbrace{\begin{bmatrix} h_1 \\ h_2 \end{bmatrix}}_{=\mathbf{h}} + \underbrace{\begin{bmatrix} n[1] \\ n[2] \end{bmatrix}}_{=\mathbf{n}}. \tag{5.65}$$

The data symbols should be embedded into the matrix $\mathbf{X}$, where each column represents the signals transmitted at a specific antenna (the space dimension), and each row contains the signals transmitted simultaneously (the time dimension). We want to send the two data symbols $\bar{x}[1]$ and $\bar{x}[2]$ over the two considered time instances. Ideally, we would send one after the other using MRT with $\mathbf{X} = [\bar{x}[1], \ \bar{x}[2]]^{\mathrm{T}} \frac{\mathbf{h}^{\mathrm{H}}}{\|\mathbf{h}\|}$ but this requires channel knowledge at the transmitter. Alamouti proposed a way to achieve a similar result without having to know the channel by embedding $\bar{x}[1]$ and $\bar{x}[2]$ into $\mathbf{X}$ as

$$\mathbf{X} = \frac{1}{\sqrt{2}} \begin{bmatrix} \bar{x}[1] & \bar{x}[2] \\ -\bar{x}^*[2] & \bar{x}^*[1] \end{bmatrix}. \tag{5.66}$$

The scaling factor $1/\sqrt{2}$ in front of the matrix ensures that the transmit power at each time instance equals the average power of the data symbols. This signal matrix does not depend on the channel realization $\mathbf{h}$. Each column in $\mathbf{X}$ represents what one of the antennas transmits over two different time

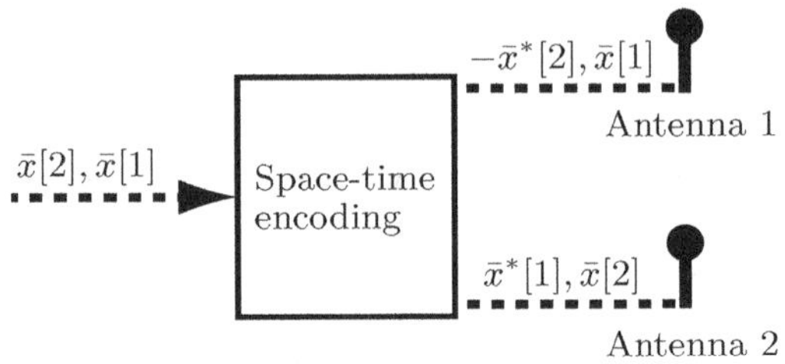

Figure 5.14: The Alamouti space-time encoding takes a sequence of two data symbols $\bar{x}[1], \bar{x}[2]$ and transmits them over two antennas according to (5.66).

instances. In the first instance, the first antenna sends the first symbol, and the second one sends the second symbol. Next, the antennas transmit the opposite symbols with complex conjugates and a minus sign on one of the antennas. The operation of taking a block of data symbols and mapping them to the antennas over a time block is called *space-time encoding*. Figure 5.14 illustrates the encoding operation for the Alamouti code.

The pattern of how the data symbols are mapped to the antennas in $\mathbf{X}$ is carefully designed so that (5.65) can be written as

$$\mathbf{y} = \mathbf{X}\mathbf{h} + \mathbf{n} = \frac{1}{\sqrt{2}} \begin{bmatrix} \bar{x}[1] & \bar{x}[2] \\ -\bar{x}^*[2] & \bar{x}^*[1] \end{bmatrix} \begin{bmatrix} h_1 \\ h_2 \end{bmatrix} + \mathbf{n}$$
$$= \frac{1}{\sqrt{2}} \begin{bmatrix} h_1\bar{x}[1] + h_2\bar{x}[2] \\ -h_1\bar{x}^*[2] + h_2\bar{x}^*[1] \end{bmatrix} + \mathbf{n}. \tag{5.67}$$

If we take the conjugate of the second row in (5.67), we obtain

$$\underbrace{\begin{bmatrix} y[1] \\ y^*[2] \end{bmatrix}}_{=\bar{\mathbf{y}}} = \frac{1}{\sqrt{2}} \begin{bmatrix} h_1\bar{x}[1] + h_2\bar{x}[2] \\ -h_1^*\bar{x}[2] + h_2^*\bar{x}[1] \end{bmatrix} + \underbrace{\begin{bmatrix} n[1] \\ n^*[2] \end{bmatrix}}_{=\bar{\mathbf{n}}}$$
$$= \underbrace{\frac{1}{\sqrt{2}} \begin{bmatrix} h_1 & h_2 \\ h_2^* & -h_1^* \end{bmatrix}}_{=\bar{\mathbf{H}}} \begin{bmatrix} \bar{x}[1] \\ \bar{x}[2] \end{bmatrix} + \bar{\mathbf{n}}. \tag{5.68}$$

By comparing (5.68) with (3.56), we notice that it has the same form as a 2×2 MIMO system with the channel matrix $\bar{\mathbf{H}}$. In fact, the Alamouti code has been selected so that this matrix has orthogonal columns. This implies that the receiver observes the two signals in two different orthogonal dimensions of the vector space so that the signals can be distinguished without mutual interference. The SVD $\bar{\mathbf{H}} = \bar{\mathbf{U}}\bar{\boldsymbol{\Sigma}}\bar{\mathbf{V}}^{\mathrm{H}}$ of the channel matrix in (5.68) has the simple form

$$\bar{\mathbf{H}} = \underbrace{\frac{1}{\|\mathbf{h}\|} \begin{bmatrix} h_1 & h_2 \\ h_2^* & -h_1^* \end{bmatrix}}_{=\bar{\mathbf{U}}} \underbrace{\begin{bmatrix} \frac{\|\mathbf{h}\|}{\sqrt{2}} & 0 \\ 0 & \frac{\|\mathbf{h}\|}{\sqrt{2}} \end{bmatrix}}_{=\bar{\boldsymbol{\Sigma}}} \underbrace{\begin{bmatrix} 1 & 0 \\ 0 & 1 \end{bmatrix}}_{=\bar{\mathbf{V}}^{\mathrm{H}}}, \tag{5.69}$$

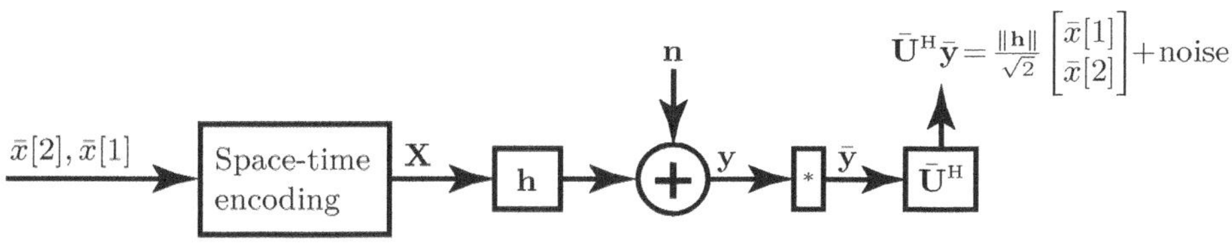

Figure 5.15: A block diagram of the transmission and reception using the Alamouti code. The transmitter maps two data symbols into a transmission block $\mathbf{X}$, as detailed in Figure 5.14. The receiver conjugates the second received signal to obtain $\bar{\mathbf{y}}$ in (5.68). It then acts as a MIMO receiver that uses the left singular vectors from (5.69) to decouple the two transmitted symbols.

where both singular values are equal to $\|\mathbf{h}\|/\sqrt{2}$. If we multiply $\bar{\mathbf{y}}$ in (5.68) with $\bar{\mathbf{U}}^{\mathrm{H}}$ from the left, we decouple the reception into two parallel channels:

$$\bar{\mathbf{U}}^{\mathrm{H}}\bar{\mathbf{y}} = \frac{\|\mathbf{h}\|}{\sqrt{2}} \begin{bmatrix} \bar{x}[1] \\ \bar{x}[2] \end{bmatrix} + \bar{\mathbf{U}}^{\mathrm{H}}\bar{\mathbf{n}}, \tag{5.70}$$

where $\bar{\mathbf{U}}^{\mathrm{H}}\bar{\mathbf{n}} \sim \mathcal{N}_{\mathbb{C}}(\mathbf{0}, N_0\mathbf{I}_2)$. Since the signal values are equally large, it is optimal to allocate the transmit power equally between $\bar{x}[1]$ and $\bar{x}[2]$:

$$\begin{bmatrix} \bar{x}[1] \\ \bar{x}[2] \end{bmatrix} \sim \mathcal{N}_{\mathbb{C}}(\mathbf{0}, q\mathbf{I}_2). \tag{5.71}$$

The transmitter sends these signals without utilizing the channel coefficients since $\bar{\mathbf{V}}$ in (5.70) is an identity matrix. The block diagram shown in Figure 5.15 summarizes the space-time encoding and decoding. Two data symbols are mapped to $\mathbf{X}$, which is then sent over the channel over two different time instances to obtain $\mathbf{y} = \mathbf{X}\mathbf{h} + \mathbf{n}$. At the receiver, the second entry of $\mathbf{y}$ is conjugated to obtain $\bar{\mathbf{y}}$, which is then multiplied by $\bar{\mathbf{U}}^{\mathrm{H}}$ that originates from the SVD in (5.69). This decouples the transmission into two parallel channels, each having the channel coefficient $\|\mathbf{h}\|/\sqrt{2}$ and independent additive noise with variance N_0. The data can be encoded and decoded separately over these channels; thus, no non-linear processing is required.

Note that each symbol is transmitted with the power q since the total power over two channel instances is $2q$, and it should be equally distributed over $\bar{x}[1]$ and $\bar{x}[2]$. Hence, for a given channel realization $\mathbf{h}$, it follows from (3.75) that the (conditional) capacity over the two time instances is

$$C_{\mathbf{h}} = 2\log_2\left(1 + \frac{q\|\mathbf{h}\|^2}{2N_0}\right) \quad \text{bit per two symbols.} \tag{5.72}$$

Since we are used to expressing the capacity in bit/symbol, it is more convenient to rewrite (5.72) as

$$C_{\mathbf{h}} = \log_2\left(1 + \frac{q\|\mathbf{h}\|^2}{2N_0}\right) \quad \text{bit/symbol.} \tag{5.73}$$

If we compare (5.73) with the SIMO case in (5.50), we notice that the only difference is that we only get half the SNR in the MISO case. This is because the signals are transmitted isotropically instead of being beamformed towards the receiver; thus, the beamforming gain is lost. If we consider i.i.d. Rayleigh fading with $\mathbf{h} \sim \mathcal{N}_{\mathbb{C}}(\mathbf{0}, \beta \mathbf{I}_2)$, then the average SNR in (5.73) becomes $\mathbb{E}\{\frac{q\|\mathbf{h}\|^2}{2N_0}\} = q\beta/N_0$, which does not depend on the number of antennas. This is the way to verify that there is no beamforming gain.

Although the transmitter can construct $\mathbf{X}$ without knowing the channel, it cannot compute $C_{\mathbf{h}}$, so it does not know how much data to encode into $\bar{x}[1], \bar{x}[2]$. If the transmitter selects the rate R, the outage probability with i.i.d. Rayleigh fading can be computed as

$$
\begin{aligned}
P_{\text{out}}(R) &= \Pr\left\{ R > \log_2\left(1 + \frac{q\|\mathbf{h}\|^2}{2N_0}\right)\right\} \\
&= 1 - e^{-\frac{2(2^R-1)}{\text{SNR}}} \sum_{m=0}^{1} \frac{\left(\frac{2(2^R-1)}{\text{SNR}}\right)^m}{m!}
\end{aligned} \tag{5.74}
$$

by following the same integration-by-parts approach as in (5.53). The average SNR is still defined as $\text{SNR} = \frac{q\beta}{N_0}$, but only half of this value is achieved when communicating in this way. Moreover, the compact upper bound

$$
P_{\text{out}}(R) \leq \left(\frac{2N_0\left(2^R - 1\right)}{q\beta}\right)^2 \frac{1}{2!} = \frac{1}{2}\left(\frac{2\left(2^R - 1\right)}{\text{SNR}}\right)^2 \tag{5.75}
$$

can be obtained by following the same steps as in (5.54)-(5.56). Recall from Section 5.3.2 that this bound is tight at high SNRs. We can see in (5.75) that the outage probability reduces as SNR^{-2}. This means the Alamouti code achieves a diversity gain of order $M = 2$, the same maximum diversity order as in the SIMO case with the matching number of antennas.

Figure 5.16 shows the outage probability for a channel with i.i.d. Rayleigh fading and the desired rate $R = 1$ bit/symbol. We compare a SISO system ($M = 1$) with the receive diversity obtained by a SIMO system ($M = 2$) and the transmit diversity obtained by a MISO system ($M = 2$) using the Alamouti code. The diversity gains are clearly visible: The outage probabilities decay as SNR^{-M}. This demonstrates that the diversity orders obtained by transmit and receive diversity are the same. However, there is a 3 dB gap between the curves. This is because receive diversity also gives rise to a beamforming gain that doubles the SNR, corresponding to a 3 dB improvement. This can be observed mathematically by comparing (5.75) with (5.56) for $M = 2$, where the only difference is that the SNR is divided by two in the MISO case.

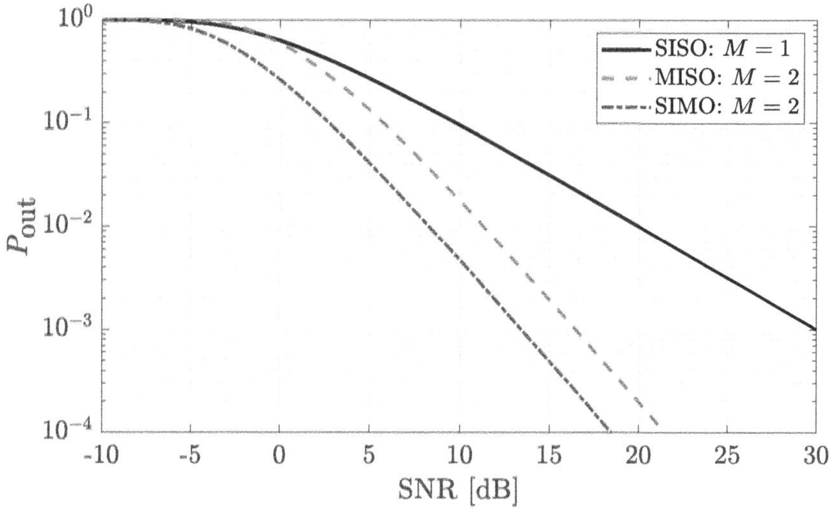

Figure 5.16: The outage probability of MISO and SIMO systems with $M = 2$ and i.i.d. Rayleigh fading is compared with the corresponding SISO system. The rate is $R = 1$ bit/symbol. The MISO and SIMO systems achieve the same diversity order, but the MISO system has a 3 dB worse SNR since it cannot obtain a beamforming gain.

Example 5.8. Show that $\mathbf{XX}^{\mathrm{H}}$ is a scaled identity matrix when using the Alamouti code. How is $\mathbb{E}\{\mathrm{tr}(\mathbf{XX}^{\mathrm{H}})\}$ related to the transmit power?

We can compute the matrix product using (5.66), which leads to

$$\mathbf{XX}^{\mathrm{H}} = \frac{1}{2}\begin{bmatrix} \bar{x}[1] & \bar{x}[2] \\ -\bar{x}^*[2] & \bar{x}^*[1] \end{bmatrix}\begin{bmatrix} \bar{x}^*[1] & -\bar{x}[2] \\ \bar{x}^*[2] & \bar{x}[1] \end{bmatrix} = \frac{1}{2}\left(|\bar{x}[1]|^2 + |\bar{x}[2]|^2\right)\mathbf{I}_2. \quad (5.76)$$

This is a scaled identity matrix, which implies that the rows of $\mathbf{X}$ are orthogonal. All STBCs that satisfy this condition are called *orthogonal* and share the property that the receiver can separate the transmitted signals without interference, as was the case for the Alamouti code. The scaling factor in (5.76) ensures that $\mathrm{tr}(\mathbf{XX}^{\mathrm{H}}) = |\bar{x}[1]|^2 + |\bar{x}[2]|^2$, which implies that $\mathbb{E}\{\mathrm{tr}(\mathbf{XX}^{\mathrm{H}})\} = 2q$ so that the total power of one block equals the power q per symbol times the length of the block.

The Alamouti code is designed for transmitting two data symbols over $M = 2$ antennas, but there are orthogonal STBCs crafted for larger numbers of transmit antennas and more symbols. The code design is nontrivial and has attracted much research attention over the past decades, starting with [37]. There are three main design parameters: i) the number of transmit antennas, ii) the number of time instances the code is transmitted over, and iii) how many data symbols are embedded. It is also essential to ensure that the symbols are assigned an equal fraction of the transmit power and obtain the maximum diversity order. When having $M > 2$ transmit antennas, only codes that use more time instances than there are symbols exist, and it can be proved that the fraction cannot surpass 3/4 [62]. The Alamouti code is the only code with as many symbols as there are time instances.[7]

We will conclude this section by describing an orthogonal code from [63] designed for $M = 4$ antennas, which we will refer to as the *Ganesan code* since Ganesan and Stoica proposed it. The code transmits the 3 data symbols $\bar{x}[1], \bar{x}[2], \bar{x}[3]$ over 4 time instances, leading to the *coding rate* $n_{\mathrm{r}} = 3/4$. The code matrix is

$$\mathbf{X} = \frac{1}{\sqrt{3}} \begin{bmatrix} \bar{x}[1] & 0 & \bar{x}[2] & -\bar{x}[3] \\ 0 & \bar{x}[1] & \bar{x}^*[3] & \bar{x}^*[2] \\ -\bar{x}^*[2] & -\bar{x}[3] & \bar{x}^*[1] & 0 \\ \bar{x}^*[3] & -\bar{x}[2] & 0 & \bar{x}^*[1] \end{bmatrix} \tag{5.77}$$

and has the property $\mathbf{X}\mathbf{X}^{\mathrm{H}} = \frac{1}{3}\left(|\bar{x}[1]|^2 + |\bar{x}[2]|^2 + |\bar{x}[3]|^2\right)\mathbf{I}_4$ that is expected from Example 5.8. The scaling factor in front of the matrix ensures that the transmit power at each time instance equals the average power of the data symbols. Each row of $\mathbf{X}$ has the same norm; thus, the transmit power is divided equally over time. We also notice that each symbol is transmitted from each antenna, which is a prerequisite for achieving maximum diversity. The received signal over the four different time instances can be expressed as

$$\underbrace{\begin{bmatrix} y[1] \\ y[2] \\ y[3] \\ y[4] \end{bmatrix}}_{=\mathbf{y}} = \mathbf{X}\mathbf{h} + \underbrace{\begin{bmatrix} n[1] \\ n[2] \\ n[3] \\ n[4] \end{bmatrix}}_{=\mathbf{n}}, \tag{5.78}$$

where $\mathbf{h} = [h_1, h_2, h_3, h_4]^{\mathrm{T}}$ is the channel response, $y[l]$ is the received signal at time l, and $n[l]$ is receiver noise at time l, for $l = 1, \ldots, 4$. Since the data symbols appear in (5.77) both with and without complex conjugates, it is convenient for decoding purposes to extend the system model in (5.78) also to include the conjugates of the received signals:

$$\underbrace{\begin{bmatrix} \mathbf{y} \\ \mathbf{y}^* \end{bmatrix}}_{=\bar{\mathbf{y}}} = \begin{bmatrix} \mathbf{X}\mathbf{h} \\ \mathbf{X}^*\mathbf{h}^* \end{bmatrix} + \underbrace{\begin{bmatrix} \mathbf{n} \\ \mathbf{n}^* \end{bmatrix}}_{=\bar{\mathbf{n}}} = \underbrace{\frac{1}{\sqrt{3}} \begin{bmatrix} h_1 & h_3 & -h_4 & 0 & 0 & 0 \\ h_2 & 0 & 0 & 0 & h_4 & h_3 \\ 0 & 0 & -h_2 & h_3 & -h_1 & 0 \\ 0 & -h_2 & 0 & h_4 & 0 & h_1 \\ 0 & 0 & 0 & h_1^* & h_3^* & -h_4^* \\ 0 & h_4^* & h_3^* & h_2^* & 0 & 0 \\ h_3^* & -h_1^* & 0 & 0 & 0 & -h_2^* \\ h_4^* & 0 & h_1^* & 0 & -h_2^* & 0 \end{bmatrix}}_{=\bar{\mathbf{H}}} \underbrace{\begin{bmatrix} \bar{x}[1] \\ \bar{x}[2] \\ \bar{x}[3] \\ \bar{x}^*[1] \\ \bar{x}^*[2] \\ \bar{x}^*[3] \end{bmatrix}}_{=\bar{\mathbf{x}}} + \bar{\mathbf{n}}. \tag{5.79}$$

[7]There exist non-orthogonal STBCs that contain as many data symbols as there are time slots, but these require more complicated receiver processing to deal with the resulting interference; we refer to [62] for details.

If we compare (5.79) with (3.56), we notice that it has the same form as an 8×6 MIMO system with the channel matrix $\bar{\mathbf{H}}$. The critical difference is that the last three symbols in $\bar{\mathbf{x}}$ are complex conjugates of the first three symbols, so they carry no extra information. All the columns of $\bar{\mathbf{H}}$ are mutually orthogonal and have norms equal to $\|\mathbf{h}\|/\sqrt{3}$, thanks to how the code matrix in (5.77) was designed. This implies that $\bar{\mathbf{H}}^{\mathrm{H}}\bar{\mathbf{H}} = \frac{\|\mathbf{h}\|^2}{3}\mathbf{I}_6$. Hence, if we multiply $\bar{\mathbf{y}}$ with $\frac{\sqrt{3}}{\|\mathbf{h}\|}\bar{\mathbf{H}}^{\mathrm{H}}$ (which has unit-norm rows) from the left, we obtain

$$\frac{\sqrt{3}}{\|\mathbf{h}\|}\bar{\mathbf{H}}^{\mathrm{H}}\bar{\mathbf{y}} = \frac{\|\mathbf{h}\|}{\sqrt{3}}\begin{bmatrix} \bar{x}[1] \\ \bar{x}[2] \\ \bar{x}[3] \\ \bar{x}^*[1] \\ \bar{x}^*[2] \\ \bar{x}^*[3] \end{bmatrix} + \frac{\sqrt{3}}{\|\mathbf{h}\|}\bar{\mathbf{H}}^{\mathrm{H}}\bar{\mathbf{n}}. \tag{5.80}$$

We notice that the data symbols $\bar{x}[1], \bar{x}[2], \bar{x}[3]$ are received separately in the first three entries of $\frac{\sqrt{3}}{\|\mathbf{h}\|}\bar{\mathbf{H}}^{\mathrm{H}}$ and exhibit a common channel coefficient of $\|\mathbf{h}\|/\sqrt{3}$. It can also be shown that the first three entries of the noise term in (5.80) are independent and have variance N_0, thanks to the normalization factor $\frac{\sqrt{3}}{\|\mathbf{h}\|}$ and that we never used a noise variable and its complex conjugate in the same expression. Since the channel gain is the same for all three symbols, it is optimal for the transmitter to allocate the power equally between them:

$$\begin{bmatrix} \bar{x}[1] \\ \bar{x}[2] \\ \bar{x}[3] \end{bmatrix} \sim \mathcal{N}_{\mathbb{C}}(\mathbf{0}, q\mathbf{I}_3). \tag{5.81}$$

It follows that $\mathbb{E}\{\mathrm{tr}(\mathbf{X}\mathbf{X}^{\mathrm{H}})\} = \frac{1}{3}(\mathbb{E}\{|\bar{x}[1]|^2 + |\bar{x}[2]|^2 + |\bar{x}[3]|^2\})\mathrm{tr}(\mathbf{I}_4) = 4q$, which is the power q times the length of the block. For a given channel realization $\mathbf{h}$, it follows from (3.75) that the (conditional) capacity is

$$C_{\mathbf{h}} = \frac{3}{4}\log_2\left(1 + \frac{q\|\mathbf{h}\|^2}{3N_0}\right) \quad \text{bit per symbol.} \tag{5.82}$$

The word "symbol" in the unit refers to the transmitted symbols, not the data symbols. Since we transmit a block of four symbols to transfer three data symbols, the coding rate $n_{\mathrm{r}} = 3/4$ appears as a pre-log factor in (5.82). Compared to the (conditional) capacity in the SIMO case in (5.50), we notice that the Ganesan code only gets a third of the SNR. There are two contributing factors: the lack of beamforming reduces the SNR by $1/M = 1/4$ while spreading three signals over four time slots increases the SNR by $1/n_{\mathrm{r}} = 4/3$. The combined effect is $1/(Mn_{\mathrm{r}}) = 1/3$.

The main reason to use an STBC is to achieve the maximum diversity order the channel can offer. The outage probability for a given rate R and i.i.d. Rayleigh fading can be computed as

$$P_{\text{out}}(R) = \Pr\left\{R > \frac{3}{4}\log_2\left(1+\frac{q\|\mathbf{h}\|^2}{3N_0}\right)\right\}$$
$$= 1 - e^{-\frac{3\left(2^{4R/3}-1\right)}{\text{SNR}}}\sum_{m=0}^{3}\frac{\left(\frac{3\left(2^{4R/3}-1\right)}{\text{SNR}}\right)^m}{m!} \tag{5.83}$$

by following the approach in (5.53). To enable comparison with the SISO case, the average SNR is still defined as $\text{SNR} = \frac{q\beta}{N_0}$, although only a third of it is achieved. An upper bound that is tight at high SNRs is obtained as

$$P_{\text{out}}(R) \leq \left(\frac{3\left(2^{4R/3}-1\right)}{\text{SNR}}\right)^4 \frac{1}{4!} \tag{5.84}$$

by following the same steps as in (5.54)-(5.56). We can see in (5.84) that the outage probability reduces as SNR^{-4}, which implies that the diversity order is $M = 4$, which is the same as in the SIMO case with four antennas.

Example 5.9. What diversity order is achieved by a repetition scheme where the same signal $x \sim \mathcal{N}_{\mathbb{C}}(0, q)$ is transmitted sequentially from M antennas over M time instances, using only one antenna at a time?

The received signal with this repetition scheme can be expressed as

$$\underbrace{\begin{bmatrix} y[1] \\ \vdots \\ y[M] \end{bmatrix}}_{=\mathbf{y}} = \underbrace{\begin{bmatrix} h_1 \\ \vdots \\ h_M \end{bmatrix}}_{=\mathbf{h}} x + \underbrace{\begin{bmatrix} n[1] \\ \vdots \\ n[M] \end{bmatrix}}_{=\mathbf{n}}, \tag{5.85}$$

if we transmit the signal from antenna m at time instance m to obtain the received signal $y[m]$. The system model in (5.85) has the same form as the SIMO system in (3.14). Hence, the (conditional) capacity is obtained from (3.22) as $C_{\mathbf{h}} = \frac{1}{M}\log_2(1+\frac{q\|\mathbf{h}\|^2}{N_0})$, where the pre-log factor $1/M$ represents that the same symbol is repeated over M time instances. Assuming i.i.d. Rayleigh fading, we can follow (5.54)-(5.56) to upper bound the outage probability as

$$P_{\text{out}}(R) = \Pr\left\{R > \frac{1}{M}\log_2\left(1+\frac{q\|\mathbf{h}\|^2}{N_0}\right)\right\} \leq \left(\frac{2^{MR}-1}{\text{SNR}}\right)^M \frac{1}{M!}. \tag{5.86}$$

The diversity order is M since the outage probability reduces as SNR^{-M}, which is the maximum value. The drawback is the inefficient use of time resources; transmitting one symbol over M time instances leads to a coding rate of only $1/M$. For this reason, the outage probability is proportional to $(2^{MR}-1)^M$ instead of $(2^R-1)^M$ as in the SIMO case in (5.56).

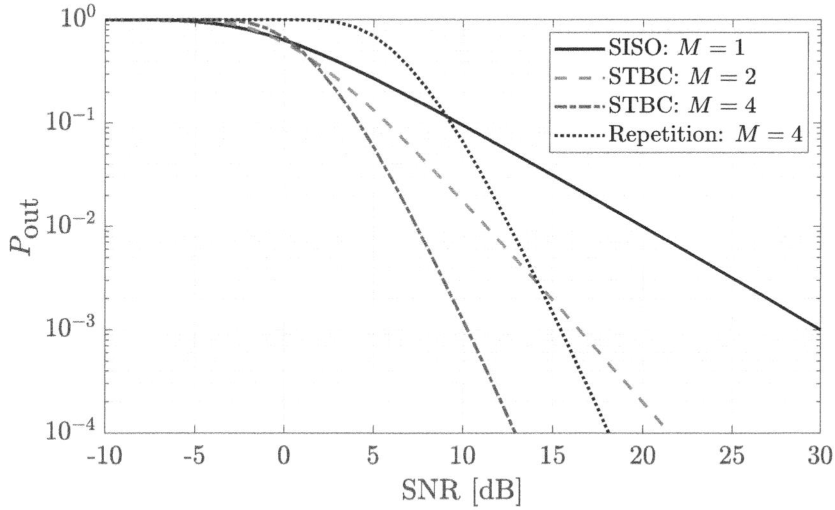

Figure 5.17: The outage probability of MISO systems with $M = 2$ or $M = 4$ antennas and i.i.d. Rayleigh fading is compared with the corresponding SISO system. The rate is $R = 1$ bit/symbol. The STBCs achieve diversity orders equal to the number of antennas and outperform the repetition scheme for the same number of antennas.

Figure 5.17 shows how the outage probability varies with the SNR when the rate is $R = 1$ bit/symbol and there is i.i.d. Rayleigh fading. We compare a SISO system ($M = 1$) with three schemes that achieve transmit diversity over MISO channels: The Alamouti code in (5.66) with $M = 2$, the Ganesan code in (5.77) with $M = 4$, and the repetition scheme from Example 5.9 with $M = 4$. The SISO system achieves the lowest outage probability when the SNR is very low, while the benefit of diversity becomes evident at medium to high SNR. Although the Alamouti code has a coding rate of $n_{\mathrm{r}} = 1$ and the Ganesan code only has $n_{\mathrm{r}} = 3/4$, the latter achieves a larger diversity order, which leads to a lower outage probability for SNRs above 1 dB. The repetition scheme's inefficiency is evident from the wide performance gap to the Ganesan code that uses the same number of antennas. However, it outperforms the Alamouti code at high SNRs thanks to the larger diversity order.

In conclusion, diversity is of utmost importance to achieve low outage probabilities, and it can be achieved at the transmitter side using STBCs.

5.3.4 Joint Transmit and Receive Diversity in MIMO Systems

When both the transmitter and receiver are equipped with multiple antennas, even better reliability against fading can be achieved through the simultaneous use of transmit and receive diversity. The MIMO channel matrix $\mathbf{H} \in \mathbb{C}^{M \times K}$ contains MK entries and, under i.i.d. Rayleigh fading, it can provide a diversity order up to MK. To this end, we must design a transmission scheme where the channel coefficient becomes proportional to $\|\mathbf{H}\|_{\mathrm{F}}$, which is the Frobenius norm of the matrix defined as follows.

Definition 5.1. The *Frobenius norm* of the matrix $\mathbf{H} \in \mathbb{C}^{M \times K}$ is defined as

$$\|\mathbf{H}\|_{\mathrm{F}} = \sqrt{\sum_{m=1}^{M} \sum_{k=1}^{K} |h_{m,k}|^2}, \tag{5.87}$$

where $h_{m,k}$ denotes the entry at the mth row in the kth column.

The Frobenius norm is a natural matrix extension of the Euclidean vector norm $\|\mathbf{h}\| = \sqrt{\sum_{m=1}^{M} |h_m|^2}$ since it also adds up the squared magnitudes of the entries. The subscript "F" is used in this book for clarity because alternative matrix norms are commonly used for matrix analysis. In this section, we will denote the kth column of $\mathbf{H}$ as $\mathbf{h}_k$, so that $\mathbf{H} = [\mathbf{h}_1, \ldots, \mathbf{h}_K]$.

The Frobenius norm is closely related to the trace and singular values $s_1, \ldots, s_r$ of $\mathbf{H}$ because the following two properties hold:

$$\|\mathbf{H}\|_{\mathrm{F}}^2 = \mathrm{tr}\left(\mathbf{H}^{\mathrm{H}}\mathbf{H}\right) = \sum_{k=1}^{K} \|\mathbf{h}_k\|^2, \tag{5.88}$$

$$\|\mathbf{H}\|_{\mathrm{F}}^2 = \sum_{k=1}^{r} s_k^2. \tag{5.89}$$

Under i.i.d. Rayleigh fading with $h_{m,k} \sim \mathcal{N}_{\mathbb{C}}(0, \beta)$, $\|\mathbf{H}\|_{\mathrm{F}}^2$ has the same distribution as the squared norm of a SIMO/MISO channel with MK antennas. Hence, the squared Frobenius norm has the scaled $\chi^2(2MK)$-distribution that was introduced in Section 2.2.5, which has the PDF

$$f_{\|\mathbf{H}\|_{\mathrm{F}}^2}(x) = \frac{x^{MK-1} e^{-\frac{x}{\beta}}}{\beta^{MK}(MK-1)!}, \quad \text{for } x \geq 0. \tag{5.90}$$

A straightforward way to achieve the maximum diversity order is to utilize the repetition scheme from Example 5.9. In this case, the same signal is transmitted over the K transmit antennas over K time instances, using only one transmit antenna at a time, while all M receive antennas are used continuously. The received signal $\mathbf{y}[k] \in \mathbb{C}^M$ at time instance k can be expressed as

$$\mathbf{y}[k] = \mathbf{h}_k x + \mathbf{n}[k], \quad k = 1, \ldots, K, \tag{5.91}$$

where x is the data symbol and $\mathbf{n}[k] \sim \mathcal{N}_{\mathbb{C}}(\mathbf{0}, N_0 \mathbf{I}_M)$ is the independent receiver noise. As the same symbol is repeated, the complete received signal can be expressed as

$$\begin{bmatrix} \mathbf{y}[1] \\ \vdots \\ \mathbf{y}[K] \end{bmatrix} = \begin{bmatrix} \mathbf{h}_1 \\ \vdots \\ \mathbf{h}_K \end{bmatrix} x + \begin{bmatrix} \mathbf{n}[1] \\ \vdots \\ \mathbf{n}[K] \end{bmatrix}, \tag{5.92}$$

which looks like a SIMO channel with an MK-dimensional channel vector containing all the columns of $\mathbf{H}$. It then follows from (3.22) that the (conditional) capacity of this channel is

$$C_{\mathbf{H}} = \frac{1}{K}\log_2\left(1+\frac{q\sum_{k=1}^{K}\|\mathbf{h}_k\|^2}{N_0}\right) = \frac{1}{K}\log_2\left(1+\frac{q\|\mathbf{H}\|_{\mathrm{F}}^2}{N_0}\right), \tag{5.93}$$

where the pre-log factor $1/K$ represents that the same data symbol is repeated over K time slots. The last expression follows from (5.88). If the transmitter selects the data rate R, then the outage probability can be computed as

$$\begin{aligned} P_{\text{out}}(R) &= \Pr\left\{R > \frac{1}{K}\log_2\left(1+\frac{q\|\mathbf{H}\|_{\mathrm{F}}^2}{N_0}\right)\right\} \\ &= 1 - e^{-\frac{2^{KR}-1}{\mathrm{SNR}}}\sum_{m=0}^{MK-1}\frac{\left(\frac{2^{KR}-1}{\mathrm{SNR}}\right)^m}{m!} \end{aligned} \tag{5.94}$$

by following the same approach as in (5.53) and defining the SNR as earlier. The diversity order becomes particularly visible in the upper bound

$$P_{\text{out}}(R) \leq \left(\frac{2^{KR}-1}{\mathrm{SNR}}\right)^{MK}\frac{1}{(MK)!} \tag{5.95}$$

that is obtained through the same steps as in (5.54)-(5.56). This expression shows that the outage probability reduces with an increasing SNR as SNR^{-MK} at high SNRs, where the bound is tight. Hence, the diversity order is MK.

The same diversity gain can be achieved using STBCs, which can be designed to outperform the repetition scheme for every given SNR value. The repetition scheme can be viewed as an inefficient STBC achieving maximum diversity but with an unnecessarily low coding rate of $n_{\mathrm{r}} = 1/K$. While each STBC is designed for a particular number of transmit antennas, they can be directly applied along with any number of receive antennas. For example, the Alamouti and Ganesan codes use $K = 2$ and $K = 4$ transmit antennas, respectively, while the number of receive antennas M can be arbitrary.

In the MISO case, the received signal in (5.67) with the Alamouti code has the form $\mathbf{y} = \mathbf{X}\mathbf{h} + \mathbf{n} \in \mathbb{C}^K$. In the MIMO case, the received signal $\mathbf{y}_m \in \mathbb{C}^K$ at receive antenna m can instead be expressed as

$$\mathbf{y}_m = \mathbf{X}\vec{\mathbf{h}}_m + \mathbf{n}_m, \quad m = 1, \ldots, M, \tag{5.96}$$

where $\mathbf{n}_m \sim \mathcal{N}_{\mathbb{C}}(\mathbf{0}, N_0\mathbf{I}_K)$ is the receiver noise and $\vec{\mathbf{h}}_m \in \mathbb{C}^K$ denotes the mth row of the channel matrix (the arrow notation points out that rows are horizontal):

$$\mathbf{H} = \begin{bmatrix} \vec{\mathbf{h}}_1^{\mathrm{T}} \\ \vdots \\ \vec{\mathbf{h}}_M^{\mathrm{T}} \end{bmatrix}. \tag{5.97}$$

By processing the received signal as described in Section 5.3.3, the counterpart to (5.70) for the mth receive antenna is the processed received signal

$$\frac{\|\vec{\mathbf{h}}_m\|}{\sqrt{2}}\begin{bmatrix}\bar{x}[1]\\ \bar{x}[2]\end{bmatrix}+\begin{bmatrix}\mathcal{N}_{\mathbb{C}}(0,N_0)\\ \mathcal{N}_{\mathbb{C}}(0,N_0)\end{bmatrix}. \tag{5.98}$$

At this stage, the receiver can use MRC to combine the signals over the M receive antennas, which will result in a summation of the channel gains:

$$\sum_{m=1}^{M}\left|\frac{\|\vec{\mathbf{h}}_m\|}{\sqrt{2}}\right|^2=\frac{\sum_{m=1}^{M}\|\vec{\mathbf{h}}_m\|^2}{2}=\frac{\|\mathbf{H}\|_{\mathrm{F}}^2}{2}. \tag{5.99}$$

Hence, we can reuse the outage probability expressions from earlier in this chapter but replace $\|\mathbf{h}\|^2$ with $\|\mathbf{H}\|_{\mathrm{F}}^2$. The diversity order with the Alamouti code increases from 2 to $2M$, and the outage probability in (5.74) becomes

$$P_{\mathrm{out}}(R)=\Pr\left\{R>\log_2\left(1+\frac{q\|\mathbf{H}\|_{\mathrm{F}}^2}{2N_0}\right)\right\}=1-e^{-\frac{2(2^R-1)}{\mathrm{SNR}}}\sum_{m=0}^{2M-1}\frac{\left(\frac{2(2^R-1)}{\mathrm{SNR}}\right)^m}{m!}. \tag{5.100}$$

The same approach of combining the signals over the M received signals can be used along with the Ganesan code, in which case the diversity order increases from 4 to $4M$, and the outage probability in (5.83) generalizes to

$$\begin{aligned}P_{\mathrm{out}}(R)&=\Pr\left\{R>\frac{3}{4}\log_2\left(1+\frac{q\|\mathbf{H}\|_{\mathrm{F}}^2}{3N_0}\right)\right\}\\&=1-e^{-\frac{3(2^{4R/3}-1)}{\mathrm{SNR}}}\sum_{m=0}^{4M-1}\frac{\left(\frac{3(2^{4R/3}-1)}{\mathrm{SNR}}\right)^m}{m!}.\end{aligned} \tag{5.101}$$

Example 5.10. What is the outage probability if the transmitter sends an independent data symbol with equal power from each antenna?

If the transmitter sends $\mathbf{x}=[x_1,\ldots,x_K]^{\mathrm{T}}\sim\mathcal{N}_{\mathbb{C}}(\mathbf{0},\frac{q}{K}\mathbf{I}_K)$, the receiver can decode the data streams sequentially as described in Section 3.4.3 with $\mathbf{P}=\mathbf{I}_K$ and $\mathbf{Q}=\frac{q}{K}\mathbf{I}_K$. The achievable data rate for x_i is stated in (3.115) as $\log_2(1+\frac{q}{KN_0}\mathbf{h}_i^{\mathrm{H}}\mathbf{C}_{i+1}^{-1}\mathbf{h}_i)$, where $\mathbf{C}_{i+1}=\mathbf{I}_M+\sum_{k=i+1}^{K}\frac{q}{KN_0}\mathbf{h}_k\mathbf{h}_k^{\mathrm{H}}$ and $\mathbf{h}_k$ is the kth column of $\mathbf{H}$. To reach the total data rate R, the transmitter can use the rate R/K for each stream. The resulting outage probability is

$$P_{\mathrm{out}}(R)=\Pr\left\{\frac{R}{K}>\min_{i\in\{1,\ldots,K\}}\log_2\left(1+\frac{q}{KN_0}\mathbf{h}_i^{\mathrm{H}}\mathbf{C}_{i+1}^{-1}\mathbf{h}_i\right)\right\} \tag{5.102}$$

because an outage occurs if at least one of the K streams does not support the rate R/K. Since each signal is sent from one antenna to M receive antennas, the diversity order is M instead of MK. Hence, this transmission scheme sacrifices diversity to achieve the maximum multiplexing gain of $\min(M,K)$.

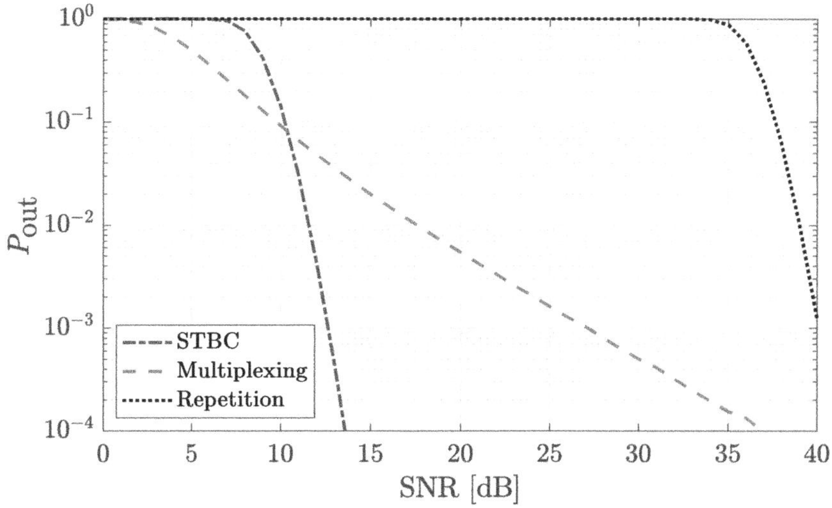

Figure 5.18: The outage probability of MIMO systems with $M = K = 4$ antennas and i.i.d. Rayleigh fading. The rate is $R = 4$ bit/symbol, and three schemes are compared. The STBC (Ganesan code) and repetition scheme achieve the maximum diversity order of 16; however, the former is more efficient thanks to the higher coding rate. Spatial multiplexing of four streams leads to lower outage probability at low and medium SNRs, but the reduced diversity order makes it less efficient than the STBC at higher SNRs.

Figure 5.18 exemplifies how the outage probabilities vary with the SNR in a MIMO system with $M = K = 4$ antennas. The data rate is $R = 4$ bit/symbol, which is selected to be larger than in previous examples since the system is now capable of spatial multiplexing. We compare the use of an STBC (Ganesan code) with spatial multiplexing of four parallel signals as in (5.102) and the repetition scheme in (5.94). Spatial multiplexing can more easily achieve large rates at moderate SNRs, which results in the lowest outage probability for SNRs below 10.3 dB. However, this scheme only achieves diversity order 4 since each symbol is transmitted from a single antenna and received by $M = 4$ antennas. The STBC gives a much lower outage probability at higher SNRs because it achieves the maximum diversity order of $MK = 16$. This is why the corresponding curve is extremely steep when it begins decaying. The repetition scheme also achieves the maximum diversity order. However, the curve is shifted by 27 dB to the right since it only transmits one symbol per four time instances, which is very inefficient compared to the Ganesan code.

This concluding example is a reminder that there are two conflicting design goals in slow-fading scenarios: Achieving a high data rate R and maintaining a low outage probability $P_{\text{out}}(R)$. In practice, the acceptable outage probability ϵ might be a given design parameter, and then the remaining goal is to achieve the ϵ-outage capacity. In the MIMO case, the preferred choice between spatial multiplexing and STBCs, or something in between, depends on ϵ. We refer to [26, Ch. 9] and [64] for a deeper theoretical framework for analyzing the diversity-multiplexing tradeoff.

5.4 Capacity Concept with Fast Fading

We now shift the attention to fast fading, where many channel realizations occur during the signal transmission. We begin by revisiting the memoryless SISO channel in (2.130), where the received signal at the time instance l is

$$y[l] = h[l] \cdot x[l] + n[l], \tag{5.103}$$

while $x[l]$ is the transmitted signal and $n[l] \sim \mathcal{N}_{\mathbb{C}}(0, N_0)$ is noise. The important new property in this section is that the channel coefficient $h[l]$ is time-dependent. For notational convenience, we will first consider the scenario where the channel takes a new independent realization at every time instance. The fading is so fast that the channel varies at the symbol rate but is constant within each symbol transmission interval. Later in this chapter, we will extend the analysis to the case where the channel is constant over a finite block of time instances before a new independent realization occurs.

The channel can be viewed as a continuous-time random process (see Section 2.2.7) from which we take samples at the symbol rate to obtain the sequence of channel realizations: $h[1], h[2], \ldots$. We assume the random process has zero mean and is stationary, which implies that each sample has the same statistics. We further assume that $h[l]$ is independent for each l and the variance is denoted by $\mathbb{E}\{|h[l]|^2\} = \beta$. The receiver knows the channel realizations perfectly, while the transmitter only knows the channel statistics. The fast channel variations lead to massive diversity since many fading realizations are observed, which makes the capacity analysis much different from the slow-fading scenario; for example, we will demonstrate that outage-free communication can be achieved without CSI at the transmitter.

The capacity of a non-fading channel is $C = \max_{f_{\mathsf{x}}(x)}(\mathcal{H}(y) - \mathcal{H}(y|x))$ bit/symbol, as defined in (2.133). To understand its operational meaning, we need to consider the transmission of a packet containing L data symbols: $x[1], \ldots, x[L]$. If each symbol is encoded to represent C bits, then the error probability in the decoding at the receiver goes to zero as $L \to \infty$.

The original capacity expression has no time index since the channel response was assumed constant throughout the communication. Hence, we need to derive a different expression that can be applied to a fast-fading channel. The starting point is to transmit a packet of length L with the received signal given by (5.103) for $l = 1, \ldots, L$, which is affected by the independent fading realizations $h[1], \ldots, h[L]$. We let $f_{\mathsf{x}[1],\ldots,\mathsf{x}[L]}(x[1], \ldots, x[L])$ denote the joint PDF of the L data symbols $x[1], \ldots, x[L]$. The capacity can then be generalized as [65, Ch. 4]

$$C = \lim_{L \to \infty} \max_{f_{\mathsf{x}[1],\ldots,\mathsf{x}[L]}(x[1],\ldots,x[L])} R_L, \tag{5.104}$$

where the average data rate over L time instances is

$$R_L = \frac{1}{L}\Big(\mathcal{H}(y[1],\dots,y[L]|h[1],\dots,h[L]) \\ - \mathcal{H}(y[1],\dots,y[L]|x[1],\dots,x[L],h[1],\dots,h[L])\Big) \tag{5.105}$$

and the differential entropies are conditioned on the L channel realizations. Since the channel is memoryless and the channel realizations are independent, the differential entropies in (5.105) can be decoupled using the chain rule in (2.136) as

$$\mathcal{H}(y[1],\dots,y[L]|h[1],\dots,h[L]) \leq \sum_{l=1}^{L} \mathcal{H}(y[l]|h[l]), \tag{5.106}$$

$$\mathcal{H}(y[1],\dots,y[L]|x[1],\dots,x[L],h[1],\dots,h[L]) = \sum_{l=1}^{L} \mathcal{H}(y[l]|x[l],h[l]), \tag{5.107}$$

where the upper bound in (5.106) is achieved when the symbols $x[1],\dots,x[L]$ are designed to be independent so that the received signals $y[l]$ are conditionally independent (given $h[l]$) for $l = 1,\dots,L$. Since the capacity in (5.104) is achieved by maximizing R_L with respect to the symbol distribution $f_{\mathsf{x}[1],\dots,\mathsf{x}[L]}(x[1],\dots,x[L])$, we should let the L signals be independent so the upper bound is achieved. By inserting these expressions into (5.104) and (5.105), we obtain

$$\begin{aligned} C &= \lim_{L\to\infty} \frac{1}{L} \sum_{l=1}^{L} \max_{f_{\mathsf{x}[l]}(x[l])} \left(\mathcal{H}(y[l]|h[l]) - \mathcal{H}(y[l]|x[l],h[l])\right) \\ &= \lim_{L\to\infty} \frac{1}{L} \sum_{l=1}^{L} \log_2\left(1 + \frac{q|h[l]|^2}{N_0}\right). \end{aligned} \tag{5.108}$$

In the last step, we utilized the capacity result in Corollary 2.1 to conclude that $x[l] \sim \mathcal{N}_{\mathbb{C}}(0,q)$ is the optimal symbol distribution, which leads to an expression having the familiar $\log_2(1+\mathrm{SNR})$ structure.

It remains to compute the limit in (5.108). We notice that this expression is the sample average of the data rate $\log_2(1+\frac{q|h[l]|^2}{N_0})$. Since the channel realizations are independent, the rate realizations are also independent. The limit of the sample average of independent and identically distributed realizations can be computed using Lemma 2.4, which is known as the law of large numbers. As long as the random data rates have finite variance, we can use this lemma to establish that

$$C = \lim_{L\to\infty} \frac{1}{L} \sum_{l=1}^{L} \log_2\left(1 + \frac{q|h[l]|^2}{N_0}\right) = \mathbb{E}\left\{\log_2\left(1 + \frac{q|h|^2}{N_0}\right)\right\}, \tag{5.109}$$

where the sample average converges to the statistical mean. This is the mean of the conditional capacity C_h in (5.37) with respect to the fading channel h, which is the only random variable in the expression. The expression in (5.109) holds for any practical fading distribution because the variance is always finite.[8] However, we will mainly analyze the Rayleigh fading case where $h \sim \mathcal{N}_{\mathbb{C}}(0, \beta)$. We summarize the capacity result as follows.

Corollary 5.1. Consider the discrete memoryless fast-fading channel with input $x[l] \in \mathbb{C}$ and output $y[l] \in \mathbb{C}$ given by

$$y[l] = h[l] \cdot x[l] + n[l], \tag{5.110}$$

where $n[l] \sim \mathcal{N}_{\mathbb{C}}(0, N_0)$ is independent noise. Suppose the input distribution is feasible whenever the symbol power satisfies $\mathbb{E}\{|x[l]|^2\} \leq q$. Furthermore, suppose the channel $h[l]$ takes independent and identically distributed realizations at every time instance l from a distribution with finite variance. If the channel realizations are known at the output, the channel capacity is

$$C = \mathbb{E}\left\{\log_2\left(1 + \frac{q|h|^2}{N_0}\right)\right\} \quad \text{bit/symbol} \tag{5.111}$$

and is achieved when $x[l] \sim \mathcal{N}_{\mathbb{C}}(0, q)$ and independent for each l.

The term *ergodic* is used in statistics to describe random processes for which the time average approaches the statistical mean. This property was used in (5.109) to obtain the capacity as the mean value of $\log_2(1 + q|h|^2/N_0)$. The proof was based on Lemma 2.4 (the law of large numbers) and utilized the assumption that the temporal channel realizations $h[1], h[2], \ldots$ are mutually independent. This is a sufficient but unnecessarily strong assumption. The same capacity expression is obtained when the channel realizations are samples from any ergodic random process, which generally features a weak temporal correlation that vanishes with time. For example, suppose the receiver moves at the speed v along a straight line in a rich multipath environment, and the data symbol l is transmitted at time $t = l/B$, where B is the bandwidth. It follows from (5.33) that the temporal correlation between the channel coefficients $h[1]$ and $h[l]$ is

$$\mathbb{E}\left\{h[1]h^*[l]\right\} = \beta \operatorname{sinc}\left(\frac{2v(l-1)}{\lambda B}\right), \tag{5.112}$$

which goes to zero as $l \to \infty$. A weak law of large numbers can be established under these conditions [66, Ex. 254]. The truly necessary condition is that

[8]The magnitude of the channel coefficient is upper bounded as $|h| \leq 1$ in practice because we cannot receive more power than was transmitted. This implies that the variance of h can also not be larger than 1.

all possible channel realizations are obtained over time, according to the underlying stationary statistical distribution, which ergodicity implies.

The capacity in (5.111) is called the *ergodic capacity* for the aforementioned reasons and this term is used to distinguish it from the conventional capacity of non-fading channels. Since the ergodic capacity in (5.111) is a deterministic constant that can be computed by the transmitter using only statistical knowledge (i.e., the distributions of the channel and noise), the transmitter can deduce how to encode the data without knowing the channel realizations. Hence, unlike the slow-fading scenario, the transmitter does not need to guess which data rate the channel supports. By encoding the data based on the ergodic capacity value, we can achieve reliable (outage-free) communications as in the non-fading channels considered in previous chapters.

Example 5.11. Compute the ergodic capacity for a Rayleigh fading channel, using the exponential integral function $E_1(x) = \int_1^{\infty} e^{-xw}/w \, \partial w$.

We consider the channel distribution $h \sim \mathcal{N}_{\mathbb{C}}(0, \beta)$. The squared magnitude of a complex Gaussian random variable is exponentially distributed (see Section 2.2.5), thus the ergodic capacity in (5.111) can be expressed as

$$C = \mathbb{E}\{\log_2(1+z)\}, \tag{5.113}$$

where $z = \frac{q|h|^2}{N_0} \sim \text{Exp}(1/\text{SNR})$ and the average SNR is defined as $\text{SNR} = \frac{q\beta}{N_0}$. By using this distribution, we can compute the mean value in (5.113) as

$$\begin{aligned} C &= \int_0^{\infty} \log_2(1+z) \frac{1}{\text{SNR}} e^{-\frac{z}{\text{SNR}}} \partial z = e^{\frac{1}{\text{SNR}}} \int_1^{\infty} \log_2(w) \frac{1}{\text{SNR}} e^{-\frac{w}{\text{SNR}}} \partial w \\ &= \log_2(e) e^{\frac{1}{\text{SNR}}} \int_1^{\infty} \frac{1}{w} e^{-\frac{w}{\text{SNR}}} \partial w = \log_2(e) e^{\frac{1}{\text{SNR}}} E_1\left(\frac{1}{\text{SNR}}\right), \end{aligned} \tag{5.114}$$

where the second equality follows from a variable change to $w = 1 + z$, and the third equality follows from an integration-by-parts approach (where some terms are omitted since they equal zero). The final expression utilizes the definition of the exponential integral function $E_1(x)$. This is an established analytical function that is implemented in many software libraries.

We can get further analytical insights by bounding the function as $\frac{1}{2}\ln(1 + 2/x) < e^x E_1(x) < \ln(1 + 1/x)$ [67, §5.1.20], which implies that

$$\frac{1}{2}\log_2(1 + 2\,\text{SNR}) < C < \log_2(1 + \text{SNR}). \tag{5.115}$$

This chain of inequalities shows that the ergodic capacity is smaller than the capacity $\log_2(1 + \text{SNR})$ of the corresponding non-fading channel, but the relative loss cannot surpass 1/2 due to the structure of the lower bound.

The closed-form ergodic capacity expression for Rayleigh fading in (5.114) enables efficient numerical evaluation but is not amenable to analysis. We will therefore continue the comparison between a Rayleigh fading channel $h \sim \mathcal{N}_{\mathbb{C}}(0,\beta)$ and the capacity $\log_2(1+\frac{q\beta}{N_0})$ of the corresponding non-fading channel by starting from the mean value expression in (5.111). At low SNR, we can use the approximation in (3.2) to obtain that the capacity difference is

$$\mathbb{E}\left\{\log_2\left(1+\frac{q|h|^2}{N_0}\right)\right\} - \log_2\left(1+\frac{q\beta}{N_0}\right) \approx \log_2(e)\frac{q\mathbb{E}\{|h|^2\}}{N_0} - \log_2(e)\frac{q\beta}{N_0} = 0. \tag{5.116}$$

Consequently, there is no capacity loss from having a fading channel when the SNR is low. The reason is that the capacity is then an approximately linear function of $|h|^2$; the realizations where $|h|^2$ is below the average β are fully compensated by the realizations that are above the average. The fading sometimes makes the channel stronger and sometimes weaker, but it behaves like a non-fading channel on average.

The situation is more troublesome at high SNR, where the capacity difference (in bit/symbol) is

$$\begin{aligned}&\mathbb{E}\left\{\log_2\left(1+\frac{q|h|^2}{N_0}\right)\right\} - \log_2\left(1+\frac{q\beta}{N_0}\right)\\ &\approx \mathbb{E}\left\{\log_2\left(\frac{q|h|^2}{N_0}\right)\right\} - \log_2\left(\frac{q\beta}{N_0}\right) = \mathbb{E}\left\{\log_2\left(\frac{|h|^2}{\beta}\right)\right\} \approx -0.83,\end{aligned} \tag{5.117}$$

where the first approximation follows from (3.3) and the second approximation is obtained by computing the mean value numerically and presenting it with two significant digits. These results imply a negligible capacity loss in Rayleigh fading channels at low SNRs, while the loss approaches 0.83 bit/symbol at high SNRs. The reason for this loss is that the capacity grows slower and slower with the SNR since it is a logarithmic function of it. Hence, the realizations of $|h|^2$ that are below the average β incur a larger rate degradation than the realizations of $|h|^2$ that are above the average improve the rate.

These differences are illustrated in Figure 5.19, where the ergodic capacity of a Rayleigh fading channel is compared with the capacity of a non-fading channel when $\mathrm{SNR} = \frac{q\beta}{N_0}$ is the same. The performance difference is negligible when the SNR is below -10 dB, but then it begins to grow. The high-SNR gap of -0.83 bit/symbol is approximately achieved when the SNR is 30 dB. In summary, fading reduces communication performance, particularly at high SNR, where we want communication systems to operate. Fortunately, this adverse effect can be mitigated using multiple antennas, as shown next.

5.4.1 Ergodic Capacity of i.i.d. Rayleigh Fading SIMO Channels

The ergodic capacity is the mean value of the data rate achieved for a single channel realization. For a SIMO channel with M antennas, we can therefore

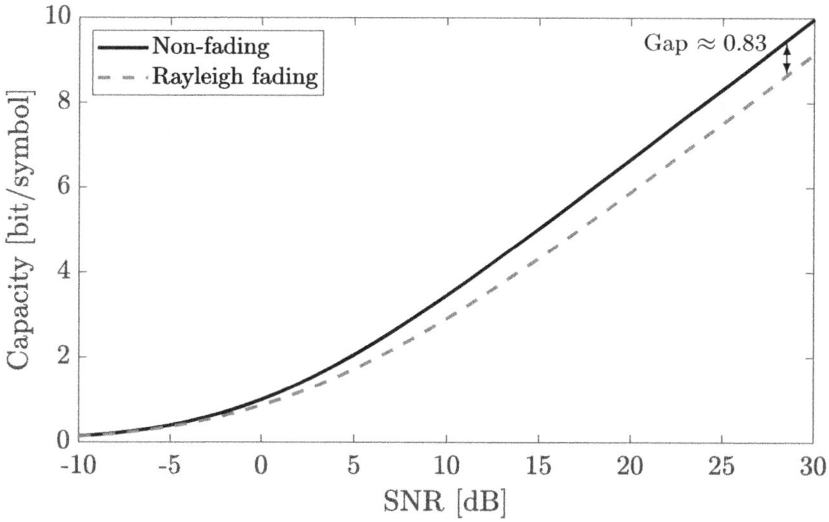

Figure 5.19: The ergodic capacity of a Rayleigh fading SISO system for a varying SNR is compared with the corresponding capacity of a non-fading channel. There is no gap at low SNR, while the capacity loss of having a fading channel approaches 0.83 bit/symbol at high SNR.

generalize the SISO ergodic capacity in (5.111) to the SIMO ergodic capacity

$$C = \mathbb{E}\left\{\log_2\left(1 + \frac{q\|\mathbf{h}\|^2}{N_0}\right)\right\} \quad \text{bit/symbol}. \tag{5.118}$$

This is the mean value of the conditional capacity $C_{\mathbf{h}}$ in (5.50). The expression in (5.118) holds for any channel distribution (with bounded variance). We consider an i.i.d. Rayleigh fading channel: $\mathbf{h} \sim \mathcal{N}_{\mathbb{C}}(\mathbf{0}, \beta\mathbf{I}_M)$. It is then possible to compute (5.118) exactly, following the approach in Example 5.11, but the expression is complicated and provides little insights.[9] We will instead compute lower and upper bounds on the ergodic capacity and compare them with the capacity $\log_2(1 + M\text{SNR})$ of the corresponding non-fading SIMO channel, for which $\|\mathbf{h}\|^2 = \beta M$ and we still use the definition $\text{SNR} = \frac{q\beta}{N_0}$. To this end, we will make use of a few mathematical properties.

Definition 5.2. A twice-differentiable scalar function $f(x)$ is said to be *convex* if $\frac{\partial^2}{\partial x^2} f(x) \geq 0$ for all x, while it is *concave* if $\frac{\partial^2}{\partial x^2} f(x) \leq 0$.

The graph of a convex function is shaped like a cup, ∪, in the sense that the line segment between any two points on the graph is above the graph. By contrast, the graph of a concave function is shaped like a cap, ∩, and has the opposite property. The expectation of a convex or concave function behaves differently, as shown by the following result, called *Jensen's inequality*.

[9]We refer to [1, Lemma B.15] for the complete result and derivation.

Lemma 5.1. Consider a scalar random variable x and scalar function $f(x)$. If $f(x)$ is a convex function, then

$$f(\mathbb{E}\{x\}) \leq \mathbb{E}\{f(x)\}. \tag{5.119}$$

If $f(x)$ is a concave function, then

$$f(\mathbb{E}\{x\}) \geq \mathbb{E}\{f(x)\}. \tag{5.120}$$

If we set $x = \|\mathbf{h}\|^2$, we can notice that $f(x) = \log_2(1 + \frac{qx}{N_0})$ is a concave function of x. Hence, we can use (5.120) to conclude that

$$\mathbb{E}\left\{\log_2\left(1 + \frac{q\|\mathbf{h}\|^2}{N_0}\right)\right\} \leq \log_2\left(1 + \frac{q\mathbb{E}\{\|\mathbf{h}\|^2\}}{N_0}\right) = \log_2\left(1 + \frac{qM\beta}{N_0}\right)$$
$$= \log_2(1 + M\mathrm{SNR}), \tag{5.121}$$

where we utilized that $\mathbb{E}\{\|\mathbf{h}\|^2\} = M\beta$. This upper bound coincides with the capacity of a non-fading LOS channel with the same average SNR. Hence, the ergodic capacity can never be larger than the capacity of a corresponding LOS channel.

To determine how much smaller the ergodic capacity can be, we can utilize Jensen's inequality again. This time we set $x = 1/\|\mathbf{h}\|^2$ and notice that $f(x) = \log_2(1 + \frac{1}{x}\frac{q}{N_0})$ is a convex function of x. It follows from (5.119) that

$$\mathbb{E}\left\{\log_2\left(1 + \frac{q\|\mathbf{h}\|^2}{N_0}\right)\right\} \geq \log_2\left(1 + \frac{q}{N_0\mathbb{E}\left\{\frac{1}{\|\mathbf{h}\|^2}\right\}}\right) = \log_2\left(1 + \frac{q(M-1)\beta}{N_0}\right)$$
$$= \log_2(1 + (M-1)\mathrm{SNR}), \tag{5.122}$$

where the first equality utilizes that

$$\mathbb{E}\left\{\frac{1}{\|\mathbf{h}\|^2}\right\} = \int_0^\infty \frac{1}{x} f_{\|\mathbf{h}\|^2}(x)\partial x = \int_0^\infty \frac{1}{x}\frac{x^{M-1}e^{-\frac{x}{\beta}}}{\beta^M(M-1)!}\partial x$$
$$= \frac{1}{(M-1)\beta}\underbrace{\int_0^\infty \frac{x^{M-2}e^{-\frac{x}{\beta}}}{\beta^{M-1}(M-2)!}\partial x}_{=1} = \frac{1}{(M-1)\beta}. \tag{5.123}$$

This mean value is computed by using the PDF in (5.51) of the scaled $\chi^2(2M)$-distribution, and the last equality follows by recognizing that the integral over the PDF of the scaled $\chi^2(2(M-1))$-distribution is one.

In summary, we have derived the following chain of inequalities:

$$\log_2(1+(M-1)\mathrm{SNR}) \leq \mathbb{E}\left\{\log_2\left(1 + \frac{q\|\mathbf{h}\|^2}{N_0}\right)\right\} \leq \log_2(1+M\mathrm{SNR}). \tag{5.124}$$

The gap between the lower and upper bounds can be substantial when M is small (as seen in the SISO case) but reduces as M increases. This is a consequence of the spatial receive diversity; as shown in (5.58), the relative variations in $\|\mathbf{h}\|^2$ reduce the more antennas are used. In the context of ergodic capacities, it is common to call it *channel hardening* [68]. This means that the influence of the random channel variations on the capacity disappears as M increases if MRC is used at the receiver. The fading still exists, and the entries of the channel vector $\mathbf{h}$ vary rapidly. However, the variations (partially) average out when the M independent random variables are combined in the SNR-maximizing way.

Example 5.12. What is the PDF of $C_{\mathbf{h}} = \log_2\left(1 + \frac{q\|\mathbf{h}\|^2}{N_0}\right)$ if $\mathbf{h} \sim \mathcal{N}_{\mathbb{C}}(\mathbf{0}, \beta\mathbf{I}_M)$?

The PDF $f_{\|\mathbf{h}\|^2}(x)$ of $\|\mathbf{h}\|^2$ is given in (5.51) and can be used to derive the PDF $f_{C_{\mathbf{h}}}(z)$ of $C_{\mathbf{h}}$. The connection between $\|\mathbf{h}\|^2$ and $C_{\mathbf{h}}$ is most easily exposed through the CDFs but, as stated in (2.101), the PDF is the derivative of the CDF. Hence, we can compute the desired PDF as

$$\begin{aligned}
f_{C_{\mathbf{h}}}(z) &= \frac{\partial}{\partial z}\Pr\{C_{\mathbf{h}} \leq z\} = \frac{\partial}{\partial z}\Pr\left\{\log_2\left(1 + \frac{q\|\mathbf{h}\|^2}{N_0}\right) \leq z\right\} \\
&= \frac{\partial}{\partial z}\Pr\left\{\|\mathbf{h}\|^2 \leq \frac{N_0}{q}\left(2^z - 1\right)\right\} \\
&= f_{\|\mathbf{h}\|^2}\left(\frac{N_0}{q}\left(2^z - 1\right)\right)\frac{\partial}{\partial z}\frac{N_0}{q}\left(2^z - 1\right) \\
&= \frac{(2^z - 1)^{M-1} e^{-\frac{2^z - 1}{\mathrm{SNR}}}}{\mathrm{SNR}^M (M-1)!} 2^z \ln(2), \quad \text{for } z \geq 0,
\end{aligned} \tag{5.125}$$

where the chain rule is used when computing the derivative and $\mathrm{SNR} = \frac{q\beta}{N_0}$ in the last step. This expression was used in Figure 5.10 for $M = 1$.

The ergodic capacity is $\mathbb{E}\{C_{\mathbf{h}}\}$ and its value depends on the PDF of $C_{\mathbf{h}}$, which is given in (5.125) for i.i.d. Rayleigh fading. Figure 5.20 shows this PDF with $M = 1$, $M = 8$, or $M = 32$ antennas. The probability mass is shifted to larger values as M increases, which results in a larger mean value (i.e., larger ergodic capacity). The mass also becomes concentrated in a smaller interval, which leads to a larger peak value of the PDF because the area under each curve is 1. It is this phenomenon that is called channel hardening.

Figure 5.21 shows the ergodic capacity of an i.i.d. Rayleigh fading channel with $\mathrm{SNR} = 10\,\mathrm{dB}$ and a varying number of antennas. The exact value is compared with the lower and upper bounds. We notice a huge gap between the curves at $M = 1$, where the lower bound is zero. However, for $M \geq 5$, the difference between the lower and upper bounds is tiny. More importantly, the ergodic capacity is close to the upper bound, representing the capacity of a

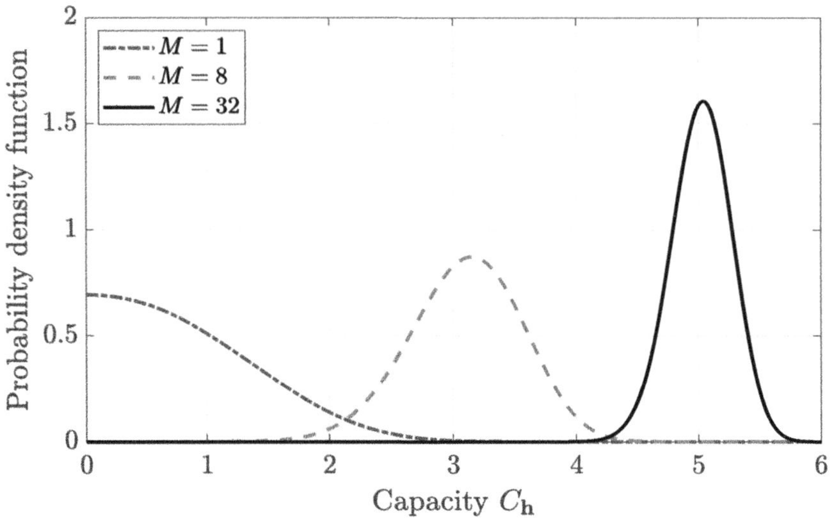

Figure 5.20: The PDFs of the (conditional) capacity $C_{\mathbf{h}}$ in (5.125) for SNR = 1 and $M = 1$, $M = 8$, or $M = 32$ antennas. As M increases, the probability mass shifts to the right and is concentrated in a smaller interval.

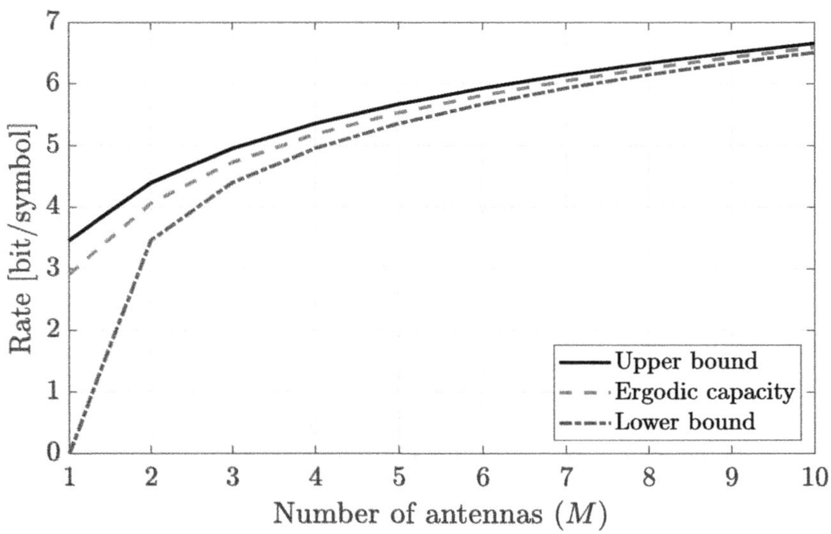

Figure 5.21: The ergodic SIMO capacity of an i.i.d. Rayleigh fading is shown for SNR = 10 dB and a varying number of antennas. It is compared with the lower and upper bounds in (5.124), where the upper bound corresponds to the capacity of a non-fading LOS channel.

non-fading LOS channel with the same SNR. Hence, the rate loss incurred by having a fading channel is relatively small when the receiver is equipped with a handful or more antennas. Hence, the spatial receive diversity makes the performance of communication systems more robust to channel fading.

5.4.2 Ergodic Capacity of i.i.d. Rayleigh Fading MIMO Channels

We now consider the MIMO setup where the transmitter has K antennas and the receiver has M antennas so that the fast-fading channel is described by the $M \times K$ channel matrix $\mathbf{H}$. The random realization of this channel matrix changes at every time instance. As earlier in this chapter, the receiver knows the realization of $\mathbf{H}$ but not the transmitter, which only knows the channel statistics. Hence, the transmitter must encode its data signal $\mathbf{x} = [x_1, \ldots, x_K]^{\mathrm{T}}$ in a way that is independent of $\mathbf{H}$, which implies that we cannot create multiple parallel channels using the SVD as in Section 3.4. Instead, we must follow the approach with arbitrary precoding from Section 3.4.3. Suppose the transmitted signal is $\mathbf{x} \sim \mathcal{N}_{\mathbb{C}}(\mathbf{0}, \mathbf{R}_x)$ for some arbitrary choice of the covariance matrix $\mathbf{R}_x$.[10] For a given realization of $\mathbf{H}$, we concluded earlier in (3.106) that an achievable (conditional) data rate is

$$\log_2\left(\det\left(\mathbf{I}_M + \frac{1}{N_0}\mathbf{H}\mathbf{R}_x\mathbf{H}^{\mathrm{H}}\right)\right). \tag{5.126}$$

By considering a transmission that takes place over infinitely many time instances with independent channel realizations, an ergodic achievable rate is

$$\mathbb{E}\left\{\log_2\left(\det\left(\mathbf{I}_M + \frac{1}{N_0}\mathbf{H}\mathbf{R}_x\mathbf{H}^{\mathrm{H}}\right)\right)\right\}, \tag{5.127}$$

where the mean value is computed with respect to the channel matrix $\mathbf{H}$. The argumentation for this result is the same as earlier in this chapter; the channel realizations are samples from a stationary and ergodic random process; thus, the time-average rate equals the statistical mean.

The capacity is the maximum achievable data rate. Although the transmitter does not know the channel realizations, it can compute the ergodic rate in (5.127) and adapt its choice of covariance matrix $\mathbf{R}_x$ to maximize it. We notice that the kth diagonal entry of $\mathbf{R}_x$ is $\mathbb{E}\{|x_k|^2\}$ and recall that $\mathrm{tr}(\mathbf{R}_x)$ is the sum of the diagonal entries. We want the signal power

$$\mathbb{E}\left\{\|\mathbf{x}\|^2\right\} = \sum_{k=1}^{K} \mathbb{E}\left\{|x_k|^2\right\} = \mathrm{tr}(\mathbf{R}_x) \tag{5.128}$$

to equal the maximum symbol power q. Hence, the ergodic capacity is

$$C = \max_{\mathbf{R}_x \in \mathbb{C}^{K\times K}:\, \mathrm{tr}(\mathbf{R}_x)=q} \mathbb{E}\left\{\log_2\left(\det\left(\mathbf{I}_M + \frac{1}{N_0}\mathbf{H}\mathbf{R}_x\mathbf{H}^{\mathrm{H}}\right)\right)\right\}, \tag{5.129}$$

where we need to find the positive semi-definite covariance matrix $\mathbf{R}_x$ that maximizes the ergodic rate under the power constraint $\mathrm{tr}(\mathbf{R}_x) = q$.

[10]This covariance matrix can be factorized as $\mathbf{R}_x = \mathbf{P}\mathbf{Q}\mathbf{P}^{\mathrm{H}}$ for some precoding matrix $\mathbf{P}$ and diagonal power allocation matrix $\mathbf{Q}$, following the definitions made in Section 3.4.3. However, this specific structure is not needed in this section.

The optimal covariance matrix depends on the distribution of the channel matrix. If we consider i.i.d. Rayleigh fading, meaning that all entries of $\mathbf{H}$ are independent complex Gaussian distributed with the zero mean and identical variance, then the optimal covariance matrix has the simple form

$$\mathbf{R}_x = \frac{q}{K}\mathbf{I}_K. \tag{5.130}$$

This means the transmitter should send one independent data symbol from each of its K antennas and divide the power q equally between them. We will outline the proof, but refer to [31] for the precise details.

For any choice of the covariance matrix $\mathbf{R}_x$, we can express its eigendecomposition as $\mathbf{R}_x = \mathbf{U}\mathbf{D}\mathbf{U}^{\mathrm{H}}$, where $\mathbf{U}$ is a unitary matrix and $\mathbf{D}$ is a diagonal matrix. The term $\mathbf{H}\mathbf{R}_x\mathbf{H}^{\mathrm{H}}$ in the ergodic capacity in (5.129) can therefore be expressed as $\mathbf{H}\mathbf{U}\mathbf{D}\mathbf{U}^{\mathrm{H}}\mathbf{H}^{\mathrm{H}} = (\mathbf{H}\mathbf{U})\mathbf{D}(\mathbf{H}\mathbf{U})^{\mathrm{H}}$. Since the entries of $\mathbf{H}$ are independent $\mathcal{N}_{\mathbb{C}}(0,\beta)$-distributed, the entries of $\mathbf{H}\mathbf{U}$ have the same distribution (see Example 2.10). Hence, when computing the mean with respect to the channel in (5.129), it is sufficient to consider diagonal covariance matrices $\mathbf{R}_x = \mathbf{D}$ because the choice of the unitary matrix makes no difference.

As all transmit antennas experience channels with the same distribution, it should not matter which antenna is assigned a specific amount of the total power; we can always reorder the antennas and get the same result. In particular, it can be proved that $\mathbb{E}\{\log_2(\det(\mathbf{I}_M + \frac{1}{N_0}\mathbf{H}\mathbf{D}\mathbf{H}^{\mathrm{H}}))\}$ is a jointly concave and symmetric function of the diagonal entries of $\mathbf{D}$, which implies that the maximum is achieved when all the entries are the same.

In summary, the ergodic capacity with i.i.d. Rayleigh fading is obtained by substituting (5.130) into (5.129):

$$C = \mathbb{E}\left\{\log_2\left(\det\left(\mathbf{I}_M + \frac{q}{KN_0}\mathbf{H}\mathbf{H}^{\mathrm{H}}\right)\right)\right\}. \tag{5.131}$$

We note that this result holds even if $K > M$, which means we transmit more signals than the receiver has antennas. This is not an issue since the transmitted data signals are encoded to enable decoding at the receiver side using the SIC procedure, even if the MIMO channel cannot be divided into parallel channels when the transmitter is unaware of the channel realizations. The multiplexing gain is, however, limited to $r = \min(M, K)$ because this is the maximum number of non-zero eigenvalues of $\mathbf{H}\mathbf{H}^{\mathrm{H}}$. If we denote these eigenvalues by $\lambda_1, \ldots, \lambda_r$, then (5.131) can also be expressed as

$$C = \mathbb{E}\left\{\log_2\left(\prod_{m=1}^{r}\left(1 + \frac{q}{KN_0}\lambda_m\right)\right)\right\} = \mathbb{E}\left\{\sum_{m=1}^{r}\log_2\left(1 + \frac{q}{KN_0}\lambda_m\right)\right\}, \tag{5.132}$$

by utilizing the fact that the determinant is the product of the eigenvalues.

The mean value in (5.132) can be computed in closed form at low SNR.

By utilizing the low-SNR approximation in (3.2), we can rewrite (5.132) as

$$C \approx \mathbb{E}\left\{\sum_{m=1}^{r} \log_2(e)\frac{q}{KN_0}\lambda_m\right\} = \log_2(e)\frac{q}{KN_0}\mathbb{E}\left\{\mathrm{tr}(\mathbf{H}\mathbf{H}^{\mathrm{H}})\right\}$$
$$= \log_2(e)\frac{q}{KN_0}MK\beta = \log_2(e)M\mathrm{SNR}. \tag{5.133}$$

The final result follows from the fact that the trace is defined as $\mathrm{tr}(\mathbf{H}\mathbf{H}^{\mathrm{H}}) = \sum_{m=1}^{r}\lambda_m$ and by computing the mean value $\mathbb{E}\{\mathbf{H}\mathbf{H}^{\mathrm{H}}\} = K\beta\mathbf{I}_M$ by using the assumption of i.i.d. Rayleigh fading. We can notice in (5.133) that the capacity is proportional to the number of receive antennas at low SNR (i.e., a receive beamforming gain), while the number of transmit antennas makes no difference. Hence, in the absence of CSI at the transmitter, multiple transmit antennas are only helpful in achieving multiplexing gains.

Example 5.13. How does the ergodic MIMO capacity with i.i.d. Rayleigh fading behave when $K \to \infty$, while M is fixed?

To answer this question, we need to determine the limit of $\frac{1}{K}\mathbf{H}\mathbf{H}^{\mathrm{H}}$ as $K \to \infty$. The mth diagonal entry of this matrix is $\frac{1}{K}\sum_{k=1}^{K}|h_{m,k}|^2$, which is the sample average of K i.i.d. random variables. The limit follows from the law of large numbers (Lemma 2.4) and equals the mean β of the individual terms $|h_{m,k}|^2$. Similarly, the (m,l)th entry is $\frac{1}{K}\sum_{k=1}^{K}h_{m,k}h_{l,k}^*$, where all terms are independent and have zero mean when $m \neq l$. It follows from the law of large numbers that the off-diagonal entries converge to zero. We have thereby proved that $\frac{1}{K}\mathbf{H}\mathbf{H}^{\mathrm{H}} \to \beta\mathbf{I}_M$. By substituting this result into (5.131), we obtain the asymptotic ergodic MIMO capacity

$$C = \log_2\left(\det\left(\mathbf{I}_M + \frac{q}{N_0}\beta\mathbf{I}_M\right)\right) = M\log_2(1+\mathrm{SNR}), \tag{5.134}$$

where the mean value is removed since the randomness vanishes as $K \to \infty$. This ergodic capacity is M times larger than the SISO capacity of the corresponding non-fading channel. The multiplexing gain is $\min(M,K) = M$.

We have now covered the ergodic capacities of SISO, SIMO, and MIMO channels. What remains is to consider the MISO channel, which is a special case of the MIMO channel with $M = 1$ receive antenna. In that special case, the channel matrix can be written as $\mathbf{H} = \mathbf{h}^{\mathrm{T}} = [h_1, \ldots, h_K]$. Hence, the ergodic capacity in (5.131) reduces to

$$C = \mathbb{E}\left\{\log_2\left(\det\left(\mathbf{I}_1 + \frac{q}{KN_0}\mathbf{h}^{\mathrm{T}}\mathbf{h}^*\right)\right)\right\} = \mathbb{E}\left\{\log_2\left(1 + \frac{q}{KN_0}\|\mathbf{h}\|^2\right)\right\}. \tag{5.135}$$

This capacity expression resembles the ergodic SIMO capacity in (5.118), but with a crucial difference: Instead of getting the sum of the channel gains

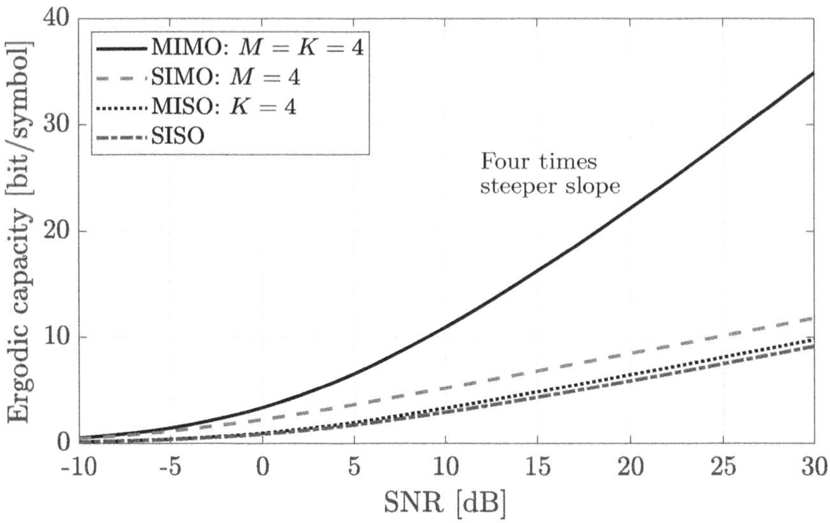

Figure 5.22: The ergodic capacity over i.i.d. Rayleigh fading channels. We compare a SISO channel with a SIMO channel with $M = 4$ antennas, a MISO channel with $K = 4$ antennas, and a MIMO channel with $M = K = 4$ antennas.

$\|\mathbf{h}\|^2 = \sum_{m=1}^{M} |h_m|^2$ of the receive antennas, we get the average $\|\mathbf{h}\|^2/K = \sum_{k=1}^{K} |h_k|^2/K$ of the channel gains among the transmit antennas. The division by K represents the absence of a beamforming gain in the MISO case, as we previously observed in slow fading. One way to interpret this is that the transmitter must spread its power over all K dimensions in $\mathbb{C}^K$ to ensure it reaches the receiver, even if only one randomly selected dimension leads to the receiver. By sending K independent signals from the antennas, the MISO channel achieves a diversity gain that provides channel hardening, making the setup preferable over a SISO channel.

We previously presented a duality result in Corollary 3.4, which stated that the capacity is the same in both directions of a MIMO channel if the ratio between the total transmit power and noise variance is the same. This result was obtained for a non-fading channel known at both the transmitter and the receiver. Considering the ergodic capacity in (5.131), the same result can only be obtained in the symmetric case of $M = K$, where there are equally many antennas in both directions. In contrast, we observed that the ergodic capacities of MISO and SIMO systems are very different because we can only obtain a beamforming gain at the receiver side.

Figure 5.22 shows the ergodic capacity as a function of the SNR in the case of i.i.d. Rayleigh fading channels. We compare four setups: a SISO case, a MISO case with $K = 4$ transmit antennas, a SIMO case with $M = 4$ receive antennas, and a MIMO case with $M = K = 4$ antennas. Note that all four cases can be computed using the MIMO expression in (5.131), but the mean value must be computed numerically. The results resemble those in

Figure 3.15 for non-fading channels, except that the SIMO and MISO cases are not equal anymore. The SISO case gives the lowest ergodic capacity. The MISO case is slightly better than the SISO case at high SNR, thanks to the diversity gain, but the performance gap is just around 2 dB. The SIMO case gives a curve having roughly the same shape as in the SISO setup, but it is shifted to the left thanks to both the beamforming gain and the diversity gain. The gap to the SISO curve is around 8 dB at high SNR, of which 6 dB is the beamforming gain and the remaining 2 dB is the diversity gain (same as in the MISO case). Note that the relative gain of having multiple receive antennas is greater for fading channels than for non-fading channels, for which there is only a beamforming gain. However, in absolute numbers, the ergodic capacities in Figure 5.22 are always smaller than the corresponding capacities of the non-fading channels in Figure 3.15. The highest capacity is achieved in the MIMO case, where a multiplexing gain is achieved since the transmitter sends four signals. At high SNR, the capacity curve grows as $M \log_2(\text{SNR})$, roughly M times faster than in the SISO case.

5.5 Block Fading and Channel Estimation

The capacity analysis in this book has thus far relied on the assumption that the receiver knows the channel perfectly. This is well motivated when the channel takes a single realization throughout the communication because the capacity is (by definition) achieved by transmitting very long data packets. Section 4.2.4 exemplified how a preamble containing a known pilot sequence can be appended to the packet to enable perfect channel estimation while still being negligibly small compared to the payload part that contains data. This argument holds in both scenarios with deterministic LOS channels (Chapter 4) and slow fading (Section 5.3), but not under fast fading. When the channel varies rapidly, we need to transmit new pilot sequences at the same pace as the channel changes, which has two main consequences. Firstly, the fraction of symbols spent on pilots instead of data is non-negligible. Secondly, the limited pilot sequence length leads to non-zero estimation errors, so perfect channel knowledge is no longer achieved. In this section, we will quantify the channel estimation errors and their detrimental impact on the ergodic capacity, following a methodology originating from [69], [70].

We return to the block fading model introduced when describing Figure 5.9. In this model, we treat the fading channel as being piecewise constant over short time intervals, but on every occasion the channel changes, the new realization is generated independently. An interval with a constant channel realization is called a *coherence block*. We let L_{c} denote the number of symbols that can be transmitted within each block. We must divide the symbols between transmitting a known pilot sequence that enables channel estimation at the receiver and unknown payload data that the receiver can decode using

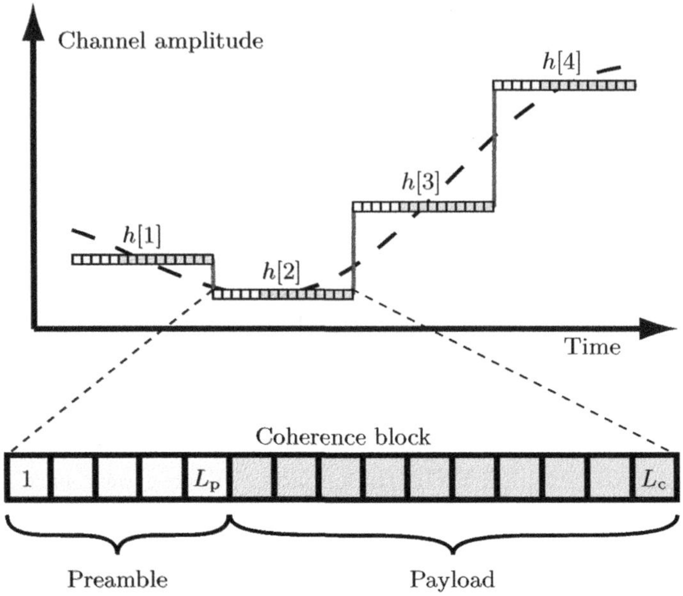

Figure 5.23: A block-fading channel has a piecewise constant channel response. Each such segment is called a coherence block. This is an abstraction of the dashed continuously varying channel. L_c symbols can be transmitted within each coherence block, of which L_p symbols are used to transmit a preamble containing a pilot sequence that enables channel estimation. The rest contains a payload with $L_c - L_p$ data symbols.

the acquired CSI. The resulting transmission protocol is repeated in each coherence block but for a new independent channel realization, as illustrated in Figure 5.23. This repetition makes it sufficient to study the operation of a single coherence block with an arbitrary random channel realization. There are two phases of each coherence block:

- Preamble: A known pilot sequence of length L_p symbols is transmitted.
- Payload: $L_c - L_p$ data symbols are transmitted.

We will analyze these phases in detail in the following sections.

5.5.1 Pilot-Based Channel Estimation

We begin by considering the channel estimation enabled by pilot transmission in a SIMO scenario where the channel vector $\mathbf{h} \in \mathbb{C}^M$ is subject to i.i.d. Rayleigh fading: $\mathbf{h} \sim \mathcal{N}_{\mathbb{C}}(\mathbf{0}, \beta \mathbf{I}_M)$. We select the known pilot of length $L_p = 1$ that equals $\sqrt{q}$ and uses the maximum symbol power. The received signal then becomes

$$\mathbf{y} = \mathbf{h}\sqrt{q} + \mathbf{n}, \tag{5.136}$$

where $\mathbf{n} \sim \mathcal{N}_{\mathbb{C}}(\mathbf{0}, N_0\mathbf{I}_M)$ is the receiver noise. Since the channel and noise are independent random variables, it follows that $\mathbf{y} \sim \mathcal{N}_{\mathbb{C}}(\mathbf{0}, (\beta q + N_0)\mathbf{I}_M)$.

The mth received signal in (5.136) has the form $y_m = h_m\sqrt{q} + n_m$, where the unknown channel realization $h_m \sim \mathcal{N}_{\mathbb{C}}(0, \beta)$ is observed in additive noise. Since the antennas' channel coefficients are independently distributed, we can estimate them separately. The received signal has the same structure as in Lemma 2.11, which implies that the MMSE estimate of h_m given the observation y_m is

$$\hat{h}_m = \frac{\beta\sqrt{q}}{\beta q + N_0} y_m, \quad m = 1, \ldots, M. \tag{5.137}$$

Among all conceivable ways of transforming y_m into a guess of h_m, (5.137) is the option that minimizes the average squared estimation error $\mathbb{E}\{|h_m - \hat{h}_m|^2\}$. The minimal value is given by (2.156) as

$$\text{MSE}_h = \mathbb{E}\left\{\left|h_m - \hat{h}_m\right|^2\right\} = \frac{\beta N_0}{\beta q + N_0}. \tag{5.138}$$

We can collect all the MMSE estimates of the channel coefficients in the vector form $\hat{\mathbf{h}} \in \mathbb{C}^M$ as

$$\hat{\mathbf{h}} = \begin{bmatrix} \hat{h}_1 \\ \vdots \\ \hat{h}_M \end{bmatrix} = \frac{\beta\sqrt{q}}{\beta q + N_0}\mathbf{y}. \tag{5.139}$$

This random vector also takes a new realization in every coherence block since we repeat the estimation once per block. By using the aforementioned distribution of $\mathbf{y}$, we notice that

$$\begin{aligned} \hat{\mathbf{h}} \sim \mathcal{N}_{\mathbb{C}}&\left(\mathbf{0}, \left|\frac{\beta\sqrt{q}}{\beta q + N_0}\right|^2 (\beta q + N_0)\mathbf{I}_M\right) = \mathcal{N}_{\mathbb{C}}\left(\mathbf{0}, \frac{\beta^2 q}{\beta q + N_0}\mathbf{I}_M\right) \\ &= \mathcal{N}_{\mathbb{C}}\left(\mathbf{0}, (\beta - \text{MSE}_h)\mathbf{I}_M\right). \end{aligned} \tag{5.140}$$

We will denote the estimation error as $\tilde{\mathbf{h}} = \mathbf{h} - \hat{\mathbf{h}}$ and each entry has a variance that equals the MSE, such that

$$\tilde{\mathbf{h}} \sim \mathcal{N}_{\mathbb{C}}\left(\mathbf{0}, \text{MSE}_h\mathbf{I}_M\right). \tag{5.141}$$

The entries of the estimate $\hat{\mathbf{h}}$ and the estimation error $\tilde{\mathbf{h}}$ have variances that add up to the variance of the entries of the original channel $\mathbf{h}$ since $(\beta - \text{MSE}_h) + \text{MSE}_h = \beta$. Hence, the channel estimation effectively splits the true channel into one known part $\hat{\mathbf{h}}$ with the reduced variance $\beta - \text{MSE}_h$ and an unknown part $\tilde{\mathbf{h}}$ with the remaining variance MSE_h. Having an estimate with a large variance is good because we want the known part of the channel to be strong. An accurate estimate is characterized by a small MSE, implying that the estimate is likely near the true channel realization.

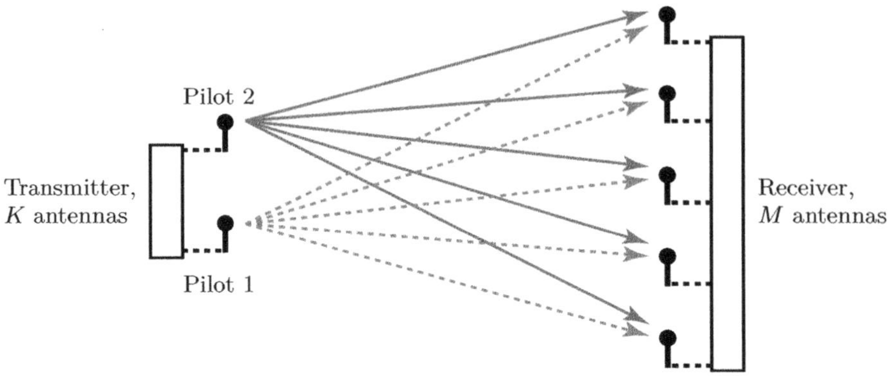

Figure 5.24: In a MIMO setup with K transmit antennas, we need to transmit K different pilots to enable estimation of the entire channel matrix $\mathbf{H}$ on the receiver side. The receiver can be equipped with any number of antennas, M, since these listen to the same pilots.

The procedure mentioned above enables the estimation of the entire SIMO channel vector $\mathbf{h}$ by transmitting only a single pilot symbol (i.e., $L_{\mathrm{p}} = 1$). This procedure works regardless of how many antennas the receiver has and resembles how public speeches are carried out: any number of people can listen simultaneously to the same speaker. However, the audience members need to take turns when asking questions to the speaker, which can become cumbersome when the audience is large. Analogously, the number of transmit antennas determines how many pilots must be transmitted to estimate the entire channel, not the number of receive antennas. We need $L_{\mathrm{p}} = K$ pilots if there are K transmit antennas.

We now switch focus to a MIMO channel with K transmit antennas and M receive antennas. It can be represented by the channel matrix $\mathbf{H} = [\mathbf{h}_1, \ldots, \mathbf{h}_K] \in \mathbb{C}^{M \times K}$, which contains K columns denoted as $\mathbf{h}_k \in \mathbb{C}^M$ for $k = 1, \ldots, K$. Suppose we transmit $L_{\mathrm{p}} = K$ pilots so that $\sqrt{q}$ is transmitted from antenna k at symbol time k. Figure 5.24 illustrates such a setup with $K = 2$ transmit antennas and $M = 5$ receive antennas, in which case we need to transmit two pilots. The received signal at symbol time k becomes

$$\mathbf{y}_k = \mathbf{h}_k \sqrt{q} + \mathbf{n}_k, \quad k = 1, \ldots, K, \tag{5.142}$$

where $\mathbf{n}_k \sim \mathcal{N}_{\mathbb{C}}(\mathbf{0}, N_0 \mathbf{I}_M)$ is the receiver noise.

The received signal in (5.142) has the same form as in (5.136) for the SIMO case. Hence, we can effectively divide the MIMO estimation problem into K SIMO estimation subproblems. Suppose $\mathbf{H}$ is modeled by i.i.d. Rayleigh fading with the entries independently distributed as $h_{m,k} \sim \mathcal{N}_{\mathbb{C}}(0, \beta)$, for $m = 1, \ldots, M$ and $k = 1, \ldots, K$. It then follows from the previous derivations that the MMSE estimate is a matrix $\hat{\mathbf{H}}$ and the estimation error is a matrix

$\tilde{\mathbf{H}} = \mathbf{H} - \hat{\mathbf{H}}$ with entries independently distributed as

$$\hat{h}_{m,k} \sim \mathcal{N}_{\mathbb{C}}(0, \beta - \mathrm{MSE}_h), \tag{5.143}$$

$$\tilde{h}_{m,k} \sim \mathcal{N}_{\mathbb{C}}(0, \mathrm{MSE}_h). \tag{5.144}$$

The receiver obtains the channel estimates, which is aligned with our standing assumption in this chapter that the receiver has CSI but not the transmitter. To provide the transmitter with CSI, we need to transmit in the reverse direction, either by feeding back the estimate or sending pilots also in that direction. This is done in some practical systems but not in this chapter.

5.5.2 Ergodic Rate with Imperfect CSI at the Receiver

We will now consider how the channel estimates from the last section can be used for signal detection and, particularly, how the achievable data rate is affected by the fact that the receiver has imperfect CSI. We consider the transmission of a packet that spans L coherence blocks and will let $L \to \infty$ to characterize the ergodic capacity. To this end, we revisit the memoryless SISO channel in (2.130). The received signal at an arbitrary time instance in coherence block l can be expressed as

$$y[l] = \left(\hat{h}[l] + \tilde{h}[l]\right) x[l] + n[l], \quad l = 1, \dots, L, \tag{5.145}$$

where $\hat{h}[l]$ is the MMSE estimated part of the channel response obtained by sending $L_{\mathrm{p}} = 1$ pilot and $\tilde{h}[l] \sim \mathcal{N}_{\mathbb{C}}(0, \mathrm{MSE}_h)$ is the independent estimation error. Similar to (5.104), the capacity is given by

$$C = \lim_{L \to \infty} \max_{f_{\mathsf{x}[1],\dots,\mathsf{x}[L]}(x[1],\dots,x[L])} \left(1 - \frac{1}{L_{\mathrm{c}}}\right) R_L, \tag{5.146}$$

where $1 - L_{\mathrm{p}}/L_{\mathrm{c}} = 1 - 1/L_{\mathrm{c}}$ is the fraction of each coherence block used for data transmission (while the fraction $1/L_{\mathrm{c}}$ is used for pilots) and the average data rate over the L coherence blocks is

$$\begin{aligned} R_L &= \frac{1}{L}\left(\mathcal{H}\left(x[1],\dots,x[L]\right) - \mathcal{H}\left(x[1],\dots,x[L] \middle| y[1],\dots,y[L],\hat{h}[1],\dots,\hat{h}[L]\right)\right) \\ &= \frac{1}{L}\sum_{l=1}^{L} \mathcal{H}\left(x[l]\right) - \mathcal{H}\left(x[l] \middle| y[l],\hat{h}[l]\right). \end{aligned} \tag{5.147}$$

Note that we used the alternative mutual information expression in (2.137), which equals the entropy of the transmitted signal minus the remaining entropy given the information known at the receiver (which is $y[l]$ and $\hat{h}[l]$ in this case). This expression is easier to evaluate under imperfect CSI. The block-fading assumption implies that independent realizations of $\hat{h}[l], \tilde{h}[l]$ are drawn in each coherence block from a stationary and ergodic random process. Hence, we can use the law of large numbers (Lemma 2.4) to compute the limit in (5.146) as

$$C = \max_{f_{\mathsf{x}}(x)} \left(1 - \frac{1}{L_{\mathrm{c}}}\right) \mathbb{E}\left\{\mathcal{H}(x) - \mathcal{H}(x|y, \hat{h})\right\}, \tag{5.148}$$

where the mean value is computed with respect to the channel estimate realization $\hat{h}$ and the maximum is computed concerning all signal distributions $f_{\mathsf{x}}(x)$ that have a symbol power of q. There is no easy way to compute the exact capacity when the receiver has imperfect CSI, but this remains one of the open problems in information theory. Therefore, we will derive a tight lower bound by following an approach from [69], [70]. To prepare for that derivation, we begin by considering the estimation of the data signal x.

Example 5.14. What is the LMMSE estimate of x when y in (5.145) is observed and $\hat{h}$ is known? What is the resulting conditional MSE?

The LMMSE estimator has the form $\hat{x} = ay$, where a is selected to minimize the MSE by satisfying the orthogonality principle $\mathbb{E}\{\tilde{x}y^*|\hat{h}\} = 0$ with $\tilde{x} = x - \hat{x}$ being the estimation error. This condition can be expanded as

$$0 = \mathbb{E}\left\{\tilde{x}y^*|\hat{h}\right\} = \mathbb{E}\left\{(x - ay)\, y^*|\hat{h}\right\} = \mathbb{E}\left\{xy^*|\hat{h}\right\} - a\mathbb{E}\left\{|y|^2|\hat{h}\right\}. \tag{5.149}$$

By solving for a in (5.149), we obtain

$$\begin{aligned} a &= \frac{\mathbb{E}\left\{xy^*|\hat{h}\right\}}{\mathbb{E}\left\{|y|^2|\hat{h}\right\}} = \frac{\mathbb{E}\left\{x\left((\hat{h}+\tilde{h})x + n\right)^*|\hat{h}\right\}}{\mathbb{E}\left\{\left|(\hat{h}+\tilde{h})x+n\right|^2|\hat{h}\right\}} \\ &= \frac{\mathbb{E}\left\{|x|^2\right\}\left(\hat{h} + \mathbb{E}\{\tilde{h}\}\right)^* + \mathbb{E}\{x\}\mathbb{E}\{n^*\}}{\mathbb{E}\left\{|x|^2\right\}\left(|\hat{h}|^2 + \mathbb{E}\left\{|\tilde{h}|^2\right\}\right) + \mathbb{E}\left\{|n|^2\right\}} = \frac{q\hat{h}^*}{q|\hat{h}|^2 + q\mathrm{MSE}_h + N_0} \end{aligned} \tag{5.150}$$

by using the fact that x, $\tilde{h}$, and n are uncorrelated and have zero mean (conditioned on $\hat{h}$). The resulting conditional MSE for the given value of $\hat{h}$ is

$$\begin{aligned} \mathrm{MSE}_{x|\hat{h}} &= \mathbb{E}\left\{|\tilde{x}|^2|\hat{h}\right\} = \mathbb{E}\left\{\tilde{x}x^*|\hat{h}\right\} - a^*\underbrace{\mathbb{E}\left\{\tilde{x}y^*|\hat{h}\right\}}_{=0} = \mathbb{E}\{|x|^2\} - a\underbrace{\mathbb{E}\left\{yx^*|\hat{h}\right\}}_{=q\hat{h}} \\ &= q - \frac{q^2|\hat{h}|^2}{q|\hat{h}|^2 + q\mathrm{MSE}_h + N_0} = \frac{q^2\mathrm{MSE}_h + qN_0}{q|\hat{h}|^2 + q\mathrm{MSE}_h + N_0}, \end{aligned} \tag{5.151}$$

where we used the orthogonality principle, reused computations from (5.150), and finally utilized that $\mathbb{E}\{|x|^2\} = q$.

We will make two (potentially) suboptimal assumptions to characterize the capacity in closed form. The first assumption is that $x \sim \mathcal{N}_{\mathbb{C}}(0, q)$. This is the optimal signal distribution when the receiver has perfect CSI, but not necessarily in our scenario with imperfect CSI. Under this assumption, we can use Lemma 2.9 to compute the first term in (5.148) as

$$\mathcal{H}(x) = \log_2(e\pi q). \tag{5.152}$$

The second suboptimal assumption is that the receiver computes an LMMSE estimate of x based on its available observations $y, \hat{h}$ and uses the resulting conditional MSE in (5.151) to upper bound the second term in (5.148) as

$$\begin{aligned} \mathcal{H}\left(x \middle| y, \hat{h}\right) &= \mathcal{H}\left(x - \hat{x} \middle| y, \hat{h}\right) \\ &\leq \mathcal{H}(x - \hat{x} | \hat{h}) \\ &\leq \log_2\left(e\pi \mathrm{MSE}_{x|\hat{h}}\right). \end{aligned} \tag{5.153}$$

The equality in (5.153) follows from subtracting the LMMSE estimate $\hat{x}$ obtained in Example 5.14 from x. This can be done without changing the differential entropy since $\hat{x}$ is deterministic given y and $\hat{h}$. We then obtain the first upper bound by removing the knowledge of y since the conditioning on a random variable cannot increase the entropy. The second upper bound follows from Lemma 2.9 since the differential entropy is maximized by a complex Gaussian distribution with the same variance as that of $x - \hat{x}$ (conditioned on the realization $\hat{h}$), which was given in (5.151).

By utilizing (5.152) and (5.153), we can obtain a lower bound on the ergodic capacity in (5.148) as

$$\begin{aligned} C &\geq \left(1 - \frac{1}{L_c}\right) \mathbb{E}\left\{\mathcal{H}(x) - \mathcal{H}\left(x \middle| y, \hat{h}\right)\right\} \\ &\geq \left(1 - \frac{1}{L_c}\right) \mathbb{E}\left\{\log_2\left(\frac{q}{\mathrm{MSE}_{x|\hat{h}}}\right)\right\} \\ &= \left(1 - \frac{1}{L_c}\right) \mathbb{E}\left\{\log_2\left(\frac{q^2|\hat{h}|^2 + q^2\mathrm{MSE}_h + qN_0}{q^2\mathrm{MSE}_h + qN_0}\right)\right\} \\ &= \left(1 - \frac{1}{L_c}\right) \mathbb{E}\left\{\log_2\left(1 + \frac{q|\hat{h}|^2}{q\mathrm{MSE}_h + N_0}\right)\right\}. \end{aligned} \tag{5.154}$$

The first lower bound follows from assuming that the transmitter uses a suboptimal signal distribution. The second lower bound follows from (5.153), which assumes that the receiver decodes the signal in a suboptimal way. We have now proved the following result.

Corollary 5.2. Consider the discrete memoryless block-fading channel with input $x[l] \in \mathbb{C}$ and output $y[l] \in \mathbb{C}$ given by

$$y[l] = \left(\hat{h}[l] + \tilde{h}[l]\right) x[l] + n[l], \tag{5.155}$$

where $n[l] \sim \mathcal{N}_{\mathbb{C}}(0, N_0)$ is independent noise. Suppose the input distribution is feasible whenever the symbol power satisfies $\mathbb{E}\{|x[l]|^2\} \leq q$. Furthermore, suppose the channel $h[l] = \hat{h}[l] + \tilde{h}[l]$ takes independent and identically distributed realizations in each coherence block from a distribution with finite variance and that a fraction $1/L_c$ of each block is used for pilots. If the channel estimate $\hat{h}[l]$ is known at the output while the channel estimation error $\tilde{h}[l] \sim \mathcal{N}_{\mathbb{C}}(0, \mathrm{MSE}_h)$ is unknown and independent, the channel capacity can be lower bounded as

$$C \geq \left(1 - \frac{1}{L_c}\right) \mathbb{E}\left\{\log_2\left(1 + \frac{q\left|\hat{h}\right|^2}{q\mathrm{MSE}_h + N_0}\right)\right\} \quad \text{bit/symbol}, \tag{5.156}$$

where the bound is achieved when $x[l] \sim \mathcal{N}_{\mathbb{C}}(0, q)$ and independent for each l.

The lower bound in (5.156) has a familiar form $(1-1/L_c)\mathbb{E}\{\log_2(1+\mathrm{SINR})\}$, where $q|\hat{h}|^2/(q\mathrm{MSE}_h + N_0)$ acts as the instantaneous SINR. If we compare the new expression with the ergodic capacity $\mathbb{E}\{\log_2(1+q|h|^2/N_0)\}$ in (5.111) for a fast-fading channel with perfect CSI, we can notice three key differences. Firstly, the pre-log factor $(1 - 1/L_c)$ in (5.156) accounts for the transmission resources spent on channel estimation in each coherence block. This makes the new expression more realistic since CSI cannot be acquired for free. Secondly, the channel response h in (5.111) is replaced by the channel estimate $\hat{h}$ in (5.156). As the variance of these coefficients determines the average channel gain, it has effectively reduced from β to $\beta - \mathrm{MSE}_h$. Finally, the noise variance N_0 has been replaced with $q\mathrm{MSE}_h + N_0$, which also contains a penalty term determined by the imperfect CSI. Its variance matches that of the signal component $\tilde{h}x$ received over the unknown portion of the channel, which is a noise-like perturbation that is worst-case modeled as complex Gaussian noise when computing the capacity bound.

We will now generalize the analysis by considering a SIMO setup with M antennas at the receiver, in which case the received signal $\mathbf{y} \in \mathbb{C}^M$ in an arbitrary coherence block can be expressed as

$$\mathbf{y} = \left(\hat{\mathbf{h}} + \tilde{\mathbf{h}}\right) x + \mathbf{n}, \tag{5.157}$$

where $\hat{\mathbf{h}} \in \mathbb{C}^M$ is the known channel estimate in (5.140) obtained using $L_p = 1$ pilot, $\tilde{\mathbf{h}} \sim \mathcal{N}_{\mathbb{C}}(\mathbf{0}, \mathrm{MSE}_h \mathbf{I}_M)$ is the independent unknown estimation error in

(5.141), and $\mathbf{n} \sim \mathcal{N}_{\mathbb{C}}(\mathbf{0}, N_0\mathbf{I}_M)$ is the receiver noise. We can notice from the SISO case in (5.154) that the ergodic capacity can be lower bounded as

$$C \geq \left(1 - \frac{1}{L_c}\right) \mathbb{E}\left\{\log_2\left(\frac{q}{\mathrm{MSE}_{x|\hat{\mathbf{h}}}}\right)\right\}, \tag{5.158}$$

where q is the transmit power and $\mathrm{MSE}_{x|\hat{\mathbf{h}}}$ is the conditional MSE when computing the LMMSE estimate of x given the received signal $\mathbf{y}$ and channel estimate $\hat{\mathbf{h}}$. This is a lower bound because it relies on the suboptimal assumptions of a Gaussian codebook $x \sim \mathcal{N}_{\mathbb{C}}(0, q)$ and that LMMSE processing is used for signal detection at the receiver, but at least we know that the rate can be achieved in this particular way. The LMMSE estimate of x has the form $\hat{x} = \mathbf{w}^{\mathrm{H}}\mathbf{y}$, where $\mathbf{w} \in \mathbb{C}^M$ is receive combining vector. We can determine the LMMSE combining vector using the orthogonality principle (as in Example 5.14), but we will instead follow the approach from Example 3.4 to directly compute the MSE as

$$\begin{aligned}
\mathbb{E}\left\{|x-\hat{x}|^2|\hat{\mathbf{h}}\right\} &= \mathbb{E}\left\{\left|x\left(1-\mathbf{w}^{\mathrm{H}}\left(\hat{\mathbf{h}}+\tilde{\mathbf{h}}\right)\right)-\mathbf{w}^{\mathrm{H}}\mathbf{n}\right|^2|\hat{\mathbf{h}}\right\} \\
&\overset{(a)}{=} \mathbb{E}\left\{|x|^2\right\}\mathbb{E}\left\{\left|1-\mathbf{w}^{\mathrm{H}}\left(\hat{\mathbf{h}}+\tilde{\mathbf{h}}\right)\right|^2|\hat{\mathbf{h}}\right\}+\mathbb{E}\left\{|\mathbf{w}^{\mathrm{H}}\mathbf{n}|^2|\hat{\mathbf{h}}\right\} \\
&= q\left(1+\mathbf{w}^{\mathrm{H}}\left(\hat{\mathbf{h}}\hat{\mathbf{h}}^{\mathrm{H}}+\mathrm{MSE}_h\mathbf{I}_M\right)\mathbf{w}-\mathbf{w}^{\mathrm{H}}\hat{\mathbf{h}}-\hat{\mathbf{h}}^{\mathrm{H}}\mathbf{w}\right)+\mathbf{w}^{\mathrm{H}}N_0\mathbf{I}_M\mathbf{w} \\
&= q+\mathbf{w}^{\mathrm{H}}\underbrace{\left(q\hat{\mathbf{h}}\hat{\mathbf{h}}^{\mathrm{H}}+(q\mathrm{MSE}_h+N_0)\mathbf{I}_M\right)}_{=\mathbf{B}}\mathbf{w}-\mathbf{w}^{\mathrm{H}}\underbrace{q\hat{\mathbf{h}}}_{=\mathbf{a}}-\underbrace{q\hat{\mathbf{h}}^{\mathrm{H}}}_{=\mathbf{a}^{\mathrm{H}}}\mathbf{w},
\end{aligned} \tag{5.159}$$

where (a) follows from utilizing the (conditional) uncorrelation $\mathbb{E}\{x\mathbf{n}^{\mathrm{H}}|\hat{\mathbf{h}}\} = \mathbf{0}$ between the signal x and the noise $\mathbf{n}$ to remove some of the terms. We stress that the combining vector was treated as deterministic when $\hat{\mathbf{h}}$ is known, but we want to find the optimal way that it depends on the estimate. By utilizing the notation $\mathbf{a}$ and $\mathbf{B}$ that was introduced in (5.159), we can write the MSE as

$$\begin{aligned}
\mathbb{E}\left\{|x-\hat{x}|^2|\hat{\mathbf{h}}\right\} &= q+\mathbf{w}^{\mathrm{H}}\mathbf{B}\mathbf{w}-\mathbf{w}^{\mathrm{H}}\mathbf{a}-\mathbf{a}^{\mathrm{H}}\mathbf{w} \\
&= q-\mathbf{a}^{\mathrm{H}}\mathbf{B}^{-1}\mathbf{a}+\left(\mathbf{w}-\mathbf{B}^{-1}\mathbf{a}\right)^{\mathrm{H}}\mathbf{B}\left(\mathbf{w}-\mathbf{B}^{-1}\mathbf{a}\right) \\
&\geq q-\mathbf{a}^{\mathrm{H}}\mathbf{B}^{-1}\mathbf{a} = \mathrm{MSE}_{x|\hat{\mathbf{h}}},
\end{aligned} \tag{5.160}$$

where we complete the squares[11] and then notice that the quadratic form in the last term attains its minimum value of zero if $\mathbf{w} = \mathbf{B}^{-1}\mathbf{a}$. This is the

[11]We utilize the fact that $(\mathbf{w}-\mathbf{B}^{-1}\mathbf{a})^{\mathrm{H}}\mathbf{B}(\mathbf{w}-\mathbf{B}^{-1}\mathbf{a}) = \mathbf{a}^{\mathrm{H}}\mathbf{B}^{-1}\mathbf{a}+\mathbf{w}^{\mathrm{H}}\mathbf{B}\mathbf{w}-\mathbf{w}^{\mathrm{H}}\mathbf{a}-\mathbf{a}^{\mathrm{H}}\mathbf{w}$ to gather all the terms that depend on $\mathbf{w}$ in a quadratic form. The missing term $\mathbf{a}^{\mathrm{H}}\mathbf{B}^{-1}\mathbf{a}$ must be subtracted when doing that. This is the matrix algebra equivalent of completing the squares in a scalar quadratic equation.

LMMSE receive combining vector, and it can be rewritten using (2.49) as

$$\begin{aligned}\mathbf{w} &= \mathbf{B}^{-1}\mathbf{a} \\ &= \left(q\hat{\mathbf{h}}\hat{\mathbf{h}}^{\mathrm{H}} + (q\mathrm{MSE}_h + N_0)\mathbf{I}_M\right)^{-1} q\hat{\mathbf{h}} \\ &= \frac{q}{q\|\hat{\mathbf{h}}\|^2 + q\mathrm{MSE}_h + N_0}\hat{\mathbf{h}}. \end{aligned} \tag{5.161}$$

This LMMSE combining vector is the counterpart to MRC under imperfect CSI because it projects the received signal into the direction of the channel estimate. The minimum MSE in (5.160) can now be expressed in a concise way using (5.161) as

$$\begin{aligned}\mathrm{MSE}_{x|\hat{\mathbf{h}}} &= q - \mathbf{a}^{\mathrm{H}}\mathbf{B}^{-1}\mathbf{a} = q - \frac{q^2\|\hat{\mathbf{h}}\|^2}{q\|\hat{\mathbf{h}}\|^2 + q\mathrm{MSE}_h + N_0} \\ &= \frac{q(q\mathrm{MSE}_h + N_0)}{q\|\hat{\mathbf{h}}\|^2 + q\mathrm{MSE}_h + N_0}. \end{aligned} \tag{5.162}$$

We can finally compute the capacity lower bound in (5.158) as

$$\begin{aligned} C &\geq \left(1 - \frac{1}{L_{\mathrm{c}}}\right)\mathbb{E}\left\{\log_2\left(q\frac{q\|\hat{\mathbf{h}}\|^2 + q\mathrm{MSE}_h + N_0}{q(q\mathrm{MSE}_h + N_0)}\right)\right\} \\ &= \left(1 - \frac{1}{L_{\mathrm{c}}}\right)\mathbb{E}\left\{\log_2\left(1 + \frac{q\|\hat{\mathbf{h}}\|^2}{q\mathrm{MSE}_h + N_0}\right)\right\}. \end{aligned} \tag{5.163}$$

This is the natural SIMO extension of the SISO capacity bound in (5.156), which has the same pre-log factor $(1 - 1/L_{\mathrm{c}})$. There is one key difference: the channel gain in the numerator is $\|\hat{\mathbf{h}}\|^2$ in the SIMO case instead of $|\hat{h}|^2$. Since these terms have the means $M(\beta - \mathrm{MSE}_h)$ and $\beta - \mathrm{MSE}_h$, respectively, we can conclude that a beamforming gain proportional to M is achievable in the SIMO setup despite the imperfect CSI. Notably, the interference term caused by the CSI imperfection remains equal to $q\mathrm{MSE}_h$, independently of the number of receive antennas. This is remarkable because the total variance of the signal received over the unknown portion of the channel is $qM\mathrm{MSE}_h$. Since $\tilde{\mathbf{h}} \sim \mathcal{N}_{\mathbb{C}}(\mathbf{0}, \mathrm{MSE}_h\mathbf{I}_M)$, that power is uniformly distributed over all M receiver dimensions and, thus, only a fraction $1/M$ of it appears on average in the dimension utilized by the LMMSE combining. This is the same phenomenon that makes the noise power independent of the number of receive antennas in the SNR expression, even if the total noise power in the receiver hardware is proportional to M. In conclusion, a SIMO receiver becomes increasingly robust to CSI imperfections as the number of antennas increases because only the desired signal power increases with M.

Example 5.15. What is the relative SNR loss caused by imperfect CSI? What happens to this loss when the SNR is high?

If we define a random vector $\mathbf{e} \sim \mathcal{N}_{\mathbb{C}}(\mathbf{0}, \mathbf{I}_M)$ with unit-variance complex Gaussian entries, we can notice that the instantaneous SNR in (5.163) satisfies

$$\frac{q\|\hat{\mathbf{h}}\|^2}{q\mathrm{MSE}_h + N_0} \sim \frac{q(\beta - \mathrm{MSE}_h)}{q\mathrm{MSE}_h + N_0}\|\mathbf{e}\|^2, \tag{5.164}$$

where the expressions are equally distributed since $\beta - \mathrm{MSE}_h$ is the variance of each entry in $\hat{\mathbf{h}}$. Similarly, the instantaneous SNR in (5.118) with perfect CSI satisfies

$$\frac{q\|\mathbf{h}\|^2}{N_0} \sim \frac{q\beta}{N_0}\|\mathbf{e}\|^2. \tag{5.165}$$

The relative SNR loss caused by the CSI imperfections becomes

$$\mathrm{Loss} = \frac{\frac{q(\beta - \mathrm{MSE}_h)}{q\mathrm{MSE}_h + N_0}}{\frac{q\beta}{N_0}} = \frac{N_0\left(1 - \mathrm{MSE}_h/\beta\right)}{q\mathrm{MSE}_h + N_0}. \tag{5.166}$$

The relative loss at high SNRs can be obtained by considering the asymptotic limit where $q \to \infty$. The MSE in (5.138) has the limit

$$\mathrm{MSE}_h = \frac{\beta N_0}{\beta q + N_0} \to 0, \tag{5.167}$$

but the convergence to zero is slow, so the interference term $q\mathrm{MSE}_h$ has the non-zero limit

$$q\mathrm{MSE}_h = \frac{q\beta N_0}{\beta q + N_0} \to \frac{\beta N_0}{\beta} = N_0. \tag{5.168}$$

Hence, the relative SNR loss in (5.166) has the asymptotic limit

$$\mathrm{Loss} \to \frac{N_0\left(1 - 0/\beta\right)}{N_0 + N_0} = \frac{1}{2}. \tag{5.169}$$

In conclusion, there is a 3 dB loss in SNR in the capacity expression in the high-SNR regime, but the capacity anyway grows unboundedly with q.

The ergodic capacity in the MIMO scenario with K transmit antennas and M receive antennas can be lower bounded similarly to the SIMO scenario. We recall from Section 3.4.3 that the received signal $\mathbf{y} \in \mathbb{C}^M$ when using an arbitrary precoding matrix $\mathbf{P} = [\mathbf{p}_1, \ldots, \mathbf{p}_K] \in \mathbb{C}^{K \times K}$ with unit-norm columns is

$$\mathbf{y} = \left(\hat{\mathbf{H}} + \tilde{\mathbf{H}}\right)\mathbf{P}\bar{\mathbf{x}} + \mathbf{n} = \sum_{k=1}^{K} \hat{\mathbf{H}}\mathbf{p}_k\bar{x}_k + \underbrace{\sum_{k=1}^{K} \tilde{\mathbf{H}}\mathbf{p}_k\bar{x}_k + \mathbf{n}}_{=\epsilon}, \tag{5.170}$$

where $\bar{\mathbf{x}} = [\bar{x}_1, \ldots, \bar{x}_K]^{\mathrm{T}} \sim \mathcal{N}_{\mathbb{C}}(\mathbf{0}, \mathbf{Q})$ contains the independent data symbols and $\mathbf{Q} = \mathrm{diag}(q_1, \ldots, q_K)$ is a diagonal power allocation matrix. The coefficients in $\mathbf{Q}$ should be selected to satisfy $\sum_{k=1}^{K} q_k = q$ so that the maximum symbol power is used. The new properties in this section are the block fading where $\hat{\mathbf{H}} \in \mathbb{C}^{M \times K}$ is the known channel estimate with i.i.d. entries distributed according to (5.143) as $\hat{h}_{m,k} \sim \mathcal{N}_{\mathbb{C}}(0, \beta - \mathrm{MSE}_h)$, while $\tilde{\mathbf{H}}$ is the estimation error with i.i.d. entries distributed according to (5.144) as $\tilde{h}_{m,k} \sim \mathcal{N}_{\mathbb{C}}(0, \mathrm{MSE}_h)$. The estimate is obtained by transmitting K pilots, while $L_{\mathrm{c}} - K$ symbols per coherence block are used for data transmission. The received signal in (5.170) can be divided into the known first term, where $\hat{\mathbf{H}}$ acts as the channel matrix, and the unknown term $\boldsymbol{\epsilon}$, which we know from earlier in this section will act as extra noise. This term has (conditional) zero mean $\mathbb{E}\{\boldsymbol{\epsilon}|\hat{\mathbf{H}}\} = \mathbf{0}$ since the data symbols have zero mean, while the conditional covariance matrix can be computed as

$$\mathbb{E}\left\{\boldsymbol{\epsilon}\boldsymbol{\epsilon}^{\mathrm{H}}\middle|\hat{\mathbf{H}}\right\} = \sum_{k=1}^{K} q_k \mathbb{E}\left\{\tilde{\mathbf{H}}\mathbf{p}_k\mathbf{p}_k^{\mathrm{H}}\tilde{\mathbf{H}}^{\mathrm{H}}\middle|\hat{\mathbf{H}}\right\} = \sum_{k=1}^{K} q_k \mathrm{MSE}_h \|\mathbf{p}_k\|^2 \mathbf{I}_M = q\mathrm{MSE}_h \mathbf{I}_M, \tag{5.171}$$

where the first equality follows from the independence of the data symbols and the second equality follows from the fact that $\tilde{\mathbf{H}}\mathbf{p}_k \sim \mathcal{N}_{\mathbb{C}}(\mathbf{0}, \mathrm{MSE}_h \|\mathbf{p}_k\|^2 \mathbf{I}_M)$, which is independent of $\hat{\mathbf{H}}$. The last equality utilizes that $\|\mathbf{p}_k\| = 1$ and $\sum_{k=1}^{K} q_k = q$ by assumption.

We demonstrated in Section 3.4.3 that the maximum achievable data rate can be achieved through the LMMSE-SIC procedure, where the K transmitted signals are decoded sequentially while treating all other signals as noise. In the block-fading scenario, the decoding of each signal will be subject to the extra noise vector $\boldsymbol{\epsilon}$, which cannot be removed at the receiver since $\tilde{\mathbf{H}}$ is unknown. Hence, based on the previous results in this section, we conclude that the variance $q\mathrm{MSE}_h$ of the entries in $\boldsymbol{\epsilon}$ can be added to the receiver noise $\mathbf{n}$. An achievable ergodic rate during the data transmission is then obtained by generalizing (3.106) as

$$\mathbb{E}\left\{\log_2\left(\det\left(\mathbf{I}_M + \frac{1}{q\mathrm{MSE}_h + N_0}\hat{\mathbf{H}}\mathbf{P}\mathbf{Q}\mathbf{P}^{\mathrm{H}}\hat{\mathbf{H}}^{\mathrm{H}}\right)\right)\right\}, \tag{5.172}$$

where the mean value is computed with respect to the random channel estimates in different coherence blocks. Since the channel estimate features i.i.d. fading, we further know from Section 5.4.2 that $\mathbf{R}_x = \mathbf{P}\mathbf{Q}\mathbf{P}^{\mathrm{H}} = \frac{q}{K}\mathbf{I}_K$ maximizes the expression in (5.172) when the transmitter is unaware of the channel. In summary, a capacity lower bound in the MIMO case is

$$C \geq \left(1 - \frac{K}{L_{\mathrm{c}}}\right)\mathbb{E}\left\{\log_2\left(\det\left(\mathbf{I}_M + \frac{q/K}{q\mathrm{MSE}_h + N_0}\hat{\mathbf{H}}\hat{\mathbf{H}}^{\mathrm{H}}\right)\right)\right\}, \tag{5.173}$$

where we also accounted for the fact that a fraction K/L_{c} of each coherence block is used for transmitting pilots. This capacity bound shares all the

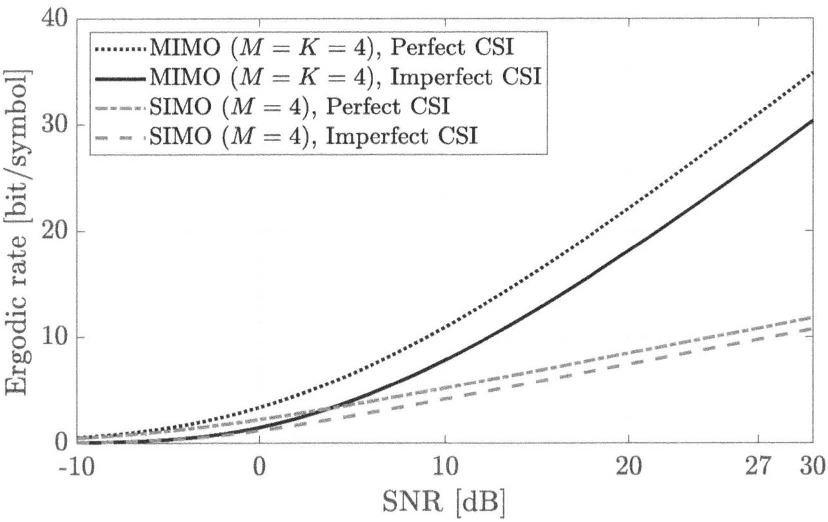

Figure 5.25: The ergodic rate over i.i.d. Rayleigh fading channels, which are either modeled as fast-fading with perfect CSI at the receiver or block-fading with imperfect CSI. We consider a SIMO channel with $M = 4$ antennas and a MIMO channel with $M = K = 4$ antennas.

essential features with the previous bounds in this section. There is a pre-log factor $(1 - K/L_c)$, the true channel matrix $\mathbf{H}$ is replaced with the estimated channel matrix $\hat{\mathbf{H}}$, and the estimation errors result in an SNR loss created by the extra noise variance $q\text{MSE}_h$. Hence, we can achieve the same multiplexing and beamforming gains under imperfect CSI as with perfect CSI but starting from a worse situation with a smaller pre-log factor and a relative SNR loss.

Figure 5.25 compares the ergodic capacity with perfect CSI at the receiver (as in Section 5.4.2) with the lower bounds obtained in this section for block-fading with imperfect CSI. We assume coherence blocks with $L_c = 200$ symbols and consider a MIMO setup with $M = K = 4$ antennas and a SIMO setup with $M = 4$ receive antennas. The figure shows the ergodic rates as functions of the average SNR $\frac{q\beta}{N_0}$. The pre-log factor is almost one in the considered setups, so the main impact of the imperfect CSI is the relative SNR loss in the rate expressions. This loss results in the shift of the rate curves to the right by approximately 3 dB in the high-SNR regime, as predicted in Example 5.15. This loss can, for instance, be identified by comparing the rates achieved at 27 and 30 dB, which are nearly the same. The shift of the curves otherwise confirms that the same beamforming and multiplexing gains are achieved with perfect and imperfect CSI; thus, MIMO systems can operate effectively also when the receiver has imperfect channel knowledge.

5.5.3 Ergodic Rate with Imperfect CSI Available Everywhere

We have derived ergodic rates achievable in a block-fading scenario where the receiver obtains CSI through pilot transmission, but the transmitter is unaware of the channel realization. This section considers the scenario where the transmitter somehow gains access to the same channel estimate, denoted by $\hat{\mathbf{H}}$ in the MIMO scenario with K transmit antennas and M receive antennas. This CSI can be utilized to optimize the transmission, particularly the precoding. We return to the achievable rate expression in (5.172) and optimize the precoding matrix $\mathbf{P}$ and power allocation matrix $\mathbf{Q}$:

$$R = \mathbb{E}\left\{ \max_{\substack{\mathbf{P}: \|\mathbf{p}_k\|=1, k=1,\ldots,K \\ q_1 \geq 0,\ldots,q_K \geq 0, \sum_{k=1}^{K} q_k = q}} \log_2 \left(\det \left(\mathbf{I}_M + \frac{1}{q\mathrm{MSE}_h + N_0} \hat{\mathbf{H}}\mathbf{P}\mathbf{Q}\mathbf{P}^{\mathrm{H}}\hat{\mathbf{H}}^{\mathrm{H}} \right) \right) \right\}. \tag{5.174}$$

The capacity is lower bounded as $C \geq (1 - K/L_{\mathrm{c}})R$ when including the pre-log factor caused by transmitting K pilots per coherence block. The key difference from the previous section is that the precoding optimization is done inside the mean value once per coherence block. This optimization problem coincides with the problem considered in Section 3.4 for deterministic channel matrices. More precisely, let $\hat{\mathbf{H}} = \mathbf{U}\boldsymbol{\Sigma}\mathbf{V}^{\mathrm{H}}$ denote the SVD of the channel estimate, where $s_1, \ldots, s_r$ denote the $r = \min(M, K)$ non-zero singular values.[12] It follows that $\mathbf{P} = \mathbf{V}$ is the rate-maximizing precoding, while $q_1, \ldots, q_K$ should be selected based on the water-filling power allocation. By utilizing Theorem 3.1, we can rewrite (5.174) as

$$R = \mathbb{E}\left\{ \sum_{k=1}^{r} \log_2 \left(1 + \frac{q_k^{\mathrm{opt}} s_k^2}{q\mathrm{MSE}_h + N_0} \right) \right\}, \tag{5.175}$$

where the power allocation in the coherence block with the singular value realization $s_1, \ldots, s_r$ is

$$q_k^{\mathrm{opt}} = \max\left(\mu - \frac{q\mathrm{MSE}_h + N_0}{s_k^2}, 0 \right), \quad k = 1, \ldots, r, \tag{5.176}$$

and the variable μ is selected to make $\sum_{k=1}^{r} q_k^{\mathrm{opt}} = q$. The water-filling is also affected by the imperfect CSI since the noise variance is increased by $q\mathrm{MSE}_h$. The singular values have the same distribution as those of the true channel matrix $\mathbf{H}$, except that the variance is reduced by a factor $(\beta - \mathrm{MSE}_h)/\beta$.

The gain in ergodic rate from having CSI at the transmitter can be quantified under i.i.d. Rayleigh fading by computing the ratio between the rate in (5.175) with CSI at the transmitter and the rate in (5.173) without CSI at the transmitter. The difference in the ergodic rates is generated by whether the precoding is based on the SVD of the estimated channel matrix or not. Figure 5.26 shows the relative rate gain as a function of the SNR for different numbers of transmit and receive antennas. We notice that having CSI at the

[12] The maximum number of non-zero singular values is $\min(M, K)$ and all these values will be non-zero with probability 1 under i.i.d. Rayleigh fading, as explained in Example 5.4.

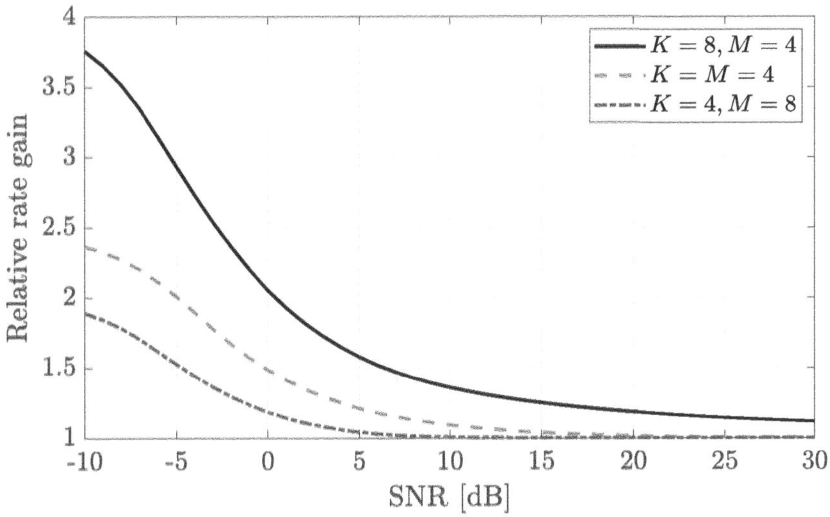

Figure 5.26: The relative gain in ergodic rate from having CSI at the transmitter, which is computed as the ratio between the rate in (5.175) with CSI and the rate in (5.173) without CSI. We consider i.i.d. Rayleigh fading MIMO channels with different numbers of transmit and receive antennas.

transmitter is primarily useful at lower SNRs, where transmit beamforming gains can be achieved by only transmitting in the estimated channel's strongest direction(s). The gain is larger when the transmitter has more antennas than the receiver (solid black line) and smaller when the receiver has more antennas than the transmitter (blue dash-dotted line). The benefit of having CSI at the transmitter vanishes asymptotically, except if $K > M$ when it remains vital for the transmitter to concentrate the transmit power in the signal dimensions that reach the receiver. In conclusion, feeding back channel estimates to the transmitter has clear benefits, particularly when the SNR is low, so transmit beamforming gains are more valuable than multiplexing gains.

5.6 Sparse Multipath Propagation and Dual Polarization

The i.i.d. Rayleigh fading model was derived in Section 5.1.2 by considering the deployment of half-wavelength-spaced ULAs in an isotropic rich multipath environment. The statistical independence between the entries of the channel vector/matrix simplifies the analysis of slow and fast fading channels, but it is generally not an accurate model of practical channels. Multiple factors can break the independence: other array geometries than ULAs, the use of directive or dual-polarized antennas, and non-isotropic multipath environments. While the system designer can control the former two factors, the propagation environment is essentially given by nature, and its non-isotropic features become particularly evident as the number of antennas increases and the wavelength shrinks. Therefore, we will develop a model for the channel fading

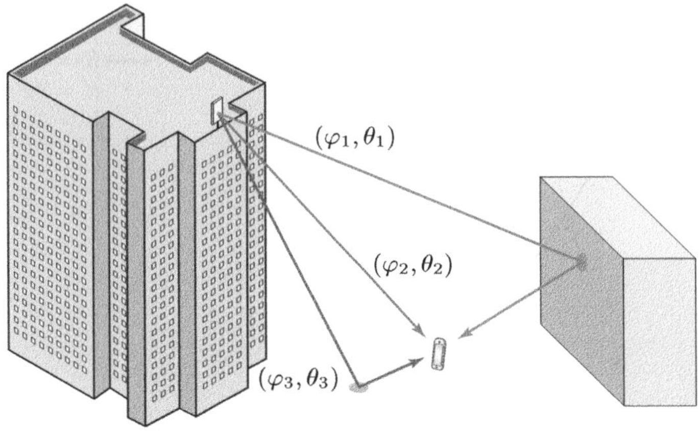

Figure 5.27: Illustration of a sparse multipath propagation environment where only signal components that leave the transmitter in some distinct angular directions (φ_i, θ_i) reach the receiver. The figure shows three such directions. The same setup was considered in Figure 2.14.

distribution that can be utilized in more realistic propagation environments. We call it *sparse multipath propagation* to distinguish it from the rich multipath propagation assumption made previously in this chapter. We will first consider single-polarized antennas and then dual-polarized antennas.

We begin by considering the MISO channel in Figure 5.27 where a base station equipped with M antennas transmits to a single-antenna user. Three propagation paths are indicated in the figure: one direct path and two reflected/scattered paths. Each path is associated with a particular azimuth angle φ_i and elevation angle θ_i, representing the direction of the path as seen from the transmitter. We let L denote the total number of propagation paths in this section. When the reflecting objects and receiver are in the far-field of the transmitter, we can utilize the array response vector $\mathbf{a}(\varphi, \theta) \in \mathbb{C}^M$ of the transmitter array to model each propagation path. A methodology for computing array response vectors for any specific array geometries was provided in Section 4.5. In this section, we will treat it as an arbitrary vector that might include antenna gains. The ith propagation path is associated with a signal attenuation $\alpha_i \in [0, 1]$ and a phase-shift $\psi_i \in [-\pi, \pi)$, where we utilize the same notation as in Section 5.1. The components of the radiated signal that reach the receiver over the different paths are superimposed. The channel vector can then be expressed as

$$\mathbf{h} = \sum_{i=1}^{L} \alpha_i e^{-\mathrm{j}\psi_i} \mathbf{a}(\varphi_i, \theta_i). \tag{5.177}$$

Since the array response vectors assign different phase-shifts to different antennas, the summation in (5.177) will lead to different complex numbers for different antennas. The signals emitted from some transmit antennas might be superimposed constructively over the multipath channel, while the signals from

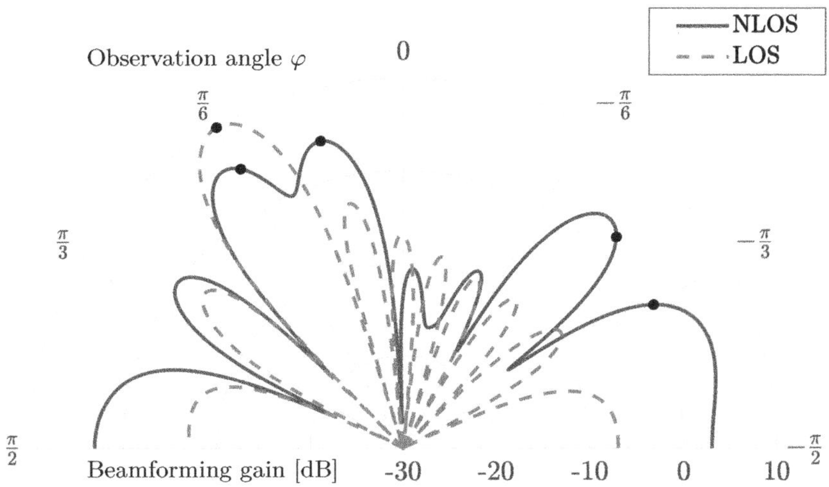

Figure 5.28: The beamforming gain that is observed in different directions φ when a ULA with $M = 10$ antennas transmits using MRT. The dashed curve considers the LOS case with a single propagation path in the direction $\varphi_1 = \pi/6$. The solid curve considers the NLOS case with $L = 4$ equally strong paths in the directions $\varphi_1 = \pi/6$, $\varphi_2 = \pi/12$, $\varphi_3 = -\pi/4$, and $\varphi_4 = -\pi/3$ (and phase-shifts are progressively shifted by $\pi/3$). These angles are marked with stars. The beam pattern becomes increasingly complex and ceases to have a distinct angular directivity when the number of paths increases.

other antennas might superimpose destructively. Hence, the transmitter should allocate its power differently over the antennas to maximize the SNR. The use of MRT with the precoding vector $\mathbf{p} = \mathbf{h}^*/\|\mathbf{h}\|$ finds the SNR-maximizing way of utilizing the multipath propagation environment. The radiated signal is a superposition of beams focused in the directions of the L different paths since $\mathbf{h}$ is a linear combination of $\mathbf{a}(\varphi_i, \theta_i)$ for $i = 1, \ldots, L$.

Whenever the channel contains multiple paths with distinctly different angles, the radiated signal generated by MRT will no longer look like a beam pointing in a single angular direction. Figure 5.28 illustrates the angular beam pattern when transmitting in a single direction (dashed curve) and when the channel consists of $L = 4$ paths (solid curve). The beam pattern is more complex in the latter case, but it has four peaks in roughly the same directions as those leading to the objects creating the multipaths (those directions are marked with stars). The beam pattern is a superposition of angular beams focused precisely toward these objects, but when their main beams and side-lobes interact, the combined angular beam pattern is smeared out. As more paths are added to the channel vector, as is typically the case in practice, the radiation pattern will look increasingly complex and lack a distinct angular directivity. The main point is that one should not expect the transmitted signal in multiple antenna communications to look like an angular beam except in the special case of free-space LOS propagation considered in Chapter 4. The only goal of precoding is to radiate the same signal from all antennas so that they add constructively at the receiver location.

5.6.1 Clustered Multipath Propagation

A key characteristic of sparse multipath propagation is that a limited number of physical objects creates the propagation paths. These objects are located in distinctly different angular directions, as seen from the transmitter and receiver. We will call each such object a *multipath cluster* and let N_{cl} denote the number of clusters. We consider an NLOS channel, so the radiated signal can only reach the receiver via one of these clusters. A multipath cluster can give rise to many propagation paths, but they are all associated with approximately the same pair of azimuth and elevation angles. This definition requires the cluster to span only a tiny angular interval from the transmitter's perspective; however, the physical size depends on how far away the cluster is and how many antennas the transmitter has. Recall that the array cannot resolve the angular differences between paths when these are smaller than the half-power beamwidth.

Example 5.16. What is the half-power angular beamwidth when using a ULA with $M = 10$ antennas and $\Delta = \lambda/2$? How physically large can a multipath cluster be if it is located 10 or 100 meters away?

The half-power beamwidth was considered in Example 4.7, where the approximate formula $1.772/M$ radians was derived. This becomes $b = 0.1772$ radians or $10.15°$ with $M = 10$ antennas. An angular interval b radians wide becomes $2d\tan(b/2)$ meters wide at a distance d from the transmitter. The width becomes $1.78\,\text{m}$ if $d = 10\,\text{m}$ and $17.8\,\text{m}$ if $d = 100\,\text{m}$; thus, a single multipath cluster can be physically large, particularly when it is far from the antenna array. The green and blue circles in Figure 5.27 might represent different multipath clusters. It can be buildings, cars, mountains, etc.

We let (φ_i, θ_i) denote the common angles associated with all the paths generated by the ith cluster. Moreover, we assume each cluster gives rise to N_{path} different paths, each having a different attenuation $\alpha_{i,n}$ and phase-shift $\psi_{i,n}$, for $n = 1, \ldots, N_{\text{path}}$. We then have a total of $L = N_{\text{cl}}N_{\text{path}}$ propagation paths, but some share the same angles. Hence, we can reformulate (5.177) as

$$\mathbf{h} = \sum_{i=1}^{N_{\text{cl}}} \left(\sum_{n=1}^{N_{\text{path}}} \alpha_{i,n} e^{-\mathrm{j}\psi_{i,n}} \right) \mathbf{a}(\varphi_i, \theta_i). \tag{5.178}$$

We can model this as a complex Gaussian distributed channel vector if N_{path} is large, but it will not result in what we previously called i.i.d. Rayleigh fading. To demonstrate this, suppose the phase-shifts $\psi_{i,n} \sim U[-\pi, \pi)$ are independent and uniformly distributed random variables and that the channel attenuations $\alpha_{i,n}$ within any cluster i are independent and identically distributed random variables with an average channel gain denoted by

$$\mathbb{E}\left\{\alpha_{i,n}^2\right\} = \frac{\beta_i}{N_{\text{path}}}. \tag{5.179}$$

This implies that $\sum_{n=1}^{N_{\text{path}}} \mathbb{E}\{\alpha_{i,n}^2\} = \beta_i$ irrespectively of the number of paths, thus, the parameter $\beta_i \in [0,1]$ determines the average channel gain of the entire cluster i. It follows from the central limit theorem in Lemma 2.6 that

$$\sum_{n=1}^{N_{\text{path}}} \alpha_{i,n} e^{-\mathrm{j}\psi_{i,n}} \to \mathcal{N}_{\mathbb{C}}(0, \beta_i) \tag{5.180}$$

in probability as $N_{\text{path}} \to \infty$. Note that the normalization by N_{path} in (5.179) is essential to keep the variance β_i constant as the number of paths increases; otherwise, the summation would diverge instead of converge to a Gaussian distribution. However, this is only a mathematical technicality because we recall from Figure 5.4 that the convergence to the Gaussian distribution is approximately achieved for $N_{\text{path}} \geq 5$ if all the attenuations are equally large.

We let $c_i \sim \mathcal{N}_{\mathbb{C}}(0, \beta_i)$ denote the fading variables in (5.180) for $i = 1, \ldots, N_{\text{cl}}$, and notice that these are independent random variables. When there are many paths per cluster, we can therefore rewrite (5.178) as

$$\mathbf{h} = \sum_{i=1}^{N_{\text{cl}}} c_i \mathbf{a}(\varphi_i, \theta_i). \tag{5.181}$$

We call this scenario *clustered rich multipath propagation* where the word "rich" signifies that there are many paths, so we get Rayleigh fading. However, these paths are not isotropically distributed over the angular domain but are confined to a limited number of multipath clusters with distinct angles.

The channel response $\mathbf{h}$ in (5.181) is a linear combination of the N_{cl} array response vectors $\mathbf{a}(\varphi_i, \theta_i)$ using the coefficients c_i that are complex Gaussian random distributed. Hence, the channel has the random distribution

$$\mathbf{h} \sim \mathcal{N}_{\mathbb{C}}(\mathbf{0}, \mathbf{R}_h), \tag{5.182}$$

where the covariance matrix can be computed as

$$\mathbf{R}_h = \mathbb{E}\{\mathbf{h}\mathbf{h}^{\mathrm{H}}\} = \sum_{i=1}^{N_{\text{cl}}} \sum_{n=1}^{N_{\text{cl}}} \mathbb{E}\{c_i c_n^*\} \mathbf{a}(\varphi_i, \theta_i) \mathbf{a}^{\mathrm{H}}(\varphi_n, \theta_n) = \sum_{i=1}^{N_{\text{cl}}} \beta_i \mathbf{a}(\varphi_i, \theta_i) \mathbf{a}^{\mathrm{H}}(\varphi_i, \theta_i). \tag{5.183}$$

The last step in (5.183) follows from utilizing that $\mathbb{E}\{c_i c_n^*\} = 0$ when $i \neq n$ since the variables are independent and have zero means. We will refer to $\mathbf{R}_h$ as the *spatial correlation matrix* since it describes how the channel coefficients at different spatial locations (i.e., antenna locations) are correlated.

Clustered rich multipath propagation gives rise to *spatially correlated Rayleigh fading* since each entry of $\mathbf{h}$ has a magnitude that is Rayleigh distributed, but the entries are statistically correlated. Practical channels generally feature spatially correlated fading. The level of correlation can vary depending on the number of multipath clusters and their angular locations.

One way to measure the level of correlation is by inspecting the eigenvalue spread of $\mathbf{R}_h$; a large spread represents a high correlation, while a small spread represents a low correlation. Note that all the eigenvalues are equal when considering i.i.d. Rayleigh fading with $\mathbf{R}_h = \beta \mathbf{I}_M$. The MISO channel model in (5.181)–(5.183) can also be utilized for SIMO channels by interchanging the roles of the transmitter and receiver.

Instead of having a single angle pair (φ_i, θ_i) per cluster, we can associate each one with a limited but continuous range of angles. For example, a typical cluster is an object that the antenna array observes over a limited but continuous range of angles. It then makes sense to replace the summation in (5.183) with an integral expression. This can be done as

$$\mathbf{R}_h = \beta \int_{-\pi}^{\pi} \int_{-\pi/2}^{\pi/2} f_{\varphi,\theta}(\varphi, \theta) \mathbf{a}(\varphi, \theta) \mathbf{a}^{\mathrm{H}}(\varphi, \theta) \partial\theta \partial\varphi, \tag{5.184}$$

where $f_{\varphi,\theta}(\varphi, \theta)$ is the angular density function of the multipath components and β represents the average channel gain over all clusters. The former function describes how the multipath components are distributed over the angular domain. The function is normalized such that $\int_{-\pi}^{\pi} \int_{-\pi/2}^{\pi/2} f_{\varphi,\theta}(\varphi, \theta) \partial\theta \partial\varphi = 1$. The covariance model in (5.184) is a continuous generalization of the discrete model in (5.183) because we can obtain the latter one by selecting $\beta = \sum_{i=1}^{N_{\mathrm{cl}}} \beta_i$ and $f_{\varphi,\theta}(\varphi, \theta) = \sum_{i=1}^{N_{\mathrm{cl}}} \frac{\beta_i}{\beta} \delta(\varphi - \varphi_i) \delta(\theta - \theta_i)$. Even if there is a finite number of clusters, the continuous model is more realistic since the abrupt Dirac delta function $\delta(\cdot)$ can be replaced with something smoother.

Example 5.17. Suppose there is only one multipath cluster centered around (φ_1, θ_1), but it spans a horizontal angular window of length $2\Delta_\varphi$ and a vertical angular window of length $2\Delta_\theta$. What will be the spatial correlation matrix $\mathbf{R}_h$ if the multipath components are uniformly distributed?

Under these conditions, the angular density function should be constant over the specified intervals. This is achieved by

$$f_{\varphi,\theta}(\varphi, \theta) = \begin{cases} \frac{1}{4\Delta_\varphi \Delta_\theta}, & \text{if } |\varphi - \varphi_1| \leq \Delta_\varphi, |\theta - \theta_1| \leq \Delta_\theta, \\ 0, & \text{otherwise}. \end{cases} \tag{5.185}$$

By substituting this into (5.184), we obtain the spatial correlation matrix

$$\mathbf{R}_h = \frac{\beta}{4\Delta_\varphi \Delta_\theta} \int_{\varphi_1 - \Delta_\varphi}^{\varphi_1 + \Delta_\varphi} \int_{\theta_1 - \Delta_\theta}^{\theta_1 + \Delta_\theta} \mathbf{a}(\varphi, \theta) \mathbf{a}^{\mathrm{H}}(\varphi, \theta) \partial\theta \partial\varphi. \tag{5.186}$$

This model originates from [71] and is known as the *one-ring model* since it appears in the hypothetical scenario where all the multipath components are on a ring-shaped object around the single-antenna device.

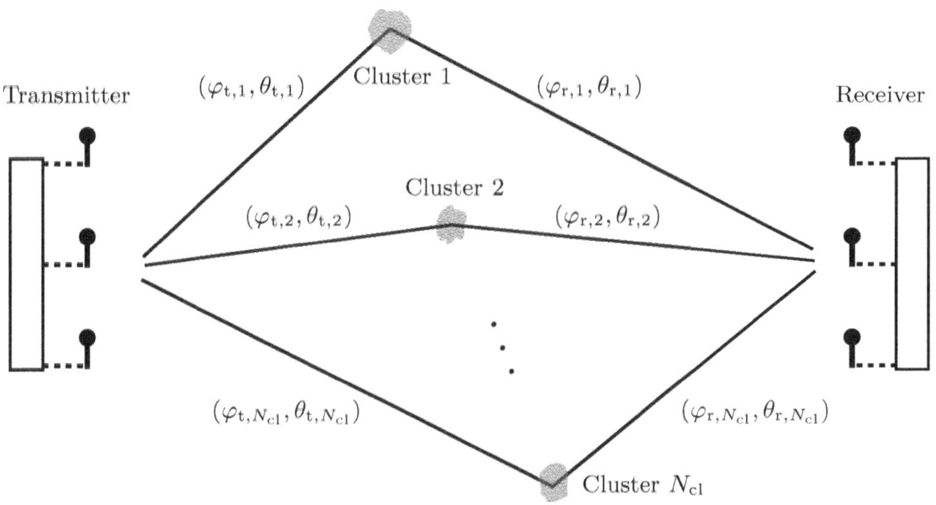

Figure 5.29: Illustration of a sparse multipath propagation environment where only signal components that depart the transmitter in some distinct angular directions $(\varphi_{t,i}, \theta_{t,i})$ will reach the receiver, and only arrive from some distinct angular directions $(\varphi_{r,i}, \theta_{r,i})$. The figure shows N_{cl} such cluster directions.

We can extend the propagation model to capture a point-to-point MIMO channel with N_{cl} clusters between the transmitter and receiver. Cluster i is located in the direction $(\varphi_{t,i}, \theta_{t,i})$ seen from the transmitter and in the direction $(\varphi_{r,i}, \theta_{r,i})$ seen from the receiver. This setup is illustrated in Figure 5.29. Let $\mathbf{a}_K(\varphi, \theta) \in \mathbb{C}^K$ denote the array response vector of the ULA at the transmitter and $\mathbf{a}_M(\varphi, \theta) \in \mathbb{C}^M$ denote the array response vector of the ULA at the receiver, which can both be modeled as in (4.120). If isotropic antennas are utilized, then the channel matrix $\mathbf{H} \in \mathbb{C}^{M \times K}$ can be expressed as

$$\mathbf{H} = \sum_{i=1}^{N_{cl}} c_i \mathbf{a}_M(\varphi_{r,i}, \theta_{r,i}) \mathbf{a}_K^{T}(\varphi_{t,i}, \theta_{t,i}), \tag{5.187}$$

where $c_i \sim \mathcal{N}_{\mathbb{C}}(0, \beta_i)$ is an independent random variable that models the rich multipath within the ith cluster. The channel matrix consists of a superposition of N_{cl} components, each determined by the angular directions of the multipath cluster via the array response vectors, and having the average channel gain $\beta_i \in [0, 1]$. The rank of this channel matrix is determined by the number of antennas, the number of multipath clusters, and the angular locations of these clusters. If the multipath clusters are well separated in the angular domain, the summation of N_{cl} paths in (5.187) implies that the rank could be N_{cl}. However, the rank is also upper bounded by $\min(M, K)$ since $\mathbf{H}$ only has that many singular values. Hence, the maximum channel rank is $\min(M, K, N_{cl})$.

Suppose the transmitter uses directive antennas with the gain function $G_t(\varphi, \theta)$ and the receiver uses directive antennas with the gain function

$G_{\mathrm{r}}(\varphi,\theta)$, where the angles are defined as for the respective array response vectors. In line with (4.148), we can extend the channel model in (5.187) as

$$\mathbf{H} = \sum_{i=1}^{N_{\mathrm{cl}}} c_i \sqrt{G_{\mathrm{t}}(\varphi_{\mathrm{t},i},\theta_{\mathrm{t},i})G_{\mathrm{r}}(\varphi_{\mathrm{r},i},\theta_{\mathrm{r},i})}\mathbf{a}_M(\varphi_{\mathrm{r},i},\theta_{\mathrm{r},i})\mathbf{a}_K^{\mathrm{T}}(\varphi_{\mathrm{t},i},\theta_{\mathrm{t},i}). \tag{5.188}$$

Since the multipath clusters are located in different directions, they will be associated with different gains $G_{\mathrm{t}}(\varphi_{\mathrm{t},i},\theta_{\mathrm{t},i})G_{\mathrm{r}}(\varphi_{\mathrm{r},i},\theta_{\mathrm{r},i})$. However, the variance β_i was already assumed to be cluster-specific, so we can simplify the notation by absorbing the antenna gains into these variables. Hence, we can define

$$\bar{c}_i = c_i\sqrt{G_{\mathrm{t}}(\varphi_{\mathrm{t},i},\theta_{\mathrm{t},i})G_{\mathrm{r}}(\varphi_{\mathrm{r},i},\theta_{\mathrm{r},i})} \sim \mathcal{N}_{\mathbb{C}}\left(0,\bar{\beta}_i\right) \tag{5.189}$$

where $\bar{\beta}_i = \beta_i G_{\mathrm{t}}(\varphi_{\mathrm{t},i},\theta_{\mathrm{t},i})G_{\mathrm{r}}(\varphi_{\mathrm{r},i},\theta_{\mathrm{r},i})$ and then use the original channel model in (5.187) with $\bar{c}_i$ instead of c_i.

Example 5.18. How does the Rician fading distribution extend to the clustered rich multipath propagation scenario?

Rician fading was introduced in Example 5.2 to model channels where there exists an LOS path in addition to the many NLOS paths. The LOS path has a particular set of departure angles $(\varphi_{\mathrm{t},0},\theta_{\mathrm{t},0})$ at the transmitter and arrival angles $(\varphi_{\mathrm{r},0},\theta_{\mathrm{r},0})$ at the receiver. This is not a cluster since there is only one path. If we add this path to (5.187), we obtain

$$\mathbf{H} = \alpha_0 e^{-\mathrm{j}\psi_0}\mathbf{a}_M(\varphi_{\mathrm{r},0},\theta_{\mathrm{r},0})\mathbf{a}_K^{\mathrm{T}}(\varphi_{\mathrm{t},0},\theta_{\mathrm{t},0}) + \sum_{i=1}^{N_{\mathrm{cl}}} c_i\mathbf{a}_M(\varphi_{\mathrm{r},i},\theta_{\mathrm{r},i})\mathbf{a}_K^{\mathrm{T}}(\varphi_{\mathrm{t},i},\theta_{\mathrm{t},i}), \tag{5.190}$$

where $\alpha_0 \in [0,1]$ models the attenuation and $\psi_0 \sim U[-\pi,\pi)$ models the phase-shift of the LOS path. Each entry $h_{m,k}$ of this matrix has a magnitude with the Rician distribution, thereof the name Rician fading.

When using this model, it is common to let $\beta = \mathbb{E}\{|h_{m,k}|^2\} = \alpha_0^2 + \sum_{i=1}^{N_{\mathrm{cl}}}\beta_i$ denote the average gain of each channel coefficient. One can then define the so-called κ-factor determining how the average gain is divided between the LOS and NLOS paths: $\kappa = \alpha_0^2/\sum_{i=1}^{N_{\mathrm{cl}}}\beta_i$. Using this notation, we can generate random MIMO channel realizations as

$$\begin{aligned}\mathbf{H} =& \sqrt{\frac{\kappa}{\kappa+1}}\sqrt{\beta}e^{-\mathrm{j}\cdot U[-\pi,\pi)}\mathbf{a}_M(\varphi_{\mathrm{r},0},\theta_{\mathrm{r},0})\mathbf{a}_K^{\mathrm{T}}(\varphi_{\mathrm{t},0},\theta_{\mathrm{t},0}) \\ &+ \sum_{i=1}^{N_{\mathrm{cl}}}\mathcal{N}_{\mathbb{C}}(0,\beta_i)\mathbf{a}_M(\varphi_{\mathrm{r},i},\theta_{\mathrm{r},i})\mathbf{a}_K^{\mathrm{T}}(\varphi_{\mathrm{t},i},\theta_{\mathrm{t},i}),\end{aligned} \tag{5.191}$$

where the phase of the LOS path and the Rayleigh fading of each cluster are the sources of randomness.

5.6.2 Beamspace Representation

Propagation channels can be sparse in the angular domain in the sense that there is a small number of multipath clusters. Such sparsity is particularly evident when communicating in the mmWave and THz bands because the wireless signals then interact less favorably with objects in the propagation environment. One example is provided in Figure 5.30, where the signal with the higher frequency is more damped when transmitted through a blocking object (e.g., propagating through a wall). Although the world between the transmitter and receiver is the same regardless of the signal frequency, the number of impactful multipath components can change drastically. The strength of the LOS and specularly reflected paths are almost wavelength-independent, while paths that interact with multiple objects virtually disappear at higher frequencies—thereby leaving only a few dominant paths. The resulting angular sparsity is not visible in the MISO channel vector $\mathbf{h}$ in (5.181) where all the entries have roughly the same magnitude because each antenna reaches the receiver with nearly the same power. However, the sparsity can be extracted by transforming the channel vector from the antenna domain (where each entry represents a physical antenna) to the angular domain (where each entry represents an angular interval). The angular domain representation is nowadays known as the *beamspace* [72], but was initially called the *virtual channel representation* [73] and has also been named the *Weichselberger model* due to Weichselberger's seminal work [74] that generalizes the model and highlights its connections to beamforming and multiplexing.

We recall from Section 4.3.3 that the columns of the DFT matrix $\mathbf{F}_M$ in (2.198) generate a grid of orthogonal beams, which spans all angular directions when using a half-wavelength-spaced ULA with the aperture length $D = M\frac{\lambda}{2}$. Hence, it acts as an orthogonal basis for the channel, where each basis vector represents a specific angular interval. We can denote the nth column of $\mathbf{F}_M$ as

$$\begin{aligned}\mathbf{f}_{M,n} &= \frac{1}{\sqrt{M}}\begin{bmatrix} 1 \\ e^{-\mathrm{j}\pi(n-1)\frac{2}{M}} \\ e^{-\mathrm{j}\pi 2(n-1)\frac{2}{M}} \\ \vdots \\ e^{-\mathrm{j}\pi(M-1)(n-1)\frac{2}{M}} \end{bmatrix} \\ &= \begin{cases} \frac{1}{\sqrt{M}}\mathbf{a}_M\left(\arcsin\left(\frac{2(n-1)}{M}\right), 0\right), & n = 1,\dots,\lfloor\frac{M}{2}\rfloor + 1, \\ \frac{1}{\sqrt{M}}\mathbf{a}_M\left(\arcsin\left(\frac{2(n-1)}{M} - 2\right), 0\right), & n = \lfloor\frac{M}{2}\rfloor + 2,\dots,M, \end{cases} \end{aligned} \tag{5.192}$$

where $\mathbf{a}_M(\varphi,\theta) \in \mathbb{C}^M$ denotes the array response vector in (4.120) for $\Delta = \lambda/2$. We can express this relation in short form as

$$\mathbf{f}_{M,n} = \frac{1}{\sqrt{M}}\mathbf{a}_M\left(\arcsin\left(\left[\frac{2(n-1)}{M}\right]_{-1:1}\right), 0\right), \quad n = 1,\dots,M, \tag{5.193}$$

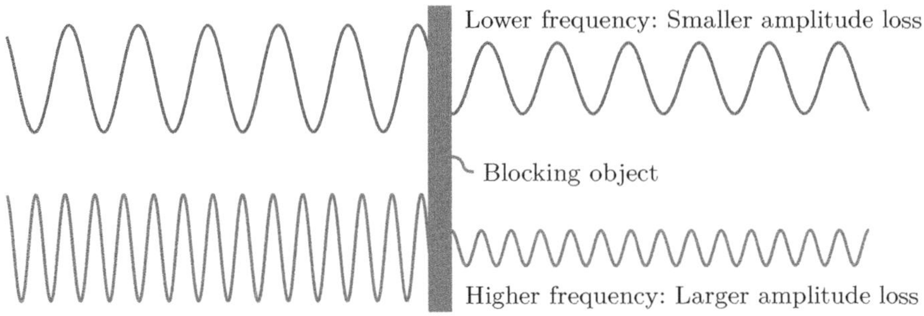

Figure 5.30: Signals with a higher frequency are subject to a larger amplitude loss when transmitted through a blocking object.

where $[\cdot]_{-1:1}$ wraps the argument within the range $(-1, 1]$ and is defined as

$$[x]_{-1:1} = \begin{cases} x, & x \leq 1, \\ x - 2, & x > 1, \end{cases} \tag{5.194}$$

If we multiply the channel vector in (5.181) with the conjugate transpose of the DFT matrix, we obtain the channel's beamspace representation

$$\breve{\mathbf{h}} = \mathbf{F}_M^{\mathrm{H}} \mathbf{h} = \sum_{i=1}^{N_{\mathrm{cl}}} c_i \begin{bmatrix} \mathbf{f}_{M,1}^{\mathrm{H}} \mathbf{a}(\varphi_i, \theta_i) \\ \vdots \\ \mathbf{f}_{M,M}^{\mathrm{H}} \mathbf{a}(\varphi_i, \theta_i) \end{bmatrix}, \tag{5.195}$$

where each cluster only contributes to one or a few of the entries.

Figure 5.31 illustrates this transformation in a scenario with $M = 10$ antennas and $N_{\mathrm{cl}} = 4$ multipath clusters, each located in one of the DFT beam directions. The DFT beams are numbered from 1 to 10 and are the same as in Figure 4.19(a). Each beam is associated with the interval where it provides the largest beamforming gain. We recall that the main beams partially overlap, so this is an approximate division of the angular domain. As the DFT beams in (5.193) have equally-spaced sine values of their azimuth angles, they represent equally-spaced spatial frequencies. Beam n is centered around the spatial frequency $[2(n-1)/M]_{-1:1}/\lambda$ and covers the interval between $[(2(n-1)-1)/M]_{-1:1}/\lambda$ and $[(2(n-1)+1)/M]_{-1:1}/\lambda$, which has the width $\frac{2}{M\lambda} = \frac{1}{D}$ that is inversely proportional to the aperture length D. Consequently, the angular spacing is wider in the end-fire directions (i.e., $\pm\pi/2$) than in the broadside direction. In the propagation scenario illustrated in Figure 5.31, the beamspace representation $\breve{\mathbf{h}}$ of the channel has $N_{\mathrm{cl}} = 4$ non-zero entries illustrated by the colored boxes, each associated with one of the four clusters. The other six entries of the vector are zero (white boxes). This illustrates how the angular sparsity created by the small number of clusters can be exposed by transforming the channel vector to the beamspace.

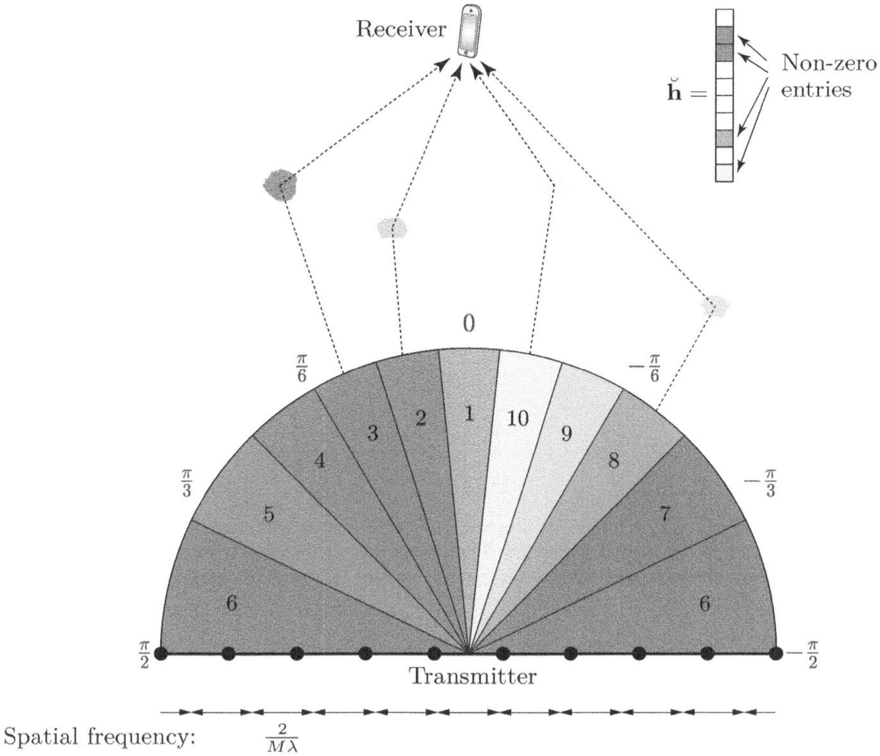

Figure 5.31: A transmitter equipped with a half-wavelength-spaced ULA with $M = 10$ antennas communicates with a single-antenna receiver. The angular domain is divided into M intervals that match the DFT beams in Figure 4.19(a). Each covers an interval of length $2/(M\lambda)$ in terms of spatial frequencies. The beamspace representation $\breve{\mathbf{h}} = \mathbf{F}_M^{\mathrm{H}}\mathbf{h}$ of the channel vector has $N_{\mathrm{cl}} = 4$ non-zero entries, each generated by a multipath cluster located in an angular direction that coincides with one of the DFT beams. Note that the size of the transmitter is exaggerated compared to the propagation distances in this figure.

To shed further light on the beamspace representation, we return to the NLOS example in Figure 5.28, where the channel contains four paths with distinctly different angles. If we transform this channel to the beamspace, we obtain 10 channel components, one per DFT beam direction. Figure 5.32 shows the relative strength of these channel components (in the decibel scale). There are six large and four small components; thus, angular sparsity is also prevalent in this scenario. However, the path directions do not precisely match the directions of the DFT beams, which is why several paths are smeared out over multiple beam directions. Moreover, none of the channel components are precisely zero. This is not because there are weak signals arriving from all directions but due to the side-lobes that show up almost everywhere; when we look for signals in a particular beam direction, we can pick up signals from a very different direction through a side-lobe. The behavior in this example represents what we will experience in practical scenarios with angular sparsity.

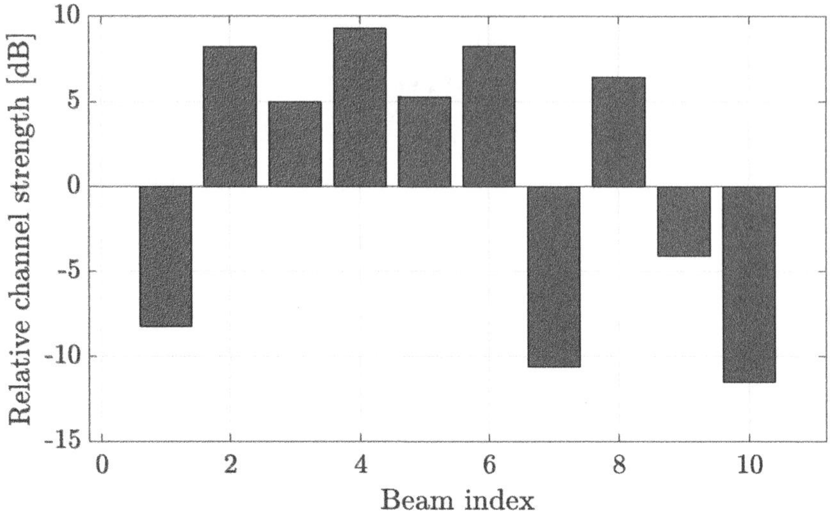

Figure 5.32: The strength of the 10 channel components when the NLOS channel from Figure 5.28 is transformed to the beamspace. The channel consists of four propagation paths but has six large channel components since some paths are smeared out over multiple DFT beams.

Example 5.19. Suppose there are $N_{\text{cl}} = M$ clusters that are equally spaced in terms of spatial frequency, such that $\mathbf{a}(\varphi_i, \theta_i) = \sqrt{M}\mathbf{f}_{M,i}$ for $i = 1, \ldots, M$. What is the spatial correlation matrix $\mathbf{R}_h$?

Under the given assumptions, we can express the spatial correlation matrix in (5.183) as

$$\mathbf{R}_h = \sum_{i=1}^{N_{\text{cl}}} \beta_i \mathbf{a}(\varphi_i, \theta_i)\mathbf{a}^{\text{H}}(\varphi_i, \theta_i) = \sum_{i=1}^{M} \beta_i M \mathbf{f}_{M,i}\mathbf{f}_{M,i}^{\text{H}} = \mathbf{F}_M \mathbf{B} \mathbf{F}_M^{\text{H}}, \tag{5.196}$$

where $\mathbf{B} = \text{diag}(M\beta_1, \ldots, M\beta_M)$. The last expression is the eigendecomposition of the spatial correlation matrix; thus, the columns of the DFT matrix are the eigenvectors and each associated eigenvalue $M\beta_i$ is the total average channel gain from the cluster i to all the antennas.

When using two half-wavelength-spaced ULAs, the MIMO channel matrix in (5.187) can also be transformed to the beamspace by multiplying by DFT matrices of matching dimensions from the left and the right:

$$\check{\mathbf{H}} = \mathbf{F}_M^{\text{H}} \mathbf{H} \mathbf{F}_K^* = \sum_{i=1}^{N_{\text{cl}}} c_i \times$$
$$\begin{bmatrix} \mathbf{f}_{M,1}^{\text{H}} \mathbf{a}_M(\varphi_{\text{r},i}, \theta_{\text{r},i}) \mathbf{a}_K^{\text{T}}(\varphi_{\text{t},i}, \theta_{\text{t},i}) \mathbf{f}_{K,1}^* & \ldots & \mathbf{f}_{M,1}^{\text{H}} \mathbf{a}_M(\varphi_{\text{r},i}, \theta_{\text{r},i}) \mathbf{a}_K^{\text{T}}(\varphi_{\text{t},i}, \theta_{\text{t},i}) \mathbf{f}_{K,K}^* \\ \vdots & \ddots & \vdots \\ \mathbf{f}_{M,M}^{\text{H}} \mathbf{a}_M(\varphi_{\text{r},i}, \theta_{\text{r},i}) \mathbf{a}_K^{\text{T}}(\varphi_{\text{t},i}, \theta_{\text{t},i}) \mathbf{f}_{K,1}^* & \ldots & \mathbf{f}_{M,M}^{\text{H}} \mathbf{a}_M(\varphi_{\text{r},i}, \theta_{\text{r},i}) \mathbf{a}_K^{\text{T}}(\varphi_{\text{t},i}, \theta_{\text{t},i}) \mathbf{f}_{K,K}^* \end{bmatrix}. \tag{5.197}$$

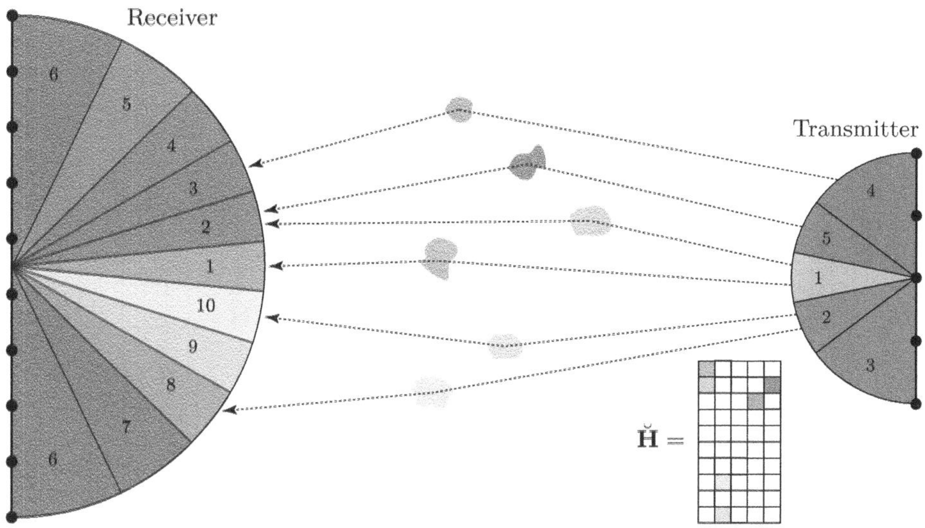

Figure 5.33: A transmitter equipped with a half-wavelength-spaced ULA with $K = 5$ antennas communicates with a receiver with a half-wavelength-spaced ULA with $M = 10$ antennas. The beamspace representation $\breve{\mathbf{H}} = \mathbf{F}_M^{\mathrm{H}}\mathbf{H}\mathbf{F}_K^*$ of the channel matrix has $N_{\mathrm{cl}} = 6$ non-zero entries, each generated by a multipath cluster located in an angular direction that coincides with one of the DFT beams at each side. The rank of the channel matrix is 4, which is the number of linearly independent columns in $\breve{\mathbf{H}}$. Note that the transmitter and receiver sizes are exaggerated compared to the propagation distances.

Each column of the transformed matrix represents a viable angular transmission direction seen from the transmitter, while each row represents a viable angular reception direction. Each of the N_{cl} multipath clusters will appear in one matrix entry or possibly a few neighboring entries. Figure 5.33 shows an example where a transmitter with $K = 5$ antennas communicates with a receiver with $M = 10$ antennas. There are $N_{\mathrm{cl}} = 6$ clusters that connect the transmitter and receiver, and the non-zero entries of $\breve{\mathbf{H}}$ are illustrated with coloring in the figure. The rank r of the channel matrix is essential to determine how many data streams can be spatially multiplexed over the channel. As the multiplication with unitary matrices (e.g., DFT matrices) does not change the rank, we can utilize the beamspace representation when determining the rank. The rank is the maximum number of linearly independent columns (or rows) of the matrix. In this example, there are four linearly independent columns and one empty column; thus, the rank and multiplexing gain are $r = 4$.

The channel rank is determined by how many dimensions the transmitter can reach the receiver through (i.e., the number of non-zero columns of $\breve{\mathbf{H}}$) and how many dimensions the receiver can hear the transmitter through (i.e., the number of non-zero rows of $\breve{\mathbf{H}}$). Figure 5.34 shows three examples of such beamspace matrices. Case (a) is a rich multipath environment with clusters in all directions, resulting in all entries of $\breve{\mathbf{H}}$ being non-zero. This channel

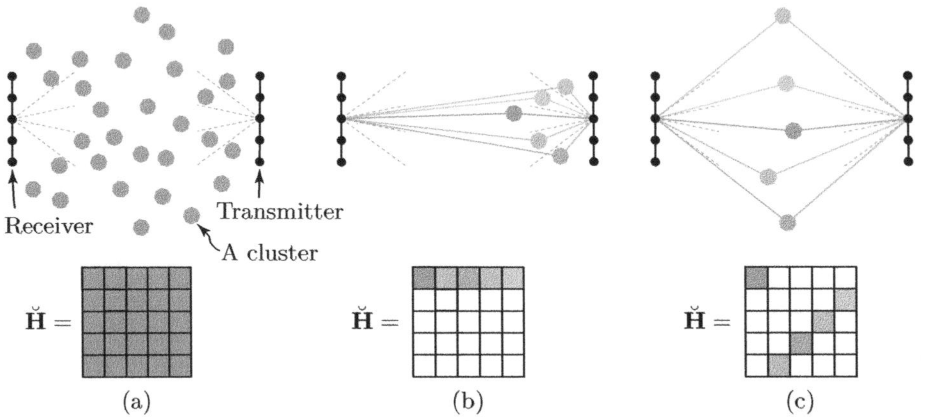

Figure 5.34: Three examples of MIMO channels represented in the beamspace. There are different numbers of multipath clusters in these examples. The clusters are distributed differently over the DFT beams, resulting in channels with different ranks and numbers of random coefficients. There are $K = 5$ transmit antennas and $M = 5$ receive antennas in all the examples. The white entries of the matrix $\check{\mathbf{H}}$ are approximately zero.

has the full rank. The MK sources of randomness make it likely that all the singular values are of comparable sizes, making this channel well-suited for spatial multiplexing. Case (b) has many clusters around the transmitter, but the receiver observes all of them through a single DFT beam. This could happen when a transmitting user device is surrounded by scattering objects while the receiving base station is elevated far above them and sees them all from roughly the same direction. The rank of $\check{\mathbf{H}}$ is 1, so this MIMO channel only provides beamforming gains. Case (c) has a small number of clusters, but these have well-separated angles that make $\check{\mathbf{H}}$ diagonal, so the channel has full rank. Since there are fewer sources of randomness than in Case (a), there might be significant variations in the singular values. This setup resembles the MIMO channel illustrated in Figure 3.16, where each singular value is associated with a single cluster.

The rank of a MIMO channel is fundamentally limited by the aperture lengths at the transmitter and receiver, along with the sizes and locations of the multipath clusters. The DFT beams are equally spaced in the spatial frequency domain from $-1/\lambda$ to $1/\lambda$. When the ULAs are half-wavelength-spaced, each beam at the transmitter covers a spatial frequency interval of length $\frac{2}{K\lambda}$ while each beam at the receiver covers an interval of length $\frac{2}{M\lambda}$. The rank is determined by how many such intervals are covered by the multipath clusters, which can be interpreted as the *spatial bandwidth* of the channel.

Suppose the transmitter is connected to the receiver through $N_{\mathrm{t,cl}}$ multipath clusters visible at the transmitter in non-overlapping angular directions. We will let the clusters have arbitrary sizes; thus, integrals are required to obtain the resulting channel covariance matrices. In particular, we let cluster

n extend from the angle-of-departure $\varphi_{\mathrm{t},n}^{\mathrm{start}}$ to $\varphi_{\mathrm{t},n}^{\mathrm{end}}$ so that it covers a range of spatial frequencies of length

$$\Omega_{\mathrm{t},n} = \frac{\left|\sin(\varphi_{\mathrm{t},n}^{\mathrm{end}}) - \sin(\varphi_{\mathrm{t},n}^{\mathrm{start}})\right|}{\lambda}. \tag{5.198}$$

The total range of spatial frequencies that the multipath clusters cover is $\sum_{n=1}^{N_{\mathrm{t,cl}}} \Omega_{\mathrm{t},n} \in [0, 2/\lambda]$, which represents the channel's spatial bandwidth from the transmitter perspective. If the spatial bandwidth is divided equally between the beamspace dimensions at the transmitter, the number of non-zero columns in $\breve{\mathbf{H}}$ will be approximately

$$\frac{\sum_{n=1}^{N_{\mathrm{t,cl}}} \Omega_{\mathrm{t},n}}{\frac{2}{K\lambda}} = \frac{K\lambda}{2} \sum_{n=1}^{N_{\mathrm{t,cl}}} \Omega_{\mathrm{t},n} = D_{\mathrm{t}} \sum_{n=1}^{N_{\mathrm{t,cl}}} \Omega_{\mathrm{t},n}, \tag{5.199}$$

where $D_{\mathrm{t}} = K\Delta = \frac{K\lambda}{2}$ denotes the aperture length of the transmitter. We divided by $2/(K\lambda)$ since this is the spatial frequency interval represented by each column. The maximum value in (5.199) is $D_{\mathrm{t}}\frac{2}{\lambda} = K$, which equals the number of transmit antennas and, thereby, the number of columns of $\breve{\mathbf{H}}$.[13]

Similarly, suppose the receiver is connected to the transmitter through $N_{\mathrm{r,cl}}$ multipath clusters visible at the receiver in non-overlapping angular directions. Cluster n extends from the angle-of-arrival $\varphi_{\mathrm{r},n}^{\mathrm{start}}$ to $\varphi_{\mathrm{r},n}^{\mathrm{end}}$ so that it covers a range of spatial frequencies of length

$$\Omega_{\mathrm{r},n} = \frac{\left|\sin(\varphi_{\mathrm{r},n}^{\mathrm{end}}) - \sin(\varphi_{\mathrm{r},n}^{\mathrm{start}})\right|}{\lambda}. \tag{5.200}$$

The total range of spatial frequencies excited by the multipath clusters is $\sum_{n=1}^{N_{\mathrm{r,cl}}} \Omega_{\mathrm{r},n} \in [0, 2/\lambda]$, which represents the channel's spatial bandwidth from the receiver perspective. If the spatial bandwidth is divided equally between the beamspace dimensions at the receiver, the number of non-zero rows in $\breve{\mathbf{H}}$ will be approximately

$$\frac{\sum_{n=1}^{N_{\mathrm{r,cl}}} \Omega_{\mathrm{r},n}}{\frac{2}{M\lambda}} = \frac{M\lambda}{2} \sum_{n=1}^{N_{\mathrm{r,cl}}} \Omega_{\mathrm{r},n} = D_{\mathrm{r}} \sum_{n=1}^{N_{\mathrm{r,cl}}} \Omega_{\mathrm{r},n}, \tag{5.201}$$

where $D_{\mathrm{r}} = M\Delta = \frac{M\lambda}{2}$ denotes the aperture length of the receiver.

These principles are illustrated in Figure 5.35 for a setup with $N_{\mathrm{t,cl}} = N_{\mathrm{r,cl}} = 4$ visible multipath clusters at the transmitter and receiver. The widths of the colored intervals at the transmitter and receiver show the ranges of spatial frequencies $\Omega_{\mathrm{t},n}$ and $\Omega_{\mathrm{r},n}$ that the respective multipath clusters are covering, for $n = 1, \ldots, 4$. The same cluster can cover a vastly different spatial frequency range at the transmitter compared to the receiver, depending on its

[13]The maximum value is, for example, achieved when there is a single cluster covering all angles from $\varphi_{\mathrm{t},1}^{\mathrm{start}} = -\pi/2$ to $\varphi_{\mathrm{t},1}^{\mathrm{end}} = \pi/2$ so that $\Omega_{\mathrm{t},1} = \left|\sin(\varphi_{\mathrm{t},1}^{\mathrm{end}}) - \sin(\varphi_{\mathrm{t},1}^{\mathrm{start}})\right| / \lambda = 2/\lambda$.

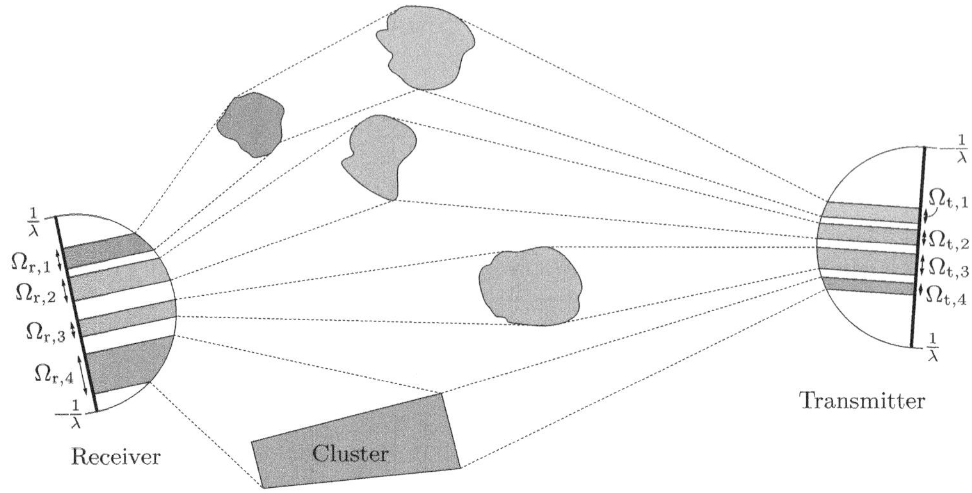

Figure 5.35: The rank of a MIMO channel matrix with half-wavelength-spaced ULAs is approximately determined by (5.202), which depends on the aperture lengths at the transmitter and receiver, as well as the widths of the spatial frequency ranges $\Omega_{t,n}$ and $\Omega_{r,n}$ that the multipath clusters are covering at the transmitter and receiver, respectively.

distance and orientation with regards to each of them. As the channel rank is the minimum number of non-zero columns and rows, we obtain

$$\mathrm{rank}(\mathbf{H}) = \mathrm{rank}(\breve{\mathbf{H}}) \approx \min\left(D_t \sum_{n=1}^{N_{t,cl}} \Omega_{t,n}, D_r \sum_{n=1}^{N_{r,cl}} \Omega_{r,n} \right), \tag{5.202}$$

which is an approximation because the rank must be integer-valued and the edges of the clusters might be divided between multiple matrix entries, which could slightly increase the rank. On the other hand, the rank only specifies the number of non-zero singular values of the channel matrix but does not guarantee that they are of comparable size. The latter depends on the relative strengths of the signals traveling through the respective multipath clusters.

The MIMO rank analysis can be extended to half-wavelength-spaced UPAs, which can resolve spatial frequencies both horizontally and vertically, as discussed in Section 4.5.3. Previously in this section, we noticed that a half-wavelength-spaced ULA with the aperture length $D = M\frac{\lambda}{2}$ can achieve a maximum rank of $M = D\frac{2}{\lambda}$, where $2/\lambda$ is the maximum range of spatial frequencies in one dimension. The UPA has a horizontal aperture length $M_H\frac{2}{\lambda}$ that can be used to resolve horizontal spatial frequencies and a vertical aperture length $M_V\frac{2}{\lambda}$ that can be used to resolve vertical spatial frequencies. By decoupling these dimensions, one might expect from the previous analysis that the maximum channel rank is $M_H M_V = \mathsf{Area} \cdot \frac{4}{\lambda^2}$, where $\mathsf{Area} = M_H\frac{\lambda}{2}M_V\frac{\lambda}{2}$ denotes the UPA's aperture area. However, this is incorrect because only some combinations of horizontal and vertical frequencies can coexist. The possible combinations lie within the circle with diameter $2/\lambda$ illustrated in Figure 4.42, while the incorrect decoupling argument above considered all combinations in a square with side length $2/\lambda$. The relative area difference between these geometrical shapes is $\pi(1/\lambda)^2/(2/\lambda)^2 = \pi/4$. Consequently, the actual maximum MIMO channel rank that the UPA can support is $\mathsf{Area} \cdot \frac{\pi}{\lambda^2}$ when $\mathsf{Area}\,\mathrm{m}^2$ is the aperture area. In other words, each segment with area λ^2 can add (approximately) π to the channel rank. If two such half-wavelength-spaced UPAs are placed in a realistic environment, the MIMO channel rank is determined by how many spatial frequencies are excited by the multipath clusters; that is, which fractions of the circle with spatial frequencies in Figure 4.42 are excited. The rank becomes approximately $\mathsf{Area} \cdot \min(\Omega_{\mathrm{t}}, \Omega_{\mathrm{r}})$, where $\Omega_{\mathrm{t}}, \Omega_{\mathrm{r}} \in [0, \frac{\pi}{\lambda^2}]$ denote the total areas of the parts of the circle that are excited at the transmitter and receiver, respectively. A precise derivation requires more extensive mathematical notation, so we refer to [75]–[77] for such details, which can also be used to analyze the MIMO channel rank arbitrarily shaped antenna arrays.

The beamspace representation for a half-wavelength-spaced 9×9 UPA is illustrated in Figure 5.36. We begin by revisiting the NLOS setup from Figure 5.28 with four propagation paths having distinct azimuth and elevation angles. Figure 5.36(a) shows the real part of the wave impinging on the array. The UPA samples the wave at the marked antenna locations. The resulting channel can be turned into the beamspace by taking a 2D-DFT of the channel coefficients, which results in the 2D spatial frequency spectrum shown in Figure 5.36(c). There are four peaks, which match the number of paths. It was implicitly assumed in this example that the paths gave rise to plane waves. However, the beamspace analysis can be applied regardless of the wavefront. Figure 5.36(b) shows the real part of the impinging wave emitted from a transmitter at a short distance of 8λ in the broadside direction. There are large circle-shaped phase variations over the array, which is typical for spherical waves. When turning this near-field channel into the beamspace, we obtain the 2D spatial frequency spectrum shown in Figure 5.36(d). The spectrum contains a range of spatial frequencies centered around zero, while a plane wave would only contain the zero-valued frequency. Since the 2D-IDFT recreates the channel using the discrete spatial frequencies shown in the figure, a spherical wave can be represented as a summation of multiple plane waves. A formal connection can be made using the Weyl identity [78]. The bottom line is that any channel matrix can be represented in the beamspace. The maximum channel rank result holds even in the presence of spherical waves, which are summations of many plane waves.

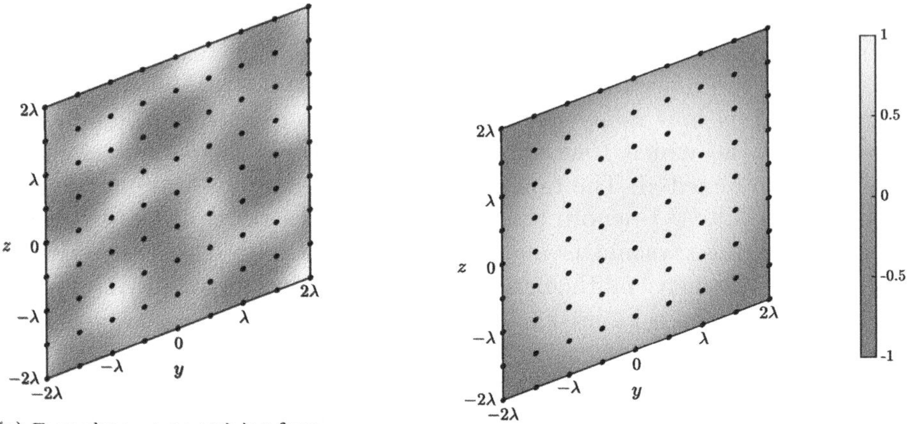

(a) Four plane waves arriving from $(\varphi_1, \theta_1) = (\pi/6, \pi/6)$, $(\varphi_2, \theta_2) = (\pi/12, -\pi/4)$, $(\varphi_3, \theta_3) = (-\pi/4, 0)$, $(\varphi_4, \theta_4) = (-\pi/3, \pi/3)$.

(b) One spherical wave from a transmitter at a distance 8λ in the broadside direction.

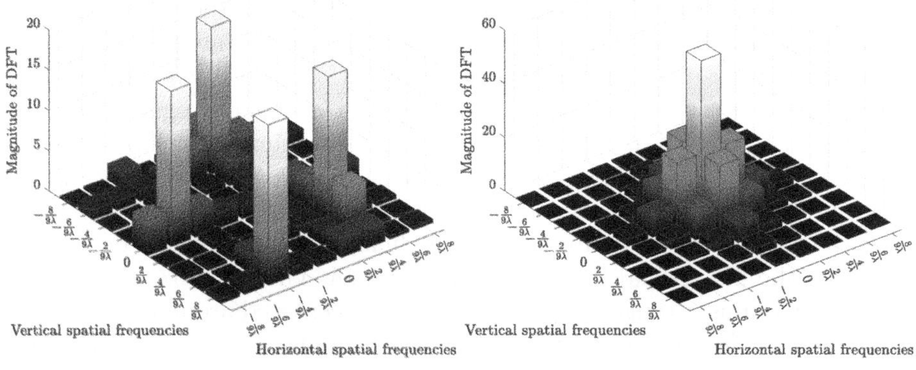

(c) 2D-DFT of the channel from (a).

(d) 2D-DFT of the channel from (b).

Figure 5.36: The wave that impinges on a UPA can have many different shapes. The real parts of two waves are shown in (a) and (b) for different kinds of propagation channels. The observed horizontal and vertical spatial frequencies differ for these channels. When the UPA has 9×9 half-wavelength-spaced antennas, the spatial frequencies shown in (c) and (d) are observable.

Example 5.20. How does the wavelength impact the rank of the channel matrix if the multipath clusters remain the same?

The rank expression in (5.202) for a ULA can be expressed as

$$\min\left(\frac{D_{\mathrm{t}}}{\lambda}\sum_{n=1}^{N_{\mathrm{t,cl}}}\left|\sin(\varphi_{\mathrm{t},n}^{\mathrm{end}})-\sin(\varphi_{\mathrm{t},n}^{\mathrm{start}})\right|,\frac{D_{\mathrm{r}}}{\lambda}\sum_{n=1}^{N_{\mathrm{r,cl}}}\left|\sin(\varphi_{\mathrm{r},n}^{\mathrm{end}})-\sin(\varphi_{\mathrm{r},n}^{\mathrm{start}})\right|\right), \tag{5.203}$$

which depends on the wavelength λ through the normalized aperture lengths $D_{\lambda,\mathrm{t}} = \frac{D_{\mathrm{t}}}{\lambda}$ and $D_{\lambda,\mathrm{r}} = \frac{D_{\mathrm{r}}}{\lambda}$ of the transmitter and receiver, respectively. If we keep the aperture lengths D_{t} and D_{r} constant (i.e., constant array sizes in meters), the rank is inversely proportional to the wavelength. Hence, we obtain a larger channel rank as the wavelength shrinks (e.g., from the low-band to the mmWave band). To achieve this, we need a larger number of antennas since the antenna spacing $\lambda/2$ is also wavelength-dependent, which explains why the spatial resolution improves so that the same clusters generate more channel dimensions. If we instead keep the normalized aperture lengths $D_{\lambda,\mathrm{t}}$ and $D_{\lambda,\mathrm{r}}$ fixed, then the rank becomes independent of the wavelength. This is achieved by using a fixed number of half-wavelength-spaced antennas.

An i.i.d. Rayleigh fading channel matrix obtained using half-wavelength-spaced ULAs can also be transformed to the beamspace. The entries of $\breve{\mathbf{H}}$ will remain independent and identically distributed because the multiplications with the unitary DFT matrices in (5.197) do not change the distribution. This demonstrates how the multipath components are uniformly distributed over all angular directions instead of clustered. The maximum diversity order of a MIMO channel equals the number of distinguishable sources of independent randomness, which is MK under i.i.d. Rayleigh fading and (approximately) equal to the number of non-zero entries of $\breve{\mathbf{H}}$ under clustered scattering. In Figure 5.34, Case (a) can represent i.i.d. Rayleigh fading if all the entries are identically distributed. In this case, the maximum diversity order is $MK = 25$. The channel matrices in Case (b) and Case (c) have 5 non-zero entries; thus, the maximum diversity order is 5. We need to use transmit diversity in Case (b) since the sources of randomness are only distinguishable from the transmitter's viewpoint. In contrast, it is sufficient to exploit receive diversity in Case (c), while the transmitter can send the same signal in each DFT beam direction. The diversity order generally equals the number of (large) non-zero entries of the beamspace channel matrix. The diversity gain increases with the number of antennas (with $\Delta = \lambda/2$) since the improved spatial resolution will divide a multipath cluster between multiple matrix entries in the beamspace representation, thereby creating additional distinguishable sources of randomness.

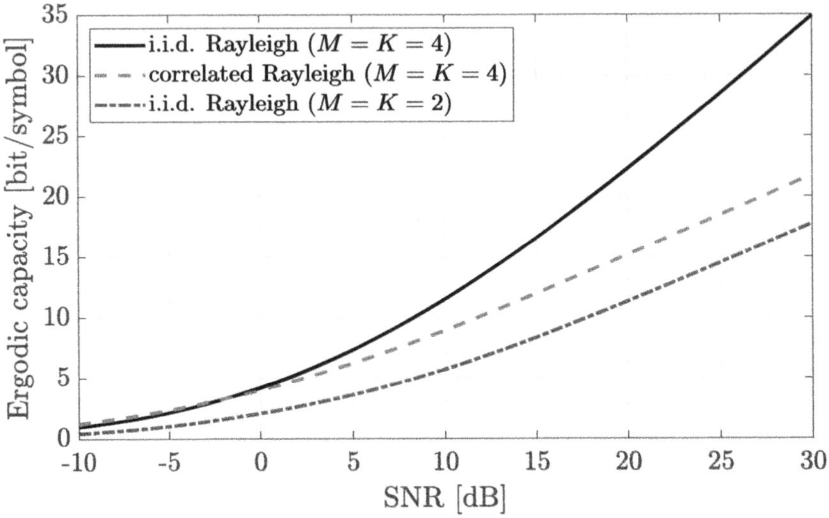

Figure 5.37: The ergodic rate as a function of the SNR with either i.i.d. Rayleigh fading or correlated fading, where the beamspace representation of the channel matrix only contains a non-zero block of size 2×2 (i.e., the rank is 2). The correlation reduces the multiplexing gain compared to i.i.d. Rayleigh fading, but a beamforming gain is still achieved.

The capacity with clustered scattering can be evaluated similarly to the cases with i.i.d. Rayleigh fading. In particular, the ergodic capacity expression in (5.129) can be applied when the receiver has perfect CSI while the transmitter has no CSI. However, the optimal covariance matrix $\mathbf{R}_x$ of the transmitted signal will not be a scaled identity matrix but will depend on the clusters seen from the transmitter. For example, in a MISO scenario in which the channel distribution $\mathbf{h} \sim \mathcal{N}_{\mathbb{C}}(\mathbf{0}, \mathbf{R}_h)$ is given in (5.182), it can be shown that the optimal covariance matrix $\mathbf{R}_x$ has the same eigenvectors as $\mathbf{R}_h$ and allocates power between these dimensions based on how large their eigenvalues are. We refer to [79], [80] for further the precise details. When the transmitter also has perfect CSI, we can compute the maximum rate for a single channel realization using Theorem 3.1 (i.e., transmitting in the directions of the right singular vectors and using the water-filling power allocation) and then compute the mean value over the fading channel.

Figure 5.37 shows the ergodic capacity achieved with $M = K = 4$ antennas when the transmitter and receiver have perfect CSI. A channel with i.i.d. Rayleigh fading is compared with a correlated channel where $\breve{\mathbf{H}}$ contains a 2×2 non-zero submatrix, representing multipath clusters that cover half of the DFT beams. The channel matrix in the correlated scenario has the rank $r = 2$ while the rank is $r = 4$ with i.i.d. Rayleigh fading, which implies that the largest singular value is generally larger under correlated fading. This results in a slightly higher capacity at low SNRs, but the benefit is lost at higher SNRs where the larger multiplexing gain leads to a faster capacity growth under i.i.d. Rayleigh fading. The capacity with $M = K = 2$ and i.i.d. Rayleigh fading is shown as a reference. It achieves the same multiplexing gain as in the correlated scenario, but the capacity curve is shifted to the right by 6 dB since the beamforming gain is 4 times smaller. In general, clustered multipath propagation has a detrimental impact on the ergodic capacity compared to i.i.d. fading, but spatial correlation is a naturally occurring characteristic that must be considered when modeling practical channels.

5.6.3 Fading Channels with Dual-Polarized Antennas

The i.i.d. Rayleigh fading model was derived earlier in this chapter under the implicit assumption of using single-polarized antenna arrays. That kind of fading cannot be achieved with dual-polarized antennas because the channel between a transmit antenna and a receive antenna is generally stronger if their polarization matches compared to if they have opposite polarizations; hence, the fading is not identically distributed. The channel model derived in Section 4.6.3 for LOS MIMO channels with dual-polarized antennas can be readily generalized for NLOS channels with clustered multipaths. That model was constructed for a scenario with $K/2$ dual-polarized transmit antennas and $M/2$ dual-polarized receive antennas. The antennas were numbered so that transmit antennas $1, \dots, K/2$ and receive antennas $1, \dots, M/2$ have matching polarization, while the same holds for transmit antennas $K/2+1, \dots, K$ and receive antennas $M/2+1, \dots, M$.

Suppose there are N_{cl} multipath clusters between the transmitter and receiver, of which cluster i is located in the direction $(\varphi_{\mathrm{t},i}, \theta_{\mathrm{t},i})$ seen from the transmitter and in the direction $(\varphi_{\mathrm{r},i}, \theta_{\mathrm{r},i})$ seen from the receiver. This is the same notation as in (5.187). The channel component through each cluster can be modeled similarly to (4.176) but with additional Rayleigh fading coefficients that model the random amplitudes and phases. We obtain the channel matrix

$$\mathbf{H} = \sum_{i=1}^{N_{\mathrm{cl}}} \begin{bmatrix} \sqrt{1-\kappa} c_{i,1,1} & \sqrt{\kappa} c_{i,1,2} \\ \sqrt{\kappa} c_{i,2,1} & \sqrt{1-\kappa} c_{i,2,2} \end{bmatrix} \otimes \left(\mathbf{a}_{M/2}(\varphi_{\mathrm{r},i}, \theta_{\mathrm{r},i}) \mathbf{a}_{K/2}^{\mathrm{T}}(\varphi_{\mathrm{t},i}, \theta_{\mathrm{t},i}) \right), \tag{5.204}$$

where $c_{i,1,1}, c_{i,1,2}, c_{i,2,1}, c_{i,2,2} \sim \mathcal{N}_{\mathbb{C}}(0, \beta_i)$ are independent Rayleigh fading coefficients and $\beta_i \in [0, 1]$ models the average channel gain of cluster i. The channel has a limited XPD in the sense that there is leakage between the orthogonal polarizations characterized by the parameter κ. The diagonal entries $\sqrt{1-\kappa} c_{i,1,1}, \sqrt{1-\kappa} c_{i,2,2}$ characterize the signal propagation that maintains its polarization, while the off-diagonal entries $\sqrt{\kappa} c_{i,1,2}, \sqrt{\kappa} c_{i,2,1}$ characterize the leakage between the polarizations. We recall that $\kappa = 0$ represents perfect discrimination, while $\kappa = 0.5$ is the worst-case situation. In the LOS scenario considered in Section 4.6.3, the limited XPD was caused by imperfect isolation within the transmitter and receiver hardware. The situation is different in

NLOS scenarios. Each reflection/scattering in a multipath environment can shift the wave's polarization, which creates further leakage that is factored into the parameter κ in the considered model; that is, κ is likely larger in NLOS setups than in LOS setups.

The primary implication of adding a second polarization is that it provides extra sources of randomness since the two polarizations fade independently, even if the XPD creates an additional kind of spatial correlation.[14] Hence, dual polarization can double the channel matrix's rank in a propagation environment with few multipath clusters. It can also double the diversity order (or even quadruple it, thanks to the imperfect XPD). Polarization has been utilized since the early days of mobile communications [56], [58] to enhance performance and reliability.

We have only considered single- and dual-polarized antennas so far. As wireless signals propagate in three dimensions, up to three mutually orthogonal linear polarization dimensions exist, which represent the x, y, and z axes in a coordinate system. We can only utilize two of these in LOS communications because the polarization must be perpendicular to the direction the waves propagate to the receiver. However, NLOS channels allow waves to follow widely different paths from the transmitter to the receiver. The signal can leave the transmitter in any direction and reach the receiver from any direction, and objects in the environment might allow a signal with any polarization to (partially) maintain that polarization when reaching the receiver. It is possible to build tri-polarized antennas, for example, with one antenna pointing along each axis in the coordinate system. The channel measurements reported in [82], [83] confirm the viability of building tri-polarized MIMO communication systems. However, the improvement in data rate is slight compared to having optimally rotated dual-polarized antennas. Hence, the main benefit is that one can keep a consistent rate regardless of how the user device is rotated, while the challenge is to integrate tri-polarized antennas into the form factor of a base station and device.

[14]Many measurements show that there exists a slight correlation between the fading realizations $c_{i,1,1}, c_{i,1,2}, c_{i,2,1}, c_{i,2,2}$, which is not captured by the model used in this chapter. There can also exist imbalances between the variances of the two polarization dimensions since the multipath distributions generally differ horizontally and vertically. The latter effect can be reduced by using slanted polarization [56], [57]. We refer to [81] for further details and ways to enrich the MIMO channel model to include such characteristics.

5.7 Exercises

Exercise 5.1. We used the central limit theorem to motivate that a SISO system has the channel response $h \sim \mathcal{N}_{\mathbb{C}}(0, \beta)$ in a rich multipath environment. In this exercise, we will revisit the technical conditions that underpin that $h = \sum_{i=1}^{L} \alpha_i e^{-\mathrm{j}\psi_i} \to \mathcal{N}_{\mathbb{C}}(0, \beta)$ when the number of paths L is very large. We assume that the path attenuations α_i are independent and identically distributed, the phases $\psi_i \sim U[-\pi, \pi)$ are independent, and $\beta > 0$ is a constant.

(a) Suppose the path gains satisfy $\mathbb{E}\{\alpha_i^2\} = \beta/L^2$, for $i = 1, \dots, L$. What is the variance of h when $L \to \infty$? Is h complex Gaussian distributed?

(b) Suppose the path gains satisfy $\mathbb{E}\{\alpha_i^2\} = \beta/L$, for $i = 1, \dots, L$. What is the variance of h when $L \to \infty$? Is h complex Gaussian distributed?

(c) Suppose the path gains satisfy $\mathbb{E}\{\alpha_i^2\} = \beta$, for $i = 1, \dots, L$. What is the variance of h when $L \to \infty$? Is h complex Gaussian distributed?

Exercise 5.2. The spatial correlation expression in (5.24) is derived for isotropic scattering, for which the multipath components are uniformly distributed over the unit sphere as stated in (5.17). When other distributions are used, one can compute the spatial correlation differently. One such example is the Clarke model, where the joint PDF of the azimuth and elevation angles is

$$f_{\varphi,\theta}(\varphi, \theta) = \frac{1}{2\pi^2}, \quad -\pi \leq \varphi < \pi, \quad -\frac{\pi}{2} \leq \theta \leq \frac{\pi}{2}. \tag{5.205}$$

(a) For a ULA located along the z-axis with the channel response in (5.15), obtain correlation $\mathbb{E}\{h_m h_n^*\}$ between the channel realizations at antenna m and n.

(b) Express the correlation in (a) using the zeroth-order Bessel function of the first kind, defined as

$$J_0(x) = \frac{1}{\pi} \int_{-\pi/2}^{\pi/2} e^{-\mathrm{j}x \sin\theta} \partial\theta. \tag{5.206}$$

Exercise 5.3. Consider the setup in Figure 5.7 for which the SISO channel coefficient is given in (5.29) as a function of time: $h(t) = 2\alpha \cos(2\pi \frac{\upsilon t}{\lambda})$. What is the coherence time if we define it as the time it takes to move from a peak to losing half the received power?

Exercise 5.4. Consider the SISO channel in (5.35) with slow fading, where $y[l] = h \cdot x[l] + n[l]$. Suppose the channel coefficient h is a realization of a random variable that is zero with probability p and one with probability $1 - p$.

(a) What is the outage probability of this channel? Express the answer as a function of the desired rate R.

(b) Suppose we instead have two receive antennas that observe independent channel realizations, each with the mentioned distribution. What is the outage probability for the desired rate R?

Exercise 5.5. Consider the SISO channel in (5.35) with slow fading, where $y[l] = h \cdot x[l] + n[l]$. The channel coefficient has the uniform distribution $h \sim U[-1, 1]$.

(a) Derive the outage probability of this channel. Express the answer as a function of the desired rate $R \geq 0$.

(b) Suppose we have M receive antennas and these observe *the same* channel realization h. What is the outage probability in this case?

(c) Derive expressions for the ϵ-outage capacities for the setups in (a) and (b). Sketch a graph of the expressions for $M = 1$, $M = 4$, and $M = 10$ with ϵ on the horizontal axis and the ϵ-outage capacity on the vertical axis for $q/N_0 = 1$.

Exercise 5.6. Consider the SISO channel in (5.35) with slow fading, where $y[l] = h \cdot x[l] + n[l]$. Compute the outage probability and ϵ-outage capacity for the following fading distributions.

(a) The channel gain $|h|^2$ has the PDF

$$f_{|h|^2}(x) = \begin{cases} 2(1-x), & 0 \leq x \leq 1, \\ 0, & \text{otherwise}. \end{cases} \tag{5.207}$$

(b) The channel gain $|h|^2$ has the PDF

$$f_{|h|^2}(x) = \begin{cases} \frac{3\sqrt{x}}{2}, & 0 \leq x \leq 1, \\ 0, & \text{otherwise}. \end{cases} \tag{5.208}$$

Exercise 5.7. Consider a SIMO system with M antennas.

(a) The receiver has a hardware limitation that only allows it to use one of its antennas at a time, known as *antenna selection*. Which antenna should the receiver select to maximize the rate for given realizations of $h_1, \ldots, h_M$? Provide an expression for the maximum rate.

(b) Suppose the channel is subject to slow i.i.d. Rayleigh fading and only the receiver knows the channel realization. Formulate the outage probability when communicating at a given rate R. Hint: Use the identity

$$\Pr\left\{\max\{|h_1|^2, \ldots, |h_M|^2\} < x\right\} = \Pr\left\{|h_1|^2 < x, \ldots, |h_M|^2 < x\right\}.$$

(c) Compare the outage probability of the antenna selection scheme with that of MRC given in (5.52). Which one is larger?

(d) Derive the high-SNR slope of the outage probability curve when using antenna selection. Compare the results with the high-SNR slope achieved with MRC. Hint: Use the approximation $e^{-x} \approx 1 - x$, which is tight when x is small.

Exercise 5.8. The diversity order of a channel can be defined as

$$\lim_{\text{SNR}\to\infty} \frac{-\ln(P_{\text{out}}(R))}{\ln(\text{SNR})}, \tag{5.209}$$

where SNR denotes the SNR and the outage probability $P_{\text{out}}(R)$ is a function of the SNR for any fixed value of R. Use the exact expression of the outage probability in (5.53) for a SIMO channel and verify that the diversity order is M according to this definition. Hint: Use the identity $e^x = \sum_{m=0}^{\infty} \frac{x^m}{m!}$ and L'Hôpital's rule.

Exercise 5.9. Consider a MISO system with $M = 2$ antennas and slow fading, where the receiver knows the channel but not the transmitter. The channel coefficients are distributed as $h_1, h_2 \sim \mathcal{N}_{\mathbb{C}}(0, \beta)$ but are *correlated* in the sense that $h_1 = h_2^*$ for every channel realization. Use the Alamouti code for transmission and compute the outage probability, both the exact value and an upper bound that exposes the diversity order. Compare the diversity order with what is achieved when h_1 and h_2 are independent. Hint: The expression in (5.73) holds for any channel distribution, so the solution can start from that point.

Exercise 5.10. One popular definition of channel hardening is that the fading SIMO/MISO channel vector $\mathbf{h} \in \mathbb{C}^M$ must satisfy

$$\frac{\|\mathbf{h}\|^2}{\mathbb{E}\{\|\mathbf{h}\|^2\}} \to 1 \quad \text{as } M \to \infty. \tag{5.210}$$

This means that all realizations of $\|\mathbf{h}\|^2$ will be close to the mean value $\mathbb{E}\{\|\mathbf{h}\|^2\}$ when there are many antennas. The convergence in (5.210) can be proved in a mean-squared sense by computing the variance of $\frac{\|\mathbf{h}\|^2}{\mathbb{E}\{\|\mathbf{h}\|^2\}}$ and show that it goes to zero as $M \to \infty$. Follow this approach to prove that an i.i.d. Rayleigh fading channel provides channel hardening.

Exercise 5.11. Consider a slow-fading SISO channel with a repetition scheme where the same signal $x \sim \mathcal{N}_{\mathbb{C}}(0, q)$ is transmitted over L time slots.

(a) Compute the conditional capacity for a given realization of the channel coefficient h.

(b) Compare the capacity obtained in (a) with a conventional slow-fading SISO channel without a repetition scheme. What value of L maximizes the capacity? Hint: Use the inequality $\frac{x}{x+1} < \ln(1+x)$, for $x > 0$.

(c) Obtain a low-SNR approximation of the capacity in (a). How does it depend on L?

Exercise 5.12. Consider MISO and SIMO channels with slow-fading and M transmit and receive antennas, respectively, under i.i.d. Rayleigh fading. Only the receiver knows the channel realization. Due to hardware limitations, we can only adjust the phase of the precoding/combining vectors, so MRT/MRC is not possible. This is called *equal gain beamforming*.

(a) Consider the MISO case and show that the full transmit diversity order can be achieved using a repetition scheme where the same signal is repeated using M different orthogonal beams.

(b) Consider the SIMO case and propose a way to achieve the full receive diversity order.

Exercise 5.13. Consider a MISO channel with slow fading and i.i.d. Rayleigh fading. There are $M = 2$ transmit antennas and the Alamouti code is used.

(a) Use the low-SNR approximation of the conditional capacity to compute a low-SNR approximation of the outage probability expression.

(b) Compare the result in (a) with the low-SNR approximation of the outage probability for the corresponding SISO channel (without the Alamouti code). Which one is better in terms of outage performance at low SNR? Hint: Use $e^x > x + 1$, for $x > 0$.

Exercise 5.14. Consider a SIMO system with M antennas. The symbol power is q and the noise power spectral density is N_0.

(a) Suppose the channel vector is $\mathbf{h} = [1, \ldots, 1]^{\mathrm{T}}$. What is the capacity of the channel? What is performance gain compared to a corresponding SISO system with $h = 1$? Exemplify the performance gain at low and high SNRs. What is this kind of performance gain called?

(b) Suppose the channel vector $\mathbf{h}$ is instead subject to i.i.d. Rayleigh fading, so that $\mathbf{h} \sim \mathcal{N}_{\mathbb{C}}(\mathbf{0}, \mathbf{I}_M)$. Consider a fast-fading scenario where the receiver knows the channel but not the transmitter. What is the capacity of the channel? What are the performance gains compared to a corresponding fast-fading SISO channel with $h \sim \mathcal{N}_{\mathbb{C}}(0, 1)$?

(c) Compare the capacities in (a) and (b). Which one is the largest? What happens with the performance difference as $M \to \infty$? Hint: Use the channel hardening property in (5.210).

Exercise 5.15. Consider the SISO channel with fast fading in (5.103), where the received signal at the time instance l is $y[l] = h[l] \cdot x[l] + n[l]$. The channel coefficient $h[l]$ is a binary random number taking the realization 1 with probability p and 0 with probability $1 - p$. A new independent realization of $h[l]$ is drawn in each time instance l and the receiver knows the realization.

(a) What is the ergodic capacity of this SISO channel?

(b) Consider a fast-fading SIMO channel with M antennas and the channel vector $\mathbf{h}[l] = [h[l], \ldots, h[l]]^{\mathrm{T}}$. This vector takes a new independent realization at every time instance, but all the entries in the vector are always mutually identical. What is the ergodic capacity of this channel?

(c) Consider a fast-fading SIMO channel with M antennas. The channel coefficients are independent and identically distributed according to the distribution specified above. What is the ergodic capacity in this case? Hint: Use that $\|\mathbf{h}\|^2$ is the summation of the independent Bernoulli random variables $|h_m|^2$ and write the expectation using the binomial sum formula.

Exercise 5.16. Consider an i.i.d. Rayleigh fading MIMO channel with fast fading. The received signal at the time instance l is $\mathbf{y}[l] = \mathbf{H}[l]\mathbf{x}[l] + \mathbf{n}[l]$, where the noise is colored with the distribution $\mathbf{n}[l] \sim \mathcal{N}_{\mathbb{C}}(\mathbf{0}, N_0\mathbf{C})$ and $\mathbf{C}$ is a non-singular matrix. The channel takes an independent realization at each time instance, and only the receiver knows the realization. Determine the ergodic capacity. Hint: Use whitening.

Exercise 5.17. The fast-fading MIMO capacity is expressed in (5.132) in terms of the non-zero eigenvalues of $\mathbf{H}\mathbf{H}^{\mathrm{H}}$. Use a high-SNR approximation to express the ergodic capacity in terms of mean values involving those eigenvalues.

Exercise 5.18. Consider an i.i.d. Rayleigh fading MIMO channel with block fading. We must estimate the channel coefficients in each block to perform spatial multiplexing. Suppose L_c is the length of the coherence block and that M, K are larger than L_c. The pilot length L_p is a variable that must be smaller than L_c.

(a) Which multiplexing gain can we achieve for a given value of L_p? What is the pre-log factor in the ergodic capacity expression, which includes the multiplexing gain and the penalty from the pilot transmission?

(b) Which value of L_p maximizes the pre-log factor in (a)? What is the multiplexing gain in this case?

Exercise 5.19. Consider a MISO channel with N_{cl} multipath clusters, which has the channel vector

$$\mathbf{h} = \sum_{i=1}^{N_{\text{cl}}} c_i \mathbf{a}(\varphi_i, \theta_i), \tag{5.211}$$

where $c_1, \ldots, c_{N_{\text{cl}}}$ are i.i.d. $\mathcal{N}_{\mathbb{C}}(0, \beta/N_{\text{cl}})$ random variables. The transmitter has a ULA deployed along the y-axis with $\Delta = \lambda/2$ antenna spacing. The angles (φ_i, θ_i) are deterministic and different for every i (i.e., $\mathbf{a}(\varphi_i, \theta_i)$ is a different vector for every i).

(a) Determine if this channel provides channel hardening by following the approach from Exercise 5.10 for a fixed number $N_{\text{cl}} < \infty$ of multipath clusters. Hint: Use the trace property given in (2.52).

(b) Determine if this channel provides channel hardening when $N_{\text{cl}} \to \infty$. Hint: It holds that $\mathbb{V}\text{ar}\{\|\mathbf{h}\|^2\} = \text{tr}(\mathbf{R}^2)$ if $\mathbf{h} \sim \mathcal{N}_{\mathbb{C}}(\mathbf{0}, \mathbf{R})$ for any covariance matrix $\mathbf{R}$. Utilize beamwidth-like expressions to prove that $\frac{1}{M}|\mathbf{a}^{\text{H}}(\varphi_i, \theta_i)\mathbf{a}(\varphi_k, \theta_k)| \to 0$ as $M \to \infty$ when k and i are different.

Exercise 5.20. Consider two half-wavelength-spaced ULAs with 4 antennas deployed to be parallel. What is the minimum number of multipath clusters needed to achieve a full-rank channel matrix if all the paths in a cluster have the same angle? Suggest angle values for these clusters to achieve full rank. Hint: Use the DFT matrix.

Exercise 5.21. A half-wavelength-spaced ULA with M antennas is deployed along the y-axis. Consider the one-ring model from Example 5.17 with $\Delta_\theta = 0$, so that the multipath components are uniformly distributed in the azimuth plane in an interval of width $2\Delta_\varphi > 0$.

(a) Generalize this one-ring model to have N_{cl} non-overlapping clusters in the azimuth plane. Cluster i is centered around the azimuth angle φ_i, for $i = 1, \ldots, N_{\text{cl}}$, and has the average channel gain β_i. Hence, the angular density function for cluster i is

$$f_i(\varphi) = \begin{cases} \frac{1}{2\Delta_\varphi}, & \text{if } |\varphi - \varphi_i| \leq \Delta_\varphi, \\ 0, & \text{otherwise}. \end{cases} \tag{5.212}$$

Derive an expression for the spatial correlation matrix $\mathbf{R}_h$.

(b) Use the width of the clusters to approximate the rank of the matrix $\mathbf{R}_h$.

Exercise 5.22. The *Kronecker model* is a classic model for spatially correlated Rayleigh fading MIMO channels. The channel matrix is then given as

$$\mathbf{H} = \mathbf{R}_{\mathrm{r}}^{1/2}\mathbf{W}\left(\mathbf{R}_{\mathrm{t}}^{1/2}\right)^{\mathrm{T}}, \tag{5.213}$$

where the entries of $\mathbf{W} \in \mathbb{C}^{M \times K}$ are i.i.d. $\mathcal{N}_{\mathbb{C}}(0, \beta)$ random variables. The normalized spatial correlation matrices $\mathbf{R}_{\mathrm{r}}$ and $\mathbf{R}_{\mathrm{t}}$ have unit diagonal entries and characterize the spatial correlation among the channel realization at the receive and transmit antennas, respectively. Let the eigendecomposition of $\mathbf{R}_{\mathrm{r}}$ and $\mathbf{R}_{\mathrm{t}}$ be denoted as $\mathbf{R}_{\mathrm{r}} = \mathbf{U}_{\mathrm{r}}\mathbf{D}_{\mathrm{r}}\mathbf{U}_{\mathrm{r}}^{\mathrm{H}}$ and $\mathbf{R}_{\mathrm{t}} = \mathbf{U}_{\mathrm{t}}\mathbf{D}_{\mathrm{t}}\mathbf{U}_{\mathrm{t}}^{\mathrm{H}}$, respectively. The eigenvalues $\lambda_{\mathrm{r},1}, \ldots, \lambda_{\mathrm{r},M}$ and $\lambda_{\mathrm{t},1}, \ldots, \lambda_{\mathrm{t},K}$ are located along the diagonals of $\mathbf{D}_{\mathrm{r}}$ and $\mathbf{D}_{\mathrm{t}}$, respectively, in descending order.

(a) Compute $\breve{\mathbf{H}} = \mathbf{U}_{\mathrm{r}}^{\mathrm{H}}\mathbf{H}\mathbf{U}_{\mathrm{t}}^{*}$ and simplify the expression to show that its entries are independently distributed. This is a beamspace matrix similar to $\breve{\mathbf{H}} = \mathbf{F}_{M}^{\mathrm{H}}\mathbf{H}\mathbf{F}_{K}^{*}$ in (5.197).

(b) How do the variances vary between the entries of $\breve{\mathbf{H}}$? Can any beamspace matrix be expressed using the Kronecker model?

Exercise 5.23. Consider an antenna array in an isotropic rich scattering propagation environment where the multipath components cover all angular dimensions uniformly.

(a) Consider a 2-antenna ULA with $\Delta = \lambda/2$. What is the spatial correlation matrix? Hint: Use the correlation expression in (5.24).

(b) Consider a 2×2 UPA with $\Delta = \lambda/2$ vertical and horizontal spacing. What is the spatial correlation matrix? Hint: The coordinate system can be rotated arbitrarily, so the expression in (5.23) can be used when considering any pair of antennas.

(c) For which kinds of arrays will isotropic rich scattering imply i.i.d. fading?

Chapter 6

Capacity of Multi-User MIMO Channels

In this chapter, we will characterize the communication performance over multi-user MIMO channels, also known as point-to-multipoint and multipoint-to-point MIMO channels. We begin by explaining why the capacity gains achieved by point-to-point MIMO are limited in many practical scenarios, as a motivation for identifying an alternative way to use multiple antennas. The focus will then be on serving multiple users connecting to the same wireless system, which raises the question of whether the users should be assigned orthogonal or non-orthogonal transmission resources. To answer this, we will extend the capacity concept to the multi-user setting and discover how the use of multiple antennas radically changes the situation. We will adapt the precoding, combining, and power allocation schemes from previous chapters to maximize performance in the multi-user context. The tradeoff between non-linear and linear signal processing methods will finally be explored.

6.1 A Practical Issue with Point-to-Point MIMO Systems

In the previous two chapters, we have derived the capacity of point-to-point MIMO channels in both LOS and NLOS scenarios. The largest capacity improvement from having multiple antennas is achieved through the multiplexing gain. If we have M receive antennas and K transmit antennas, the capacity ideally becomes $\min(M, K)$ times larger than in a corresponding SISO channel. This can lead to a huge performance improvement if the channel satisfies two properties:

1. The channel matrix $\mathbf{H}$ has $\min(M, K)$ singular values of similar size;
2. The SNR is high.

Unfortunately, these properties seldom appear at the same time in practice. When the SNR is large, it is often because the channel contains one dominant propagation path (e.g., a LOS path) while the remaining paths are substantially weaker. Hence, $\mathbf{H}$ has only one or two large singular values (depending on whether single-polarized or dual-polarized antennas are considered), while the remaining ones are much smaller and potentially zero-valued regardless of how many antennas are deployed. In NLOS scenarios with isotropic rich scattering, the channel matrix will instead have full rank, and the singular value variations are quite small. However, the SNR is relatively low since a large fraction of the transmitted power disappears due to the multipath propagation; thus, only beamforming gains might be practically useful. Hence, LOS and NLOS propagation typically provide the opposite conditions of what is preferable from a theoretical perspective. At low SNRs, we would prefer a low-rank channel to make the most out of the beamforming gain, while we want a high-rank channel with many similar singular values to make the most out of the multiplexing gain at high SNRs.

Figure 6.1 exemplifies these issues by considering a point-to-point MIMO channel with $K = 4$ transmit antennas and $M = 4$ receive antennas. We compare an NLOS case with i.i.d. Rayleigh fading, where the ergodic capacity is computed using (5.131), and a single-polarized LOS case where the capacity is computed according to (4.96). The capacity of a non-fading SISO channel is shown as a reference. The use of multiple antennas provides the most significant capacity gains compared to the SISO case when considering NLOS channels with high SNR and LOS channels with low SNR. However, these events are unlikely to happen in practice. The figure indicates two more likely events: LOS with high SNR and NLOS with low SNR. In both cases, there are clear gains compared to the SISO channel, but they are still modest compared to what could be achieved in those SNR ranges.

In summary, an ideal point-to-point MIMO system operates at high SNR and achieves a multiplexing gain. However, in practice, we are likely to mainly achieve beamforming (and diversity) gains either because the SNR is low or the channel has a low rank. Reality can be slightly better than was illustrated in Figure 6.1 because LOS channels can contain a few strong reflected paths useful for spatial multiplexing, while NLOS channels feature clustered multipath propagation where a few directions provide better SNRs. Yet, the nature of signal propagation seems to hinder the point-to-point MIMO from reaching its "full capacity", except in short-range NLOS scenarios.

Another practical issue with having multiple antennas at both the transmitter and receiver is that one of them is usually a handheld user device. The number of antennas that can fit into such a device is limited for aesthetic reasons, particularly in the low-band and mid-band spectrum. This observation should not be interpreted as beamforming gains being pointless; on the contrary, practical cellular networks are designed and deployed to make good use of them. Recall from Section 3.1 that point-to-point SISO systems can either be power-limited (i.e., operate at low SNR) or bandwidth-limited (i.e., operate at high SNR). The capacity of a power-limited SISO system can

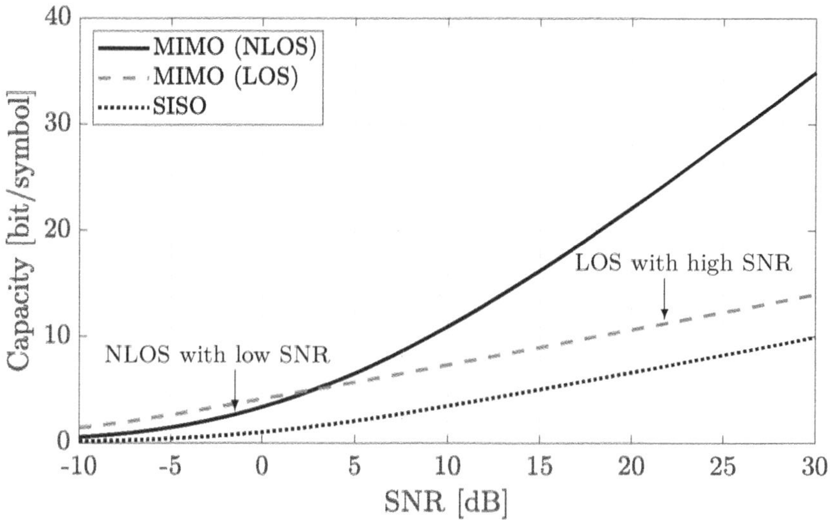

Figure 6.1: A comparison of the capacities of different MIMO channels with $M = K = 4$ antennas. There is i.i.d. Rayleigh fading in the NLOS case, while there is a rank-one channel in the LOS case. Since practical channels with high SNRs are often LOS channels, while NLOS channels experience low SNRs, the large potential capacity gains of point-to-point MIMO channels over SISO channels are hard to achieve in practice.

be greatly improved by adding antennas to achieve a beamforming gain that increases the SNR. This is practically relevant in systems operating over large distances (i.e., with a small channel gain per antenna) and/or using large bandwidths (e.g., in the high-band spectrum). Beamforming is probably a prerequisite for systems operating in mmWave and THz bands because we need similar aperture lengths as in the lower bands to achieve similar SNRs, which calls for using antenna arrays. However, beamforming gains have a limited impact on the bandwidth-limited SISO systems' capacity; thus, adding antenna arrays to such systems might only be worthwhile if we can achieve multiplexing gains.

An early indication of how to achieve multiplexing gains also in LOS scenarios was provided in Figure 4.28, where a ULA transmits to a receiver equipped with distributed antennas. Strictly speaking, this is not a point-to-point MIMO system but a point-to-multipoint MIMO system because the receive antennas were located at multiple geographically distributed points. While a user device will only exist at one point, deploying base stations at different points and letting them cooperate to serve a user is practically viable. This is called *coordinated multipoint* transmission [52] or *Cell-free MIMO* [2]. We refer to those references for further details since this chapter considers a different scenario: a base station at one point communicates with multiple user devices, each located at a geographically different point. This is known as *multi-user MIMO* communications.

6.2 Capacity Definition in Uplink and Downlink

In the rest of this chapter, we consider a setup where a base station deployed at a fixed location serves K user devices. This could represent an entire communication system or a single cell in a larger cellular network with many base stations that serve different geographical regions (cells). The users can move around in the coverage area; thus, the base station must adapt its transmissions to the current set of users. There are two relevant directions of communication. The transmission from the base station to the user devices is called the *downlink*, inspired by the fact that base stations are typically deployed at elevated locations and transmit down toward the users. Similarly, the transmission from the user devices to the base station is called the *uplink*. These communication directions are also known as the *forward link* and *reverse link*, respectively, especially in contexts where the down/up analogy is not applicable (e.g., when a ground-based base station serves flying objects). In information theory, the downlink of a multi-user system is known as the *broadcast channel*, while the uplink is called the *multiple access channel* [42]. The downlink and uplink are illustrated in Figure 6.2. The base station is shown as equipped with an antenna array capable of directing beams toward each user device, a feature we will explore later in this chapter. If there are NLOS channels to the users, the radiated signals will not look like angular beams, as discussed in Section 5.6. The user devices radiate signals (almost) isotropically, but only the parts directed toward the base station are indicated. Both the downlink and uplink will be analyzed in this chapter.

The downlink is a point-to-multipoint system where we transmit from one point (the base station location) to multiple points (the K user locations). It resembles the point-to-multipoint example in Figure 4.28, where a transmitter communicated with a receiver equipped with distributed antennas. However, two fundamental properties make the downlink setup different from an operational perspective. Firstly, each user only has access to its own received signal and not those at antennas belonging to other user devices. Secondly, the users want to access different data and are not interested in the data intended for others. Hence, each user measures its performance in terms of its individual channel capacity. In a system with K users, there are K different capacities to consider. This makes the system design more complicated and we will develop a theory in this chapter to manage it.

Let R_k bit/s be the variable denoting an achievable data rate of user k. From previous chapters, we know how to characterize the user's capacity when the user is alone in the system; for example, Theorem 3.1 gives the capacity when the base station and user operate as a point-to-point MIMO system. We will denote that capacity value by $C_k^{\rm su}$ bit/s, where su indicates that this is the *single-user capacity*. Hence, we know that the range of achievable data rates for user k is

$$0 \leq R_k \leq C_k^{\rm su}. \tag{6.1}$$

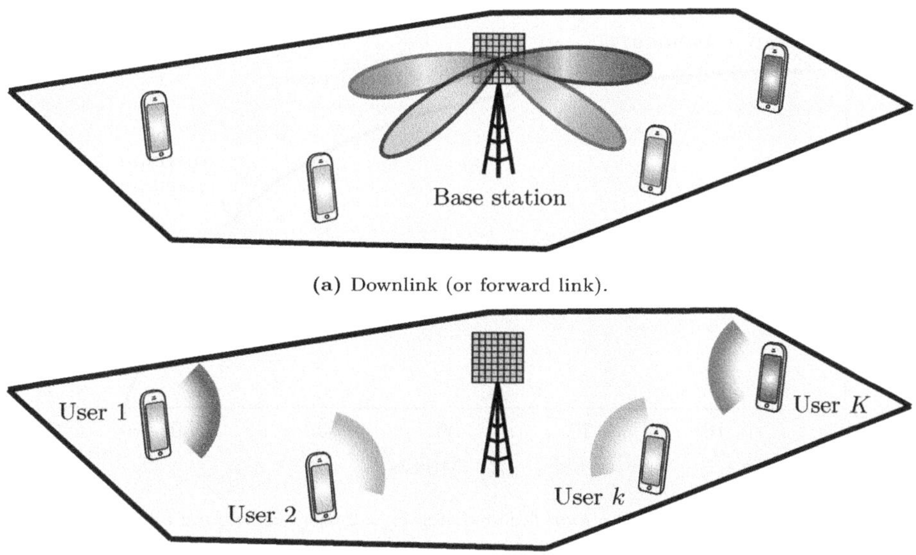

(a) Downlink (or forward link).

(b) Uplink (or reverse link).

Figure 6.2: In a cellular network, a fixed-location base station serves mobile user devices in a given coverage area called a cell. It can transmit beams toward the users in the downlink and receive signals from the users in the uplink, either simultaneously or sequentially.

It is usually impossible for two users to achieve their single-user capacities simultaneously because they share transmission resources, namely the time, frequency, transmit power, and spatial dimensions. Hence, a tradeoff exists between the performance different users can achieve, which must be modeled and dealt with in the system design. We will characterize this tradeoff in different scenarios but begin by describing the framework that can quantify it: the rate region. This is a set $\mathcal{R} \subset \mathbb{R}^K$ containing all the combinations of rates $(R_1, \ldots, R_K)$ that are simultaneously achievable in a given system (i.e., for the given channel conditions and transmission resources).

Figure 6.3 illustrates a rate region for a setup with $K = 2$ users, where the yellow-shaded region shows all the combinations/tuples of rates (R_1, R_2) simultaneously achievable. This includes the single-user capacity points $(C_1^{\mathrm{su}}, 0)$ and $(0, C_2^{\mathrm{su}})$. It also includes many different tradeoffs between these points, where one user reduces its capacity to allow the other user to increase its capacity. If a point (R_1, R_2) is inside the region, then any other point (R_1', R_2') that satisfies $0 \leq R_1' \leq R_1$ and $0 \leq R_2' \leq R_2$ is also inside the region. The intuition is that we can always purposely reduce the users' data rates and still obtain an achievable system operation. However, the interesting question is: how can we simultaneously make the rates as large as possible? The points on the *Pareto boundary*, which is the curved portion of the outer boundary, are of particular interest because these points are such that the rate cannot be improved for any user without deteriorating the rate for at least one other

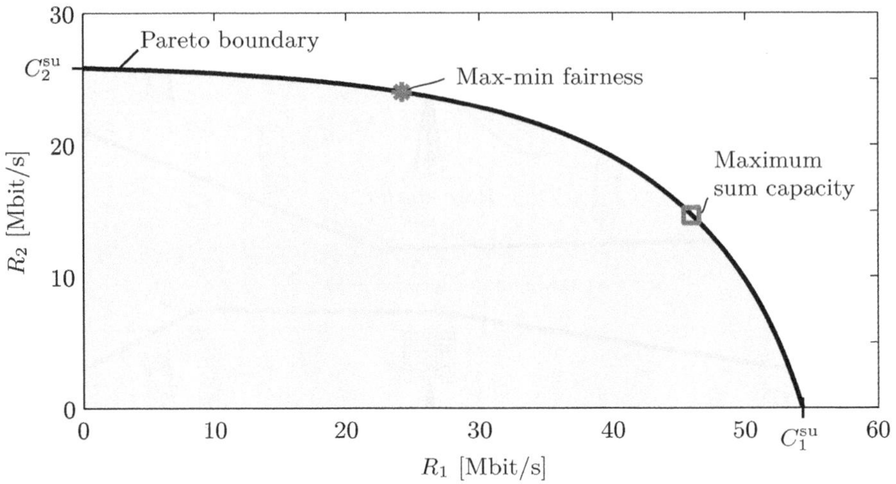

Figure 6.3: Example of a rate region (shaded) for $K = 2$ users containing all the rate points (R_1, R_2) that are simultaneously achievable in a multi-user system. The points on the Pareto boundary are the ones of practical interest. The points that give maximum sum capacity and max-min fairness (i.e., the minimum rate among the users is maximized) are indicated.

user. This stands in contrast to the points in the interior of the rate region, where we can improve the rates for some users without reducing the rates of other users. The Pareto boundary is formally defined as follows.

Definition 6.1. The Pareto boundary $\partial\mathcal{R}$ of the K-dimensional rate region $\mathcal{R}$ consists of all points $(R_1, \ldots, R_K) \in \mathcal{R}$ for which there does not exist any $(R'_1, \ldots, R'_K) \in \mathcal{R} \setminus \{(R_1, \ldots, R_K)\}$ with $R'_k \geq R_k$ for $k = 1, \ldots, K$.

Since there are K user rates, but we can only operate the system in one way, there is no *objectively* optimal way of operating a multi-user system. The Pareto boundary is the closest characterization of optimality that we can obtain because any point $(R_1, \ldots, R_K) \in \mathcal{R}$ that is not on the Pareto boundary is suboptimal in the sense that there exist other rate points $(R'_1, \ldots, R'_K) \in \partial\mathcal{R}$ that are better or at least as good for every user. However, there are generally infinitely many points on the Pareto boundary. Hence, when designing the system, we need to make a *subjective* tradeoff between the rates achieved by the different users. Each point on the Pareto boundary represents one *Pareto optimal* tradeoff between the K user rates, but they are mutually unordered.

To address this issue stringently, the system designer can select a utility function $u(R_1, \ldots, R_K)$ that takes any rate point $(R_1, \ldots, R_K)$ and provides a real scalar number representing the preference of that point; a larger value represents larger preference. Since a larger rate should be desirable for the system, the function should be increasing or non-decreasing with respect to all the rates. The choice of function will impose a subjective ordering of the points in the rate region, leading to that we can now identify an operating point as the optimum with respect to the selected utility function. The corresponding optimization problem can be stated as

$$\underset{(R_1,\ldots,R_K)\in\mathcal{R}}{\text{maximize}} \quad u(R_1,\ldots,R_K). \tag{6.2}$$

This is called a *resource allocation problem* since the goal is determining how the transmission resources (i.e., the time, frequency, power, and spatial resources) are allocated to the K users. Depending on the utility and scenario, there might be one or multiple solutions to (6.2). We will exemplify two ways of selecting the utility function based on very different design principles.

6.2.1 Max-Min Fairness

The first example focuses on delivering equal rates to all users and making that common rate value as large as possible. This utility function can be defined as

$$u(R_1,\ldots,R_K) = \min_{k\in\{1,\ldots,K\}} R_k, \tag{6.3}$$

where the preference value is the lowest rate among all the users. This is an increasing function of all the rates, but it is not strictly increasing because only improving the lowest rate will increase the function value. Hence, it can also be called a non-decreasing function of the individual users' rates. Substituting this utility into (6.2) results in the *max-min fairness* problem

$$\underset{(R_1,\ldots,R_K)\in\mathcal{R}}{\text{maximize}} \quad \min_{k\in\{1,\ldots,K\}} R_k. \tag{6.4}$$

By solving this problem, we will identify a point $(R_1,\ldots,R_K) \in \mathcal{R}$ that satisfies $R_1 = R_2 = \ldots = R_K$ since the utility gives no incentive to assign a larger rate to any user than to the other ones.[1] This condition is the equation of a line that passes through the origin and has the slope $+1$ in all dimensions. When illustrated in two dimensions, as in Figure 6.4, this line has a 45° angle to both axes. Solving the optimization problem in (6.4) entails identifying the point on this line that provides the largest rate values (i.e., is furthest from the origin) but belongs to the rate region. Hence, the optimum is the intersection point between the line and the Pareto boundary, as illustrated in Figure 6.4. The optimal rate is denoted by R_{mmf} in the figure and satisfies $R_1 = R_2 = R_{\text{mmf}}$ and $(R_{\text{mmf}}, R_{\text{mmf}}) \in \partial\mathcal{R}$. Hence, it is relatively easy to identify the optimum by searching along the given line. The practical challenge is that it can be computationally complex to compute the region and to find

[1]There exist special cases where there are multiple solutions to the max-min fairness problem. All points provide the same max-min rate value, but some points provide higher rates for a subset of the users. This happens when the Pareto boundary only constitutes a subset of the outer boundary because some segments of the outer boundary are parallel to some of the axes. An example of this is shown later in Figure 6.8.

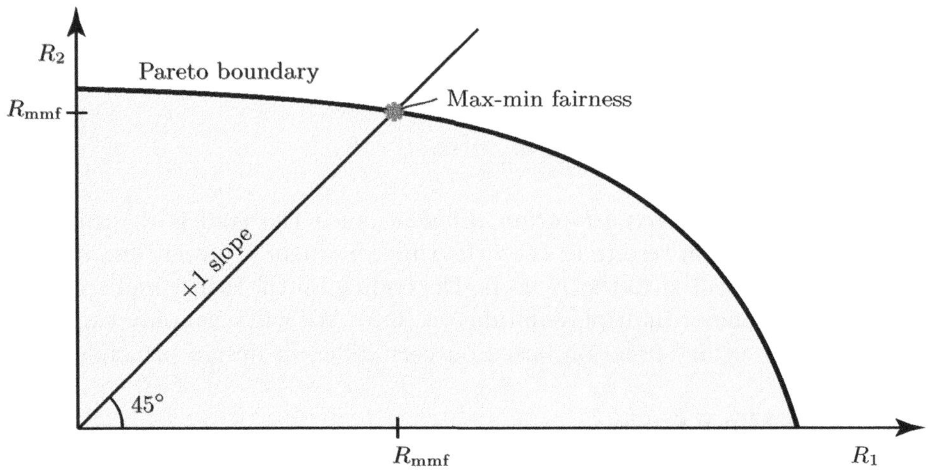

Figure 6.4: Example of a rate region for $K = 2$ users containing all the rate points (R_1, R_2) that are simultaneously achievable. If the max-min utility optimization in (6.4) is used to find the preferred operating point, the optimum (marked by a red star) is the intersection between the Pareto boundary and a line through the origin with a slope of $+1$. At the optimum, the users will have equal rates, denoted by R_{mmf}.

how the transmission resources should be allocated to achieve a certain point in the region. Several detailed examples of how to characterize rate regions will be provided later in this chapter.

The max-min utility results in the *egalitarian solution* to resource allocation, which builds on the principle that all users have equal rights, in this case referring to the right to equal communication performance. When the users have widely different channel qualities (i.e., widely different single-user capacities), users with weak channel gains will achieve a larger fraction of their single-user capacities than users with strong channel gains. This principle can be observed in Figure 6.4, where user 1 has a stronger channel than user 2 but gets the same rate R_{mmf}. One can argue whether that is a fair decision, but it reinforces the point that resource allocation decisions are always subjective.

6.2.2 Maximum Sum Rate

The max-min fairness problem focuses on achieving short-term fairness by allocating the transmission resources to give the users equal rates at every time instance, for the current set of active users and their current channel conditions. This can lead to the undesired side-effect that adding a single user with weak channel conditions to the system will throttle the performance of all other users. An alternative approach is to assume that the users will move around in the same coverage area over time and thereby switch between being the one with a strong channel gain and the one with a weak channel gain. To achieve long-term fairness, it is preferable to transmit as many bits per second as possible at every time instance, irrespective of how the sum rate is currently divided between the users. On average, as the users move around in the cell, they will be allocated an equal share of the long-term average rate.

Based on this logic, we should maximize the sum of the users' rates, represented by the utility function

$$u(R_1, \ldots, R_K) = \sum_{k=1}^{K} R_k. \tag{6.5}$$

This is a strictly increasing function of all the user rates. Substituting this utility into (6.2) results in the *sum-rate maximization* problem

$$\underset{(R_1,\ldots,R_K)\in\mathcal{R}}{\text{maximize}} \quad \sum_{k=1}^{K} R_k. \tag{6.6}$$

By solving this problem, we will identify a point $(R_1, \ldots, R_K) \in \mathcal{R}$ that satisfies $R_1 + R_2 + \ldots + R_K = R_{\text{sr}}$ for the largest possible sum-rate value R_{sr}. Using linear algebra terminology, this condition is the equation of a hyperplane of dimension $K-1$. When illustrated for $K=2$ users, as in Figure 6.5, it becomes the equation $R_2 = R_{\text{sr}} - R_1$ of a line with the slope -1. It intersects the two axes at $(R_{\text{sr}}, 0)$ and $(0, R_{\text{sr}})$, and the line has an angle 45° to both axes. From the geometric perspective, the challenge in sum-rate maximization is to find the value R_{sr} so that the line touches the Pareto boundary without entering the region's interior. The intersection point(s) are the sum-rate maximizing solution(s) to the resource allocation problem. Finding such a point is relatively easy in two dimensions, but as K increases, it entails moving around a $(K-1)$-dimensional hyperplane to find when it intersects a K-dimensional rate region; this can be as computationally complicated as it sounds. Hence, sum-rate maximization is generally a computationally complex problem to solve, but there exists a wealth of algorithms [84], [85].

The sum-rate utility results in the *utilitarian solution* to resource allocation, which builds on the principle that an efficient system produces as much value (or goods) as possible using the given resources, in this case, measured in bits transferred per second. The allocation of the value between the users is not part of the utility function. Hence, when the users have widely different channel qualities (i.e., widely different single-user capacities), users with strong channels will achieve larger rates than users with weak channels. This principle is illustrated in Figure 6.5, where user 1 has a stronger channel gain than user 2. As noted earlier, the short-term differences in rates will average out if the users move around the same coverage area in the same way.[2]

[2]In practice, the users of a communication system will likely move around in the coverage area according to different distributions and spend a large fraction of time in their respective homes and workplaces. Moreover, different data services might be of importance at different locations. In summary, selecting an appropriate utility function can be very challenging. One possible solution is to consider the weighted sum-rate $u(R_1, \ldots, R_K) = \sum_{k=1}^{K} \omega_k R_k$, where the weights $\omega_k \geq 0$ are tuned depending on the users' locations, requested data service, and recent rates to maximize the users' perceived quality-of-service [86], [87].

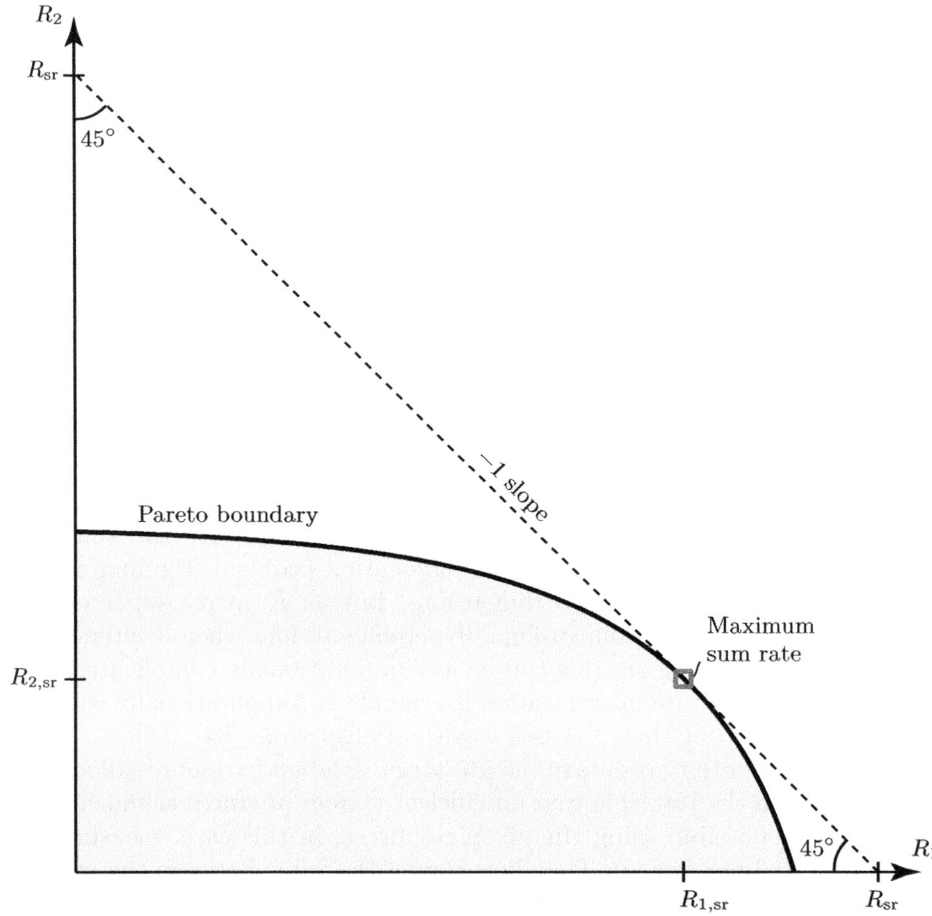

Figure 6.5: Example of a rate region for $K = 2$ users containing all the rate points (R_1, R_2) that are simultaneously achievable. If the sum-rate optimization in (6.6) is used to find the preferred operating point, the optimum (marked by a red square) is the intersection between the Pareto boundary and a line with a slope of -1 slope. All the points on this line provide $R_1 + R_2 = R_{\rm sr}$ but only one point $(R_{1,\rm sr}, R_{2,\rm sr}) \in \partial\mathcal{R}$ can be achieved by the system.

6.3 Uplink Communications

We will now consider different ways to operate the uplink of a multi-user system in terms of different resource allocation solutions and the number of antennas at the base station. Since many different rate expressions will be presented and compared, we define the function

$$C(x) = B \log_2 (1 + x) . \tag{6.7}$$

This is the capacity of a discrete memoryless channel with bandwidth B and SNR x, and we recall from Chapter 3 that the capacity has this form in SISO, SIMO, and MISO scenarios. We will use this concise notation to explore how different uplink solutions provide different rate values $C(x)$ that differ in the effective SNR x that is attained. We will compare three types of operation: orthogonal and non-orthogonal multiple access, and multi-user MIMO.

6.3.1 Orthogonal Multiple Access

We begin by considering the classical setup where a single-antenna base station receives signals from K single-antenna user devices that share a communication channel with bandwidth B Hz. The channel gain of user k is denoted by $\beta_k \in [0, 1]$, for $k = 1, \ldots, K$. It then follows from (2.146) that the single-user capacity of user k is

$$C_k^{\text{su}} = C\left(\frac{P\beta_k}{BN_0}\right) = B \log_2 \left(1 + \frac{P\beta_k}{BN_0}\right) \quad \text{bit/s}, \tag{6.8}$$

where P is the maximum transmit power of the user. The transmission resources involved in this multi-user system are time, frequency, and power. Each user has a separate power amplifier and maximum transmit power P; thus, the resources that can be divided between the users are time and frequency. In this section, we consider *orthogonal multiple access (OMA)*, where the users are assigned orthogonal time-frequency resources.

We begin by considering *frequency-division multiple access (FDMA)*, which is an OMA scheme where the users are assigned non-overlapping fractions of the bandwidth B. We let $\xi_k \in [0, 1]$ denote the bandwidth fraction allocated to user k, for $k = 1, \ldots, K$. These fractions can be selected arbitrarily under the constraint $\xi_1 + \xi_2 + \ldots + \xi_K \leq 1$ so that each bandwidth portion is assigned to at most one user. All users transmit continuously over their assigned bands; thus, user k experiences a point-to-point system with bandwidth $\xi_k B$. By replacing B with $\xi_k B$ in (6.8), the data rate of user k becomes

$$R_k(\xi_k) = \xi_k C\left(\frac{P\beta_k}{\xi_k BN_0}\right) = \xi_k B \log_2 \left(1 + \frac{P\beta_k}{\xi_k BN_0}\right) \quad \text{bit/s}, \tag{6.9}$$

where the notation $R_k(\xi_k)$ emphasizes that the rate is a function of the fraction of the total bandwidth assigned to the user. It can be shown (by computing the first-order derivative) that $R_k(\xi_k)$ is an increasing function of ξ_k, which agrees with the intuition that using as much bandwidth as possible is preferable. It is crucial to notice that only the pre-log factor in (6.9) increases with ξ_k, while the SNR $\frac{P\beta_k}{\xi_k BN_0}$ decreases with ξ_k. This highlights that we can increase the SNR in the communication by concentrating the transmit power in a narrower frequency band, where there is less noise.

Based on the rate expression in (6.9), we can define the rate region as

$$\mathcal{R} = \left\{(R_1(\xi_1), \ldots, R_K(\xi_K)) : \text{for } \xi_1, \ldots, \xi_K \geq 0, \xi_1 + \ldots + \xi_K \leq 1\right\}, \tag{6.10}$$

which is the continuous set of all points $(R_1(\xi_1), \ldots, R_K(\xi_K))$ that can be obtained by dividing the bandwidth between the users in different ways. While the rate of user k in (6.9) is an increasing function of the fraction ξ_k assigned to this user, it is independent of how the remaining bandwidth fraction $1 - \xi_k$ is assigned to the other users. Hence, only the constraint $\xi_1 + \xi_2 + \ldots + \xi_K \leq 1$ creates a tradeoff between the users' rates. Whenever there is equality in this constraint, we will obtain a Pareto optimal point because then the only way to increase the rate of user k is to reallocate bandwidth from another user to user k (e.g., increasing ξ_k and simultaneously decreasing ξ_i by an equal amount, for some $i \neq k$). The Pareto boundary is thus characterized by replacing the inequality with equality in (6.10):

$$\partial\mathcal{R} = \left\{(R_1(\xi_1), \ldots, R_K(\xi_K)) : \text{for } \xi_1, \ldots, \xi_K \geq 0, \xi_1 + \ldots + \xi_K = 1\right\}. \tag{6.11}$$

The rate region is exemplified in Figure 6.6 for a setup with $K = 2$ users and $B = 10\,\text{MHz}$. In Figure 6.6(a), the two users have equal channel quality, represented by $\frac{P\beta_1}{BN_0} = \frac{P\beta_2}{BN_0} = 10$. The rate region is then symmetric, which implies that max-min fairness is achieved at the sum-rate maximizing operating point. In this example, that operating point is $(R_1, R_2) = (22.0, 22.0)\,\text{Mbit/s}$. The tradeoff created by dividing the bandwidth between the two users results in a curved Pareto boundary. In Figure 6.6(b), the users have different channel qualities, represented by $\frac{P\beta_1}{BN_0} = 10$ and $\frac{P\beta_2}{BN_0} = 5$. The rate region remains curved but is now asymmetric. Max-min fairness is achieved at $(19.1, 19.1)\,\text{Mbit/s}$, while the sum rate is maximized at $(26.7, 13.3)\,\text{Mbit/s}$. The maximum sum rate is $40\,\text{Mbit/s}$, while the max-min fairness achieves the sum rate of $38.2\,\text{Mbit/s}$. The regions were generated numerically using (6.11).

Example 6.1. How does FDMA behave when the bandwidth is very large?

When $B \to \infty$, the rate in (6.9) can be approximated using (3.2) as

$$R_k(\xi_k) \approx \xi_k B \log_2(e) \frac{P\beta_k}{\xi_k BN_0} = \log_2(e) \frac{P\beta_k}{N_0}. \tag{6.12}$$

This expression is independent of the fractions $\xi_1, \ldots, \xi_K$, which shows that bandwidth allocation is easy when spectrum is abundant.

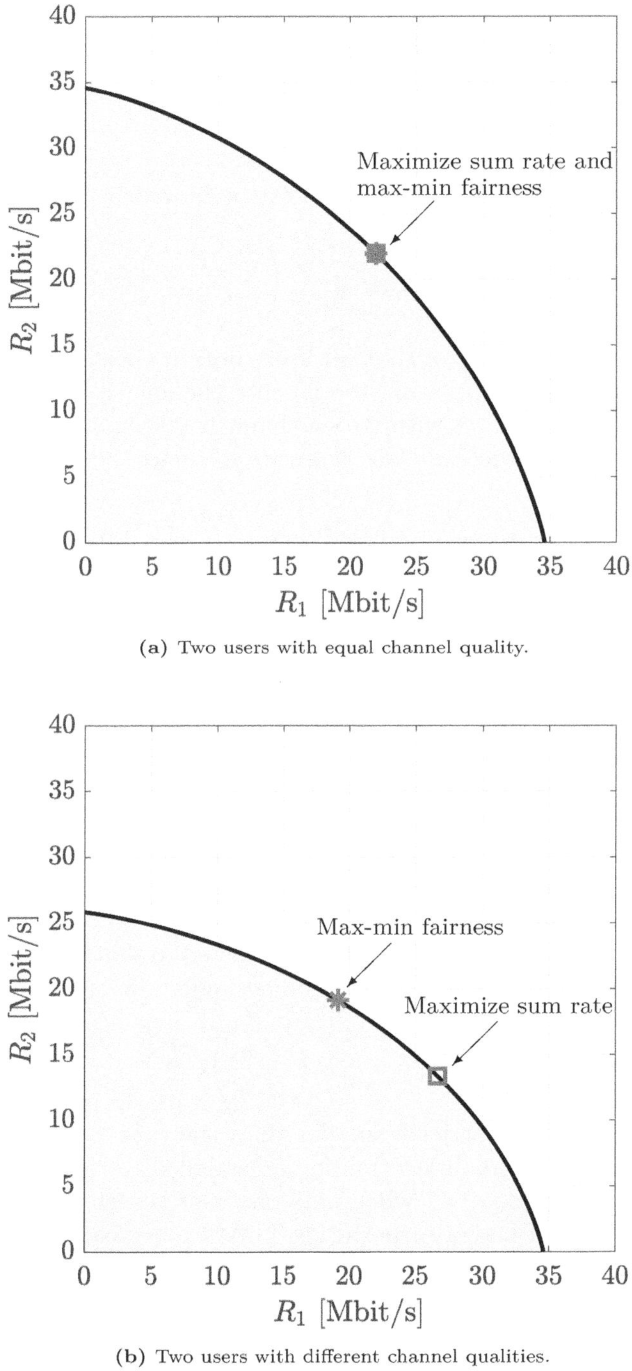

Figure 6.6: Examples of the uplink rate regions for $K = 2$ users when using orthogonal multiple access based on FDMA.

For a given finite amount of bandwidth, one can prove (by differentiation) that the maximum sum rate is achieved by selecting the weights proportionally to the channel gains:

$$\xi_k = \frac{\beta_k}{\sum_{i=1}^{K} \beta_i}. \tag{6.13}$$

By substituting this value into (6.9), the rate achieved by user k is

$$R_k\left(\frac{\beta_k}{\sum_{i=1}^{K}\beta_i}\right) = \frac{\beta_k}{\sum_{i=1}^{K}\beta_i} C\left(\sum_{i=1}^{K} \frac{P\beta_i}{BN_0}\right). \quad \text{bit/s} \tag{6.14}$$

The bandwidth is allocated so that all users obtain the same SNR value to prevent the system from operating too far into the logarithmic regime of the rate function. However, users with strong channels obtain a larger rate thanks to a larger bandwidth fraction. The sum rate becomes

$$\sum_{k=1}^{K} R_k\left(\frac{\beta_k}{\sum_{i=1}^{K}\beta_i}\right) = C\left(\sum_{i=1}^{K} \frac{P\beta_i}{BN_0}\right) \quad \text{bit/s.} \tag{6.15}$$

Interestingly, this is the same rate as one would get over a point-to-point MISO channel with K antennas, the channel vector $\mathbf{h} = [\sqrt{\beta_1}, \ldots, \sqrt{\beta_K}]^{\mathrm{T}}$, and a total transmit power of P. However, in the FDMA scenario, each user transmits from a single antenna with power P, so the total transmit power is KP. The reason for the increased power in the multi-user system, compared to the MISO system, is that the users transmit different signals in orthogonal frequency bands; thus, there is no beamforming gain.

FDMA is not the only OMA scheme. Another option is *time-division multiple access (TDMA)*, where the users take turns transmitting over the entire bandwidth. Suppose user k is active for a fraction of time denoted by $\xi_k \in [0,1]$, for $k = 1, \ldots, K$, which has been selected so that $\xi_1 + \xi_2 + \ldots + \xi_K \leq 1$. The user will then achieve a fraction ξ_k of its single-user capacity, represented by the rate

$$\xi_k C_k^{\mathrm{su}} = \xi_k C\left(\frac{P\beta_k}{BN_0}\right). \tag{6.16}$$

Interestingly, this rate is strictly smaller than the rate $\xi_k C(\frac{P\beta_k}{\xi_k BN_0})$ in (6.9) that is achieved by FDMA, if the fraction ξ_k assigned to the user is the same (equality is achieved for $\xi_k = 1$ when only one user is served). It might seem counterintuitive that FDMA outperforms TDMA since each user is assigned the same fraction of the total time-frequency resources in both cases. The reason behind this result is that the power amplifier is turned on and off in TDMA; thus, even if the instantaneous transmit power is P when the user is transmitting, the time-average transmit power is reduced to $\xi_k P$. This explains why the SNR is ξ_k times smaller when using TDMA than FDMA.

Example 6.2. Can the time resources be divided orthogonally between the users without turning off the power amplifiers?

Yes, this can be achieved by letting the users repeat their data symbols in a way that allows the receiver to separate the transmissions. For example, in a setup with $K = 2$ users and $\xi_1 = \xi_2 = 1/2$, the users can be assigned the orthogonal vectors $[1, 1]^{\mathrm{T}}$ and $[1, -1]^{\mathrm{T}}$ where all entries have unit magnitude. The users multiply each data symbol with their respective vectors and transmit the result as two consecutive symbols in time. The base station can undo the operation by multiplying its received signal over the two consecutive symbol times with the respective vectors. The SNR is improved by a factor $1/\xi_k = 2$ (compared to TDMA) by the repeated transmission, while the orthogonality ensures that there is no interference between the users. Since the multiplication with the vectors spreads out each data symbol over time, the vectors are called *spreading sequences*. This example represents a third type of OMA scheme and is a special case of the general concept of *code-division multiple access (CDMA)*. While CDMA is a remedy to the SNR issue that TDMA suffers from, it will not outperform FDMA and limits which values of ξ_k can be selected to match with spreading sequences. Hence, based on its performance and flexibility, FDMA remains the preferred option among the OMA schemes. We refer to [26] for further details on CDMA and its extensions.

6.3.2 Non-Orthogonal Multiple Access

The previous section demonstrated how a base station can serve multiple users by dividing the time-frequency resources between them in an orthogonal manner; for example, each portion of the frequency band can be assigned to one user. The reason for the orthogonal resource division is to avoid interference. Still, such a protective system design might not be optimal for maximizing our utility function (e.g., max-min fairness or maximum sum rate). An alternative solution is *non-orthogonal multiple access (NOMA)*, where the K users share the same time-frequency resources, and the interference is instead managed by signal processing. In this section, we will show that the rate region obtained by NOMA is larger than the region achieved by FDMA (and other OMA schemes). In fact, it is the largest rate region that can be obtained in the considered setup, called the *capacity region*.

For brevity, we will describe the concept of NOMA in the case of $K = 2$ users that share a bandwidth of B Hz. Each user has a maximum power P and transmits with some power $P_k^{\mathrm{ul}} \in [0, P]$, for $k = 1, 2$. We consider a discrete memoryless multiple access channel where the two users transmit simultaneously, as illustrated in Figure 6.7. The received signal is

$$y[l] = h_1 x_1[l] + h_2 x_2[l] + n[l], \tag{6.17}$$

where $x_k[l]$ is the input signal from user k at the discrete time l and the energy

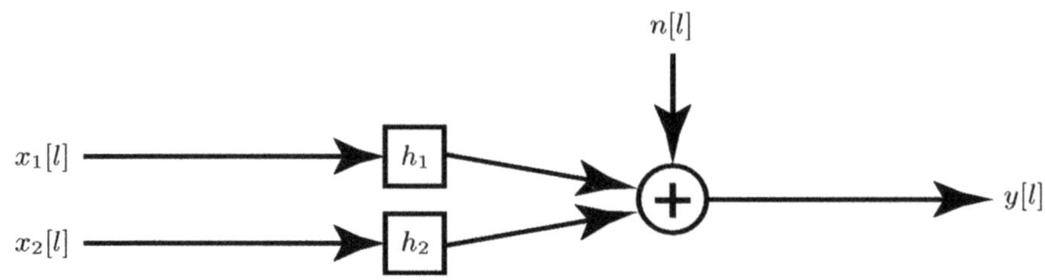

Figure 6.7: A discrete memoryless multiple access channel with $K = 2$ users. The two input signals are $x_1[l]$ and $x_2[l]$, where l is a discrete-time index. The output is $y[l] = h_1 \cdot x_1[l] + h_2 \cdot x_2[l] + n[l]$, where h_1, h_2 are the channel responses and $n[l]$ is the independent complex Gaussian receiver noise.

per symbol is P_k^{ul}/B (because there are B symbols per second). The complex channel response from user k is denoted by h_k and assumed deterministic, so it might be a LOS channel. The magnitude square of the channel is denoted as $\beta_k = |h_k|^2$. Moreover, $n[l] \sim \mathcal{N}_{\mathbb{C}}(0, N_0)$ is the independent receiver noise. There are two input signals $x_1[l], x_2[l]$ but only one output signal $y[l]$. It is nevertheless possible for the receiver to extract data from both signals if the data is appropriately encoded. Note that (6.17) is an extension of the discrete memoryless channel in (2.130) to the case where two users transmit simultaneously and therefore interfere with each other.

Suppose the input signals are Gaussian distributed: $x_k[l] \sim \mathcal{N}_{\mathbb{C}}(0, P_k^{\mathrm{ul}}/B)$ for $k = 1, 2$. This is the optimal input distribution in the point-to-point case and can be proved to be optimal also for multiple access channels. We refer to [26, Appendix B.9] for details. If the receiver focuses on user 1, the received signal in (6.17) can be rewritten as

$$y[l] = h_1 x_1[l] + n_1'[l], \tag{6.18}$$

where $n_1'[l] = h_2 x_2[l] + n[l] \sim \mathcal{N}_{\mathbb{C}}(0, P_2^{\mathrm{ul}}\beta_2/B + N_0)$ is an independent complex Gaussian distributed variable. It is not conventional noise since it consists of both an interfering signal and receiver noise. However, from the perspective of decoding the signal from user 1, it takes the role of an *effective noise* term distributed in the same way as receiver noise (apart from the larger variance). Hence, it follows from Corollary 2.1 that an achievable rate of user 1 is

$$R_1 = C\left(\frac{P_1^{\mathrm{ul}}\beta_1}{P_2^{\mathrm{ul}}\beta_2 + BN_0}\right) \quad \text{bit/s}, \tag{6.19}$$

where we utilized the fact that $\beta_1 = |h_1|^2$ and $\beta_2 = |h_2|^2$. The term $\frac{P_1^{\mathrm{ul}}\beta_1}{P_2^{\mathrm{ul}}\beta_2 + BN_0}$ is the SINR, but here it takes the role as an *effective SNR*; that is, user 1 achieves the same rate as in a point-to-point SISO channel with the SNR value $\frac{P_1^{\mathrm{ul}}\beta_1}{P_2^{\mathrm{ul}}\beta_2 + BN_0}$. This also means that the data can be encoded and decoded identically, which aligns with our previous assumption of having Gaussian distributed data symbols, which achieve the capacity of point-to-point channels.

Once the receiver has decoded the signal sequence $\{x_1[l]\}$ from user 1, it can subtract it from the original received signal in (6.17) as

$$y[l] - h_1 x_1[l] = h_2 x_2[l] + n[l]. \tag{6.20}$$

The result looks like the received signal of a conventional SISO channel with the additive receiver noise $n[l]$ but no interference. Hence, the achievable rate is

$$R_2 = C\left(\frac{P_2^{\mathrm{ul}}\beta_2}{BN_0}\right) \quad \text{bit/s}. \tag{6.21}$$

Interestingly, user 2 achieves the same rate as if it had been assigned the entire bandwidth in OMA. This is enabled by the procedure of first decoding the signal from user 1 and then subtracting it from the received signal. We followed a similar procedure in Section 3.4.3 to sequentially decode the transmitted streams over a point-to-point MIMO channel, in which case we called it *successive interference cancellation (SIC).* We will use the same terminology here and recall that it is a non-linear receiver processing scheme because we must decode one signal sequence entirely before subtracting interference.

It follows from (6.19) and (6.21) that the sum rate is

$$\begin{aligned} R_1 + R_2 &= B\log_2\left(1 + \frac{P_1^{\mathrm{ul}}\beta_1}{P_2^{\mathrm{ul}}\beta_2 + BN_0}\right) + B\log_2\left(1 + \frac{P_2^{\mathrm{ul}}\beta_2}{BN_0}\right) \\ &= B\log_2\left(\frac{P_1^{\mathrm{ul}}\beta_1 + P_2^{\mathrm{ul}}\beta_2 + BN_0}{P_2^{\mathrm{ul}}\beta_2 + BN_0}\right) + B\log_2\left(\frac{P_2^{\mathrm{ul}}\beta_2 + BN_0}{BN_0}\right) \\ &= B\log_2\left(1 + \frac{P_1^{\mathrm{ul}}\beta_1 + P_2^{\mathrm{ul}}\beta_2}{BN_0}\right) = C\left(\frac{P_1^{\mathrm{ul}}\beta_1 + P_2^{\mathrm{ul}}\beta_2}{BN_0}\right). \end{aligned} \tag{6.22}$$

We can notice from (6.22) that the sum rate is an increasing function of both P_1^{ul} and P_2^{ul}, thus it is maximized when both users transmit at their maximum power P. This implies that any tuple of achievable rates (R_1, R_2) must satisfy

$$R_1 + R_2 \leq C\left(\frac{P\beta_1 + P\beta_2}{BN_0}\right). \tag{6.23}$$

The sum-rate expression is symmetric with respect to the two users; thus, the same sum rate can be achieved if the receiver first decodes the signal from user 2, subtracts that signal from the originally received signal, and finally decodes the signal from user 1. However, in the latter case, it is user 1 that achieves the same rate as if assigned the entire bandwidth in OMA mode. In

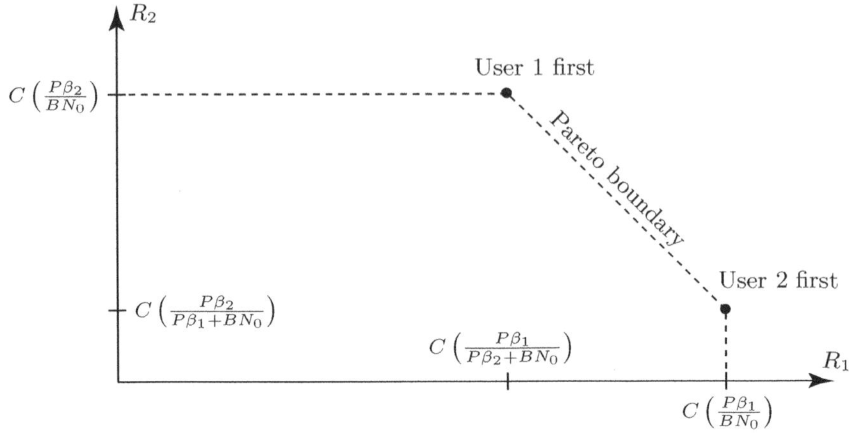

Figure 6.8: When using NOMA, the uplink rate region with $K = 2$ users is characterized by the three line segments shown in the figure. The Pareto boundary is the diagonal line segment between the operating points in (6.24) and (6.25), marked with filled circles, which are achieved by decoding the signals from one user first and subtracting its interference before decoding the signals from the other user.

summary, we know that we can achieve the points

$$(R_1, R_2) = \left(C\left(\frac{P\beta_1}{P\beta_2 + BN_0}\right), C\left(\frac{P\beta_2}{BN_0}\right)\right), \tag{6.24}$$

$$(R_1, R_2) = \left(C\left(\frac{P\beta_1}{BN_0}\right), C\left(\frac{P\beta_2}{P\beta_1 + BN_0}\right)\right). \tag{6.25}$$

These points are marked with filled circles in Figure 6.8, and it is indicated that the first point is achieved by decoding the signal from user 1 first, while the second point is achieved by decoding the signal from user 2 first. By switching between operating at these different points over time, a procedure called *time-sharing*, we can achieve any point on the dashed line segment drawn between the two points. This is the Pareto boundary of the rate region, and it is a segment of the line defined by the maximum sum rate equation $R_1 + R_2 = C(\frac{P\beta_1 + P\beta_2}{BN_0})$.

It is also possible to achieve any point for which the entries are strictly smaller than the points on the Pareto boundary; that is, any point between the axes and the dashed line segments in the figure. The vertical segment is a portion of the line defined by $R_1 = C(\frac{P\beta_1}{BN_0})$, while the horizontal segment is a portion of the line defined by $R_2 = C(\frac{P\beta_2}{BN_0})$. Hence, the rate region is the pentagon determined by these three lines and can be characterized as

$$\mathcal{R} = \Bigg\{(R_1, R_2): \quad 0 \leq R_1 \leq C\left(\frac{P\beta_1}{BN_0}\right), \quad 0 \leq R_2 \leq C\left(\frac{P\beta_2}{BN_0}\right), \\ R_1 + R_2 \leq C\left(\frac{P\beta_1 + P\beta_2}{BN_0}\right)\Bigg\}. \tag{6.26}$$

Any point that satisfies the three equations in (6.26) belongs to the rate region. We have demonstrated the achievability of this rate region by SIC and time-sharing. It can also be proved that no other NOMA scheme can achieve a larger rate region; we refer to [42, Ch. 14] for details. The Pareto boundary is obtained when there is equality in the third equation of (6.26) so that the maximum sum rate is achieved:

$$\partial\mathcal{R} = \left\{(R_1, R_2): \quad 0 \leq R_1 \leq C\left(\frac{P\beta_1}{BN_0}\right), \quad 0 \leq R_2 \leq C\left(\frac{P\beta_2}{BN_0}\right), \right.$$
$$\left. R_1 + R_2 = C\left(\frac{P\beta_1 + P\beta_2}{BN_0}\right)\right\}. \tag{6.27}$$

Example 6.3. Compare the sum rate in (6.23) with that of a point-to-point MISO channel with the channel vector $\mathbf{h} = [\sqrt{\beta_1}, \sqrt{\beta_2}]^{\mathrm{T}}$.

The MISO channel capacity is given in (3.47) and by substituting $q = P/B$ into the expression, we obtain

$$B\log_2\left(1 + \frac{P\|\mathbf{h}\|^2}{BN_0}\right) = C\left(\frac{P\beta_1 + P\beta_2}{BN_0}\right) \quad \text{bit/s}. \tag{6.28}$$

This is the same as the sum rate in (6.23), but the latter is achieved using a total transmit power of $2P$ instead of P because there are two users. The NOMA setup is instead mathematically identical to a MISO system that uses suboptimal precoding where each antenna transmits different data. We first came across SIC in Section 3.4.3 when analyzing how to decode the received data with arbitrary precoding. Using the notation from that section, the precoding matrix is $\mathbf{P} = \mathbf{I}_2$ and the power allocation matrix is $\mathbf{Q} = \mathrm{diag}(P/B, P/B)$.

To compare the rate regions attained by the orthogonal and non-orthogonal types of multiple access, we will continue the example from Figure 6.6(b). Recall that the two users have different channel qualities: $\frac{P\beta_1}{BN_0} = 10$ and $\frac{P\beta_2}{BN_0} = 5$. Figure 6.9 shows the rate regions obtained with OMA/FDMA and NOMA. We notice that NOMA achieves a larger rate region, containing all the operating points OMA achieves and some additional points along the Pareto boundary. When using NOMA, all the points on the Pareto boundary maximize the sum rate and represent different ways of allocating the sum rate between the users. One point on the Pareto boundary is also optimal in the max-min fairness sense; thus, we can maximize both utility functions simultaneously when using NOMA. Interestingly, the maximum sum rate is 40 Mbit/s for both FDMA and NOMA. While this specific value depends on the simulation assumptions, the equivalence is not unique to this example but can be noticed by comparing the maximum sum rate expression for FDMA in (6.15) with

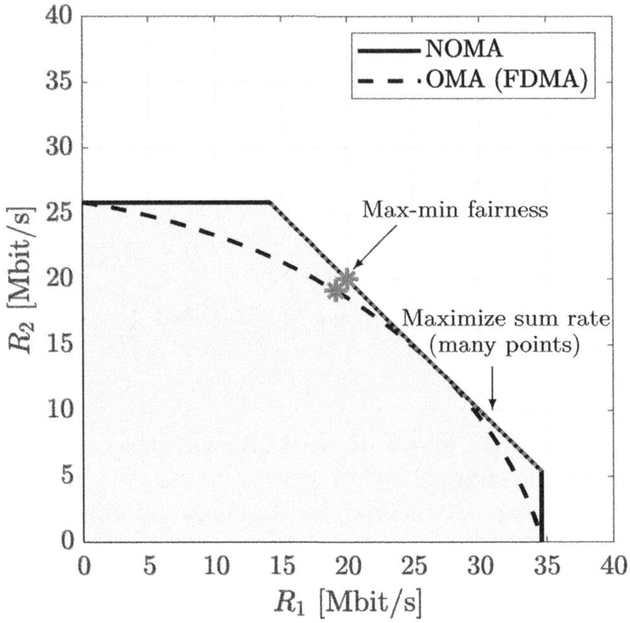

Figure 6.9: Example of uplink rate regions for $K = 2$ users with different channel qualities when using either NOMA or OMA based on FDMA. This is a continuation of the example from Figure 6.6(b).

the corresponding expression for NOMA in (6.23), which are identical. There is, however, a rate difference when it comes to max-min fairness, which is achieved at (20, 20) Mbit/s with NOMA and at (19.1, 19.1) Mbit/s when using orthogonal access. Hence, if the system is designed for max-min fairness, both users can achieve a 5% higher rate when using NOMA.

This example indicates the benefit that NOMA provides over OMA: the maximum sum rate value is the same but can be allocated between the users in a variety of different ways, while FDMA only achieves it using one specific rate division. For example, NOMA allows for the max-min fairness and sum-rate utilities to be maximized simultaneously; however, there is generally a tradeoff between these performance targets when considering OMA. If the users have widely different channel gains, the max-min fairness point with NOMA might be at a corner point of the Pareto boundary. At this point, the user with the weakest channel gain achieves its single-user capacity by being decoded last, while other users might achieve higher rates than that.

The rate region with NOMA for $K \geq 2$ users can be formulated and achieved similarly to what was described earlier in this section. Recall that three equations characterize the rate region in (6.26) in the two-user case: each user's rate must be lower than or equal to the respective single-user capacity, and the sum rate is upper bounded by the capacity of a point-to-point MISO channel with the same transmit power P. When extending this to the K-user case, there will be $2^K - 1$ equations, each describing how the sum rate of a certain subset of the users is upper bounded by the point-to-point MISO channel capacity with a channel vector containing the users' channel coefficients.[3] More precisely, the rate region can be characterized as follows (where the time index l has been omitted for brevity) [42, Sec. 14.3.5].

Theorem 6.1. Consider a K-user discrete memoryless multiple access channel with the inputs $x_1, \ldots, x_K \in \mathbb{C}$ and the output $y \in \mathbb{C}$ given by

$$y = \sum_{k=1}^{K} h_k x_k + n, \tag{6.29}$$

where $n \sim \mathcal{N}_{\mathbb{C}}(0, N_0)$ is independent noise and $h_1, \ldots, h_K \in \mathbb{C}$ are constant channel coefficients known at the output. Suppose the input distributions are feasible whenever $\mathbb{E}\{|x_k|^2\} \leq P/B$, where P is the transmit power and B is the bandwidth (and symbol rate). If R_k denotes the rate of user k and $\beta_k = |h_k|^2$, the capacity region is given by

$$\mathcal{R} = \left\{ (R_1, \ldots, R_K) : 0 \leq \sum_{k \in \mathcal{K}} R_k \leq C\left(\sum_{k \in \mathcal{K}} \frac{P\beta_k}{BN_0} \right) \text{ for all } \mathcal{K} \subset \{1, \ldots, K\} \right\}. \tag{6.30}$$

Notice that $\mathcal{K}$ in (6.30) denotes a subset of the indices of the K users, and there are $2^K - 1$ different non-empty subsets to consider. For example, if $K = 3$, the seven subsets are $\{1\}$, $\{2\}$, $\{3\}$, $\{1, 2\}$, $\{2, 3\}$, $\{1, 3\}$, and $\{1, 2, 3\}$.

Figure 6.10 exemplifies the rate region achieved with NOMA for $K = 3$ users, with $\frac{P\beta_1}{BN_0} = 10$, $\frac{P\beta_2}{BN_0} = 5$, $\frac{P\beta_3}{BN_0} = 2.5$, and $B = 10\,\text{MHz}$. The Pareto boundary is the area enclosed by the solid line segments. The six corner points are achieved by letting the users transmit at maximum power and then decode their signals sequentially in different orders using SIC. Other points on the Pareto boundary can be achieved by time-sharing between operating at the different corner points. All the points on the Pareto boundary achieve the same sum rate

$$R_1 + \ldots + R_K = C\left(\sum_{i=1}^{K} \frac{P\beta_i}{BN_0} \right), \tag{6.31}$$

which is also the same as the maximum sum rate in (6.15) achieved by FDMA. As stated earlier, the key difference is that NOMA can divide the sum rate between the users in multiple ways, while FDMA can not.

[3] K of these subsets correspond to the single-user capacity bounds since each of those subsets includes a single user. The number of subsets that include k users is $\binom{K}{k}$. When we add all the subsets for $k = 1, \ldots, K$, it follows from the binomial theorem that there are $2^K - 1$ subsets.

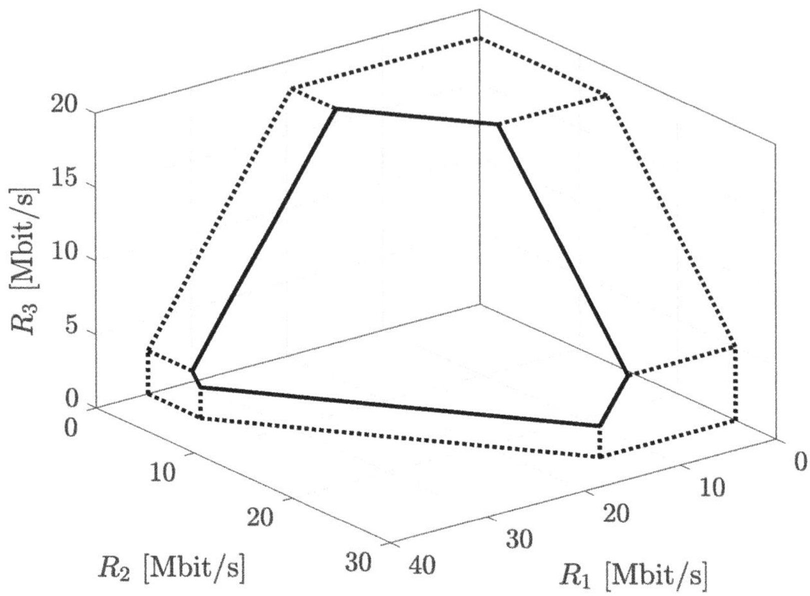

Figure 6.10: Example of an uplink rate region for $K = 3$ users when using NOMA.

6.3.3 Uplink Multi-User MIMO with Non-Linear Processing

The underlying reason that NOMA cannot increase the sum rate compared with OMA is that the base station is only equipped with a single antenna; thus, it can only distinguish one signal dimension per received symbol. Different access schemes can allocate different fractions of this dimension to different users but not create additional signal dimensions. It is instructive to compare multiple access schemes with the operation of point-to-point channels. As mentioned in Example 6.3, the multiple access system model is mathematically indistinguishable from a MISO channel where the K transmit antennas are sending different messages (instead of using MRT) and the receiver has $M = 1$ antenna. The achievable rate for such a setup is given by (3.106) and can be shown to coincide with the sum rate achieved by FDMA and NOMA. Since the base station is only equipped with a single antenna, the multiplexing gain is $\min(M, K) = M = 1$, which is another way to quantify that the users share one signal dimension. However, this analogy reveals a potential solution to the dimensionality bottleneck: if the base station would be equipped with M antennas, for some $M \geq K$, the maximum multiplexing gain becomes $\min(M, K) = K$. In that case, the sum rate can possibly be improved by serving multiple users simultaneously over the entire bandwidth—the more users the better, as long as $K \leq M$. This is the essence of *multi-user MIMO*. Note that the MIMO terminology is utilized even when each user device only has a single antenna because the multiple inputs are the multiple transmitting users, and the multiple outputs are the multiple receive antennas.

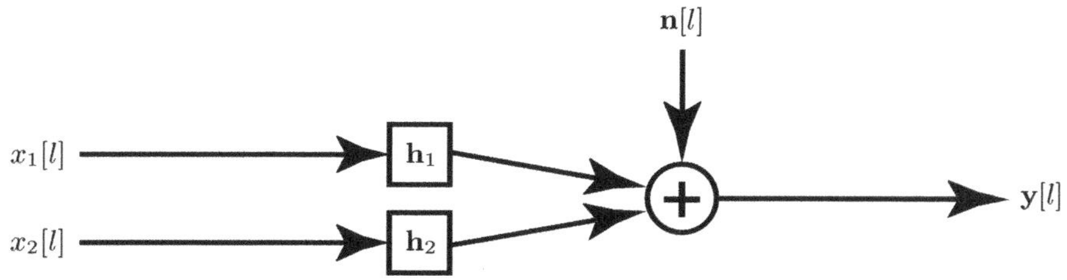

Figure 6.11: A discrete memoryless uplink multi-user MIMO channel with $K = 2$ single-antenna users and M receive antennas. The two input signals are $x_1[l]$ and $x_2[l]$, where l is a discrete-time index. The output is $\mathbf{y}[l] = \mathbf{h}_1 \cdot x_1[l] + \mathbf{h}_2 \cdot x_2[l] + \mathbf{n}[l]$, where $\mathbf{h}_1, \mathbf{h}_2$ are the channel vectors and $\mathbf{n}[l]$ is the independent complex Gaussian receiver noise.

We begin by considering a discrete memoryless channel with $K = 2$ single-antenna user devices and a receiving base station equipped with $M \geq 2$ antennas. Both users transmit simultaneously over a bandwidth of B Hz and their transmit powers are $P_k^{\mathrm{ul}} \in [0, P]$, for $k = 1, 2$, where P is the maximum power. The received signal $\mathbf{y}[l] \in \mathbb{C}^M$ at the discrete time l is

$$\mathbf{y}[l] = \mathbf{h}_1 x_1[l] + \mathbf{h}_2 x_2[l] + \mathbf{n}[l], \tag{6.32}$$

where $x_k[l]$ is the input signal from user k, for $k = 1, 2$. The energy per symbol is P_k^{ul}/B and we assume Gaussian codebooks, such that $x_k[l] \sim \mathcal{N}_{\mathbb{C}}(0, P_k^{\mathrm{ul}}/B)$. The channel vector from user k is denoted by $\mathbf{h}_k \in \mathbb{C}^M$, while $\mathbf{n}[l] \sim \mathcal{N}_{\mathbb{C}}(\mathbf{0}, N_0 \mathbf{I}_M)$ is independent receiver noise. A block diagram of this uplink multi-user MIMO channel is provided in Figure 6.11.

We will characterize the rate region by following the same non-linear receiver processing as in the case of NOMA, namely SIC. If the receiver focuses on user 1, the received signal in (6.32) can be rewritten as

$$\mathbf{y}[l] = \mathbf{h}_1 x_1[l] + \mathbf{n}_1'[l], \tag{6.33}$$

where $\mathbf{n}_1'[l] = \mathbf{h}_2 x_2[l] + \mathbf{n}[l] \sim \mathcal{N}_{\mathbb{C}}(\mathbf{0}, \frac{P_2^{\mathrm{ul}}}{B} \mathbf{h}_2 \mathbf{h}_2^{\mathrm{H}} + N_0 \mathbf{I}_M)$ is an independent complex Gaussian distributed variable. This effective noise term contains both an interfering signal and receiver noise. Since the covariance matrix $\frac{P_2^{\mathrm{ul}}}{B} \mathbf{h}_2 \mathbf{h}_2^{\mathrm{H}} + N_0 \mathbf{I}_M$ has non-zero off-diagonal entries, the effective noise is colored. As described in Section 2.2.4, colored noise can be whitened by multiplying with the inverse square root of the covariance matrix of the noise:

$$\left(\frac{P_2^{\mathrm{ul}}}{B} \mathbf{h}_2 \mathbf{h}_2^{\mathrm{H}} + N_0 \mathbf{I}_M \right)^{-1/2} \mathbf{y}[l] = \left(\frac{P_2^{\mathrm{ul}}}{B} \mathbf{h}_2 \mathbf{h}_2^{\mathrm{H}} + N_0 \mathbf{I}_M \right)^{-1/2} \mathbf{h}_1 x_1[l] + \mathbf{n}_1''[l], \tag{6.34}$$

where the new effective noise $\mathbf{n}_1''[l] = (\frac{P_2^{\mathrm{ul}}}{B} \mathbf{h}_2 \mathbf{h}_2^{\mathrm{H}} + N_0 \mathbf{I}_M)^{-1/2} \mathbf{n}_1'[l] \sim \mathcal{N}_{\mathbb{C}}(\mathbf{0}, \mathbf{I}_M)$ is spatially white. We notice that (6.34) is the system model of a SIMO channel of the type considered in Corollary 3.1, but the physical channel vector $\mathbf{h}_1$ to user 1 is replaced by the *effective channel vector* $(\frac{P_2^{\mathrm{ul}}}{B}\mathbf{h}_2\mathbf{h}_2^{\mathrm{H}} + N_0\mathbf{I}_M)^{-1/2}\mathbf{h}_1$. Hence, the achievable rate of user 1 is

$$\begin{aligned} R_1 &= B\log_2\left(1 + \frac{P_1^{\mathrm{ul}}}{B}\left\|\left(\frac{P_2^{\mathrm{ul}}}{B}\mathbf{h}_2\mathbf{h}_2^{\mathrm{H}} + N_0\mathbf{I}_M\right)^{-1/2}\mathbf{h}_1\right\|^2\right) \\ &= B\log_2\left(1 + P_1^{\mathrm{ul}}\mathbf{h}_1^{\mathrm{H}}\left(P_2^{\mathrm{ul}}\mathbf{h}_2\mathbf{h}_2^{\mathrm{H}} + BN_0\mathbf{I}_M\right)^{-1}\mathbf{h}_1\right) \\ &= C\left(P_1^{\mathrm{ul}}\mathbf{h}_1^{\mathrm{H}}\left(P_2^{\mathrm{ul}}\mathbf{h}_2\mathbf{h}_2^{\mathrm{H}} + BN_0\mathbf{I}_M\right)^{-1}\mathbf{h}_1\right) \quad \text{bit/s.} \end{aligned} \tag{6.35}$$

This rate is achieved by applying an MRC vector based on the effective channel to the whitened received signal in (6.34). Instead of carrying out the whitening and MRC as two separate steps, the combining vector

$$\begin{aligned} \mathbf{w}_1 &= \left(\frac{P_2^{\mathrm{ul}}}{B}\mathbf{h}_2\mathbf{h}_2^{\mathrm{H}} + N_0\mathbf{I}_M\right)^{-1/2}\left(\frac{P_2^{\mathrm{ul}}}{B}\mathbf{h}_2\mathbf{h}_2^{\mathrm{H}} + N_0\mathbf{I}_M\right)^{-1/2}\mathbf{h}_1 \\ &= \left(\frac{P_2^{\mathrm{ul}}}{B}\mathbf{h}_2\mathbf{h}_2^{\mathrm{H}} + N_0\mathbf{I}_M\right)^{-1}\mathbf{h}_1 \end{aligned} \tag{6.36}$$

can be applied to the original received signal in (6.33). Since the receiver computes the inner product $\mathbf{w}_1^{\mathrm{H}}\mathbf{y}[l]$, receive combining is a *linear processing* scheme. We call this LMMSE combining since it can be shown similar to Example 3.4 that $\hat{x}_1[l] = \frac{P_1^{\mathrm{ul}}}{P_1^{\mathrm{ul}}\mathbf{w}_1^{\mathrm{H}}\mathbf{h}_1 + B}\mathbf{w}_1^{\mathrm{H}}\mathbf{y}[l]$ is the LMMSE estimate of $x_1[l]$.

Once the receiver has decoded the signal sequence $\{x_1[l]\}$ from user 1, it can subtract it from the original received signal in (6.32) as

$$\mathbf{y}[l] - \mathbf{h}_1 x_1[l] = \mathbf{h}_2 x_2[l] + \mathbf{n}[l]. \tag{6.37}$$

This resembles the received signal of a conventional SIMO channel with white receiver noise; thus, the achievable rate is equal to the single-user capacity of user 2:

$$R_2 = C\left(\frac{P_2^{\mathrm{ul}}}{BN_0}\|\mathbf{h}_2\|^2\right) \quad \text{bit/s.} \tag{6.38}$$

This rate is achieved by applying an MRC vector $\mathbf{w}_2 = \mathbf{h}_2/\|\mathbf{h}_2\|$ to the received signal in (6.37) after the interference cancellation. Notice that the receiver processing related to user 2 is non-linear since we are not only computing an inner product between the received signal and a combining vector but also subtracting interference caused by the decoded signal from user 1.

The sum rate can be computed by adding (6.35) to (6.38), but some lengthy matrix algebra of the kind in Section 3.4.3 is required to simplify the expression. However, we can take a shortcut by interpreting (6.32) as a point-to-point MIMO channel by writing the received signal as

$$\mathbf{y}[l] = \underbrace{\begin{bmatrix}\mathbf{h}_1 & \mathbf{h}_2\end{bmatrix}}_{=\mathbf{H}} \underbrace{\mathbf{I}_2}_{=\mathbf{P}} \underbrace{\begin{bmatrix} x_1[l] \\ x_2[l] \end{bmatrix}}_{=\bar{\mathbf{x}}[l]} + \mathbf{n}[l], \tag{6.39}$$

where $\mathbf{P} = \mathbf{I}_2$ is the precoding matrix and the signal vector $\bar{\mathbf{x}}[l] \sim \mathcal{N}_{\mathbb{C}}(\mathbf{0}, \mathbf{Q})$ has the covariance matrix $\mathbf{Q} = \mathrm{diag}(P_1^{\mathrm{ul}}/B, P_2^{\mathrm{ul}}/B)$. The diagonal precoding matrix indicates that the two users transmit independently encoded signals, as required in a multiple access channel. It then follows from (3.106) that the sum rate is

$$R_1 + R_2 = B \log_2 \left(\det \left(\mathbf{I}_M + \frac{P_1^{\mathrm{ul}}}{BN_0} \mathbf{h}_1 \mathbf{h}_1^{\mathrm{H}} + \frac{P_2^{\mathrm{ul}}}{BN_0} \mathbf{h}_2 \mathbf{h}_2^{\mathrm{H}} \right) \right). \tag{6.40}$$

The expression in (6.40) is an increasing function of both P_1^{ul} and P_2^{ul}, thus it is maximized when both users transmit at their maximum power P. This implies that any tuple of achievable rates (R_1, R_2) must satisfy

$$R_1 + R_2 \leq B \log_2 \left(\det \left(\mathbf{I}_M + \frac{P}{BN_0} \mathbf{h}_1 \mathbf{h}_1^{\mathrm{H}} + \frac{P}{BN_0} \mathbf{h}_2 \mathbf{h}_2^{\mathrm{H}} \right) \right). \tag{6.41}$$

The sum-rate expression is symmetric with respect to the two users, indicating that there are multiple ways of achieving it. The procedure of decoding the signal from user 1 first and removing its interference before decoding the signal from user 2 is only one of these ways. By exchanging the roles of the two users, another operating point can be achieved. These points are marked with filled circles in Figure 6.12 and given by

$$(R_1, R_2) = \left(C \left(P \mathbf{h}_1^{\mathrm{H}} \left(P \mathbf{h}_2 \mathbf{h}_2^{\mathrm{H}} + BN_0 \mathbf{I}_M \right)^{-1} \mathbf{h}_1 \right), C \left(\frac{P}{BN_0} \|\mathbf{h}_2\|^2 \right) \right), \tag{6.42}$$

$$(R_1, R_2) = \left(C \left(\frac{P}{BN_0} \|\mathbf{h}_1\|^2 \right), C \left(P \mathbf{h}_2^{\mathrm{H}} \left(P \mathbf{h}_1 \mathbf{h}_1^{\mathrm{H}} + BN_0 \mathbf{I}_M \right)^{-1} \mathbf{h}_2 \right) \right). \tag{6.43}$$

The line segment between these points is the Pareto boundary. The pentagon structure of the rate region is clearly the same as with NOMA, but the rate points are computed differently; in fact, NOMA is the special case of multi-user MIMO obtained with $M = 1$. The complete rate region can be defined as

$$\begin{aligned} \mathcal{R} = \Bigg\{ (R_1, R_2) : 0 \leq R_1 \leq C \left(\frac{P}{BN_0} \|\mathbf{h}_1\|^2 \right), \quad 0 \leq R_2 \leq C \left(\frac{P}{BN_0} \|\mathbf{h}_2\|^2 \right), \\ R_1 + R_2 \leq B \log_2 \left(\det \left(\mathbf{I}_M + \frac{P}{BN_0} \mathbf{h}_1 \mathbf{h}_1^{\mathrm{H}} + \frac{P}{BN_0} \mathbf{h}_2 \mathbf{h}_2^{\mathrm{H}} \right) \right) \Bigg\}. \end{aligned} \tag{6.44}$$

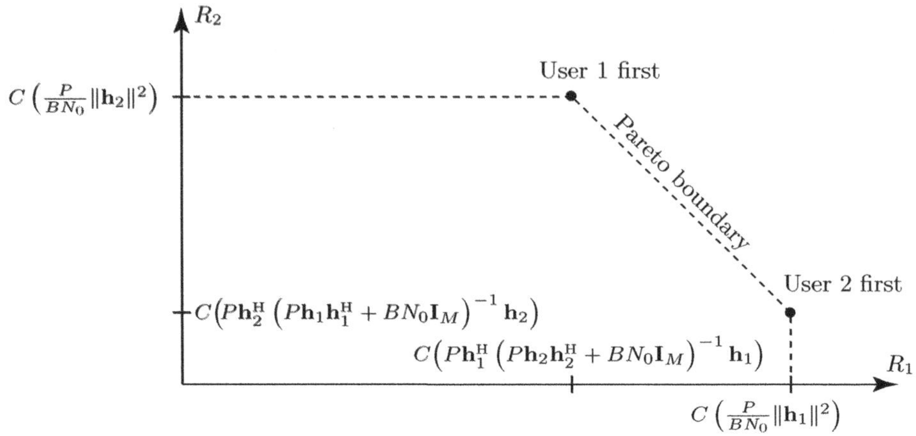

Figure 6.12: The uplink rate region with $K = 2$ users is characterized by three line segments when using multi-user MIMO. The Pareto boundary is the diagonal line segment between the operating points in (6.42) and (6.43), marked with filled circles, which are achieved by decoding the signals from one user first and then subtracting its interference before decoding the signals from the other user.

Any point that satisfies the three equations in (6.44) belongs to the rate region. The Pareto boundary is obtained when there is equality in the third equation so that the maximum sum rate is achieved:

$$\partial\mathcal{R} = \Bigg\{(R_1, R_2) : 0 \leq R_1 \leq C\left(\frac{P}{BN_0}\|\mathbf{h}_1\|^2\right), \quad 0 \leq R_2 \leq C\left(\frac{P}{BN_0}\|\mathbf{h}_2\|^2\right), \\ R_1 + R_2 = B\log_2\left(\det\left(\mathbf{I}_M + \frac{P}{BN_0}\mathbf{h}_1\mathbf{h}_1^{\mathrm{H}} + \frac{P}{BN_0}\mathbf{h}_2\mathbf{h}_2^{\mathrm{H}}\right)\right)\Bigg\}. \tag{6.45}$$

The rate region and the Pareto boundary are illustrated in Figure 6.12.

These results can be extended to the general case $K \geq 2$ by following the same approach as in the NOMA case. The critical point to notice is that the equations defining the rate region are considering each non-empty subset of the K users and specifying that their sum rate should be lower than or equal to the corresponding rate achieved by point-to-point MIMO where the considered users transmit independent signals at their maximum power P. Even the single-user rates have this structure, which can be noticed from that

$$C\left(\frac{P}{BN_0}\|\mathbf{h}_k\|^2\right) = B\log_2\left(1 + \frac{P}{BN_0}\|\mathbf{h}_k\|^2\right) \\ = B\log_2\left(\det\left(\mathbf{I}_M + \frac{P}{BN_0}\mathbf{h}_k\mathbf{h}_k^{\mathrm{H}}\right)\right). \tag{6.46}$$

We obtain the following general result regarding the capacity region of uplink multi-user MIMO.

Theorem 6.2. Consider a K-user discrete memoryless uplink multi-user MIMO channel with the inputs $x_1, \ldots, x_K \in \mathbb{C}$ and the output $\mathbf{y} \in \mathbb{C}^M$ given by

$$\mathbf{y} = \sum_{k=1}^{K} \mathbf{h}_k x_k + \mathbf{n}, \tag{6.47}$$

where $\mathbf{n} \sim \mathcal{N}_{\mathbb{C}}(\mathbf{0}, N_0 \mathbf{I}_M)$ is independent noise and $\mathbf{h}_1, \ldots, \mathbf{h}_K \in \mathbb{C}^M$ are constant channel vectors known at the output. Suppose the input distributions are feasible whenever $\mathbb{E}\{|x_k|^2\} \leq P/B$, where P is the transmit power and B is the bandwidth (and symbol rate). If R_k denotes the rate achieved by user k, then the capacity region is given by

$$\mathcal{R} = \Bigg\{ (R_1, \ldots, R_K) : 0 \leq \sum_{k \in \mathcal{K}} R_k \leq B \log_2 \left(\det \left(\mathbf{I}_M + \sum_{k \in \mathcal{K}} \frac{P}{BN_0} \mathbf{h}_k \mathbf{h}_k^{\mathrm{H}} \right) \right) \\ \text{for all } \mathcal{K} \subset \{1, \ldots, K\} \Bigg\}. \tag{6.48}$$

By considering (6.48) with $\mathcal{K} = \{1, \ldots, K\}$, we obtain

$$\begin{aligned} \sum_{k=1}^{K} R_k &\leq B \log_2 \left(\det \left(\mathbf{I}_M + \sum_{k=1}^{K} \frac{P}{BN_0} \mathbf{h}_k \mathbf{h}_k^{\mathrm{H}} \right) \right) \\ &= B \log_2 \left(\det \left(\mathbf{I}_M + \frac{P}{BN_0} \mathbf{H}\mathbf{H}^{\mathrm{H}} \right) \right), \end{aligned} \tag{6.49}$$

where the upper bound is the sum capacity. The last expression is obtained using the notation $\mathbf{H} = [\mathbf{h}_1, \ldots, \mathbf{h}_K]$ and is identical to the rate expression in (3.106) for a point-to-point MIMO system if each antenna transmits a different signal using the precoding matrix $\mathbf{P} = \mathbf{I}_K$ and $\mathbf{Q} = \mathrm{diag}(\frac{P}{B}, \ldots, \frac{P}{B})$. This means that, from a total bit rate perspective, uplink multi-user MIMO is like a point-to-point MIMO system where the transmit antennas are not collaborating. However, the channel modeling will be very different.

To demonstrate how the use of multiple base station antennas affects the shape of the rate region, we will continue the example with $K = 2$ users from Figure 6.9. Recall that the users have different channel qualities: $\frac{P\beta_1}{BN_0} = 10$ and $\frac{P\beta_2}{BN_0} = 5$. Figure 6.13 shows the rate regions that multi-user MIMO achieves with $M = 2$, $M = 4$, and $M = 8$ antennas, as well as $M = 1$, which represents the previously considered NOMA setup. We assume the base station has a ULA with half-wavelength antenna spacing. We use the LOS channel model from (4.23) and let the users be located in two different azimuth angle directions: $\varphi_1 = -\pi/20$ and $\varphi_2 = \pi/20$ (i.e., there is a 18° angular spacing). As the number of antennas increases, the beamforming gain increases the single-user capacities, thus pushing the horizontal and vertical

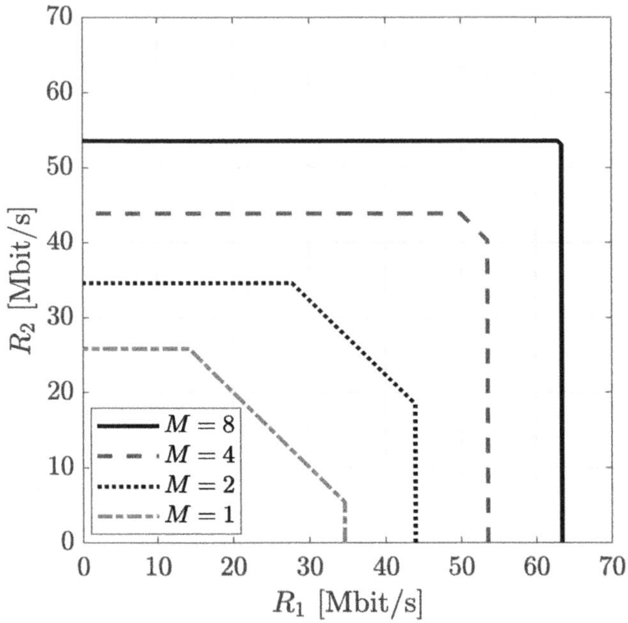

Figure 6.13: Examples of uplink rate regions for $K = 2$ users with multi-user MIMO and a varying number of antennas M, where $M = 1$ corresponds to NOMA. This is a continuation of the example in Figure 6.9.

lines toward larger numbers. Moreover, the increased multiplexing gain from 1 to 2 makes it possible to deal with the inter-user interference so that the sum rate increases with a faster slope. This can be seen from the fact that the diagonal part of the region becomes shorter the more antennas are used. This eventually implies that both users can simultaneously achieve rates almost equal to their respective single-user capacities. In conclusion, this benefit from adding antennas (i.e., increasing M) continues even if the full multiplexing gain is achieved already at $M = 2$.

To understand the reason for this result, we can take a closer look at the effective SNR in (6.42) that user 1 achieves when it is decoding its signal first:

$$P\mathbf{h}_1^{\mathrm{H}} \left(P\mathbf{h}_2\mathbf{h}_2^{\mathrm{H}} + BN_0\mathbf{I}_M\right)^{-1} \mathbf{h}_1 = P\mathbf{h}_1^{\mathrm{H}} \left(\frac{1}{BN_0}\mathbf{I}_M - \frac{\mathbf{h}_2\mathbf{h}_2^{\mathrm{H}}}{(BN_0)^2\left(\frac{1}{P} + \frac{\|\mathbf{h}_2\|^2}{BN_0}\right)} \right) \mathbf{h}_1$$

$$= \underbrace{\frac{P}{BN_0}\|\mathbf{h}_1\|^2}_{\text{Single-user SNR}} \underbrace{\left(1 - \frac{P|\mathbf{h}_2^{\mathrm{H}}\mathbf{h}_1|^2}{\|\mathbf{h}_1\|^2(BN_0 + P\|\mathbf{h}_2\|^2)}\right)}_{\text{Reduction due to interference}}, \tag{6.50}$$

where the first equality follows from Lemma 2.3.[4] The last expression reveals that the effective SNR consists of two factors. The first factor is $\frac{P}{BN_0}\|\mathbf{h}_1\|^2$,

[4]The matrix inversion lemma is utilized with $\mathbf{A} = BN_0\mathbf{I}_M$, $\mathbf{B} = \mathbf{h}_2$, $\mathbf{C} = P$, and $\mathbf{D} = \mathbf{h}_2^{\mathrm{H}}$.

which is the SNR expression that appears in the single-user capacity expression. The second factor also depends on the channel $\mathbf{h}_2$ of the interfering user and determines the performance reduction caused by inter-user interference. This factor takes a value between 0 and 1, representing what fraction of the single-user SNR is achieved under interference. This factor usually approaches 1 as we increase the number of antennas, thereby making the diagonal part of the rate region shorter and shorter, as observed in Figure 6.13.

Example 6.4. Show that the second factor in (6.50) goes to 1 as the number of antennas $M \to \infty$ in the setup considered in Figure 6.13.

In the simulated scenario with a ULA and LOS channel conditions, we have $\mathbf{h}_1 = \sqrt{\beta_1}\mathbf{a}_M(-\pi/8)$ and $\mathbf{h}_2 = \sqrt{\beta_2}\mathbf{a}_M(\pi/8)$ using the array response vector expression in (4.49). We can utilize (4.50) and (4.52) to rewrite the second factor in (6.50) as

$$1 - \frac{P|\mathbf{h}_2^{\mathrm{H}}\mathbf{h}_1|^2}{\|\mathbf{h}_1\|^2(BN_0 + P\|\mathbf{h}_2\|^2)} = 1 - \frac{P\beta_1\beta_2}{M\beta_1(BN_0 + PM\beta_2)} \frac{\sin^2\left(M\pi\sin(\frac{\pi}{8})\right)}{\sin^2\left(\pi\sin(\frac{\pi}{8})\right)}$$
$$\geq 1 - \frac{1}{M^2}\frac{1}{\sin^2\left(\pi\sin(\frac{\pi}{8})\right)}, \tag{6.51}$$

where the lower bound is obtained by replacing $\sin^2\left(M\pi\sin(\frac{\pi}{8})\right)$ with 1 and BN_0 with 0, which are two operations that result in subtracting a larger term. The lower bound goes to 1 when $M \to \infty$. The mathematical explanation is that the directions $\mathbf{h}_1/\|\mathbf{h}_1\|$ and $\mathbf{h}_2/\|\mathbf{h}_2\|$ of the channel vectors become increasingly orthogonal as more antennas are added to the ULA, making it easier to suppress interference without sacrificing much of the desired signal. The physical explanation is that the beamwidth shrinks with M.

Interference is the performance limiting factor when the SNR is high, while it drowns in the noise when the SNR is low. The last term in (6.50) can be upper bounded as

$$\frac{P|\mathbf{h}_2^{\mathrm{H}}\mathbf{h}_1|^2}{\|\mathbf{h}_1\|^2(BN_0 + P\|\mathbf{h}_2\|^2)} \leq \frac{|\mathbf{h}_2^{\mathrm{H}}\mathbf{h}_1|^2}{\|\mathbf{h}_1\|^2\|\mathbf{h}_2\|^2}, \tag{6.52}$$

where equality is achieved at high SNR where $\frac{P}{BN_0} \to \infty$. Propagation environments where this term vanishes when using many antennas are said to provide *favorable propagation* [88] because the effective SNR then approaches the single-user SNR. This property can be formalized as follows.

Definition 6.2. The pair of channels $\mathbf{h}_1, \mathbf{h}_2 \in \mathbb{C}^M$ is said to provide *favorable propagation* if

$$\frac{|\mathbf{h}_2^{\mathrm{H}}\mathbf{h}_1|}{\|\mathbf{h}_1\|\|\mathbf{h}_2\|} \to 0 \quad \text{as } M \to \infty. \tag{6.53}$$

This definition aims to evaluate if a given channel model has the desired property that the performance loss from interference reduces gradually as we add more antennas. However, taking the limit, $M \to \infty$, is just a mathematical curiosity and should not be interpreted literally. In practice, we typically only need 10-100 antennas to make the impact of interference negligibly small for most kinds of channels. Moreover, most channel models considered in this book were derived under a far-field assumption, which will eventually be invalidated as the aperture length increases (i.e., the Fraunhofer distance grows with M). A precise analysis is more complicated but can be found in [89]. In summary, the favorable propagation property simply says: interference is easy to suppress if we have many antennas.

6.3.4 Uplink Multi-User MIMO with Linear Processing

The last two sections demonstrated how SIC could be utilized in NOMA and multi-user MIMO to achieve Pareto optimal operating points. Unfortunately, this non-linear processing scheme has some practical drawbacks. Firstly, the sequential decoding of the users' signals leads to a decoding delay that grows proportionally to the number of users. Secondly, practical data packets have a finite length and, thus, a non-zero probability of decoding errors. When one user's data is decoded incorrectly, the interference cancellation will fail, which implies that the users whose data are decoded later in the sequence get more interference rather than less. This most likely leads to further decoding errors, which is called error propagation. Thirdly, the individual users' data rates must be selected jointly based on the decoding order. Hence, if one user experiences a sudden change in channel conditions, all user rates must be updated accordingly. If we omit the SIC step instead, Pareto optimality cannot be ensured, and most users will experience a rate reduction. However, the practical benefits are that the receiver can now decode all the users' data simultaneously (e.g., using a multi-core processor), decoding errors for one user will not cause decoding errors for other users, and the rates can be selected independently for the different users based on only their individual SINR. In this section, we will analyze multi-user MIMO with such *linear processing*, where each user's signal is essentially decoded as if it is the first to be decoded. In particular, we will investigate under what conditions the performance loss is slight when omitting the SIC procedure.

We consider a discrete memoryless channel with K single-antenna user devices and a receiving base station equipped with $M \geq 2$ antennas. Setups with $M \geq K$ are particularly important, but the performance analysis does not require that assumption. The users transmit simultaneously over a bandwidth of B Hz and their transmit powers are $P_k^{\mathrm{ul}} \in [0, P]$, for $k = 1, \ldots, K$, where P is the maximum power. The received signal $\mathbf{y}[l] \in \mathbb{C}^M$ at the discrete time l is

$$\mathbf{y}[l] = \sum_{k=1}^{K} \mathbf{h}_k x_k[l] + \mathbf{n}[l], \tag{6.54}$$

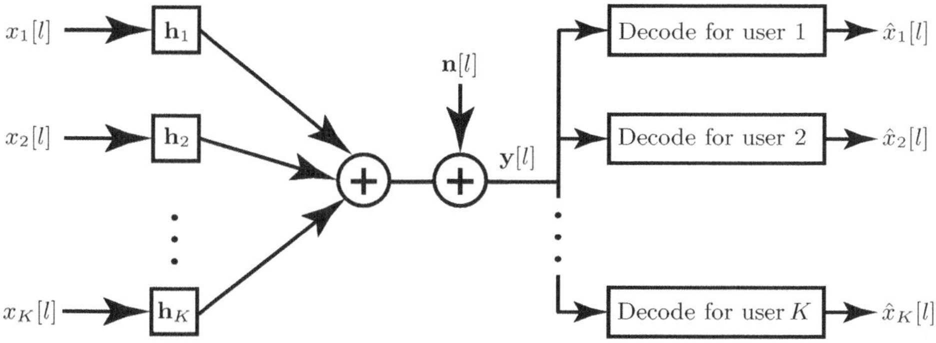

Figure 6.14: A discrete memoryless uplink multi-user MIMO channel with the inputs $x_k[l]$ for $k = 1, \ldots, K$ and the output $\mathbf{y}[l] = \sum_{k=1}^{K} \mathbf{h}_k x_k[l] + \mathbf{n}[l] \in \mathbb{C}^M$, where l is a discrete-time index, $\mathbf{h}_k \in \mathbb{C}^M$ is the channel vector from user k, and $\mathbf{n}[l]$ is the independent Gaussian receiver noise. The user signals are decoded in parallel by treating interference as noise.

where $x_k[l]$ is the input signal from user k, for $k = 1, \ldots, K$. The energy per symbol is P_k^{ul}/B. We assume the use of Gaussian codebooks, such that $x_k[l] \sim \mathcal{N}_{\mathbb{C}}(0, P_k^{\text{ul}}/B)$. The channel vector from user k is denoted by $\mathbf{h}_k \in \mathbb{C}^M$, while $\mathbf{n}[l] \sim \mathcal{N}_{\mathbb{C}}(\mathbf{0}, N_0\mathbf{I}_M)$ is the independent receiver noise. The key difference is how the receiver processing will be carried out, namely, each user's data is decoded separately and, potentially, in parallel. This setup is the same as in the previous section and is illustrated in Figure 6.14.

Since the channel is memoryless, we can remove the time indices from (6.54). However, we must remember that the channel vectors are constant, while the signals and noise are random variables that take new independent realizations at every time instance. When considering an arbitrary user k at an arbitrary time, the received signal in (6.54) can be rewritten as

$$\mathbf{y} = \underbrace{\mathbf{h}_k x_k}_{\text{Desired signal}} + \underbrace{\sum_{i=1, i\neq k}^{K} \mathbf{h}_i x_i}_{\text{Interference}} + \mathbf{n} = \mathbf{h}_k x_k + \mathbf{n}_k' \tag{6.55}$$

where $\mathbf{n}_k' = \sum_{i=1, i\neq k}^{K} \mathbf{h}_i x_i + \mathbf{n} \sim \mathcal{N}_{\mathbb{C}}(\mathbf{0}, \mathbf{C}_k)$ is a colored noise term with the non-diagonal covariance matrix

$$\mathbf{C}_k = \mathbb{E}\{\mathbf{n}_k'(\mathbf{n}_k')^{\text{H}}\} = \sum_{i=1, i\neq k}^{K} \frac{P_i^{\text{ul}}}{B} \mathbf{h}_i \mathbf{h}_i^{\text{H}} + N_0 \mathbf{I}_M. \tag{6.56}$$

Hence, if we refrain from decoding the other users' interfering signals but treat them as extra noise, we can view (6.55) as a SIMO channel with the additive colored noise $\mathbf{n}_k'$. We recall from Section 3.2 that we can obtain an estimate $\hat{x}_k$ of x_k by projecting $\mathbf{y}$ onto a scalar value using a receive combining vector. This is a type of linear processing. In this section, we denote the receive combining vector associated with user k as $\mathbf{w}_k \in \mathbb{C}^M$ and we then obtain

$$\hat{x}_k = \mathbf{w}_k^{\mathrm{H}}\mathbf{y} = \mathbf{w}_k^{\mathrm{H}}\mathbf{h}_k x_k + \mathbf{w}_k^{\mathrm{H}}\mathbf{n}_k'. \tag{6.57}$$

We notice that (6.57) is effectively a memoryless SISO channel of the kind in (2.130) with the received signal $y = \hat{x}_k$, the effective channel $h = \mathbf{w}_k^{\mathrm{H}}\mathbf{h}_k$, and the processed noise $n = \mathbf{w}_k^{\mathrm{H}}\mathbf{n}_k' \sim \mathcal{N}_{\mathbb{C}}(0, \mathbf{w}_k^{\mathrm{H}}\mathbf{C}_k\mathbf{w}_k)$. Hence, it follows from Corollary 2.1 that an achievable data rate (in bit/s) is

$$R_k = B\log_2\left(1 + \frac{\frac{P_k^{\mathrm{ul}}}{B}|\mathbf{w}_k^{\mathrm{H}}\mathbf{h}_k|^2}{\mathbf{w}_k^{\mathrm{H}}\mathbf{C}_k\mathbf{w}_k}\right) = C\left(\frac{\frac{P_k^{\mathrm{ul}}}{B}|\mathbf{w}_k^{\mathrm{H}}\mathbf{h}_k|^2}{\sum_{i=1,i\neq k}^{K}\frac{P_i^{\mathrm{ul}}}{B}|\mathbf{w}_k^{\mathrm{H}}\mathbf{h}_i|^2 + N_0\|\mathbf{w}_k\|^2}\right). \tag{6.58}$$

This rate depends on the selection of the receive combining vector $\mathbf{w}_k$, which appears in the effective SNR term $\frac{P_k^{\mathrm{ul}}}{B}|\mathbf{w}_k^{\mathrm{H}}\mathbf{h}_k|^2/(\mathbf{w}_k^{\mathrm{H}}\mathbf{C}_k\mathbf{w}_k)$. More precisely, the direction of the combining vector will determine the SNR, while the length of the vector is immaterial since it affects the numerator and denominator equally. When having white noise, as in Section 3.2, the direction of the receive combining vector will not affect the noise variance $\mathbf{w}_k^{\mathrm{H}}\mathbf{C}_k\mathbf{w}_k$ since $\mathbf{C}_k$ is a scaled identity matrix. In that case, the SNR is maximized by selecting $\mathbf{w}_k$ as a vector parallel to the channel $\mathbf{h}_k$, which we previously called MRC. The situation changes when having the colored noise $\mathbf{n}_k'$ because then the noise (or rather the interference) is stronger in some directions and weaker in others. For example, if we compute an eigendecomposition of $\mathbf{C}_k$, eigenvectors associated with large eigenvalues represent strong directions, and eigenvectors associated with small eigenvalues represent weaker directions. Hence, the receive combining that maximizes the effective SNR must balance maximizing the numerator and minimizing the denominator.

To identify the combining vector that maximizes the effective SNR in (6.58), we can divide the receive combining vector into two parts: one part that performs whitening of the noise and one that performs receive combining after the whitening. Recall from the previous section that the whitening of the colored noise is achieved by multiplying the received signal with $\mathbf{C}_k^{-1/2}$, thus we set

$$\mathbf{w}_k = \mathbf{C}_k^{-1/2}\bar{\mathbf{w}}_k \tag{6.59}$$

where $\bar{\mathbf{w}}_k \in \mathbb{C}^M$ is the effective combining vector after the whitening. There is a one-to-one mapping between $\mathbf{w}_k$ and $\bar{\mathbf{w}}_k$, so we can make this assumption without risking any loss-of-optimality. By substituting (6.59) into (6.58), we obtain

$$R_k = B\log_2\left(1 + \frac{\frac{P_k^{\mathrm{ul}}}{B}|\bar{\mathbf{w}}_k^{\mathrm{H}}\mathbf{C}_k^{-1/2}\mathbf{h}_k|^2}{\bar{\mathbf{w}}_k^{\mathrm{H}}\mathbf{C}_k^{-1/2}\mathbf{C}_k\mathbf{C}_k^{-1/2}\bar{\mathbf{w}}_k}\right) = B\log_2\left(1 + \frac{\frac{P_k^{\mathrm{ul}}}{B}|\bar{\mathbf{w}}_k^{\mathrm{H}}\mathbf{C}_k^{-1/2}\mathbf{h}_k|^2}{\|\bar{\mathbf{w}}_k\|^2}\right). \tag{6.60}$$

We can now notice that the variance of the whitened noise only depends on the squared norm $\|\bar{\mathbf{w}}_k\|^2$ and not on the direction of $\bar{\mathbf{w}}_k$. Hence, we can maximize the effective SNR in (6.60) by making $\bar{\mathbf{w}}_k$ parallel to the effective channel $\mathbf{C}_k^{-1/2}\mathbf{h}_k$; that is, applying MRC to the effective channel with $\bar{\mathbf{w}}_k = \mathbf{C}_k^{-1/2}\mathbf{h}_k$. In conclusion, the effective SNR is maximized by selecting the receive combining vector as

$$\mathbf{w}_k = \underbrace{\mathbf{C}_k^{-1/2}}_{\text{Whitening}} \underbrace{\mathbf{C}_k^{-1/2}\mathbf{h}_k}_{\text{MRC}} = \mathbf{C}_k^{-1}\mathbf{h}_k$$
$$= \left(\sum_{i=1, i\neq k}^{K} \frac{P_i^{\text{ul}}}{B} \mathbf{h}_i \mathbf{h}_i^{\text{H}} + N_0 \mathbf{I}_M \right)^{-1} \mathbf{h}_k. \tag{6.61}$$

By substituting this vector into the rate expression in (6.58), we obtain

$$R_k = B\log_2\left(1 + \frac{\frac{P_k^{\text{ul}}}{B}|\mathbf{h}_k^{\text{H}}\mathbf{C}_k^{-1}\mathbf{h}_k|^2}{\mathbf{h}_k^{\text{H}}\mathbf{C}_k^{-1}\mathbf{C}_k\mathbf{C}_k^{-1}\mathbf{h}_k}\right) = B\log_2\left(1 + \frac{P_k^{\text{ul}}}{B}\mathbf{h}_k^{\text{H}}\mathbf{C}_k^{-1}\mathbf{h}_k\right). \tag{6.62}$$

This is a generalization of (6.35) to the case with an arbitrary number of users, which decode their signals separately by treating all interfering signals as colored noise. We call this a linear processing scheme since the receiver only computes the inner product between the received signal $\mathbf{y}$ and the combining vector $\mathbf{w}_k$ before decoding the data.

The rate-maximizing receive combining vector in (6.61) is referred to as LMMSE combining because we can also derive it by looking for the vector that minimizes the MSE $\mathbb{E}\{|x_k - \hat{x}_k|^2\}$ between the transmitted signal and the estimate in (6.57). Such a problem was solved in Example 3.4, except that there were no user indices in that case. By substituting $q = P_k^{\text{ul}}/B$, $\mathbf{h} = \mathbf{h}_k$, and $\mathbf{C} = \mathbf{C}_k$ into (3.34), we obtain the MSE-minimizing combining vector

$$\mathbf{w}_k = \frac{P_k^{\text{ul}}}{B}\left(\frac{P_k^{\text{ul}}}{B}\mathbf{h}_k\mathbf{h}_k^{\text{H}} + \mathbf{C}_k\right)^{-1}\mathbf{h}_k = \frac{P_k^{\text{ul}}}{P_k^{\text{ul}}\mathbf{h}_k^{\text{H}}\mathbf{C}_k^{-1}\mathbf{h}_k + B}\mathbf{C}_k^{-1}\mathbf{h}_k, \tag{6.63}$$

which is equal to (6.61) except for the extra scaling factor $P_k^{\text{ul}}/(P_k^{\text{ul}}\mathbf{h}_k^{\text{H}}\mathbf{C}_k^{-1}\mathbf{h}_k + B)$. Strictly speaking, only the combining vector in (6.63) minimizes the MSE; however, both expressions are commonly referred to as LMMSE combining. The reason is that both vectors maximize the rate because the SINR only depends on the direction of the combining vector, not on its length. The rate expression originates from a mutual information expression that ignores the scaling because it implicitly assumes an optimal decoder, which will scale the received signal on its own, thereby compensating for whatever undesired scaling has been applied earlier. The preferred scaling depends on the decoding algorithm but is likely similar to the *true* LMMSE combining in (6.63).

We can summarize the results with linear processing as follows.

Corollary 6.1. Consider the discrete memoryless uplink multi-user MIMO channel in Figure 6.14 where the K users have the respective inputs $x_k \in \mathbb{C}$, $k = 1, \dots, K$, and the output $\mathbf{y} \in \mathbb{C}^M$ given by

$$\mathbf{y} = \sum_{k=1}^{K} \mathbf{h}_k x_k + \mathbf{n}, \tag{6.64}$$

where $\mathbf{n} \sim \mathcal{N}_{\mathbb{C}}(\mathbf{0}, N_0 \mathbf{I}_M)$ is independent noise. Suppose $\mathbf{h}_k \in \mathbb{C}^M$ is a constant vector known at the output, for $k = 1, \dots, K$. If each input signal is independently distributed as $x_k \sim \mathcal{N}_{\mathbb{C}}(0, P_k^{\mathrm{ul}}/B)$ and decoded separately by treating inter-user interference as noise, the largest achievable rate for user k is

$$R_k = B \log_2 \left(1 + P_k^{\mathrm{ul}} \mathbf{h}_k^{\mathrm{H}} \left(\sum_{i=1, i \neq k}^{K} P_i^{\mathrm{ul}} \mathbf{h}_i \mathbf{h}_i^{\mathrm{H}} + B N_0 \mathbf{I}_M \right)^{-1} \mathbf{h}_k \right) \tag{6.65}$$

and is achieved by using LMMSE combining.

While it is possible to operate a multi-user MIMO system without utilizing SIC, the pertinent question is: how large is the performance loss? The achievable rate region with linear processing can be characterized by considering all the rate tuples $(R_1, \dots, R_K)$ that can be obtained for different selections of the transmit powers:

$$\mathcal{R} = \Bigg\{ (R_1, \dots, R_K) \colon R_k = B \log_2 \left(1 + P_k^{\mathrm{ul}} \mathbf{h}_k^{\mathrm{H}} \left(\sum_{i=1, i \neq k}^{K} P_i^{\mathrm{ul}} \mathbf{h}_i \mathbf{h}_i^{\mathrm{H}} + B N_0 \mathbf{I}_M \right)^{-1} \mathbf{h}_k \right)$$
$$\text{for } k = 1, \dots, K, \text{ for some } P_1^{\mathrm{ul}}, \dots, P_K^{\mathrm{ul}} \in [0, P] \Bigg\}. \tag{6.66}$$

To compare this rate region with the one achieved with non-linear processing, we continue the example from Figure 6.13. Recall that we considered $K = 2$ users with different channel qualities: $\frac{P\beta_1}{BN_0} = 10$ and $\frac{P\beta_2}{BN_0} = 5$. Figures 6.15(a) and 6.15(b) show the rate regions obtained with $M = 4$ and $M = 8$ antennas, respectively. The regions called "non-linear" are achieved using SIC and are the same as those in Figure 6.13, while the regions called "linear" are computed using (6.66). The boundary points are obtained by assigning the maximum power P to one of the users and varying the other user's power from 0 to P. The corner point in the middle of the boundary is achieved by $P_1^{\mathrm{ul}} = P_2^{\mathrm{ul}} = P$. As expected, linear processing results in a smaller region than non-linear processing, but the difference reduces as we increase the number of antennas. For example, the loss in sum rate from using linear processing is 4% with $M = 4$ but only 0.4% with $M = 8$. The explanation is

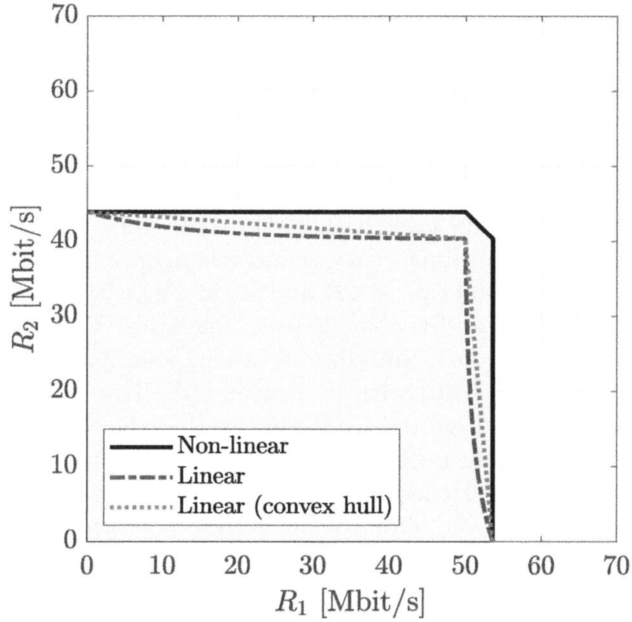

(a) $M = 4$ antennas.

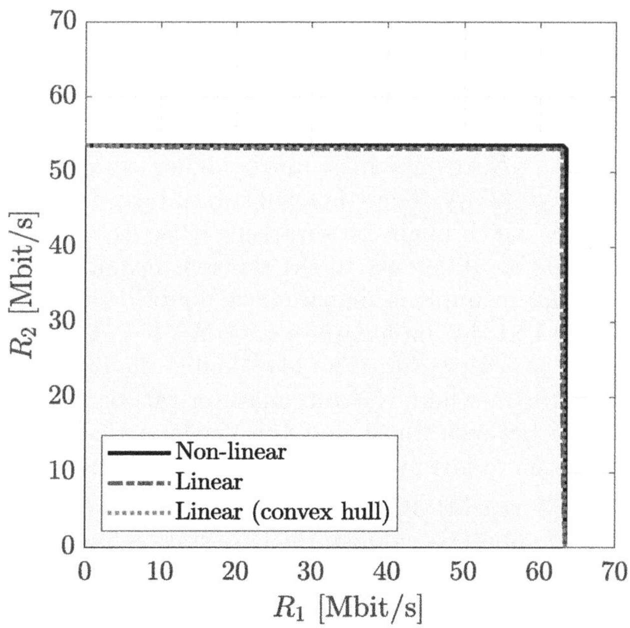

(b) $M = 8$ antennas.

Figure 6.15: Examples of uplink rate regions with $K = 2$ users when multi-user MIMO is used with either non-linear or linear processing. The region obtained in (6.66) is called "linear" and its convex hull is also shown. This is a continuation of the example from Figure 6.13.

the favorable propagation property discussed in the previous section; that is, the directions of the users' channel vectors become increasingly different as M grows, making it possible to suppress interference using LMMSE combining without sacrificing much of the desired signal power.

All the rate regions we have considered earlier in this chapter are convex sets, meaning that if we pick any two operating points in the region and draw a line between them, the line stays within the region. The pentagon shape in the two-user case with non-linear processing was achieved by drawing lines between the operating points in (6.42) and (6.43), which are achieved with different decoding orders, and the single-user capacities. This procedure was called time-sharing. If we allow time-sharing when using linear processing, we can replace the region in (6.66) with its convex hull. This way of expanding the region is also shown in Figure 6.15. We notice that the benefit of switching between different operating points over time shrinks as more antennas are added to the base station. Hence, having base stations with many antennas will increase the sum rate and simplify the system operation.

To further elaborate on these properties, Figure 6.16 shows the sum rate in a setup with $K = 4$ users when the SNR is varied. The base station is equipped with a ULA with half-wavelength-spaced antennas, and the users have equal SNRs but have different azimuth angles-of-arrivals: $-\pi/16, -\pi/32, 0, +\pi/24$. We compare the sum rates achieved with multi-user MIMO with non-linear and linear processing, as well as OMA/FDMA, where each user is allocated a quarter of the bandwidth. We will not specify the bandwidth in this example but plot the sum rate in bit/symbol to keep it general. Figure 6.16(a) shows the sum rate with $M = 10$ base station antennas. We notice that the multiplexing gain of $\min(M, K) = K$ results in a much higher sum rate when using multi-user MIMO than OMA. There is a substantial gap between linear and non-linear processing, which might be surprising since $M = 10$ antennas only resulted in a 3% sum-rate difference in the previous example. The reason for the broader gap in this example is that we have doubled the number of users. In Figure 6.15(b) we had the antenna-user ratio $M/K = 4$, and now we only have $M/K = 10/4 = 2.5$. However, if we also double the number of antennas, we obtain Figure 6.16(b), where the antenna-user ratio is 5. We notice that the performance gap between linear and non-linear processing is once again negligible, thanks to more favorable propagation that limits interference.

In summary, multi-user MIMO with linear processing performs almost the same as its non-linear counterpart when the base station has around five times more antennas than the number of single-antenna users. This operating regime is often called *Massive MIMO*. A typical 5G NR mid-band configuration is $M = 64$ and $1 \leq K \leq 16$ (depending on the traffic load), which results in antenna-user ratios of 4 to 64, for which linear processing works well. We refer to the textbook [1] for a deeper theoretical analysis of Massive MIMO, focusing on cellular networks with inter-cell interference and fading channels.

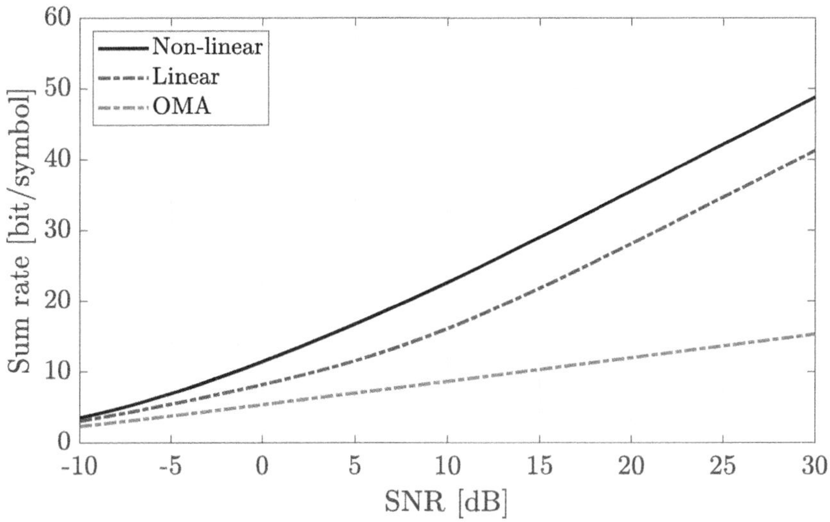

(a) $M = 10$ antennas.

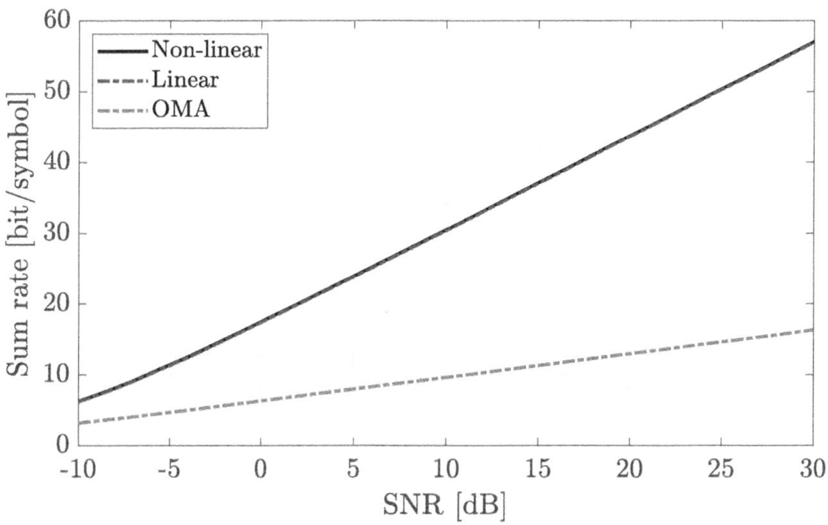

(b) $M = 20$ antennas.

Figure 6.16: The sum rate in a multi-user MIMO system with $K = 4$ users and either non-linear or linear processing. All the users have the same SNR and LOS channels with different azimuth angles: $-\pi/16, -\pi/32, 0, +\pi/24$. OMA/FDMA, where the users are allocated equal fractions of the bandwidth, is shown as a reference and does not provide any multiplexing gain.

Example 6.5. What are the ergodic rates in a multi-user MIMO system with fast-fading channels, linear processing, and perfect CSI at the receiver?

The ergodic capacity of SIMO channels with white noise was studied in Section 5.4.1, where it was concluded that it is the mean value of the conditional capacity achieved for a given channel realization. By following that principle, in a fast-fading multi-user scenario where $\mathbf{h}_1, \ldots, \mathbf{h}_K$ are realizations from stationary and ergodic random processes, the ergodic rate of user k is obtained by taking the mean value of (6.65):

$$R_k = B\,\mathbb{E}\left\{ \log_2 \left(1 + P_k^{\text{ul}} \mathbf{h}_k^{\text{H}} \left(\sum_{i=1, i\neq k}^{K} P_i^{\text{ul}} \mathbf{h}_i \mathbf{h}_i^{\text{H}} + B N_0 \mathbf{I}_M \right)^{-1} \mathbf{h}_k \right) \right\}. \quad (6.67)$$

Apart from the mean value, the rate region can be defined similarly.

6.3.5 Alternative Linear Uplink Processing Schemes

Although LMMSE combining is the rate-maximizing linear processing scheme, other schemes are commonly considered within the area of multi-user MIMO. There are situations where MRC works almost equally well as LMMSE combining, and there are other situations where a scheme called *zero-forcing (ZF)* is nearly optimal. These situations are connected with the SNR at which the system operates. If we consider the direction of the LMMSE combining vector in (6.63), we notice that

$$\frac{\mathbf{w}_k}{\|\mathbf{w}_k\|} = \frac{\left(\sum_{i=1, i\neq k}^{K} \frac{P_i^{\text{ul}}}{B} \mathbf{h}_i \mathbf{h}_i^{\text{H}} + N_0 \mathbf{I}_M \right)^{-1} \mathbf{h}_k}{\left\| \left(\sum_{i=1, i\neq k}^{K} \frac{P_i^{\text{ul}}}{B} \mathbf{h}_i \mathbf{h}_i^{\text{H}} + N_0 \mathbf{I}_M \right)^{-1} \mathbf{h}_k \right\|} \to \frac{\left(N_0 \mathbf{I}_M \right)^{-1} \mathbf{h}_k}{\left\| \left(N_0 \mathbf{I}_M \right)^{-1} \mathbf{h}_k \right\|} = \frac{\mathbf{h}_k}{\|\mathbf{h}_k\|}$$

as $P_1^{\text{ul}}, \ldots, P_K^{\text{ul}} \to 0$. This is the same direction as when using MRC, which proves that LMMSE combining turns into MRC when all the users experience low SNRs. The intuition behind this result is that the inter-user interference will be much weaker than the noise in this situation; thus, every user is experiencing a SIMO channel with only receiver noise. If we substitute MRC with $\mathbf{w}_k^{\text{MRC}} = \mathbf{h}_k / \|\mathbf{h}_k\|$ into the general rate expression in (6.58), we obtain

$$R_k^{\text{MRC}} = B \log_2 \left(1 + \frac{\frac{P_k^{\text{ul}}}{B} \|\mathbf{h}_k\|^2}{\sum_{i=1, i\neq k}^{K} \frac{P_i^{\text{ul}}}{B} \frac{|\mathbf{h}_i^{\text{H}} \mathbf{h}_k|^2}{\|\mathbf{h}_k\|^2} + N_0} \right). \quad (6.68)$$

This is the achievable rate when using MRC in a multi-user MIMO system. The interference term $\frac{|\mathbf{h}_i^{\text{H}} \mathbf{h}_k|^2}{\|\mathbf{h}_k\|^2}$ in the denominator resembles the expression that appeared in the favorable propagation definition in (6.53); thus, MRC is also considered to work well in situations with very many antennas [90].

To study the high-SNR regime, we will utilize the channel matrix notation $\mathbf{H} = [\mathbf{h}_1, \ldots, \mathbf{h}_K] \in \mathbb{C}^{M \times K}$ and the diagonal matrix $\mathbf{Q} = \mathrm{diag}(\frac{P_1^{\mathrm{ul}}}{B}, \ldots, \frac{P_K^{\mathrm{ul}}}{B}) \in \mathbb{C}^{K \times K}$ containing the transmit powers. We assume that $\mathbf{H}^{\mathrm{H}}\mathbf{H} \in \mathbb{C}^{K \times K}$ is invertible, which is generally the case if $M \geq K$ and the users are at physically different locations so that their channel vectors become linearly independent. Suppose we gather all the LMMSE combining vectors from (6.63) as the columns of a combining matrix $\mathbf{W} = [\mathbf{w}_1, \ldots, \mathbf{w}_K] \in \mathbb{C}^{M \times K}$. By noticing that $\sum_{i=1}^{K} \frac{P_i^{\mathrm{ul}}}{B} \mathbf{h}_i \mathbf{h}_i^{\mathrm{H}} = \mathbf{H}\mathbf{Q}\mathbf{H}^{\mathrm{H}}$, we can express this matrix as

$$
\begin{aligned}
\mathbf{W} &= \left(\mathbf{H}\mathbf{Q}\mathbf{H}^{\mathrm{H}} + N_0 \mathbf{I}_M\right)^{-1} \mathbf{H}\mathbf{Q} \\
&= \mathbf{H}\mathbf{Q}\left(\mathbf{H}^{\mathrm{H}}\mathbf{H}\mathbf{Q} + N_0 \mathbf{I}_K\right)^{-1} = \mathbf{H}\mathbf{Q}\mathbf{Q}^{-1}\left(\mathbf{H}^{\mathrm{H}}\mathbf{H} + N_0 \mathbf{Q}^{-1}\right)^{-1} \\
&\to \mathbf{H}\left(\mathbf{H}^{\mathrm{H}}\mathbf{H}\right)^{-1} = \mathbf{W}^{\mathrm{ZF}}
\end{aligned} \tag{6.69}
$$

as $P_1^{\mathrm{ul}}, \ldots, P_K^{\mathrm{ul}} \to \infty$ since then $N_0 \mathbf{Q}^{-1} \to \mathbf{0}$ (a matrix with only zeros). The second equality in (6.69) follows from the matrix identity in (2.50).[5] The resulting combining scheme is called ZF because all the interference terms become zero when using it. This can be seen from the fact that $(\mathbf{W}^{\mathrm{ZF}})^{\mathrm{H}}\mathbf{H} = \left(\mathbf{H}^{\mathrm{H}}\mathbf{H}\right)^{-1} \mathbf{H}^{\mathrm{H}}\mathbf{H} = \mathbf{I}_K$, which implies that

$$
(\mathbf{w}_k^{\mathrm{ZF}})^{\mathrm{H}} \mathbf{h}_i = \begin{cases} 1 & \text{if } k = i, \\ 0 & \text{if } k \neq i. \end{cases} \tag{6.70}
$$

LMMSE combining turns into ZF at high SNR because the receiver noise becomes negligibly small under these conditions; thus, the rate is maximized by removing all interference. The interference affecting user k exists in the $(K-1)$-dimensional subspace of $\mathbb{C}^M$ spanned by the channel vectors $\mathbf{h}_1, \ldots, \mathbf{h}_{k-1}, \mathbf{h}_{k+1}, \ldots, \mathbf{h}_K$ of the $K-1$ other users. This subspace is removed from the received signal when using ZF combining because $\mathbf{w}_k$ is selected orthogonally to it. If we substitute the ZF combining vector into the general rate expression in (6.58), we obtain

$$
R_k^{\mathrm{ZF}} = B \log_2 \left(1 + \frac{P_k^{\mathrm{ul}}}{B N_0 \left[\left(\mathbf{H}^{\mathrm{H}}\mathbf{H}\right)^{-1} \right]_{kk}} \right) \tag{6.71}
$$

where the interference terms vanish thanks to (6.70) and $[(\mathbf{H}^{\mathrm{H}}\mathbf{H})^{-1}]_{kk}$ is the kth diagonal entry of $(\mathbf{H}^{\mathrm{H}}\mathbf{H})^{-1}$. This term is obtained by utilizing the fact that $\|\mathbf{w}_k^{\mathrm{ZF}}\|^2 = [(\mathbf{W}^{\mathrm{ZF}})^{\mathrm{H}}\mathbf{W}^{\mathrm{ZF}}]_{kk} = [(\mathbf{H}^{\mathrm{H}}\mathbf{H})^{-1}]_{kk}$ when using ZF. All users should transmit at maximum power when ZF is used since there is no interference.

There are other ways to achieve a high SNR than to increase the transmit powers, as was done in (6.69). In particular, we can increase the beamforming

[5]This matrix identity is applied with $\mathbf{A} = \mathbf{H}\mathbf{Q}/N_0$ and $\mathbf{B} = \mathbf{H}^{\mathrm{H}}$.

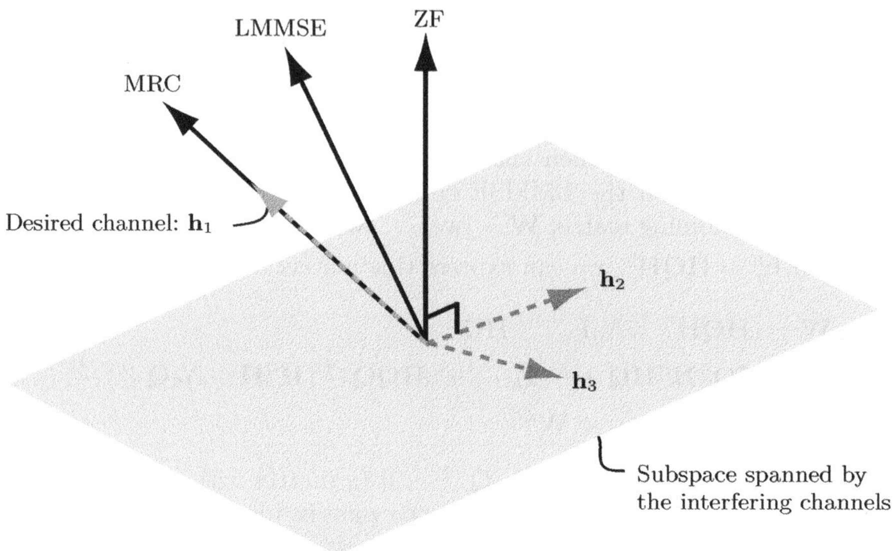

Figure 6.17: A geometrical interpretation of LMMSE, ZF, and MRC in a scenario with $K = 3$ users and $M = 3$ antennas. The focus is on user 1, while users 2 and 3 cause interference.

gain by adding more antennas. This will reduce the transmit power needed to enter the high-SNR regime. Hence, in the Massive MIMO regime where the base stations are equipped with many antennas compared to the number of users, ZF will likely perform similarly to LMMSE combining.

Figure 6.17 provides a geometrical interpretation of LMMSE, ZF, and MRC in a scenario with $K = 3$ users and $M = 3$ antennas. The channel vectors $\mathbf{h}_1, \mathbf{h}_2, \mathbf{h}_3$ point in three different directions in the three-dimensional vector space $\mathbb{C}^3$. We focus on receiving the signal from user 1; thus, all the interference will exist in the subspace spanned by $\mathbf{h}_2$ and $\mathbf{h}_3$. This is the red-shaded plane in the figure. The desired signal is received along the dimension spanned by the channel vector $\mathbf{h}_1$. If MRC is used, the combining vector is parallel with $\mathbf{h}_1$ to maximize the received signal power. If ZF is used, the combining vector is selected orthogonally to the subspace spanned by the $\mathbf{h}_2$ and $\mathbf{h}_3$. If $M > K$, there are $M - K + 1 > 1$ dimensions free from interference, and ZF will collect the received signal power from all of them. LMMSE combining is a vector between the two extremes (MRC and ZF) and will move between them depending on the SNR.

Figure 6.18 illustrates the low and high SNR behaviors of LMMSE combining by continuing the example with $K = 4$ and $M = 10$ from Figure 6.16(a). We notice that MRC provides the same sum rate as LMMSE combining at low SNRs, where the system performance is noise-limited. In contrast, ZF provides the same sum rate as LMMSE combining at high SNRs, where the system performance is interference-limited. There is a large gap between LMMSE combining and the other schemes at intermediate SNRs.

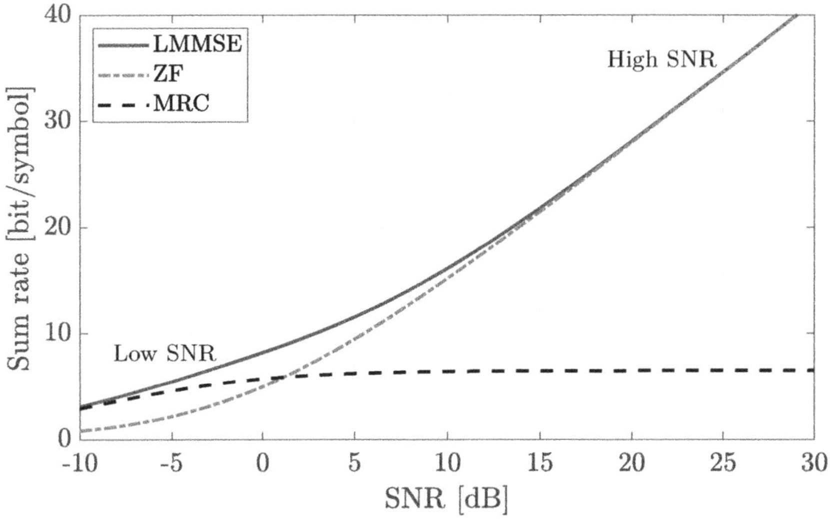

Figure 6.18: The uplink sum rate in a multi-user MIMO system with $K = 4$ users, $M = 10$ antennas, and different linear receive combining schemes. All users have the same SNR.

In practical systems, some users might experience high SNRs while other users simultaneously experience low SNRs; thus, it is preferable to utilize LMMSE combining to identify the rate-maximizing tradeoff between interference and noise suppression automatically. Nevertheless, there is a vast literature on rate analysis for MRC and ZF, mainly focused on the respective asymptotic regimes where these methods are optimal. The reason is that the rate expressions obtained with these schemes are analytically simpler than the ones obtained with LMMSE combining (e.g., there is no matrix inverse) and more amenable to mathematical analysis and extraction of insights.

Example 6.6. When ZF combining is used, the effective SNR in (6.71) is proportional to $1/[(\mathbf{H}^{\mathrm{H}}\mathbf{H})^{-1}]_{kk}$. How is this term distributed if the user channels are subject to i.i.d. Rayleigh fading: $\mathbf{h}_k \sim \mathcal{N}_{\mathbb{C}}(\mathbf{0}_M, \beta_k \mathbf{I}_M)$?

The channel gain after combining is $|(\mathbf{w}_k^{\mathrm{ZF}})^{\mathrm{H}}\mathbf{h}_k|^2/\|\mathbf{w}_k^{\mathrm{ZF}}\|^2 = 1/[(\mathbf{H}^{\mathrm{H}}\mathbf{H})^{-1}]_{kk}$. It is hard to analyze it directly, so we start from the ZF principle in Figure 6.17: ZF projects $\mathbf{h}_k$ orthogonally to the interfering channels. We can create a unitary matrix $\mathbf{A}_k = [\mathbf{A}_k^{\mathrm{interf}}, \mathbf{A}_k^{\mathrm{free}}]$ in which the $K-1$ columns of $\mathbf{A}_k^{\mathrm{interf}} \in \mathbb{C}^{M\times(K-1)}$ is an orthonormal basis of the subspace spanned by the $K-1$ interfering channels. The columns of $\mathbf{A}_k^{\mathrm{free}} \in \mathbb{C}^{M\times(M-K+1)}$ span the remaining $M-(K-1)$ interference-free dimensions. ZF combining reduces the user channel to $(\mathbf{A}_k^{\mathrm{free}})^{\mathrm{H}}\mathbf{h}_k \sim \mathcal{N}_{\mathbb{C}}(\mathbf{0}_{M-K+1}, \beta_k \mathbf{I}_{M-K+1})$. Hence, $1/[(\mathbf{H}^{\mathrm{H}}\mathbf{H})^{-1}]_{kk} = \|(\mathbf{A}_k^{\mathrm{free}})^{\mathrm{H}}\mathbf{h}_k\|^2$, which has a scaled $\chi^2(2(M-K+1))$-distribution with the PDF $f(x) = \frac{x^{M-K}e^{-\frac{x}{\beta_k}}}{\beta_k^{M-K+1}(M-K)!}$ and mean value $\beta_k(M-K+1)$. This channel behaves the same as if we would remove $K-1$ antennas to cancel interference.

6.3.6 Power Control for Max-Min Fairness

The uplink transmit power coefficients $P_k^{\mathrm{ul}} \in [0, P]$, for $k = 1, \dots, K$ can be selected to maximize a specific utility function. This is known as *power control* since it entails controlling the transmit power each user uses to achieve the preferred balance between their capacities. In this section, we consider power control for max-min fairness. We will introduce an efficient fixed-point algorithm that obtains the power coefficients that maximize the utility function

$$u(R_1, \dots, R_K) = \min_{k \in \{1, \dots, K\}} R_k. \tag{6.72}$$

The max-min fairness problem was formulated in (6.4) as

$$\underset{(R_1, \dots, R_K) \in \mathcal{R}}{\text{maximize}} \quad \min_{k \in \{1, \dots, K\}} R_k. \tag{6.73}$$

The achievable rate region $\mathcal{R}$ depends on the adopted receive combining scheme. When LMMSE combining is used, $\mathcal{R}$ is given by (6.66). However, for any linear processing scheme, we can express the rate region in the generic form

$$\mathcal{R} = \Big\{ (R_1, \dots, R_K) \colon R_k = B \log_2\Big(1 + \mathrm{SINR}_k(P_1^{\mathrm{ul}}, \dots, P_K^{\mathrm{ul}})\Big) \text{ for } k = 1, \dots, K, \\ \text{for some } P_1^{\mathrm{ul}}, \dots, P_K^{\mathrm{ul}} \in [0, P] \Big\}, \tag{6.74}$$

where the SINR for each user is a function of the transmit power coefficients $P_1^{\mathrm{ul}}, \dots, P_K^{\mathrm{ul}}$. The SINR of user k obtained by LMMSE combining is given in (6.65) as

$$\mathrm{SINR}_k^{\mathrm{LMMSE}}(P_1^{\mathrm{ul}}, \dots, P_K^{\mathrm{ul}}) = P_k^{\mathrm{ul}} \mathbf{h}_k^{\mathrm{H}} \left(\sum_{i=1, i \neq k}^{K} P_i^{\mathrm{ul}} \mathbf{h}_i \mathbf{h}_i^{\mathrm{H}} + B N_0 \mathbf{I}_M \right)^{-1} \mathbf{h}_k. \tag{6.75}$$

Similarly, the SINRs of user k when using MRC or ZF are given in (6.68) and (6.71), respectively, as

$$\mathrm{SINR}_k^{\mathrm{MRC}}(P_1^{\mathrm{ul}}, \dots, P_K^{\mathrm{ul}}) = \frac{P_k^{\mathrm{ul}} \|\mathbf{h}_k\|^2}{\sum_{i=1, i \neq k}^{K} P_i^{\mathrm{ul}} \frac{|\mathbf{h}_i^{\mathrm{H}} \mathbf{h}_k|^2}{\|\mathbf{h}_k\|^2} + B N_0}, \tag{6.76}$$

$$\mathrm{SINR}_k^{\mathrm{ZF}}(P_1^{\mathrm{ul}}, \dots, P_K^{\mathrm{ul}}) = \frac{P_k^{\mathrm{ul}}}{B N_0 \left[(\mathbf{H}^{\mathrm{H}} \mathbf{H})^{-1} \right]_{kk}}. \tag{6.77}$$

Maximizing the minimum rate is equivalent to maximizing the minimum SINR among the users. Hence, the max-min fairness problem in (6.73) can be expressed for uplink multi-user MIMO with linear processing as

$$\begin{aligned} &\underset{P_1^{\mathrm{ul}}, \dots, P_K^{\mathrm{ul}}}{\text{maximize}} \quad \min_{k \in \{1, \dots, K\}} \mathrm{SINR}_k(P_1^{\mathrm{ul}}, \dots, P_K^{\mathrm{ul}}) \\ &\text{subject to} \quad P_k^{\mathrm{ul}} \in [0, P], \quad \text{for } k = 1, \dots, K. \end{aligned} \tag{6.78}$$

Since the lowest SINR determines the utility, there is no incentive to provide any user with a larger SINR than the others. This property is crucial in devising Algorithm 6.1 that finds the optimal solution. The algorithm starts from arbitrarily selected non-zero power coefficients $P_k^{\mathrm{ul}} \in (0, P]$ and sets a solution accuracy $\epsilon > 0$. In Step 3, each user that achieves an SINR larger than the current minimum SINR reduces its transmit power. Next, in Step 4, all the power coefficients are scaled so that at least one user transmits at maximum power. These steps are repeated iteratively until a stopping criterion is satisfied. The difference between the maximum and minimum SINRs among the users goes to zero asymptotically; thus, the stopping criterion in Step 2 identifies when the difference becomes smaller than ϵ. The algorithm usually converges in less than ten iterations because Step 2 is a so-called fixed-point iteration that rapidly reduces the range of power values to consider.

The algorithm has been stated as if any SINR expression can be utilized, but certain technical conditions must be satisfied; we refer to [91, Lem. 1, Th. 1] for the specific details. These conditions are satisfied when using LMMSE combining or MRC, in which case convergence to the optimal solution to (6.78) is guaranteed. The algorithm builds on Perron-Frobenius theory and interference functions covered in the textbooks [92], [93].

Example 6.7. Consider max-min fairness power control along with ZF combining. Show that an optimal solution is to use full power for all users. Find the corresponding max-min fair rate.

The SINR expression in (6.77) with ZF is $P_k^{\mathrm{ul}}/(BN_0[(\mathbf{H}^{\mathrm{H}}\mathbf{H})^{-1}]_{kk})$. The SINR of user k is an increasing function of P_k^{ul}, but unaffected by the powers used by other users. Hence, the only way to improve a user's rate is to increase its power, and it can be done without degrading for anyone. Hence, using full power $P_k^{\mathrm{ul}} = P$ for all users is one solution to the max-min fairness problem. The corresponding max-min fair rate is the minimum rate among the users:

$$r^{\text{max-min fair}} = \min_{k \in \{1,\dots,K\}} B \log_2 \left(1 + \frac{P}{BN_0 \left[(\mathbf{H}^{\mathrm{H}}\mathbf{H})^{-1} \right]_{kk}} \right). \tag{6.79}$$

The users typically get different rates when using full power and only the user with the largest value of $[(\mathbf{H}^{\mathrm{H}}\mathbf{H})^{-1}]_{kk}$ achieves exactly $r^{\text{max-min fair}}$, while the others achieve larger rates. We can alternatively ensure that all users get exactly the rate $r^{\text{max-min fair}}$ by selecting the transmit powers as

$$P_k^{\mathrm{ul}} = \frac{[(\mathbf{H}^{\mathrm{H}}\mathbf{H})^{-1}]_{kk}}{\min_{i \in \{1,\dots,K\}} [(\mathbf{H}^{\mathrm{H}}\mathbf{H})^{-1}]_{ii}} P. \tag{6.80}$$

The non-uniqueness of the solution to the max-min fairness problem is why Algorithm 6.1 might not converge when using ZF combining.

Algorithm 6.1 Solution to the max-min fairness problem in (6.78).

Initialization: Select arbitrary $P_k^{\text{ul}} \in (0, P]$, for $k = 1, \ldots, K$, and the solution accuracy $\epsilon > 0$

while $\max\limits_{i \in \{1,\ldots,K\}} \text{SINR}_i\left(P_1^{\text{ul}}, \ldots, P_K^{\text{ul}}\right) - \min\limits_{i \in \{1,\ldots,K\}} \text{SINR}_i\left(P_1^{\text{ul}}, \ldots, P_K^{\text{ul}}\right) > \epsilon$ **do**

$\quad P_k^{\text{ul}} \leftarrow \frac{\min\limits_{i \in \{1,\ldots,K\}} \text{SINR}_i\left(P_1^{\text{ul}},\ldots,P_K^{\text{ul}}\right)}{\text{SINR}_k\left(P_1^{\text{ul}},\ldots,P_K^{\text{ul}}\right)} P_k^{\text{ul}}$, for $k = 1, \ldots, K$

$\quad P_k^{\text{ul}} \leftarrow \frac{P}{\max\limits_{i \in \{1,\ldots,K\}} P_i^{\text{ul}}} P_k^{\text{ul}}$, for $k = 1, \ldots, K$

end while

Output: $P_1^{\text{ul}}, \ldots, P_K^{\text{ul}}$

Figure 6.19 demonstrates the max-min fairness solution obtained by Algorithm 6.1 in a system with $K = 4$ users. The setup is the same as in Figure 6.16(a) and Figure 6.18. Each user achieves an SNR of 10 dB when using full power. In Figure 6.19(a), we show how the rates obtained by the four users vary with the iterations of the algorithm for $M = 6$ antennas and LMMSE combining. During the initial iterations, there are substantial rate variations between the users. However, as the algorithm proceeds, the four rates converge to a common value: the max-min fairness solution. The minimum rate among the users is gradually improved, but the convergence is not monotonic because a power reduction for some users will improve the rates of other users. In this example, 6–8 iterations are sufficient for convergence, but similar behavior can also be expected in other scenarios.

Figure 6.19(b) shows the minimum rate among the $K = 4$ users for different numbers of antennas M. Both LMMSE combining and MRC are considered. In addition to the max-min fairness solutions obtained by Algorithm 6.1, the minimum rates achieved when using full power at every user are shown as references. In all the considered cases, the minimum rate increases with M, which highlights how the communication performance is improved by using more antennas. As expected, the max-min fairness power control provides larger minimum rates than full-power transmission for both combining schemes. With LMMSE combining, the gap between the max-min fairness solution and full-power transmission reduces with an increasing number of antennas and diminishes asymptotically. The reason is that LMMSE combining resembles ZF combining when M is larger, and Example 6.7 demonstrated that all users should then use full power.

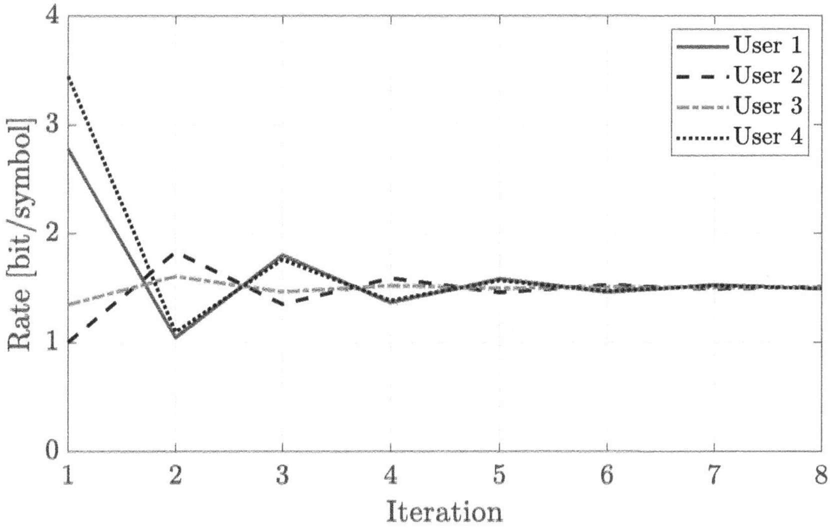

(a) The rates achieved by the four users at different iterations.

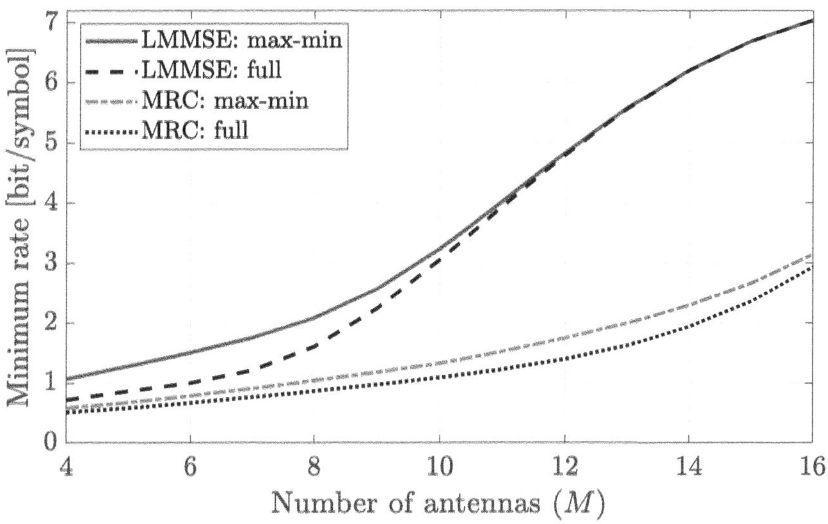

(b) The minimum rate versus the number of antennas M.

Figure 6.19: The max-min fairness solution obtained by Algorithm 6.1 with $K = 4$ users, using the setup from Figure 6.16(a). All the users have the same SNR of 10 dB when using full power. In (a), the rates of the four users during the algorithm's iterations are shown when LMMSE combining and $M = 6$ antennas are used. In (b), the minimum rate among the users is shown for a varying number of antennas M when using LMMSE and MRC combining. The minimum rate obtained using full power $P_k^{\text{ul}} = P$ for each user is shown as a reference.

6.4 Downlink Communications

There are several ways to operate the downlink of a multi-user system. The channel gains are identical in uplink and downlink, but the resource allocation solutions differ for two main reasons. Firstly, the base station can divide its power flexibly between the users in the downlink, while each user has an individual power budget in the uplink. Secondly, interference has a different impact since each user receives it through the same channel as its desired signal in the downlink, while it is received through different user channels in the uplink. In line with the uplink analysis in Section 6.3, we will consider three types of downlink operation: OMA, NOMA, and multi-user MIMO.

6.4.1 Orthogonal Multiple Access

We begin by considering the scenario where a single-antenna base station transmits to K single-antenna user devices over a shared communication channel with the total bandwidth B Hz. We will use FDMA, the OMA scheme where the bandwidth is divided orthogonally between the users. We let $\xi_k \in [0,1]$ denote the bandwidth fraction allocated to user k, for $k = 1,\ldots,K$, and recall that these fractions can be selected arbitrarily as long as $\xi_1 + \xi_2 + \ldots + \xi_K \leq 1$. Power allocation is another design dimension. The base station has a maximum transmit power denoted by P, which it can divide freely between the users. We let $P_k^{\mathrm{dl}} \in [0,P]$ denote the power allocated to user k, for $k = 1,\ldots,K$, and notice that these powers can be selected arbitrarily under the constraint $P_1^{\mathrm{dl}} + P_2^{\mathrm{dl}} + \ldots + P_K^{\mathrm{dl}} \leq P$. The channel gain of user k is denoted by $\beta_k \in [0,1]$. We assume the users are ordered such that $\beta_1 \geq \beta_2 \geq \ldots \geq \beta_K \geq 0$, which can be done without loss of generality.

Under these assumptions, user k experiences a point-to-point system with signal power P_k^{dl} and bandwidth $\xi_k B$, so the data rate in (6.8) becomes

$$R_k\left(\xi_k, P_k^{\mathrm{dl}}\right) = \xi_k C\left(\frac{P_k^{\mathrm{dl}}\beta_k}{\xi_k B N_0}\right) = \xi_k B \log_2\left(1 + \frac{P_k^{\mathrm{dl}}\beta_k}{\xi_k B N_0}\right) \quad \text{bit/s}, \tag{6.81}$$

where the notation $R_k(\xi_k, P_k^{\mathrm{dl}})$ emphasizes that the rate is a function of the bandwidth and power allocated to the user. It is a strictly increasing function of both variables (as can be shown by computing the first-order derivatives), which shows the fundamental conflict in resource allocation: If we increase a specific user's rate by allocating more power or bandwidth, we must take this power/bandwidth from other users that will experience rate reductions.

Based on the rate expression in (6.81), we can define the rate region as

$$\mathcal{R} = \Bigg\{\left(R_1(\xi_1, P_1^{\mathrm{dl}}),\ldots,R_K(\xi_K, P_K^{\mathrm{dl}})\right) : R_k(\xi_k, P_k^{\mathrm{dl}}) = \xi_k B \log_2\left(1 + \frac{P_k^{\mathrm{dl}}\beta_k}{\xi_k B N_0}\right),$$
$$\text{for } \xi_k, P_k^{\mathrm{dl}} \geq 0, k = 1,\ldots,K,\ \xi_1 + \ldots + \xi_K \leq 1, P_1^{\mathrm{dl}} + \ldots + P_K^{\mathrm{dl}} \leq P\Bigg\}, \tag{6.82}$$

which is the set of all points $(R_1(\xi_1, P_1^{\text{dl}}), \ldots, R_K(\xi_K, P_K^{\text{dl}}))$ that can be achieved by dividing the bandwidth and power between the users in different permissible ways. There must be equality in the last two constraints to reach the Pareto boundary. However, this is not a sufficient condition because some points at the rate region's interior also use all bandwidth and power. This situation differs from the uplink, where the bandwidth fractions are the only parameters varied in the rate region. For the downlink, even if a point uses all the bandwidth/power, improving a user's rate might be possible without sacrificing others by jointly changing bandwidth and power values. For simulation purposes, the Pareto boundary can be identified by generating many points that belong to the rate region and calculating their convex hull (i.e., the smallest convex outer boundary that encloses all the points).[6]

Suppose we want to achieve the maximum sum rate. For any given set of transmit powers $P_1^{\text{dl}}, \ldots, P_K^{\text{dl}}$, one can prove (by differentiation) that the sum rate is maximized by selecting the bandwidth fractions proportionally to the users' received signal powers:

$$\xi_k = \frac{P_k^{\text{dl}} \beta_k}{\sum_{i=1}^{K} P_i^{\text{dl}} \beta_i}. \tag{6.83}$$

By substituting this value into (6.81), the rate achieved by user k becomes

$$R_k\left(\frac{P_k^{\text{dl}} \beta_k}{\sum_{i=1}^{K} P_i^{\text{dl}} \beta_i}, P_k^{\text{dl}}\right) = \frac{P_k^{\text{dl}} \beta_k}{\sum_{i=1}^{K} P_i^{\text{dl}} \beta_i} C\left(\sum_{i=1}^{K} \frac{P_i^{\text{dl}} \beta_i}{B N_0}\right) \quad \text{bit/s} \tag{6.84}$$

and the sum rate becomes

$$\sum_{k=1}^{K} R_k\left(\frac{P_k^{\text{dl}} \beta_k}{\sum_{i=1}^{K} P_i^{\text{dl}} \beta_i}, P_k^{\text{dl}}\right) = C\left(\sum_{i=1}^{K} \frac{P_i^{\text{dl}} \beta_i}{B N_0}\right) \quad \text{bit/s}. \tag{6.85}$$

We can further maximize this expression by selecting the power allocation. The $C(\cdot)$ function is increasing with its argument $\sum_{i=1}^{K} P_i^{\text{dl}} \beta_i / (B N_0)$. The expression $\sum_{i=1}^{K} P_i^{\text{dl}} \beta_i$ is maximized by assigning all power to the user with the largest β_i, which is user 1 based on the assumed user ordering. Hence, the sum rate is maximized by $P_1^{\text{dl}} = P$ and $P_i^{\text{dl}} = 0$ for $i = 2, \ldots, K$. The maximum sum rate equals the single-user capacity $C_1^{\text{su}} = B \log_2(1 + P\beta_1/(B N_0))$ of user 1. It is beneficial from a sum-rate perspective only to serve one user, while any attempt to serve multiple users will result in a sum rate reduction (except if the served users have identical channel gains). Nevertheless, this must be done in practical multi-user systems because everyone must be served to some extent. This issue is different from the uplink, for which we noticed in Section 6.3.1 that the sum-rate-maximizing solution with FDMA is to split

[6] A close approximation of the Pareto boundary is obtained by assuming that the power is allocated proportionally to the bandwidth: $P_k^{\text{dl}} = P\xi_k$. The approximate boundary is then generated by varying $\xi_1, \ldots, \xi_K$ under the condition that $\xi_1 + \xi_2 + \ldots + \xi_K = 1$.

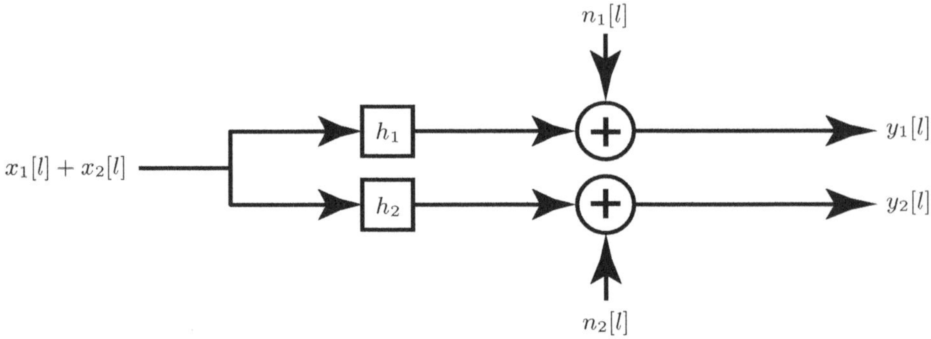

Figure 6.20: A discrete memoryless broadcast channel with $K = 2$ users and l denoting the discrete-time index. The input signal is the superposition of $x_1[l]$ and $x_2[l]$, designated for user 1 and user 2, respectively. The output at user k is $y_k[l] = h_k(x_1[l] + x_2[l]) + n_k[l]$, where h_k is the channel response and $n_k[l]$ is the independent complex Gaussian receiver noise, for $k = 1, 2$.

the bandwidth proportionally to the channel gains, which results in non-zero rates for everyone. The reason for the difference is that the base station can divide the power arbitrarily between the users in the downlink, while each user has a fixed power budget in the uplink, so they must all be served to use all the available transmit power.

6.4.2 Non-Orthogonal Multiple Access

We can achieve a larger rate region by letting the users share all time-frequency resources instead of dividing them orthogonally using FDMA. This was demonstrated in Section 6.3.2 for the uplink, while this section considers the downlink counterpart. A downlink setup with inter-user interference is called a broadcast channel, and the transmission scheme is called NOMA.

We will first describe the downlink NOMA scheme in the case of $K = 2$ users that share a bandwidth of B Hz. The base station divides its maximum transmit power P between the users so that $P_k^{\mathrm{dl}} \in [0, P]$ is the power assigned to user k, for $k = 1, 2$. We consider a discrete memoryless broadcast channel where the base station transmits simultaneously to both users, as illustrated in Figure 6.20. The received signal at user k at the discrete time l is

$$y_k[l] = h_k \left(x_1[l] + x_2[l]\right) + n_k[l], \tag{6.86}$$

where $x_1[l]$ is the input signal designated for user 1, $x_2[l]$ is the input signal meant for user 2, and $n_k[l] \sim \mathcal{N}_{\mathbb{C}}(0, N_0)$ is the independent receiver noise. The complex channel response to user k is denoted as h_k and is assumed to be deterministic. Its magnitude square is denoted as $\beta_k = |h_k|^2$.

Each user receives a superposition of both signals, but despite the mutual interference, it is possible to extract data if it is encoded correctly. Suppose the two input signals are independently complex Gaussian distributed so that $x_k[l] \sim \mathcal{N}_{\mathbb{C}}(0, P_k^{\mathrm{dl}}/B)$ for $k = 1, 2$, where the symbol power is P_k^{dl}/B. We further assume that the users are ordered such that $\beta_1 \geq \beta_2$. The received signal at user 2 in (6.86) can be expressed as

$$y_2[l] = h_2 x_2[l] + n_2'[l], \tag{6.87}$$

where $n_2'[l] = h_2 x_1[l] + n_2[l] \sim \mathcal{N}_{\mathbb{C}}(0, P_1^{\mathrm{dl}}\beta_2/B + N_0)$ is an effective noise term that is independent complex Gaussian distributed. Hence, it follows from Corollary 2.1 that an achievable rate of user 2 is

$$R_2 = C\left(\frac{P_2^{\mathrm{dl}}\beta_2}{P_1^{\mathrm{dl}}\beta_2 + BN_0}\right) \quad \text{bit/s}. \tag{6.88}$$

This expression resembles the ones obtained in the uplink, but an essential difference is that the interference term $P_1^{\mathrm{dl}}\beta_2$ depends on the user's own channel gain β_2 and not the channel gain β_1 of the other user. The reason is that the interfering downlink signal arrives through the same channel from the base station as the desired signal.

If user 1 is informed of the channel coding used by user 2, it can try to decode the signal designated for user 2. The received signal in (6.86) can then be expressed as

$$y_1[l] = h_1 x_2[l] + n_1'[l], \tag{6.89}$$

where $n_1'[l] = h_1 x_1[l] + n_1[l] \sim \mathcal{N}_{\mathbb{C}}(0, P_1^{\mathrm{dl}}\beta_1/B + N_0)$ is the effective noise term. Hence, an achievable rate is

$$C\left(\frac{P_2^{\mathrm{dl}}\beta_1}{P_1^{\mathrm{dl}}\beta_1 + BN_0}\right) \quad \text{bit/s}. \tag{6.90}$$

This value is larger than or equal to (6.88) because we assumed that $\beta_1 \geq \beta_2$. Consequently, any rate achievable for user 2 is also achievable for user 1, in the sense that it can also decode it successfully. Although user 1 is not interested in the actual data designated for user 2 (the data can even be encrypted so only user 2 can extract its original meaning), it can decode it to apply SIC. By subtracting the decoded signal sequence $\{x_2[l]\}$ from the original received signal in (6.86), user 1 obtains

$$y_1[l] - h_1 x_2[l] = h_1 x_1[l] + n_1[l]. \tag{6.91}$$

This is a conventional SISO channel without interference, so the achievable rate for user 1 regarding its designated signal is

$$R_1 = C\left(\frac{P_1^{\mathrm{dl}}\beta_1}{BN_0}\right) \quad \text{bit/s}. \tag{6.92}$$

This is the same rate as if the user was assigned the entire bandwidth in OMA, except that the power P_1^{dl} might be smaller than the maximum transmit power, depending on the base station's selected power allocation. Since the

SIC procedure is utilized, we have obtained a non-linear receiver processing scheme where we both scale the received signal before decoding and subtract interference from previously decoded signals. When we utilized SIC for uplink NOMA in Section 6.3.2, we could order the users arbitrarily. The situation is different in the downlink because the user with the strongest channel can decode signals encoded for users with weaker channels, but not vice versa. This makes the rate region easier to characterize since all points are obtained by varying the power allocation:

$$\mathcal{R} = \Bigg\{ (R_1, R_2) : 0 \leq R_1 \leq C\left(\frac{P_1^{\mathrm{dl}}\beta_1}{BN_0}\right), 0 \leq R_2 \leq C\left(\frac{P_2^{\mathrm{dl}}\beta_2}{P_1^{\mathrm{dl}}\beta_2 + BN_0}\right), \\ \text{for some } P_1^{\mathrm{dl}}, P_2^{\mathrm{dl}} \geq 0, P_1^{\mathrm{dl}} + P_2^{\mathrm{dl}} \leq P \Bigg\}. \tag{6.93}$$

The Pareto boundary is obtained whenever the maximum power is utilized:

$$\partial\mathcal{R} = \left\{ \left(C\left(\frac{P_1^{\mathrm{dl}}\beta_1}{BN_0}\right), C\left(\frac{P_2^{\mathrm{dl}}\beta_2}{P_1^{\mathrm{dl}}\beta_2 + BN_0}\right) \right) : P_1^{\mathrm{dl}}, P_2^{\mathrm{dl}} \geq 0, P_1^{\mathrm{dl}} + P_2^{\mathrm{dl}} = P \right\}.$$

It follows from (6.88) and (6.92) that the sum rate with NOMA is

$$\begin{aligned} R_1 + R_2 &= B\log_2\left(1 + \frac{P_1^{\mathrm{dl}}\beta_1}{BN_0}\right) + B\log_2\left(1 + \frac{P_2^{\mathrm{dl}}\beta_2}{P_1^{\mathrm{dl}}\beta_2 + BN_0}\right) \\ &= B\log_2\left(\frac{P_1^{\mathrm{dl}}\beta_1 + BN_0}{BN_0}\right) + B\log_2\left(\frac{(P_1^{\mathrm{dl}} + P_2^{\mathrm{dl}})\beta_2 + BN_0}{P_1^{\mathrm{dl}}\beta_2 + BN_0}\right) \\ &= B\log_2\left(1 + \frac{(P_1^{\mathrm{dl}} + P_2^{\mathrm{dl}})\beta_2}{BN_0}\right) + B\log_2\left(\frac{P_1^{\mathrm{dl}}\beta_1 + BN_0}{P_1^{\mathrm{dl}}\beta_2 + BN_0}\right), \end{aligned} \tag{6.94}$$

where the last equality follows from swapping the numerators between the two logarithms. The first term depends on $P_1^{\mathrm{dl}} + P_2^{\mathrm{dl}}$, but not on how the power is allocated between the users, so it is maximized when using the maximum power P. The second term is an increasing function of P_1^{dl} but independent of P_2^{dl}, thus the sum rate is maximized by setting $P_1^{\mathrm{dl}} = P$ and $P_2^{\mathrm{dl}} = 0$. The resulting maximum sum rate is $B\log_2(1 + P\beta_1/(BN_0)) = C_1^{\mathrm{su}}$, which is the single-user capacity of user 1. This is the same maximum value as with FDMA; thus, NOMA cannot improve the sum rate compared to OMA and reduces to single-user transmission at the optimal point. The rate region in (6.93) is anyway larger than the region in (6.82) obtained by FDMA.

The rate regions with NOMA and FDMA are exemplified in Figure 6.21(a) with $B = 10\,\text{MHz}$. This is the downlink counterpart to the two-user scenario considered in Figure 6.9. The two users have unequal channel qualities that become $\frac{P\beta_1}{2BN_0} = 10$ and $\frac{P\beta_2}{2BN_0} = 5$ if an equal power allocation of $P/2$ per user is used. The rate region with FDMA has a nearly linear Pareto boundary.

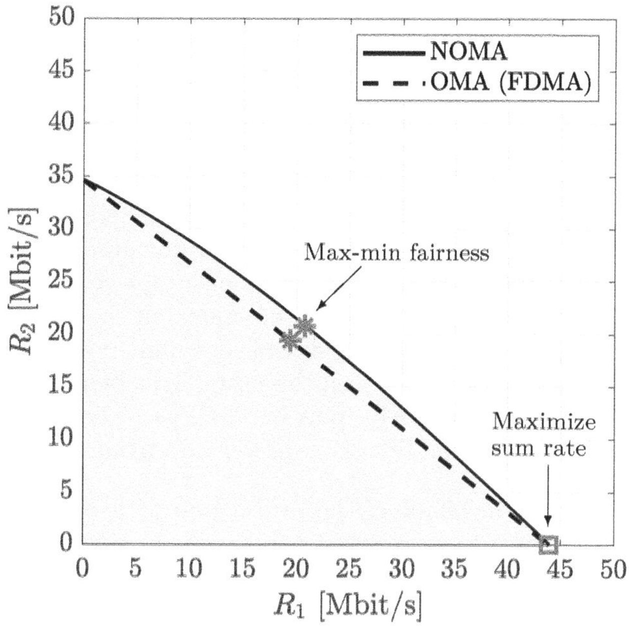

(a) Comparison of downlink NOMA with OMA based on FDMA.

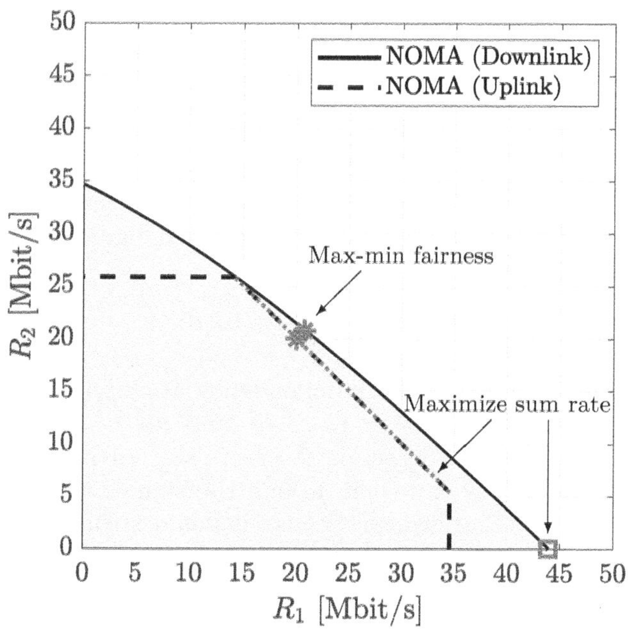

(b) Comparison of downlink NOMA and uplink NOMA.

Figure 6.21: Example of downlink rate regions for $K = 2$ users with different channel qualities. This is a continuation of the example in Figure 6.9.

The boundary with NOMA is more curved and results in a slightly larger region, demonstrating that it is easier to balance user performance when using NOMA. The sum rate is maximized at the single-user capacity point $(43.9, 0)$ Mbit/s that both access schemes can achieve. Max-min fairness is achieved at $(20.7, 20.7)$ Mbit/s with NOMA and at $(19.4, 19.4)$ Mbit/s with FDMA, so NOMA increases this rate point by 7%. In other words, fairness is achieved with a smaller sum-rate reduction when using NOMA.

Figure 6.21(b) compares the rate region with downlink NOMA and the rate region with uplink NOMA, which was previously shown in Figure 6.9. The channel gains and total transmit power are the same, but the downlink region is nevertheless larger. This is thanks to the flexible downlink power allocation that can divide the power unequally between the users. This benefit is particularly large close to the single-user capacity points. The Pareto boundaries intersect in one point, achieved by equal downlink power allocation.

Example 6.8. What is the shape of the rate region with NOMA if $\beta_1 = \beta_2$?

The users could have been ordered arbitrarily in this situation, which implies that the rate region is symmetric. The sum rate in (6.94) reduces to

$$R_1 + R_2 = B \log_2 \left(1 + \frac{(P_1^{\mathrm{dl}} + P_2^{\mathrm{dl}})\beta_2}{BN_0} \right) \leq B \log_2 \left(1 + \frac{P\beta_2}{BN_0} \right), \tag{6.95}$$

where the upper bound is achieved for any power allocation that satisfies $P_1^{\mathrm{dl}} + P_2^{\mathrm{dl}} = P$. Hence, the Pareto boundary is the straight line between the single-user points $(C_1^{\mathrm{su}}, 0)$ and $(0, C_2^{\mathrm{su}})$. Any point on that line is achieved by one specific selection of P_1^{dl} and $P_2^{\mathrm{dl}} = P - P_1^{\mathrm{dl}}$.

The rate region with NOMA for $K \geq 2$ users can be derived by following the same principles as with two users. The received signal at user k at the discrete time l is

$$y_k[l] = h_k \sum_{i=1}^{K} x_i[l] + n_k[l], \tag{6.96}$$

where $x_i[l] \sim \mathcal{N}_{\mathbb{C}}(0, P_i^{\mathrm{dl}}/B)$ is the independent data signal transmitted to user i with the power P_i^{dl}, for $i = 1, \ldots, K$, and $n_k[l] \sim \mathcal{N}_{\mathbb{C}}(0, N_0)$ is the independent receiver noise. We assume the users are ordered such that $\beta_1 \geq \beta_2 \geq \ldots \geq \beta_K \geq 0$. User k can then decode the signals intended for users $k+1, \ldots, K$ (in descending order) because it has a stronger channel than them. By subtracting those interfering signals before decoding its desired signal, user k will only be exposed to interference from users $1, \ldots, k-1$:

$$y_k[l] - h_k \sum_{i=k+1}^{K} x_i[l] = h_k x_k[l] + \underbrace{h_k \sum_{i=1}^{k-1} x_i[l] + n_k[l]}_{\sim \mathcal{N}_{\mathbb{C}}\left(0, \sum_{i=1}^{k-1} P_i^{\mathrm{dl}} \beta_k / B + N_0\right)} . \tag{6.97}$$

By treating the last terms as an effective noise term, an achievable rate for user k becomes

$$R_k = C\left(\frac{P_k^{\mathrm{dl}}\beta_k}{\sum_{i=1}^{k-1} P_i^{\mathrm{dl}}\beta_k + BN_0}\right) \quad \text{bit/s}. \tag{6.98}$$

Theorem 6.3. Consider a K-user discrete memoryless broadcast channel with the input $x_1 + \ldots + x_K \in \mathbb{C}$ and the outputs $y_1, \ldots, y_K \in \mathbb{C}$ given by

$$y_k = h_k \sum_{i=1}^{K} x_i + n_k, \quad k = 1, \ldots, K, \tag{6.99}$$

where $n_k \sim \mathcal{N}_{\mathbb{C}}(0, N_0)$ is independent noise and $h_1, \ldots, h_K \in \mathbb{C}$ are constant channel coefficients known at the output. Suppose the input distributions are feasible whenever $\mathbb{E}\{|x_k|^2\} \leq P_k^{\mathrm{dl}}/B$, where the transmit powers $P_1^{\mathrm{dl}}, \ldots, P_K^{\mathrm{dl}} \geq 0$ satisfy $P_1^{\mathrm{dl}} + \ldots + P_K^{\mathrm{dl}} \leq P$, P denotes the maximum transmit power, and B is the bandwidth (and symbol rate). If R_k denotes the rate of user k and $\beta_k = |h_k|^2$, the capacity region is given by

$$\mathcal{R} = \Bigg\{(R_1, \ldots, R_K) : 0 \leq R_k \leq C\left(\frac{P_k^{\mathrm{dl}}\beta_k}{\sum_{i=1}^{k-1} P_i^{\mathrm{dl}}\beta_k + BN_0}\right) \\ \text{for some } P_1^{\mathrm{dl}}, \ldots, P_K^{\mathrm{dl}} \geq 0, P_1^{\mathrm{dl}} + \ldots + P_K^{\mathrm{dl}} \leq P\Bigg\}. \tag{6.100}$$

Figure 6.22 exemplifies the rate region achieved with NOMA for $K = 3$ users. The considered setup is a downlink counterpart of Figure 6.10 with $\frac{P\beta_1}{3BN_0} = 10$, $\frac{P\beta_2}{3BN_0} = 5$, $\frac{P\beta_3}{3BN_0} = 2.5$, and $B = 10\,\mathrm{MHz}$. The shape of the outer boundary is indicated by a collection of curved lines that lie on the Pareto boundary. The lines are generated by fixing one of the transmit powers and then varying the others such that $P_1^{\mathrm{dl}} + P_2^{\mathrm{dl}} + P_3^{\mathrm{dl}} = P$. Each point on the boundary represents a specific tradeoff between the users' performance. The maximum sum rate is achieved at the single-user capacity point $(C_1^{\mathrm{su}}, 0, 0)$.

6.4.3 Downlink Multi-User MIMO with Non-Linear Processing

The reason that NOMA cannot increase the sum rate compared with FDMA is that all the transmitted signals propagate in the same way because the base station utilizes a single antenna. We have $M = 1$ transmit antenna and a total of K receive antennas, so the multiplexing gain of the corresponding point-to-point MIMO channel is $\min(M, K) = M = 1$. This means that the sum rate in the NOMA setup cannot surpass the capacity of the point-to-point

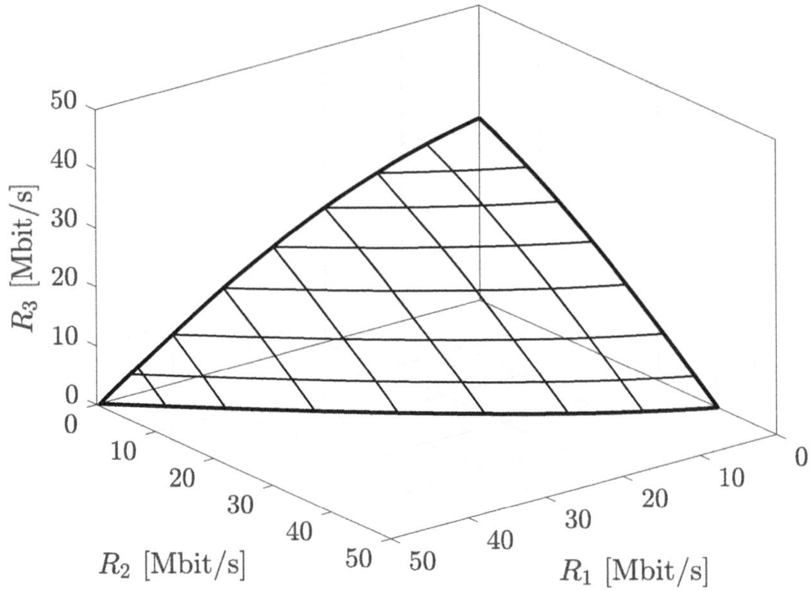

Figure 6.22: Example of a downlink rate region for $K = 3$ users when using NOMA.

MIMO setup.[7] Hence, it is unsurprising that the sum capacity with NOMA is limited by what can be achieved when serving one user at a time using FDMA. To serve K users efficiently in the downlink, the base station should have at least $M \geq K$ antennas so that a multiplexing gain of $\min(M, K) = K$ is theoretically achievable. This will enable the base station to send the K signals with substantially different spatial directivity so that the inter-user interference can be managed through clever processing/precoding at the transmitter side. This brings us to a downlink multi-user MIMO setup, where the multiple inputs are the M transmit antennas and the multiple outputs are the K user antennas; that is, the MIMO terminology is utilized even if each user device only has a single antenna.

We begin by considering a discrete memoryless channel with a transmitting base station equipped with $M \geq 2$ antennas and $K = 2$ receiving single-antenna user devices. Both users are served simultaneously over a bandwidth of B Hz. The maximum transmit power P is divided between the users, such that $P_k^{\text{dl}} \in [0, P]$ is the power allocated to user k and $P_1^{\text{dl}} + P_2^{\text{dl}} \leq P$. Moreover, each user is assigned a unit-norm precoding vector $\mathbf{p}_k \in \mathbb{C}^M$. The received signal $y_k[l] \in \mathbb{C}$ at user k at the discrete time l is

$$y_k[l] = \mathbf{h}_k^{\mathrm{T}} \left(\mathbf{p}_1 x_1[l] + \mathbf{p}_2 x_2[l] \right) + n_k[l], \tag{6.101}$$

[7] The essential difference between the broadcast channel and point-to-point MIMO channel is that the receiving users cannot decode signals cooperatively in the former setup, which is a restriction that can only lower the sum capacity.

where $x_i[l]$ is the date signal designated for user i, for $i = 1, 2$. The symbol power is P_i^{dl}/B and we assume Gaussian codebooks: $x_i[l] \sim \mathcal{N}_{\mathbb{C}}(0, P_i^{\text{dl}}/B)$. The channel vector to user k is denoted by $\mathbf{h}_k \in \mathbb{C}^M$, while $n_k[l] \sim \mathcal{N}_{\mathbb{C}}(0, N_0)$ is independent receiver noise. This channel is illustrated in Figure 6.23.

The uplink multi-user MIMO capacity region was obtained in Section 6.3.3 by utilizing two processing components: SIC and LMMSE combining. The capacity-achieving downlink operation has counterparts to these components, but another interference cancellation technique must replace SIC. In the downlink NOMA scenario considered in the last section, the users were ordered based on their channel gains as $|h_1|^2 \geq |h_2|^2$, and we noticed that the first user can always decode the second user's signal thanks to its stronger channel. This enabled SIC to be used to achieve the capacity. However, the same principle cannot be applied with $M \geq 2$ antennas because even if we order the users so that $\|\mathbf{h}_1\|^2 \geq \|\mathbf{h}_2\|^2$, the precoding vectors can be selected so that neither user can decode the other user's signal.[8]

Example 6.9. Consider a scenario with $M = 2$ antennas where the channels are $\mathbf{h}_1 = [1, 1]^{\mathrm{T}}$ and $\mathbf{h}_2 = [1, 0]^{\mathrm{T}}$. Suppose that $\frac{P_1^{\text{dl}}}{BN_0} = \frac{P_2^{\text{dl}}}{BN_0} = 1$ and MRT is used for precoding. Can user 1 decode the signal meant for user 2?

If user 2 treats the interfering signal as noise, it achieves the rate

$$\log_2\left(1 + \frac{P_2^{\text{dl}}|\mathbf{h}_2^{\mathrm{T}}\mathbf{p}_2|^2}{P_1^{\text{dl}}|\mathbf{h}_2^{\mathrm{T}}\mathbf{p}_1|^2 + BN_0}\right) = \log_2\left(1 + \frac{1}{0.5+1}\right) \approx 0.74\,\text{bit/symbol}, \tag{6.102}$$

because $\mathbf{p}_1 = \frac{1}{\sqrt{2}}\mathbf{h}_1^*$ and $\mathbf{p}_2 = \mathbf{h}_2^*$ when MRT is used. User 1 can only decode this signal if it is encoded at a rate that is lower than or equal to

$$\log_2\left(1 + \frac{P_2^{\text{dl}}|\mathbf{h}_1^{\mathrm{T}}\mathbf{p}_2|^2}{P_1^{\text{dl}}|\mathbf{h}_1^{\mathrm{T}}\mathbf{p}_1|^2 + BN_0}\right) = \log_2\left(1 + \frac{1}{2+1}\right) \approx 0.42\,\text{bit/symbol}. \tag{6.103}$$

Since $0.42 < 0.74$, user 1 cannot decode the signal designated for user 2, even if it has a stronger channel (i.e., $\|\mathbf{h}_1\|^2 = 2$ and $\|\mathbf{h}_2\|^2 = 1$). A similar computation will show that user 2 cannot decode the signal designated for user 1, so none of them can apply SIC. The multi-antenna precoding creates this effect because it reduces inter-user interference compared to the single-antenna case considered in NOMA. Another contributing factor to this result is that there is no unique ordering of vectors from strong to weak.

Instead of relying on interference cancellation at the receiving users, the base station can arrange a kind of interference subtraction before the data

[8]One can create special cases, called degraded broadcast channels, where the SIC procedure from Section 6.4.2 can also be utilized with multiple transmit antennas. One example is when $\mathbf{h}_1$ and $\mathbf{h}_2$ are equal except for a scaling factor. However, these cases are unlikely to occur in practical scenarios.

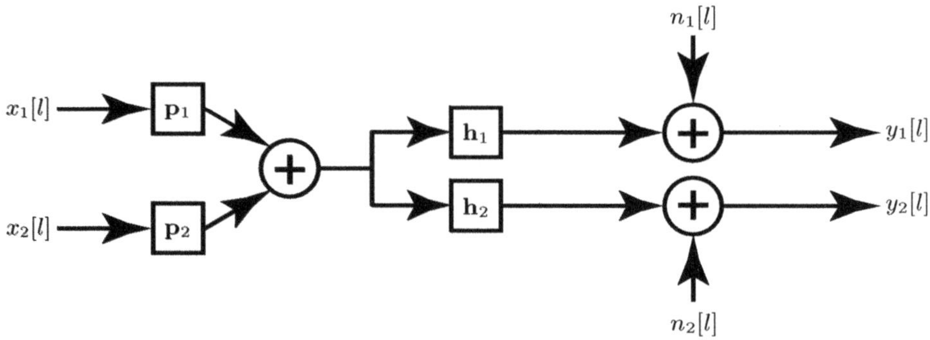

Figure 6.23: A discrete memoryless downlink multi-user MIMO channel with M transmit antennas and $K = 2$ receiving single-antenna users. The two input signals are $x_1[l]$ and $x_2[l]$, where l is a discrete-time index. The output at user k is $y_k[l] = \mathbf{h}_k^{\mathrm{T}}(\mathbf{p}_1 x_1[l] + \mathbf{p}_2 x_2[l]) + n_k[l]$ for $k = 1, 2$, where $\mathbf{h}_1, \mathbf{h}_2$ are the channel vectors, $\mathbf{p}_1, \mathbf{p}_2$ are the precoding vectors, and $n_1[l], n_2[l]$ are the independent complex Gaussian receiver noise terms.

signals leave its antennas. This approach builds on an information-theoretic result from [94], which considers the SISO channel shown in Figure 6.24, which has the unique characteristic that an extra interfering signal ι is added to the received signal. Suppose this interfering signal is random but known to the transmitter. In that case, the channel capacity is the same as if the interference was not there—even if the receiver is unaware of the realization of the interference.

Theorem 6.4. Consider the discrete memoryless channel in Figure 6.24 with input $x \in \mathbb{C}$ and output $y \in \mathbb{C}$ given by

$$y = h \cdot x + \iota + n, \tag{6.104}$$

where $n \sim \mathcal{N}_{\mathbb{C}}(0, N_0)$ is independent noise and $\iota \sim \mathcal{N}_{\mathbb{C}}(0, P_\iota)$ is an interfering signal that is only known at the transmitter. Suppose the input distribution is feasible whenever the symbol power satisfies $\mathbb{E}\{|x|^2\} \leq q$ and $h \in \mathbb{C}$ is a known constant. The channel capacity is

$$C = \log_2\left(1 + \frac{q|h|^2}{N_0}\right) \quad \text{bit/symbol} \tag{6.105}$$

and is achieved when the input is distributed as $x \sim \mathcal{N}_{\mathbb{C}}(0, q)$.

This somewhat surprising result is known as *dirty paper coding (DPC)* due to Max Costa's analogy in [94] between the proposed transmission scheme and how one can write a message on a paper that contains dirt spots. The paper represents the channel and the dirt is the interference that the transmitter/writer knows beforehand. The transmitter can write the message by adding ink so that the combination of ink and dirt becomes a message that

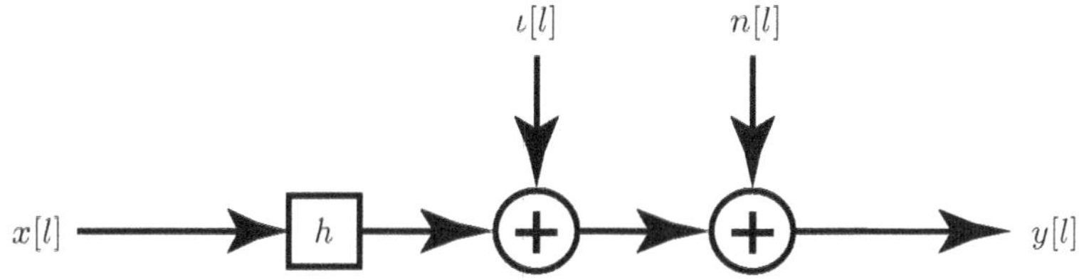

Figure 6.24: A discrete memoryless SISO channel with input $x[l]$ and output $y[l] = h \cdot x[l] + \iota[l] + n[l]$, where l is a discrete-time index, h is the channel response, $\iota[l]$ is an interfering signal that is known at the transmitter, and $n[l]$ is the independent Gaussian receiver noise.

the receiver/reader can understand without distinguishing ink from dirt. The main point is to utilize the dirt, not combat it.

We will not detail the proof of Theorem 6.4, which can be found in [94], or convey the precise implementation details, but only describe the fundamental principles of DPC. The transmitter and receiver take the codebook that could have been used to achieve the capacity in the absence of the interfering signal ι and augment it. Figure 6.25 illustrates this augmentation, where the black points around the origin represent the original Gaussian codebook. There are six copies of this codebook, highlighted with different colors, that are sufficiently far away not to overlap but sufficiently close not to create holes in between. For a given data symbol x and (normalized) interfering signal $\frac{\iota}{h}$, the transmitter determines which copy of x is closest to $\frac{\iota}{h}$ in the complex plane. We denote the closest copy by $\tilde{x}$, as illustrated in the figure. Instead of transmitting $\tilde{x}$ directly, we transmit $a = \tilde{x} - \frac{\iota}{h}$, because the receiver will then observe $ha + \iota + n = h\tilde{x} + n$ which is free from interference (but still contains noise). To make DPC efficient, the distance between the copied codebooks must be selected precisely so that $a \sim \mathcal{N}_{\mathbb{C}}(0, q)$. Due to the augmentation, the receiver has many more potential constellation points to consider during signal detection, which increases the decoding complexity, but the design does not create extra decoding errors since the copies are relatively far apart.

In the remainder of this section, we will utilize DPC to characterize the capacity region. When there are multiple users, their data signals must be encoded sequentially because DPC can only eliminate interference from already encoded signals. When there are $K = 2$ users, the user whose signal is encoded first must treat interference as noise, while the other user can benefit from DPC to get an interference-free transmission. The iterative procedure makes this a non-linear processing scheme. For every choice of precoding vectors and encoding order, we can generate the outer boundary of the rate region by varying the power allocation so that $P_1^{\text{dl}} + P_2^{\text{dl}} \leq P$. The capacity region will then be the union of all these rate regions; however, it is computationally hard to generate the region in this way. Searching over different encoding orders and power allocations is manageable, as shown earlier in this chapter,

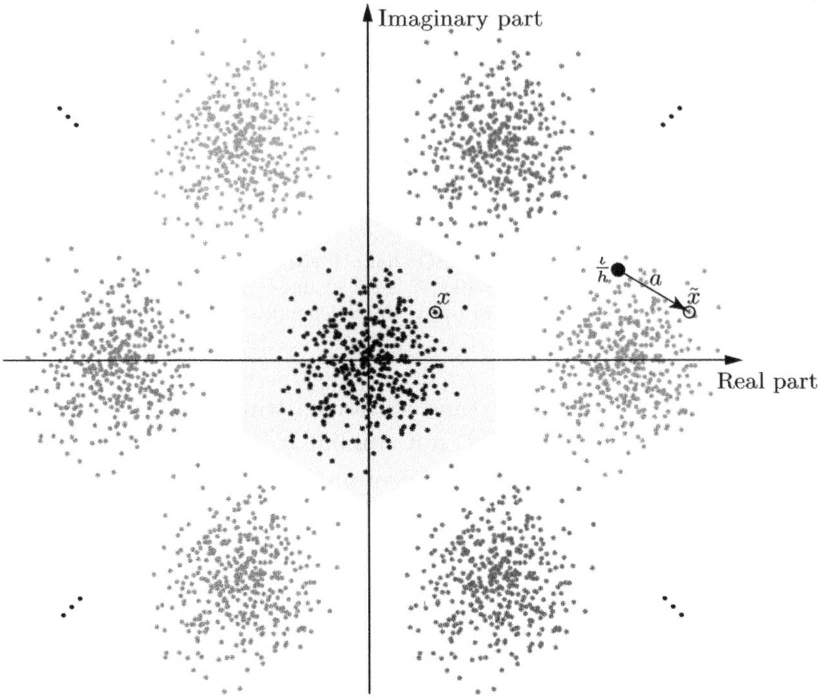

Figure 6.25: Illustration of the main principle of DPC. The original codebook is augmented with the colored codeword copies to fill the complex plane. When the data symbol x is to be transmitted, and the interfering signal is ι, the transmitter identifies the copy of x closest to ι/h, denoted by $\tilde{x}$. The transmitted signal is then selected as $a = \tilde{x} - \iota/h$ so that the summation of a and the interfering signal becomes $\tilde{x}$. The interference thereby becomes invisible to the receiver.

but there are too many ways of selecting the precoding vectors. Hence, we will look deeper into this issue to identify the optimal precoding when the other parameters have been selected.

If the signal designated for user 1 is encoded first, the user's received signal in (6.101) will be affected by interference from the signal meant for user 2. By treating this signal as additional noise with the power $\frac{P_2^{\text{dl}}}{B}|\mathbf{h}_1^{\text{T}}\mathbf{p}_2|^2$, the achievable rate becomes

$$R_1 = C\left(\frac{P_1^{\text{dl}}|\mathbf{h}_1^{\text{T}}\mathbf{p}_1|^2}{P_2^{\text{dl}}|\mathbf{h}_1^{\text{T}}\mathbf{p}_2|^2 + BN_0}\right). \tag{6.106}$$

When the signal to user 1 has been determined, the transmission to user 2 can be encoded using DPC. It then follows from Theorem 6.4 that the achievable rate will be the same as in the absence of interference; that is, as if the received signal was $y_2[l] = \mathbf{h}_2^{\mathrm{T}}\mathbf{p}_2 x_2[l] + n_2[l]$. The resulting rate for user 2 is

$$R_2 = C\left(\frac{P_2^{\mathrm{dl}}|\mathbf{h}_2^{\mathrm{T}}\mathbf{p}_2|^2}{BN_0}\right). \tag{6.107}$$

The rates in (6.106) and (6.107) can be computed for any precoding vectors. It is challenging to select the precoding because $\mathbf{p}_2$ affects the numerator of the SINR of user 2 and the denominator of the SINR of user 1. Hence, we must make a tradeoff between maximizing the received signal power of user 2 and limiting the interference caused to user 1. This is a crucial difference from the uplink, where each combining vector only affected its designated user and could be optimized without making tradeoffs. Interestingly, a mathematical connection between the uplink and downlink can be utilized to identify the optimal precoding. This is known as the *uplink-downlink duality.*

Consider the dual uplink scenario where the same two users send their signals to the base station using the transmit powers P_1^{ul} and P_2^{ul}, respectively. The channel vectors $\mathbf{h}_1, \mathbf{h}_2$ are the same as before, and the receive combining vectors are selected based on the precoding vectors as $\mathbf{w}_1 = \mathbf{p}_1^*$ and $\mathbf{w}_2 = \mathbf{p}_2^*$. If the signal from user 2 is decoded first and SIC is utilized to remove its interference before decoding the signal from user 1, the same approach as in Section 6.3.3 can be used to compute the achievable uplink rates as

$$R_1^{\mathrm{ul}} = C\left(\frac{P_1^{\mathrm{ul}}|\mathbf{p}_1^{\mathrm{T}}\mathbf{h}_1|^2}{\|\mathbf{p}_1\|^2 BN_0}\right) = C\left(\frac{P_1^{\mathrm{ul}}|\mathbf{p}_1^{\mathrm{T}}\mathbf{h}_1|^2}{BN_0}\right), \tag{6.108}$$

$$R_2^{\mathrm{ul}} = C\left(\frac{P_2^{\mathrm{ul}}|\mathbf{p}_2^{\mathrm{T}}\mathbf{h}_2|^2}{P_1^{\mathrm{ul}}|\mathbf{p}_2^{\mathrm{T}}\mathbf{h}_1|^2 + \|\mathbf{p}_2\|^2 BN_0}\right) = C\left(\frac{P_2^{\mathrm{ul}}|\mathbf{p}_2^{\mathrm{T}}\mathbf{h}_2|^2}{P_1^{\mathrm{ul}}|\mathbf{p}_2^{\mathrm{T}}\mathbf{h}_1|^2 + BN_0}\right), \tag{6.109}$$

where we simplified the expressions by utilizing that the precoding vectors have unit norm. Suppose we want user k to achieve a specific rate $C(\gamma_k)$ in both the uplink and downlink, for $k = 1, 2$, where $\gamma_k \geq 0$ denotes the desired SINR value. We can find the downlink transmit powers that lead to these rates by solving the equation

$$\left.\begin{aligned} \frac{P_1^{\mathrm{dl}}|\mathbf{h}_1^{\mathrm{T}}\mathbf{p}_1|^2}{P_2^{\mathrm{dl}}|\mathbf{h}_1^{\mathrm{T}}\mathbf{p}_2|^2 + BN_0} &= \gamma_1 \\ \frac{P_2^{\mathrm{dl}}|\mathbf{h}_2^{\mathrm{T}}\mathbf{p}_2|^2}{BN_0} &= \gamma_2 \end{aligned}\right\} \Rightarrow \underbrace{\begin{bmatrix} \frac{|\mathbf{h}_1^{\mathrm{T}}\mathbf{p}_1|^2}{\gamma_1 BN_0} & -\frac{|\mathbf{h}_1^{\mathrm{T}}\mathbf{p}_2|^2}{BN_0} \\ 0 & \frac{|\mathbf{h}_2^{\mathrm{T}}\mathbf{p}_2|^2}{\gamma_2 BN_0} \end{bmatrix}}_{=\boldsymbol{\Gamma}} \begin{bmatrix} P_1^{\mathrm{dl}} \\ P_2^{\mathrm{dl}} \end{bmatrix} = \begin{bmatrix} 1 \\ 1 \end{bmatrix}. \tag{6.110}$$

This is a linear system of equations, so the solution is $[P_1^{\mathrm{dl}}, P_2^{\mathrm{dl}}]^{\mathrm{T}} = \boldsymbol{\Gamma}^{-1}[1, 1]^{\mathrm{T}}$ if the matrix $\boldsymbol{\Gamma}$ is invertible.[9] The corresponding equations for the dual uplink

[9] The transmit powers must be positive and this condition is only satisfied for some invertible matrices $\boldsymbol{\Gamma}$, which showcases that some rate combinations can never be achieved. The necessary and sufficient condition is that $\mathbf{I} - \boldsymbol{\Gamma}_{\mathrm{diag}}^{-1}\boldsymbol{\Gamma}$ has eigenvalues in the range $(-1, 1)$ [95], where $\boldsymbol{\Gamma}_{\mathrm{diag}}$ is the diagonal matrix having the same diagonal entries as $\boldsymbol{\Gamma}$. In this book, we only want to establish that rates achievable in the uplink with non-zero power coefficients are also achievable in the downlink with non-zero power coefficients and vice versa, which implies that the condition is automatically satisfied.

transmit powers $P_1^{\mathrm{ul}}, P_2^{\mathrm{ul}}$ are

$$\left.\begin{aligned} \frac{P_1^{\mathrm{ul}}|\mathbf{p}_1^{\mathrm{T}}\mathbf{h}_1|^2}{BN_0} &= \gamma_1 \\ \frac{P_2^{\mathrm{ul}}|\mathbf{p}_2^{\mathrm{T}}\mathbf{h}_2|^2}{P_1^{\mathrm{ul}}|\mathbf{p}_2^{\mathrm{T}}\mathbf{h}_1|^2+BN_0} &= \gamma_2 \end{aligned}\right\} \Rightarrow \underbrace{\begin{bmatrix} \frac{|\mathbf{p}_1^{\mathrm{T}}\mathbf{h}_1|^2}{\gamma_1 BN_0} & 0 \\ -\frac{|\mathbf{p}_2^{\mathrm{T}}\mathbf{h}_1|^2}{BN_0} & \frac{|\mathbf{p}_2^{\mathrm{T}}\mathbf{h}_2|^2}{\gamma_2 BN_0} \end{bmatrix}}_{=\boldsymbol{\Gamma}^{\mathrm{T}}} \begin{bmatrix} P_1^{\mathrm{ul}} \\ P_2^{\mathrm{ul}} \end{bmatrix} = \begin{bmatrix} 1 \\ 1 \end{bmatrix}. \tag{6.111}$$

The only difference from the downlink is that the uplink equation system contains $\boldsymbol{\Gamma}^{\mathrm{T}}$ instead of $\boldsymbol{\Gamma}$. This showcases that the downlink and uplink SINR expressions contain the same kind of interference terms but at different places: the interference term $|\mathbf{p}_2^{\mathrm{T}}\mathbf{h}_1|^2 = |\mathbf{h}_1^{\mathrm{T}}\mathbf{p}_2|^2$ affects user 1 in the downlink and user 2 in the uplink. Due to this asymmetry, the values of $P_1^{\mathrm{dl}}, P_2^{\mathrm{dl}}$ that deliver the desired downlink rates are generally different from the values of $P_1^{\mathrm{ul}}, P_2^{\mathrm{ul}}$ that deliver the same uplink rates. However, the values are tightly related because

$$P_1^{\mathrm{dl}} + P_2^{\mathrm{dl}} = \begin{bmatrix} 1 \\ 1 \end{bmatrix}^{\mathrm{T}} \begin{bmatrix} P_1^{\mathrm{dl}} \\ P_2^{\mathrm{dl}} \end{bmatrix} = \begin{bmatrix} P_1^{\mathrm{ul}} \\ P_2^{\mathrm{ul}} \end{bmatrix}^{\mathrm{T}} \boldsymbol{\Gamma} \begin{bmatrix} P_1^{\mathrm{dl}} \\ P_2^{\mathrm{dl}} \end{bmatrix} = \begin{bmatrix} P_1^{\mathrm{ul}} \\ P_2^{\mathrm{ul}} \end{bmatrix}^{\mathrm{T}} \begin{bmatrix} 1 \\ 1 \end{bmatrix} = P_1^{\mathrm{ul}} + P_2^{\mathrm{ul}}, \tag{6.112}$$

where the second and third equalities follow from (6.110) and (6.111), respectively. Hence, when the precoding and combining vectors are identical, the same data rates are achievable in the downlink and uplink using the same total transmit power but allocating it differently between the users. If the uplink powers are known, the corresponding downlink powers can be computed as

$$\begin{bmatrix} P_1^{\mathrm{dl}} \\ P_2^{\mathrm{dl}} \end{bmatrix} = \boldsymbol{\Gamma}^{-1}\boldsymbol{\Gamma}^{\mathrm{T}} \begin{bmatrix} P_1^{\mathrm{ul}} \\ P_2^{\mathrm{ul}} \end{bmatrix}. \tag{6.113}$$

This result also has implications for the precoding selection. We know from Section 6.3.3 that the uplink rate in (6.108) is maximized by $\mathbf{p}_1 = \mathbf{h}_1^*/\|\mathbf{h}_1\|$, which is MRC. Moreover, we know that the uplink rate in (6.109) is maximized by the LMMSE combining in (6.36). By revising the notation, including complex conjugates, and normalizing the expression to have unit norm, we obtain the SINR-maximizing precoding vector

$$\mathbf{p}_2 = \frac{\left(P_1^{\mathrm{ul}}\mathbf{h}_1^*\mathbf{h}_1^{\mathrm{T}} + BN_0\mathbf{I}_M\right)^{-1}\mathbf{h}_2^*}{\left\|\left(P_1^{\mathrm{ul}}\mathbf{h}_1^*\mathbf{h}_1^{\mathrm{T}} + BN_0\mathbf{I}_M\right)^{-1}\mathbf{h}_2^*\right\|}. \tag{6.114}$$

The uplink-downlink duality dictates that the same rate points $(C(\gamma_1), C(\gamma_2))$ are achievable in both uplink and downlink; thus, if some points can only be reached by the optimal uplink combining, we must use the complex conjugates of the same vectors for downlink precoding to reach those points. We have thereby established a mechanism to transform optimal uplink combining vectors into downlink precoding vectors that are optimal for reaching the same rate point, thereby resolving the complicated tradeoff. The intuition is that the spatial direction (in the M-dimensional vector space) in which the base station should listen to the uplink signal from user k is the same as it should transmit back to user k in the downlink.

The uplink rate region with two users was characterized in (6.44) for the case when the users use the same transmit power. When we instead assign arbitrary uplink powers $P_1^{\text{ul}}, P_2^{\text{ul}}$, we can generalize the expression as

$$\mathcal{R}^{\text{ul}}_{P_1^{\text{ul}},P_2^{\text{ul}}} = \Bigg\{ (R_1, R_2) : 0 \leq R_1 \leq C\left(\frac{P_1^{\text{ul}}\|\mathbf{h}_1\|^2}{BN_0}\right), 0 \leq R_2 \leq C\left(\frac{P_2^{\text{ul}}\|\mathbf{h}_2\|^2}{BN_0}\right), \\ R_1 + R_2 \leq B\log_2\left(\det\left(\mathbf{I}_M + \frac{P_1^{\text{ul}}}{BN_0}\mathbf{h}_1\mathbf{h}_1^{\text{H}} + \frac{P_2^{\text{ul}}}{BN_0}\mathbf{h}_2\mathbf{h}_2^{\text{H}}\right)\right)\Bigg\}. \tag{6.115}$$

This scenario is typically called the *virtual dual uplink* since we cannot allocate the total uplink power arbitrarily between the users in practice. However, the power allocation feature exists in the downlink, and the uplink-downlink duality connects the downlink scenario to these hypothetical/virtual dual uplink scenarios with arbitrary power splits between the users. In particular, the downlink rate region that is achievable in multi-user MIMO setups with DPC is the union of all conceivable virtual uplink rate regions:

$$\mathcal{R} = \bigcup_{P_1^{\text{ul}},P_2^{\text{ul}}:\, P_1^{\text{ul}}+P_2^{\text{ul}}\leq P} \mathcal{R}^{\text{ul}}_{P_1^{\text{ul}},P_2^{\text{ul}}}. \tag{6.116}$$

This is the largest achievable rate region of the downlink multi-user MIMO channel, which is formally proved in [96], [97], so we call it the capacity region.

The downlink rate region with $M = 4$ and $B = 10\,\text{MHz}$ is exemplified in Figure 6.26, as a continuation to the NOMA scenario in Figure 6.21 with the LOS channel model in (4.23) where the UEs have the azimuth angles $\varphi_1 = -\pi/20$ and $\varphi_2 = \pi/20$. The users have unequal channel qualities that become $\frac{P\beta_1}{2BN_0} = 10$ and $\frac{P\beta_2}{2BN_0} = 5$ under an equal power allocation of $P/2$ per user. Nine virtual uplink regions, obtained by different power splits, are illustrated using blue-dotted lines. These regions have the pentagonal shape typical for the uplink. The downlink region is the union of all such virtual uplink regions; thus, it is larger than any given uplink region thanks to the ability to allocate downlink power arbitrarily between users. The Pareto boundary has a smoother shape where each point is obtained from one specific uplink region, following from the uplink-downlink duality. Suppose we start from a point in the virtual uplink. In that case, we can obtain the corresponding downlink transmission method by using the same combining vectors as precoding vectors, transforming the uplink powers into downlink powers using (6.113), and encoding the downlink signals using DPC in the opposite order as the uplink signals are decoded using SIC.

The operating points that provide the maximum sum rate and max-min fairness are indicated in Figure 6.26, and there are multiple points in the

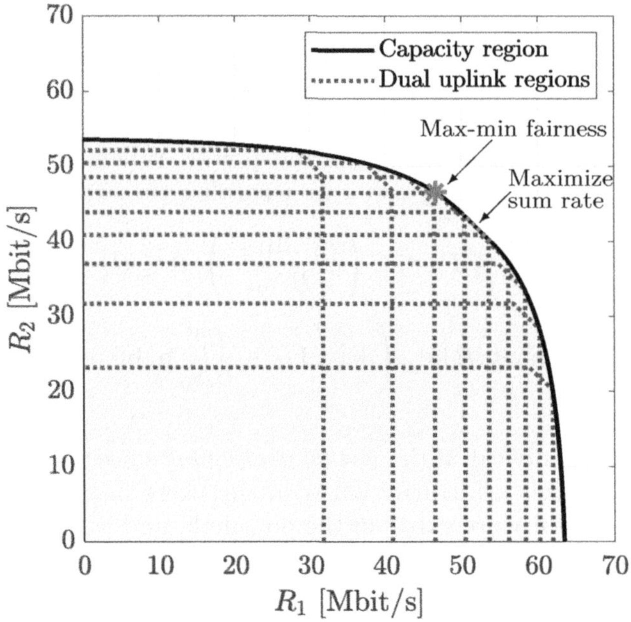

Figure 6.26: Example of the downlink rate region for $K = 2$ users with multi-user MIMO, DPC, and $M = 4$ antennas. The region is the union of all possible dual uplink rate regions with different transmit power divisions between the users. This is a continuation of the example in Figure 6.21. Many red-marked points achieve the maximum sum rate, while a single point (red star) gives max-min fairness.

former case. The sum rate with the optimal receive combining in the virtual uplink can be computed using (6.43) with arbitrary uplink powers $P_1^{\text{ul}}, P_2^{\text{ul}}$ as

$$R_1^{\text{ul}} + R_2^{\text{ul}} = C\left(\frac{P_1^{\text{ul}}\|\mathbf{h}_1\|^2}{BN_0}\right) + C\left(P_2^{\text{ul}}\mathbf{h}_2^{\text{H}}\left(P_1^{\text{ul}}\mathbf{h}_1\mathbf{h}_1^{\text{H}} + BN_0\mathbf{I}_M\right)^{-1}\mathbf{h}_2\right)$$
$$= B\log_2\left(\det\left(\mathbf{I}_M + \frac{P_1^{\text{ul}}}{BN_0}\mathbf{h}_1\mathbf{h}_1^{\text{H}} + \frac{P_2^{\text{ul}}}{BN_0}\mathbf{h}_2\mathbf{h}_2^{\text{H}}\right)\right). \quad (6.117)$$

This expression is symmetric with respect to the user indices, which means that the same sum rate can be achieved irrespective of which user signal is decoded first in the virtual uplink (i.e., encoded last in the downlink). The maximum downlink sum rate is obtained by maximizing this expression with respect to the uplink powers [98]:

$$\underset{P_1^{\text{ul}},P_2^{\text{ul}}:P_1^{\text{ul}}+P_2^{\text{ul}}\leq P}{\text{maximize}} \quad B\log_2\left(\det\left(\mathbf{I}_M + \frac{P_1^{\text{ul}}}{BN_0}\mathbf{h}_1\mathbf{h}_1^{\text{H}} + \frac{P_2^{\text{ul}}}{BN_0}\mathbf{h}_2\mathbf{h}_2^{\text{H}}\right)\right). \quad (6.118)$$

There is no closed-form solution to this problem, but the objective function is a concave function of the power variables. Hence, this is a convex optimization problem that can be solved using general-purpose convex solvers [99].

The derivation of the rate region can be extended to the general case of $K \geq 2$. The uplink-downlink duality is then generalized to consider a virtual uplink scenario of the same kind as in Theorem 6.2 but with different powers among the users. The rate region is the union of all such virtual uplink regions. Each point in the downlink region is achieved by encoding the user signals sequentially and applying DPC to protect each user from interference from the previously encoded signals. We can summarize the result as follows.

Theorem 6.5. Consider a K-user discrete memoryless downlink multi-user MIMO channel with the input $\mathbf{p}_1 x_1 + \ldots + \mathbf{p}_K x_K \in \mathbb{C}^M$, where $x_k \in \mathbb{C}$ is the input signal designated for user k and $\mathbf{p}_k \in \mathbb{C}^M$ is the corresponding unit-norm precoding vector. The outputs $y_1, \ldots, y_K \in \mathbb{C}$ at the users are given by

$$y_k = \mathbf{h}_k^{\mathrm{T}} \sum_{i=1}^{K} \mathbf{p}_i x_i + n_k, \quad k = 1, \ldots, K, \tag{6.119}$$

where $n_k \sim \mathcal{N}_{\mathbb{C}}(0, N_0)$ is independent noise and $\mathbf{h}_1, \ldots, \mathbf{h}_K \in \mathbb{C}^M$ are constant channel vectors known at the output. Suppose the input distributions are feasible whenever $\mathbb{E}\{|x_k|^2\} \leq P_k^{\mathrm{dl}}/B$, where the transmit powers $P_1^{\mathrm{dl}}, \ldots, P_K^{\mathrm{dl}} \geq 0$ satisfy $P_1^{\mathrm{dl}} + \ldots + P_K^{\mathrm{dl}} \leq P$, P denotes the maximum transmit power, and B is the bandwidth (and symbol rate). The capacity region is given by

$$\mathcal{R} = \bigcup_{\substack{P_1^{\mathrm{ul}}, \ldots, P_K^{\mathrm{ul}}: \\ P_1^{\mathrm{ul}} + \ldots + P_K^{\mathrm{ul}} \leq P}} \mathcal{R}^{\mathrm{ul}}_{P_1^{\mathrm{ul}}, \ldots, P_K^{\mathrm{ul}}}, \tag{6.120}$$

which is the union of the virtual uplink regions

$$\mathcal{R}^{\mathrm{ul}}_{P_1^{\mathrm{ul}}, \ldots, P_K^{\mathrm{ul}}} = \Bigg\{ (R_1, \ldots, R_K) : \sum_{k \in \mathcal{K}} R_k \leq B \log_2 \left(\det \left(\mathbf{I}_M + \sum_{k \in \mathcal{K}} \frac{P_k^{\mathrm{ul}}}{B N_0} \mathbf{h}_k \mathbf{h}_k^{\mathrm{H}} \right) \right)$$
$$\text{for all } \mathcal{K} \subset \{1, \ldots, K\}, \ R_k \geq 0 \text{ for all } k \Bigg\}. \tag{6.121}$$

The parameterization in (6.121) reveals the expression for the sum rate, obtained with $\mathcal{K} = \{1, \ldots, K\}$. Hence, we can maximize the sum rate as

$$\underset{\substack{P_1^{\mathrm{ul}}, \ldots, P_K^{\mathrm{ul}} \\ P_1^{\mathrm{ul}} + \ldots + P_K^{\mathrm{ul}} \leq P}}{\text{maximize}} \quad B \log_2 \left(\det \left(\mathbf{I}_M + \sum_{k=1}^{K} \frac{P_k^{\mathrm{ul}}}{B N_0} \mathbf{h}_k \mathbf{h}_k^{\mathrm{H}} \right) \right). \tag{6.122}$$

This is a generalization of the two-user problem in (6.118), and it remains to be a convex optimization problem that lacks a closed-form solution [98], but can be solved efficiently using any software for solving such problems.

The benefit of increasing the number of antennas is illustrated in Figure 6.27 by revisiting the example from Figures 6.21 and 6.26. The rate regions with

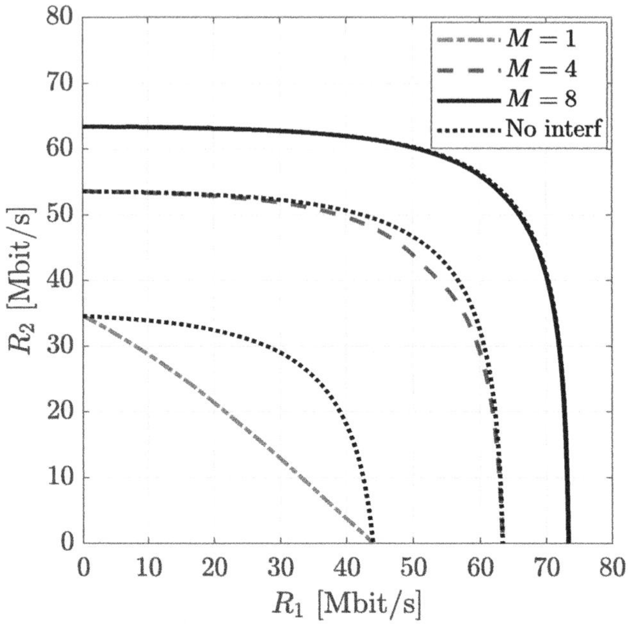

Figure 6.27: Examples of downlink rate regions for $K = 2$ users with multi-user MIMO and a varying number of antennas M, where $M = 1$ corresponds to NOMA. This is a continuation of the example in Figures 6.21 and 6.26. The dotted curves represent the hypothetical cases where inter-user interference is neglected.

NOMA (i.e., $M = 1$) and multi-user MIMO with $M = 4$ or $M = 8$ are compared. As the beamforming gain increases with M, the single-user capacity points are shifted towards larger values along the two axes. The Pareto boundaries also become increasingly curved, demonstrating how interference becomes less of an issue thanks to favorable propagation, and the sum rate is substantially larger than the single-user capacities. The dotted curves illustrate the three hypothetical rate regions obtained without inter-user interference to emphasize this effect further. The difference between the hypothetical interference-free case and the actual Pareto boundary is large for $M = 1$, but tiny for $M = 8$. The reason for the curvature in the interference-free cases is the need to divide the total transmit power between the users.

6.4.4 Downlink Multi-User MIMO with Linear Processing

The last section demonstrated how DPC could be utilized in downlink multi-user MIMO to achieve all Pareto optimal operating points in the rate region. Unfortunately, this non-linear encoding scheme has some practical drawbacks. Firstly, the sequential encoding of the users' signals leads to an encoding delay that grows proportionally to the number of users, and the encoding complexity per user signal is also increased. Secondly, the decoding complexity at the user devices is increased as the codebook is augmented with many codebook copies. Thirdly, the individual users' data rates must be selected jointly based on the encoding order, which makes the operation less flexible. Finally, the transmitter might have imperfect channel knowledge in practice, so the interference can only be partially removed. In this section, we will analyze downlink multi-user MIMO without DPC, where each user is subject to interference from all other users and treats it as additional noise. This is referred to as *linear processing* and requires the precoding and power allocation to be fine-tuned to suppress inter-user interference further.

We consider a K-user downlink multi-user MIMO channel of the kind defined in Theorem 6.5. The received signal at user k is

$$y_k = \mathbf{h}_k^{\mathrm{T}} \sum_{i=1}^{K} \mathbf{p}_i x_i + n_k, \quad k = 1, \ldots, K, \tag{6.123}$$

where $x_i \sim \mathcal{N}_{\mathbb{C}}(0, P_i^{\mathrm{dl}}/B)$ is the data signal sent to user i, $P_i^{\mathrm{dl}} \geq 0$ is the allocated transmit power, $\mathbf{p}_i \in \mathbb{C}^M$ is the associated unit-norm precoding vector, and $n_k \sim \mathcal{N}_{\mathbb{C}}(0, N_0)$ is independent noise. The complete received signal $\mathbf{y} = [y_1, \ldots, y_K]^{\mathrm{T}}$ for all users is expressed as

$$\mathbf{y} = \mathbf{H}^{\mathrm{T}}\mathbf{P}\mathbf{x} + \mathbf{n}, \tag{6.124}$$

where $\mathbf{H} = [\mathbf{h}_1, \ldots, \mathbf{h}_K] \in \mathbb{C}^{M \times K}$ is the channel matrix, $\mathbf{P} = [\mathbf{p}_1, \ldots, \mathbf{p}_K] \in \mathbb{C}^{M \times K}$ is the precoding matrix with unit-norm columns, $\mathbf{x} = [x_1, \ldots, x_K]^{\mathrm{T}}$ contains all the data signals, and $\mathbf{n} = [n_1, \ldots, n_K]^{\mathrm{T}}$ contains the noise.

Under these conditions, the sum of the interfering signals at user k is

$$\sum_{i=1, i\neq k}^{K} \mathbf{h}_k^{\mathrm{T}} \mathbf{p}_i x_i \sim \mathcal{N}_{\mathbb{C}}\left(0, \sum_{i=1, i\neq k}^{K} \frac{P_i^{\mathrm{dl}}}{B} |\mathbf{h}_k^{\mathrm{T}} \mathbf{p}_i|^2\right), \tag{6.125}$$

which has the same distribution as receiver noise but a different variance. By treating the interference as additional noise in the signal decoding, it follows from Corollary 2.1 that user k can achieve the downlink rate

$$R_k = C\left(\frac{P_k^{\mathrm{dl}} |\mathbf{h}_k^{\mathrm{T}} \mathbf{p}_k|^2}{\sum_{i=1, i\neq k}^{K} P_i^{\mathrm{dl}} |\mathbf{h}_k^{\mathrm{T}} \mathbf{p}_i|^2 + BN_0}\right). \tag{6.126}$$

The same decoding algorithm as in a single-user system can be utilized since DPC is not used. It is instructive to compare this rate to the uplink rate expression in (6.58), under the assumption that the receive combining vectors are selected based on the precoding vectors as $\mathbf{w}_k = \mathbf{p}_k^*$. The uplink rate can then becomes

$$R_k^{\mathrm{ul}} = C\left(\frac{P_k^{\mathrm{ul}} |\mathbf{p}_k^{\mathrm{T}} \mathbf{h}_k|^2}{\sum_{i=1, i\neq k}^{K} P_i^{\mathrm{ul}} |\mathbf{p}_k^{\mathrm{T}} \mathbf{h}_i|^2 + BN_0}\right). \tag{6.127}$$

There is a striking similarity between (6.126) and (6.127), where the only differences are the power coefficients and that the indices are switched in the interference terms so that $|\mathbf{h}_k^{\mathrm{T}}\mathbf{p}_i|^2$ in the downlink becomes $|\mathbf{p}_k^{\mathrm{T}}\mathbf{h}_i|^2$ in the uplink. This is another instance of the uplink-downlink duality but for systems with linear processing. By following the same approach as in the previous section, one can prove that any combination $(R_1, \dots, R_K)$ of user rates that is achievable in the downlink is also achievable in the uplink by selecting the combining vectors as $\mathbf{w}_k = \mathbf{p}_k^*$ and using the same total transmit power but allocating it differently between the users. The duality results with linear processing can be traced back to [100], [101]. Since the uplink powers cannot be distributed freely between the users, the duality holds between the downlink scenario and a virtual uplink scenario that allows for power reallocation between users. Hence, the downlink rate region with linear processing is obtained from the uplink region in (6.66) by changing the constraint for how uplink powers are allocated:

$$\mathcal{R} = \Bigg\{ (R_1, \dots, R_K) : R_k = B \log_2 \Bigg(1 + P_k^{\mathrm{ul}} \mathbf{h}_k^{\mathrm{H}} \Bigg(\sum_{i=1, i \neq k}^{K} P_i^{\mathrm{ul}} \mathbf{h}_i \mathbf{h}_i^{\mathrm{H}} + B N_0 \mathbf{I}_M \Bigg)^{-1} \mathbf{h}_k \Bigg)$$
$$\text{for } k = 1, \dots, K, \text{ for some } P_1^{\mathrm{ul}}, \dots, P_K^{\mathrm{ul}} \geq 0 \text{ satisfying } \sum_{k=1}^{K} P_k^{\mathrm{ul}} \leq P \Bigg\}. \tag{6.128}$$

The rate expression and transmit power terms in (6.128) originate from the corresponding uplink scenario, so how to achieve each specific rate in the downlink is not apparent. Before taking a closer look at that, we will compare (6.128) with the capacity region in Theorem 6.5, obtained with DPC.

Figure 6.28 compares the downlink rate regions achieved with non-linear processing (using DPC) and linear processing. We continue the example with $K = 2$ users considered in many previous figures, such as Figure 6.26. The rate regions with $M = 4$ and $M = 8$ antennas are shown in Figures 6.28(a) and 6.28(b), respectively. The boundary points with linear processing are obtained from the parameterization in (6.128) by considering all combinations of virtual uplink powers that satisfy $P_1^{\mathrm{ul}} + P_2^{\mathrm{ul}} = P$. Linear processing results in a smaller region than non-linear processing, but the difference reduces as we increase the number of antennas, just as in the uplink. The loss in sum rate from linear processing is 4% with $M = 4$ but only 0.4% with $M = 8$, roughly the same as in the uplink. From a mathematical perspective, the channel vectors become more easily distinguishable as the number of dimensions in the vector space increases, which makes it possible to find precoding vectors that avoid causing inter-user interference without sacrificing much of the signal strength at the intended receiver. This is an instance of the favorable propagation property specified in Definition 6.2. We notice that the rate region obtained with linear processing is not a convex set but has a slightly

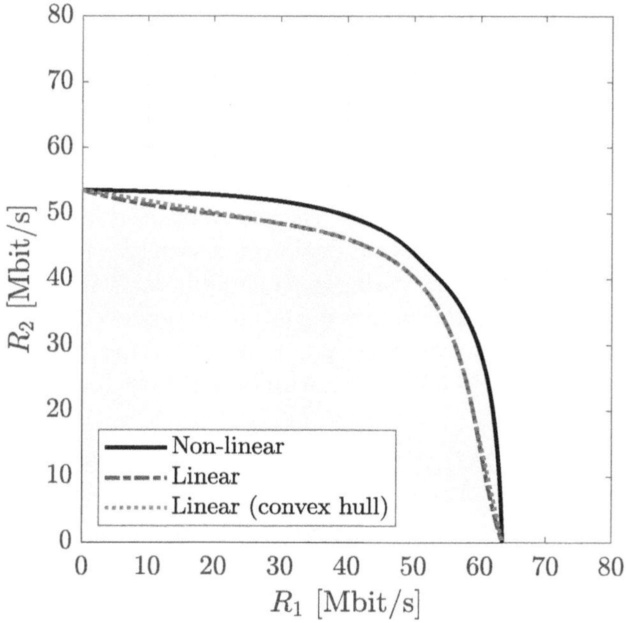

(a) $M = 4$ antennas.

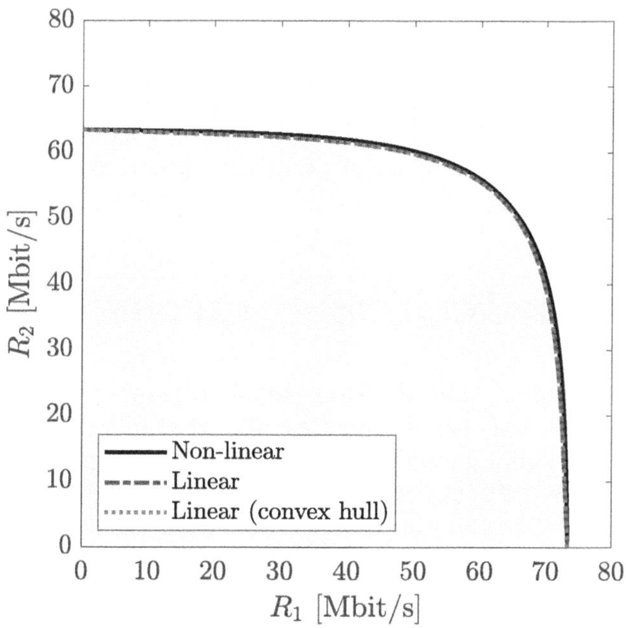

(b) $M = 8$ antennas.

Figure 6.28: Examples of downlink rate regions with $K = 2$ users when multi-user MIMO is used with either non-linear or linear processing. The region obtained in (6.128) is called "linear" and its convex hull is also shown. This is a continuation of the example considered in Figure 6.26.

curvy outer boundary. The convex hull of the region is also shown in the figure, and it is achieved by the time-sharing procedure described earlier in the chapter, where we switch between two operating points to achieve points on the straight line in between. The region's size can be slightly increased by time-sharing when $M = 4$, while the benefit is unnoticeable when $M = 8$, thanks to the more favorable propagation.

Each operating point in the rate region characterization in (6.128) is obtained from a corresponding virtual dual uplink scenario. LMMSE combining is the optimal linear receiver processing in the uplink; thus, the uplink-downlink duality implies that the same operating point is achieved by some kind of *LMMSE precoding* because we need $\mathbf{p}_k = \mathbf{w}_k^*$. Starting from the LMMSE combining expression in (6.63) and normalizing it to have unit norm, we obtain

$$\mathbf{p}_k = \frac{\left(\sum_{i=1}^{K} \frac{P_i^{\text{ul}}}{B} \mathbf{h}_i^* \mathbf{h}_i^{\text{T}} + N_0 \mathbf{I}_M\right)^{-1} \mathbf{h}_k^*}{\left\| \left(\sum_{i=1}^{K} \frac{P_i^{\text{ul}}}{B} \mathbf{h}_i^* \mathbf{h}_i^{\text{T}} + N_0 \mathbf{I}_M\right)^{-1} \mathbf{h}_k^* \right\|}, \tag{6.129}$$

after removing common scaling factors from the numerator and denominator. We can express the precoding matrix $\mathbf{P} = [\mathbf{p}_1, \dots, \mathbf{p}_K]$ for all users as

$$\mathbf{P} = \left(\mathbf{H}^* \tilde{\mathbf{Q}} \mathbf{H}^{\text{T}} + N_0 \mathbf{I}_M\right)^{-1} \mathbf{H}^* \mathbf{Z}, \tag{6.130}$$

using the channel matrix $\mathbf{H}$ and a diagonal matrix with uplink powers divided by the bandwidth: $\tilde{\mathbf{Q}} = \text{diag}(\frac{P_1^{\text{ul}}}{B}, \dots, \frac{P_K^{\text{ul}}}{B}) \in \mathbb{C}^{K \times K}$. The matrix $\mathbf{Z}$ ensures that each column of $\mathbf{P}$ has unit norm by being selected as

$$\mathbf{Z} = \text{diag}\left(\frac{1}{\left\| \left(\mathbf{H}^* \tilde{\mathbf{Q}} \mathbf{H}^{\text{T}} + N_0 \mathbf{I}_M\right)^{-1} \mathbf{h}_1^* \right\|}, \dots, \frac{1}{\left\| \left(\mathbf{H}^* \tilde{\mathbf{Q}} \mathbf{H}^{\text{T}} + N_0 \mathbf{I}_M\right)^{-1} \mathbf{h}_K^* \right\|} \right). \tag{6.131}$$

The duality also implies that the same total transmit power is needed in uplink and downlink but usually must be allocated differently between the users. For the given uplink powers, $P_1^{\text{ul}}, \dots, P_K^{\text{ul}}$, we can compute the resulting uplink SINR values $\gamma_1, \dots, \gamma_K$ using (6.127). If we equate the downlink SINR expressions in (6.126) to the same values, we obtain the equations

$$\frac{P_k^{\text{dl}} |\mathbf{h}_k^{\text{T}} \mathbf{p}_k|^2}{\sum\limits_{i=1, i \neq k}^{K} P_i^{\text{dl}} |\mathbf{h}_k^{\text{T}} \mathbf{p}_i|^2 + B N_0} = \gamma_k \rightarrow \frac{P_k^{\text{dl}}}{\gamma_k B N_0} |\mathbf{h}_k^{\text{T}} \mathbf{p}_k|^2 - \sum_{i=1, i \neq k}^{K} \frac{P_i^{\text{dl}}}{B N_0} |\mathbf{h}_k^{\text{T}} \mathbf{p}_i|^2 = 1 \tag{6.132}$$

for $k = 1, \dots, K$. These are K linear equations of the K downlink transmit powers $P_1^{\rm dl}, \ldots, P_K^{\rm dl}$, thus, we obtain the downlink powers by solving them as

$$\begin{bmatrix} P_1^{\rm dl} \\ \vdots \\ P_K^{\rm dl} \end{bmatrix} = \mathbf{\Gamma}^{-1} \begin{bmatrix} 1 \\ \vdots \\ 1 \end{bmatrix}, \quad \text{where } [\mathbf{\Gamma}]_{ki} = \begin{cases} \frac{|\mathbf{h}_k^{\rm T}\mathbf{p}_k|^2}{\gamma_k B N_0} & \text{if } k = i, \\ -\frac{|\mathbf{h}_k^{\rm T}\mathbf{p}_i|^2}{B N_0} & \text{if } k \neq i, \end{cases} \tag{6.133}$$

contains all the equation coefficients and $[\mathbf{\Gamma}]_{ki}$ denotes the (k, i)th entry of the $K \times K$ matrix $\mathbf{\Gamma}$. We now have a way to map any point in the downlink region, which was parameterized in (6.128) based on the dual uplink powers, to the downlink precoding vectors and power allocation that achieves it.

6.4.5 Alternative Linear Downlink Processing Schemes

Despite the uplink-downlink duality, selecting preferable downlink precoding and power allocation can be challenging in practice. The duality holds between the downlink and the virtual uplink, where we are allowed to allocate power freely between the users. Hence, even if we design the actual uplink operation optimally in some sense (e.g., using the max-min fairness power control described in Section 6.3.6), the corresponding downlink operation obtained through (6.130) and (6.133) might not achieve a point on the Pareto boundary of the downlink rate region. It is unlikely that the specific uplink power division enforced by the maximum power per user in the uplink will happen to be optimal in the downlink. For this reason, the duality result is typically interpreted more loosely as the following rule of thumb [1]: the base station should transmit to a user in the downlink in roughly the same direction as it obtains a strong uplink SINR through receive combining. In other words, a combining vector that works well in the uplink also works well as a precoding vector in the downlink but might not be optimal.

A key challenge when selecting the downlink processing is that the power allocation and precoding selection are intertwined in a complex way through the mapping between power coefficients in the downlink and the virtual uplink. A common approximate solution is to replace the uplink powers with heuristically selected coefficients in the precoding expression and then optimize the downlink power allocation separately. If we replace each uplink coefficient $P_k^{\rm ul}$ in the LMMSE precoding vector in (6.129) by the downlink coefficient $P_k^{\rm dl}$, we obtain

$$\mathbf{p}_k^{\rm TWF} = \frac{\left(\sum_{i=1}^{K} \frac{P_i^{\rm dl}}{B} \mathbf{h}_i^* \mathbf{h}_i^{\rm T} + N_0 \mathbf{I}_M\right)^{-1} \mathbf{h}_k^*}{\left\| \left(\sum_{i=1}^{K} \frac{P_i^{\rm dl}}{B} \mathbf{h}_i^* \mathbf{h}_i^{\rm T} + N_0 \mathbf{I}_M\right)^{-1} \mathbf{h}_k^* \right\|}. \tag{6.134}$$

This alternative design was called the *transmit Wiener filter (TWF)* in [102], where it was motivated by minimizing the MSE between the transmitted signal vector and the scaled received signal vector at the users. It was also proposed in [103], [104] as the precoding that maximizes the *signal-to-leakage-and-noise ratio (SLNR)* obtained by replacing the downlink interference term for a given user by a sum of how much interference the user leaks to other users.

Another way of simplifying the precoding expression is to assume equal power allocation in the virtual uplink. By substituting $P_k^{\mathrm{ul}} = P/K$ into (6.130) and moving the power and bandwidth terms to the noise term, we obtain

$$\begin{aligned}\mathbf{P}^{\mathrm{RZF}} &= \left(\mathbf{H}^*\mathbf{H}^{\mathrm{T}} + \frac{KBN_0}{P}\mathbf{I}_M\right)^{-1}\mathbf{H}^*\mathbf{Z}^{\mathrm{RZF}} \\ &= \mathbf{H}^*\left(\mathbf{H}^{\mathrm{T}}\mathbf{H}^* + \frac{KBN_0}{P}\mathbf{I}_K\right)^{-1}\mathbf{Z}^{\mathrm{RZF}} \qquad (6.135)\end{aligned}$$

where the second equality follows from the matrix identity in (2.50) and

$$\mathbf{Z}^{\mathrm{RZF}} = \mathrm{diag}\left(\frac{1}{\|(\mathbf{H}^*\mathbf{H}^{\mathrm{T}}+\frac{KBN_0}{P}\mathbf{I}_M)^{-1}\mathbf{h}_1^*\|}, \ldots, \frac{1}{\|(\mathbf{H}^*\mathbf{H}^{\mathrm{T}}+\frac{KBN_0}{P}\mathbf{I}_M)^{-1}\mathbf{h}_K^*\|}\right). \qquad (6.136)$$

This is often referred to as *regularized zero-forcing (RZF)* because the expression resembles that of ZF in (6.69), but the inverse of $\mathbf{H}^{\mathrm{T}}\mathbf{H}^*$ has been regularized by adding a scaled identity matrix. Regularization is a classical way to enhance numerical stability in linear algebra algorithms, but here, it determines how strong the interference is compared to the noise. By considering the high-SNR limit $P \to \infty$, (6.135) converges to *ZF precoding*

$$\mathbf{P}^{\mathrm{RZF}} \to \mathbf{P}^{\mathrm{ZF}} = \mathbf{H}^*\left(\mathbf{H}^{\mathrm{T}}\mathbf{H}^*\right)^{-1}\mathbf{Z}^{\mathrm{ZF}}. \qquad (6.137)$$

ZF precoding has the property that $\mathbf{H}^{\mathrm{T}}\mathbf{P}^{\mathrm{ZF}} = \mathbf{H}^{\mathrm{T}}\mathbf{H}^*(\mathbf{H}^{\mathrm{T}}\mathbf{H}^*)^{-1}\mathbf{Z}^{\mathrm{ZF}} = \mathbf{Z}^{\mathrm{ZF}}$, so the impact of the channel matrix appears to vanish from the received signal expression in (6.124). However, the channel still impacts the selection of the matrix $\mathbf{Z}^{\mathrm{ZF}}$ that normalizes the columns of the precoding matrix. We need one-valued diagonal entries of $(\mathbf{P}^{\mathrm{ZF}})^{\mathrm{H}}\mathbf{P}^{\mathrm{ZF}}$, which can be expressed as

$$\left(\mathbf{H}^*\left(\mathbf{H}^{\mathrm{T}}\mathbf{H}^*\right)^{-1}\mathbf{Z}^{\mathrm{ZF}}\right)^{\mathrm{H}}\mathbf{H}^*\left(\mathbf{H}^{\mathrm{T}}\mathbf{H}^*\right)^{-1}\mathbf{Z}^{\mathrm{ZF}} = \left(\mathbf{Z}^{\mathrm{ZF}}\right)^{\mathrm{H}}\left(\mathbf{H}^{\mathrm{T}}\mathbf{H}^*\right)^{-1}\mathbf{Z}^{\mathrm{ZF}}. \qquad (6.138)$$

We need $\mathbf{Z}^{\mathrm{ZF}} = \mathrm{diag}(1/\sqrt{[(\mathbf{H}^{\mathrm{T}}\mathbf{H}^*)^{-1}]_{11}}, \ldots, 1/\sqrt{[(\mathbf{H}^{\mathrm{T}}\mathbf{H}^*)^{-1}]_{KK}})$ to make the diagonal entries equal to one. If we substitute the ZF precoding matrix into the general rate expression in (6.126), we obtain the simplified expression

$$R_k^{\mathrm{ZF}} = B\log_2\left(1 + \frac{P_k^{\mathrm{dl}}}{BN_0\left[\left(\mathbf{H}^{\mathrm{T}}\mathbf{H}^*\right)^{-1}\right]_{kk}}\right). \qquad (6.139)$$

This expression contains no interference since ZF precoding leads to a beamformed transmission that creates nulls at all the co-users. The SNR-term contains the factor $1/[(\mathbf{H}^{\mathrm{T}}\mathbf{H}^*)^{-1}]_{kk}$ that determines how strong the remaining channel to user k is when the precoding has been restricted to cause no interference. It is no surprise that RZF turns into ZF at high SNR because the interference will then dominate over the noise.

Example 6.10. How should the transmit power be allocated to maximize the sum rate or achieve max-min fairness when ZF precoding is utilized?

Power allocation optimization is relatively simple when using ZF precoding, thanks to the lack of interference. The sum of the rates in (6.139) is

$$\sum_{k=1}^{K} B \log_2 \left(1 + \frac{P_k^{\mathrm{dl}} s_k^2}{N_0} \right), \tag{6.140}$$

using the notation $s_k^2 = 1/(B[(\mathbf{H}^{\mathrm{T}}\mathbf{H}^*)^{-1}]_{kk})$, which is the summation over K parallel user channels. It has the same form as the rate expression in (3.67) for a point-to-point MIMO channel, in which case the parallel channels were created using the SVD and the sum rate was maximized by water-filling power allocation. Hence, the corresponding way of maximizing (6.140) under the total power constraint $\sum_{k=1}^{K} P_k^{\mathrm{dl}} \leq P$ is to use the transmit power

$$P_k^{\mathrm{dl,sum\text{-}rate}} = \max \left(\mu - BN_0 \left[(\mathbf{H}^{\mathrm{T}}\mathbf{H}^*)^{-1} \right]_{kk}, 0 \right), \quad k = 1, \ldots, K, \tag{6.141}$$

where the variable μ is selected to make $\sum_{k=1}^{K} P_k^{\mathrm{dl,sum\text{-}rate}} = P$.

Max-min fairness is achieved by giving all users the same SINR value and maximizing that common value. The SINR in (6.139) becomes $c/(BN_0)$ for all users if $P_k^{\mathrm{dl}} = c[(\mathbf{H}^{\mathrm{T}}\mathbf{H}^*)^{-1}]_{kk}$ for $k = 1, \ldots, K$. This common SINR is maximized by making the scaling factor c as large as possible while complying with the sum power constraint. The maximum is achieved for $c = P/(\sum_{i=1}^{K} [(\mathbf{H}^{\mathrm{T}}\mathbf{H}^*)^{-1}]_{ii})$ for which all the available power is used; thus, the max-min fairness power allocation is

$$P_k^{\mathrm{dl,max\text{-}min}} = P \frac{[(\mathbf{H}^{\mathrm{T}}\mathbf{H}^*)^{-1}]_{kk}}{\sum_{i=1}^{K} [(\mathbf{H}^{\mathrm{T}}\mathbf{H}^*)^{-1}]_{ii}}, \quad k = 1, \ldots, K. \tag{6.142}$$

To analyze the low-SNR regime, we can return to the precoding vector expression in (6.134) and let $P_1^{\mathrm{dl}}, \ldots, P_K^{\mathrm{dl}} \to 0$, which leads to

$$\mathbf{p}_k^{\mathrm{TWF}} \to \frac{\left(\sum_{i=1}^{K} \frac{0}{B} \mathbf{h}_i^* \mathbf{h}_i^{\mathrm{T}} + N_0 \mathbf{I}_M \right)^{-1} \mathbf{h}_k^*}{\left\| \left(\sum_{i=1}^{K} \frac{0}{B} \mathbf{h}_i^* \mathbf{h}_i^{\mathrm{T}} + N_0 \mathbf{I}_M \right)^{-1} \mathbf{h}_k^* \right\|} = \frac{\mathbf{h}_k^*}{\|\mathbf{h}_k^*\|} = \mathbf{p}_k^{\mathrm{MRT}}. \tag{6.143}$$

We recognize this as the expression for MRT precoding from (3.44), which we recall will maximize the SNR in the absence of interference. It also maximizes the SINR in the multi-user setting when the interference is negligibly weak compared to the noise. If we substitute the MRT vector into the general rate expression in (6.126), we obtain

$$R_k^{\mathrm{MRT}} = B \log_2 \left(1 + \frac{P_k^{\mathrm{dl}} \|\mathbf{h}_k\|^2}{\sum_{i=1, i \neq k}^{K} P_i^{\mathrm{dl}} \frac{|\mathbf{h}_k^{\mathrm{T}} \mathbf{h}_i^*|^2}{\|\mathbf{h}_i\|^2} + BN_0} \right). \tag{6.144}$$

To compare the mentioned precoding schemes, Figure 6.29 shows the sum rate in the downlink counterpart to Figure 6.16. There are $K = 4$ users with equal channel strengths, and the SNR value in the figure represents what is achieved with equal power allocation. The base station is equipped with a ULA with half-wavelength-spaced antennas, and the users have LOS channels with different azimuth angles-of-arrivals: $-\pi/16, -\pi/32, 0, +\pi/24$. We compare multi-user MIMO with non-linear processing (using DPC) and linear processing with LMMSE precoding, both based on the virtual uplink power allocations that maximize the sum rate. The sum rates with RZF and ZF using equal power allocation and the sum-rate-maximizing OMA scheme are also shown. The case of $M = 10$ antennas is considered in Figure 6.29(a) and reveals substantial differences between the curves. All the multi-user MIMO schemes have the same slope at high SNRs, demonstrating that they reach the same multiplexing gain of $\min(M, K) = K$. However, there is a substantial gap between the non-linear and linear processing schemes, which showcases the benefit of removing interference using DPC. All the considered linear schemes perform identically at high SNRs, as expected from the fact that they all converge to ZF in that regime. At lower SNRs, the optimal LMMSE precoding is better than the simplified RZF precoding and much better than ZF. The OMA curve outperforms ZF at low SNRs, although it has a four times smaller slope as only a single user is served at a time. Hence, if one must choose between the simplified RZF and ZF schemes, then RZF is preferred since it works reasonably well at all SNRs.

The number of antennas is increased to $M = 20$ in Figure 6.29(b), and then all the multi-user MIMO schemes provide indistinguishable performance. The antenna-user ratio is $M/4 = 5$. The same kind of behavior was observed in the uplink: linear processing is nearly optimal when the base station has around five times more antennas than the number of single-antenna users. This is the Massive MIMO operating regime for which 5G NR systems (in the mid-band) are designed by having $M = 64$ antennas and serving $1 \leq K \leq 16$ users, depending on the traffic load. These systems are purposely designed not to need complex non-linear processing, and the sizeable antenna-user ratio gives robustness to various practical imperfections, such as imperfect channel knowledge and hardware imperfections; see [1] for further details.

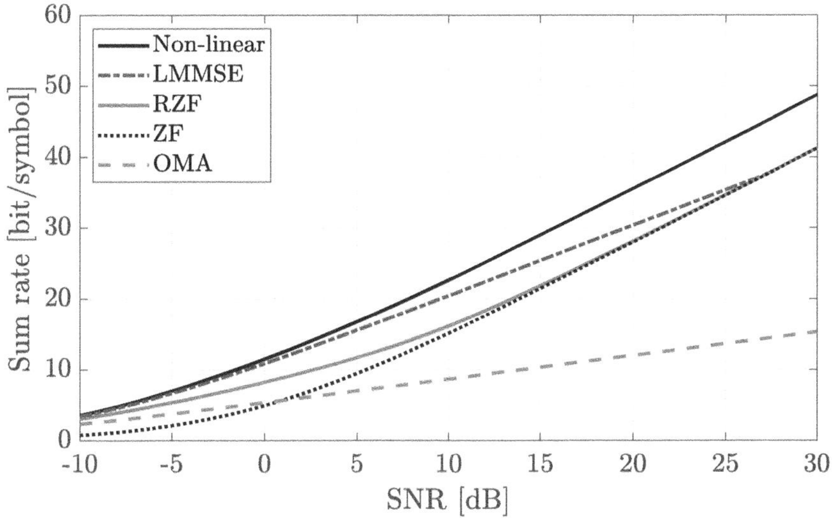

(a) $M = 10$ antennas.

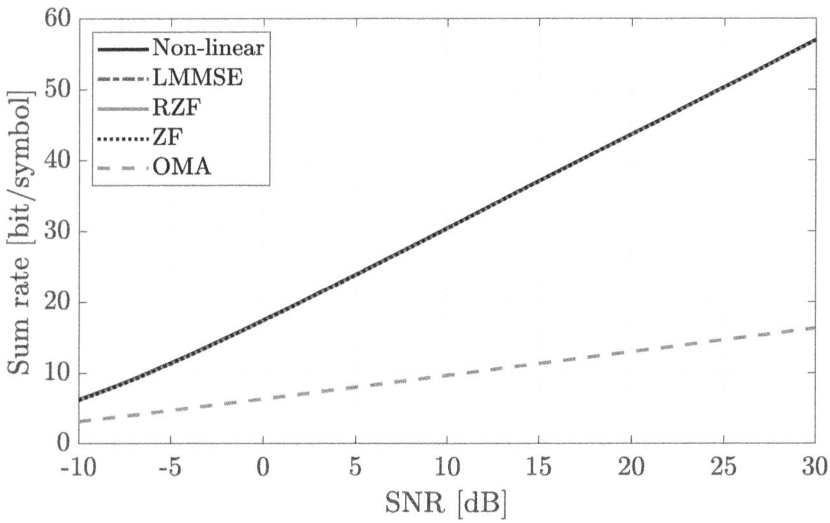

(b) $M = 20$ antennas.

Figure 6.29: The downlink sum rate in a multi-user MIMO system with $K = 4$ users and either non-linear or linear processing. All the users have the same SNR if equal power allocation is used. The non-linear and LMMSE processing curves are obtained using sum-rate maximizing power allocation, while RZF and ZF use equal power allocation. OMA, where only one user is served, is shown as a reference and does not provide any multiplexing gain.

Example 6.11. Can we reduce the gap between DPC-based processing and LMMSE precoding without using DPC?

Yes, one way to increase the sum rate is to use the *rate-splitting* technique [105]. The core idea is to transmit an additional data signal x_{c} using the power $P_{\mathrm{c}}^{\mathrm{dl}}$ and the precoding vector $\mathbf{p}_{\mathrm{c}}$. This signal contains a collection of data for everyone and is decoded by all users while treating other interfering signals as noise. The common signal can be encoded at the rate

$$R_{\mathrm{c}} = \min_{k \in \{1,\ldots,K\}} C\left(\frac{P_{\mathrm{c}}^{\mathrm{dl}}|\mathbf{h}_k^{\mathrm{T}}\mathbf{p}_{\mathrm{c}}|^2}{\sum_{i=1}^{K} P_i^{\mathrm{dl}}|\mathbf{h}_k^{\mathrm{T}}\mathbf{p}_i|^2 + BN_0}\right), \tag{6.145}$$

where the minimization over the user indices allows all users to decode it. The data contained in the common signal is divided between the users, but any user k can remove the entire common signal from its received signal before decoding x_k as described earlier in this section. The sum rate is therefore

$$R_{\mathrm{c}} + \sum_{k=1}^{K} C\left(\frac{P_k^{\mathrm{dl}}|\mathbf{h}_k^{\mathrm{T}}\mathbf{p}_k|^2}{\sum_{i=1, i\neq k}^{K} P_i^{\mathrm{dl}}|\mathbf{h}_k^{\mathrm{T}}\mathbf{p}_i|^2 + BN_0}\right), \tag{6.146}$$

and can be maximized with respect to the precoding vectors $\mathbf{p}_{\mathrm{c}}, \mathbf{p}_1, \ldots, \mathbf{p}_K$ and transmit power coefficients $P_{\mathrm{c}}^{\mathrm{dl}}, P_1^{\mathrm{dl}}, \ldots, P_K^{\mathrm{dl}}$, which must satisfy the constraint $P_{\mathrm{c}}^{\mathrm{dl}} + \sum_{k=1}^{K} P_k^{\mathrm{dl}} \leq P$. The term rate-splitting refers to how each user's data rate is split into a "public" part contained in x_{c} and a "private" part contained in x_k. With an informed design, communication with rate-splitting cannot be worse than linear precoding since that is a special case obtained by setting $P_{\mathrm{c}}^{\mathrm{dl}} = 0$. On the other hand, it relies on SIC, which has the many practical downsides described earlier in this chapter. Moreover, it can only give a noticeable improvement in scenarios such as Figure 6.29(a), where there is a substantial gap between DPC-based processing and LMMSE precoding. The most attractive gains might exist in situations with limited channel knowledge, which are beyond the scope of this book.

6.4.6 Power Allocation for Max-Min Fairness

Once the linear precoding scheme is determined, the downlink transmit power coefficients $P_1^{\mathrm{dl}}, \ldots, P_K^{\mathrm{dl}} \geq 0$ can be selected to maximize a specific utility function, under the constraint $\sum_{k=1}^{K} P_k^{\mathrm{dl}} \leq P$. This is known as *power allocation* since it entails distributing the available transmit power among the users to achieve the desired balance among their achieved rates. In this section, we consider power allocation for max-min fairness. We will introduce the downlink counterpart to the efficient fixed-point algorithm previously given in Algorithm 6.1 for the uplink. Hence, we aim to find the power coefficients that achieve the solution to the max-min fairness problem

$$\underset{(R_1,\ldots,R_K)\in\mathcal{R}}{\text{maximize}} \quad \min_{k\in\{1,\ldots,K\}} R_k, \tag{6.147}$$

where the downlink rate region $\mathcal{R}$ depends on the adopted linear precoding scheme. For any such scheme, it can be expressed in the generic form

$$\mathcal{R} = \Bigg\{(R_1,\ldots,R_K) : R_k = B\log_2\left(1+\mathrm{SINR}_k(P_1^{\mathrm{dl}},\ldots,P_K^{\mathrm{dl}})\right)$$
$$\text{for } k=1,\ldots,K, \text{ for some } P_1^{\mathrm{dl}},\ldots,P_K^{\mathrm{dl}} \geq 0 \text{ satisfying } \sum_{k=1}^{K} P_k^{\mathrm{dl}} \leq P\Bigg\}, \tag{6.148}$$

where the SINR for each user is a function of the transmit power coefficients $P_1^{\mathrm{dl}},\ldots,P_K^{\mathrm{dl}}$. When ZF precoding is used, the downlink power allocation that achieves max-min fairness was already derived in Example 6.10. When any other fixed normalized precoding vectors $\mathbf{p}_1,\ldots,\mathbf{p}_K$ that are independent of the downlink power coefficients (e.g., RZF precoding in (6.135) or MRT) are used, the SINR of user k can be expressed using (6.126) as

$$\mathrm{SINR}_k(P_1^{\mathrm{dl}},\ldots,P_K^{\mathrm{dl}}) = \frac{P_k^{\mathrm{dl}}|\mathbf{h}_k^{\mathrm{T}}\mathbf{p}_k|^2}{\sum_{i=1,i\neq k}^{K} P_i^{\mathrm{dl}}|\mathbf{h}_k^{\mathrm{T}}\mathbf{p}_i|^2 + BN_0}, \tag{6.149}$$

where the numerators and denominators are linear functions of the downlink power coefficients P_k^{dl}, for $k=1,\ldots,K$.

Since maximizing the minimum rate is equivalent to maximizing the minimum SINR value among the users, (6.147) can be expressed for fixed precoding vectors as

$$\begin{aligned} &\underset{P_1^{\mathrm{dl}},\ldots,P_K^{\mathrm{dl}}\geq 0}{\text{maximize}} \quad \min_{k\in\{1,\ldots,K\}} \mathrm{SINR}_k(P_1^{\mathrm{dl}},\ldots,P_K^{\mathrm{dl}}) \\ &\text{subject to} \quad \sum_{k=1}^{K} P_k^{\mathrm{dl}} \leq P. \end{aligned} \tag{6.150}$$

Algorithm 6.2 states a fixed-point iteration that finds the optimal solution. The algorithm starts from arbitrarily selected non-zero power coefficients $P_k^{\mathrm{dl}} \in (0,P]$ and sets a solution accuracy $\epsilon > 0$. As in the uplink counterpart in Algorithm 6.1, each user that achieves an SINR larger than the minimum SINR is assigned a reduced transmit power in Step 3. Next, in Step 4, all the power coefficients are scaled so that the total transmit power equals the maximum value of P. In fact, it can be proved that the optimal power allocation must use all the available power. The process continues iteratively until a stopping criterion is satisfied. The difference between the maximum and minimum SINRs among the users gradually diminishes, and the stopping

Algorithm 6.2 Solution to the max-min fairness problem in (6.150).

1: **Initialization:** Select arbitrary $P_k^{\text{dl}} \in (0, P]$, for $k = 1, \ldots, K$, and the solution accuracy $\epsilon > 0$

2: **while** $\max\limits_{i \in \{1,\ldots,K\}} \text{SINR}_i\left(P_1^{\text{dl}}, \ldots, P_K^{\text{dl}}\right) - \min\limits_{i \in \{1,\ldots,K\}} \text{SINR}_i\left(P_1^{\text{dl}}, \ldots, P_K^{\text{dl}}\right) > \epsilon$ **do**

3: $\quad P_k^{\text{dl}} \leftarrow \frac{\min\limits_{i \in \{1,\ldots,K\}} \text{SINR}_i\left(P_1^{\text{dl}},\ldots,P_K^{\text{dl}}\right)}{\text{SINR}_k\left(P_1^{\text{dl}},\ldots,P_K^{\text{dl}}\right)} P_k^{\text{dl}}$, for $k = 1, \ldots, K$

4: $\quad P_k^{\text{dl}} \leftarrow \frac{P}{\sum_{i=1}^{K} P_i^{\text{dl}}} P_k^{\text{dl}}$, for $k = 1, \ldots, K$

5: **end while**

6: **Output:** $P_1^{\text{dl}}, \ldots, P_K^{\text{dl}}$

criterion in Step 2 determines when the difference becomes less than ϵ. As in the uplink, the algorithm usually converges in fewer than ten iterations.

The convergence to the optimal solution to the max-min fairness problem is guaranteed if certain technical conditions are satisfied [91, Lem. 1, Th. 1], which is the case when the downlink SINR has the generic form in (6.149).

Figure 6.30 demonstrates the max-min fairness solution obtained by Algorithm 6.2 in a system with $K = 4$ users. The setup is the same as in Figure 6.29(a) and each user achieves an SNR of 10 dB if equal power allocation is used. Figure 6.30(a) shows the variations in the rates obtained by the four users throughout the algorithm's iterations when using $M = 6$ antennas and RZF precoding. Initially, there are significant rate discrepancies among the users because the initial equal power allocation is suboptimal. However, as the algorithm progresses, the rates of all four users gradually converge to a common value, representing the max-min fairness solution. The minimum rate among the users experiences gradual enhancement; however, the convergence behavior is not strictly monotonic because reducing the power for some users can improve the rates of other users after the power normalization.

Figure 6.30(b) demonstrates the minimum rate among the $K = 4$ users as the number of antennas M increases, considering both RZF and MRT precoding. In addition to the max-min fairness solutions obtained by Algorithm 6.2, the minimum rates achieved by equal power allocation among the users are shown as references. For both RZF and MRT, employing max-min fair power allocation increases the minimum rate as M grows, indicating improved communication performance with a greater number of antennas. As expected, the max-min fairness power allocation yields higher minimum rates than equal power allocation, regardless of the precoding scheme employed. However, a non-monotonic trend is observed when increasing M and using RZF precoding with equal power allocation. This peculiarity arises since the power is not allocated based on the interference levels generated by the precoding.

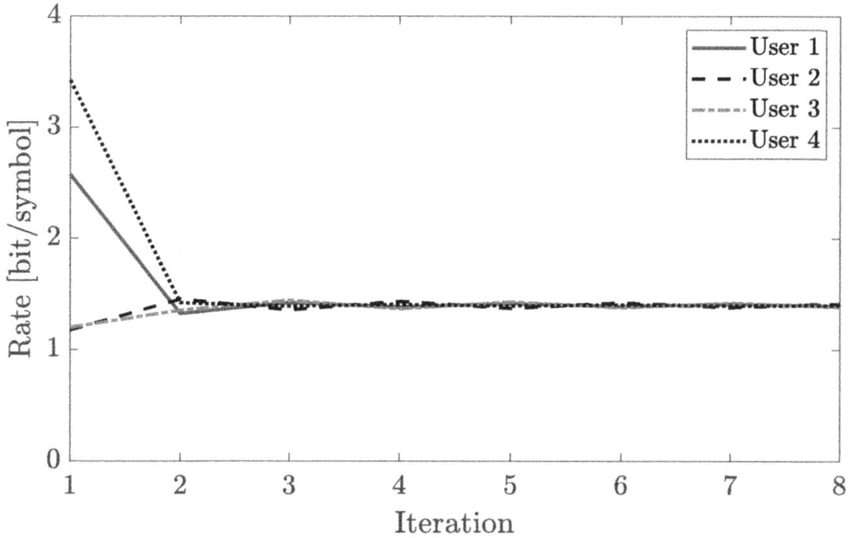

(a) The rates achieved by the four users at different iterations.

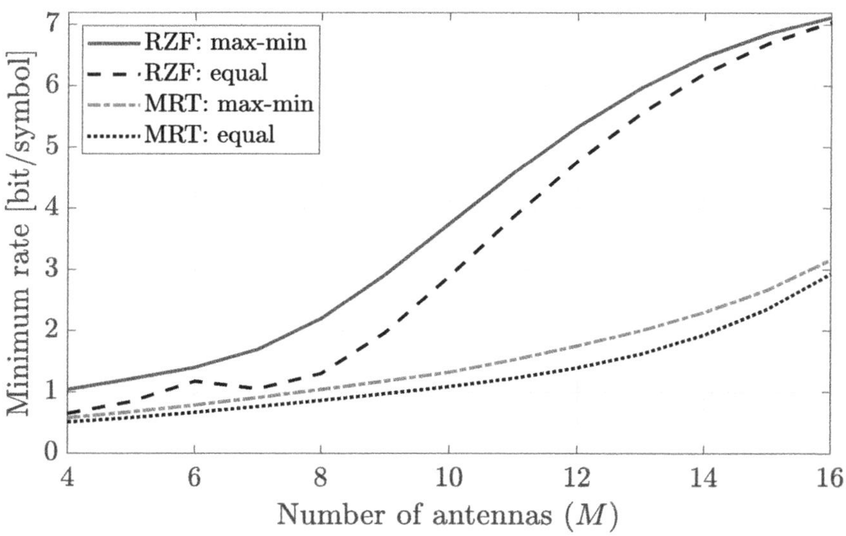

(b) The minimum rate versus the number of antennas M.

Figure 6.30: The max-min fairness solution obtained by Algorithm 6.2 with $K = 4$ users, in the same setup as in Figure 6.29(a). All the users have the same SNR of 10 dB when using equal power allocation. In (a), the rates of the four users during the fixed-point iterations are shown when RZF precoding and $M = 6$ antennas are used. In (b), the minimum rate among the users is shown for a varying number of antennas M when using RZF and MRT precoding. The minimum rate obtained using equal power allocation is shown as a reference.

6.5 Exercises

Exercise 6.1. The uplink rate region of a multi-user MIMO system with $K = 2$ is

$$\mathcal{R} = \left\{ (R_1, R_2) : R_1, R_2 \geq 0, \frac{R_1}{2} + R_2 \leq 10\,\text{Mbit/s} \right\}. \tag{6.151}$$

(a) Find the expression for the Pareto boundary.

(b) Find the maximum achievable rate of the second user if $R_1 \geq 15\,\text{Mbit/s}$ is required.

Exercise 6.2. The Pareto boundary of the uplink rate region for a multi-user MIMO system with $K = 2$ is

$$\partial\mathcal{R} = \left\{ (R_1, R_2) : R_1, R_2 \geq 0, R_1^2 + 2R_2 = 48\,\text{Mbit/s} \right\}. \tag{6.152}$$

(a) Find the max-min fairness point on the Pareto boundary.

(b) Find the maximum sum-rate point on the Pareto boundary.

(c) Find the point on the Pareto boundary that maximizes the weighted sum rate $3R_1 + R_2$.

Exercise 6.3. The bandwidth allocation that maximizes the uplink sum rate with FDMA is stated in (6.13). Derive this expression by maximizing $\sum_{k=1}^{K} \xi_k B \log_2(1 + \frac{P\beta_k}{\xi_k B N_0})$ with respect to $\xi_k \geq 0$, for $k = 1, \ldots, K$, under the condition $\xi_1 + \ldots + \xi_K = 1$.

Exercise 6.4. Consider the uplink multi-user MIMO channel in Theorem 6.2 with $M = 4$ base station antennas, $K = 2$ users, and $B = 10\,\text{MHz}$.

(a) Suppose $\frac{P}{BN_0} = \frac{3}{4}$, $\mathbf{h}_1 = [1, 1, 1, 1]^{\mathrm{T}}$, and $\mathbf{h}_2 = [1, -1, 1, -1]^{\mathrm{T}}$. Sketch the capacity region and explain its shape.

(b) Suppose $\frac{P}{BN_0} = \frac{3}{4}$, $\mathbf{h}_1 = [1, 1, 1, 1]^{\mathrm{T}}$, and $\mathbf{h}_2 = [1, 1, 1, 1]^{\mathrm{T}}$. Sketch the capacity region and explain its shape.

(c) Suppose $\frac{P}{BN_0} = 1$, $\mathbf{h}_1 = [1, 1, \ldots]^{\mathrm{T}}$, and $\mathbf{h}_1 = [1, -1, 1, -1, \ldots]^{\mathrm{T}}$. For which values of M is the sum rate greater or equal to $100\,\text{Mbit/s}$?

Exercise 6.5. Consider the uplink rate region in Figure 6.10 with NOMA and $K = 3$.

(a) Which user data decoding order is needed to operate at the top-left corner of the Pareto boundary? Is time-sharing required?

(b) Which user data decoding order is needed to operate at the top-right corner of the Pareto boundary? Is time-sharing required?

(c) How can we achieve an arbitrary point on the line between the two top corners of the Pareto boundary?

Exercise 6.6. Prove that at least one user must use maximum uplink power when achieving the max-min fairness solution with any linear receive combining scheme that gives $|\mathbf{w}_k^{\mathrm{H}}\mathbf{h}_k|^2 > 0$ for all users. Hint: Use a proof-by-contradiction approach.

Exercise 6.7. Consider an uplink multi-user MIMO system with linear processing. Show that the optimal receive combining is a linear combination of the channels: $\mathbf{w}_k = \mathbf{H}\breve{\mathbf{w}}_k$ for some $\breve{\mathbf{w}}_k \in \mathbb{C}^K$ for $k = 1, \ldots, K$. Hint: An arbitrary combining vector can be expressed as $\mathbf{w}_k = \mathbf{H}\breve{\mathbf{w}}_k + \dot{\mathbf{w}}_k$, where $\dot{\mathbf{w}}_k \in \mathbb{C}^M$ is orthogonal to the channel vectors, i.e., $\mathbf{H}^{\mathrm{H}}\dot{\mathbf{w}}_k = \mathbf{0}$. It is sufficient to prove that picking $\dot{\mathbf{w}}_k = \mathbf{0}$ does not reduce the rates.

Exercise 6.8. Consider an uplink multi-user MIMO system with linear processing and $K = 2$. Show that the optimal combining vector $\mathbf{w}_k$ for user k, for $k = 1, 2$, is a linear combination of the MRC and ZF combining vectors:

$$\mathbf{w}_k = \alpha_k^{\mathrm{MRC}} \mathbf{w}_k^{\mathrm{MRC}} + \alpha_k^{\mathrm{ZF}} \mathbf{w}_k^{\mathrm{ZF}} \quad \text{for some } \alpha_k^{\mathrm{MRC}}, \alpha_k^{\mathrm{ZF}} \in \mathbb{C}. \tag{6.153}$$

Hint: Use the result from Exercise 6.7 to express the optimal receive combining vector as $\mathbf{w}_k = \alpha_{k,1}\mathbf{h}_1 + \alpha_{k,2}\mathbf{h}_2$ for some values of $\alpha_{k,1}, \alpha_{k,2} \in \mathbb{C}$. Show that for any $\alpha_{k,1}, \alpha_{k,2}$, one can find $\alpha_k^{\mathrm{MRC}}, \alpha_k^{\mathrm{ZF}} \in \mathbb{C}$ so that $\mathbf{w}_k = \alpha_k^{\mathrm{MRC}} \mathbf{w}_k^{\mathrm{MRC}} + \alpha_k^{\mathrm{ZF}} \mathbf{w}_k^{\mathrm{ZF}}$ holds.

Exercise 6.9. The ZF combining matrix was defined as $\mathbf{W}^{\mathrm{ZF}} = \mathbf{H}\left(\mathbf{H}^{\mathrm{H}}\mathbf{H}\right)^{-1}$ in (6.69), which leads to the rate expression in (6.71). Alternatively, the ZF vector can be interpreted as an orthogonal projection of the desired channel vector onto the null space of the interfering channels. By following this approach in a system with $K = 2$ users, the ZF combining vector of user 1 becomes

$$\mathbf{w}_1^{\text{ZF-alternative}} = \left(\mathbf{I}_M - \frac{\mathbf{h}_2}{\|\mathbf{h}_2\|}\frac{\mathbf{h}_2^{\mathrm{H}}}{\|\mathbf{h}_2\|}\right)\mathbf{h}_1, \tag{6.154}$$

which is the orthogonal projection of $\mathbf{h}_1$ onto the null space of the other user's channel $\mathbf{h}_2$. Show that user 1 achieves the rate in (6.71) when using $\mathbf{w}_1^{\text{ZF-alternative}}$.

Exercise 6.10. Consider downlink communication to $K = 2$ users from an M-antenna base station, where M is an even number. The channels to the users are decomposed as $\mathbf{h}_1 = [\mathbf{h}_{1,1}^{\mathrm{T}}\, \mathbf{h}_{1,2}^{\mathrm{T}}]^{\mathrm{T}}$ and $\mathbf{h}_2 = [\mathbf{h}_{2,1}^{\mathrm{T}}\, \mathbf{h}_{2,2}^{\mathrm{T}}]^{\mathrm{T}}$, respectively, where $\mathbf{h}_{1,1} \in \mathbb{C}^{M/2}$ and $\mathbf{h}_{2,1} \in \mathbb{C}^{M/2}$ correspond to the channels from the first $M/2$ antennas of the base station to the users. Similarly, $\mathbf{h}_{1,2} \in \mathbb{C}^{M/2}$ and $\mathbf{h}_{2,2} \in \mathbb{C}^{M/2}$ are the channels from the last $M/2$ antennas to the users. Suppose the channels are orthogonal in the sense that $\mathbf{h}_{1,1}^{\mathrm{H}}\mathbf{h}_{2,1} = 0$ and $\mathbf{h}_{1,2}^{\mathrm{H}}\mathbf{h}_{2,2} = 0$. Moreover, it holds that $\|\mathbf{h}_{1,1}\|^2 = \|\mathbf{h}_{1,2}\|^2 = \|\mathbf{h}_{2,1}\|^2 = \|\mathbf{h}_{2,2}\|^2 = M\beta/2$, where β is the common channel gain.

(a) Suppose the first user is served only by the first $M/2$ antennas, and the second user is served only by the last $M/2$ antennas. What are the rates of the users if MRT and equal power allocation are used? What is the sum rate?

(b) Suppose all the antennas are used for serving both users with MRT precoding. What are the rates of the users with equal power allocation? What is the sum rate? Compare the results with those obtained in (a) when each antenna is assigned to a single user.

Exercise 6.11. Consider a base station with a ULA with $M = 4$ antennas and half-wavelength antenna spacing. There are free-space LOS channels with zero elevation angle to all K users.

(a) Suppose we transmit with MRT to a user with the channel gain β_1 located in the azimuth angular direction φ_1. The transmit power is denoted by P. What is the power of the received interfering signal at another user, located in some other azimuth direction φ_2 and having the channel gain β_2?

(b) How should the angles φ_1 and φ_2 in (a) be related to have zero interference?

(c) Find a set of four user angles $\varphi_1, \varphi_2, \varphi_3, \varphi_4$ so that we can transmit to all the users using MRT without causing any interference.

(d) Suppose the four user angles are all different but do not satisfy the condition derived in (c). Suggest a precoding matrix that removes the interference.

Exercise 6.12. A telecom operator divides its customers into two categories: i) standard and ii) premium. It promises that a premium user will always get four times higher SINR than any standard user. Consider the downlink of a multi-user MIMO with some arbitrary fixed linear precoding. Suppose that the K_{p} users with the indices $k = 1, \ldots, K_{\mathrm{p}}$ are premium users while the remaining $K - K_{\mathrm{p}}$ users are standard users with the indices $k = K_{\mathrm{p}} + 1, \ldots, K$.

(a) For an arbitrary transmit precoding scheme, design a fixed-point algorithm that obtains the optimal solution to the problem

$$\begin{aligned} \underset{P_1^{\mathrm{dl}},\ldots,P_K^{\mathrm{dl}} \geq 0}{\text{maximize}} \quad & \overline{\mathrm{SINR}} \\ \text{subject to} \quad & \mathrm{SINR}_k \geq 4\overline{\mathrm{SINR}}, \quad k = 1, \ldots, K_{\mathrm{p}}, \\ & \mathrm{SINR}_k \geq \overline{\mathrm{SINR}}, \quad k = K_{\mathrm{p}} + 1, \ldots, K, \\ & \sum_{k=1}^{K} P_k^{\mathrm{dl}} \leq P. \end{aligned} \tag{6.155}$$

(b) Suppose ZF precoding is utilized. Find a closed-form solution to the power allocation problem in (6.155).

Exercise 6.13. Consider a base station with a ULA with M antennas and half-wavelength antenna spacing. Free-space LOS channels and $K = 2$ users are considered in the uplink. Suppose the users have equal channel gains $\beta_1 = \beta_2 = \beta$ and transmit with maximum power: $P_1^{\mathrm{ul}} = P_2^{\mathrm{ul}} = P$. Moreover, assume $B = 10\,\mathrm{MHz}$ and $\frac{P\beta}{BN_0} = 1$. The users are located in the azimuth angle directions $\varphi_1 = 0$ and $\varphi_2 = \pi/8$, while the elevation angles are zero.

(a) For $M = 4$, compute the sum rate achieved with FDMA using MRC with the optimal bandwidth allocation.

(b) For $M = 4$, compute the sum rate achieved with multi-user MIMO based on MRC. Compare the result with that of FDMA from (a).

(c) Increase the number of base station antennas to $M = 8$. Compute the maximum sum rate achieved with FDMA.

(d) For $M = 8$, compute the sum rate achieved with multi-user MIMO based on MRC. Compare the result with that of FDMA from (c). Is the gap between FDMA and multi-user MIMO increasing with the number of antennas?

Exercise 6.14. Consider uplink multi-user MIMO with fast-fading channels, linear processing, and perfect CSI at the receiver.

(a) What are the ergodic rate expressions when using MRC and ZF combining?

(b) Assume i.i.d. Rayleigh fading. Compute closed-form lower bounds on the ergodic rates using Jensen's inequality from Lemma 5.1. How do the resulting expressions depend on M? Hint: Apply Jensen's inequality to the convex function $f(x) = \log_2(1 + x^{-1})$, $x > 0$. Use that $\mathbb{E}\{\frac{1}{\|\mathbf{h}_k\|^2}\} = \frac{1}{\beta_k(M-1)}$ and $\mathbb{E}\left\{[(\mathbf{H}^{\mathrm{H}}\mathbf{H})^{-1}]_{kk}\right\} = \frac{1}{\beta_k(M-K)}$ for i.i.d. Rayleigh fading channels [3, App. B.3].

(c) Simplify the lower bounds from (b) by assuming the same channel gain β and transmission with maximum power P for all users. What happens to the ratio of the lower bounds achieved with ZF and MRC as $M \to \infty$? What happens to their difference as $M \to \infty$?

Exercise 6.15. Consider downlink multi-user MIMO with fast-fading channels, linear processing, and perfect CSI at the receiver.

(a) What are the ergodic rates when using MRT and ZF precoding?

(b) Assume i.i.d. Rayleigh fading. Compute closed-form lower bounds on the ergodic rates using Jensen's inequality from Lemma 5.1. How do the resulting expressions depend on M? Hint: Apply Jensen's inequality to the convex function $f(x) = \log_2(1+x^{-1})$, $x > 0$. Use that $\mathbb{E}\left\{\frac{1}{\|\mathbf{h}_k\|^2}\right\} = \frac{1}{\beta_k(M-1)}$, $\mathbb{E}\left\{[(\mathbf{H}^{\mathrm{H}}\mathbf{H})^{-1}]_{kk}\right\} = \frac{1}{\beta_k(M-K)}$, and $\mathbb{E}\left\{\frac{\mathbf{h}_i^*\mathbf{h}_i^{\mathrm{T}}}{\|\mathbf{h}_i\|^2}\right\} = \frac{1}{M}\mathbf{I}_M$ for i.i.d. Rayleigh fading channels [3, App. B.3].

(c) Simplify the lower bounds from (b) by assuming the same channel gain β and equal power allocation $P_k^{\mathrm{dl}} = P/K$ among the users. What happens to the ratio of the lower bounds achieved with ZF combining and MRT as $M \to \infty$? What happens to their difference as $M \to \infty$?

Exercise 6.16. Consider an uplink multi-user MIMO system with $K = 2$ users and block-fading channels with inputs $x_1[l] \in \mathbb{C}$ and $x_2[l] \in \mathbb{C}$, for $l = 1, \ldots, L_{\mathrm{c}}$, where L_{c} is the number of symbols transmitted in each coherence block. The two users send simultaneous pilot sequences that span the initial $L_{\mathrm{p}} = 2$ symbols of each coherence block. For the base station to distinguish between the users' channels, the pilot sequences are selected as

$$\boldsymbol{\phi}_1 = \begin{bmatrix} x_1[1] \\ x_1[2] \end{bmatrix} = \sqrt{\frac{P}{B}} \begin{bmatrix} 1 \\ 1 \end{bmatrix}, \quad \boldsymbol{\phi}_2 = \begin{bmatrix} x_2[1] \\ x_2[2] \end{bmatrix} = \sqrt{\frac{P}{B}} \begin{bmatrix} 1 \\ -1 \end{bmatrix}, \tag{6.156}$$

which are orthogonal vectors since $\boldsymbol{\phi}_1^{\mathrm{H}}\boldsymbol{\phi}_2 = 0$. During the pilot transmission phase, the maximum uplink power P is used by both users. The received signal at the initial two time instances is

$$\begin{bmatrix} \mathbf{y}[1] & \mathbf{y}[2] \end{bmatrix} = \mathbf{h}_1\boldsymbol{\phi}_1^{\mathrm{T}} + \mathbf{h}_2\boldsymbol{\phi}_2^{\mathrm{T}} + \begin{bmatrix} \mathbf{n}[1] & \mathbf{n}[2] \end{bmatrix}, \tag{6.157}$$

where $\mathbf{n}[l] \sim \mathcal{N}_{\mathbb{C}}(\mathbf{0}, N_0\mathbf{I}_M)$ is the independent receiver noise. During the $L_{\mathrm{c}} - L_{\mathrm{p}} = L_{\mathrm{c}} - 2$ remaining symbols of each coherence block, the received signal is

$$\mathbf{y}[l] = \mathbf{h}_1 x_1[l] + \mathbf{h}_2 x_2[l] + \mathbf{n}[l], \quad l = 3, \ldots, L_{\mathrm{c}}, \tag{6.158}$$

where $x_1[l] \sim \mathcal{N}_{\mathbb{C}}(0, P_1^{\mathrm{ul}}/B)$ and $x_2[l] \sim \mathcal{N}_{\mathbb{C}}(0, P_2^{\mathrm{ul}}/B)$.

(a) Compute the MMSE estimates of $\mathbf{h}_1, \mathbf{h}_2$ based on the received signal in (6.157), assuming that $\mathbf{h}_k \sim \mathcal{N}_{\mathbb{C}}(\mathbf{0}, \beta_k\mathbf{I}_M)$, for $k = 1, 2$. Hint: Multiply with $\frac{\boldsymbol{\phi}_1^*}{\|\boldsymbol{\phi}_1\|}$ and $\frac{\boldsymbol{\phi}_2^*}{\|\boldsymbol{\phi}_2\|}$ from the right-hand side in (6.157) to obtain two interference-free received signals. You can then follow the approach from (5.137).

(b) Suppose the base station applies MRC to the received signals in (6.158) based on the estimated channels: $\mathbf{w}_k = \frac{\hat{\mathbf{h}}_k}{\|\hat{\mathbf{h}}_k\|}$, for $k = 1, 2$. Obtain the ergodic rate of user k, for $k = 1, 2$, by treating the channel estimation error and the interference as noise. Hint: Use Corollary 5.2.

Exercise 6.17. Consider uplink multi-user MIMO with i.i.d. Rayleigh slow-fading channels, ZF combining, and perfect CSI at the receiver.

(a) Show that the outage probability $P_{\text{out},k}(R_k)$ when the rate R_k [bit/s] is used for user k can be expressed involving a $(M-K+1)$-dimensional vector $\breve{\mathbf{h}}_k \sim \mathcal{N}_{\mathbb{C}}(\mathbf{0}, \beta_k \mathbf{I}_{M-K+1})$. This is equivalent to showing that each user experiences an interference-free channel with $M-K+1$ *degrees of freedom*. Hint: Use the result from Example 6.6 with $\breve{\mathbf{h}}_k = (\mathbf{A}_k^{\text{free}})^{\text{H}} \mathbf{h}_k$.

(b) Obtain an upper bound on the outage probability $P_{\text{out},k}(R_k)$ using the bound from (5.54), and find the diversity order.

Exercise 6.18. Consider a multi-user MIMO where each user has N antennas, and perfect CSI is available everywhere.

(a) Consider the uplink and suppose that user k uses a specific precoding matrix $\mathbf{P}_k^{\text{ul}} \in \mathbb{C}^{N \times N}$, which has unit-norm columns and is known at the base station. The transmitted signal is generated as $\mathbf{P}_k^{\text{ul}} \bar{\mathbf{x}}_k^{\text{ul}}$ where the data vector is distributed as $\bar{\mathbf{x}}_k^{\text{ul}} \sim \mathcal{N}_{\mathbb{C}}(\mathbf{0}, \mathbf{Q}_k^{\text{ul}})$, for $k = 1, \ldots, K$. The covariance matrix $\mathbf{Q}_k^{\text{ul}} \in \mathbb{C}^{N \times N}$ is the diagonal power allocation matrix with $\text{tr}\left(\mathbf{Q}_k^{\text{ul}}\right)$ being the user's total transmit power. The received signal at the base station is

$$\mathbf{y}^{\text{ul}} = \sum_{k=1}^{K} \mathbf{H}_k \mathbf{P}_k^{\text{ul}} \bar{\mathbf{x}}_k^{\text{ul}} + \mathbf{n}^{\text{ul}}, \tag{6.159}$$

where $\mathbf{H}_k \in \mathbb{C}^{M \times N}$ is the channel vector from user k to the base station and $\mathbf{n}^{\text{ul}} \sim \mathcal{N}_{\mathbb{C}}(0, N_0 \mathbf{I}_M)$ is the independent receiver noise. What is the achievable data rate for user k if the interference from other users is treated as colored noise (i.e., linear receiver processing is used)?

(b) Consider the downlink and suppose that the base station uses a specific precoding matrix $\mathbf{P}_k^{\text{dl}} \in \mathbb{C}^{M \times M}$ for user k, for $k = 1, \ldots, K$, which has unit-norm columns and is known at the users. The transmitted signal is generated as $\sum_{i=1}^{K} \mathbf{P}_i^{\text{dl}} \bar{\mathbf{x}}_i^{\text{dl}}$. The data vector is distributed as $\bar{\mathbf{x}}_i^{\text{dl}} \sim \mathcal{N}_{\mathbb{C}}(\mathbf{0}, \mathbf{Q}_i^{\text{dl}})$, where $\mathbf{Q}_i^{\text{dl}} \in \mathbb{C}^{M \times M}$ is the diagonal power allocation matrix with $\sum_{i=1}^{K} \text{tr}\left(\mathbf{Q}_i^{\text{dl}}\right)$ being the total transmit power. The received signal at user k is

$$\mathbf{y}_k^{\text{dl}} = \mathbf{H}_k^{\text{T}} \sum_{i=1}^{K} \mathbf{P}_i^{\text{dl}} \bar{\mathbf{x}}_i^{\text{dl}} + \mathbf{n}_k^{\text{dl}}, \tag{6.160}$$

where $\mathbf{n}_k^{\text{dl}} \sim \mathcal{N}_{\mathbb{C}}(0, N_0 \mathbf{I}_N)$ is the independent receiver noise. What is the achievable data rate for user k if the interference from signals meant for other users is treated as colored noise (i.e., linear receiver processing is used)?

Chapter 7

Wideband MIMO Channels and Practical Aspects

Practical communication systems utilize vast bandwidths to the extent that the channel coefficients vary over it, which might result in inter-symbol interference. In this chapter, we extend the previously developed MIMO theory to handle these situations. We will first show how multicarrier modulation appears as the natural transmission method when dealing with inter-symbol interference. We then derive the resulting multicarrier MIMO capacity and describe how the subcarrier channels depend on the multipath clusters. Next, we discuss practical hardware implementation of precoding and combining, and when the typical digital architecture can be simplified into an analog or hybrid architecture. Finally, we will exemplify two practical MIMO implementations and elaborate on different MIMO-related terminologies and their meanings.

7.1 Basics of Multicarrier Modulation

The analysis and algorithmic development in previous chapters were based on the discrete memoryless channel model derived in Section 2.3.4. To reach that model, we made the *narrowband signal assumption*, which essentially means that the time interval $1/B$ between two transmitted symbols is much larger than the delay spread, which is the variation in delay between the fastest and slowest propagation paths in the propagation environment. Under that condition, delayed copies of the previous symbols will not interfere with the currently transmitted symbol. One can get an intuitive sense of this phenomenon by listening to acoustic waves. When we hear speech or music, the waves will be reflected on various objects before reaching the listener. In a normal-sized room, the delay spread of acoustic waves is smaller than 50 ms, giving rise to the reverberation effect where each distinct sound becomes less sharp but still apprehensible and sometimes perceived as more pleasant to the ears. In contrast, in a large room or outdoor environment with a delay spread larger than 50 ms, there can be distinct echoes that disturb the listening experience. When there are echoes of this kind, the acoustic channel is said to have a *memory*. The same physical principles apply to radio waves, but the bandwidth and propagation speed are entirely different.

Many wireless communication systems designed for broadband connectivity use more bandwidth than permitted under the narrowband signal assumption. Hence, we want to design systems that function irrespective of whether the environment has a long or short delay spread. To model such *wideband channels*, we return to the received signal $y[l]$ in (2.128) at symbol time l, which was obtained before making the narrowband signal assumption:

$$y[l] = \sum_{k=-\infty}^{\infty} h[l-k]x[k] + n[l], \tag{7.1}$$

where $x[k]$ is the transmitted symbol at time k, $n[l] \sim \mathcal{N}_{\mathbb{C}}(0, N_0)$ is the additive receiver noise, and the communication channel is represented by the coefficients

$$h[k] = \sum_{i=1}^{L} \alpha_i e^{-\mathrm{j}2\pi f_c(\tau_i - \eta)} \mathrm{sinc}(k + B(\eta - \tau_i)). \tag{7.2}$$

These coefficients describe L propagation paths for which path i has the attenuation α_i and the delay τ_i, while η is the sampling delay at the receiver. The important thing in this chapter will not be the exact channel model in (7.2) but the general structure in (7.1). The received signal $y[l]$ contains a weighted summation of many transmitted symbols $\{x[k]\}$. The copy of $x[k]$ received at time l is multiplied by the weight denoted by $h[l-k]$.

The sinc-function appears in (7.2) because it was utilized in Section 2.3.2 as the pulse $p(t)$ in the PAM transmission and for bandpass receiver filtering that removes noise outside the signal band. This function satisfies the transmission design requirements from that section while requiring the minimum bandwidth. However, the downside is that it has a long time duration around its peak value, spanning both forward and backward in time. Strictly speaking, $\mathrm{sinc}(Bt)$ has an infinite duration, but 90% of its energy is in the interval $t \in [-1/B, 1/B]$ and 99% in the interval $t \in [-8/B, 8/B]$. We will refer to the latter as the *effective time duration* of the pulse, and the fact that it is much larger than the symbol time is important when characterizing the channel coefficients in (7.2). Recall that (7.2) is obtained in (2.126) by sampling the function

$$(p * g * p)(t) = \sum_{i=1}^{L} \alpha_i e^{-\mathrm{j}2\pi f_c(\tau_i - \eta)} \mathrm{sinc}(B(t + \eta - \tau_i)) \tag{7.3}$$

at the time instance $t = \frac{k}{B}$ where k is an integer. This function is illustrated in Figure 7.1 for $L = 3$ paths with amplitudes and delays specified in the legend. The three path components in (7.3) are shown individually, and all take the form of an attenuated and delayed sinc-function. The summation of these components results in the dotted curve that represents $(p * g * p)(t)$. By

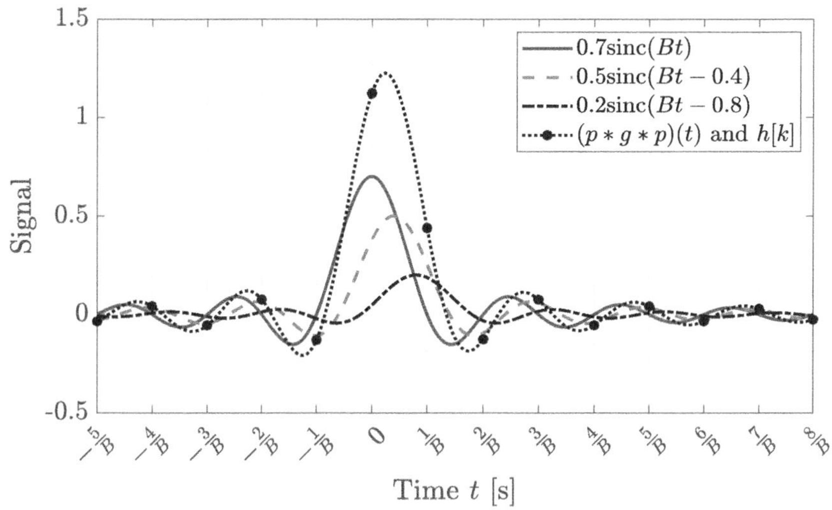

Figure 7.1: Example of a channel with $L = 3$ distinct paths with different amplitudes and delays (the complex phase-shifts are neglected). The summation of these paths results in $(p * g * p)(t)$ in (7.3). By sampling this signal at k/B, where k is an integer, we get the channel coefficients $h[k]$ defined in (7.2).

taking samples of this function at time instances k/B, where k is an integer, we get the channel coefficients $h[k]$ in (7.2). Interestingly, these coefficients are non-zero for both positive and negative values of k, but the oscillations become smaller as $|k|$ increases. The fact that $h[k]$ can be non-zero for negative indices should not be interpreted as having an unrealistic non-causal system but highlights that real pulse functions start long before they reach their peak values. Importantly, the three paths give rise to much more than three non-zero channel coefficients due to the pulse's long effective time duration.

Practical systems mitigate this effect by using pulses with a shorter effective time duration than the sinc-function, represented by a faster decay around the peak value. However, all feasible pulses have a non-zero effective time duration, so this issue cannot be fully alleviated.[1] Hence, even if the sampling delay is selected as $\eta = \min_i \tau_i$ to match with the peak of the fastest propagation path (as was done in Figure 7.1), there will be $h[k] \neq 0$ for negative values of k. To achieve a causal discrete-time system model, we should instead select η to take the first sample of the received signal at the beginning of the pulse

[1]The pulse $p(t)$ must satisfy the Nyquist criterion, which for a given symbol rate B requires that $p(k/B) = 0$ for non-zero integers k and results in a signal bandwidth that is larger than B. In theory, we could minimize the effective time duration by using a rectangle-shaped pulse that is only non-zero in the interval $t \in [-1/(2B), 1/(2B)]$, but it will have a huge bandwidth (the Fourier transform is a sinc-function). A common practical choice is the so-called root-raised-cosine pulse, for which one can conveniently control the tradeoff between the effective time duration and the excess bandwidth compared to B (required by the sinc-pulse). For example, with 25% excess bandwidth, 99% of the energy is contained in the interval $t \in [-2/B, 2/B]$.

that arrives through the fastest propagation path. The number of samples should be selected to take the last sample at the end of the pulse that arrives through the slowest propagation path. The relevant parameters are then the delay spread

$$\tau_{\text{spread}} = \max_{i \in \{1,\ldots,L\}} \tau_i - \min_{i \in \{1,\ldots,L\}} \tau_i \tag{7.4}$$

of the channel and the integer number of periods N_{pulse} for which the pulse takes values that cannot be approximated as zero; that is, N_{pulse} is the smallest even[2] integer so that $(p * p)(t) = \text{sinc}(Bt) \approx 0$ for $|t| \geq N_{\text{pulse}}/(2B)$. Since N_{pulse} and τ_{spread} are finite in practice, we can describe the channel using a finite number of channel coefficients $h[k]$ that we will denote as $T+1$ in the remainder of this chapter. If we select the sampling delay as

$$\eta = \min_{i \in \{1,\ldots,L\}} \tau_i - \frac{N_{\text{pulse}} - 2}{2B}, \tag{7.5}$$

then the fastest path in (7.2), with the smallest τ_i, will contain the time-shifted pulse $\text{sinc}(k - \frac{N_{\text{pulse}}}{2} + 1)$, which can be approximated as zero for all $k < 0$. Since all other propagation paths are slower, we can conclude that $h[k] \approx 0$ for all $k < 0$ in (7.1). Moreover, the slowest path (with the largest τ_i) will contain the time-shifted pulse $\text{sinc}(k - B\tau_{\text{spread}} - \frac{N_{\text{pulse}}}{2} + 1)$, which can be approximated as zero for $k \geq B\tau_{\text{spread}} + N_{\text{pulse}} - 1$. Hence, the channel coefficient with the largest time index that we need to consider in (7.1) is $h[T]$ with

$$T = \lfloor B\tau_{\text{spread}} \rfloor + N_{\text{pulse}} - 1, \tag{7.6}$$

where $\lfloor \cdot \rfloor$ truncates its argument to the nearest smaller integer.

In summary, when selecting the sampling delay as in (7.5), the summation in (7.1) will approximately end at $k = l$ and contain $T+1$ terms:

$$\begin{aligned} y[l] &= \sum_{k=l-T}^{l} h[l-k]x[k] + n[l] \\ &= \sum_{\ell=0}^{T} h[\ell]x[l-\ell] + n[l], \end{aligned} \tag{7.7}$$

where the equality follows from changing the summation index from k to $\ell = l - k$. We notice that the channel now behaves as a causal FIR filter of order T with the non-zero coefficients $h[0], \ldots, h[T]$ as the impulse response. These coefficients are the discrete-time representation of the channel and can be computed based on the physical channel using (7.2) and (7.5).

[2]The considered sinc-pulse is symmetric around its peak value in the time domain; thus, we should consider the same number of periods before and after the peak value. Since the fastest path is typically the strongest one, it is essential to take samples when the pulse received over that path reaches its peak value to make the corresponding path as strong as possible.

The discrete-time system model in (7.7) describes a dispersive channel with a memory of T previous symbols; that is, the received signal $y[l]$ contains not only the currently transmitted signal $x[l]$ but also inter-symbol interference from $x[l-1], \dots, x[l-T]$. There are multiple ways of dealing with interference. We can remove the interference by "transmitting" T zero-valued symbols after each data symbol so that the inter-symbol interference becomes zero. This approach will reduce the symbol rate from B to $B/(T+1)$ and is more-or-less equivalent to the narrowband signal assumption since we effectively reduce the signal bandwidth to alleviate inter-symbol interference. Another option is to design a digital receiver filter that inverts the operation of the FIR filter of the channel. This is known as single-carrier transmission. In this chapter, we will focus on a third option: divide the bandwidth into multiple frequency subcarriers that each can be modeled as a memoryless channel.

7.1.1 Orthogonal Frequency-Division Multiplexing (OFDM)

If a narrowband signal can be transmitted over a small piece of bandwidth B_{narrow} without generating inter-symbol interference, then it must be possible to take a larger bandwidth B, divide it equally into B/B_{narrow} pieces with bandwidth B_{narrow}, and transmit separate narrowband signals in each of them. *Orthogonal frequency-division multiplexing (OFDM)* is a way to implement this procedure without requiring a strict bandwidth division or separate hardware components for each piece of bandwidth. OFDM has become the standard digital transmission method in WiFi, LTE, NR, and many other standards.

The main characteristic of OFDM is that the transmitted time-domain symbols $\{x[k]\}$ in (7.7) are not equal to the data symbols, but they are instead designed to convey different data over different parts of the frequency band. To achieve this, we would like to transform the wideband channel in (7.7) into the frequency domain using the DFT that was defined in Section 2.8. In this section, we will show that this is the optimal way of operating under the assumption that the time-domain signal has a block-wise cyclic structure.

Suppose we want to transmit a block of S symbols, called $\chi[0], \dots, \chi[S-1]$, over the channel in (7.7). For any given value of T, determined by the propagation environment, we can always select $S > T$ since we are the ones designing the communication protocol. Since the channel has a memory of T previous symbols, we must control what was transmitted at the previous T symbol times before time 0. In particular, we will append a *cyclic prefix* to obtain the following cyclic sequence of length $S+T$:

$$x[k] = \begin{cases} \chi[k] & k = 0, \dots, S-1, \\ \chi[k+S] & k = -T, \dots, -1. \end{cases} \tag{7.8}$$

This procedure of creating one transmission block is illustrated in Figure 7.2, where the complete transmitted signal consists of $\{x[k] : k = -T, \dots, S-1\}$.

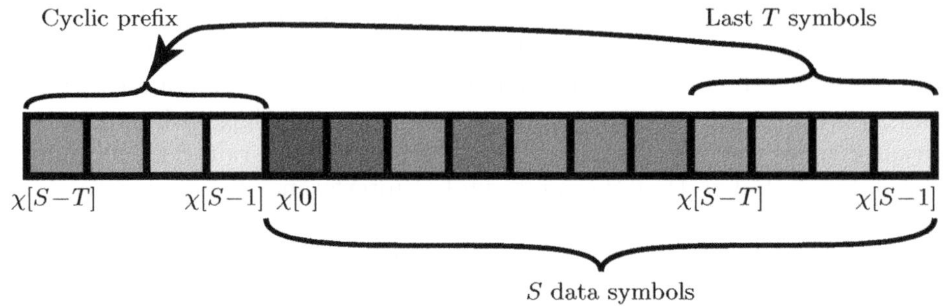

Figure 7.2: Each block in an OFDM transmission consists of S data symbols and a cyclic prefix containing the last T symbols.

Since we added the last T symbols as a prefix, we can interpret the received signal in (7.7) as the cyclic convolution

$$y[l] = \sum_{\ell=0}^{T} h[\ell]x[l-\ell] + n[l] = \sum_{\ell=0}^{T} h[\ell]\chi[(l-\ell)_{\mathrm{mod}\, S}] + n[l], \quad l = 0, \dots, S-1, \tag{7.9}$$

between the input signal sequence $\{\chi[s] : s = 0, \dots, S-1\}$ and the sequence $\{h[\ell] : \ell = 0, \dots, T\}$ with the channel taps, plus the independent noise $n[l] \sim \mathcal{N}_{\mathbb{C}}(0, N_0)$.[3] The cyclic convolution and its properties were previously discussed in Section 2.8.2. Thanks to the cyclic prefix, we can write the relationship between the S received signals and S transmitted signals in matrix-vector form as

$$\underbrace{\begin{bmatrix} y[0] \\ \vdots \\ y[S-1] \end{bmatrix}}_{=\mathbf{y}} = \mathbf{C}_h \underbrace{\begin{bmatrix} \chi[0] \\ \vdots \\ \chi[S-1] \end{bmatrix}}_{=\boldsymbol{\chi}} + \underbrace{\begin{bmatrix} n[0] \\ \vdots \\ n[S-1] \end{bmatrix}}_{=\mathbf{n}}, \tag{7.10}$$

where the channel is represented by the $S \times S$ circulant matrix

$$\mathbf{C}_h = \begin{bmatrix} h[0] & h[S-1] & \dots & h[2] & h[1] \\ h[1] & h[0] & h[S-1] & \dots & h[2] \\ \vdots & h[1] & h[0] & \ddots & \vdots \\ h[S-2] & \ddots & \ddots & \ddots & h[S-1] \\ h[S-1] & h[S-2] & \dots & h[1] & h[0] \end{bmatrix}, \tag{7.11}$$

[3]In principle, we could also consider the previously received signals $y[-T], \dots, y[-1]$ that contain a combination of the signals in the cyclic prefix and signals that were transmitted even earlier in time, but these received signals are normally discarded in OFDM since they contain interference from even earlier signals that are generally unknown. Even if these are previous data symbols for which estimates are available at the receiver, error propagation effects can be created if we rely on them for decoding the new block of symbols.

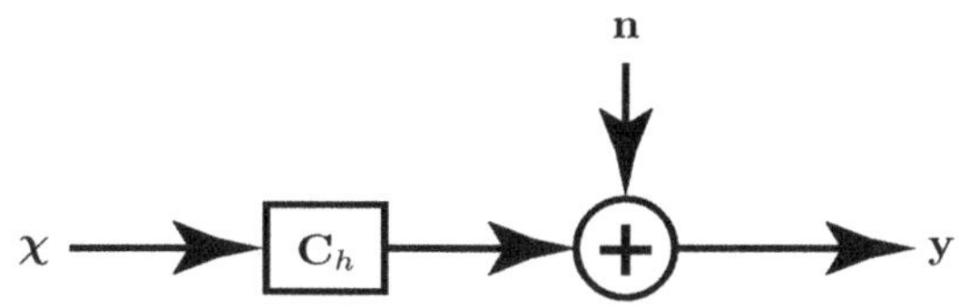

Figure 7.3: The operation of an OFDM system is divided into blocks of S symbols (plus a cyclic prefix). The transmission in a block can be expressed as a discrete memoryless MIMO channel with vector input $\boldsymbol{\chi} \in \mathbb{C}^S$ and vector output $\mathbf{y} \in \mathbb{C}^S$. The channel matrix $\mathbf{C}_h$ in (7.11) is circulant and the independent noise vector $\mathbf{n}$ is complex Gaussian distributed.

which contains the FIR filter taps $h[0], \ldots, h[T]$ that have been padded with the zero-valued taps $h[T+1] = \ldots = h[S-1] = 0$ when $S-1 > T$ for notational convenience.

Interestingly, there is a mathematical equivalence between (7.10) and the system model of a point-to-point MIMO channel with S inputs, S outputs, and the channel matrix $\mathbf{C}_h$. We can write (7.10) in the familiar MIMO-like form

$$\mathbf{y} = \mathbf{C}_h \boldsymbol{\chi} + \mathbf{n} \tag{7.12}$$

and Figure 7.3 shows the corresponding block diagram. We recall from Section 3.4 that the capacity of such a channel is achieved by diagonalizing the channel matrix, thereby creating S parallel memoryless subchannels. Since the channel matrix $\mathbf{C}_h$ of an OFDM system is a circulant matrix, its eigendecomposition has a simple form that was derived in Section 2.8.2:

$$\mathbf{C}_h = \mathbf{F}_S^{\mathrm{H}} \mathbf{D}_{\bar{h}} \mathbf{F}_S, \tag{7.13}$$

where $\mathbf{F}_S$ is the DFT matrix defined in (2.198) and $\mathbf{D}_{\bar{h}}$ is the diagonal matrix

$$\mathbf{D}_{\bar{h}} = \begin{bmatrix} \bar{h}[0] & 0 & \ldots & 0 \\ 0 & \bar{h}[1] & \ddots & \vdots \\ \vdots & \ddots & \ddots & 0 \\ 0 & \ldots & 0 & \bar{h}[S-1] \end{bmatrix} \tag{7.14}$$

containing the frequency response of the FIR filter. It is computed as

$$\bar{h}[\nu] = \sum_{\ell=0}^{T} h[\ell] e^{-\mathrm{j}2\pi\ell\nu/S}, \quad \text{for } \nu = 0, \ldots, S-1. \tag{7.15}$$

Suppose we let the transmitter generate the time-domain signal sequence $\boldsymbol{\chi}$ as

$$\boldsymbol{\chi} = \mathbf{F}_S^{\mathrm{H}} \bar{\boldsymbol{\chi}} \tag{7.16}$$

for some data-bearing vector $\bar{\boldsymbol{\chi}} \in \mathbb{C}^S$. If the receiver multiplies the received signal sequence $\mathbf{y}$ with the DFT matrix as $\mathbf{F}_S \mathbf{y}$, it will obtain

$$\begin{aligned}\bar{\mathbf{y}} = \mathbf{F}_S\mathbf{y} = \mathbf{F}_S(\mathbf{C}_h\mathbf{F}_S^{\mathrm{H}}\bar{\chi} + \mathbf{n}) &= \underbrace{\mathbf{F}_S\mathbf{F}_S^{\mathrm{H}}}_{=\mathbf{I}_S}\mathbf{D}_{\bar{h}}\underbrace{\mathbf{F}_S\mathbf{F}_S^{\mathrm{H}}}_{=\mathbf{I}_S}\bar{\chi} + \mathbf{F}_S\mathbf{n} \\ &= \mathbf{D}_{\bar{h}}\bar{\chi} + \bar{\mathbf{n}},\end{aligned} \tag{7.17}$$

which has the same form as a MIMO channel with the diagonal channel matrix $\mathbf{D}_{\bar{h}}$ and the rotated noise vector

$$\bar{\mathbf{n}} = \begin{bmatrix} \bar{n}[0] \\ \vdots \\ \bar{n}[S-1] \end{bmatrix} = \mathbf{F}_S\mathbf{n} \sim \mathcal{N}_{\mathbb{C}}(\mathbf{0}, N_0\mathbf{I}_S). \tag{7.18}$$

To obtain this result, we have utilized the eigendecomposition in (7.13) and the fact that the DFT matrix $\mathbf{F}_S$ is unitary. The latter property makes $\mathbf{F}_S\mathbf{F}_S^{\mathrm{H}} = \mathbf{I}_S$ and ensures that the rotated noise vector $\bar{\mathbf{n}}$ contains independent entries with the same variance as $\mathbf{n}$. The transmitter and receiver processing that creates the S parallel SISO channels is summarized in Figure 7.4(a).

We used the eigendecomposition to diagonalize the channel matrix $\mathbf{C}_h$, while the SVD was used for the same purpose in Section 3.4. These decompositions are closely related but differ in whether the diagonal matrix is real or complex.[4] The eigendecomposition has a simpler form but only exists for square matrices (as in this section), while the SVD always exists.

Example 7.1. Compute the frequency responses with $S = 4$ subcarriers for the following channels. The first channel has $h_1[0] = 1$ but $h_1[\ell] = 0$ for $\ell \neq 0$. The second channel has $h_2[0] = h_2[1] = 1$, while $h_2[\ell] = 0$ for any other ℓ.

The frequency responses of these channels are obtained from (7.15) as

$$\bar{h}_1[\nu] = h_1[0]e^{-\mathrm{j}2\pi\cdot 0\cdot\nu/4} = 1, \tag{7.19}$$

$$\begin{aligned}\bar{h}_2[\nu] &= h_2[0]e^{-\mathrm{j}2\pi\cdot 0\cdot\nu/4} + h_2[1]e^{-\mathrm{j}2\pi\cdot 1\cdot\nu/4} = 1 + e^{-\mathrm{j}\pi\nu/2} \\ &= 2e^{-\mathrm{j}\pi\nu/4}\cos(\pi\nu/4),\end{aligned} \tag{7.20}$$

for $\nu = 0, \ldots, 3$, where the last equality follows from Euler's formula.

The magnitude of the first channel's frequency response is 1 on all subcarriers, so this channel is frequency-flat. This is a consequence of only having a single tap. On the other hand, the magnitude of the second channel's frequency response is $2|\cos(\pi\nu/4)|$, which results in the values $|\bar{h}_2[0]| = 2$, $|\bar{h}_2[1]| = \sqrt{2}$, $|\bar{h}_2[2]| = 0$, $|\bar{h}_2[3]| = \sqrt{2}$ on the different subcarriers. This channel has frequency-varying characteristics since the two taps superimpose differently between the subcarriers.

[4]The SVD $\mathbf{U}\boldsymbol{\Sigma}\mathbf{V}^{\mathrm{H}}$ of $\mathbf{C}_h$ has the matrix $\boldsymbol{\Sigma} = \mathrm{diag}(|\bar{h}[0]|, \ldots, |\bar{h}[S-1]|)$ with singular values, which are the magnitudes of the corresponding entries of $\mathbf{D}_{\bar{h}}$ in the eigendecomposition. The unitary matrices can be selected as $\mathbf{U} = \mathbf{F}_S^{\mathrm{H}}\mathbf{D}_{\bar{h}}\boldsymbol{\Sigma}^{-1}$ and $\mathbf{V} = \mathbf{F}_S^{\mathrm{H}}$.

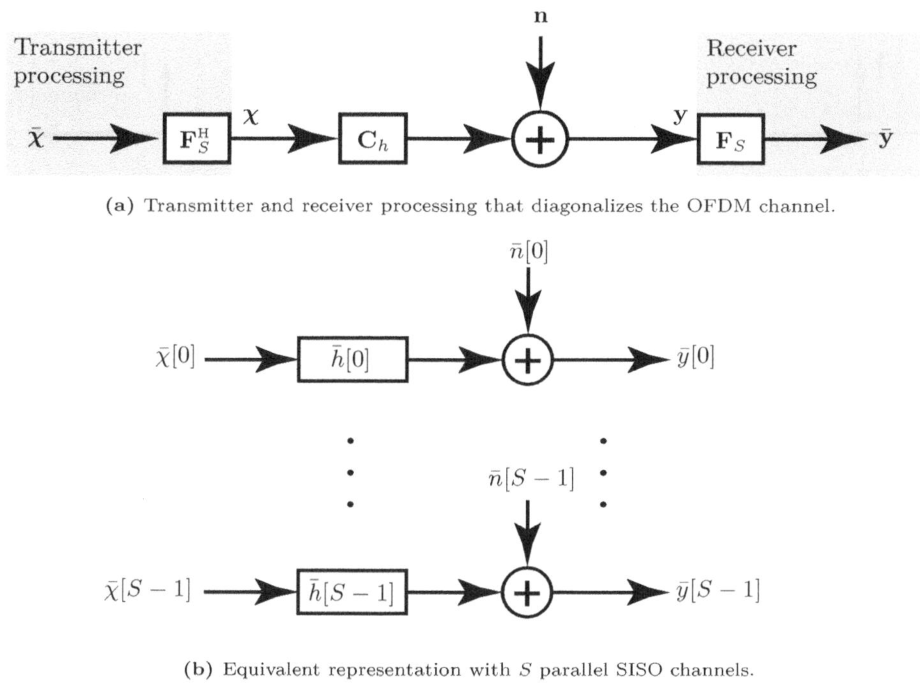

Figure 7.4: The transmission of an S-length block in an OFDM system can be represented as a MIMO channel where the channel matrix has the eigendecomposition $\mathbf{C}_h = \mathbf{F}_S^{\mathrm{H}}\mathbf{D}_{\bar{h}}\mathbf{F}_S$. Hence, the transmitter and receiver can process the signals using the $S \times S$ DFT matrix as shown in (a) to achieve S parallel SISO channels as shown in (b).

We have now derived the system operation generally referred to as OFDM. The reason for calling it orthogonal frequency-division multiplexing is that we multiplex the S data symbols in $\bar{\chi}$ using the frequency domain. More precisely, we generate the transmitted sequence χ of time-domain symbols using the IDFT as $\chi = \mathbf{F}_S^{\mathrm{H}}\bar{\chi}$, which implies that $\bar{\chi}$ is the frequency-domain representation of the transmitted signal. Similarly, the receiver obtains the received signals $\mathbf{y}$ in the time domain and computes its DFT $\bar{\mathbf{y}} = \mathbf{F}_S\mathbf{y}$. We thereby obtain S parallel (orthogonal) discrete memoryless channels

$$\bar{y}[\nu] = \bar{h}[\nu]\bar{\chi}[\nu] + \bar{n}[\nu], \quad \text{for } \nu = 0, \dots, S-1,$$

as illustrated in Figure 7.4(b). We call these *subcarriers* since OFDM divides the wideband channel into S equally spaced subchannels in the frequency domain. The frequency value of a given subcarrier depends on how we measure frequencies. Subcarrier ν utilizes the normalized frequency ν/S, but since we use a symbol rate equal to the bandwidth B, this corresponds to the unnormalized frequency $\nu B/S$ in the complex baseband. Moreover, as described in Section 2.8.1, the DFT is a periodic function where each normalized frequency has aliases at $\nu/S + n$ for any integer n. This ambiguity is due to the sampling

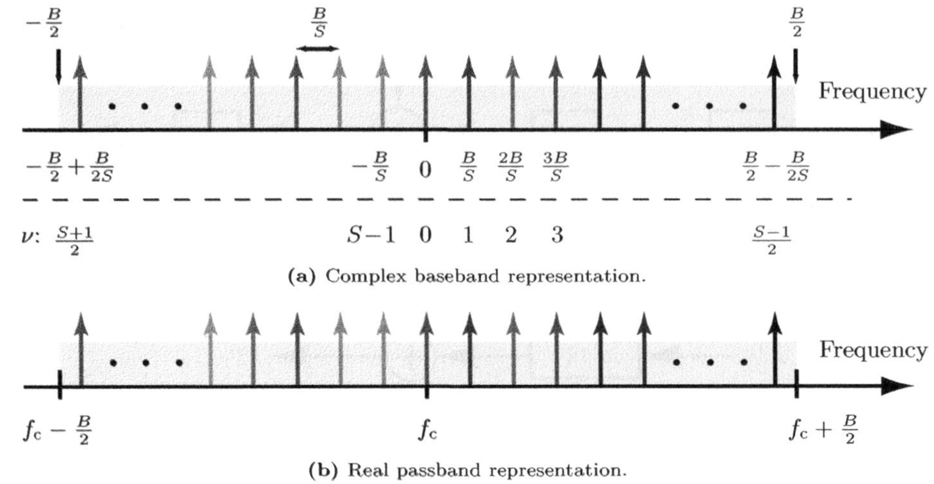

Figure 7.5: The S subcarriers in an OFDM system are equally spaced over the bandwidth B and centered around the carrier frequency, which is 0 in the complex baseband representation shown in (a) and f_c in the real passband representation shown in (b). The subcarrier index ν counts subcarriers from the center towards the right and then continues from left to center.

and implies that many different continuous-time frequencies can give rise to the same discrete-time frequency. Since we analyze the complex baseband representation of a passband signal and take samples at the symbol rate, we know from Figure 2.9 that the actual frequencies occur in the interval from $-B/2$ to $B/2$. Each subcarrier ν only has one alias in that range; hence, its true frequency is

$$\begin{cases} \frac{\nu B}{S}, & \text{if } 0 \leq \nu < \frac{S}{2}, \\ \frac{(\nu - S)B}{S}, & \text{if } \frac{S}{2} \leq \nu < S, \end{cases} \tag{7.21}$$

which is aligned with the symmetric range of positive and negative normalized frequencies shown in Figure 2.29. Figure 7.5(a) illustrates the location of the subcarriers along the frequency axis and which subcarrier index ν gives rise to each of them. The figure considers the case when S is odd, while the outermost frequency values change slightly when S is even. We can multiply (2.207) by B to obtain a list of all the subcarrier frequencies in the complex baseband:

$$\begin{aligned} &\left\{ \frac{B\lceil \frac{S}{2} \rceil}{S} - B, \ldots, -\frac{B}{S}, 0, \frac{B}{S}, \ldots, \frac{B\lceil \frac{S}{2} \rceil - B}{S} \right\} \\ &= \begin{cases} -\frac{B}{2}, \ldots, \frac{B}{2} - \frac{B}{S}, & \text{if } S \text{ is even}, \\ -\frac{B}{2} + \frac{B}{2S}, \ldots, \frac{B}{2} - \frac{B}{2S}, & \text{if } S \text{ is odd}. \end{cases} \end{aligned} \tag{7.22}$$

The separation B/S between two adjacent subcarrier frequencies is called the *subcarrier spacing*. The theory for OFDM is developed in the complex baseband, but the physical communications occur in a passband centered around some carrier frequency f_c. We can obtain the corresponding subcarrier frequencies by shifting the entire spectrum to be centered around that frequency. The resulting real passband representation is illustrated in Figure 7.5(b), which shows the positive subcarrier frequencies around $+f_c$ (there is also a copy around $-f_c$, as illustrated in Figure 2.9).

When transmitting a large amount of data, the OFDM system operation is divided into many consecutive blocks, each managed as described above. Each block is called an *OFDM symbol*. The structure of an OFDM symbol is illustrated in Figure 7.6, which shows both the time- and frequency-domain representations. Since we transmit $T + S$ time-domain symbols with a symbol rate of B Hz, the total time duration of an OFDM symbol is $(T + S)/B$ seconds. The OFDM symbol spans the entire bandwidth B, as shown in Figure 7.6(a). Since each time-domain symbol has a duration of $1/B$ seconds and a bandwidth of B Hz, it covers an area of $\frac{1}{B}B = 1$ in the time-frequency plane. This unit area is dimensionless but is sometimes called one *complex degree of freedom* because it represents the minimum component from which time-frequency signals can be created. Just as any molecule is made of a group of atoms, any communication signal is made from a group of complex degrees of freedom.

Figure 7.6(b) shows how $\bar{\chi}[0], \ldots, \bar{\chi}[S-1]$ represent the transmitted signals over S subcarriers. The subcarriers are equally spaced over the frequency domain, each utilizing a bandwidth of B/S Hz. Since an OFDM symbol has a time duration of $(T + S)/B$ seconds, each subcarrier covers an area of

$$\frac{T+S}{B}\frac{B}{S} = 1 + \frac{T}{S} \text{ degrees of freedom} \tag{7.23}$$

in the time-frequency plane. This is larger than the unit area of a time-domain symbol because each OFDM symbol consists of a sequence of $T + S$ time-domain symbols, of which T symbols are sacrificed in the cyclic prefix to remove inter-symbol interference. However, if we select $S \gg T$, then $1 + \frac{T}{S} \approx 1$ so that the loss is small in relative terms.

The complete transmitter and receiver implementations of OFDM are illustrated in Figures 7.7(a) and (b), respectively. The transmitter first encodes data into the S symbols $\bar{\boldsymbol{\chi}} = [\bar{\chi}[0], \ldots, \bar{\chi}[S-1]]^{\mathrm{T}}$. It then computes the IDFT to obtain $\boldsymbol{\chi} = [\chi[0], \ldots, \chi[S-1]]^{\mathrm{T}} = \mathbf{F}_S^{\mathrm{H}}\bar{\boldsymbol{\chi}}$. The transmitter then appends the cyclic prefix to obtain a sequence $\chi[S-T], \ldots, \chi[S-1], \chi[0], \ldots, \chi[S-1]$ of $T + S$ time-domain symbols, which are transmitted serially over the communication channel. The receiver stores a sequence of $T + S$ time-domain symbols $y[-T], \ldots, y[S-1]$ but discards the cyclic prefix to obtain $\mathbf{y} = [y[0], \ldots, y[S-1]]^{\mathrm{T}}$. It then computes the frequency-domain signals $\bar{\mathbf{y}} = [\bar{y}[0], \ldots, \bar{y}[S-1]]^{\mathrm{T}} = \mathbf{F}_S\mathbf{y}$ using the DFT.

The IDFT $\boldsymbol{\chi} = \mathbf{F}_S^{\mathrm{H}}\bar{\boldsymbol{\chi}}$ at the transmitter and DFT $\bar{\mathbf{y}} = \mathbf{F}_S\mathbf{y}$ at the receiver are obtained as matrix-vector multiplications. The multiplication between an $S \times S$ matrix and an S-length vector generally requires the computation of

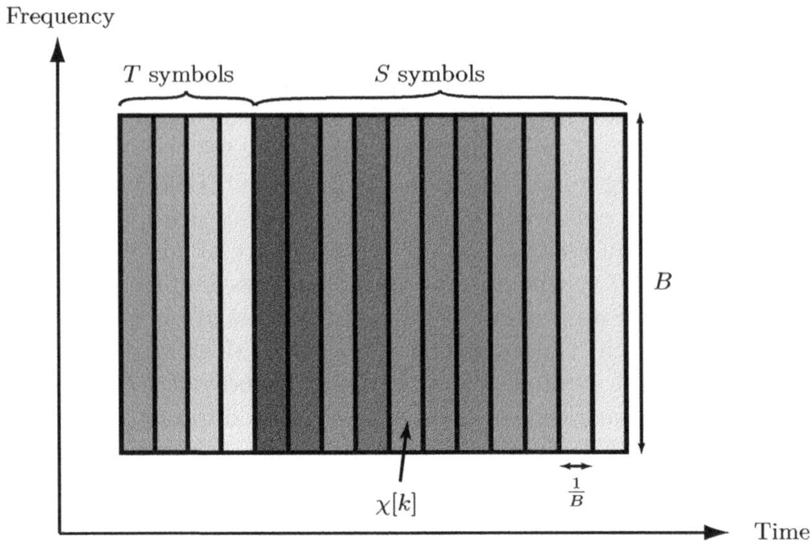

(a) Time-domain representation of an OFDM symbol.

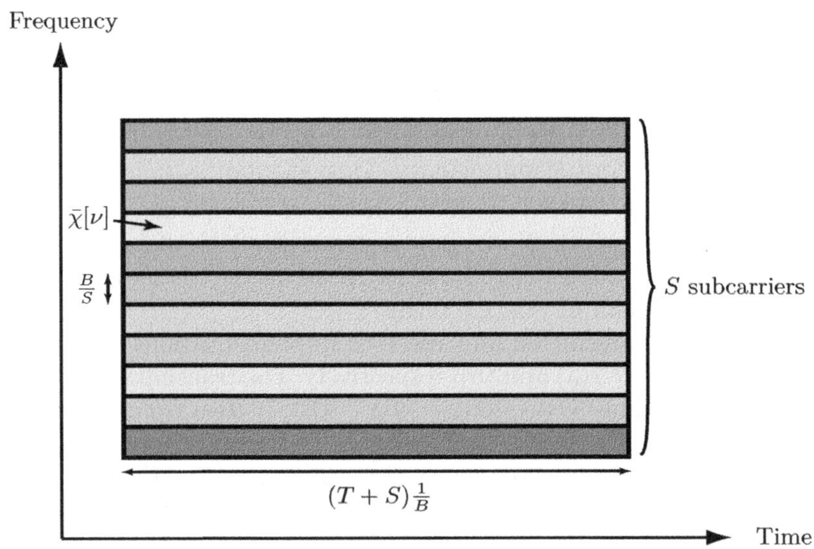

(b) Frequency-domain representation of an OFDM symbol.

Figure 7.6: OFDM systems divide the transmission into blocks called OFDM symbols, which span the entire bandwidth B and $T + S$ time-domain symbols. This block is utilized to generate S memoryless subcarrier channels in the frequency domain.

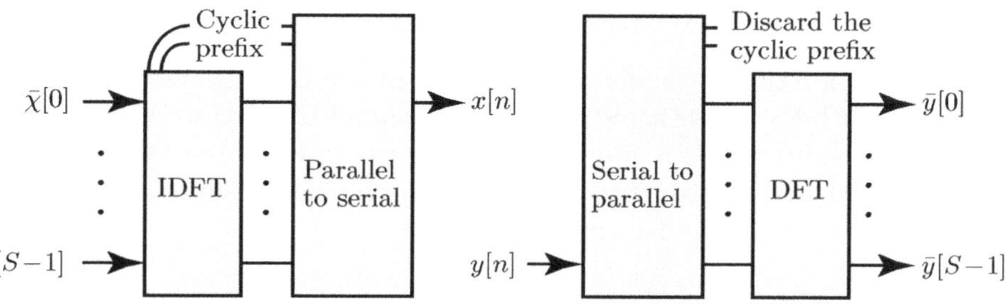

(a) Transmitter. (b) Receiver.

Figure 7.7: Block diagrams of the transmitter and receiver in an OFDM system.

S^2 multiplications and $S(S-1)$ additions, but the DFT matrix has a special structure with repeated entries that can be utilized to lower the computational complexity. In particular, there is a classical algorithm called the *fast Fourier transform* [106] that computes the DFT or IDFT using a number of arithmetic operations proportional to $S\log_2(S)$ instead of S^2. This fast implementation is typically used in practical systems.

Example 7.2. The OFDM symbols in 4G LTE and 5G NR use the subcarrier spacing $B/S = 15\,\text{kHz}$, irrespectively of the bandwidth; thus, the block length S grows proportionally to the bandwidth B. The cyclic prefix is selected to have the time duration 4.69 µs, which specifies a particular largest admissible delay spread and corresponds to $T \approx B \cdot 4.69 \cdot 10^{-6}$. How many complex degrees of freedom does each subcarrier utilize?

We can compute the complex degrees of freedom directly using (7.23) as

$$1 + \frac{T}{S} \approx 1 + 4.69 \cdot 10^{-6}\frac{B}{S} = 1 + 4.69 \cdot 10^{-6} \cdot 15 \cdot 10^3 \approx 1.07. \tag{7.24}$$

This indicates that the cyclic prefix increases the utilization of signal resources by 7%, which is the price to pay for dealing with inter-symbol interference.

There are alternative OFDM configurations in 5G NR [107], including an extended cyclic prefix option that can be selected to manage larger delay spreads and increased subcarrier spacings (by a factor of 2^n for $n \in \{1,2,3,4\}$) to handle latency-critical services and small-cell deployments with low delay spread. In the latter case, the cyclic prefix is shortened accordingly to maintain the same number of degrees of freedom per subcarrier.

There are other multicarrier modulation schemes than OFDM, and some alleviate the cyclic prefix to increase the resource efficiency; however, this can only increase the capacity by 7%. A more important reason to avoid OFDM is that the IDFT operation creates a time-domain signal with relatively large power variations, which makes it hard to build efficient power amplifiers. Hence, some low-power communication systems use other modulation schemes.

7.1.2 Capacity of SISO-OFDM Channels

We will now determine the channel capacity of the OFDM system in (7.17), which we will refer to as the *SISO-OFDM channel* because we have a single-antenna transmitter and a single-antenna receiver. By using DFT matrices for transmitter and receiver processing, as illustrated in Figure 7.4, we create the S memoryless subcarrier channels

$$\bar{y}[\nu] = \bar{h}[\nu]\bar{\chi}[\nu] + \bar{n}[\nu], \quad \text{for } \nu = 0, \dots, S-1. \tag{7.25}$$

Suppose we use the symbol power q_ν when sending the data symbol $\bar{\chi}[\nu]$ at subcarrier ν; that is, $\mathbb{E}\{|\bar{\chi}[\nu]|^2\} = q_\nu$. We can then utilize Corollary 2.1 to conclude that the resulting data rate at subcarrier ν is

$$\log_2\left(1 + \frac{q_\nu |\bar{h}[\nu]|^2}{N_0}\right) \quad \text{bit per subcarrier symbol.} \tag{7.26}$$

This rate is achieved when the data symbol is distributed as $\bar{\chi}[\nu] \sim \mathcal{N}_{\mathbb{C}}(0, q_\nu)$. The accumulated data rate within one OFDM symbol is the summation of (7.26) for all S subcarriers:

$$\sum_{\nu=0}^{S-1} \log_2\left(1 + \frac{q_\nu |\bar{h}[\nu]|^2}{N_0}\right) \quad \text{bit per OFDM symbol.} \tag{7.27}$$

Since each OFDM symbol has a time duration of $(T+S)/B$ seconds, we can equivalently express (7.27) as

$$\frac{B}{T+S} \sum_{\nu=0}^{S-1} \log_2\left(1 + \frac{q_\nu |\bar{h}[\nu]|^2}{N_0}\right) \quad \text{bit/s.} \tag{7.28}$$

This expression is *almost* the bandwidth B multiplied by the average rate of the S subcarriers, but we are dividing by $T+S$ instead of S, which is the price to pay for the cyclic prefix. We have referred to (7.26)–(7.28) as data rates, not the capacities, because we initially assumed arbitrary symbol powers $q_0, \dots, q_{S-1}$ on the subcarriers. Since the channel capacity is the maximum data rate, it can be obtained by maximizing (7.27) with respect to all permissible ways of selecting these power parameters. We used q in previous chapters to denote the maximum symbol power in the time domain. The corresponding requirement in the OFDM case is that $\mathbb{E}\{|\chi[s]|^2\} \leq q$ for the time-domain symbols, for $s = 0, \dots, S-1$. We can utilize the definition $\chi[s] = \frac{1}{\sqrt{S}} \sum_{\nu=0}^{S-1} \bar{\chi}[\nu] e^{\mathrm{j}2\pi s\nu/S}$ of the IDFT to connect this requirement to the data symbols $\bar{\chi}[0], \dots, \bar{\chi}[S-1]$ that are transmitted in the frequency domain:

$$\begin{aligned}
\mathbb{E}\left\{|\chi[s]|^2\right\} &= \mathbb{E}\left\{\left|\frac{1}{\sqrt{S}} \sum_{\nu=0}^{S-1} \bar{\chi}[\nu] e^{\mathrm{j}2\pi s\nu/S}\right|^2\right\} \\
&= \frac{1}{S} \sum_{\nu=0}^{S-1} \mathbb{E}\left\{\left|\bar{\chi}[\nu] e^{\mathrm{j}2\pi s\nu/S}\right|^2\right\} = \frac{1}{S} \sum_{\nu=0}^{S-1} q_\nu,
\end{aligned} \tag{7.29}$$

where we utilized the fact that the data symbols are independent and have zero mean when achieving the aforementioned rates. The conclusion is that we can select the (non-negative) symbol powers $q_0, \ldots, q_{S-1}$ at the different subcarriers freely under the power constraint

$$\frac{1}{S}\sum_{\nu=0}^{S-1} q_\nu \leq q. \tag{7.30}$$

This constraint says that the average power over the subcarriers should equal the power per time-domain symbol. Another way to phrase it is that the sum power of the S subcarriers should be smaller or equal to the power of S time-domain symbols: $\sum_{\nu=0}^{S-1} q_\nu \leq qS$. The capacity of the OFDM channel (in bit per OFDM symbol) is therefore obtained by maximizing the sum rate of S memoryless channels under a sum power constraint:

$$C = \max_{\substack{q_0 \geq 0, \ldots, q_{S-1} \geq 0: \\ \sum_{\nu=0}^{S-1} q_\nu = qS}} \sum_{\nu=0}^{S-1} \log_2\left(1 + \frac{q_\nu |\bar{h}[\nu]|^2}{N_0}\right). \tag{7.31}$$

Apart from a somewhat different notation, this is precisely what we did when considering the point-to-point MIMO capacity in Theorem 3.1. The optimal solution was obtained by the water-filling power allocation:

$$q_\nu^{\text{opt}} = \max\left(\mu - \frac{N_0}{|\bar{h}[\nu]|^2}, 0\right), \quad \nu = 0, \ldots, S-1, \tag{7.32}$$

where the variable μ is selected to make $\sum_{\nu=0}^{S-1} q_\nu = qS$.

When we previously utilized water-filling to achieve the MIMO channel capacity, we divided the power between different spatial dimensions. It is common that a few spatial dimensions are much stronger than the other dimensions since more power reaches the receiver when transmitting towards some specific multipath clusters. This can result in only allocating power to a subset of the subchannels, particularly at low SNRs or when considering LOS channels. In contrast, the S subcarrier channels $\bar{h}[0], \ldots, \bar{h}[S-1]$ are often of similar strength because they are all created as linear combinations of the same channel taps, as can be seen from (7.15). Hence, except in very low SNR scenarios, we will allocate power to all subcarriers. When that happens, we can utilize (3.72) to identify the optimal value of μ in (7.32):

$$\mu = q + \frac{1}{S}\sum_{\nu=0}^{S-1} \frac{N_0}{|\bar{h}[\nu]|^2}. \tag{7.33}$$

The channel capacity of a SISO-OFDM system can be summarized as follows.

Theorem 7.1. Consider the SISO-OFDM system in Figure 7.3 with input $\boldsymbol{\chi} \in \mathbb{C}^S$ and output $\mathbf{y} \in \mathbb{C}^S$ given by

$$\mathbf{y} = \mathbf{C}_h \boldsymbol{\chi} + \mathbf{n}, \tag{7.34}$$

where $\mathbf{n} \sim \mathcal{N}_{\mathbb{C}}(\mathbf{0}, N_0 \mathbf{I}_S)$ is independent noise. Suppose the input distribution is feasible whenever the symbol power satisfies $\mathbb{E}\{\|\boldsymbol{\chi}\|^2\} \leq qS$. The channel matrix $\mathbf{C}_h$ has the eigenvalues $\bar{h}[0], \ldots, \bar{h}[S-1]$ given by (7.15) and the corresponding eigenvectors are columns of the IDFT matrix $\mathbf{F}_S^{\mathrm{H}}$. If the channel matrix is constant and known at the input and output, the channel capacity is

$$C = \frac{B}{T+S} \sum_{\nu=0}^{S-1} \log_2 \left(1 + \frac{q_\nu^{\mathrm{opt}} |\bar{h}[\nu]|^2}{N_0}\right) \quad \text{bit/s}, \tag{7.35}$$

where T is the length of the cyclic prefix,

$$q_\nu^{\mathrm{opt}} = \max\left(\mu - \frac{N_0}{|\bar{h}[\nu]|^2}, 0\right), \quad \nu = 0, \ldots, S-1, \tag{7.36}$$

and the variable μ is selected to make $\sum_{\nu=0}^{S-1} q_\nu^{\mathrm{opt}} = qS$.

The capacity is achieved by the input distribution $\boldsymbol{\chi} \sim \mathcal{N}_{\mathbb{C}}(\mathbf{0}, \mathbf{F}_S^{\mathrm{H}} \mathbf{Q}^{\mathrm{opt}} \mathbf{F}_S)$, where $\mathbf{Q}^{\mathrm{opt}} = \mathrm{diag}(q_0^{\mathrm{opt}}, \ldots, q_{S-1}^{\mathrm{opt}})$ is an $S \times S$ diagonal matrix.

In summary, OFDM is the capacity-achieving way to communicate over wideband SISO channels under the assumption that a cyclic prefix is appended to the data transmission. We observed this by rewriting the transmission of a block of S symbols into a MIMO-like matrix form and showing that the resulting channel matrix $\mathbf{C}_h$ is diagonalized by DFT and IDFT operations at the receiver and transmitter, respectively. We then obtain S parallel subcarrier channels, similar to Section 3.4, and achieve the capacity by dividing the power between them using water-filling.

7.2 Capacity of MIMO-OFDM Channels

We will now extend the capacity analysis from the last section to cover OFDM systems with multiple antennas at both the transmitter and the receiver. When there are M receive antennas, each of the received signals can be modeled using an FIR channel filter as in (7.7), but with the essential difference that signals are received simultaneously from K transmit antennas. Hence, the received signal on antenna m at time l can be expressed as

$$y_m[l] = \sum_{\ell=0}^{T} \sum_{k=1}^{K} h_{m,k}[\ell] x_k[l-\ell] + n_m[l], \tag{7.37}$$

where $h_{m,k}[0], \ldots, h_{m,k}[T]$ are the channel coefficients between receive antenna m and transmit antenna k, $x_k[l]$ is the transmitted signal from antenna k at time l, and $n_m[l] \sim \mathcal{N}_{\mathbb{C}}(0, N_0)$ is the independent receiver noise. This is a messy system model because the received signal at a given time instance l depends on the signals transmitted from K antennas at $T+1$ time instances. However, we can resolve this inter-symbol interference as in the SISO case.

Suppose a T-length cyclic prefix is applied in accordance to (7.8), then the collection of received signals at antenna m in a block containing S time-domain symbols can be expressed as

$$\underbrace{\begin{bmatrix} y_m[0] \\ \vdots \\ y_m[S-1] \end{bmatrix}}_{=\mathbf{y}_m} = \sum_{k=1}^{K} \mathbf{C}_{h_{m,k}} \underbrace{\begin{bmatrix} \chi_k[0] \\ \vdots \\ \chi_k[S-1] \end{bmatrix}}_{=\boldsymbol{\chi}_k} + \underbrace{\begin{bmatrix} n_m[0] \\ \vdots \\ n_m[S-1] \end{bmatrix}}_{=\mathbf{n}_m}, \tag{7.38}$$

where $\boldsymbol{\chi}_k \in \mathbb{C}^S$ is the signal sequence transmitted from antenna k, $\mathbf{n}_m \sim \mathcal{N}_{\mathbb{C}}(\mathbf{0}, N_0\mathbf{I}_S)$ is the receiver noise, and the channel between receive antenna m and transmit antenna k is represented by the $S \times S$ circulant matrix

$$\mathbf{C}_{h_{m,k}} = \begin{bmatrix} h_{m,k}[0] & h_{m,k}[S-1] & \ldots & h_{m,k}[2] & h_{m,k}[1] \\ h_{m,k}[1] & h_{m,k}[0] & h_{m,k}[S-1] & \ldots & h_{m,k}[2] \\ \vdots & h_{m,k}[1] & h_{m,k}[0] & \ddots & \vdots \\ h_{m,k}[S-2] & \ddots & \ddots & \ddots & h_{m,k}[S-1] \\ h_{m,k}[S-1] & h_{m,k}[S-2] & \ldots & h_{m,k}[1] & h_{m,k}[0] \end{bmatrix}. \tag{7.39}$$

This matrix has the same shape as (7.11), which implies that its eigenvectors also coincide with the columns of the IDFT matrix $\mathbf{F}_S^{\mathrm{H}}$ in (2.198). In particular, the eigendecomposition of $\mathbf{C}_{h_{m,k}}$ is

$$\mathbf{C}_{h_{m,k}} = \mathbf{F}_S^{\mathrm{H}} \mathbf{D}_{\bar{h}_{m,k}} \mathbf{F}_S, \tag{7.40}$$

where the diagonal matrix $\mathbf{D}_{\bar{h}_{m,k}} = \mathrm{diag}(\bar{h}_{m,k}[0], \ldots, \bar{h}_{m,k}[S-1])$ contains the frequency response coefficients of the FIR filter that describes the channel:

$$\bar{h}_{m,k}[\nu] = \sum_{\ell=0}^{T} h_{m,k}[\ell] e^{-\mathrm{j}2\pi\ell\nu/S}, \quad \nu = 0, \ldots, S-1. \tag{7.41}$$

This implies that we can diagonalize the matrix $\mathbf{C}_{h_{m,k}}$ by considering signals transmitted and received in the frequency domain instead of the time domain. If we express the DFT of the transmitted signal at antenna k as $\bar{\boldsymbol{\chi}}_k = \mathbf{F}_S \boldsymbol{\chi}_k$, we can write the DFT $\bar{\mathbf{y}}_m = \mathbf{F}_S \mathbf{y}_m$ of the received signal in (7.38) as

$$\bar{\mathbf{y}}_m = \begin{bmatrix} \bar{y}_m[0] \\ \vdots \\ \bar{y}_m[S-1] \end{bmatrix} = \mathbf{F}_S \left(\sum_{k=1}^{K} \mathbf{C}_{h_{m,k}} \mathbf{F}_S^{\mathrm{H}} \bar{\boldsymbol{\chi}}_k + \mathbf{n}_m \right) = \sum_{k=1}^{K} \mathbf{D}_{\bar{h}_{m,k}} \bar{\boldsymbol{\chi}}_k + \bar{\mathbf{n}}_m, \tag{7.42}$$

where we utilized (7.40) and denoted the noise vector in the frequency domain as $\bar{\mathbf{n}}_m = [\bar{n}_m[0], \ldots, \bar{n}_m[S-1]]^{\mathrm{T}} = \mathbf{F}_S \mathbf{n}_m \sim \mathcal{N}_{\mathbb{C}}(\mathbf{0}, N_0 \mathbf{I}_S)$. We notice that the νth entry in $\bar{\mathbf{y}}_m$ is independent of the other entries in the sense of only depending on variables with the same index. Hence, we can separately describe the signals received over the M receive antennas at subcarrier ν as

$$\underbrace{\begin{bmatrix} \bar{y}_1[\nu] \\ \vdots \\ \bar{y}_M[\nu] \end{bmatrix}}_{=\bar{\mathbf{y}}[\nu]} = \underbrace{\begin{bmatrix} \bar{h}_{1,1}[\nu] & \ldots & \bar{h}_{1,K}[\nu] \\ \vdots & \ddots & \vdots \\ \bar{h}_{M,1}[\nu] & \ldots & \bar{h}_{M,K}[\nu] \end{bmatrix}}_{=\bar{\mathbf{H}}[\nu]} \underbrace{\begin{bmatrix} \bar{\chi}_1[\nu] \\ \vdots \\ \bar{\chi}_K[\nu] \end{bmatrix}}_{=\bar{\boldsymbol{\chi}}[\nu]} + \underbrace{\begin{bmatrix} \bar{n}_1[\nu] \\ \vdots \\ \bar{n}_M[\nu] \end{bmatrix}}_{=\bar{\mathbf{n}}[\nu]}, \tag{7.43}$$

which we can write in short form as

$$\bar{\mathbf{y}}[\nu] = \bar{\mathbf{H}}[\nu]\bar{\boldsymbol{\chi}}[\nu] + \bar{\mathbf{n}}[\nu], \quad \nu = 0, \ldots, S-1. \tag{7.44}$$

This looks precisely like a MIMO channel of the kind considered in Section 3.4, but it is based on the frequency-domain channel matrix $\bar{\mathbf{H}}[\nu] \in \mathbb{C}^{M \times K}$ that is a weighted sum of the channel matrices at the different channel taps:

$$\bar{\mathbf{H}}[\nu] = \sum_{\ell=0}^{T} \mathbf{H}[\ell] e^{-\mathrm{j}2\pi\ell\nu/S}, \quad \nu = 0, \ldots, S-1, \tag{7.45}$$

where $\mathbf{H}[\ell] \in \mathbb{C}^{M \times K}$ is the time-domain channel matrix at the tap with index ℓ, whose (m,k)th entry is $h_{m,k}[\ell]$. The matrices $\bar{\mathbf{H}}[0], \ldots, \bar{\mathbf{H}}[S-1]$ is the frequency response of the considered MIMO channel.

Thanks to the cyclic prefix, we managed to rewrite the system model in (7.37) for one S-length block with inter-symbol interference into the S separate subcarrier channels in (7.44). This is the model of a MIMO-OFDM system and is summarized in Figure 7.8. The subcarriers are mutually independent in the sense of depending on different signal vectors $\bar{\boldsymbol{\chi}}[\nu]$ and independent noise terms $\bar{\mathbf{n}}[\nu]$. The only thing that couples them is the power budget of the transmitter: the total energy per block is limited to qS. This power constraint can be expressed in different ways:

$$\sum_{k=1}^{K} \mathbb{E}\{\|\boldsymbol{\chi}_k\|^2\} = \sum_{k=1}^{K} \mathbb{E}\{\|\bar{\boldsymbol{\chi}}_k\|^2\} = \sum_{\nu=0}^{S-1} \mathbb{E}\{\|\bar{\boldsymbol{\chi}}[\nu]\|^2\} \leq qS. \tag{7.46}$$

The first and second summations consider the time-domain and frequency-domain signals at the K antennas, respectively. The third summation considers the frequency-domain signals at the S subcarriers. It showcases that the power limit applies to the average sum of the symbol powers $\|\bar{\boldsymbol{\chi}}[\nu]\|^2$ per subcarrier. We know from Theorem 3.1 that the capacity of a point-to-point MIMO channel is achieved by transmitting in the right singular vector directions and applying water-filling power allocation. The same can be done in the OFDM case, with the only exception that water-filling is carried out by considering all subcarriers and their respective spatial dimensions.

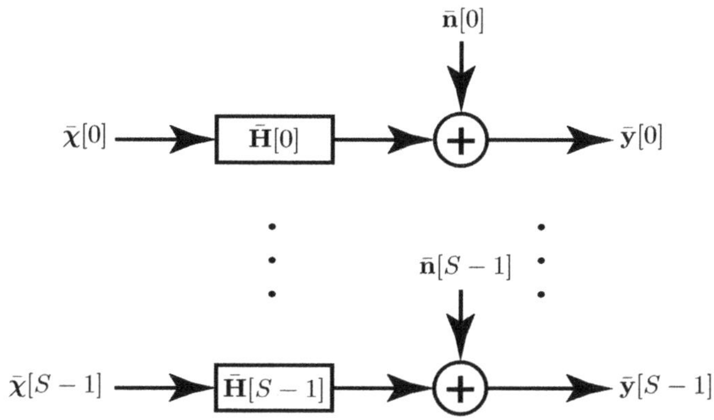

Figure 7.8: A MIMO-OFDM system can be represented as S parallel MIMO channels.

Theorem 7.2. Consider the point-to-point MIMO-OFDM system in Figure 7.8, where subcarrier ν has the input $\bar{\chi}[\nu] \in \mathbb{C}^K$ and output $\bar{\mathbf{y}}[\nu] \in \mathbb{C}^M$ given by

$$\bar{\mathbf{y}}[\nu] = \bar{\mathbf{H}}[\nu]\bar{\chi}[\nu] + \bar{\mathbf{n}}[\nu], \quad \nu = 0, \ldots, S-1, \tag{7.47}$$

where $\bar{\mathbf{n}}[\nu] \sim \mathcal{N}_{\mathbb{C}}(\mathbf{0}, N_0\mathbf{I}_M)$ is independent noise. Suppose the input distribution is feasible whenever $\sum_{\nu=0}^{S-1} \mathbb{E}\{\|\bar{\chi}[\nu]\|^2\} \leq qS$. The channel matrices $\bar{\mathbf{H}}[\nu]$ are constant and known at the transmitter and receiver. Let the r_ν non-zero singular values of $\bar{\mathbf{H}}[\nu]$ be denoted as $s_{\nu,1}, \ldots, s_{\nu,r_\nu}$. The channel capacity is

$$C = \frac{B}{T+S} \sum_{\nu=0}^{S-1} \sum_{k=1}^{r_\nu} \log_2\left(1 + \frac{q_{\nu,k}^{\text{opt}} s_{\nu,k}^2}{N_0}\right) \quad \text{bit/s}, \tag{7.48}$$

where T is the length of the cyclic prefix,

$$q_{\nu,k}^{\text{opt}} = \max\left(\mu - \frac{N_0}{s_{\nu,k}^2}, 0\right), \quad \nu = 0, \ldots, S-1, \quad k = 1, \ldots, r_\nu, \tag{7.49}$$

and the variable μ is selected to make $\sum_{\nu=0}^{S-1} \sum_{k=1}^{r_\nu} q_{\nu,k}^{\text{opt}} = qS$.

The capacity is achieved by the input distribution $\bar{\chi}[\nu] \sim \mathcal{N}_{\mathbb{C}}(\mathbf{0}, \mathbf{V}_\nu \mathbf{Q}_\nu^{\text{opt}} \mathbf{V}_\nu^{\text{H}})$, where $\mathbf{Q}_\nu^{\text{opt}} = \text{diag}(q_{\nu,1}^{\text{opt}}, \ldots, q_{\nu,r_\nu}^{\text{opt}}, 0, \ldots, 0)$ is a $K \times K$ diagonal matrix and $\mathbf{V}_\nu$ contains the ordered right singular vectors of $\bar{\mathbf{H}}[\nu]$.

If we would instead use an arbitrary precoding matrix $\mathbf{P}_\nu$ and diagonal power allocation matrix $\mathbf{Q}_\nu$ on subcarrier ν, the resulting achievable rate can be expressed similarly to (3.106) as

$$C = \frac{B}{T+S} \sum_{\nu=0}^{S-1} \log_2\left(\det\left(\mathbf{I}_M + \frac{1}{N_0}\bar{\mathbf{H}}[\nu]\mathbf{P}_\nu\mathbf{Q}_\nu\mathbf{P}_\nu^{\text{H}}\bar{\mathbf{H}}^{\text{H}}[\nu]\right)\right) \quad \text{bit/s}. \tag{7.50}$$

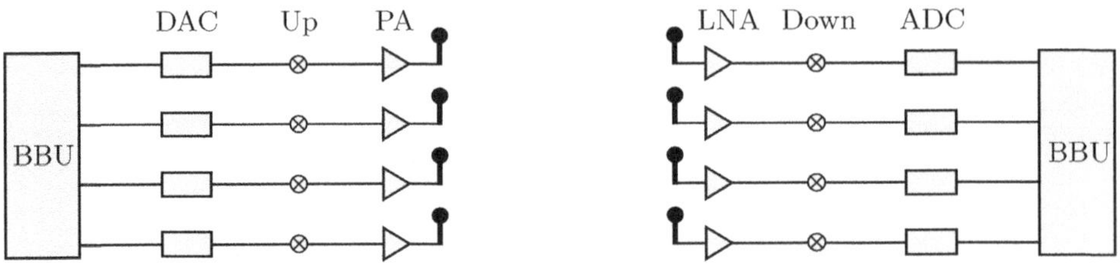

(a) Transmitter. (b) Receiver.

Figure 7.9: Block diagrams of the main components between the baseband unit and the antennas when using the digital beamforming architecture with $K = M = 4$ antennas.

7.2.1 Digital Beamforming Architecture

The theory and algorithms in this and previous chapters were developed using discrete-time complex baseband signals and channel models. There is a direct mapping between these models and the real continuous-time passband models used for practical communications, which was described in Section 2.3.1. In practice, this transformation is done by a sequence of hardware components at the transmitter and receiver. Figure 7.9 exemplifies the *digital beamforming architecture*, where each antenna has a dedicated chain of components between itself and the baseband processor [108]. This versatile architecture is capable of implementing all the features considered in this book.

At the transmitter, the discrete-time OFDM signal sequence is generated in the baseband unit (BBU) and then converted to an analog baseband signal using a digital-to-analog converter (DAC). The signal is then up-converted to the passband through multiplication with a sinusoidal carrier frequency signal generated by a local oscillator. The passband signal is then fed to a power amplifier (PA) that greatly increases the power before the signal reaches the antenna, which radiates it as an electromagnetic wave. Each antenna has a dedicated branch in Figure 7.9(a) with a DAC, up-converter, and PA.

The receiver performs similar processing but in the opposite order. The receive antenna converts the incoming wave into an electric current that is typically very weak and, therefore, fed to a low-noise amplifier (LNA) for immediate amplification. Next, the signal is down-converted to the baseband through multiplication with a sinusoidal carrier frequency signal and lowpass filtering. Finally, the signal is sampled using an analog-to-digital converter (ADC), and the output signal sequence reaches the BBU. Each antenna has a dedicated branch in Figure 7.9(b) with an LNA, down-converter, and ADC.

This is a high-level description of the digital beamforming architecture, which highlights the essential processing blocks. In practice, there are also bandpass filters next to the amplifiers to reject out-of-band distortion. Sometimes, the down-conversion is done in two stages: from the carrier to an intermediate frequency, where the bandpass filtering is done more conveniently and then converted to the baseband. The number of components of each kind is directly proportional to the number of antennas.

7.3 Clustered Multipath Propagation and Hybrid Beamforming

The clustered rich multipath propagation model was introduced in Section 5.6.1 to determine the MIMO channel matrix in an environment with N_{cl} clusters that scatter signals from the transmitter to the receiver. Cluster $i \in \{1, \ldots, N_{\mathrm{cl}}\}$ is located in the direction $(\varphi_{\mathrm{t},i}, \theta_{\mathrm{t},i})$ seen from the transmitter and in the direction $(\varphi_{\mathrm{r},i}, \theta_{\mathrm{r},i})$ seen from the receiver. In this section, we will extend this model to the OFDM case and explore what kind of hardware implementation is necessary to achieve the MIMO capacity.

We consider a ULA with K antennas at the transmitter and a ULA with M antennas at the receiver. The array response vectors are denoted as $\mathbf{a}_K(\varphi, \theta) \in \mathbb{C}^K$ and $\mathbf{a}_M(\varphi, \theta) \in \mathbb{C}^M$, respectively, and can be modeled as in (4.120). Each multipath cluster contains a large number of paths with varying delays, which are spread out so that the cluster might contribute to multiple channel taps. Hence, the channel matrix $\mathbf{H}[\ell] \in \mathbb{C}^{M \times K}$ at tap ℓ is modeled as

$$\mathbf{H}[\ell] = \sum_{i=1}^{N_{\mathrm{cl}}} c_i[\ell] \mathbf{a}_M(\varphi_{\mathrm{r},i}, \theta_{\mathrm{r},i}) \mathbf{a}_K^{\mathrm{T}}(\varphi_{\mathrm{t},i}, \theta_{\mathrm{t},i}), \quad \ell = 0, \ldots, T, \tag{7.51}$$

which is a generalization of (5.187) where the new channel coefficient $c_i[\ell] \sim \mathcal{N}_{\mathbb{C}}(0, \beta_i[\ell])$ depends on the tap index. The shape of the sequence $\beta_i[0], \ldots, \beta_i[T]$ of variances is known as the *power-delay profile* and is characterized by the cluster arrival time (i.e., the first tap index with a non-zero variance) and how the power decays with time as the propagation distance increases. The channel model in (7.51) is commonly considered in the analysis of mmWave channels [109], which typically contain fewer clusters than in the low-band and mid-band because of the greater penetration losses and negligible diffraction in those bands. We refer to [110], [111] for further motivations of this model, which is also appropriate for sub-THz bands [13].

Example 7.3. In the Saleh-Valenzuela model from [112], the power-delay profile is determined by the power-delay coefficients $\Gamma > 0$ and $\gamma > 0$ for the clusters and individual paths, respectively. Each cluster i is associated with a discrete arrival time t_i, and the variances of the respective channel coefficients are given by

$$\beta_i[\ell] = \begin{cases} 0, & \text{if } \ell \in \{0, \ldots, t_i - 1\}, \\ \beta_0 e^{-\ell/\Gamma} e^{-(\ell - t_i)/\gamma}, & \text{if } \ell \in \{t_i, \ldots, T\}. \end{cases} \tag{7.52}$$

The factor $\beta_0 e^{-\ell/\Gamma}$ describes how all channel gain coefficients decay exponentially with time since the waves spread out, while the factor $e^{-(\ell - t_i)/\gamma}$ determines how much weaker the slower paths in a cluster are compared to the quickest path. This model was originally proposed for SISO channels but is commonly used with the clustered MIMO channel model in (7.51).

If we substitute the time-domain channel matrices in (7.51) into (7.45), we obtain the channel matrices at each of the S OFDM subcarriers:

$$\bar{\mathbf{H}}[\nu] = \sum_{i=1}^{N_{\text{cl}}} \left(\sum_{\ell=0}^{T} c_i[\ell] e^{-\text{j}2\pi\ell\nu/S} \right) \mathbf{a}_M(\varphi_{\text{r},i}, \theta_{\text{r},i}) \mathbf{a}_K^{\text{T}}(\varphi_{\text{t},i}, \theta_{\text{t},i}), \quad \nu = 0, \dots, S-1. \tag{7.53}$$

Each matrix is a weighted sum of all clusters, with the weights being the S-length DFT of the sequence $c_i[0], \dots, c_i[T]$ of time-domain channel coefficients. The weights vary with the subcarrier index, which creates frequency-dependent fading variations depending on whether the channel coefficients superimpose constructively or destructively. However, the large-scale geometric channel properties, such as the number of clusters, their angular directions, and average strength, are the same for all subcarriers. These large-scale properties primarily determine the rank of the channel matrix. For example, the rank of each matrix in (7.53) is upper bounded by $\min(M, K, N_{\text{cl}})$, which is equal to N_{cl} when the clusters have well-separated angles from both the transmitter's and receiver's perspective (and $\min(M, K) \geq N_{\text{cl}}$); see Figure 5.34(c) for an example. In situations where the number of clusters is small compared to the number of antennas so that the channel matrix has at most rank N_{cl}, a simplified hardware architecture is sufficient to achieve the channel capacity. These practical aspects will be the focus of the remainder of this chapter.

Example 7.4. Suppose the channel coefficients are distributed as $c_i[\ell] \sim \mathcal{N}_{\mathbb{C}}(0, \beta_i[\ell])$ and independent across the clusters i and tap indices ℓ. What is the average squared Frobenius norm of the channel matrix in (7.53)? How does it depend on the subcarrier index?

The average squared Frobenius norm at subcarrier $\nu \in \{0, \dots, S-1\}$ is

$$\begin{aligned} \mathbb{E}\{\|\bar{\mathbf{H}}[\nu]\|_{\text{F}}^2\} &= \mathbb{E}\left\{ \left| \sum_{i=1}^{N_{\text{cl}}} \left(\sum_{\ell=0}^{T} c_i[\ell] e^{-\text{j}2\pi\ell\nu/S} \right) \right|^2 \right\} MK \\ &= MK \sum_{i=1}^{N_{\text{cl}}} \sum_{\ell=0}^{T} \mathbb{E}\left\{ |c_i[\ell]|^2 \right\} = MK \sum_{i=1}^{N_{\text{cl}}} \sum_{\ell=0}^{T} \beta_i[\ell], \end{aligned} \tag{7.54}$$

which is independent of the subcarrier index. The channel realizations will be different between subcarriers so that some are stronger than others momentarily, but all subcarriers are equally good statistically.

In practice, the channel coefficients of different clusters are independent since they involve different physical paths. However, the channel taps of a given cluster can be slightly correlated since the pulse functions have a non-zero effective time duration, so each physical path affects multiple taps.

7.3.1 One Dominant Cluster: Analog Beamforming is Sufficient

There are propagation scenarios where one of the clusters is significantly stronger than the others, for example, because it provides specular reflection while all other clusters provide diffuse scattering. Moreover, if there is a LOS path, it is typically much stronger than the scattered paths and can be modeled similarly to a cluster but with a deterministic $c_i[\ell]$ and a short power-delay profile.[5] In this section, we assume that $i = 1$ is the dominant cluster. The general subcarrier channel matrix in (7.53) can then be approximated as

$$\bar{\mathbf{H}}[\nu] \approx \left(\sum_{\ell=0}^{T} c_1[\ell] e^{-\mathrm{j}2\pi\ell\nu/S} \right) \mathbf{a}_M(\varphi_{\mathrm{r},1}, \theta_{\mathrm{r},1}) \mathbf{a}_K^{\mathrm{T}}(\varphi_{\mathrm{t},1}, \theta_{\mathrm{t},1}), \quad \nu = 0, \dots, S-1. \tag{7.55}$$

This is an approximately rank-one matrix with $\sqrt{MK} | \sum_{\ell=0}^{T} c_1[\ell] e^{-\mathrm{j}2\pi\ell\nu/S} |$ being the only non-zero singular value. This value varies with the subcarrier index, ν, but the eigenvectors remain the same. At every subcarrier, it is optimal for the transmitter to apply MRT with $\mathbf{p}_1 = \mathbf{a}_K^*(\varphi_{\mathrm{t},1}, \theta_{\mathrm{t},1})/\sqrt{K}$ and for the receiver to use MRC with $\mathbf{w}_1 = \mathbf{a}_M(\varphi_{\mathrm{r},1}, \theta_{\mathrm{r},1})/\sqrt{M}$. The resulting effective SISO channel on subcarrier ν is

$$\begin{aligned} \mathbf{w}_1^{\mathrm{H}} \bar{\mathbf{H}}[\nu] \mathbf{p}_1 &\approx \left(\sum_{\ell=0}^{T} c_1[\ell] e^{-\mathrm{j}2\pi\ell\nu/S} \right) \frac{\|\mathbf{a}_M(\varphi_{\mathrm{r},1}, \theta_{\mathrm{r},1})\|^2}{\sqrt{M}} \frac{\|\mathbf{a}_K(\varphi_{\mathrm{t},1}, \theta_{\mathrm{t},1})\|^2}{\sqrt{K}} \\ &= \left(\sum_{\ell=0}^{T} c_1[\ell] e^{-\mathrm{j}2\pi\ell\nu/S} \right) \sqrt{MK}. \end{aligned} \tag{7.56}$$

Thanks to MRT and MRC, the amplitude of each channel tap is increased by a factor $\sqrt{MK}$; thus, the maximum beamforming gain of MK is achieved in this setup. Since we only transmit one signal per subcarrier and use the same precoding/combining vectors on all subcarriers, we can implement the transmitter and receiver using a simpler architecture than the digital beamforming architecture illustrated in Figure 7.9.

The simplified *analog beamforming architecture* is shown in Figure 7.10. The transmitter only generates one OFDM signal sequence in the BBU and uses a DAC and up-converter to transform it into an analog passband signal centered at the carrier frequency. This signal is then divided into K branches, one per antenna. Each branch contains a phase shifter (PS) and a PA, which is sufficient to implement the multiplication with the phase-shift and amplitude of one of the entries in $\mathbf{p}_1$. This is called analog beamforming (or a phased array) because the beamforming operation is implemented in the analog part of the transmitter, in contrast to the digital baseband as in the digital beamforming architecture. There are multiple ways to implement PSs. If

[5]The addition of a LOS path to the clustered multipath model was previously considered in Example 5.18. Although there is only a single LOS path, it can contribute to multiple taps since the sinc-pulse has a long time duration.

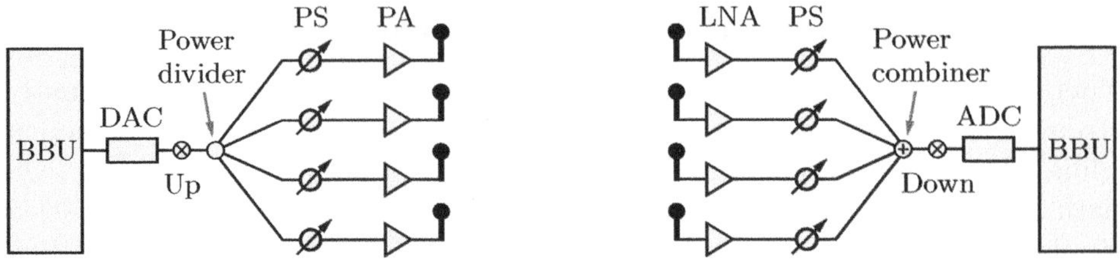

(a) Transmitter. (b) Receiver.

Figure 7.10: Block diagrams of the main components between the baseband unit and the antennas when using the analog beamforming architecture with $K = M = 4$ antennas. The phase shifters at the transmitter are used to implement a precoding vector common to all subcarriers, while the phase shifters at the receiver implement a combining vector common to all subcarriers.

a few predefined phase-shift values are sufficient to choose between (which restricts the selection of $\mathbf{p}_1$), then the circuit can contain transmission lines of different lengths, each causing a propagation delay that matches one of those phase-shifts. This kind of digital PS circuit controls the phase by switching between which transmission line the signal propagates through. There are also analog PS circuits that can control the phase continuously, for example, by controlling a voltage that determines which phase-shift the circuit imposes on the signal. Each PS causes a relative power loss of a few dBs when shifting the phase, called an insertion loss. There are similar losses in the power divider. Hence, to minimize the total power dissipation in the transmitter (and the need for cooling), the signals are not amplified until right before the antennas, which is why the PAs are placed after the PSs in Figure 7.10. In principle, the PAs could operate at different powers, but this feature is not needed in the LOS and single-cluster scenarios that analog beamforming is meant for.

The receiver carries out nearly the same operations but in the opposite order. The real passband signal received at a specific antenna is amplified by an LNA and then phase-shifted using a PS unit. The M phase-shifted received signals are then added to obtain a combined signal that is down-converted and sampled by an ADC. The LNA is placed before the PS since the received signals to the antennas are typically much weaker than the minimum input power that a PS can handle. Since the received signal power can vary by many tens of dB depending on the propagation conditions, the amplification level in the LNA must be dynamically adjusted to maintain an almost constant output power. This feature is called *automatic gain control* and is implemented as a feedback loop between the amplifier output and the regulating circuit. This is done in both analog and digital architectures.

The analog beamforming architecture is tailored to propagation scenarios with a single dominant cluster (or LOS path). We can thereby reduce the number of converters (i.e., DAC, ADC, up/down converters) since all antennas share these, but it comes at the expense of adding PSs and power dividers/combiners. We can obtain beamforming gains and spatial diversity gains in this way but no multiplexing gains. If the channel contains more than one strong cluster, then the achievable beamforming gain is less than what can be reached using digital beamforming, and the achievable rate can be much below the capacity since spatial multiplexing cannot be used.

Figure 7.11 exemplifies the effective channel gain that can be achieved on different subcarriers by beamforming in different angular directions. There are $K = 5$ transmit antennas and $S = 200$ subcarriers. The channel gains are normalized so that the maximum is 0 dB. There is $N_{\text{cl}} = 1$ cluster in Figure 7.11(a), and it spans one channel tap and is seen from the azimuth angle $0°$. The channel gain is the same on all subcarriers, and the beam pattern with its main beam and side-lobes is seen over the angles. This is a situation where analog beamforming is capacity-achieving.

Figure 7.11(b) considers a case with $N_{\text{cl}} = 3$ clusters, which are located in the angular directions $0°, 25°, -35°$. The clusters have equally strong channel gains but different time delays, so they appear in three different channel taps. Hence, the clusters interact to create channel variations between the subcarriers, known as *frequency-selective fading.* There is no angular direction that simultaneously maximizes the channel gain on all subcarriers, but we will have to vary the precoding over the subcarriers and utilize MRT. Analog beamforming cannot achieve the maximum beamforming gain in this case and can also not utilize the three clusters for spatial multiplexing. The next section will determine a simplified hardware architecture tailored to the case with $\min(M, K) > N_{\text{cl}} > 1$.

The precoding and combining are applied to the time-domain passband signals when using analog beamforming, while it is done in the frequency-domain in digital beamforming. The equivalence between these implementations can be established mathematically. We begin by taking the IDFT of the MIMO-OFDM received signal in (7.44), which is given at time instance l as

$$
\begin{aligned}
\mathbf{y}[l] &= \frac{1}{\sqrt{S}} \sum_{\nu=0}^{S-1} \bar{\mathbf{y}}[\nu] e^{\mathrm{j}2\pi l\nu/S} \\
&= \frac{1}{\sqrt{S}} \sum_{\nu=0}^{S-1} \bar{\mathbf{H}}[\nu] \bar{\boldsymbol{\chi}}[\nu] e^{\mathrm{j}2\pi l\nu/S} + \underbrace{\frac{1}{\sqrt{S}} \sum_{\nu=0}^{S-1} \bar{\mathbf{n}}[\nu] e^{\mathrm{j}2\pi l\nu/S}}_{=\mathbf{n}[l]} \\
&= \sum_{\ell=0}^{T} \mathbf{H}[\ell] \boldsymbol{\chi}[(l-\ell)_{\operatorname{mod} S}] + \mathbf{n}[l], \quad l = 0, \dots, S-1,
\end{aligned}
\tag{7.57}
$$

where we used the cyclic convolution theorem from Lemma 2.15. By inserting the time-domain channel $\mathbf{H}[\ell]$ from (7.51) with $N_{\text{cl}} = 1$ cluster into (7.57),

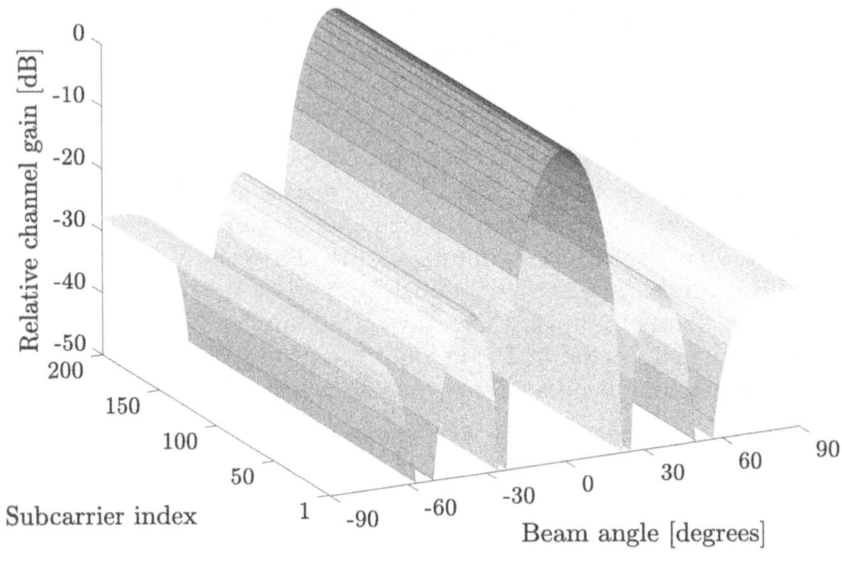

(a) One multipath cluster appearing in one tap and seen in the direction 0°.

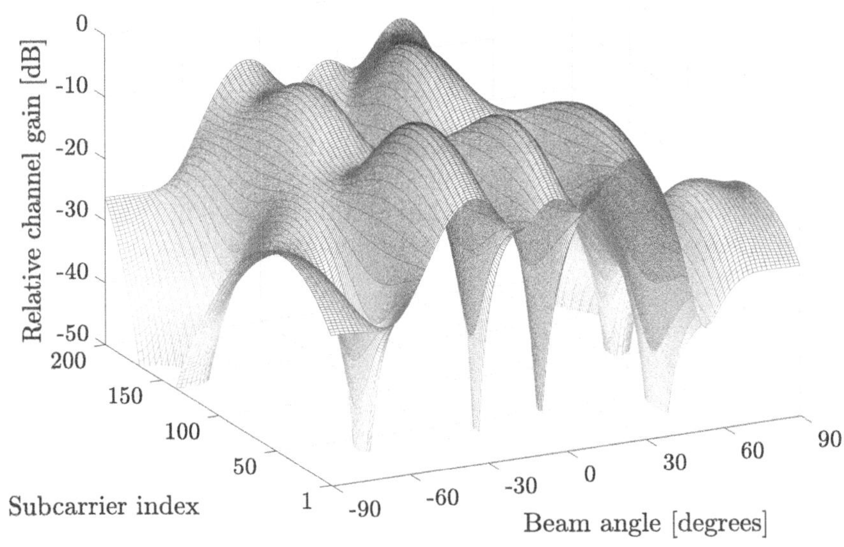

(b) Three multipath clusters appearing in different taps and seen in the directions $0^\circ, 25^\circ, -35^\circ$.

Figure 7.11: The effective channel gain (i.e., the squared norm of the inner product between the channel and precoding vector) can vary with the subcarrier index and beam angle (assuming that the precoding vector is an array response vector). The variations' size depends on the number of multipath clusters and their respective time delays. There are $K = 5$ antennas and $S = 200$ subcarriers, but different numbers of clusters in (a) and (b).

we obtain

$$\mathbf{y}[l] = \sum_{\ell=0}^{T} c_1[\ell]\mathbf{a}_M(\varphi_{\mathrm{r},1}, \theta_{\mathrm{r},1})\mathbf{a}_K^{\mathrm{T}}(\varphi_{\mathrm{t},1}, \theta_{\mathrm{t},1})\boldsymbol{\chi}[(l-\ell)_{\mathrm{mod}\, S}] + \mathbf{n}[l]. \tag{7.58}$$

At every time instance $l \in \{0, \dots, S-1\}$, we maximize the received power by applying the precoding vector $\mathbf{p}_1 = \mathbf{a}_K^*(\varphi_{\mathrm{t},1}, \theta_{\mathrm{t},1})/\sqrt{K}$ and combining vector $\mathbf{w}_1 = \mathbf{a}_M(\varphi_{\mathrm{r},1}, \theta_{\mathrm{r},1})/\sqrt{M}$. By writing the transmitted signal with time-domain precoding as $\boldsymbol{\chi}[l] = \mathbf{p}_1\chi[l]$, and applying receive combining to $\mathbf{y}[l]$ as $y[l] = \mathbf{w}_1^{\mathrm{H}}\mathbf{y}[l]$, we obtain the equivalent SISO-OFDM system

$$y[l] = \sum_{\ell=0}^{T} c_1[\ell]\sqrt{MK}\chi[(l-\ell)_{\mathrm{mod}\, S}] + n[l], \tag{7.59}$$

where $n[l] = \mathbf{w}_1^{\mathrm{H}}\mathbf{n}[l]$ is the noise. Every tap is scaled by a factor $\sqrt{MK}$, precisely as in (7.56) that was derived based on frequency-domain precoding/combining. Hence, the frequency-domain representation of the effective SISO channel in (7.59) is the same as in (7.56). We stress that time-domain precoding/combining leads to using the same precoding/combining vectors on all subcarriers, so the equivalence only holds when this is our goal (i.e., when having one dominant cluster).[6]

7.3.2 A Few Dominant Clusters: Hybrid Beamforming is Sufficient

A more general scenario where the hardware architecture can also be simplified is when the number of clusters N_{cl} is any number smaller than $\min(M, K)$. In this case, the rank of the channel matrices in (7.53) equals N_{cl} (or could possibly be even smaller), and this is the maximum number of parallel data streams that need to be transmitted and received per subcarrier. We can express the channel matrix on subcarrier ν as

$$\bar{\mathbf{H}}[\nu] = \mathbf{A}_{\mathrm{r}}\mathbf{D}[\nu]\mathbf{A}_{\mathrm{t}}^{\mathrm{H}} \tag{7.60}$$

by using the matrix notation

$$\mathbf{A}_{\mathrm{r}} = \left[\mathbf{a}_M(\varphi_{\mathrm{r},1}, \theta_{\mathrm{r},1}) \quad \dots \quad \mathbf{a}_M(\varphi_{\mathrm{r},N_{\mathrm{cl}}}, \theta_{\mathrm{r},N_{\mathrm{cl}}})\right] \in \mathbb{C}^{M \times N_{\mathrm{cl}}}, \tag{7.61}$$

$$\mathbf{D}[\nu] = \mathrm{diag}\left(\sum_{\ell=0}^{T} c_1[\ell]e^{-\mathrm{j}2\pi\ell\nu/S}, \dots, \sum_{\ell=0}^{T} c_{N_{\mathrm{cl}}}[\ell]e^{-\mathrm{j}2\pi\ell\nu/S}\right) \in \mathbb{C}^{N_{\mathrm{cl}} \times N_{\mathrm{cl}}}, \tag{7.62}$$

$$\mathbf{A}_{\mathrm{t}} = \left[\mathbf{a}_K^*(\varphi_{\mathrm{t},1}, \theta_{\mathrm{t},1}) \quad \dots \quad \mathbf{a}_K^*(\varphi_{\mathrm{t},N_{\mathrm{cl}}}, \theta_{\mathrm{t},N_{\mathrm{cl}}})\right] \in \mathbb{C}^{K \times N_{\mathrm{cl}}}. \tag{7.63}$$

The $N_{\mathrm{cl}} \times N_{\mathrm{cl}}$ diagonal matrix $\mathbf{D}[\nu]$ varies between the subcarriers. In contrast, the matrices $\mathbf{A}_{\mathrm{t}}, \mathbf{A}_{\mathrm{r}}$ with array response vectors remain the same since these

[6]In principle, any linear frequency-domain precoding/combining that is constant over an OFDM symbol can alternatively be implemented using time-domain filters, but it generally requires more complex impulse responses than what can be implemented using PSs.

describe the cluster geometry. When $N_{\text{cl}} < \min(M, K)$, these matrices expand the channel dimension from $N_{\text{cl}} \times N_{\text{cl}}$ in $\mathbf{D}[\nu]$ to $M \times K$ in $\bar{\mathbf{H}}[\nu]$.

If the transmitter uses a precoding vector that is not within the span of $\mathbf{A}_{\text{t}}$ (i.e., it cannot be written as a linear combination of its columns), then the unspanned component cannot reach the receiver. This would be a waste of signal power; thus, any transmit precoding matrix $\mathbf{P}[\nu]$ of practical interest on subcarrier ν can be expressed

$$\mathbf{P}[\nu] = \mathbf{A}_{\text{t}}\mathbf{P}_{\text{BB}}[\nu], \tag{7.64}$$

where $\mathbf{P}_{\text{BB}}[\nu] \in \mathbb{C}^{N_{\text{cl}} \times N_{\text{cl}}}$ is the subcarrier-unique part with a dimension that matches with the channel's rank. The subscript indicates that this part of the precoding matrix must be generated in the baseband (BB) before the IDFT is used to generate the time-domain OFDM signal sequence.

Similarly, if the receiver uses a receive combining vector $\mathbf{w}$ that is not within the span of $\mathbf{A}_{\text{r}}$, it will try to extract signals from dimensions in $\mathbb{C}^M$ where the channel matrix can never place any signal components. Hence, any combining matrix $\mathbf{W}[\nu]$ of practical interest on subcarrier ν can be expressed

$$\mathbf{W}[\nu] = \mathbf{A}_{\text{r}}\mathbf{W}_{\text{BB}}[\nu], \tag{7.65}$$

where $\mathbf{W}_{\text{BB}}[\nu] \in \mathbb{C}^{N_{\text{cl}} \times N_{\text{cl}}}$ is the subcarrier-unique part of reduced dimension.

Suppose we generate the transmitted signal on subcarrier ν as $\bar{\boldsymbol{\chi}}[\nu] = \mathbf{P}[\nu]\bar{\bar{\boldsymbol{\chi}}}[\nu]$, where $\bar{\bar{\boldsymbol{\chi}}}[\nu] \in \mathbb{C}^{N_{\text{cl}}}$ is the data signal. By applying the combining matrix in (7.65) to the received signal in (7.47), we obtain

$$\mathbf{W}^{\text{H}}[\nu]\bar{\mathbf{y}}[\nu] = \mathbf{W}_{\text{BB}}^{\text{H}}[\nu] \underbrace{\mathbf{A}_{\text{r}}^{\text{H}}\bar{\mathbf{H}}[\nu]\mathbf{A}_{\text{t}}}_{\text{Analog domain}} \mathbf{P}_{\text{BB}}[\nu]\bar{\bar{\boldsymbol{\chi}}}[\nu] + \underbrace{\mathbf{W}_{\text{BB}}^{\text{H}}[\nu]\mathbf{A}_{\text{r}}^{\text{H}}\bar{\mathbf{n}}[\nu]}_{\text{Effective noise}}. \tag{7.66}$$

The pre-processing by $\mathbf{A}_{\text{t}}$ at the transmitter and post-processing by $\mathbf{A}_{\text{r}}$ can be implemented in the analog domain in the transceiver hardware since these matrices are common to all subcarriers. Since these matrices contain array response vectors, each entry represents a phase-shift that can be implemented using a PS, as in the previous case of analog beamforming.

Figure 7.12 illustrates a possible hardware architecture for the case of $M = K = 4$ and $N_{\text{cl}} = 2$. The transmitter generates $N_{\text{cl}} = 2$ OFDM signals in the BBU, each representing one of the entries of $\mathbf{P}_{\text{BB}}[\nu]\bar{\bar{\boldsymbol{\chi}}}[\nu] \in \mathbb{C}^2$ for $\nu = 0, \ldots, S-1$. Each OFDM signal is transformed into an analog passband signal using a DAC and up-converter, and then multiplied with the respective columns of $\mathbf{A}_{\text{t}} \in \mathbb{C}^{4\times 2}$ by using the upper and lower collection of PSs, respectively. Each collection contains $K = 4$ PSs. The phase-shifted signals are then sent to the respective antennas, where they are added up before being amplified and radiated.

The opposite procedure is carried out at the receiver side, where the received signal at any given antenna is first amplified by an LNA. The signal is then divided into two parts that are sent to different sets of PSs, representing

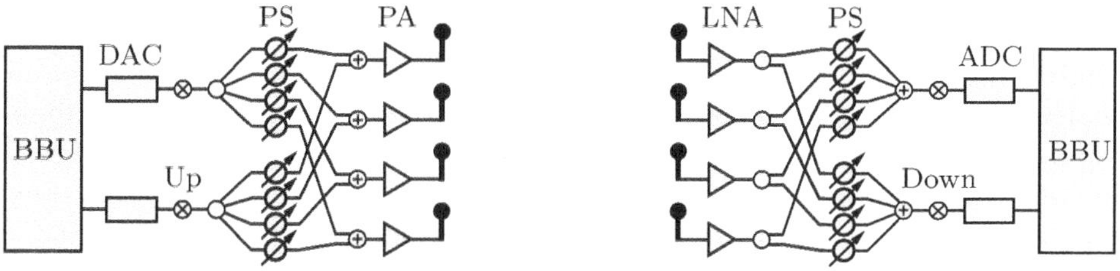

(a) Transmitter. (b) Receiver.

Figure 7.12: Block diagrams of the main components between the baseband unit and the antennas when using the hybrid beamforming architecture with $K = M = 4$ antennas. The power dividers and combiners are illustrated with circles as in Figure 7.10, but not labeled with text to avoid clutter.

the different rows of $\mathbf{A}_{\mathrm{r}}^{\mathrm{H}}$. The signals are then added up within each branch, down-converted, and sampled to obtain two output signals in the BBU. The effective noise in (7.66) has the covariance matrix

$$N_0 \mathbf{W}_{\mathrm{BB}}^{\mathrm{H}}[\nu]\mathbf{A}_{\mathrm{r}}^{\mathrm{H}}\mathbf{A}_{\mathrm{r}}\mathbf{W}_{\mathrm{BB}}[\nu], \tag{7.67}$$

which should preferably equal $N_0\mathbf{I}_{N_{\mathrm{cl}}}$ so the noise is white. This condition is satisfied by combining matrices of the kind $\mathbf{W}_{\mathrm{BB}}[\nu] = (\mathbf{A}_{\mathrm{r}}^{\mathrm{H}}\mathbf{A}_{\mathrm{r}})^{-1/2}\mathbf{U}_{\mathrm{BB}}[\nu]$, where $\mathbf{U}_{\mathrm{BB}}[\nu] \in \mathbb{C}^{N_{\mathrm{cl}} \times N_{\mathrm{cl}}}$ can be any unitary matrix.

We have described an instance of the *hybrid analog-digital beamforming architecture* [72], [113]. The name indicates that the precoding and combining operations are divided between the analog and digital domains. One could view this as a generalized architecture since $N_{\mathrm{cl}} = 1$ results in analog beamforming and $N_{\mathrm{cl}} = M = K$ is equivalent to digital beamforming. However, the reality is more complicated because N_{cl} is a variable that changes with the propagation environment where the transmitter/receiver is utilized while the hardware architecture must remain fixed. Hence, it is more suitable to decouple these variables and let N_{RF} denote the number of radio-frequency (RF) signals generated in the transmitter's BBU and sampled at the receiver; that is, the number of DACs/ADCs and up/down converters. The hybrid architecture is sufficient to achieve the MIMO capacity if $N_{\mathrm{cl}} \leq N_{\mathrm{RF}}$. The following table summarizes the number of hardware components needed in the transmitter (or receiver by replacing K with M):

Component	Digital	Hybrid	Analog
Converters	K	N_{RF}	1
Phase shifters	0	$N_{\mathrm{RF}}K$	K
Power amplifiers	K	K	K

The choice between these architectures requires making tradeoffs. The number of converters (i.e., ADC/DAC, up/down) can be reduced by going from a digital to a hybrid or analog architecture, but at the expense of requiring PSs and being unable to achieve the capacity when $N_{\text{cl}} > N_{\text{RF}}$. Digital beamforming is predominant in systems operating in the low-band and mid-band, while analog/hybrid beamforming is common in the high-band. A general trend seems to be that the frequency range for which digital architectures are practically feasible gradually increases. However, this does not change the fact that some propagation environments (e.g., LOS-dominant scenarios) do not require the extra capabilities the digital architecture provides.

When using an arbitrary value of N_{RF}, we can denote the hybrid precoding matrix on subcarrier ν as $\mathbf{P}[\nu] = \mathbf{P}_{\text{RF}}\mathbf{P}_{\text{BB}}[\nu]$, where $\mathbf{P}_{\text{RF}} \in \mathbb{C}^{K \times N_{\text{RF}}}$ is the analog part and the subcarrier-specific digital part is $\mathbf{P}_{\text{BB}}[\nu] \in \mathbb{C}^{N_{\text{RF}} \times N_{\text{RF}}}$. The latter part can be further factorized as $\mathbf{P}_{\text{BB}}[\nu] = (\mathbf{P}_{\text{RF}}^{\text{H}}\mathbf{P}_{\text{RF}})^{-1/2}\mathbf{V}_{\text{BB}}[\nu]$ where $\mathbf{V}_{\text{BB}}[\nu] \in \mathbb{C}^{N_{\text{RF}} \times N_{\text{RF}}}$ is the effective precoding matrix that the transmitter can freely select because it has the same power as $\mathbf{P}[\nu]$:

$$\|\mathbf{P}[\nu]\|_{\text{F}}^2 = \|\mathbf{V}_{\text{BB}}[\nu]\|_{\text{F}}^2, \tag{7.68}$$

where $\|\cdot\|_{\text{F}}$ denotes the Frobenius norm defined in (5.87).

Similarly, the combining matrix is denoted as $\mathbf{W}[\nu] = \mathbf{W}_{\text{RF}}\mathbf{W}_{\text{BB}}[\nu]$, where $\mathbf{W}_{\text{RF}} \in \mathbb{C}^{M \times N_{\text{RF}}}$ is the analog part and $\mathbf{W}_{\text{BB}}[\nu] = (\mathbf{W}_{\text{RF}}^{\text{H}}\mathbf{W}_{\text{RF}})^{-1/2}\mathbf{U}_{\text{BB}}[\nu] \in \mathbb{C}^{N_{\text{RF}} \times N_{\text{RF}}}$ is the digital part with $\mathbf{U}_{\text{BB}}[\nu]$ being a unitary matrix.

Using this notation, the received signal in (7.66) can be reformulated as

$$\mathbf{W}^{\text{H}}[\nu]\bar{\mathbf{y}}[\nu] = \mathbf{U}_{\text{BB}}^{\text{H}}[\nu]\breve{\mathbf{H}}[\nu]\mathbf{V}_{\text{BB}}[\nu]\bar{\bar{\boldsymbol{\chi}}}[\nu] + \underbrace{\mathbf{W}_{\text{BB}}^{\text{H}}[\nu]\mathbf{W}_{\text{RF}}^{\text{H}}\bar{\mathbf{n}}[\nu]}_{\sim \mathcal{N}_{\mathbb{C}}(\mathbf{0}, N_0\mathbf{I}_{N_{\text{RF}}})}, \tag{7.69}$$

where the data signal is $\bar{\bar{\boldsymbol{\chi}}}[\nu] \sim \mathcal{N}_{\mathbb{C}}(\mathbf{0}, \mathbf{Q}[\nu])$ and $\mathbf{Q}[\nu]$ is the diagonal matrix with power coefficients. The effective channel matrix is denoted by $\breve{\mathbf{H}}[\nu] \in \mathbb{C}^{N_{\text{RF}} \times N_{\text{RF}}}$ and defined as

$$\breve{\mathbf{H}}[\nu] = (\mathbf{W}_{\text{RF}}^{\text{H}}\mathbf{W}_{\text{RF}})^{-1/2}\mathbf{W}_{\text{RF}}^{\text{H}}\bar{\mathbf{H}}[\nu]\mathbf{P}_{\text{RF}}(\mathbf{P}_{\text{RF}}^{\text{H}}\mathbf{P}_{\text{RF}})^{-1/2}. \tag{7.70}$$

It follows from (3.106) that an achievable rate (in bit per subcarrier symbol) is

$$\log_2\left(\det\left(\mathbf{I}_{N_{\text{RF}}} + \frac{1}{N_0}\breve{\mathbf{H}}[\nu]\mathbf{V}_{\text{BB}}[\nu]\mathbf{Q}[\nu]\mathbf{V}_{\text{BB}}^{\text{H}}[\nu]\breve{\mathbf{H}}^{\text{H}}[\nu]\right)\right). \tag{7.71}$$

This rate can be maximized by selecting $\mathbf{V}_{\text{BB}}[\nu]$ as the right singular vectors of $\breve{\mathbf{H}}[\nu]$ and $\mathbf{Q}[\nu]$ according to water-filling power allocation.

The expression in (7.70) demonstrates how the analog parts of the precoding and combining matrices transform the $M \times K$ channel matrix $\bar{\mathbf{H}}[\nu]$ into an effective channel matrix $\breve{\mathbf{H}}[\nu]$ with the reduced dimensions $N_{\text{RF}} \times N_{\text{RF}}$. This limits the maximum multiplexing gain to N_{RF} and reduces the maximum achievable rate if the original channel matrix had a higher rank.

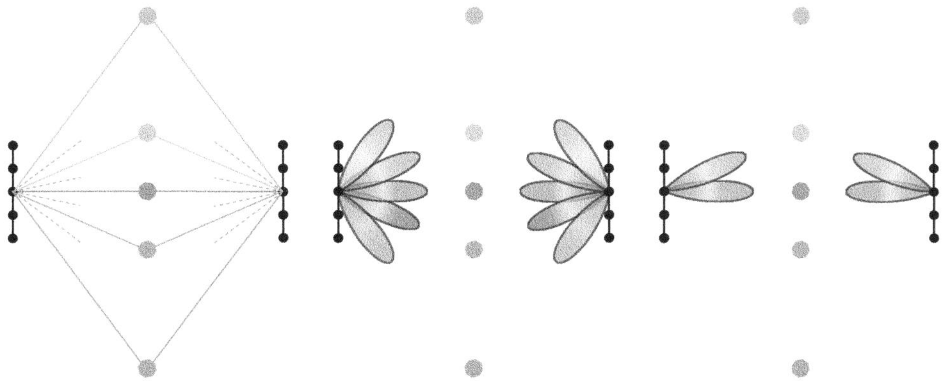

(a) Multipath clusters. (b) Digital beamforming. (c) Hybrid beamforming.

Figure 7.13: A sparse multipath propagation scenario with $M = K = 5$ antennas and $N_{\text{cl}} = 5$ clusters distributed in angle to match perfectly with the beamspace representation, as illustrated in (a). The transmitter is to the right, the receiver to the left, and the dotted lines show the boundaries between the angular intervals considered in the beamspace representation. The digital beamforming architecture achieves the MIMO capacity by transmitting/receiving one beam per cluster as in (b). The hybrid beamforming architecture can only transmit as many beams as there are RF inputs/outputs, which is $N_{\text{RF}} = 2$ in (c).

Example 7.5. Consider a hybrid ULA architecture with $M = K$ antennas and N_{RF} RF inputs/outputs. What rate can be achieved over the channel in (7.47) if each of $\mathbf{A}_{\text{t}}$ and $\mathbf{A}_{\text{r}}$ contains $N_{\text{cl}} \geq N_{\text{RF}}$ columns from the scaled DFT matrix $\sqrt{M}\mathbf{F}_M$, and $c_n[\ell] = \sqrt{\beta/N_{\text{cl}}}$ if $\ell = n$ and $c_n[\ell] = 0$ otherwise?

The columns of $\mathbf{A}_{\text{t}}$ are orthogonal, and the same holds for $\mathbf{A}_{\text{r}}$. Hence, we can only use $N_{\text{RF}} \leq N_{\text{cl}}$ clusters. Since all clusters are equally strong, we can select $\mathbf{P}_{\text{RF}}$ as the first N_{RF} columns of $\mathbf{A}_{\text{t}}$ and $\mathbf{W}_{\text{RF}}$ as the first N_{RF} columns of $\mathbf{A}_{\text{r}}$ without loss of optimality. As these columns originate from the DFT matrix, the channel can be represented in the beamspace, and the hybrid transmitter/receiver will point beams directly toward N_{RF} of the N_{cl} clusters, as illustrated in Figure 7.13(c) for $N_{\text{RF}} = 2$ and $N_{\text{cl}} = 5$. The effective channel matrix in (7.70) simplifies to

$$\breve{\bar{\mathbf{H}}}[\nu] = \sqrt{\frac{\beta}{N_{\text{cl}}}}\sqrt{MK}\text{diag}\left(e^{-\text{j}2\pi\nu/S}, \ldots, e^{-\text{j}2\pi N_{\text{RF}}\nu/S}\right) \tag{7.72}$$

since $\mathbf{P}_{\text{RF}}^{\text{H}}\mathbf{P}_{\text{RF}} = \mathbf{W}_{\text{RF}}^{\text{H}}\mathbf{W}_{\text{RF}} = M\mathbf{I}_{N_{\text{RF}}}$ and $\mathbf{W}_{\text{RF}}^{\text{H}}\mathbf{A}_{\text{r}} = \mathbf{P}_{\text{RF}}^{\text{H}}\mathbf{A}_{\text{t}} = [M\mathbf{I}_{N_{\text{RF}}}, \mathbf{0}]$. All N_{RF} singular values of $\breve{\bar{\mathbf{H}}}[\nu]$ equals $\sqrt{\beta/N_{\text{cl}}}M$, which turns water-filling into equal power allocation of q/N_{RF} and $\mathbf{V}_{\text{BB}}[\nu]\mathbf{Q}[\nu]\mathbf{V}_{\text{BB}}^{\text{H}}[\nu] = q/N_{\text{RF}}\mathbf{I}_{N_{\text{RF}}}$. Hence, the rate in (7.71) becomes

$$N_{\text{RF}}\log_2\left(1 + \frac{q\beta}{N_0}\frac{MK}{N_{\text{cl}}N_{\text{RF}}}\right) \quad \text{bit per subcarrier symbol.} \tag{7.73}$$

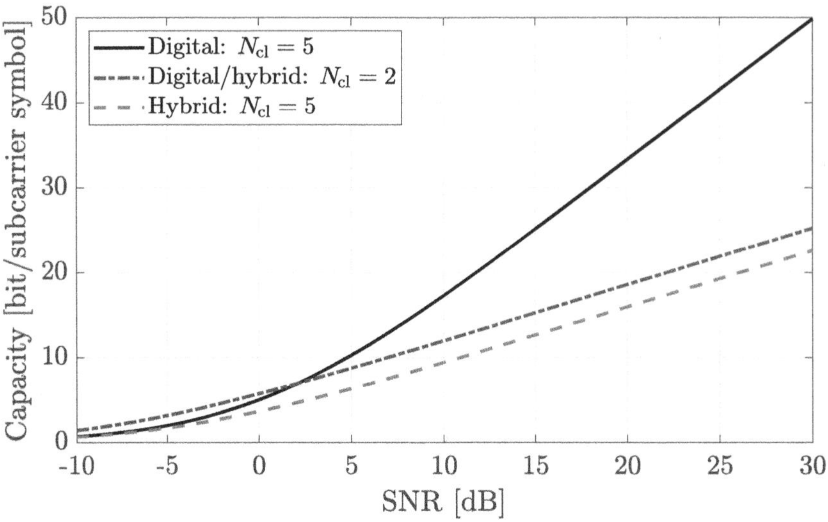

Figure 7.14: The capacity per OFDM subcarrier symbol achieved in a setup with $M = K = 5$ antennas and either $N_{\text{cl}} = 2$ or $N_{\text{cl}} = 5$ multipath clusters. The digital beamforming architecture is compared with the hybrid architecture with $N_{\text{RF}} = 2$. The architectures achieve different capacities when $N_{\text{cl}} > N_{\text{RF}}$.

The simplified rate expression in (7.73) for hybrid beamforming showcases the essential limitations of this architecture. The multiplexing gain is N_{RF}, even if the channel supports a larger multiplexing gain N_{cl}. The full beamforming gain MK is achieved but divided by N_{cl} since the total channel gain β is equally distributed between that many clusters. It is also divided by N_{RF} since the total transmit power is divided into that many pieces when performing spatial multiplexing. The channel capacity in the considered setup is $N_{\text{cl}} \log_2(1 + \text{SNR}MK/N_{\text{cl}}^2)$, where $\text{SNR} = q\beta/N_0$, and can be achieved using the digital beamforming architecture as shown in Figure 7.13(b). The main difference is the multiplexing gain that is different if $N_{\text{cl}} > N_{\text{RF}}$.

Figure 7.14 compares the rates achieved by the digital and hybrid architectures with $M = K = 5$. We consider the same setup as in Example 7.5 with $N_{\text{RF}} = 2$ and $N_{\text{cl}} \in \{2, 5\}$. The digital and hybrid architectures provide exactly the same rate when $N_{\text{cl}} = N_{\text{RF}} = 2$. By contrast, there is a large gap at high SNRs when $N_{\text{cl}} = 5$ because the hybrid architecture achieves a multiplexing gain of 2 instead of 5. This result confirms that hybrid beamforming is only a suitable alternative in propagation environments with a small number of clusters, not more than N_{RF}.

The selection of the analog precoding/combining matrices $\mathbf{P}_{\text{RF}}, \mathbf{W}_{\text{RF}}$ is easy when the clusters are located in orthogonal angular directions, as in Example 7.5 where they match with the DFT beam directions. This situation is unlikely to arise in practice, which makes the selection more challenging. For example, we might have $N_{\text{cl}} > N_{\text{RF}}$, but the clusters are unevenly distributed over the angles so that we can capture most of the signal power using N_{RF} carefully selected beam directions. There is a vast literature on this topic [114], where the main principle is that we want to retain as much as possible of the average channel gain over the subcarriers when applying analog precoding/combining. Specifically, it holds that

$$\frac{1}{S}\sum_{\nu=0}^{S-1} \|\mathbf{W}_{\mathrm{RF}}^{\mathrm{H}}\bar{\mathbf{H}}[\nu]\mathbf{P}_{\mathrm{RF}}\|_{\mathrm{F}}^{2} \leq \frac{1}{S}\sum_{\nu=0}^{S-1} \|\bar{\mathbf{H}}[\nu]\|_{\mathrm{F}}^{2} \tag{7.74}$$

since we lose some channel dimensions when using hybrid beamforming. Intuitively, we want to make the gap between the two expressions in (7.74) small. Hence, we want to maximize

$$\frac{1}{S}\sum_{\nu=0}^{S-1} \|\mathbf{W}_{\mathrm{RF}}^{\mathrm{H}}\bar{\mathbf{H}}[\nu]\mathbf{P}_{\mathrm{RF}}\|_{\mathrm{F}}^{2} = \frac{1}{S}\sum_{\nu=0}^{S-1} \mathrm{tr}(\mathbf{P}_{\mathrm{RF}}^{\mathrm{H}} \underbrace{\bar{\mathbf{H}}^{\mathrm{H}}[\nu]\mathbf{W}_{\mathrm{RF}}\mathbf{W}_{\mathrm{RF}}^{\mathrm{H}}\bar{\mathbf{H}}[\nu]}_{\approx \bar{\mathbf{H}}^{\mathrm{H}}[\nu]\bar{\mathbf{H}}[\nu]} \mathbf{P}_{\mathrm{RF}}) \tag{7.75}$$

$$\approx \mathrm{tr}\left(\mathbf{P}_{\mathrm{RF}}^{\mathrm{H}}\left(\frac{1}{S}\sum_{\nu=0}^{S-1} \bar{\mathbf{H}}^{\mathrm{H}}[\nu]\bar{\mathbf{H}}[\nu]\right)\mathbf{P}_{\mathrm{RF}}\right), \tag{7.76}$$

where the approximation is motivated by having a receiver that can capture all the signal power in the N_{RF}-dimensional subspace where the transmitted signals exist. Based on this approximation, it follows that the analog precoding matrix $\mathbf{P}_{\mathrm{RF}}$ should be selected based on the average channel matrix expression $\frac{1}{S}\sum_{\nu=0}^{S-1} \bar{\mathbf{H}}^{\mathrm{H}}[\nu]\bar{\mathbf{H}}[\nu] \in \mathbb{C}^{K\times K}$. More precisely, it should use the N_{RF} strongest dimensions of this matrix, which are spanned by the eigenvectors associated with its N_{RF} largest eigenvalues. Since these eigenvectors generally have entries with varying magnitudes that cannot be implemented using PSs, a transformation step is required; we refer to [115] for the precise details. When $\mathbf{P}_{\mathrm{RF}}$ has been selected, one can further argue that $\mathbf{W}_{\mathrm{RF}}$ should be selected to contain the eigenvectors corresponding to the N_{RF} strongest eigenvalues of $\frac{1}{S}\sum_{\nu=0}^{S-1} \bar{\mathbf{H}}[\nu]\mathbf{P}_{\mathrm{RF}}\mathbf{P}_{\mathrm{RF}}^{\mathrm{H}}\bar{\mathbf{H}}^{\mathrm{H}}[\nu]$, but this can also only be done approximately. When the analog precoding/combining matrices have been selected, the digital precoding/combining matrices are computed separately on each subcarrier based on the SVD of $\breve{\mathbf{H}}[\nu]$ in (7.70), and the water-filling power allocation is finally performed over all the subcarriers.

7.3.3 Beam-Squint Effect

In the clustered multipath propagation model, the channel taps in (7.51) depend on the array response vectors of the ULAs at the transmitter and receiver. The array response vector expression was initially derived in Section 4.2.1 under two assumptions: far-field propagation and frequency flatness. The latter condition can be invalidated when the bandwidth B is very large because the array response expression depends on the wavelength, and it varies with the frequency. This can lead to issues when using analog beamforming.

To analyze this phenomenon in detail, we revisit the array response expression in (4.120). We let $\lambda_\text{c} = c/f_\text{c}$ denote the wavelength at the carrier frequency and assume an antenna spacing of $\Delta = \lambda_\text{c}/2$. The array response vector for a signal with frequency f that arrives from the azimuth angle φ and elevation angle θ then becomes

$$\mathbf{a}_M(\varphi,\theta,f) = \begin{bmatrix} 1 \\ e^{-\mathrm{j}2\pi\frac{(\lambda_\text{c}/2)\sin(\varphi)\cos(\theta)}{c/f}} \\ \vdots \\ e^{-\mathrm{j}2\pi\frac{(M-1)(\lambda_\text{c}/2)\sin(\varphi)\cos(\theta)}{c/f}} \end{bmatrix} = \begin{bmatrix} 1 \\ e^{-\mathrm{j}\pi\frac{f}{f_\text{c}}\sin(\varphi)\cos(\theta)} \\ \vdots \\ e^{-\mathrm{j}\pi(M-1)\frac{f}{f_\text{c}}\sin(\varphi)\cos(\theta)} \end{bmatrix}. \tag{7.77}$$

This is the same expression as in (4.120) when $f = f_\text{c}$. When considering OFDM, we are interested in frequencies that can be expressed as $f = f_\text{c} + \frac{\nu B}{S}$ for subcarrier indices in the range $\nu \in [-S/2, S/2]$, where S is the number of subcarriers. By substituting this into (7.77), we obtain

$$\begin{aligned}\mathbf{a}_M\left(\varphi,\theta,f_\text{c}+\frac{\nu B}{S}\right) &= \begin{bmatrix} 1 \\ e^{-\mathrm{j}\pi\frac{f_\text{c}+\frac{\nu B}{S}}{f_\text{c}}\sin(\varphi)\cos(\theta)} \\ \vdots \\ e^{-\mathrm{j}\pi(M-1)\frac{f_\text{c}+\frac{\nu B}{S}}{f_\text{c}}\sin(\varphi)\cos(\theta)} \end{bmatrix} \\ &= \mathrm{diag}\left(b^0[\nu],\ldots,b^{M-1}[\nu]\right)\underbrace{\begin{bmatrix} 1 \\ e^{-\mathrm{j}\pi\sin(\varphi)\cos(\theta)} \\ \vdots \\ e^{-\mathrm{j}\pi(M-1)\sin(\varphi)\cos(\theta)} \end{bmatrix}}_{=\mathbf{a}_M(\varphi,\theta,f_\text{c})},\end{aligned} \tag{7.78}$$

where $b[\nu] = e^{-\mathrm{j}\pi\frac{\nu B}{S f_\text{c}}\sin(\varphi)\cos(\theta)}$. The last term is the conventional array response vector for a half-wavelength-spaced ULA, while the diagonal matrix shifts the phases of the entries depending on the subcarrier index.

The frequency-dependent array response affects the subcarrier channels in an OFDM system. For example, the $M \times K$ MIMO channel matrix on subcarrier ν in (7.53) must be revised as

$$\bar{\mathbf{H}}[\nu] = \sum_{i=1}^{N_\text{cl}}\left(\sum_{\ell=0}^{T} c_i[\ell]e^{-\mathrm{j}2\pi\ell\nu/S}\right)\mathbf{a}_M\left(\varphi_{\text{r},i},\theta_{\text{r},i},f_\text{c}+\frac{\nu B}{S}\right)\mathbf{a}_K^\mathrm{T}\left(\varphi_{\text{t},i},\theta_{\text{t},i},f_\text{c}+\frac{\nu B}{S}\right), \tag{7.79}$$

where the two array response vectors now depend on the subcarrier index. We recall that the beamwidth depends on the aperture length compared to the wavelength. Since the physical aperture length is constant in a practical array, the relative length varies over the signal bandwidth, so the beamwidth shrinks or grows. This is a minor issue when using the digital beamforming architecture because we can then adapt the precoding/combining to the channel conditions on each subcarrier. The frequency dependence is more problematic when using the analog beamforming architecture because it can lead to the *beam-squint* effect. To showcase the phenomenon, we consider a SIMO channel with $N_{\mathrm{cl}} = 1$ cluster where the channel vector on subcarrier ν is

$$\bar{\mathbf{h}}[\nu] = \left(\sum_{\ell=0}^{T} c[\ell] e^{-\mathrm{j}2\pi\ell\nu/S}\right) \mathbf{a}_M\left(\varphi, \theta, f_{\mathrm{c}} + \frac{\nu B}{S}\right). \tag{7.80}$$

Suppose the receiver is built using the analog beamforming architecture and applies the MRC vector $\mathbf{w} = \mathbf{a}_M(\varphi, \theta, f_{\mathrm{c}})/\sqrt{M}$ designed for the carrier frequency f_{c}. The effective channel on subcarrier ν becomes

$$\begin{aligned} \mathbf{w}^{\mathrm{H}} \bar{\mathbf{h}}[\nu] &= \left(\sum_{\ell=0}^{T} c[\ell] e^{-\mathrm{j}2\pi\ell\nu/S}\right) \frac{\mathbf{a}_M^{\mathrm{H}}(\varphi, \theta, f_{\mathrm{c}}) \mathrm{diag}(b^0[\nu], \ldots, b^{M-1}[\nu]) \mathbf{a}_M(\varphi, \theta, f_{\mathrm{c}})}{\sqrt{M}} \\ &= \left(\sum_{\ell=0}^{T} c[\ell] e^{-\mathrm{j}2\pi\ell\nu/S}\right) \frac{1}{\sqrt{M}} \sum_{m=1}^{M} b^{m-1}[\nu], \end{aligned} \tag{7.81}$$

where we utilized the expression in (7.78). The term in parenthesis is obtained also in the single-antenna case, while the rest is due to having multiple antennas. Hence, the beamforming gain on subcarrier ν is $|\frac{1}{\sqrt{M}} \sum_{m=1}^{M} b^{m-1}[\nu]|^2$ and becomes M if $b[\nu] = 1$, as we normally expect when using MRC. This property holds at the center subcarrier with $\nu = 0$ or when the transmitter is located in a direction where $\sin(\varphi)\cos(\theta) = 0$ (e.g., $\varphi = 0$). In other cases, we get

$$\begin{aligned} \left| \frac{1}{\sqrt{M}} \sum_{m=1}^{M} b^{m-1}[\nu] \right|^2 &= \frac{1}{M} \left| \sum_{m=1}^{M} e^{-\mathrm{j}\pi(m-1)\frac{\nu B}{S f_{\mathrm{c}}} \sin(\varphi)\cos(\theta)} \right|^2 \\ &= \frac{1}{M} \frac{\sin^2\left(M \frac{\nu B}{S f_{\mathrm{c}}} \frac{\pi \sin(\varphi)\cos(\theta)}{2}\right)}{\sin^2\left(\frac{\nu B}{S f_{\mathrm{c}}} \frac{\pi \sin(\varphi)\cos(\theta)}{2}\right)}, \end{aligned} \tag{7.82}$$

where the last equality follows in the same way as the beamwidth calculation in (4.52). This is generally a decreasing function of the magnitude $|\nu|$ of the subcarrier index but can sometimes oscillate. The function also depends on the number of antennas and ratio B/f_{c} of how large the bandwidth is compared to the carrier frequency.

Figure 7.15 shows the beamforming gain in (7.82) at different subcarriers with indices $\nu \in [-S/2, S/2]$, where S is the total number of subcarriers. We consider a setup with $M = 20$ antennas, $B/f_{\mathrm{c}} = 0.1$, and transmitters located in the azimuth plane in three directions: $\varphi \in \{0, \pi/6, \pi/3\}$. The beamforming gain is the same on all subcarriers if $\varphi = 0$, but for all other directions,

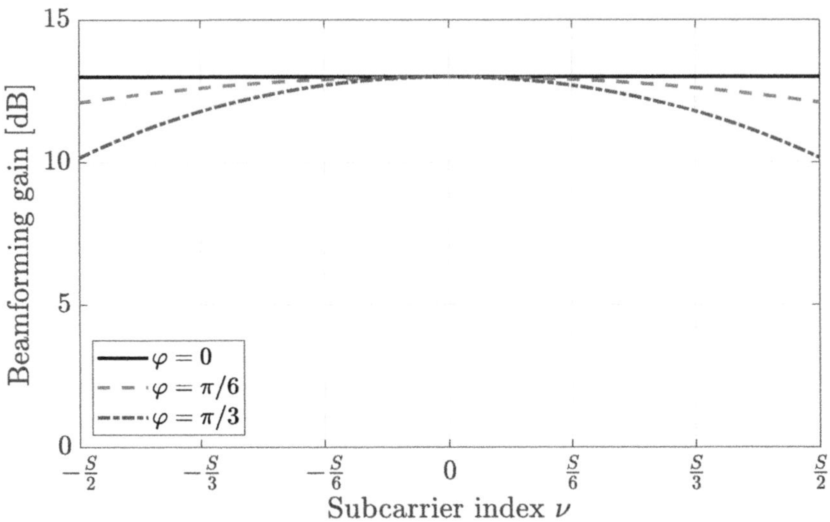

Figure 7.15: The beamforming gain in (7.82) achieved at different subcarriers in a setup with $M = 20$, $B/f_c = 0.1$, and $\theta = 0$.

the gain is reduced as $|\nu|$ increases. The reason is that MRC is supposed to compensate for the phase differences between adjacent antennas, and these vary substantially between subcarriers when the bandwidth is large compared to the carrier frequency. The figure shows that there can be several dB of gain losses at the edge of the band. The results shown in this figure could appear in the mid-band if $f_c = 3$ GHz and $B = 300$ MHz, or in the mmWave band if $f_c = 30$ GHz and $B = 3$ GHz. Those specific bandwidth values might be larger than what practical systems use; it is more typical to consider a third of it so that $B/f_c = 0.1/3$. In that case, the gain losses observed in the middle third of the figure (i.e., $\nu \in [-S/6, S/6]$) should be anticipated in analog beamforming systems.

The noun "squint" is used to describe a mismatch in the directions that a person's eyes are pointing. A similar directional mismatch causes beam-squint. Figure 7.16 shows the beamforming gain

$$\frac{1}{M}\left|\mathbf{a}_M^{\mathrm{H}}\left(\varphi, 0, f_c\right)\mathbf{a}_M\left(\pi/3, 0, f_c + \frac{\nu B}{S}\right)\right|^2 \tag{7.83}$$

obtained using MRC vectors with different observation angles φ when the true signal arrives from the angle $\pi/3$. We still consider $M = 20$ and $B/f_c = 0.1$. The beam pattern has its peak at the correct angle $\varphi = \pi/3$ at the center frequency ($\nu = 0$), but when we consider subcarriers further from the center, the pattern is shifted outwards; that is, the beam is not pointing in the direction we expect it to do. With analog beamforming, we would use $\varphi = \pi/3$ on all subcarriers, leading to the gain losses observed previously since the actual beam direction changes.

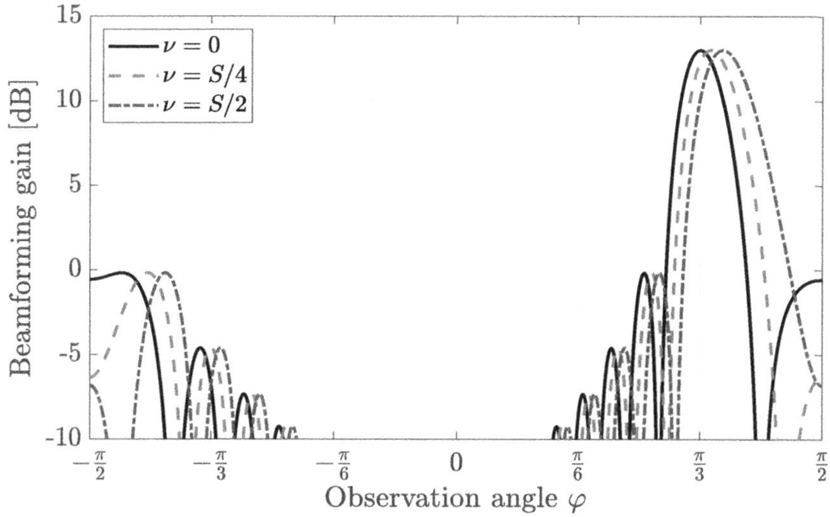

Figure 7.16: The beamforming gain in (7.83) obtained using MRC vectors with different angles φ. The pattern is shown for different subcarrier indices when the signal arrives from the azimuth angle $\pi/3$, $M = 20$, and $B/f_c = 0.1$.

The beam-squint effect is present in the analog beamforming architecture and limits how large bandwidths can be used effectively. When a signal arrives from the angle φ, the propagation delay d_m/c at receive antenna m is frequency-independent and only depends on the propagation distance d_m and speed of light c. However, the phase-shift $2\pi d_m/\lambda$ is not since the wavelength λ depends on the frequency. Conventional PSs nevertheless assign the same phase-shift to the entire signal band, giving rise to the described beam-squint. An implementation solution is to replace the PSs with more complex *true time delay (TTD)* units that assign the same delay to all frequencies; this alleviates the beam-squint effect when the signal is transmitted/received in a single direction. However, it does not address the general limitation of analog beamforming when it comes to multipath propagation.

7.4 Practical Implementations and Terminology

MIMO communication technology has existed for decades, but in the 5G era, it switched from being an optional high-end feature to becoming mainstream. It is utilized in both mid-band and mmWave deployments, at both base stations and user devices. In this section, we will take a look at two specific implementations, highlight some practical design characteristics, and shed light on a few ambiguities that exist in academic and industrial terminology.

Figure 7.17 shows a mmWave transmitter designed for the 28 GHz band. It consists of 16 single-polarized antenna elements, arranged on a 4×4 square grid. The horizontal and vertical element spacings are $\lambda/2 \approx 5.3$ mm, so this is a critically spaced array. Four RF inputs are visible at the bottom of the

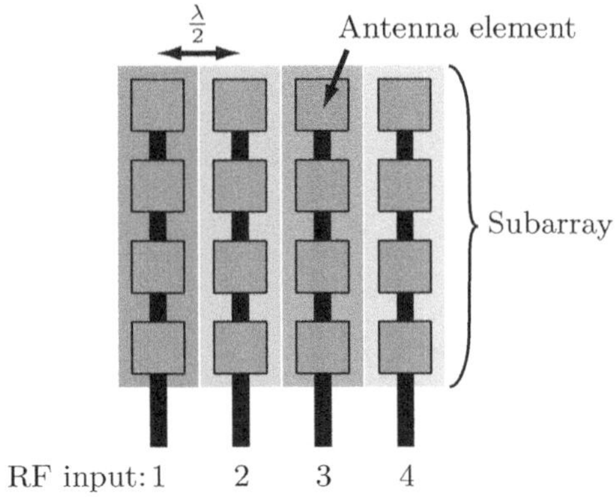

Figure 7.17: The photo shows the antenna array in the TMYTEK Developer Kit from 2022, which is designed for the 28 GHz band. It consists of 16 antenna elements, which are arranged into subarrays. Each column is a subarray that shares an RF input and, therefore, always transmits the same signal. Hence, from a MIMO communication perspective, this is a horizontal ULA with directive antennas, and it uses an analog beamforming architecture.

photo, and these are connected to the antenna elements so that elements in the same column are always sending the same signal. The set of elements that share the same RF input is called a *subarray* in the industry; however, from the perspective of this book, each column corresponds to a single antenna. Hence, this is a horizontal ULA with half-wavelength spacing, but with directive antennas implemented using subarrays consisting of a few elements. Each individual element has a 3 dBi directivity gain while each antenna has a $3 + 10\log_{10}(4) = 9$ dBi gain. The extra 6 dB comes from the fixed "beamforming" gain obtained when feeding four elements with the same signal. The consequence of this design is that the array has a limited beamwidth both horizontally and vertically, but it can only control the beamforming in the horizontal plane. This is sufficient when the transmitter and prospective receivers are located in roughly the same plane (e.g., a person carrying a device in a room or along a street). The four RF inputs are connected to individual PSs located behind the antenna arrays; thus, this device uses the analog beamforming architecture.

Figure 7.18 shows a base station array for the 3.5 GHz band, and it has 32 RF input/output signals, which is referred to as 32T32R. This kind of product is marketed as "Massive MIMO". If we look inside the box, it contains 64 dual-polarized antenna elements arranged on a 8×8 grid, so the total number of elements is 128. These elements are arranged into subarrays, each consisting of four vertically stacked elements having the same polarization. Hence, using the terminology of this book, we are considering a UPA with 8 dual-polarized antennas per row and 2 dual-polarized antennas per column. Each dual-polarized antenna uses $\pm 45°$ polarizations, which are illustrated using red and blue colors in the figure. This subarray arrangement gives the maximum beamforming capability in the horizontal plane for the given 8×8

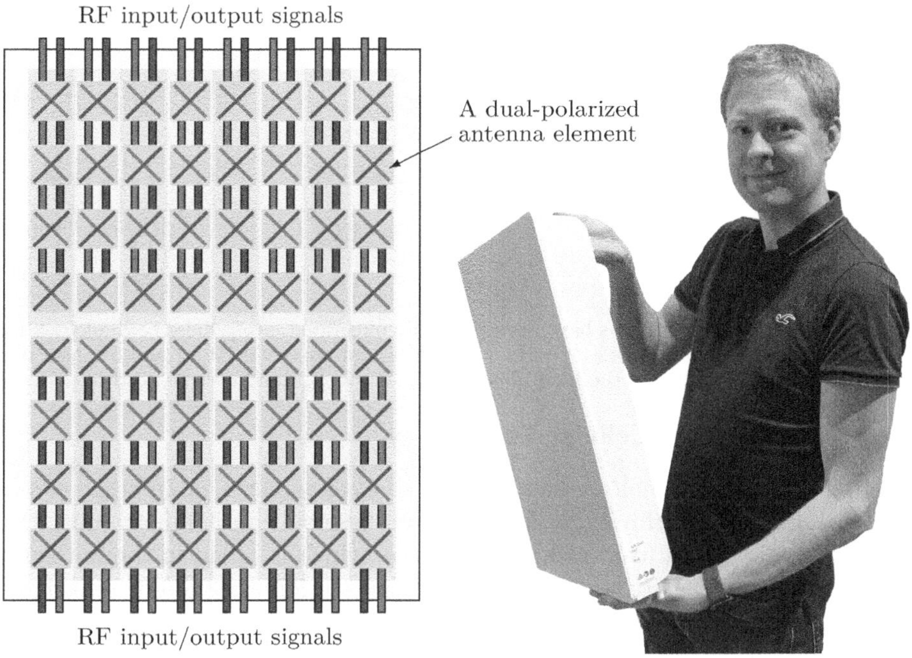

Figure 7.18: The photo shows the Ericsson AIR 3268 base station from 2021. It is designed for the 3.5 GHz band and has 32 antennas (16 dual-polarized antennas) and RF/BBU hardware integrated into the box, following the digital architecture. Each antenna is designed as a subarray with four vertically stacked elements having the same polarization. The array is dual-polarized and each element location contains two elements with orthogonal polarization ($\pm 45°$). This product supports a bandwidth of 200 MHz, a total transmit power of 200 W, and passive cooling.

element grid. However, it has a limited ability to change the vertical beam directivity. A product of this kind is meant for deployments in geographical areas with low-rise buildings, where the base station sees all the users and multipath clusters from roughly the same elevation angle, so there is no need for drastically changing the vertical beam directivity. There are other base station products with the same number of elements but more RF inputs/outputs, each connected to PAs/LNAs, DACs/ADCs, filters, etc. These products are thicker, heavier, and more expensive, but are capable of spatial multiplexing of users on different floors in high-rise buildings. This particular product weighs 12 kg and is implemented using the digital architecture, and all the components are integrated into a box with the dimensions 0.5×0.7 m. It is clear that the word "massive" refers to the number of antennas, not the weight or size. This base station array is rectangular, although the dual-polarized elements are deployed on an 8×8 grid. The reason is that the horizontal element spacing is $\lambda/2$, while the vertical element spacing is 0.7λ. The latter is a sparsely spaced array configuration that reduces the vertical beamwidth at the expense of occasionally creating grating lobes, but these point into the sky, where they cause no interference to users on the ground.

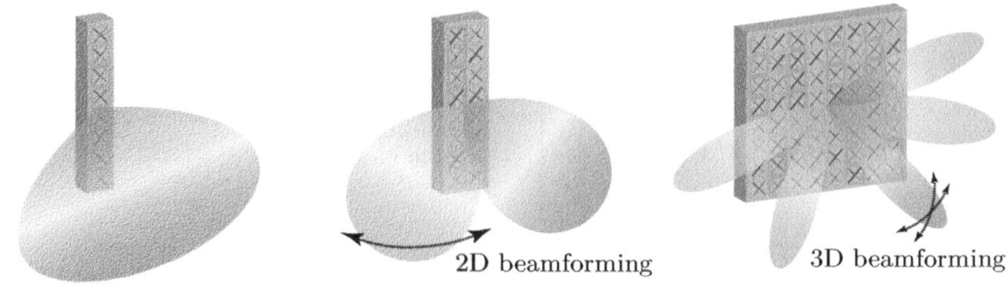

(a) 1G–3G (b) 4G (c) 5G

Figure 7.19: Multiple antennas have been gradually integrated into cellular technology. Fixed beams are used in 1G-3G, while 4G uses horizontal dual-polarized ULAs and 5G uses dual-polarized UPAs. Subarrays of the kind illustrated in Figure 7.18 are used in all three cases.

7.4.1 Evolution of Cellular Technology

The base station technology has thus far evolved in three main steps toward integrating MIMO technology. A traditional base station is illustrated in Figure 7.19(a) and has a fixed radiation pattern, which is broad in the horizontal plane but relatively narrow vertically. The base station can thereby aim signals toward the ground and cover a 120° sector where the intended users reside. This radiation pattern is typically achieved using a single subarray with multiple vertically stacked antenna elements, resulting in fixed beamforming. The base station might have dual-polarized antennas, which enables polarization diversity. 1G-3G technology featured such base stations.

Figure 7.19(b) illustrates how basic MIMO features were enabled in 4G by deploying two traditional dual-polarized antennas next to each other horizontally. The horizontal beamwidth can then be halved compared to Figure 7.19(a) and the beamforming gain doubled. The typical array configuration is a dual-polarized horizontal ULA where each antenna consists of a subarray with multiple vertically stacked antenna elements. The directivity can only be adapted in the horizontal plane, which is called 2D beamforming. The 4G standard supports basic spatial multiplexing and diversity features.

Figure 7.19(c) shows a typical 5G base station configuration with a UPA that enables both horizontal and vertical beamforming, so-called 3D beamforming. The illustrated configuration is the same as in Figure 7.18, which contains subarrays because many telecom operators want the beamforming gain provided by having many antenna elements but save costs by reducing the number of RF components. The 5G MIMO implementation is called Massive MIMO and supports beamforming, diversity, and spatial multiplexing. One reason that 5G can support many more antennas than in the past is that all the components in the digital beamforming architecture in Figure 7.9, except the BBU, nowadays can be integrated into a single box. In previous generations, each chain required separate boxes, which made MIMO bulky and heavy. 5G base stations for mmWave frequencies can be similar to the 4G example, except that each subarray is an analog beamforming array.

7.4.2 MIMO-Related Terminology

The history of multiple antenna communications spans more than a century, and several ambiguities in the terminology have appeared along the way. One reason is that different people prefer different terms for roughly the same concepts. Another reason is that the MIMO functionalities and use cases have expanded with time, which raises the question of whether one should generalize existing terms to cover these changes or make up new terms. We have selected a particular terminology in this book and tried to define its meaning rigorously, but in this section, we will describe additional terms and briefly explain what different meanings they might have.

Antenna port: The mapping between physical antenna elements and what we call "antennas" in the baseband processing can be rather complicated and implementation-specific, as exemplified by the subarrays in Figures 7.17 and 7.18. Therefore, 3GPP uses the term *antenna port* to refer to what is perceived as an antenna in the BBU; in other words, a typical MIMO channel in this book has K antenna ports at the transmitter and M antenna ports at the receiver. How these "logical" antennas are mapped to physical antenna elements needs not to be standardized, as long as we have a mechanism to obtain the corresponding channel matrix $\mathbf{H}$. It is even possible for a practical MIMO system to vary its number of antenna ports with time, by changing how large groups of elements constitute a subarray with a common logical antenna port. The traffic and device capabilities might trigger these changes.

Beamsteering: This refers to the mechanism of varying the angular direction of the beam transmitted from an antenna array. This can be achieved using either the analog, hybrid, or digital architecture. The term is usually used when the beam direction is changed over time to cover different geographical regions, but without aiming the beam at a known user location. This feature is used for broadcasting common messages over different parts of the coverage area (as discussed in Section 4.3.3) or for scanning an area in radar applications.

Beamforming, precoding, combining: This refers to the tuning of amplitudes and phases in antenna arrays to achieve directional signal transmission and reception that maximize communication performance. Beamforming was originally considered in LOS scenarios, where the optimal design creates beams that point in specific angular directions leading to the intended receivers. When the concept is applied in NLOS scenarios, MRT instead results in sending a signal with no apparent angular directivity. These two cases are illustrated in Figure 7.20. Some people use the beamforming term also in NLOS scenarios, while others prefer to call it *generalized beamforming* to highlight that the transmission has an entirely different physical shape than in LOS scenarios. In this book, we avoid using the beamforming term in NLOS scenarios to limit the risk of confusion. Instead, we have used the generalized terms transmit precoding and receive combining. The SNR gain that is obtained when focusing signals using multiple antennas is called beamforming

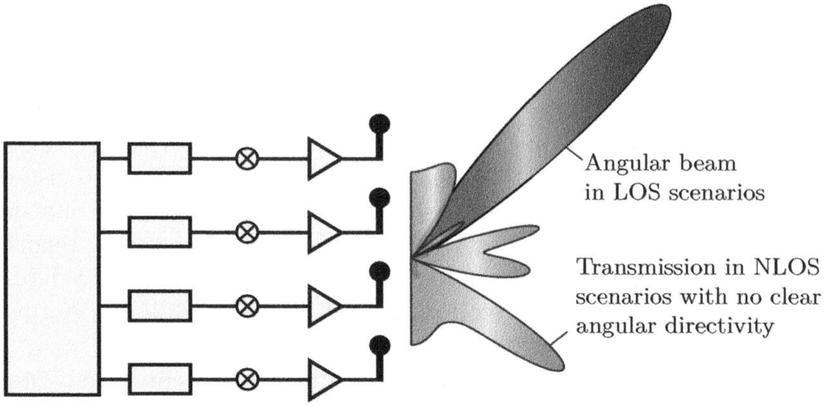

Figure 7.20: The SNR-maximizing transmission takes a different physical shape in LOS and NLOS scenarios because it is based on the channel. Some people use "beamforming" to refer to both cases, while some people use the terms precoding or generalized beamforming in the NLOS case.

gain, array gain, aperture gain, or power gain.

Multi-stream or multi-user beamforming: When spatial multiplexing is used in point-to-point MIMO scenarios, each signal is transmitted and received using a different "beam", which might not point in specific angular directions. This is sometimes called multi-stream or multi-layer beamforming, to extend the classical beamforming terminology further. In this book, we have instead used the precoding and combining terms, and we let the power allocation (e.g., water-filling) determine how many parallel signals are transmitted and received. The signals can be called streams or layers. Similarly, some people refer to the transmission and reception in multi-user MIMO scenarios as multi-user beamforming. Furthermore, it happens that the term precoding is viewed as a combination of beamforming (selection of the signal direction) and power allocation (distribution of power between different beams).

Full-dimensional and three-dimensional beamforming: This refers to beamforming using UPAs or other array geometries that can control the beam directivity both horizontally and vertically. The industry introduced the term to highlight this new feature in their product lines because the first multiple antenna features in 3G and 4G systems used horizontal ULAs only capable of beamforming in the two-dimensional horizontal plane.

Block-level and symbol-level precoding: The previous chapters described block-level precoding, where a fixed set of precoding vectors is used for a block of data symbols. Specifically, the transmitted signal was expressed as $\sum_{i=1}^{K} \mathbf{p}_i x_i$, where the value of the symbol x_i changes at every time instance based on the user data, while the precoding vector $\mathbf{p}_i$ only depends on the channels and is fixed for as long as the channels are. This structure is capacity-achieving when an infinitely large block of data is transmitted, but other options can be considered when transmitting a finite data block in practice. In symbol-level precoding [116], the precoding vectors change at every time instance, based on which data symbols will be transmitted. Instead of sending signals through fixed beams, the transmission is optimized so that each receiver observes a signal that is seemingly interference-free and as close to the transmitted constellation point as possible. The shape of the decision regions for the constellation points is exploited to accept interference when it will not increase the decoding error probability. The downside with symbol-level precoding is that the precoding optimization is computationally complex and must be redone at every symbol time instance.

Spatial layers: The parallel data streams that are spatially multiplexed to one or multiple devices are called spatial layers in 3GPP standards. In theory, the maximum number of spatial layers $r = \min(M, K)$ is determined by the number of antenna ports, but it can be smaller in practice. The number of orthogonal pilot sequences is predefined by the standard and manifests the maximum number of spatial layers, because we need a mechanism to estimate each column of the effective channel matrix $\mathbf{HP}$ that is obtained when applying the precoding matrix. Hence, once the standard has been defined, we can build base stations and devices with arbitrarily many antenna elements and antenna ports but the maximum number of spatial layers remains fixed. On the other hand, standards are revised when needed to utilize new functionalities, so adding more antenna ports and supporting more spatial layers typically come hand-in-hand.

Null-steering, MMSE, and other linear precoding schemes: The optimal linear precoding vectors are given by (6.129), but they depend on the virtual uplink power coefficients that are generally challenging to compute for a given performance metric (e.g., maximum sum rate). The TWF, RZF, and ZF schemes were described in Section 6.4.5 as simplifications of the optimal precoding. One can find many other heuristic/simplified precoding schemes in the literature [85, Remark 3.2], having names such as null-steering, SLNR precoding, multi-cell MMSE precoding, minimum-variance distortionless response (MVDR) precoding, and virtual SINR beamforming. These schemes are motivated through (slightly) different heuristic arguments, which are often connected to the uplink-downlink duality. Nevertheless, they usually perform roughly the same and are nearly optimal, so it can be puzzling that there are many names for almost the same thing. There are fewer alternative uplink schemes since MMSE combining is optimal for any performance metric. However, the MMSE scheme has alternative names, such as MVDR beamforming and interference-rejection combining.

Holographic MIMO: This term is used to describe densely spaced antenna arrays with antenna spacings much smaller than $\lambda/2$. The small spacing leads to spatial oversampling and mutual coupling effects. The latter can be exploited to achieve superdirectivity with beamforming gains that can be larger than usual in specific directions. The holographic terminology indicates that a densely spaced array can be implemented by having a surface

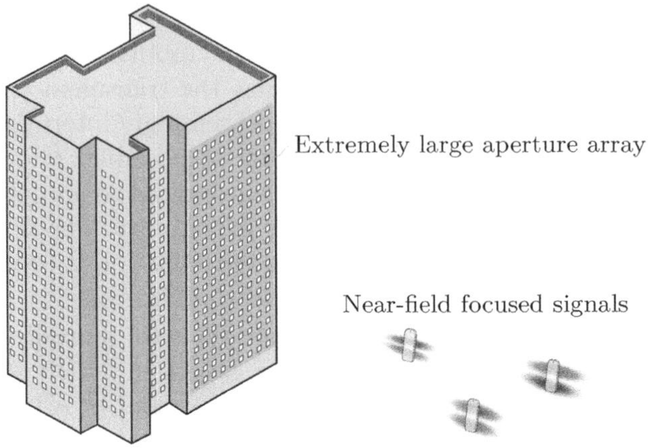

Figure 7.21: An ELAA consists of many antennas distributed over a huge aperture area, here exemplified as the facade of a building. The antenna spacing can be larger than in conventional arrays (e.g., one antenna per window) since the goal is to achieve tiny "beams" that have a finite depth and can be focused on individual user devices.

with a specific impedance pattern (i.e., the hologram) that is illuminated by a reference wave from a nearby antenna to generate an emitted wave [117], [118]. Each desired wavefront corresponds to a specific hologram that can be synthesized if the surface contains a dense grid of dielectric microstructures. The term *large intelligent surface* has also been used for similar purposes [76]. There exist commercial metamaterial antennas inspired by the holographic principle for both terrestrial and satellite communications, but as a way to implement the analog beamforming architecture without traditional PSs.

Extremely large aperture array (ELAA): This term was coined in [118] to refer to antenna arrays with an aperture size that is very large compared to the wavelength. Different from holographic MIMO, the antenna spacing might be larger than in conventional arrays. The motivation for the vast aperture is that the prospective receivers will be in its radiative near-field, where the propagation phenomena differ from the conventional far-field models. In particular, "beams" can both be focused in angle and depth, thereby creating an elliptical region with a strong beamforming gain around the intended receiver [119]. This could enable spatial multiplexing of very many devices and data streams per device as a way to manage more traffic without requiring more bandwidth. An example skyscraper deployment is illustrated in Figure 7.21. Another option is to deploy circular arrays in the radiative near-field, which is called orbital angular momentum (OAM) because it results in helical beams [120]. The MIMO capacity expressions from earlier chapters can still be utilized in these cases, but the far-field approximations cannot be used when computing the channel matrix $\mathbf{H}$. Indications of the propagation phenomena that appear in the radiative near-field were provided in Sections 4.4.2 and 4.4.3, which showed how high-rank channel matrices can be achieved in LOS conditions when using distributed or large arrays.

7.5 Exercises

Exercise 7.1. Consider a SISO-OFDM channel with $T+1$ taps, S subcarriers, and $h[\ell]=1$ for all $\ell \in \{0,\ldots,T\}$. The subcarrier spacing is $B/S = 15\,\text{kHz}$.

(a) Compute the subcarrier channels $\bar{h}[\nu]$, for $\nu \in \{0,\ldots,S-1\}$.

(b) Express the channel gains $|\bar{h}[\nu]|^2$, for $\nu \in \{0,\ldots,S-1\}$ in terms of sinusoidal functions. Hint: Rewrite the expression using (4.52).

(c) Assume that $T=3$ and $S=32$. The maximum channel gain in (b) is obtained at $\nu=0$. The first-null coherence bandwidth can be measured as $2(B/S)\nu^\star$, where $\nu^\star$ is the smallest subcarrier index in $\{1,\ldots,31\}$ for which $\bar{h}[\nu]=0$. This is the frequency interval (in Hz) between two nulls. What is the coherence bandwidth in this setup?

(d) What is the first-null coherence bandwidth if $T=7$ and $S=32$? Is it smaller or larger than in (c)?

Exercise 7.2. Prove the identity in (7.68): $\|\mathbf{P}[\nu]\|_{\text{F}}^2 = \|\mathbf{V}_{\text{BB}}[\nu]\|_{\text{F}}^2$.

Exercise 7.3. Suppose the pulse used in the PAM is selected such that

$$(p * p)(t) = \begin{cases} 1 & \text{for } |t| \leq 1/B, \\ 3 - 2B|t| & \text{for } 1/B < |t| \leq 1.5/B, \\ 0 & \text{otherwise.} \end{cases} \tag{7.84}$$

Recall that $(p * p)(t)$ appears in (2.126) when computing the coefficients of a multipath channel. Consider a channel with three propagation paths having the lengths 30 m, 45 m, and 108 m, respectively. The bandwidth is $B = 20\,\text{MHz}$ and the carrier frequency is $f_\text{c} = 3\,\text{GHz}$.

(a) What is the delay spread τ_{spread}?

(b) Determine the sampling delay η according to (7.5).

(c) Compute the channel taps $h[\ell]$, for $\ell \in 0,\ldots,T$, by sampling (2.126) using η from (b). The attenuations α_1, α_2, and α_3 have arbitrary values.

Exercise 7.4. The subcarrier spacing in 5G NR can either be 15, 30, or 60 kHz. Consider an OFDM setup with $S = 4000$ subcarriers, $\tau_{\text{spread}} = 4\,\mu\text{s}$, and $T \approx B\tau_{\text{spread}}$.

(a) Which of the three subcarrier spacings minimizes the OFDM symbol duration while ensuring that the cyclic prefix does not increase the signal resource utilization (i.e., the complex degrees of freedom) by more than 20%?

(b) What is the total bandwidth when using the subcarrier spacing from (a)?

Exercise 7.5. Consider a SISO-OFDM channel with $T+1=4$ taps and $S=32$ subcarriers. Each channel tap features independent and identical Rayleigh fading: $h[\ell] \sim \mathcal{N}_{\mathbb{C}}(0,\beta/4)$.

(a) Compute the correlation between the frequency-domain channel coefficients at two different subcarriers ν and ν'.

(b) How does the squared magnitude of the correlation vary as the difference $|\nu - \nu'|$ increases?

Exercise 7.6. Consider a SISO-OFDM channel with S subcarriers and $T+1$ channel taps. We would like to estimate the channels on all subcarriers, and therefore, the deterministic symbol $\sqrt{q}$ is transmitted on all subcarriers. The received signal on subcarrier $\nu \in \{0, S-1\}$ is

$$\bar{y}[\nu] = \sqrt{q}\bar{h}[\nu] + \bar{n}[\nu], \tag{7.85}$$

where $\bar{h}[\nu]$ is given in (7.15) and $\bar{n}[\nu] \sim \mathcal{N}_{\mathbb{C}}(0, N_0)$ is independent receiver noise. We will follow the estimation methodology from Section 4.2.4.

(a) Suppose the channel on subcarrier ν is estimated as $\hat{\bar{h}}[\nu] = \bar{y}[\nu]/\sqrt{q}$. What is the variance of the estimation error $\bar{h}[\nu] - \hat{\bar{h}}[\nu]$? What is the total error variance across the S subcarriers?

(b) The S subcarrier channels $\bar{\mathbf{h}} = [\bar{h}[0], \ldots, \bar{h}[S-1]]^{\mathrm{T}}$ are determined by the $T+1$ channel taps $\mathbf{h} = [h[0], \ldots, h[T]]^{\mathrm{T}}$ as

$$\bar{\mathbf{h}} = \mathbf{F}_{S,T+1}\mathbf{h}, \tag{7.86}$$

where $\mathbf{F}_{S,T+1} = \mathbb{C}^{S\times(T+1)}$ contains the first $T+1$ columns of the DFT matrix $\mathbf{F}_S$. Suppose we estimate the time-domain channel taps as

$$\hat{\mathbf{h}} = \frac{1}{\sqrt{q}}\mathbf{F}_{S,T+1}^{\mathrm{H}}[\bar{y}[0], \ldots, \bar{y}[S-1]]^{\mathrm{T}} \tag{7.87}$$

and then transform it to an estimate of $\bar{\mathbf{h}}$ as $\hat{\bar{\mathbf{h}}} = \mathbf{F}_{S,T+1}\hat{\mathbf{h}}$. What is the total error variance across the S subcarriers? Hint: $\mathbf{F}_{S,T+1}\mathbf{F}_{S,T+1}^{\mathrm{H}}\bar{\mathbf{h}} = \bar{\mathbf{h}}$.

(c) Suppose $S = 2000$ and $T+1 = 20$. How large is the difference between the total error variances in (a) and (b)? Explain the difference.

Exercise 7.7. Consider the dual-polarized array shown in Figure 7.18 and assume it consists of isotropic antenna elements.

(a) What are the horizontal and vertical first-null beamwidths (in radians) in the broadside direction?

(b) Consider a receiver located 50 m from the array in the broadside direction. How wide is the beam in meters in the horizontal and vertical directions?

(c) Consider another 8×2 UPA consisting of isotropic antennas with no subarrays. The horizontal antenna spacing is 0.5λ and the vertical antenna spacing is 0.7λ; thus, the array aperture is smaller than in Figure 7.18. What are the horizontal and vertical first-null beamwidths of this array in the broadside direction? Compare the beamwidths and the maximum beamforming gain with those obtained by the original array in Figure 7.18.

Exercise 7.8. Consider a hybrid ULA architecture with $M = K$ antennas and N_{RF} RF inputs/outputs. What rate can be achieved over the channel in (7.47) if each of $\mathbf{A}_{\mathrm{t}}$ and $\mathbf{A}_{\mathrm{r}}$ contain $N_{\mathrm{cl}} \geq N_{\mathrm{RF}}$ columns from the scaled DFT matrix $\sqrt{M}\mathbf{F}_M$, and $c_i[\ell] = \sqrt{\beta_0}e^{-\ell/\Gamma}$ if $\ell = i-1$ and $c_i[\ell] = 0$ otherwise, where $\beta_0 > 0$ is a constant and $\Gamma > 0$ specifies power-decay behavior? Assume that N_{RF} data streams are transmitted with equal power allocation.

Exercise 7.9. Consider the channel in (7.60) with $N_{\text{cl}} = M = K$. Suppose $\mathbf{A}_{\text{r}}$ and $\mathbf{A}_{\text{t}}$ are two DFT matrices. Each cluster only appears in one specific channel tap, such that $c_i[\ell] = \sqrt{\beta_i}$ for $\ell + 1 = i$, but $c_i[\ell] = 0$ otherwise. The channel gains $\beta_1, \ldots, \beta_{N_{\text{cl}}}$ will be treated as variables that can be selected freely under the constraint $\sum_{i=1}^{N_{\text{cl}}} \beta_i = \beta$, where β is the total channel gain.

(a) For a given value of β, which selection of $\beta_1, \ldots, \beta_{N_{\text{cl}}}$ maximizes the achievable rate at low SNRs? Answer this question for three different architectures: analog beamforming, hybrid beamforming with $N_{\text{RF}} < N_{\text{cl}}$ RF inputs/outputs, and digital beamforming.

(b) Repeat (a) but consider the achievable rate at high SNRs.

Exercise 7.10. Consider a MIMO-OFDM channel with $T + 1 = 2$ channel taps, $S \geq 2$ subcarriers, and $M = K = 2$ antennas. The channel matrices are

$$\mathbf{H}[0] = \sqrt{\beta}\begin{bmatrix} 1 & 1 \\ 1 & 1 \end{bmatrix}, \quad \mathbf{H}[1] = \sqrt{\beta}\begin{bmatrix} 1 & -1 \\ -1 & 1 \end{bmatrix}. \tag{7.88}$$

(a) Compute the MIMO-OFDM capacity in terms of bit per OFDM symbol.

(b) What is the capacity-achieving input distribution on subcarrier ν?

Exercise 7.11. Consider a Rician fading MIMO-OFDM channel with $T + 1 = 2$ taps. The first tap is the LOS path and the second tap is an i.i.d. fading matrix. Using the κ-factor notation from Example 5.2, the two taps of this channel are defined as

$$\mathbf{H}[0] = \sqrt{\frac{\kappa}{\kappa+1}}\sqrt{\beta}e^{-\mathrm{j}\psi}\mathbf{a}_M(\varphi_r, \theta_r)\mathbf{a}_K^{\mathrm{T}}(\varphi_t, \theta_t), \quad \mathbf{H}[1] = \sqrt{\frac{1}{\kappa+1}}\sqrt{\beta}\mathbf{H}_{\text{iid}}, \tag{7.89}$$

where the entries of $\mathbf{H}_{\text{iid}} \in \mathbb{C}^{M \times K}$ are i.i.d. $\mathcal{N}_{\mathbb{C}}(0, 1)$-distributed. Suppose an analog beamforming architecture is used, and the transmit precoding and receive combining are based on the LOS path in $\mathbf{H}[0]$. What fraction of the total channel gain $MK\beta$ will be received on the average? Is it an increasing or decreasing function of the κ-factor?

Exercise 7.12. When using analog beamforming and large bandwidth, the beam-squint effect can change the beam direction at the edges of the signal bandwidth. Consider the beamforming gain in (7.83) with $B/f_{\text{c}} = 0.1$.

(a) At what observation angle φ is the gain maximized? Hint: The answer is an expression that depends on ν/S.

(b) Does the answer in (a) depend on M?

(c) How many degrees is the beam shifted if $\nu/S = 1/2$?

Exercise 7.13. The approximation $\bar{\mathbf{H}}^{\mathrm{H}}[\nu]\mathbf{W}_{\text{RF}}\mathbf{W}_{\text{RF}}^{\mathrm{H}}\bar{\mathbf{H}}[\nu] \approx \bar{\mathbf{H}}^{\mathrm{H}}[\nu]\bar{\mathbf{H}}[\nu]$ is used in (7.75). Quantify the approximation error by computing the squared Frobenius norm of the difference

$$\bar{\mathbf{H}}^{\mathrm{H}}[\nu]\mathbf{W}_{\text{RF}}\mathbf{W}_{\text{RF}}^{\mathrm{H}}\bar{\mathbf{H}}[\nu] - \bar{\mathbf{H}}^{\mathrm{H}}[\nu]\bar{\mathbf{H}}[\nu]. \tag{7.90}$$

Assume that $\mathbf{W}_{\text{RF}}$ contains the first N_{RF} columns of $\mathbf{U}$, which comes from the SVD $\bar{\mathbf{H}}[\nu] = \mathbf{U}\mathbf{\Sigma}\mathbf{V}^{\mathrm{H}}$ of the channel matrix on subcarrier ν.

Chapter 8

Localization and Sensing with MIMO

The previous chapters considered different forms of MIMO communications. Apart from communications, antenna arrays are also used for classical radar applications such as direction-of-arrival (DOA) estimation, target detection, localization, velocity estimation, etc. These topics are covered under the umbrella of *(sensor/radar) array signal processing* [51], [121], [122]. Communication and radar technologies have evolved along separate paths for many years, requiring different physical equipment and separate deployments. A commonality is that progressively more antennas/sensors have been utilized to exploit the spatial dimension further to improve the respective performance metrics. As the radio hardware becomes more versatile and software-defined, it is desirable to use the same physical network equipment for multiple applications, including communication, localization, and sensing. This design paradigm is called *integrated sensing and communication (ISAC)* [123] and can enable cost savings and new innovative use cases but also require fundamental design tradeoffs. Since existing wireless communication networks feature wide-area coverage, it is a suitable platform to evolve into an ISAC system. The integration can take place by sharing the hardware and/or waveforms.

Sensing refers to radar-like applications that aim to obtain spatial knowledge of the physical environment by transmitting known signals and observing their reflections on various objects. Typical sensing applications are target detection, target range and velocity estimation, and target tracking. Localization refers to determining the map coordinates of an object. This chapter covers the following fundamental applications: 1) far-field DOA estimation, 2) localization, and 3) target detection. The aim is to analyze how having multiple antennas helps carry out the respective tasks.

8.1 Direction-of-Arrival Estimation

In this section, we consider DOA estimation, where the goal is to determine the angular directions (φ, θ) of multiple waves that impinge on an antenna array. We consider the free-space LOS channel model, developed in Chapter 4. The transmitters/sources that radiate the waves are assumed to be in the far-field of the receiver array, and we will use the narrowband signal assumption from Section 2.3.4. In other words, we assume the maximum difference in the propagation delay over the array is much shorter than the symbol time as in (4.7). This results in frequency flatness and the system model in (4.9). We note that estimation of the angles (φ, θ) is the first step towards solving a LOS *localization* problem, where the physical locations of the sources are determined by also estimating what distance each signal has traveled.

We consider a receiver array with M antennas that has K single-antenna radiating sources (transmitters) in its far-field. The signal radiated by source k impinges on the array as a planar wave from some angular direction (φ_k, θ_k), and the common channel gain to all the receive antennas is denoted by $\beta_k \geq 0$. The location of the receive antenna m is denoted by $\mathbf{u}_m \in \mathbb{R}^3$, as in Chapter 4. It follows from (4.113) that the array response vector for source k is

$$\mathbf{a}(\varphi_k, \theta_k) = \begin{bmatrix} e^{\mathrm{j}\frac{2\pi}{\lambda}\mathbf{u}_1^{\mathrm{T}}\boldsymbol{\rho}_k} \\ e^{\mathrm{j}\frac{2\pi}{\lambda}\mathbf{u}_2^{\mathrm{T}}\boldsymbol{\rho}_k} \\ \vdots \\ e^{\mathrm{j}\frac{2\pi}{\lambda}\mathbf{u}_M^{\mathrm{T}}\boldsymbol{\rho}_k} \end{bmatrix}, \tag{8.1}$$

where $\boldsymbol{\rho}_k$ the unit-length vector that points from the origin to source k:

$$\boldsymbol{\rho}_k = \begin{bmatrix} \cos(\varphi_k)\cos(\theta_k) \\ \sin(\varphi_k)\cos(\theta_k) \\ \sin(\theta_k) \end{bmatrix}. \tag{8.2}$$

The received signal $\mathbf{y}[l] \in \mathbb{C}^M$ at integer sample index l can be expressed as

$$\mathbf{y}[l] = \sum_{k=1}^{K} \sqrt{\beta_k} e^{-\mathrm{j}\psi_k} \mathbf{a}(\varphi_k, \theta_k) x_k[l] + \mathbf{n}[l], \tag{8.3}$$

where $x_k[l]$ is the baseband equivalent of the signal emitted by source k, that is sampled at time index l and ψ_k is the phase-shift introduced along the respective propagation path. The signals $x_k[l]$ have zero mean and variance P_k and might contain data because they are random and unknown to the receiver. The independent receiver noise is distributed as $\mathbf{n}[l] \sim \mathcal{N}_{\mathbb{C}}(\mathbf{0}, \sigma^2 \mathbf{I}_M)$. We assume the source signals $x_k[l]$ and $x_i[l]$ are independent for $k \neq i$.

The DOA estimation problem is to estimate (φ_k, θ_k), for $k = 1, \ldots, K$, using the received signals $\mathbf{y}[l]$, for $l = 1, \ldots, L$. We assume the channel gains, phase-shifts, and array response vectors are constant during these consecutive L samples. Exploiting multiple samples is useful to increase the estimation accuracy by improving the SNR and averaging out randomness in the source signals. We assume the number of sources, K, is known.[1] We further assume that $K < M$, which is required by some of the algorithms we will describe.

[1]There are algorithms that detect the number of sources; see [122] for details.

The DOA estimation algorithms can be classified under two main branches:

- Non-parametric (model-free) methods;
- Parametric (model-based) methods.

The non-parametric methods assume that the characteristics of the signals $x_k[l]$ are unknown. However, the structure of the array response vector in (8.1) is assumed to be known because it is only based on the array geometry. On the other hand, the parametric methods utilize the statistics of the input signals $x_k[l]$ in addition to the structure of the array response vector. Since they exploit the specific system model parameters, they generally perform better than the non-parametric methods. In the following, we will first cover two non-parametric beamforming methods for DOA estimation and then describe a parametric subspace-based method that exploits the noise subspace.

8.1.1 Conventional Non-Parametric Beamforming Method

The beamforming methods considered for DOA estimation in this chapter are non-parametric. They are sometimes called *spectral-based* since they construct a DFT-like spatial spectrum that shows how much power is received from different angles (φ, θ). The main peaks of that spectrum are the DOA estimates, but there will also be ripples created by side-lobes from receive beamforming.

We recall that there are K sources whose angular directions are to be estimated. In beamforming techniques, a receive combining vector $\mathbf{w} \in \mathbb{C}^M$ is applied to all the received signals in (8.3): $\mathbf{w}^{\mathrm{H}}\mathbf{y}[l]$, for $l = 1, \ldots, L$. Then, the average squared magnitude of the combined signals is computed as

$$P(\mathbf{w}) = \frac{1}{L}\sum_{l=1}^{L} |\mathbf{w}^{\mathrm{H}}\mathbf{y}[l]|^2 = \mathbf{w}^{\mathrm{H}} \underbrace{\left(\frac{1}{L}\sum_{l=1}^{L} \mathbf{y}[l]\mathbf{y}^{\mathrm{H}}[l]\right)}_{=\hat{\mathbf{R}}_L} \mathbf{w} = \mathbf{w}^{\mathrm{H}}\hat{\mathbf{R}}_L\mathbf{w}. \tag{8.4}$$

We recognize $\hat{\mathbf{R}}_L$ as the unbiased sample average estimator (i.e., a matrix generalization of (2.171)) of the correlation matrix of $\mathbf{y}[l]$, which is defined as

$$\mathbf{R} = \mathbb{E}\left\{\mathbf{y}[l]\mathbf{y}^{\mathrm{H}}[l]\right\}. \tag{8.5}$$

The randomness in $\mathbf{y}[l]$ is assumed independent across the L samples. Hence, the sample average correlation matrix $\hat{\mathbf{R}}_L$ approaches its statistical mean, $\mathbf{R}$, when the number of samples L goes to infinity, as previously discussed in Section 2.6.1. This implies that $P(\mathbf{w})$ in (8.4) is the sample estimate of $\mathbb{E}\{|\mathbf{w}^{\mathrm{H}}\mathbf{y}[l]|^2\}$, which is the average power of the signal obtained when the receive combining vector $\mathbf{w}$ is applied.

Suppose one of the true DOAs is (φ_k, θ_k). We can maximize $|\mathbf{w}^{\mathrm{H}}\mathbf{a}(\varphi_k, \theta_k)|^2$ (among all unit-norm combining vectors) by selecting $\mathbf{w}$ equal to $\mathbf{a}(\varphi_k, \theta_k)$. This corresponds to an MRC receiver, and we recall from Figure 4.8 that

MRC is a spatial bandpass filter that only provides a large power value if the angle used for MRC matches the angle of the incoming signal. Based on this principle, in the *conventional beamforming* method, we select the combining vector $\mathbf{w}(\varphi, \theta)$ as a function of (φ, θ) to match the array response vector:

$$\mathbf{w}(\varphi, \theta) = \mathbf{a}(\varphi, \theta). \tag{8.6}$$

Therefore, $P(\mathbf{w}(\varphi, \theta))$ is an estimate of how much power is received from the direction (φ, θ), and estimating the DOAs corresponds to finding the K peaks of the function $P(\mathbf{w}(\varphi, \theta))$. The peaks will be clearly noticeable when the SNR is high and/or the number of symbols L is sufficiently large.

Inserting (8.6) into the power spectrum in (8.4), the DOA estimates $(\hat{\varphi}_k, \hat{\theta}_k)$, for $k = 1, \ldots, K$, are obtained as the K highest peaks of the function

$$P_{\text{conv}}(\varphi, \theta) = \frac{1}{L} \sum_{l=1}^{L} |\mathbf{a}^{\text{H}}(\varphi, \theta)\mathbf{y}[l]|^2 = \mathbf{a}^{\text{H}}(\varphi, \theta)\hat{\mathbf{R}}_L \mathbf{a}(\varphi, \theta). \tag{8.7}$$

This method can be applied when having any array geometry. The only requirement is that the array response vector is known for any angle pair (φ, θ), which implies that the antennas must be phase-synchronized.

Suppose we have a ULA with M antennas, and the K sources are in the same horizontal plane as the array. The array response vector is then given in (4.74) as

$$\mathbf{a}(\varphi) = \begin{bmatrix} 1 \\ e^{-\text{j}2\pi \frac{\Delta \sin(\varphi)}{\lambda}} \\ e^{-\text{j}2\pi \frac{2\Delta \sin(\varphi)}{\lambda}} \\ \vdots \\ e^{-\text{j}2\pi \frac{(M-1)\Delta \sin(\varphi)}{\lambda}} \end{bmatrix}, \tag{8.8}$$

which is only a function of the azimuth angle. In this case, the DOA estimation problem turns into estimating the azimuth angles $\varphi_1, \ldots, \varphi_K$. The power spectrum in (8.7) whose K highest peaks are the DOA estimates simplifies to

$$P_{\text{conv}}(\varphi) = \frac{1}{L} \sum_{l=1}^{L} |\mathbf{a}^{\text{H}}(\varphi)\mathbf{y}[l]|^2 = \frac{1}{L} \sum_{l=1}^{L} \left| \sum_{m=1}^{M} e^{\text{j}2\pi \frac{(m-1)\Delta \sin(\varphi)}{\lambda}} y_m[l] \right|^2. \tag{8.9}$$

Consider DOA estimation of a single source ($K = 1$) located at the DOA angle $\varphi = \pi/6$ in the same plane as the receiver. The receiver has a ULA with $M = 2$ or $M = 10$ antennas and $\Delta = \lambda/2$ spacing. Figure 8.1 shows the normalized power spectrum obtained with 0 dB SNR and $L = 25$ time samples. The normalization ensures that the peak value on each curve is 0 dB.[2] The

[2]The normalization is done by dividing the power spectrum $P_{\text{conv}}(\varphi)$ by its maximum value $P_{\max} = \max_{\varphi} P_{\text{conv}}(\varphi)$. It becomes easier to compare power spectra obtained with different SNRs and different numbers of antennas when applying the normalization.

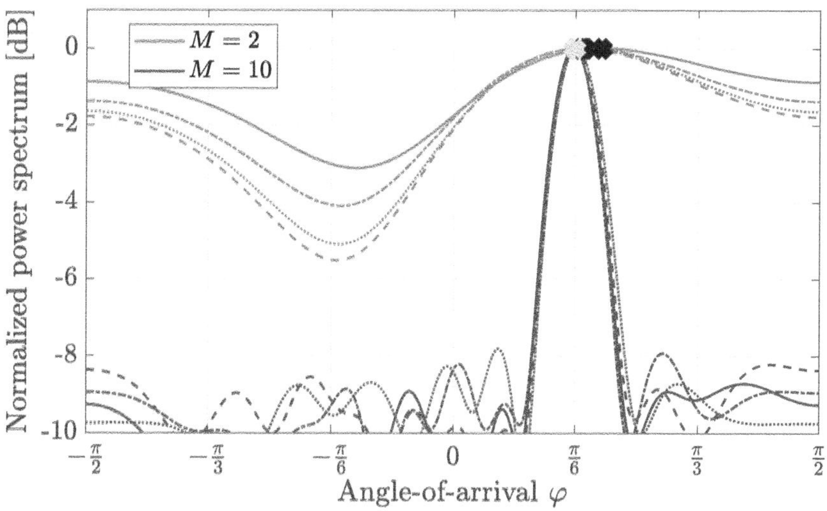

Figure 8.1: The normalized power spectrum obtained with four random realizations of the source signals and noise with an M-antenna ULA and conventional beamforming. There is one source, and its DOA is $\varphi = \pi/6$. $L = 25$ time samples are used to find the generate the power spectrum. The respective DOA estimates are the horizontal values at the peaks of the curves. The peaks are marked with black and yellow crosses for $M = 2$ and $M = 10$, respectively.

shape of the estimated power spectrum is affected by the random source signals and noise samples, which are all Gaussian distributed. We show four curves with different random realizations in the figure to showcase the variations one can expect. The black and yellow crosses denote the peak values on the curves with $M = 2$ and $M = 10$, respectively. The corresponding angle value is the DOA estimate $\hat{\varphi}$. The curves resemble beam patterns (recall the terminology in Figure 4.14) with narrower main beams and smaller side-lobes with $M = 10$ antennas compared to $M = 2$. This results in more accurate angle estimates with $M = 10$, in the sense that the yellow crosses are very close to the true DOA angle. The randomness shifts the curves and, particularly, modifies the side-lobes. However, the system becomes more robust to randomness when there are more antennas, thanks to the higher *spatial resolution* (i.e., smaller beamwidth) and larger beamforming gain.

In Figure 8.2, we show the MSE between the true DOA $\pi/6$ and the estimate obtained (in radians) using conventional beamforming. The setup is the same as in Figure 8.1, except that we vary the number of samples, L, on the horizontal axis and consider two different SNR values: 0 dB and 10 dB. As expected, the lowest MSE is achieved using the most antennas and having the highest SNR. All four curves show that increasing the number of samples improves the DOA estimation quality. This happens because the sample average estimator $\hat{\mathbf{R}}_L$ in (8.4) approaches the true correlation matrix $\mathbf{R} = \mathbb{E}\{\mathbf{y}[l]\mathbf{y}^{\mathrm{H}}[l]\}$ as $L \to \infty$, which progressively makes the power spectrum less dependent on the random signals and noise.

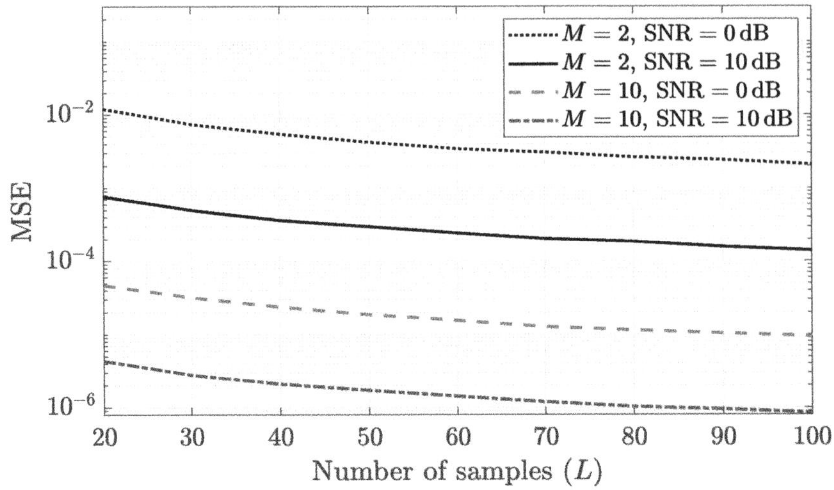

Figure 8.2: The MSE of the DOA estimation with conventional beamforming. A ULA is considered with either $M = 2$ or $M = 10$ antennas and SNR $= 0$ or SNR $= 10$ dB. There is a single source with the DOA $\varphi_1 = \pi/6$.

The estimation performance can be studied analytically in the limit $L \to \infty$, where the power spectrum $P_{\text{conv}}(\varphi)$ with the ULA has the limit

$$P_{\text{conv}}(\varphi) \to \overline{P}_{\text{conv}}(\varphi) = \mathbf{a}^{\text{H}}(\varphi)\mathbf{R}\mathbf{a}(\varphi). \tag{8.10}$$

There is $K = 1$ source that sends the zero-mean signal $x_1[l]$ with power P_1. The correlation matrix $\mathbf{R}$ of the received signal in (8.3) can then be computed as

$$\mathbf{R} = \mathbb{E}\{\mathbf{y}[l]\mathbf{y}^{\text{H}}[l]\} = P_1\beta_1\mathbf{a}(\varphi_1)\mathbf{a}^{\text{H}}(\varphi_1) + \sigma^2\mathbf{I}_M. \tag{8.11}$$

By inserting this expression into the right-hand side of (8.10), we obtain

$$\begin{aligned}
\overline{P}_{\text{conv}}(\varphi) &= \mathbf{a}^{\text{H}}(\varphi)\left(P_1\beta_1\mathbf{a}(\varphi_1)\mathbf{a}^{\text{H}}(\varphi_1) + \sigma^2\mathbf{I}_M\right)\mathbf{a}(\varphi) \\
&= P_1\beta_1\left|\mathbf{a}^{\text{H}}(\varphi)\mathbf{a}(\varphi_1)\right|^2 + \sigma^2\|\mathbf{a}(\varphi)\|^2 \\
&\leq P_1\beta_1\|\mathbf{a}(\varphi)\|^2\|\mathbf{a}(\varphi_1)\|^2 + \sigma^2 M \\
&= P_1\beta_1 M^2 + \sigma^2 M,
\end{aligned} \tag{8.12}$$

where we utilized the Cauchy-Schwartz inequality from (2.18) and that array response vectors satisfy $\|\mathbf{a}(\varphi)\|^2 = M$. The inequality is only satisfied with equality when $\mathbf{a}(\varphi)$ and $\mathbf{a}(\varphi_1)$ are parallel vectors, which happens for $\varphi = \varphi_1$ and $\varphi = \pi - \varphi_1$ when using a ULA with half-wavelength spacing (recall the mirror ambiguity from Figure 4.7). If the ULA is deployed so that only sources in the range $\varphi \in [-\pi/2, \pi/2]$ can occur, $\varphi = \varphi_1$ is the unique maximum of the asymptotic power spectrum $\overline{P}_{\text{conv}}(\varphi)$. Since this asymptotic DOA estimate is exact, the conventional beamforming method is a consistent DOA estimator.

Example 8.1. Is the power spectrum in (8.9) related to the Fourier transform?

By introducing the variable $\nu = -M\Delta \sin(\varphi)/\lambda$, we can express the power spectrum in (8.9) as

$$P_{\text{conv}}(\varphi) = \frac{1}{L}\sum_{l=1}^{L}\Bigg|\underbrace{\sum_{m=0}^{M-1} y_{m+1}[l]e^{-\mathrm{j}2\pi m\nu/M}}_{=\sqrt{M}\mathcal{F}_d\{y_{m+1}[l]\}}\Bigg|^2. \tag{8.13}$$

If ν is an integer, we can recognize the term inside the magnitude square as $\sqrt{M}$ times the DFT of the M-length spatial sequence $y_1[l], \ldots, y_M[l]$, based on the definition in (2.195). Since the DFT is applied to the antenna index domain instead of the time domain, we call this the spatial DFT, and ν is the normalized spatial frequency. The power spectrum is the average of these spatial DFTs with respect to the time samples $l = 1, \ldots, L$. Different from the classical DFT that only considers the normalized frequencies $\nu = 0, 1, \ldots, M-1$, we consider a real-valued spatial frequency variable $\nu = -M\Delta \sin(\varphi)/\lambda$ because we want to evaluate the power spectrum $P_{\text{conv}}(\varphi)$ for any search direction $\varphi \in [-\pi/2, \pi/2]$ to find its peaks. Hence, the Fourier transform appearing in this context is the spatial counterpart to the discrete-time Fourier transform (DTFT), defined as the DFT but with real-valued frequencies ν. We might call it the *discrete-space Fourier transform (DSFT).*

Thus far, we have considered a half-wavelength-spaced ULA to avoid spatial undersampling. As discussed in Section 4.3.4, grating lobes appear in directions other than the main beam's direction when Δ is greater than $\lambda/2$ in a ULA. Grating lobes can be acceptable in communications because the total interference level is unaffected; instead of sending interference to places close to the intended receiver when $\Delta = \lambda/2$, the same amount is sent somewhere else when $\Delta > \lambda/2$. The issue is more severe in DOA estimation since the grating lobes make the ULA unable to distinguish between some widely different directions. To showcase this phenomenon, we consider the same setup as in Figure 8.1 but increase the antenna spacing to $\Delta = \lambda$ in Figure 8.3. The power spectrum now has two equally tall peaks: one at the true DOA $\varphi = \pi/6$ and another at $\varphi = -\pi/6$. The estimator cannot determine which one is the true DOA because the array response vectors $\mathbf{a}(\pi/6)$ and $\mathbf{a}(-\pi/6)$ are equal, as can be seen by computing (8.8) with $\Delta = \lambda$:

$$\mathbf{a}(\pi/6) = \begin{bmatrix} 1 \\ e^{-\mathrm{j}2\pi\frac{\lambda\sin(\pi/6)}{\lambda}} \\ e^{-\mathrm{j}2\pi\frac{2\lambda\sin(\pi/6)}{\lambda}} \\ \vdots \\ e^{-\mathrm{j}2\pi\frac{(M-1)\lambda\sin(\pi/6)}{\lambda}} \end{bmatrix} = \begin{bmatrix} 1 \\ -1 \\ 1 \\ \vdots \\ (-1)^{M-1} \end{bmatrix} = \begin{bmatrix} 1 \\ e^{-\mathrm{j}2\pi\frac{\lambda\sin(-\pi/6)}{\lambda}} \\ e^{-\mathrm{j}2\pi\frac{2\lambda\sin(-\pi/6)}{\lambda}} \\ \vdots \\ e^{-\mathrm{j}2\pi\frac{(M-1)\lambda\sin(-\pi/6)}{\lambda}} \end{bmatrix}. \tag{8.14}$$

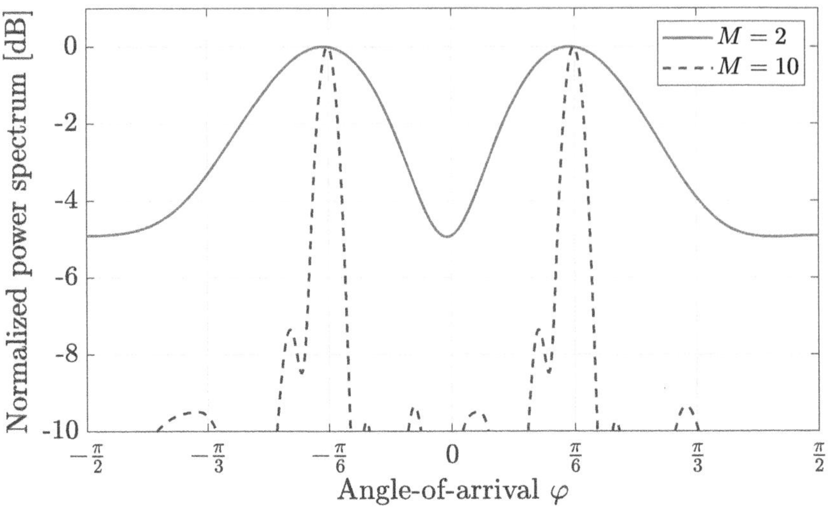

Figure 8.3: The normalized power spectrum in the same setup as in Figure 8.1 but for a single random realization and with the antenna spacing $\Delta = \lambda$ instead of $\Delta = \lambda/2$. The large spacing results in two indistinguishable peaks in the power spectrum: One at the correct angle $\pi/6$ and a grating lobe at $-\pi/6$.

It is the spatial undersampling (i.e., aliasing) that creates the grating lobe at $\varphi = -\pi/6$, and the ambiguity remains when L, M, or the SNR goes to infinity. The beamforming method cannot be consistent with such ambiguity; thus, one should only use ULAs with $\Delta \leq \lambda/2$ for DOA estimation.

Example 8.2. Consider a ULA with antenna spacing $\Delta = 2\lambda/3$ and a source with the DOA $\varphi = \pi/6$. Is there any ambiguity in the DOA estimator? Can it be resolved by increasing the number of antennas?

For the given spacing, the mth entry of the array response vector $\mathbf{a}(\pi/6)$ is

$$e^{-\mathrm{j}2\pi\frac{2\lambda/3\cdot(m-1)\sin(\pi/6)}{\lambda}} = e^{-\mathrm{j}\frac{2\pi}{3}(m-1)}. \tag{8.15}$$

If we can find another angle $\varphi \in [-\pi/2, \pi/2]$ for which the mth entry of $\mathbf{a}(\varphi)$ coincides with (8.15), we have a grating lobe at that angle and this results in DOA ambiguity. To check for such an angle, we equate $e^{-\mathrm{j}\frac{2\pi}{3}(m-1)}$ to the mth element of $\mathbf{a}(\varphi)$:

$$e^{-\mathrm{j}2\pi\frac{2\lambda/3\cdot(m-1)\sin(\varphi)}{\lambda}} = e^{-\mathrm{j}\frac{2\pi}{3}(m-1)} \Rightarrow \frac{4\pi}{3}\sin(\varphi) = \frac{2\pi}{3} + 2\pi \cdot n \tag{8.16}$$

for some integer n. The equality is satisfied for $n = 0$ and $n = -1$. For $n = 0$, we obtain $\varphi = \pi/6$, which is the true DOA. For $n = -1$, we obtain $\varphi = -\pi/2$ as another solution, so there is a grating lobe at that angle. This ambiguity cannot be resolved by changing the number of antennas.

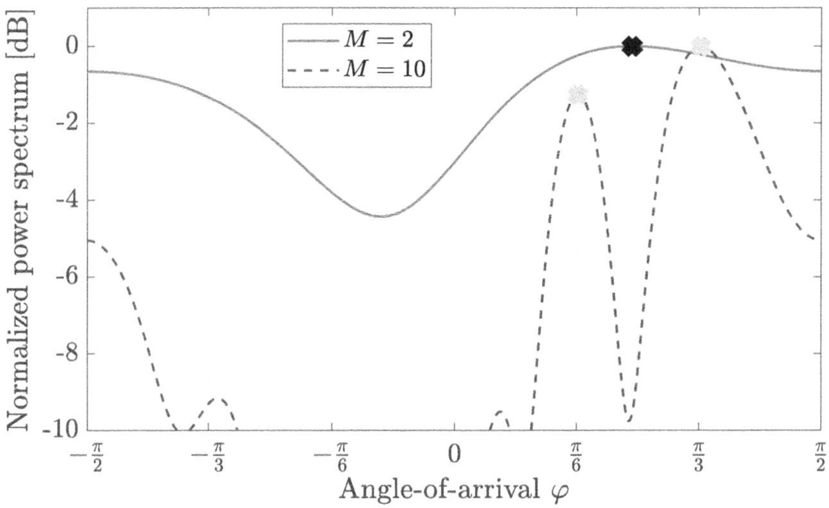

Figure 8.4: The normalized power spectrum with an M-antenna ULA and conventional beamforming for a single random realization. There are $K = 2$ sources with the DOAs $\varphi_1 = \pi/6$ and $\varphi_2 = \pi/3$. $L = 25$ time samples are used to compute the power spectrum. The peaks of the power spectrum are the DOA estimates. The black and yellow crosses denote the peaks with $M = 2$ and $M = 10$, respectively.

Next, we consider $K = 2$ sources located in the same horizontal plane as the ULA. The sources have the DOAs $\varphi_1 = \pi/6$ and $\varphi_2 = \pi/3$. The ULA has $M = 2$ or $M = 10$ antennas with $\Delta = \lambda/2$ spacing. Figure 8.4 shows the normalized power spectrum obtained using $L = 25$ time samples where the SNR is 0 dB. We show the spectrum for a single random signal/noise realization for each value of M. In this case, the DOA estimates $\hat{\varphi}_1, \hat{\varphi}_2$ should be the two tallest power spectrum peaks. When $M = 2$, there is only a single peak, which is located between the true DOAs and marked with a black cross. This ULA cannot distinguish between the two sources using its small number of antennas. This effect can be explained following the beamwidth discussion in Section 4.3.2. Suppose that the ULA points its beamforming toward one of the sources in the receiver processing. If the other source is located within the main beam (i.e., closer than the first nulls), it will disturb the angle estimation. The receiver observes constructive interference of the signals from both sources, which makes the power spectrum look as if there were only one source. As the number of antennas increases, the beamwidth becomes narrower, and we can observe two distinct peaks in the power spectrum. This can be seen in the case of $M = 10$, where the peak values are marked with yellow crosses. To get a rough idea of how many antennas are needed to resolve two sources, we can use the approximation in (4.62) of the distance between the beam direction and first null: $2/M$ radians. If the angular separation of the sources is larger than this, we can expect them to be distinguishable in

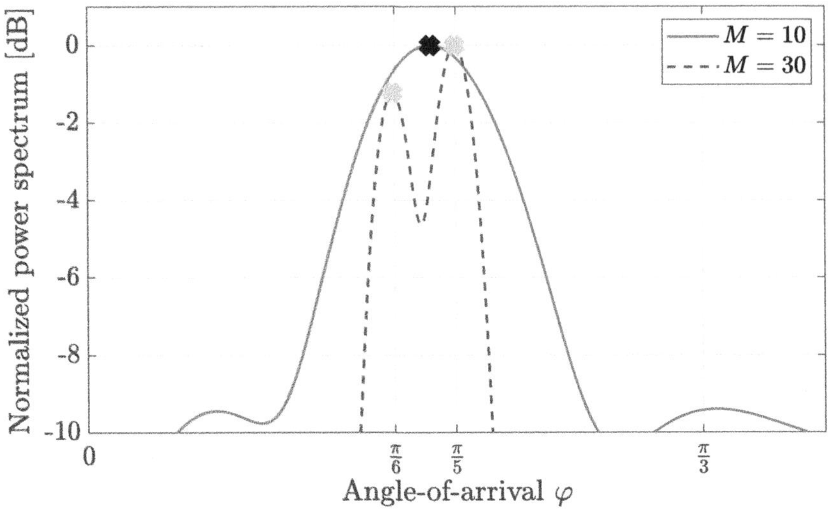

Figure 8.5: The normalized power spectrum with an M-antenna ULA and conventional beamforming for a single random realization. There are $K = 2$ sources with the DOAs $\varphi_1 = \pi/6$ and $\varphi_2 = \pi/5$. $L = 25$ time samples are used to compute the power spectrum. The peaks of the power spectrum are the DOA estimates. The black and yellow crosses denote the peaks with $M = 10$ and $M = 30$, respectively.

the power spectrum (when the SNR or L is sufficiently large). Hence, this is a measure of the array's spatial resolutions. In Figure 8.4, the angle difference between $\varphi_1 = \pi/6$ and $\varphi_2 = \pi/3$ is $\varphi_2 - \varphi_1 = \pi/6 \approx 0.52$ rad. When using $M = 2$ antennas, we need the spacing to be greater than $2/M = 1$ rad to ensure that two distinct peaks are visible in the spectrum. The corresponding minimum separation is $2/M = 0.2$ rad when $M = 10$, which is sufficient to clearly distinguish the sources, as seen in the figure.

We will now change the DOA of the second source to $\varphi_2 = \pi/5$, which reduces the angular separation to $\varphi_2 - \varphi_1 = \pi/30 \approx 0.1$ rad. Figure 8.5 shows the normalized power spectrum for this scenario with either $M = 10$ or $M = 30$ antennas. In this case, we cannot resolve the sources using 10 antennas, but we only observe a single peak marked with a black cross. However, we can separate the sources with $M = 30$ antennas because $2/M = 2/30 \approx 0.07$, which is smaller than 0.1. The two peaks in the power spectrum are marked with yellow crosses and are close to the true DOAs.

The above principles also apply to cases with $K > 2$ sources. The numbering of the sources is arbitrary in the system model. The conventional beamforming method finds K DOA estimates (when the spatial resolution is sufficient) but cannot determine how the sources were numbered. Further information regarding the transmitted signals is required to distinguish between sources, which is assumed unavailable when using non-parametric methods.

8.1.2 Non-Parametric Capon Beamforming

The conventional beamforming method works very well for DOA estimation in the single-source scenario. However, several modifications exist to enhance the resolution in multi-source scenarios. The general idea is to go beyond array response vectors and use other combining vectors $\mathbf{w}$ that make it easier to distinguish the sources. This is reminiscent of how MRC can be replaced by LMMSE combining in the uplink of multi-user MIMO to suppress inter-user interference and thereby achieve a higher data rate.

One important technique is *Capon beamforming*, named after its originator Jack Capon [124]. This technique is also known as *minimum-variance distortionless response (MVDR)* beamforming. As the latter name indicates, when inspecting a specific direction (φ, θ), we should use the beamforming vector $\mathbf{w}$ that minimizes the variance of the received signal while not distorting the signal that arrives from the intended direction. The variance is the received power $P(\mathbf{w}) = \mathbf{w}^{\mathrm{H}}\hat{\mathbf{R}}_L\mathbf{w}$ defined in (8.4), while $\mathbf{w}^{\mathrm{H}}\mathbf{a}(\varphi, \theta) = 1$ is required not to distort the impinging wave coming from the direction (φ, θ). We find the Capon beamforming by solving the optimization problem

$$\begin{aligned} \underset{\mathbf{w}\in\mathbb{C}^M}{\text{minimize}} \quad & \mathbf{w}^{\mathrm{H}}\hat{\mathbf{R}}_L\mathbf{w} \\ \text{subject to} \quad & \mathbf{w}^{\mathrm{H}}\mathbf{a}(\varphi, \theta) = 1. \end{aligned} \tag{8.17}$$

When there are $L \geq M$ received signal samples, the estimate $\hat{\mathbf{R}}_L$ of the correlation matrix is almost always non-singular due to the noise. By defining $\overline{\mathbf{w}} = \hat{\mathbf{R}}_L^{1/2}\mathbf{w}$ as a new optimization variable, the problem in (8.17) can be rewritten (by utilizing the invertibility of $\hat{\mathbf{R}}_L$) as

$$\begin{aligned} \underset{\overline{\mathbf{w}}\in\mathbb{C}^M}{\text{minimize}} \quad & \overline{\mathbf{w}}^{\mathrm{H}}\overline{\mathbf{w}} \\ \text{subject to} \quad & \overline{\mathbf{w}}^{\mathrm{H}}\hat{\mathbf{R}}_L^{-1/2}\mathbf{a}(\varphi, \theta) = 1. \end{aligned} \tag{8.18}$$

The vector $\overline{\mathbf{w}} = \hat{\mathbf{R}}_L^{-1/2}\mathbf{a}(\varphi, \theta)/\|\hat{\mathbf{R}}_L^{-1/2}\mathbf{a}(\varphi, \theta)\|^2$ gives equality in the constraint and has the minimum norm among all potential solutions because it is parallel to $\hat{\mathbf{R}}_L^{-1/2}\mathbf{a}(\varphi, \theta)$. Hence, this is the solution to (8.18). The corresponding solution to (8.17) is

$$\mathbf{w} = \hat{\mathbf{R}}_L^{-1/2}\overline{\mathbf{w}} = \frac{\hat{\mathbf{R}}_L^{-1}\mathbf{a}(\varphi, \theta)}{\mathbf{a}^{\mathrm{H}}(\varphi, \theta)\hat{\mathbf{R}}_L^{-1}\mathbf{a}(\varphi, \theta)}, \tag{8.19}$$

which is called Capon/MVDR beamforming. If we insert this vector into the general power spectrum expression in (8.4), we obtain the *Capon spectrum*

$$\begin{aligned} P_{\text{Capon}}(\varphi, \theta) &= \frac{\mathbf{a}^{\mathrm{H}}(\varphi, \theta)\hat{\mathbf{R}}_L^{-1}\hat{\mathbf{R}}_L\hat{\mathbf{R}}_L^{-1}\mathbf{a}(\varphi, \theta)}{\left(\mathbf{a}^{\mathrm{H}}(\varphi, \theta)\hat{\mathbf{R}}_L^{-1}\mathbf{a}(\varphi, \theta)\right)^2} = \frac{\mathbf{a}^{\mathrm{H}}(\varphi, \theta)\hat{\mathbf{R}}_L^{-1}\mathbf{a}(\varphi, \theta)}{\left(\mathbf{a}^{\mathrm{H}}(\varphi, \theta)\hat{\mathbf{R}}_L^{-1}\mathbf{a}(\varphi, \theta)\right)^2} \\ &= \frac{1}{\mathbf{a}^{\mathrm{H}}(\varphi, \theta)\hat{\mathbf{R}}_L^{-1}\mathbf{a}(\varphi, \theta)}. \end{aligned} \tag{8.20}$$

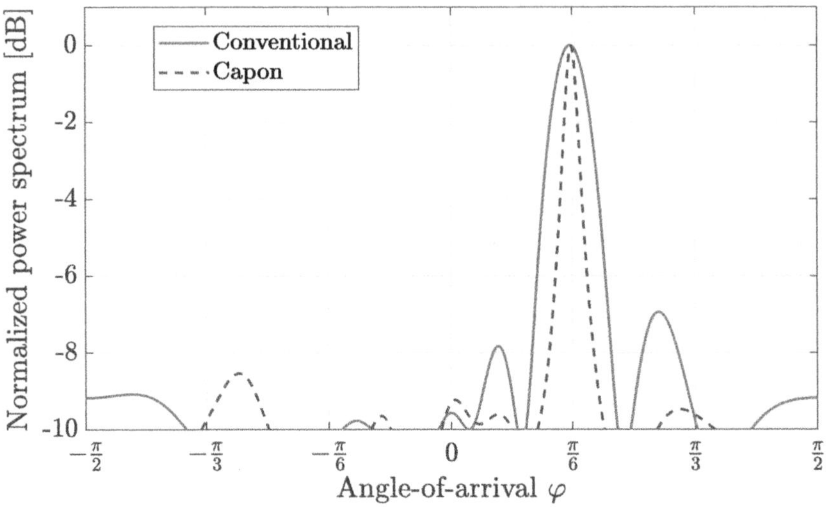

Figure 8.6: The normalized power spectrum with conventional and Capon beamforming for a single random realization. A ULA with $M = 10$ antennas and $\Delta = \lambda/2$ is considered. There is a single source with the DOA $\varphi = \pi/6$.

The DOA estimates $\{\hat{\varphi}_k, \hat{\theta}_k\}$, for $k = 1, \ldots, K$, are obtained as the K highest peaks of the Capon spectrum.

Figure 8.6 compares the normalized power spectrum of conventional and Capon beamforming for a single random realization. A ULA with $M = 10$ antennas and $\Delta = \lambda/2$ is considered. There is a single source with the DOA $\varphi = \pi/6$, $L = 25$ samples are used, and the SNR is 0 dB. The figure shows that the beamwidth of the Capon beamformer is narrower; thus, it has a higher spatial resolution. The price to pay is that the peak of the power spectrum can be shifted more from the true DOA when Capon beamforming is used.

The consequence of the larger deviation of the peak is highlighted in Figure 8.7, which shows the MSE of the DOA estimates with conventional and Capon beamforming for the same setup as in the last figure. This time, we vary the number of samples L and consider two SNR values: 0 dB and 10 dB. Conventional beamforming provides a smaller MSE than Capon beamforming in this single-source scenario when the number of samples is low. However, as L increases, the MSE gap diminishes.

The bottom line is that Capon beamforming is unnecessary in the single-source scenario. However, it is designed to deal with situations with multiple sources, where the improved spatial resolution can help resolve closely spaced sources for which conventional beamforming fails. An example of this is provided in Figure 8.8, where we consider $K = 2$ sources with the DOAs $\varphi_1 = \pi/6$ and $\varphi_2 = \pi/5$. A ULA with $M = 20$ antennas and $\Delta = \lambda/2$ is considered. Figure 8.8 shows the normalized power spectrum of conventional and Capon beamforming for a single random realization. The spectrum with

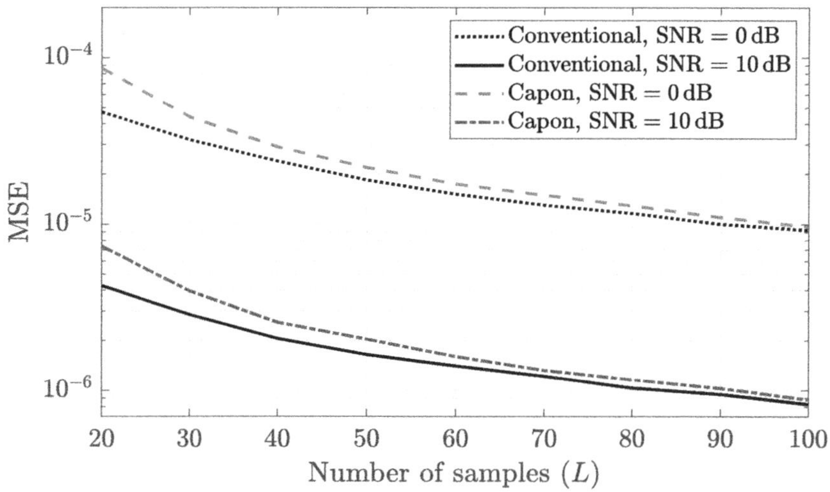

Figure 8.7: The MSE of DOA estimation with conventional and Capon beamforming as a function of the number of samples used to compute the power spectra. A ULA with $M = 10$ antennas and $\Delta = \lambda/2$ is considered. There is a single source with the DOA $\varphi = \pi/6$, and the SNR is varied.

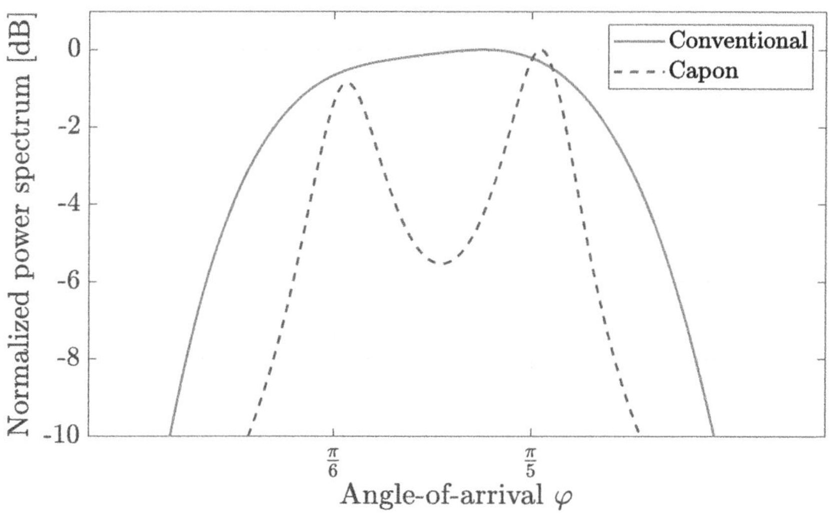

Figure 8.8: The normalized power spectrum with conventional and Capon beamforming for a single random realization. A ULA with $M = 20$ antennas and $\Delta = \lambda/2$ antenna spacing is considered. There are $K = 2$ sources with the DOAs $\varphi_1 = \pi/6$ and $\varphi_2 = \pi/5$. The peaks of the power spectrum are the DOA estimates.

conventional beamforming only has one peak, so it cannot resolve the two sources. On the other hand, the Capon spectrum has two clearly distinguishable peaks thanks to its increased spatial resolution. The locations of the peaks are slightly biased/shifted compared to the true DOAs, but Capon beamforming can at least provide two decent DOA estimates in this challenging setup.

Example 8.3. Prove that Capon beamforming is a consistent DOA estimator when there is a single source, $\Delta = \lambda/2$, and $\varphi \in [-\pi/2, \pi/2]$ is of interest.

An estimator is consistent if the estimation error vanishes asymptotically. When $L \to \infty$, it follows that $\hat{\mathbf{R}}_L \to \mathbf{R}$ and the Capon spectrum approaches

$$P_{\text{Capon}}(\varphi, \theta) \to \overline{P}_{\text{Capon}}(\varphi, \theta) = \frac{1}{\mathbf{a}^{\mathrm{H}}(\varphi, \theta)\mathbf{R}^{-1}\mathbf{a}(\varphi, \theta)}, \tag{8.21}$$

where the correlation matrix $\mathbf{R}$ is given by (8.11) for the single-source case. Using the matrix inversion lemma from Lemma 2.3, $\mathbf{R}^{-1}$ can be expressed as

$$\begin{aligned}\mathbf{R}^{-1} &= \left(P_1\beta_1\mathbf{a}(\varphi_1, \theta_1)\mathbf{a}^{\mathrm{H}}(\varphi_1, \theta_1) + \sigma^2\mathbf{I}_M\right)^{-1} \\ &= \sigma^{-2}\mathbf{I}_M - \frac{\sigma^{-2}P_1\beta_1}{\sigma^2 + P_1\beta_1 M}\mathbf{a}(\varphi_1, \theta_1)\mathbf{a}^{\mathrm{H}}(\varphi_1, \theta_1). \end{aligned} \tag{8.22}$$

Hence, $\mathbf{a}^{\mathrm{H}}(\varphi, \theta)\mathbf{R}^{-1}\mathbf{a}(\varphi, \theta)$ in the denominator of (8.21) becomes

$$\begin{aligned}&\sigma^{-2}\mathbf{a}^{\mathrm{H}}(\varphi, \theta)\mathbf{I}_M\mathbf{a}(\varphi, \theta) - \frac{\sigma^{-2}P_1\beta_1}{\sigma^2 + P_1\beta_1 M}\left|\mathbf{a}^{\mathrm{H}}(\varphi, \theta)\mathbf{a}(\varphi_1, \theta_1)\right|^2 \\ &= \sigma^{-2}M - \frac{\sigma^{-2}P_1\beta_1}{\sigma^2 + P_1\beta_1 M}\left|\mathbf{a}^{\mathrm{H}}(\varphi, \theta)\mathbf{a}(\varphi_1, \theta_1)\right|^2. \end{aligned} \tag{8.23}$$

Inserting this expression into the right-hand side of (8.21), we obtain

$$\overline{P}_{\text{Capon}}(\varphi, \theta) = \frac{1}{\sigma^{-2}M - \frac{\sigma^{-2}P_1\beta_1}{\sigma^2 + P_1\beta_1 M}\left|\mathbf{a}^{\mathrm{H}}(\varphi, \theta)\mathbf{a}(\varphi_1, \theta_1)\right|^2}. \tag{8.24}$$

The asymptotic Capon spectrum is maximized when $|\mathbf{a}^{\mathrm{H}}(\varphi, \theta)\mathbf{a}(\varphi_1, \theta_1)|^2$ is maximized, which according to the Cauchy-Schwartz inequality in (2.18) only happens if $\mathbf{a}(\varphi, \theta) = \mathbf{a}(\varphi_1, \theta_1)$. This equation has only one solution $\varphi \in [-\pi/2, \pi/2]$; thus, Capon beamforming is a consistent DOA estimator.

The DOA estimation performance changes if the sources transmit correlated signals. We will consider a ULA with $M = 3$ antennas and $\Delta = \lambda/2$ to exemplify this. There are $K = 2$ sources with the DOAs $\varphi_1 = 0$ and $\varphi_2 = \pi/6$. The corresponding array response vectors can be computed using (8.8) as

$$\mathbf{a}(\varphi_1 = 0) = \begin{bmatrix} 1 \\ e^{-\mathrm{j}\pi\sin(0)} \\ e^{-\mathrm{j}\pi 2\sin(0)} \end{bmatrix} = \begin{bmatrix} 1 \\ 1 \\ 1 \end{bmatrix}, \quad \mathbf{a}(\varphi_2 = \pi/6) = \begin{bmatrix} 1 \\ e^{-\mathrm{j}\pi\sin(\pi/6)} \\ e^{-\mathrm{j}\pi 2\sin(\pi/6)} \end{bmatrix} = \begin{bmatrix} 1 \\ -\mathrm{j} \\ -1 \end{bmatrix}. \tag{8.25}$$

Suppose the random source signals $x_1[l], x_2[l]$ in (8.3) are always equal except for a phase-shift: $x_1[l] = x_2[l]e^{\mathrm{j}\phi}$. Such sources are called *coherent*. This scenario can happen when the same beacon signal is broadcasted from two sources or when the signal from one source is reflected on two objects before reaching the receiver. For simplicity, suppose that $\beta_1 = \beta_2 = \beta$, $\psi_1 = \psi_2 = 0$, and $\phi = 0$. The received signal in (8.3) then becomes

$$\begin{aligned}\mathbf{y}[l] &= \sqrt{\beta_1}\mathbf{a}(\varphi_1)x_1[l] + \sqrt{\beta_2}\mathbf{a}_2(\varphi_2)x_2[l] + \mathbf{n}[l] \\ &= \sqrt{\beta}\underbrace{\left(\mathbf{a}(0) + \mathbf{a}(\pi/6)\right)}_{=\bar{\mathbf{a}}}x[l] + \mathbf{n}[l]. \end{aligned} \tag{8.26}$$

Hence, when the source signals are coherent, the received signal is the same as if there were a single source with the effective array response vector

$$\bar{\mathbf{a}} = \begin{bmatrix}1\\1\\1\end{bmatrix} + \begin{bmatrix}1\\-\mathrm{j}\\-1\end{bmatrix} = \begin{bmatrix}2\\1-\mathrm{j}\\0\end{bmatrix}. \tag{8.27}$$

The asymptotic power spectra in (8.12) and (8.24) for conventional and Capon beamforming, respectively, are proportional to $|\mathbf{a}^{\mathrm{H}}(\varphi)\bar{\mathbf{a}}|$. To obtain consistent estimates, we expect the power spectra to have their peaks at $\varphi = \varphi_1 = 0$ and $\varphi = \varphi_2 = \pi/6$. However, this is not the case because $|\mathbf{a}^{\mathrm{H}}(0)\bar{\mathbf{a}}| \approx 3.16$ and $|\mathbf{a}^{\mathrm{H}}(\pi/6)\bar{\mathbf{a}}| \approx 1.41$, while $|\mathbf{a}^{\mathrm{H}}(\arcsin(1/4))\bar{\mathbf{a}}| \approx 3.41$ gives a larger value. This specific angle gives $\mathbf{a}(\arcsin(1/4)) = [1, (1-\mathrm{j})/\sqrt{2}, -\mathrm{j}]^{\mathrm{T}}$, which resembles (8.27). Hence, the conventional and Capon beamforming methods are not consistent DOA estimators when the sources are coherent.

Example 8.4. Consider K sources that transmit independent data signals with power P and suppose the noise variance is σ^2. How is Capon beamforming related to LMMSE combining in this case?

In this setup, the Capon beamforming expression in (8.19) has the limit

$$\mathbf{w} \to c\mathbf{R}^{-1}\mathbf{a}(\varphi,\theta) = c\left(\sum_{k=1}^{K} P\beta_k \mathbf{a}(\varphi_k,\theta_k)\mathbf{a}^{\mathrm{H}}(\varphi_k,\theta_k) + \sigma^2\mathbf{I}_M\right)^{-1}\mathbf{a}(\varphi,\theta) \tag{8.28}$$

as $L \to \infty$, where $c = 1/(\mathbf{a}^{\mathrm{H}}(\varphi,\theta)\hat{\mathbf{R}}_L^{-1}\mathbf{a}(\varphi,\theta))$ is a scalar. When inspecting the kth source direction by setting $\varphi = \varphi_k$ and $\theta = \theta_k$, the limit in (8.28) coincides with the LMMSE combining vector in (6.63) for an uplink multi-user MIMO system where the users transmit with power P and have the LOS channels $\mathbf{h}_k = \mathbf{a}(\varphi_k,\theta_k)$ for $k = 1,\ldots,K$. The only difference between the two expressions is the scalar c, which is selected to get a distortionless signal in Capon beamforming, while it is picked to minimize the MSE in LMMSE combining. The vital difference is the application: LMMSE combining is implemented to receive uplink data signals when the channels are known, while Capon beamforming aims at estimating the channel parameters without knowing the signals. However, the similarities between the system models imply that Capon beamforming is an approximate form of LMMSE combining.

In summary, conventional and Capon beamforming are consistent DOA estimators when there is a single source, if the array deployment causes no DOA ambiguity. The main beam with Capon beamforming can be slightly shifted; thus, it requires more samples to be as accurate as conventional beamforming when there is a single source. On the other hand, Capon beamforming has a higher spatial resolution, and when multiple sources have close DOAs, it can resolve sources that conventional beamforming cannot. There is a limit on how closely spaced sources these methods can distinguish for a given number of antennas. Correlation between the source signals will reduce the accuracy of the DOA estimates. These are the main reasons for developing more advanced methods that exploit the source and signal statistics. Later in this chapter, we will present a subspace-based technique belonging to that category.

8.1.3 Joint Azimuth and Elevation DOA Estimation

The Capon spectrum in (8.20) can be applied with arbitrary array geometries and source locations, but all the previous simulation examples have considered ULAs and sources with zero elevation angles. In this section, we will have a closer look at how the theory can be used to jointly estimate the azimuth and elevation angles of the sources.

Figure 8.9 shows the normalized power spectrum with Capon beamforming for a single random realization containing $L = 25$ time samples. A ULA is considered with $M = 16$ antennas and $\Delta = \lambda/2$. There is a single source with the azimuth and elevation DOAs $\varphi = \pi/4$ and $\theta = -\pi/4$, respectively. The SNR is 0 dB. The figure shows infinitely many peak values along the yellow arc in the azimuth-elevation plane. Hence, there is a DOA estimation ambiguity when using a ULA to simultaneously estimate the azimuth and elevation angles. The true source location is marked with a green circle and is on the arc, but we cannot distinguish it from the other points. This reason can be identified by analyzing the array response vector of the source:

$$\mathbf{a}(\pi/4, -\pi/4) = \begin{bmatrix} 1 \\ e^{-\mathrm{j}\pi \sin(\pi/4)\cos(-\pi/4)} \\ \vdots \\ e^{-\mathrm{j}\pi(M-1)\sin(\pi/4)\cos(-\pi/4)} \end{bmatrix} = \begin{bmatrix} 1 \\ e^{-\mathrm{j}\pi/2} \\ \vdots \\ e^{-\mathrm{j}\pi(M-1)/2} \end{bmatrix}. \tag{8.29}$$

The same array response vector can be obtained for $\varphi = \pi/6$ and $\theta = 0$:

$$\mathbf{a}(\pi/6, 0) = \begin{bmatrix} 1 \\ e^{-\mathrm{j}\pi \sin(\pi/6)\cos(0)} \\ \vdots \\ e^{-\mathrm{j}\pi(M-1)\sin(\pi/6)\cos(0)} \end{bmatrix} = \begin{bmatrix} 1 \\ e^{-\mathrm{j}\pi/2} \\ \vdots \\ e^{-\mathrm{j}\pi(M-1)/2} \end{bmatrix}. \tag{8.30}$$

Suppose we somehow know the elevation angle to the source (as in previous examples where it was zero). In that case, we only need to look for the peak

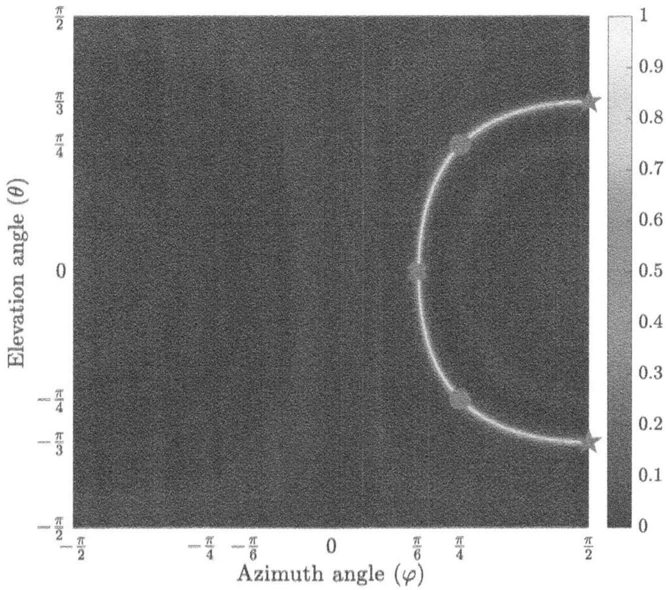

Figure 8.9: The normalized 2D power spectrum of DOA estimation for a ULA with $M = 16$ antennas, $L = 25$ samples, 0 dB SNR, and Capon beamforming. There is one source with the azimuth and elevation DOAs $\varphi = \pi/4$ and $\theta = -\pi/4$, indicated by the green circle. The color shows the spectrum value. It has infinitely many peaks along the yellow arc, which results in ambiguity. The correct point is only found if the receiver somehow knows the correct elevation DOA. The red cross, red stars, and red circle show the alternative DOA estimates obtained if the receiver knows that the elevation DOA is $\theta = 0$, $\theta = \pm\pi/3$, or $\theta = \pi/4$, respectively.

value along the corresponding horizontal line in Figure 8.9, and there is only a single yellow peak on that line. For example, if we know that $\theta = -\pi/4$, we will find the correct DOA estimate. However, if we incorrectly believe that $\theta = 0$, the red cross at $\varphi = \pi/6$ denotes the unique but incorrect solution we will get. Similarly, if we incorrectly believe that the correct elevation DOA is $\theta = \pm\pi/3$, we will obtain $\varphi = \pi/2$ as the azimuth DOA estimate because it also gives the same array response vector as in (8.29):

$$\mathbf{a}(\pi/2, \pm\pi/3) = \begin{bmatrix} 1 \\ e^{-\mathrm{j}\pi \sin(\pi/2)\cos(\pm\pi/3)} \\ \vdots \\ e^{-\mathrm{j}\pi(M-1)\sin(\pi/2)\cos(\pm\pi/3)} \end{bmatrix} = \begin{bmatrix} 1 \\ e^{-\mathrm{j}\pi/2} \\ \vdots \\ e^{-\mathrm{j}\pi(M-1)/2} \end{bmatrix}. \tag{8.31}$$

The red stars in the figure show these DOA estimates.

On the other hand, it is not enough to know that $\varphi = \pi/4$ is the correct azimuth angle because there are two peaks on the corresponding vertical line, resulting in an ambiguity between $\theta = \pm\pi/4$ (marked with circles). The bottom line is that a ULA cannot estimate the azimuth and elevation DOAs jointly, but we must know the elevation angle to find the correct DOA.

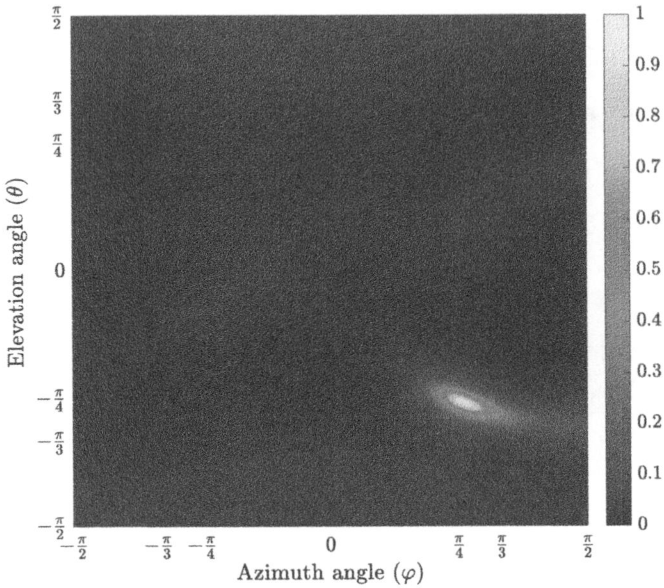

Figure 8.10: The normalized power spectrum of DOA estimation for a UPA with $M_{\mathrm{H}} = M_{\mathrm{V}} = 4$, $\Delta = \lambda/2$, and Capon beamforming. The parameters are otherwise the same as in the ULA case in Figure 8.9. Unlike that case, a single peak is located at the true DOA when using a UPA.

One way to resolve the ambiguity is to use a two-dimensional array, capable of 3D beamforming, to resolve signals both horizontally and vertically. We will exemplify this feature by considering a UPA with $M_{\mathrm{H}} = 4$ horizontal antennas, $M_{\mathrm{V}} = 4$ vertical antennas, and $\Delta = \lambda/2$. The total number of antennas is the same as in the ULA, and all other parameters are unchanged. Figure 8.10 shows the normalized power spectrum of Capon beamforming for a single random realization. In this case, there is only a single peak, and it is located at the true azimuth and elevation DOAs (i.e., $\varphi = \pi/4$ and $\theta = -\pi/4$). This implies that switching from a ULA to a UPA resolves the angular ambiguity issue. The enabling factor is that the array response vector can be expressed using (4.128) as

$$\mathbf{a}_{4,4}(\pi/4, -\pi/4) = \mathbf{a}_4(-\pi/4, 0) \otimes \mathbf{a}_4(\pi/4, -\pi/4), \tag{8.32}$$

which is the Kronecker product of the array responses of two 4-antenna ULAs. When considering the ULA earlier in this section, we noticed that multiple DOA pairs give rise to the same vector as in the second factor in (8.32). These are all the values (φ, θ) that give $\sin(\varphi)\cos(\theta) = \sin(\pi/4)\cos(-\pi/4) = 1/2$. If we pick the wrong angle pair, it will give the wrong vector in the first factor in (8.32). Hence, the array response vector is unique, and the UPA can provide a consistent estimate of both the azimuth and elevation angle. The only necessary condition is that the antenna spacing satisfies $\Delta \leq \lambda/2$ and that we only consider azimuth angles on one side of the array: $\varphi \in [-\pi/2, \pi/2]$.

Example 8.5. Consider a UPA with $M_{\mathrm{H}} = M_{\mathrm{V}} = 2$ and the antenna spacing $\Delta = \lambda$. If there is a single source with the azimuth and elevation DOAs $\varphi_1 = \pi/4$ and $\theta_1 = -\pi/4$, is there a unique peak in the Capon spectrum?

Following (4.128), the array response vector is

$$\mathbf{a}_{2,2}(\varphi,\theta) = \begin{bmatrix} 1 \\ e^{-\mathrm{j}2\pi\sin(\theta)} \end{bmatrix} \otimes \begin{bmatrix} 1 \\ e^{-\mathrm{j}2\pi\sin(\varphi)\cos(\theta)} \end{bmatrix}. \tag{8.33}$$

If there is an extra peak at some angle pair $(\tilde{\varphi},\tilde{\theta})$ in the Capon spectrum, then both factors in (8.33) must be the same as for the true source angles. We begin by comparing the first factors, which are equal if

$$-2\pi\sin(-\pi/4) = \sqrt{2}\pi = -2\pi\sin(\tilde{\theta}) + 2\pi n_1 \ \Rightarrow \ \sin(\tilde{\theta}) = n_1 - \frac{1}{\sqrt{2}} \tag{8.34}$$

for any integer n_1. Two elevation angles satisfy this condition: the true DOA $\tilde{\theta}_1 = -\pi/4$ (for $n_1 = 0$) and the extra solution (for $n_1 = 1$)

$$\tilde{\theta}_2 = \arcsin\left(1 - \frac{1}{\sqrt{2}}\right) \approx 0.297\,\mathrm{rad}. \tag{8.35}$$

For any given value of $\tilde{\theta}$, the second factor in (8.33) is the same as for the source if

$$\begin{aligned} &-2\pi\sin(\pi/4)\cos(-\pi/4) = -\pi = -2\pi\sin(\tilde{\varphi})\cos(\tilde{\theta}) + 2\pi n_2 \\ &\Rightarrow \ \sin(\tilde{\varphi})\cos(\tilde{\theta}) = n_2 + \frac{1}{2} \ \Rightarrow \ \tilde{\varphi} = \arcsin\left(\frac{n_2 + 1/2}{\cos(\tilde{\theta})}\right) \end{aligned} \tag{8.36}$$

for any integer n_2. This equation has the two solutions: $\tilde{\varphi}_{1,1} = \pi/4$ (for $n_2 = 0$) and $\tilde{\varphi}_{1,2} = -\pi/4$ (for $n_2 = -1$) when $\tilde{\theta}_1 = -\pi/4$ is considered. For $\tilde{\theta}_2$, we obtain the additionalsolutions

$$\tilde{\varphi}_{2,1} = \arcsin\left(\frac{1/2}{\cos(\tilde{\theta}_2)}\right) \approx 0.55\,\mathrm{rad}, \quad \tilde{\varphi}_{2,2} = \arcsin\left(\frac{-1/2}{\cos(\tilde{\theta}_2)}\right) \approx -0.55\,\mathrm{rad}. \tag{8.37}$$

Hence, the power spectrum has the four peaks $(\tilde{\varphi}_{1,1},\tilde{\theta}_1)$, $(\tilde{\varphi}_{1,2},\tilde{\theta}_1)$, $(\tilde{\varphi}_{2,1},\tilde{\theta}_2)$, and $(\tilde{\varphi}_{2,2},\tilde{\theta}_2)$. The reason for not having a unique peak is the large antenna spacing of $\Delta = \lambda$, which creates one grating lobe in the azimuth plane and one in the elevation plane. The latter one has a grating lobe on its own.

8.1.4 Parametric Subspace-Based Methods

Subspace-based methods can provide better DOA estimation accuracy than beamforming methods by exploiting further information regarding the source signals. Similar to beamforming methods, they exploit the estimate $\hat{\mathbf{R}}_L$ of the received signal's correlation matrix. As the name "subspace-based" suggests, these methods rely on explicitly separating the eigendecomposition of $\hat{\mathbf{R}}_L$ into *signal* and *noise* subspaces [51]. *MUltiple SIgnal Classification (MUSIC)* [125], [126] and *Estimation of Signal Parameters by Rotational Invariance Techniques (ESPRIT)* [127], [128] are two classic subspace-based DOA estimation methods. The former method exploits the noise subspace, which is spanned by the eigenvectors of the smallest eigenvalues of $\hat{\mathbf{R}}_L$, while the latter technique uses the signal subspace spanned by the eigenvectors of the largest eigenvalues. In this section, we will describe the basic form of the MUSIC algorithm and compare it to Capon beamforming. We refer to the textbook [129] for a detailed description of ESPRIT and variations on MUSIC.

We revisit the signal model in (8.3), where the received signal at time l is

$$\mathbf{y}[l] = \sum_{k=1}^{K} \sqrt{\beta_k} e^{-\mathrm{j}\psi_k} \mathbf{a}(\varphi_k, \theta_k) x_k[l] + \mathbf{n}[l]. \tag{8.38}$$

We assume the number of sources is smaller than the number of antennas (i.e., $K < M$) and define the vector $\mathbf{p}[l] = [\sqrt{\beta_1} e^{-\mathrm{j}\psi_1} x_1[l], \ldots, \sqrt{\beta_K} e^{-\mathrm{j}\psi_K} x_K[l]]^{\mathrm{T}}$ containing the received signals at the first antenna. If we denote its correlation matrix as $\mathbf{P} = \mathbb{E}\{\mathbf{p}[l]\mathbf{p}^{\mathrm{H}}[l]\}$, the correlation matrix of $\mathbf{y}[l]$ can be expressed as

$$\mathbf{R} = \mathbb{E}\left\{\mathbf{y}[l]\mathbf{y}^{\mathrm{H}}[l]\right\} = \mathbf{A}\mathbf{P}\mathbf{A}^{\mathrm{H}} + \sigma^2 \mathbf{I}_M, \tag{8.39}$$

where $\mathbf{A} \in \mathbb{C}^{M \times K}$ contains the array response vectors of the sources as its columns:

$$\mathbf{A} = \begin{bmatrix} \mathbf{a}(\varphi_1, \theta_1) & \ldots & \mathbf{a}(\varphi_K, \theta_K) \end{bmatrix}. \tag{8.40}$$

The eigendecomposition of the positive semi-definite Hermitian matrix $\mathbf{A}\mathbf{P}\mathbf{A}^{\mathrm{H}}$ in (8.39) always exists and can be expressed as

$$\mathbf{A}\mathbf{P}\mathbf{A}^{\mathrm{H}} = \mathbf{U}\mathbf{D}\mathbf{U}^{\mathrm{H}}, \tag{8.41}$$

where the diagonal entries of $\mathbf{D}$ contain the real-valued positive eigenvalues in decreasing order and the columns of $\mathbf{U}$ are the corresponding unit-length eigenvectors. Adding a scaled identity matrix to $\mathbf{U}\mathbf{D}\mathbf{U}^{\mathrm{H}}$ preserves the eigenvectors but increases all the eigenvalues (see Example 2.7). Hence, the eigendecomposition of $\mathbf{R}$ in (8.39) is

$$\mathbf{R} = \mathbf{A}\mathbf{P}\mathbf{A}^{\mathrm{H}} + \sigma^2 \mathbf{I}_M = \mathbf{U}\left(\mathbf{D} + \sigma^2 \mathbf{I}_M\right)\mathbf{U}^{\mathrm{H}}. \tag{8.42}$$

If the matrix $\mathbf{A}\mathbf{P}\mathbf{A}^{\mathrm{H}}$ has rank of r, then r eigenvalues of $\mathbf{R}$ are strictly greater than σ^2 and the remaining $M - r$ eigenvalues are exactly σ^2. Since we index the eigenvalues in decreasing order, we can decompose $\mathbf{U}$ as $\mathbf{U} = [\mathbf{U}_{\mathrm{s}}, \mathbf{U}_{\mathrm{n}}]$, where $\mathbf{U}_{\mathrm{s}} \in \mathbb{C}^{M \times r}$ contains the eigenvectors corresponding to the non-zero eigenvalues of $\mathbf{APA}^{\mathrm{H}}$. These r eigenvectors span the signal subspace of $\mathbf{R}$, as the subscript s indicates. This subspace contains all the received signals and additive noise. On the other hand, the columns of $\mathbf{U}_{\mathrm{n}} \in \mathbb{C}^{M \times (M-r)}$ contain the eigenvectors corresponding to the zero-valued eigenvalues of $\mathbf{APA}^{\mathrm{H}}$. These $M - r$ eigenvectors span the noise subspace of $\mathbf{R}$, as the subscript n indicates. This subspace only contains noise with variance σ^2.

Since the eigenvectors in $\mathbf{U}_{\mathrm{n}}$ correspond to the zero-valued eigenvalues of $\mathbf{APA}^{\mathrm{H}}$, we have the relation

$$\mathbf{APA}^{\mathrm{H}}\mathbf{U}_{\mathrm{n}} = \mathbf{0}. \tag{8.43}$$

From linear algebra, we know that if $\mathbf{AP} \in \mathbb{C}^{M \times K}$ has the full rank of K (recall the assumption $K < M$), then (8.43) implies

$$\begin{aligned} \mathbf{A}^{\mathrm{H}}\mathbf{U}_{\mathrm{n}} = \mathbf{0} \quad &\Rightarrow \quad \mathbf{a}^{\mathrm{H}}(\varphi_k, \theta_k)\mathbf{U}_{\mathrm{n}} = \mathbf{0}, \quad k = 1, \dots, K \\ &\Rightarrow \quad \mathbf{a}^{\mathrm{H}}(\varphi_k, \theta_k)\mathbf{U}_{\mathrm{n}}\mathbf{U}_{\mathrm{n}}^{\mathrm{H}}\mathbf{a}(\varphi_k, \theta_k) = 0, \quad k = 1, \dots, K. \end{aligned} \tag{8.44}$$

The rank of $\mathbf{AP}$ is equal to the rank of $\mathbf{APA}^{\mathrm{H}}$. To achieve full rank, we need both $\mathbf{A}$ and $\mathbf{P}$ to have full rank. The correlation matrix $\mathbf{P}$ is non-singular when the source signals are not fully correlated (coherent). Secondly, the matrix $\mathbf{A}$ has full rank if and only if the K array response vectors $\mathbf{a}(\varphi_k, \theta_k)$ are linearly independent. When the second condition is satisfied, the array is said to be *unambigious*, which enables unique DOA estimates [51]. This is a necessary condition for the existence of a consistent estimator, but in non-asymptotic cases, there might nevertheless be multiple peaks in the spectrum even if there is only a single source, and the DOA estimate might be erroneous. The unambiguity is a usual assumption valid for the most commonly used arrays. The following lemma presents the conditions for a ULA.

Lemma 8.1. The array response vectors $\mathbf{a}(\varphi_k, \theta_k)$ for $k = 1, \dots, K$, where $K \leq M$, are linearly independent for a horizontal ULA with $\Delta \leq \lambda/2$ if the K DOAs result in distinctly different values of $\sin(\varphi_k)\cos(\theta_k)$.

Suppose that the source angles satisfy $\theta_k = 0$ and $\varphi_k \in [-\pi/2, \pi/2]$, for $k = 1, \dots, K$. If the K azimuth angles φ_k are different, then Lemma 8.1 implies that the array response vectors $\mathbf{a}(\varphi_k, \theta_k)$ are linearly independent when using a ULA with $\Delta \leq \lambda/2$.

If we know $\mathbf{U}_{\mathrm{n}}$ and the array is unambiguous, we can find the DOA angles of the sources by searching for K linearly independent array response vectors that give equality in (8.44). The MUSIC algorithm builds on this principle but deals with the situation where $\mathbf{U}_{\mathrm{n}}$ is estimated from the received signals.

Under the assumption that $\mathbf{APA}^{\mathrm{H}}$ has full rank (i.e., $r = K$), the MUSIC algorithm estimates the DOA angles by first constructing the sample average estimator of $\mathbf{R}$ using L samples as

$$\hat{\mathbf{R}}_L = \frac{1}{L}\sum_{l=1}^{L} \mathbf{y}[l]\mathbf{y}^{\mathrm{H}}[l]. \tag{8.45}$$

We then compute the eigendecomposition of $\hat{\mathbf{R}}_L$ and let $\hat{\mathbf{U}}_{\mathrm{n}} \in \mathbb{C}^{M\times(M-K)}$ be the matrix whose columns are the unit-length eigenvectors corresponding to the $M-K$ smallest eigenvalues. Inspired by the fact that $\mathbf{a}^{\mathrm{H}}(\varphi,\theta)\mathbf{U}_{\mathrm{n}}\mathbf{U}_{\mathrm{n}}^{\mathrm{H}}\mathbf{a}(\varphi,\theta) = 0$ when considering the DOA of a source, we define the MUSIC spectrum as

$$P_{\mathrm{MUSIC}}(\varphi,\theta) = \frac{1}{\mathbf{a}^{\mathrm{H}}(\varphi,\theta)\hat{\mathbf{U}}_{\mathrm{n}}\hat{\mathbf{U}}_{\mathrm{n}}^{\mathrm{H}}\mathbf{a}(\varphi,\theta)} \tag{8.46}$$

for azimuth angles $\varphi \in [-\pi/2,\pi/2]$ and elevation angles $\theta \in [-\pi/2,\pi/2]$. The denominator is nearly zero when the angle (φ,θ) is close to a source, which will generate a peak in the spectrum. If $\hat{\mathbf{U}}_{\mathrm{n}}$ is exactly equal to $\mathbf{U}_{\mathrm{n}}$ (i.e., $\hat{\mathbf{R}}_L = \mathbf{R}$), then the MUSIC spectrum is infinite at the true DOAs. Since we only have access to the estimate $\hat{\mathbf{U}}_{\mathrm{n}}$, the peak values and locations are approximations. When K is known, the K tallest peaks of the MUSIC spectrum are declared as the DOA estimates. When K is unknown, the MUSIC algorithm can also detect the number of sources by inspecting the eigenvalues of $\hat{\mathbf{R}}_L$. By comparing them with a threshold, we can determine how many are substantially larger than σ^2 and use this value as the estimate of K. We then proceed by identifying the K tallest peaks of the MUSIC spectrum.

In Figure 8.11, we show the normalized power spectra using either Capon beamforming or the MUSIC algorithm for DOA estimation. A ULA is considered with $M = 50$ antennas and $\Delta = \lambda/2$. There are $K = 2$ sources with the azimuth DOAs $\varphi_1 = \pi/6$ and $\varphi_2 = \pi/5$, respectively. The elevation angles are zero. The source signals are independent and Gaussian distributed. The transmit power and channel gains are the same, and the common SNR is 0 dB. We use $L = 100$ samples in Figure 8.11(a). In this case, both Capon beamforming and the MUSIC algorithm have peaks around the true DOAs, although the peak values are not exactly centered at the true values, so the DOA estimates are not exact. However, the resolution of MUSIC is superior since the main beams are narrower. When we decrease the number of samples to $L = 50$ in Figure 8.11(b), we see that the performance of Capon beamforming deteriorates, while MUSIC still performs roughly the same. Since the MUSIC algorithm explicitly exploits the eigenstructure of $\mathbf{R}$ by only using the estimated noise subspace when constructing the power spectrum, it generally provides higher resolution than beamforming methods. The difference is particularly large when L is small; thus, MUSIC is said to be more sample-efficient than the beamforming methods.

In Figure 8.12, we consider the same setup as in 8.11(b), but reduce the number of antennas to $M = 10$. In this scenario, neither MUSIC nor Capon beamforming can provide useful DOA estimates. Although MUSIC generally

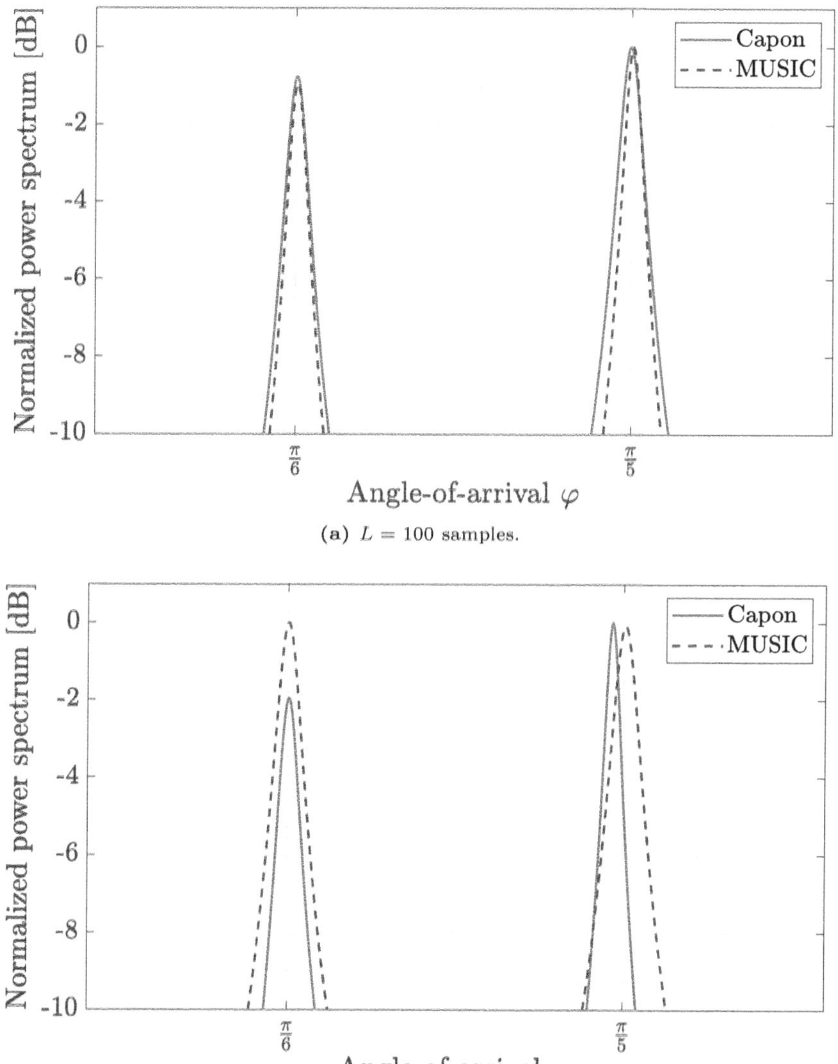

(b) $L = 50$ samples.

Figure 8.11: The normalized power spectrum for a single random realization. Capon beamforming and the MUSIC algorithm are used for DOA estimation using a ULA with $M = 50$ and $\Delta = \lambda/2$. There are $K = 2$ sources with the DOAs $\varphi_1 = \pi/6$ and $\varphi_2 = \pi/5$, respectively. Different numbers of samples are considered when generating the spectra.

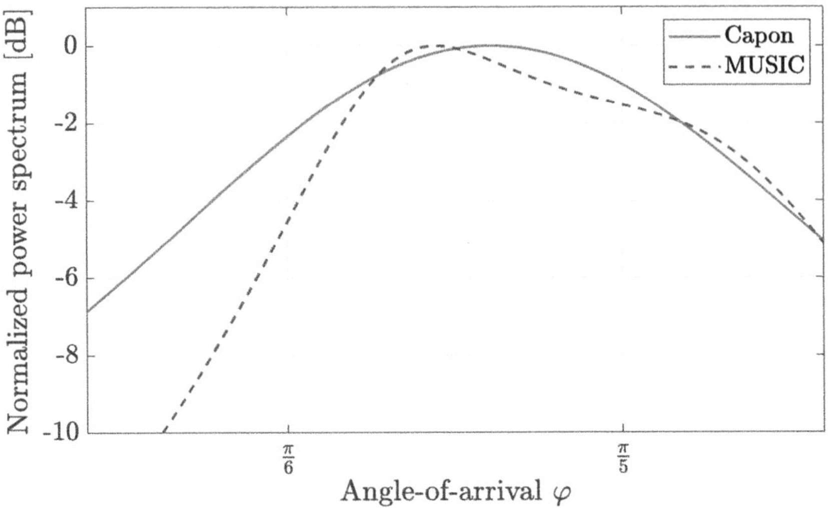

Figure 8.12: The normalized power spectrum with Capon beamforming and the MUSIC algorithm in the same setup as in 8.11(b) but with $M = 10$ antennas.

provides higher estimation accuracy than the beamforming methods, the number of antennas limits the spatial resolution. The MUSIC algorithm also fails if the sources are closely located, compared to the beamwidth.

Example 8.6. Consider DOA estimation with $K = 2$ fully correlated sources. What is the correlation matrix $\mathbf{P}$? What is the rank of $\mathbf{APA}^{\mathrm{H}}$?

For $K = 2$ sources, we have $\mathbf{p}[l] = [\sqrt{\beta_1}e^{-\mathrm{j}\psi_1}x_1[l], \sqrt{\beta_2}e^{-\mathrm{j}\psi_2}x_2[l]]^{\mathrm{T}}$. When the sources are fully correlated, their correlation coefficient has a magnitude of one. Assuming the correlation coefficient is 1 (real-valued) and $\psi_1 = \psi_2 = 0$ for notational convenience, we have $\mathbb{E}\{x_1[l]x_2^*[l]\} = \sqrt{P_1P_2}$. The correlation matrix $\mathbf{P} = \mathbb{E}\{\mathbf{p}[l]\mathbf{p}^{\mathrm{H}}[l]\}$ then becomes

$$\mathbf{P} = \begin{bmatrix} \beta_1 P_1 & \sqrt{\beta_1\beta_2P_1P_2} \\ \sqrt{\beta_1\beta_2P_1P_2} & \beta_2 P_2 \end{bmatrix} = \begin{bmatrix} \sqrt{\beta_1P_1} \\ \sqrt{\beta_2P_2} \end{bmatrix} \begin{bmatrix} \sqrt{\beta_1P_1} & \sqrt{\beta_2P_2} \end{bmatrix}, \tag{8.47}$$

which has rank one since it can be decomposed as the outer product of two vectors. Irrespective of the rank of $\mathbf{A}$, the rank of $\mathbf{APA}^{\mathrm{H}}$ is also one because

$$\begin{aligned} \mathbf{APA}^{\mathrm{H}} &= \begin{bmatrix} \mathbf{a}(\varphi_1,\theta_1) & \mathbf{a}(\varphi_2,\theta_2) \end{bmatrix} \begin{bmatrix} \sqrt{\beta_1P_1} \\ \sqrt{\beta_2P_2} \end{bmatrix} \begin{bmatrix} \sqrt{\beta_1P_1} & \sqrt{\beta_2P_2} \end{bmatrix} \begin{bmatrix} \mathbf{a}^{\mathrm{H}}(\varphi_1,\theta_1) \\ \mathbf{a}^{\mathrm{H}}(\varphi_2,\theta_2) \end{bmatrix} \\ &= \left(\sqrt{\beta_1P_1}\mathbf{a}(\varphi_1,\theta_1) + \sqrt{\beta_2P_2}\mathbf{a}(\varphi_2,\theta_2)\right)\left(\sqrt{\beta_1P_1}\mathbf{a}(\varphi_1,\theta_1) + \sqrt{\beta_2P_2}\mathbf{a}(\varphi_2,\theta_2)\right)^{\mathrm{H}} \end{aligned} \tag{8.48}$$

is the outer product of two vectors. When the correlation coefficient is smaller than 1, $\mathbf{P}$ has rank 2, so rank deficiency only occurs with full correlation.

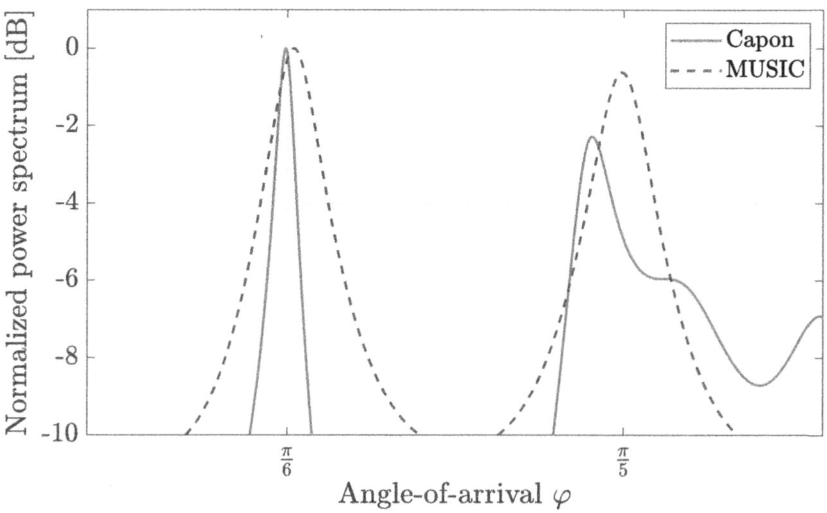

(a) The correlation coefficient is 0.9.

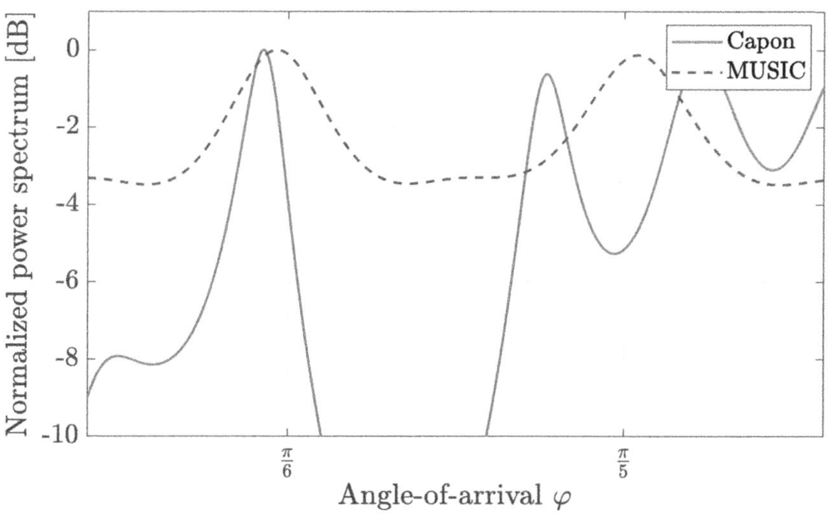

(b) The correlation coefficient is 1 (i.e., fully correlated sources).

Figure 8.13: The normalized power spectrum in the same setup as in 8.11(b) except that the source signals are either highly correlated or fully correlated.

We have seen previously in Section 8.1.2 that statistical correlation between the source signals can degrade the DOA estimation accuracy when using Capon beamforming. To further explore this phenomenon, Figure 8.13 considers the same setup as in 8.11(b), but now the source signals are correlated with Gaussian distributions. The correlation coefficient is 0.9 in Figure 8.13(a), whereas it is 1 in Figure 8.13(b). Capon beamforming cannot provide accurate DOA estimates in any of these cases, but it gets even worse when the sources are fully correlated. In contrast, the MUSIC algorithm is relatively robust to source correlation. The rank of $\mathbf{APA}^{\mathrm{H}}$ is two when the correlation coefficient is 0.9, and MUSIC proves peaks around the true DOAs. When the sources are fully correlated, the peaks are slightly shifted since the rank of $\mathbf{APA}^{\mathrm{H}}$ drops to 1 but remains fairly accurate. This demonstrates that subspace-based methods can handle source correlation relatively efficiently.

Despite the better resolution, the MUSIC algorithm cannot jointly estimate the azimuth and elevation DOA angles when a ULA is utilized. Hence, it is required to use a two-dimensional array (e.g., a UPA) capable of 3D beamforming to solve the general DOA estimation problem.

Example 8.7. Consider a UPA with $M_{\mathrm{H}} > 1$ horizontal and $M_{\mathrm{V}} > 1$ vertical antennas with the spacing $\Delta \leq \lambda/2$. Show that the array response vectors $\mathbf{a}_{M_{\mathrm{H}},M_{\mathrm{V}}}(\varphi_k, \theta_k)$, for $k = 1, 2$, are linearly independent for any combination of $\varphi_1, \theta_1, \varphi_2, \theta_2 \in [-\pi/2, \pi/2]$, except if both $\varphi_1 = \varphi_2$ and $\theta_1 = \theta_2$.

The UPA array response vector is given in (4.128) as

$$\mathbf{a}_{M_{\mathrm{H}},M_{\mathrm{V}}}(\varphi, \theta) = \mathbf{a}_{M_{\mathrm{V}}}(\theta, 0) \otimes \mathbf{a}_{M_{\mathrm{H}}}(\varphi, \theta), \tag{8.49}$$

which is the Kronecker product of two array response vectors for ULAs. For $\mathbf{a}_{M_{\mathrm{H}},M_{\mathrm{V}}}(\varphi_1, \theta_1)$ and $\mathbf{a}_{M_{\mathrm{H}},M_{\mathrm{V}}}(\varphi_2, \theta_2)$ to be linearly dependent, both factors in the Kronecker product must be equal. We know from Lemma 8.1 that $\mathbf{a}_{M_{\mathrm{V}}}(\theta_1, 0)$ and $\mathbf{a}_{M_{\mathrm{V}}}(\theta_2, 0)$ are only linearly dependent if $\sin(\theta_1)\cos(0) = \sin(\theta_2)\cos(0)$. The only solution in the range $[-\pi/2, \pi/2]$ is $\theta_1 = \theta_2$. Hence, linear dependence requires the elevation angles to be equal.

The same lemma says that $\mathbf{a}_{M_{\mathrm{H}}}(\varphi_k, \theta_k)$, for $k = 1, 2$, are only linearly dependent when $\sin(\varphi_k)\cos(\theta_k)$ has the same value for both sources. Since we already know that $\theta_1 = \theta_2$ is required for linear dependence, this implies that we further need $\sin(\varphi_1) = \sin(\varphi_2)$. The only solution in the range $[-\pi/2, \pi/2]$ is $\varphi_1 = \varphi_2$. Hence, a UPA can uniquely identify sources located in different directions thanks to its ability to resolve sources in the elevation angle domain.

In Figure 8.14, we show the normalized 2D power spectrum obtained with either Capon beamforming or the MUSIC algorithm when using a UPA with $M_{\mathrm{H}} = 10$, $M_{\mathrm{V}} = 5$, and $\Delta = \lambda/2$. There are $K = 4$ sources located at the intersection points of the red dashed lines; that is, at the DOA azimuth and elevation angle pairs $(\pi/20, \pi/20)$, $(\pi/20, -\pi/20)$, $(-\pi/20, \pi/20)$, and

$(-\pi/20, -\pi/20)$. $L = 50$ time samples are used to compute the power spectra and the SNR is 0 dB. The source signals are independent and Gaussian distributed. By comparing the peaks of the Capon and MUSIC spectra, we note that MUSIC is more accurate and gives peaks close to the true DOA locations. Although the four DOAs share the same azimuth or elevation angles pairwise, the MUSIC algorithm can resolve these similar sources using a UPA.

The MUSIC spectrum in (8.46) is generated under the assumption that there are K sources by using the eigenvectors corresponding to the $M - K$ smallest eigenvalues of $\hat{\mathbf{R}}_L$. We will now look at the impact of wrongly estimating the number of sources. We consider the same setup as in Figure 8.14 but only consider the MUSIC algorithm. There are $K = 4$ sources but $\hat{\mathbf{U}}_{\mathrm{n}}$ is constructed by incorrectly assuming $\hat{K} = 3$ sources in Figure 8.15(a) and $\hat{K} = 10$ sources in Figure 8.15(b). When we underestimate the number of sources, we effectively treat one dimension of the signal space as a part of the noise subspace. Since that dimension generally contains components from all four source signals (except in the special case where the array response vectors are mutually orthogonal), the result is that we lose the ability to estimate the DOAs of all the sources. On the other hand, the MUSIC algorithm is much more robust to overestimating the number of sources. When ten sources are assumed, the dimension of the noise subspace is reduced from 47 to 40, but it remains orthogonal to the signal space, so the peaks of the spectrum appear roughly at the correct locations. Hence, it is better first to overestimate the number of sources and then refine the estimate if the spectrum contains fewer peaks. If we know that $\hat{K}$ might overestimate K, we need an extra step in the algorithm to determine how many peaks to consider as source estimates.

In summary, the subspace-based MUSIC algorithm provides higher DOA estimation accuracy than the beamforming methods. It is relatively robust to source signal correlation and can be used with an unknown number of sources. There exist modified versions of the MUSIC algorithm that are even better at managing signal correlation [129]. Other than that, there are more advanced parametric methods that further exploit the structure of the source signals for better accuracy [51]. Although the theoretical development of the MUSIC algorithm relies on the rank of $\mathbf{P}$, this matrix is not explicitly considered when generating the MUSIC spectrum in (8.46). The correlation matrix estimate $\hat{\mathbf{R}}_L$ and the corresponding noise subspace $\hat{\mathbf{U}}_{\mathrm{n}}$ are used instead. There also exist parametric ML methods (i.e., extensions of the method described in Section 4.2.5) that exploit further knowledge of the source signal's characteristics for enhanced estimation [51]. The more is known about the source signals, the higher DOA estimation accuracy can be achieved, but the computational complexity might also grow. In all the considered methods, we need to evaluate the power spectra for a dense grid of discrete angles to identify the peaks, which is especially complex in the 2D case.

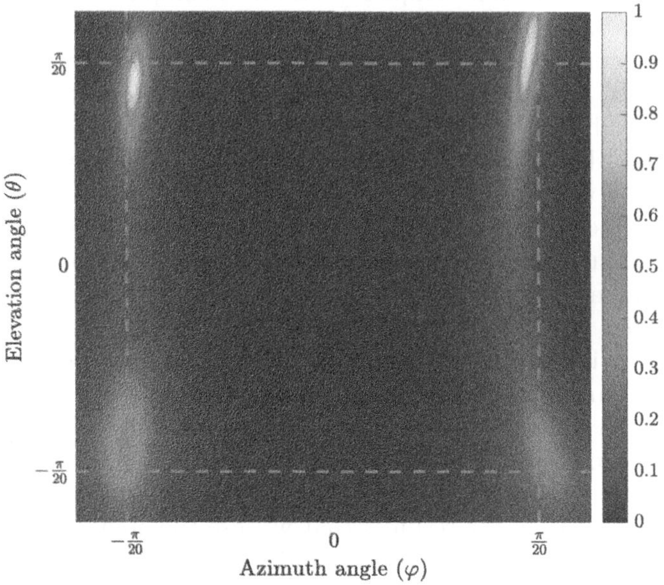

(a) 2D spectrum with Capon beamforming.

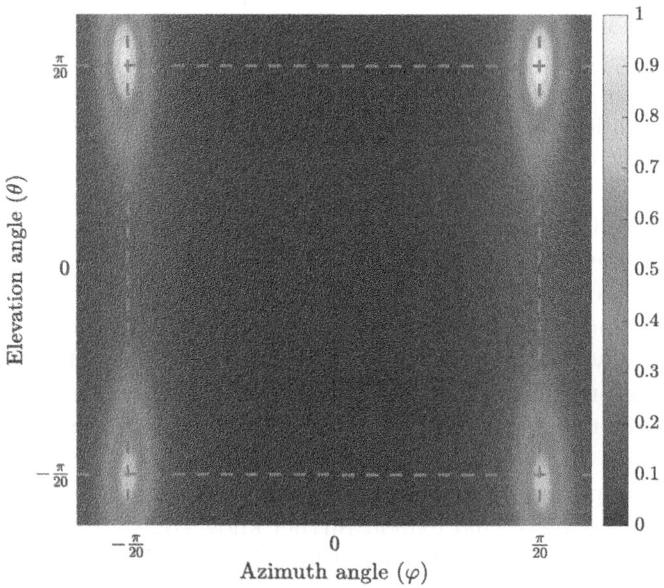

(b) 2D spectrum with the MUSIC algorithm.

Figure 8.14: The normalized power spectrum for a single random realization when using a UPA with $M_{\mathrm{H}} = 10$, $M_{\mathrm{V}} = 5$, and $\Delta = \lambda/2$. There are $K = 4$ sources located at the intersection points of the red dashed lines: $(\pi/20, \pi/20)$, $(\pi/20, -\pi/20)$, $(-\pi/20, \pi/20)$, and $(-\pi/20, -\pi/20)$. $L = 50$ time samples are used to compute the power spectra. Capon beamforming is compared with the MUSIC algorithm.

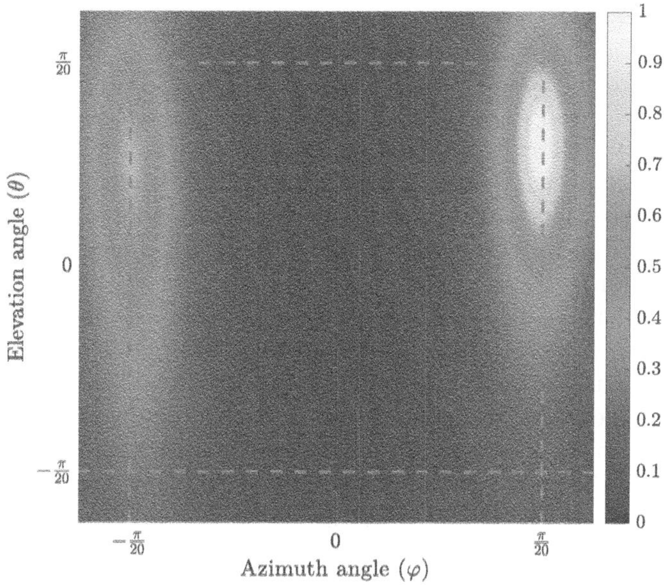

(a) Power spectrum when believing there are $\hat{K} = 3$ sources (too few).

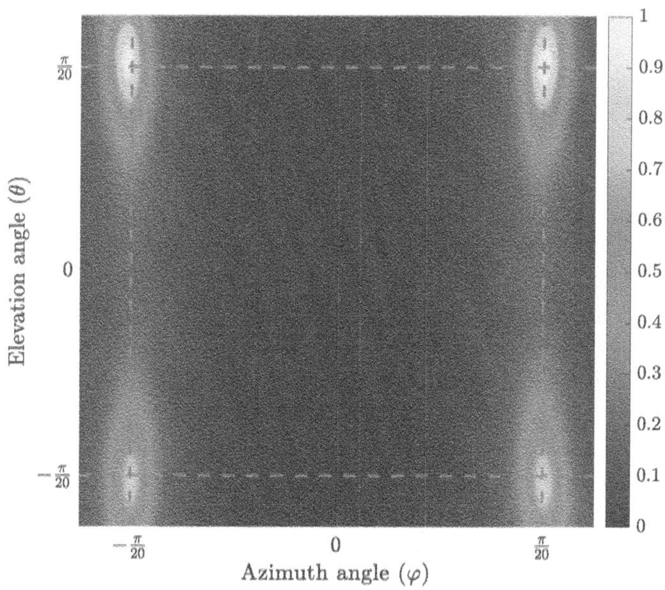

(b) Power spectrum when believing there are $\hat{K} = 10$ sources (too many).

Figure 8.15: The normalized power spectrum obtained by the MUSIC algorithm in the same setup as in Figure 8.14. There are $K = 4$ sources, but the noise subspace is constructed by the eigenvectors corresponding to the $M - \hat{K}$ smallest eigenvalues. The number of sources is presumed to either be $\hat{K} = 3$ or $\hat{K} = 10$.

8.2 Localization

Source localization or simply *localization* is an extensively studied topic, where the aim is to estimate the unknown location of a source node, object, or person by using the measured data from multiple other sensors that have known locations [130]. We will call the object of interest (with an unknown location) the *target node* and other sensors that collect measurements the *receivers*. The location refers to a point in a selected coordinate system [131], such as a 2D location in $\mathbb{R}^2$ or a 3D location in $\mathbb{R}^3$. The origin is at an arbitrary but predefined location.

We will consider so-called cooperative localization, where the measurements collected at M receivers are fused to estimate the target node's location. When the target transmits a signal, the receivers constitute a distributed receive antenna array. The signal propagates over an M-dimensional SIMO channel, but the goal is not to estimate its M complex coefficients (as in previous chapters) but only to extract the location. To this end, each receiver can measure the *time-of-arrival (TOA)* of the transmitted signal. If the target node and the receivers have synchronized clocks, the propagation delays to the respective receivers can be computed by knowing the time the signal was transmitted.[3] In a LOS scenario, these measurements can be used to deduce the respective distances to the target node by multiplying the delay by the speed of light. The distance measurements can be combined with the known locations of the receivers to extract the target location. If the receivers are synchronized but the transmission time is uncertain, they can compare their TOA measurements instead and determine the *time-difference-of-arrival (TDOA)*. This scenario is of practical interest because it is hard to maintain precise synchronization between a mobile target node at an unknown location and a network of receivers. On the other hand, cables can be drawn between the fixed receivers to enable sharing of measurements and synchronization. When each receiver is equipped with multiple antennas, the receivers can individually estimate their DOA from the target node. By combining these DOA measurements with the known receiver locations, the target node's location can be precisely estimated. Many practical systems use hybrid localization methods that fuse different kinds of physical measurements (e.g., angles, signal strengths, inertial sensor measures, different radio systems, etc.) so their respective weaknesses can be counteracted. In this section, we only cover the fundamentals of localization. We begin by exemplifying the basic principles of TOA-based localization and then cover the details of the TDOA- and DOA-based localization techniques.

[3]Alternatively, the round-trip delay can be measured by sending a signal from a receiver to the target node, which immediately sends it back [131]. Half the round-trip delay plus the initial transmission time can then be treated as the TOA. This procedure does not require clock synchronization but must be repeated M times when there are M receivers, making it inefficient for implementing cooperative localization.

8.2.1 TOA-Based Localization

We will focus on 2D localization for notational convenience. Hence, the aim is to estimate the $(x, y) \in \mathbb{R}^2$ coordinates of the target node using M receivers that are distributed over the azimuth plane. A setup of this kind is shown in Figure 8.16(a), where the target node is denoted by a red star and located at the (unknown) coordinate $(100, 0)$ m. There are $M = 3$ receivers shown as blue squares at the known coordinates $(-100, 0)$, $(0, 100)$, and $(0, -100)$. Note that the target and receivers are equally spaced on a circle with a 100 m radius centered at the origin. We assume there are free-space LOS channels from the target node to each receiver.

If a signal is transmitted by the target node at time 0 (or any other known time instance), the TOA at receiver m becomes

$$t_m = \frac{\sqrt{(x_m - x)^2 + (y_m - y)^2}}{c}, \tag{8.50}$$

where c is the speed of light and (x_m, y_m) denotes the 2D coordinates of receiver m, for $m = 1, \ldots, M$. Suppose the TOAs are measured perfectly. Receive m can then compute the corresponding propagation distance

$$d_m = t_m c = \sqrt{(x_m - x)^2 + (y_m - y)^2} \tag{8.51}$$

and knows that the target node is located somewhere on a circle around receiver m with radius d_m. Figure 8.16(a) shows these circles for the $M = 3$ receivers. The three circles only intersect at the precise location of the target node; thus, three distance measurements are sufficient to uniquely estimate the location, which is known as *trilateration*. However, if we remove one of the receivers, the remaining two circles intersect at two locations, which creates ambiguity. In conclusion, at least $M = 3$ TOAs must be measured to find the 2D target location in the noise-free case.

In practice, the location estimate will be imperfect due to TOA measurement errors. The receiver noise creates an upper limit on the TOA measurement accuracy for a given bandwidth and SNR. Other error sources are multipath propagation (in addition to the LOS path) and synchronization mismatches. Suppose the total errors can be modeled as additive Gaussian noise so that the noisy distance measured at receiver m is

$$r_m = d_m + n_m = \sqrt{(x_m - x)^2 + (y_m - y)^2} + n_m, \quad m = 1, \ldots, M, \tag{8.52}$$

where $n_m \sim \mathcal{N}(0, \sigma_{\mathrm{d}}^2)$. The variance σ_{d}^2 depends on the wireless technology, bandwidth, carrier frequency, range, etc. The aim of TOA-based localization is then to estimate the target location (x, y) as accurately as possible based on the noisy measurements r_m, for $m = 1, \ldots, M$. In Figure 8.16(b), we consider the same setup as in Figure 8.16(a), but the receivers only know the noisy measurements $r_1, \ldots, r_M$ and the variance σ_{d}^2. Based on this information, each

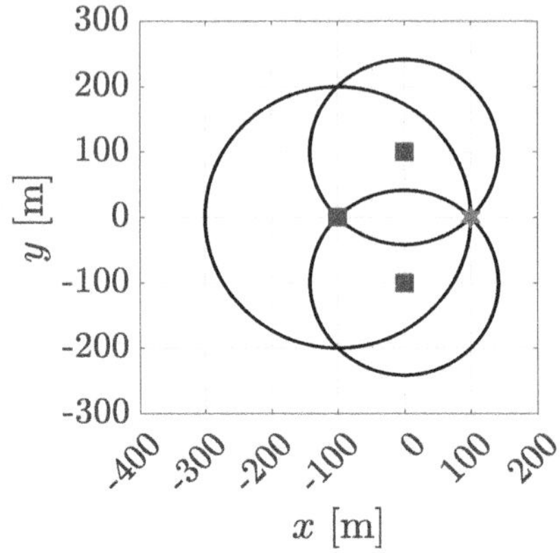

(a) $M = 3$ receivers and noise-free measurements.

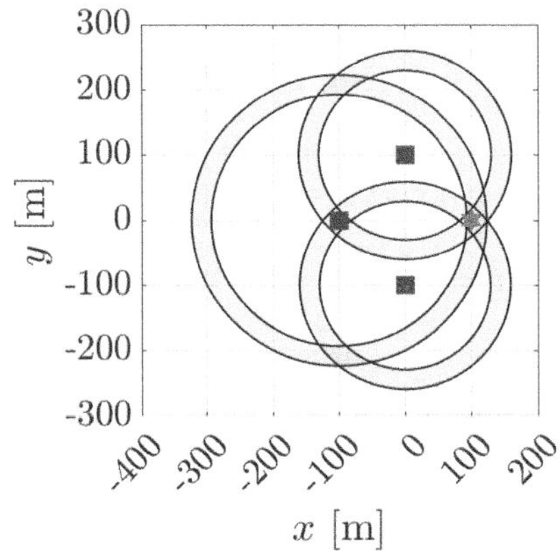

(b) $M = 3$ receivers and noisy measurements.

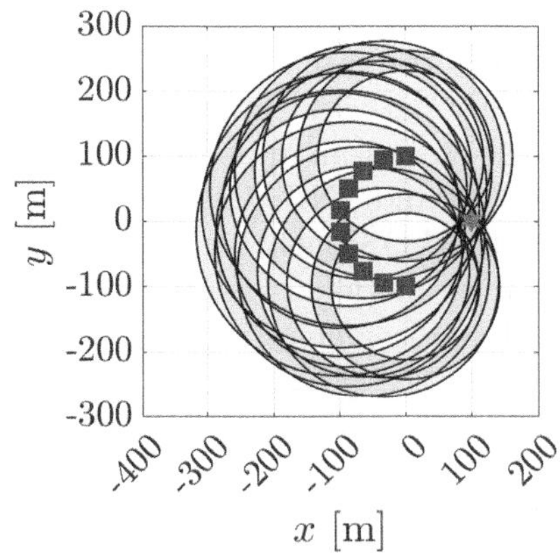

(c) $M = 10$ receivers and noisy measurements.

Figure 8.16: Example of TOA-based localization in the azimuth plane with M receivers and a single target node. The location of the target node is indicated by a red star, and the locations of the receivers are shown as blue squares. The circle (or yellow annulus between two circles) indicates where each receiver believes the target is located. The intersection points/regions can be used to estimate the target's location.

receiver can construct a confidence interval for the true value of d_m. If we plot the lower and upper interval limits as two circles, the annulus between them contains the likely locations of the target node.

In Figure 8.16(b), we consider the noise standard deviation $\sigma_{\mathrm{d}} = 5\,\mathrm{m}$ and construct our confidence intervals to contain three standard deviations: $d_m \in [r_m - 15, r_m + 15]$. Hence, the confidence interval for receiver m is represented by the yellow annulus between an inner circle around the receiver with radius $r_m - 15$ and an outer circle with radius $r_m + 15$. We can confidently say that the target node is located somewhere in the overlapping area between the $M = 3$ annuluses. We notice that the red star is in this area, but there is always an uncertainty in the location estimation when measurement errors occur. If the errors were smaller, each annulus shrinks, which improves the localization accuracy since the overlapping area also shrinks. Another way to improve the accuracy (for a fixed noise variance) is to fuse the measurements from more receivers. To see this impact visually, we consider $M = 10$ receivers in Figure 8.16(c) and distribute them uniformly on the left half of a circle centered at the origin with radius 100 m. The confidence intervals are generated as before, and we can be certain that the receiver is located in the area where all the ten annuluses intersect. This area shrinks with an increased number of receivers as the confidence intervals point in different directions and thereby have less overlap.

8.2.2 TDOA-Based Localization

As mentioned earlier, TOA-based localization requires clock synchronization between the target node and all the receivers to turn the TOA measurements into distance measurements. In practice, it is desirable to alleviate the need for the target node to be precisely synchronized with the receiver because that is hard to achieve when the location is unknown and there is only a wireless connection to it. Even a tiny clock bias of 1 µs can lead to a bias of 300 m in the distance measurement because the speed of light is immense. In this section, we consider TDOA-based localization that does not rely on target node synchronization but only requires that the receivers have a common reference clock. The target node is assumed to transmit a signal at some unknown time δ, according to the receivers' clock. The TOA at receiver m (in the absence of noise) is then changed from (8.50) to

$$t_m = \frac{d_m}{c} + \delta = \frac{\sqrt{(x_m - x)^2 + (y_m - y)^2}}{c} + \delta, \quad m = 1, \ldots, M. \tag{8.53}$$

To remove the unknown δ from these equations, in TDOA-based cooperative localization, we compute the differences between the TOAs measured at different receivers. In particular, we pick a reference receiver and give it the index 1. The TDOA between receivers m and 1 is $t_m - t_1$, and becomes independent of δ. If we can measure this TDOA perfectly, the corresponding distance difference can be computed as

$$\begin{aligned} d_{m,1} &= (t_m - t_1)c \\ &= \sqrt{(x_m - x)^2 + (y_m - y)^2} - \sqrt{(x_1 - x)^2 + (y_1 - y)^2}. \end{aligned} \tag{8.54}$$

For a given measurement value $d_{m,1}$ and known receiver locations (x_1, y_1) and (x_m, y_m), the equation (8.54) defines one branch of a hyperbola with respect to (x, y) in the 2D Cartesian coordinate system.[4] This bowl-like curve identifies all potential target locations that would give rise to the measured TDOA. In Figure 8.17(a), we revisit the setup from Figure 8.16(a) with $M = 3$ receivers. We let the receiver located at $(0, 100)$ m have the index 1 and be used as the reference for the TDOAs. By knowing the distance differences $d_{2,1}$ and $d_{3,1}$, and the receiver locations, we can draw two hyperbola branches. One of the curves is straight while the other is bent, and they intersect at one point: the target node location $(x, y) = (100, 0)$ m. In this noise-free case, we notice that at least $M = 3$ receivers are needed for unambiguous 2D localization based on TDOAs. This is the same as for TOA-based localization.

TDOA-based localization can be utilized even if the distance measurements are noisy. Similarly to (8.52), we let $n_m \sim \mathcal{N}(0, \sigma_{\mathrm{d}}^2)$ denote the independent additive noise at receiver m. The $M-1$ noisy distance difference measurements are then given as

$$\begin{aligned} r_{m,1} &= d_m - d_1 + \underbrace{n_m - n_1}_{=n_{m,1}} \\ &= \sqrt{(x_m - x)^2 + (y_m - y)^2} - \sqrt{(x_1 - x)^2 + (y_1 - y)^2} + n_{m,1}, \end{aligned} \tag{8.55}$$

for $m = 2, \ldots, M$. This equation with respect to (x, y) also defines one branch of a hyperbola, but we cannot draw it due to the noise. We would like to have an equation of the kind $d_{m,1} = \sqrt{(x_m - x)^2 + (y_m - y)^2} - \sqrt{(x_1 - x)^2 + (y_1 - y)^2}$ as in (8.54). However, the term $d_{m,1}$ is replaced by $r_{m,1} - n_{m,1}$ in (8.55) where the collective noise realization $n_{m,1} \sim \mathcal{N}(0, 2\sigma_{\mathrm{d}}^2)$ is unknown. Since the measurement value $r_{m,1}$ and the noise distribution are known, we can compute a confidence interval for the value of $d_{m,1}$ and use its limits to draw two hyperbola branches. We can then be confident that the target node is located in between these curves.

In Figure 8.17(b), we consider the same setup as in Figure 8.17(a) but perform localization based on the noisy measurements $r_{m,1}$ for $m = 2, 3$. We assume the noise has the standard deviation $\sigma_{\mathrm{d}} = 5$ m, which implies that the collective noise realization $n_{m,1}$ has the standard deviation $5\sqrt{2}$ m.

[4]A hyperbola is the curve obtained when a double-cone is cut by a plane. The general equation is $|\sqrt{(x_m - x)^2 + (y_m - y)^2} - \sqrt{(x_1 - x)^2 + (y_1 - y)^2}| = d_{m,1}$, where (x_1, y_1) and (x_m, y_m) are the two focal points and $d_{m,1} > 0$ is a constant. A hyperbola contains two branches, which are two unconnected bent curves. Only one of these branches remains when the absolute value is removed as in (8.54).

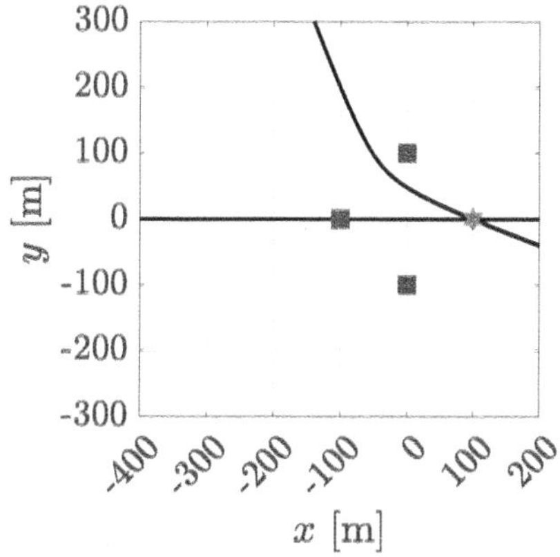

(a) $M = 3$ receivers and noise-free measurements.

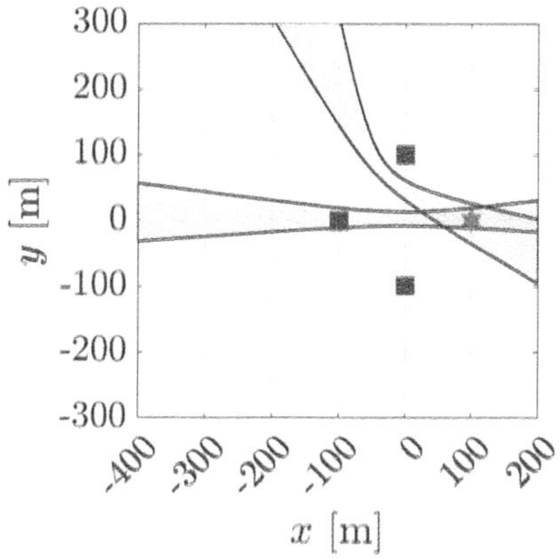

(b) $M = 3$ receivers and noisy measurements.

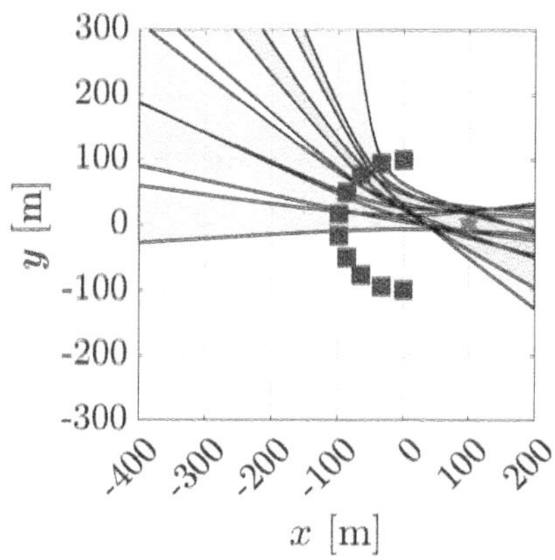

(c) $M = 10$ receivers and noisy measurements.

Figure 8.17: Example of TDOA-based localization in the azimuth plane with M receivers and a single target node. The location of the target node is indicated by a red star, and the locations of the receivers are shown as blue squares. The hyperbola branch (or yellow regions between two branches) indicates where each receiver believes the target is located. The intersection points/regions can be used to estimate the target's location.

We construct our confidence intervals to contain three standard deviations: $d_{m,1} \in [r_{m,1} - 15\sqrt{2}, r_{m,1} + 15\sqrt{2}]$. The hyperbola branches obtained using the lower and upper limits of this interval are shown as curves in the figure, and the area in between is marked in yellow. We can be confident that the target node (red star) is located somewhere in the region where the two yellow areas intersect. This is also the case, but the intersection is pretty large—particularly compared to Figure 8.16(b), where we considered the same setup but with TOA-based localization. Hence, the price to pay for not having a clock-synchronized target node is reduced localization accuracy.

We increase the number of receivers to $M = 10$ in Figure 8.17(c). There are now $M - 1 = 9$ yellow regions to consider, and their intersection region determines where the target node might be. The localization accuracy increases monotonically with the number of receivers. Since all the receivers in this example are located on the left-hand side of the target node, the intersection region has a long tail toward the right. This can be dealt with in practice by surrounding the potential target location with receivers.

The yellow confidence areas in Figure 8.17 indicate where the target node might be, but some points in the areas are more likely than others. This statistical information can be utilized to obtain a specific localization estimate $(\hat{x}, \hat{y})$. Unfortunately, there is no simple closed-form solution to this estimation problem because the equations are nonlinear and the noise terms $n_{2,1}, \ldots, n_{M,1}$ are correlated. Several algorithms have been developed to tackle this problem [131]. One approach is to compute the ML estimate of (x, y) given the noisy observations $r_{m,1}$, for $m = 2, \ldots, M$. In this case, it is convenient to define the distance measurement vector $\mathbf{r} = [r_{2,1}, \ldots, r_{M,1}]^{\mathrm{T}} \in \mathbb{R}^{M-1}$, the noise vector $\mathbf{n} = [n_{2,1}, \ldots, n_{M,1}] \in \mathbb{R}^{M-1}$, and the theoretical distance difference vector function $\bar{\mathbf{d}}(x, y) = [\bar{d}_{2,1}(x, y), \ldots, \bar{d}_{M,1}(x, y)]^{\mathrm{T}} \in \mathbb{R}^{M-1}$, where the distance difference for receiver m is given by the function

$$\bar{d}_{m,1}(x, y) = \sqrt{(x_m - x)^2 + (y_m - y)^2} - \sqrt{(x_1 - x)^2 + (y_1 - y)^2}. \tag{8.56}$$

This function computes what the distance difference would be for a specific guess (x, y) of the target node location. The ML estimation approach assumes that the target node's unknown location (x, y) is deterministic. The received signal $\mathbf{r} = \bar{\mathbf{d}}(x, y) + \mathbf{n}$ then has the real Gaussian multivariate distribution $\mathcal{N}(\bar{\mathbf{d}}(x, y), \mathbf{C})$, where $\mathbf{C} = \mathbb{E}\{\mathbf{n}\mathbf{n}^{\mathrm{T}}\}$ is the covariance matrix of the noise vector. We know from (2.87) that the PDF of $\mathbf{r}$ is

$$f_{\mathbf{r}}(\mathbf{r}) = \frac{1}{(2\pi)^{\frac{M-1}{2}} \sqrt{\det(\mathbf{C})}} e^{-\frac{1}{2}\left(\mathbf{r} - \bar{\mathbf{d}}(x,y)\right)^{\mathrm{T}} \mathbf{C}^{-1} \left(\mathbf{r} - \bar{\mathbf{d}}(x,y)\right)}. \tag{8.57}$$

We recall that the measurement errors n_m in (8.55) were assumed to be independent and identically distributed as $n_m \sim \mathcal{N}(0, \sigma_{\mathrm{d}}^2)$. This implies that the $(m-1)$th diagonal entry of $\mathbf{C}$ can be computed as

$$\mathbb{E}\left\{n_{m,1}^2\right\} = \mathbb{E}\left\{(n_m - n_1)^2\right\} = \mathbb{E}\left\{n_m^2\right\} + \mathbb{E}\left\{n_1^2\right\} = 2\sigma_{\mathrm{d}}^2, \tag{8.58}$$

for $m = 2, \ldots, M$. The $(m-1, i-1)$th off-diagonal entry of $\mathbf{C}$ is given as

$$\mathbb{E}\{n_{m,1}n_{i,1}\} = \mathbb{E}\{(n_m - n_1)(n_i - n_1)\} = \mathbb{E}\{n_1^2\} = \sigma_{\mathrm{d}}^2. \tag{8.59}$$

The noise covariance matrix is

$$\mathbf{C} = \begin{bmatrix} 2\sigma_{\mathrm{d}}^2 & \sigma_{\mathrm{d}}^2 & \cdots & \sigma_{\mathrm{d}}^2 \\ \sigma_{\mathrm{d}}^2 & 2\sigma_{\mathrm{d}}^2 & \ddots & \vdots \\ \vdots & \ddots & \ddots & \sigma_{\mathrm{d}}^2 \\ \sigma_{\mathrm{d}}^2 & \cdots & \sigma_{\mathrm{d}}^2 & 2\sigma_{\mathrm{d}}^2 \end{bmatrix} \tag{8.60}$$

and it is non-diagonal since the noise at the reference receiver affects all the TDOAs. The ML estimates of x and y are the values that jointly maximize (8.57), which is equivalent to maximizing the argument of the exponential function or minimizing $(\mathbf{r} - \bar{\mathbf{d}}(x,y))^{\mathrm{T}}\mathbf{C}^{-1}(\mathbf{r} - \bar{\mathbf{d}}(x,y))$. Therefore, the ML estimates are obtained by solving the problem

$$(\hat{x}, \hat{y}) = \underset{(x,y)}{\arg\min} \quad \left(\mathbf{r} - \bar{\mathbf{d}}(x,y)\right)^{\mathrm{T}} \mathbf{C}^{-1} \left(\mathbf{r} - \bar{\mathbf{d}}(x,y)\right). \tag{8.61}$$

The objective function to be minimized in (8.61) is not convex with respect to (x, y). This makes it computationally expensive to find the solution, for example, by evaluating the objective function on a dense grid of potential (x, y)-values and picking the best of them. The complexity can be managed using an iterative gradient descent algorithm, but it might not converge to the global optimum. If sufficient computational resources can be assigned to solve the ML estimation problem, it will provide better accuracy than other methods; however, alternative lower-complexity methods exist [131].

The estimation accuracy can be evaluated using the root MSE (RMSE) of the distance, which is defined as

$$\mathrm{RMSE} = \sqrt{\mathbb{E}\left\{(x-\hat{x})^2 + (y-\hat{y})^2\right\}}, \tag{8.62}$$

where the expectation is computed with respect to the measurement noise.

The RMSE is shown in Figure 8.18 for the same setup as in Figure 8.17(c), but with a varying number of receivers. The location estimate $(\hat{x}, \hat{y})$ is obtained by minimizing the objective in (8.61) using a gradient-descent algorithm. We consider two values of the noise standard deviation: $\sigma_{\mathrm{d}} = 10\,\mathrm{m}$ and $\sigma_{\mathrm{d}} = 5\,\mathrm{m}$. This figure shows that increasing the number of receivers leads to improved localization accuracy. This effect is particularly noticeable up until 15 receivers, after which the RMSE decays more slowly because the extra receivers are placed next to existing ones on the edge of the same half circle. The noise standard deviation greatly impacts the localization accuracy, both for a given number of receivers and when considering the saturation level that is approached when M is large.

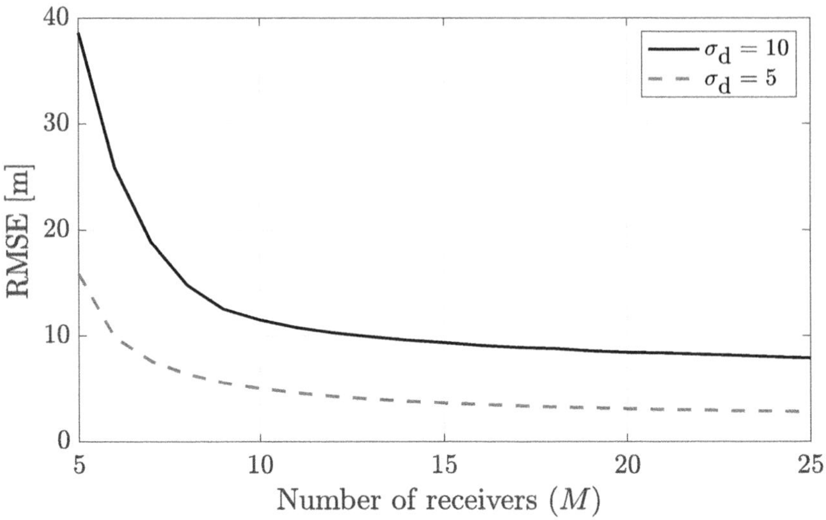

Figure 8.18: The RMSE of localization in (8.62) with respect to the number of receivers. The TDOA-based location estimate $(\hat{x}, \hat{y})$ is obtained by minimizing the ML objective function in (8.61) using a gradient-descent algorithm. The same setup is considered as in Figure 8.17(c), where the receivers are located along the edge of a half circle.

Example 8.8. The TOA measurement errors limit the accuracy of TOA- and TDOA-based localization methods. How are these measurements made?

The TOA is measured by sending a known signal from the target node with some time duration T, carrier frequency f_c, and bandwidth B. The receiver correlates the received noisy signal with different time-delayed versions of the transmitted signal to determine which delay matches the most with the observation. The peak of the resulting crosscorrelation function is the TOA estimate. The variance of the TOA measurement depends on the mentioned parameters and the SNR. In particular, the variance in a free-space LOS channel can be lower bounded as [130, Eq. (5)]

$$\mathbb{V}\text{ar}\{\text{TOA}\} \geq \frac{1}{8\pi^2 BTf_c^2\text{SNR}} \tag{8.63}$$

when $f_c \gg B$. The TOA measurement accuracy improves as the bandwidth and carrier frequency increase. Since new wireless systems designed for high-capacity communications progressively use higher carrier frequencies to make more bandwidth available, the TOA/TDOA-based localization accuracy can gradually improve if localization features are integrated into these systems. For example, a shift from a mid-band system with $f_c = 3\,\text{GHz}$ and $B = 100\,\text{MHz}$ to a high-band system with $f_c = 30\,\text{GHz}$ and $B = 500\,\text{MHz}$ will result in a 500 times lower TOA measurement variance, if all other parameters are unchanged.

8.2.3 DOA-Based Localization

In TOA- and TDOA-based localization, the measurements are taken in the time domain, and we assumed that each receiver provides a single TOA measurement. When the receiver is equipped with multiple antennas, each one can measure a TOA. When the target node is in the far-field of the receiver, the TOA is approximately equal at all the receive antennas, but there are noticeable phase-shift differences that enable DOA estimation using the methods described in Section 8.1.[5] In DOA-based localization, also called AOA-based localization, each of the M receivers uses its multiple antennas to estimate the DOA from the target node. To explain the basics of DOA-based localization, we consider 2D localization, where the target node and all the receiver arrays are located in the azimuth plane. The DOA is then represented by an azimuth angle, which for a target node at the location (x, y) and receiver m at (x_m, y_m) with $x > x_m$ becomes[6]

$$\varphi_m = \arctan\left(\frac{y - y_m}{x - x_m}\right), \quad m = 1, \dots, M. \tag{8.64}$$

If the value of φ_m is measured perfectly and the receiver location is known, we can treat (8.64) as an equation with respect to (x, y). In particular, the relation can be rearranged as $y = \tan(\varphi_m)x + y_m - \tan(\varphi_m)x_m$, which is the equation of a straight line in the 2D Cartesian coordinate system.

In Figure 8.19(a), we revisit the localization scenario from Figure 8.16(a) and Figure 8.17(a). By measuring the three angles $\varphi_1, \varphi_2, \varphi_3 \in [-\pi/2, \pi/2]$ in (8.64), we can draw three straight lines. Each line starts from the respective receiver location and extends towards the positive x-axis direction since we assume $x > x_m$. These lines intersect at one point: the target node location. This is the only intersection point in the figure because any two non-identical lines can intersect at most once. Hence, having two multiple antenna receivers is sufficient for unambiguous 2D localization in the noise-free case if $\varphi_1 \neq \varphi_2$. This principle is known as *triangulation* because the two lines plus the line between the receivers define a triangle. Since we know the length of one side of the triangle (between the receivers) and two angles (to the target), we can compute anything related to this triangle—including the target location.

In practice, the DOA estimates will be subject to measurement errors. Suppose we can model the estimate as

$$r_m = \varphi_m + n_m, \tag{8.65}$$

[5]When the receiver is in the radiative near-field of the target node, the TOA differences over the receiver array are so large that range estimation is also possible—similar to when having M distributed receivers. We refer to [132] for further details.

[6]For notational convenience, we assume that $x > x_m$ so that the angle to the target node is between $-\pi/2$ and $\pi/2$, and can be obtained using the arctan function. For $x < x_m$, we must add $\pm\pi$ to (8.64) to get the correct angle. If the receiver array is subject to mirror-like ambiguity, as illustrated in Figure 4.7 for ULAs, this ambiguity must also be resolved. This can, for instance, be done using rough TDOA estimation that determines which side the target is at.

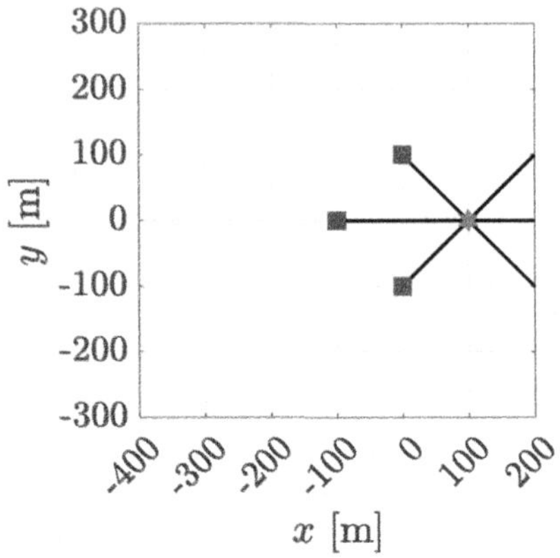

(a) $M = 3$ receivers and noise-free measurements.

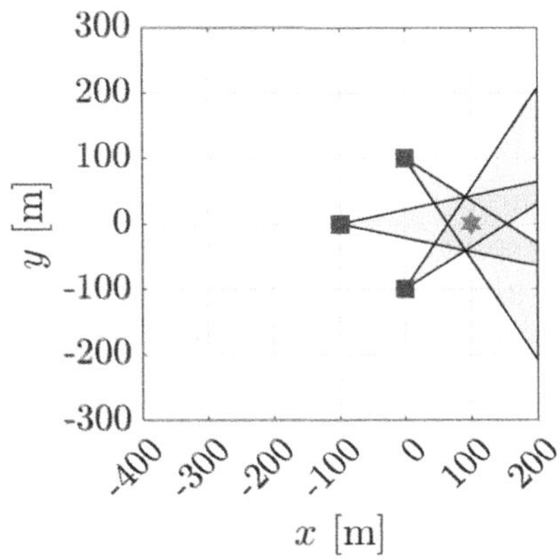

(b) $M = 3$ receivers and noisy measurements.

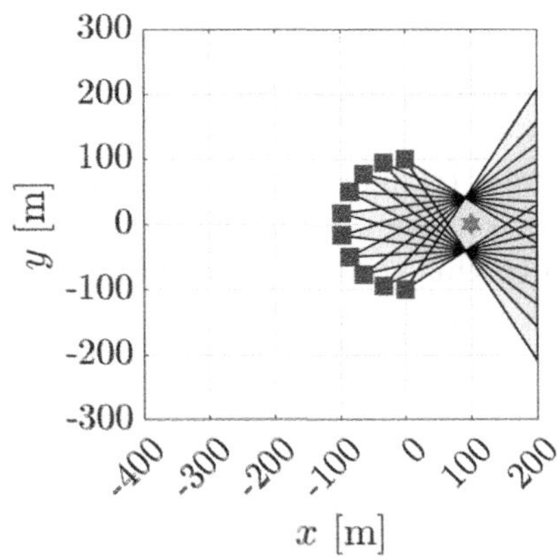

(c) $M = 10$ receivers and noisy measurements.

Figure 8.19: Example of DOA-based localization in the azimuth plane with M multi-antenna receivers and a single target node. The location of the target node is indicated by a red star, and the locations of the receivers are shown as blue squares. The straight line (or yellow areas between two lines) indicates where each receiver believes the target is located. The intersection points/regions can be used to estimate the target's location.

which is the true DOA from (8.64) plus a Gaussian random noise realization $n_m \sim \mathcal{N}(0, \sigma_\varphi^2)$. The noise is independent between the receivers, but we let the variance σ_φ^2 be the same for simplicity. The noise variance will depend on the number of antennas and SNR. It also depends on the wavelength because we get better angular resolution when the wavelength shrinks (for a given physical length of the array), so the measurement noise will be reduced. Based on the measured received signal r_m, we know that the true DOA is $\varphi_m = r_m - n_m$. Although the noise realization is unknown, we can use this relation to deduce a confidence interval for the DOA. By considering the lower and upper limits of this interval, we can draw two lines that start at receiver m and point in slightly different directions. We can then be confident that the target node is located somewhere between these lines.

In Figure 8.19(b), we consider the same localization setup as in Figure 8.19(a), but with noisy angle measurements with the standard deviation $\sigma_\varphi = 4°$. We construct our confidence interval as $\varphi_m \in [r_m - 12°, r_m + 12°]$ by considering three standard deviations. In the figure, we show the straight lines obtained using the lower and upper limits of this interval, and the area in between is yellow. There are three such yellow areas whose intersection region specifies where the receiver must be located. The target node (red star) is located in this area, which is relatively small because the three receivers observe the target from very different angles, but it would be even smaller if σ_φ was reduced. In Figure 8.19(c), we increase the number of receivers to $M = 10$ by adding extra receivers on the edge of the half-circle where the original receivers are located. There are many more yellow areas in this case, but their intersection region remains roughly the same as in Figure 8.19(b) because the new receivers cover angular directions between the previous ones. We need to deploy receivers that observe the target from the right-hand side or reduce the noise variance to get even higher estimation accuracy.

Since the measurement error is Gaussian distributed, the true DOA is more likely to be at the center of the confidence interval than at the edges. We can identify the most likely target location among those in the intersection region of the yellow areas. This would be the ML estimate $(\hat{x}, \hat{y})$ of the target node location. To formulate the ML estimation problem, we first introduce suitable notation: the measurement vector $\mathbf{r} = [r_1, \dots, r_M]^{\mathrm{T}} \in \mathbb{R}^M$, the noise vector $\mathbf{n} = [n_1, \dots, n_M] \in \mathbb{R}^M$, and the theoretical azimuth DOA vector function $\bar{\boldsymbol{\varphi}}(x, y) = [\bar{\varphi}_1(x, y), \dots, \bar{\varphi}_M(x, y)]^{\mathrm{T}} \in \mathbb{R}^M$, where the DOA at receiver m is given by the function

$$\bar{\varphi}_m(x, y) = \arctan\left(\frac{y - y_m}{x - x_m}\right). \tag{8.66}$$

This function computes what the DOA angle would be for a specific guess (x, y) of the target node location.

The ML estimation approach assumes that the target node's unknown location (x, y) is deterministic. The received signal $\mathbf{r} = \bar{\boldsymbol{\varphi}}(x, y) + \mathbf{n}$ is distributed

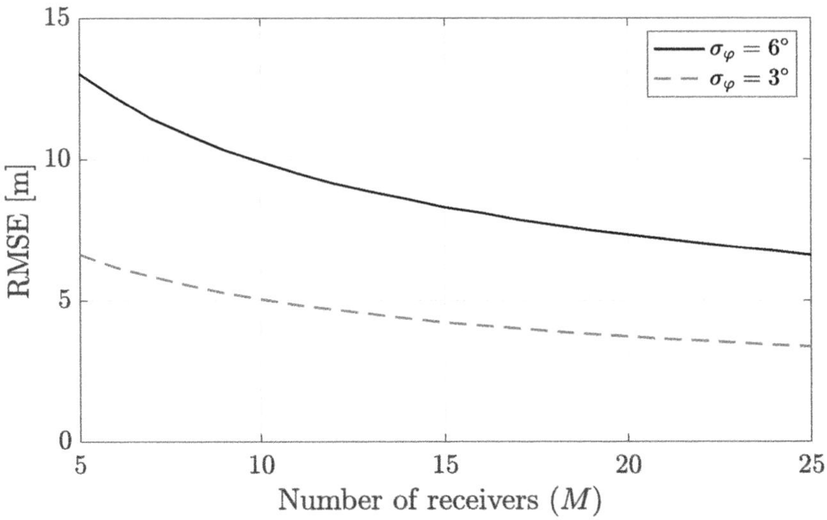

Figure 8.20: The RMSE of localization in (8.62) with respect to the number of receivers. The DOA-based location estimate $(\hat{x}, \hat{y})$ is obtained by minimizing the ML objective function in (8.68) using a gradient-descent algorithm. The same setup is considered as in Figure 8.19(c), where the receivers are located along the edge of a half circle.

according to the real Gaussian multivariate distribution $\mathcal{N}(\bar{\boldsymbol{\varphi}}(x, y), \mathbf{C})$ where $\mathbf{C} = \mathbb{E}\{\mathbf{n}\mathbf{n}^{\mathrm{T}}\}$ is the covariance matrix of $\mathbf{n}$. We know from (2.87) that the PDF of $\mathbf{r}$ is

$$f_{\mathbf{r}}(\mathbf{r}) = \frac{1}{(2\pi)^{\frac{M}{2}}\sqrt{\det(\mathbf{C})}} e^{-\frac{1}{2}(\mathbf{r}-\bar{\boldsymbol{\varphi}}(x,y))^{\mathrm{T}}\mathbf{C}^{-1}(\mathbf{r}-\bar{\boldsymbol{\varphi}}(x,y))}. \tag{8.67}$$

The ML estimates of x and y are the values that jointly maximize (8.67), which is equivalent to maximizing the argument of the exponential function or minimizing $(\mathbf{r} - \bar{\boldsymbol{\varphi}}(x, y))^{\mathrm{T}}\mathbf{C}^{-1}(\mathbf{r} - \bar{\boldsymbol{\varphi}}(x, y))$. Therefore, the ML estimates are obtained by solving the problem

$$(\hat{x}, \hat{y}) = \underset{(x,y)}{\arg\min} \quad (\mathbf{r} - \bar{\boldsymbol{\varphi}}(x, y))^{\mathrm{T}} \mathbf{C}^{-1} (\mathbf{r} - \bar{\boldsymbol{\varphi}}(x, y)). \tag{8.68}$$

We have previously assumed that $\mathbf{C} = \sigma_\varphi^2 \mathbf{I}_M$, but this problem can be solved with arbitrary noise covariance matrices (e.g., when some receivers have more accurate measurements than others). The main issue is that $\bar{\varphi}_m(x, y)$ is a nonlinear function of x and y, which makes it computationally complicated to compute the solution to (8.68). As in the case of TDOA-based localization, we can find the solution to a predefined accuracy by evaluating the objective function on a dense grid of potential (x, y)-values and picking the best of these points. Using an iterative gradient descent algorithm leads to a more tractable complexity, but convergence to the global optimum is not guaranteed.

In Figure 8.20, we plot the RMSE of the distance in (8.62) with respect to the number of DOA receivers. The location estimate $(\hat{x}, \hat{y})$ is obtained by minimizing the objective in (8.68) using a gradient-descent algorithm. We consider the same setup as in Figure 8.19(c), except that we consider a varying number of receivers and two different standard deviations of the measurement noise: $\sigma_\varphi = 6°$ and $\sigma_\varphi = 3°$. The figure shows that the RMSE decreases consistently with an increasing M, so having more receivers lead to better localization accuracy. We previously noticed in Figure 8.19 that the intersection region was nearly the same with $M = 3$ and $M = 10$ receivers. However, the probability distribution within the region becomes more favorable as M increases, which makes the ML estimate more accurate. Moreover, the noise variance greatly impacts the localization accuracy; if the standard deviation is cut in half, so is the RMSE.

Example 8.9. We have seen that $M = 3$ receivers are sufficient to estimate the target's 2D location unambiguously with TOA- and TDOA-based methods, while $M = 2$ is sufficient in DOA-based localization. How many receivers are needed for 3D localization?

The M circles determined by the TOA measurements in noise-free TOA-based 2D localization turn into M spheres in the 3D coordinate system. In the noise-free case, there will be a unique intersection point (x, y, z) if there are at least $M = 4$ spheres. In noise-free TDOA localization, the TDOA measurements define $M - 1$ hyperboloids in the 3D coordinate system. We need at least $M - 1 = 3$ hyperboloids to get a unique intersection point; thus, at least $M = 4$ receivers are needed for unambiguous localization with these two methods [133]. Hence, we can get away with the same number of receivers regardless of whether the target node is synchronized with the receivers or not. The TOA-based method will, however, provide more accurate location estimates in the noisy case.

If each of the M receivers can estimate its azimuth and elevation DOA from the target node without noise, these measurements will define M lines in the 3D coordinate system. Since two non-identical lines can only intersect at one point, $M = 2$ receivers are sufficient for unambiguous 3D localization. This is the same triangulation principle as in the 2D case. Note that the receivers need two-dimensional arrays (e.g., UPAs) to estimate both the azimuth and elevation angles. If each receiver is instead equipped with a horizontal ULA, then there will be an ambiguity in the azimuth-elevation plane, as exemplified in Figure 8.9. In such a case, each DOA measurement defines a surface in the 3D coordinate system, and we need at least $M = 3$ receivers to locate the target node unambiguously.

DOA-based localization requires multiple antennas, unlike the TOA- and TDOA-based approaches that only require a single antenna. These methods build on different principles by measuring angles and ranges, respectively, and can be combined for even higher estimation accuracy.

8.3 Target Detection

The methods described thus far in this chapter rely on the target node actively transmitting a signal so that physical parameters (e.g., time delays, angles, and location) can be estimated by a wireless system equipped with receive antennas. In radar applications, the target is instead passive, so the wireless system must both transmit signals and receive them. Radar is originally an abbreviation of *radio detection and ranging*; thus, its first aim is to detect targets, and its second aim is to estimate physical parameters such as the range. We will focus on the detection part in this section because the previous section described the fundamental principles for parameter estimation.

Target detection is the core problem of detecting whether there is an object of interest at a particular location by sending known signal pulses toward that target location. A receiver located near the transmitter (or at another predefined location) listens to the noisy echoes of the transmitted signal, which might be reflected off the target of interest. If there is no target, the received signal in a free-space LOS scenario contains only noise. On the other hand, if there is a target, the attenuated reflected signal is received along with the noise. The task of the detector is to determine whether there is a target or not by processing the received signal and exploiting prior information regarding the signal characteristics. There are two events in target detection:

- There is no target;
- The target exists.

The binary hypothesis testing framework outlined in Section 2.7 is commonly used for target detection. In hypothesis testing, the absence of the target represents the null hypothesis $\mathcal{H}_0$, whereas the alternative hypothesis $\mathcal{H}_1$ corresponds to the existence of the target. The detection method should take the reflection properties of the target into account. Intuitively, it is easier to detect an object if it is large, made of reflecting material, or happens to focus its reflected signal toward the receiver. When a planar wavefront impinges on the object from a specific angle, the reflected wave will have a complicated shape determined by the object's physical characteristics, as illustrated in Figure 8.21. The receiver only observes the signal component that is reflected toward it; thus, we can quantify the reflection using a single number σ_{RCS} called the *radar cross section (RCS)*. Using antenna terminology, the RCS is the effective area of the object when facing the transmitter multiplied by the antenna gain toward the receiver for the reflected wave, which makes it measured in m^2. Suppose the power flux density of the impinging wave (i.e., the power of the electromagnetic field at the target location per unit area) is Q, measured in W/m^2. The reflected power by the target is then

$$P = Q\sigma_{\text{RCS}}. \tag{8.69}$$

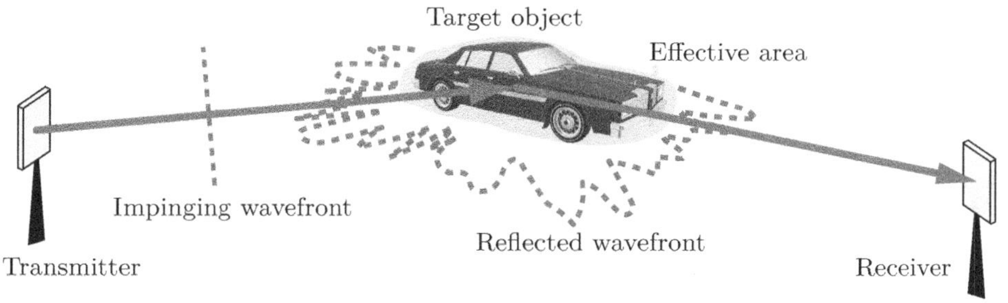

Figure 8.21: The RCS σ_{RCS} [m^2] quantifies how the target object reflects an impinging signal from the transmitter toward the receiver. It can be interpreted as the effective area of the object toward the transmitter multiplied by the antenna gain achieved by the reflected wavefront in the receiver direction. The RCS depends on the object's physical properties and the location/rotation of the transmitter, receiver, and object.

The RCS is the cumulative effect of the diffuse/specular reflection at different parts of the target. The value fluctuates as the target moves and is rotated because the effective area toward the transmitter and the antenna gain toward the receiver are angle-dependent. Several approaches exist in the radar literature to statistically characterize the reflection of a target [134]. One key factor that creates modeling differences is the fluctuation frequency. The so-called *Swerling models* developed by Peter Swerling in the 1950s [134]–[136] take into account different fluctuating conditions and use different probability distributions. In this chapter, we will outline the basic target detection methods for two such models: Swerling 1 and Swerling 2.

Apart from the target reflection, the numbers of the transmitters and receivers, and their locations, also affect the detection problem and solution method. In Figure 8.22, we illustrate three categories of setups used for target detection. Each category can also be used for radar and sensing applications other than target detection. The basic setup is *mono-static sensing*, shown in Figure 8.22(a), where the transmitter and receiver are co-located. In this figure, the solid lines represent the radiated signal from the transmitter(s) toward the target, and the dashed lines represent the received signals at the receiver(s) after being reflected by the target. The antenna array is typically divided into two parts, where one is used for transmission and the other for reception. Figure 8.22(b) shows a *bi-static sensing* setup where the transmitter and the receiver are at different locations, thereby viewing the target from different angles. The RCS will be different in the bi-static and mono-static cases because the angles from the transmitter and receiver determine the RCS. The detection performance can be improved using multiple transmitters and receivers, which operate in mono-static or bi-static sensing mode. Figure 8.22(c) illustrates the corresponding *multi-static sensing* case. For example, the operation represented by purple lines is mono-static, whereas the red lines represent a bi-static setup. It is also possible to exploit the received signals at multiple receivers for detection, as shown by the green lines.

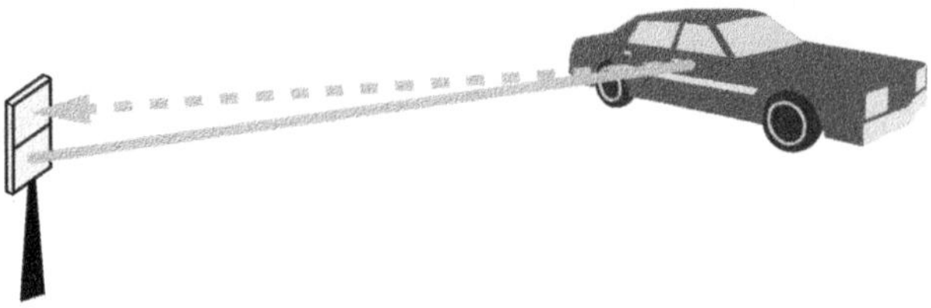

(a) Mono-static sensing for target detection.

(b) Bi-static sensing for target detection.

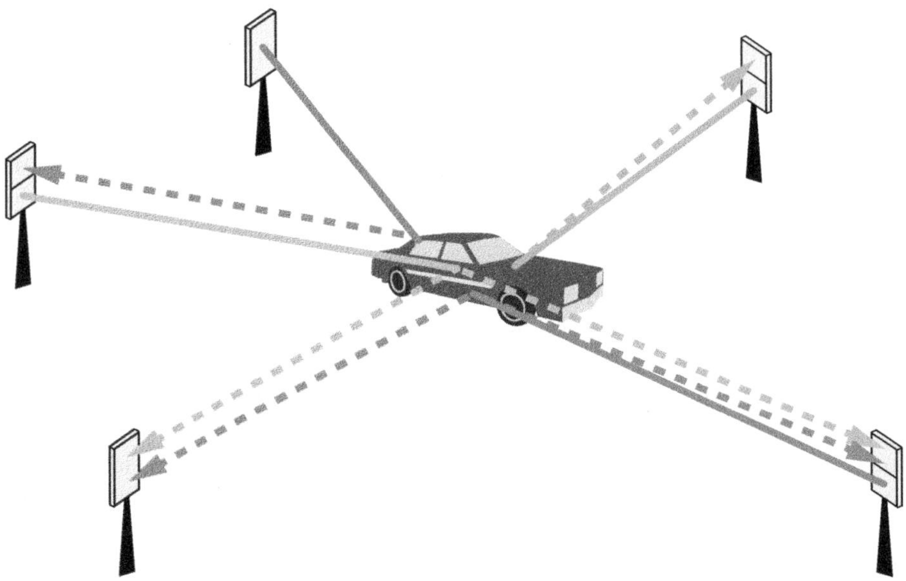

(c) Multi-static sensing for target detection.

Figure 8.22: Three categories of sensing systems are illustrated: mono-static, bi-static, and multi-static. Each system can be used for target detection but also other sensing applications. The solid lines represent the transmitted signal from each transmitter toward the potential target location. The dashed lines represent the received signal at each receiver after being reflected by the target object.

Another alternative is combining the mono-static and bi-static setups, as the blue lines show. Each of these propagation paths experiences a different RCS value for the same object due to the different transmission/reception angles. The primary purpose of multi-static sensing is to exploit spatial diversity because the RCS value can be very small for some angles and transmit-receiver pairs but likely not for all combinations simultaneously.

8.3.1 Radar Range Equation

We will now derive the radar range equation, which describes the average received power when a signal with a specific power is transmitted toward and reflected by the target object. The received power depends on the transmit power, frequency, antenna gains, distances to the target, and the RCS. We begin by considering the radar range equation for the bi-static sensing case in Figure 8.22(b). Initially, we assume a single-antenna transmitter that sends a signal with power P_{t} and has the antenna gain function $G_{\mathrm{t}}(\varphi_{\mathrm{t}}, \theta_{\mathrm{t}})$, where the angles $(\varphi_{\mathrm{t}}, \theta_{\mathrm{t}})$ lead from the transmitter to the target. In a free-space LOS propagation scenario with the distance d_{t} to the target, the power flux density at the target location will be

$$Q = \frac{P_{\mathrm{t}} G_{\mathrm{t}}(\varphi_{\mathrm{t}}, \theta_{\mathrm{t}})}{4\pi d_{\mathrm{t}}^2} \quad \mathrm{W/m^2} \tag{8.70}$$

because the power is divided over a sphere with surface area $4\pi d_{\mathrm{t}}^2$.

The RCS of the target is denoted σ_{RCS} in m^2. Practical RCS values can vary immensely; thus, the decibel scale is often used when specifying them. By taking one square meter as the reference value, the RCS can be reported in decibel-of-square-meter (dBsm) as $10\log_{10}\left(\frac{\sigma_{\mathrm{RCS}}}{1\,\mathrm{m}^2}\right)$. Measured values from -50 dBsm (insects) to 60 dBsm (large ships, aircraft carriers) are reported in [134]. The RCS value is not always proportional to the size of the object; for example, the typical RCS value of a small truck is 20 dBsm while it is only 8 dBsm for a large fighter aircraft and even smaller for stealth aircrafts [137]. We will now determine how the RCS value affects the received signal power. For a given value of σ_{RCS}, the *effective isotropic reflected power* from the target towards the receiver is $Q\sigma_{\mathrm{RCS}}$. We use the term "effective isotropic" similar to how the EIRP concept was defined in Section 4.5.5: the reflected power emitted towards the receiver is the same as if the target had an isotropic antenna that transmits with power $Q\sigma_{\mathrm{RCS}}$. The total reflected power can be entirely different because an object typically does not reflect power isotropically, but σ_{RCS} depends on the angles that lead to the transmitter and receiver.

Suppose the receiver is also equipped with a single antenna and has the antenna gain function $G_{\mathrm{r}}(\varphi_{\mathrm{r}}, \theta_{\mathrm{r}})$, where the angles $(\varphi_{\mathrm{r}}, \theta_{\mathrm{r}})$ lead from the receiver to the target. In a free-space LOS propagation scenario with the distance d_{r} from the target to the receiver, the channel gain from an isotropic transmitter (effective isotropic reflector in this case) to the receiver is given by (1.40) as $\beta = \frac{\lambda^2}{(4\pi d_r)^2} G_r(\varphi_r, \theta_r)$. Hence, the received signal power is

$$P_r = Q\sigma_{RCS}\beta = \frac{P_t G_t(\varphi_t, \theta_t)}{4\pi d_t^2} \sigma_{RCS} \frac{\lambda^2}{(4\pi d_r)^2} G_r(\varphi_r, \theta_r)$$
$$= \frac{P_t G_t(\varphi_t, \theta_t) G_r(\varphi_r, \theta_r) \lambda^2 \sigma_{RCS}}{(4\pi)^3 d_t^2 d_r^2}. \tag{8.71}$$

This is known as the radar range equation and applies to a bi-static setup. The received power is proportional to the RCS and will later be used to distinguish the signal from the noise. An object with a small RCS provides a smaller SNR and is, therefore, more challenging to detect.

We can use (8.71) to determine the mono-static radar range equation. Since the angles and distances are now the same to and from the target, it holds that $\varphi_t = \varphi_r = \varphi$, $\theta_t = \theta_r = \theta$, and $d_t = d_r = d$. Inserting these values without subscripts into (8.71), we obtain the radar range equation for the mono-static case as

$$P_r = \frac{P_t G_t(\varphi, \theta) G_r(\varphi, \theta) \lambda^2 \sigma_{RCS}}{(4\pi)^3 d^4}. \tag{8.72}$$

One crucial difference from the bi-static case is that the RCS σ_{RCS} depends on the location and orientation of two nodes instead of three.

Example 8.10. Suppose an SNR of $-10\,\text{dB}$ is needed to detect the target. What is the smallest RCS that enables target detection if $P_t = 10\,\text{W}$, $G_t(\varphi, \theta) = G_r(\varphi, \theta) = 2$, $\lambda = 0.01\,\text{m}$ (i.e., $f = 30\,\text{GHz}$), $B = 100\,\text{MHz}$, $d = 100\,\text{m}$, and $N_0 = 10^{-20.4}\,\text{W/Hz}$?

By substituting the given values into (8.72) and dividing by the noise variance $N_0 B = 10^{-20.4+8} = 10^{-12.4}\,\text{W}$, we obtain the SNR as

$$\text{SNR} = \frac{P_r}{N_0 B} = \frac{10 \cdot 2^2 \cdot 0.01^2 \sigma_{RCS}}{(4\pi)^3 \cdot 100^4 \cdot 10^{-12.4}}. \tag{8.73}$$

We can now solve the equation $\text{SNR} \geq -10\,\text{dB}$ for σ_{RCS} to obtain that the RCS should be at least

$$\sigma_{RCS} \geq 0.1 \frac{(4\pi)^3 \cdot 10^{-4.4}}{4 \cdot 10^{-3}} \approx 1.98 \approx 2.96 \text{ dBsm}. \tag{8.74}$$

The bi-static received power in (8.71) is proportional to the squared wavelength, which implies that it reduces when the carrier frequency is increased if the antenna gain functions and RCS are fixed. However, we can also rewrite the received power in terms of the effective areas $A_t(\varphi_t, \theta_t) = \frac{\lambda^2}{4\pi} G_t(\varphi_t, \theta_t)$ and $A_r(\varphi_r, \theta_r) = \frac{\lambda^2}{4\pi} G_r(\varphi_r, \theta_r)$ of the transmitter and receiver. In this case, (8.71) becomes

$$P_{\mathrm{r}} = \frac{P_{\mathrm{t}} A_{\mathrm{t}}(\varphi_{\mathrm{t}}, \theta_{\mathrm{t}}) A_{\mathrm{r}}(\varphi_{\mathrm{r}}, \theta_{\mathrm{r}}) \sigma_{\mathrm{RCS}}}{4\pi \lambda^2 d_{\mathrm{t}}^2 d_{\mathrm{r}}^2}. \tag{8.75}$$

This expression is inversely proportional to the squared wavelength, which implies that it increases when the carrier frequency is increased if the effective antenna areas and RCS are constant. Hence, target detection can become easier at higher frequencies, particularly if antenna arrays are utilized to achieve large effective areas toward the target.

The radar range equation can be easily extended to manage the case where the transmitter is equipped with K antennas, whereas the receiver has M antennas. When inspecting whether a target exists at a specific location, the transmitter can apply MRT precoding towards the prospective target location, while the receiver can apply MRC. We then achieve a combined beamforming gain of MK over a LOS channel, as shown in Section 4.4. The radar range equation in (8.71) for the bi-static setup is generalized by multiplying with the beamforming gain, which results in

$$P_{\mathrm{r}} = \frac{P_{\mathrm{t}} G_{\mathrm{t}}(\varphi_{\mathrm{t}}, \theta_{\mathrm{t}}) G_{\mathrm{r}}(\varphi_{\mathrm{r}}, \theta_{\mathrm{r}}) M K \lambda^2 \sigma_{\mathrm{RCS}}}{(4\pi)^3 d_{\mathrm{t}}^2 d_{\mathrm{r}}^2}. \tag{8.76}$$

MRT focuses the transmission in a specific direction. If the target location is unknown (e.g., we want to detect if a vehicle exists somewhere on the road), the transmitter must scan for the target by sending beamformed signals in different directions. The orthogonal DFT beams described in Section 4.3.3 can be used to cover all dimensions, but a denser grid of non-orthogonal beams can also be used to ensure that nearly the maximum beamforming gain is achieved in any potential target direction. This kind of radar sweeping is often presented as a circle with a rotating beam in movies, and the detected targets show up as dots. Conventional radar systems perform mechanical beamforming by rotating the array instead of using electrical beamforming.

Example 8.11. Consider the mono-static setup from Example 8.10. What is the minimum RCS value that a detectable target can have if the number of antennas at the transmitter and receiver is $M = K = 4$?

Due to beamforming gain of $MK = 4 \cdot 4 = 16$, the SNR is improved by a factor of 16 compared to the single-antenna case in (8.73). If we solve the equation $\mathrm{SNR} \geq -10\,\mathrm{dB}$ for σ_{RCS}, we obtain

$$\sigma_{\mathrm{RCS}} \geq \frac{0.1}{16} \frac{(4\pi)^3 \cdot 10^{-4.4}}{4 \cdot 10^{-3}} \approx 0.12 \approx -9.1 \quad \mathrm{dBsm}. \tag{8.77}$$

A target with 16 times smaller RCS can be detected thanks to the beamforming gain. Even smaller targets can be found by using more antennas.

In the remainder of this chapter, we will consider the Swerling 1 and Swerling 2 target models, in which the target consists of many small diffuse reflectors that contribute to the overall effective RCS. Similar to the derivation of the Rayleigh fading channel in Section 5.1.1, the independent random-like phase-variations across the many reflectors give rise to a complex Gaussian coefficient in the complex baseband: $c_{\mathrm{RCS}} \sim \mathcal{N}_{\mathbb{C}}(0, \sigma_{\mathrm{RCS}})$. In fact, the target behaves as a multipath cluster when interacting with wireless signals. The magnitude $|c_{\mathrm{RCS}}|$ has a Rayleigh distribution, while the RCS realization $|c_{\mathrm{RCS}}|^2$ has an exponential distribution that satisfies $\mathbb{E}\{|c_{\mathrm{RCS}}|^2\} = \sigma_{\mathrm{RCS}}$. Hence, we will now treat σ_{RCS} as the average RCS value and c_{RCS} as the random realization. In analogy with the slow fading case in Chapter 5, in the Swerling 1 target model, the RCS realization is assumed to be fixed throughout the signal transmission interval used for target detection. When the target's RCS fluctuates more rapidly, the Swerling 2 target model can be used, where the RCS takes a new independent realization for each transmitted symbol. The latter is the radar counterpart of the fast fading in communication channels. We will analyze the target detection problem for each of these models.

8.3.2 Target Detection with the Swerling 1 Target Model

In the Swerling 1 model, the target's RCS is assumed to fluctuate slowly, so it is fixed throughout the L received symbols collected for target detection. If the target exists at the analyzed location, the received signal power is $P_{\mathrm{r}}|c_{\mathrm{RCS}}|^2$, where P_{r} is the average power given by the radar range equation in (8.76) and $c_{\mathrm{RCS}} \sim \mathcal{N}_{\mathbb{C}}(0, 1)$ models the randomness. Note that, unlike the last section, c_{RCS} has unit variance because the average RCS σ_{RCS} is now included in P_{r} for notational convenience. We assume that a constant symbol "1" is transmitted during the L transmissions without loss of generality. Hence, the corresponding binary hypothesis test is

$$\mathcal{H}_0 \quad : \quad y[l] = n[l], \quad l = 1, \ldots, L, \tag{8.78}$$

$$\mathcal{H}_1 \quad : \quad y[l] = \sqrt{P_{\mathrm{r}}} c_{\mathrm{RCS}} + n[l], \quad l = 1, \ldots, L, \tag{8.79}$$

where the additive noise samples $n[l] \sim \mathcal{N}_{\mathbb{C}}(0, \sigma^2)$ are independent.

In radar applications, there is typically no prior knowledge of the hypothesis probabilities. Hence, the Neyman-Pearson detector from Section 2.7 can be used to maximize the detection probability P_{D} for a desired value $P_{\mathrm{FA}} = \alpha$ of the false alarm probability. The Neyman-Pearson detector, which is optimal in this sense, was presented in Lemma 2.14. It states that $\mathcal{H}_1$ is selected if

$$\frac{f_{\mathbf{y}|\mathcal{H}_1}(\mathbf{y}|\mathcal{H}_1)}{f_{\mathbf{y}|\mathcal{H}_0}(\mathbf{y}|\mathcal{H}_0)} \geq \gamma, \tag{8.80}$$

where the threshold $\gamma \geq 0$ will later be selected so that $P_{\mathrm{FA}} = \alpha$. To particularize the Neyman-Pearson detector for the hypothesis test in (8.78)-(8.79), we collect all the received samples in a vector $\mathbf{y} = [y[1], \ldots, y[L]]^{\mathrm{T}} \in \mathbb{C}^L$, and define the noise vector $\mathbf{n} = [n[1], \ldots, n[L]]^{\mathrm{T}} \in \mathbb{C}^L$. By letting $\mathbf{1}_L$ denote the L-dimensional vector with only ones, the received signal vector in (8.79) under the hypothesis $\mathcal{H}_1$ can be expressed as

$$\mathbf{y} = \sqrt{P_{\mathrm{r}}}\mathbf{1}_L c_{\mathrm{RCS}} + \mathbf{n} \sim \mathcal{N}_{\mathbb{C}}(\mathbf{0}, P_{\mathrm{r}}\mathbf{1}_L\mathbf{1}_L^{\mathrm{H}} + \sigma^2\mathbf{I}_L). \tag{8.81}$$

On the other hand, we have $\mathbf{y} \sim \mathcal{N}_{\mathbb{C}}(\mathbf{0}, \sigma^2\mathbf{I}_L)$ when the hypothesis $\mathcal{H}_0$ is true. We can use the PDF of a complex Gaussian vector in (2.85) to evaluate the likelihood ratio in (8.80) as

$$\gamma \leq \frac{f_{\mathbf{y}|\mathcal{H}_1}(\mathbf{y}|\mathcal{H}_1)}{f_{\mathbf{y}|\mathcal{H}_0}(\mathbf{y}|\mathcal{H}_0)} = \frac{\frac{1}{\pi^L \det\left(P_{\mathrm{r}}\mathbf{1}_L\mathbf{1}_L^{\mathrm{H}}+\sigma^2\mathbf{I}_L\right)} e^{-\mathbf{y}^{\mathrm{H}}\left(P_{\mathrm{r}}\mathbf{1}_L\mathbf{1}_L^{\mathrm{H}}+\sigma^2\mathbf{I}_L\right)^{-1}\mathbf{y}}}{\frac{1}{\pi^L \det(\sigma^2\mathbf{I}_L)} e^{-\mathbf{y}^{\mathrm{H}}(\sigma^2\mathbf{I}_L)^{-1}\mathbf{y}}}. \tag{8.82}$$

Using the fact that $\ln(\gamma)$ is a monotonically increasing function for $\gamma \geq 0$, the Neyman-Pearson detector in (8.82) decides on the hypothesis $\mathcal{H}_1$ if

$$\sigma^{-2}\mathbf{y}^{\mathrm{H}}\mathbf{y} - \mathbf{y}^{\mathrm{H}}\left(P_{\mathrm{r}}\mathbf{1}_L\mathbf{1}_L^{\mathrm{H}} + \sigma^2\mathbf{I}_L\right)^{-1}\mathbf{y} \geq \ln(\gamma) - \ln(b), \tag{8.83}$$

where the constant $b = \det\left(\sigma^2\mathbf{I}_L\right) / \det\left(P_{\mathrm{r}}\mathbf{1}_L\mathbf{1}_L^{\mathrm{H}} + \sigma^2\mathbf{I}_L\right)$ is independent of the received signal $\mathbf{y}$. Using the rank-one update formula in (2.48), we have

$$\left(P_{\mathrm{r}}\mathbf{1}_L\mathbf{1}_L^{\mathrm{H}} + \sigma^2\mathbf{I}_L\right)^{-1} = \sigma^{-2}\mathbf{I}_L - \frac{P_{\mathrm{r}}\sigma^{-4}}{1 + P_{\mathrm{r}}L\sigma^{-2}}\mathbf{1}_L\mathbf{1}_L^{\mathrm{H}}. \tag{8.84}$$

Inserting this result into (8.83), the detector decides on $\mathcal{H}_1$ if

$$|\mathbf{1}_L^{\mathrm{H}}\mathbf{y}|^2 \geq \underbrace{\frac{(1 + P_{\mathrm{r}}L\sigma^{-2})(\ln(\gamma) - \ln(b))}{P_{\mathrm{r}}\sigma^{-4}}}_{=\gamma'}, \tag{8.85}$$

where γ' is the revised threshold variable that must be selected so that $P_{\mathrm{FA}} = \alpha$. We have $\mathbf{1}_L^{\mathrm{H}}\mathbf{y} \sim \mathcal{N}_{\mathbb{C}}(0, L\sigma^2)$ if hypothesis $\mathcal{H}_0$ is true, which implies that $|\mathbf{1}_L^{\mathrm{H}}\mathbf{y}|^2 \sim \mathrm{Exp}(1/(L\sigma^2))$. Hence, we can compute the threshold using (2.91) as

$$P_{\mathrm{FA}} = \alpha = \int_{|\mathbf{1}_L^{\mathrm{H}}\mathbf{y}|^2 \geq \gamma'} f_{\mathbf{y}|\mathcal{H}_0}(\mathbf{y}|\mathcal{H}_0)\,\partial\mathbf{y} = \int_{\gamma'}^{\infty} \frac{1}{L\sigma^2} e^{-\frac{z}{L\sigma^2}}\,\partial z = e^{-\frac{\gamma'}{L\sigma^2}}$$
$$\Rightarrow \quad \gamma' = L\sigma^2\ln(\alpha^{-1}). \tag{8.86}$$

This threshold is inversely proportional to the specified false alarm probability α. If α is reduced, the threshold γ' increases but the detection probability

$$P_{\mathrm{D}} = \int_{|\mathbf{1}_L^{\mathrm{H}}\mathbf{y}|^2 \geq \gamma' = L\sigma^2\ln(\alpha^{-1})} f_{\mathbf{y}|\mathcal{H}_1}(\mathbf{y}|\mathcal{H}_1)\,\partial\mathbf{y} \tag{8.87}$$

becomes smaller since we integrate the PDF over a smaller set of values. This result highlights a fundamental tradeoff in target detection: a large detection probability is associated with a large false alarm probability, and vice versa.

The term $|\mathbf{1}_L^{\mathrm{H}}\mathbf{y}|^2$ in (8.85) is called the *sufficient statistics* for target detection because it is only this variable that must be measured and compared to the threshold γ' to implement the Neyman-Pearson detector, and it determines the detection probability in (8.87). Hence, the optimal receiver processing for target detection coherently combines the L received signal as $\mathbf{1}_L^{\mathrm{H}}\mathbf{y}$ and then compares its power (i.e., its squared magnitude) to the predefined threshold γ'. We note that the realization of c_{RCS} is unknown, but coherent combining is achievable anyway because the realization is the same for all the L received symbols. In particular, under hypothesis $\mathcal{H}_1$, it holds that

$$\mathbb{E}\left\{|\mathbf{1}_L^{\mathrm{H}}\mathbf{y}|^2\right\} = \mathbf{1}_L^{\mathrm{H}}\left(P_{\mathrm{r}}\mathbf{1}_L\mathbf{1}_L^{\mathrm{H}} + \sigma^2\mathbf{I}_L\right)\mathbf{1}_L = L\left(P_{\mathrm{r}}L + \sigma^2\right). \tag{8.88}$$

The average effective SNR is $P_{\mathrm{r}}L/\sigma^2$, which increases proportionally to L. We also recall from (8.76) that P_{r} is proportional to the beamforming gain MK.

To exemplify the Neyman-Pearson detector for solving the binary hypothesis test with the Swerling 1 target model, we consider the false alarm probability $P_{\mathrm{FA}} = \alpha = 10^{-3}$. Figure 8.23 shows the resulting detection probability, P_{D}, versus the single-antenna SNR, which is computed by dividing the received power at a single antenna in (8.71) by the noise power $\sigma^2 = BN_0$. We consider a symmetric setup where both the transmitter and receiver have M antennas (i.e., $K = M$). Hence, the effective SNR is obtained by multiplying the single-antenna SNR at the horizontal axis by the beamforming gain M^2. We compare three setups: i) $M = 1$ antenna and $L = 10$ symbols; ii) $M = 10$ antennas and $L = 10$ symbols; and iii) $M = 10$ antennas and $L = 100$ symbols. We notice that the detection probability improves with the SNR in all three cases, which is logical since target detection revolves around distinguishing signals from noise. The three curves have identical shapes but are shifted horizontally. The solid black curve is furthest to the right since it has the fewest antennas and symbols. The dashed red curve is shifted 20 dB to the left because it has $M = 10$ antennas instead of one, which results in a beamforming gain of $M^2 = 100 = 20$ dB. When the single-antenna SNR is -10 dB, P_{D} increases from 0.03 to 0.93 when $M = 1$ is increased to $M = 10$; thus, the use of multiple antennas can make a huge difference. When $M = 10$, an additional performance improvement can be achieved by increasing the number of symbols. When going from $L = 10$ to $L = 100$, the total received power is increased by a factor of 10 thanks to the coherent combining. This explains why the dash-dotted blue curve is shifted 10 dB to the left compared to the red curve. Hence, it is possible to obtain reasonable detection probability values at very low SNR values by utilizing many antennas or symbols. In practice, there is a limit on how large L can be made before the realization of c_{RCS} changes due to target movement.

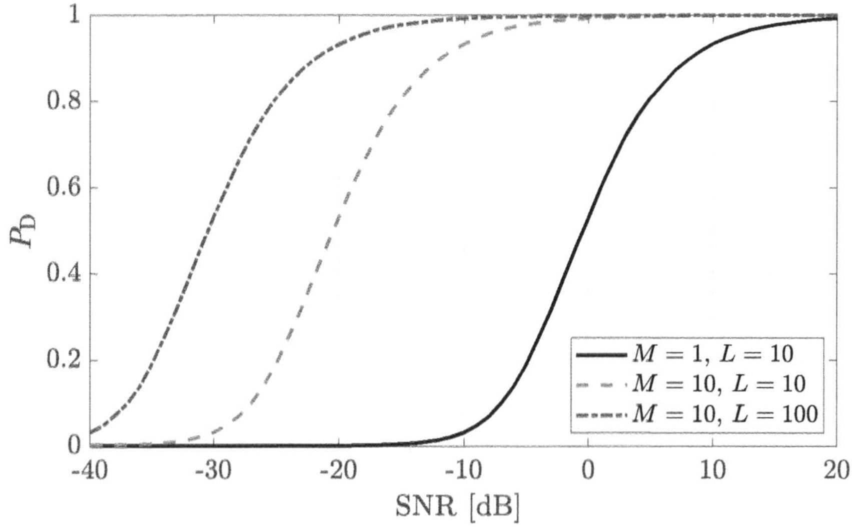

Figure 8.23: The detection probability for different numbers of transmit/receive antennas and received symbols with respect to the single-antenna SNR for the Swerling 1 model.

Example 8.12. We assumed that the transmitter sends the constant symbol "1" throughout the L symbol times when formulating the hypothesis test in (8.78)-(8.79). What changes in the Neyman-Pearson detector if the transmitted signal is $\mathbf{x} = [x[1], \dots, x[L]]^{\mathrm{T}} \in \mathbb{C}^L$, which is known at the receiver?

The new received signal vector can be expressed as $\mathbf{y} = \sqrt{P_{\mathrm{r}}}\mathbf{x}c_{\mathrm{RCS}} + \mathbf{n}$ under the hypothesis $\mathcal{H}_1$. This vector is distributed as $\mathbf{y} \sim \mathcal{N}_{\mathbb{C}}(\mathbf{0}, P_{\mathrm{r}}\mathbf{x}\mathbf{x}^{\mathrm{H}} + \sigma^2\mathbf{I}_L)$, while it still holds that $\mathbf{y} \sim \mathcal{N}_{\mathbb{C}}(\mathbf{0}, \sigma^2\mathbf{I}_L)$ when the hypothesis $\mathcal{H}_0$ is true. Following similar steps as in (8.82)-(8.85), we end up with a Neyman-Pearson detector that decides on the hypothesis $\mathcal{H}_1$ if $|\mathbf{x}^{\mathrm{H}}\mathbf{y}|^2 \geq \gamma'$, where the threshold γ' is selected to have the desired value $P_{\mathrm{FA}} = \alpha$ as

$$
\begin{aligned}
P_{\mathrm{FA}} = \alpha = \int_{|\mathbf{x}^{\mathrm{H}}\mathbf{y}|^2 \geq \gamma'} f_{\mathbf{y}|\mathcal{H}_0}(\mathbf{y}|\mathcal{H}_0)\,\partial\mathbf{y} = \int_{\gamma'}^{\infty} \frac{1}{\|\mathbf{x}\|^2\sigma^2} e^{-\frac{z}{\|\mathbf{x}\|^2\sigma^2}}\,\partial z = e^{-\frac{\gamma'}{\|\mathbf{x}\|^2\sigma^2}} \\
\Rightarrow \quad \gamma' = \|\mathbf{x}\|^2\sigma^2 \ln(\alpha^{-1}). \qquad (8.89)
\end{aligned}
$$

The optimal detector combines the received signal as $\mathbf{x}^{\mathrm{H}}\mathbf{y}$, where each received signal $y[l]$ is multiplied by $x^*[l]$ before being summed up. The multiplication aligns the L signals in phase, and if the magnitudes $|x[1]|, \dots, |x[L]|$ are varying, it also weighs them to maximize the SNR according to the MRC principle. Finally, the detector compares $|\mathbf{x}^{\mathrm{H}}\mathbf{y}|^2$ with the threshold $\gamma' = \|\mathbf{x}\|^2\sigma^2\ln(\alpha^{-1})$. The average received power is $\mathbb{E}\{|\mathbf{x}^{\mathrm{H}}\mathbf{y}|^2\} = P_{\mathrm{r}}\|\mathbf{x}\|^4 + \sigma^2\|\mathbf{x}\|^2$, which shows that it is the value $\|\mathbf{x}\|^2$ that matters not the individual symbols. This is why $\mathbf{x} = \mathbf{1}_L$ works equally well as any other sequence that satisfies $\|\mathbf{x}\|^2 = L$.

8.3.3 Target Detection with the Swerling 2 Target Model

In the Swerling 2 model, the target's RCS is assumed to fluctuate so rapidly that it takes a new independent realization at each symbol time. The realization at time l is denoted by $c_{\text{RCS}}[l] \sim \mathcal{N}_{\mathbb{C}}(0,1)$. We assume that L received signals are collected for target detection and that the constant symbol "1" is transmitted during all of them, as in the previous section. The corresponding binary hypothesis test is

$$\mathcal{H}_0 \quad : \quad y[l] = n[l], \quad l = 1, \ldots, L, \tag{8.90}$$

$$\mathcal{H}_1 \quad : \quad y[l] = \sqrt{P_{\text{r}}} c_{\text{RCS}}[l] + n[l], \quad l = 1, \ldots, L, \tag{8.91}$$

where P_{r} is the average received power reflected through the target, which can be computed using the radar range equation in (8.76). The noise $n[l] \sim \mathcal{N}_{\mathbb{C}}(0, \sigma^2)$ and channel coefficients $c_{\text{RCS}}[l]$ are independent.

We will now particularize the Neyman-Pearson detector for this scenario, where the channel coefficients fluctuate. To prepare for this, we define the vectors $\mathbf{y} = [y[1], \ldots, y[L]] \in \mathbb{C}^L$, $\mathbf{c}_{\text{RCS}} = [c_{\text{RCS}}[1], \ldots, c_{\text{RCS}}[L]]^{\text{T}} \in \mathbb{C}^L$, and $\mathbf{n} = [n[1], \ldots, n[L]]^{\text{T}} \in \mathbb{C}^L$. We note that $\mathbf{c}_{\text{RCS}} \sim \mathcal{N}_{\mathbb{C}}(\mathbf{0}, \mathbf{I}_L)$ and $\mathbf{n} \sim \mathcal{N}_{\mathbb{C}}(\mathbf{0}, \sigma^2 \mathbf{I}_L)$. When the null hypothesis $\mathcal{H}_0$ is correct, the received signal vector becomes $\mathbf{y} = \mathbf{n} \sim \mathcal{N}_{\mathbb{C}}(\mathbf{0}, \sigma^2 \mathbf{I}_L)$. When the hypothesis $\mathcal{H}_1$ is true, the received signal instead becomes $\mathbf{y} = \sqrt{P_{\text{r}}} \mathbf{c}_{\text{RCS}} + \mathbf{n} \sim \mathcal{N}_{\mathbb{C}}(\mathbf{0}, (P_{\text{r}} + \sigma^2)\mathbf{I}_L)$. We can use the PDF of a complex Gaussian vector in (2.85) to evaluate the likelihood ratio in (2.191) from Lemma 2.14 as

$$\gamma \leq \frac{f_{\mathbf{y}|\mathcal{H}_1}(\mathbf{y}|\mathcal{H}_1)}{f_{\mathbf{y}|\mathcal{H}_0}(\mathbf{y}|\mathcal{H}_0)} = \frac{\frac{1}{\pi^L \det((P_{\text{r}}+\sigma^2)\mathbf{I}_L)} e^{-\mathbf{y}^{\text{H}}\left((P_{\text{r}}+\sigma^2)\mathbf{I}_L\right)^{-1}\mathbf{y}}}{\frac{1}{\pi^L \det(\sigma^2 \mathbf{I}_L)} e^{-\mathbf{y}^{\text{H}}(\sigma^2 \mathbf{I}_L)^{-1}\mathbf{y}}}. \tag{8.92}$$

Using the fact that $\ln(\gamma)$ is a monotonically increasing function for $\gamma \geq 0$, the Neyman-Pearson detector in (8.92) decides on the hypothesis $\mathcal{H}_1$ if

$$\frac{\mathbf{y}^{\text{H}}\mathbf{y}}{\sigma^2} - \frac{\mathbf{y}^{\text{H}}\mathbf{y}}{P_{\text{r}} + \sigma^2} \geq \ln(\gamma) - \ln(b), \tag{8.93}$$

where the constant $b = \det\left(\sigma^2 \mathbf{I}_L\right) / \det\left((P_{\text{r}} + \sigma^2)\mathbf{I}_L\right) = \left(\sigma^2/(P_{\text{r}} + \sigma^2)\right)^L$ is independent of the received signal $\mathbf{y}$. By arranging the terms in (8.93), we can express the condition for selecting hypothesis $\mathcal{H}_1$ as

$$\mathbf{y}^{\text{H}}\mathbf{y} \geq \underbrace{\frac{\sigma^2(P_{\text{r}} + \sigma^2)(\ln(\gamma) - \ln(b))}{P_{\text{r}}}}_{=\gamma'}, \tag{8.94}$$

where γ' is the revised threshold variable that must be selected so that

$$P_{\text{FA}} = \alpha = \int_{\mathbf{y}^{\text{H}}\mathbf{y} \geq \gamma'} f_{\mathbf{y}|\mathcal{H}_0}(\mathbf{y}|\mathcal{H}_0)\, \partial\mathbf{y} = \int_{\gamma'}^{\infty} \frac{z^{L-1} e^{-\frac{z}{\sigma^2}}}{(\sigma^2)^L (L-1)!} \partial z, \tag{8.95}$$

where the last equality follows from (2.99) because $\mathbf{y}^{\mathrm{H}}\mathbf{y} = \|\mathbf{y}\|^2$ has a scaled χ^2-distribution under the hypothesis $\mathcal{H}_0$. The integral can be computed using the incomplete gamma function, but it lacks a closed-form inverse, so (8.95) must be solved numerically.

The term $\mathbf{y}^{\mathrm{H}}\mathbf{y} = \sum_{l=1}^{L} |y[l]|^2$ in (8.94) is the sufficient statistics for target detection in this scenario. Hence, the optimal receiver processing for target detection adds up the powers of the individual received signals $y[l]$ and compares the result to the predefined threshold γ'. This approach differs from the detector derived for the Swerling 1 target model. The reason is that the channel coefficient $c_{\mathrm{RCS}}[l]$ takes a new unknown realization at every time instant, so the receiver cannot coherently combine the signals. One way to quantify the difference is to compute the total power of the received signal:

$$\mathbb{E}\left\{\|\mathbf{y}\|^2\right\} = \mathrm{tr}\left((P_{\mathrm{r}} + \sigma^2)\mathbf{I}_L\right) = L(P_{\mathrm{r}} + \sigma^2). \tag{8.96}$$

The average effective SNR is P_{r}/σ^2, which is independent of L. This is different from (8.88) where an L times larger SNR value was achieved with the Swerling 1 target model, thanks to the coherent combining at the receiver. Fortunately, the term P_{r} remains proportional to the beamforming gain MK, so we still benefit from having multiple antennas because the RCS realization is the same for all antennas.

For a selected threshold γ', the detection probability P_{D} is given as

$$P_{\mathrm{D}} = \int_{\mathbf{y}^{\mathrm{H}}\mathbf{y} \geq \gamma'} f_{\mathbf{y}|\mathcal{H}_1}(\mathbf{y}|\mathcal{H}_1)\, \partial \mathbf{y}. \tag{8.97}$$

To exemplify the Neyman-Pearson detector for solving the binary hypothesis test with the Swerling 2 target model, we consider the false alarm probability $P_{\mathrm{FA}} = \alpha = 10^{-3}$. Figure 8.24 shows the detection probability, P_{D}, versus the single-antenna SNR, which is computed as in Figure 8.23. We consider a symmetric setup where both the transmitter and receiver have M antennas (i.e., $K = M$). Hence, the effective SNR is obtained by multiplying the single-antenna SNR at the horizontal axis by the beamforming gain M^2. There are three curves, which represent different numbers of antennas M and symbols L. As expected, the detection probability improves as the SNR increases. When the number of antennas increases from $M = 1$ to $M = 10$ (i.e., from solid black to dashed red), a beamforming gain of $M^2 = 100 = 20\,\mathrm{dB}$ is achieved. This shifts the detection probability curve to the left by 20 dB. This can make an immense difference: when the single-antenna SNR is $-10\,\mathrm{dB}$, P_{D} increases from almost zero to almost one. If we increase the number of symbols from $L = 10$ to $L = 100$, the detection probability curve is further shifted to the left, but the gain is much less than 10 dB, even if we receive 10 times more power. It might come as a surprise that the curve is shifted at all because we observed in (8.96) that the average SNR is independent of L. Although the receive combining does not provide any coherent power gain,

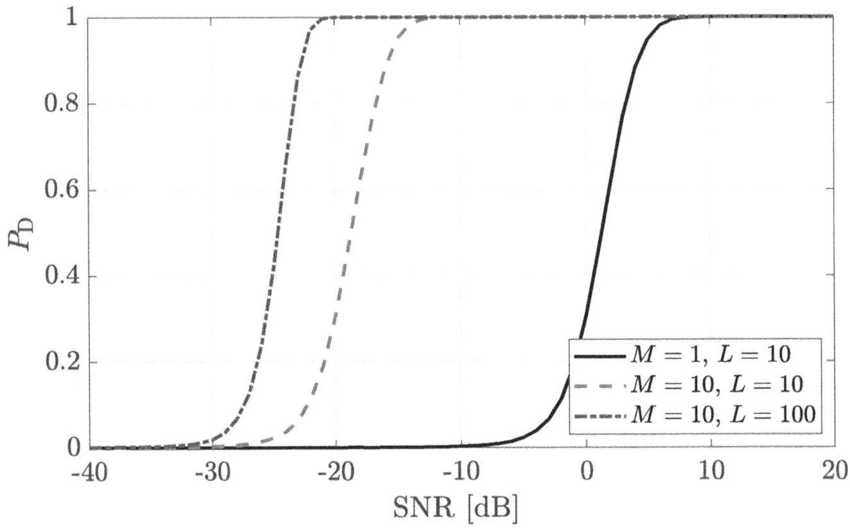

Figure 8.24: The detection probability for different numbers of transmit/receive antennas and received symbols with respect to the single-antenna SNR for the Swerling 2 model.

we achieve a time diversity gain that makes the distribution of $\|\mathbf{y}\|^2$ more confined around its mean when L is increased. Such diversity is beneficial when we try to reach small probabilities, such as $P_{\text{FA}} = 10^{-3}$.

By comparing Figure 8.24 with the Swerling 1 counterpart in Figure 8.23, we can notice two main things. Firstly, the SNR values that give 0.5 (i.e., the median) are shifted to the right in Swerling 2, so the increased randomness generally leads to performance degradation. Secondly, the detection probability curves are steeper with Swerling 2 due to the time diversity that suppresses the channel's randomness. When the SNR is low, the power gain brought by coherent combining with Swerling 1 is preferable over the diversity gain. However, the diversity gain obtained in Swerling 2 dominates the loss due to non-coherent combining at high SNR, where the noise level is already much smaller than the average signal level. Hence, Swerling 2 provides better performance than Swerling 1 in these situations.

The choice of the RCS model clearly impacts the target detection performance. The Swerling 1 model is suitable when the target is approximately static during the transmission time, while the Swerling 2 is suitable for highly mobile targets. One could also create an intermediate block-fading-like model where the RCS realization is constant for multiple symbols but not the entire transmission time. There are further Swerling models where the RCS parameter has a different distribution than complex Gaussian [134]–[136]. As shown in Figure 5.4, the Gaussian distribution appears when there are at least five equally strong scattering objects on the target object, but some targets might have a shape that is not well modeled like that.

Example 8.13. Consider a multi-static target detection setup with a single transmitter and two spatially separated receivers. Derive the sufficient statistics of the Neyman-Pearson detector with the Swerling 2 target model.

In this scenario, we can express the binary hypothesis test as

$$\mathcal{H}_0 \quad : \quad y_1[l] = n_1[l], \quad y_2[l] = n_2[l], \; l = 1, \dots, L, \tag{8.98}$$

$$\mathcal{H}_1 \quad : \quad y_1[l] = \sqrt{P_{\mathrm{r},1}} c_1[l] + n_1[l], \; y_2[l] = \sqrt{P_{\mathrm{r},2}} c_2[l] + n_2[l], \; l = 1, \dots, L, \tag{8.99}$$

where $c_1[l] \sim \mathcal{N}_{\mathbb{C}}(0,1)$ and $c_2[l] \sim \mathcal{N}_{\mathbb{C}}(0,1)$ are the independent random RCS coefficients, while $P_{\mathrm{r},1}$ and $P_{\mathrm{r},2}$ denote the average received powers at the two receivers. These might be different since the average RCS depends on the receivers' angles to the target. The noise samples $n_1[l] \sim \mathcal{N}_{\mathbb{C}}(0, \sigma^2)$ and $n_2[l] \sim \mathcal{N}_{\mathbb{C}}(0, \sigma^2)$ are independent since the receivers are spatially separated.

We define $\mathbf{y}_m = [y_m[1], \dots, y_m[L]]^{\mathrm{T}} \in \mathbb{C}^L$, $\mathbf{c}_m = [c_m[1], \dots, c_m[L]]^{\mathrm{T}} \in \mathbb{C}^L$, and $\mathbf{n}_m = [n_m[1], \dots, n_m[L]]^{\mathrm{T}} \in \mathbb{C}^L$, for $m = 1, 2$. We can now note that $\mathbf{c}_m \sim \mathcal{N}_{\mathbb{C}}(\mathbf{0}, \mathbf{I}_L)$ and $\mathbf{n}_m \sim \mathcal{N}_{\mathbb{C}}(\mathbf{0}, \sigma^2 \mathbf{I}_L)$. Since $\mathbf{y}_1$ and $\mathbf{y}_2$ are independent under both hypotheses, we construct the likelihood ratio in (2.191) as

$$\gamma \leq \frac{f_{\mathbf{y}_1,\mathbf{y}_2|\mathcal{H}_1}(\mathbf{y}_1,\mathbf{y}_2|\mathcal{H}_1)}{f_{\mathbf{y}_1,\mathbf{y}_2|\mathcal{H}_0}(\mathbf{y}_1,\mathbf{y}_2|\mathcal{H}_0)} = \frac{\frac{1}{\pi^{2L}(P_{\mathrm{r},1}+\sigma^2)^L (P_{\mathrm{r},2}+\sigma^2)^L} e^{-\frac{\mathbf{y}_1^{\mathrm{H}}\mathbf{y}_1}{P_{\mathrm{r},1}+\sigma^2}} e^{-\frac{\mathbf{y}_2^{\mathrm{H}}\mathbf{y}_2}{P_{\mathrm{r},2}+\sigma^2}}}{\frac{1}{\pi^{2L}\sigma^{2L}} e^{-\frac{\mathbf{y}_1^{\mathrm{H}}\mathbf{y}_1}{\sigma^2}} e^{-\frac{\mathbf{y}_2^{\mathrm{H}}\mathbf{y}_2}{\sigma^2}}}. \tag{8.100}$$

Taking the logarithm of both sides and omitting the constant coefficients, the sufficient statistics for the Neyman-Pearson detector can be expressed as

$$\left(\frac{1}{\sigma^2} - \frac{1}{P_{\mathrm{r},1} + \sigma^2} \right) \mathbf{y}_1^{\mathrm{H}} \mathbf{y}_1 + \left(\frac{1}{\sigma^2} - \frac{1}{P_{\mathrm{r},2} + \sigma^2} \right) \mathbf{y}_2^{\mathrm{H}} \mathbf{y}_2. \tag{8.101}$$

This is a weighted sum of the sufficient statistics the receivers would use in the single-receiver case. The receiver that experiences the largest received power uses the largest weight, but both receivers are useful.

8.3.4 Different Types of Radar Antenna Arrays

The radar technology dates back to Christian Hülsmeyer, who filed a patent in 1904 on a system that uses electromagnetic waves to detect metallic objects [138], and it was demonstrated for target detection at sea to avoid ship collisions. The technology was not utilized at scale until the Second World War, which is when the United States Navy introduced the radar abbreviation. A classical radar consists of a highly directive antenna that is mechanically rotated over time to scan different angular directions sequentially. The *passive electronically scanned array (PESA)* technology appeared in the 1960s based on the analog beamforming architecture, previously illustrated in Figure 7.10. The directivity is controlled by electrical beamforming in PESA radars, which enables faster scanning and flexibility in which directions are considered than mechanical beamforming. These features are particularly useful for target tracking. Some PESA radars can emit/receive multiple beams simultaneously, which resembles the hybrid beamforming architecture in Figure 7.12.

The most capable radars use the digital beamforming architecture, where each antenna is directly connected to the digital baseband as in Figure 7.9. This is called the *active electronically scanned array (AESA)* technology and enables simultaneous beamforming in different directions at different frequencies. Practical implementations began in the 1990s, but the higher implementation cost has thus far led to AESA radars primarily being used in mission-critical military applications where many targets must be simultaneously detected, localized, tracked, and potentially attacked. This might change when MIMO communication systems evolve into ISAC systems, where the digital architecture required for high-capacity MIMO communications is also used for sensing applications. For this reason, the theory described in this chapter presumes the use of the digital architecture.

The term *MIMO radar* has been used for decades [139], [140] and created some controversy [141] because not all MIMO communication features are helpful for radars. For example, the ergodic capacity in (5.131) over a point-to-point MIMO channel is achieved by spreading many independent data signals in different directions, and the sum capacity of a multi-user MIMO channel is achieved by sending many simultaneous signals even if this reduces the capacity of individual streams and users. In radar applications, the accuracy of individual sensing tasks might be more important than the ability to spatially multiplex many sensing tasks if the latter comes with reduced accuracy. The pragmatic view is that MIMO radar theory [142] describes how to operate AESA radars in different situations, which sometimes results in the same functionality as a PESA radar—similar to how beamforming of one signal is capacity-achieving in point-to-point MIMO systems that have low SNR. In other situations, AESA radars can benefit from simultaneous detection of multiple targets, higher spatial resolution, flexible interference suppression, and different directivity at different frequencies [118].

An additional way to improve spatial resolution is to utilize a synthetic aperture created by moving the antennas during the measurement period. If the deployment location is fixed, the antennas can be moved around at that location. If the radar is deployed on a satellite that travels around the Earth, a synthetic aperture is created even if the antennas are fixed at the satellite. In any case, by combining the measurements made at different times, the resolution of radar sensing becomes identical to using a physical array that simultaneously has antennas at all the measurement locations.

8.3.5 Integrated Sensing and Communication

The term ISAC is used to describe network deployments that are jointly designed for sensing and communication applications [123], in contrast to how communication networks and radar systems have been developed and deployed independently in the past. The fusing of these technologies became particularly interesting when communication systems began to use mmWave bands, which is the spectrum range traditionally used for radar [143]. Apart from cost savings, a dual-functional network might provide performance benefits to the different applications by sharing information between them, and new joint radar communication services might arise [144].

The integration can come at different levels of which three are illustrated in Figure 8.25. At the first level, shown in Figure 8.25(a), the deployment sites for communication networks are reused for deploying radar transceivers, but the systems are otherwise independent: they use different hardware and frequency bands. At the second level, shown in Figure 8.25(b), the same transceiver hardware is used for both applications, but they use orthogonal signal resources. The network can either switch between sensing and communication over time or use non-overlapping frequency bands, which are sufficiently similar so the same hardware components can be used for dual purposes. The benefit of this approach is that the signal waveforms can be optimized for the respective applications without making tradeoffs. At the third level, shown in Figure 8.25(c), the same time-frequency resources are used for both sensing and communication purposes. The benefit of this approach is that more signal resources are available for both applications, while the drawback is interference and signal transmissions that are not optimized for dual purposes.

The theory for sensing provided in this chapter directly applies to the first two integration levels, while the third level gives rise to different system models. A basic mono-static level-three ISAC setup is illustrated in Figure 8.25(c), where the base station transmits a communication signal to a data-receiving user but also listens to the reflection of the data signal from the target. Since the transmitter knows the data signal, it can be used for target detection; we recall from Example 8.12 that any signal with a specified average power works equally well for that purpose. However, the user prefers data transmission with MRT precoding, while the target detection probability is maximized if the signal is beamformed towards the potential target location. This is an example of the inherent tradeoff between sensing and communication, which materializes in conflicting precoding designs in this basic scenario. We refer to [145] for a more profound overview of ISAC, also known as joint communication and sensing.

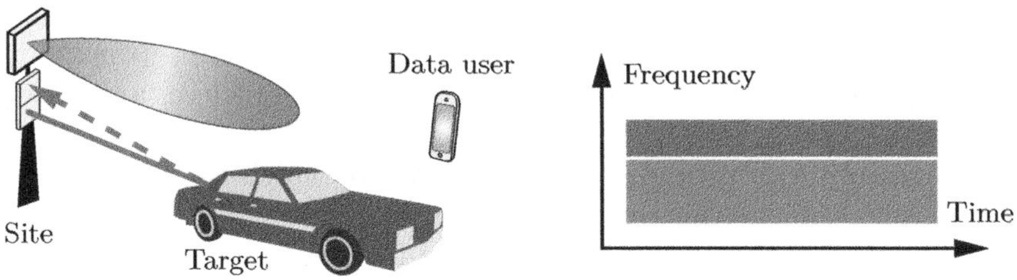

(a) First integration level: Site-sharing but separate hardware and frequency bands.

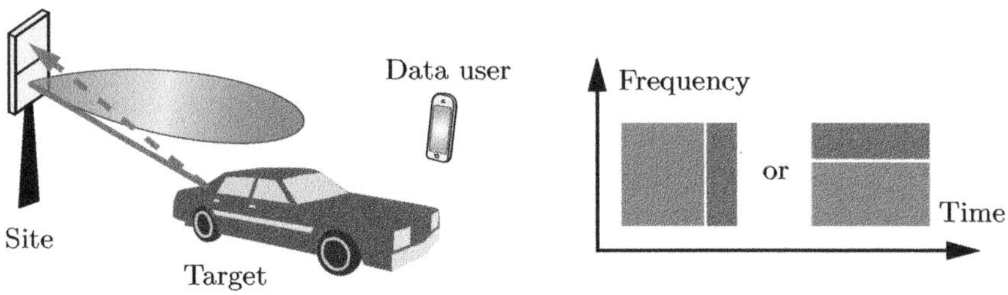

(b) Second integration level: Hardware-sharing but orthogonal time/frequency signals.

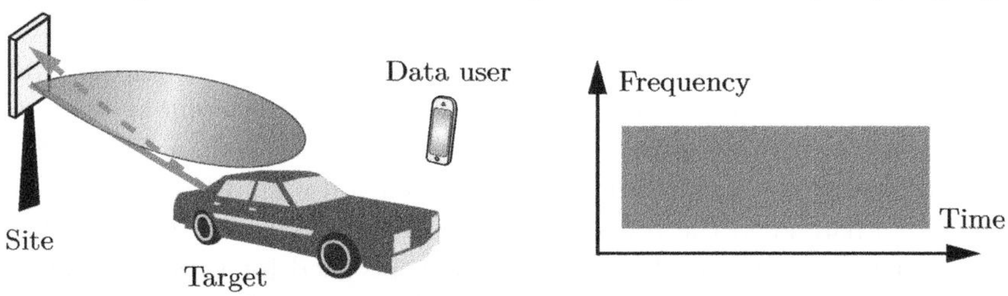

(c) Third integration level: Hardware- and signal-sharing for sensing/communication.

Figure 8.25: Example of three integration levels for sensing and communication. Different hardware, technology, and spectrum are used in (a), but the site location is shared. The same hardware is used in (b), but the time/frequency resources differ. The same hardware and resources are used for both applications in (c).

8.4 Exercises

Exercise 8.1. When deriving the Capon spectrum in (8.20), it is implicitly assumed that the sample average estimate $\hat{\mathbf{R}}_L$ in (8.4) is an invertible matrix. In this exercise, we will analyze the opposite case when $\hat{\mathbf{R}}_L$ is rank-deficient. By defining $\mathbf{Y}_L = [\mathbf{y}[1], \ldots, \mathbf{y}[L]] \in \mathbb{C}^{M \times L}$, we can write $\hat{\mathbf{R}}_L = \mathbf{Y}_L \mathbf{Y}_L^{\mathrm{H}}/L$. Let the SVD of $\mathbf{Y}_L$ be denoted as $\mathbf{Y}_L = \mathbf{U}\boldsymbol{\Sigma}\mathbf{V}^{\mathrm{H}}$. Then, the eigendecomposition of $\hat{\mathbf{R}}_L$ can be expressed as $\hat{\mathbf{R}}_L = \mathbf{U}\boldsymbol{\Sigma}\mathbf{V}^{\mathrm{H}}\mathbf{V}\boldsymbol{\Sigma}^{\mathrm{T}}\mathbf{U}^{\mathrm{H}}/L = \mathbf{U}\left(\boldsymbol{\Sigma}\boldsymbol{\Sigma}^{\mathrm{T}}/L\right)\mathbf{U}^{\mathrm{H}}$.

(a) Show that if $L < M$, $\hat{\mathbf{R}}_L$ is rank-deficient. Hint: You can use the SVD of $\mathbf{Y}_L$ to compute the number of non-zero eigenvalues of $\hat{\mathbf{R}}_L$.

(b) Assume $L < M$ so that $\mathbf{Y}_L$ has L positive singular values. The left singular vector matrix can be factorized as $\mathbf{U} = [\bar{\mathbf{U}}, \tilde{\mathbf{U}}]$, where $\bar{\mathbf{U}} \in \mathbb{C}^{M \times L}$ corresponds to the strictly positive singular values (in decreasing order) and $\tilde{\mathbf{U}} \in \mathbb{C}^{M \times (M-L)}$ corresponds to the zero singular values. We can express any array response vector as $\mathbf{a}(\varphi, \theta) = \bar{\mathbf{U}}\bar{\mathbf{x}} + \tilde{\mathbf{U}}\tilde{\mathbf{x}}$ in terms of the vectors $\bar{\mathbf{x}} = \bar{\mathbf{U}}^{\mathrm{H}}\mathbf{a}(\varphi, \theta)$ and $\tilde{\mathbf{x}} = \tilde{\mathbf{U}}^{\mathrm{H}}\mathbf{a}(\varphi, \theta)$. Show that the objective function of the Capon spectrum in (8.17) becomes zero for $\mathbf{a}(\varphi, \theta)$ that satisfies $\tilde{\mathbf{x}} = \tilde{\mathbf{U}}^{\mathrm{H}}\mathbf{a}(\varphi, \theta) \neq \mathbf{0}$.

(c) According to (b), the Capon spectrum becomes zero when $\tilde{\mathbf{x}} = \tilde{\mathbf{U}}^{\mathrm{H}}\mathbf{a}(\varphi, \theta) \neq \mathbf{0}$, regardless of the value of $\bar{\mathbf{x}} = \bar{\mathbf{U}}^{\mathrm{H}}\mathbf{a}(\varphi, \theta)$. A more noise-robust version of the Capon spectrum that differentiates between the power of $\bar{\mathbf{x}}$ for different array response vectors can be constructed by so-called *diagonal loading*. In this method, a regularization term $\epsilon\mathbf{I}_M$ with a small $\epsilon > 0$ is added to $\hat{\mathbf{R}}_L$ to make it invertible, and the modified Capon spectrum is obtained as

$$P(\varphi, \theta) = \frac{1}{\mathbf{a}^{\mathrm{H}}(\varphi, \theta)\left(\hat{\mathbf{R}}_L + \epsilon\mathbf{I}_M\right)^{-1}\mathbf{a}(\varphi, \theta)}. \tag{8.102}$$

Assuming $L < M$ and that $\mathbf{Y}_L$ has the singular values $s_1 \geq \ldots \geq s_L > 0$, express the value of the Capon spectrum for an arbitrary $\mathbf{a}(\varphi, \theta) = \bar{\mathbf{U}}\bar{\mathbf{x}} + \tilde{\mathbf{U}}\tilde{\mathbf{x}} = \mathbf{U}[\bar{\mathbf{x}}^{\mathrm{T}}, \tilde{\mathbf{x}}^{\mathrm{T}}]^{\mathrm{T}}$ in terms of $\bar{\mathbf{x}}$ and $\tilde{\mathbf{x}}$. Does the spectrum value differ for the array response vectors that satisfy $\tilde{\mathbf{x}} = \tilde{\mathbf{U}}^{\mathrm{H}}\mathbf{a}(\varphi, \theta) \neq \mathbf{0}$? Hint: Use the relation

$$\begin{aligned}\left(\hat{\mathbf{R}}_L + \epsilon\mathbf{I}_M\right)^{-1} &= \left(\mathbf{U}\left(\boldsymbol{\Sigma}\boldsymbol{\Sigma}^{\mathrm{T}}/L\right)\mathbf{U}^{\mathrm{H}} + \epsilon\mathbf{U}\mathbf{U}^{\mathrm{H}}\right)^{-1} \\ &= \mathbf{U}\left(\boldsymbol{\Sigma}\boldsymbol{\Sigma}^{\mathrm{T}}/L + \epsilon\mathbf{I}_M\right)^{-1}\mathbf{U}^{\mathrm{H}}.\end{aligned} \tag{8.103}$$

Exercise 8.2. When generating the MUSIC spectrum in (8.46), we need to create a grid of angles and evaluate the value of the spectrum at each grid point. Hence, the accuracy of the DOA estimation highly depends on the grid resolution. Although having a dense grid for better accuracy is good, one major drawback of the original MUSIC algorithm, called *spectral MUSIC*, is the high computational complexity. A modified version of the MUSIC algorithm that avoids the grid search is called *root MUSIC* [146].

(a) Define the complex variable $z = e^{-\mathrm{j}2\pi\frac{\Delta\sin(\varphi)}{\lambda}}$ and express the array response vector for the ULA in (8.8) as a function $\mathbf{a}(z)$ of z.

(b) Suppose that there are K sources. Show that the DOA estimates $\hat{\varphi}_k$ can be obtained from the angular positions $-2\pi\frac{\Delta\sin(\hat{\varphi}_k)}{\lambda}$ of the K complex roots, which are closest to the unit circle and appear in pairs of reciprocal, of the equation $\mathbf{a}^{\mathrm{T}}(z^{-1})\hat{\mathbf{U}}_{\mathrm{n}}\hat{\mathbf{U}}_{\mathrm{n}}^{\mathrm{H}}\mathbf{a}(z) = 0$.

Exercise 8.3. Consider a DOA estimation problem in the 2D plane with the elevation angle $\theta = 0$. There are $M = 3$ antennas in a triangular array with the antenna positions $(0,0)$, $(\Delta,0)$, and $(0,\Delta)$.

(a) If $\Delta = \lambda/2$, can we unambiguously estimate the DOA for all the angles $\varphi \in [0, 2\pi)$? If not, what are the azimuth angles that create ambiguity?

(b) If $\Delta = \lambda/2 - \epsilon$ for some arbitrary $0 < \epsilon < \lambda/2$, can we unambiguously estimate the DOA for all the angles $\varphi \in [0, 2\pi)$? If not, what are the azimuth angles that create ambiguity?

Exercise 8.4. Non-linear least squares (NLS) is a parametric DOA estimation method, where the DOA estimates are obtained as the angles that minimize the norm square of the difference between the received noisy signals and the noise-free part in (8.3):

$$\sum_{l=1}^{L} \left\| \mathbf{y}[l] - \sum_{k=1}^{K} \sqrt{\beta_k}\mathbf{a}(\varphi_k, \theta_k)x_k[l] \right\|^2 = \sum_{l=1}^{L} \|\mathbf{y}[l] - \mathbf{A}\mathbf{p}[l]\|^2, \tag{8.104}$$

where $\mathbf{A} = [\mathbf{a}(\varphi_1, \theta_1), \ldots, \mathbf{a}(\varphi_K, \theta_K)]$ and $\mathbf{p}[l] = [\sqrt{\beta_1}x_1[l], \ldots, \sqrt{\beta_K}x_K[l]]$. The unknown source signals $\mathbf{p}[l]$ are treated as deterministic in the NLS method. Assume the rank of $\mathbf{A}$ equals K.

(a) Find the vectors $\mathbf{p}[l]$ that minimize (8.104) by expressing the objective function as a quadratic function of $\mathbf{p}[l]$, for $l = 1, \ldots, L$.

(b) Insert the optimal $\mathbf{p}[l]$ found in (a) into the objective function in (8.104) and show that the DOA estimates are found as

$$\{(\hat{\varphi}_k, \hat{\theta}_k)\}_{k=1}^{K} = \underset{\{\varphi_k, \theta_k\}_{k=1}^{K}}{\arg\max} \sum_{l=1}^{L} \mathbf{y}^{\mathrm{H}}[l]\mathbf{A}\left(\mathbf{A}^{\mathrm{H}}\mathbf{A}\right)^{-1}\mathbf{A}^{\mathrm{H}}\mathbf{y}[l]. \tag{8.105}$$

(c) Show that the NLS method becomes equivalent to conventional beamforming if $K = 1$.

Exercise 8.5. Consider the received signal given in (8.38) for DOA estimation, which can be expressed as

$$\mathbf{y}[l] = \mathbf{A}\mathbf{p}[l] + \mathbf{n}[l], \tag{8.106}$$

where $K < M$, $\mathbf{A} = [\mathbf{a}(\varphi_1, \theta_1), \ldots, \mathbf{a}(\varphi_K, \theta_K)]$ and $\mathbf{p}[l] = [\sqrt{\beta_1}x_1[l], \ldots, \sqrt{\beta_K}x_K[l]]$.

(a) Suppose the noise signal is colored with an invertible covariance matrix $\mathbf{C}$, i.e., $\mathbf{n}[l] \sim \mathcal{N}_{\mathbb{C}}(\mathbf{0}, \mathbf{C})$. To apply the MUSIC algorithm, we must first whiten the signals $\mathbf{y}[l]$. What is the resulting MUSIC spectrum?

(b) In practice, mutual coupling can occur due to interaction between closely spaced antennas in an array. There exist array calibration methods that can mitigate these effects, but there will be residual calibration errors. Suppose the received signal can be modeled as [147]

$$\mathbf{y}[l] = \mathbf{M}\mathbf{A}\mathbf{p}[l] + \mathbf{n}[l], \tag{8.107}$$

where $\mathbf{M} \in \mathbb{C}^{M \times M}$ is a non-singular matrix. If $\mathbf{n}[l] \sim \mathcal{N}_{\mathbb{C}}(\mathbf{0}, \sigma^2\mathbf{I}_M)$, obtain the MUSIC spectrum for this signal model.

Exercise 8.6. In this exercise, we consider mixed TDOA/DOA localization. A single target node is located at (x, y) and M receivers. The 2D coordinates of receiver m is denoted by (x_m, y_m). Suppose the receivers with the indices $2, \ldots, \overline{M}$, where $\overline{M} < M$, provide TDOA measurements with respect to the reference receiver 1. The remaining receivers with the indices $\overline{M}+1, \ldots, M$ provide DOA estimates. Let $\mathbf{r}_{\text{TDOA}} = [r_{2,1}, \ldots, r_{\overline{M},1}]^{\text{T}}$ and $\mathbf{r}_{\text{DOA}} = [r_{\overline{M}+1}, \ldots, r_M]^{\text{T}}$ be the noisy distance measurements obtained with TDOA and DOA measurements, respectively. The respective measurement noise is denoted by $\mathbf{n}_{\text{TDOA}} = [n_{2,1}, \ldots, n_{\overline{M},1}]^{\text{T}} \sim \mathcal{N}_{\mathbb{C}}(\mathbf{0}, \mathbf{C}_{\text{TDOA}})$ and $\mathbf{n}_{\text{DOA}} = [n_{\overline{M}+1}, \ldots, n_M]^{\text{T}} \sim \mathcal{N}_{\mathbb{C}}(\mathbf{0}, \mathbf{C}_{\text{DOA}})$.

(a) Derive the ML cost function for the mixed TDOA/DOA localization, assuming that the measurement noises are independent.

(b) What is the minimum number of receivers for unambiguous 2D localization under the condition $2 \leq \overline{M} < M$ if the azimuth and elevation DOAs can be estimated separately?

(c) What is the minimum number of receivers for unambiguous 3D localization under the condition $2 \leq \overline{M} < M$ if the receivers that estimate DOA have ULAs?

Exercise 8.7. Consider 2D TOA-based localization with a single target node and M receivers with the received signals given in (8.52). One approach is to arrange the equations in (8.52) to obtain a linear relation with zero-mean additive noise. The LS solution can then be obtained in closed form. The aim of this exercise is to obtain a relation in the form of $\mathbf{b} = \mathbf{A}\mathbf{z} + \mathbf{w}$, where $\mathbf{z}$ consists of unknown variables, $\mathbf{b}$ and $\mathbf{A}$ are fixed, and $\mathbf{w}$ has zero-mean noise entries. Given such a model, the LS solution is obtained as $\hat{\mathbf{z}} = \left(\mathbf{A}^{\text{T}}\mathbf{A}\right)^{-1}\mathbf{A}^{\text{T}}\mathbf{b}$ if $\mathbf{A}^{\text{T}}\mathbf{A}$ is invertible, and we can obtain $(\hat{x}, \hat{y})$ using $\hat{\mathbf{z}}$.

(a) Let us define $\mathbf{z} = [x, y, x^2 + y^2]^{\text{T}} \in \mathbb{R}^3$. Find the matrix $\mathbf{A} \in \mathbb{R}^{M \times 3}$ that contains only the known receiver locations (x_m, y_m) and constants. Obtain also the observation vector $\mathbf{b}$ and noise vector $\mathbf{w}$. Hint: Take the squares on both sides of

$$r_m = \sqrt{(x_m - x)^2 + (y_m - y)^2} + n_m, \quad m = 1, \ldots, M, \tag{8.108}$$

where $n_m \sim \mathcal{N}(0, \sigma_{\text{d}}^2)$.

(b) Do the entries of $\mathbf{w}$ have zero mean? If not, under what conditions can we approximate it as a zero-mean vector?

Exercise 8.8. One of the main contributors to the reduced localization accuracy is NLOS paths between the target node and some of the receivers due to the blockage of the LOS path. In TOA, the effect of NLOS paths can be modeled as a positive bias to the true TOAs with much larger power than the measurement errors. Suppose there is a single target node located at (x, y) and M receivers. The 2D coordinates of receiver m is denoted by (x_m, y_m). Suppose the receivers with the indices $1, \ldots, \overline{M}$, where $\overline{M} < M$, have a blocked LOS towards the target node, and the NLOS bias is modeled as an exponential random variable $b_m \sim \text{Exp}(1/\sigma_b^2)$ with the PDF from (2.91). For the other receivers $\overline{M}+1, \ldots, M$, the relations in (8.52) are valid. The distance measurements for this TOA-based localization setup can be expressed as

$$r_m = \sqrt{(x_m - x)^2 + (y_m - y)^2} + b_m, \quad m = 1, \ldots, \overline{M}, \tag{8.109}$$

$$r_m = \sqrt{(x_m - x)^2 + (y_m - y)^2} + n_m, \quad m = \overline{M}+1, \ldots, M, \tag{8.110}$$

where we have omitted the measurement noise n_m for $m = 1, \ldots, \overline{M}$ since the NLOS bias is much stronger. Assuming that $n_m \sim \mathcal{N}(0, \sigma_{\text{d}}^2)$ and all b_m and n_m are mutually independent, derive the cost function to be minimized for ML estimation of (x, y).

Exercise 8.9. Consider target detection with the binary hypothesis test

$$\mathcal{H}_0 \quad : \quad y[l] = n[l], \quad l = 1, \ldots, L, \tag{8.111}$$

$$\mathcal{H}_1 \quad : \quad y[l] = \sqrt{P_{\mathrm{r}}} c_{\mathrm{RCS}} + n[l], \quad l = 1, \ldots, L, \tag{8.112}$$

where the RCS coefficient c_{RCS} is constant and known at the receiver (called the Swerling 0 model). Derive the sufficient statistics for the Neyman-Pearson detector when the noise samples $n[l] \sim \mathcal{N}_{\mathbb{C}}(0, \sigma^2)$ are independent.

Exercise 8.10. Consider the sufficient statistics $\left|\mathbf{1}_L^{\mathrm{H}}\mathbf{y}\right|^2$ in (8.85) of the Neyman-Pearson detector with the Swerling 1 target model. The SNR of the coherently combined signal $\mathbf{1}_L^{\mathrm{H}}\mathbf{y}$ under the target existence determines the detection performance of the radar detector.

(a) Let ϱ denote the average single-antenna SNR of the considered radar channel. Under the target existence, express the SNR of the coherently combined signal $\mathbf{1}_L^{\mathrm{H}}\mathbf{y}$ for a given number of transmit/receive antennas M and the number of coherently combined symbols L.

(b) Suppose the average power consumption of the system is

$$L\frac{P}{0.25} + LM \cdot 1 + \bar{L}M \cdot 1 \quad \mathrm{W}, \tag{8.113}$$

where the first term includes a power amplifier efficiency of 25%. The second term models that each transmit antenna consumes 1 W and is turned on for L symbols. The third term models that each receive antenna consumes 1 W and must be active for a fixed window of $\bar{L}$ symbols to capture the reflected signal. Which combination of L and M minimizes the average power consumption in (8.113) while guaranteeing an SNR of at least 10 dB for $\mathbf{1}_L^{\mathrm{H}}\mathbf{y}$ if $\varrho = -10$ dB, $P = 10$ W, and $\bar{L} = 100$?

Exercise 8.11. Consider a mono-static setup with a single transmit/receive antenna for a target detection task. Suppose the propagation between the transmitter/receiver and the potential target is modeled using the radar range equation. Moreover, assume that a target having RCS of 0 dBsm is detectable at a distance of 100 m when $L = 1$. It is desired to detect a smaller target with an RCS of -10 dBsm at a distance of 200 m. The antenna gains are assumed to be fixed in this exercise.

(a) How many transmit/receive antennas M are needed to achieve the given task without changing any other parameters?

(b) If the target follows the Swerling 1 model, how many symbols L are needed to achieve the given task without changing any other parameters?

Exercise 8.12. Consider a multi-static target detection setup with a single transmitter and two spatially separated receivers. Assuming the target reflection follows the Swerling 1 model, we can express the binary hypothesis test as

$$\mathcal{H}_0 \quad : \quad y_1[l] = n_1[l], \quad y_2[l] = n_2[l], \; l = 1, \ldots, L, \tag{8.114}$$

$$\mathcal{H}_1 \quad : \quad y_1[l] = \sqrt{P_{\mathrm{r},1}} c_1 + n_1[l], \; y_2[l] = \sqrt{P_{\mathrm{r},2}} c_2 + n_2[l], \quad l = 1, \ldots, L, \tag{8.115}$$

where $c_1 \sim \mathcal{N}_{\mathbb{C}}(0, 1)$ and $c_2 \sim \mathcal{N}_{\mathbb{C}}(0, 1)$ are the random RCSs of the target towards the receiver 1 and 2, respectively. Similarly, the subscript in the other symbols refers to the receiver index. The noise samples have the distributions $n_1[l] \sim \mathcal{N}_{\mathbb{C}}(0, \sigma^2)$ and $n_2[l] \sim \mathcal{N}_{\mathbb{C}}(0, \sigma^2)$. Derive the sufficient statistics in the Neyman-Pearson detector if all the random variables are independent.

Exercise 8.13. In the hypothesis test stated in (8.90)-(8.91) for the Swerling 2 target model, we have assumed that the transmitter sends the constant symbol "1" throughout the L symbol times. What are the sufficient statistics of the Neyman-Pearson detector if the transmitter instead sends $\mathbf{x} = [x[1], \dots, x[L]]^{\mathrm{T}} \in \mathbb{C}^L$, which is deterministic and known at the receiver?

Exercise 8.14. Consider target detection setup with the binary hypothesis test

$$\mathcal{H}_0 \quad : \quad y[l] = n[l], \quad l = 1, \dots, L, \tag{8.116}$$

$$\mathcal{H}_1 \quad : \quad y[l] = \sqrt{P_{\mathrm{r}}} c_{\mathrm{RCS}}[l] + n[l], \quad l = 1, \dots, L, \tag{8.117}$$

where the noise is colored such that $\mathbf{n} = [n[1], \dots, n[L]]^{\mathrm{T}} \sim \mathcal{N}_{\mathbb{C}}(\mathbf{0}, \mathbf{C})$ with an invertible covariance matrix $\mathbf{C}$.

(a) Consider the Swerling 1 target model with $c_{\mathrm{RCS}}[l] = c_{\mathrm{RCS}} \sim \mathcal{N}_{\mathbb{C}}(0, 1)$, for $l = 1, \dots, L$. Derive the sufficient statistics of the Neyman-Pearson detector. Interpret the result.

(b) Consider the Swerling 2 target model with independent $c_{\mathrm{RCS}}[l] \sim \mathcal{N}_{\mathbb{C}}(0, 1)$, for $l = 1, \dots, L$. Derive the sufficient statistics of the Neyman-Pearson detector. Interpret the result.

Exercise 8.15. Consider a mono-static ISAC transceiver with K transmit and K receive antennas. The transceiver sends data to a single user. At the same time, it wants to detect the existence of a target at a specific location. The channels from the ISAC transceiver to the data user and the target location are denoted as $\mathbf{h}_{\mathrm{u}} \in \mathbb{C}^K$ and $\mathbf{h}_{\mathrm{t}} \in \mathbb{C}^K$, respectively. The channel from the target location to the receiver is $\mathbf{h}_{\mathrm{r}} \in \mathbb{C}^K$. Suppose $\mathbf{p}s$ is transmitted where $s \sim \mathcal{N}_{\mathbb{C}}(0, P)$ is the data signal and $\mathbf{p} \in \mathbb{C}^K$ is the unit-norm precoding vector. The target detection is done based on the received signal for one time instance at the ISAC receiver, which is reflected by the target. A receive combining vector $\mathbf{w} \in \mathbb{C}^K$ is applied to the received signal. The RCS variance is σ_{RCS}. Suppose that the receiver noises at the user and ISAC receiver are both zero-mean and have the variance σ^2.

(a) What is the SNR of the received signal at the ISAC transceiver if the target exists?

(b) What is the combining vector $\mathbf{w}$ that maximizes the sensing SNR in (a)?

(c) Suppose that $P\|\mathbf{h}_{\mathrm{u}}\|^2/\sigma^2 = 9$ and the user requires an SNR of 1. How should the precoding vector $\mathbf{p}$ be selected to maximize the sensing SNR subject to the user SNR constraint if $\mathbf{h}_{\mathrm{t}}^{\mathrm{H}}\mathbf{h}_{\mathrm{u}} = 0$?

Chapter 9

Reconfigurable Surfaces

The previous chapters have demonstrated how the signal strength can be increased by equipping the transmitter and receiver with multiple antennas used for precoding and combining. Unfortunately, the MIMO technology can hardly turn a weak channel into a strong one; if the channel gain β is tiny, then $MK\beta$ will remain mediocre. Communication systems that operate under NLOS conditions rely heavily on reflections by various surfaces for the signals to reach the intended receivers. This might lead to multiple propagation paths but often immense signal losses along these paths, particularly in the mmWave and THz bands. Figure 9.1 illustrates such a scenario, where the NLOS receiver can only be reached by beamforming towards a building that reflects the signal. Unfortunately, the building in this example is rotated such that the signal is mainly reflected away from the receiver, following the solid arrow. Can we change the reflection properties somehow so the signal bends around the corner and follows the dashed arrow instead? Yes, this can be achieved using *reconfigurable surfaces*, which is the topic of this chapter.

This chapter will explore how the reflection properties can be dynamically tuned using reconfigurable surfaces to aid the communication between a transmitter and receiver. We begin by explaining how reflections can be interpreted using the beamforming characteristics from previous chapters and how reconfigurable surfaces can control these characteristics. We will then analyze how these surfaces can be configured to maximize the capacity of narrowband and wideband SISO channels and MIMO channels.

9.1 Basic Physics of Reflecting Surfaces

There are two primary categories of reflections: specular and diffuse. These categories are illustrated in Figure 9.2 and represent the extremes in how a plane wave can interact with a reflecting object. In the specular case, the reflected wave remains planar but changes its direction. If angles are measured counterclockwise with $0°$ being the broadside direction, then a wave with

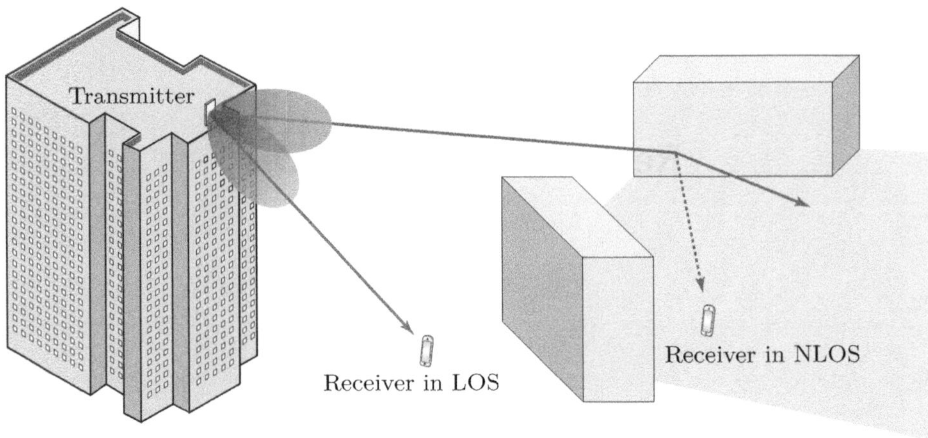

Figure 9.1: A downlink scenario where the transmitter can easily reach the LOS receiver. By contrast, the NLOS receiver has rather weak channel conditions since the wall reflection directs the signal along the solid arrow. If a reconfigurable surface is deployed on that building, the signal can be reflected following the dashed arrow instead.

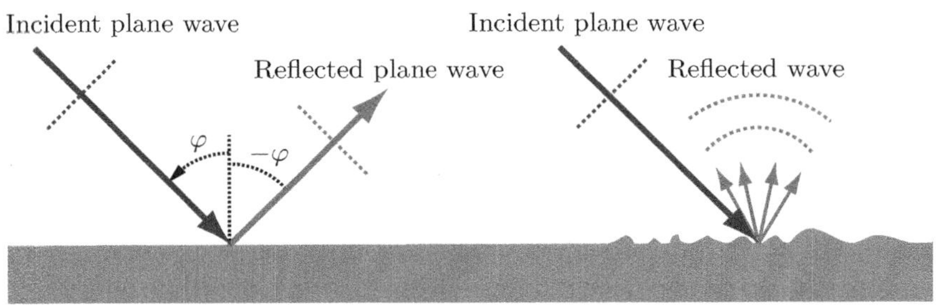

(a) Specular reflection. (b) Diffuse reflection/scattering.

Figure 9.2: A plane wave that reaches an object can be reflected in different ways, with specular reflection and diffuse reflection/scattering being the two extremes.

the incident angle φ is reflecting having the outgoing angle $-\varphi$. This is a consequence of *Snell's law* of refraction, often considered in optics [148]. By contrast, in the diffuse scattering case, the reflected wave has a spherical shape with no particular directivity; thus, the wave's energy is further diffused over the propagation environments.

These categories might seem familiar because we constantly observe how visible light interacts with objects around us to create specular reflections on smooth surfaces (e.g., mirrors that provide an undistorted image) and diffuse reflections on rough surfaces (e.g., white walls that spread the light through the room). The smoothness level is measured compared to the wavelength and size of the object. Firstly, the object must be many wavelengths wide to have the chance to provide (approximately) specular reflection. Secondly, the surface roughness must be small compared to the wavelength. Hence, a large object that is smooth enough to provide specular reflection for radio waves might be too rough to provide that for visible light. On the other hand, a perfectly smooth but physically small object might be a specular reflector for visible light but be too small to act that way for radio waves since these have a roughly 10^5 times larger wavelength.[1] In fact, the physics that underpins the specular reflection case assumes an infinitely large surface. If a finite-sized mirror approximately provides specular reflection for visible light, it must be 10^5 times larger to give the same approximation accuracy for radio waves. It is not only specular reflection that is an idealization, but ideal diffuse scattering that is uniform in all directions (as in Figure 9.2) is also unlikely to occur in wireless channels; even a rough object has a specific geometry that affects the reflected wave's shape.

The properties of the reflected signal from a finite-sized flat object can be derived using similar methods as in Chapter 4, where we studied antenna arrays. Figure 9.3(a) shows a plane wave impinging on a surface from the angle φ. We will measure the resulting phase-shifts at three points on the surface, which are selected as a ULA with the separation Δ. If we use the left-most point as the phase reference, then the second point observes a phase-shift of $2\pi\frac{\Delta\sin(\varphi)}{\lambda}$ and the third point observes a phase-shift of $2\pi\frac{2\Delta\sin(\varphi)}{\lambda}$. These phases are obtained from the wave needing to travel the additional distances $\Delta\sin(\varphi)$ and $2\Delta\sin(\varphi)$ to reach these points. In general, we obtain the relative phase-shifts at M points in a ULA configuration with separation Δ from the array response vector in (4.19):

$$\mathbf{a}(\varphi) = \begin{bmatrix} 1 \\ e^{-\mathrm{j}2\pi\frac{\Delta\sin(\varphi)}{\lambda}} \\ e^{-\mathrm{j}2\pi\frac{2\Delta\sin(\varphi)}{\lambda}} \\ \vdots \\ e^{-\mathrm{j}2\pi\frac{(M-1)\Delta\sin(\varphi)}{\lambda}} \end{bmatrix} \in \mathbb{C}^M. \tag{9.1}$$

If the same M points retransmit the signal isotropically with the mentioned phase-shifts, we obtain the situation illustrated in Figure 9.3(b). The signal is beamformed using the precoding vector $\mathbf{p} = \mathbf{a}(\varphi)$, which can be expressed as

$$\mathbf{p} = \begin{bmatrix} 1 \\ e^{-\mathrm{j}2\pi\frac{\Delta\sin(\varphi)}{\lambda}} \\ e^{-\mathrm{j}2\pi\frac{2\Delta\sin(\varphi)}{\lambda}} \\ \vdots \\ e^{-\mathrm{j}2\pi\frac{(M-1)\Delta\sin(\varphi)}{\lambda}} \end{bmatrix} = \begin{bmatrix} 1 \\ e^{-\mathrm{j}2\pi\frac{\Delta\sin(-\varphi)}{\lambda}} \\ e^{-\mathrm{j}2\pi\frac{2\Delta\sin(-\varphi)}{\lambda}} \\ \vdots \\ e^{-\mathrm{j}2\pi\frac{(M-1)\Delta\sin(-\varphi)}{\lambda}} \end{bmatrix}^* = \mathbf{a}^*(-\varphi) \tag{9.2}$$

since $\sin(\varphi) = -\sin(-\varphi)$. We recognize this as the MRT vector (without a normalization factor) for transmission in the angular direction $-\varphi$; thus, this

[1]This number is obtained by comparing green light having the carrier frequency 600 THz with a wireless communication signal at 6 GHz.

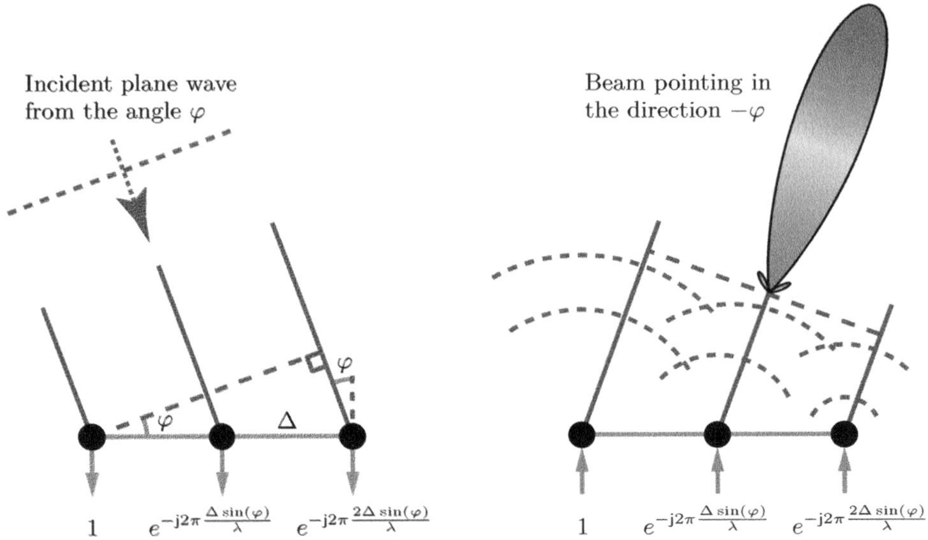

(a) Phase-shifts in signal reception. (b) Transmission with the same phases.

Figure 9.3: When a plane wave impinges on a plane surface from the direction φ, the phase-shifts over the surface can be computed by viewing it as a ULA as in (a). If the same points on the surface retransmit the signal with the phase-shifts obtained from (a), then a beam will be formed in the opposite direction $-\varphi$ as shown in (b).

is where the retransmitted beam is pointing. This observation is essential to determining the shape of the reflected signal from a plane finite-sized surface and is an instance of the *Huygens-Fresnel principle*. This principle says that when a wavefront interacts with an object, every point on that object can be viewed as a new source that emits spherical waves (isotropically). These waves' constructive/destructive combinations determine the new wavefront [149]. The reason that a plane wave that arrives from the angle φ is beamformed with the angle $-\varphi$ is that the distance to a far-away receiver in that direction is identical through all the elements in the ULA shown in Figure 9.3; thus, there will be constructive interference in that direction. The reflected wavefront can be determined using the methodology for computing beam patterns developed in Chapter 4.

9.1.1 Beam Pattern from a Reflecting Surface

We will now compute the angular shape of the reflected signal from the two-dimensional surface illustrated in Figure 9.4, which is deployed in the yz-plane. We denote its horizontal length as L_{H} and vertical length as L_{V} to follow the notation from the analysis of UPAs in Section 4.5.3. The considered surface is a homogeneous perfect electric conductor (PEC), but we will first treat it as a UPA by (hypothetically) cutting it into many tiny pieces that each has the

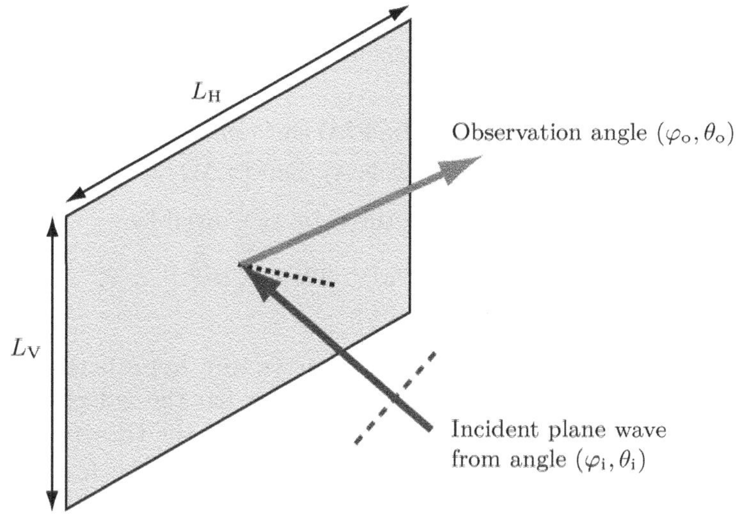

Figure 9.4: A plane wave impinges on a flat homogenous PEC surface in the yz-plane of size $L_{\mathrm{H}} \times L_{\mathrm{V}}$ meters. The channel gain observed in different observation directions $(\varphi_{\mathrm{o}}, \theta_{\mathrm{o}})$ can be computed similarly to the beam pattern of a UPA, which was characterized in Chapter 4.

physical dimension $\Delta \times \Delta$ for some $\Delta \leq \lambda/4$. Hence, the horizontal/vertical antenna spacing is Δ. Each antenna has an area Δ^2 that is smaller than that of an isotropic antenna, implying that it also has an approximately isotropic radiation pattern. The number of horizontal and vertical antennas can be computed as $N_{\mathrm{H}} = L_{\mathrm{H}}/\Delta$ and $N_{\mathrm{V}} = L_{\mathrm{V}}/\Delta$, respectively. We will use this notation to determine the beam pattern and then let $\Delta \to 0$ so that the UPA is made of asymptotically many small antennas that we will call *atoms*. The considered setup is shown in Figure 9.4.

If a plane wave impinges on the surface from the azimuth angle $\varphi_{\mathrm{i}} \in [-\pi/2, \pi/2]$ and elevation angle $\theta_{\mathrm{i}} \in [-\pi/2, \pi/2]$, then the relative phase-shifts among the atoms are given by the respective entries of the array response vector $\mathbf{a}_{N_{\mathrm{H}},N_{\mathrm{V}}}(\varphi_{\mathrm{i}}, \theta_{\mathrm{i}})$ in (4.128). By following the Huygens-Fresnel principle, we can obtain the reflected signal by considering transmission using the precoding vector

$$\mathbf{p} = \mathbf{a}_{N_{\mathrm{H}},N_{\mathrm{V}}}(\varphi_{\mathrm{i}}, \theta_{\mathrm{i}}) = \mathbf{a}^*_{N_{\mathrm{H}},N_{\mathrm{V}}}(-\varphi_{\mathrm{i}}, -\theta_{\mathrm{i}}), \tag{9.3}$$

which corresponds to MRT (without power normalization) in the direction $(-\varphi_{\mathrm{i}}, -\theta_{\mathrm{i}})$. The beamforming gain that is observed in an arbitrary observation direction, represented by the azimuth angle $\varphi_{\mathrm{o}} \in [-\pi/2, \pi/2]$ and elevation angle $\theta_{\mathrm{o}} \in [-\pi/2, \pi/2]$, is obtained by multiplying with the array response vector $\mathbf{a}^{\mathrm{T}}_{N_{\mathrm{H}},N_{\mathrm{V}}}(\varphi_{\mathrm{o}}, \theta_{\mathrm{o}})$ representing the channel in that direction:

$$\begin{aligned} B(\varphi_{\mathrm{o}}, \theta_{\mathrm{o}}) &= \left|\mathbf{a}^{\mathrm{T}}_{N_{\mathrm{H}},N_{\mathrm{V}}}(\varphi_{\mathrm{o}}, \theta_{\mathrm{o}})\mathbf{p}\right|^2 \\ &= \left|\mathbf{a}^{\mathrm{T}}_{N_{\mathrm{H}},N_{\mathrm{V}}}(\varphi_{\mathrm{o}}, \theta_{\mathrm{o}})\mathbf{a}^*_{N_{\mathrm{H}},N_{\mathrm{V}}}(-\varphi_{\mathrm{i}}, -\theta_{\mathrm{i}})\right|^2. \end{aligned} \tag{9.4}$$

A compact formula for this kind of expression was previously derived in Section 4.5.3, using a slightly different notation. In our case, (4.139) turns into

$$B(\varphi_o,\theta_o) = \frac{\sin^2(\pi L_H\Phi/\lambda)}{\sin^2(\pi\Delta\Phi/\lambda)}\frac{\sin^2(\pi L_V\Omega/\lambda)}{\sin^2(\pi\Delta\Omega/\lambda)}, \tag{9.5}$$

where the impact of the angles is captured by the variables

$$\Phi = \sin(\varphi_o)\cos(\theta_o) + \sin(\varphi_i)\cos(\theta_i), \tag{9.6}$$

$$\Omega = \sin(\theta_o) + \sin(\theta_i). \tag{9.7}$$

Suppose the channel gain from the transmitter to a hypothetical isotropic antenna inside the surface is β_t. Each of the atoms has the (effective) area Δ^2 and will experience the channel gain $\beta_t\Delta^2/A_{iso}$ from the transmitter, where $A_{iso} = \frac{\lambda^2}{4\pi}$ is the area of an isotropic antenna. This is the fraction of the transmitted power that reaches a single atom and will be reflected by it. Similarly, suppose the channel gain from the hypothetical isotropic antenna to a receiver in the observation direction is β_r. Each atom with area Δ^2 will then experience the channel gain $\beta_r\Delta^2/A_{iso}$, which is the propagation loss from an atom to the receiver. In conclusion, the end-to-end channel gain from the transmitter to the receiver via the reflecting surface is

$$\begin{aligned}\beta &= \beta_t\frac{\Delta^2}{A_{iso}}\beta_r\frac{\Delta^2}{A_{iso}}B(\varphi_o,\theta_o)\\ &= \beta_t\beta_r\frac{\Delta^4}{A_{iso}^2}\frac{\sin^2(\pi L_H\Phi/\lambda)}{\sin^2(\pi\Delta\Phi/\lambda)}\frac{\sin^2(\pi L_V\Omega/\lambda)}{\sin^2(\pi\Delta\Omega/\lambda)}\\ &\approx \beta_t\beta_r\frac{\Delta^4}{A_{iso}^2}\frac{\sin^2(\pi L_H\Phi/\lambda)}{(\pi\Delta\Phi/\lambda)^2}\frac{\sin^2(\pi L_V\Omega/\lambda)}{(\pi\Delta\Omega/\lambda)^2}\\ &= \beta_t\beta_r\frac{L_H^2L_V^2}{A_{iso}^2}\text{sinc}^2\left(\frac{L_H\Phi}{\lambda}\right)\text{sinc}^2\left(\frac{L_V\Omega}{\lambda}\right),\end{aligned} \tag{9.8}$$

where the approximation utilizes the fact that $\sin(x) \approx x$ when $x \approx 0$ and is tight when $\Delta \to 0$. The last equality identifies the sinc-function expression. The expression in (9.8) shows how the channel gain depends on the angles and captures all the essential channel properties, except for polarization. The potential polarization mismatch between the transmitter and receiver can be included in β_t and β_r but are also angle-dependent [150].

The largest channel gain is obtained in (9.8) when $\Phi = \Omega = 0$. By inspecting (9.6) and (9.7), we can conclude that the maximum is obtained for the observation angles $\varphi_o = -\varphi_i$ and $\theta_o = -\theta_i$, which is expected from Snell's law and the previous discussion. If the reflected signal were a plane wave, the channel gain would be zero in all other directions, which is not the case. Instead, the angular gain variations are the same as for a UPA with the same physical size and can be analyzed as in Section 4.5.3.

Example 9.1. What are the reflected signal's first-null horizontal and vertical beamwidths when a plane wave impinges from $\varphi_{\mathrm{i}} = \theta_{\mathrm{i}} = 0$?

The first nulls appear when the argument of one of the sinc-functions in (9.8) is ± 1. In the horizontal plane (i.e., $\theta_{\mathrm{o}} = 0$), this happens for $\varphi_{\mathrm{o}} = \pm \arcsin(\lambda/L_{\mathrm{H}}) \approx \lambda/L_{\mathrm{H}}$. Hence, the first-null horizontal beamwidth is approximately $2\lambda/L_{\mathrm{H}}$. It follows from the same computation that the first-null vertical beamwidth is approximately $2\lambda/L_{\mathrm{V}}$. The beamwidths are proportional to the wavelength, which demonstrates how a PEC surface of a given physical size can give an extremely narrow beamwidth for visible light but a relatively wide beamwidth for radio spectrum.

As the surface's lengths $L_{\mathrm{H}}, L_{\mathrm{V}}$ grow large, for a given wavelength, the beamwidths approach zero. This implies the reflected signal will be a plane wave with zero beamwidth in the asymptotic limit. This corresponds to ideal specular reflection where the incident plane wave only changes direction.

The maximum gain value in (9.8) can be factorized as

$$\beta = \beta_{\mathrm{t}} \cdot \underbrace{\frac{L_{\mathrm{H}} L_{\mathrm{V}}}{A_{\mathrm{iso}}}}_{=\text{Aperture gain for reception}} \cdot \beta_{\mathrm{r}} \cdot \underbrace{\frac{L_{\mathrm{H}} L_{\mathrm{V}}}{A_{\mathrm{iso}}}}_{=\text{Beamforming gain for retransmission}}. \tag{9.9}$$

Recall that $L_{\mathrm{H}} L_{\mathrm{V}}$ is the total area of the surface. The first term in (9.9) is the channel gain from the transmitter to an isotropic-antenna-sized receiver surface, while the second term $L_{\mathrm{H}} L_{\mathrm{V}}/A_{\mathrm{iso}}$ determines how much larger aperture the surface has. Hence, we will call the second term the *aperture gain*, but it could also be interpreted as a receive beamforming gain. This part of the expression models the fact that a surface collects an amount of power from the impinging plane wave that is proportional to its area. The third term in (9.9) is the channel gain from an isotropic-antenna-sized transmitter surface to the receiver, while the fourth term $L_{\mathrm{H}} L_{\mathrm{V}}/A_{\mathrm{iso}}$ determines the transmit *beamforming gain* delivered by the surface. This part of the expression highlights how a big surface can beamform/reflect the signal with a narrower beamwidth and a power concentration in the main beam proportional to its area.

Figure 9.5 shows the end-to-end channel gain in (9.8) observed in different angle directions φ_{o} in the azimuth plane (where $\theta_{\mathrm{o}} = 0$) when a plane wave impinges from the direction $\varphi_{\mathrm{i}} = \pi/6$, $\theta_{\mathrm{i}} = 0$. We consider a square surface with the side lengths $L = L_{\mathrm{H}} = L_{\mathrm{V}} \in \{4\lambda, 16\lambda\}$ and a propagation scenario with $\beta_{\mathrm{t}}\beta_{\mathrm{r}} = 10^{-8}$. The figure shows how the reflected signals are beams pointing in the direction $\varphi_{\mathrm{o}} = -\pi/6 = -\varphi_{\mathrm{i}}$, as expected. The horizontal beamwidth shrinks slightly when the surface increases in size, but the more dominant effect is the increased channel gain, which grows quadratically with the surface area. Hence, when each side increases by a factor of 4, the channel gain grows by $4^4 = 24\,\mathrm{dB}$. This is the combination of the aperture gain and the transmit beamforming gain.

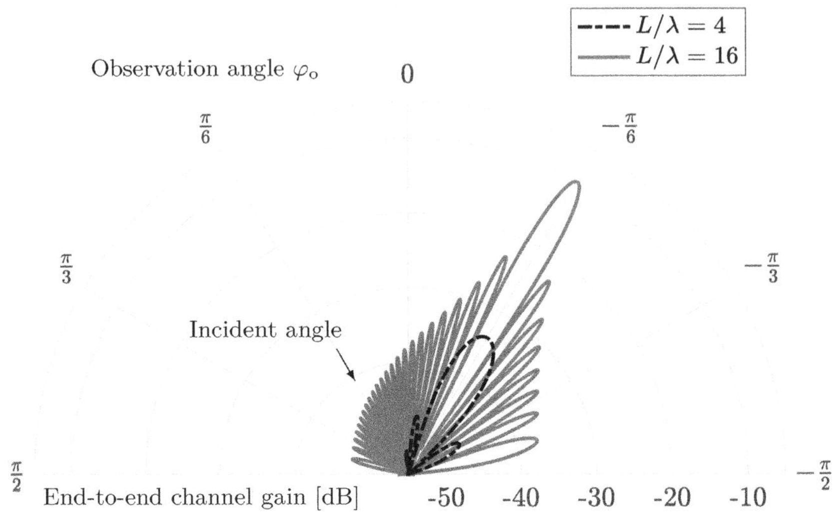

Figure 9.5: The end-to-end channel gain in (9.8) from a transmitter in the direction $\varphi_{\mathrm{i}} = \pi/6$ via a reflecting PEC surface to a receiver in a varying observation angle direction φ_{o}. The channel gain depends on the surface's size $L \times L$ and the observation angle.

Example 9.2. How does the end-to-end channel gain in (9.8) relate to the radar range equation in (8.71)?

The connection between these expressions can be identified by using the notation $\beta_{\mathrm{t}} = G_{\mathrm{t}}(\varphi_{\mathrm{t}}, \theta_{\mathrm{t}})\frac{\lambda^2}{(4\pi d_{\mathrm{t}})^2}$ and $\beta_{\mathrm{r}} = G_{\mathrm{r}}(\varphi_{\mathrm{r}}, \theta_{\mathrm{r}})\frac{\lambda^2}{(4\pi d_r)^2}$ for the channel gains of the LOS paths to and from the radar target, respectively. These channel gains depend on the propagation distances and antenna gains at the transmitter and receiver. The received power in (8.71) can then be expressed as

$$P_{\mathrm{r}} = P_{\mathrm{t}}\beta_{\mathrm{t}}\beta_{\mathrm{r}}\frac{\sigma_{\mathrm{RCS}}}{A_{\mathrm{iso}}}. \tag{9.10}$$

The corresponding received power when using the reflecting surface is $P_{\mathrm{t}}\beta$ so by comparing (9.10) with (9.8), we can identify the RCS of the surface as

$$\sigma_{\mathrm{RCS}} = \frac{L_{\mathrm{H}}^2 L_{\mathrm{V}}^2}{A_{\mathrm{iso}}}\mathrm{sinc}^2\left(\frac{L_{\mathrm{H}}\Phi}{\lambda}\right)\mathrm{sinc}^2\left(\frac{L_{\mathrm{V}}\Omega}{\lambda}\right). \tag{9.11}$$

This expression characterizes how the RCS depends on the incident and observation angles through Φ and Ω. The largest value is achieved when $\Phi = \Omega = 0$, but we can also achieve zero RCS if $\Phi = \pm\lambda/L_{\mathrm{H}}$ or $\Omega = \pm\lambda/L_{\mathrm{V}}$.

If the surface would be rotated randomly with respect to the transmitter and receiver, then Φ and Ω are random, and so is the angle-dependent RCS σ_{RCS}. This is the core principle that leads to randomness in radar sensing.

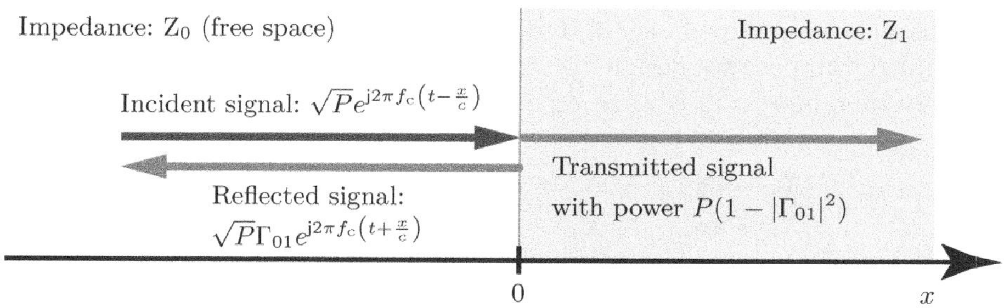

Figure 9.6: When the plane wave $\sqrt{P}e^{\mathrm{j}2\pi f_c(t-\frac{x}{c})}$ reaches the boundary between two mediums, a fraction of it is reflected backward and the remaining is transmitted into the new medium. The reflection coefficient Γ_{01} determines these behaviors.

9.1.2 Reconfigurable Reflection from Heterogeneous Surfaces

The previous section analyzed the reflection from a homogeneous surface, which reflects all the power of the incident wave as a beam pointing in the specular reflection direction. The situation is different when the surface is heterogeneous. We need the *reflection coefficient* to study that scenario.

When a sinusoidal wave reaches the boundary between two mediums, their respective characteristic impedances determine what fraction of the signal is reflected back versus transmitted into the new medium. We let Z_0 denote the impedance of the first medium (e.g., free space) and Z_1 denote the impedance of the second medium (e.g., the surface). The reflection coefficient can then be computed as [151, Sec. 1.7]

$$\Gamma_{01} = \frac{\mathrm{Z}_1 - \mathrm{Z}_0}{\mathrm{Z}_1 + \mathrm{Z}_0} \tag{9.12}$$

and is the reflected signal divided by the incident signal at the boundary between the mediums. The reflection coefficient can be complex, in which case $\arg(\Gamma_{01})$ represents the phase-shift incurred to the signal before it is reflected. This scenario is illustrated in Figure 9.6, where the first medium is free space (vacuum) for which the speed of light has been denoted c earlier in this book. A fraction $|\Gamma_{01}|^2$ of the power is reflected, while the remaining fraction $1-|\Gamma_{01}|^2$ is transmitted into the new medium and might be absorbed by it. We will focus on the reflected signal.

A homogenous surface has a constant impedance Z_1, which results in a reduced power by a factor of $|\Gamma_{01}|^2 \in [0,1]$, but otherwise, the same reflection behavior as in the previous section. However, suppose the surface is divided into N small units that are structurally similar but have heterogeneous electrical properties. We call these *metaatoms* and each has a specific impedance Z_n for $n = 1,\ldots,N$. The corresponding reflection coefficients then become

$$\Gamma_{0n} = \frac{\mathrm{Z}_n - \mathrm{Z}_0}{\mathrm{Z}_n + \mathrm{Z}_0}. \tag{9.13}$$

The characteristic impedance of free space is $\mathrm{Z}_0 \approx 376.7 \approx 120\pi$ ohm, so it is real-valued. Suppose we design the metaatom using a reactance element with a purely imaginary impedance $\mathrm{Z}_n = \mathrm{j}X_n$ for some $X_n \in \mathbb{R}$. It then follows that

$$|\Gamma_{0n}| = \left|\frac{\mathrm{j}X_n - \mathrm{Z}_0}{\mathrm{j}X_n + \mathrm{Z}_0}\right| = \frac{\sqrt{X_n^2 + \mathrm{Z}_0^2}}{\sqrt{X_n^2 + \mathrm{Z}_0^2}} = 1, \tag{9.14}$$

$$\arg(\Gamma_{0n}) = \arg\left(\frac{\mathrm{j}X_n - \mathrm{Z}_0}{\mathrm{j}X_n + \mathrm{Z}_0}\right) = \begin{cases} \pi - 2\arctan\left(\frac{X_n}{\mathrm{Z}_0}\right), & \text{if } X_n \geq 0, \\ -\pi - 2\arctan\left(\frac{X_n}{\mathrm{Z}_0}\right), & \text{if } X_n < 0. \end{cases} \tag{9.15}$$

Such a metaatom will reflect all the incident power since $|\Gamma_{0n}| = 1$ and it causes a phase-shift $\psi_n = \arg(\Gamma_{0n})$ that can be continuously tuned between $-\pi$ and π by varying X_n. Such tuning can be achieved by configuring a capacitor that determines the capacitive part of the reactance, which can be implemented using a varactor diode.[2] This feature is the key to designing reconfigurable surfaces that can shape the reflected signals.

We will now return to the reflection example in Figure 9.4 that analyzed a homogeneous PEC surface. Suppose that surface is replaced by the one shown in Figure 9.7, which consists of $N = N_\mathrm{H}N_\mathrm{V}$ metaatoms that have varying impedance values corresponding to phase-shifts between $-\pi$ and π. Each color represents a specific phase value, and we let $\psi_n \in [-\pi, \pi)$ denote the phase-shift incurred by the nth metaatom. If the incident plane wave arrives from the angular direction $(\varphi_\mathrm{i}, \theta_\mathrm{i})$, then the incident phase-shifts over the surface are given by the array response vector $\mathbf{a}_{N_\mathrm{H},N_\mathrm{V}}(\varphi_\mathrm{i}, \theta_\mathrm{i})$. Each metaatom then adjusts its local incident phase value by ψ_n; therefore, the phase profile of the retransmitted/reflected signals is given by the precoding vector

$$\mathbf{p} = \mathbf{D}_\psi \mathbf{a}_{N_\mathrm{H},N_\mathrm{V}}(\varphi_\mathrm{i}, \theta_\mathrm{i}) = \mathbf{D}_\psi \mathbf{a}^*_{N_\mathrm{H},N_\mathrm{V}}(-\varphi_\mathrm{i}, -\theta_\mathrm{i}), \tag{9.16}$$

where the surface's phase adjustments are applied using the diagonal *reflection matrix*

$$\mathbf{D}_\psi = \mathrm{diag}\left(e^{\mathrm{j}\psi_1}, \ldots, e^{\mathrm{j}\psi_N}\right). \tag{9.17}$$

The beamforming gain expression in (9.4) can then be updated as

$$\begin{aligned} B(\varphi_\mathrm{o}, \theta_\mathrm{o}) &= \left|\mathbf{a}^\mathrm{T}_{N_\mathrm{H},N_\mathrm{V}}(\varphi_\mathrm{o}, \theta_\mathrm{o})\mathbf{p}\right|^2 \\ &= \left|\mathbf{a}^\mathrm{T}_{N_\mathrm{H},N_\mathrm{V}}(\varphi_\mathrm{o}, \theta_\mathrm{o})\mathbf{D}_\psi \mathbf{a}^*_{N_\mathrm{H},N_\mathrm{V}}(-\varphi_\mathrm{i}, -\theta_\mathrm{i})\right|^2. \end{aligned} \tag{9.18}$$

Since each entry of an array response vector is a complex exponential entirely determined by a phase value, we can turn $\mathbf{p}$ into any array response vector of our choice by selecting $\mathbf{D}_\psi$ accordingly. We can thereby control the main direction of the reflected beam. We can also generate precoding vectors that are not array response vectors if we happen to prefer that.

[2]A capacitor adds a positive value to X_n and an inductor adds a negative value. If the metaatom is a circuit consisting of both fixed inductive elements and variable capacitive elements, then we can control X_n over a range of both positive and negative values, resulting in the range of positive and negative phase-shifts shown in Figure 9.17.

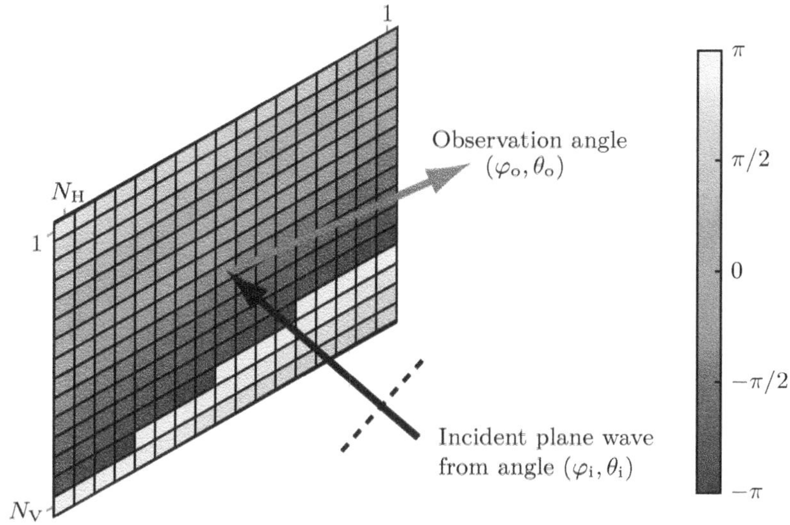

Figure 9.7: A plane wave impinges on a flat surface in the yz-plane that consists of $N_H \times N_V$ reflecting metaatoms with heterogeneous properties. The metaatoms' varying impedances cause different phase-shifts between $-\pi$ and π, as exemplified using colors. This enables the surface to control the channel gain in the observation directions (φ_o, θ_o).

Example 9.3. How should the surface's phase-shifts be selected to point the reflected beam in a specific desired direction (φ_d, θ_d)?

The reflected beam will point in that direction if $\mathbf{p} = \mathbf{a}^*_{N_H,N_V}(\varphi_d, \theta_d)$. By equating (9.16) to this value, we obtain the relation

$$\mathbf{D}_\psi \underbrace{\mathbf{a}^*_{N_H,N_V}(-\varphi_i, -\theta_i)}_{=[a_{i,1},\ldots,a_{i,N}]^T} = \underbrace{\mathbf{a}^*_{N_H,N_V}(\varphi_d, \theta_d)}_{=[a_{d,1},\ldots,a_{d,N}]^T}. \tag{9.19}$$

The nth entry can be expressed as $e^{j\psi_n} a_{i,n} = a_{d,n}$, which holds if $\psi_n = \arg(a_{d,n}/a_{i,n})$. Hence, each metaatom compensates for the phase difference between the desired array response and the actual array response of the incident wave. Since the phase varies gradually in both vectors, the phase profile of the surface will also vary gradually, as exemplified in Figure 9.7.

Another way to control the direction of the reflected beam would be to rotate the surface mechanically, but greater flexibility is achieved by the electrical implementation described above. A similar discussion was made in relation to Example 4.22, which compared the mechanical and electrical downtilt of an antenna array. When considering reflecting surfaces, we seek a way to deploy them on building facades, as illustrated in Figure 9.1, to point the reflection angle toward the receiving user without mechanical rotations. Using the radar sensing terminology from Section 8.3, we want to configure the electrical properties of the surface to achieve the largest possible RCS in the direction leading to the receiver.

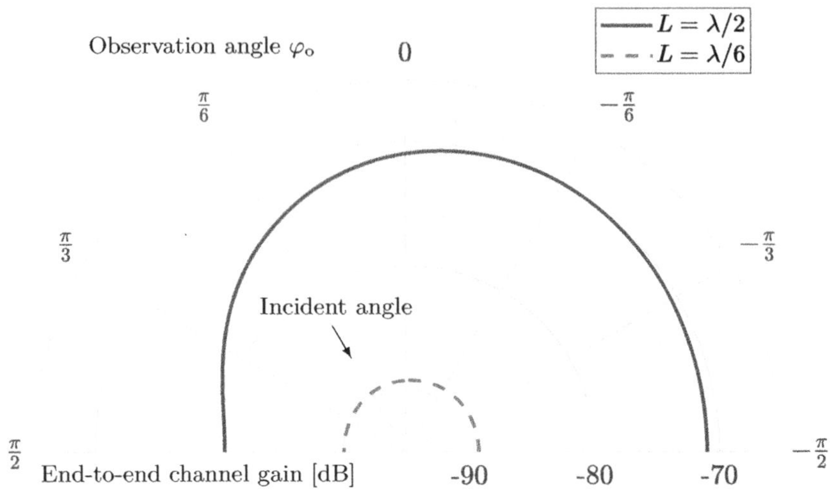

Figure 9.8: The end-to-end channel gain in (9.8) from a transmitter in the direction $\varphi_i = \pi/6$ via a reflecting surface of size $L \times L$ to a receiver in a varying observation angle direction φ_o. The smaller the reflector is, the more circular/isotropic its radiation pattern is.

We use the term "metaatom" when referring to each controllable piece of the surface to signify that they are tiny compared to the wavelength. This is because a small object provides approximately diffuse reflection with no preferred directivity, even if it is flat. Figure 9.8 shows the end-to-end channel gain in (9.8) observed in different azimuth angle directions φ_o when a plane wave impinges from the direction $\varphi_i = \pi/6$, $\theta_i = 0$. The setup is the same as in Figure 9.5, but now the surface dimensions are $L \times L$ with $L \in \{\lambda/6, \lambda/2\}$. Both sizes result in a reflected signal that is spread over all angles, even back toward the transmitter, but the radiation pattern becomes more circular (i.e., closer to isotropic) as the size shrinks. For the reconfigurable surface to fully steer the direction of the reflected signal, it should be made of many tiny controllable pieces that each lack a preferable directivity but can be used to jointly beamform the reflected signal where we want it to go.

9.1.3 Terminology and Implementation Aspects

Reconfigurable surfaces are often associated with metamaterials, which are engineered materials typically containing sub-wavelength-sized structures that create a heterogeneous impedance profile over the surface. These structures are typically referred to as metaatoms, which is why we have already adopted that terminology in this chapter. The engineered material concept was first utilized in communications to design static *reflectarrays* with a fixed reflection matrix $\mathbf{D}_\psi$ that was not a scaled identity matrix. This results in an anomalous reflection angle that differs from Snell's law [152]. This was followed by

Figure 9.9: The photo shows the reconfigurable surface prototype from [159], which is designed for the 5.8 GHz band. It consists of $N = 1100$ metaatoms, arranged as a UPA with $N_{\mathrm{V}} = 20$ rows and $N_{\mathrm{H}} = 55$ metaatoms per row. The impedance of each metaatom is controlled using two varactor diodes, which enable phase control of the reflected signals over a range of 240°.

reconfigurable reflectarrays that can electrically tune the reflection matrix [153]. The purpose was to build transmitters/receivers consisting of a single antenna pointing towards the reflectarray that controls the beam direction. This is an alternative implementation of the analog beamforming architecture, discussed in Section 7.3.1, that is particularly used for satellite communications and radars but has also been commercialized for mmWave transceivers.

The alternative concept of deploying reconfigurable surfaces in the propagation environment to relay signals between a transmitter and receiver gained traction in the late 2010s. This concept has been called *software-controlled metasurfaces* [154], *reconfigurable intelligent surfaces (RIS)* [155], *intelligent reflecting surfaces (IRS)* [156], and *reconfigurable intelligent metasurfaces* [157]. In this book, we will call them reconfigurable surfaces. There is a wealth of implementation challenges and details that go beyond the purpose of this book. We refer to [158] for a review of software-controlled metasurfaces designed for everything from the low-band to the infrared frequency range. While it is possible to build reconfigurable surfaces that can vary the phase-shifts continuously using varactor diodes, many designs use PIN diodes that can be switched on and off to shift between a discrete set of phase values.

A reconfigurable surface prototype from [159] with $N = 1100$ metaatoms is shown in Figure 9.9. Each metaatom contains two metallic patches connected to varactor diodes, which are controlled by an external bias voltage to tune the impedance. This enables the prototype to select phases $\psi_n \in [-120°, 120°]$ and the corresponding power loss varies slightly with the phase but satisfies $1 - |\Gamma_{0n}|^2 < -3\,\mathrm{dB}$. The indoor and outdoor measurements presented in [159] verify that the prototype can change the angle of the reflected beam as described above. In conclusion, reconfigurable surfaces can be implemented, and the remainder of this chapter will analyze how they can aid communication and radar systems.

9.2 Narrowband Communication using Reconfigurable Surfaces

We will now analyze how a reconfigurable surface can be tuned to aid a point-to-point communication system. We begin by considering a narrowband SISO channel between a single-antenna transmitter and a single-antenna receiver. The received signal was stated in (2.144) as

$$y = h \cdot x + n, \tag{9.20}$$

where h is the channel coefficient, $x \sim \mathcal{N}_{\mathbb{C}}(0, q)$ is the capacity-achieving transmit signal, and $n \sim \mathcal{N}_{\mathbb{C}}(0, N_0)$ is independent noise. According to Corollary 2.1, the capacity of such a channel can be expressed as

$$C = \log_2\left(1 + \frac{q|h|^2}{N_0}\right) \quad \text{bit/symbol.} \tag{9.21}$$

When a reconfigurable surface is deployed in the propagation environment, it affects how the channel coefficient h is modeled. Suppose the surface consists of N metaatoms that reflect all the incident power with the controllable phase-shifts ψ_n for $n = 1, \ldots, N$. This setup is illustrated in Figure 9.10. In general, the end-to-end channel can be modeled as

$$h = h_{\mathrm{s}} + \sum_{n=1}^{N} h_{\mathrm{r},n} e^{\mathrm{j}\psi_n} h_{\mathrm{t},n}, \tag{9.22}$$

where the *static channel* $h_{\mathrm{s}} \in \mathbb{C}$ includes all propagation paths unaffected by the surface. The propagation path via metaatom n is described by the channel coefficient $h_{\mathrm{t},n} \in \mathbb{C}$ from the transmitter to the metaatom, the phase-shift $e^{\mathrm{j}\psi_n}$ incurred by the metaatom, and the channel coefficient $h_{\mathrm{r},n} \in \mathbb{C}$ from the metaatom to the receiver. These three coefficients are multiplied together following Section 9.1.1. Since waves that travel through different paths are superimposed at the receive antenna, the channel coefficients are added up as in (9.22). We can express (9.22) in the matrix/vector form

$$h = h_{\mathrm{s}} + \mathbf{h}_{\mathrm{r}}^{\mathrm{T}} \mathbf{D}_{\psi} \mathbf{h}_{\mathrm{t}} \tag{9.23}$$

by introducing the notation

$$\mathbf{h}_{\mathrm{t}} = \begin{bmatrix} h_{\mathrm{t},1} \\ \vdots \\ h_{\mathrm{t},N} \end{bmatrix}, \quad \mathbf{h}_{\mathrm{r}} = \begin{bmatrix} h_{\mathrm{r},1} \\ \vdots \\ h_{\mathrm{r},N} \end{bmatrix}, \tag{9.24}$$

and recalling the reflection matrix notation $\mathbf{D}_{\psi} = \mathrm{diag}\left(e^{\mathrm{j}\psi_1}, \ldots, e^{\mathrm{j}\psi_N}\right)$ from (9.17). Using this notation, the capacity in (9.21) for a given value of $\mathbf{D}_{\psi}$ becomes

$$\log_2\left(1 + \frac{q|h_{\mathrm{s}} + \mathbf{h}_{\mathrm{r}}^{\mathrm{T}} \mathbf{D}_{\psi} \mathbf{h}_{\mathrm{t}}|^2}{N_0}\right). \tag{9.25}$$

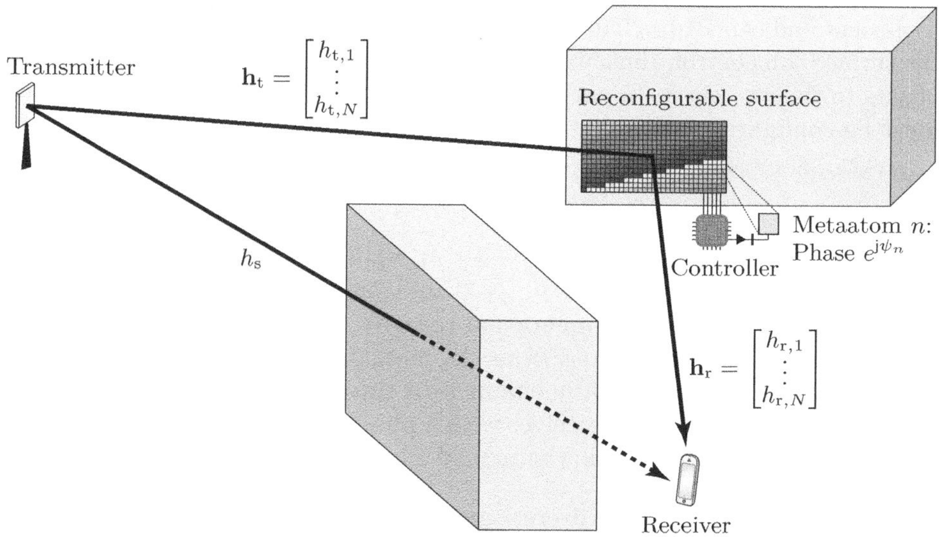

Figure 9.10: An example of a narrowband SISO channel aided by a reconfigurable surface with N metaatoms. The SIMO channel from the single-antenna transmitter to the surface is denoted by $\mathbf{h}_{\mathrm{t}}$ and the MISO channel from the surface to the single-antenna receiver is denoted by $\mathbf{h}_{\mathrm{r}}$. Each metaatom incurs a tunable phase-shift $e^{\mathrm{j}\psi_n}$ and the static channel that does not involve the surface is denoted as h_{s}. The end-to-end channel coefficient becomes $h = h_{\mathrm{s}} + \mathbf{h}_{\mathrm{r}}^{\mathrm{T}}\mathbf{D}_{\psi}\mathbf{h}_{\mathrm{t}}$, where $\mathbf{D}_{\psi} = \mathrm{diag}(e^{\mathrm{j}\psi_1}, \ldots, e^{\mathrm{j}\psi_N})$ is the reflection matrix.

We can aid the system by identifying the matrix that maximizes this capacity expression. In particular, we want to maximize

$$
\begin{aligned}
|h_{\mathrm{s}} + \mathbf{h}_{\mathrm{r}}^{\mathrm{T}}\mathbf{D}_{\psi}\mathbf{h}_{\mathrm{t}}|^2 &= \left| \begin{bmatrix} \sqrt{|h_{\mathrm{s}}|} \\ \sqrt{|h_{\mathrm{r},1}h_{\mathrm{t},1}|} \\ \vdots \\ \sqrt{|h_{\mathrm{r},N}h_{\mathrm{t},N}|} \end{bmatrix}^{\mathrm{H}} \begin{bmatrix} \sqrt{|h_{\mathrm{s}}|}e^{\mathrm{j}\arg(h_{\mathrm{s}})} \\ \sqrt{|h_{\mathrm{r},1}h_{\mathrm{t},1}|}e^{\mathrm{j}(\arg(h_{\mathrm{r},1}h_{\mathrm{t},1})+\psi_1)} \\ \vdots \\ \sqrt{|h_{\mathrm{r},N}h_{\mathrm{t},N}|}e^{\mathrm{j}(\arg(h_{\mathrm{r},N}h_{\mathrm{t},N})+\psi_N)} \end{bmatrix} \right|^2 \\
&\leq \left\| \begin{bmatrix} \sqrt{|h_{\mathrm{s}}|} \\ \sqrt{|h_{\mathrm{r},1}h_{\mathrm{t},1}|} \\ \vdots \\ \sqrt{|h_{\mathrm{r},N}h_{\mathrm{t},N}|} \end{bmatrix} \right\|^2 \left\| \begin{bmatrix} \sqrt{|h_{\mathrm{s}}|}e^{\mathrm{j}\arg(h_{\mathrm{s}})} \\ \sqrt{|h_{\mathrm{r},1}h_{\mathrm{t},1}|}e^{\mathrm{j}(\arg(h_{\mathrm{r},1}h_{\mathrm{t},1})+\psi_1)} \\ \vdots \\ \sqrt{|h_{\mathrm{r},N}h_{\mathrm{t},N}|}e^{\mathrm{j}(\arg(h_{\mathrm{r},N}h_{\mathrm{t},N})+\psi_N)} \end{bmatrix} \right\|^2 \\
&= \left(|h_{\mathrm{s}}| + \sum_{n=1}^{N} |h_{\mathrm{r},n}h_{\mathrm{t},n}| \right)^2 \qquad (9.26)
\end{aligned}
$$

where the second row follows from the Cauchy-Schwartz inequality in (2.18). The upper bound in that inequality is achieved if and only if the two vectors are equal, except for a scaling factor. The first entries of the two vectors differ by the phase-shift $e^{\mathrm{j}\arg(h_{\mathrm{s}})}$, which is determined by the static channel coefficient and cannot be changed. The other entries differ by a phase-shift that the surface can control; thus, we can attain the upper bound by matching their phases to the first entry. In particular, the entry corresponding to metaatom n must be configured such that

$$e^{\mathrm{j}\arg(h_{\mathrm{r},n}h_{\mathrm{t},n})}e^{\mathrm{j}\psi_n} = e^{\mathrm{j}\arg(h_{\mathrm{s}})} \quad \Rightarrow \quad \psi_n = \arg(h_{\mathrm{s}}) - \arg(h_{\mathrm{r},n}h_{\mathrm{t},n}) + 2\pi k_n \tag{9.27}$$

for the integer k_n that gives that $\psi_n \in [-\pi, \pi)$. This solution is also obtained as $\psi_n = \arg\left(h_{\mathrm{s}}/(h_{\mathrm{r},n}h_{\mathrm{t},n})\right)$ if $h_{\mathrm{s}} \neq 0$ and $h_{\mathrm{r},n}h_{\mathrm{t},n} \neq 0$. We notice that the capacity-maximizing configuration removes the phase-shift $\arg(h_{\mathrm{r},n}h_{\mathrm{t},n})$ created by the channels to and from the metaatom and replaces it with the phase-shift $\arg(h_{\mathrm{s}})$ of the static channel. In this way, the signals that the N metaatoms reflect reach the receiver with a phase that matches the signal that propagates through the static channel. We have proved the following result.

Corollary 9.1. Consider a discrete memoryless SISO channel aided by a reconfigurable surface, for which the channel coefficient is

$$h = h_{\mathrm{s}} + \sum_{n=1}^{N} h_{\mathrm{r},n}e^{\mathrm{j}\psi_n}h_{\mathrm{t},n}. \tag{9.28}$$

The channel capacity is maximized by configuring the surface as $\psi_n = \arg(h_{\mathrm{s}}) - \arg(h_{\mathrm{r},n}h_{\mathrm{t},n}) + 2\pi k_n$, where k_n is the integer that gives $\psi_n \in [-\pi, \pi)$, for $n = 1, \ldots, N$. This results in the capacity

$$C = \log_2\left(1 + \frac{q\left(|h_{\mathrm{s}}| + \sum_{n=1}^{N}|h_{\mathrm{r},n}h_{\mathrm{t},n}|\right)^2}{N_0}\right) \quad \text{bit/symbol.} \tag{9.29}$$

The maximum end-to-end channel gain is the squared sum of the amplitudes $|h_{\mathrm{s}}|$ and $|h_{\mathrm{r},n}h_{\mathrm{t},n}|$ for $n = 1, \ldots, N$. If we define the *effective channel vector*

$$\check{\mathbf{h}} = \begin{bmatrix} h_{\mathrm{s}} \\ h_{\mathrm{r},1}h_{\mathrm{t},1} \\ \vdots \\ h_{\mathrm{r},N}h_{\mathrm{t},N} \end{bmatrix}, \tag{9.30}$$

we can alternatively express the end-to-end channel gain using the 1-norm[3] as

$$\left(|h_{\mathrm{s}}| + \sum_{n=1}^{N}|h_{\mathrm{r},n}h_{\mathrm{t},n}|\right)^2 = \|\check{\mathbf{h}}\|_1^2. \tag{9.31}$$

[3]The 1-norm is defined for an arbitrary vector $\mathbf{x} \in \mathbb{C}^M$ as $\|\mathbf{x}\|_1 = \sum_{m=1}^{M}|x_m|$. It is also known as the Manhattan norm since it adds up the distances in the M dimensions as if one has to travel along straight perpendicular streets on a map.

Example 9.4. Suppose the N reflected paths have the same propagation losses: $|h_{t,n}|^2 = \beta_t$ and $|h_{r,n}|^2 = \beta_r$ for $n = 1, \ldots, N$. How does the end-to-end channel gain behave when the channel gain $|h_s|^2 = \beta_s$ of the static path is either relatively weak or strong?

The contribution from the metaatoms to (9.29) is $\sum_{n=1}^{N} |h_{r,n} h_{t,n}| = N\sqrt{\beta_r \beta_t}$ under these assumptions. Hence, the end-to-end channel gain becomes

$$\left(\sqrt{\beta_s} + N\sqrt{\beta_r \beta_t}\right)^2 \approx \begin{cases} N^2 \beta_r \beta_t, & \text{if } \beta_s \text{ is small,} \\ \beta_s, & \text{if } \beta_s \text{ is large,} \end{cases} \tag{9.32}$$

where "small" means that $\beta_s \ll N^2 \beta_r \beta_t$ and "large" means that $\beta_s \gg N^2 \beta_r \beta_t$. In the former case, when the vast majority of the received power comes from the surface, the end-to-end channel gain is proportional to $N^2 \beta_r \beta_t$. This term grows quadratically with the number of metaatoms, thanks to an aperture gain of N and a transmit beamforming gain of N. When the static path is relatively strong (i.e., $\beta_s \gg N^2 \beta_r \beta_t$), the reconfigurable surface barely affects the end-to-end channel gain, which is approximately equal to β_s.

The conclusion is that physically large reconfigurable surfaces are much more effective than small surfaces, thanks to the quadratic scaling law. This is important because the N^2 factor is multiplied by $\beta_r \beta_t$, which is the product of two channel gains that both can be very small numbers.

We will now use the exact expression on the left-hand side of (9.32) to demonstrate how the number of metaatoms affects communication performance. Figure 9.11 shows the capacity as a function of the number of metaatoms when $\beta_t = -80$ dB, $\beta_r = -60$ dB, and $q/N_0 = 100$ dB. We compare two types of static paths: $\beta_s = -80$ dB (strong) and $\beta_s = -110$ dB (weak). When the static path is weak, the capacity is nearly zero for $N = 0$ but grows rapidly as metaatoms are added to the surface. In this case, the N^2 SNR growth from (9.32) dominates. By contrast, when the static path is strong, the capacity is already quite high for $N = 0$, and a huge surface is needed before it has a noticeable impact on the capacity. The relative strength of the propagation path provided by the surface matters, not how much power it provides in an absolute sense. This indicates that reconfigurable surfaces are particularly valuable in deployment scenarios with weak static paths, where even a small surface makes a significant difference.

While Section 9.1 considered reflections in LOS scenarios, we have not assumed any specific channel model in this section. By optimizing the reflection matrix, we can find the phase-shift profile of a surface with fixed dimensions that maximizes the received signal power in a given propagation environment. We could mechanically bend and deform a homogeneous surface to obtain the corresponding physical shape, but instead, a reconfigurable surface synthesizes that shape using a heterogeneous impedance pattern. This

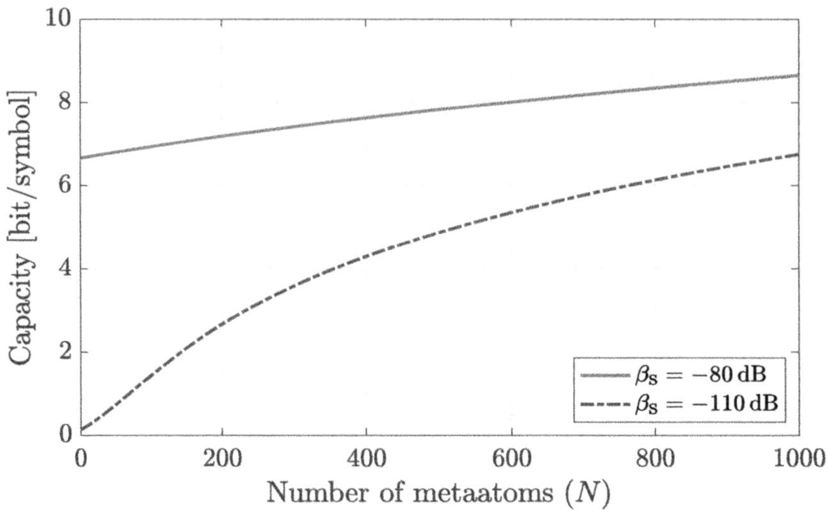

Figure 9.11: The capacity of a SISO channel that is aided by a reconfigurable surface. The capacity increases with the number of metaatoms, and the relative improvement is particularly large when the channel gain β_S of the static path is weak.

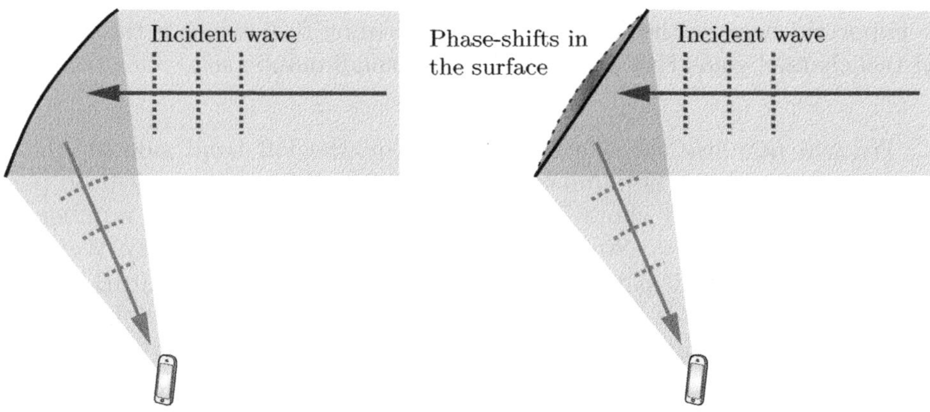

(a) Reflection from a bent surface.
(b) Reflection from a reconfigurable surface.

Figure 9.12: The shape and electric properties of the reflecting surface jointly determine the reflection direction. In (a), a homogenous surface is bent to reflect the incident wave toward the receiver. In (b), a flat reconfigurable surface has a phase-shift profile that achieves the same result by adding extra phase-shifts to signals in the center compared to the edges.

enables us to reconfigure the reflection properties rapidly when the environment or transmitter/receiver locations change. Figure 9.12 illustrates this principle. When the wave reaches the surface from the right, a parabolically bent surface will reflect the signal toward the indicated receiver, as shown in Figure 9.12(a). A flat, reconfigurable surface can synthesize the same reflection behavior by adding extra phase-shifts to the wave components reflected at its center compared to its edges. This is illustrated in Figure 9.12(b), where the coloring behind the surface represents the phase-shifts using the same scale as in Figure 9.7. The phase-shifts are negative in this scenario since they represent extra delays incurred by the metaatoms to synthesize a bent surface.

9.2.1 Line-of-Sight Channel Modeling and Surface Placement

The channel gain can be computed precisely in free-space LOS propagation using the formulas provided in Section 1.1.4. Suppose the distance from the transmitter to the surface is d_{t} and the distance from the surface to the receiver is d_{r}. If the transmitter has the antenna gain $G_{\mathrm{t}}(\varphi_{\mathrm{t}}, \theta_{\mathrm{t}})$ towards the surface and each small metaatom has the area A_{m} (i.e., the antenna gain is $\frac{4\pi}{\lambda^2}A_{\mathrm{m}}$), it follows from (1.40) that the channel gain between them is

$$\beta_{\mathrm{t}} = \frac{\lambda^2}{(4\pi d_{\mathrm{t}})^2} G_{\mathrm{t}}(\varphi_{\mathrm{t}}, \theta_{\mathrm{t}}) \frac{4\pi}{\lambda^2} A_{\mathrm{m}} = \frac{G_{\mathrm{t}}(\varphi_{\mathrm{t}}, \theta_{\mathrm{t}}) A_{\mathrm{m}}}{4\pi d_{\mathrm{t}}^2}. \tag{9.33}$$

Similarly, if the receiver has the antenna gain $G_{\mathrm{r}}(\varphi_{\mathrm{r}}, \theta_{\mathrm{r}})$ towards the surface, then the channel gain between them is

$$\beta_{\mathrm{r}} = \frac{\lambda^2}{(4\pi d_{\mathrm{r}})^2} G_{\mathrm{r}}(\varphi_{\mathrm{r}}, \theta_{\mathrm{r}}) \frac{4\pi}{\lambda^2} A_{\mathrm{m}} = \frac{G_{\mathrm{r}}(\varphi_{\mathrm{r}}, \theta_{\mathrm{r}}) A_{\mathrm{m}}}{4\pi d_{\mathrm{r}}^2}. \tag{9.34}$$

When the propagation path via the surface dominates over the static path, it follows from (9.32) that the end-to-end channel gain can be expressed as

$$N^2 \beta_{\mathrm{r}} \beta_{\mathrm{t}} = N^2 \frac{G_{\mathrm{t}}(\varphi_{\mathrm{t}}, \theta_{\mathrm{t}}) A_{\mathrm{m}}}{4\pi d_{\mathrm{t}}^2} \frac{G_{\mathrm{r}}(\varphi_{\mathrm{r}}, \theta_{\mathrm{r}}) A_{\mathrm{m}}}{4\pi d_{\mathrm{r}}^2} = \frac{N^2 G_{\mathrm{t}}(\varphi_{\mathrm{t}}, \theta_{\mathrm{t}}) G_{\mathrm{r}}(\varphi_{\mathrm{r}}, \theta_{\mathrm{r}}) A_{\mathrm{m}}^2}{(4\pi d_{\mathrm{t}} d_{\mathrm{r}})^2}, \tag{9.35}$$

where we utilized the gain expressions in (9.33) and (9.34). There are many squares in (9.35) because the end-to-end channel gain is the product of two conventional channel gain expressions. First, the metaatom's area is squared because it appears in both gain expressions. It is also multiplied by N^2, which implies that it is the total area NA_{m} of the surface that determines the end-to-end gain. Second, the squared propagation distances d_{t}^2 and d_{r}^2 appear in the expression since the signal power attenuates inversely proportional to them in free-space propagation. The distances are also multiplied together.

In NLOS propagation scenarios, any channel model could be used for the individual channels. The only important aspect is to account for the small area A_{m} of each metaatom, which results in a gain $\frac{4\pi}{\lambda^2}A_{\mathrm{m}}$. This value is smaller than one since we consider sufficiently small-sized metaatoms to be able to capture and retransmit power almost isotropically.

It is not only the number of metaatoms that determines the end-to-end channel gain but also where the reconfigurable surface is deployed. Ideally, the transmitter and receiver should have LOS channels to the surface because these are generally stronger than NLOS channels. When multiple deployment locations satisfy that condition, further characteristics can be considered. The following example highlights one key property.

Example 9.5. Suppose a transmitter and a receiver with isotropic antennas are located in the same horizontal plane. They are both 10 m from a long wall but 100 m apart from each other. There is no static path between them, but they can see any point on the wall. Where on the wall should a reconfigurable surface be deployed to maximize the end-to-end channel gain?

We let N denote the number of metaatoms, A_{m} denote the area per metaatom, and $\sqrt{10^2 + d_{\mathrm{w}}^2}$ be the distance between the transmitter and the surface, where $d_{\mathrm{w}} \in [0, 100]$ m is the distance along the wall. The distance between the surface and the receiver is then given as $\sqrt{10^2 + (100 - d_{\mathrm{w}})^2}$. Since we have LOS channels and isotropic transmit and receive antennas, the end-to-end channel gain can be expressed using (9.35) as

$$N^2 \beta_{\mathrm{r}} \beta_{\mathrm{t}} = N^2 \frac{A_{\mathrm{m}}}{4\pi(10^2 + (100 - d_{\mathrm{w}})^2)} \frac{A_{\mathrm{m}}}{4\pi(10^2 + d_{\mathrm{w}}^2)}. \tag{9.36}$$

The deployment location affects the term $(10^2 + (100 - d_{\mathrm{w}})^2)(10^2 + d_{\mathrm{w}}^2)$ in the denominator. This term has the first-order derivative

$$\begin{aligned} \frac{\partial}{\partial d_{\mathrm{w}}} \left(10^2 + (100 - d_{\mathrm{w}})^2\right) \left(10^2 + d_{\mathrm{w}}^2\right) &= 4d_{\mathrm{w}}^3 - 600 d_{\mathrm{w}}^2 + 20400 d_{\mathrm{w}} - 20000 \\ &= 4(d_{\mathrm{w}} - 50)\left(d_{\mathrm{w}}^2 - 100 d_{\mathrm{w}} + 100\right), \end{aligned} \tag{9.37}$$

which has the roots $d_{\mathrm{w}} = 50$ m, $d_{\mathrm{w}} = 50 - 20\sqrt{6} \approx 1$ m, and $d_{\mathrm{w}} = 50 + 20\sqrt{6} \approx 99$ m. The former value is a maximum and the latter values are two minima, as can be proved by checking the signs of the second-order derivative. Hence, the channel gain is minimized when the surface is deployed in the middle and maximized when it is close to the transmitter or receiver.

The conclusion from this example is that we should look for all deployment locations where the surface has LOS conditions to both the base station and prospective user locations. Among these locations, we should pick the one that is closest to either of them since this maximizes the channel gain. This insight motivates the holographic MIMO architecture, mentioned briefly in Section 7.4.2, where a metasurface is deployed as a part of the base station to create an analog beamforming architecture with small antenna spacing. In this chapter, the reconfigurable surface is meant to be decoupled from the transmitter and receiver, but it should preferably be quite near one of them.

Figure 9.13 shows the end-to-end channel gain in (9.36) for $N = 200$ metaatoms that each has the area $A_{\mathrm{m}} = (\lambda/4)^2$ where $\lambda = 0.1$ m is the wavelength. As expected from the example above, the maximum channel gain is achieved close to either the transmitter or receiver, while the minimum value is obtained in the middle. The difference is around 8 dB in this example, which is substantial but not huge compared to the fact that all the considered channel gains are at the order of -100 dB.

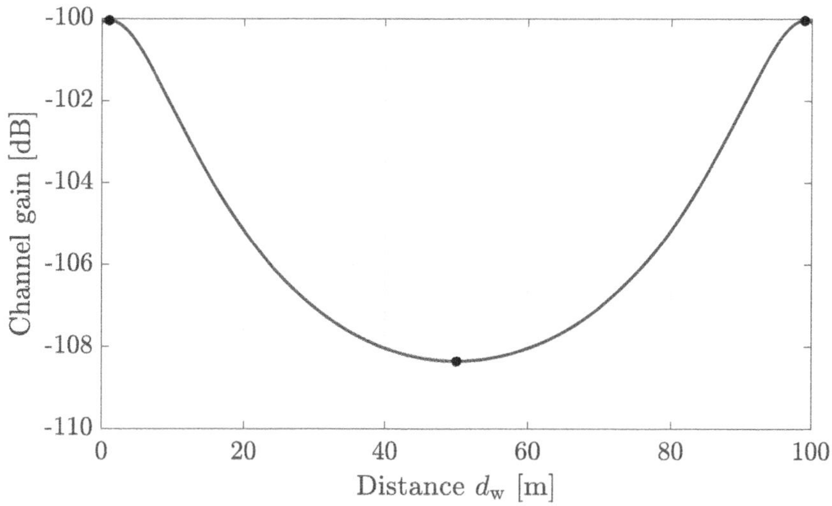

Figure 9.13: The end-to-end channel gain in (9.36) depending on how far the reconfigurable surface is from the transmitter, when the total distance between the transmitter and receiver is 100 m. The stars show the maximum and minimum values, which were derived in Example 9.5.

9.2.2 Acquiring Channel State Information and Feedback Signaling

The reconfigurable surface requires CSI to compute the capacity-maximizing reflection matrix. Specifically, (9.27) shows that it must know the phase of the static channel $\arg(h_s)$ and the phases $\arg(h_{r,n}h_{t,n})$ of the paths through each of the N metaatoms. It is sufficient to know the phase of the *cascaded channel coefficient* $h_{r,n}h_{t,n}$, while the individual characteristics of $h_{r,n}$ and $h_{t,n}$ are unimportant. These $N+1$ real-valued phase coefficients can be estimated by sending a pilot sequence similar to the point-to-point scenario described in Section 4.2.4. The precise details are somewhat different since the reconfigurable surface can only reflect signals, not measure them. To describe the procedure, we begin by factorizing the end-to-end channel in (9.22) as

$$h = h_s + \sum_{n=1}^{N} h_{r,n} e^{j\psi_n} h_{t,n} = \underbrace{\begin{bmatrix} 1 \\ e^{j\psi_1} \\ \vdots \\ e^{j\psi_N} \end{bmatrix}^{\mathrm{T}}}_{=\boldsymbol{\psi}^{\mathrm{T}}} \underbrace{\begin{bmatrix} h_s \\ h_{r,1}h_{t,1} \\ \vdots \\ h_{r,N}h_{t,N} \end{bmatrix}}_{=\check{\mathbf{h}}}, \tag{9.38}$$

where $\boldsymbol{\psi} \in \mathbb{C}^{N+1}$ contains all the configurable phase-shifts (including a 1 for the static path) and $\check{\mathbf{h}} \in \mathbb{C}^{N+1}$ contains all the necessary channel coefficients. This vector was previously defined in (9.30). Every time a signal is transmitted over the channel, it will experience the scalar channel coefficient h obtained as the inner product between the channel vector $\check{\mathbf{h}}$ and the complex conjugate of the phase-shift vector $\boldsymbol{\psi}$. Hence, only the one-dimensional part of the

$(N+1)$-dimensional channel vector that is aligned with $\boldsymbol{\psi}^*$ is utilized to reflect the signal toward the receiver, while the remaining dimensions are invisible to the receiver. For the receiver to observe the entire channel vector, we must send multiple pilot signals and vary the phase-shift configuration vector to explore all the dimensions of $\mathbb{C}^{N+1}$ where $\check{\mathbf{h}}$ might have a component.

By following the notation from Section 4.2.4, we consider the transmission of a preamble of length L_{p} designed to enable channel estimation. Specifically, a constant pilot sequence $x[l] = \sqrt{q}$ is transmitted for $l = 1, \dots, L_{\mathrm{p}}$ and we reflect it using a sequence of different configuration vectors: $\boldsymbol{\psi}[1], \dots, \boldsymbol{\psi}[L_{\mathrm{p}}] \in \mathbb{C}^{N+1}$. The received signal at time instance l can then be expressed as

$$y[l] = \boldsymbol{\psi}^{\mathrm{T}}[l]\check{\mathbf{h}}\sqrt{q} + n[l], \quad l = 1, \dots, L_{\mathrm{p}}, \tag{9.39}$$

but it is convenient to write it in matrix/vector form as

$$\underbrace{\begin{bmatrix} y[1] \\ \vdots \\ y[L_{\mathrm{p}}] \end{bmatrix}}_{=\check{\mathbf{y}}} = \underbrace{\begin{bmatrix} \boldsymbol{\psi}[1] & \dots & \boldsymbol{\psi}[L_{\mathrm{p}}] \end{bmatrix}^{\mathrm{T}}}_{=\boldsymbol{\Psi}} \check{\mathbf{h}}\sqrt{q} + \underbrace{\begin{bmatrix} n[1] \\ \vdots \\ n[L_{\mathrm{p}}] \end{bmatrix}}_{=\check{\mathbf{n}}}. \tag{9.40}$$

If the channel vector $\check{\mathbf{h}}$ is treated as deterministic but unknown, the PDF of $\check{\mathbf{y}} \in \mathbb{C}^{L_{\mathrm{p}}}$ can be expressed using (2.80) as

$$f_{\check{\mathbf{y}}}(\check{\mathbf{y}}) = \frac{1}{(\pi N_0)^{L_{\mathrm{p}}}} e^{-\frac{\|\check{\mathbf{y}} - \boldsymbol{\Psi}\check{\mathbf{h}}\sqrt{q}\|^2}{N_0}} \tag{9.41}$$

because $\check{\mathbf{y}} - \boldsymbol{\Psi}\check{\mathbf{h}}\sqrt{q} \sim \mathcal{N}_{\mathbb{C}}(\mathbf{0}, N_0\mathbf{I}_{L_{\mathrm{p}}})$. The ML estimate $\widehat{\check{\mathbf{h}}}$ of $\check{\mathbf{h}}$ is the vector that maximizes the PDF, which corresponds to minimizing the squared norm expression in its exponent. By equating the argument of the norm to zero, we obtain

$$\check{\mathbf{y}} - \boldsymbol{\Psi}\widehat{\check{\mathbf{h}}}\sqrt{q} = \mathbf{0} \quad \Rightarrow \quad \widehat{\check{\mathbf{h}}} = \frac{1}{\sqrt{q}}\boldsymbol{\Psi}^{-1}\check{\mathbf{y}} \tag{9.42}$$

if the matrix $\boldsymbol{\Psi} \in \mathbb{C}^{L_{\mathrm{p}} \times (N+1)}$ is invertible. This requires that $L_{\mathrm{p}} = N+1$ since only square matrices are invertible. Moreover, we need to find an invertible matrix of that size that satisfies two conditions: all entries of the first column are equal to 1, and all other entries are complex exponentials that can be implemented using the phase-shifting ability of a reconfigurable surface. Both properties are satisfied by the DFT matrix in (2.198) if it is scaled properly:

$$\boldsymbol{\Psi} = \sqrt{N+1}\mathbf{F}_{N+1}. \tag{9.43}$$

For this particular choice, the ML estimate in (9.42) can be expressed as

$$\widehat{\check{\mathbf{h}}} = \frac{1}{\sqrt{q}}\boldsymbol{\Psi}^{-1}\left(\boldsymbol{\Psi}\check{\mathbf{h}}\sqrt{q} + \mathbf{n}\right) = \check{\mathbf{h}} + \frac{1}{\sqrt{q(N+1)}}\mathbf{F}_{N+1}^{\mathrm{H}}\mathbf{n}, \tag{9.44}$$

which is the true channel vector plus a scaled noise term with i.i.d. entries distributed as $\mathcal{N}_{\mathbb{C}}(0, N_0/(q(N+1)))$. The estimation error vanishes as $q \to \infty$ as expected from a well-crafted estimator.

Example 9.6. Can an ML estimate of $\check{\mathbf{h}}$ be computed if $L_\mathrm{p} \neq N+1$?

Yes, we can always do our best to maximize the PDF in (9.41) even if the performance varies. If $L_\mathrm{p} > N+1$, we can pick $\mathbf{\Psi}$ as a full-rank matrix. We still want to solve the linear system of equations $\check{\mathbf{y}} - \mathbf{\Psi}\check{\mathbf{h}}\sqrt{q} = \mathbf{0}$ with respect to $\check{\mathbf{h}}$. This system is overdetermined and might lack a solution. However, the channel vector can only be observed in the subspace spanned by the columns of $\mathbf{\Psi}$; thus, we can project the equation to that subspace and then solve it:

$$\mathbf{\Psi}^\mathrm{H}\check{\mathbf{y}} - \mathbf{\Psi}^\mathrm{H}\mathbf{\Psi}\widehat{\check{\mathbf{h}}}\sqrt{q} = \mathbf{0} \quad \Rightarrow \quad \widehat{\check{\mathbf{h}}} = \frac{1}{\sqrt{q}}(\mathbf{\Psi}^\mathrm{H}\mathbf{\Psi})^{-1}\mathbf{\Psi}^\mathrm{H}\check{\mathbf{y}} = \check{\mathbf{h}} + \frac{1}{\sqrt{q}}(\mathbf{\Psi}^\mathrm{H}\mathbf{\Psi})^{-1}\mathbf{\Psi}^\mathrm{H}\mathbf{n}. \tag{9.45}$$

The matrix $(\mathbf{\Psi}^\mathrm{H}\mathbf{\Psi})^{-1}\mathbf{\Psi}^\mathrm{H}$ is called the left pseudo-inverse of $\mathbf{\Psi}$. The estimate is more precise than with $L_\mathrm{p} = N+1$ since it builds on more observations.

If $L_\mathrm{p} < N+1$, the linear system of equations $\check{\mathbf{y}} - \mathbf{\Psi}\check{\mathbf{h}}\sqrt{q} = \mathbf{0}$ is underdetermined and has many solutions. Instead of picking an arbitrary solution, which leads to estimation errors that remain as $q \to \infty$, it can be desirable to reformulate the entire problem to reduce the number of unknowns. We can group N_s adjacent metaatoms together into a subarray that must use the same phase-shift value, motivated by the fact that the optimal phase-shift pattern often varies slowly over the surface. A single channel coefficient can then represent the cascaded channel through one subarray. Hence, in the reformulated estimation problem, detailed in [160], there are only $N/N_\mathrm{s}+1$ unknown coefficients. For any $L_\mathrm{p} \geq 2$, we can pick the subarray size N_s such that $L_\mathrm{p} \geq N/N_\mathrm{s}+1$ to avoid an underdetermined estimation problem.

The ML estimate is computed at the receiver, not the reconfigurable surface that needs it. A possible solution is that the receiver (e.g., the base station) computes the estimate, then determines the desirable configuration by putting the estimates into (9.27), and finally feeds this information back to the surface. This procedure is illustrated in Figure 9.14. The feedback link requires the reconfigurable surface to be equipped with a transceiver.

We will now compare the capacity-maximizing configuration (based on perfect CSI) with the capacities obtained when the reconfigurable surface is tuned based on the ML estimate (imperfect CSI) and when random phase-shifts uniformly distributed in $[-\pi, \pi)$ are used. Since randomness affects h in the latter cases, we present the average capacity values in Figure 9.15. These values are computed assuming the receiver knows h perfectly during data transmission, so the randomness only affects how the surface is configured. We consider $N = 200$ metaatoms, $\beta_\mathrm{t} = -80\,\mathrm{dB}$, $\beta_\mathrm{r} = -60\,\mathrm{dB}$, and $\beta_\mathrm{s} = -110\,\mathrm{dB}$ (as in Figure 9.11). The SNR shown on the horizontal axis is defined based on the static path as $q\beta_\mathrm{s}/N_0$. The estimation accuracy is low when the SNR is small; thus, random phase-shifts give the same capacity as when the ML

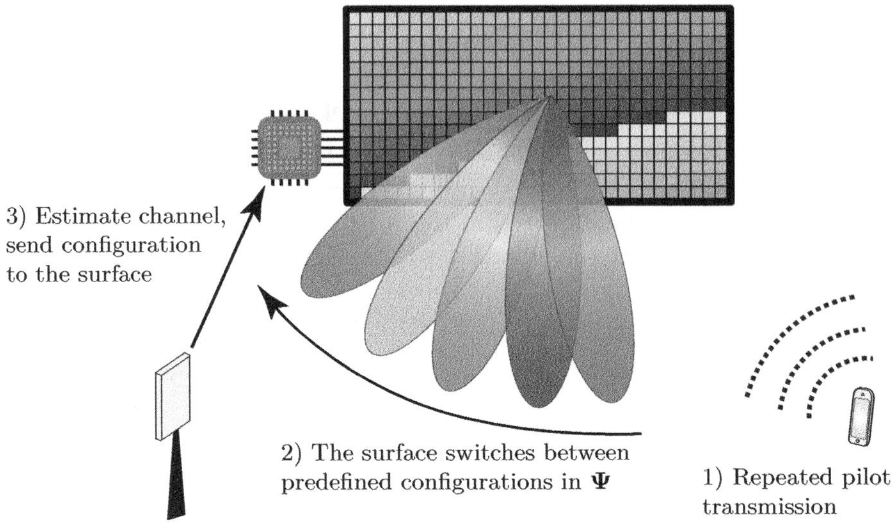

Figure 9.14: A reconfigurable surface can be configured by letting the transmitter repeat a pilot transmission while the surface reflects it using different predefined configurations. The receiver then computes an estimate of the channel vector and uses it to compute the desirable configuration. This information is then sent to the surface using a feedback link.

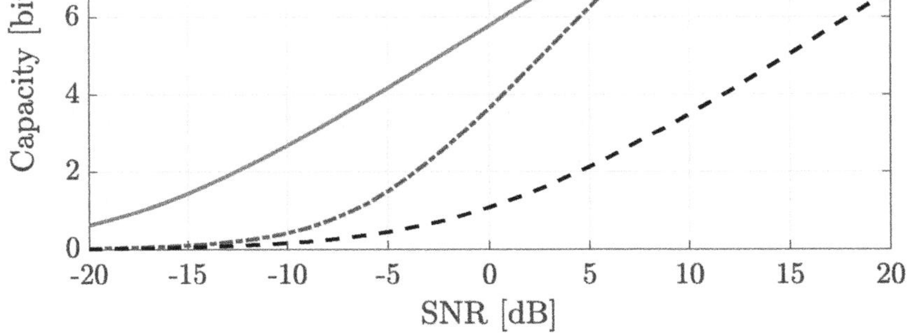

Figure 9.15: The capacity as a function of the SNR, considering a SISO channel aided by a reconfigurable surface with $N = 200$ metaatoms. Three different configurations are compared: the capacity-maximizing one based on perfect CSI, one based on the ML estimator (imperfect CSI), and one using random phase-shifts.

estimate is used for selecting the phases. As the SNR increases, the imperfect CSI curve improves rapidly, thanks to the better estimation accuracy, and converges to the perfect CSI curve. The gap to the case with random phase-shifts is then large. It is important to note that the SNR per metaatom, $q\beta_r\beta_t/N_0$, is 30 dB smaller than what is shown on the horizontal axis; on the other hand, the 201-length pilot sequence increases the SNR during the channel estimation by $10\log_{10}(201) \approx 23$ dB. In conclusion, assessing what SNR value is small versus large in this context is complicated.

The ML estimator derived and discussed above is non-parametric, which means that we look for any conceivable channel vector in $\mathbb{C}^{N+1}$. When the ML estimation framework was previously applied in Section 4.2.5, we restricted the search to LOS channels that are parametrized by the angle-of-arrival to a multi-antenna receiver. A similar parametric ML estimator can be developed when the channels to and from the surface are array response vectors, but we refer to [161] for the precise details.

9.3 Wideband Communication using Reconfigurable Surfaces

In this section, we will analyze how reconfigurable surfaces can be utilized to enhance communication over wideband channels. To ensure we capture the essential new characteristics, we must revisit how practical continuous passband channels were transformed into discrete complex baseband channels in Section 2.3. The previous analysis considered the general setup in Figure 9.16(a), where a passband signal $z_p(t)$ is transmitted over a wireless channel and $v_p(t)$ denotes the filtered version that reaches the receiver before noise is added to it. The channel was described by the impulse response $g_p(t)$, and it depends on the propagation environment that the reconfigurable surface can control. Hence, we will now denote the impulse response as $g_{p;\boldsymbol{\psi}}(t)$, where the vector $\boldsymbol{\psi}$ represents the surface configuration.

The end-to-end channel impulse response $g_{p;\boldsymbol{\psi}}(t)$ is the sum of the impulse responses of the different propagation paths. We begin by defining the impulse response $g_{s,p}(t)$ of the static LTI channel that the signal propagates over in the absence of the reconfigurable surface. The transmitted signal $z_p(t)$ also propagates to each of the N metaatoms in the reconfigurable surface through a separate LTI channel represented by an arbitrary impulse response $g_{t,n,p}(t)$, for $n = 1, \ldots, N$. When the signal reaches metaatom n, it will be filtered by its internal circuitry and then reradiated. The filtering happens in the analog domain and will be modeled as an LTI filter. We denote the impulse response as $\vartheta_{n,p;\psi_n}(t)$ and stress that it is reconfigurable in the sense that it depends on an external stimulus represented by the variable ψ_n from the vector $\boldsymbol{\psi}$. In other words, we can choose from a set of possible impulse responses by selecting ψ_n. To be consistent with the LTI assumption, only one value of ψ_n can be used during the considered signal transmission.

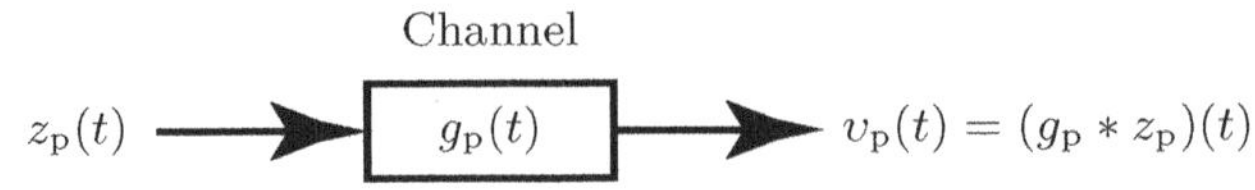

(a) The general relation between the transmitted and received passband signals.

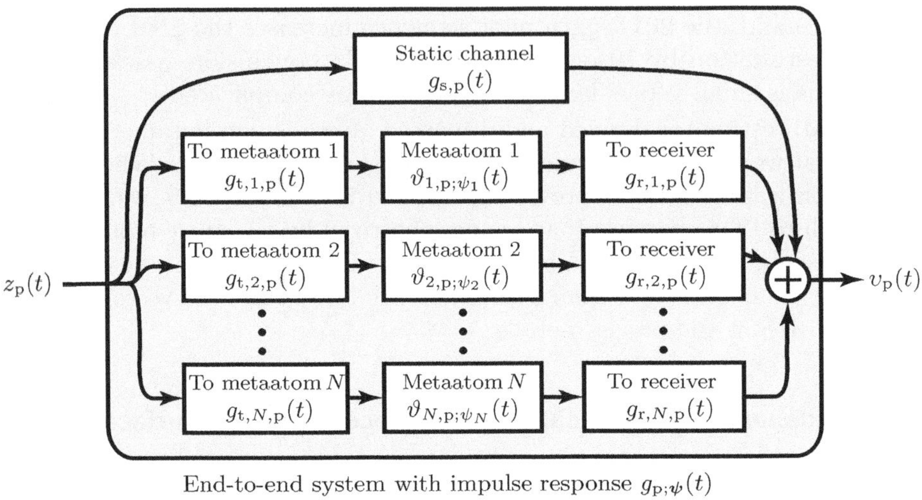

(b) A detailed channel description when utilizing a reconfigurable surface.

Figure 9.16: The received passband signal $v_p(t)$ is the convolution between the transmitted signal $z_p(t)$ and the channel impulse response $g_p(t)$, as shown in (a). When utilizing a reconfigurable surface with N metaatoms, the impulse response is the superposition/addition of the impulse response $g_{s,p}(t)$ of the static channel and the N controllable impulse responses of the channels via each of the N metaatoms, as shown in (b).

When the signal is reradiated from metaatom n, it propagates to the receiver over yet another LTI channel with an arbitrary impulse response $g_{r,n,p}(t)$. Since the transmitted signal propagates via metaatom n to the receiver through a cascade of three LTI filters, the joint impulse response is the convolution of their impulse responses: $(g_{r,n,p} * \vartheta_{n,p;\psi_n} * g_{t,n,p})(t)$. We thereby obtain the input-output relation illustrated in Figure 9.16(b):

$$\begin{aligned} v_p(t) &= (g_{s,p} * z_p)(t) + \sum_{n=1}^{N} (g_{r,n,p} * \vartheta_{n,p;\psi_n} * g_{t,n,p} * z_p)(t) \\ &= \Bigg(\underbrace{\left[g_{s,p} + \sum_{n=1}^{N} g_{r,n,p} * \vartheta_{n,p;\psi_n} * g_{t,n,p} \right]}_{=g_{p;\psi}} * z_p \Bigg)(t), \end{aligned} \tag{9.46}$$

where we identify the impulse response of the end-to-end system as

$$g_{p;\psi}(t) = g_{s,p}(t) + \sum_{n=1}^{N} (g_{r,n,p} * \vartheta_{n,p;\psi_n} * g_{t,n,p})(t). \tag{9.47}$$

We recall from (2.116)-(2.117) that filtering of a passband signal can be represented by filtering the equivalent complex-baseband signal using impulse responses that are downshifted. By applying this principle to each component of $g_{\text{p};\psi}(t)$, the complex baseband representation of (9.46) becomes

$$v(t) = (g_\text{s} * z)(t) + \sum_{n=1}^{N} (g_{\text{r},n} * \vartheta_{n;\psi_n} * g_{\text{t},n} * z)(t), \tag{9.48}$$

where the impulse responses of the downshifted channels and filters are defined as

$$g_\text{s}(t) = g_{\text{s,p}}(t)e^{-\text{j}2\pi f_\text{c} t}, \tag{9.49}$$
$$g_{\text{t},n}(t) = g_{\text{t},n,\text{p}}(t)e^{-\text{j}2\pi f_\text{c} t}, \tag{9.50}$$
$$g_{\text{r},n}(t) = g_{\text{r},n,\text{p}}(t)e^{-\text{j}2\pi f_\text{c} t}, \tag{9.51}$$
$$\vartheta_{n;\psi_n}(t) = \vartheta_{n,\text{p};\psi_n}(t)e^{-\text{j}2\pi f_\text{c} t}. \tag{9.52}$$

The end-to-end channel has the impulse response

$$g_{\psi}(t) = g_\text{s}(t) + \sum_{n=1}^{N} (g_{\text{r},n} * \vartheta_{n;\psi_n} * g_{\text{t},n})(t). \tag{9.53}$$

We notice that the convolution of a chain of impulse responses in the passband becomes the convolution of the corresponding chain of complex-baseband impulse responses. This property seems natural but is actually a feature of the definitions previously made in Section 2.3.1. We considered a passband signal that is sent over a channel with arbitrary frequency support and defined how the signal and channel/filter are transformed to the baseband differently. This can be called the *pseudo-baseband representation* since the channel is not a baseband filter, but the output is a baseband signal since the input is a baseband signal. By contrast, many other textbooks consider a stricter complex-baseband representation, where each channel is a passband filter that is transformed to the baseband identically to the signal. That definition is less practical since wireless channels are not passband filters but support signals of any frequency. More importantly, it gives rise to extra scaling factors, and these multiply when considering convolutions of filters, making the stricter model inappropriate when studying reconfigurable surfaces.

When a discrete sequence of data symbols $x[k]$ is transmitted using PAM, bandpass filtered at the receiver, and sampled on the symbol rate, the resulting received signal sequence $y[l]$ was expressed in (7.7) as

$$y[l] = \sum_{\ell=0}^{T} h_{\psi}[\ell]x[l-\ell] + n[l], \tag{9.54}$$

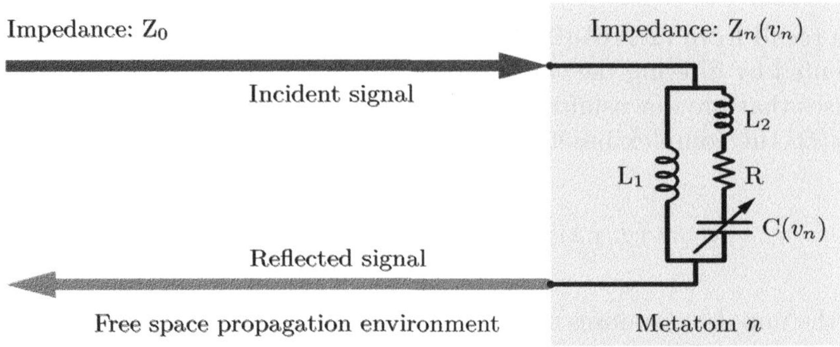

Figure 9.17: An example of a lumped-element model of a metaatom containing two inductors $(\mathrm{L}_1, \mathrm{L}_2)$, one resistor (R), and a varactor with a controllable capacitance $\mathrm{C}(v_n)$.

where the $T+1$ discrete-time channel coefficients $h_{\boldsymbol{\psi}}[0], \ldots, h_{\boldsymbol{\psi}}[T]$ are computed based on the end-to-end channel model in (9.53) as

$$\begin{aligned} h_{\boldsymbol{\psi}}[\ell] &= (p * g_{\boldsymbol{\psi}} * p)\,(t)\,\Big|_{t=\ell/B} \\ &= (p * g_{\mathrm{s}} * p)(t)\Big|_{t=\ell/B} + \sum_{n=1}^{N} (p * g_{\mathrm{r},n} * \vartheta_{n;\psi_n} * g_{\mathrm{t},n} * p)(t)\Big|_{t=\ell/B}. \end{aligned} \quad (9.55)$$

Conventional propagation models can be used for the impulse responses $g_{\mathrm{s}}(t)$, $g_{\mathrm{t},n}(t)$, $g_{\mathrm{r},n}(t)$ of the wireless channels, by accounting for the effective areas of antennas and metaatoms. To compute an expression of (9.55), we must also characterize the impulse response $\vartheta_{n;\psi_n}(t)$ of a metaatom.

9.3.1 Impulse Response of a Metaatom

We will showcase a basic model of the impulse response $\vartheta_{n;\psi_n}(t)$ of a metaatom by analyzing a practical implementation. Figure 9.17 shows a lumped-element model of metaatom n containing two parallel branches where the first contains an inductor with inductance L_1 and the second contains a series with an inductor with inductance L_2, a resistor with resistance R, and a varactor with a capacitance $\mathrm{C}(v_n)$ controlled by the bias voltage v_n. This parallel resonance circuit is a simplified version of the metaatom design in [162] and was considered for reconfigurable surfaces in [163]. Using circuit theory methods, the impedance of the metaatom can be shown to be

$$\mathrm{Z}_n(v_n) = \frac{\mathrm{j}2\pi f \mathrm{L}_1 \left(\mathrm{j}2\pi f \mathrm{L}_2 + \mathrm{R} + \frac{1}{\mathrm{j}2\pi f \mathrm{C}(v_n)}\right)}{\mathrm{j}2\pi f \mathrm{L}_1 + \left(\mathrm{j}2\pi f \mathrm{L}_2 + \mathrm{R} + \frac{1}{\mathrm{j}2\pi f \mathrm{C}(v_n)}\right)}, \quad (9.56)$$

for a signal with the frequency f. We can compute the frequency-dependent reflection coefficient by substituting this expression into (9.13).

Figure 9.18 shows the frequency response of the metaatom for frequencies around an intended carrier frequency of $f_{\mathrm{c}} = 3\,\mathrm{GHz}$. The parameters of the lumped-element model in Figure 9.17 are $L_1 = 2.5$ nH, $L_2 = 0.7$ nH, $Z_0 = 377$ ohm, and $R = 1$ ohm. Since the reflection coefficient is complex, the phase and amplitude responses are presented in (a) and (b), respectively. There are curves in Figure 9.18(a) for four different capacitance values obtained by controlling the bias voltage of the varactor. The specific values have been selected to give the phase-shifts $3\pi/4, \pi/4, -\pi/4, -3\pi/4$ at the carrier frequency. The phase begins close to $+\pi$ on all the curves because the reradiated electric field is inverted (upside down). The phase variations are large when considering the GHz range, which is natural for all filters. The simplest representation of reflection would be a pure time delay τ, which has the impulse response $\delta(t-\tau)$ and frequency response $e^{-\mathrm{j}2\pi f\tau}$ with a phase that varies linearly with the frequency. Since the curves are approximately linear next to the carrier frequency, this model is suitable if the signal bandwidth B is limited to a few hundred MHz. Outside this range, the phase response curves have non-linear shapes that will distort the signal in the time domain; thus, a metaatom has a limited useful bandwidth range. As long as the system uses a smaller bandwidth, the reflected signal will be undistorted, and the propagation channels will determine whether the communication system is narrowband or wideband. The four curves have roughly the same shape but are shifted in the frequency domain; it is this shift that the varactor controls.

The amplitude responses are shown in Figure 9.18(b) for the same capacitance values, but different resistances: $R = 1$ ohm and $R = 0$ ohm. The theoretical maximum amplitude response is 0 dB since the metaatom is a passive circuit that reflects the signal without amplification.[4] All the signal power is reflected when the resistance is negligible, while there are a few dB of amplitude losses when the resistance is non-zero. In the latter case, the amplitude loss is also frequency-dependent. The loss is largest at the frequency where the phase response is zero due to resonance in the circuit. While building metaatoms with minimal reflection losses is desirable, we should keep in mind that a few dB is minor compared to the propagation losses over wireless channels that are typically at the order of 100 dB.

In summary, an ideal metaatom design has no amplitude losses and a linear phase within the signal band. Hence, its impulse response can be expressed as $\vartheta_{n,\mathrm{p};\psi_n}(t) = \delta(t-\tau_{\psi_n})$ in the passband, which results in the downshifted filter

$$\vartheta_{n;\psi_n}(t) = \delta(t-\tau_{\psi_n})e^{-\mathrm{j}2\pi f_c t}, \tag{9.57}$$

where the controllable delay is denoted by τ_{ψ_n}. We will soon show that (9.57) results in a phase-shift of $\psi_n = -2\pi f_c \tau_{\psi_n}$ in the system model; thus, the phase-shift caused by a metaatom is controlled by tuning the reflection delay.

[4]The passiveness is a key feature since active components (e.g., an amplifier) add noise to the reflected signal, which is not the case when using reconfigurable surfaces. Retransmitting devices that contain amplifiers are normally referred to as repeaters or relays and must be studied using different system models than in this chapter.

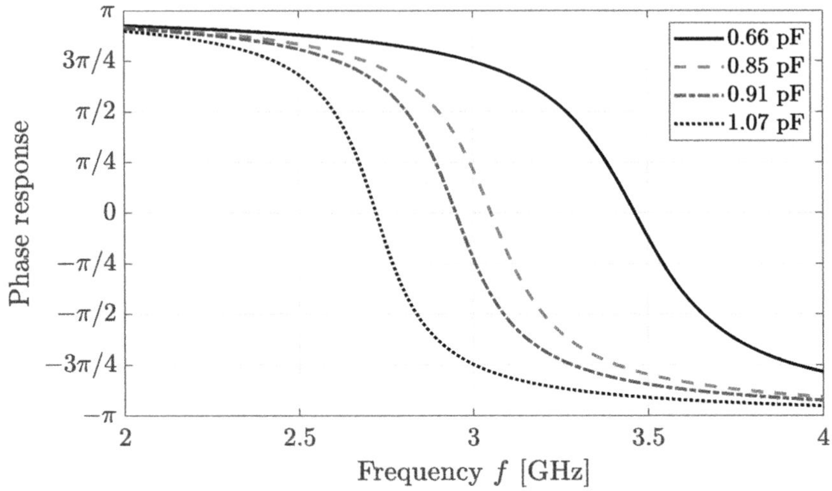

(a) Phase response for different frequencies when R = 1 ohm.

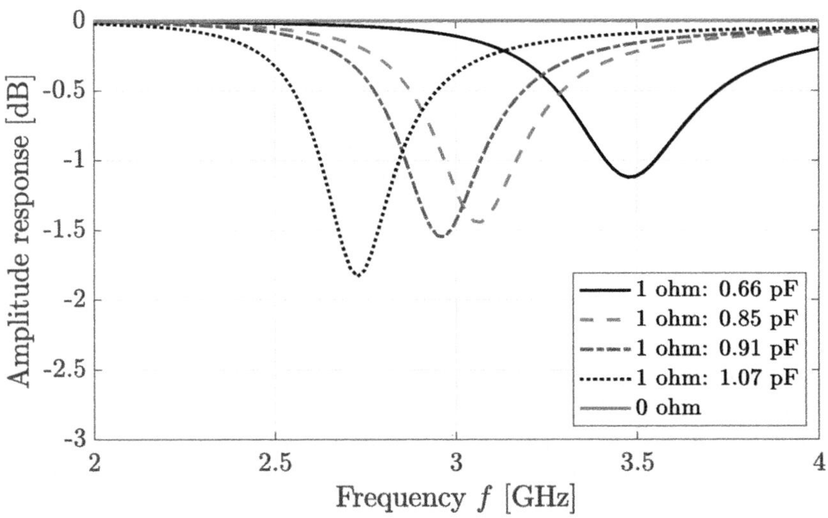

(b) Amplitude response for different frequencies.

Figure 9.18: The frequency response when a metaatom with the impedance in (9.56) reflects a signal in free space. The curves are obtained for different capacitances of the varactor, which are selected to give the phase-shifts $3\pi/4, \pi/4, -\pi/4, -3\pi/4$ at 3 GHz. The phase response is shown in (a) and the the amplitude response is shown in (b).

9.3.2 OFDM-Based Channel Model with a Reconfigurable Surface

In this section, we will determine the channel coefficients in an OFDM system aided by a reconfigurable surface. We can use the basic channel model from (2.124) to express a static channel with L_{s} propagation paths as

$$g_{\mathrm{s,p}}(t) = \sum_{i=1}^{L_{\mathrm{s}}} \alpha_{\mathrm{s},i}\delta(t+\eta-\tau_{\mathrm{s},i}) \quad \Rightarrow \quad g_{\mathrm{s}}(t) = \sum_{i=1}^{L_{\mathrm{s}}} \alpha_{\mathrm{s},i}e^{-\mathrm{j}2\pi f_{\mathrm{c}}t}\delta(t+\eta-\tau_{\mathrm{s},i}), \tag{9.58}$$

where $\alpha_{\mathrm{s},i} \in [0,1]$ is the attenuation and $\tau_{\mathrm{s},i} \geq 0$ is the delay of path i, for $i = 1,\ldots,L_{\mathrm{s}}$. We recall that η denotes the receiver's clock delay, which ensures that the receiver takes samples when the signal reaches it and not when the signal leaves the transmitter. This parameter must be selected as described in Section 7.1 to achieve the causal FIR filter representation in (9.54). We denote the number of propagation paths between the transmitter and reconfigurable surface as L_{t} and between the surface and receiver as L_{r}. Similarly to (9.58), we can then model the impulse responses to and from metaatom n as $g_{\mathrm{t},n,\mathrm{p}}(t) = \sum_{i=1}^{L_{\mathrm{t}}} \alpha_{\mathrm{t},n,i}\delta(t-\tau_{\mathrm{t},n,i})$ and $g_{\mathrm{r},n,\mathrm{p}}(t) = \sum_{j=1}^{L_{\mathrm{r}}} \alpha_{\mathrm{r},n,j}\delta(t+\eta-\tau_{\mathrm{r},n,j})$. These can be expressed in the complex baseband as

$$g_{\mathrm{t},n}(t) = \sum_{i=1}^{L_{\mathrm{t}}} \alpha_{\mathrm{t},n,i}e^{-\mathrm{j}2\pi f_{\mathrm{c}}t}\delta(t-\tau_{\mathrm{t},n,i}), \tag{9.59}$$

$$g_{\mathrm{r},n}(t) = \sum_{j=1}^{L_{\mathrm{r}}} \alpha_{\mathrm{r},n,j}e^{-\mathrm{j}2\pi f_{\mathrm{c}}t}\delta(t+\eta-\tau_{\mathrm{r},n,j}), \tag{9.60}$$

where $\alpha_{\mathrm{t},n,i}, \alpha_{\mathrm{r},n,j} \in [0,1]$ are the attenuations and $\tau_{\mathrm{t},n,i}, \tau_{\mathrm{r},n,j} \geq 0$ are the propagation delays. Note that the receiver's timing delay η is only included in the channels that lead to the receiver. By substituting the metaatom's impulse response in (9.57) and the channel impulse responses in (9.58)–(9.60) into (9.55), the channel coefficients can be computed as[5]

$$\begin{aligned} h_{\boldsymbol{\psi}}[\ell] &= \sum_{i=1}^{L_{\mathrm{s}}} \alpha_{\mathrm{s},i}e^{-\mathrm{j}2\pi f_{\mathrm{c}}(\tau_{\mathrm{s},i}-\eta)}\mathrm{sinc}(\ell+B(\eta-\tau_{\mathrm{s},i}))+ \\ &\sum_{n=1}^{N}\sum_{j=1}^{L_{\mathrm{r}}}\sum_{i=1}^{L_{\mathrm{t}}} \alpha_{\mathrm{r},n,j}\alpha_{\mathrm{t},n,i}e^{-\mathrm{j}2\pi f_{\mathrm{c}}(\tau_{\mathrm{r},n,j}+\tau_{\mathrm{t},n,i}+\tau_{\psi_n}-\eta)}\mathrm{sinc}(\ell+B(\eta-\tau_{\mathrm{r},n,j}-\tau_{\mathrm{t},n,i})) \end{aligned} \tag{9.61}$$

for $\ell = 0,\ldots,T$, by utilizing the facts that $(p*p)(t) = \mathrm{sinc}(Bt)$ and that the convolution between $\mathrm{sinc}(Bt)$ and $e^{-\mathrm{j}2\pi f_{\mathrm{c}}t}\delta(t-\tau)$ is $\mathrm{sinc}(B(t-\tau))e^{-\mathrm{j}2\pi f_{\mathrm{c}}\tau}$.

[5]The last sinc-term in (9.61) becomes $\mathrm{sinc}(\ell+B(\eta-\tau_{\mathrm{r},n,j}-\tau_{\mathrm{t},n,i}-\tau_{\psi_n}))$ but the term containing τ_{ψ_n} can be dropped since the delay caused by the reflection is negligible compared to the symbol time $1/B$, in the sense that $\tau_{\psi_n}/(1/B) = B\tau_{\psi_n} \approx 0$. The metaatom nevertheless creates a noticeable phase-shift since $f_{\mathrm{c}} \gg B$.

We notice that (9.61) contains L_{s} static paths from the transmitter to the receiver and $NL_{\mathrm{t}}L_{\mathrm{r}}$ paths involving the reconfigurable surface. Each of the latter paths has an attenuation $\alpha_{\mathrm{r},n,j}\alpha_{\mathrm{t},n,i}$ that is the product of the attenuation from the transmitter to the metaatom and from the metaatom to the receiver. Each such path is also associated with a phase-shift $e^{-\mathrm{j}2\pi f_{\mathrm{c}}(\tau_{\mathrm{r},n,j}+\tau_{\mathrm{t},n,i}+\tau_{\psi_n}-\eta)}$ containing the accumulated delays. The sinc-function determines how the signal energy carried by the path is divided between the $T+1$ channel taps.

When the reconfigurable surface is in the far-field of the transmitter, receiver, and other objects in the propagation environment, we can represent the channel using array response vectors. We assume that the surface is a UPA with N_{V} rows and N_{H} metaatoms per row. The ith incident path to the surface can be associated with an angle pair $(\varphi_{\mathrm{i},i}, \theta_{\mathrm{i},i})$, measured from the broadside direction of the surface, and the jth outgoing path can be associated with an angle pair $(\varphi_{\mathrm{o},j}, \theta_{\mathrm{o},j})$. If we gather the N phase-shifts related to such a path, they match with the array response vector expression in (4.128):

$$\mathbf{a}_{N_{\mathrm{H}},N_{\mathrm{V}}}(\varphi_{\mathrm{i},i}, \theta_{\mathrm{i},i}) = \begin{bmatrix} 1 \\ e^{-\mathrm{j}2\pi f_{\mathrm{c}}(\tau_{\mathrm{t},2,i}-\tau_{\mathrm{t},1,i})} \\ \vdots \\ e^{-\mathrm{j}2\pi f_{\mathrm{c}}(\tau_{\mathrm{t},N,i}-\tau_{\mathrm{t},1,i})} \end{bmatrix}, \tag{9.62}$$

$$\mathbf{a}_{N_{\mathrm{H}},N_{\mathrm{V}}}(\varphi_{\mathrm{o},j}, \theta_{\mathrm{o},j}) = \begin{bmatrix} 1 \\ e^{-\mathrm{j}2\pi f_{\mathrm{c}}(\tau_{\mathrm{r},2,j}-\tau_{\mathrm{r},1,j})} \\ \vdots \\ e^{-\mathrm{j}2\pi f_{\mathrm{c}}(\tau_{\mathrm{r},N,j}-\tau_{\mathrm{r},1,j})} \end{bmatrix}. \tag{9.63}$$

Furthermore, the attenuation is the same for all the metaatoms: $\alpha_{\mathrm{t},n,i} = \alpha_{\mathrm{t},1,i}$ and $\alpha_{\mathrm{r},n,j} = \alpha_{\mathrm{r},1,j}$ for all n, where we take the first metaatom as the reference. For a given path, the delay variations across the surface are negligible in the sense that $B\tau_{\mathrm{r},n,j} \approx B\tau_{\mathrm{r},1,j}$ and $B\tau_{\mathrm{t},n,i} \approx B\tau_{\mathrm{t},1,i}$, for all n. We can utilize these properties to rewrite the channel coefficients in (9.61) as

$$h_{\psi}[\ell] = c_{\mathrm{s}}[\ell] + \sum_{j=1}^{L_{\mathrm{r}}}\sum_{i=1}^{L_{\mathrm{t}}} c_{i,j}[\ell]\mathbf{a}_{N_{\mathrm{H}},N_{\mathrm{V}}}^{\mathrm{T}}(\varphi_{\mathrm{o},j}, \theta_{\mathrm{o},j})\mathbf{D}_{\psi}\mathbf{a}_{N_{\mathrm{H}},N_{\mathrm{V}}}(\varphi_{\mathrm{i},i}, \theta_{\mathrm{i},i}), \tag{9.64}$$

where $\mathbf{D}_{\psi} = \mathrm{diag}(e^{\mathrm{j}\psi_1}, \ldots, e^{\mathrm{j}\psi_N})$ contains the controllable phase-shifts

$$\psi_n = -2\pi f_{\mathrm{c}}\tau_{\psi_n} \tag{9.65}$$

created by each of the metaatoms, and the channel coefficients that depend on the tap index are gathered in the sequences

$$c_{\mathrm{s}}[\ell] = \sum_{i=1}^{L_{\mathrm{s}}} \alpha_{\mathrm{s},i}e^{-\mathrm{j}2\pi f_{\mathrm{c}}(\tau_{\mathrm{s},i}-\eta)}\mathrm{sinc}(\ell + B(\eta - \tau_{\mathrm{s},i})), \tag{9.66}$$

$$c_{i,j}[\ell] = \alpha_{\mathrm{r},1,j}\alpha_{\mathrm{t},1,i}e^{-\mathrm{j}2\pi f_{\mathrm{c}}(\tau_{\mathrm{r},1,j}+\tau_{\mathrm{t},1,i}-\eta)}\mathrm{sinc}(\ell + B(\eta - \tau_{\mathrm{r},1,j} - \tau_{\mathrm{t},1,i})). \tag{9.67}$$

The channel coefficient at the ℓth tap has the same structure as the narrowband channel expression in (9.23), but the fact that there are multiple taps calls for a different kind of signal transmission. If we apply OFDM with S data symbols per block and a cyclic prefix of length T, it follows from (7.25) that we obtain S memoryless subcarriers of the kind

$$\bar{y}[\nu] = \bar{h}_{\boldsymbol{\psi}}[\nu]\bar{\chi}[\nu] + \bar{n}[\nu], \quad \text{for } \nu = 0, \dots, S-1, \tag{9.68}$$

with the reconfigurable frequency-domain channel coefficients

$$\bar{h}_{\boldsymbol{\psi}}[\nu] = \bar{c}_{\mathrm{s}}[\nu] + \sum_{j=1}^{L_{\mathrm{r}}} \sum_{i=1}^{L_{\mathrm{t}}} \bar{c}_{i,j}[\nu] \mathbf{a}_{N_{\mathrm{H}},N_{\mathrm{V}}}^{\mathrm{T}}(\varphi_{\mathrm{o},j}, \theta_{\mathrm{o},j}) \mathbf{D}_{\boldsymbol{\psi}} \mathbf{a}_{N_{\mathrm{H}},N_{\mathrm{V}}}(\varphi_{\mathrm{i},i}, \theta_{\mathrm{i},i}) \tag{9.69}$$

that depend on the DFTs of the time-domain channel coefficients:

$$\bar{c}_{\mathrm{s}}[\nu] = \sum_{\ell=0}^{T} c_{\mathrm{s}}[\ell] e^{-\mathrm{j}2\pi\ell\nu/S}, \quad \nu = 0, \dots, S-1, \tag{9.70}$$

$$\bar{c}_{i,j}[\nu] = \sum_{\ell=0}^{T} c_{i,j}[\ell] e^{-\mathrm{j}2\pi\ell\nu/S}, \quad \nu = 0, \dots, S-1. \tag{9.71}$$

The expression in (9.69) might seem complicated but can be expressed as

$$\bar{h}_{\boldsymbol{\psi}}[\nu] = \underbrace{\begin{bmatrix} 1 \\ e^{\mathrm{j}\psi_1} \\ \vdots \\ e^{\mathrm{j}\psi_N} \end{bmatrix}^{\mathrm{T}}}_{=\boldsymbol{\psi}^{\mathrm{T}}} \underbrace{\begin{bmatrix} \bar{c}_{\mathrm{s}}[\nu] \\ \sum\limits_{j=1}^{L_{\mathrm{r}}} \sum\limits_{i=1}^{L_{\mathrm{t}}} \bar{c}_{i,j}[\nu] \mathbf{a}_{N_{\mathrm{H}},N_{\mathrm{V}}}(\varphi_{\mathrm{o},j}, \theta_{\mathrm{o},j}) \odot \mathbf{a}_{N_{\mathrm{H}},N_{\mathrm{V}}}(\varphi_{\mathrm{i},i}, \theta_{\mathrm{i},i}) \end{bmatrix}}_{=\check{\mathbf{h}}[\nu]}, \tag{9.72}$$

where $\odot$ denotes the entry-wise product between two vectors. The expression $\bar{h}_{\boldsymbol{\psi}}[\nu] = \boldsymbol{\psi}^{\mathrm{T}}\check{\mathbf{h}}[\nu]$ is the simplest we can obtain for the subcarrier channel coefficient when using a reconfigurable surface. It is the inner product between $\boldsymbol{\psi}^*$ and $\check{\mathbf{h}}[\nu]$, where the former vector depends on the surface configuration, while the latter fully characterizes the channel on subcarrier ν.

9.3.3 Wideband Capacity Maximization

The capacity of a SISO-OFDM system was presented in Theorem 7.1 with arbitrary but static channel coefficients. For a given surface configuration $\boldsymbol{\psi}$, as defined in (9.72), the capacity becomes

$$C = \frac{B}{T+S} \sum_{\nu=0}^{S-1} \log_2 \left(1 + \frac{q_{\nu}^{\mathrm{opt}} \left| \boldsymbol{\psi}^{\mathrm{T}} \check{\mathbf{h}}[\nu] \right|^2}{N_0} \right) \quad \text{bit/s}, \tag{9.73}$$

where T is the length of the cyclic prefix,

$$q_\nu^{\text{opt}} = \max\left(\mu - \frac{N_0}{\left|\boldsymbol{\psi}^{\mathrm{T}}\check{\mathbf{h}}[\nu]\right|^2}, 0\right), \quad \nu = 0, \ldots, S-1, \tag{9.74}$$

and the variable μ is selected to make $\sum_{\nu=0}^{S-1} q_\nu^{\text{opt}} = qS$.

The capacity value in (9.73) depends on the configuration $\boldsymbol{\psi}$. Corollary 9.1 showed which configuration maximizes the SNR in the narrowband case, which also leads to the maximum capacity in that case. The solution was to select $\boldsymbol{\psi}$ so that the $N+1$ terms in the inner product $\boldsymbol{\psi}^{\mathrm{T}}\check{\mathbf{h}}$, where $\check{\mathbf{h}}$ was defined in (9.38), get the same phase. The optimization task is more challenging in the wideband OFDM scenario because there are S different SNRs among the S subcarriers. The SNR on subcarrier ν is proportional to $|\boldsymbol{\psi}^{\mathrm{T}}\check{\mathbf{h}}[\nu]|^2$, where the channel vector $\check{\mathbf{h}}[\nu]$ is subcarrier-dependent while the surface configuration vector $\boldsymbol{\psi}$ is not. This resembles the analog beamforming situation in Section 7.3.1, where the same beamforming vector had to be used on all subcarriers. The wideband capacity maximization problem is computationally challenging but is partially addressed in [164], [165]. In this section, we will cover a suboptimal but effective way to configure the surface in these situations.

The optimal surface configuration in the narrowband case maximizes the channel gain. We can aim to do the same in the wideband case by maximizing the total channel gain over all subcarriers:

$$\sum_{\nu=0}^{S-1}\left|\bar{h}_{\boldsymbol{\psi}}[\nu]\right|^2 = \sum_{\nu=0}^{S-1}\left|\boldsymbol{\psi}^{\mathrm{T}}\check{\mathbf{h}}[\nu]\right|^2 = \boldsymbol{\psi}^{\mathrm{H}}\underbrace{\left(\sum_{\nu=0}^{S-1}\check{\mathbf{h}}^*[\nu]\check{\mathbf{h}}^{\mathrm{T}}[\nu]\right)}_{=\mathbf{A}}\boldsymbol{\psi}. \tag{9.75}$$

If we could pick $\boldsymbol{\psi}$ as any unit-norm vector, this quadratic form would be maximized by selecting $\boldsymbol{\psi}$ as the dominant eigenvector of $\mathbf{A} = \sum_{\nu=0}^{S-1}\check{\mathbf{h}}^*[\nu]\check{\mathbf{h}}^{\mathrm{T}}[\nu]$ (associated with the largest positive eigenvalue). However, we have a stricter constraint on the configuration vector because the first entry must be 1, and the remaining ones must have unit magnitude. There is no simple way to maximize (9.75) under this constraint, but an efficient iterative algorithm was proposed in [166]. The starting point is the *power iteration method* [167], which finds the dominant eigenvector of $\mathbf{A}$ by the iterative computation

$$\mathbf{w}_{i+1} = \frac{\mathbf{A}\mathbf{w}_i}{\|\mathbf{A}\mathbf{w}_i\|}, \quad i = 0, 1, \ldots, \tag{9.76}$$

which is initialized from arbitrary non-zero vector $\mathbf{w}_0 \in \mathbb{C}^{N+1}$. In each iteration, the multiplication $\mathbf{A}\mathbf{w}_i$ amplifies the component of $\mathbf{w}_i$ that is aligned with the dominant eigenvector relative to all other components. The convergence speed of the power iteration method depends on how much larger the largest eigenvalue is compared to the second largest eigenvalue.

A modified power iteration is described in Algorithm 9.1, which is initialized from $\boldsymbol{\psi}_0 = [1, \ldots, 1]^{\mathrm{T}}$, where the surface is not changing the phases. In each iteration, the computation in (9.76) is made using $\mathbf{w}_i = \boldsymbol{\psi}_i$. The result is used to determine the next surface configuration $\boldsymbol{\psi}_{i+1}$ by only keeping the phases of the entries of $\mathbf{w}_{i+1}$ while replacing their magnitudes by 1. Since the same value is obtained in (9.75) for $\boldsymbol{\psi}$ and $e^{-\mathrm{j}\phi}\boldsymbol{\psi}$, for any common phase-shift ϕ, we can shift the phase of all the entries so that the first entry in $\boldsymbol{\psi}_{i+1}$ becomes 1. This shift is necessary since the phase of the static channel cannot be modified. Note that $[\mathbf{w}]_n$ denotes the nth entry of $\mathbf{w}$ in the algorithm.

Example 9.7. Suppose there is only one path to and from the reconfigurable surface (i.e., $L_{\mathrm{t}} = L_{\mathrm{r}} = 1$). Which phase-shift configuration maximizes (9.75)?

Under these conditions, the channel vector in (9.72) simplifies to

$$\check{\mathbf{h}}[\nu] = \begin{bmatrix} \bar{c}_{\mathrm{s}}[\nu] \\ \bar{c}[\nu]\mathbf{a} \end{bmatrix}, \tag{9.77}$$

where $\mathbf{a} = \mathbf{a}_{N_{\mathrm{H}},N_{\mathrm{V}}}(\varphi_{\mathrm{o}}, \theta_{\mathrm{o}}) \odot \mathbf{a}_{N_{\mathrm{H}},N_{\mathrm{V}}}(\varphi_{\mathrm{i}}, \theta_{\mathrm{i}})$ and we dropped the path indices. Hence, the matrix $\mathbf{A}$ in (9.75) can be expressed as

$$\begin{aligned} \mathbf{A} &= \sum_{\nu=0}^{S-1} \check{\mathbf{h}}^*[\nu]\check{\mathbf{h}}^{\mathrm{T}}[\nu] \\ &= \sum_{\nu=0}^{S-1} \begin{bmatrix} |\bar{c}_{\mathrm{s}}[\nu]|^2 & \bar{c}_{\mathrm{s}}^*[\nu]\bar{c}[\nu]\mathbf{a}^{\mathrm{T}} \\ \bar{c}_{\mathrm{s}}[\nu]\bar{c}^*[\nu]\mathbf{a}^* & |\bar{c}[\nu]|^2\mathbf{a}^*\mathbf{a}^{\mathrm{T}} \end{bmatrix} = \begin{bmatrix} b_{\mathrm{ss}} & b_{\mathrm{s}}\mathbf{a}^{\mathrm{T}} \\ b_{\mathrm{s}}^*\mathbf{a}^* & b\mathbf{a}^*\mathbf{a}^{\mathrm{T}} \end{bmatrix}, \end{aligned} \tag{9.78}$$

where $b_{\mathrm{ss}} = \sum_{\nu=0}^{S-1} |\bar{c}_{\mathrm{s}}[\nu]|^2 \geq 0$, $b = \sum_{\nu=0}^{S-1} |\bar{c}[\nu]|^2 \geq 0$, and $b_{\mathrm{s}} = \sum_{\nu=0}^{S-1} \bar{c}_{\mathrm{s}}^*[\nu]\bar{c}[\nu]$. If the phase-shift configuration is expressed as $\boldsymbol{\psi} = [1, \mathbf{w}^{\mathrm{T}}]^{\mathrm{T}}$, then

$$\begin{aligned} \boldsymbol{\psi}^{\mathrm{H}}\mathbf{A}\boldsymbol{\psi} &= b_{\mathrm{ss}} + b_{\mathrm{s}}\mathbf{a}^{\mathrm{T}}\mathbf{w} + b_{\mathrm{s}}^*\mathbf{w}^{\mathrm{H}}\mathbf{a}^* + b\mathbf{w}^{\mathrm{H}}\mathbf{a}^*\mathbf{a}^{\mathrm{T}}\mathbf{w} \\ &= b_{\mathrm{ss}} + 2\Re\{b_{\mathrm{s}}\mathbf{a}^{\mathrm{T}}\mathbf{w}\} + b|\mathbf{a}^{\mathrm{T}}\mathbf{w}|^2. \end{aligned} \tag{9.79}$$

Among all vectors that satisfy $\|\mathbf{w}\|^2 = N$, the third term is maximized when $\mathbf{w} = e^{-\mathrm{j}\phi}\mathbf{a}^*$ for any ϕ, while the second term is maximized similarly but only for $\phi = \arg(b_{\mathrm{s}})$. The resulting solution $\mathbf{w} = e^{-\mathrm{j}\arg(b_{\mathrm{s}})}\mathbf{a}^*$ is achievable with a reconfigurable surface since $\mathbf{a}$ is a vector with phase-shifts obtained from two array response vectors. Hence, the configuration that maximizes the total channel gain is parallel to the complex conjugate of the element-wise product $\mathbf{a}$ of the array response vectors to/from the surface and has an additional common phase-shift $\arg(\sum_{\nu=0}^{S-1} \bar{c}_{\mathrm{s}}^*[\nu]\bar{c}[\nu])$ that aligns the static and controllable paths to the extent possible in an OFDM system.

Algorithm 9.1 Constrained power iteration to maximize (9.75).

Initialization: Select $\boldsymbol{\psi}_0 = [1, \ldots, 1]^{\mathrm{T}}$ and number of iterations L
for $i = 0, \ldots, L-1$ **do**
  $\mathbf{w}_{i+1} \leftarrow \frac{\mathbf{A}\boldsymbol{\psi}_i}{\|\mathbf{A}\boldsymbol{\psi}_i\|}$
  $\phi \leftarrow \arg\left([\mathbf{w}_{i+1}]_1\right)$
  $\boldsymbol{\psi}_{i+1} = \left[1, e^{\mathrm{j}(\arg([\mathbf{w}_{i+1}]_2)-\phi)}, \ldots, e^{\mathrm{j}(\arg([\mathbf{w}_{i+1}]_{N+1})-\phi)}\right]^{\mathrm{T}}$
end for
Output: $\boldsymbol{\psi}_L$

To evaluate the effectiveness of the power iteration method in general propagation scenarios, we can compare the capacity that it achieves with an upper bound. Suppose we could select a different value of $\boldsymbol{\psi}$ on each subcarrier. We could then simultaneously maximize the channel gains of all subcarriers by following the approach in (9.30)–(9.31) in the narrowband case. The resulting upper bound on the capacity can be expressed as

$$C \leq \frac{B}{T+S} \sum_{\nu=0}^{S-1} \log_2 \left(1 + \frac{q_\nu \|\check{\mathbf{h}}[\nu]\|_1^2}{N_0}\right), \tag{9.80}$$

where $\|\cdot\|_1$ denotes the 1-norm and $q_\nu = \max(\mu - N_0/\|\check{\mathbf{h}}[\nu]\|_1^2, 0)$, for $\nu = 0, \ldots, S-1$, with μ selected to make $\sum_{\nu=0}^{S-1} q_\nu = qS$. The upper bound is only achieved with equality in the unlikely event that the same surface configuration happens to maximize the channel gains on all subcarriers simultaneously.

Figure 9.19 shows how the capacity varies with the bandwidth in a scenario of the kind illustrated in Figure 9.10. Specifically, a reconfigurable surface is deployed along the yz-plane with its reflective side facing the positive x-axis and its center at $(0, 0, 0)$ m. The base station and the user are located at $(40, -200, 0)$ m and $(20, 0, 0)$ m, respectively. We assume LOS channels with multiple reflected paths to and from the surface, while the static channel is of NLOS nature. The surface has the size 0.5×0.5 m, which for a carrier frequency of 3 GHz corresponds to $N = 400$ metaatoms that each has the dimension $\lambda/4 \times \lambda/4$. The channels are modeled similarly to the 3GPP channel model in [168], and the capacity is averaged over random realizations of the multipath components' characteristics. The capacity in (9.73) is shown in Figure 9.19 as a function of the bandwidth B, assuming that the transmit power grows proportionally to the bandwidth so that the signal power spectral density is 1 W per MHz. The subcarrier spacing is 150 kHz, so the number of subcarriers increases with B as well as the number of channel taps.

The dashed-dotted curve in Figure 9.19 uses the power iteration method, and it provides 96–99% of the upper bound from (9.80). This method will find a configuration that takes the signal power from the strongest incident direction

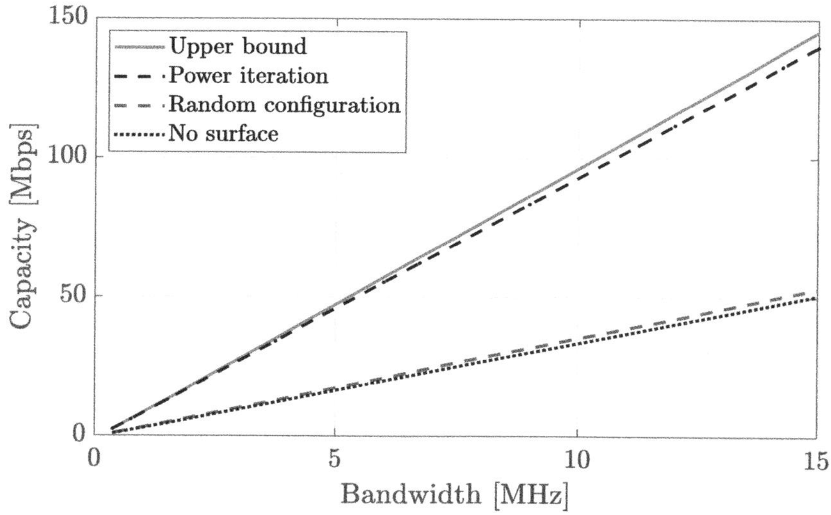

Figure 9.19: The capacity achieved over a wideband OFDM channel grows proportionally to the bandwidth and can be improved using a reconfigurable surface. The capacity achieved when configuring the surface using the power iteration in Algorithm 9.1 is compared with the upper bound in (9.80), the use of a random configuration, and the removal of the surface.

and reflects it in the direction that maximizes the received signal power. Since there are LOS channels to and from the surface, this is approximately equal to taking the signal from the LOS path and beamform it along the LOS path to the receiver. The performance gap grows with B due to the increased frequency-selectivity, but since the LOS paths to/from the surface are stronger than the scattered paths, it is possible to find a single surface configuration that works well over the entire band. This is reminiscent of how the analog beamforming architecture can provide rates close to the capacity in LOS-dominant scenarios. A part of the gap between the power iteration method and the upper bound can be closed using a more advanced configuration algorithm (examples are given in [164], [165]), but at the expense of increased computational complexity.

The importance of properly configuring the surface is also illustrated in Figure 9.19. The dashed curve considers the average capacity over random phase-shift configurations with independent uniformly distributed phases from $[-\pi, \pi)$, while the dotted curve considers the absence of a surface (i.e., it is replaced by an absorbing material). There is barely any difference between these curves, but there is a huge performance gap compared to the power iteration method. Hence, deploying a reflecting surface in this setup is only meaningful if it is configured to beamform the signal toward the receiver. The capacity is increased by 2–3 times when doing that, which results in large bit rate differences when many MHz of bandwidth are used.

9.4 MIMO Applications of Reconfigurable Surfaces

Reconfigurable surfaces can also be used to enhance MIMO channels, but there are limitations to what can be achieved since all signals reflected by a particular metaatom are phase-shifted identically. To shed light on the fundamentals using simple notation, we return to the narrowband case in this section and consider three different scenarios: point-to-point MIMO and multi-user MIMO communications, as well as multi-antenna target detection.

9.4.1 Enhanced Point-to-Point MIMO Communication

In the point-to-point SISO case considered previously in this chapter, the end-to-end channel coefficient was expressed in (9.23) as $h = h_{\mathrm{s}} + \mathbf{h}_{\mathrm{r}}^{\mathrm{T}}\mathbf{D}_{\psi}\mathbf{h}_{\mathrm{t}}$. In a point-to-point MIMO scenario with K transmit antennas and M receive antennas, the channel matrix $\mathbf{H} \in \mathbb{C}^{M\times K}$ can be expressed similarly as

$$\mathbf{H} = \mathbf{H}_{\mathrm{s}} + \mathbf{H}_{\mathrm{r}}\mathbf{D}_{\psi}\mathbf{H}_{\mathrm{t}}, \tag{9.81}$$

where $\mathbf{H}_{\mathrm{s}} \in \mathbb{C}^{M\times K}$ is the static channel, $\mathbf{H}_{\mathrm{t}} \in \mathbb{C}^{N\times K}$ is the channel from the transmitter to the surface, and $\mathbf{H}_{\mathrm{r}} \in \mathbb{C}^{M\times N}$ is the channel from the surface to the receiver. These channel matrices can be modeled just as any other MIMO channels because the propagation to/from the surface is the same as if the array of metaatoms were an array of antennas.

The reflection matrix does not change in the MIMO case but is defined as in (9.17): $\mathbf{D}_{\psi} = \mathrm{diag}\left(e^{\mathrm{j}\psi_1}, \dots, e^{\mathrm{j}\psi_N}\right)$. This matrix can adjust how the matrices $\mathbf{H}_{\mathrm{r}}$ and $\mathbf{H}_{\mathrm{t}}$ are multiplied in (9.81), but the flexibility is limited since it contains N controllable phase parameters while there are respectively MN and KN coefficients in the channel matrices. These numbers were equal in the SISO case with $M = K = 1$, while there are many more channel coefficients than controllable parameters in the MIMO case.

When matrices are multiplied, the rank of the resulting matrix is always smaller or equal to the minimum rank of the individual matrices. The reflection matrix always has full rank. However, the rank of $\mathbf{H}_{\mathrm{r}}\mathbf{D}_{\psi}\mathbf{H}_{\mathrm{t}}$ cannot surpass the minimum rank of the channel matrices $\mathbf{H}_{\mathrm{r}}$ and $\mathbf{H}_{\mathrm{t}}$. This implies that the reconfigurable surface cannot improve the channel rank in any dramatic fashion. However, it can be configured to improve specific singular values, match the strongest channel dimensions from the two channel matrices, and ensure that the static and configurable terms in (9.81) fit well together. Since many possibilities exist, we should identify the surface configuration that maximizes the MIMO channel capacity. In this section, we will first provide some geometrical insights into what can be achieved and then derive a general algorithm that iteratively refines the configuration to increase capacity.

Example 9.8. Suppose the surface is deployed to have far-field LOS channels to both the transmitter and receiver. How should the surface be configured to maximize the total end-to-end channel gain $\|\mathbf{H}\|_{\mathrm{F}}^2$?

The matrices $\mathbf{H}_{\mathrm{r}}$ and $\mathbf{H}_{\mathrm{t}}$ have rank one under these conditions, as explained in Section 4.4.1. They could be expressed as outer products of array response vectors, but for notational convenience, we will express them as

$$\mathbf{H}_{\mathrm{r}} = \mathbf{a}_{\mathrm{r}}\mathbf{b}_{\mathrm{r}}^{\mathrm{T}}, \quad \mathbf{H}_{\mathrm{t}} = \mathbf{a}_{\mathrm{t}}\mathbf{b}_{\mathrm{t}}^{\mathrm{T}}, \tag{9.82}$$

where $\mathbf{a}_{\mathrm{r}} \in \mathbb{C}^M$, $\mathbf{b}_{\mathrm{r}} = [b_{\mathrm{r},1}, \ldots, b_{\mathrm{r},N}]^{\mathrm{T}} \in \mathbb{C}^N$, $\mathbf{a}_{\mathrm{t}} = [a_{\mathrm{t},1}, \ldots, a_{\mathrm{t},N}]^{\mathrm{T}} \in \mathbb{C}^N$, and $\mathbf{b}_{\mathrm{t}} \in \mathbb{C}^K$ are vectors. We can then simplify (9.81) as

$$\mathbf{H} = \mathbf{H}_{\mathrm{s}} + \mathbf{a}_{\mathrm{r}} \underbrace{\mathbf{b}_{\mathrm{r}}^{\mathrm{T}}\mathbf{D}_{\psi}\mathbf{a}_{\mathrm{t}}}_{=\alpha} \mathbf{b}_{\mathrm{t}}^{\mathrm{T}}, \tag{9.83}$$

where $\alpha = \mathbf{b}_{\mathrm{r}}^{\mathrm{T}}\mathbf{D}_{\psi}\mathbf{a}_{\mathrm{t}} \in \mathbb{C}$ is a scalar. The second term adds the rank-one matrix $\mathbf{a}_{\mathrm{r}}\mathbf{b}_{\mathrm{t}}^{\mathrm{T}}$ to the static channel with a scaling that is determined by the reflection matrix. The total end-to-end channel gain can be rewritten using (5.88) as

$$\begin{aligned}
\|\mathbf{H}\|_{\mathrm{F}}^2 &= \mathrm{tr}\left(\mathbf{H}^{\mathrm{H}}\mathbf{H}\right) \\
&= \mathrm{tr}\left(\mathbf{H}_{\mathrm{s}}^{\mathrm{H}}\mathbf{H}_{\mathrm{s}}\right) + |\alpha|^2\mathrm{tr}\left(\mathbf{b}_{\mathrm{t}}^{*}\mathbf{a}_{\mathrm{r}}^{\mathrm{H}}\mathbf{a}_{\mathrm{r}}\mathbf{b}_{\mathrm{t}}^{\mathrm{T}}\right) + \mathrm{tr}\left(\alpha\mathbf{H}_{\mathrm{s}}^{\mathrm{H}}\mathbf{a}_{\mathrm{r}}\mathbf{b}_{\mathrm{t}}^{\mathrm{T}}\right) + \mathrm{tr}\left(\alpha^*\mathbf{b}_{\mathrm{t}}^{*}\mathbf{a}_{\mathrm{r}}^{\mathrm{H}}\mathbf{H}_{\mathrm{s}}\right) \\
&= \|\mathbf{H}_{\mathrm{s}}\|_{\mathrm{F}}^2 + |\alpha|^2\|\mathbf{a}_{\mathrm{r}}\|^2\|\mathbf{b}_{\mathrm{t}}\|^2 + 2\Re\left(\alpha\mathbf{b}_{\mathrm{t}}^{\mathrm{T}}\mathbf{H}_{\mathrm{s}}^{\mathrm{H}}\mathbf{a}_{\mathrm{r}}\right),
\end{aligned} \tag{9.84}$$

where the last equality follows from (2.52) that states how one can shift the order of matrices in the trace function. The final expression is maximized when $|\alpha|$ takes its largest possible value while its phase makes the last term positive. This happens when $\psi_n = -\arg(\mathbf{b}_{\mathrm{t}}^{\mathrm{T}}\mathbf{H}_{\mathrm{s}}^{\mathrm{H}}\mathbf{a}_{\mathrm{r}}) - \arg(b_{\mathrm{r},n}a_{\mathrm{t},n}) + 2\pi k_n$, where k_n is the integer that gives $\psi_n \in [-\pi, \pi)$, for $n = 1, \ldots, N$. When the static channel is weak, maximizing $\|\mathbf{H}\|_{\mathrm{F}}^2$ is approximately equivalent to maximizing the channel capacity because there will only be one strong singular value, which is amplified using the surface.

As noted earlier in this chapter, deploying the reconfigurable surface to have LOS channels to the transmitter and receiver is desirable. However, in contrast to the example, the channel matrices will not have rank one in practice due to multipath propagation. This makes it harder to compute the optimal phase-shift configuration directly. The design principle remains the same, as can be showcased using the beamspace representation. Suppose the transmitter and receiver are equipped with half-wavelength-spaced ULAs. As explained in Section 5.6.2, we can then transform the channel matrix to the beamspace by multiplying by DFT matrices from the left and the right:

$$\breve{\mathbf{H}} = \mathbf{F}_M^{\mathrm{H}}\mathbf{H}\mathbf{F}_K^{*} = \mathbf{F}_M^{\mathrm{H}}\mathbf{H}_{\mathrm{s}}\mathbf{F}_K^{*} + \mathbf{F}_M^{\mathrm{H}}\mathbf{H}_{\mathrm{r}}\mathbf{D}_{\psi}\mathbf{H}_{\mathrm{t}}\mathbf{F}_K^{*}. \tag{9.85}$$

The first term represents all the static multipath clusters, while the second

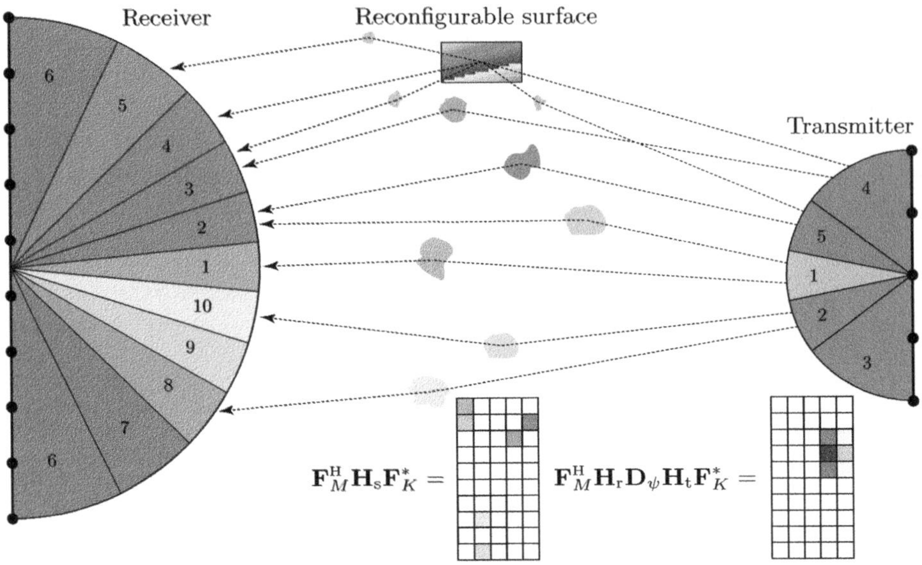

Figure 9.20: A transmitter with a half-wavelength-spaced ULA with $K = 5$ antennas communicates with a receiver having a half-wavelength-spaced ULA with $M = 10$ antennas. The communication is aided by a reconfigurable surface that adds extra paths to the MIMO channel. In the beamspace representation, these extra paths are concentrated around a few angular directions since the surface mostly interacts with multipath components in its vicinity. This is a continuation of Figure 5.33 where no reconfigurable surfaces existed. Note that the transmitter and receiver sizes are exaggerated compared to the propagation distances.

term represents the propagation paths affected by the reconfigurable surface. Figure 9.20 exemplifies how these matrices depend on the angular geometry and is a continuation of Figure 5.33, which considered the same setup without a reconfigurable surface. Each of the six static multipath clusters contributes to the entry of $\mathbf{F}_M^{\mathrm{H}}\mathbf{H}_{\mathrm{s}}\mathbf{F}_K^*$ with the matching color. A white entry means its value is nearly zero because no propagation path connects that pair of transmit/receive directions. Since we consider deployment with LOS to the transmitter and receiver, the reconfigurable surface mainly contributes to one entry, determined by its physical location. However, it could slightly affect a few neighboring entries using the multipath clusters around it, as illustrated by the brightness of the coloring (brighter means smaller).

The rank of $\mathbf{H}$ is the same as the rank of $\breve{\mathbf{H}}$. The static channel matrix describes the contributions from all multipath clusters that create paths between the transmitter and the receiver. Clusters seen from very different angles barely interact and contribute to different singular values in $\breve{\mathbf{H}}$. The reconfigurable surface can increase the channel rank by creating new non-zero entries. Since the surface is deployed at a specific location, it will primarily interact with the propagation environment in its vicinity. Hence, all the new propagation paths it creates have similar angles from the transmitter's and receiver's perspectives.

This implies that we should expect the surface to mainly contribute to one or a few singular values in $\breve{\mathbf{H}}$. The reconfigurable surface can raise the channel capacity by selecting a good phase-shift configuration, but the increase will resemble the SISO case since only one channel dimension is improved. Multiple reconfigurable surfaces deployed at physically diverse locations are generally required to enhance multiple singular values substantially. An analogy can be made with the analog/hybrid beamforming considered in Sections 7.3.1 and 7.3.2: a single surface is like an analog beamforming architecture that can only receive/transmit one signal, while multiple surfaces are like the hybrid beamforming architecture that can jointly receive/transmit as many signals as there are surfaces. While this is generally true, there are specific scenarios where the surface contributes to many singular values [169]. This happens when the arrays and surface are very large compared to the wavelength so that spherical wavefronts can be utilized to achieve high-rank channels via the surface. The enabling factor is the same as in Section 4.4.3, where we showed how to achieve full-rank LOS channels by making the antenna spacing sufficiently large compared to the propagation distance. Similarly, one can make the reconfigurable surface so large compared to the propagation distances that it acts as multiple surfaces.

For a given MIMO channel matrix, the capacity is achieved by transmitting along the right singular vectors and allocating power using water-filling (see Theorem 3.1). If we modify the channel matrix by refining the reflection matrix $\mathbf{D}_{\psi}$, the singular vectors and values will change, and so will the capacity-achieving precoding. We will describe an algorithm from [170] that progressively improves the MIMO capacity by iterating between updating the precoding/water-filling and reflection matrix. We recall from (3.100) that the capacity for a given channel matrix $\mathbf{H}$ can be expressed as

$$C = \log_2\left(\det\left(\mathbf{I}_M + \frac{1}{N_0}\mathbf{H}\mathbf{V}\mathbf{Q}^{\mathrm{opt}}\mathbf{V}^{\mathrm{H}}\mathbf{H}^{\mathrm{H}}\right)\right). \tag{9.86}$$

We introduce the notation $\mathbf{H}_{\mathrm{r}} = [\mathbf{h}_{\mathrm{r},1}, \ldots, \mathbf{h}_{\mathrm{r},N}]$ and $\mathbf{H}_{\mathrm{t}} = [\vec{\mathbf{h}}_{\mathrm{t},1}, \ldots, \vec{\mathbf{h}}_{\mathrm{t},N}]^{\mathrm{T}}$, where $\vec{\mathbf{h}}_{\mathrm{t},n}^{\mathrm{T}}$ is the nth row and the arrow notation points out that rows are horizontal. We can then rewrite the channel matrix in (9.81) as

$$\mathbf{H} = \mathbf{H}_{\mathrm{s}} + \mathbf{H}_{\mathrm{r}}\mathbf{D}_{\psi}\mathbf{H}_{\mathrm{t}} = \mathbf{H}_{\mathrm{s}} + \sum_{i=1}^{N}\mathbf{h}_{\mathrm{r},i}e^{\mathrm{j}\psi_i}\vec{\mathbf{h}}_{\mathrm{t},i}^{\mathrm{T}} = \mathbf{H}_n + e^{\mathrm{j}\psi_n}\mathbf{h}_{\mathrm{r},n}\vec{\mathbf{h}}_{\mathrm{t},n}^{\mathrm{T}}, \tag{9.87}$$

where $\mathbf{H}_n = \mathbf{H}_{\mathrm{s}} + \sum_{i=1, i\neq n}^{N}\mathbf{h}_{\mathrm{r},i}e^{\mathrm{j}\psi_i}\vec{\mathbf{h}}_{\mathrm{t},i}^{\mathrm{T}}$ contains all the terms except the one involving ψ_n. We want to reconfigure this phase-shift to increase the capacity when all other parameters are fixed. By substituting (9.87) into (9.86), we can express the capacity as

$$\begin{aligned}
&\log_2\left(\det\left(\mathbf{I}_M + \frac{1}{N_0}\left(\mathbf{H}_n + e^{\mathrm{j}\psi_n}\mathbf{h}_{\mathrm{r},n}\vec{\mathbf{h}}_{\mathrm{t},n}^{\mathrm{T}}\right)\mathbf{V}\mathbf{Q}^{\mathrm{opt}}\mathbf{V}^{\mathrm{H}}\left(\mathbf{H}_n + e^{\mathrm{j}\psi_n}\mathbf{h}_{\mathrm{r},n}\vec{\mathbf{h}}_{\mathrm{t},n}^{\mathrm{T}}\right)^{\mathrm{H}}\right)\right)\\
&= \log_2\left(\det\left(\mathbf{A}_n + e^{\mathrm{j}\psi_n}\mathbf{h}_{\mathrm{r},n}\mathbf{b}_n^{\mathrm{H}} + e^{-\mathrm{j}\psi_n}\mathbf{b}_n\mathbf{h}_{\mathrm{r},n}^{\mathrm{H}}\right)\right),\\
&= \log_2\left(\det\left(\mathbf{A}_n\right)\right) + \log_2\left(\det\left(\mathbf{I}_M + e^{\mathrm{j}\psi_n}\mathbf{A}_n^{-1}\mathbf{h}_{\mathrm{r},n}\mathbf{b}_n^{\mathrm{H}} + e^{-\mathrm{j}\psi_n}\mathbf{A}_n^{-1}\mathbf{b}_n\mathbf{h}_{\mathrm{r},n}^{\mathrm{H}}\right)\right),
\end{aligned} \tag{9.88}$$

where the terms that are independent of ψ_n are included in

$$\mathbf{A}_n = \mathbf{I}_M + \frac{1}{N_0}\mathbf{H}_n\mathbf{V}\mathbf{Q}^{\mathrm{opt}}\mathbf{V}^{\mathrm{H}}\mathbf{H}_n^{\mathrm{H}} + \frac{1}{N_0}\mathbf{h}_{\mathrm{r},n}\vec{\mathbf{h}}_{\mathrm{t},n}^{\mathrm{T}}\mathbf{V}\mathbf{Q}^{\mathrm{opt}}\mathbf{V}^{\mathrm{H}}\vec{\mathbf{h}}_{\mathrm{t},n}^{*}\mathbf{h}_{\mathrm{r},n}^{\mathrm{H}}, \tag{9.89}$$

$$\mathbf{b}_n = \frac{1}{N_0}\mathbf{H}_n\mathbf{V}\mathbf{Q}^{\mathrm{opt}}\mathbf{V}^{\mathrm{H}}\vec{\mathbf{h}}_{\mathrm{t},n}^{*}. \tag{9.90}$$

Only the determinant in the second term of (9.88) depends on the phase-shift. This determinant can be computed as

$$\begin{aligned}
&\det\left(\mathbf{I}_M + \begin{bmatrix}\mathbf{A}_n^{-1}\mathbf{h}_{\mathrm{r},n} & \mathbf{A}_n^{-1}\mathbf{b}_n\end{bmatrix}\begin{bmatrix}e^{\mathrm{j}\psi_n}\mathbf{b}_n^{\mathrm{H}}\\ e^{-\mathrm{j}\psi_n}\mathbf{h}_{\mathrm{r},n}^{\mathrm{H}}\end{bmatrix}\right)\\
&= \det\left(\mathbf{I}_2 + \begin{bmatrix}e^{\mathrm{j}\psi_n}\mathbf{b}_n^{\mathrm{H}}\\ e^{-\mathrm{j}\psi_n}\mathbf{h}_{\mathrm{r},n}^{\mathrm{H}}\end{bmatrix}\begin{bmatrix}\mathbf{A}_n^{-1}\mathbf{h}_{\mathrm{r},n} & \mathbf{A}_n^{-1}\mathbf{b}_n\end{bmatrix}\right)\\
&= (1 + e^{\mathrm{j}\psi_n}\mathbf{b}_n^{\mathrm{H}}\mathbf{A}_n^{-1}\mathbf{h}_{\mathrm{r},n})(1 + e^{-\mathrm{j}\psi_n}\mathbf{h}_{\mathrm{r},n}^{\mathrm{H}}\mathbf{A}_n^{-1}\mathbf{b}_n) - \mathbf{b}_n^{\mathrm{H}}\mathbf{A}_n^{-1}\mathbf{b}_n\mathbf{h}_{\mathrm{r},n}^{\mathrm{H}}\mathbf{A}_n^{-1}\mathbf{h}_{\mathrm{r},n}\\
&= e^{\mathrm{j}\psi_n}\mathbf{b}_n^{\mathrm{H}}\mathbf{A}_n^{-1}\mathbf{h}_{\mathrm{r},n} + e^{-\mathrm{j}\psi_n}\mathbf{h}_{\mathrm{r},n}^{\mathrm{H}}\mathbf{A}_n^{-1}\mathbf{b}_n + \text{constants},
\end{aligned} \tag{9.91}$$

where the first equality follows from Sylvester's determinant theorem in (2.53), and we then compute the determinant for the resulting 2×2 matrix. The final expression in (9.91) is maximized when the first two terms are positive, which is achieved by

$$\psi_n = -\arg(\mathbf{b}_n^{\mathrm{H}}\mathbf{A}_n^{-1}\mathbf{h}_{\mathrm{r},n}). \tag{9.92}$$

We can utilize this result to obtain the iterative procedure described in Algorithm 9.2. The algorithm begins by computing the capacity-achieving signal covariance matrix $\mathbf{V}\mathbf{Q}^{\mathrm{opt}}\mathbf{V}^{\mathrm{H}}$ for an initial set of phase-shifts. It then refines the N phase-shifts sequentially using (9.92). When this is done, the capacity-achieving signal covariance matrix is recomputed for the new channel matrix obtained with the new reflection matrix, and the procedure is repeated L times. Each step in the algorithm either improves the achievable rate or keeps it fixed because we can always choose not to modify the phase. The rate will eventually converge when no further changes are beneficial. However, a consequence of sequential optimization is that the algorithm might not converge to the best possible configuration but only a locally optimal solution where one cannot further increase the capacity unless multiple phases are

Algorithm 9.2 Reconfigurable surface configuration for point-to-point MIMO capacity maximization.

Initialization: Set $\psi_1, \ldots, \psi_N$ randomly and select the number of iterations L
for $i = 1, \ldots, L$ **do**
  Compute the capacity-achieving covariance matrix $\mathbf{V}\mathbf{Q}^{\mathrm{opt}}\mathbf{V}^{\mathrm{H}}$ for the channel matrix in (9.81) with $\mathbf{D}_{\psi} = \mathrm{diag}(e^{\mathrm{j}\psi_1}, \ldots, e^{\mathrm{j}\psi_N})$
  for $n = 1, \ldots, N$ **do**
    Compute $\mathbf{A}_n$ in (9.89) and $\mathbf{b}_n$ in (9.90) for fixed $\psi_1, \ldots, \psi_N$
    $\psi_n \leftarrow -\arg(\mathbf{b}_n^{\mathrm{H}} \mathbf{A}_n^{-1} \mathbf{h}_{\mathrm{r},n})$
  end for
end for
Output: $\psi_1, \ldots, \psi_N$

updated simultaneously.[6] The channel matrices must be estimated before running Algorithm 9.2. No extra wireless signaling is required while running the algorithm, which can be executed at the receiver. Hence, the procedure shown in Figure 9.14 can still be followed: the transmitter sends pilots while the surface switches between predefined configurations, and then the receiver computes the preferred configuration and sends it to the surface.

Figure 9.21 shows how the capacity is improved with the iteration index from Algorithm 9.2 in a point-to-point MIMO scenario with $M = K \in \{1, 2, 4, 8\}$ antennas and $N = 100$ metaatoms. The SNR of the static path is $0\,$dB and $\mathbf{H}_{\mathrm{s}}$ has i.i.d. Rayleigh fading entries. The channel matrices $\mathbf{H}_{\mathrm{r}}$ and $\mathbf{H}_{\mathrm{t}}$ via the surface are subject to Rician fading with the κ-factor $\kappa = 10$ and the NLOS part having an i.i.d. Rayleigh fading distribution (see Example 5.18). The cascaded path via a single metaatom has the SNR $-10\,$dB. The results are averaged over many channel realizations.

The iteration index 0 in Figure 9.21 represents the initial state when the phase-shifts are uniformly distributed between 0 and 2π, thereby approximating diffuse scattering. The first iteration of Algorithm 9.2 leads to a substantial capacity improvement, while only minor improvements occur in the subsequent iterations. The vertical gaps between the curves grow when comparing the first and last points on the curves. This shows that a system with more antennas benefits slightly more from having a well-configured surface, but the difference is small because the surface mainly contributes to one singular value. We use the optimal configuration from Corollary 9.1 when considering the SISO case with $M = K = 1$, which is why that curve does not vary with

[6] The initial phase-shifts determine which configuration that Algorithm 9.2 converges to. A simple way to explore if better solutions exist is to consider multiple random phase-shift initializations and compare the capacity values they converge to.

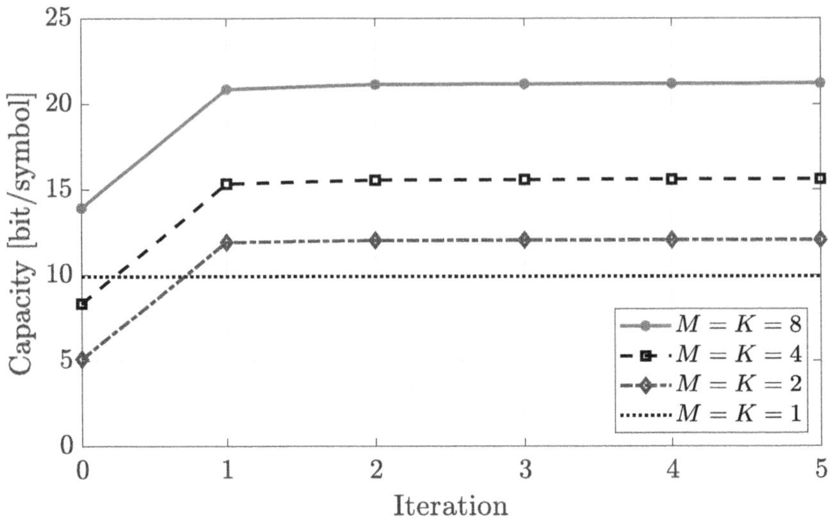

Figure 9.21: The point-to-point MIMO capacity as a function of the iteration index when running Algorithm 9.2 to iteratively select the phase-shifts of the reconfigurable surface to increase the capacity. The index 0 represents the initial random configuration. The optimal configuration is directly used in the SISO case ($M = K = 1$) since it is known in closed form.

the iteration index. Interestingly, the optimal SISO configuration leads to a higher capacity than the initial capacity in the 2×2 and 4×4 MIMO setups, reiterating the importance of correctly configuring the surface. In summary, the deployment of a reconfigurable surface can greatly improve the capacity of a point-to-point MIMO system.

9.4.2 Enhanced Multi-User MIMO Communication

A reconfigurable surface can also improve the communication performance over multi-user MIMO channels. As explained in Chapter 6, the precoding and combining differ substantially from the point-to-point case because users are not collaborating in the signal processing and measure their capacity separately. Nevertheless, the sum capacity expression in multi-user MIMO resembles the capacity expression in point-to-point MIMO, which implies that we can use similar algorithms to optimize the reconfigurable surface.

The uplink sum capacity in a multi-user MIMO system with K single-antenna users and M antennas at the base station is given in (6.49) as

$$\log_2\left(\det\left(\mathbf{I}_M + \frac{q}{N_0}\mathbf{H}\mathbf{H}^{\mathrm{H}}\right)\right) \quad \text{bit/symbol,} \tag{9.93}$$

where $\mathbf{H} = [\mathbf{h}_1, \dots, \mathbf{h}_K] \in \mathbb{C}^{M \times K}$ is the channel matrix and $q = P/B$ is the signal energy per symbol. When the uplink is aided by a reconfigurable surface with N metaatoms, the channel $\mathbf{h}_k \in \mathbb{C}^M$ from user k can be expressed as

$$\mathbf{h}_k = \mathbf{h}_{\mathrm{s},k} + \mathbf{H}_\mathrm{r}\mathbf{D}_\psi \mathbf{h}_{\mathrm{t},k}, \tag{9.94}$$

where $\mathbf{h}_{\mathrm{s},k} \in \mathbb{C}^M$ is the static channel, $\mathbf{h}_{\mathrm{t},k} \in \mathbb{C}^N$ is the channel from the user to the surface, and $\mathbf{H}_\mathrm{r} \in \mathbb{C}^{M\times N}$ is the channel from the surface to the receiver. This is the SIMO counterpart to the channel model in (9.81). The term $\mathbf{H}_\mathrm{r}\mathbf{D}_\psi \mathbf{h}_{\mathrm{t},k}$ in (9.94) can be viewed as the projection of the user-specific channel vector $\mathbf{h}_{\mathrm{t},k}$ onto the span of the matrix $\mathbf{H}_\mathrm{r}\mathbf{D}_\psi$, which is the same for all users but controllable using the reflection matrix $\mathbf{D}_\psi = \mathrm{diag}(e^{\mathrm{j}\psi_1}, \ldots, e^{\mathrm{j}\psi_N})$. The combined channel matrix of all users becomes

$$\mathbf{H} = [\mathbf{h}_1, \ldots, \mathbf{h}_K] = \underbrace{[\mathbf{h}_{\mathrm{s},1}, \ldots, \mathbf{h}_{\mathrm{s},K}]}_{=\mathbf{H}_\mathrm{s}} + \mathbf{H}_\mathrm{r}\mathbf{D}_\psi \underbrace{[\mathbf{h}_{\mathrm{t},1}, \ldots, \mathbf{h}_{\mathrm{t},K}]}_{=\mathbf{H}_\mathrm{t}}, \tag{9.95}$$

which has the same form $\mathbf{H} = \mathbf{H}_\mathrm{s} + \mathbf{H}_\mathrm{r}\mathbf{D}_\psi\mathbf{H}_\mathrm{t}$ as in the point-to-point MIMO case. In particular, the reflection matrix enters into the equation identically.

Example 9.9. Suppose the surface is deployed to have a far-field LOS channel to the base station. How can it modify the user channels in this case?

The matrix $\mathbf{H}_\mathrm{r}$ has rank one under these conditions, as explained in Section 4.4.1, and can be expressed as $\mathbf{H}_\mathrm{r} = \mathbf{a}_\mathrm{r}\mathbf{b}_\mathrm{r}^\mathrm{T}$ for some vectors $\mathbf{a}_\mathrm{r} \in \mathbb{C}^M$ and $\mathbf{b}_\mathrm{r} \in \mathbb{C}^N$. The channel of user k in (9.94) then becomes

$$\mathbf{h}_k = \mathbf{h}_{\mathrm{s},k} + \mathbf{a}_\mathrm{r} \underbrace{\mathbf{b}_\mathrm{r}^\mathrm{T}\mathbf{D}_\psi \mathbf{h}_{\mathrm{t},k}}_{=\alpha_k}. \tag{9.96}$$

This implies that the surface adds a component $\alpha_k \mathbf{a}_\mathrm{r}$ to the static channel vector, where only the complex scaling factor α_k can be controlled and depends on the user index. In case the static channels are blocked (i.e., $\mathbf{h}_{\mathrm{s},k} = \mathbf{0}$ for all k), the K channel vectors are parallel. We cannot suppress interference under such circumstances; thus, FDMA achieves the same sum capacity as multi-user MIMO in this case. In conclusion, the reconfigurable surface cannot enable multi-user MIMO communications on its own, but it can improve performance by making the channels $\mathbf{h}_k$ more diverse than the original static channels by adding the components $\alpha_k \mathbf{a}_\mathrm{r}$.

Since there are K user capacities to consider in multi-user MIMO systems, different phase-shift configurations are preferred for different users. In other words, the reconfigurable surface bends the shape of the capacity region, and there is typically no configuration that results in a region that is larger than all other achievable regions in all user dimensions. In this section, we will concentrate on maximizing the sum capacity in (9.93). Similarly to the last section, we will develop an iterative algorithm that updates one of the N phase-shifts at a time to increase the capacity. When refining the phase ψ_n of metaatom n, it is convenient to express the channel matrix $\mathbf{H} = \mathbf{H}_\mathrm{s} + \mathbf{H}_\mathrm{r}\mathbf{D}_\psi\mathbf{H}_\mathrm{t}$ in (9.95) as

$$\mathbf{H} = \mathbf{H}_\mathrm{s} + \sum_{i=1}^{N} \mathbf{h}_{\mathrm{r},i} e^{\mathrm{j}\psi_i} \vec{\mathbf{h}}_{\mathrm{t},i}^{\mathrm{T}} = \underbrace{\mathbf{H}_\mathrm{s} + \sum_{i=1, i\neq n}^{N} \mathbf{h}_{\mathrm{r},i} e^{\mathrm{j}\psi_i} \vec{\mathbf{h}}_{\mathrm{t},i}^{\mathrm{T}}}_{=\mathbf{H}_n} + e^{\mathrm{j}\psi_n} \mathbf{h}_{\mathrm{r},n} \vec{\mathbf{h}}_{\mathrm{t},n}^{\mathrm{T}}, \tag{9.97}$$

using the notation $\mathbf{H}_\mathrm{r} = [\mathbf{h}_{\mathrm{r},1}, \ldots, \mathbf{h}_{\mathrm{r},N}]$ and $\mathbf{H}_\mathrm{t} = [\vec{\mathbf{h}}_{\mathrm{t},1}, \ldots, \vec{\mathbf{h}}_{\mathrm{t},N}]^\mathrm{T}$. Note that $\vec{\mathbf{h}}_{\mathrm{t},n}^{\mathrm{T}}$ is the nth row of $\mathbf{H}_\mathrm{t}$ and differs from the user channel vector $\mathbf{h}_{\mathrm{t},k}$ appearing as the kth column of the matrix. By substituting (9.97) into (9.93), we can express the sum capacity as

$$\begin{aligned}
&\log_2\left(\det\left(\mathbf{I}_M + \frac{q}{N_0}\left(\mathbf{H}_n + e^{\mathrm{j}\psi_n}\mathbf{h}_{\mathrm{r},n}\vec{\mathbf{h}}_{\mathrm{t},n}^{\mathrm{T}}\right)\left(\mathbf{H}_n + e^{\mathrm{j}\psi_n}\mathbf{h}_{\mathrm{r},n}\vec{\mathbf{h}}_{\mathrm{t},n}^{\mathrm{T}}\right)^{\mathrm{H}}\right)\right) \\
&= \log_2\left(\det\left(\mathbf{A}_n + e^{\mathrm{j}\psi_n}\mathbf{h}_{\mathrm{r},n}\mathbf{b}_n^{\mathrm{H}} + e^{-\mathrm{j}\psi_n}\mathbf{b}_n\mathbf{h}_{\mathrm{r},n}^{\mathrm{H}}\right)\right), \\
&= \log_2\left(\det\left(\mathbf{A}_n\right)\right) + \log_2\left(\det\left(\mathbf{I}_M + e^{\mathrm{j}\psi_n}\mathbf{A}_n^{-1}\mathbf{h}_{\mathrm{r},n}\mathbf{b}_n^{\mathrm{H}} + e^{-\mathrm{j}\psi_n}\mathbf{A}_n^{-1}\mathbf{b}_n\mathbf{h}_{\mathrm{r},n}^{\mathrm{H}}\right)\right),
\end{aligned} \tag{9.98}$$

where the terms that are independent of ψ_n are included in

$$\mathbf{b}_n = \frac{q}{N_0}\mathbf{H}_n\vec{\mathbf{h}}_{\mathrm{t},n}^{*}, \quad \mathbf{A}_n = \mathbf{I}_M + \frac{q}{N_0}\mathbf{H}_n\mathbf{H}_n^{\mathrm{H}} + \frac{q}{N_0}\mathbf{h}_{\mathrm{r},n}\vec{\mathbf{h}}_{\mathrm{t},n}^{\mathrm{T}}\vec{\mathbf{h}}_{\mathrm{t},n}^{*}\mathbf{h}_{\mathrm{r},n}^{\mathrm{H}}. \tag{9.99}$$

It remains to select the phase-shift to maximize the second determinant in (9.98), and this problem has the same form as in (9.91) of the point-to-point MIMO case. Hence, the optimal phase is obtained from (9.92) as

$$\psi_n = -\arg(\mathbf{b}_n^{\mathrm{H}}\mathbf{A}_n^{-1}\mathbf{h}_{\mathrm{r},n}), \tag{9.100}$$

but using the expressions for $\mathbf{b}_n$ and $\mathbf{A}_n$ defined above. By sequentially updating the N phases using (9.100), we obtain Algorithm 9.3. This algorithm resembles Algorithm 9.2 for the point-to-point MIMO case, but a key difference is that the precoding is not updated in multi-user MIMO because the sum capacity is always achieved when the users transmit their signals using maximum power. Each step in the algorithm either improves the sum capacity or keeps it fixed because we can always choose not to modify the phase; thus, the sum capacity gradually increases and converges to a final value. We let L denote the predefined number of iterations to consider, but the algorithm can also be terminated earlier when the sum capacity has not been improved much from one iteration to the next. Although the sum capacity improves monotonically, there is no guarantee that the algorithm will converge to the best conceivable configuration because the variables are optimized sequentially rather than jointly.

Figure 9.22 shows how the sum capacity of an uplink multi-user MIMO system increases with the number of metaatoms. There are $K = 4$ users,

Algorithm 9.3 Reconfigurable surface configuration for uplink multi-user MIMO sum capacity maximization.

Initialization: Set $\psi_1, \dots, \psi_N$ randomly and select the number of iterations L
for $i = 1, \dots, L$ **do**
    for $n = 1, \dots, N$ **do**
        Compute $\mathbf{A}_n$ and $\mathbf{b}_n$ in (9.99) using current $\psi_1, \dots, \psi_N$
        $\psi_n \leftarrow -\arg(\mathbf{b}_n^{\mathrm{H}} \mathbf{A}_n^{-1} \mathbf{h}_{\mathrm{r},n})$
    end for
end for
Output: $\psi_1, \dots, \psi_N$

$M = 10$ receive antennas, and the static channel is modeled as in Figure 6.16. The channel matrix $\mathbf{H}_{\mathrm{r}}$ between the base station and reconfigurable surface is modeled as Rician fading with the κ-factor $\kappa = 10$ and the channels between the users and surface are subject to i.i.d. Rayleigh fading. SNR of the static path is 0 dB, while the cascaded path via a single metaatom has the SNR −20 dB. The results are averaged over many channel realizations. We notice that the sum capacity grows rapidly with the number of metaatoms when Algorithm 9.3 is used; hence, deploying the reconfigurable surface in this particular setup makes a great difference. $L = 1$ iteration of the algorithm is sufficient to outperform the initial configuration with random phase-shifts. Further capacity improvements are achieved by running $L = 5$ iterations of the algorithm, especially when there are many metaatoms to configure.

We have focused on the uplink thus far, but the results are also useful for the downlink because the uplink-downlink duality implies that we can achieve the same user rates in both directions. Hence, if the surface is configured to provide a high uplink sum capacity, we can achieve the same downlink sum rate using the same power without changing the surface configuration. However, this will generally not be the downlink sum capacity because we might have a different total downlink transmit power and can allocate it arbitrarily between the users. The downlink sum capacity was stated in (6.122) as the problem of maximizing the sum rate in the virtual uplink with respect to the virtual uplink powers, and it can be solved efficiently using convex optimization tools. It is straightforward to devise an iterative algorithm that switches between solving (6.122) for given user channels and enhancing the channels using Algorithm 9.3 for given virtual uplink powers.

The uplink sum capacity requires SIC, while the downlink sum capacity is achieved using DPC. It is easier to implement uplink and downlink multi-user MIMO systems with linear signal processing, but unfortunately, it comes at the price of more complex parameter optimization problems. For example, the

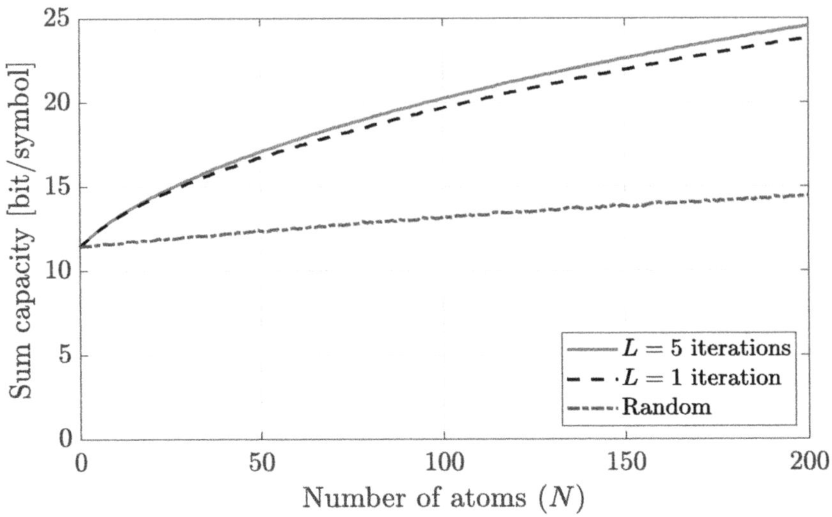

Figure 9.22: The sum capacity of an uplink multi-user MIMO system that is aided by a reconfigurable surface with a varying number of metaatoms. The phase-shift configuration is either selected randomly or by running Algorithm 9.3 with $L = 1$ or $L = 5$ iterations.

uplink sum capacity is achieved when all users transmit with their maximum power, while the maximum uplink sum rate with linear combining requires power control optimization (as exemplified in Section 6.3.6). Similarly, the parameters required to achieve the downlink sum capacity are obtained by solving the convex optimization problem in (6.122), while the linear precoding that maximizes the downlink sum rate can only be computed using high-complexity global optimization algorithms [84], [85]. The structure of the rate expressions causes increased complexity and makes phase-shift optimization more complicated when a reconfigurable surface supports a multi-user MIMO system that employs linear processing. We refer to [155], [171], [172] for further details and solutions to these problems. The bottom line is that reconfigurable surfaces can improve the user rates in multi-user MIMO systems, and many algorithms for phase-shift optimization can be developed for various utility functions and kinds of signal processing for data transmission and reception.

9.4.3 Enhanced Target Detection

A reconfigurable surface can also be used to improve the wireless channel properties for sensing applications [173], [174], particularly to increase the SNR and reliability. To exemplify this, we will consider a mono-static target detection scenario, where a multi-antenna radar system must determine whether a target exists at a specific location or not. A reconfigurable surface is deployed in the same area, and there are free-space LOS channels between the different locations, as illustrated in Figure 9.23. The radar transceiver

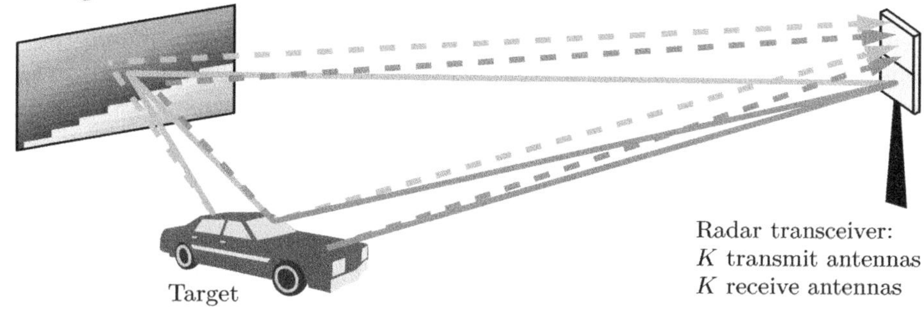

Figure 9.23: A radar transceiver with K transmit antennas and K receive antennas wants to detect the presence of a target with assistance from a reconfigurable surface with N metaatoms. There are two paths from the transmitter to the target and two paths from the target to the receiver, resulting in four propagation paths. Solid lines represent paths leading to the target and dashed lines are paths leading back to the receiver.

has K transmit antennas and K receive antennas, which are symmetrically arranged to achieve identical array response vectors. The surface consists of N metaatoms.

A predefined radar signal $\mathbf{p}x$ is transmitted using the precoding vector $\mathbf{p} \in \mathbb{C}^K$, and it reaches the target location in two ways: through the direct LOS path or via the reflection by the surface. If the target exists, it will reflect the signals, and these can reach the receiver either through the direct LOS path or via reflection by the surface. This gives rise to a total of four propagation paths from the transmitter to the receiver. We let $\mathbf{h}_{\mathrm{s}} \in \mathbb{C}^K$ denote the static LOS channel between the transmitter and target location. Furthermore, the cascaded channel from the radar to the target via the reconfigurable surface is represented by the vector

$$\mathbf{h}_{\mathrm{c}} = \mathbf{a}_{\mathrm{t}} \mathbf{b}_{\mathrm{t}}^{\mathrm{T}} \mathbf{D}_{\psi} \mathbf{h}_{\mathrm{r}}, \tag{9.101}$$

where $\mathbf{a}_{\mathrm{t}} \mathbf{b}_{\mathrm{t}}^{\mathrm{T}} \in \mathbb{C}^{K \times N}$ is the rank-one LOS channel matrix between the radar transceiver and surface, $\mathbf{D}_{\psi} \in \mathbb{C}^{N \times N}$ is the reflection matrix, and $\mathbf{h}_{\mathrm{r}} \in \mathbb{C}^N$ is the channel between the surface and target. For notational simplicity, we will not include any channel gains in $\mathbf{h}_{\mathrm{s}}$, $\mathbf{a}_{\mathrm{t}}$, $\mathbf{b}_{\mathrm{t}}$, and $\mathbf{h}_{\mathrm{r}}$ but model them separately. Hence, these are four array response vectors that describe the LOS propagation between the different locations, which implies that the squared norm of each vector equals the number of entries it has.

If the target exists, the effective end-to-end channel to the receiver is

$$\mathbf{h} = \Big(\underbrace{c_1 \mathbf{h}_{\mathrm{s}} \mathbf{h}_{\mathrm{s}}^{\mathrm{T}}}_{\text{LOS path}} + \underbrace{c_2 \mathbf{h}_{\mathrm{c}} \mathbf{h}_{\mathrm{c}}^{\mathrm{T}}}_{\text{Via surface}} + \underbrace{c_3 \mathbf{h}_{\mathrm{s}} \mathbf{h}_{\mathrm{c}}^{\mathrm{T}} + c_3 \mathbf{h}_{\mathrm{c}} \mathbf{h}_{\mathrm{s}}^{\mathrm{T}}}_{\text{Mix of LOS and surface paths}} \Big) \mathbf{p}, \tag{9.102}$$

where we included the precoding vector and $c_1 \sim \mathcal{N}_{\mathbb{C}}(0, \beta_1)$, $c_2 \sim \mathcal{N}_{\mathbb{C}}(0, \beta_2)$, and $c_3 \sim \mathcal{N}_{\mathbb{C}}(0, \beta_3)$ are three independent RCS realizations for the target, which include the channel gains as well. Multiple realizations are required because we consider signals reaching and leaving the target in different directions. However, the coefficient c_3 appears twice due to channel reciprocity, which implies that the RCS is the same when the signal propagates from the transmitter to the target and back via the surface, and when the signal travels in the opposite direction. There are four terms in (9.102) representing the four propagation paths from the transmitter to the receiver. The first term is the direct reflection by the target that would also happen in the absence of the surface. The second term is the path that reaches the target via the surface and then goes back to the receiver in the same way. The third term is the path that reaches the target via the surface and then is reflected through the LOS path, while the fourth term takes the opposite direction. The variances $\beta_1, \beta_2, \beta_3$ are generally different because they include the multiplication of the channel gains between the different locations that the radar signal passes on its way from the transmitter to the receiver. We can expect that $\beta_1 > \beta_3 > \beta_2$ since the LOS path is typically stronger than the path via a single metaatom; however, with an appropriate surface configuration, the combined effect of the N metaatoms can make a large difference for target detection.

We assume that the signal $\sqrt{P}$ is transmitted, denote the received signal by $\mathbf{y} \in \mathbb{C}^K$, and formulate the binary hypothesis test

$$\mathcal{H}_0 \quad : \quad \mathbf{y} = \mathbf{n}, \tag{9.103}$$

$$\mathcal{H}_1 \quad : \quad \mathbf{y} = \sqrt{P}\mathbf{h} + \mathbf{n}, \tag{9.104}$$

where $\mathbf{n} \sim \mathcal{N}_{\mathbb{C}}(\mathbf{0}, \sigma^2 \mathbf{I}_K)$ is the additive noise vector.

If the hypothesis $\mathcal{H}_1$ is true, the channel covariance matrix is

$$\begin{aligned}\mathbf{R} = \mathbb{E}\{\mathbf{h}\mathbf{h}^{\mathrm{H}}\} &= \beta_1 |\mathbf{h}_{\mathrm{s}}^{\mathrm{T}}\mathbf{p}|^2 \mathbf{h}_{\mathrm{s}}\mathbf{h}_{\mathrm{s}}^{\mathrm{H}} + \beta_2 |\mathbf{h}_{\mathrm{c}}^{\mathrm{T}}\mathbf{p}|^2 \mathbf{h}_{\mathrm{c}}\mathbf{h}_{\mathrm{c}}^{\mathrm{H}} \\ &+ \beta_3 \big((\mathbf{h}_{\mathrm{c}}^{\mathrm{T}}\mathbf{p})\mathbf{h}_{\mathrm{s}} + (\mathbf{h}_{\mathrm{s}}^{\mathrm{T}}\mathbf{p})\mathbf{h}_{\mathrm{c}} \big) \big((\mathbf{h}_{\mathrm{c}}^{\mathrm{T}}\mathbf{p})\mathbf{h}_{\mathrm{s}} + (\mathbf{h}_{\mathrm{s}}^{\mathrm{T}}\mathbf{p})\mathbf{h}_{\mathrm{c}} \big)^{\mathrm{H}}. \end{aligned} \tag{9.105}$$

This matrix consists of three terms, where the first term only utilizes the LOS path while the remaining two terms are created thanks to the reconfigurable surface. Each of the terms has rank one because they are outer products of vectors, but all terms are spanned by $\mathbf{h}_{\mathrm{s}}$ and $\mathbf{h}_{\mathrm{c}}$ so $\mathbf{R}$ has rank two (if $K \geq 2$).

The precoding and surface configuration can be selected to optimize this covariance matrix. The reflection matrix $\mathbf{D}_{\psi}$ only affects the norm of the cascaded channel vector $\mathbf{h}_{\mathrm{c}}$ in (9.101) since $\mathbf{b}_{\mathrm{t}}^{\mathrm{T}}\mathbf{D}_{\psi}\mathbf{h}_{\mathrm{r}}$ is a scalar. We showed in Section 9.2 that the magnitude of this term is maximized by (9.27), where the phase-shifts ensure that we sum up N phase-aligned terms. Since $\mathbf{h}_{\mathrm{r}}$ and $\mathbf{b}_{\mathrm{t}}$ are array response vectors where each entry has unit magnitude, it follows that $\mathbf{b}_{\mathrm{t}}^{\mathrm{T}}\mathbf{D}_{\psi}\mathbf{h}_{\mathrm{r}} = N$ when using the optimal configuration.

The precoding vector should be a linear combination of $\mathbf{h}_{\mathrm{s}}$ and $\mathbf{h}_{\mathrm{c}} = N\mathbf{a}_{\mathrm{t}}$ since these are the two transmission directions that lead to the target. To maximize the average SNR, we can select the precoding vector that maximizes

$$\begin{aligned}\mathbb{E}\left\{\|\mathbf{h}\|^2\right\} &= \mathrm{tr}(\mathbf{R}) \\ &= \beta_1|\mathbf{h}_{\mathrm{s}}^{\mathrm{T}}\mathbf{p}|^2\|\mathbf{h}_{\mathrm{s}}\|^2 + \beta_2|\mathbf{h}_{\mathrm{c}}^{\mathrm{T}}\mathbf{p}|^2\|\mathbf{h}_{\mathrm{c}}\|^2 + \beta_3\|(\mathbf{h}_{\mathrm{c}}^{\mathrm{T}}\mathbf{p})\mathbf{h}_{\mathrm{s}}+(\mathbf{h}_{\mathrm{s}}^{\mathrm{T}}\mathbf{p})\mathbf{h}_{\mathrm{c}}\|^2 \\ &= \mathbf{p}^{\mathrm{H}}\Big((\beta_1\|\mathbf{h}_{\mathrm{s}}\|^2 + \beta_3\|\mathbf{h}_{\mathrm{c}}\|^2)\mathbf{h}_{\mathrm{s}}^{*}\mathbf{h}_{\mathrm{s}}^{\mathrm{T}} + (\beta_2\|\mathbf{h}_{\mathrm{c}}\|^2 + \beta_3\|\mathbf{h}_{\mathrm{s}}\|^2)\mathbf{h}_{\mathrm{c}}^{*}\mathbf{h}_{\mathrm{c}}^{\mathrm{T}} \\ &\quad + \beta_3(\mathbf{h}_{\mathrm{s}}^{\mathrm{H}}\mathbf{h}_{\mathrm{c}})\mathbf{h}_{\mathrm{c}}^{*}\mathbf{h}_{\mathrm{s}}^{\mathrm{T}} + \beta_3(\mathbf{h}_{\mathrm{c}}^{\mathrm{H}}\mathbf{h}_{\mathrm{s}})\mathbf{h}_{\mathrm{s}}^{*}\mathbf{h}_{\mathrm{c}}^{\mathrm{T}}\Big)\mathbf{p}. \end{aligned} \tag{9.106}$$

This is a quadratic form with respect to the precoding vector and with a Hermitian matrix in the middle. Hence, it is maximized when $\mathbf{p}$ is selected as the unit-length eigenvector associated with the largest eigenvalue.

With the optimized precoding vector and surface configuration described above, the covariance matrix $\mathbf{R}$ will take a particular value that we will denote as $\bar{\mathbf{R}}$. Based on this matrix, we can derive the Neyman-Pearson detector that gives a desired false alarm probability $P_{\mathrm{FA}} = \alpha$ following the approach from Section 8.3.2. In particular, $\mathbf{y} \sim \mathcal{N}_{\mathbb{C}}(\mathbf{0}, \sigma^2\mathbf{I}_K)$ under the hypothesis $\mathcal{H}_0$ and $\mathbf{y} \sim \mathcal{N}_{\mathbb{C}}(\mathbf{0}, P\bar{\mathbf{R}} + \sigma^2\mathbf{I}_K)$ under the hypothesis $\mathcal{H}_1$. Lemma 2.14 says that we should decide on the hypothesis $\mathcal{H}_1$ if

$$\gamma \leq \frac{f_{\mathbf{y}|\mathcal{H}_1}(\mathbf{y}|\mathcal{H}_1)}{f_{\mathbf{y}|\mathcal{H}_0}(\mathbf{y}|\mathcal{H}_0)} = \frac{\frac{1}{\pi^K \det\left(P\bar{\mathbf{R}}+\sigma^2\mathbf{I}_K\right)}e^{-\mathbf{y}^{\mathrm{H}}\left(P\bar{\mathbf{R}}+\sigma^2\mathbf{I}_K\right)^{-1}\mathbf{y}}}{\frac{1}{\pi^K \det(\sigma^2\mathbf{I}_K)}e^{-\mathbf{y}^{\mathrm{H}}(\sigma^2\mathbf{I}_K)^{-1}\mathbf{y}}}. \tag{9.107}$$

We can rewrite this condition by using the fact that $\ln(\gamma)$ is a monotonically increasing function for $\gamma \geq 0$:

$$\ln(\gamma) - \ln(b) \leq \sigma^{-2}\mathbf{y}^{\mathrm{H}}\mathbf{y} - \mathbf{y}^{\mathrm{H}}\left(P\bar{\mathbf{R}} + \sigma^2\mathbf{I}_K\right)^{-1}\mathbf{y}, \tag{9.108}$$

where the constant $b = \det(\sigma^2\mathbf{I}_K)/\det(P\bar{\mathbf{R}} + \sigma^2\mathbf{I}_K)$ is independent of the received signal $\mathbf{y}$. Hence, the Neyman-Pearson detector decides on $\mathcal{H}_1$ if

$$\|\mathbf{y}\|^2 - \mathbf{y}^{\mathrm{H}}\left(\frac{P}{\sigma^2}\bar{\mathbf{R}} + \mathbf{I}_K\right)^{-1}\mathbf{y} \geq \underbrace{\sigma^2\left(\ln(\gamma) - \ln(b)\right)}_{=\gamma'}, \tag{9.109}$$

where γ' is the revised threshold variable that must be selected so that

$$P_{\mathrm{FA}} = \alpha = \int_{\|\mathbf{y}\|^2-\mathbf{y}^{\mathrm{H}}\left(\frac{P}{\sigma^2}\bar{\mathbf{R}}+\mathbf{I}_K\right)^{-1}\mathbf{y}\geq\gamma'} f_{\mathbf{y}|\mathcal{H}_0}(\mathbf{y}|\mathcal{H}_0)\,\partial\mathbf{y}. \tag{9.110}$$

The sufficient statistics for target detection is $\|\mathbf{y}\|^2 - \mathbf{y}^{\mathrm{H}}\left(\frac{P}{\sigma^2}\bar{\mathbf{R}} + \mathbf{I}_K\right)^{-1}\mathbf{y}$ and is affected by the precoding and surface configuration through $\bar{\mathbf{R}}$.

Figure 9.24 shows the detection probability, P_{D}, versus the reference SNR obtained if the radar transceiver has a single antenna and there is no reconfigurable surface. The Neyman-Pearson detector is used with the false

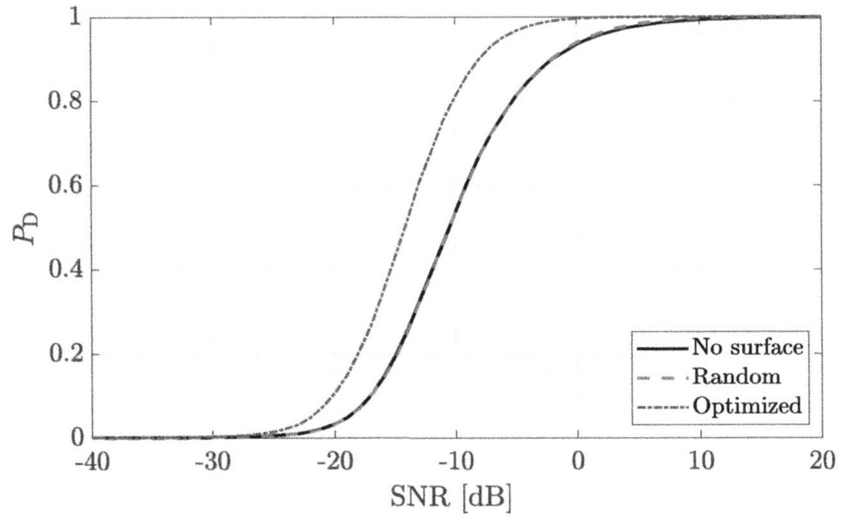

Figure 9.24: The detection probability with respect to the reference SNR in a setup with or without a reconfigurable surface. The surface either has a random phase-shift configuration or is optimized to maximize the received power.

alarm probability $P_{\mathrm{FA}} = \alpha = 10^{-3}$. The radar transceiver is equipped with a half-wavelength-spaced ULA with $K = 10$ transmit and receive antennas. The target location is in the direction $\varphi = 0$ seen from the transceiver, while a reconfigurable surface with $N = 100$ elements is seen in the direction $\varphi = \pi/6$. We let $\beta_1 = \beta_2 N^4 = \beta_3 N^2$ so all the propagation paths are equally strong when the surface is optimally configured. The solid black curve shows the detection performance without the surface, in which case MRT is the optimal precoding. The dashed red curve is obtained when the reconfigurable surface is added to the setup, but it has a random configuration and the precoding still points the signal directly toward the target. The detection probability is improved, but the effect is negligible since the extra paths are weak. It is when the precoding and surface configuration are jointly optimized that we can observe large improvements. The dash-dotted blue curve represents this case and is shifted by roughly 4 dB to the left, compared to the original black curve. This is explained by the fact that total received power $P\mathrm{tr}(\mathbf{R})$ is increased by 3.9 dB. The blue curve is steeper thanks to the spatial diversity gain obtained by having three random RCS coefficients instead of one.

Beyond this basic example, there are many other MIMO radar system configurations and more complex propagation channels where a reconfigurable surface can enhance detection performance. The expected gains are created by ensuring that a larger fraction of the transmitted power reaches the target location and is then reflected toward the receivers, as well as by creating extra spatially distinguishable paths that provide diversity against randomness and improved spatial resolution. We refer to [175], [176] for further details.

9.5 Exercises

Exercise 9.1. Consider the setup from Example 9.4 with equal propagation losses to all elements, such that the end-to-end channel gain is $(\sqrt{\beta_s} + N\sqrt{\beta_r\beta_t})^2$. How many metaatoms N are required for the reconfigurable surface to double the received power, compared to the case of $N = 0$?

(a) Answer the question when $\beta_s = \beta_t = \beta_r = 10^{-8}$.

(b) Answer the question when $\beta_s = 10^{-10}$, $\beta_t = 10^{-8}$, and $\beta_r = 10^{-6}$.

Exercise 9.2. The end-to-end channel in (9.28) becomes $h = \sum_{n=1}^{N} h_{r,n} e^{j\psi_n} h_{t,n}$ if the static channel is totally blocked. Suppose the channels are equally strong to/from all metaatoms: $h_{r,n} = \sqrt{\beta_r}$ and $h_{t,n} = \sqrt{\beta_t}$, $n = 1, \ldots, N$.

(a) What is the average channel gain $\mathbb{E}\{|h|^2\}$ if the phase-shifts ψ_n are selected as independent random variables that are uniformly distributed between 0 and 2π?

(b) What will $|h|^2$ become if the phase-shifts are selected to maximize it?

(c) Compare the results in (a) and (b). What kind of gain is missing in (a)?

Exercise 9.3. The end-to-end channel gain in (9.25) becomes $|\mathbf{h}_r^T \mathbf{D}_\psi \mathbf{h}_t|^2$ if the static channel is totally blocked by some objects (e.g., at high frequencies). Suppose $\mathbf{h}_t \sim \mathcal{N}_{\mathbb{C}}(\mathbf{0}, \beta_t \mathbf{I}_N)$ and $\mathbf{h}_r \sim \mathcal{N}_{\mathbb{C}}(\mathbf{0}, \beta_r \mathbf{I}_N)$ and they are independent.

(a) What is the average channel gain with a static surface with $\mathbf{D}_\psi = \mathbf{I}_N$?

(b) What is the average channel gain with a reconfigurable surface that is configured to maximize the channel gain? Hint: $\mathbb{E}\{|h_{t,n}|\} = \sqrt{\beta_t}\sqrt{\pi/4}$.

Exercise 9.4. The LOS end-to-end channel gain is stated in (9.35) when the static channel is negligible. Suppose the transmitter and receiver are equipped with isotropic antennas, while each metaatom has the effective area $A_m = (\lambda/4)^2$.

(a) Determine an expression of the end-to-end channel gain when the distance between the transmitter and the surface is d_t and the distance between the surface and the receiver is d_r.

(b) How many metaatoms are needed to achieve an end-to-end channel gain of 10^{-9} if the wavelength is $\lambda = 0.1$ m (i.e., 3 GHz), $d_t = 50$ m, and $d_r = 2$ m? How large is the total area NA_m of the surface?

(c) How many metaatoms are needed to achieve the same channel gain as in (b) when $\lambda = 0.01$ m (i.e., 30 GHz). How large is the total area NA_m of the surface?

Exercise 9.5. Suppose the reconfigurable surface can turn off specific metaatoms so they absorb all incident signal energy instead of reflecting anything.

(a) Use this feature to estimate each of the cascaded channels $h_{r,n}h_{t,n}$ sequentially while the remaining $N-1$ metaatoms are turned off. Follow the ML estimation framework in Section 9.2.2 and assume that $h_s = 0$.

(b) Suppose all metaatoms are turned on during the ML estimation and simplify the ML estimator in (9.42) for the case when $h_s = 0$.

(c) Show that the ML estimate can be expressed as $\widehat{\check{\mathbf{h}}} = \check{\mathbf{h}} +$ effective noise in both (a) and (b). Compare the variances of the noise terms. Is it preferable to turn metaatoms on/off during the channel estimation? Explain the result.

Exercise 9.6. Consider a reconfigurable surface designed as a uniform planar array with N_{H} columns and N_{V} metaatoms per column. Suppose a plane wave impinges from the direction φ_{i} in the azimuth plane and should be reflected towards a user in the direction φ_{o} in the azimuth plane (i.e., $\theta_{\mathrm{i}} = \theta_{\mathrm{o}} = 0$).

(a) Prove that the cascaded channel coefficient $h_{\mathrm{r},n}h_{\mathrm{t},n}$ is equal for all the N_{V} metaatoms located in the same column.

(b) Based on the property proved in (a), if we deploy the reconfigurable surface in an environment where signals only propagate in the azimuth plane, we can reduce the number of phase-shift variables from $N_{\mathrm{H}}N_{\mathrm{V}}$ to N_{H}. This is achieved by assigning the same phase-shift to metaatoms in the same column. Write up the corresponding end-to-end channel and factorize it similarly to (9.38).

(c) Write up a new ML estimator that utilizes the new factorization from (b). Show that the minimum pilot length is now $L_{\mathrm{p}} = N_{\mathrm{H}} + 1$.

(d) Suppose the same total energy $(N+1)q$ is utilized for pilot transmission with the new ML estimator as with one considered in (9.44). How much smaller total variance will the scaled noise term have with the new estimator?

Exercise 9.7. A classic way of extending wireless coverage (e.g., into tunnels) is to use a repeater that picks up the signal using one antenna and immediately retransmits an amplified version using another antenna. In this exercise, it will be compared with a reconfigurable surface in the same deployment scenario.

(a) The received signal at the repeater is $y_1 = \sqrt{\beta_{\mathrm{t}}}x_1 + n_1$ and the received signal at the receiver is $y_2 = \sqrt{\beta_{\mathrm{r}}}x_2 + n_2$, where $n_1, n_2 \sim \mathcal{N}_{\mathbb{C}}(0, N_0)$. Suppose the data signal $x_1 \sim \mathcal{N}_{\mathbb{C}}(0, q_1)$ is transmitted and that the repeater sends $x_2 = \sqrt{a}y_1$, where a is the amplification gain. How should a be selected to ensure that $\mathbb{E}\{|x_2|^2\} = q_2$?

(b) What is the SNR at the receiver when using the repeater with the amplification gain obtained in (a)?

(c) If a reconfigurable surface is used in the same scenario, the SNR would be $qN^2\beta_{\mathrm{t}}\beta_{\mathrm{r}}/N_0$. Derive an expression for how many metaatoms N are required to achieve a larger SNR with the surface than with the repeater.

(d) Compute the number of metaatoms in (c) if $\beta_{\mathrm{t}} = \beta_{\mathrm{r}} = 10^{-8}$ and $q/N_0 = 10^8$. To make the total transmit power the same in both setups, we let $q_1 = q_2 = q/2$.

Exercise 9.8. The end-to-end channel gain in (9.31) with an optimal phase-shift configuration is $\|\check{\mathbf{h}}\|_1^2$, where the 1-norm is used.

(a) A MISO channel with the same channel vector achieves the channel gain $\|\check{\mathbf{h}}\|^2$, where the Euclidean norm (2-norm) is used. Which of the two squared norms is the largest? Under which conditions are they equal?

(b) The phase-shift vector $\boldsymbol{\psi}$ acts as a beamforming vector with $\|\boldsymbol{\psi}\|^2 = N+1$. If MRT is used with a precoding vector that has the same squared norm, what will be the resulting channel gain? Is it larger than $\|\check{\mathbf{h}}\|_1^2$?

Exercise 9.9. Derive (9.61) from (9.55) step-by-step by utilizing the two properties stated after the equation. Hint: Use the commutative and associative properties of the convolution. The commutative property states that $(f*g)(t) = (g*f)(t)$. The associative property of the convolution is $(f*g*h)(t) = (f*g)*(h)(t) = (f)*(g*h)(t)$.

Exercise 9.10. Consider the reflection coefficient Γ_{0n} when the metaatom's impedance Z_n is given by (9.56).

(a) What does Γ_{0n} converge to as $f \to 0$? What is its amplitude and phase?

(b) What does Γ_{0n} converge to as $f \to \infty$? What is its amplitude and phase?

(c) Show that $|\Gamma_{0n}| = 1$ for all f if $\mathrm{R} = 0$.

Exercise 9.11. If a known pilot signal is transmitted on all subcarriers, the ML estimator described in Section 9.2.2 can be applied to separately estimate each of the channel vectors $\check{\mathbf{h}}[0], \ldots, \check{\mathbf{h}}[S-1]$. However, this is unnecessary because adjacent subcarriers have similar channel vectors.

(a) Show that $h_{\boldsymbol{\psi}}[\ell]$ in (9.64) can be expressed as $\boldsymbol{\psi}^{\mathrm{T}}\mathbf{h}[\ell]$, where $\boldsymbol{\psi} = [1, e^{\mathrm{j}\psi_1}, \ldots, e^{\mathrm{j}\psi_N}]^{\mathrm{T}}$.

(b) Relate $h_{\boldsymbol{\psi}}[0], \ldots, h_{\boldsymbol{\psi}}[T]$ to $\check{\mathbf{h}}[0], \ldots, \check{\mathbf{h}}[S-1]$ using a DFT matrix.

(c) Use the property in (b) to determine an ML estimator of $\mathbf{h}[0], \ldots, \mathbf{h}[T]$, based on the received signals from pilot transmission on $T+1$ subcarriers.

Exercise 9.12. Consider a SIMO channel aided by a reconfigurable surface where the channel vector is

$$\mathbf{h} = \mathbf{h}_{\mathrm{s}} + \mathbf{H}_{\mathrm{r}}\mathbf{D}_{\psi}\mathbf{h}_{\mathrm{t}}. \tag{9.111}$$

Suppose there is a free-space LOS channel $\mathbf{H}_{\mathrm{r}} = \mathbf{a}_{\mathrm{r}}\mathbf{b}_{\mathrm{r}}^{\mathrm{T}}$ between the surface and receiver, where $\mathbf{a}_{\mathrm{r}}, \mathbf{b}_{\mathrm{r}}$ are vectors. Determine the surface configuration that maximizes the capacity.

Exercise 9.13. Consider a point-to-point MIMO system with $M = K$ where the channel matrix $\mathbf{H} \in \mathbb{C}^{M \times M}$ has full rank. The system operates at high SNR so that equal power allocation is optimal.

(a) Prove that the high-SNR capacity is upper bounded by $M \log_2(1 + \frac{q\|\mathbf{H}\|_F^2}{M^2 N_0})$. Under what conditions on $\mathbf{H}$ is the upper bound achieved? Hint: Use the inequality of arithmetic and geometric means from Lemma 3.2.

(b) The channel contains a reconfigurable surface, so the channel matrix is modeled according to (9.81) as $\mathbf{H} = \mathbf{H}_{\mathrm{s}} + \mathbf{H}_{\mathrm{r}}\mathbf{D}_{\psi}\mathbf{H}_{\mathrm{t}}$. How should the surface be configured to maximize the high-SNR capacity if $\mathbf{H}_{\mathrm{s}}$, $\mathbf{H}_{\mathrm{r}}$, and $\mathbf{H}_{\mathrm{t}}$ are rank-one matrices?

Exercise 9.14. Propose an algorithm for downlink sum capacity maximization that switches between solving (6.122) for given user channels and updating the phase-shifts similarly to Algorithm 9.3.

Appendix

Mathematical Notation

Upper-case boldface letters are used to denote matrices (e.g., $\mathbf{X}, \mathbf{Y}$), while column vectors are denoted with lower-case boldface letters (e.g., $\mathbf{x}, \mathbf{y}$). Scalars are denoted by lower/upper-case italic letters (e.g., x, y, X, Y) and sets by calligraphic letters (e.g., $\mathcal{X}, \mathcal{Y}$).

The following general mathematical notations are used:

$\mathbb{R}$	The space of real-valued numbers
$\mathbb{C}$	The space of complex-valued numbers
$\mathbb{R}^N$	The space of real-valued N-dimensional vectors
$\mathbb{C}^N$	The space of complex-valued N-dimensional vectors
$\mathbb{C}^{N \times M}$	The set of complex-valued $N \times M$ matrices
$\mathcal{A} = \{a_1, \ldots, a_N\}$	A set with the members $a_1, \ldots, a_N$
$x \in \mathcal{A}$	x is a member of the set $\mathcal{A}$
$x \notin \mathcal{A}$	x is not a member of the set $\mathcal{A}$
$\mathcal{A} \subset \mathcal{B}$	$\mathcal{A}$ is a subset of $\mathcal{B}$
$\{(R_1, R_2) : \text{cond}\}$	The set of all (R_1, R_2) that satisfy the condition
$[\mathbf{x}]_i$	The ith entry of a vector $\mathbf{x}$
$[\mathbf{X}]_{ij}$	The (i, j)th entry of a matrix $\mathbf{X}$
$\text{diag}(d_1, \ldots, d_N)$	Diagonal matrix with $d_1, \ldots, d_N$ on the diagonal
$\mathbf{I}_M$	The $M \times M$ identity matrix
$\mathbf{0}$	A matrix with only zeros with matching size
$\mathbf{X}^*$	The entry-wise complex conjugate of $\mathbf{X}$
$\mathbf{X}^{\mathrm{T}}$	The transpose of $\mathbf{X}$
$\mathbf{X}^{\mathrm{H}}$	The conjugate/Hermitian transpose of $\mathbf{X}$
$\mathbf{X}^{-1}$	The inverse of a square matrix $\mathbf{X}$
$\mathbf{X}^{1/2}$	The square root of a square matrix $\mathbf{X}$
$\text{tr}(\mathbf{X})$	Trace of a square matrix $\mathbf{X}$
$\det(\mathbf{X})$	Determinant of a square matrix $\mathbf{X}$
$\mathbf{x} \odot \mathbf{y}$	Entry-wise (Hadamard) product of $\mathbf{x}$, $\mathbf{y}$
$\mathbf{X} \otimes \mathbf{Y}$	Kronecker product of $\mathbf{X}$ and $\mathbf{Y}$
$\|\mathbf{x}\|$	The Euclidean norm $\|\mathbf{x}\| = \sqrt{\sum_i \lvert[\mathbf{x}]_i\rvert^2}$ of $\mathbf{x}$
$\|\mathbf{X}\|_{\mathrm{F}}$	The Frobenius norm of $\mathbf{X}$, defined in (5.87)
$\Re(x), \Im(x)$	Real part and imaginary part of x
j	The imaginary unit $\sqrt{-1}$

$\sqrt{x}$	The square root of x
$\sqrt[n]{x}$	The nth root $x^{1/n}$ of $x > 0$
$\|x\|$	Magnitude (or absolute value) of a scalar x
$\arg(x)$	The phase in $[-\pi, \pi)$ of complex number x
$\lfloor x \rfloor$	Closest integer smaller or equal to x
$\lceil x \rceil$	Closest integer greater or equal to x
$n!$	The factorial function for positive integers n
e	Euler's number ($e \approx 2.71828$)
$\min(x, y)$	The minimum of x and y
$\max(x, y)$	The maximum of x and y
$\bmod S$	Modulo operation (the remainder after division by S)
$[x]_{-1:1}$	Wraps x within the range $(-1, 1)$, see (5.194)
$\log_a(x)$	The logarithm of x using the base $a > 0$
$\ln(x)$	The natural logarithm of x (base e)
$\sin(x), \cos(x)$	The sine and cosine functions of x
$\tan(x)$	The tangent function of x
$\arcsin(x)$	The inverse sine function
$\arctan(x)$	The inverse tangent function
$e^{\mathrm{j}x}$	The complex exponential function of x
$\mathrm{sinc}(x)$	The sinc function $\mathrm{sinc}(x) = \sin(\pi x)/(\pi x)$
$\delta(t)$	The Dirac delta function
$(f * g)(t)$	Convolution of the continuous functions $f(t), g(t)$
$(f * g)[k]$	Linear convolution of the discrete sequences $f[k], g[k]$
$(f \circledast g)[k]$	Cyclic convolution of the discrete sequences $f[k], g[k]$
$\mathcal{F}\{a(t)\}$	Fourier transform of the continuous function $a(t)$
$\mathcal{F}_{\mathrm{d}}\{\chi[s]\}$	DFT of the discrete sequence $\chi[s]$
$\sim$	Means "distributed as"
$\mathcal{N}(\mu, \sigma^2)$	Gaussian distribution with mean μ and variance σ^2
$\mathcal{N}_{\mathbb{C}}(\mathbf{0}, \mathbf{R})$	Circularly symmetric complex Gaussian distribution with zero mean and covariance matrix $\mathbf{R}$
$\mathrm{Rayleigh}(\sigma)$	Rayleigh distribution with the scale parameter $\sigma \geq 0$
$\mathrm{Rice}(\nu, \sigma)$	Rician distribution with the parameters $\nu, \sigma \geq 0$
$\mathrm{Exp}(x)$	Exponential distribution with the rate $x > 0$
$\chi^2(N)$	Chi-squared distribution with N degrees of freedom
$U[a, b]$	Uniform distribution between a and b
$\mathbb{E}\{x\}$	The mean of a random variable x
$\mathbb{V}\mathrm{ar}\{x\}$	The variance of a random variable x
$\mathrm{Cov}\{\mathbf{x}\}$	The covariance matrix of a random vector $\mathbf{x}$
$\Pr\{\mathrm{cond}\}$	The probability that the condition "cond" is satisfied
$\mathcal{H}(y)$	The differential entropy of y, see (2.134)
$\mathcal{H}(y\|x)$	The conditional differential entropy, see (2.135)
$\mathcal{I}(x; y)$	The mutual information between x and y
$\mathcal{H}_0, \mathcal{H}_1$	The null hypothesis and alternative hypothesis

Specific Notation

Many variables are used in the different chapters and some names are used for multiple purposes. The following is a non-exhaustive list of such notation:

α_i	Attenuation of path i
$\mathbf{a}(\varphi), \mathbf{a}_M(\varphi)$	Array response vector of a ULA in 2D
$\mathbf{a}_M(\varphi, \theta)$	Array response vector of a ULA in 3D
$\mathbf{a}_{M_{\mathrm{H}},M_{\mathrm{V}}}(\varphi, \theta)$	Array response vector of a UPA
A_{iso}	Area of an isotropic antenna [m^2]
A_{m}	Area of a metaatom [m^2]
$A(\varphi, \theta)$	Effective area function of an antenna [m^2]
B	The signal bandwidth [Hz] and symbol rate [symbol/s]
β	The channel gain
c	The speed of light in free space (vacuum) [m/s]
C	The capacity of a channel [bit/s] or [bit/symbol]
C_ϵ	The ϵ-outage capacity of a channel [bit/s] or [bit/symbol]
C_k^{su}	The single-user capacity [bit/s]
$C(x)$	Capacity function in (6.7) [bit/s]
$d, d_m, d_i, d_{\mathrm{t}}, d_{\mathrm{r}}$	The propagation distance (for paths or to antennas)
D, D_λ	The aperture length and normalized length $D_\lambda = D/\lambda$
$\mathbf{D}$	A diagonal matrix, often from the SVD
$\mathbf{D}_\psi$	Reflection matrix in (9.17) of a reconfigurable surface
Δ, Δ_λ	The antenna spacing and normalized spacing $\Delta_\lambda = \Delta/\lambda$
η	The sampling delay at the receiver [s]
f	A frequency variable [Hz]
f_{c}	Carrier frequency of the signal [Hz]
$f_{\mathsf{x}}(x)$	The PDF of a random variable x
$F_{\mathsf{x}}(x)$	The CDF of a random variable x
$\mathbf{F}_S$	The $S \times S$ DFT matrix defined in (2.198)
$G(\varphi, \theta)$	Antenna gain function
h	A scalar channel coefficient
$\mathbf{h}$	A SIMO/MISO channel vector
$\mathbf{H}$	A MIMO channel matrix
$\hat{h}, \hat{\mathbf{h}}, \hat{\mathbf{H}}$	Estimates of a channel scalar/vector/matrix
$\tilde{h}, \tilde{\mathbf{h}}, \tilde{\mathbf{H}}$	Estimation errors of a channel scalar/vector/matrix
$\breve{\mathbf{H}}$	Beamspace representation of the channel matrix
$i(m), j(m)$	Horizontal/vertical indices in a UPA, see (4.124)–(4.125)
K, M	Number of transmit or receive antennas
κ	XPD-related variable, see (4.171)
λ, λ_m	Wavelength [m] or an eigenvalue of a matrix
L	Number of variables in different contexts
$L_{\mathrm{V}}, L_{\mathrm{H}}$	Vertical and horizontal length of a UPA
$L_{\mathrm{c}}, L_{\mathrm{p}}$	Length of a coherence block and pilot length [symbols]

$M_{\mathrm{V}}, M_{\mathrm{H}}$	Number of antennas per column and row in a UPA
MSE_x	The MSE of an estimate of x
N	Number of metaatoms in a reconfigurable surface
N_0	Noise power spectral density [W/Hz]
$N_{\mathrm{cl}}, N_{\mathrm{path}}$	Number of clusters/paths in a multipath channel
N_{RF}	Number of RF inputs/outputs in hybrid beamforming
$p(t)$	The pulse used in PAM, often $p(t) = \sqrt{B}\mathrm{sinc}(Bt)$
$\mathbf{p}$	A transmit precoding vector
$\mathbf{P}, \mathbf{P}_{\mathrm{BB}}, \mathbf{P}_{\mathrm{RF}}$	Arbitrary digital and hybrid precoding matrices
P, P_{t}	Transmitted signal power [W]
P_{r}	Received signal power [W]
$P_k^{\mathrm{ul}}, P_k^{\mathrm{dl}}$	Transmit power in uplink/downlink
P_{D}	The correct detection probability
P_{FA}	The false alarm probability
P_{M}	The missing probability
$P_{\mathrm{conv}}(\varphi, \theta)$	Power spectrum with conventional beamforming
$P_{\mathrm{Capon}}(\varphi, \theta)$	Power spectrum with Capon beamforming
$P_{\mathrm{out}}(R)$	The outage probability given the rate R
$\varphi, \varphi_{\mathrm{t}}, \varphi_{\mathrm{r}}, \varphi_{\mathrm{i}}, \varphi_{\mathrm{o}}$	Azimuth angle
$\varphi_{\mathrm{beam}}, \theta_{\mathrm{beam}}$	The beam direction
ψ_n	Phase-shift of metaatom n
R, R_k	Achievable rates [bit/s] or [bit/symbol]
$\mathbf{R}_h$	The covariance matrix of a random vector $\mathbf{h}$
$\mathcal{R}, \partial\mathcal{R}$	Rate region and its Pareto boundary
q	The symbol power $q = P/B$ [Joule]
q_k	The symbol power assigned to the kth channel
$\mathbf{Q}$	Diagonal matrix with $q_1, \ldots, q_K$
r	The rank of a matrix
S	Number of subcarriers in OFDM
$\mathbf{\Sigma}$	Diagonal matrix with singular values
s_k	The kth singular value in the SVD of a matrix
σ^2	Variance of the noise [W] or of another variable
σ_{RCS}	The RCS of a target object
SNR	SNR variable
t	A time variable [s]
τ_i	Propagation delay of path i [s]
$\theta, \theta_{\mathrm{t}}, \theta_{\mathrm{r}}, \theta_{\mathrm{i}}, \theta_{\mathrm{o}}$	Elevation angle
T	The memory of an FIR filter [symbols]
T_{c}	Channel coherence time [s]
v	Speed of movement [m/s]
$\mathbf{U}, \mathbf{V}$	Unitary matrices, often obtained from the SVD
$\mathbf{w}$	A receive combining vector
$\mathbf{W}, \mathbf{W}_{\mathrm{BB}}, \mathbf{W}_{\mathrm{RF}}$	Arbitrary digital and hybrid combining matrices

Abbreviations

The following acronyms and abbreviations are used in this book:

2D	two-dimensional
3D	three-dimensional
3GPP	3rd generation partnership project (an organization)
5G,4G,3G,2G	fifth/fourth/third/second generation
ADC	analog-to-digital converter
AESA	active electronically scanned array
AWGN	additive white Gaussian noise
BB	baseband
BBU	baseband unit
CDF	cumulative distribution function
CDMA	code-division multiple access
CDMA2000	name of a CDMA-based 3G standard
CSI	channel state information
DAC	digital-to-analog converter
dBi	decibels referenced to an isotropic antenna
dBm	decibels referenced to 1 mW
DFT	discrete Fourier transform
dl	downlink
DOA	direction-of-arrival
DPC	dirty paper coding
DSFT	discrete-space Fourier transform
DTFT	discrete-time Fourier transform
eCDF	empirical cumulative distribution function
EIRP	effective isotropic radiated power
ELAA	extremely large aperture array
ESPRIT	estimation of signal parameters by rotational invariance techniques
EV-DO	Evolution-data optimized (4G standard)
FDMA	frequency-division multiple access
FIR	finite impulse response
GSM	Global system for mobile communications (2G standard)
i.i.d.	independent and identically distributed
IDFT	inverse DFT
IEEE	Institute of electrical and electronics engineers
IRS	intelligent reflecting surfaces
IS-95	Interim standard 95 (2G standard)
ISAC	integrated sensing and communication
ITU	International telecommunication union
LDPC	low-density parity-check
LMMSE	linear MMSE

LNA	low-noise amplifier
LOS	line-of-sight
LTE	Long-term evolution (4G standard)
LTI	linear time-invariant
MCS	modulation and coding scheme
MIMO	multiple-input multiple-output
MISO	multiple-input single-output
ML	maximum likelihood
mmf	max-min fairness
MMSE	minimum mean-squared error
mmWave	millimeter-wave
MRC	maximum-ratio combining
MRT	maximum-ratio transmission
MSE	mean-squared error
MUSIC	multiple signal classification
MVDR	minimum-variance distortionless response
NFC	Near-field communication (a wireless standard)
NLOS	non-LOS
NMSE	normalized MSE
NOMA	non-orthogonal multiple access
NR	New radio (5G standard)
OAM	orbital angular momentum
OFDM	orthogonal frequency-division multiplexing
OMA	orthogonal multiple access
opt	optimal
PA	power amplifier
PAM	pulse-amplitude modulation
PDF	probability density function
PEC	perfect electric conductor
PESA	passive electronically scanned array
PS	phase shifter
PSS	primary synchronization signal
QAM	quadrature amplitude modulation
RCS	radar cross section
RF	radio-frequency
RIS	reconfigurable intelligent surfaces
RMSE	root MSE
RZF	regularized zero-forcing
SDMA	space-division multiple access
SIC	successive interference cancellation
SIMO	single-input multiple-output
SINR	signal-to-interference-plus-noise ratio
SISO	single-input single-output

SLNR	signal-to-leakage-and-noise ratio
SNR	signal-to-noise ratio
sr	sum rate
STBC	space-time block code
su	single user
SVD	singular-value decomposition
TDMA	time-division multiple access
TDOA	time-difference-of-arrival
TOA	time-of-arrival
TTD	true time delay
TWF	transmit Wiener filter
ul	uplink
ULA	uniform linear array
UMi	urban microcell
UMTS	Universal mobile telecommunications system (3G standard)
UPA	uniform planar array
WiFi	Trademark used for WLAN
WLAN	wireless local area network
XPD	cross-polar discrimination
ZF	zero-forcing

References

[1] E. Björnson, J. Hoydis, and L. Sanguinetti, "Massive MIMO networks: Spectral, energy, and hardware efficiency", *Foundations and Trends in Signal Processing*, vol. 11, no. 3-4, 2017, pp. 154–655.

[2] Ö. T. Demir, E. Björnson, and L. Sanguinetti, "Foundations of user-centric cell-free massive MIMO", *Foundations and Trends in Signal Processing*, vol. 14, no. 3-4, 2021, pp. 162–472.

[3] T. L. Marzetta, E. G. Larsson, H. Yang, and H. Q. Ngo, *Fundamentals of Massive MIMO*. Cambridge University Press, 2016.

[4] A. Ghosh, J. Zhang, J. G. Andrews, and R. Muhamed, *Fundamentals of LTE*. Prentice Hall, 2010.

[5] S. Parkvall, E. Dahlman, A. Furuskär, and M. Frenne, "NR: The new 5G radio access technology", *IEEE Communications Standards Magazine*, vol. 1, no. 4, 2017, pp. 24–30.

[6] 3GPP, *Further advancements for E-UTRA physical layer aspects (Release 9)*. 3GPP TS 36.814, Mar. 2017.

[7] 3GPP, *NR; Base Station (BS) radio transmission and reception (Release 17)*. 3GPP TS 38.104, Dec. 2020.

[8] C. A. Balanis, *Antenna theory: analysis and design*. John Wiley & Sons, 2015.

[9] H. T. Friis, "A note on a simple transmission formula", *IRE*, vol. 34, no. 5, 1946, pp. 254–256.

[10] Ericsson, *Ericsson mobility report*, Jun. 2018. [Online]. Available: http://www.ericsson.com/mobility-report.

[11] ITU, "Radio regulations: Articles", Tech. Rep., 2020. [Online]. Available: http://handle.itu.int/11.1002/pub/814b0c44-en.

[12] IEEE, "IEEE standard letter designations for radar-frequency bands", *Std 521-2019 (Revision of IEEE Std 521-2002)*, 2020, pp. 1–15.

[13] T. S. Rappaport, Y. Xing, O. Kanhere, S. Ju, A. Madanayake, S. Mandal, A. Alkhateeb, and G. C. Trichopoulos, "Wireless communications and applications above 100 GHz: Opportunities and challenges for 6G and beyond", *IEEE Access*, vol. 7, 2019, pp. 78 729–78 757.

[14] K. T. Selvan and R. Janaswamy, "Fraunhofer and Fresnel distances: Unified derivation for aperture antennas", *IEEE Antennas Propag. Mag.*, vol. 59, no. 4, 2017, pp. 12–15.

[15] E. F. W. Alexanderson, "Transatlantic radio communication", *Trans. American Institute of Electrical Engineers*, vol. 38, no. 2, 1919, pp. 1269–1285.

[16] P. Bondyopadhyay, "The first application of array antenna", in *Proc. IEEE International Conference on Phased Array Systems and Technology*, pp. 29–32, 2000.

[17] K. F. Braun, "Electrical oscillations and wireless telegraphy", *Nobel Lecture*, Dec. 1909, pp. 226–245.

[18] E. F. W. Alexanderson, *Antenna*, US Patent, 1920.

[19] S. Anderson, U. Forssen, J. Karlsson, T. Witzschel, P. Fischer, and A. Krug, "Ericsson/Mannesmann GSM field-trials with adaptive antennas", in *IEE Colloquium on Advanced TDMA Techniques and Applications*, 1996.

[20] M. Nilsson, B. Lindmark, M. Ahlberg, M. Larsson, and C. Beckman, "Measurements of the spatio-temporal polarization characteristics of a radio channel at 1800 MHz", in *IEEE VTC*, vol. 1, pp. 386–391, 1999.

[21] J. H. Winters, "Optimum combining for indoor radio systems with multiple users", *IEEE Trans. Commun.*, vol. 35, no. 11, 1987, pp. 1222–1230.

[22] S. C. Swales, M. A. Beach, D. J. Edwards, and J. P. McGeehan, "The performance enhancement of multibeam adaptive base-station antennas for cellular land mobile radio systems", *IEEE Trans. Veh. Technol.*, vol. 39, no. 1, 1990, pp. 56–67.

[23] S. Anderson, M. Millnert, M. Viberg, and B. Wahlberg, "An adaptive array for mobile communication systems", *IEEE Trans. Veh. Technol.*, vol. 40, no. 1, 1991, pp. 230–236.

[24] R. H. Roy and B. Ottersten, *Spatial division multiple access wireless communication systems*, US Patent, 1991.

[25] Y. Tsuji and Y. Tada, *Transmit phase control system of synchronization burst for SDMA/TDMA satellite communication system*, US Patent, 1974.

[26] D. Tse and P. Viswanath, *Fundamentals of wireless communications*. Cambridge University Press, 2005.

[27] Qualcomm, *802.11ac MU-MIMO: Bridging the MIMO gap in Wi-Fi*, Jan. 2015.

[28] G. Raleigh and J. Cioffi, "Spatio-temporal coding for wireless communication", *IEEE Trans. Commun.*, vol. 46, no. 3, 1998, pp. 357–366.

[29] G. J. Foschini and M. J. Gans, "On limits of wireless communications in a fading environment when using multiple antennas", *Wireless Personal Commun.*, vol. 6, no. 3, 1998, pp. 311–335.

[30] P. W. Wolniansky, G. J. Foschini, G. D. Golden, and R. A. Valenzuela, "V-BLAST: An architecture for realizing very high data rates over the rich-scattering wireless channel", *URSI Int'l Symp. Signals, Systems, and Electronics (ISSSE)*, 1998, pp. 295–300.

[31] E. Telatar, "Capacity of multi-antenna Gaussian channels", *European Trans. Telecom.*, vol. 10, no. 6, 1999, pp. 585–595.

[32] H. O. Peterson, H. H. Beverage, and J. B. Moore, "Diversity telephone receiving system of R.C.A. communications, Inc.", *IRE*, vol. 19, no. 4, 1931, pp. 562–584.

[33] H. T. Friis and C. B. Feldman, "A multiple unit steerable antenna for short-wave reception", *IRE*, vol. 25, no. 7, 1937, pp. 841–917.

[34] D. G. Brennan, "Linear diversity combining techniques", *IRE*, vol. 43, no. 6, 1959, pp. 1975–1102.

[35] A. Wittneben, "Basestation modulation diversity for digital simulcast", in *IEEE VTC*, pp. 848–853, 1991.

[36] S. M. Alamouti, "A simple transmit diversity technique for wireless communications", *IEEE J. Sel. Areas Commun.*, vol. 16, no. 8, 1998, pp. 1451–1458.

[37] V. Tarokh, H. Jafarkhani, and A. Calderbank, "Space-time block codes from orthogonal designs", *IEEE Trans. Inf. Theory*, vol. 45, no. 5, 1999, pp. 1456–1467.

[38] C. E. Shannon, "Communication in the presence of noise", *IRE*, vol. 37, no. 1, 1949, pp. 10–21.

[39] J. G. Proakis and M. Salehi, *Digital Communications*, 5th edition. McGraw-Hill, 2008.

[40] C. E. Shannon, "A mathematical theory of communication", *Bell System Technical Journal*, vol. 27, 1948, pp. 379–423, 623–656.

[41] E. Björnson, E. G. Larsson, and T. L. Marzetta, "Massive MIMO: Ten myths and one critical question", *IEEE Commun. Mag.*, vol. 54, no. 2, Feb. 2016, pp. 114–123.

[42] T. M. Cover and J. A. Thomas, *Elements of information theory*. Wiley, 1991.

[43] 3GPP, *NR; Physical layer procedures for data (Release 15)*. 3GPP TS 38.214, Oct. 2018.

[44] S. M. Kay, *Fundamentals of statistical signal processing: Estimation theory*. Prentice Hall, 1993.

[45] M. Jeruchim, "Techniques for estimating the bit error rate in the simulation of digital communication systems", *IEEE J. Sel. Areas Commun.*, vol. 2, no. 1, 1984, pp. 153–170.

[46] B. Mazzeo and M. Rice, "On monte carlo simulation of the bit error rate", in *IEEE International Conference on Communications (ICC)*, pp. 1–5, 2011.

[47] J. B. S. Haldane, "On a method of estimating frequencies", *Biometrika*, vol. 33, no. 3, 1945, pp. 222–225.

[48] S. Kay, *Fundamentals of Statistical Signal Processing: Detection theory*, ser. Fundamentals of Statistical Signal Processing. Prentice-Hall, 1993.

[49] R. A. Horn and C. R. Johnson, *Matrix analysis*. Cambridge University Press, 1999.

[50] A. Lozano, A. Tulino, and S. Verdú, "Optimum power allocation for parallel Gaussian channels with arbitrary input distributions", *IEEE Trans. Inf. Theory*, vol. 52, no. 7, 2006, pp. 3033–3051.

[51] H. Krim and M. Viberg, "Two decades of array signal processing research: The parametric approach", *IEEE Signal Process. Mag.*, vol. 13, no. 4, 1996, pp. 67–94.

[52] R. Irmer, H. Droste, P. Marsch, M. Grieger, G. Fettweis, S. Brueck, H.-P. Mayer, L. Thiele, and V. Jungnickel, "Coordinated multipoint: Concepts, performance, and field trial results", *IEEE Commun. Mag.*, vol. 49, no. 2, 2011, pp. 102–111.

[53] E. Torkildson, U. Madhow, and M. Rodwell, "Indoor millimeter wave mimo: Feasibility and performance", *IEEE Trans. Wireless Commun.*, vol. 10, no. 12, 2011, pp. 4150–4160.

[54] C. Polk, "Optical Fresnel-zone gain of a rectangular aperture", *IRE Trans. Antennas Propag.*, vol. 4, no. 1, 1956, pp. 65–69.

[55] T. E. Commission, "On amending decision 2008/411/ec on the harmonisation of the 3 400-3 800 mhz frequency band for terrestrial systems capable of providing electronic communications services in the community", *Official Journal of the European Union*, vol. 276, May 2014, pp. 18–35.

[56] W. Lee and Y. Yeh, "Polarization diversity system for mobile radio", *IEEE Trans. Commun.*, vol. 20, no. 5, 1972, pp. 912–923.

[57] R. Visoz and E. Bejjani, "Matched filter bound for multichannel diversity over frequency-selective Rayleigh-fading mobile channels", *IEEE Trans. Veh. Technol.*, vol. 49, no. 5, 2000, pp. 1832–1845.

[58] N. Amitay and J. Salz, "Linear equalization theory in digital data transmission over dually polarized fading radio channels", *Bell Labs Tech. J.*, vol. 63, no. 10, 1984, pp. 2215–2259.

[59] R. Nabar, H. Bolcskei, V. Erceg, D. Gesbert, and A. Paulraj, "Performance of multi-antenna signaling techniques in the presence of polarization diversity", *IEEE Trans. Signal Process.*, vol. 50, no. 10, 2002, pp. 2553–2562.

[60] M. Coldrey, "Modeling and capacity of polarized MIMO channels", in *IEEE VTC-Spring*, IEEE, pp. 440–444, 2008.

[61] T. S. Rappaport, *Wireless Communications: Principles and Practice*, second. Upper Saddle, NJ: Prentice Hall, 2002.

[62] E. Larsson and P. Stoica, *Space-time block coding for wireless communications.* Cambridge University Press, 2003.

[63] G. Ganesan and P. Stoica, "Space-time block codes: A maximum SNR approach", *IEEE Trans. Inf. Theory*, vol. 47, no. 4, 2001, pp. 1650–1656.

[64] A. Lozano and N. Jindal, "Transmit diversity vs. spatial multiplexing in modern MIMO systems", *IEEE Trans. Wireless Commun.*, vol. 9, no. 1, 2010, pp. 186–197.

[65] R. G. Gallager, *Information Theory and Reliable Communication.* Wiley, 1968.

[66] T. Cacoullos, *Exercises in Probability*, ser. Problem Books in Mathematics. New York, NY: Springer New York, 1989.

[67] M. Abramowitz and I. Stegun, *Handbook of mathematical functions.* Dover Publications, 1965.

[68] B. M. Hochwald, T. L. Marzetta, and V. Tarokh, "Multiple-antenna channel hardening and its implications for rate feedback and scheduling", *IEEE Trans. Inf. Theory*, vol. 60, no. 9, 2004, pp. 1893–1909.

[69] M. Medard, "The effect upon channel capacity in wireless communications of perfect and imperfect knowledge of the channel", *IEEE Trans. Inf. Theory*, vol. 46, no. 3, 2000, pp. 933–946.

[70] A. Sabharwal, E. Erkip, and B. Aazhang, "On channel state information in multiple antenna block fading channels", in *IEEE Symposium on Information Theory (ISIT)*, pp. 116–119, 2000.

[71] J. Salz and J. H. Winters, "Effect of fading correlation on adaptive arrays in digital mobile radio", *IEEE Trans. Veh. Technol.*, vol. 43, no. 4, 1994, pp. 1049–1057.

[72] J. Brady, N. Behdad, and A. M. Sayeed, "Beamspace MIMO for millimeter-wave communications: System architecture, modeling, analysis, and measurements", *IEEE Trans. Antennas Propag.*, vol. 61, no. 7, 2013, pp. 3814–3827.

[73] A. M. Sayeed, "Deconstructing multiantenna fading channels", *IEEE Trans. Signal Process.*, vol. 50, no. 10, 2002, pp. 2563–2579.

[74] W. Weichselberger, M. Herdin, H. Özcelik, and E. Bonek, "A stochastic MIMO channel model with joint correlation of both link ends", *IEEE Trans. Wireless Commun.*, vol. 5, no. 1, 2006, pp. 90–100.

[75] A. S. Y. Poon, R. W. Brodersen, and D. N. C. Tse, "Degrees of freedom in multiple-antenna channels: A signal space approach", *IEEE Trans. Inf. Theory*, vol. 51, no. 2, 2005, pp. 523–536.

[76] S. Hu, F. Rusek, and O. Edfors, "Beyond massive MIMO: The potential of data transmission with large intelligent surfaces", *IEEE Trans. Signal Process.*, vol. 66, no. 10, 2018, pp. 2746–2758.

[77] A. Pizzo, T. L. Marzetta, and L. Sanguinetti, "Degrees of freedom of holographic MIMO channels", in *IEEE International Workshop on Signal Processing Advances in Wireless Communications (SPAWC)*, pp. 1–5, 2020.

[78] W. Chew, "A quick way to approximate a sommerfeld-weyl-type integral (antenna far-field radiation)", *IEEE Trans. Antennas Propag.*, vol. 36, no. 11, 1988, pp. 1654–1657.

[79] E. Visotsky and U. Madhow, "Space-time transmit precoding with imperfect feedback", *IEEE Trans. Inf. Theory*, vol. 47, no. 6, 2001, pp. 2632–2639.

[80] E. Jorswieck and H. Boche, "Optimal transmission strategies and impact of correlation in multiantenna systems with different types of channel state information", *IEEE Trans. Signal Process.*, vol. 52, no. 12, 2004, pp. 3440–3453.

[81] C. Oestges, B. Clerckx, M. Guillaud, and M. Debbah, "Dual-polarized wireless communications: From propagation models to system performance evaluation", *IEEE Transactions on Wireless Communications*, vol. 7, no. 10, 2008, pp. 4019–4031.

[82] N. Das, T. Inoue, T. Taniguchi, and Y. Karasawa, "An experiment on MIMO system having three orthogonal polarization diversity branches in multipath-rich environment", in *IEEE VTC-Fall*, vol. 2, pp. 1528–1532, 2004.

[83] F. Quitin, C. Oestges, A. Panahandeh, F. Horlin, and P. De Doncker, "Tri-polarized MIMO systems in real-world channels: Channel investigation and performance analysis", *Physical Commun.*, vol. 5, no. 4, 2012, pp. 308–316.

[84] P. Weeraddana, M. Codreanu, M. Latva-aho, A. Ephremides, and C. Fischione, "Weighted sum-rate maximization in wireless networks: A review", *Foundations and Trends in Networking*, vol. 6, no. 1-2, 2012, pp. 1–163.

[85] E. Björnson and E. Jorswieck, "Optimal resource allocation in coordinated multi-cell systems", *Foundations and Trends in Communications and Information Theory*, vol. 9, no. 2-3, 2013, pp. 113–381.

[86] L. Georgiadis, M. J. Neely, and L. Tassiulas, "Resource Allocation and Cross-Layer Control in Wireless Networks", *Foundations and Trends in Networking*, vol. 1, no. 1, 2006, pp. 1–144.

[87] Z. Chen, E. Björnson, and E. G. Larsson, "Dynamic resource allocation in co-located and cell-free massive MIMO", *IEEE Trans. Green Commun. Netw.*, vol. 4, no. 1, 2020, pp. 209–220.

[88] F. Rusek, D. Persson, B. K. Lau, E. G. Larsson, T. L. Marzetta, O. Edfors, and F. Tufvesson, "Scaling up MIMO: Opportunities and challenges with very large arrays", *IEEE Signal Process. Mag.*, vol. 30, no. 1, 2013, pp. 40–60.

[89] A. de Jesus Torres, L. Sanguinetti, and E. Björnson, "Near- and far-field communications with large intelligent surfaces", in *Asilomar Conference on Signals, Systems, and Computers*, pp. 564–568, 2020.

[90] T. L. Marzetta, "Noncooperative cellular wireless with unlimited numbers of base station antennas", *IEEE Trans. Wireless Commun.*, vol. 9, no. 11, 2010, pp. 3590–3600.

[91] Y.-W. P. Hong, C. W. Tan, L. Zheng, C. Hsieh, and C. Lee, "A unified framework for wireless max-min utility optimization with general monotonic constraints", in *IEEE Int. Conf. Comp. Commun. (INFOCOM)*, pp. 2076–2084, 2014.

[92] M. Schubert and H. Boche, "QoS-based resource allocation and transceiver optimization", *Foundations and Trends in Communications and Information Theory*, vol. 2, no. 6, 2005, pp. 383–529.

[93] C. W. Tan, "Wireless network optimization by Perron-Frobenius theory", *Foundations and Trends in Networking*, vol. 9, no. 2-3, 2015, pp. 107–218.

[94] M. Costa, "Writing on dirty paper", *IEEE Trans. Inf. Theory*, vol. 29, no. 3, 1983, pp. 439–441.

[95] S. U. Pillai, T. Suel, and S. Cha, "The Perron-Frobenius theorem: Some of its applications", *IEEE Signal Process. Mag.*, vol. 22, no. 2, 2005, pp. 62–75.

[96] G. Caire and S. Shamai, "On the achievable throughput of a multiantenna Gaussian broadcast channel", *IEEE Trans. Inf. Theory*, vol. 49, no. 7, 2003, pp. 1691–1706.

[97] S. Vishwanath, N. Jindal, and A. Goldsmith, "Duality, achievable rates, and sum-rate capacity of Gaussian MIMO broadcast channels", *IEEE Trans. Inf. Theory*, vol. 49, no. 10, 2003, pp. 2658–2668.

[98] P. Viswanath and D. N. C. Tse, "Sum capacity of the vector Gaussian broadcast channel and uplink-downlink duality", *IEEE Trans. Inf. Theory*, vol. 49, no. 8, 2003, pp. 1912–1921.

[99] M. Grant and S. Boyd, *CVX: Matlab software for disciplined convex programming, version 2.2*, http://cvxr.com/cvx, Jan. 2020.

[100] J. Zander, "Performance of optimum transmitter power control in cellular radio systems", *IEEE Trans. Veh. Technol.*, vol. 41, no. 1, 1992, pp. 57–62.

[101] H. Boche and M. Schubert, "A general duality theory for uplink and downlink beamforming", in *IEEE VTC-Fall*, pp. 87–91, 2002.

[102] M. Joham, W. Utschick, and J. Nossek, "Linear transmit processing in MIMO communications systems", *IEEE Trans. Signal Process.*, vol. 53, no. 8, 2005, pp. 2700–2712.

[103] P. Zetterberg and B. Ottersten, "The spectrum efficiency of a base station antenna array system for spatially selective transmission", *IEEE Trans. Veh. Technol.*, vol. 44, no. 3, 1995, pp. 651–660.

[104] M. Sadek, A. Tarighat, and A. Sayed, "A leakage-based precoding scheme for downlink multi-user MIMO channels", *IEEE Trans. Wireless Commun.*, vol. 6, no. 5, 2007, pp. 1711–1721.

[105] Y. Mao, O. Dizdar, B. Clerckx, R. Schober, P. Popovski, and H. V. Poor, "Rate-splitting multiple access: Fundamentals, survey, and future research trends", *IEEE Commun. Surveys Tuts.*, vol. 24, no. 4, 2022, pp. 2073–2126.

[106] E. O. Brigham, *The Fast Fourier Transform and Its Applications.* Pearson, 1988.

[107] A. A. Zaidi, R. Baldemair, M. Andersson, S. Faxér, V. Molés-Cases, and Z. Wang, "Designing for the future: The 5G NR physical layer", *Ericsson Review*, no. 7, 2017.

[108] H. Steyskal, "Digital beamforming antennas - an introduction", *Microwave Journal*, vol. 30, 1987, pp. 107–124.

[109] R. W. Heath, N. Gonzalez-Prelcic, S. Rangan, W. Roh, and A. M. Sayeed, "An overview of signal processing techniques for millimeter wave MIMO systems", *IEEE J. Sel. Topics Signal Process.*, vol. 10, no. 3, 2016, pp. 436–453.

[110] M. Stege, J. Jelitto, M. Bronzel, and G. Fettweis, "A multiple input-multiple output channel model for simulation of Tx- and Rx-diversity wireless systems", in *IEEE VTC-Fall*, vol. 2, pp. 833–839, 2000.

[111] A. Molisch, "A generic model for MIMO wireless propagation channels in macro- and microcells", *IEEE Trans. Signal Process.*, vol. 52, no. 1, 2004, pp. 61–71.

[112] A. A. Saleh and R. A. Valenzuela, "A statistical model for indoor multipath propagation", *IEEE J. Sel. Areas Commun.*, vol. 5, no. 2, 1987, pp. 128–137.

[113] X. Zhang, A. Molisch, and S.-Y. Kung, "Variable-phase-shift-based RF-baseband codesign for MIMO antenna selection", *IEEE Trans. Signal Process.*, vol. 53, no. 11, 2005, pp. 4091–4103.

[114] I. Ahmed, H. Khammari, A. Shahid, A. Musa, K. S. Kim, E. De Poorter, and I. Moerman, "A survey on hybrid beamforming techniques in 5G: Architecture and system model perspectives", *IEEE Commun. Surveys & Tutorials*, vol. 20, no. 4, 2018, pp. 3060–3097.

[115] F. Sohrabi and W. Yu, "Hybrid analog and digital beamforming for mmWave OFDM large-scale antenna arrays", *IEEE J. Sel. Areas Commun.*, vol. 35, no. 7, 2017, pp. 1432–1443.

[116] A. Li, D. Spano, J. Krivochiza, S. Domouchtsidis, C. G. Tsinos, C. Masouros, S. Chatzinotas, Y. Li, B. Vucetic, and B. Ottersten, "A tutorial on interference exploitation via symbol-level precoding: Overview, state-of-the-art and future directions", *IEEE Communications Surveys & Tutorials*, vol. 22, no. 2, 2020, pp. 796–839.

[117] P. Checcacci, V. Russo, and A. Scheggi, "Holographic antennas", *IEEE Trans. Antennas Propag.*, vol. 18, no. 6, 1970, pp. 811–813.

[118] E. Björnson, L. Sanguinetti, H. Wymeersch, J. Hoydis, and T. L. Marzetta, "Massive MIMO is a reality—What is next? Five promising research directions for antenna arrays", *Digital Signal Processing*, vol. 94, 2019, pp. 3–20.

[119] P. Ramezani and E. Björnson, "Fundamentals of 6G communications and networking", in ch. Near-Field Beamforming and Multiplexing Using Extremely Large Aperture Arrays, Springer: Cambridge University Press, 2023.

[120] O. Edfors and A. J. Johansson, "Is orbital angular momentum (OAM) based radio communication an unexploited area?", *IEEE Trans. Antennas Propag.*, vol. 60, no. 2, 2012, pp. 1126–1131.

[121] A. L. Swindlehurst and P. Stoica, "Maximum likelihood methods in radar array signal processing", *Proc. IEEE*, vol. 86, no. 2, 1998, pp. 421–441.

[122] S. S. Haykin, J. Litva, and T. J. Shepherd, *Radar Array Processing*. Springer-Verlag, 1993.

[123] F. Liu, Y. Cui, C. Masouros, J. Xu, T. X. Han, Y. C. Eldar, and S. Buzzi, "Integrated sensing and communications: Toward dual-functional wireless networks for 6G and beyond", *IEEE J. Sel. Areas Commun.*, vol. 40, no. 6, 2022, pp. 1728–1767.

[124] J. Capon, "High-resolution frequency-wavenumber spectrum analysis", *Proceedings of the IEEE*, vol. 57, no. 8, 1969, pp. 1408–1418.

[125] R. Schmidt, "Multiple emitter location and signal parameter estimation", *IEEE Transactions on Antennas and Propagation*, vol. 34, no. 3, 1986, pp. 276–280.

[126] G. Bienvenu, "Influence of the spatial coherence of the background noise on high resolution passive methods", in *IEEE International Conference on Acoustics, Speech and Signal Processing (ICASSP)*, vol. 4, pp. 306–309, 1979.

[127] A. Paulraj, R. Roy, and T. Kailath, "A subspace rotation approach to signal parameter estimation", *Proc. IEEE*, vol. 74, no. 7, 1986, pp. 1044–1046.

[128] R. Roy and T. Kailath, "ESPRIT-estimation of signal parameters via rotational invariance techniques", *IEEE Trans. Acoust., Speech, Signal Process.*, vol. 37, no. 7, 1989, pp. 984–995.

[129] P. Stoica and R. L. Moses, *Spectral analysis of signals*. Pearson Prentice Hall, 2005.

[130] N. Patwari, J. N. Ash, S. Kyperountas, A. O. Hero, R. L. Moses, and N. S. Correal, "Locating the nodes: Cooperative localization in wireless sensor networks", *IEEE Signal Process. Mag.*, vol. 22, no. 4, 2005, pp. 54–69.

[131] R. Zekavat and R. M. Buehrer, *Handbook of position location: Theory, practice and advances*. John Wiley & Sons, 2011.

[132] B. Friedlander, "Localization of signals in the near-field of an antenna array", *IEEE Trans. Signal Process.*, vol. 67, no. 15, 2019, pp. 3885–3893.

[133] S. Sand, A. Dammann, and C. Mensing, *Positioning in wireless communications systems*. John Wiley & Sons, 2014.

[134] M. A. Richards, J. Scheer, W. A. Holm, and W. L. Melvin, *Principles of modern radar*, vol. 1. SciTech Publishing, 2010.

[135] P. Swerling, "Probability of detection for fluctuating targets", *ASTIA RM-1217*, 80638 Mar. 1954.

[136] M. I. Skolnik, *Introduction to radar systems*. McGraw-Hill, 1980.

[137] M. A. Richards, *Fundamentals of radar signal processing*. McGraw-Hill Education, 2014.

[138] C. Hülsmeyer, *Wireless transmitting and receiving mechanism for electric wawes*, US Patent, 1906.

[139] D. W. Bliss and K. W. Forsythe, "Multiple-input multiple-output (MIMO) radar and imaging: Degrees of freedom and resolution", in *Asilomar Conference on Signals, Systems and Computers*, pp. 54–59, 2003.

[140] E. Fishler, A. Haimovich, R. Blum, D. Chizhik, L. Cimini, and R. Valenzuela, "MIMO radar: An idea whose time has come", in *IEEE Radar Conference*, pp. 71–78, 2004.

[141] F. Daum and J. Huang, "MIMO radar: Snake oil or good idea?", *IEEE Aerosp. Electron. Syst. Mag.*, vol. 24, no. 5, 2009, pp. 8–12.

[142] J. Li and P. Stoica, *MIMO Radar Signal Processing.* Wiley, 2009.

[143] C. Sturm and W. Wiesbeck, "Waveform design and signal processing aspects for fusion of wireless communications and radar sensing", *Proc. IEEE*, vol. 99, no. 7, 2011, pp. 1236–1259.

[144] K. V. Mishra, M. Bhavani Shankar, V. Koivunen, B. Ottersten, and S. A. Vorobyov, "Toward millimeter-wave joint radar communications: A signal processing perspective", *IEEE Signal Process. Mag.*, vol. 36, no. 5, 2019, pp. 100–114.

[145] J. A. Zhang, F. Liu, C. Masouros, R. W. Heath, Z. Feng, L. Zheng, and A. Petropulu, "An overview of signal processing techniques for joint communication and radar sensing", *IEEE J. Sel. Topics Signal Process.*, vol. 15, no. 6, 2021, pp. 1295–1315.

[146] A. Barabell, "Improving the resolution performance of eigenstructure-based direction-finding algorithms", in *IEEE International Conference on Acoustics, Speech and Signal Processing (ICASSP)*, vol. 8, pp. 336–339, 1983.

[147] B. C. Ng and C. M. S. See, "Sensor-array calibration using a maximum-likelihood approach", *IEEE Trans. Antennas Propag.*, vol. 44, no. 6, 1996, pp. 827–835.

[148] P. Khare and A. Swarup, *Engineering Physics Fundamentals and Modern Applications.* Laxmi Publications, Ltd., 2008.

[149] S. A. Schelkunoff, "Some equivalence theorems of electromagnetics and their application to radiation problems", *Bell System Technical Journal*, vol. 15, no. 1, 1936, pp. 92–112.

[150] M. Najafi, V. Jamali, R. Schober, and H. V. Poor, "Physics-based modeling and scalable optimization of large intelligent reflecting surfaces", *IEEE Trans. Commun.*, vol. 69, no. 4, 2021, pp. 2673–2691.

[151] D. Pozar, *Microwave Engineering*, 4th edition. Wiley, 2011.

[152] D. Berry, R. Malech, and W. Kennedy, "The reflectarray antenna", *IEEE Trans. Antennas Propag.*, vol. 11, no. 6, 1963, pp. 645–651.

[153] S. V. Hum and J. Perruisseau-Carrier, "Reconfigurable reflectarrays and array lenses for dynamic antenna beam control: A review", *IEEE Trans. Antennas Propag.*, vol. 62, no. 1, 2014, pp. 183–198.

[154] C. Liaskos, S. Nie, A. Tsioliaridou, A. Pitsillides, S. Ioannidis, and I. Akyildiz, "A new wireless communication paradigm through software-controlled metasurfaces", *IEEE Commun. Mag.*, vol. 56, no. 9, 2018, pp. 162–169.

[155] C. Huang, A. Zappone, G. C. Alexandropoulos, M. Debbah, and C. Yuen, "Reconfigurable intelligent surfaces for energy efficiency in wireless communication", *IEEE Trans. Wireless Commun.*, vol. 18, no. 8, 2019, pp. 4157–4170.

[156] Q. Wu and R. Zhang, "Towards smart and reconfigurable environment: Intelligent reflecting surface aided wireless network", *IEEE Commun. Mag.*, vol. 58, no. 1, 2020, pp. 106–112.

[157] M. D. Renzo, M. Debbah, D.-T. Phan-Huy, A. Zappone, M.-S. Alouini, C. Yuen, V. Sciancalepore, G. C. Alexandropoulos, J. Hoydis, H. Gacanin, J. de Rosny, A. Bounceur, G. Lerosey, and M. Fink, "Smart radio environments empowered by reconfigurable AI meta-surfaces: An idea whose time has come", *EURASIP J. Wirel. Commun. Netw.*, no. 129, 2019.

[158] O. Tsilipakos, A. C. Tasolamprou, A. Pitilakis, F. Liu, X. Wang, M. S. Mirmoosa, D. C. Tzarouchis, S. Abadal, H. Taghvaee, C. Liaskos, A. Tsioliaridou, J. Georgiou, A. Cabellos-Aparicio, E. Alarcon, S. Ioannidis, A. Pitsillides, I. F. Akyildiz, N. V. Kantartzis, E. N. Economou, C. M. Soukoulis, M. Kafesaki, and S. Tretyakov, "Toward intelligent metasurfaces: The progress from globally tunable metasurfaces to software-defined metasurfaces with an embedded network of controllers", *Advanced Optical Materials*, no. 2000783, 2020.

[159] X. Pei, H. Yin, L. Tan, L. Cao, Z. Li, K. Wang, K. Zhang, and E. Björnson, "RIS-aided wireless communications: Prototyping, adaptive beamforming, and indoor/outdoor field trials", *IEEE Trans. Commun.*, vol. 69, no. 12, 2021, pp. 8627–8640.

[160] A. Enqvist, Ö. T. Demir, C. Cavdar, and E. Björnson, "Optimizing reconfigurable intelligent surfaces for short transmissions: How detailed configurations can be afforded?", *IEEE Trans. Wireless Commun.*, 2023. DOI: 10.1109/TWC.2023.3307605.

[161] M. Haghshenas, P. Ramezani, and E. Björnson, "Efficient LOS channel estimation for RIS-aided communications under non-stationary mobility", in *IEEE International Conference on Communication (ICC)*, 2023.

[162] B. Zhu, J. Zhao, and Y. Feng, "Active impedance metasurface with full 360 reflection phase tuning", *Scientific Reports*, vol. 3, no. 3059, 2013.

[163] S. Abeywickrama, R. Zhang, Q. Wu, and C. Yuen, "Intelligent reflecting surface: Practical phase shift model and beamforming optimization", *IEEE Trans. Commun.*, vol. 68, no. 9, 2020, pp. 5849–5863.

[164] Y. Yang, B. Zheng, S. Zhang, and R. Zhang, "Intelligent reflecting surface meets OFDM: Protocol design and rate maximization", *IEEE Trans. Commun.*, vol. 68, no. 7, 2020, pp. 4522–4535.

[165] S. Lin, B. Zheng, G. C. Alexandropoulos, M. Wen, F. Chen, and S. Mumtaz, "Adaptive transmission for reconfigurable intelligent surface-assisted OFDM wireless communications", *IEEE J. Sel. Areas Commun.*, vol. 38, no. 11, 2020, pp. 2653–2665.

[166] E. Björnson, "Optimizing a binary intelligent reflecting surface for OFDM communications under mutual coupling", in *International ITG Workshop on Smart Antennas (WSA)*, pp. 1–6, 2021.

[167] O. Axelsson, *Iterative solution methods*. Cambridge University Press, 1996.

[168] *Spatial channel model for Multiple Input Multiple Output (MIMO) simulations (Release 16)*. 3GPP TS 25.996, Jul. 2020.

[169] H. Do, N. Lee, and A. Lozano, "Line-of-sight MIMO via intelligent reflecting surface", *IEEE Trans. Wireless Commun.*, vol. 22, no. 6, 2023, pp. 4215–4231.

[170] S. Zhang and R. Zhang, "Capacity characterization for intelligent reflecting surface aided MIMO communication", *IEEE J. Sel. Areas Commun.*, vol. 38, no. 8, 2020, pp. 1823–1838.

[171] Q. Wu and R. Zhang, "Intelligent reflecting surface enhanced wireless network via joint active and passive beamforming", *IEEE Trans. Wireless Commun.*, vol. 18, no. 11, 2019, pp. 5394–5409.

[172] Y. Liu, J. Zhao, M. Li, and Q. Wu, "Intelligent reflecting surface aided MISO uplink communication network: Feasibility and power minimization for perfect and imperfect CSI", *IEEE Trans. Commun.*, vol. 69, no. 3, 2021, pp. 1975–1989.

[173] R. Liu, M. Li, H. Luo, Q. Liu, and A. L. Swindlehurst, "Integrated sensing and communication with reconfigurable intelligent surfaces: Opportunities, applications, and future directions", *IEEE Wireless Commun.*, vol. 30, no. 1, 2023, pp. 50–57.

[174] A. M. Elbir, K. V. Mishra, M. R. B. Shankar, and S. Chatzinotas, "The rise of intelligent reflecting surfaces in integrated sensing and communications paradigms", *IEEE Netw.*, 2023. DOI: 10.1109/MNET.128.2200446.

[175] S. Buzzi, E. Grossi, M. Lops, and L. Venturino, "Foundations of MIMO radar detection aided by reconfigurable intelligent surfaces", *IEEE Trans. Signal Process.*, vol. 70, 2022, pp. 1749–1763.

[176] H. Zhang, H. Zhang, B. Di, K. Bian, Z. Han, and L. Song, "Metaradar: Multi-target detection for reconfigurable intelligent surface aided radar systems", *IEEE Trans. Wireless Commun.*, vol. 21, no. 9, 2022, pp. 6994–7010.

Index

About the Authors

Dr. Emil Björnson is a Professor of Wireless Communication at the KTH Royal Institute of Technology, Stockholm, Sweden. He is an IEEE Fellow, Digital Futures Fellow, and Wallenberg Academy Fellow. He received his M.S. degree from Lund University, Sweden, in 2007 and his Ph.D. degree from KTH in 2011. He is a co-host of the podcast and YouTube channel called Wireless Future. He has previously authored the textbooks *Optimal Resource Allocation in Coordinated Multi-Cell Systems* (2013), *Massive MIMO Networks: Spectral, Energy, and Hardware Efficiency* (2017), and *Foundations of User-Centric Cell-Free Massive MIMO* (2021). He has published a large amount of simulation code on GitHub related to these books and selected scientific papers.

His research focuses on multi-antenna communications, reconfigurable intelligent surfaces, energy-efficient communication, and radio resource management, using methods from communication theory, signal processing, and machine learning. His work on these topics received the 2018 and 2022 IEEE Marconi Prize Paper Awards, the 2019 EURASIP Early Career Award, the 2019 IEEE Fred W. Ellersick Prize, the 2019 IEEE Signal Processing Magazine Best Column Award, the 2020 Pierre-Simon Laplace Early Career Technical Achievement Award, the 2020 CTTC Early Achievement Award, the 2021 IEEE ComSoc RCC Early Achievement Award, and the 2023 IEEE Communications Society Outstanding Paper Award.

Dr. Özlem Tuğfe Demir is an Assistant Professor of Electrical and Electronics Engineering at TOBB University of Economics and Technology, Ankara, Türkiye. She received her B.S., M.S., and Ph.D. degrees in Electrical and Electronics Engineering from Middle East Technical University, Ankara, Türkiye, in 2012, 2014, and 2018, respectively. She was a Postdoctoral Researcher at Linköping University, Sweden, in 2019–2020 and at the KTH Royal Institute of Technology, Sweden, in 2021–2022. She has previously co-authored the textbook *Foundations of User-Centric Cell-Free Massive MIMO* (2021). She is an Associate Editor of the IEEE Transactions on Wireless Communications.

Her research interests are on signal processing and optimization in wireless communications, massive MIMO, cell-free massive MIMO, beyond 5G multiple antenna technologies, reconfigurable intelligent surfaces, near-field communications, machine learning for communications, mobile data analysis, and green mobile networks.

Printed and bound by CPI Group (UK) Ltd, Croydon, CR0 4YY

15/09/2026

14962365-0008